CUMULATIVE SUBJECT INDEX

VOLUMES 1–75 (including Revised Series Volumes 1–31)

PART III: O–Z

HANDBOOK OF
CLINICAL NEUROLOGY

Editors

PIERRE J. VINKEN   GEORGE W. BRUYN

Executive Editor

KENNETH ELLISON DAVIS

VOLUME 78

ELSEVIER

AMSTERDAM • LONDON • NEW YORK • OXFORD • PARIS •
SHANNON • TOKYO

CUMULATIVE SUBJECT INDEX

VOLUMES 1–75 (including Revised Series Volumes 1–31)

PART III: O–Z

Editors

PIERRE J. VINKEN   GEORGE W. BRUYN

In collaboration with

WILLEKE VAN OCKENBURG

REVISED SERIES 34

ELSEVIER

AMSTERDAM • LONDON • NEW YORK • OXFORD • PARIS •
SHANNON • TOKYO

ELSEVIER SCIENCE B.V.
Sara Burgerhartstraat 25
P.O. Box 211, 1000 AE Amsterdam, The Netherlands

First edition 2002

Library of Congress Cataloging in Publication Data
A catalog record from the Library of Congress has been applied for.

ISBN: 0-444-50919-4 (Part I: A–E)
       0-444-50918-6 (Part II: F–N)
       0-444-50917-8 (Part III: O–Z)
ISSN: 0072-9752

⊗ The paper used in this publication meets the requirements of ANSI/NISO Z39.48-1992 (Permanence of Paper). Printed in The Netherlands.

Cumulative Subject Index
to Volumes 1–75 (including Revised Series 1–31)
Part III: O–Z

In collaboration with W. van Ockenburg

Oasthouse urine disease
  methionine malabsorption syndrome, 29/128,
    42/600
Oat cell carcinoma
  late cerebellar atrophy, 60/585
  paraneoplastic polyneuropathy, 51/466
Obersteiner, H., 1/10, 27
Obersteiner layer, *see* Cerebellar external granular
    layer
Obesity
  acrocephalosyndactyly type II, 38/422, 43/324
  adiposis dolorosa, 43/57
  Alström-Hallgren syndrome, 13/460, 22/501
  arterial hypertension, 54/206
  Bardet-Biedl syndrome, 13/383, 22/536, 43/233
  benign intracranial hypertension, 42/740
  cardiorespiratory syndrome, *see* Pickwickian
    syndrome
  central alveolar hypoventilation, 63/464
  central sleep apnea, 63/461
  cerebrovascular disease, 53/17
  Cushing syndrome, 2/449
  daytime hypersomnia, 45/134
  decompression myelopathy, 61/224
  diencephalic lesion, 2/449
  familial spastic paraplegia, 22/425
  Fröhlich syndrome, 2/449
  hypertension, 75/477
  hypoventilation syndrome, *see* Pickwickian
    syndrome
  meningeal leukemia, 63/341
  mental deficiency, 43/273-275, 288

Morgagni-Stewart-Morel syndrome, 43/411
  narcolepsy, 2/448
  obstructive sleep apnea, 63/454
  pickwickian syndrome, 3/96, 38/299, 63/455
  pituitary dwarfism type I, 42/612
  pituitary dwarfism type III, 42/615
  Prader-Labhart-Willi syndrome, 31/322, 324,
    40/336, 43/463, 46/93, 50/590
  pseudohypoparathyroidism, 42/621
  sensorineural deafness, 42/391
  sleep apnea, 75/477
  snoring, 63/450
  spina bifida, 50/495
  Summit syndrome, 43/325, 372-374
  Weiss-Alström syndrome, 13/463
Obesity cardiorespiratory syndrome, *see*
    Pickwickian syndrome
Obesity hypoventilation syndrome, *see* Pickwickian
    syndrome
Obex plugging
  hydromyelia, 50/431
  syringomyelia, 50/458
Object agnosia
  visual, *see* Visual object agnosia
Object apraxia
  developmental dyspraxia, 4/459
Object localization
  space perception, 45/411
Object recognition
  agnosia, 45/333-347
  perception classification, 45/339
Oblique amyotrophy

Hirayama disease, 59/108
  spinal muscular atrophy, 59/108
Obliterating brachiocephalic arteritis, *see* Takayasu
  disease
Obliterative brachiocephalic arteriopathy
  (Ash-Upmark), *see* Takayasu disease
Obsessional neurosis
  lobotomy, 45/27
Obstetrical paralysis
  maternal, *see* Maternal obstetrical paralysis
Obstructions
  congenital, *see* Congenital obstruction
  duodenal, *see* Duodenal obstruction
  intestinal pseudo, *see* Intestinal
    pseudo-obstruction
  pseudo, *see* Pseudo-obstruction
  superior vena cava, *see* Superior vena cava
    obstruction
  thoracic duct, *see* Thoracic duct obstruction
  venous drainage, *see* Venous drainage obstruction
Obstructive apnea syndrome
  chronic respiratory insufficiency, 63/413
  hypoventilation, 63/413
  hypoventilation syndrome, 63/413
  respiratory disease, 63/413
Obstructive hydrocephalus
  acquired immune deficiency syndrome, 56/495
  Arnold-Chiari malformation type I, 46/325
  cerebellar infarction, 53/383
  computer assisted tomography, 57/164
  Dandy-Walker syndrome, 46/325, 50/323
  etiology, 46/324
  head injury, 57/164
  hydrocephalic dementia, 46/324
  neurocysticercosis, 16/225, 35/315, 52/530
  pneumococcal meningitis, 52/43
  posterior fossa tumor, 18/406
  skull, 16/94
  type, 16/99
Obstructive liver disease
  neuropathology, 63/278
  vitamin E deficiency, 63/278
Obstructive pulmonary disease
  chronic, *see* Chronic obstructive pulmonary
    disease
  eosinophilia myalgia syndrome, 63/376
Obstructive respiratory tract disorder
  chronic, *see* Chronic obstructive respiratory tract
    disorder
Obstructive sleep apnea
  acromegaly, 63/454
  adenotonsillar hypertrophy, 63/454
  airway size, 63/453

alcohol, 63/454
alveolar hypoventilation, 63/455
amnesia, 63/454
arterial carbon dioxide tension, 63/455
arterial hypertension, 63/456
arterial oxygen tension, 63/455
blood pressure, 63/456
cardiac output, 63/457
continuous positive airway pressure, 63/457
diagnosis, 63/457
dysphoria, 63/453
ECG, 63/457
edema, 63/454
esophageal reflux, 63/453
fluoxetine, 63/460
headache, 63/454
heart rate, 63/456
heart rhythm, 63/456
hyoid bone, 63/454
hypercapnia, 63/455
hypnagogic hallucination, 63/453
hypopharynx, 63/452
hypoxia, 63/455
irritability, 63/453
macroglossia, 63/454
mortality rate, 63/457
muscular hypotonia, 63/452
myxedema, 63/454
nasal pressure, 63/453
non-REM sleep, 63/453
obesity, 63/454
oxyhemoglobin desaturation, 63/455
palate, 63/452
pathophysiology, 63/452
pharyngeal airway, 63/452
pharyngeal pressure, 63/452
pickwickian syndrome, 63/455
polycythemia, 63/454
polysomnography, 63/457
prevalence, 63/453
prosthetic device, 63/459
protriptyline, 63/460
pulmonary hypertension, 63/456
REM sleep, 63/458
sexual impotence, 63/454
sleep disorder, 63/449
sleep phase, 63/453
somnolence, 63/453
surgery, 63/458
symptom, 63/453
tracheostomy, 63/457
uvulopalatopharyngoplasty, 63/459
voluminous tongue, 63/454

weight loss, 63/458
Obtundation
ethylene glycol intoxication, 64/123
hypercalcemia, 63/561
hypocalcemia, 63/558
hypophosphatemia, 63/564
Obturator nerve
compression neuropathy, 7/314, 51/108
Howship-Romberg sign, 2/40
topographical diagnosis, 2/40
Obturator nerve injury
sacroiliac fracture dislocation, 25/473
Obturator neuropathy
aorta aneurysm, 63/49
Obturator neurotomy
spastic paraplegia, 61/369
spasticity treatment, 61/369
spinal cord injury, 61/369
spinal spasticity, 61/369
transverse spinal cord lesion, 61/369
Occipital acalculia
secondary acalculia, 4/188
Occipital artery
lateral, see Lateral occipital artery
parieto, see Parieto-occipital artery
temporal, see Temporal occipital artery
Occipital bone
basal part hypoplasia, 32/14
development, 30/210, 32/6
malformation, 32/14
occipital dysplasia, 32/14, 20-22
posterior fossa hematoma, 57/157, 254
skull fracture, 57/254
Occipital bone shortening
basilar impression, 32/14
craniovertebral region, 32/14
hypoplasia, 32/14
measurement, 32/14
platybasia, 32/14
Occipital cephalocele
aqueduct stenosis, 50/101
Arnold-Chiari malformation type III, 30/211,
50/101
associated abnormality, 50/101
cerebellar dysplasia, 50/101
Dandy-Walker syndrome, 50/101
diastematomyelia, 50/101
hydrocephalus, 50/101
incidence, 50/99
intelligence quotient, 50/101
Klippel-Feil syndrome, 50/101
nerve tissue heterotopia, 50/102
radiation, 50/100

sex ratio, 50/99
spina bifida, 50/101
surgical repair, 50/101
Occipital cerebral vein
ascending, see Ascending cerebral vein
Occipital condyle syndrome
brain metastasis, 71/612
Occipital condylus
anterior, see Anterior occipital condyle
hypoplasia, 32/22
skull base metastasis, 69/128
third, see Third occipital condylus
Occipital diploic vein
anatomy, 11/55
Occipital dysplasia
achondroplasia, 50/395
Arnold-Chiari malformation type II, 50/404
atlas, 32/25
basilar impression, 32/16-20, 50/399
bone abnormality, 50/388
foramen magnum, 50/395
form, 50/388
hypoplasia, 32/14, 20-22
occipital bone, 32/14, 20-22
platybasia, 32/15
Occipital encephalocele
aqueduct stenosis, 42/26
Arnold-Chiari malformation type III, 42/26
basilar impression, 32/60
cerebellar agenesis, 50/177
computer assisted tomography, 50/100
corpus callosum agenesis, 42/26
Dandy-Walker syndrome, 42/26
Gruber syndrome, 14/505, 43/391
heterotopia, 42/26
Klippel-Feil syndrome, 32/60, 42/26
micrencephaly, 42/26
morphologic data, 42/26
prognosis, 50/101
sonography, 50/100
spina bifida, 42/26
trigeminal nerve agenesis, 50/214
Occipital headache
anxiety, 5/371
cervical vertebral column injury, 61/33
continuous, 5/371
muscle contraction, 5/371
nuchal rigidity, 5/371, 48/137
subarachnoid hemorrhage, 48/137
vomiting, 48/137
Occipital horn
occipital lobe tumor, 17/343
Occipital infarction

splenio-occipital alexia, 45/444
temporal arteritis, 48/315
Occipital ischemia
brain cortex, 12/3
Occipital lobe
anatomy, 2/641
body scheme disorder, 2/668
brain infarction, 54/404
cerebral dominance, 4/5, 303
cerebrovascular disease, 55/118
cortex layer, 2/642
elementary hallucination, 45/367
hallucination, 45/364-368
palinacousia, 45/367
phosphene, 45/364
posterior alexia, 45/438
symptomatology, 2/671
temporal crescent, 2/597
topographical memory, 2/613
traumatic encephalopathy, 3/69
visual field defect, 2/588, 17/314-317, 319
visual perseveration, 45/364-366
Occipital lobe epilepsy, 2/646-648, 662, 15/43
benign childhood epilepsy with occipital
paroxysm, 73/109
calcification, 73/110
cause, 73/107
classification, 72/11
EEG, 73/112
lactic acidosis, 73/110
MELAS syndrome, 73/110
metamorphopsia, 2/647
migraine, 73/109
mitochondrial disease, 73/110
neuroimaging, 73/112
pathology, 72/117
pharmacology, 73/112
seizure spread, 73/108
surgery, 73/114
survey, 73/107
visual aura, 73/108
Occipital lobe infarction
cortical blindness, 55/119
visual hallucination, 55/146
Occipital lobe lesion
acalculia, 4/188
accommodation disorder, 2/667
agnostic alexia, 4/95
agraphia, 4/5
alexia, 4/5, 125
Balint syndrome, 45/410
cerebral dominance, 4/303
color agnosia, 4/5, 300

constructional apraxia, 45/500
dissociated visual perception, 2/584
hallucination, 4/330
homonymous hemianopia, 2/672
left side, 4/5
memory disorder, 3/283
metamorphopsia, 2/662
monocular polyopia, 2/663, 17/331
right side, 4/5
spatial agraphia, 4/147
spatial orientation disorder, 2/665
visual agnosia, 2/541, 4/20-22
Occipital lobe resection, see Occipital lobectomy
Occipital lobe syndrome, 45/49-58
amnesic color anomia, 45/52
anosognosia, 45/51
Anton syndrome, 45/50
color agnosia, 45/52
color aphasia, 45/52
color naming defect, 45/52
cortical blindness, 45/49-51
EEG, 45/57
epilepsy, 45/57
hallucination, 45/56-58
homonymous hemiachromatopsia, 2/650
illusion, 45/56-58
macular sparing, 45/51
metamorphopsia, 2/670
migraine, 45/57
oculomotor disease, 45/52
optic allesthesia, 2/664
prosopagnosia, 45/53
visual agnosia, 45/53-56
Wernicke-Korsakoff syndrome, 45/51
Occipital lobe tumor, 37/310-343
abnormal visual clarity, 17/331
agnostic alexia, 17/327
alexia, 17/327, 45/436
anatomy, 17/310, 311
autotopagnosia, 17/329
Balint syndrome, 17/329
blindness, 17/326
campimetry, 17/318-320
central sparing, 17/316, 319
cerebellar sign, 17/333
color anomia, 67/156
color blindness, 17/325
color sense, 17/324
completion phenomenon, 17/342
conjugate gaze, 17/324
cortical blindness, 17/326
cranial nerve, 17/336
deafness, 17/337

diagnosis, 17/332, 336, 342
epilepsy, 17/322
extrapyramidal disorder, 17/338
eye fixation, 17/324, 329
false localization sign, 2/672, 17/332
Gerstmann syndrome, 17/325
hallucination, 67/154
hemianopia, 17/314-316, 319, 337
hemiparesis, 17/335
hemiplegia, 17/335
homonymous hemianopia, 2/672
incidence, 17/312, 313
intelligence, 17/334
localization, 17/339
macular sparing, 17/316, 319
macular splitting, 17/320
meiosis, 17/336
mental disorder, 17/334
metamorphopsia, 17/323, 326, 330
monocular polyopia, 2/663, 17/331
motion perception, 17/331
motor disorder, 17/335
motor weakness, 17/335
mydriasis, 17/336
nystagmus, 17/335
occipital horn, 17/343
oculomotor tracking, 67/155
optic allesthesia, 17/331
polyopia, 17/331
prognosis, 17/314
prosopagnosia, 17/327
ptosis, 17/336
seizure, 67/154
simultanagnosia, 17/327
spatial localization, 17/329
spatial orientation disorder, 17/329
spelling dyslexia, 17/328
temporary blindness, 17/326
time sense, 17/331
topography, 17/311
transtentorial herniation, 17/333
vertical axis sparing, 17/321
visual agnosia, 17/325, 327, 331, 67/156
visual defect, 67/154
visual dreaming, 17/328
visual field, 17/314-317, 319, 339-342
visual hallucination, 17/322, 331
visual memory, 17/326, 328
visual perseveration, 17/331
visual recognition, 17/327
Occipital lobectomy
alexia, 45/436, 438
homonymous hemianopia, 2/671

macular sparing, 2/580
macular splitting, 2/580-582, 671
optokinetic nystagmus, 2/670
Occipital nerve
anatomy, 5/368, 369
greater, see Greater occipital nerve
headache, 5/110
lesion, 7/326, 335
minor, see Lesser occipital nerve
Occipital neuralgia, 5/368-374
acute, 5/370
anatomy, 5/368
cervical arthrosis, 5/373
cervical spondylarthrosis, 5/373
chronic, 5/370
clinical aspect, 5/369
head posture, 19/39
Kuhlendahl syndrome, 5/373
neoplasm, 19/39
paroxysmal pain, 5/277, 369
sternocleidomastoid muscle, 19/39
subacute, 5/370
surgery, 5/402
Occipital pain
referred, 5/277, 278
Occipital paroxysm
idiopathic partial epilepsy, 73/18
Occipital pug
fibrous dysplasia, 14/168
Occipital reversal
computer assisted tomography, 46/484
schizophrenia, 46/484
Occipital sinus
dura mater, 30/424
petro, see Petro-occipital sinus
radioanatomy, 11/112
topography, 11/112
Occipital synostosis
atlanto, see Atlanto-occipital synostosis
Occipital vein
descending medial, see Descending medial
occipital vein
Occipital vertebrae
embryonic anlage, 32/6
Occipitoatlantal joint
movement, 61/5
Occipitoatlantoaxial complex
instability, 61/6
ligament, 61/6
radiography, 61/6
Occipitoaxial injury
anatomy, 25/229
dens, 25/229

Occipitofrontal circumference
  macrencephaly, 42/41, 43
Occipitoparietal artery
  anatomy, 11/21
Occipitoparietal region
  astereopsis, 45/16
  Balint syndrome, 45/15
Occipitoparietal tumor
  color anomia, 45/52
Occipitotemporal syndrome
  mesial, *see* Mesial occipitotemporal syndrome
Occlusion
  anterior choroidal artery, *see* Anterior choroidal
    artery occlusion
  anridromic, *see* Antidromic occlusion
  arterial, *see* Artery occlusion
  basilar artery, 53/386
  brain artery, *see* Brain artery occlusion
  brain collateral blood flow, 11/170
  carotid artery, *see* Carotid artery occlusion
  central retinal artery, *see* Central retinal artery
    occlusion
  chronic superior mesenteric artery, *see* Chronic
    superior mesenteric artery occlusion
  internal carotid artery, 5/126
  intracranial venous, *see* Intracranial venous
    occlusion
  mesenteric artery, *see* Mesenteric artery occlusion
  middle cerebral artery, *see* Middle cerebral artery
    occlusion
  multiple intracranial arterial, *see* Multiple
    intracranial arterial occlusion
  posterior groove, *see* Posterior groove occlusion
  posterior inferior cerebellar artery, *see* Posterior
    inferior cerebellar artery occlusion
  retinal artery, *see* Retinal artery occlusion
  subclavian artery, *see* Subclavian artery occlusion
  venous, *see* Venous occlusion
  venous sinus, *see* Venous sinus occlusion
  Willis circle, 42/749-751
Occlusive aortoiliac disease
  cauda equina intermittent claudication, 12/535
  diagnostic pitfalls, 12/535
  painful, 12/535
Occlusive disease
  cerebrovascular, *see* Cerebrovascular occlusive
    disease
  vertebral artery, 12/1
Occlusive vasculopathy
  diamorphine addiction, 55/518
  homocystinuria, 55/328
  sickle cell anemia, 55/509
Occupational cramp, *see* Occupational neurosis

Occupational neurosis
  anatomic classification, 1/279
  feedback, 1/290
  muscle cramp, 1/290
  writers cramp, 1/290, 7/338-340
Occupational palsy
  hereditary multiple recurrent mononeuropathy,
    51/560
Occupational Raynaud phenomenon
  features, 51/136
  mechanism, 51/136
Occupational spasm, *see* Occupational neurosis
Occupational therapy
  hyperventilation syndrome, 38/330, 63/443
  lead intoxication, 64/432
  neuropathy, 8/402
  paraplegia, 26/233
  Pick disease, 46/244
  rehabilitation, 41/481, 55/241
  school, 41/481
  spinal cord injury, 26/233, 61/457
Octamethyl pyrophosphoramide
  neurotoxicity, 37/549
  neurotoxin, 37/549, 64/152, 155
  organophosphate, 64/152
  organophosphate intoxication, 64/152, 155
Octamethyl pyrophosphoramide intoxication
  chemical classification, 37/549
  organophosphorus compound, 37/549
Octamethyl pyrophosphorotetramide intoxication,
    see Octamethyl pyrophosphoramide intoxication
Octavus nerve
  cervical spondylosis, 7/454
Octopoda intoxication
  artificial respiration, 37/62
  cephalotoxin, 37/62
  convulsion, 37/62
  dopamine, 37/62
  flaccid paralysis, 37/62
  glycoprotein, 37/62
  Hapalochlaena lunulata, 37/62
  hapalochlaena maculosa, 37/62
  histamine, 37/62
  octopamine, 37/62
  octopus, 37/62
  serotonin, 37/62
  treatment, 37/62
  vomiting, 37/62
Octopus intoxication
  envenomation, 37/62
  venom, 37/62
Octreotide
  acromegaly, 68/353

autonomic failure, 75/725
orthostatic hypotension, 63/164
thyrotropic hormone secreting adenoma, 68/357
Ocular abnormality
chondrodystrophic myotonia, 40/328
congenital heart disease, 63/11
hereditary amyloid polyneuropathy type 1, 51/421
myotonic dystrophy, 40/512
Patau syndrome, 31/504, 513, 50/556, 559
Rubinstein-Taybi syndrome, 31/282, 43/234
Wernicke-Korsakoff syndrome, 28/247-249
Ocular accommodation
diphtheria, 51/183
Ocular apraxia, *see* Oculomotor apraxia
Ocular ataxia
Balint syndrome, 45/15, 408, 55/144
etiology, 45/408
latent, 1/314
parietal lobe syndrome, 45/73
pseudoataxia, 1/310
reaching disorder, 45/408
visual fixation, 1/311
Ocular bobbing
brain embolism, 53/213
brain infarction, 53/213
cardiac arrest, 63/215
central pontine myelinolysis, 47/588
cerebrovascular disease, 53/213
extrapyramidal disorder, 2/345
intracranial hemorrhage, 53/213
lesion site, 53/213
pontine hemorrhage, 12/43, 57
pontine infarction, 53/390
transient ischemic attack, 53/213
vertebrobasilar system syndrome, 53/390
Ocular bradykinesia
doll head eye phenomenon, 1/599
eye fixation, 1/599
Ocular convergence, *see* Eye convergence
Ocular dominance column
original finding, 2/542
Ocular dyskinesia
eye movement, 1/584
Ocular dysmetria
cerebellar tumor, 16/326
definition, 21/528
Gerstmann-Sträussler-Scheinker disease, 56/553
internuclear ophthalmoplegia, 2/329
Ocular fixation, *see* Eye fixation
Ocular flutter
cerebellar tumor, 16/326
opsoclonus, 21/528
posterior fossa disease, 60/657

Ocular hypertelorism, *see* Hypertelorism
Ocular imbalance
see also Abducent nerve injury, Oculomotor nerve
injury *and* Trochlear nerve injury
convergence weakness, 24/75
definition, 24/73
divergence paralysis, 24/75
enophthalmos, 24/91
exophthalmos, 24/89
eye convergence, 24/75
history, 24/73
internuclear ophthalmoplegia, 24/75
midbrain injury, 24/79
nasomaxillary fracture, 24/97
ocular muscle injury, 24/87
orbital fracture, 24/89
orbital hemorrhage, 24/87, 89
Parinaud syndrome, 24/74
skew deviation, 24/75
Ocular motility
metabolic encephalopathy, 69/400
Ocular movement, *see* Eye movement
Ocular muscle, *see* Eye muscle
Ocular muscle disorder
optovestibular, *see* Optovestibular ocular muscle
disorder
Ocular muscle injury
ocular imbalance, 24/87
Ocular myasthenia gravis
misdiagnosis, 2/323
progressive external ophthalmoplegia, 62/302
Ocular myoclonia
Mollaret triangle lesion, 2/344
Ocular myopathy
see also Progressive external ophthalmoplegia
curare sensitive, *see* Curare sensitive ocular
myopathy
hereditary motor and sensory neuropathy type I,
21/285
hereditary motor and sensory neuropathy type II,
21/285
muscular dystrophy, 1/612
ophthalmoplegia, 22/178
progressive bulbar palsy, 22/140
progressive external ophthalmoplegia, 22/186,
41/215, 62/288, 291
progressive external ophthalmoplegia
classification, 22/179, 62/290
Ocular myositis
see also Orbital myositis
cause, 2/323
classification, 22/179
Ocular pain

carotid artery stenosis, 55/111
Charlin neuralgia, 48/484
cluster headache, 48/241
nasociliary nerve, 48/484
temporal arteritis, 48/312
Ocular periodicity
eye movement, 1/585
Ocular plethysmography
cerebrovascular occlusive disease, 11/228
Ocular pseudoparesis
Brown syndrome, 2/327
Ocular pursuit
Charlevoix-Saguenay spastic ataxia, 60/657
hereditary ataxia, 60/657
olivopontocerebellar atrophy (Wadia-Swami),
60/657
Ocular reflex
vestibular, *see* Vestibulo-ocular reflex
Ocular retraction syndrome, *see* Acquired retraction
syndrome *and* Stilling-Türk-Duane syndrome
Ocular sign
chronic respiratory insufficiency, 38/290
hereditary motor and sensory neuropathy type I,
21/284
hereditary motor and sensory neuropathy type II,
21/284
metabolic encephalopathy, 69/400
olivopontocerebellar atrophy (Wadia-Swami),
60/492
Parkinson dementia, 49/172
Refsum disease, 66/489
respiratory encephalopathy, 38/290
sympathetic nervous disorder, 25/382
Ocular spasm
anatomic classification, 1/279
postencephalitic parkinsonism, 1/291, 6/748
Ocular tonography
cerebrovascular occlusive disease, 11/228
Oculoacousticocerebral degeneration, *see* Norrie
disease
Oculoauriculovertebral dysplasia, *see* Goldenhar
syndrome
Oculoauriculovertebral syndrome, *see* Goldenhar
syndrome
Oculocephalic reflex
brain stem death, 57/458
midbrain syndrome, 24/585
progressive supranuclear palsy, 22/219, 221,
49/241
Oculocerebral angiomatosis, *see* Brégeat syndrome
Oculocerebrocutaneous syndrome
athetosis, 43/444
Babinski sign, 43/444

hyperreflexia, 43/444
hypopigmentation, 43/444
mental deficiency, 43/444
microphthalmia, 43/444
Oculocerebrofacial syndrome
chorea, 43/280
mental deficiency, 43/280
microcornea, 43/280
micrognathia, 43/280
Oculocerebrorenal syndrome, *see* Lowe syndrome
Oculocraniosomatic neuromuscular disease, *see*
Progressive external ophthalmoplegia
Oculocutaneous albinism
autosomal dominant, 43/6
black lock, 43/6
brown, 43/6
Chédiak-Higashi syndrome, 14/783, 42/536
Forsius-Eriksson syndrome, 42/398
Hermansky-Pudlak syndrome, 43/5
heterochromia iridis, 42/404
mental deficiency, 43/3
myopia, 42/405
nystagmus, 42/404, 43/3-7
pachydermia, 43/5
photophobia, 42/404, 43/3-7
red reflex, 43/4
rufous, *see* Rufous oculocutaneous albinism
sensorineural deafness, 42/404, 43/3, 6
squamous carcinoma, 43/5
strabismus, 43/4, 7
tyrosinase negative, 43/4
tyrosinase positive, 43/5
visual acuity, 43/3
yellow mutant, 43/7
Oculocutaneous telangiectasia
progressive, *see* Progressive oculocutaneous
telangiectasia
Oculocutaneous tyrosinosis
Oregon type, 42/581
Richner-Hanhart syndrome, 42/582
Oculodentodigital dysplasia, *see*
Oculodento-osseous dysplasia
Oculodento-osseous dysplasia
cataract, 43/445
intracerebral calcification, 43/445
mental deficiency, 43/445
microphthalmia, 43/445
Oculodermal depigmentation
Vogt-Koyanagi-Harada syndrome, 56/612
Oculodermal melanocytosis, *see* Ota nevus
Oculofacial dystonia, *see* Progressive supranuclear
palsy
Oculofacial palsy, *see* Möbius syndrome

Oculofaciocervical dystonia, *see* Progressive
  supranuclear palsy
Oculogyric crisis
  see also Gaze spasm
  anatomic classification, 1/279
  blepharospasm, 1/583
  diazoxide intoxication, 37/433, 65/446
  extrapyramidal disorder, 2/344
  heterocyclic antidepressant, 65/319
  intestinal pseudo-obstruction, 60/650
  neuroleptic dystonia, 65/281
  nifedipine, 63/194
  postencephalitic, 1/284
  postencephalitic parkinsonism, 6/189, 748
  treatment, 6/220
  verapamil, 63/194
Oculomandibular reflex
  pure reflex, 1/410
Oculomandibular visual direction synergy
  synkinesis, 1/410
Oculomandibulodyscephaly, *see* Hallermann-Streiff
  syndrome
Oculomotor apraxia
  ataxia telangiectasia, 14/267, 283, 60/360, 387
  ataxic, *see* Ataxic oculomotor apraxia
  Balint syndrome, 45/15
  bilateral occipital lobe infarction, 53/413
  cardiac arrest, 63/216
  cerebellar ataxia, 42/329
  choreoathetosis, 60/656
  congenital, *see* Cogan syndrome type I
  constructional apraxia, 45/492
  developmental dyspraxia, 4/459
  eye movement, 1/605, 14/653
  gaze paralysis, 2/332
  Huntington chorea, 6/308, 49/280
  hysteria, 2/332
  operculum syndrome, 2/777, 781
  posterior cerebral artery syndrome, 53/413
  progressive supranuclear palsy, 49/243
  reading disability, 42/329
  spatial agnosia, 2/606
  Wieacker-Wolff syndrome, 60/656
Oculomotor degeneration
  hereditary spinodentatonigro, *see* Hereditary
    spinodentatonigro-oculomotor degeneration
Oculomotor disease
  metabolic encephalopathy, 45/123
  occipital lobe syndrome, 45/52
  optokinetic nystagmus, 45/52
Oculomotor disorder
  agnosia, 3/226
  Alzheimer disease, 46/252

doll head eye phenomenon, 2/306
Foville syndrome, 2/301, 304, 315-318
hereditary motor and sensory neuropathy type I,
  21/285
hereditary motor and sensory neuropathy type II,
  21/285
Parkinson dementia, 46/375
psychosis, 46/498
schizophrenia, 46/498, 505
Shy-Drager syndrome, 22/141
Stilling-Türk-Duane syndrome, 2/326
Oculomotor nerve
  see also Stilling-Türk-Duane syndrome
  autonomic fiber, 2/296, 74/401
  autonomic nervous system, 2/107
  Benedikt syndrome, 2/276, 298
  clinical examination, 57/136
  clinical features, 2/292
  course, 2/295
  diabetes mellitus, 27/117
  diagnosis, 2/292
  embryology, 30/396
  eye movement, 1/576, 583
  headache, 48/275
  intracranial hypertension, 16/124, 132
  lesion, 2/292
  migraine, 5/80
  neurinoma, 68/536
  nuclear lesion, 2/300-304
  osteitis deformans (Paget), 70/13
  parasympathetic fiber, 2/107
  pontine infarction, 12/16, 22
  ptosis, 57/136
  pupil, 1/623
  root lesion, 2/298-300
  sphenoidal sinus syndrome, 2/296
  subarachnoid hemorrhage, 55/17
  temporal arteritis, 55/342
  topographical anatomy, 2/295
  topography, 16/123
Oculomotor nerve abnormality
  see also Congenital ptosis, Marcus Gunn
    phenomenon *and* Stilling-Türk-Duane
    syndrome
  congenital ptosis, 30/401
  Marcus Gunn phenomenon, 30/401
  Stilling-Türk-Duane syndrome, 30/403
Oculomotor nerve agenesis
  congenital ptosis, 50/212
  cranial nerve agenesis, 50/212
  Marcus Gunn phenomenon, 50/212
  Möbius syndrome, 50/212
  nuclear aplasia, 50/212

oculopharyngeal muscular dystrophy, 50/212
Oculomotor nerve injury
  see also Ocular imbalance and Oculomotor
    paralysis
  aberrant regeneration, 24/64, 83
  acute subdural hematoma, 24/63
  anatomy, 24/59
  ciliary ganglion, 24/61
  clinical examination, 24/61, 80
  cruciate paralysis, 24/63
  delayed, 24/81, 84
  Edinger-Westphal nucleus, 24/59
  epidural hematoma, 24/63, 268
  fascicular lesion, 24/78
  head injury, 24/61
  history, 24/59
  incidence, 24/76
  interpeduncular syndrome, 2/298
  mechanism, 24/61, 81
  missile injury, 24/61
  nuclear lesion, 24/78
  orbital injury, 24/79, 82, 86
  pathology, 24/81
  penetrating, 24/61
  physiology, 24/59
  prognosis, 24/83
  sign, 24/60, 62, 80, 82
  site, 24/60
  skull fracture, 24/61, 81
  sympathetic innervation, 24/60
  symptom, 24/60, 62, 80, 82
  topical diagnosis, 24/80
  transtentorial herniation, 24/61, 64
  treatment, 24/64
Oculomotor neuropathy
  diabetes mellitus, 70/143
Oculomotor nucleus
  anatomy, 2/301
  cyclic oculomotor paralysis, 2/304
  familial amyotrophic lateral sclerosis, 59/246
  Fazio-Londe disease, 22/181
  opticocochleodentate atrophy, 22/174
  progressive bulbar palsy, 59/220
Oculomotor paralysis
  see also Oculomotor nerve injury,
    Ophthalmoplegia and Progressive external
    ophthalmoplegia
  amyotrophic lateral sclerosis, 22/295
  anisocoria, 74/410
  aqueduct stenosis, 50/311
  Benedikt syndrome, 1/187, 6/162
  brain lacunar infarction, 54/244
  cause, 2/319-322

concomitant strabismus, 2/291
cyclic, see Cyclic oculomotor paralysis
eyelid, 1/614
hereditary olivopontocerebellar atrophy (Menzel),
  21/443
involved muscles, 1/609-611
localization, 2/286-357
migraine, 48/156
neuromyelitis optica, 9/428
nuclear form, 54/245
ophthalmoplegia totalis, 2/293
ophthalmoplegic migraine, 48/132
orbital disease, 2/319-322
Parinaud syndrome, 2/304
prenuclear form, 54/245
pseudo-Von Gräfe sign, 2/293, 319
ptosis, 2/292
pupil, 1/623, 74/411
pure motor hemiplegia, 54/244
recurrent, see Recurrent oculomotor paralysis
scarlet fever, 7/487
spastic lid retraction, 1/614
spinopontine degeneration, 21/390
superior orbital fissure syndrome, 33/180
temporal arteritis, 48/313
terminology, 2/287, 288
Tolosa-Hunt syndrome, 31/147
transtentorial herniation, 1/559
traumatic carotid cavernous sinus fistula, 57/352,
  361
Oculomotor syndrome
  conjugate gaze, 2/278
Oculomotor system
  cerebrovascular disease, 55/120
  coordination, 2/274
  neuroanatomy, 2/274
Oculomotor tracking
  occipital lobe tumor, 67/155
Oculonasal syndrome, see Charlin neuralgia
Oculonuclear artery syndrome
  supply area, 2/304
Oculo-orbital angiomatosis, see Brégeat syndrome
Oculo-otocerebrorenal syndrome
  secondary pigmentary retinal degeneration,
    60/721
Oculopharyngeal amyotrophy
  progressive spinal muscular atrophy, 22/12
Oculopharyngeal muscular dystrophy
  anterior horn syndrome, 43/101
  areflexia, 43/100
  clinical features, 40/302, 60/49, 62/296
  congenital ptosis, 50/212
  definition, 62/295

differential diagnosis, 40/520
dysphagia, 43/100
electrophysiologic study, 60/50
esophagoscopy, 60/52
facial paralysis, 43/101
8-10 nm filament, 62/351
French-Canadian family, 22/185
genetics, 62/296
histochemistry, 40/47, 54
histopathology, 62/296
IgA, 43/101
IgG, 43/101
IgM, 43/101
inclusion body myositis, 62/297
Kearns-Sayre-Daroff-Shy syndrome, 40/302
laboratory investigation, 60/50
muscle biopsy, 60/51
muscle tissue culture, 62/102
myasthenia gravis, 62/415
myotonic dystrophy, 62/240
nerve conduction velocity, 43/101
nuclear inclusion, 62/296
oculomotor nerve agenesis, 50/212
pathogenesis, 60/52
pathologic reaction, 40/253
premature senescence, 40/199
progressive bulbar palsy, 59/375
progressive external ophthalmoplegia, 22/179,
    187, 206, 40/302, 41/215, 43/136, 60/49-52,
    62/295
progressive external ophthalmoplegia
    classification, 22/179, 62/290
ptosis, 43/100
rimmed muscle fiber vacuole, 40/253, 62/297
sporadic progressive external ophthalmoplegia
    with ragged red fiber, 62/508
Tomé-Fardeau body, 62/103
treatment, 60/52
Oculopharyngeal neuromuscular disease
bulbar paralysis, 22/140
Oculorenocerebellar syndrome
secondary pigmentary retinal degeneration,
    60/721
Oculosphygmography
cerebrovascular occlusive disease, 11/228
Oculosympathetic paralysis
carotid dissecting aneurysm, 53/40
cluster headache, 48/335
headache, 48/337
paratrigeminal syndrome, 48/332
Oculosympathetic pathway
autonomic nervous system, 74/409
headache, 48/331

paratrigeminal syndrome, 48/331
Oculovertebral syndrome
Goldenhar syndrome, 31/248
mental deficiency, 31/248
microphthalmia, 31/248
Oculovestibular reflex
brain concussion, 23/429
brain death, 24/719, 764
brain stem death, 57/458
general, 24/585
Oddi sphincter
bile flow, 75/640
Odontoid, see Dens
Odontoid agenesis, see Dens agenesis
Odontoid aplasia
glycosaminoglycanosis type IV, 66/304
Odontoid fracture
axis, 25/326
base, 25/325
cervical vertebral column injury, 61/63
displacement, 25/131
epiphysiolysis, 25/320
instability, 25/280
management, 61/63
mechanism, 25/130
nonfatal result, 26/290
nonunited, 25/280
process, 25/127, 280, 315-326
radiology, 61/518
surgical treatment, 61/63
symptom, 25/232
tomography, 25/323
Odontoid peg
compression, 71/23
Odor
see also Body odor
arousal, 74/365
autonomic nervous system, 74/367
cutaneous blood flow, 74/367
diphtheria, 65/237
disulfiram intoxication, 37/321
emotional memory, 74/371
heart rate, 74/367
human, 74/375
organophosphate intoxication, 64/170
phenylketonuria, 29/30, 42/610
relaxation, 74/365
respiratory frequency, 74/367
skin potential, 74/367
skin resistance, 74/367
skin temperature, 74/367
sweat, 74/375
sweat gland, 74/375

vaginal secretion, 74/378
Odor blindness, *see* Specific anosmia
Odorant
   see also Body odor
   arousal, 74/364
   autonomic nervous system, 74/364, 366
   autonomic nervous system response, 74/370
   basic, 74/372
   decision tree, 74/370
   emotion, 74/370, 372
   facial mimic, 74/368
   pleasant, 74/367
   relaxation, 74/365
   unpleasant, 74/367
   vomeronasal organ, 74/376
OFD I syndrome, *see* Orofaciodigital syndrome
   type I
OFD II syndrome, *see* Mohr syndrome
Ofloxacin
   benign intracranial hypertension, 67/111
Ogilvie syndrome
   autonomic nervous system, 75/633
Oguchi disease
   cataract, 13/262
   histology, 13/260
   mental deficiency, 13/262
   Mizuo phenomenon, 13/261
   myopia, 13/262
   optic disc, 13/261
Ohtahara syndrome, *see* Early infantile epileptic
   encephalopathy
Ohton, *see* Dimethylthiambutene
Oil red O staining
   lipid, 9/25, 28
   skeletal muscle, 40/4
Okadaic acid
   diarrheic shellfish poisoning, 65/165
   dinophysiotoxin, 65/149
   marine toxin intoxication, 65/149
   neurotoxin, 65/142
   pectenotoxin, 65/149
   protein phosphatase inhibition, 65/150
   structure, 65/150
   toxic dose, 65/149
   tumor promotor, 65/150
   yessotoxin, 65/149
OKT3
   adverse effect, 69/498
   agitation, 63/536
   akinesia, 63/536
   asterixis, 63/536
   ataxia, 63/536
   CSF pleocytosis, 63/536

delirium, 63/536
epilepsy, 63/536
headache, 63/536
immunosuppressive therapy, 65/550
meningism, 63/536
meningitis, 63/536
meningitis serosa, 63/536
paranoia, 63/536
psychosis, 63/536
renal transplantation, 63/534, 536
OKT3 encephalopathy
   symptom, 63/536
OKT3 intoxication
   epilepsy, 65/560
   meningitis, 65/560
Old age
   astrocytoma, 18/18
   autonomic nervous system, 74/3
   bladder, 75/669
   sympathetic nervous disorder, 2/121
Oldberg, E., 1/1, 24
O'Leary, J.L., 1/36
Oleic acid
   biochemistry, 66/35
   chemistry, 10/235
   migraine, 48/95
   myopathy, 40/158
   plasma, 48/96
Olfaction
   see also Anosmia *and* Smell
   anatomy, 24/2
   arousal, 74/364
   autonomic nervous system, 74/363, 366
   emotion, 74/363
   humidity, 74/363
   mechanism, 24/3
   menstrual cycle, 74/377
   migraine, 48/162
   mother-infant, 74/378
   olfactory cortex, 24/2, 8
   pathway, 24/2, 8, 20
   pheromonal perception, 74/380
   physiology, 24/2
   primary odor, 24/3
   regeneration, 24/11
   sex difference, 74/380
   sexual activity, 74/377
   steroid, 74/380
   stimulation, 74/364
   temperature, 74/363
   test, 24/4
Olfactogenital dysplasia, *see* Kallmann syndrome
Olfactory agnosia, 4/14

Olfactory bulb
  anosmia, 24/8
  basal ganglion tier II, 49/21
  homocystinuria, 55/330
  human, 74/380
  Rubinstein-Taybi syndrome, 30/472
  ventral globus pallidus, 49/20
  ventral neostriatum, 49/20
Olfactory bulb agenesis
  corpus callosum, 50/178
  epidemiology, 30/150
  mental deficiency, 43/256
  trigonocephaly, 43/486
Olfactory bulb hypoplasia
  homocystinuria, 55/330
Olfactory fissure syndrome
  anterior ethmoidal nerve, 5/217
  treatment, 5/217
Olfactory hallucination
  diagnostic symptom, 4/14
  dreamy state, 2/702
  temporal lobe tumor, 17/286
Olfactory illusion
  epileptic, 15/95
  hyperosmia, 73/72
  parosmia, 73/72
  temporal lobe epilepsy, 73/72
Olfactory nerve
  abnormality, *see* Orbital hypertelorism
  acute amebic meningoencephalitis, 52/323, 326
  anosmia, 2/52
  diabetes mellitus, 27/118
  embryology, 30/395
  Foster Kennedy syndrome, 2/53, 570
  granulomatous amebic encephalitis, 52/325
  localization, 2/52-54
  Naegleria fowleri, 52/323
  neurinoma, 68/536
  neuroamebiasis, 52/314
  orbital apex syndrome, 2/570
  osteitis deformans (Paget), 70/12
  primary amebic meningoencephalitis, 52/316
  radiation injury, 7/397
Olfactory nerve aplasia
  Edwards syndrome, 50/562
Olfactory neuroblastoma
  adult, 8/472
  epilepsy, 72/180
  nose, 8/472
Olfactory stimulation
  epilepsy, 15/447
Olfactory system
  stress, 74/313

  vomeronasal organ, 74/376
Oligoastrocyte
  terminology, 18/86
Oligoclonal banding
  progressive systemic sclerosis, 51/448
Oligoclonal IgG band
  CSF, *see* CSF oligoclonal IgG
Oligodactyly
  De Lange syndrome, 50/276
Oligodendroblastoma, *see* Oligodendroglioma
Oligodendrocyte
  autoantigen, 9/505
  butyrylcholinesterase, 9/14
  cell to cell coupling, 47/12
  cuprizone intoxication, 64/17
  cytology, 18/84
  damage, 47/511
  demyelination, 47/551
  encephalitogenic myelin basic protein, 9/511
  globoid cell leukodystrophy, 10/77
  lead, 64/437
  lead intoxication, 64/437
  Marchiafava-Bignami disease, 47/557
  multiple sclerosis, 47/213, 222-230, 241-244
  multiple sclerosis plaque, 9/38
  nerve regeneration, 61/481
  neuropil, 47/1, 7, 12
  nuclear alteration, 47/508
  oligodendroglial cytoplasm, 47/6
  oxidative enzyme, 9/628
  Pelizaeus-Merzbacher disease, 66/560
  pentose shunt, 9/14
  plaque, 9/274-279
  progressive multifocal leukoencephalopathy,
    9/486, 34/310, 39/524, 46/388, 47/508, 511-513,
    56/514
  radiation injury, 23/646
  radiation myelopathy, 26/93, 61/205
  respiration, 9/624
Oligodendrocyte antibody
  multiple sclerosis, 47/109
Oligodendrocyte hyperplasia
  multiple sclerosis, 47/224, 244
Oligodendrocyte inclusion body
  cuprizone, 64/17
  homidium bromide, 64/17
  homidium bromide intoxication, 64/17
Oligodendrocytoma, *see* Oligodendroglioma
Oligodendroglia cell, *see* Oligodendrocyte
Oligodendroglial cytoplasm
  oligodendrocyte, 47/6
Oligodendroglioma, 18/81-102
  see also Optic chiasm glioma

age, 18/90
alkylating agent, 68/132
anatomic growth, 18/83
arachnoidal metastasis, 16/13
biopsy, 18/87, 67/227
blood vessel, 18/87
brain angiography, 18/95
brain biopsy, 16/713, 715-717
brain scintigraphy, 18/99
brain tumor, 18/502
calcification, 18/86
cerebellar medulloblastoma, 18/178
chemosensitivity, 68/132
chemotherapy, 68/129
chromosome, 67/50
chromosome abnormality, 67/45
classification, 16/4, 7, 18/5, 68/12
clinical features, 68/124
congenital, 31/47
CSF cytology, 16/396
differential diagnosis, 18/12, 20/333
distribution, 67/130
DNA ploidy, 68/127
EEG, 18/91
epidemiology, 68/123
epilepsy, 18/20, 91, 72/180, 357
frontal lobe, 18/82
glioblastoma, 72/357
grade, 68/12-16, 126
hemorrhage, 68/95
histogenesis, 67/10
histology, 18/84, 20/333, 68/125
history, 18/81
honeycomb structure, 16/13, 22, 20/333
imaging, 67/169, 172
infant, 42/733
macroscopic aspect, 18/82
malignancy, 16/22, 18/88
malignant change, 18/16
marker, 68/125
metastasis, 72/357
midline, 18/82
mixed glioma, 68/127
mode of onset, 18/90
molecular basis, 68/127
multiple neurofibromatosis, 43/35
neuroastrocytoma, 72/357
optic, 42/733
optic nerve tumor, 17/355, 359
optic pathway tumor, 68/74
PEG, 18/95
peritumoral zone, 18/87
plain roentgenogram, 18/92

pleomorphic xanthoastrocytoma, 72/357
pleomorphism, 20/335
polymorphism, 16/14, 22
pons, 42/332
postoperative radiotherapy, 18/100
preferential site, 18/82
prevalence, 18/90
prognosis, 68/128
pseudopalisading, 16/14
radiologic features, 68/125
radionucleotide diagnosis, 18/99
radiosensitivity, 18/100
radiotherapy, 18/38, 502, 68/128, 130
recurrence, 18/88, 101
retractile tumor, 18/95
Ringertz classification, 18/5
Rosenthal fiber, 20/335
satellitosis, 16/12
schwannoid, *see* Schwannoid oligodendroglioma
spinal, *see* Spinal oligodendroglioma
spinal cord, 18/83
spinal cord compression, 19/361
spinal ependymoma, 20/362
spinal glioma, 19/359, 361, 20/332
spread, 18/82
subtentorial, 18/83
supratentorial brain tumor, 18/317
surgery, 18/99, 68/128
survey, 68/123
survival, 18/101
symptom, 18/90-92
synonym, 20/333
thalamic tumor, 68/65
tissue culture, 16/38, 17/65, 18/84
treatment, 18/99, 68/128
ultrastructure, 18/86
Oligodendrogliomatosis
  meningeal, *see* Meningeal oligodendrogliomatosis
Oligodendroma, *see* Oligodendroglioma
Oligohydramnios
  arthrogryposis multiplex congenita, 42/74
  caudal aplasia, 50/513
  prune belly syndrome, 43/467, 50/515
Oligophrenia, *see* Mental deficiency
Oligophrenic cerebellolental degeneration
  hereditary, *see* Marinesco-Sjögren syndrome
Oligosaccharide
  degradation, 66/63
Oligosaccharidosis
  blindness, 66/329
  bone marrow transplantation, 66/329
  classification, 66/378
  demyelination, 66/329

genotype, 66/329
glycoprotein, 66/329
glycoprotein degradation, 66/62
hepatosplenomegaly, 66/329
Oligosymptomatic hyperthyroidism
Tolosa-Hunt syndrome, 48/303
Olisthesis, *see* Anterolisthesis *and* Retrolisthesis
Olivary hypertrophy
inferior, *see* Inferior olivary hypertrophy
Olivary nucleus
cerebellar vermis agenesis, 42/5
hereditary cerebello-olivary atrophy (Holmes),
21/407
inferior, *see* Inferior olivary nucleus
Olivary nucleus heterotopia
inferior, *see* Inferior olivary nucleus heterotopia
Olive
inferior, *see* Inferior olive
Olivecrona, H., 1/12, 17
Oliver-McFarlane syndrome
secondary pigmentary retinal degeneration,
60/721
Olivocerebellar atrophy
hereditary cerebello-olivary atrophy (Holmes),
21/18
striatal lesion, 6/669
striatonigral degeneration, 6/699
thalamic degeneration, 2/483
thalamic lesion, 2/489
Olivocochlear tract
hypopigmentation, 43/43
Olivopontocerebellar atrophy
amyotrophic lateral sclerosis, 22/371, 59/411
autonomic dysfunction, 49/206, 63/146
autonomic nervous system, 22/265
biochemical study, 60/552, 559
bulbar paralysis, 60/658
Carter-Sukavajana syndrome, 21/411
cerebellar ataxia, 42/161-163
cerebellar atrophy, 42/161
cerebello-olivary atrophy, 60/571
classification, 21/20
clinical features, 60/552
congenital, *see* Congenital olivopontocerebellar
atrophy
dementia, 21/397, 59/411
dentatorubropallidoluysian atrophy, 21/521, 526,
49/443
dominant, *see* Dominant olivopontocerebellar
atrophy
dyssynergia cerebellaris myoclonica, 60/599
enzyme deficiency polyneuropathy, 51/388
familial spastic paraplegia, 21/400

Ferguson-Critchley syndrome, 59/322
forgotten respiration, 63/488
Friedreich ataxia, 21/21, 353, 22/172
glutamate dehydrogenase, 60/553
glutamate dehydrogenase deficiency, 49/108, 130,
209, 60/551-566
glutamic acid receptor, 60/563
Greenfield classification, 21/397, 42/161
Hallervorden-Spatz syndrome, 49/401
hastening phenomenon, 49/68
hereditary sensory and autonomic neuropathy type
I, 21/11
hereditary spastic ataxia, 21/369, 60/463
history, 60/551
Huntington chorea, 49/293
insomnia, 49/245
Joseph-Machado disease, 60/476
Kii Peninsula amyotrophic lateral sclerosis,
22/371
Konigsmark, 21/397, 42/161
Menzel type, *see* Hereditary olivopontocerebellar
atrophy (Menzel)
mixed spinocerebellar type, 21/416
multiple neuronal system degeneration, 59/137,
75/161
myoclonus, 60/599
neuropathologic study, 60/556-559
nonhereditary, *see* Olivopontocerebellar atrophy
(Dejerine-Thomas)
nuclear magnetic resonance, 60/673
nystagmus, 42/161
ophthalmoplegia, 22/172-174
optic atrophy, 42/161
orthostatic hypotension, 2/121, 43/66, 63/146
pallidal degeneration, 49/454
pallidoluysian atrophy, 22/172
pallidoluysian degeneration, 49/454
pallidoluysiodentate degeneration, 49/452, 456
pallidoluysionigral degeneration, 49/446
Parkinson disease, 49/94, 108, 130, 206
parkinsonism, 21/397, 526, 49/130, 206
pathogenesis, 60/560
progressive dysautonomia, 59/137, 141
progressive external ophthalmoplegia, 43/136,
62/304
Purkinje cell loss, 42/161
recessive, *see* Recessive olivopontocerebellar
atrophy
respiratory failure, 49/208
retinal degeneration, *see* Olivopontocerebellar
atrophy with retinal degeneration
rigidity, 49/69
secondary pigmentary retinal degeneration,

13/305, 60/735
sensory loss, 42/161
Shy-Drager syndrome, 49/211
slow eye movement, 42/187
spinopontine degeneration, 21/397
sporadic, *see* Olivopontocerebellar atrophy
   (Dejerine-Thomas)
striatal lesion, 6/669
striatonigral degeneration, 49/205, 208-211,
   60/540
supranuclear ophthalmoplegia, 60/657
thalamic lesion, 2/489
tremor, 21/397, 455
type I, *see* Hereditary olivopontocerebellar
   atrophy (Menzel)
type II, *see* Olivopontocerebellar atrophy
   (Fickler-Winkler)
type III, 42/162
type IV, *see* Olivopontocerebellar atrophy
   (Schut-Haymaker)
type V, 43/163
X-linked recessive ataxia, 60/638
Olivopontocerebellar atrophy (Dejerine-Thomas),
   21/415-429, 60/511-516, 519-535
age at onset, 21/424, 60/514
Alzheimer disease, 60/516
amyotrophic lateral sclerosis, 60/531
animal model, 21/418
arcuate nucleus, 21/418
ataxia, 21/425
basal ganglion, 21/423
biochemical study, 60/513
brain cortex, 21/423
brain stem, 21/420
cause, 21/418
cerebellum, 21/421
chorea, 60/663
choreoathetosis, 21/426
classification, 21/415-417
clinical diagnosis, 60/520
clinical features, 21/424, 60/514, 521
computer assisted tomography, 60/514-516, 519,
   522-528, 671
Creutzfeldt-Jakob disease, 21/428, 60/534
CSF, 21/427
dementia, 21/426, 60/665
depression, 60/665
description, 60/511
diagnosis, 21/427
differential diagnosis, 21/428, 60/514, 528
distal amyotrophy, 60/659
duration of illness, 21/424
dysarthria, 21/425

dyssynergia cerebellaris myoclonica, 60/532
dystonia, 60/664
earlier report, 21/415
ECG, 21/427
EEG, 21/427
electrophysiologic study, 60/514
epidemiology, 60/521
Essick cell band, 21/418
external ophthalmoplegia, 60/656
fasciculation, 60/659
Friedreich ataxia, 60/531
gait, 21/424
history, 21/417, 60/511
hypoesthesia, 60/661
inferior olivary nucleus, 21/418
irradiation encephalopathy, 60/533
lateral reticular nucleus, 21/418
Marie-Foix-Alajouanine disease, 21/416
mesencephalon, 21/425
multiple sclerosis, 60/516, 533
neoplasm, 60/516
nerve conduction velocity, 60/670
neuropathology, 21/419-424, 60/512, 516
neuropathy, 60/660
nigral inclusion body, 21/423
nosology, 60/512
nuclear magnetic resonance, 60/519, 522
optic atrophy, 60/655
orthostatic hypotension, 60/667
Parkinson disease, 60/516, 528
parkinsonism, 21/425
pathogenesis, 21/418
PEG, 21/427
pontine nucleus, 21/418
ponto-olivary atrophy, 21/418
posterior fossa lesion, 60/516
progressive external ophthalmoplegia, 60/656
progressive supranuclear palsy, 49/246, 60/530
pyramidal sign, 21/426
Ramsay Hunt syndrome, 60/532
Refsum disease, 8/23, 60/661
Shy-Drager syndrome, 60/530
spastic dysphonia, 21/425
sphincter disturbance, 21/427
spinal cord, 21/424
static tremor, 21/425
striatonigral degeneration, 21/417, 60/528
thalamus, 21/425
treatment, 21/429
tremor, 60/661
vertical gaze palsy, 60/656
Olivopontocerebellar atrophy (Fickler-Winkler)
   autosomal recessive, 42/162

hereditary spinocerebellar degeneration, 21/18,
    42/162
  rare type, 42/162
Olivopontocerebellar atrophy (Menzel), *see*
    Hereditary olivopontocerebellar atrophy
    (Menzel)
Olivopontocerebellar atrophy (Schut-Haymaker)
  see also Hereditary olivopontocerebellar atrophy
  computer assisted tomography, 60/671
  cranial nerve, 42/162
  depression, 60/665
  genetic linkage, 42/163
  hereditary spinocerebellar degeneration, 21/18
  HLA antigen, 42/162
  hypoesthesia, 60/661
  nuclear magnetic resonance, 60/673
  optic atrophy, 60/655
  positron emission tomography, 60/673
Olivopontocerebellar atrophy variant, 21/451-456
  amyotrophy, 21/452, 455
  athetosis, 21/452
  autosomal dominant, 21/454
  Behr disease, 21/454
  blindness, 21/453
  Carpenter-Schumacher family, 21/454
  chorea, 21/453
  choreoathetosis, 21/456
  clinical features, 21/454
  congenital retinal blindness, 21/454
  cranial neuropathy, 21/455
  dementia, 21/452, 455
  external ophthalmoplegia, 21/452
  genetics, 21/454
  grimacing, 21/452-454, 456
  internuclear ophthalmoplegia, 21/453
  Leber optic atrophy, 21/454
  macular degeneration, 21/452
  muscular atrophy, 22/171
  myoclonus, 21/452
  nystagmus, 21/455, 42/161
  ophthalmoplegia, 21/452
  optic atrophy, 21/451-453
  parkinsonism, 21/453
  pathologic laughing and crying, 21/456
  secondary pigmentary retinal degeneration,
    21/452
  spinopontine degeneration, 22/172
  unilateral case, 21/452
  unilateral type, 21/455
Olivopontocerebellar atrophy (Wadia-Swami),
    60/491-503
  see also Hereditary olivopontocerebellar atrophy
  action myoclonus, 60/494

  age at onset, 60/494
  areflexia, 60/660
  basilar impression, 60/497
  bulbar paralysis, 60/494
  cataract, 60/653
  cerebellar disorder, 60/492
  choreoathetosis, 60/663
  chronic multiple sclerosis, 60/497
  clinical features, 60/492
  clinicopathologic correlation, 60/495-497
  computer assisted tomography, 60/499, 671
  course, 60/494
  dementia, 60/494, 665
  differential diagnosis, 60/497
  distal amyotrophy, 60/494, 659
  distal hypoesthesia, 60/494
  dyssynergia cerebellaris myoclonica, 60/665
  dystonia, 60/494
  electrophysiologic study, 60/500-503
  epidemiology, 60/494
  facial weakness, 60/494
  hereditary cerebello-olivary atrophy (Holmes),
    60/497
  hereditary olivopontocerebellar atrophy, 21/443
  history, 60/491
  hypoesthesia, 60/661
  inheritance mode, 60/494
  kyphoscoliosis, 60/669
  macular degeneration, 60/494
  muscle stretch reflex, 60/493
  myoclonus, 60/662
  narcolepsy, 60/494
  neoplasm, 60/497
  nerve biopsy, 60/501
  nerve conduction velocity, 60/670
  neuropathology, 60/495
  neuropathy, 60/660
  nuclear magnetic resonance, 60/499
  ocular pursuit, 60/657
  ocular sign, 60/492
  oculography, 60/499, 502
  optic atrophy, 60/494, 655
  orthostatic hypotension, 60/494, 667
  ota nevus, 60/667
  pes cavus, 60/669
  Ramsay Hunt syndrome, 60/497
  Shy-Drager syndrome, 60/497
  treatment, 60/503
  vertical gaze palsy, 60/656
Olivopontocerebellar atrophy with retinal
    degeneration, 60/505-508
  internuclear ophthalmoplegia, 60/507, 657
  nystagmus, 60/507

ocular symptom, 60/507
ophthalmoplegia, 60/507
optic atrophy, 60/507
optokinetic nystagmus, 60/507
ptosis, 60/507
retinal pigment dispersion, 60/507
secondary pigmentary retinal degeneration,
    60/653
tremor, 60/661
visual acuity loss, 60/507
Olivopontocerebellar atrophy with slow eye
    movement, *see* Olivopontocerebellar atrophy
    (Wadia-Swami)
Olivopontocerebellar degeneration, *see*
    Olivopontocerebellar atrophy
Ollier disease
    fibrous dysplasia, 14/192
    Maffucci-Kast syndrome, 43/29
Ollier-Klippel-Trénaunay syndrome, *see*
    Klippel-Trénaunay syndrome
Olympics for the Paralysed
    rehabilitative sport, 26/528
Omega oxidation
    phytanic acid, 27/525
    Refsum disease, 13/314, 27/520
Omega oxidation block
    Refsum disease, 51/387
Ommaya reservoir
    Burkitt lymphoma, 63/352
    Candida meningitis, 52/404
    coccidioidomycosis, 35/451, 52/422
    hydrocephalus, 30/552
    meningeal leukemia, 63/342
    staphylococcal meningitis, 52/73
OMPA intoxication, *see* Octamethyl
    pyrophosphoramide intoxication
Omphalocele
    anencephaly, 50/84
    caudal aplasia, 50/512
    Patau syndrome, 31/504, 507, 50/558, 560
    renal agenesis syndrome, 50/512
Omsk hemorrhagic fever
    acute viral encephalitis, 34/77
Onchocerciasis
    filariasis, 35/161
    secondary pigmentary retinal degeneration,
        13/218
Oncogene
    brain metastasis, 69/108
    retrovirus, 56/17
    tumor cell, 67/18
Oncogenic virus
    brain tumor, 17/4

brain tumor carcinogenesis, 17/4
    simian virus 40, 18/132
Oncovirus
    retrovirus, 56/529
Ondine curse, *see* Forgotten respiration
One and one half syndrome
    brain lacunar infarction, 54/246
    pontine lacuna, 54/245
    symptom, 53/389
One eyed vertebra
    radiologic finding, 20/22
    spinal tumor, 20/22
Onion bulb formation
    carcinogenic agent, 21/252
    chronic inflammatory demyelinating
        polyradiculoneuropathy, 51/537
    chronic renal disease, 21/252
    Cockayne-Neill-Dingwall disease, 51/399
    diabetes mellitus, 21/252
    diabetic polyneuropathy, 21/163, 51/511
    diphtheria antitoxin, 21/252
    diphtheria toxin, 21/252
    experimental allergic neuritis, 21/252
    hereditary motor and sensory neuropathy type I,
        42/340
    hypertrophic interstitial neuropathy, 42/317, 51/71
    lead intoxication, 21/252
    neurofibromatosis, 21/252
    Refsum disease, 8/23, 21/9, 12, 233, 240, 248,
        251, 254, 257, 51/385
    Roussy-Lévy syndrome, 21/177
    Rud syndrome, 51/398
    ultrastructure, 21/154-158
Onion skin pattern, *see* Trigeminal descending
    spinal tract lesion
Onset zone
    definition, 73/386
Onuf nucleus
    autonomic nervous system, 74/70
    orthostatic hypotension, 63/146
    striatonigral degeneration, 49/209
Onufrowicz nucleus
    familial amyotrophic lateral sclerosis, 59/246
Onychodystrophy
    hereditary deafness, 50/218
    sensorineural deafness, 42/384-386
Onychotrichodysplasia
    mental deficiency, 43/289
O'nyong-nyong virus
    alphavirus, 56/12
Opacity
    corneal, *see* Corneal opacity
    hereditary vitreoretinal degeneration, 13/178

lenticular, *see* Lenticular opacity
vitreous, *see* Vitreous opacity
Opalski cell
    acquired hepatocerebral degeneration, 6/280,
      46/338
    features, 6/291
    Wilson disease, 6/268, 289
Opalski syndrome
    central motor disorder, 1/189
    hemiplegia, 1/189
    pyramidal tract lesion, 1/189
Open heart surgery
    see also Cardiac surgery
    anoxic ischemic leukoencephalopathy, 9/578
    brain arteriolar dilatation, 63/175
    brain blood flow, 63/175
    brain capillary dilatation, 63/175
    brain embolism, 55/168, 63/175
    brain infarction, 53/165
    brain ischemia, 53/165
    brain microembolism, 63/175
    brain perfusion pressure, 63/188
    calcific embolus, 63/187
    cerebellar infarction, 63/8, 187
    coma, 63/187
    encephalopathy, 63/187
    iatrogenic brain infarction, 53/165
    iatrogenic neurological disease, 63/175
    neurologic complication, 63/175
    neurologic complication prophylaxis, 63/188
    postoperative dysrhythmia, 63/186
    prior stroke, 63/186
    prior stroke significance, 63/188
    pump time *vs* stroke risk, 63/186
    regional brain blood flow, 63/175
    stroke, 63/8
    stroke risk, 63/186
Operculum
    anatomy, 2/776
    blood supply, 2/782
    connection, 2/776
    function, 2/776
    parietal, *see* Parietal operculum
    pars opercularis, 2/776
    pars orbitalis, 2/776
    pars triangularis, 2/776
Operculum hypoplasia
    Down syndrome, 50/531
Operculum syndrome, 2/776-782
    see also Pseudobulbar paralysis
    anarthria, 2/781
    anterior type, 2/781
    Bell sign, 2/781

brain infarction, 53/359
brain lacunar infarction, 54/243
cause, 2/781
conjugate deviation, 2/781
dysmasesia, 2/779
dysphagia, 2/779
hemiplegie à bascule, 2/781
middle cerebral artery syndrome, 53/359
oculomotor apraxia, 2/777, 781
pathologic laughing and crying, 2/780
posterior type, 2/781
pseudobulbar paralysis, 2/777, 779
pseudoperipheral facial palsy, 53/359
symptomatology, 2/779
Ophthalmia
    sympathetic, *see* Sympathetic ophthalmia
Ophthalmic artery
    anastomosis, 53/296
    anatomy, 53/296
    blood supply, 48/317
    brain aneurysm, 12/91
    branch, 53/295
    carotid artery anastomosis, 12/308
    carotid artery stenosis, 54/2
    carotid cavernous fistula, 24/403, 425
    carotid rete mirabile, 12/308-311
    congenital aneurysm, 31/153
    migraine, 48/135
    radioanatomy, 11/69
    sympathetic innervation, 5/335
    temporal arteritis, 11/133
Ophthalmic nerve
    Gradenigo syndrome, 2/91, 313
    paratrigeminal syndrome, 48/334
Ophthalmic vein
    anatomy, 11/53
    inferior, *see* Inferior ophthalmic vein
    superior, *see* Superior ophthalmic vein
Ophthalmodynamometry, 11/223-228
    apparatus, 11/225
    brain embolism, 53/219
    brain infarction, 53/219
    carotid artery patency, 11/210
    carotid artery thrombosis, 26/45
    cerebrovascular disease, 53/219
    complication, 11/227
    contraindication, 11/227
    direct method, 11/226
    history, 11/224
    indication, 11/225
    indirect method, 11/227
    interpretation, 11/227
    intracranial hemorrhage, 53/219

20

papilledema, 16/273
suction, *see* Suction ophthalmodynamometry
technique, 11/225
theory, 11/224
transient ischemic attack, 53/219, 268
Ophthalmo-otoencephalica
heredopathia, *see* Heredopathia
ophthalmo-otoencephalica
Ophthalmoparesis, *see* Ophthalmoplegia
Ophthalmopathy, *see* Thyroid associated
ophthalmopathy
Ophthalmoplegia
see also Oculomotor paralysis *and* Progressive
external ophthalmoplegia
Aarskog syndrome, 43/314
abiotrophic external, *see* Abiotrophic external
ophthalmoplegia
abnormal muscle mitochondria, 43/137
adducted thumb syndrome, 43/331
aminoaciduria, 43/125-127
amnesic shellfish poisoning, 65/167
amyotrophic lateral sclerosis, 59/232, 411
aqueduct stenosis, 42/57
bacterial endocarditis, 52/299
Barnard-Scholz syndrome, 22/471
Bassen-Kornzweig syndrome, 22/191, 51/327,
394
Batten disease, 10/663
Bielschowsky relapsing alternating, *see*
Bielschowsky relapsing alternating
ophthalmoplegia
carotid cavernous fistula, 24/415
cause, 2/294
cavernous sinus thrombosis, 54/398
cephalic tetanus, 7/476
cerebellar ataxia, 42/132
chloroma, 63/343
chronic progressive external, *see* Chronic
progressive external ophthalmoplegia
complex I deficiency, 62/504, 66/422
complex IV deficiency, 66/423
complicated migraine, 5/68, 42/747
congenital, *see* Congenital ophthalmoplegia
congenital acetylcholine receptor deficiency/short
channel opentime syndrome, 62/437
congenital ptosis, 62/297
cryptococcosis, 52/431
curare sensitive, 43/135
dementia, 59/411
dental abnormality, 43/141
diabetes mellitus, 27/117
diabetic neuropathy, 27/117
dialysis, 63/528

dialysis encephalopathy, 63/528
dysphagia, 43/138
dysthyroid, *see* Dysthyroid ophthalmoplegia
encephalitis lethargica, 22/180
Epstein-Barr virus, 56/253
exophthalmos, 2/320
external, *see* External ophthalmoplegia
extrinsic, 2/293
facial paralysis, 43/138, 145
familial intermittent, *see* Familial intermittent
ophthalmoplegia
familial relapsing, *see* Familial relapsing
ophthalmoplegia
familial spastic paraplegia, *see* Ferguson-Critchley
syndrome
Fazio-Londe disease, 42/95
Ferguson-Critchley syndrome, 22/191, 471
Fisher syndrome, 22/180
Friedreich ataxia, 2/326, 13/304, 21/347, 355,
22/191
gold encephalopathy, 51/307
Guillain-Barré syndrome, 22/180, 51/245
headache, 65/178
heart disease, 43/141
hereditary cerebellar ataxia, 21/575, 22/191
hereditary cranial nerve palsy, 60/43
hereditary motor and sensory neuropathy type I,
21/285, 308, 42/339
hereditary motor and sensory neuropathy type II,
21/285, 308, 42/339
hereditary spastic ataxia, 21/383
herpes zoster, 56/231
Horner syndrome, 22/209
hypogonadism, 43/138-140
infantile hypotonia, 43/125-127
infantile spinal muscular atrophy, 22/180
internuclear, *see* Internuclear ophthalmoplegia
intrinsic factor, 2/293
Joseph-Machado disease, 42/262
juvenile Gaucher disease, 10/307
Kearns-Sayre-Daroff-Shy syndrome, 22/179,
38/224, 229, 43/142
late cerebellar atrophy, 60/583
malnutrition, 28/20
maple syrup urine disease, 42/599
masseter muscle, 43/138
median longitudinal fasciculus syndrome, 42/321
mesencephalic syndrome, 2/276
metabolic encephalopathy, 69/400
metachromatic leukodystrophy, 42/493
Miller Fisher syndrome, 51/246
miosis, 43/145
mitochondrial disease, 40/33, 43/137

mitochondrial myopathy, 41/214
Möbius syndrome, 22/179, 43/134, 136
mucormycosis, 35/551, 52/471
multiple sclerosis, 9/167
muscle weakness, 40/298
muscular dystrophy, 22/180, 43/139
myopathy, 43/125-127
myopia, 43/141
myotonic dystrophy, 22/206
myotubular myopathy, 43/113
nemaline myopathy, 62/342
neural disease, 22/190
neurovisceral storage disease, 42/604-606
nigrospinodentate degeneration, 21/401, 22/158,
    164, 170
nuclear, see Nuclear ophthalmoplegia
nuclear progressive, see Progressive external
    ophthalmoplegia
ocular myopathy, 22/178
olivopontocerebellar atrophy, 22/172-174
olivopontocerebellar atrophy variant, 21/452
olivopontocerebellar atrophy with retinal
    degeneration, 60/507
painful, see Painful ophthalmoplegia
pallidoluysionigral degeneration, 49/456
partial, 2/293
phenytoin intoxication, 65/500
polyneuropathy, 22/194
progressive, see Progressive ophthalmoplegia
progressive external, see Progressive external
    ophthalmoplegia
progressive rubella panencephalitis, 56/409
progressive supranuclear palsy, 22/192, 219,
    42/259, 46/301, 49/240
pseudo, see Oculomotor apraxia
ptosis, 22/185, 43/138, 145
pyruvate decarboxylase deficiency, 62/502
pyruvate dehydrogenase complex deficiency,
    66/416
Refsum disease, 21/211, 22/195, 60/228, 66/489
retinal degeneration, 43/141
secondary pigmentary retinal degeneration,
    22/191-193, 310-314
sensorineural deafness, 42/605, 43/138
snake envenomation, 65/178
spastic paraplegia, 22/191, 471
speech disorder, 43/139
sporadic progressive external ophthalmoplegia
    with ragged red fiber, 62/508
subacute necrotizing encephalomyelopathy,
    42/588
supranuclear, see Supranuclear ophthalmoplegia
synkinesis, 2/294

temporal arteritis, 5/120
tetanus, 7/476
Tolosa-Hunt syndrome, 48/296, 299, 302
traumatic carotid cavernous sinus fistula, 57/352
trichinosis, 52/567
tuberculous meningitis, 52/202
vertical supranuclear, see Vertical supranuclear
    ophthalmoplegia
vincristine polyneuropathy, 51/299
visceral myopathy, 51/493
Wegener granulomatosis, 71/177
Wernicke encephalopathy, 28/17
Whipple disease, 52/137
Wohlfart-Kugelberg-Welander disease, 22/70, 73,
    180, 192
Ophthalmoplegia plus syndrome, see Progressive
    external ophthalmoplegia
Ophthalmoplegia totalis
  nystagmus, 43/145
  oculomotor paralysis, 2/293
Ophthalmoplegic migraine
  aneurysm site, 12/89
  child, 42/748, 48/37
  classification, 48/6, 141
  clinical picture, 5/72
  differential diagnosis, 8/228
  diplopia, 48/132
  features, 48/131
  hereditary cranial nerve palsy, 60/42
  history, 5/69-72
  Melkersson-Rosenthal syndrome, 8/228
  oculomotor paralysis, 48/132
  orbital myositis, 62/301
  pathology, 5/73
  pathophysiology, 5/73
  posterior cerebral artery, 2/295
  progressive external ophthalmoplegia, 2/324
  ptosis, 48/132
  recurrent oculomotor paralysis, 1/611
  strabismus, 48/132
  symptom schema, 48/156
  Tolosa-Hunt syndrome, 42/748, 48/302
Ophthalmoscopy
  altitudinal hemianopia, 2/563
  aplasia papillae N. opticorum, 22/537
  bacterial endocarditis, 52/299
  brain embolism, 53/218
  brain infarction, 53/218
  cerebrovascular disease, 11/212, 53/218
  diabetes mellitus, 11/214
  direct ophthalmoscopy, 13/3
  embolism, 11/214
  indirect ophthalmoscopy, 13/4

intracranial hemorrhage, 53/218
neuroretinal degeneration, 13/3-9
retinovestibular hereditary spastic ataxia, 21/386
slit lamp, 13/5
subarachnoid hemorrhage, 11/214
systemic lupus erythematosus, 11/215
tapetal reflex, 13/155
transient ischemic attack, 53/218
uremia, 11/214
venous occlusion, 11/213
Opiate
   alfentanil, 65/350
   alphaprodine, 37/366, 65/350
   anorexia nervosa, 46/587
   apomorphine, 65/350
   brain infarction, 55/517
   buprenorphine, 65/350
   butorphanol, 65/350
   codeine, 65/350
   cyclazocine, 65/350
   dextropropoxyphene, 65/350
   dezocine, 65/350
   diamorphine, 65/350
   dihydrocodeine, 65/350
   diphenoxylate, 65/350
   endogenous, *see* Endogenous opioid
   epilepsy, 72/234
   etorphine, 65/350
   fentanyl, 65/350
   haloxon, 65/350
   hydrocodone, 65/350
   hydromorphone, 65/350
   levorphanol, 37/366, 65/350
   meptazinol, 65/350
   mesocorticolimbic reward circuit, 65/350
   methadone, 37/366, 65/350
   morphine, 65/350
   nalbuphine, 65/350
   nalorphine, 37/366
   naloxone, 37/365, 65/350
   naltrexone, 37/365, 65/350
   neurotoxin, 37/365, 379, 65/349
   oxycodone, 65/350
   oxymorphone, 65/350
   pentazocine, 65/350
   pethidine, 65/350
   pharmacology, 65/350
   profadol, 65/350
   propiram, 65/350
   sufentanil, 65/350
Opiate addiction
   drug dependence, 65/349
   epidemiology, 65/351

myelopathy, 65/356
opiate agonist, 65/349
opiate antagonist, 65/349
physical dependence, 65/349
psychic dependence, 65/349
term definition, 65/349
tolerance, 65/349
toxic myopathy, 62/598
Opiate binding
   Parkinson disease, 75/183
Opiate intoxication, 37/365-394, 65/349-358
   abscess, 37/393
   amblyopia, 37/386
   apnea, 65/353
   brachial plexus neuritis, 37/388
   brachial plexus neuropathy, 65/356
   brain infarction, 65/355
   chemistry, 37/365-367
   coma, 65/353
   complication, 37/379, 65/355
   compression neuropathy, 65/356
   crush syndrome, 37/391
   dependence, 37/374-376, 65/349, 354
   endocarditis, 65/355
   fatality rate, 65/355
   fungal infection, 65/355
   hepatotoxicity, 37/393, 65/355
   history, 65/350
   human immunodeficiency virus infection, 65/356
   infection, 65/355
   infectious hepatitis, 65/355
   lumbar plexitis, 37/388
   malaria, 37/393, 65/355
   meningitis, 65/355
   1-methyl-4-phenyl-1,2,3,6-tetrahydropyridine, 65/357
   miosis, 65/353
   mononeuritis, 37/389
   mycotic aneurysm, 37/391, 65/355
   myelitis, 37/387
   myoglobinuria, 65/357
   myopathy, 37/390
   naloxone, 65/353
   neonatal withdrawal toxicity, 37/378, 65/354
   nerve lesion, 37/387
   neurologic complication, 37/379, 65/355
   neuronal activity, 37/372
   neuropathy, 65/356
   neurophysiology, 37/374
   neurotransmitter, 37/368
   opium war, 65/350
   osteomyelitis, 37/391
   Papaver somniferum, 65/349

pharmacology, 65/350
polyneuropathy, 37/389, 65/356
pulmonary edema, 65/353
pulmonary hypertension, 65/355
receptor, 37/369-371
renal disease, 65/355
rhabdomyolysis, 37/389
social implication, 37/393
spongiform encephalopathy, 65/357
subarachnoid hemorrhage, 65/355
tetanus, 37/393, 65/355
thrombophlebitis, 65/355
tolerance, 37/374-376, 65/349
treatment, 65/353
tripelennamine, 65/357
vasculopathy, 65/356
vertebral osteomyelitis, 65/355
withdrawal syndrome, 65/354
wound botulism, 65/355
Opiate peptide
alcohol intoxication, 64/114
migraine, 48/99
Opiate receptor
delta type, 57/83, 65/350
kappa type, 57/83, 65/350
mu type, 57/83, 65/350
pain, 45/238
substantia gelatinosa, 45/230
thalamus, 45/236
Opiate withdrawal syndrome
neonate, 65/355
symptom, 65/354
Opioid, see Opiate
Opioid analgesic agent
autonomic nervous system, 75/501
classification, 69/40
receptor selectivity, 69/40
Opioid peptide, see Opiate peptide
Opisthotonos
ablation, 2/402
animal experiment, 2/401
Arnold-Chiari malformation type II, 50/410
ascariasis, 35/359
astasia, 2/394-399
bilirubin encephalopathy, 27/421
cerebellar anterior lobe, 2/401
decaborane intoxication, 64/360
decerebration, 6/149
globoid cell leukodystrophy, 10/303, 29/356,
42/489
Hallervorden-Spatz syndrome, 66/715
hypernatremia, 63/554
hypocalcemia, 63/559

hyponatremia, 63/545
infantile Gaucher disease, 42/438
kernicterus, 42/583
Lesch-Nyhan syndrome, 42/153
lissencephaly, 30/486, 42/40
meningitis, 1/542
mesencephalic syndrome, 2/283
3,4-methylenedioxyamphetamine intoxication,
65/366
neuroleptic dystonia, 65/281
pallidonigral degeneration, 42/242
pentaborane intoxication, 64/360
Reye syndrome, 27/363, 65/116
snake envenomation, 65/180
strychnine intoxication, 41/305
sulfite oxidase deficiency, 29/121
tetanus, 52/230, 65/219
tetany, 41/301, 63/559
tuberculous meningitis, 52/201
Opitz-Kaveggia syndrome, see FG syndrome
Opium, see Opiate
Opium war
opiate intoxication, 65/350
Papaver somniferum, 65/350
Oppenheim, H., 1/8, 11, 13
Oppenheim-Scholz-Morel syndrome, see Dyshoric
angiopathy
Oppenheim sign
Babinski sign, 1/184, 250
Opportunistic infection
acquired immune deficiency syndrome, 56/490,
507, 513
acquired immune deficiency syndrome dementia,
56/18
immunocompromised host, 56/472
nervous system, 71/261
renal transplantation, 63/533
Opsanus intoxication
family, 37/78
venomous spine, 37/78
Opsoclonia, see Opsoclonus
Opsoclonus, 34/611-615
acute cerebellar ataxia, 34/613, 629
breast cancer, 60/657
carcinoma, 34/614
carcinomatous cerebellar degeneration, 21/492
child, 34/613
chlordecone intoxication, 64/204
dentatorubropallidoluysian atrophy, 21/520, 528,
49/440
encephalitis, 34/613, 60/657
extrapyramidal disorder, 2/344
furor bulborum, 2/344

gaze apraxia, 45/406
Joseph-Machado disease, 42/262
lung cancer, 60/657
manifestation, 34/611
multiple sclerosis, 47/133
myoclonic epilepsy, 38/582
neuroblastoma, 21/574, 34/613, 60/657
ocular flutter, 21/528
opticocochleodentate degeneration, 21/538
organochlorine insecticide intoxication, 64/204
palatal myoclonus, 34/612
paraneoplastic syndrome, 69/361
posterior fossa tumor, 18/407
thyroid cancer, 60/657
uterus cancer, 60/657
visual sensory system, 2/534
Whipple disease, 52/137
Opsoclonus myoclonus
paraneoplastic syndrome, 71/691, 694
Opsoclonus myoclonus ataxia syndrome
acute cerebellar ataxia, 34/629-632
Opsoclonus myoclonus syndrome
Coxsackie virus B3, 56/330
paraneoplastic syndrome, 69/361, 364
Opsonin
experimental allergic encephalomyelitis, 9/520
gram-negative bacillary meningitis, 52/105
Opsonization
viral infection, 56/58
Optic allesthesia
migraine, 48/162
occipital lobe syndrome, 2/664
occipital lobe tumor, 17/331
Optic aphasia
neuropsychology, 45/13
Optic ataxia, see Ocular ataxia
Optic atrophy
acrocephalosyndactyly type I, 13/81, 43/318
acrodysostosis, 43/291
Addison disease, 13/81
adrenoleukodystrophy, 10/132, 42/483, 51/384,
60/655
adult GM1 gangliosidosis, 60/654
Aicardi syndrome, 42/696
alaninuria, 42/516
alcohol, 13/61
amyotrophic lateral sclerosis, 13/97, 22/138
anatomy, 2/518
anencephaly, 42/12
angioid streak, 13/41
angiostrongyliasis, 52/555
ape, 13/56
arsenic intoxication, 13/63

arteriosclerosis, 13/58
associated disease, 13/39-42, 68
ataxia, 13/39, 42/117, 406, 412-417
ataxia telangiectasia, 60/654
autosomal dominant, see Autosomal dominant
optic atrophy
autosomal dominant infantile, see Autosomal
dominant infantile optic atrophy
autosomal dominant pseudoglaucoma, 13/120
autosomal recessive, see Autosomal recessive
optic atrophy
autosomal recessive congenital, 13/91
Bassen-Kornzweig syndrome, 13/40
Batten disease, 42/461, 463
Behr disease, 13/42, 81, 98
Biemond syndrome, 21/380, 42/164, 60/654
bilateral, 13/56
blood loss, 13/60
bone disease, 13/40, 68
bone membranous lipodystrophy, 42/280
brain aneurysm, 13/69
Bruyn-Went disease, 13/42, 22/454
bulbar paralysis, 22/138
burn trauma, 13/64
cachexia, 13/64
carotid artery occlusion, 13/83
carotid cavernous fistula, 24/416, 422
cataract, 42/406
cause, 13/52
cerebrovascular disease, 55/114
chloramphenicol intoxication, 13/63
chloroquine, 65/484
cinchonism, 65/485
classification, 13/45
Cockayne-Neill-Dingwall disease, 10/182, 13/40,
43/350
complex IV deficiency, 66/423
compression, 13/69
congenital, see Congenital retinal blindness
congenital autonomic dysfunction, 60/18
congenital recessive, see Congenital recessive
optic atrophy
congenital retinal blindness, 13/38, 83
Conradi-Hünermann syndrome, 43/345
corpus callosum agenesis, 42/8, 50/162
cranial dysostosis (Franceschetti), 13/82
craniodiaphyseal dysplasia, 43/354
craniofacial dysostosis (Crouzon), 13/70, 81,
43/359
craniometaphyseal dysplasia, 43/362-364
craniosynostosis, 50/126
Creutzfeldt-Jakob disease, 56/545
cyanide intoxication, 65/30

cyanogenic glycoside intoxication, 36/520-522
cytomegalovirus, 56/268
Dandy-Walker syndrome, 50/328
De Lange syndrome, 43/246
deaf mutism, 13/41
deafness, 13/40, 22/138, 60/660
deficiency neuropathy, 51/326, 328
diabetes insipidus, 42/411
diabetes mellitus, 13/39, 60, 64, 42/411
diagnosis, 13/47
diaphyseal dysplasia, 43/376
differential diagnosis, 13/72-81
diffuse sclerosis, 13/83
Divry-Van Bogaert syndrome, 14/514
dominant infantile, *see* Dominant infantile optic
    atrophy
Dumon-Radermecker syndrome, 42/477
dysosteosclerosis, 43/393
Ehlers-Danlos syndrome, 13/41
elliptocytosis, 13/40
ERG, 13/49
external ophthalmoplegia, 13/40
familial amaurotic idiocy, 13/39
familial amyotrophic lateral sclerosis, 22/138
familial spastic paraplegia, 21/16, 22/425, 427,
    472, 42/409
Ferguson-Critchley syndrome, 42/142, 59/319
fibrous dysplasia, 14/170
filix mas intoxication, 13/63
fluorescein angiography, 13/48
foramen lacerum syndrome (Jefferson), 13/82
Foster Kennedy syndrome, 2/53, 13/58, 68, 82
Friedreich ataxia, 13/39, 82, 21/325, 334, 42/143,
    409, 51/388, 60/309, 654
Friedreich ataxia like Sylvester syndrome, 60/654
Gaucher disease, 13/39
Gerstmann-Sträussler-Scheinker disease, 56/553
globoid cell leukodystrophy, 10/302, 350, 13/83,
    29/356
glycosaminoglycanosis type I, 13/40
glycosaminoglycanosis type IS, 13/40
glycosaminoglycanosis type II, 13/40
glycosaminoglycanosis type III, 10/342, 13/40
GM2 gangliosidosis, 13/83, 42/433
GM2 gangliosidosis type I, 42/433
GM2 gangliosidosis type III, 42/436
Goltz-Gorlin syndrome, 13/82, 14/113
gyrate atrophy (Fuchs), 13/82
Hallervorden-Spatz syndrome, 6/626, 13/39,
    49/401, 66/712
Hallgren syndrome, 13/40
hereditary ataxia, 13/39, 300, 305
hereditary cerebellar ataxia, 42/117

hereditary cerebellar ataxia (Marie), 13/83
hereditary congenital, *see* Hereditary congenital
    optic atrophy
hereditary deafness, 50/218
hereditary hearing loss, 22/499, 504-508
hereditary motor and sensory neuropathy type I,
    13/39, 81, 21/284, 308, 22/507, 42/117, 339, 409
hereditary motor and sensory neuropathy type II,
    13/39, 81, 21/284, 308, 22/507, 42/117, 339, 409
hereditary motor and sensory neuropathy variant,
    60/245
hereditary muscular atrophy, 13/39
hereditary olivopontocerebellar atrophy (Menzel),
    21/443
hereditary optic atrophy (Leber), *see* Congenital
    retinal blindness
hereditary periodic ataxia, 21/569
hereditary sensory and autonomic neuropathy type
    I, 13/42
hereditary spastic ataxia, 21/365, 367, 60/462, 654
hereditary type, 13/65
heredodegeneration, 13/68
histology, 13/51
homocystinuria, 13/42
Hooft-Bruens syndrome, 42/508
Hooft disease, 13/41
hydrocephalus, 13/69
hypertrophic interstitial neuropathy, 13/82, 42/317
hypertrophic polyneuropathy, 13/82
hypoesthesia, 60/661
hypogonadal cerebellar ataxia, 21/468, 476
hypophyseal tumor, 13/69
incontinentia pigmenti, 13/42, 81, 14/12, 108,
    43/23
infantile, *see* Infantile optic atrophy
infantile neuroaxonal dystrophy, 6/627, 13/83,
    42/229, 49/402
infantile optic glioma, 42/733
infantile spinal muscular atrophy, 59/70
injury, 13/70, 24/38
isoniazid intoxication, 65/480
isoniazid polyneuropathy, 51/297
juvenile, *see* Juvenile optic atrophy
Kjer, *see* Infantile optic atrophy
Klinefelter syndrome, 13/40
lactation, 13/64
late cerebellar atrophy, 60/655
Laurence-Moon-Bardet-Biedl syndrome, 13/40,
    42
lead intoxication, 13/63
Leber, *see* Leber optic atrophy
Leber hereditary optic neuropathy, 62/509
leukodystrophy, 60/655

Lundberg syndrome, 13/91
Marfan syndrome, 13/41
Marinesco-Sjögren syndrome, 60/654
MERRF syndrome, 62/509
methanol, 13/62
methanol intoxication, 36/353, 64/99
midfrequency hearing loss, 42/414
Minamata disease, 7/642, 64/415
multiple mitochondrial DNA deletion, 62/510
multiple myeloma, 63/393
multiple sclerosis, 13/56-58, 81, 42/495
myotonic dystrophy, 13/40, 42
neuroblastoma, 42/758
neurofibromatosis, 14/151
neurofibromatosis type I, 13/69
neurology, 13/44-88
neuronal ceroid lipofuscinosis, 42/461, 463
neuropathy, 60/660
neurosyphilis, 33/365, 52/277
olivopontocerebellar atrophy, 42/161
olivopontocerebellar atrophy (Dejerine-Thomas), 60/655
olivopontocerebellar atrophy (Schut-Haymaker), 60/655
olivopontocerebellar atrophy variant, 21/451-453
olivopontocerebellar atrophy (Wadia-Swami), 60/494, 655
olivopontocerebellar atrophy with retinal degeneration, 60/507
optic chiasm syndrome, 13/81
optic nerve coloboma, 42/407
optic nerve tumor, 17/352
optic neuritis, 13/65
optic tract lesion, 2/577
opticochiasmatic arachnoiditis, 13/65
orbital cellulitis, 33/179
organotin intoxication, 64/136
osteitis deformans (Paget), 13/83, 38/364
osteopetrosis, 13/68, 81, 43/456
oxycephaly, 13/69
Paine syndrome, 43/433, 60/293
pallidoluysionigral degeneration, 49/447
papilledema, 16/289
parathyroid dysfunction, 13/41
Pelizaeus-Merzbacher disease, 13/39, 83
pernicious anemia, 42/609
petrosphenoid syndrome (Jaccodo), 13/82
pheniprazine intoxication, 13/63
Pierre-Robin syndrome, 13/41
poliodystrophy, 60/655
polyarteritis nodosa, 55/356
polyneuropathy, 60/661
porencephaly, 30/685, 50/359

posterior column ataxia, 42/164
postrubella embryopathy, 13/82
pregnancy, 13/64
progressive bulbar palsy, 22/108, 138
progressive external ophthalmoplegia, 13/37, 82, 62/305
progressive muscular dystrophy, 13/40
progressive myoclonus epilepsy, 13/39
progressive paralysis, 13/53-56
pseudoxanthoma elasticum, 13/41
psoriasis, 13/40
Pyle disease, 13/83, 22/493, 60/774
pyramidal sign, 42/412-414
pyruvate decarboxylase deficiency, 62/502
pyruvate dehydrogenase complex deficiency, 66/416
pyruvate dehydrogenase deficiency, 29/212
18q partial monosomy, 43/532
quinine intoxication, 13/63
recessive, see Recessive optic atrophy
recessive infantile, see Recessive infantile optic atrophy
Refsum disease, 8/22, 13/40, 21/195, 51/385, 60/229, 66/489
renal disease, 13/40
Rosenberg-Chutorian syndrome, 22/505, 42/418, 60/654
rubella virus, 56/408
Rubinstein-Taybi syndrome, 13/83, 31/282
Schnabel cavern, 13/52
sclerosteosis, 31/258, 43/475
sensorineural deafness, 42/386, 411, 414
sickle cell anemia, 13/41
Sjögren-Larsson syndrome, 13/41
spastic ataxia, 42/165
spastic diplegia, 42/415
spastic paraplegia, 13/39, 22/434, 42/177
sphenoid syndrome, 13/82
sphenoid tumor, 13/82
spinal muscular atrophy, 59/93, 373
spinocerebellar ataxia, 42/181
spinocerebellar degeneration, 13/97
spino-olivocerebellar degeneration, 60/655
Stargardt disease, 13/83
Stickler syndrome, 13/41
Stilling-Türk-Duane syndrome, 13/83
subacute myelo-optic neuropathy, 51/296
subacute necrotizing encephalomyelopathy, 10/209, 51/388, 60/654
subacute sclerosing panencephalitis, 56/419
subarachnoid hemorrhage, 12/139
syndrome part, 13/81-84
syphilis, 33/366

systemic lupus erythematosus, 55/377
tabes dorsalis, 13/53-56, 19/86, 52/280
Takayasu disease, 13/60, 83
talipes, 42/406, 412-414, 417
temporal arteritis, 13/60, 82
tetraplegia, 13/40
tobacco, 13/61
trichopoliodystrophy, 14/784
tropical ataxic neuropathy, 65/29
true porencephaly, 30/685
tuberculosis, 51/186
tuberculous meningitis, 52/202
tuberous sclerosis, 13/81, 14/346
tumor, 13/68
Usher syndrome, 13/40
Van Buchem disease, 13/81, 31/258, 38/408, 43/410
Van den Bosch syndrome, 13/41
vascular, 13/58-61
vincristine polyneuropathy, 51/299
visual function, 13/49
vitamin B$_{12}$, 28/175
Vogt-Koyanagi-Harada syndrome, 13/83
Weiss-Alström syndrome, 13/40
Werner syndrome, 13/40
X-linked, *see* X-linked optic atrophy
Optic bundle
accessory, *see* Accessory optic bundle
Optic canal
see also Optic foramen
anatomy, 2/519
Optic canal fracture
bilateral, 24/40
head injury, 24/39
optic chiasm injury, 24/44
optic nerve injury, 24/39-41
radiology, 24/39
skull fracture, 24/44
surgery, 24/42
type, 24/39
Optic chiasm, 2/524-529
accessory optic bundle, 2/529
astrocytoma, 68/80
brain injury, 24/30, 42
corpus callosum agenesis, 2/528
development, 2/527
ethambutol polyneuropathy, 51/295
fiber type, 2/528
meningeal leukemia, 63/341
multiple sclerosis, 47/55
pituitary adenoma, 17/385, 388
pituitary tumor, 68/344
retinotopic organization, 2/551

spongioblastoma, *see* Oligodendroglioma
topography, 2/526
variation, 2/524
visual fiber, 2/550-554
visual field defect, 2/570
Optic chiasm agenesis
anophthalmia, 42/14
Optic chiasm compression
aneurysm, 68/75
arachnoid cyst, 68/75
arteriovenous malformation, 68/75
cavernous hemangioma, 68/75
chordoma, 68/75
craniopharyngioma, 68/75
epidermoid, 68/75
fibrillary astrocytoma, 68/75
germinoma, 68/75
hamartoma, 68/75
hemangioblastoma, 68/75
histiocytosis, 68/75
malignant astrocytoma, 68/75
meningioma, 68/75
metastasis, 68/75
neuroastrocytoma, 68/75
pituitary adenoma, 68/75
sarcoid, 68/75
spongioblastoma, 68/75
tuberculoma, 68/75
Optic chiasm glioma
see also Oligodendroglioma
neurofibromatosis type I, 14/33
terminology, 18/83
Optic chiasm injury
bitemporal hemianopia, 24/43
brain injury, 24/30, 42
central vision, 24/44
cranial nerve, 24/44
optic canal fracture, 24/44
pathogenesis, 24/44
seesaw nystagmus, 1/605, 16/323, 24/44
surgery, 24/45
traumatic blindness, 24/28, 44
treatment, 24/45
vascular supply, 24/30
Optic chiasm syndrome
acute onset, 2/576
bitemporal hemianopia, 18/354
definition, 2/574
false localization, 2/576
optic atrophy, 13/81
papillary atrophy, 18/354
vision loss, 18/354
Optic chiasm tumor

28

treatment, 68/80
Optic decussation
  supra, *see* Supraoptic decussation
Optic disc
  acute ischemic papillopathy, 48/312
  angioid streak, 13/74
  blood supply, 48/317
  Cloquet canal, 13/81
  neurofibromatosis, 14/638
  Oguchi disease, 13/261
  Takayasu disease, 55/335
  temporal arteritis, 48/312
  tilted, *see* Tilted optic disc
Optic disc dysplasia
  amaurosis fugax, 55/108
Optic disc hypoplasia
  congenital heart disease, 63/11
Optic disc pit
  basal encephalocele, 50/106
  basal meningoencephalocele, 50/106
Optic dysplasia
  septal, *see* Septo-optic dysplasia
Optic foramen
  see also Optic canal
  callus, 13/70
  fracture, 13/69
  measurement, 2/520
  roentgenology, 2/520
Optic glioma, *see* Optic nerve glioma
Optic nerve
  arteritis, 48/319
  binasal hemianopia, 2/572-574
  blood supply, 48/317
  brain aneurysm, 12/90
  brain injury, 24/27
  centrocecal scotoma, 2/567
  chiasm, 2/550
  developmental anatomy, 2/528
  diabetes mellitus, 27/118
  fiber size, 2/549
  fiber type, 2/521
  glycosaminoglycanosis type VII, 66/311
  Langerhans cell histiocytosis, 70/60
  meningeal leukemia, 63/341
  methanol intoxication, 36/353, 357, 64/101
  multiple sclerosis, 47/134-137
  neurofibromatosis, 14/630
  orbital apex syndrome, 2/570
  osteitis deformans (Paget), 70/12
  osteopetrosis, 70/16
  pellagra, 7/573
  pial septum, 2/549, 553
  quantitative histology, 2/522, 524

  retinotopic organization, 2/548
  Schnabel cavern, 13/78
  schwannoid oligodendroglioma, 18/83
  sheath, 2/521
  sincipital encephalocele, 50/103
  subacute combined spinal cord degeneration, 28/173-178
  suprachiasmatic nucleus, 74/480
  temporal arteritis, 48/312, 319, 326
  thyroid disease, 70/90
  visual fiber, 2/548
  Wegener granulomatosis, 71/177
Optic nerve abnormality
  basal encephalocele, 50/106
  basal meningoencephalocele, 50/106
Optic nerve agenesis
  anencephaly, 50/211
  anophthalmia, 42/14
  cyclopia, 50/211
  encephalocele, 50/211
  hypopituitarism, 50/211
  juvenile diabetes mellitus, 50/211
  lissencephaly, 50/211
  measles, 50/211
  microphthalmia, 50/211
  porencephaly, 50/211
  prevalence, 50/211
Optic nerve aplasia
  animal, 30/400
  associated malformation, 30/400
  clinical features, 30/400
  Edwards syndrome, 50/562
  epidemiology, 30/400
  familial incidence, 30/400
  pathogenesis, 30/400
Optic nerve atrophy, *see* Optic atrophy
Optic nerve coloboma
  Aicardi syndrome, 50/165
  encephalomeningocele, 42/27
  Goltz-Gorlin syndrome, 43/20
  optic atrophy, 42/407
Optic nerve degeneration
  cisplatin intoxication, 65/536
Optic nerve demyelination
  methanol intoxication, 64/101
Optic nerve glioma
  see also Neurofibromatosis type I *and* Optic nerve tumor
  benign hamartoma, 68/75
  brain tumor, 18/503
  clinical features, 17/350, 18/345
  diagnosis, 18/345
  growth hormone, 42/734

incidence, 50/367
infant, *see* Infantile optic glioma
J-shaped sella, 14/151, 18/345
meningoblastoma, 14/630
neurofibromatosis, 50/365, 367
neurofibromatosis type I, 14/134, 142, 147, 151,
    18/344
orbital tumor, 17/202, 18/339
pathology, 18/343
prognosis, 18/346
radiotherapy, 18/503
Russell syndrome, 16/349
skull base tumor, 17/196
supratentorial brain tumor, 18/343-347
treatment, 18/346, 68/78, 81
Optic nerve infarction
    temporal arteritis, 48/317
Optic nerve injury
    see also Traumatic blindness
    anatomy, 24/29
    blindness, 24/27
    brain injury, 24/27
    clinical features, 24/37
    ERG, 24/37
    facial injury, 24/36
    facial nerve injury, 24/36
    hemorrhage, 24/41
    incidence, 24/32, 36, 52
    indirect, 24/36
    intrabulbar, 24/35
    intracanalicular, 24/36, 40
    ischemic necrosis, 24/40
    optic canal fracture, 24/39-41
    orbital injury, 24/35
    pathogenesis, 24/40
    prognosis, 24/38
    result, 24/42
    shearing lesion, 24/40
    superior orbital fissure syndrome, 2/569
    surgery, 24/39, 42
    traumatic blindness, 24/27
    treatment, 24/42
    vascular supply, 24/30
    visual evoked response, 24/37
    visual type, 2/568
Optic nerve meningoblastoma
    features, 14/631
Optic nerve neurinoma, *see* Optic nerve glioma
Optic nerve pilocytic astrocytoma, *see* Sarcoid
Optic nerve spongioblastoma, *see*
    Oligodendroglioma
Optic nerve tumor
    see also Optic nerve glioma

astrocytoma, 17/350, 354, 356, 359
clinical features, 17/350-352
glioblastoma multiforme, 17/353
hemangioma, 17/365
leukemia, 17/371
melanocytoma, 17/366
meningioma, 17/362
nerve head, 17/366, 369
nerve trunk, 17/372
occurrence, 17/353
oligodendroglioma, 17/355, 359
optic atrophy, 17/352
pathology, 17/353, 357
treatment, 17/361
tuberous sclerosis, 17/353
Optic neuritis
    acute, *see* Anterior ischemic optic neuropathy
    angiostrongyliasis, 52/549
    canine distemper, 9/683
    carcinoma, 38/658-660
    chloramphenicol intoxication, 65/487
    clinical features, 9/172
    disulfiram polyneuropathy, 51/309
    Epstein-Barr virus, 56/253
    ethambutol polyneuropathy, 51/294
    features, 13/64
    Hodgkin disease, 39/54
    leptospirosis, 7/489
    lindane intoxication, 65/479
    lithium intoxication, 64/296
    malignancy, 38/658-660
    malignant lymphoma, 39/54
    multiple sclerosis, 9/195, 47/54, 69, 171
    neurobrucellosis, 33/316, 52/588
    optic atrophy, 13/65
    optic pathway tumor, 68/74
    papilledema, 16/288
    pattern shift visual evoked response, 47/135
    pellagra, 46/399
    pernicious anemia, 38/22
    retinal nerve fiber, 47/54
    retrobulbar, *see* Retrobulbar optic neuritis
    rubella virus, 56/407
    Sjögren syndrome, 71/74
    syphilis, 52/281
    trichinosis, 52/567
    tuberculous meningitis, 52/202
    varicella zoster virus, 56/229
    visual evoked response, 47/54
    Wegener granulomatosis, 39/345, 51/451
Optic neurofibromatosis
    geniculocalcarine tract, 68/79
Optic neuromyelitis, *see* Neuromyelitis optica

30

Optic neuropathy
  acute ischemia, *see* Acute ischemic optic
    neuropathy
  allergic granulomatous angiitis, 63/383
  amiodarone, 63/190
  amiodarone intoxication, 65/459
  aspergillosis, 52/380
  bromoureide intoxication, 36/306
  carbon disulfide, 64/8
  cerebrotendinous xanthomatosis, 60/167
  chloramphenicol intoxication, 65/486
  chloroquine, 65/484
  cisplatin, 65/529
  clioquinol, 64/8
  cyanogenic glycoside, 64/8
  cyproterone acetate intoxication, 65/538
  cytarabine intoxication, 65/533
  dazzling test, 9/174
  deficiency neuropathy, 51/321
  disulfiram, 64/8
  ethambutol, 51/294, 64/8, 65/481
  ethambutol polyneuropathy, 51/295
  ischemic, *see* Ischemic optic neuropathy
  isoniazid intoxication, 65/480
  lead encephalopathy, 64/436
  lead intoxication, 64/436
  lindane intoxication, 65/479
  lomustine, 65/528
  methanol, 64/8
  methanol intoxication, 64/101
  multiple sclerosis, 9/172, 176
  organic bromide intoxication, 36/306
  quinine, 64/8
  radiation induced, 67/340
  Sjögren syndrome, 71/85
  streptomycin, 65/483
  streptomycin intoxication, 65/483
  tamoxifen, 65/529
  tamoxifen intoxication, 65/538
  thallium, 64/8
  tropical ataxic neuropathy, 51/322
  vitamin B_2 deficiency, 51/332
  vitamin B_{12} deficiency, 70/382
Optic nucleus
  supra, *see* Supraoptic nucleus
Optic pathway
  acetonylacetone intoxication, 64/89
  diencephalon, 68/69
  pathology, 68/77
  topical diagnosis, 2/561
  vascular supply, 2/560
  visual disorder, 68/76
Optic pathway tumor

  epidemiology, 68/74
  hemangioblastoma, 68/74
  lymphoma, 68/74
  malignant astrocytoma, 68/74
  metastasis, 68/74
  neuroastrocytoma, 68/74
  neurofibromatosis, 68/75
  oligodendroglioma, 68/74
  optic neuritis, 68/74
  optic sheath meningioma, 68/74
  sarcoid, 68/74
  spongioblastoma, 68/77
  symptom, 68/76
  tuberculoma, 68/74
Optic radiation, *see* Geniculocalcarine tract
Optic recess
  blood-brain barrier, 56/80
Optic sheath meningioma, 68/412
  optic pathway tumor, 68/74
Optic stimulation
  epilepsy, 15/442
Optic tract
  see also Geniculocalcarine tract *and* Visual
    pathway
  acrylamide intoxication, 64/74
  anatomy, 2/529
  anophthalmia, 2/528
  cerebrovascular disease, 55/115
  ethambutol polyneuropathy, 51/295
  hereditary motor and sensory neuropathy type I,
    21/303
  hereditary motor and sensory neuropathy type II,
    21/303
  opticocochleodentate degeneration, 21/542
  retinotopic organization, 2/554
  suprachiasmatic nucleus, 74/481
  visual fiber, 2/554-557
  visual field defect, 2/576
Optic tract injury
  brain injury, 24/30, 45
  clinical features, 24/45
  homonymous hemianopia, 24/45
  vascular supply, 24/30
Optic tract lesion
  anisocoria, 2/577
  hemianopic reaction, 2/577
  optic atrophy, 2/577
  optokinetic nystagmus, 2/577
Optically evoked response, *see* Visual evoked
  response
Opticoacoustic nerve degeneration
  polyneuritis, 42/418
  polyneuropathy, 42/418

Opticochiasmatic arachnoiditis
  familial spinal arachnoiditis, 42/107
  optic atrophy, 13/65
Opticocochleodentate atrophy
  oculomotor nucleus, 22/174
Opticocochleodentate degeneration, 21/535-551
  age at onset, 60/751
  argentophilic plaque, 21/543
  associated lesion, 60/755
  Babinski sign, 42/241
  blindness, 21/538, 42/241, 60/752
  capillary basement membrane, 21/548
  cerebellar cortex, 21/546-548
  choreoathetosis, 21/538, 60/754
  clinical features, 21/536, 60/751
  cochlear system, 21/542
  contracture, 42/241
  course, 21/541
  deafness, 21/538, 60/657, 752
  dentatofugal system, 21/544
  differential diagnosis, 60/757
  EEG, 21/542, 60/754
  epidemiology, 60/751
  epilepsy, 42/241, 60/754
  Friedreich ataxia, 60/757
  general pathology, 60/754
  genetics, 21/535
  GM2 gangliosidosis type I, 21/550
  hereditary hearing loss, 22/508
  hippocampus, 21/548
  history, 60/751
  hyperkinesia, 21/540
  hyperreflexia, 42/241
  laboratory data, 21/541, 60/754
  medial lemniscus, 21/545
  mental deficiency, 21/540, 22/508, 42/241
  microcephaly, 21/538
  motor disturbance, 60/752
  muscle biopsy, 60/754
  neuropathology, 21/542, 60/754
  opsoclonus, 21/538
  optic tract, 21/542
  presenting sign, 21/536
  psychomotor development, 60/751
  pyramidal tract, 21/545, 42/241
  Rosenberg-Chutorian syndrome, 60/758
  seizure, 21/541
  sensorineural deafness, 42/241
  spastic tetraplegia, 21/540
  status epilepticus, 42/241
  thalamus, 21/547, 42/241
  Wolfram syndrome, 60/758
  X-linked syndrome, 60/758

Opticohypophyseal tract
  supra, see Supraopticohypophyseal tract
Opticoseptal dysplasia, see Septo-optic dysplasia
Optochine, see Ethylhydrocupreine
Optokinetic nystagmus
  after-nystagmus, 1/577
  Bárány drum, 2/368
  brain stem pathway, 2/368
  cerebral pathway, 2/369
  congenital nystagmus, 42/158-161
  cortical area, 2/370
  cortical blindness, 45/50
  defect, 1/587
  examination method, 2/333
  gaze, 2/653
  method, 1/576
  middle cerebral artery infarction, 54/162
  multiple sclerosis, 47/133
  occipital lobectomy, 2/670
  oculomotor disease, 45/52
  olivopontocerebellar atrophy with retinal
    degeneration, 60/507
  optic tract lesion, 2/577
  parietal lobe lesion, 2/588
  parietal lobe syndrome, 45/69
  parietotemporo-occipital area, 54/165
  progressive supranuclear palsy, 22/219, 221,
    49/241
  significance, 2/370
  sleep, 2/533
  supranuclear mechanism, 2/368
  temporal lobe lesion, 2/707
  test, 16/319
  type, 16/323
  vestibular function, 2/706
  vestibular syndrome, 2/367
  visual cortex injury, 24/47
Optomotor ataxia
  convergence spasm, 2/344
  visual attention, 2/340
  visual fixation, 2/340
Optomotor cortex, see Parastriate cortex
Optomotor system
  vestibular system, 16/331
Optovestibular ocular muscle disorder
  localization, 2/342
Ora serrata
  anatomy, 2/507
Oral contraception
  antithrombin, 55/525
  aorta thrombosis, 63/41
  behavior disorder, 46/599
  brain hemorrhage, 55/524

brain infarction, 53/164, 55/524
brain ischemia, 53/164
brain lacunar infarction, 54/240
brain vasculitis, 55/525
deep vein thrombosis, 55/524
depression, 28/131, 46/599
endothelial proliferation, 55/525
epilepsy, 73/372
estrogen, 55/525
ethinylestradiol, 55/525
migraine, 55/524
myocardial infarction, 55/525
neurofibromatosis type II, 50/370
neuropathy, 7/520
pulmonary embolism, 55/524
smoking, 55/525
subarachnoid hemorrhage, 55/524
systemic brain infarction, 11/457
Oral contraceptive agent
    autosomal dominant hypoalphalipoproteinemia,
        60/136
    biological rhythm, 74/469
    brain infarction, 53/36, 55/524, 63/332
    cavernous sinus thrombosis, 52/173
    cerebrovascular disease, 11/199, 53/18, 36
    coagulopathy, 63/332
    dopaminergic nerve cell, 49/49
    drug induced systemic lupus erythematosus,
        16/222
    headache, 48/201, 435
    melatonin, 74/503
    migraine, 48/201, 435, 53/36
    multiple sclerosis, 47/153
    sickle cell trait, 55/466
    smoking, 53/36
    Sneddon syndrome, 55/404
    venous sinus occlusion, 54/426
    vertebral artery occlusion, 53/379
    vertebrobasilar system syndrome, 53/379
    vitamin B6, 28/131
Oral contraceptive chorea, see Contraceptive drug
    induced chorea
Oral grasping
    Klüver-Bucy syndrome, 3/350, 355, 45/257-260
    temporal lobe epilepsy, 45/259
Oral lesion
    acrocephalosyndactyly type I, 31/232
    acrodysostosis, 31/230
    Behçet syndrome, 34/480
    Coffin-Lowry syndrome, 31/245
    Coffin-Siris syndrome, 31/240
    craniofacial dysostosis (Crouzon), 31/234
    Goldenhar syndrome, 31/247

    Gruber syndrome, 31/243
    Hallermann-Streiff syndrome, 31/249
    incontinentia pigmenti, 31/242
    orofaciodigital syndrome type I, 31/251
Oral tendency, see Oral grasping
Orbicularis oculi reflex, see Blink reflex
Orbicularis oris reflex, see Lip reflex
Orbit
    ciliary ganglion, 74/402
    cirsoid aneurysm, 14/651
    paratrigeminal syndrome, 48/333
    skull base tumor, 17/176
    sympathetic nerve fiber, 48/333
Orbital abscess
    orbital cellulitis, 33/179
Orbital apex syndrome
    aspergillosis, 52/380
    brain arteriovenous malformation, 31/200
    cavernous sinus syndrome, 48/301
    central scotoma, 2/322
    clinical picture, 2/101
    definition, 2/86
    diplopia, 2/322
    migraine, 2/322
    mucormycosis, 35/544, 52/471
    nerve compression, 2/570
    nonrecurrent nonhereditary multiple cranial
        neuropathy, 51/571
    olfactory nerve, 2/570
    optic nerve, 2/570
    ptosis, 2/322
    skull base tumor, 17/181, 68/468
    superior orbital fissure syndrome, 2/296
    symptom, 2/90, 322
    Tolosa-Hunt syndrome, 48/291, 300
    Wegener granulomatosis, 51/450
Orbital arteritis
    temporal arteritis, 48/318
Orbital artery, see Frontobasilar artery
Orbital cellulitis
    chemosis, 33/179
    corneal ulceration, 33/179
    erysipelas, 33/179
    exophthalmos, 33/179
    optic atrophy, 33/179
    orbital abscess, 33/179
    papilledema, 33/179
    sinusitis, 33/179
    temporal arteritis, 48/312
Orbital disease
    oculomotor paralysis, 2/319-322
Orbital encephalocele
    sphenoid wing aplasia, 50/144

Orbital fissure syndrome
  superior, *see* Superior orbital fissure syndrome
Orbital floor syndrome
  neoplasm, 2/88
  pterygopalatine fossa syndrome, 2/88
Orbital fracture
  see also Orbital injury
  blowout fracture, 24/99, 57/137
  classification, 24/92
  contrast study, 24/95
  diplopia, 24/93
  etiology, 24/92
  examination, 24/92
  Hess chart, 24/93
  nasomaxillary fracture, 24/97
  ocular imbalance, 24/89
  orbital roof fracture, 23/381
  radiology, 24/94
  sign, 24/92
  superior, *see* Superior orbital fracture
  zygomatic bone fracture, 23/375, 24/95
Orbital frontal cortex
  ECG change, 63/235
Orbital hemorrhage
  ocular imbalance, 24/87, 89
Orbital hypertelorism, 30/235-251
  acrocephalosyndactyly type V, 50/123
  brain defect, 30/246, 247
  canthal index, 30/238
  cheilopalatoschisis, 30/247
  chromosomal abnormality, 30/249
  classification, 30/242, 246-251
  clinical syndrome, 30/243-250
  cloverleaf skull syndrome, 30/245
  definition, 30/235
  diagnosis, 30/238
  embryology, 30/242
  epidemiology, 30/154
  etiology, 30/243
  intercanthal distance, 30/239-241
  interorbital distance, 30/239
  interpupillary distance, 30/239, 241
  iris dysplasia, 43/417
  measurement, 30/238
  median facial malformation, 30/246
  mental deficiency, 30/248, 43/257, 439
  nomenclature, 30/238
  ocular defect, 30/247
  sensorineural deafness, 43/417
  sex organ malformation, 30/249
  skeletal deformity, 30/248
  skin malformation, 30/248
  skull dysplasia, 30/246

  teeth defect, 30/246
  treatment, 30/251
Orbital hypotelorism
  holoprosencephaly, 30/431, 446-450, 42/32,
    50/233
Orbital injury
  see also Orbital fracture
  abducent nerve injury, 24/86
  blowout fracture, 24/35, 88, 99
  dislocated bone fragment, 24/88
  emphysema, 24/90
  enophthalmos, 24/91
  eyeball displacement, 24/89
  foreign body, 24/88
  globus fixus, 24/88
  head injury, 24/87
  hemorrhage, 24/87
  intrabulbar optic nerve, 24/35
  intracanalicular optic nerve, 24/36
  intraorbital optic nerve, 24/35
  missile injury, 23/520
  oculomotor nerve injury, 24/79, 82, 86
  optic nerve injury, 24/35
  Stilling-Türk-Duane syndrome, 24/89
  traumatic exophthalmos, 24/89
  trochlear nerve injury, 24/86
Orbital myositis
  see also Ocular myositis
  clinical features, 62/299
  computer assisted tomography, 62/299, 305
  definition, 62/298
  differential diagnosis, 62/298, 305
  familial relapsing ophthalmoplegia, 62/302
  histopathology, 62/299
  nuclear magnetic resonance, 62/299
  ophthalmoplegic migraine, 62/301
  progressive external ophthalmoplegia, 22/190,
    62/298
  progressive external ophthalmoplegia
    classification, 62/290
  T3, 62/300
  thyroid associated ophthalmopathy, 62/298
  thyrotropic hormone, 62/300
  Tolosa-Hunt syndrome, 62/301
Orbital neuralgia
  supra, *see* Supraorbital neuralgia
Orbital neurofibroma
  neurofibromatosis type I, 50/367
Orbital pain
  carotid artery stenosis, 55/111
  cluster headache, 48/241, 270
  Gradenigo syndrome, 2/313
  sphenopalatine neuralgia, 48/480

temporal arteritis, 48/314
Tolosa-Hunt syndrome, 48/291
Orbital phlebography
cavernous sinus thrombosis, 52/174
Tolosa-Hunt syndrome, 48/299
Orbital pseudotumor
Tolosa-Hunt syndrome, 48/302
Wegener granulomatosis, 51/450
Orbital syndrome
brain metastasis, 71/612
Orbital tumor
aneurysmal bone cyst, 17/202
angioma, 17/202
benign tumor, 17/176
dermoid cyst, 17/202
encephalocele, 17/202
epithelioma, 17/177
fibroma, 17/177
lymphangioma, 17/177
malignant tumor, 17/177
meningioma, 17/176
meningocele, 17/202
neurofibroma, 17/176, 202
optic nerve glioma, 17/202, 18/339
sarcoma, 17/177
skull base tumor, 17/201
Orbital vein
anatomy, 11/61
radioanatomy, 11/113
topography, 11/113
Orbitostenosis
craniosynostosis, 50/125
Orbitothalamencephalic angiomatosis, *see* Brégeat
syndrome
Orbivirus
Colorado tick fever, 56/11
RNA virus, 56/10
Orchidalgia
lumbar vertebral canal stenosis, 20/689
Orchitis
Freund adjuvant, 9/506
mumps, 9/545, 56/129, 429
Orcin reaction, *see* Orcinol reaction
Orcinol reaction
glycolipid, 9/26
Orciprenaline
hyperkalemic periodic paralysis, 62/472
myotonia, 62/276
Orellanine
see also Mushroom toxin
chemical structure, 65/38
Orellanine intoxication
nephrotoxicity, 65/38

time delay, 65/38
Orf virus
neurotropic virus, 56/5
Organ failure
multiple, *see* Multiple organ failure
Organ transplantation
acquired toxoplasmosis, 52/353
adenovirus, 56/283
brain injury, 24/813
cytomegalovirus infection, 56/263, 265
death, 24/813
immunocompromised host, 56/472
nocardiosis, 52/445
primary CNS lymphoma, 56/255
progressive multifocal leukoencephalopathy,
47/503-506
viral infection, 56/472
Organelle
axon, *see* Axon organelle
Organic acid
CSF, 66/646
Reye syndrome, 56/152
Organic acid metabolism
N-acetylaspartic aciduria, 66/639, 641
acidosis, 66/641
agranulocytosis, 66/641
athetosis, 66/641
biotinidase deficiency, 66/641
cardiomyopathy, 66/641
cerebellar atrophy, 66/641
cerebral palsy, 66/641
coma, 66/641
congenital brain malformation, 66/641
diagnosis, 66/646
dystonia, 66/641
frontotemporal atrophy, 66/641
fumarate hydratase deficiency, 66/639
germ cell cyst, 66/641
glutaryl-coenzyme A dehydrogenase deficiency,
66/639, 641
hematoma, 66/641
hemolytic anemia, 66/641
hepatosplenomegaly, 66/641
D-2-hydroxyglutaric aciduria, 66/639, 641
L-2-hydroxyglutaric aciduria, 66/639, 641
3-hydroxy-3-methylglutaric aciduria, 66/641
hyperammonemia, 66/641
hypertriglyceridemia, 66/641
hyperuricemia, 66/641
hypoglycemia, 66/641
hypotonia, 66/641
increased anion gap, 66/641
infant, 66/640

ketonuria, 66/641
ketosis, 66/641
lactic acidosis, 66/641
leukodystrophy, 66/641
macrencephaly, 66/641
macrocephaly, 66/639, 641
malonyl-coenzyme A decarboxylase deficiency, 66/639
metabolic acidosis, 66/641
metabolic stroke, 66/639, 641
3-methylglutaconyl coenzyme A hydratase deficiency, 66/641
mevalonate kinase deficiency, 66/639
multiple acyl-coenzyme A dehydrogenase deficiency, 66/641
myeloproliferative disease, 66/641
myoclonic, 66/641
myoclonus, 66/641
myopathy, 66/641
newborn, 66/640
nuclear magnetic resonance, 66/642
oxoglutarate dehydrogenase deficiency, 66/639
pancreatitis, 66/641
polyneuropathy, 66/641
progressive ataxia, 66/641
pyramidal sign, 66/641
Reye syndrome, 66/641
speech retardation, 66/641
spongiform encephalopathy, 66/641
spongy cerebral degeneration, 66/639
succinate semialdehyde dehydrogenase deficiency, 66/641
symptom, 66/640
thrombocytopenia, 66/641
Organic affective syndrome
clinicopathologic correlation, 46/427-429
depression, 46/425-429
functional psychiatry, 46/418
limbic system, 46/427-429
organic solvent intoxication, 64/41
Organic brain syndrome
akinetic mutism, 46/424
depression, 46/476
Epstein-Barr virus infection, 56/252
infectious mononucleosis, 56/252
mania, 46/476
psychosis, 46/460-462
Organic bromide intoxication
chronic syndrome, 36/305-307
clinical features, 36/304
history, 36/303
optic neuropathy, 36/306
Organic delusional syndrome

functional psychiatry, 46/418
schizophreniform psychosis, 46/420
somatoparaphrenia, 46/420
Organic lead intoxication
symptom, 36/6
Organic mental disorder
depression, 46/476
mania, 46/476
Organic mental syndrome
behavior change, 46/416
psychosis, 46/460-462
sensory extinction, 3/193
vegetative sign, 46/417
Organic mercury
neurotoxin, 36/83, 64/413
Organic mercury intoxication
see also Minamata disease
animal, 36/127
ataxia, 36/74
bulbar paralysis, 22/144
chelating agent, 46/391
chorea, 49/556
clinical features, 36/73-80, 119-127
deafness, 36/74
demyelination, 9/643
diagnosis, 36/75
dysarthria, 36/74
experimental, 36/128-135
fetal effect, 64/413
history, 36/73
Hunter-Russell syndrome, 36/73-75
late cerebellar atrophy, 60/586
late cortical cerebellar atrophy, 21/497
leukoencephalopathy, 9/643, 47/559
methylmercury source, 36/79
Minamata disease, 37/88
motoneuron disease, 22/27
neuropathologic aspect, 36/83-141
penicillamine, 36/76
prenatal, 36/135-139
prevention, 36/79
sensation, 36/74
sign, 36/74
sodium selenite, 64/425
source, 64/413
spontaneous, 36/127
symptom, 36/74
treatment, 36/76
in utero poisoning, 36/77
vision, 36/74
Organic personality syndrome
frontal lobe syndrome, 46/431
functional psychiatry, 46/418

hysteria, 46/434
multiple sclerosis, 46/434
temporal lobe epilepsy, 46/432-434
Organic psychosis
EEG, 46/493
primary alcoholic dementia, 46/341
Organic solvent
acetonylacetone, 64/82
acetonylacetone intoxication, 64/81
acrylamide, 64/82
2-butanone, 64/82
carbon disulfide, 64/82
dimethylaminopropionitrile, 64/82
3,4-dimethyl-2,5-hexanedione, 64/82
N,N-dimethylhydrazine, 64/82
hexane, 64/82
2-hexanone, 64/82
hydrazine, 64/82
3,3'-iminodipropionitrile, 64/82
leukoencephalopathy, 66/721
methyl chloride, 64/82
neurotoxin, 51/479, 64/39
toluene, 64/82
Organic solvent intoxication, 36/361-386
see also Styrene intoxication and Toluene
intoxication
absorption, 64/39
acetone, 64/42
acute effect, 64/40
attention, 64/48
autonomic polyneuropathy, 51/479
benzene, 64/39
brain atrophy, 64/43
brain blood flow, 64/44
2-butanone, 64/44
butyl acetate, 64/44
carbon disulfide, 64/40
chronic effect, 64/41
color discrimination, 64/48
confusion, 64/42
dementia, 64/41
1,2-dichloroethylene, 64/40
EEG, 64/40
EMG, 64/41
encephalopathy, 64/41
ether, 64/40
ethylene, 64/40
euphoria, 64/42
gasoline, 64/40
glue sniffing, 64/41
hallucination, 64/42
Huffer neuropathy, 36/374
leukoencephalopathy, 47/555

long-term effect, 64/43
mental depression, 64/43
metabolism, 64/39
neuropathy, 75/505
neuropsychologic data, 64/45
nitrobenzene, 64/40
organic affective syndrome, 64/41
orthostatic hypotension, 63/152
paresthesia, 64/41
pyridine, 64/40
Raleigh type, 64/41
solvent mixture, 36/374, 64/42
stoddard solvent, 64/40
styrene, 64/39
tetrachloroethane, 64/40
tetrachloroethylene, 64/40, 42
toluene, 64/39, 42
trichloroethylene, 64/40, 44
use, 64/39
visuomotor coordination, 64/48
whipped cream dispenser neuropathy, 36/374
xylene, 64/40, 44
Organic thiocyanate intoxication
action mechanism, 36/438
Organochlorine
aldrin, 36/393, 64/197
chemical formula, 36/393
chlordecone, 36/393, 64/197
chlorinated cyclodiene, 64/197, 200
chlorphenotane, 36/393, 64/197
1,1-dichloro-2,2-bis(4-chlorophenyl)ethane,
36/393, 64/197
dichlorodiphenylethane, 64/197
dicofol, 36/393, 64/197
dieldrin, 36/393, 64/197
heptachlor, 36/393, 64/197
hexachlorobenzene, 64/197
hexachlorocyclohexane, 64/197, 202
lindane, 64/197
mirex, 36/393, 64/197
neurotoxin, 36/391, 64/197
polychlorinated biphenyl, 64/206
toxic polyneuropathy, 64/197
Organochlorine insecticide intoxication, 36/391-424
aldrin, 36/411-417, 64/197
ataxia, 64/204
calcium-adenosine triphosphate, 64/200
calcium homeostasis, 64/200
calmodulin, 64/200
chloracne, 64/207
chlordane, 36/408-411
chlordecone, 36/420-423, 64/197, 203
chlordecone shake, 64/204

chloride flux, 64/202
chlorinated cyclodiene, 64/197, 200
chlorphenotane, 36/398-405, 64/197
CNS activation, 64/202
confusion, 64/199
dementia, 64/207
developmental delay, 64/207
1,1-dichloro-2,2-bis(4-chlorophenyl)ethane,
    64/197
dichlorodiphenylethane, 64/197
dicofol, 64/197
dieldrin, 36/411-417, 64/197
dysarthria, 64/204
EEG, 64/199
endosulfan, 36/418
endrin, 36/411-417
epilepsy, 64/199, 202, 204
excitatory postsynaptic potential, 64/200
fatality, 64/202
GABA blockade, 64/200
GABA receptor, 64/202
headache, 64/200
heptachlor, 36/417, 64/197
hexachlorobenzene, 64/197
hexachlorocyclohexane, 36/405-408, 64/197, 202
hyperesthesia, 64/199
hyperexcitability, 64/199
inhibitory postsynaptic potential, 64/200
insecticide, 64/198
isobenz, 36/411-417
lindane, 64/197, 202
methoxychlor, 36/405, 64/198
mirex, 36/418, 64/197, 203
myoclonia, 64/200
neuropathology, 64/204
opsoclonus, 64/204
paresthesia, 64/199
phthalate ester, 36/444
polychlorinated biphenyl, 64/206
polychlorinated biphenyl intoxication, 64/207
pyrethrum, 36/435
rotenone, 36/444-447
ryania, 36/447
Schwann cell, 64/204
sensory neuron, 64/199
sodium permeability, 64/199
stroban, 36/423
symptom, 64/199
toxaphene, 36/419
toxicity mechanism, 64/198
tremor, 64/199, 202, 204
use history, 64/198
Yucheng disease, 64/207

yusho disease, 64/207
Organochlorine pesticide intoxication
    autonomic nervous system, 75/509
Organolead
    chemistry, 64/129
    compound, 64/129
    general formula, 64/129
    neurotoxin, 64/129, 133
    tetra-alkyllead, 64/129
    tetramethyllead intoxication, 64/133
Organolead intoxication
    acute intoxication, 64/133
    adenosine triphosphate hydrolysis, 64/143
    aggression, 64/131
    animal experiment, 64/131
    anorexia, 64/133
    anxiety, 64/133
    ataxia, 64/131
    blood level, 64/130
    brain concentration, 64/130
    brain stem, 64/132
    carbon disulfide like effect, 64/133
    chemistry, 64/129
    chloride transport, 64/143
    chlorine/hydroxide ion exchange, 64/146
    delirium, 64/133
    dendritic vacuolation, 64/146
    dentate fascia, 64/132
    dentate fascia neuron, 64/146
    dimercaprol, 64/134
    edetic acid, 64/134
    epilepsy, 64/131, 144
    erythrocyte stippling, 64/134
    excitotoxic effect, 64/145
    fatal case, 64/133
    GABA neuron, 64/146
    gene, 64/129
    Golgi apparatus, 64/132
    Golgi saccule vacuolation, 64/145
    half-life, 64/130
    hallucination, 64/133
    headache, 64/133
    hemorrhagic encephalopathy, 64/133
    hippocampus, 64/131, 145
    hyperexcitability, 64/131
    hyperreflexia, 64/133
    inclusion body, 64/132
    individual susceptibility, 64/133
    insomnia, 64/133
    intracellular dense body, 64/132
    intracranial hypertension, 64/133
    liver concentration, 64/130
    metabolism, 64/130

metallic taste, 64/133
microtubular assembly, 64/143
motoneuron, 64/145
myelin, 64/132
nausea, 64/133
neuronal necrosis, 64/131
neuronal toxicity, 64/144
neuropathology, 64/131
nonlysosomal degradation, 64/145
pathomechanism, 64/134, 144
penicillamine, 64/134
perforant path, 64/146
petechial hemorrhage, 64/133
petroleum vapour, 64/133
porphobilinogen synthase, 64/134
primary neuronal change, 64/145
pyriform cortex, 64/131
radical binding, 64/143
rage, 64/131
retinal photoreceptor, 64/145
schizophrenia, 64/133
startle epilepsy, 64/131
suicide, 64/133
tetraethyllead, 64/130
tetramethyllead, 64/130
tissue distribution, 64/130
treatment, 64/134
tremor, 64/131, 134
tributyllead, 64/130
triethyl like change, 64/132
triethyllead, 64/130
trimethyllead, 64/130
trimethyltin like change, 64/133
tripropyllead, 64/130
tubulovesicular body, 64/145
Organomegaly, *see* Visceromegaly
Organophosphate
  absorption, 64/155
  acetylcholinesterase inhibition, 64/157
  acetylcholinesterase phosphorylation, 64/158
  agaricide, 64/152
  azinphos, 64/156
  bladan, 64/153
  chemical formula, 64/153
  chemistry, 37/542-551, 64/152
  chlorthiophos, 64/169
  cholinesterase, 37/541
  chronic axonal neuropathy, 51/531
  cyanazine, 64/153
  deacetylation, 64/157
  demethylation, 64/155
  demeton, 64/152, 156
  demeton sulfoxide, 64/154, 169

detoxification, 64/155
dialifos, 64/169
dialkoxy group, 64/155
dichlorvos, 64/152
diethoxy group, 64/155
dimethoate, 64/154, 169
dimethoxy group, 64/155
distal axonopathy, 64/11
dyflos, 64/152
dyflos intoxication, 64/152
ecothiopate iodide, 64/153
fungicide, 64/152
helminthiasis, 64/155
heptenophos, 64/169
hexaethyl tetraphosphate, 64/153
historic, 37/541
hydrolysis, 64/155
leaving group category, 64/153
malathion, 64/152
mercaptophos, 64/152
metabolism, 37/551, 64/155
methidathion, 64/169
metrifonate, 64/152
mixed substitute, 64/155
nematocide, 64/152
neurotoxicity measure, 64/161
neurotoxin, 37/541, 46/393, 51/531, 64/11, 151,
     161, 229
octamethyl pyrophosphoramide, 64/152
oxidative biotransformation, 64/156
oxidative dealkylation, 64/156
oxidative dearylation, 64/156
oxidative desulfuration, 64/156
paraoxon, 64/152
parathion, 8/338, 64/152, 169
parathion/demeton sulfide, 64/169
parathion methyl, 64/169
pharmacology, 37/551
phosphocreosote, 64/152
phospholine, 64/153
pyrophosphates, 64/153
sarin, 64/152
schistosomiasis, 64/155
segmental demyelination, 64/160
slow eye movement, 64/156
soman, 64/152
sulfide oxidation, 64/156
synthesis, 64/151
tabun, 64/152
tetraethyl pyrophosphate, 64/152
toxic neuromuscular junction disorder, 64/15
toxic neuropathy, 51/283, 64/11, 151
toxic polyneuropathy, 64/151

toxicology, 64/156
treatment, 37/557-559
tri-*o*-cresyl phosphate, 64/152
tricresyl phosphates, 64/152
triphenyl compound, 64/155
trithioalkyl compound, 64/155
venom X, 64/174
Organophosphate induced delayed neuropathy
axonal degeneration, 64/229
cranial nerve palsy, 64/230
intermediate syndrome, 64/230
neuropathy target esterase, 64/229
paraparesis, 64/229
Schwann cell hypertrophy, 64/229
steppage gait, 64/229
wallerian degeneration, 64/229
Organophosphate induced delayed polyneuropathy
see also Organophosphate intoxication
ataxia, 64/161
carbamate, 64/177
carbaril, 64/177
chlorpyrifos, 64/177
CNS change, 64/160
corticospinal tract, 64/175
denervation, 64/177
dichlorvos, 64/177
EMG, 64/176
fibrillation, 64/177
gingili oil, 64/176
jake, 64/175
jake leg, 64/175
jake walk, 64/175
methamidophos, 64/176
motor polyneuropathy, 64/176
muscle spasm, 64/176
muscular atrophy, 64/176
neuropathology, 64/160
neuropathy target esterase inhibition, 64/151
organophosphate intoxication, 64/151
parathion, 64/177
pathogenesis, 64/161
phosphocreosote, 64/175
spastic paraplegia, 64/176
spinal cord, 64/175
term adequacy, 64/161
m-tolylmethyl carbamate, 64/177
tri-*o*-cresyl phosphate, 64/152, 175
tri-*o*-cresyl phosphate intoxication, 64/152
Organophosphate intoxication, 37/541-559
see also Acute muscle weakness *and*
Organophosphate induced delayed
polyneuropathy
A band loss, 64/228

acetylcholine accumulation, 37/555, 64/224
acetylcholine hydrolysis, 64/157
acetylcholine intoxication, 64/160
acetylcholinesterase, 64/151
acetylcholinesterase inhibition, 37/555, 64/151,
157
acetylcholinesterase phosphorylation, 64/158
activated charcoal, 64/172
acute intoxication, 64/226
acute muscle weakness, 41/288
agaricide, 64/152
aging, 64/224
anoxia, 64/228
anticholinesterase, 64/224
antidromic backfiring, 64/163
apnea, 64/174
areflexia, 64/167
astrocytic edema, 64/228
ataxia, 37/555, 64/167
atropine, 37/555, 557, 64/170, 172, 230
autonomic nervous system, 64/162, 75/507
autonomic neuropathy, 51/284
axonal neuropathy, 64/160
azinphos, 64/156
azinphos ethyl, 64/155
azinphos methyl, 64/155
behavior, 64/226
benzodiazepine, 37/559, 64/172
bladan, 64/153
bromophos, 64/155
calcium influx, 64/228
carbofenotion, 64/155
cardiac arrest, 64/169
cardiac dysrhythmia, 64/227
cardiovascular symptom, 37/554, 556, 64/167
carejin, 64/155
cathartic, 64/172
central core myopathy, 62/333
central effect, 64/226
central respiratory depression, 37/556, 64/167,
226
central respiratory failure, 37/556, 64/167
central type, 64/224
chemical structure, 37/542-550
chemistry, 37/542, 64/152
Cheyne-Stokes respiration, 37/555, 64/167
chlormethine intoxication, 64/151
chlorphoxim, 64/155
chlorpyrifos, 64/155
chlorpyrifos methyl, 64/155
chlorthiophos, 64/169
cholinergic crisis, 51/284, 64/170
cholinesterase blocking agent, 37/554, 51/284

clinical course, 64/170
clinical features, 64/166
clinical use, 37/552
clofenvinfos, 64/155
coma, 37/555, 64/167
compound muscle action potential, 64/163
confusion, 64/226
convulsion, 37/555, 64/174
coumafos, 64/155
cranial nerve palsy, 64/230
curare, 64/165
cyanazine, 64/153
cyanide intoxication, 64/151
cytoplasmic inclusion body myopathy, 62/345
deacetylation, 64/157
death, 37/554, 556, 64/167
delayed effect, 64/227
delayed neuropathy, 51/284, 64/229
demeton, 64/152, 155
demeton methyl, 64/155
demeton sulfoxide, 64/154, 169
detoxification, 64/155
diagnosis, 64/170
dialifos, 64/155, 169
dialkylphosphoryl enzyme, 64/224
dicapthon, 64/155
dichlofenthion, 64/155
dichlorvos, 64/152, 155
dicrotophos, 64/155
dimethoate, 64/154, 169
dimpylate, 64/154
dioxation, 64/155
dyflos, 64/152, 223
dysarthria, 64/167
ecothiopate iodide, 64/153
edifenphos, 64/155
EEG, 37/554, 64/172, 227
encephalopathy, 51/284
endothion, 64/155
epilepsy, 37/555, 64/172, 174, 226
erythrocyte acetylcholinesterase, 64/171
O-ethyl S-[2-(diisopropylamino)ethyl]methyl-
   phosphonothionate, 64/223
famphur, 64/155
fasciculation, 64/168
fenclofos, 64/155
fenitrothion, 64/155
fensulfothion, 64/155
fenthion, 64/155
fonofos, 64/155
formothion, 64/155
fungicide, 64/152
gastric lavage, 64/171

glutathione transferase, 64/156
headache, 37/555, 64/167, 226
heptenophos, 64/169
hexaethyl tetraphosphate, 64/153
higher nervous function, 64/226
hippocampus, 37/555, 64/228
history, 37/541, 64/151
hypertension, 64/228
ICU therapy, 64/168
industrial plasticizer, 64/151
insecticide, 64/151
insomnia, 64/167
intermediate syndrome, 64/230
ischemia, 64/228
jodfenfos, 64/155
leaving group, 64/224
leptophos, 64/155
long-term effect, 64/227
malathion, 37/554, 64/152, 155
mechanical ventilation, 64/172
mercaptophos, 64/152
metabolism, 37/551
methamidophos, 64/155
methidathion, 64/155, 169
metrifonate, 64/152, 155
mevinphos, 64/155
midazolam, 64/173
miosis, 64/167, 170, 228
monoalkylphosphoryl enzyme, 64/224
monocrotophos, 64/155
motor end plate, 64/228
motor end plate dysfunction, 64/162
muscarinic symptom, 64/167
muscarinic type, 37/554, 64/224
muscle weakness, 37/554, 64/151
myasthenic syndrome, 64/167
myoclonia, 64/167
myopathy, 64/160, 228
naled, 64/155
nematocide, 64/152
neuromuscular block, 37/555, 558, 40/297, 304
neuron degeneration, 64/228
neuropathology, 64/160, 228
neuropathy target esterase, 64/151, 229
neuropathy target esterase inhibition, 64/160
neurotoxic esterase, 37/475, 51/284
nicotinic effect, 64/167
nicotinic type, 37/554, 64/224
nonanticholinesterase effect, 64/224
obidoxime, 37/558, 64/159, 172, 230
octamethyl pyrophosphoramide, 64/152, 155
odor, 64/170
onset, 64/167

organophosphate induced delayed polyneuropathy, 64/151
oxime, 64/158
oxydemeton methyl, 64/155
pancuronium, 64/173
paralysis, 37/554, 64/167
paraoxon, 64/152
parathion, 37/554, 558, 64/152, 155, 169
parathion/demeton sulfide, 64/169
parathion methyl, 64/155, 169
pathogenesis, 64/224
pathomechanism, 64/11
pathophysiology, 64/162
peripheral effect, 64/228
pesticide, 64/151
petroleum additive, 64/151
pharmacodynamics, 37/551
phenthoate, 64/155
phenylphosphonothioic acid
    O-ethyl-O-(4-nitrophenyl)ester, 64/155
phenytoin, 64/172
phorate, 64/155
phosalone, 64/155
phosfolan, 64/155
phosphamidon, 64/155
phosphocreosote, 64/152
phospholine, 64/153
phosphorotrithioic acid tributyl ester, 64/155
phoxim, 64/155
physostigmine, 64/230
piriform cortex, 64/228
pirimiphos methyl, 64/155
pralidoxime, 37/557, 64/159, 172, 230
pralidoxime chloride, 37/558, 64/230
pralidoxime mesilate, 64/230
pralidoxime methylsulfate, 64/230
propaphos, 64/155
pulmonary symptom, 37/554, 64/167
pyridostigmine, 64/174, 230
pyrophosphates, 64/154
receptor, 64/226
repetitive discharge, 64/165
respiratory failure, 37/555, 557, 64/169
sarin, 37/556, 64/152, 223
Schwann cell, 64/229
secondary demyelination, 64/160
serum cholinesterase, 64/170
sex ratio, 64/169
sign, 64/226
soman, 37/556, 64/152, 223
suicide attempt, 64/151
sweating, 37/554, 64/170
symptom, 64/226

tabun, 64/152, 223
tachycardia, 64/228
target organ, 64/226
targetoid fiber, 40/80
temefos, 64/155
tetraethyl pyrophosphate, 64/151, 155, 223
thalamic nucleus, 64/228
thiometon, 64/155
toxic myopathy, 62/612
toxicology, 37/553-557, 64/156
treatment, 37/557, 64/171, 230
tremor, 37/555, 64/167, 226
tributyl phosphorotrithioite, 64/155
trichloronate, 64/155
tri-o-cresyl phosphate, 37/475, 554, 64/152, 155
tricresyl phosphates, 64/152
type I syndrome, 64/168
type II syndrome, 64/168
visual impairment, 37/554, 64/167
wartime use, 64/173
Z band loss, 64/228
Organophosphorus compound intoxication
  chemistry, 37/542
  cholinesterase blocking agent, 37/541-558
  death, 37/541
  demyelination, 9/652
  ergot intoxication, 9/652
  exposure, 37/541
  headache, 48/420
  insecticide, 37/542
  irreversible inhibitor, 37/451
  neurotoxicity, 37/541
  peripheral nervous system, 9/652
Organotin
  brain edema, 64/4
  neurotoxin, 64/4, 129, 135
  toxic encephalopathy, 64/4
Organotin intoxication
  absence, 64/142
  absorption, 64/135
  acaricide like change, 64/137
  adenosine triphosphate hydrolysis, 64/143
  aggression, 64/142
  alkyl chain length, 64/135
  animal experiment, 64/135, 139
  axonal reaction, 64/140
  basal ganglion, 64/142
  biochemical effect, 64/142
  blindness, 64/138
  blood level, 64/135
  brain edema, 64/135, 138
  brain level, 64/135
  chemistry, 64/134

chloride transport, 64/143
chlorine/hydroxide ion exchange, 64/146
confusion, 64/141
convulsion, 64/139
cytochrome P450 system, 64/135
dementia, 64/142
demyelination, 64/136
dendritic vacuolation, 64/146
dentate fascia neuron, 64/146
dimethylmonoethyltin, 64/144
EEG, 64/142
epilepsy, 64/138, 141
erythrocyte affinity, 64/135
excitotoxic damage, 64/140
excitotoxic effect, 64/145
excretion, 64/135
extracellular edema, 64/136
GABA neuron, 64/146
Golgi apparatus, 64/140
headache, 64/138, 141
hexachlorophene like change, 64/137
hippocampus, 64/140, 145
inclusion body, 64/140
individual susceptibility, 64/139
insomnia, 64/141
intracranial hypertension, 64/138
kidney level, 64/135
lethargy, 64/139
limbic system, 64/140
mental derangement, 64/138
metabolism, 64/135
microtubular assembly, 64/143
myelin protein, 64/142
myelin splitting, 64/137
myelin vacuolation, 64/136
myelinotoxicity, 64/144
nausea, 64/138
nerve conduction velocity, 64/137
neuronal necrosis, 64/142
neuropathology, 64/136, 140, 142
optic atrophy, 64/136
papilledema, 64/136, 138
paraplegia, 64/135
pathomechanism, 64/142, 144
perforant path, 64/146
peripheral nerve, 64/137
photophobia, 64/138
plasma ultrafiltrate, 64/138
Purkinje cell, 64/142
pyruvate dehydrogenase complex, 64/144
radical binding, 64/143
rage, 64/142
retinal neuron, 64/137

self-mutilation, 64/139
skin contact, 64/138
Sommer sector, 64/140
sphincter control, 64/138
spinal cord edema, 64/136
spongiform encephalopathy, 64/136
spongiform myelopathy, 64/136
stalinon, 64/138
startle epilepsy, 64/139
temporal lobe cortex, 64/142
tetrasubstituted compound, 64/135
tissue toxicity, 64/129
tremor, 64/139
trialkyltin, 64/134
trimethyltin syndrome, 64/139
triphenyltin, 64/135, 139
tri-*n*-propyltin, 64/135
ultrastructure, 64/136, 140
vertigo, 64/138
Orgasm
    exertional headache, 48/380
    female sexual function, 61/333
    headache, 48/6, 378
    paraplegia, 26/456
    sexual function, 26/440, 442, 448
Oriental language
    alexia, 4/114, 45/446
Orientation
    arrow symbol, 3/210
    body, *see* Body orientation
    cerebral localization, 3/207
    delirium, 46/528
    disorder, *see* Disorientation
    geographic, 3/220
    line, *see* Line orientation
    navigational performance, 3/210
    neuropsychologic defect, 45/524
    physiologic aspect, 3/202-211
    psychological aspect, 3/202-211
    reduplication phenomenon, 3/251-256
    sensation and movement, 3/205
    sense of direction, 3/209
    spatial, 3/203
    spatial localization, 3/216
    spatial orientation, 3/212-227
    spatial scheme, 3/221
    time sense, 3/229-249
    vertigo, 3/206
Orienting reaction
    alertness, 3/156
    behavior change, 3/52
    coma, 3/73
    consciousness, 3/52

emotion, 3/322
prefrontal syndrome, 2/742
subcortical, 3/63
terminology, 3/52
Ornate butterfly cod intoxication
picture, 37/73
Ornithine carbamoyltransferase
epilepsy, 72/222
mitochondrial protein import, 66/396
Ornithine carbamoyltransferase deficiency
amino acid, 29/91
ammonia, 29/91
biochemical finding, 29/91
blood urea nitrogen, 29/92
classic type, 29/90
diagnosis, 29/94
diet, 29/90
genetics, 29/93
metabolic ataxia, 21/574, 577
neomycin, 29/93
orotic aciduria, 29/91, 249
pathology, 29/93
Rett syndrome, 29/325, 63/437
Reye syndrome, 27/363, 29/339, 56/160
treatment, 29/93
variant type, 29/91
Ornithine transcarbamyl deficiency, *see* Ornithine
carbamoyltransferase deficiency
Ornithine transcarbamylase, *see* Ornithine
carbamoyltransferase
Ornithorhynchus anatinus
mammalia intoxication, 37/94
Orocraniodigital dysostosis, *see* Juberg-Hayward
syndrome
Orodyskinesia
edentulous, *see* Edentulous orodyskinesia
Orofacial dyskinesia
see also Linguobuccofacial dyskinesia
amiodarone, 63/190
antiepileptic agent, 65/499
athetosis, 49/382
chorea-acanthocytosis, 63/281
ergometrine, 49/56
focal athetosis, 49/383
hyperactive mesolimbic dopamine activity, 49/56
hypoactive mesolimbic noradrenalin activity,
49/56
levodopa induced dyskinesia, 49/195
neuropharmacology, 49/55
Orofaciodigital syndrome
corpus callosum agenesis, 50/163
Orofaciodigital syndrome type I
alopecia, 43/447

birth incidence, 43/447
cheilopalatoschisis, 30/247, 43/447
cheiloschisis, 50/167
CNS, 31/251
corpus callosum agenesis, 50/167
differential diagnosis, 31/252
digital abnormality, 43/447
frenula abnormality, 50/167
genetics, 31/250
hydrocephalus, 30/100, 31/252, 43/448
hypertelorism, 30/247
interhemispheric cyst, 50/168
lyonization, 43/448
mental deficiency, 31/251, 43/447, 46/67
Mohr syndrome, 31/252
multilobulated tongue, 43/447
oral lesion, 31/251
skeletal abnormality, 31/251
skin lesion, 31/250
subependymal heterotopia, 50/168
tongue hamartoma, 43/447, 50/167
X-linked dominant, 50/168
Orofaciodigital syndrome type II, *see* Mohr
syndrome
Oromandibular dyskinesia
focal athetosis, 49/383
Oromandibular dystonia
levodopa induced dyskinesia, 49/195
Oropharyngeal dysphagia
disorder, 75/629
Orotate phosphoribosyltransferase deficiency
orotic aciduria, 42/608
Orotic aciduria
anemia, 42/607
citrullinemia, 29/96
hereditary, *see* Hereditary orotic aciduria
Lesch-Nyhan syndrome, 63/256
leukopenia, 42/607
megaloblastic anemia, 63/256
mental deficiency, 42/607, 63/256
ornithine carbamoyltransferase deficiency, 29/91,
249
orotate phosphoribosyltransferase deficiency,
42/608
orotidine-5'-phosphate decarboxylase deficiency,
42/608
pregnancy, 29/240
psychomotor retardation, 42/607, 63/256
Orotidine-5'-phosphate decarboxylase deficiency
orotic aciduria, 42/608
Oroya fever, *see* Verruga peruana
Orphenadrine
muscle contraction headache, 5/169

parkinsonism, 6/219
Orthochromatic leukodystrophy
    see also Sudanophilic leukodystrophy
    Addison disease, 10/114
    adrenoleukodystrophy, 10/114
    adult form, 10/29, 42/501
    associated disease, 10/106
    ataxia, 42/500
    behavior disorder, 42/501
    behavior disorder in children, 42/500
    Binswanger disease, 10/125
    brain atrophy, 10/114
    cholesterol metabolism, 42/501
    choreoathetosis, 42/500
    classification, 10/37
    clinical features, 10/108
    combined form, 10/113, 42/500
    corpus callosum agenesis, 10/114
    cutis marmorata, 10/113, 123
    dementia, 42/500
    demyelination, 42/501
    Divry-Van Bogaert syndrome, 14/72
    EEG, 42/501
    endocrine disorder, 10/114
    epilepsy, 10/123
    genetics, 10/109, 113, 42/501
    histochemistry, 10/110, 42/501
    history, 10/11
    infantile case, 10/107, 111
    leptomeningeal angiomatosis, 10/113
    lipid hexosamine, 42/501
    Löwenberg-Hill leukodystrophy, 10/106
    mental symptom, 42/501
    microcephaly, 10/114
    microgyria, 10/114
    muscular dystrophy, 10/114
    Norman disease, 10/115
    Norman type, *see* Norman leukodystrophy
    nystagmus, 42/500
    pachygyria, 10/114
    palmitic acid, 42/501
    pathology, 10/109, 113, 42/501
    Peiffer classification, 10/106
    Pelizaeus-Merzbacher disease, 10/106, 42/500
    phakomatosis, 10/113, 120-127
    prelipoid, 10/115
    progressive myoclonus epilepsy, 10/113
    pure form, 10/105, 107, 42/500
    rare combination, 10/114
    secondary pigmentary retinal degeneration, 10/113
    Seitelberger type, 10/106
    seizure, 10/123, 42/500
    sex ratio, 42/501
    simple type, 10/107-111, 42/500
    spastic paraplegia, 42/500
    spongy degeneration, 10/114
    strabismus, 42/500
    subacute necrotizing encephalomyelopathy, 10/114
    symptomatic type, 10/114
    system degeneration, 10/114
    tremor, 42/500
    urinary incontinence, 42/500
    Wolman disease, 10/114
Orthodromic sensory potential
    recording, 7/146
Orthodromic stimulation
    technique, 7/146
Orthomyxovirus
    acute viral myositis, 56/193
    features, 56/15
    influenza A virus, 56/15
    influenza B virus, 56/15
    leukoencephalitis, 56/15
    RNA virus, 56/15
Orthopedic syndrome
    diastematomyelia, 50/436
    diplomyelia, 50/436
Orthopedic trauma
    cervical spine, 70/27
    neurologic complication, 70/25
    phantom limb, 70/45
    reflex sympathetic dystrophy, 70/43
    spinal trauma, 70/27
    survey, 70/25
Orthopoxvirus
    neurotropic virus, 56/5
Orthostatic hypotension
    see also Shy-Drager syndrome
    acquired immune deficiency syndrome, 63/153
    acute intermittent porphyria, 63/150
    acute pandysautonomia, 51/475, 63/152
    adrenal insufficiency, 63/145
    adrenergic receptor stimulating agent, 63/163
    age, 43/66, 63/145
    aged, 1/473, 2/121, 63/154
    alcoholic polyneuropathy, 63/151
    alcoholism, 1/473, 63/148, 151
    aldosterone release, 63/145
    alveolar hypoventilation, 63/146
    amyloid autonomic polyneuropathy, 63/149
    amyloid polyneuropathy, 51/416
    amyloidosis, 1/472, 63/148
    amyotrophic lateral sclerosis, 63/147
    angiotensin II, 63/145
    anhidrosis, 8/344, 63/145

anterior horn syndrome, 43/66
antihypertensive agent, 63/152
arsenic intoxication, 63/152
autonomic failure, 75/114, 713
autonomic nervous system, 22/253, 43/66, 75/715
autonomic polyneuropathy, 51/486, 63/148
autonomic syncope, 22/262
baroreflex, 2/120, 63/143
baroreflex sensitivity, 63/160
bedside test, 63/157
beta adrenergic receptor blocking agent, 63/162
blood pressure response, 63/157
botulinum toxin intoxication, 63/152
brain blood flow, 11/123
brain infarction, 63/156
brain tumor, 63/155
caffeine, 63/164
catecholamine, 43/66, 74/325
cause, 63/145
ciguatoxin intoxication, 65/161
classification, 75/714
cold pressor test, 63/158
cold stimulus, 63/157
collagen vascular disease, 63/152
countermaneuver, 75/718
definition, 63/143
delayed, 75/127
desmopressin, 63/164
diabetes insipidus, 63/145
diabetes mellitus, 63/148, 70/139, 75/599
diabetic autonomic polyneuropathy, 1/472,
    22/263, 63/148
diabetic autonomic visceral neuropathy, 27/115
diabetic polyneuropathy, 1/472, 22/263, 63/148
diagnosis, 63/156
diamorphine intoxication, 65/352
differential diagnosis, 63/156
dihydroergotamine, 63/164
dipeptidyl carboxypeptidase I inhibitor
    intoxication, 65/445
dopamine β-mono-oxygenase deficiency, 59/163,
    63/148
dopamine receptor blocking agent, 63/164
drug, 75/177, 723
dysautonomia, 22/262
ephedrine, 75/723
ergot alkaloid, 63/164
Fabry disease, 63/156
fastigial nucleus, 63/232
fecal incontinence, 43/66
fludrocortisone, 75/723
galvanic skin response, 63/160
geriatric syncope, 22/263

guanethidine intoxication, 37/436
Guillain-Barré syndrome, 51/257, 63/148, 151
handgrip, 63/157
handgrip pressor response, 63/159
heart disease, 63/145
heart rate, 63/144
heart rate response, 63/157
hereditary amyloid polyneuropathy, 42/518,
    63/156
hereditary amyloid polyneuropathy type 1, 51/421
hereditary amyloidosis, 63/149
hereditary motor and sensory neuropathy type I,
    63/148
hereditary motor and sensory neuropathy type II,
    63/148
hereditary sensory and autonomic neuropathy type
    II, 63/148
hereditary sensory and autonomic neuropathy type
    III, 1/478, 21/109, 43/58, 60/7, 27
heterocyclic antidepressant, 65/316
Holmes-Adie syndrome, 22/262, 63/156
Horner syndrome, 63/145
Huntington chorea, 63/147
hydrocephalus, 22/262, 63/156
hyperbradykininemia, 43/67
hypoglycemia, 63/148
hypovolemia, 63/145
idiopathic, see Pure autonomic failure
inspiratory gasping, 63/146
irregular respiration, 63/146
juvenile spinal cord injury, 61/243
leg crossing, 75/719
leprosy, 51/218, 63/154
light headedness, 63/143
MAO inhibitor, 63/152
medulla oblongata lesion, 63/232
mercury intoxication, 63/152
mineralocorticoid, 63/161
mitral valve prolapse, 63/155
mixed connective tissue disease, 63/152
multiple myeloma polyneuropathy, 63/393
multiple neuronal system degeneration, 43/66,
    63/145, 232, 75/168
multiple sclerosis, 63/156
neuroanatomy, 63/143
neurogenic, see Neurogenic orthostatic
    hypotension
neurogenic cause, 63/145
neurotransmitter, 63/144
nonneurogenic cause, 63/156
nonpharmacologic treatment, 63/161
noradrenalin, 26/319, 63/144
nucleus tractus solitarii, 63/143

octreotide, 63/164
olivopontocerebellar atrophy, 2/121, 43/66, 63/146
olivopontocerebellar atrophy (Dejerine-Thomas), 60/667
olivopontocerebellar atrophy (Wadia-Swami), 60/494, 667
Onuf nucleus, 63/146
organic solvent intoxication, 63/152
paraneoplastic polyneuropathy, 63/152
Parkinson disease, 63/145, 147, 75/180
parkinsonism, 22/262, 43/66, 63/146
pharmacologic treatment, 63/161
pheochromocytoma, 14/252
physical countermaneuver, 75/719
physiology, 43/67, 63/143
plasma catecholamine, 26/319, 63/159
polyneuropathy, 22/262, 51/416, 63/146
porphyria, 22/263, 63/148
porphyric autonomic neuropathy, 63/150
postural blood pressure response, 63/158
postural heart rate response, 63/158
pressor drug infusion, 63/160
primary, *see* Pure autonomic failure
primary amyloidosis, 22/253, 63/149
progressive dysautonomia, 59/136, 151
progressive supranuclear palsy, 63/147
prostaglandin synthesis inhibitor, 63/162
ptosis, 63/148
pure autonomic failure, 22/231, 63/146
renal insufficiency, 63/151
renin-angiotensin system, 63/145
renin release, 63/144
rheumatoid arthritis, 63/152
secondary amyloidosis, 22/253, 63/149
selective serotonin reuptake inhibitor, 65/316
sexual impotence, 63/146
Shy-Drager syndrome, 1/474, 2/121, 43/66, 63/145
sinus dysrhythmia, 63/158
sleep apnea syndrome, 63/146
spinal cord injury, 22/262, 26/318, 61/442
spinal cord lesion, 63/153
splanchnic innervation, 63/144
squatting, 75/719
stress, 74/325
striatal lesion, 6/670
striatonigral degeneration, 63/146
stridor, 63/146
sudomotor axon reflex, 63/160
sudomotor sympathetic function, 63/160
supine, 75/725
survey, 75/114, 714

sympathetic skin response, 63/160
symptom, 43/66, 63/143, 75/169, 713
syncope, 63/143, 75/715
syringobulbia, 50/445, 63/153
syringomyelia, 63/153
systemic lupus erythematosus, 63/152
tabes dorsalis, 1/472, 2/120, 22/262
Tangier disease, 63/156
test, 74/596
test battery, 63/156
tetraplegia, 26/318, 61/442
thallium intoxication, 63/152
transverse spinal cord lesion, 22/263, 61/442, 63/153, 232
transverse spinal cord syndrome, 2/198
transverse syndrome, 2/198
treatment, 63/161
tricyclic antidepressant, 65/316
tyramine, 22/270
uremic polyneuropathy, 63/511
urinary catecholamine, 63/159
urinary incontinence, 43/66, 63/143
vagus nerve, 63/144
Valsalva maneuver, 63/157
Valsalva ratio, 63/157
various disorders, 22/262
vasoconstriction, 75/723
vasopressin, 63/145
vincristine intoxication, 63/152
visual impairment, 63/143
vitamin B12 deficiency, 51/341, 63/152
Von Hippel-Lindau disease, 14/255
Wernicke encephalopathy, 63/151
Orthostatic intolerance
    altitude, 74/326
    endurance exercise training, 74/325
    mitral valve prolapse, 74/326
    pheochromocytoma, 74/326
    stress, 74/325
Orthostatic intolerance syndrome
    stress, 74/324
Orthostatic stress
    response, 75/112
Orthostatic tremor
    task specific tremor, 49/592
Ortiiomyxovirus
    human, 56/3
    taxonomic place, 56/3
Orton theory
    developmental dyslexia, 4/380, 46/106
Os odontoideum, *see* Dens
Oscillation
    attention, 3/164

basal ganglion function, 6/93
  congenital nystagmus, 2/367
  miner nystagmus, 1/597, 2/367
  spasmus nutans, 2/367
Oscillatory myoclonus
  features, 49/619
Oscillopsia
  brain embolism, 53/207
  brain infarction, 53/207
  cerebrovascular disease, 53/207
  intracranial hemorrhage, 53/207
  syringobulbia, 50/445
  transient ischemic attack, 53/207
Osler syndrome, *see* Hereditary hemorrhagic
  telangiectasia (Osler)
Osler-Vaquez disease, *see* Polycythemia vera
Osler-Weber-Rendu disease, *see* Hereditary
  hemorrhagic telangiectasia (Osler)
Oslo, Norway
  neurology, 1/12
Osmolality
  serum definition, 63/545
  water intoxication, 63/545
Osmoregulation
  betaine, 63/555
  brain hypernatremia, 63/555
  brain hyponatremia, 63/547
  brain potassium, 63/555
  brain taurine, 63/555
  brain water content, 63/545
  glutamine, 63/555
  inositol, 63/555
  spinal cord disease, 75/581
  trimethylamine, 63/555
Osmotic agent
  brain edema, 57/214
  cytotoxic brain edema, 63/213
  dehydration, 57/215
  neurologic intensive care, 55/209
  rebound, 57/215
OsO4 staining
  lipid, 9/31
OSPOM syndrome
  Kearns-Sayre-Daroff-Shy syndrome, 62/313
  progressive external ophthalmoplegia, 62/313
  progressive external ophthalmoplegia
    classification, 62/290
Osseous dysplasia
  mental deficiency, *see* Ruvalcaba syndrome
Ossicle hypoplasia
  auditory, *see* Auditory ossicle hypoplasia
Ossiculum terminale
  site, 50/392

Ossification
  craniosynostosis, 42/20
  ependymoma, 18/120
  heterotopic, *see* Neurogenic osteoarthropathy
  styloid chain bone, 48/506
  stylomandibular ligament, 48/507
Ossification center
  dens, 25/253
  vertebra, 20/21
Ossifying fibroma
  cranial vault tumor, 17/109
  fibrous dysplasia, 14/176
  histologic variant, 14/175
  skull base tumor, 17/146
Osteitis deformans (Paget)
  alkaline phosphatase, 38/361, 43/451
  amyotrophic lateral sclerosis, 22/309
  angioid streak, 13/141, 38/364, 43/452
  basilar impression, 32/67, 38/365, 50/399, 70/14
  bilateral facial paresis, 32/76
  blindness, 43/451
  brain stem, 38/361
  calcitonin, 38/368, 43/452, 60/773
  cauda equina, 70/14
  cauda equina intermittent claudication, 38/367
  compression neuropathy, 38/361, 43/451
  cranial hyperostosis, 43/450
  cranial nerve, 38/361, 70/12
  cranial nerve palsy, 2/364
  cranial vault tumor, 17/118, 170, 38/369
  CSF flow, 38/366
  cytotoxic agent, 43/452
  deafness, 70/13
  dementia, 38/366, 46/400, 70/14
  diphosphonate, 43/452
  epidemiology, 43/451
  etidronic acid, 38/368
  facial pain, 70/13
  facial paralysis, 17/76, 38/363, 43/451, 70/13
  fibrous dysplasia, 14/193, 199
  foramen magnum, 38/361, 365, 50/396
  gait disturbance, 38/362, 366, 43/451, 46/400
  genetics, 43/452
  glucagon, 43/452
  headache, 38/362, 43/451, 48/6
  hearing loss, 22/493, 38/362, 43/451, 50/396,
    60/773
  hereditary progressive cochleovestibular atrophy,
    22/492, 60/772
  history, 38/362
  Huntington chorea, 6/316, 49/291
  hydrocephalus, 38/366, 46/400, 70/14
  hydroxyproline, 38/361, 43/451

ivory vertebra, 19/172
lumbar vertebral canal stenosis, 20/783
malignant degeneration, 17/170, 38/369, 43/451
memory loss, 38/366
mithramycin, 38/368, 43/452
nerve, 38/367
neural foramina, 70/12
neurologic aspect, 38/361-369
nonrecurrent nonhereditary multiple cranial
    neuropathy, 51/571
oculomotor nerve, 70/13
olfactory nerve, 70/12
optic atrophy, 13/83, 38/364
optic nerve, 70/12
osteoporosis, 38/361
osteosarcoma, 38/369, 43/451
pain, 19/172, 38/361, 363
papilledema, 38/364
platybasia, 32/67, 38/365
psychiatric disorder, 43/451
radiculopathy, 38/366
sarcoma, 17/118, 170
sex ratio, 17/118, 170, 43/451
skull, 17/118, 38/362-365, 43/450
skull base tumor, 17/170, 172
space occupying lesion, 17/170, 38/361
spinal cord, 38/361, 366, 70/14
spinal cord compression, 19/172, 20/782, 38/366,
    46/400
spinal nerve, 70/15
survey, 70/11
tetraplegia, 38/367
tinnitus, 38/362
treatment, 38/367-369, 43/452
trigeminal neuralgia, 38/364, 50/399
urinary incontinence, 38/366, 46/400
vertebral abnormality, 20/782, 43/450
vertebral scalloping, 14/193
vertigo, 2/362, 38/364
X-ray finding, 20/28
Osteitis fibrosa, see Fibrous dysplasia
Osteitis fibrosa cystica, see Aneurysmal bone cyst
Osteitis fibrosa disseminata, see Fibrous dysplasia
Osteitis fibrosa generalisata
    hyperthyroidism, 14/163
Osteoarthropathy
    bismuth intoxication, 64/337
    neurogenic, see Neurogenic osteoarthropathy
    neurotrophic, see Neurotrophic osteoarthropathy
Osteoarthrosis
    cervicogenic headache, 48/405
    posttraumatic syringomyelia, 61/42
    spondyloepiphyseal dysplasia tarda, 43/481

temporomandibular cephalalgia, 48/413
Osteoblastoma
    benign, see Benign osteoblastoma
    cervical spine, 67/196
    incidence, 68/530
    spinal cord, 67/195
    spinal cord compression, 19/367
    spinal cord tumor, 68/513
    vertebral column tumor, 20/26
Osteocartilaginous exostosis, see Hereditary
        multiple exostosis
Osteochondrodystrophy
    vertebral canal stenosis, 20/637
Osteochondrofibroma
    intracerebral, 18/321
Osteochondroma
    see also Hereditary multiple exostosis
    age, 19/306
    congenital spinal cord tumor, 32/355-386
    hereditary multiple exostosis, 42/326
    histology, 19/307
    prognosis, 19/308
    sex ratio, 19/307
    site, 19/306
    spinal cord compression, 19/174
    treatment, 19/308
Osteochondrosis
    spinal cord compression, 19/372
Osteodysgenesis
    multisynostotic, see Multisynostotic
        osteodysgenesis
Osteodysplasia
    arthrodento, see Arthrodento-osteodysplasia
    fibrous, see Fibrous dysplasia
    lipomembranous polycystic, see Lipomembranous
        polycystic osteodysplasia
    Melnick-Needles, see Melnick-Needles syndrome
Osteodysplasty, see Melnick-Needles syndrome
Osteodystrophy
    Albright hereditary, see
        Pseudopseudohypoparathyroidism
    renal, see Renal osteodystrophy
Osteoectasia
    hyperphosphatasia, 43/409
Osteofibrosis deformans juvenilis, see Fibrous
        dysplasia
Osteogenesis imperfecta
    basilar impression, 32/67, 43/454, 50/399
    birth incidence, 43/454
    blue sclerae, 43/453
    bone fragility, 43/453
    clinical presentation, 40/336
    dental abnormality, 43/453

Divry-Van Bogaert syndrome, 14/513
fibrous dysplasia, 14/199
heart disease, 43/453
hereditary connective tissue disease, 39/393-400
hydrocephalus, 30/100, 43/454
hypotonia, 40/336
kyphoscoliosis, 43/453
osteoporosis, 43/453
scoliosis, 50/418
sensorineural deafness, 43/453
Vrolik disease, 22/494
Osteogenesis imperfecta tarda, *see* Ekman-Lobstein
disease
Osteogenic fibroma, *see* Benign osteoblastoma
Osteogenic sarcoma, *see* Osteosarcoma
Osteogenic tumor
benign spinal, *see* Benign spinal osteogenic tumor
Osteohypertrophy
Klippel-Trénaunay syndrome, 14/13, 67, 116, 393,
43/24
Osteoid osteoma
age, 19/297
benign, 19/295
differential diagnosis, 19/297, 301
histopathology, 19/298
incidence, 68/530
pain, 19/300
sex ratio, 19/297, 300
site, 19/297
skull base tumor, 17/147
spinal cord, 67/194
spinal cord compression, 19/173, 367
spinal cord tumor, 68/513
tomography, 19/173
vertebral column, 20/25
Osteolathyrism
aminoacetonitrile, 7/644
β-aminoproprionitrile, 7/644
lathyrism, 7/644, 36/505, 507, 65/6
Osteolysis
arthrodento-osteodysplasia, 43/331
arthrogryposis, 43/85
Kuskokwim arthrogryposis, 43/85
neurogenic acro-osteolysis, 42/294
neurotrophic osteoarthropathy, 38/447, 468-470
spinal ependymoma, 20/370
vertebral, *see* Vertebral osteolysis
Osteoma
age, 19/295
cranial vault tumor, 17/107
flexion contracture, 20/827
osteoid, *see* Osteoid osteoma
sex ratio, 19/295

sinus, *see* Sinus osteoma
site, 19/295
skull base tumor, 17/138, 140, 143
vertebral, *see* Vertebral osteoma
Osteomalacia
aluminum intoxication, 64/274
basilar impression, 50/399
hereditary hypophosphatemic, *see* Hereditary
hypophosphatemic osteomalacia
hyperparathyroid myopathy, 62/538
postgastrectomy syndrome, 28/233, 51/327
proximal muscle weakness, 40/317
vitamin D, 41/247
Osteomyelitis
bacterial endocarditis, 52/299
basal, *see* Basal osteomyelitis
dental headache, 5/227, 48/392
diamorphine, 37/391
hereditary sensory and autonomic neuropathy type
IV, 60/12
intracranial infection, 24/216, 657
jaw, 5/277
neurogenic acro-osteolysis, 42/294
opiate intoxication, 37/391
skull, *see* Skull osteomyelitis
vertebral, *see* Vertebral osteomyelitis
Osteoperiostitis
syphilitic, *see* Syphilitic osteoperiostitis
Osteopetrosis, 38/371-379
see also Albers-Schönberg disease
Albers-Schönberg disease, 70/15
bone fragility, 43/456
brain atrophy, 70/16
carbonate dehydratase II deficiency, 70/17
clinical features, 38/371-374
complication, 70/16
compression neuropathy, 43/456
facial paralysis, 43/456
familial, *see* Hereditary osteopetrosis
fibrous dysplasia, 14/198
hepatosplenomegaly, 43/456
history, 38/371
inherited, 70/15
ivory vertebra, 19/171
laboratory evaluation, 38/377
macrocephaly, 30/100
malignant, 70/17
mental deficiency, 46/63
neurologic manifestation, 38/378
optic atrophy, 13/68, 81, 43/456
optic nerve, 70/16
pathology, 38/379
radiologic features, 38/374-376

sensorineural deafness, 43/456
survey, 70/11
treatment, 38/379
Osteophyte
terminology, 7/447
Osteoporosis
ADR syndrome, 42/113
alcohol intoxication, 64/113
aminoaciduria, 42/517
anticonvulsant, 70/19
arthrodento-osteodysplasia, 43/331
cardiac pharmacotherapy, 63/189
cerebro-oculofacioskeletal syndrome, 43/341
citrullinuria, 42/538
deficiency neuropathy, 51/327
diabetic neuropathy, 70/20
drug therapy, 70/19
Flynn-Aird syndrome, 42/327
foot, 70/20
glycosaminoglycanosis type IV, 66/303
homocystinuria, 42/555
juvenile spinal cord injury, 61/247
Klippel-Trénaunay syndrome, 43/30
Lowe syndrome, 42/606
lysinuric protein intolerance, 29/210
menopause, 70/19
mental deficiency, 42/75
methylmalonic aciduria, 42/601
multiple neuronal system degeneration, 42/75
multiple sclerosis, 70/20
muscular dystrophy, 41/474
myopathy, 43/124
osteitis deformans (Paget), 38/361
osteogenesis imperfecta, 43/453
phenylketonuria, 29/43, 42/611
physical rehabilitation, 61/471
postmenopausal, see Postmenopausal osteoporosis
progeria, 43/466
reflex sympathetic dystrophy, 1/513, 8/328,
    51/480
scoliosis, 50/418
senile, see Senile osteoporosis
spinal cord, 70/20
spinal cord compression, 19/365
survey, 70/11, 18
thoracolumbar vertebral column injury, 61/95
Turner syndrome, 50/544
Werner syndrome, 43/489
xanthomatosis, 70/20
Osteosarcoma
brain metastasis, 18/205
cranial vault tumor, 17/117
differential diagnosis, 19/301

fibrous dysplasia, 14/198
incidence, 68/530
ivory vertebra, 19/171
osteitis deformans (Paget), 38/369, 43/451
radiation malignancy, 20/6
skull base tumor, 17/167
spinal cord, 67/197
spinal cord compression, 19/368
spinal cord tumor, 68/523
spinal epidural tumor, 20/129
vertebral column tumor, 20/32
Osteosclerosis
sclerosteosis, 31/258
vertebral, see Vertebral osteosclerosis
Osteosclerotic myeloma
chemotherapy, 63/396
classification, 69/290
confusional state, 63/395
corticosteroid, 63/396
cranial nerve palsy, 63/395
cranial neuropathy, 63/395
cyclophosphamide, 63/396
motor polyneuropathy, 63/393
multiple myeloma polyneuropathy, 51/430, 63/393
paraproteinemia, 69/290, 303
plasmapheresis, 63/395
POEMS syndrome, 47/619, 51/430, 63/394
polyneuropathy pathogenesis, 63/395
radiotherapy, 63/395
specific sign, 63/393
survey, 71/435
treatment, 63/395
vasculopathy, 63/395
Osteotomy
talipes equinovarus, 40/365
Ostracioidei
ostraciontidae, 37/82
Ostracion lentiginosus intoxication
chemical structure, 37/82
pahutoxic, 37/82
Ostracion meleagris Shaw intoxication, see
    Ostracion lentiginosus intoxication
Ostraciontidae intoxication
chemical structure, 37/82
pahutoxic, 37/82
Ostrea gigas intoxication
abdominal pain, 37/64
anorexia, 37/64
blood clotting, 37/64
constipation, 37/64
headache, 37/64
mortality, 37/64
nausea, 37/64

petechial hemorrhage, 37/64
vomiting, 37/64
Ostrum-Furst syndrome
dysraphia, 14/111
Osuntokun syndrome
congenital auditory agnosia, 22/522
congenital indifference to pain, 22/522
Ota nevus
Klippel-Trénaunay syndrome, 43/30
olivopontocerebellar atrophy (Wadia-Swami),
60/667
sensorineural deafness, 43/37
spinocerebellar degeneration, 43/37
Sturge-Weber syndrome, 43/37
trigeminal nerve, 43/37
Ota-Sato disease, *see* Ota syndrome
Ota syndrome
clinical features, 14/120
neurocutaneous syndrome, 14/101
spinocerebellar degeneration, 14/791
Otalgia
glossopharyngeal neuralgia, 5/209
herpes zoster oticus, 56/118
intermedius neuralgia, 48/487
stylohyoid syndrome, 48/510
superior laryngeal neuralgia, 48/495
OTAN staining
lipid, 9/25, 28
myelin, 9/31
Otani tumor, *see* Glomus jugulare tumor
Otic artery
anatomy, 11/1
primitive, *see* Primitive otic artery
Otic ganglion
anatomy, 5/338, 351, 48/488
Otic sinus
pro, *see* Pro-otic sinus
Otitic herpes, *see* Herpes zoster oticus
Otitic hydrocephalus
benign intracranial hypertension, 33/171
brain phlebothrombosis, 12/425-429
headache, 5/155, 48/6
pathogenesis, 16/236
superior longitudinal sinus thrombosis, 33/171
venous sinus thrombosis, 54/431
Otitis externa
headache, 5/209
Otitis media
Abt-Letterer-Siwe disease, 38/96
bacterial meningitis, 52/3
benign paroxysmal positional nystagmus, 2/377
brain abscess, 52/144
Candida meningitis, 52/400

cavernous sinus thrombosis, 52/173
cranial epidural empyema, 52/167
cranial subdural empyema, 52/167, 170
Down syndrome, 50/526
glycosaminoglycanosis type III, 66/302
Gradenigo syndrome, 2/91, 313, 16/210
gram-negative bacillary meningitis, 52/103
headache, 5/210
miosis, 48/332
ptosis, 48/332
transverse sinus thrombosis, 52/176, 54/397
tuberculous pachymeningitis, 33/161
Turner syndrome, 50/543
Otocephaly
see also Holoprosencephaly
cyclopia, 30/442
holoprosencephaly, 30/437, 50/233
Otodental dysplasia
dental abnormality, 42/419
sensorineural deafness, 42/419
Otogenic hydrocephalus
transverse sinus thrombosis, 54/397
Otolith lesion
benign paroxysmal positional nystagmus, 2/364,
370
positional nystagmus, 2/371
Otologic disease
headache, 5/35, 208, 212, 214, 330
Otologic function
examination sheet, 2/389
Otopalatodigital syndrome
broad thumb, 43/457
cheilopalatoschisis, 43/457
dwarfism, 43/457
hypertelorism, 30/248
mental deficiency, 43/457, 46/64
pectus excavatum, 43/457
skeletal malformation, 30/248
speech, 46/30
Otorrhagia
temporal bone fracture, 24/120, 125, 185
Otorrhea
see also CSF fistula
posttraumatic meningitis, 57/317
temporal bone fracture, 24/120, 125, 185
Otosclerosis
vertigo, 2/365
Ototoxicity
alkylating agent, 67/354
aminoglycoside intoxication, 65/482
carboplatin intoxication, 65/537
chemotherapy, 67/359
cisplatin, 67/354

cisplatin intoxication, 64/358
radiation, 67/359
salazosulfapyridine, 51/309
Ouabain
brain cortex sodium, 28/447
brain infarction, 53/116
brain ischemia, 53/116
Duchenne muscular dystrophy, 40/386
Ouabain intoxication
CSF production, 37/426
Ouabain sensitive adenosine triphosphatase
Huntington chorea, 49/261
Outcome scales
Glasgow, *see* Glasgow Outcome Scale
Outcomes
head injury, *see* Head injury outcome
pediatric head injury, *see* Pediatric head injury
outcome
posttraumatic brain abscess, *see* Posttraumatic
brain abscess outcome
traumatic intracranial hematoma, *see* Traumatic
intracranial hematoma outcome
Outer mesaxon
Schwann cell, 51/9
Outer pallidal segment
dentatorubropallidoluysian atrophy, 49/440
hemiballismus, 49/370
Ovarian cyst
multiple hamartoma syndrome, 42/754
Ovary carcinoma
late cerebellar atrophy, 60/585
metastasis, 71/619
Overactivity, *see* Hyperactivity
Overeating, *see* Bulimia
Overestimation
nosoagnosic, *see* Nosoagnosic overestimation
Overlap syndrome
classification, 71/4
dermatomyositis, 62/374, 71/131
inclusion body myositis, 62/374
polyarteritis nodosa, 55/358
polymyositis, 62/374, 71/131
Oversecretion
CSF, *see* CSF oversecretion
Overstretching
cervical spinal cord, 25/261
spinal cord injury, 25/261, 268
Ovulation
seasonal rhythm, 74/490
Oxaloacetic acid
hepatic coma, 49/215
Oxalosis
ethylene glycol intoxication, 64/124

facioacoustic, *see* Facioacoustic oxalosis
vitamin B_6 intoxication, 65/572
3-Oxalylaminoalanine
excitotoxic aminoacid, 65/13
glutamic acid receptor, 65/3
Guam amyotrophic lateral sclerosis, 59/274
isoxazole propionate stimulating agent, 65/3
Kii Peninsula amyotrophic lateral sclerosis,
59/274
neurotoxin, 36/507, 64/10, 65/1
pathomechanism, 65/1
toxic myelopathy, 64/10
β-*N*-Oxalylamino-L-alanine, *see*
3-Oxalylaminoalanine
3-Oxalylaminoalanine intoxication
action site, 64/10
β-*N*-Oxalyl-L-α,β-diaminopropionic acid, *see*
3-Oxalylaminoalanine
Oxapropanium iodide
restless legs syndrome, 8/318, 51/548
Oxazepam
benzodiazepine intoxication, 65/334
chemical structure, 37/356
clinical study, 37/359
diazepam intoxication, 37/199, 357
tranquilizing agent, 37/355, 359
Oxazepam intoxication
anticonvulsant, 37/199
Oxazolidinedione
epilepsy, 15/641
Oxazolidinedione intoxication
anticonvulsant, 37/200
Oxcarbazepine
antiepileptic agent, 65/496
carbamazepine intoxication, 65/507
epilepsy, 73/361
Oxcarbazepine intoxication
hyponatremia, 65/507
water intoxication, 65/507
Oxford method
arterial baroreflex, 75/432
autonomic nervous system, 75/432
Oxford, United Kingdom
neurology, 1/7
Oxicam derivative
headache, 48/174
migraine, 48/174
Oxidation-phosphorylation
defect, 66/396, 425
Luft syndrome, 66/425
Oxidation-phosphorylation coupling defect
bulimia, 62/506
exercise intolerance, 62/506

heat intolerance, 62/506
hyperthermia, 62/506
metabolic myopathy, 62/506
mitochondrial disease, 62/506
polydipsia, 62/506
sweating, 62/506
tachycardia, 62/506
Oxidative enzyme
    muscle biopsy, 40/6, 136
    myelination, 9/628
    oligodendrocyte, 9/628
Oxidative phosphorylation
    see also Mitochondrial abnormality
    achondroplasia, 41/214
    arsenic intoxication, 64/286
    arsenic polyneuropathy, 51/267
    Duchenne muscular dystrophy, 40/379
    experimental change, 40/158-160
    impairment, 40/171
    Luft syndrome, 41/210
    metabolic myopathy, 41/216
    mitochondrial myopathy, 41/209, 216
    muscle tissue culture, 40/187
    phenothiazine, 13/211
    progressive external ophthalmoplegia, 41/215
    thyrotoxic myopathy, 41/239
    thyrotoxic periodic paralysis, 41/242
    vitamin E deficiency, 40/171
Oxidative phosphorylation disorder
    respiratory chain defect, 66/422
Oxidopamine
    akinesia, 49/67
    autonomic nervous system, 75/38
    Parkinson disease, 49/52
Oxitriptan
    dyssynergia cerebellaris myoclonica, 60/603
    eosinophilia myalgia syndrome, 64/250
    manganese intoxication, 36/229
    migraine, 48/86
    phenylketonuria, 29/25
    posthypoxic myoclonus, 49/618
    schizophrenia, 46/496
Oxoacid decarboxylase
    branched chain, see Branched chain oxoacid
        decarboxylase
Oxoacid metabolism
    branched chain, see Branched chain oxoacid
        metabolism
2-Oxoglutaramic acid
    CSF, see CSF 2-oxoglutaramic acid
    hepatic coma, 49/215
α-Oxoglutarate dehydrogenase
    tricarboxylic acid cycle, 27/16

Oxoglutarate dehydrogenase deficiency
    choreoathetosis, 66/419
    dystonia, 66/419
    encephalomyopathy, 62/503
    fumarate hydratase deficiency, 62/503
    hyporeflexia, 62/503
    hypotonia, 66/419
    infantile hypotonia, 62/503
    lethargy, 62/503
    metabolic acidosis, 66/419
    microcephaly, 62/503
    mitochondrial disease, 62/503
    muscle biopsy, 62/503
    organic acid metabolism, 66/639
    pyruvate dehydrogenase-dihydrolipoyl
        dehydrogenase, 62/503
    pyruvate dehydrogenase-dihydrolipoyl
        transacetylase, 62/502
    3 subunits, 66/419
    sudden death, 66/419
    treatment, 66/420
    tricarboxylic acid cycle, 66/654
    vision, 62/503
    vomiting, 62/503
2-Oxoglutaric acid
    deficiency neuropathy, 51/330
    dihydrolipoamide dehydrogenase, 66/417
2-Oxoglutaric aciduria, see Oxoglutarate
        dehydrogenase deficiency
2-Oxoisocaproic acid
    maple syrup urine disease, 42/598
α-Oxoisovaleric acid
    maple syrup urine disease, 42/598
α-Oxo-β-methylvaleric acid
    maple syrup urine disease, 42/598
3-Oxothiolase deficiency, see
        2-Methylacetoacetyl-coenzyme A thiolase
        deficiency
β-Oxothiolase deficiency
    see also Hyperglycinemia
    hyperammonemia, 29/106
    hyperglycinemia, 29/204
Oxprenolol
    acquired myoglobinuria, 62/577
    beta adrenergic receptor blocking agent, 65/434
    migraine, 48/180
    toxic myopathy, 62/601
Oxycephaly
    see also Craniosynostosis
    acrocephalosyndactyly type I, 13/401
    acrocephalosyndactyly type V, 50/123
    Bardet-Biedl syndrome, 13/386
    Bonnet-Dechaume-Blanc syndrome, 14/264

convulsion, 30/226
craniosynostosis, 30/154, 220, 42/20, 50/119
lissencephaly, 50/260
mental deficiency, 30/227
methotrexate syndrome, 30/122
optic atrophy, 13/69
pituitary adenoma, 17/412
Sturge-Weber syndrome, 14/512
Oxydemeton, *see* Demeton sulfoxide
Oxygen
brain diaschisis, 54/125
brain infarction, 54/121
brain injury, 57/211
brain metabolism, 54/110, 121, 125, 57/70, 211,
463
brain microcirculation, 53/79
brain stem death, 57/463
chronic migrainous neuralgia, 48/251
exercise, 74/247
intracranial pressure, 23/215
positron emission tomography, 54/110, 72/371
work, 74/247
Oxygen deficiency
acute mountain sickness, 29/539-541
carbon monoxide intoxication, 29/541-547
neuropathology, 29/541, 545
Oxygen free radical
brain edema, 67/80
Oxygen intoxication
dysbarism, 12/668
mechanism, 29/533-536
neuropathology, 29/536-538
treatment, 29/538, 539
Oxygen radical
critical illness polyneuropathy, 51/584
Oxygen therapy
azobenzene intoxication, 36/434
cluster headache, 48/231, 242
Oxygenation
brain infarction, 53/431
hyperbaric, *see* Hyperbaric oxygenation
Oxymetazoline
classification, 74/149
Oxymorphone
opiate, 65/350
Oxynervon
chemistry, 10/249
Oxyphenbutazone
chronic paroxysmal hemicrania, 48/264
Oxytetracycline
neurotoxin, 65/472
Oxytocin
adenohypophysis, 2/443

water intoxication, 64/244
Ozeretski test
handedness, 4/255

P component
amyloid, 51/414
P protein
measles virus, 56/423
P0 protein
nerve myelin, 51/27, 29-31
wallerian degeneration, 51/32
P1 protein
nerve myelin, 51/27
P2 protein
nerve myelin, 51/27
Pa Ping, *see* Endemic periodic paralysis
Pachycheilia
Melkersson-Rosenthal syndrome, 8/209
Pachydermia
hypertrophic polyneuropathy, 14/791
kerato, *see* Keratopachydermia
oculocutaneous albinism, 43/5
Pachydermoperiostosis
cutis verticis gyrata, 43/244
Pachygyria
see also Brain cortex heterotopia
brain cortex disorganization, 50/249
brain cortex heterotopia, 50/248
cerebrohepatorenal syndrome, 43/340
corpus callosum agenesis, 50/150
epilepsy, 72/109
lissencephaly, 30/485, 42/39-41, 50/258
Lowe syndrome, 42/607
mental deficiency, 46/11, 48
microcephaly, 42/46, 50/263
microgyria, 30/484
morphology, 50/247
neurofibromatosis type I, 14/37
Norman microcephalic familial leukodystrophy,
10/103, 182
orthochromatic leukodystrophy, 10/114
pathology, 30/485-487
polymicrogyria, 50/257
porencephaly, 50/356
tetraploidy, 43/567
Pachymeningitis
see also Dura mater
basal, *see* Basal pachymeningitis
chronic meningitis, 56/644
definition, 52/185
externa, *see* Spinal epidural empyema
externa purulenta, *see* Spinal epidural empyema
hypertrophic, *see* Hypertrophic pachymeningitis

interna, 33/157-161
spinal, *see* Spinal pachymeningitis
spinal cord compression, 52/188
spinal hemorrhagic, 12/567
syphilis, 33/161
syringomyelia, 61/376
tuberculous, *see* Tuberculous pachymeningitis
venous dural sinusitis, 33/155-162
Pachymeningitis cervicalis hypertrophicans
differential diagnosis, 20/339
spinal glioma, 20/339
Pachymeningitis hemorrhagica interna, *see* Chronic
subdural hematoma
Pachymicrogyria
Edwards syndrome, 50/562
Pacinian corpuscles
acrylamide intoxication, 64/74
anatomy, 1/83
proprioception, 1/93
sensation, 7/82
Pack palsy
brachial plexus injury, 2/139
neuralgic amyotrophy, 8/81, 51/174
Paclitaxel, *see* Taxol
Padua, Italy
neurology, 1/11
Paecilomyces
fungal CNS disease, 35/567, 52/489
Paecilomycosis, *see* Poikilomycosis
Paget disease, *see* Osteitis deformans (Paget)
Pagophagia
iron deficiency anemia, 55/464
PAH, *see* p-Aminohippuric acid
Pahutoxic intoxication
chemical structure, 37/82
classification, 37/78
Pain, 1/114-142
a-alphalipoproteinemia, 51/397
abdominal, *see* Abdominal pain
acupuncture, 26/496, 45/231
acute intermittent porphyria, 42/617
adiposis dolorosa, 8/347, 43/57
alpha adrenergic receptor stimulating agent, 69/47
amphetamine, 69/48
amputation, 70/44
amyloid polyneuropathy, 51/416
analgesic agent, 69/45
anesthetic agent, 69/46
angiodysgenetic necrotizing myelomalacia,
20/487
anterior tibial syndrome, 8/152, 12/538, 20/798
anterolateral cordotomy, 26/497, 45/232
anterolateral funiculus, 45/231-234

anticonvulsant, 69/47
antiemetic agent, 69/47
antipsychotic agent, 69/47
antitumor therapy, 69/49
anxiety, 1/110, 26/495
anxiolytic agent, 69/47
aorta body tumor, 8/494
apraxia, 4/223
arterial dissection, 48/287
arterial obstruction, 1/140
arthrodento-osteodysplasia, 43/332
ascending anterolateral fiber, 45/234-236
asymbolia for pain, *see* Indifference to pain
atlantoaxial dislocation, 5/373
atlas syndrome, 5/373
autonomic dysfunction, 75/329
autonomic nervous system, 74/75, 75/309
back, *see* Back pain
Bannwarth syndrome, 51/201
bee sting intoxication, 37/107
biochemistry, 1/129
bladder, 1/363, 366, 379
body scheme disorder, 4/223
bone, *see* Bone pain
brachial plexus, 51/152
bradykinin, 1/129
brain cortex, 45/237
brain stem reticular formation, 45/236
brain tumor, 69/31
Brown hyperparathyroid tumor, 19/301
calcitonin, 69/47
calcium, 75/325
cancer, 69/31, 60
carotid artery, 5/375, 48/274
carpal tunnel syndrome, 42/309
cauda equina intermittent claudication, 20/756,
32/337-339
causalgia, 26/491
cause, 69/35
central, 69/32
central hyperexcitable state, 75/320
cerebral localization, 4/223
cervical arthrosis, 5/373
cervical rib syndrome, 5/287
cervical spondylarthrosis, 5/373
cervical syndrome, 5/287
cervico-occipital, *see* Cervico-occipital pain
chemotherapy, 69/35
chordoma, 19/18
chronic muscle contraction headache, 48/344
chrysaora intoxication, 37/51
cingulotomy, 1/110
cluster headache, 48/225, 241

56

cold induced, *see* Cold induced pain
congenital indifference, *see* Congenital indifference to pain
congenital spinal cord tumor, 32/366
cortical, 1/106
couvade syndrome, 48/358
cryoglobulinemia, 63/404
cutaneous, 1/128
deafferentation, *see* Deafferentation pain
deep, *see* Deep pain
dental, 48/391
diabetes mellitus, 70/154
diabetic proximal neuropathy, 40/326
diagnosis, 69/36
diaphyseal dysplasia, 43/376
2,4-dichlorophenoxyacetic acid polyneuropathy, 7/523
diphosphonate, 69/46
disability, 8/533
discography, 19/221
dissecting aorta aneurysm, 63/46
dorsal column stimulation, 26/497
dorsal spinal root entry zone, 45/229-231
dorsal spinal root ganglion, 75/321
drug, 69/45
drug use, 26/495
dysautonomia, 22/253
dyschondrosteosis, 43/389, 391
ear, *see* Otalgia
ear lobe, 48/497
early burning pain, 26/490
early spinal pain, 26/490
echinoidea intoxication, 37/68
end zone pain, 26/492
epidural hematoma, 19/15
experimental, 1/128
extraterritorial, 75/313
eye, 5/330
Fabry disease, 51/376, 66/226
facial, *see* Facial pain
facial paralysis, 42/313
familial spinal arachnoiditis, 33/269, 42/107
fast, *see* First pain
first, *see* First pain
fracture management, 26/495
frontotemporal, 48/330
funicular, *see* Funicular pain
gasserian ganglion, 45/229
gate control system, 45/231
Gaucher disease, 42/437
Gellé syndrome, 2/241
genetics, 75/325
glomus tumor, 8/456

glucocorticoid, 69/46
Gradenigo syndrome, 2/91, 16/210
growing, *see* Growing pain
Guillain-Barré syndrome, 7/495, 51/245
headache, 26/494, 48/1, 358
heart ischemia, 1/137
hereditary amyloid polyneuropathy, 8/368, 42/518, 521
hereditary amyloid polyneuropathy type 1, 8/368, 51/420
Horner syndrome, 48/330
hydromyelia, 50/429
hyperlipoproteinemia type I, 42/443
hypnosis, 45/239
hypnotic agent, 69/47
hypokalemic periodic paralysis, 43/170
hypothalamus, 45/236
idiopathic, 69/32
incidence, 26/489
incomplete nerve lesion, *see* Incomplete lesion
indifference, *see* Indifference to pain
infection, 69/35
inhibition of painful stimuli, 3/65
intractable, *see* Intractable pain
intralaminar nucleus, 45/237
ischemic, 1/134, 140
jaw, 5/227
joint, 8/12
juvenile Gaucher disease, 42/438
kinesigenic, *see* Kinesigenic pain
kuru, 42/657
lancinating, *see* Lancinating pain
laryngeal, *see* Laryngeal pain
late chronic spinal pain, 26/493
lathyrism, 36/509, 65/7
laxative agent, 69/47
lightning, *see* Lightning pain
lobotomy, 1/110
low back, *see* Low back pain
loxosceles intoxication, 37/112
lumbar intervertebral disc prolapse, 20/584
lumbar vertebral canal stenosis, 20/756
lung tumor, 1/139
management, 26/494
mastication, 48/392
mastoid tip, 48/503
measurement, 1/130, 69/37
mechanism, 26/492
medication, 69/37
meningeal, *see* Meningeal pain
meralgia paresthetica, 19/86, 42/323
metastatic myopathy, 69/72
N-methyl-D-aspartic acid receptor, 75/323

midbrain reticular formation, 45/237
migraine, 1/140, 48/130
migrant sensory neuritis, 51/461
mononeuropathy, 69/72
Morvan fibrillary chorea, 6/355, 49/559
motoneuron disease, 59/468
movement disorder, 75/332
Muckle-Wells syndrome, 42/396
mucolipidosis type I, 51/380
multiple neuronal system degeneration, 59/138
multiple sclerosis, 47/170
muscle contraction, 48/349
muscular, *see* Myalgia
myelotomy, 26/497
myotonia fluctuans, 62/266
nasal, *see* Nasal pain
neck, *see* Neck pain
neospinothalamic tract, 45/234
nerve block, 26/496, 69/49
nerve root, *see* Nerve root pain
neuroborreliosis, 51/201
neuropathic, 69/32, 34, 73
neurophysiology, 5/282
nociceptive, 69/32, 34, 38, 48
occipital, *see* Occipital pain
ocular, *see* Ocular pain
opiate receptor, 45/238
orbital, *see* Orbital pain
osteitis deformans (Paget), 19/172, 38/361, 363
osteoid osteoma, 19/300
paleospinothalamic fiber, 45/235
paleospinothalamic tract, 45/235
paralysing sciatica, 20/585
parietal cortex, 1/106
parietal lobe, 8/196-198
parietal lobe epilepsy, 73/100
paroxysmal, *see* Paroxysmal pain
pathophysiology, 3/65
pathway, 1/114-116, 45/227-238
patient type, 5/284
pattern, 48/274
Pelger-Huët abnormality, 43/459
pellagra, 6/748
pelvic, *see* Pelvic pain
penis, *see* Penis pain
periorbital, 48/330
peripheral neurolymphomatosis, 56/181
peritoneal, 1/364
phantom limb, *see* Phantom limb pain
physiotherapy, 26/496
pituitary gland, 45/239
pleura, 1/138
plexopathy, 69/72, 85

polyneuropathy, 69/72
postdissection, 69/34
postoperative, 48/392
posttraumatic syringomyelia, 50/454, 61/381
precordial, *see* Precordial pain
prefrontal leukotomy, 1/110
proctalgia, 43/69
pseudoradicular, *see* Pseudoradicular pain
pseudothalamic, *see* Pseudothalamic pain
psychiatric, 48/358
psychoanalytic scheme, 5/285
psychogenic, *see* Psychogenic pain
radicular, *see* Radicular pain
radiculopathy, 69/72
radiology, 26/494
radiotherapy, 69/35
receptor, 1/114, 7/83
referred, *see* Referred pain
restless legs syndrome, 8/313, 42/260
rhizostomeae intoxication, 37/53
rhizotomy, 26/497
root, 26/492
sacrococcygeal chordoma, 19/289
sclerosteosis, 43/475
scorpion sting intoxication, 37/112
second, *see* Second pain
sensory mechanism, 45/227-240
sensory polyneuropathy, 8/31
sensory radicular neuropathy, 42/351
serotonin, 1/142
shooting, *see* Shooting pain
shoulder, *see* Shoulder pain
sickle cell anemia, 42/623
sinus, *see* Sinus pain
sinusitis, 5/330
somatic deep, 1/133
spasmodic torticollis, 42/264
sphenopalatine neuralgia, 5/115, 48/477
spinal, *see* Spinal pain
spinal arachnoiditis, 42/107
spinal cord injury, 26/489-499, 75/579
spinal cord intermittent claudication, 20/756,
    32/337-339
spinal cord metastasis, 69/174
spinal ependymoma, 19/360, 20/364-366
spinal meningioma, 19/68, 358, 20/200
spinal nerve root, *see* Spinal nerve root pain
spinal nerve root syndrome, *see* Spinal nerve root
    pain
spinal neurinoma, 20/250
spinal pathway, 45/234
spinal subarachnoid hemorrhage, 12/133, 173
spinal tumor, 19/54

58

spinoreticulodiencephalic fiber, 45/235
spontaneous, 1/135
stingray intoxication, 37/72
stump, *see* Stump pain
superficial, 1/81
surgery, *see* Pain surgery
survey, 1/114, 69/31, 75/310
swallowing, 48/502
sympathectomy, 26/497
sympathetic nervous system, 75/314
syringomyelia, 32/263
tabes dorsalis, 19/86, 45/230
temporal arteritis, 39/316-319, 48/312
temporomandibular cephalalgia, 48/413
temporomandibular joint, 5/316, 345
thalamic, *see* Thalamic pain
thalamic syndrome (Dejerine-Roussy), 45/235
thalamoneocortical fiber, 45/236-238
thalamotomy, 2/491
thalamus, 45/234-236
thallium intoxication, 36/245, 64/325
theory, 1/114
thoracic outlet syndrome, 51/122
thoracolumbar spine injury, 25/463
threshold, 1/92, 114, 130-132
throat, 48/502
tongue, 48/502
tonsillar fossa, 48/503
transorbital leukotomy, 1/110
treatment, 69/38
tricyclic antidepressant, 69/47
trigeminal, *see* Trigeminal pain
trigeminal neuralgia, 45/230
vagus nerve, 5/209
vegetative disorder, 1/512
vegetative response, 3/65
ventrobasal complex, 45/236
vertebral, *see* Vertebral pain
vertebral artery, 48/274
vertebral column, 32/140
vertebral metastasis, 20/418, 427, 69/61
visceral, *see* Visceral pain
vitamin B6 intoxication, 65/572
vitamin B12 deficiency, 70/377
Wohlfart-Kugelberg-Welander disease, 59/86
Pain agnosia
   dominant hemisphere lesion, 2/695
   Gerstmann syndrome, 2/691
Pain asymbolia, *see* Indifference to pain
Pain control
   diabetic polyneuropathy, 51/514
   hypothalamus, 45/238-240
   nucleus raphe magnus, 45/238-240

periaqueductal gray matter, 45/238-240
pituitary gland, 45/239
Pain insensitivity
   autonomic nervous system, 74/218
   autonomic polyneuropathy, 51/476
   autosomal dominant, *see* Autosomal dominant
      pain insensitivity
   congenital, *see* Congenital analgesia *and*
      Congenital pain insensitivity
   hereditary sensory and autonomic neuropathy type
      III, 60/7, 27
Pain mechanism
   histamine, 1/142
   migraine, 1/140
   postherpetic neuralgia, 1/119
   reflex sympathetic dystrophy, 61/123
Pain perception
   A fiber, 1/117-120
   nociceptor, 45/227-229
   substance P, 45/228
   threshold, 45/227
   trigeminal neuralgia, 45/228
Pain producing substance
   variety, 1/128
Pain reaction
   Alzheimer disease, 46/253
   brain injury, 23/230
   postherpetic neuralgia, 5/324
Pain sensory loss
   amyloid polyneuropathy, 51/416
Pain surgery
   cause, 69/35
   dorsal rhizotomy, 45/229
   paraplegia, 26/496
Pain syncope
   type, 15/824
Pain syndrome
   antebrachial, *see* Antebrachial pain syndrome
   autonomic nervous system, 75/35
   chronic regional, *see* Chronic regional pain
      syndrome
Pain tolerance
   analgesic agent, 45/228
   morphine, 45/228
   threshold, 45/228
Paine syndrome
   aminoaciduria, 30/514, 43/433
   cerebellar atrophy, 43/433
   cerebral palsy, 60/293
   CSF, 30/514, 43/432, 60/293
   EEG, 43/433, 60/293
   epilepsy, 60/293
   hereditary congenital cerebellar atrophy, 60/293

hyperreflexia, 43/433
mental deficiency, 30/514, 43/433
microcephaly, 30/509, 514, 43/432, 60/293
optic atrophy, 43/433, 60/293
psychomotor retardation, 60/293
seizure, 43/433
spastic diplegia, 30/514, 43/432
Painful
asteroidea intoxication, 37/66
centrostephanus intoxication, 37/67
diadematidae, 37/67
diadematidae intoxication, 37/67
echinothrix intoxication, 37/67
feet syndrome, *see* Burning feet syndrome
hydrozoa intoxication, 37/38
hyperalgesic diabetic neuropathy, 70/136
myogelosis, 41/385
occlusive aortoiliac disease, 12/535
penis erection, 74/560
spinal injury, 37/67
subacute myoclonic spinal neuronitis, 6/775
thallium polyneuropathy, 51/274
Painful anosognosia
thalamic lesion, 2/479
Painful dominant myotonia
chromosome 17q, 62/266
fasting, 62/266
muscle stiffness, 62/266
myotonic syndrome, 62/263
oral potassium, 62/266
Painful feet syndrome, *see* Burning feet syndrome
Painful legs moving toes syndrome
motor unit hyperactivity, 41/311
restless legs syndrome, 42/260, 51/545
Painful neuropathy
diabetic, 70/136
lymphoma, 63/350
Painful ophthalmoplegia
systemic lupus erythematosus, 55/377
Painful paresthesia
primary hyperoxaluria, 51/398
PALA, *see* Sparfosic acid
Palatal myoclonus
amyotrophic lateral sclerosis, 22/140
brain embolism, 53/211
brain infarction, 53/211
brain stem infarction, 38/584
cerebellar tumor, 16/325
cerebrotendinous xanthomatosis, 51/399, 60/167
cerebrovascular disease, 53/211
clicking, 49/614
clinical classification, 6/763
clinical features, 6/772-774

definition, 2/240
dysarthria, 38/584
dysphagia, 38/584
encephalitis, 38/584
intracranial hemorrhage, 53/211
lesion site, 6/776
multiple sclerosis, 9/181, 38/584
neuropathology, 6/774
opsoclonus, 34/612
paramedian medullary syndrome, 2/227
pontine infarction, 53/390
progressive bulbar palsy, 38/584, 59/225
syringobulbia, 38/584
transient ischemic attack, 53/211
trauma, 38/584
vertebrobasilar system syndrome, 53/390
Palatal reflex
diagnostic value, 1/241
elicitation, 2/78
Palatal weakness
diphtheria, 33/482, 51/183
Palate
Avellis syndrome, 1/189
cleft, *see* Palatoschisis
high arched, *see* High arched palate
obstructive sleep apnea, 63/452
Turner syndrome, 50/543
Palate sign (Mathews)
spinal rheumatoid arthritis, 19/337
Palatognathoschisis
hereditary sensory and autonomic neuropathy type
I, 21/77
Palatoschisis
congenital spondyloepiphyseal dysplasia, 43/478
Gruber syndrome, 14/505
hereditary sensory and autonomic neuropathy type
I, 60/9
iniencephaly, 50/131
Klinefelter variant XXXXY, 31/483
mental deficiency, 43/256, 269, 46/30
neuralgic amyotrophy, 51/175
Patau syndrome, 31/507, 43/528, 50/556
13q partial monosomy, 50/587
Smith-Lemli-Opitz syndrome, 66/582
speech, 46/30
trigeminal nerve agenesis, 50/214
trisomy 8, 50/563
trisomy 9, 50/564
trisomy 22, 50/564
Paleocerebellar syndrome
see also Archicerebellar syndrome *and*
Neocerebellar syndrome
archicerebellar syndrome, 2/417

clinical symptomatology, 2/417
extensor thrust reflex, 2/418
locomotor ataxia, 2/418
pendular reflex, 2/419
postural reflex, 2/418
Paleocerebellum
definition, 1/329
late cortical cerebellar atrophy, 21/480
Paleoneuroanatomy
survey, 74/186
Paleoneuroarthropathy
survey, 74/187
Paleoneurobiology
autonomic nervous system, 74/181
nerve, 74/181
nervous system, 74/181
survey, 74/189
Paleoneuropathology
survey, 74/187
Paleopathology
autonomic nervous system, 74/183
survey, 74/183
Paleospinothalamic tract
pain, 45/235
Paleostriatum
see also Globus pallidus
anatomy, 6/3
limbic part, 49/4
nonlimbic part, 49/4
Palfium intoxication, see Dextromoramide
Paligraphia
agraphia, 45/467
etiology, 45/468
palilalia, 45/467
progressive supranuclear palsy, 22/220
Palilalia
Alzheimer disease, 43/225
anterior cerebral artery syndrome, 53/346
brain infarction, 53/346
chorea, 43/225
Gilles de la Tourette syndrome, 6/788
intracerebral calcification, 43/225
paligraphia, 45/467
pallidoluysionigral degeneration, 49/456
Parkinson disease, 43/225
parkinsonism, 43/225
Pick disease, 46/240
progressive supranuclear palsy, 22/220
seizure, 43/226
stuttering, 46/171
supplementary motor area syndrome, 53/346
Palinacousia
hallucination, 45/367

hallucinosis, 46/562
occipital lobe, 45/367
perception, 45/367
Palinopsia, see Visual perseveration
Palitypia
parkinsonism, 45/468
Pallanesthesia
terminology, 1/96
Pallesthesia
terminology, 1/95
Pallidal atrophy
amyotrophic lateral sclerosis, 6/619
classification, 21/531
Huntington chorea, 6/645-647
Parkinson disease, 6/648
Pick disease, 6/645, 647, 46/236, 49/293
postencephalitic parkinsonism, 6/649
spasmodic torticollis, 6/638
striatonigral degeneration, 6/649
Pallidal atrophy (Hunt-Van Bogaert)
progressive, see Progressive pallidal atrophy
(Hunt-Van Bogaert)
Pallidal degeneration
age at onset, 49/454
ataxia, 49/450
bajonet posture, 49/455
Bielschowsky body, 49/459
carbamate intoxication, 64/188
carbon disulfide intoxication, 64/188
chorea, 49/450
choreoathetosis, 6/644, 49/454
classification, 6/632
clinical features, 49/454
dementia, 49/450
disease duration, 49/454
disulfiram, 64/188
dystonia, 49/454
epidemiology, 49/446
exogenous lesion, 6/632, 652
genetics, 49/446
Gerstmann-Sträussler-Scheinker disease, 49/454
Hallervorden-Spatz syndrome, 6/650
history, 6/635
hypersalivation, 49/455
infantile neuroaxonal dystrophy, 6/650
juvenile parkinsonism, 49/454
manganese encephalopathy, 49/477
myoclonic epilepsy, 6/650
nosologic group, 49/445
olivopontocerebellar atrophy, 49/454
primary type, 6/635
progressive, see Progressive pallidal degeneration
pseudo-Babinski sign, 49/455

pure, *see* Pure pallidal degeneration
spongiform encephalopathy, 6/651
spongy cerebral degeneration, 6/651
striatal involvement, 49/450, 454
striatopallidodentate calcification, 6/650
Wilson disease, 6/651
Pallidal demyelination
  Parkinson disease, 6/176
Pallidal efferent
  anatomy, 6/19-21
Pallidal lesion
  anoxia, 6/652
  globus pallidus, 6/633
  kernicterus, 6/661
  lateral pallidal segment, 6/633
  lead encephalopathy, 6/664
  manganese encephalopathy, 6/662-664
Pallidal necrosis
  Adams-Stokes syndrome, 49/466
  anemia, 49/466
  anesthesia, 6/658
  anoxia, 6/654
  anoxic leukomalacia, 47/527
  anterior choroidal artery occlusion, 49/466, 469
  arsphenamine encephalopathy, 49/479
  arsphenamine intoxication, 6/652, 49/466
  athetosis, 2/502, 49/465
  atmospheric hypoxidosis, 49/466
  barbiturate intoxication, 6/652, 49/466
  bilirubin encephalopathy, 49/465, 481
  bradykinesia, 49/465
  brain arteriosclerosis, 49/466
  brain hypoperfusion, 49/465
  brain hypoxia, 49/465
  brain ischemia, 49/465, 63/213
  carbon disulfide intoxication, 6/652, 49/466
  carbon monoxide intoxication, 6/652, 654, 656,
    660, 47/527, 49/466, 471, 473, 63/418
  cardiac arrest, 49/466, 63/213
  carotid artery occlusion, 49/466, 469
  cerebral palsy, 49/481
  cerebrovascular disease, 49/465
  clinical features, 49/465
  comparative neuropathology, 49/483
  cyanide intoxication, 6/652, 49/466
  dyskinesia, 49/465
  dystonia, 49/465
  Hallervorden-Spatz syndrome, 49/465, 481
  head injury, 49/470
  hemiplegia inflection, 49/470
  hemorrhagic shock, 49/466
  hyperbaric oxygenation, 6/660
  hypoglycemic shock, 49/479

  hypoxemia, 49/466
  hypoxic hypertensive state, 49/471
  iatrogenic nerve injury, 49/468
  intracerebral hematoma, 49/465
  intracranial hypertension, 49/465
  intracranial pressure, 49/470
  kernicterus, 49/465
  lead intoxication, 6/652, 49/466
  manganese intoxication, 6/652, 49/466
  Marchiafava-Bignami disease, 49/479
  methyl chloride intoxication, 6/652, 49/466
  morbus caeruleus, 49/466
  morphine intoxication, 6/652, 49/466, 479
  neuroleptic agent intoxication, 49/466
  newborn asphyxia, 49/466
  nitrogen inhalation, 49/466
  obscure cause, 49/482
  parkinsonism, 49/465
  pathogenesis, 49/466
  pathology, 49/471
  perinatal brain injury, 49/465
  perinatal hypoxic ischemia, 49/466
  phosphorus intoxication, 6/652, 49/466
  poliomyelitis, 49/479
  postanesthetic encephalopathy, 49/466
  pseudobulbar paralysis, 49/470
  respiratory failure, 49/466, 479
  Rh incompatibility, 49/466
  rigidity, 49/465
  shock, 49/466
  sodium nitrate intoxication, 49/479
  sodium nitrite intoxication, 6/652, 49/466
  strangulation, 49/466
  toxic encephalopathy, 49/465
  transtentorial herniation, 49/469
Pallidal posture
  bilateral lesion, 6/634
  hereditary progressive diurnal fluctuating
    dystonia, 49/532
  lesion indication, 49/78
Pallidal segment
  internal, *see* Internal pallidal segment
  lateral, *see* Lateral pallidal segment
  medial, *see* Medial pallidal segment
  outer, *see* Outer pallidal segment
Pallidal syndrome
  characteristics, 6/633
  choreoathetosis, 2/502, 6/644
  flexion dystonia, 21/526
  main symptom, 21/526
Pallidocerebello-olivary degeneration
  bulbar paralysis, 60/658
  choreoathetosis, 60/663

eunuchoidism, 60/658, 663
hypogonadism, 60/668
mental deficiency, 60/664
Pallidodentate degeneration
  evidence, 6/638
  heterogeneity, 21/531
Pallidofugal atrophy
  dentatorubropallidoluysian atrophy, 21/531
Pallidofugal fiber
  anatomy, 6/29
  system, 49/6
Pallidohabenular fiber
  lateral habenular nucleus, 49/7
Pallidoluysian atrophy
  see also Progressive pallidal atrophy (Hunt-Van
    Bogaert)
  differential diagnosis, 6/643
  dystonia, 6/642
  genetics, 6/636
  hemiballismus, 42/224
  hereditary cerebellar ataxia, 6/675
  heterogeneity, 21/531
  neuropathology, 6/636-640
  nosology, 6/674
  olivopontocerebellar atrophy, 22/172
Pallidoluysian degeneration
  choreoathetosis, 49/450
  Gerstmann-Sträussler-Scheinker disease, 49/454
  head tremor, 49/455
  olivopontocerebellar atrophy, 49/454
  psychosis, 49/455
  pure, *see* Pure pallidoluysian degeneration
  rigidity, 49/455
  spasmodic torticollis, 49/455
  striatal involvement, 49/454
Pallidoluysian system
  dentatorubropallidoluysian atrophy, 49/440
Pallidoluysiodentate degeneration
  age at onset, 49/454
  athetosis, 49/452
  chorea, 49/456
  choreoathetosis, 49/454
  dentatorubropallidoluysian atrophy, 49/452
  disease duration, 49/454
  dystonia, 49/454
  epidemiology, 49/446
  epilepsy, 49/456
  genetics, 49/446
  hemiballismus, 49/456
  hereditary, *see* Hereditary pallidoluysiodentate
    degeneration
  myoclonus, 49/452, 456
  nosologic group, 49/445

olivopontocerebellar atrophy, 49/452, 456
Pallidoluysionigral degeneration
  see also Juvenile parkinsonism
  age at onset, 49/454
  akinesia, 49/451
  amyotrophic lateral sclerosis, 49/447, 453, 456
  anemia, 6/665
  astrocytosis, 6/665
  ataxia, 49/456
  athetosis, 49/446, 455
  Bielschowsky body, 49/459
  bradykinesia, 49/446
  catatonia, 49/456
  chorea, 49/455
  choreoathetosis, 49/454
  clinical features, 49/454
  clinicopathologic correlation, 49/457
  convulsion, 6/665
  Creutzfeldt-Jakob disease, 49/459
  dementia, 49/452, 456
  demyelination, 6/665
  dentatorubropallidoluysian atrophy, 49/459
  differential diagnosis, 49/458
  disease duration, 49/454
  double athetosis, 49/459
  dystonia, 49/446, 451, 454, 456
  dystonia musculorum deformans, 49/458
  epidemiology, 49/446
  familial, *see* Familial pallidoluysionigral
    degeneration
  gait disturbance, 6/665
  genetics, 49/446
  Gerstmann-Sträussler-Scheinker disease, 49/454,
    459
  Hallervorden-Spatz syndrome, 49/458
  hepatosplenomegaly, 6/665
  hereditary, *see* Hereditary pallidoluysionigral
    degeneration
  historic aspect, 49/445
  Huntington chorea, 49/459
  intention tremor, 49/456
  Joseph-Machado disease, 49/459
  juvenile parkinsonism, 49/454, 458
  laboratory data, 49/458
  lead intoxication, 49/460
  lesion, 49/445
  manganese intoxication, 49/460
  muscle stretch reflex, 6/665
  neuronal loss, 6/665
  neuropathology, 49/447
  nuchal rigidity, 49/456
  olivopontocerebellar atrophy, 49/446
  ophthalmoplegia, 49/456

optic atrophy, 49/447
palilalia, 49/456
parkinsonism, 49/452
progressive supranuclear palsy, 49/456, 459
reticuloendothelial disorder, 6/665
reticuloendothelial system, 49/445
rigidity, 49/446, 456
sickle cell anemia, 49/483
sickle cell trait, 6/665
spasmodic torticollis, 49/455
spasticity, 6/665
spinal muscular atrophy, 49/454, 457
spinopontocerebellar degeneration, 49/457
striatonigral degeneration, 49/459
toxic encephalopathy, 49/460
treatment, 49/460
tremor, 49/446, 455
vertical gaze palsy, 49/451, 456
Wilson disease, 49/459
Pallidoluysionigrothalamic degeneration
nosologic group, 49/445, 447
nuchal rigidity, 49/456
progressive supranuclear palsy, 49/456
vertical gaze palsy, 49/456
Pallidonigral degeneration
age, 49/454
ataxia, 42/242
cause, 49/460
choreoathetosis, 49/454
clinical features, 49/454
disease duration, 49/454
epidemiology, 49/446
familial, see Familial pallidonigral degeneration
genetics, 49/446
hemiplegia, 42/242
Huntington chorea, 6/356, 49/293
Joseph-Machado disease, 49/445
juvenile parkinsonism, 49/454
multiple neuronal system degeneration, 49/445
nosologic group, 49/445
opisthotonos, 42/242
reticuloendothelial disorder, 42/242
sickle cell anemia, 49/450
spasticity, 49/456
tonic clonic seizure, 42/242, 49/456
tremor, 42/242
Pallidonigral encephalomalacia
horse, 6/665, 49/466
pallidonigral necrosis, 49/466
yellow star thistle intoxication, 6/665, 49/466, 483
Pallidonigral fiber
GABA, 49/8
substance P, 49/8
substantia nigra, 49/8
termination, 49/7
Pallidonigral necrosis
athetosis, 49/465
bilirubin encephalopathy, 49/465, 482
bradykinesia, 49/465
brain hypoperfusion, 49/465
brain hypoxia, 49/465
brain ischemia, 49/465
carbon monoxide intoxication, 49/466
cardiac circulatory failure, 49/466
cerebrovascular disease, 49/465
clinical features, 49/465
dyskinesia, 49/465
dystonia, 49/465
familial pallidonigral degeneration, 49/466
Hallervorden-Spatz syndrome, 49/465
intracerebral hematoma, 49/465
intracranial hypertension, 49/465
kernicterus, 49/465
pallidonigral encephalomalacia, 49/466
parkinsonism, 49/465
pathogenesis, 49/466
perinatal brain injury, 49/465
postanesthetic accident, 49/466
postanesthetic encephalopathy, 49/478
rigidity, 49/465
suffocation, 49/466
symmetrical encephalomalacia, 49/484
toxic encephalopathy, 49/465
yellow star thistle intoxication, 49/483
Pallidonigral pigmentary degeneration, see
    Hallervorden-Spatz syndrome
Pallidopetal fiber
anatomy, 6/30
Pallidopyramidal degeneration
age at onset, 49/454
Babinski sign, 42/244
choreoathetosis, 49/454
clinical features, 49/454
dementia, 49/456
disease duration, 49/454
dysarthria, 49/456
dystonia, 42/244, 49/454, 456
evidence, 6/640
hereditary, see Hereditary pallidopyramidal
    degeneration
hereditary dystonic paraplegia, 22/457
Huntington chorea, 6/356
hyperreflexia, 42/244
intention tremor, 49/456
Jollymore family, 22/463
juvenile parkinsonism, 49/454

mental deficiency, 42/244
nosologic group, 49/445
speech disorder in children, 42/244
striatal involvement, 49/446
tremor, 42/243
Pallidopyramidal syndrome, *see* Pallidopyramidal
degeneration
Pallidostriatal necrosis
athetosis, 49/465
bilirubin encephalopathy, 49/465
bradykinesia, 49/465
brain hypoperfusion, 49/465
brain hypoxia, 49/465
brain ischemia, 49/465
carbon monoxide intoxication, 49/466
cardiac arrest, 49/466
cerebrovascular disease, 49/465
clinical features, 49/465
dyskinesia, 49/465
dystonia, 49/465
Hallervorden-Spatz syndrome, 49/465
hypoglycemia, 49/466
intracerebral hematoma, 49/465
intracranial hypertension, 49/465
kernicterus, 49/465
parkinsonism, 49/465
pathogenesis, 49/466
perinatal brain injury, 49/465
postanesthetic complication, 49/466
posttraumatic encephalopathy, 49/466
respiratory failure, 49/466
rigidity, 6/657, 49/465
strangulation, 49/466
strangulation encephalopathy, 49/480
toxic encephalopathy, 49/465
Pallidosubthalamic fiber
subthalamic nucleus, 49/11
Pallidosubthalamic projection
choline acetyltransferase, 49/7
GABA, 49/7
glutamate decarboxylase, 49/7
subthalamic nucleus, 49/11
Pallidotegmental fiber
pedunculopontine tegmental nucleus, 49/7
Pallidothalamic pathway
hemiballismus, 49/370
Pallidothalamic projection
thalamic ventral anterior nucleus, 49/6
thalamic ventral lateral nucleus, 49/6
Pallidum, *see* Globus pallidus
Pallidum Epilepsie, *see* Infantile spasm
Pallister syndrome, *see* W syndrome
Pallor of disc

temporal, *see* Temporal disc pallor
Palmar hyperhidrosis
autonomic nervous system, 74/628
Palmar and plantar pit
differentiation, 14/585
multiple nevoid basal cell carcinoma syndrome,
14/582, 18/172
Palmitaldehyde
chemistry, 10/246
Palmitate, *see* Palmitic acid
Palmitic acid
carnitine palmitoyltransferase deficiency, 41/206
chorea-acanthocytosis, 63/286
migraine, 48/95
orthochromatic leukodystrophy, 42/501
plasma, 48/96
Palmomental reflex
Argentinian hemorrhagic fever, 56/365
central motor disorder, 1/174
definition, 1/241
significance, 1/174
Palmoplantar keratosis
corneal dystrophy, 42/581
Friedreich ataxia, 21/348
hereditary, *see* Hereditary palmoplantar keratosis
mal de Meleda, 14/786
mental deficiency, 42/581
Papillon-Lefevre form, 14/786
Richner-Hanhart syndrome, 42/581
Palpebral reflex
eyelid, 1/614
Palythoa tuberculosa intoxication
fatal, 37/55
molecular weight, 37/55
palytoxin, 37/55
Palythoa vestitus intoxication
fatal, 37/55
molecular weight, 37/55
palytoxin, 37/55
Palytoxin intoxication
convulsion, 37/55
death, 37/55
diarrhea, 37/55
dyspnea, 37/55
respiratory failure, 37/55
PAN method, *see* Perchloric acid naphthoquinone
method
Panarteritis brachiocephalica, *see* Takayasu disease
Panarteritis nodosa, *see* Polyarteritis nodosa
Panatrophy
focal cerebellar cortical, *see* Focal cerebellar
cortical panatrophy
Pancoast tumor

Horner syndrome, 2/143
radicular syndrome, 20/556
Pancorphen
multiple sclerosis, 9/388
Pancreas
ganglion, 75/615
Pancreas adenoma
tuberous sclerosis, 14/51
Pancreas transplantation
indication, 70/260
neurology, 70/249
survey, 70/249
Pancreas tumor
neuropathy, 7/607
Pancreatic cyst
Von Hippel-Lindau disease, 14/13, 242, 694,
16/31, 50/375
Pancreatic encephalopathy, 27/449-457
acute pancreatitis, 27/449-452
anoxic ischemic leukoencephalopathy, 47/538
clinical features, 27/453-455
definition, 70/259
delirium, 46/543
differential diagnosis, 27/455
hypokalemia, 27/456
incidence, 27/449
pathology, 27/450
treatment, 27/457
Pancreatic exocrine function
Pearson syndrome, 66/397, 427
Pancreatic extract, see Kallikrein
Pancreatic insufficiency
deficiency neuropathy, 51/327
vitamin B12 deficiency, 51/339
Pancreatic insulinoma
antiepileptic treatment, 14/751
Pancreatic islet cell hyperplasia
leprechaunism, 42/590
Pancreatic polypeptide
hypoglycemia, 74/624
progressive dysautonomia, 59/157
Pancreatitis
acute, see Acute pancreatitis
acute hemorrhagic, see Acute hemorrhagic
pancreatitis
clinical features, 27/449-452
diagnosis, 27/452
etiology, 27/449
hypokalemia, 27/456
intravascular consumption coagulopathy, 55/495
neuropathy, 7/603
organic acid metabolism, 66/641
pathology, 27/450

Pancreatogenic malabsorption
deficiency neuropathy, 7/603
Pancuronium
intensive care, 71/536
multiple organ failure, 71/536
organophosphate intoxication, 64/173
Pancytopenia
dyskeratosis congenita, 43/392
familial, see Fanconi syndrome
mephenytoin intoxication, 37/202
Pearson syndrome, 66/397, 427
Panda eyes
anterior cranial fossa fracture, 57/131
clinical examination, 57/131
skull fracture, 57/131
Pandysautonomia
acute, see Acute pandysautonomia
concept, 22/251
leprosy, 63/154
Pandysautonomic syndrome
acute, see Acute pandysautonomia
Panencephalitis
chronic progressive, see Chronic progressive
panencephalitis
leukodystrophy classification, 10/20
progressive rubella, see Progressive rubella
panencephalitis
rubella, see Rubella panencephalitis
rubella virus, 56/12
subacute sclerosing, see Subacute sclerosing
panencephalitis
Whipple disease, 52/139
Panencephalitis (Pette-Döring), see Subacute
sclerosing panencephalitis
Panhypopituitarism
acanthocyte, 49/332
acanthocytosis, 63/271, 290
delirium, 46/544
Sheehan syndrome, 46/397
Panhypopituitary dwarfism, see Pituitary dwarfism
type II
Panic disorder
catecholamine, 74/325
stress, 74/325
Pankration
boxing injury, 23/533
Panleukopenia virus
animal viral disease, 34/296
ataxia, 34/296
feline ataxia, 34/296
Panstrongylus megistus
Chagas disease, 75/387
Pantelakis meningocortical amyloid angiopathy, see

Brain amyloid angiopathy
Panting
spinal cord injury, 26/367, 61/282
thermoregulation, 26/367
Pantothenic acid, *see* Vitamin B5
Pantothenic acid deficiency, *see* Vitamin B5
deficiency
Papaver somniferum
delta receptor gene, 65/350
dopamine enhancement, 65/350
opiate intoxication, 65/349
opium war, 65/350
Papaverine
action mechanism, 37/455
brain infarction, 37/455, 53/421
brain vasospasm, 11/125
chemical structure, 37/455
dementia, 37/455
ischemia, 37/455
migraine, 37/456
regional brain blood flow, 37/455
tardive dyskinesia, 37/456
vasodilator agent, 37/454-456
Papaverine intoxication
exacerbation, 37/455
facial flushing, 37/456
hypertension, 37/456
liver toxicity, 37/456
vasodilatation, 37/456
Papez circuit
Alzheimer disease, 46/259
limbic system, 73/57
memory, 3/259
progressive multifocal leukoencephalopathy,
46/388
Papez-MacLean theory
emotion, 3/320
Papillary fibroelastoma
cardiac, *see* Cardiac papillary fibroelastoma
primary cardiac tumor, 63/93
Papillary medulloblastoma
pigmented, *see* Pigmented papillary
medulloblastoma
Papilledema
acute subdural hematoma, 24/280
Addison disease, 39/479
amaurosis fugax, 55/108
anatomy, 16/272
anemia, 38/27, 63/252
aneurysmal bone cyst, 20/3
anterior ischemic optic neuropathy, 55/114
arsacetin, 9/648
arsphenamine, 9/648

arterial hypertension, 63/72
basilar impression, 42/17
Behr disease, 13/88, 60/655
blind spot, 2/567
Bonnet-Dechaume-Blanc syndrome, 14/108
brain edema, 63/417
brain hypoxia, 63/417
brain tumor, 67/143
cavernous sinus thrombosis, 52/174
cerebellar infarction, 53/383
cerebellar tumor, 17/710
Chédiak-Higashi syndrome, 14/783, 60/655
chiasmal lesion, 2/574
chlormethine intoxication, 64/233
choroid plexus papilloma, 42/761
chronic subdural hematoma, 24/305
ciprofloxacin intoxication, 65/486
clinical features, 16/281
corpus callosum syndrome, 2/761
craniopharyngioma, 18/541
craniosynostosis, 30/226
Cryptococcus neoformans, 63/532
current theory, 16/274
Dandy-Walker syndrome, 50/328
diabetes mellitus, 16/286
drusen body, 16/294
endocrine candidiasis syndrome, 42/642
epidural hematoma, 24/266
foramen magnum tumor, 19/60
Foster Kennedy syndrome, 2/53
fox intoxication, 37/96
frontal lobe tumor, 17/260
glycosaminoglycanosis type II, 42/454
glycosaminoglycanosis type IIB, 60/176
Guillain-Barré syndrome, 7/501, 21/151, 51/246
headache, 48/33
hemolytic uremic syndrome, 63/324
hexachlorophene intoxication, 37/486, 65/476,
478
hypertensive encephalopathy, 11/557, 54/212,
63/72
hypertrophic interstitial neuropathy, 21/151
hypocalcemia, 63/561
hypoparathyroidism, 27/310
hypothyroidism, 42/577
incidence, 16/284
intracerebral hematoma, 24/360
intracranial hypertension, 16/133, 42/740
intramedullary spinal tumor, 19/60
intraocular pressure, 16/275
iron deficiency anemia, 63/252
ischemic optic neuropathy, 16/287
lead encephalopathy, 64/436

lead intoxication, 9/642, 64/436
lithium intoxication, 64/296
lymphocytic choriomeningitis, 9/197
malignant hypertension, 16/282
Melkersson-Rosenthal syndrome, 16/222
meningeal leukemia, 16/234, 63/341
meningeal lymphoma, 63/347
mountain sickness, 63/417
multiple myeloma, 18/256, 63/393
multiple myeloma polyneuropathy, 51/430
nalidixic acid intoxication, 65/486
neurobrucellosis, 33/316, 52/584, 586, 594
ocular cause, 16/285
ophthalmodynamometry, 16/273
optic atrophy, 16/289
optic neuritis, 16/288
orbital cause, 16/285
orbital cellulitis, 33/179
organotin intoxication, 64/136, 138
osteitis deformans (Paget), 38/364
parietal lobe tumor, 17/301
pathology, 16/279
perhexiline maleate polyneuropathy, 51/302
physiology, 16/271
pituitary adenoma, 17/391
POEMS syndrome, 47/619, 51/430, 63/394
polar bear intoxication, 37/96
pseudo, see Pseudopapilledema
pseudoneuritis, 16/290-292, 294, 296
quinolone intoxication, 65/486
research, 16/274
respiratory encephalopathy, 63/414
review, 16/270
Reye syndrome, 29/333, 31/168, 34/169, 56/151
sclerosteosis, 43/475
shunt mechanism, 16/276
spinal cord tumor, 19/38, 88
spinal sympathicoblastoma, 20/112
spinal tumor, 19/60, 88
striatopallidodentate calcification, 6/707-709,
    49/422
subarachnoid hemorrhage, 12/135, 137, 55/16
temporal lobe tumor, 17/292
thrombotic thrombocytopenic purpura, 63/324
traumatic, 16/283
traumatic venous sinus thrombosis, 24/376
tryparsamide, 9/648
tuberous sclerosis, 14/345
vascular hypertension, 16/282, 291
vasculitis, 16/287
vitamin A intoxication, 28/45, 37/96
Von Hippel-Lindau disease, 43/21, 60/655
Waldenström macroglobulinemia, 63/398

wolf intoxication, 37/96
X-linked ataxia, 60/655
X-linked recessive ataxia, 60/636
Papilloma
    choroid plexus, see Choroid plexus papilloma
    plexus, see Plexus papilloma
    pseudo, see Pseudopapilloma
Papillomacular bundle
    retinal nerve fiber, 2/548
    topographical anatomy, 2/546
Papillomavirus
    neurotropic virus, 56/9
Papillon-Léage-Psaume syndrome, see
    Orofaciodigital syndrome type I
Papillon-Léage syndrome, see Orofaciodigital
    syndrome type I
Papillopathy
    acute ischemic, see Acute ischemic papillopathy
    ischemic, see Ischemic papillopathy
Papovavirus
    acute viral myositis, 56/193
    classification, 34/7
    CSF titer, 9/327
    human, 56/2
    immunocompromised host, 56/480
    neurotropic virus, 56/9
    papillomavirus genus, 56/9
    polyomavirus genus, 56/9
    progressive multifocal leukoencephalopathy,
        9/327, 495, 34/307, 310, 313-321, 39/19,
        46/388, 47/503, 63/354
    simian virus 40, 56/9
    taxonomic place, 56/2
Papovavirus JC
    acquired immune deficiency syndrome, 56/18
Papulosis
    malignant atrophic, see Köhlmeier-Degos disease
Papyrus
    Edwin Smyth, see Edwin Smyth papyrus
Paraballism
    definition, 49/369
    frequency, 6/476
    head trauma, 49/374
Paracentral lobule
    hemiplegia, 1/186
    motor cortex, 1/210
    motor cortex function, 1/210
Paracentral lobule syndrome
    anterior cerebral artery, 53/345
    brain infarction, 53/345
    motor deficit, 53/345
    sensory symptom, 53/345
    urinary incontinence, 53/345

Paracentral scotoma
   homonymous, *see* Homonymous paracentral
      scotoma
Paracetamol
   cranial neuralgia, 5/378
   delirium, 46/540
   migraine, 5/98, 48/175
   toxic myopathy, 62/601
Parachute reflex
   childhood motor development, 4/348
Paracoccidioidal granuloma, *see*
      Paracoccidioidomycosis
Paracoccidioides brasiliensis
   cerebrovascular disease, 55/432
   meningitis, 55/432
Paracoccidioidomycosis
   see also Blastomyces brasiliensis
   amphotericin B, 35/537, 52/460
   brain granuloma, 35/532, 52/457
   clinical features, 35/533
   computer assisted tomography, 52/459
   convulsion, 35/531, 52/458
   CSF, 35/536, 52/460
   CSF culture, 35/534, 52/457
   diagnosis, 35/534-536, 52/459
   EEG, 35/536
   epidemiology, 52/455
   etiology, 35/532
   geography, 35/380, 52/456
   history, 35/531
   immunofluorescence, 35/535, 52/459
   intracranial hypertension, 35/532, 534, 52/458
   ketoconazole, 52/461
   meningitis, 35/534, 52/457
   meningoencephalitis, 35/531, 52/457
   mycosis, 35/376-379, 531, 52/455
   neurologic involvement, 35/531, 52/457
   nomenclature, 35/531, 52/385
   obliterating panangiitis, 52/458
   pathogenesis, 35/533
   pathophysiology, 52/457
   serodiagnosis, 37/534, 52/456, 459
   serologic test, 35/534
   skin test, 35/535
   spastic paraplegia, 59/436
   spinal granuloma, 52/457
   sulfamethoxazole, 35/537
   sulfonamide, 35/537
   treatment, 35/537, 52/460
   trimethoprim, 35/537
Paracolobactrum
   gram-negative bacillary meningitis, 52/118
   infantile enteric bacillary meningitis, 33/62

Paracolobactrum aeroginoides
   gram-negative bacillary meningitis, 52/118
Paracrystalline body
   giant axonal neuropathy, 60/85
Paracrystalline inclusion
   complex V deficiency, 62/506
Paradoxical contraction (Westphal)
   dystonic shortening reaction, 6/141
   hereditary progressive diurnal fluctuating
      dystonia, 49/533
   Huntington chorea, 6/310, 49/280
Paradoxical embolism
   brain embolism, 11/391, 53/161
   brain infarction, 53/161
   brain ischemia, 53/161
   cardiac valvular disease, 63/22
   pregnancy, 11/460
   systemic brain infarction, 11/441
Paradoxical eyelid retraction
   clinical significance, 1/615
Paradoxical gustolacrimal reflex, *see* Crocodile tear
Paradoxical ischemia
   diabetic polyneuropathy, 51/503
Paradoxical kinesia
   antagonist hyperreflexia, 6/136
   dystonia musculorum deformans, 6/522, 49/522
   freezing gait, 49/69
   neuropharmacology, 49/58
   parkinsonism, 6/136, 188
Paradoxical myotonia
   myotonic syndrome, 62/261
   paramyotonia congenita, 62/268
Paradoxical pupillary reaction
   dilatation, 1/623
Paradoxical reaction, *see* Paradoxical kinesia
Paradoxical reflex
   head, *see* Head paradoxical reflex
   spinal cord injury, 26/250
Paradoxical respiration
   infantile spinal muscular atrophy, 59/66
   tetraplegia, 61/262
Parafascicular nucleus
   intralamellar thalamic nuclear group, 2/472
Parafascicularis complex
   centromedian, *see* Centromedian parafascicularis
      complex
Paraflocculus
   neuroanatomy, 2/399
Paraganglioma
   see also Chemodectoma
   autopsy incidence, 42/763
   carotid body, 42/762
   carotid body tumor, 8/482

catecholamine, 42/762
ectopic, 8/496
glomus jugulare, 18/435, 42/762
glomus jugulare tumor, 18/435
hypervascular, 68/270
mutation rate, 42/763
nodosum ganglion, 42/762
paraganglion, 8/461
sex ratio, 42/763
skull base, 68/466
skull base tumor, 68/482
survey, 68/270
terminology, 8/473
vagal, *see* Vagal paraganglioma
Paraganglioma abdominale
nerve tumor, 8/495
Paraganglioma aorticum
nerve tumor, 8/494
Paraganglioma caroticum, *see* Carotid body tumor
Paraganglioma nodosum
glomus jugulare tumor, 18/493
jugular body, 8/493
nerve tumor, 8/493
Paraganglioma tympanojugulare, *see* Glomus
jugulare tumor
Paraganglion
nerve tumor, 8/472-496
paraganglioma, 8/461
Paraganglion caroticum, *see* Carotid body
Paraganglion caroticum sarcoma, *see* Carotid body
tumor
Paraganglion caroticum tumor, *see* Carotid body
tumor
Paraganglion intercaroticum, *see* Carotid body
Paraganglion tympanicum
terminology, 18/435
Paraganglionic tissue
adrenal medulla, 8/473
Paraglioma
parietal lobe tumor, 17/297
subarachnoid hemorrhage, 12/119
Paragonimiasis, 35/243-264
cerebral, *see* Cerebral paragonimiasis
CSF, 35/257, 261-263
distribution, 35/243
epidemiology, 35/215, 243
epilepsy, 72/169
helminthiasis, 35/215
history, 35/243
life cycle, 35/244-246
meningitis, 35/260-262
pathogenesis, 35/246
pathology, 35/247-251

spinal, *see* Spinal paragonimiasis
treatment, 35/259, 262, 264
Paragonimus, 35/243-246
brain abscess, 52/151
helminthiasis, 52/511
specy, 35/244-246
Paragonimus meningitis
clinical features, 35/269
CSF, 35/261
differential diagnosis, 35/262
prognosis, 35/262
treatment, 35/262
Paragrammatism
aphasia, 4/92, 45/306
Paragranuloma caroticum, *see* Carotid body tumor
Paragraphia
agraphia, 4/145, 169
fluent agraphia, 45/460
semantic, *see* Semantic paragraphia
Parainfectious encephalomyelitis, *see* Acute
postinfectious disseminated encephalomyelitis
Parainfluenza virus
acute polymyositis, 56/199
acute viral myositis, 56/195
autonomic nervous system, 74/335
cluster headache, 74/335
deafness, 56/107
multiple sclerosis, 34/438
neurologic disease, 34/344
paramyxovirus, 56/431
Reye syndrome, 27/363, 29/332, 34/169, 56/150
vertigo, 56/107
Parainfluenza virus type 1
multiple sclerosis, 9/129, 47/330, 56/431
Parainfluenza virus type 2
multiple sclerosis, 9/129
Parainfluenza virus type 3
multiple sclerosis, 47/329
Paraldehyde
antiepileptic agent, 65/497
neonatal seizure, 73/258
Paraldehyde intoxication
nausea, 37/349
vomiting, 37/349
Paralexia
alexia, 4/129, 133, 45/433, 445, 448
grapheme-phoneme, 45/448
semantic, *see* Semantic paralexia
unilateral, 45/439
visual, 45/448
visuospatial agnosia, 45/168
Parallel fiber
cerebellar stimulation, 2/412

excitatory action, 2/413
function, 2/417
Paralysie des amants, *see* Honeymoon palsy
Paralysie d'amour, *see* Saturday night palsy
Paralysie d'ivrognes
brachial plexus, 51/145
Paralysing sciatica
aneurysmal bone cyst, 20/30
diagnosis, 20/585
pain, 20/585
spinal nerve root syndrome, 20/592
Paralysis
see also Paresis
abducent nerve, *see* Abducent nerve paralysis
ADR syndrome, 42/112
anterior cervical cord syndrome, 25/270
anterior horn syndrome, 2/210
anterolateral tract syndrome, 2/210
arthropod envenomation, 65/194
arthropoda intoxication, 37/56
ascending, *see* Ascending paralysis
autonomic, *see* Autonomic paralysis
barium intoxication, 36/325
Bell, *see* Peripheral facial paralysis
brachial plexus, *see* Brachial plexus paralysis
bulb diggers, *see* Compression neuropathy
bulbar, *see* Bulbar paralysis
bulbopontine, *see* Bulbopontine paralysis
cadet, *see* Cadet palsy
carotid dissecting aneurysm, 54/272
caudal aplasia, 42/50
central, 1/175
central motor, 1/169-171
central motor disorder, 1/175-185
cerebellar ataxia, 42/129
cerebral, *see* Cerebral palsy
Chastek, *see* Chastek paralysis
chlormethine intoxication, 64/233
compression neuropathy, 21/89
congenital laryngeal abductor, *see* Congenital
laryngeal abductor paralysis
conidae intoxication, 37/60
conjugate gaze, *see* Gaze paralysis
cranial nerve, *see* Cranial nerve palsy
cruciate, *see* Cruciate paralysis
CSF paraprotein, 42/603
cyclic oculomotor, *see* Cyclic oculomotor
paralysis
cyclist, *see* Cyclist palsy
dementia, *see* General paresis of the insane
diaphragm, *see* Diaphragm paralysis
divergence, *see* Divergence paralysis
drummer, *see* Drummer palsy
endemic periodic, *see* Endemic periodic paralysis
Erb, *see* Erb paralysis
extraocular, *see* Extraocular palsy
extraocular muscle, *see* Extraocular muscle
paralysis
eye convergence, 1/607, 2/341
eye muscle, 1/608
facial, *see* Facial paralysis
familial Bell, *see* Familial Bell palsy
familial bulbopontine, *see* Familial bulbopontine
paralysis
familial periodic, *see* Familial periodic paralysis
fixation, 2/340
flaccid, *see* Flaccid paralysis
flaccid cerebral, 1/232
gaze, *see* Gaze paralysis
general, *see* General paresis
glossopharyngeal nerve, *see* Glossopharyngeal
nerve palsy
Gradenigo syndrome, 2/91, 16/210
Guillain-Barré syndrome, 7/497, 42/645
Guttmann sign, 2/190, 202
hapalochlaena lunulata intoxication, 37/62
hapalochlaena maculosa intoxication, 37/62
hemophilia A, 42/738
hereditary cerebellar, *see* Hereditary cerebellar
palsy
hereditary cranial nerve, *see* Hereditary cranial
nerve palsy
hereditary facial nerve, *see* Hereditary facial nerve
palsy
hereditary hyperkalemic periodic, *see* Hereditary
hyperkalemic periodic paralysis
hereditary nerve pressure, *see* Hereditary nerve
pressure palsy
hereditary recurrent cranial nerve, *see* Hereditary
recurrent cranial nerve palsy
honeymoon, *see* Honeymoon palsy
horizontal gaze, *see* Horizontal gaze paralysis
hyperkalemic periodic, *see* Hyperkalemic periodic
paralysis
hypoglossal, *see* Hypoglossal paralysis
hypokalemic periodic, *see* Hypokalemic periodic
paralysis
hysteria, 41/166, 43/207
idiopathic facial, *see* Idiopathic facial paralysis
infantile, *see* Poliomyelitis
ischemic, *see* Ischemic paralysis
jake, *see* Organophosphate induced delayed
polyneuropathy
Klumpke, *see* Brachial paralysis
Landry, *see* Landry paralysis
laryngeal abductor, *see* Laryngeal abductor

paralysis
lateral gaze, *see* Lateral gaze palsy
lateral popliteal nerve, *see* Lateral popliteal nerve palsy
lightning injury, 7/355, 23/696
lumbosacral plexus, *see* Lumbosacral plexus palsy
lymphocytic choriomeningitis, 9/197
maternal obstetrical, *see* Maternal obstetrical paralysis
median nerve, *see* Median nerve palsy
multiple cranial nerve, *see* Multiple cranial neuropathy
neophocaena phocaenoides intoxication, 37/95
nerve, *see* Nerve paralysis
nerve pressure, *see* Nerve pressure palsy
normokalemic periodic, *see* Normokalemic periodic paralysis
occupational, *see* Occupational palsy
ocular, *see* Oculomotor paralysis
oculomotor, *see* Oculomotor paralysis
oculosympathetic, *see* Oculosympathetic paralysis
organophosphate intoxication, 37/554, 64/167
pack, *see* Pack palsy
paralytic shellfish poisoning, 37/33
paramyotonia congenita, 40/543
paratrigeminal, *see* Paratrigeminal paralysis
paratrigeminal oculosympathetic, *see* Paratrigeminal syndrome
parietal lobe epilepsy, 73/102
periodic, *see* Periodic paralysis
peripheral neuron, 1/219
phrenic nerve, *see* Phrenic nerve palsy
platypodia granulosa intoxication, 37/56
pontine gaze, *see* Pontine gaze paralysis
porpoise intoxication, 37/95
pressure, *see* Pressure palsy
progressive bulbar, *see* Progressive bulbar palsy
progressive supranuclear, *see* Progressive supranuclear palsy
pseudoabducens nerve, *see* Pseudoabducens nerve palsy
pseudobulbar, *see* Pseudobulbar paralysis
psychic gaze, *see* Psychic gaze paralysis
quail eater disease, 41/272
rabies, *see* Rabies paralysis
radial nerve, *see* Radial nerve palsy
recurrent laryngeal nerve, *see* Recurrent laryngeal nerve palsy
recurrent oculomotor, *see* Recurrent oculomotor paralysis
renal insufficiency, 27/336
respiratory muscle, *see* Respiratory muscle paralysis
rifle sling, *see* Rifle sling palsy
rucksack, *see* Rucksack paralysis
Saturday night, *see* Saturday night palsy
sciatic nerve, *see* Sciatic nerve palsy
sea snake intoxication, 37/92
sensory, *see* Sensory paralysis
serratus anterior, *see* Serratus anterior paralysis
serum, *see* Serum paralysis
serum paraprotein, 42/603
sixth nerve, *see* Abducent nerve paralysis
sleep, *see* Sleep paralysis
spastic cerebral, *see* Spastic cerebral palsy
spina bifida, 42/54
spinal cord lesion, 2/190, 202
superior longitudinal sinus thrombosis, 54/396
superior medialis, *see* Superior medialis paralysis
supranuclear, *see* Supranuclear palsy
tardy ulnar, *see* Tardy ulnar paralysis
terminology, 1/147
tick, *see* Tick paralysis
tongue, *see* Tongue paralysis
topography, 1/164-167
tri-*o*-cresyl phosphate, *see* Tri-*o*-cresyl phosphate paralysis
trigeminal nerve, *see* Trigeminal nerve palsy
trochlear nerve, *see* Trochlear nerve paralysis
ulnar, *see* Ulnar paralysis
unilateral elevator, *see* Unilateral elevator palsy
upward gaze, *see* Upward gaze paralysis
vasomotor disorder, 7/272
vertical gaze, *see* Vertical gaze palsy
vitamin B1 intoxication, 65/572
vocal cord, *see* Vocal cord paralysis
zozymus aeneus intoxication, 37/56
Paralysis agitans, *see* Parkinson disease
Paralysis periodica paramyotonica, 43/166-168
  cold injury, 43/167
  creatine kinase, 43/167
  diplopia, 43/167
  distal muscle weakness, 43/167
  dysphagia, 43/167
  fatigue, 43/167
  lactate dehydrogenase, 43/167
  myotonia paradoxa, 43/167
  pseudohypertrophy, 43/167
Paralytic brachial neuritis
  acute, *see* Acute paralytic brachial neuritis
Paralytic disease
  acute hemorrhagic conjunctivitis, 56/329
  Coxsackie virus, 56/329
  ECHO virus, 56/329
  enterovirus, 56/327
Paralytic ileus

see also Intestinal pseudo-obstruction
electrical injury, 74/321
tetanus, 65/218
Paralytic poliomyelitis, *see* Poliomyelitis
Paralytic pontine exotropia, *see* One and one half
   syndrome
Paralytic shellfish poisoning
abdominal pain, 37/33
Alexandrium genus, 37/33, 65/151
anarthria, 65/153
anuria, 37/33, 65/153
ataxia, 37/33, 65/158
autonomic nervous system, 75/512
clam, 37/33, 65/152
death, 37/33, 65/151
death rate, 37/33, 65/153
decarbamoylsaxitoxin, 65/152
diagnosis, 65/153
diarrhea, 37/33
dinoflagellates, 65/151
dysarthria, 65/153
emerita analoga intoxication, 37/56
EMG, 65/153
epidemic, 37/33, 65/151
geographic distribution, 37/33
gonyautoxin, 65/152
headache, 37/33, 65/153
history, 65/151
incidence, 65/151
lactic acidosis, 65/153
management, 65/154
marine toxin intoxication, 37/33, 65/143, 151
muscle action potential, 65/152
mussels, 37/33, 65/152
nausea, 37/33, 65/153
neosaxitoxin, 65/151
nerve conduction, 37/35, 65/152
oyster, 37/33, 65/152
paralysis, 37/33
paresthesia, 37/33, 65/153, 158
pharmacology, 37/33, 35
protozoa intoxication, 37/33
respiratory failure, 65/153
respiratory paralysis, 37/33, 65/152
salivation, 37/33
saxitoxin, 37/33, 65/152
sodium channel, 65/152
speech arrest, 37/33
symptom, 65/153
temperature reversal, 65/158
tetrahydropurine, 65/151
tetrodotoxin, 65/152
toxic encephalopathy, 65/141

transvector, 65/152
treatment, 37/35, 65/154
vertigo, 65/153
vomiting, 37/33, 65/153
weakness, 37/33
Paralytic strabismus
concomitant strabismus, 2/291
congenital type, 2/324
differential diagnosis, 2/291
diplopia, 2/291
heterophoria, 2/291
paradoxical innervation, 2/325
Paramedian medullary syndrome
inferior olive, 2/227
palatal myoclonus, 2/227
Paramedian nucleus
dorsal, *see* Dorsal paramedian nucleus
paraventricular thalamic nuclear group, 2/472
Paramedian pontine artery
en chapelet malacia, 2/241
microangiography, 11/92
symptom, 2/241, 244
Paramedian pontine syndrome
case report, 12/18
unilateral, *see* Unilateral paramedian pontine
   syndrome
Paramethadione
serum level, 15/690
side effect, 15/713
Paramnesia
memory disorder, 3/275
reduplicative, *see* Reduplicative paramnesia
Paramyeloidosis
cerebrovascular, 11/140
Paramyoclonus
juvenile hereditary benign chorea, 49/336
Paramyoclonus multiplex (Friedreich), *see*
   Hereditary essential myoclonus
Paramyotonia
amyotrophy, 22/32
hyperkalemic periodic paralysis, 40/317, 325, 335,
   537, 544, 547, 557, 562
myotonia, 62/262
myotonic syndrome, 62/262
Paramyotonia congenita
see also Hyperkalemic periodic paralysis
age at onset, 62/267
ambient temperature, 62/267
autosomal dominant, 41/423, 62/267
case history, 40/562
classification, 40/537
clinical features, 40/324, 328, 542
cold effect, 40/542-544, 41/423, 62/267

cold injury, 43/168
differential diagnosis, 40/545, 62/267
excitation contraction coupling, 40/556
genetics, 41/423
lid lag, 40/544
morphology, 40/547
myotonia, 40/542, 62/224
myotonia paradoxa, 43/169
myotonic syndrome, 40/542, 62/263
paradoxical myotonia, 62/268
paralysis, 40/543
progressive amyotrophy, 22/32
proximal muscle weakness, 43/168
pseudohypertrophy, 43/169
serum potassium, 43/169
sodium channel gene, 62/269
speech disorder, 43/168
speech disorder in children, 43/168
treatment, 40/557
Paramyotonia congenita (Eulenberg), *see*
    Paramyotonia congenita
Paramyotonia periodic paralysis, *see* Paralysis
    periodica paramyotonica
Paramyxovirus
    acute viral myositis, 56/193
    canine distemper, 34/293
    canine distemper virus, 56/587
    definition, 34/6
    encephalitis, 56/14
    human, 56/3
    human nervous system, 34/343
    measles virus, 56/14
    morbillivirus genus, 56/14
    mumps virus, 56/14
    parainfluenza virus, 56/431
    parainfluenza virus type 1-4, 56/13
    paramyxovirus genus, 56/13
    pneumovirus genus, 56/14
    respiratory syncytial virus, 56/431
    RNA virus, 56/13
    spastic paraplegia, 59/436
    subacute sclerosing panencephalitis, 56/14
    taxonomic place, 56/3
    ultrastructure, 56/14
    viral meningitis, 56/14
Paramyxovirus infection
    human nervous system, 34/343
    measles virus, 56/417
    mumps virus, 56/430
    neurology, 56/417
Paranasal sinus
    skull tumor, 17/204
    subdural empyema, 24/219

Paraneoplastic dysautonomia
    autonomic nervous system, 75/15
Paraneoplastic encephalomyelitis, 69/364
    limbic encephalitis, 71/682
    paraneoplastic syndrome, 71/682
    therapy, 71/687
Paraneoplastic neuromyotonia
    paraneoplastic syndrome, 69/374, 380
Paraneoplastic neuropathy, 69/353, 356, 378
    prevalence, 7/554
    sensorimotor, *see* Sensorimotor paraneoplastic
        neuropathy
    therapy, 71/677
Paraneoplastic polyneuropathy
    axonal degeneration, 51/469
    classification, 51/465
    CSF, 51/467
    dorsal funiculi degeneration, 51/468
    dorsal spinal nerve root ganglion, 51/468
    EMG, 51/467
    ganglioradiculopathy, 51/466
    history, 51/465
    Hodgkin disease, 51/468
    immunologic disorder, 51/470
    incidence, 51/465
    intestinal autonomic neuropathy, 51/494
    leukemia, 51/468
    metabolic cause, 51/469
    myeloma, 51/468
    Nageotte nodule, 51/468
    neuropathology, 51/468
    non-Hodgkin lymphoma, 51/468
    nutritional deficiency, 51/469
    oat cell carcinoma, 51/466
    orthostatic hypotension, 63/152
    pathogenesis, 51/469
    relapsing peripheral neuropathy, 51/467
    remitting peripheral neuropathy, 51/467
    segmental demyelination, 51/467
    sensorimotor neuropathy, 51/466
    sensory neuropathy, 51/466
    subacute disabling neuropathy, 51/467
    terminal mild neuropathy, 51/466
    toxic cause, 51/469
    viral infection, 51/470
Paraneoplastic sensory neuropathy
    paraneoplastic syndrome, 69/374
Paraneoplastic syndrome
    acetylcholine receptor, 69/338
    acute necrotizing myopathy, 69/374, 386
    acute sensorimotor neuropathy, 69/374
    amphiphysin, 69/330, 336
    antibody, 69/354, 356, 359

antigen, 69/329, 357
anti-Hu syndrome, 71/684
anti-Yo syndrome, 71/689
autoantibody, 69/341
autonomic dysfunction, 69/356
autonomic nervous system, 75/15, 536
autonomic neuropathy, 69/374, 378
bowel, 75/650
brachial neuritis, 69/374
brachial plexopathy, 69/379
brain stem encephalitis, 69/355, 71/684
cachetic myopathy, 69/374, 387
calcium, 69/338
cancer, 69/325, 71/696, 75/533
carcinoid myopathy, 69/374, 387
carcinomatous neuromyopathy, 69/374, 386
cerebellar ataxia, 69/356
cerebellar degeneration, 69/357, 71/688, 692
cerebellar syndrome, 69/364
chronic sensorimotor neuropathy, 69/374, 380
classification, 69/320, 330, 350
cloning, 69/333
dermatomyositis, 69/374, 384, 71/673
diagnosis, 69/319
Eaton-Lambert myasthenic syndrome, 69/374,
    383, 71/673, 678
epidemiology, 69/349
expression, 69/333, 337
Foix-Alajouanine disease, 71/696
history, 71/674
hodgkin antigen, 69/330
Hodgkin disease, 39/53-56, 69/337, 71/689
Hu antibody, 69/353, 355
Hu family, 69/330
immunity, 69/340
immunohistochemistry, 69/332
incidence, 69/351
inflammatory myopathy, 71/681
intestinal pseudo-obstruction, 51/493
leukocytoclastic vasculitis, 71/678
limbic encephalitis, 69/355
lung cancer, 71/673
lymphoma, 71/696
malignant lymphoma, 39/53-56
management, 75/536
molecular biology, 69/329
motoneuron disease, 69/363, 71/693, 697
motor end plate, 69/338
myasthenia gravis, 69/374, 382
myelitis, 71/684
myoclonus, 71/691
myopathy, 69/386, 71/681
myotonia, 69/374, 380

necrotizing myopathy, 71/681
nerve, 69/373
NOVA family, 69/330
NOVA gene family, 69/333
opsoclonus, 69/361
opsoclonus myoclonus, 71/691, 694
opsoclonus myoclonus syndrome, 69/361, 364
paraneoplastic encephalomyelitis, 71/682
paraneoplastic neuromyotonia, 69/374, 380
paraneoplastic sensory neuropathy, 69/374
pathogenesis, 69/321, 351
pathogenetic factor, 69/375
polymyositis, 69/374, 384
polyradiculoneuropathy, 69/374, 379
progressive dysautonomia, 59/161
recoverin, 69/330, 335
retinopathy, 71/696
sensorimotor paraneoplastic neuropathy, 71/675
sensory neuropathy, 69/354, 364
stiff-man syndrome, 69/363, 374, 380, 71/682
subacute motor neuronopathy, 69/374, 377
subacute sensory neuropathy, 71/674
survey, 69/329, 349, 373, 71/673
synaptotagmin, 69/330, 337
The Netherlands, 69/351
treatment, 69/365
tumor, 69/375
tumor expression, 69/332
vasculitic neuropathy, 69/374, 380
Yo antigen, 69/330, 334
Paranodal area
    hereditary multiple recurrent mononeuropathy,
        51/555
Paranodal axolemma
    Ranvier node, 47/10, 51/14, 16
Paranodal demyelination
    amyloid polyneuropathy, 51/417
    eosinophilia myalgia syndrome, 64/255
    renal insufficiency polyneuropathy, 51/358
    uremic polyneuropathy, 51/359
Paranodal junction
    axon, 51/41
Paranodal tomacula
    hereditary multiple recurrent mononeuropathy,
        51/559
Paranode
    Schwann cell, see Schwann cell paranode
Paranoia
    cocaine intoxication, 65/253, 256
    consciousness, 3/119
    diffuse Lewy body disease, 75/189
    disulfiram intoxication, 37/321
    dysphrenia hemicranica, 5/79

Huntington chorea, 49/282
hypercalcemia, 63/561
OKT3, 63/536
Paranoid schizophrenia
brain ventricle enlargement, 46/482
diffuse focal sign, 46/484
EEG, 46/494
neuropsychologic assessment, 46/488
nonspecific, 46/419
psychophysiology, 46/506
Paranoid state
migraine, 48/162
Paranoid syndrome
epilepsy, 15/598
Paraosteoarthropathy, *see* Neurogenic
osteoarthropathy
Paraoxon
experimental myopathy, 40/150
organophosphate, 64/152
organophosphate intoxication, 64/152
Paraparesis
aorta coarctation, 63/4
aorta thrombosis, 63/41
arsenic intoxication, 64/288
barium intoxication, 64/354
Behçet syndrome, 56/598
cerebrotendinous xanthomatosis, 59/357
ciclosporin, 63/10
lumbar vertebral canal stenosis, 20/714
multiple sclerosis, 47/56
neurobrucellosis, 52/588
organophosphate induced delayed neuropathy,
64/229
progressive, *see* Progressive paraparesis
spastic, *see* Spastic paraparesis
tropical spastic, *see* Tropical spastic paraplegia
Parapharyngeal space syndrome
see also Multiple cranial neuropathy
glomus jugulare tumor, 18/445
multiple cranial neuropathy, 51/569
Paraphasia
agraphia, 4/169
Alzheimer disease, 46/251
amusia, 45/487
aphasia, 4/91-93, 45/17-20, 57/404
definition, 2/599
delirium, 46/531
dialysis encephalopathy, 64/277
head injury outcome, 57/404
migraine, 48/161
thalamic aphasia, 54/309
verbal, *see* Verbal paraphasia
Wernicke aphasia, 4/98

Paraphrasia
cluttering, 43/220
Paraphyseal cyst, *see* Ependymal cyst
Paraphysis
blood-brain barrier, 56/80
Paraplegia
see also Spinal cord injury
acute abdominal pain, 26/211
adjustment stage, 61/537
aged, 1/208
4-aminopyridine, 61/401
amyotrophic dystonic, *see* Amyotrophic dystonic
paraplegia
amyotrophy, 60/244
anal reflex, 61/295
animal model, 56/455, 587
ankylosing spondylitis, 38/518
anocutaneous reflex, 61/295
aorta arteriosclerosis, 61/115
aorta coarctation, 61/115, 63/4
aorta disease, 61/111
aorta injury, 61/111
aorta surgery, 61/116, 63/55
aorta thrombosis, 61/115
arteriosclerosis, 1/208
autosomal dominant dystonic, *see* Dystonic
paraplegia
axon regeneration, 2/179, 61/407
axon survival, 61/400
axonal dysfunction, 61/401
basal cardiovascular state, 61/440
Basedow, *see* Basedow paraplegia
bismuth intoxication, 36/327, 329
bladder infection, 26/208
bulbocavernosus reflex, 61/295
Burkitt lymphoma, 39/75, 63/351
cardiac arrest, 63/215
cardiovascular drug response, 61/438
cardiovascular system, 61/435
central motor disorder, 1/205-210
central origin, 1/205
cerebral, 1/208
clinical trial, 61/399
Cobb syndrome, 14/439
coitus, 61/315
congenital heart disease, 63/4
corticosteroid, 61/399
decompression myelopathy, 61/216
diamorphine myelopathy, 55/519
dissecting aorta aneurysm, 61/114, 63/37, 39, 46
distress, 61/537
dystonia, 6/520, 556
ejaculation, 61/315, 322

electrical injury, 61/195
electroejaculation, 61/323
en chapelet malacia, 2/241
epidural anesthesia, 61/137, 142
exercise induced hyperthermia, 61/285
experimental therapy, 61/399
extension, 1/206, 251
familial amyotrophic dystonic, *see* Familial
 amyotrophic dystonic paraplegia
familial spastic, *see* Familial spastic paraplegia
female sexual function, 61/331
fertility, 61/313
filariasis, 35/167, 52/516
flaccid, *see* Flaccid paraplegia
flexion, 1/206, 251, 2/182, 26/252
GABA, 61/401
gunshot sequel, 61/255
hereditary dystonic, *see* Hereditary dystonic
 paraplegia
hereditary motor and sensory neuropathy variant,
 60/244
hydromyelia, 50/429
hypoparathyroidism, 27/311
iatrogenic, *see* Iatrogenic paraplegia
iatrogenic origin, 61/116
impotentia coeundi, 61/315
intrathecal chemotherapy, 61/151
intrathecal sodium chloride, 61/152
ischemic myelopathy, 61/115, 160
K opiate receptor blocker, 61/399, 403, 406
leukemia, 63/343
life quality, 61/541
lipid peroxidation inhibitor, 61/399
male sexual function, 61/313
maturity stage, 61/537
meningeal leukemia, 63/343
mental depression, 61/537
N-methylaspartic acid receptor blocker, 61/399,
 406
methylprednisolone, 61/399, 402
monosialoganglioside, 61/399, 406
multiple myeloma, 20/11, 39/141
murine leukemia virus, 56/587
muscle stretch reflex, 1/206
myelitis, 61/161
myeloma, 20/16
myelomalacia, 61/111
naloxone, 61/399
neonatal spinal, 1/210
neurobrucellosis, 33/316, 52/587, 590
neuromyelitis optica, 9/426, 429, 47/401
occupational therapy, 26/233
organotin intoxication, 64/135

orgasm, 26/456
pain surgery, 26/496
patient adjustment, 61/537
patient personality characteristics, 61/542
penis prosthesis, 61/321
phantom limb, 4/224, 226
plantar reflex, 26/248
plasma hydrocortisone, 26/397, 402
pontine, *see* Pontine paraplegia
pontine infarction, 12/15, 23
pontine syndrome, 1/210
postpneumonectomy ischemic myelopathy,
 61/142
posttraumatic syringomyelia, 26/117, 50/454,
 61/375, 393
pressure sore, 26/463, 61/347
protirelin, 61/403, 406
psychology, 61/537
radiation myelopathy, 12/501, 61/203
rebellion stage, 61/537
recovery, 61/399
recovery substrate, 61/400
recurrent transient, *see* Recurrent transient
 paraplegia
regenerative therapy, 61/406
rehabilitation, 20/830
remyelination, 61/402
schistosomiasis, 52/538
sexual capability, 61/315
sexual function, 26/454-457
sexual intercourse, 61/313
sociology, 61/537
spastic, *see* Spastic paraplegia
spastic cerebral palsy, 42/167
spasticity, 1/206
spinal, 1/207
spinal cord abscess, 61/161
spinal cord coning, 61/161
spinal cord electrotrauma, 61/195
spinal cord hemorrhage, 61/160
spinal cord injury, 26/252
spinal epidural abscess, 61/161
spinal epidural hematoma, 61/137
spinal epidural lymphoma, 63/349
spinal nerve root pain, 26/492
spinal neurinoma, 20/251
spinal sympathicoblastoma, 20/112
Stargardt disease, 13/132
steal syndrome, 32/487
steroid antioxidant, 61/405
Stokes-Mandeville-Egerton bed, 26/221, 301, 472
striatopallidodentate calcification, 49/423
superior longitudinal sinus thrombosis, 54/396

suprapubic cystotomy, 26/409
surgical complication, 61/116
systemic hypotension, 61/114
tirilazad, 61/405
traffic injury, 25/142
traumatic aorta rupture, 61/112, 116
tropical spastic, *see* Tropical spastic paraplegia
tuberculous spondylitis, 33/241-244
urinary calculi, 61/308
urinary incontinence, 61/291
vacuum constrictor devices, 61/321
vesicoureteric reflux, 61/308
vesicourethral dysfunction, 61/291
visceral pain, 26/494
Vogt-Koyanagi-Harada syndrome, 56/621
Waldenström macroglobulinemia, 63/396
X-linked spastic, *see* X-linked spastic paraplegia
Paraplegia dolorosa
   spinal meningioma, 20/198
Paraplegia in extension
   cause, 26/252
Paraplegic deep vein thrombosis
   pulmonary embolism, 26/315, 345
Paraplégie douloureuse cancéreuse
   historic note, 20/40
Parapoxvirus
   neurotropic virus, 56/5
Paraprotein
   CSF, *see* CSF paraprotein
   neuropathy, 63/402
   serum, *see* Serum paraprotein
   urine, 42/603
Paraproteinemia
   see also Plasma cell dyscrasia
   amyloidosis, 69/290, 302
   antibody, 69/297
   benign, *see* Benign paraproteinemia
   benign monoclonal, *see* Benign monoclonal
    paraproteinemia
   biclonal gammopathy, 69/290
   brain, 16/233
   brain infarction, 53/166
   brain ischemia, 53/166
   brain thrombosis, 11/461
   carpal tunnel syndrome, 39/155-168
   Castleman disease, 69/307
   chronic demyelinating neuropathy, 47/615-620
   classification, 39/135
   clinical features, 39/137-146
   cryoglobulinemia, 63/403, 69/297, 308
   demyelinating neuropathy, 47/614-620
   diagnosis, 39/158-165
   gamma heavy chain disease, 69/290, 307

ganglioside, 69/297
glycolipid, 69/297
Guillain-Barré syndrome, 39/153
history, 39/131-135
hyperviscosity syndrome, 55/484, 71/439
Ig type, 63/398
IgA, *see* IgA paraproteinemia
IgG, *see* IgG paraproteinemia
IgG-kappa, *see* IgG-kappa paraproteinemia
IgM monoclonal gammopathy, 69/297
IgM type, 63/396
leptomeningeal involvement, 39/152
leukemia, 69/308
lymphoma, 69/308
lymphoma associated, *see* Lymphoma associated
  paraproteinemia
lytic myeloma, 69/302
meningeal invasion, 71/439
monoclonal gammopathy, 69/290
multiple myeloma, 69/290, 301
myelin lamella widening, 63/399
myelin sheath, 63/399
neurologic involvement, 39/131-168
neuropathy, 69/292, 71/441
osteosclerotic myeloma, 69/290, 303
pathology, 39/146-152, 156
plasmacytoma, 69/306
POEMS syndrome, 47/619, 51/430, 432
polyneuropathy, 63/398
prevalence, 69/291
primary systemic amyloidosis, 69/290, 300
progressive bulbar palsy, 59/225
psychosis, 20/11
recurrent nonhereditary multiple cranial
  neuropathy, 51/570
reticulosarcoma, 18/235
solitary plasmacytoma, 69/306
survey, 69/289
symptom, 71/431
syndrome, 71/437
treatment, 39/165-168
undetermined significant monoclonal
  gammopathy, 63/399
Waldenström macroglobulinemia, 69/290, 306
Paraproteinemic coma
   multiple myeloma, 18/256, 20/11
Paraproteinemic neuropathy
   antibody, 69/296
   pathogenesis, 39/153-156, 69/296
   tremor, 49/589
Paraproteinemic polyneuropathy
   benign monoclonal paraproteinemia, 51/436
   benign paraproteinemia, 51/429

heavy chain disease, 51/429
IgA paraproteinemia, 51/436
IgG paraproteinemia, 51/436
IgM, *see* IgM paraproteinemic polyneuropathy
lymphoma associated paraproteinemia, 51/429
multiple myeloma, 51/429
primary amyloidosis, 51/429
treatment, 51/439
Waldenström macroglobulinemia, 51/429
Parapsilosis
Candida, 35/417, 52/398
Paraquat
amphibian parkinsonism, 65/397
1-methyl-4-phenylpyridinium, 65/397
neurotoxicity, 51/286
rodent neonate, 65/397
tremor, 51/287
Paraquat intoxication
paraquat lung, 65/398
Purkinje cell, 65/398
substantia nigra, 65/398
Parasagittal hematoma
surgery, 24/639
Parasellar cholesteatoma
skull base tumor, 17/197
Parasellar space occupying lesion
Tolosa-Hunt syndrome, 48/302
Parasellar syndrome
brain metastasis, 71/612
Parasite
ascariasis, 35/357
CNS infection, 69/444
spinal cord compression, 19/371
Parasitic disease
African trypanosomiasis, 35/8-10
amebiasis, 35/11
aqueduct stenosis, 2/279, 50/308
cestode, 35/12
Chagas disease, 35/10
echinococcosis, 35/12
epilepsy, 72/169
filariasis, 35/11-15
larva migrans, 35/15
malaria, 35/6-8
neurocysticercosis, 35/13, 52/529
schistosomiasis, 35/14
tropical neurology, 35/1-15
Paraskepsia, *see* Dementia
Parasomnia
adult, *see* Adult parasomnia
brain death, 24/701, 711, 713
brain injury, 24/701
coma, 3/76

juvenile, *see* Juvenile parasomnia
posttraumatic brain stem lesion, 24/593
sleep, 74/558
sleep disorder, 45/140-142
Paraspinal abscess
coccidioidomycosis, 52/416
curvulariosis, 52/484
Paraspinal neurofibroma
neurofibromatosis type II, 50/370
Parastriate cortex
anatomy, 2/541, 641, 643
antagonistic area, 2/542
Balint syndrome, 55/144
connection, 2/540
corpus callosum agenesis, 50/150
hypercomplex cell, 2/542
receptive field, 2/542
splenium corporis callosi, 2/541
Parasympathetic denervation
Chagas disease, 52/347
Parasympathetic efferent fiber
test, 74/596
Parasympathetic ganglion
aged, 75/6
aging, 74/226
normal, 75/5
progressive dysautonomia, 59/143
structure, 74/41
Parasympathetic nerve
neurotransmission, 74/101
sympathetic nerve, 74/104
Parasympathetic nervous system
activation, 74/8
anatomy, 1/430, 74/8
autonomic nervous system, 74/201
congenital indifference to pain, 42/297
diagram, 22/247
sacral efferent, *see* Sacral parasympathetic efferent
spinal efferent, *see* Spinal parasympathetic efferent
Parasympathetic neuron
autonomic nervous system, 75/620
classification, 74/37
Parasympathetic system
reflex, 74/36
Parasympathomimetic agent, *see* Cholinergic receptor stimulating agent
Parathion
myopathy, 40/150
neurotoxin, 8/338, 37/546, 64/177
organophosphate, 8/338, 64/152, 169
organophosphate induced delayed polyneuropathy, 64/177

organophosphate intoxication, 37/554, 558, 64/152, 155, 169
Parathion intoxication
chemical classification, 37/546
headache, 48/420
organophosphorus compound, 37/546
Parathormone, *see* Parathyroid hormone
Parathyroid adenoma
Huntington chorea, 49/291
neurofibromatosis type I, 14/492
Parathyroid disease
neurologic manifestation, 27/283-314
survey, 70/111
vitamin D, 27/284
Parathyroid dysfunction
chorioretinal degeneration, 13/41
optic atrophy, 13/41
Parathyroid function
spinal cord injury, 26/397
Parathyroid gland
calcitonin, 27/273
epilepsy, 15/319
symptom, 70/116
Parathyroid hormone
amyotrophic lateral sclerosis, 59/177
calcitonin, 27/284
chemistry, 27/283
multiple nevoid basal cell carcinoma syndrome, 14/80
myoglobinuria, 62/556
uremic encephalopathy, 63/514-516
uremic polyneuropathy, 51/362, 63/513-515
Parathyroid hormone metabolism
intracerebral calcification, 42/534
pseudohypoparathyroidism, 42/578, 621
Paratonia
African trypanosomiasis, 52/341
kuru, 56/557
motor deficiency, 4/454
Paratrigeminal oculosympathetic paralysis, *see* Paratrigeminal syndrome
Paratrigeminal paralysis
paratrigeminal syndrome, 48/329
Paratrigeminal syndrome, 48/329-339
anatomy, 5/334
anhidrosis, 48/331
arachnoiditis, 48/337
carotid gasserian anastomosis, 5/335
cause, 48/337
cavernous sinus, 48/332, 334
chronic migrainous neuralgia, 48/250
cluster headache, 5/115, 48/226
cranial nerve, 48/336

facial pain, 48/336
facial sweating, 2/113, 48/329
features, 2/314
forehead anhidrosis, 48/334, 336
headache, 48/6
herpes zoster, 48/337
history, 48/329
Horner syndrome, 1/514, 48/329, 75/293
internal carotid artery, 48/332
lesion site, 48/334
middle ear, 48/332
miosis, 48/330, 334
nonrecurrent nonhereditary multiple cranial neuropathy, 51/571
oculosympathetic paralysis, 48/332
oculosympathetic pathway, 48/331
ophthalmic nerve, 48/334
orbit, 48/333
original, 48/330
original case, 2/92
paratrigeminal paralysis, 48/329
pathogenesis, 5/334
pericarotid syndrome, 48/329
petrositis, 48/337
posterior cavernous sinus syndrome, 2/90
ptosis, 48/330, 334
sphenopalatine neuralgia, 5/331
subarachnoid hemorrhage, 55/20
subgroup, 75/294
supraorbital nerve, 48/333
sweating, 75/293
sympathetic nerve fiber, 48/332
sympathetic pathway, 48/331
sympathetic plexus, 48/334
temporal arteritis, 48/337
Tolosa-Hunt syndrome, 48/301, 303, 337
treatment, 5/382
trigeminal ganglion, 48/332
type, 5/333
Villaret syndrome, 2/314
Wegener granulomatosis, 48/337
Paratyphoid vaccine
neurologic complication, 8/87
Paravenous retinochoroidal atrophy
pigmented, *see* Pigmented paravenous retinochoroidal atrophy
Paraventricular nucleus
anatomy, 2/434
heat stroke, 2/447
suprachiasmatic nucleus, 74/479
Paraventricular nucleus degeneration
hypophyseal diabetes insipidus, 42/543
Paraventricular thalamic nuclear group

paramedian nucleus, 2/472
paratenial nucleus, 2/472
reuniens nucleus, 2/472
Paravermal artery
    inferior, *see* Inferior paravermal artery
Paravermal zone
    rostral, *see* Rostral paravermal zone
Paravertebral soft tissue
    radiography, 19/163-166
    spinal cord compression, 19/163-166
Pareiitis granulomatosa
    Melkersson-Rosenthal syndrome, 8/216
Parenchymal hemorrhage
    brain arteriovenous malformation, 54/375
Paresis
    see also Paralysis
    Bell, *see* Peripheral facial paralysis
    bilateral facial, *see* Bilateral facial paresis
    canal, *see* Canal paresis
    chronic benign trigeminal, *see* Chronic benign
        trigeminal paresis
    cold, *see* Cold paresis
    endemic, *see* Endemic paresis
    facial, *see* Facial paralysis
    Friedreich ataxia, 21/330
    GPI, *see* General paresis of the insane
    incontinentia pigmenti, 14/85, 214
    larynx, *see* Larynx paresis
    pharynx, *see* Pharynx paresis
    pseudogeneral, *see* Pseudogeneral paresis
    Refsum disease, 60/228
    spastic, *see* Spastic paresis
    spinal ependymoma, 19/360, 20/365
Paresthesia
    acetazolamide intoxication, 37/202
    acetonylacetone intoxication, 64/86
    amputation, 70/44
    anxiety neurosis, 43/199
    atenolol, 63/195
    beriberi, 28/8
    blennorrhagic polyneuritis, 7/478
    brain edema, 63/417
    carpal tunnel syndrome, 42/309
    cheiro-oral, *see* Cheiro-oral paresthesia
    ciguatoxin intoxication, 37/83, 65/162
    cluster headache, 48/223
    complicated migraine, 5/61
    conidae intoxication, 37/60
    craniometaphyseal dysplasia, 43/362
    cryoglobulinemia, 63/404
    Eaton-Lambert myasthenic syndrome, 62/426
    eosinophilia myalgia syndrome, 63/376, 64/252,
        254

eosinophilic fasciitis, 63/385
ergot intoxication, 65/67
ethambutol intoxication, 65/481
facial, *see* Facial paresthesia
flecainide intoxication, 65/456
Flynn-Aird syndrome, 42/327
foramen magnum tumor, 19/58
frontal lobe tumor, 17/259
headache, 48/443
hereditary amyloid polyneuropathy, 8/368,
    42/518, 521
hereditary hyperkalemic periodic paralysis,
    43/152
hyperabduction syndrome, 7/441
hypereosinophilic syndrome, 63/373
hyperkalemia, 63/558
hyperkalemic periodic paralysis, 43/152
hypertrophia musculorum vera, 43/84
hyperventilation syndrome, 38/318, 63/433
hypoparathyroid myopathy, 62/541
hypophosphatemia, 63/565
ichthyohaemotix fish intoxication, 37/87
intracranial hypertension, 16/135
isoniazid neuropathy, 7/516, 533, 42/648
Köhlmeier-Degos disease, 39/437
lead polyneuropathy, 64/436
lysergide, 65/45
malignant radiculopathy, 69/73
marine toxin intoxication, 65/142, 153, 158
Melkersson-Rosenthal syndrome, 42/322
methyldopa intoxication, 37/440
migraine, 42/746, 48/156
Minamata disease, 64/413
Muckle-Wells syndrome, 42/396
multiple sclerosis, 42/496, 47/55
neophocaena phocaenoides intoxication, 37/95
neurotoxic shellfish poisoning, 65/158
organic solvent intoxication, 64/41
organochlorine insecticide intoxication, 64/199
painful, *see* Painful paresthesia
paralytic shellfish poisoning, 37/33, 65/153, 158
parietal lobe epilepsy, 73/100
parietal lobe tumor, 17/300
pellagra, 6/748, 7/573
pernicious anemia, 42/609
phantom limb, 4/225, 45/398
pheochromocytoma, 39/498, 501
pontine infarction, 2/241, 12/15
primary pontine hemorrhage, 12/39
progressive hemifacial atrophy, 31/253, 43/408
propranolol, 63/195
puffer fish intoxication, 37/80
pyrethroid intoxication, 64/218

restless legs syndrome, 8/311, 317, 27/338, 42/260, 49/621
spinal tumor, 19/57
terminology, 1/81, 89
tetany, 63/560
tetrodotoxin intoxication, 65/156
thalamus, 1/109
thalidomide, 42/660
thallium intoxication, 64/325
thoracic outlet syndrome, 51/122
tocainide intoxication, 65/455
toxic oil syndrome, 63/381
tropical ataxic neuropathy, 70/411
uremic polyneuropathy, 51/356
vitamin B6 intoxication, 65/572
vitamin B12 deficiency, 70/377
zozymus aeneus intoxication, 37/56
Paresthesia nocturna
  compression syndrome, 7/437
Paresthetic causalgic syndrome
  deficiency neuropathy, 7/636
Paresthetica
  meralgia, *see* Meralgia paresthetica
Pargos intoxication
  classification, 37/78
Pargyline
  myopathy, 40/169
Parietal acalculia
  secondary acalculia, 4/189
Parietal amyotrophy
  parietal lobe lesion, 2/684
Parietal aphasia syndrome
  dyslexia, 45/75
Parietal ataxia
  description, 1/320
  parietal lobe syndrome, 45/67
  parietal lobe tumor, 17/306
Parietal bone
  development, 30/213
Parietal cephalocele
  classification, 50/108
  computer assisted tomography, 50/108
  corpus callosum agenesis, 50/108
  Dandy-Walker syndrome, 50/108
  holoprosencephaly, 50/108
  hydrocephalus, 50/108
  incidence, 50/107
  microcephaly, 50/108
  ultrasonography, 50/108
Parietal emissary foramen
  see also Foramina parietalia permagna
  anatomy, 11/55
  foramina parietalia permagna, 30/276, 50/141

Parietal emissary vein
  anatomy, 11/53
Parietal encephalocele
  atretic cephalocele, 50/110
  computer assisted tomography, 50/108
  corpus callosum agenesis, 50/108
  Dandy-Walker syndrome, 50/108
  glial heterotopia, 50/110
  holoprosencephaly, 50/108
  hydrocephalus, 50/108
  microcephaly, 50/108
  prognosis, 50/109
  surgical correction, 50/109
  ultrasonography, 50/108
Parietal endocarditis
  neurotropic, *see* Neurotrophic parietal endocarditis
Parietal foramina
  enlarged, *see* Foramina parietalia permagna
Parietal hemianesthesia
  parietal lobe syndrome, 45/64
Parietal lobe
  alexia, 4/127
  anatomic synopsis, 2/680-682
  attention, 45/172
  blood supply, 2/681
  cerebrovascular disease, 55/117
  conduction aphasia, 45/18
  congenital heart disease, 63/2
  cytoarchitectonic asymmetry, 45/78
  distance perception, 45/362
  dysmegalopsia, 45/362
  EEG, 45/173, 72/300
  eidetic imagery, 45/360
  electrostimulation, 2/682
  hallucination, 45/358-364
  hemiasomatognosia, 45/375
  illusion, 45/361-363
  indifference to pain, 8/196-198
  lateral aspect, 2/680, 73/99
  macropsia, 45/362
  memory imagery, 45/358-360
  mesial aspect, 2/681, 73/99
  metamorphopsia, 45/361-363
  micropsia, 45/362
  motion perception, 45/363
  movement plan, 3/34
  neglect, 45/16, 172
  neglect syndrome, 45/172
  pain, 8/196-198
  parkinsonism, 45/77
  phantom sensation, 45/400
  polyopia, 45/363

schizophrenia, 46/458, 489
stereoscopy, 45/362
thought imagery, 45/358-360
visual memory, 45/359
visual perseveration, 45/360
visuospatial agnosia, 45/172
Parietal lobe epilepsy
  anatomic physiology, 73/98
  anxiety, 73/102
  asomatognosia, 73/101
  autoscopia, 73/102
  cerebrovascular disease, 73/104
  changed body shape, 73/101
  classification, 72/11
  complex partial seizure, 73/103
  cryptogenic, 73/104
  EEG, 73/103
  etiology, 73/104
  eye turning, 73/102
  genital sensation, 73/100
  gustatory, 73/102
  hallucination, 73/101
  head injury, 73/104
  illusion, 73/101
  induced seizure, 73/103
  language, 73/102
  lesion, 73/103
  numbness, 73/100
  nystagmus, 73/102
  pain, 73/100
  paralysis, 73/102
  paresthesia, 73/100
  pathology, 72/117
  perseveration, 73/101
  postural, 73/102
  somatosensory, 73/99, 101
  subtle seizure, 73/103
  sudden vague sensation, 73/100
  survey, 73/97
  symptomatic, 73/104
  vertigo, 73/101
Parietal lobe lesion
  acalculia, 4/189
  agraphia, 4/154
  apraxia, 5/50, 54-57
  autotopagnosia, 1/107
  avoidance apraxia, 4/54
  Balint syndrome, 45/410
  body scheme disorder, 2/668, 4/4, 207-240
  cerebral dominance, 4/4, 302
  congenital tactile agnosia, 4/34
  conjugate eye movement, 2/588
  constructional apraxia, 4/4, 56, 45/500

homonymous hemianopia, 2/587
interparietal syndrome, 4/233
lower quadrantanopia, 2/587
optokinetic nystagmus, 2/588
parietal amyotrophy, 2/684
phantom limb phenomena, 4/224
postural sensibility, 2/683
sensory extinction, 3/193
tactile agnosia, 4/36
tactile apraxia, 4/34
unawareness of defect, 2/588
virgin hand, 4/467
visual agnosia, 2/587, 4/20-22
Wernicke aphasia, 4/98
Parietal lobe syndrome, 2/680-698, 45/63-82
  agraphia, 45/75, 77, 80
  alexia, 45/75, 77, 81
  Alzheimer disease, 46/201
  anatomy, 45/76-79
  Anton-Babinski syndrome, 45/69
  apraxia, 45/74, 77
  autotopagnosia, 45/70
  Balint syndrome, 45/73, 77
  body scheme disorder, 45/69
  brain injury, 57/134
  Charcot-Wilbrand syndrome, 45/73
  clinical examination, 57/134
  conduction aphasia, 45/75, 77
  conscious hemiasomatognosia, 45/69
  constructional apraxia, 45/74, 80
  dementia, 45/76, 46/124
  discriminative sensibility, 2/682, 45/64-67
  dressing apraxia, 2/692, 45/74
  dysgraphia, 45/75
  dyslexia, 45/75
  epileptic focus, 45/64
  Gerstmann syndrome, 2/686, 45/69
  hemiagnosia, 45/70
  hemianopia, 45/69
  hemiasomatognosia, 45/69
  hemidepersonalization, 45/69, 375
  ideational apraxia, 45/74
  ideomotor apraxia, 45/74
  motility disorder, 45/67-69
  motor deficit, 45/67
  motor disorder, 45/67-69
  muscular atrophy, 45/67
  neuropsychology, 45/79-82
  nominal aphasia, 45/74
  ocular ataxia, 45/73
  optokinetic nystagmus, 45/69
  parietal ataxia, 45/67
  parietal hemianesthesia, 45/64

placing reaction, 45/68, 77
primary tactile agnosia, 45/66
pseudothalamic syndrome, 45/65, 78
psychic gaze paralysis, 45/73
psychosis, 45/76
sensory aphasia, 45/75
sensory extinction, 45/65
somatosensory disorder, 45/64
space perception, 45/80
spatial agnosia, 45/78, 80
spatial orientation disorder, 2/697, 45/70-74
speech disorder, 45/74
symptom, 45/64-76
tactile agnosia, 45/66
tactile asymbolia, 45/66
tactile perception, 45/79
topographical memory, 45/73
transcortical sensory aphasia, 45/75
traumatic, *see* Traumatic parietal syndrome
Verger-Dejerine syndrome, 45/65
visuospatial agnosia, 45/73, 78, 80
Parietal lobe tumor
agnosia, 17/301
aphasia, 17/305, 67/153
apraxia, 17/303, 67/153
astrocytoma, 17/297
asymptomatic, 17/306
classification, 17/297
cranial nerve, 17/301
echoEG, 17/306
EEG, 17/306
Gerstmann syndrome, 17/302
glioblastoma, 17/297
glioblastoma multiforme, 17/297
hypoesthesia, 17/300
incidence, 17/298
location, 17/298
meningioma, 17/298
mental disorder, 17/305
metastasis, 17/298
motor disorder, 17/300
papilledema, 17/301
paraglioma, 17/297
paresthesia, 17/300
parietal ataxia, 17/306
prognosis, 17/306
psychiatric disorder, 17/305
seizure, 67/152
sensory abnormality, 67/152
sensory disorder, 17/300
symptom, 17/299
tactile agnosia, 17/300
visual disorder, 67/153

Parietal lobule
inferior, *see* Inferior parietal lobule
Parietal operculum
gustatory hallucination, 45/357
Parietal permagna foramen, *see* Foramina parietalia
permagna
Parietal vein
anatomy, 11/47
posterior, *see* Posterior parietal vein
Parieto-occipital artery
radioanatomy, 11/97
topography, 11/97
Parinaud oculoglandular syndrome, *see* Parinaud
syndrome
Parinaud syndrome
aqueduct stenosis, 2/279, 50/311
bilateral pretectal site, 1/602
brain embolism, 53/212
brain infarction, 53/212
brain injury, 57/137
cat scratch disease, 34/462, 52/128
cause, 2/302
cerebrovascular disease, 53/212
clinical examination, 57/137
Collier sign, 55/124
downward gaze, 2/339
inferior colliculi, 2/339
intracranial hemorrhage, 53/212
Kii Peninsula amyotrophic lateral sclerosis,
22/371
lesion site, 53/212
medial longitudinal fasciculus, 55/124
mitral valve prolapse, 63/21
ocular imbalance, 24/74
oculomotor paralysis, 2/304
pineal tumor, 17/656
pinealoma, 16/352, 18/362, 42/766
progressive supranuclear palsy, 6/557
rostral interstitial medial longitudinal fasciculus
nucleus, 55/123
secondary pigmentary retinal degeneration,
13/304
striatonigral degeneration, 49/206
superior colliculi, 2/338
symptom, 2/279
transient ischemic attack, 53/212
trochlear nerve paralysis, 2/304
upward gaze, 2/338
vertical gaze palsy, 2/337
Paris Conference of Chromosomes, 30/92, 104,
31/353, 376, 561, 572
Paris, France
neurology, 1/7

Paris green, *see* Cupric acetoarsenite
Parkes Weber-Klippel syndrome, *see*
    Klippel-Trénaunay syndrome
Parkes Weber syndrome
    see also Klippel-Trénaunay syndrome
    Klippel-Trénaunay syndrome, 14/395-397
    segmental hypertrophy, 14/67
    vascular malformation, 14/396
Parkinson dementia, 46/371-381, 49/167-179
    see also Dementia
    abstract reasoning, 46/375
    acetylcholine, 49/178
    Alzheimer disease, 46/376, 49/178, 59/284-287
    amyotrophic lateral sclerosis, 49/167, 174, 401,
        59/284-287
    anarthria, 49/171
    arteriosclerosis, 46/376
    atypical, 49/174
    Auyu people, 49/167
    bradykinesia, 46/375, 49/171
    Chamorro people, 49/167
    choline acetyltransferase, 49/178
    clinical features, 49/170
    course, 49/174
    depression, 46/376
    diagnosis, 49/174
    dysarthria, 49/171
    dysphagia, 49/171
    epidemiology, 49/169
    etiology, 46/376-381
    familial amyotrophic lateral sclerosis, 59/236
    flexion contracture, 49/173
    foot deformity, 49/172
    gait, 49/172
    genetics, 49/170
    granulovacuolar degeneration, 49/177
    Guam, *see* Guam Parkinson dementia
    Guam amyotrophic lateral sclerosis, 22/341,
        42/70, 249-251, 59/237
    Hallervorden-Spatz syndrome, 49/401
    hand malformation, 49/172
    Hirano body, 49/177
    historic note, 46/371
    history, 49/169
    homovanillic acid, 49/174
    5-hydroxyindoleacetic acid, 49/174
    idiopathic Parkinson disease, 46/378
    IgA increase, 49/174
    IgM reduction, 49/174
    incidence, 46/372
    infection, 49/178
    intelligence, 46/375
    Jakai people, 49/167

    Japanese, 49/167
    Kii Peninsula, 49/167
    Kii Peninsula amyotrophic lateral sclerosis,
        59/237
    laboratory finding, 49/174
    language, 46/375
    levodopa, 46/373, 379
    Lewy body, 46/377, 49/178
    memory, 46/374
    Meynert nucleus basalis, 46/377
    motoneuron involvement, 49/173
    neurofibrillary, 49/176
    neurofibrillary tangle, 46/377, 49/401
    neuropathology, 49/175
    neuropsychologic study, 46/374-376
    ocular sign, 49/172
    oculomotor disorder, 46/375
    onset, 49/171
    parkinsonism, 49/401
    postencephalitic parkinsonism, 46/377
    postural deformity, 49/172
    rigidity, 49/171
    senile plaque, 49/178
    spatial orientation disorder, 46/374
    sporadic amyotrophic lateral sclerosis, 59/235
    toxin, 49/178
    trace minerals, 49/178
    treatment, 49/175
    visual discrimination, 46/374
    visuospatial ability, 46/374
    West New Guinea, 49/167
Parkinson dementia complex
    see also Amyotrophic lateral sclerosis
    Alzheimer neurofibrillary change, 22/131
    amyotrophic lateral sclerosis, 22/316, 403,
        59/282-284, 409
    annual incidence, 42/250
    Babinski sign, 42/249
    behavior disorder, 42/250
    biochemical aspect, 17/485-489
    brain atrophy, 42/250
    bulbar paralysis, 22/143
    bulbar symptom, 22/130
    Chamorro people, 6/176, 22/342
    dementia, 59/409
    epidemiology, 59/274
    granulovacuolar degeneration, 22/399, 49/250
    Guam, *see* Guam Parkinson dementia
    Hallervorden-Spatz syndrome, 6/623, 22/383, 397
    hyperreflexia, 42/249
    hypokinesia, 42/249
    Kii Peninsula amyotrophic lateral sclerosis,
        22/361, 377, 59/274

locus ceruleus degeneration, 42/250
marginal beaded fibril, 22/17
muscular atrophy, 42/250
neurofibrillary tangle, 22/223, 42/250, 49/249,
    59/282-284
neuropathology, 22/343, 383
prevalence, 21/22
progressive supranuclear palsy, 22/226
psychiatric disorder, 42/250
rigidity, 49/69
sex ratio, 42/250
sleep disorder, 42/249
speech disorder, 42/249
spinocerebellar degeneration, 21/15
striatonigral degeneration, 60/540
substantia nigra degeneration, 6/176, 42/250
tetraplegia, 42/250
Parkinson disease, 6/173-207
    see also Parkinsonism
acetylcholine, 29/446, 42/246
age at onset, 49/105
aging, 49/127
akathisia, 42/245, 49/67, 92
akinesia, 45/165, 49/65, 91
akinetic rigid syndrome, 49/108
alcoholism, 49/96
alveolar hypoventilation, 49/130
Alzheimer disease, 49/96, 126
Alzheimer type brain pathology, 49/112
anhidrosis, 59/138
animal model, 49/100
annual incidence, 42/246
antioxidant treatment, 65/393
arteriosclerosis, 6/174, 230
assessment, 49/108
associated disease, 49/126
associated disorder, 49/130
associated movement loss, 49/91
asterixis, 49/96
autonomic dysfunction, 75/178
autonomic nervous system, 75/643
autonomic symptom, 49/92, 75/179
autosomal dominant, 49/108
ballismus, 49/66, 103
basal ganglion, 49/471
benign essential tremor, 49/96, 130, 567, 569
biochemistry, 6/127, 49/118
biogenic amine, 29/469-476
bladder dysfunction, 75/178
blood, 49/120
bombesin, 49/121
Bouttier push maneuver, 49/93
bradykinesia, 49/103

brain atrophy, 6/177, 42/246
brain cortex, 6/212, 49/110
brain stem lesion, 49/113
brain tumor, 6/240
cancer incidence, 49/127
carbon monoxide intoxication, 49/108
catecholamine turnover, 49/124
catechol methyltransferase, 49/120
cause, 6/174-176
cause of death, 49/107
cell ageing, 49/124
chemical change, 49/38
cholecystokinin, 49/123
cholecystokinin octapeptide, 49/120
choline acetyltransferase, 49/67, 111
chorea, 49/197
clinical features, 49/90
clock function, 74/468
cognition, 46/372
computer assisted tomography, 49/98
corticotropin, 49/123
course, 6/195
CSF examination, 49/122
CSF 3-methoxy-4-hydroxyphenylethylene glycol,
    49/123
cyclic adenosine monophosphate, 49/123
cyclic guanine monophosphate, 49/123
cytochrome P450, 49/133
daily life activity scale, 49/108
debrisoquine, 49/133
dementia, 42/246, 46/371-381, 593, 49/112, 130
2-deoxy-2-fluoroglucose F 18, 49/98
2-deoxyglucose C 14, 49/98
depression, 46/208, 379, 477, 49/90, 108
diagnosis, 6/196-198
differential diagnosis, 6/198
threo-3,4-dihydroxyphenylserine, 49/69
DIMPEA, 6/180
disease course, 49/105
disease duration, 49/106
dopamine, 6/180, 29/469-475, 49/118, 123
dopamine β-mono-oxygenase, 42/246
dopamine receptor type D2, 49/99
dopaminergic function, 75/183
dorsal 10th nucleus, 6/191
dysautonomia, 75/178, 182
dyskinesia, 49/195
dystonia, 6/555, 49/197
early diagnosis, 49/90
EEG, 6/196
EMG, 6/195, 49/102
encephalitis lethargica, 49/89, 124
environmental factor, 49/134

enzymopathy, 21/53
epidemiology, 49/124
epilepsy, 15/336
essential tremor, 49/108
essential tremor related parkinsonism, 49/129
etiology, 49/123
extra symptom, 49/94
familial recessive tremor, 49/108
familial tremor syndrome, 49/131
familial type, 49/108
fatigue, 49/90
festination, 49/68
fine hand coordination, 49/90
flexion dystonia, 6/140, 142-144
6-fluorodopamine, 49/39
6-fluorolevodopa F 18, 49/99
free radical, 49/129
freezing phenomenon, 49/68
GABA, 29/505, 42/246, 49/123
gait disturbance, 6/189
gastrointestinal disorder, 6/195
Gegenhalten, 49/94
genetic susceptibility, 49/129, 132
genetics, 49/129
glia involvement, 49/112
global hypothesis, 49/124
globus pallidus, 6/212
glucose, 75/181
glutamate decarboxylase, 49/130
glutamate dehydrogenase, 49/130
glutamic acid, 49/123
Guam motoneuron disease, 65/22
Guam Parkinson dementia, 49/108
H-reflex, 49/70
Hallervorden-Spatz syndrome, 49/130
hastening phenomenon, 49/68
head tremor, 49/96
herbicide intoxication, 7/523, 49/136
hereditary, 6/175
hereditary mental depression, 29/515
hereditary paroxysmal kinesigenic
  choreoathetosis, 49/355
HLA antigen, 49/133
Hoehn-Yahr staging, 49/108
homoprotocatechuic acid, 49/123
homovanillic acid, 6/180, 49/120
homoveratrylamine, 49/120
4-hydroxydebrisoquine, 49/133
4-hydroxyphenylpyruvic acid, 49/120
hyperthyroidism, 49/96
hypokinesia, 6/180-182, 42/245
hypoparathyroidism, 27/308
hypothalamus, 49/110

hypothyroidism, 49/96
idiopathic, see Idiopathic Parkinson disease
incidence rate, 49/125
initiation time, 49/65
insomnia, 49/97, 245
intelligence, 6/193
interdose dyskinesia, 49/195
intestinal pseudo-obstruction, 51/494
iron, 49/36, 129
juvenile parkinsonism, 6/635, 49/108, 153, 163
kinetic melody defect, 49/92
laboratory investigation, 6/195-198
learning strategy defect, 49/92
lenticulostriate artery, 49/97
lergotrile, 37/236
leucine enkephalin, 49/120
levodopa, 42/246, 60/736
levodopa induced dyskinesia, 49/196
Lewy body, 6/177, 215, 21/54, 42/246, 49/108,
  113
Lewy body disease, 49/110
Lewy body ultrastructure, 49/115
lipofuscin, 49/117, 129
β-lipoprotein, 49/123
locus ceruleus cell number, 49/111
locus ceruleus degeneration, 6/214, 42/246,
  49/111
long loop reflex, 49/70, 102
magnetic resonance spectroscopy, 75/187
malaria, 6/244
malignant melanoma, 49/126
manganese intoxication, 6/241, 36/231, 49/94
MAO B inhibitor treatment, 65/393
mental change, 49/104
mental depression, 29/515
mercury intoxication, 6/243
metenkephalin, 49/120, 123
3-methoxytyramine, 49/120
1-methyl-4-phenylpyridinium, 49/53
1-methyl-4-phenyl-1,2,3,6-tetrahydropyridine,
  49/41, 53, 100, 108, 198
3-methylspiperone, 49/99
metoclopramide, 49/108
Meynert nucleus basalis, 49/110, 112
micrographia, 49/93
mimicking disorder, see Multiple neuronal system
  degeneration
misdiagnosis, 49/93, 129
morbidity, 49/125
mortality, 49/125
motor initiative defect, 49/91
motor planning, 49/66
movement reluctance, 6/186

movement time, 49/66, 103
multifactorial origin, 49/87
multi-infarct arteriosclerosis, 49/94
multiple brain infarction, 49/95
mummy, 74/194
muscarinic cholinergic receptor, 49/112
muscle stretch, 49/70
natural history, 49/104
neostriatal model, 49/53
neurofibrillary tangle, 49/110
neurofilament protein, 49/114
neuromelanin, 49/116, 124, 129
neuronal loss, 49/127
neuropathology, 49/108
neurotensin, 49/120
neurotransmitter, 42/246
nocturia, 59/138
nocturnal myoclonus, 49/97
Nöika-Froment cross maneuver, 49/93
noradrenalin, 29/475-477, 49/119
normal pressure hydrocephalus, 49/94-96
nosology, 49/107
nuclear magnetic resonance, 49/98
nucleus accumbens, 49/67
nystagmus, 49/94
olivopontocerebellar atrophy, 49/94, 108, 130, 206
olivopontocerebellar atrophy (Dejerine-Thomas),
    60/516, 528
opiate binding, 75/183
orthostatic hypotension, 63/145, 147, 75/180
oxidopamine, 49/52
palilalia, 43/225
pallidal atrophy, 6/648
pallidal demyelination, 6/176
parkinsonism with motoneuron disease, 49/94
pathology, 6/63, 176-178, 212-216
pathophysiology, 49/100
peak-dose dyskinesia, 49/195, 198
PEG, 6/195, 49/97
Pelizaeus-Merzbacher disease, 49/96
personality, 6/194
phenolsulfotransferase, 49/122
phenothiazine, 6/829, 49/108
physiologic tremor, 49/100
pill rolling movement, 6/183
plastic rigidity, 6/139
poliomyelitis, 6/244
polyneuropathy, 49/108
positron emission tomography, 49/35, 98, 75/181
postencephalitic type, 49/108
postsurgical ataxia, 6/206
postsurgical dementia, 6/207
postsurgical dysarthria, 6/205

postural instability, 49/92
postural tremor, 49/96
posture, 6/197
precipitating factor, 6/176
predicted movement, 49/66
premorbid personality, 49/127
prevalence, 42/246, 49/89
probenecid, 49/122
probenecid test, 49/122
prognosis, 6/194, 49/107
progressive dysautonomia, 59/134
progressive supranuclear palsy, 49/94, 108, 240
protirelin, 49/120
pure autonomic failure, 49/94
ramp movement, 49/66
rapid fatigability, 49/92
rating scale, 49/108
reaction time, 49/103
readiness potential, 49/66
recessive dominant, 49/108
regional brain blood flow, 49/97
renin, 49/122
reserpine, 49/108
resting tremor, 42/253, 49/73, 584
rigidity, 49/54, 69, 91, 102
rubral tremor, 49/96
schizophrenia, 46/447
scientific landmark, 49/88
seborrhea, 6/191, 75/179
secondary neuroaxonal dystrophy, 49/408
selection for surgery, 6/201
seleginine, 65/393
senile plaque, 42/246
sensory input defective utilization, 49/104
serotonin, 29/478, 49/119
sex ratio, 42/246, 49/125
sexual dysfunction, 75/99
Shy-Drager syndrome, 49/95, 108, 59/134
sign provoking maneuver, 49/92
signpost phenomenon, 6/184
simultaneous performance, 49/66
sine agitatione, 60/540
single photon emission computer assisted
    tomography, 75/184
Sjögren syndrome, 71/74
sleep, 74/547
sleep disorder, 75/180
slow initiation, 49/91
smoking, 49/126
somatostatin, 49/112, 120, 123
space occupying lesion, 49/108
spastic paresis, 49/96
speech disorder, 6/188, 42/245

spinocerebellonigral degeneration, 49/94
stereoencephalotomy, 6/203
stereotaxic surgery, 46/380
stooped posture, 6/189
stress, 49/124
striatal lesion, 6/670
striatonigral degeneration, 49/94, 206, 211, 60/540
striatum, 6/212
subcortical dementia, 46/312-314, 378
substance P, 49/120
substantia nigra, 6/214
substantia nigra cell number, 49/111
substantia nigra degeneration, 42/246
substantia nigra involvement, 49/108
surgical complication, 6/203-206
sympathetic ganglion, 49/110
symptom complex, 49/87
symptomatic dystonia, 6/555
symptomatic type, 49/108
syphilis, 6/243
taurine deficiency, 29/515
tetrahydrobiopterin, 49/123
thalamic ventralis intermedius nucleus, 49/599
thalamotomy, 2/491, 49/599
thermoregulation, 75/179
thyroid cancer, 27/272, 49/126
thyroid dysfunction, 49/126
trauma, 6/176, 228, 230-233, 49/108
treatment, 6/198, 65/393, 75/181
tremor, 6/161, 178-180, 196, 42/245, 49/90
tremor generation, 49/100
type A course, 49/105, 130
type B course, 49/105, 130
typhoid fever, 6/245
tyrosine hydroxylase, 49/123
unusual symptom, 49/94
urinary hesitancy, 59/138
urinary incontinence, 49/94, 59/138
urine, 49/120
uvea melanoma, 60/736
vagus nerve dorsal nucleus, 49/110
vasopressin, 49/120
ventral tegmental area, 49/110
vitamin B$_6$, 28/123-129
weight loss, 75/180
Wilson disease, 49/96
writing, 49/90
Parkinson plus syndrome, see Multiple neuronal
    system degeneration
Parkinson syndrome, see Parkinsonism
Parkinson tremor
  anatomic aspect, 6/178-180
  benign essential tremor, 49/568

carbon disulfide intoxication, 6/243
  constancy, 6/161
  dementia, 46/373
  physiologic aspect, 6/178-180
  type, 6/161
Parkinsonism
  see also Parkinson disease and Rabbit sign
  akathisia, 1/290, 6/186, 49/67
  altretamine intoxication, 65/534
  alveolar hypoventilation, 49/130
  amiodarone, 63/190
  amyotrophic lateral sclerosis, 59/231, 410
  antihypertensive agent, 63/87
  antioxidant treatment, 65/393
  atropine, 6/833-838
  Bardet-Biedl syndrome, 13/392
  behavior disorder, 42/255
  benign essential tremor, 49/567
  benzhexol hydrochloride, 6/218
  benztropine sulfonate, 6/219
  Binswanger disease, 46/319, 49/471
  biopterin, 49/159
  biperiden, 6/219
  boxing injury, 23/556
  bradykinesia, 46/376
  brain atrophy, 6/177
  brain biopsy, 10/685
  brain lacunar infarction, 54/247
  breath holding, 63/481
  bromide intoxication, 36/300
  calcium antagonist intoxication, 65/442
  captopril intoxication, 65/445
  caramiphen, 6/219
  carbon disulfide intoxication, 64/25
  carbon disulfide polyneuropathy, 49/477
  carbon monoxide intoxication, 6/242, 49/475,
      64/7, 31
  cardiac arrest, 63/216
  chorea-acanthocytosis, 63/281
  course, 6/194
  Coxsackie virus B2, 56/330
  Creutzfeldt-Jakob disease, 56/545
  CSF, 42/255
  cyanide intoxication, 6/242, 49/477, 64/232
  cycrimine, 6/219
  debrisoquine, 49/133
  dementia, 49/130, 59/410, 63/481
  dentatorubropallidoluysian atrophy, 21/521
  depression, 29/515, 49/130
  diazoxide intoxication, 37/433, 65/446
  diffuse Lewy body disease, 59/410
  diltiazem, 63/194
  dipeptidyl carboxypeptidase I inhibitor, 65/445

dipeptidyl carboxypeptidase I inhibitor intoxication, 65/445
diphenhydramine, 6/219, 828
dopamine β-mono-oxygenase, 49/69
dopaminergic intoxication, 46/592-594
drug induced, *see* Drug induced parkinsonism
drug list, 6/834
drug treatment, 6/218-226, 826-839
dysautonomia, 22/253
dystonia, 42/253
dystonia musculorum deformans, 49/524
electroshock, 6/233
enalapril intoxication, 65/446
encephalitis lethargica, 34/455
enzyme deficiency, 21/53
essential tremor related, *see* Essential tremor related parkinsonism
ethosuximide intoxication, 65/508
extrapyramidal sign, 49/187
familial amyotrophic lateral sclerosis, 22/319, 59/236, 245
familial spastic paraplegia, 22/434
Ferguson-Critchley syndrome, 59/321
filariasis, 52/516
Friedreich ataxia, 21/354, 42/252
gait, 1/346
gait disturbance, 1/346
general management, 6/221
glutamate dehydrogenase deficiency, 49/130
Guam amyotrophic lateral sclerosis-Parkinson dementia complex, 65/22
Guam motoneuron disease, 65/22
Hallervorden-Spatz syndrome, 6/624, 628, 22/403, 49/130
hereditary cerebello-olivary atrophy (Holmes), 21/410, 412
hereditary motor and sensory neuropathy type I, 21/285, 308, 42/251-253
hereditary motor and sensory neuropathy type II, 21/285, 308, 42/251-253
hereditary olivopontocerebellar atrophy (Menzel), 21/412, 442
hereditary progressive diurnal fluctuating dystonia, 49/533
Huntington chorea, 42/226
hyperkinetic agraphia, 45/465
hyperparathyroidism, 27/292
hypoparathyroidism, 27/308
hyporeflexia, 42/253
intestinal pseudo-obstruction, 51/493
juvenile, *see* Juvenile parkinsonism *and* Progressive pallidal atrophy (Hunt-Van Bogaert)
Klinefelter syndrome, 43/558

kuru, 56/557
lead encephalopathy, 49/478
levodopa, 42/253, 256
levodopa induced dyskinesia, 46/592, 49/195
lithium intoxication, 64/296
lordosis, 42/253
manganese intoxication, 49/108, 64/307
MAO B inhibitor treatment, 65/393
mask like face, 6/186
methanol intoxication, 64/100
methyldopa, 37/439, 65/448
methyldopa intoxication, 37/441
methylphenidate, 6/219
1-methyl-4-phenyl-1,2,3,6-tetrahydropyridine, 65/369
1-methyl-4-phenyl-1,2,3,6-tetrahydropyridine intoxication, 65/369
micrographia, 45/464
Minamata disease, 36/121, 64/414
multi-infarct dementia, 49/471
multiple neuronal system degeneration, 49/108, 63/146, 75/162
neurobrucellosis, 33/314, 52/587
neurofibrillary tangle, 46/274
neuroleptic, *see* Neuroleptic parkinsonism
neuroleptic agent, 6/261, 49/191
neuroleptic dyskinesia, 65/274
nifedipine intoxication, 65/442
nigrospinodentate degeneration, 21/401
nimodipine intoxication, 65/442
nonsteroid anti-inflammatory agent intoxication, 65/416
nursing care, 6/221
olivopontocerebellar atrophy, 21/397, 526, 49/130, 206
olivopontocerebellar atrophy (Dejerine-Thomas), 21/425
olivopontocerebellar atrophy variant, 21/453
orphenadrine, 6/219
orthostatic hypotension, 22/262, 43/66, 63/146
palilalia, 43/225
palitypia, 45/468
pallidal necrosis, 49/465
pallidoluysionigral degeneration, 49/452
pallidonigral necrosis, 49/465
pallidostriatal necrosis, 49/465
paradoxical kinesia, 6/136, 188
parietal lobe, 45/77
Parkinson dementia, 49/401
pathophysiology, 6/100-108
PEG, 6/177
personality, 6/194
pneumococcal meningitis, 33/38, 52/44

postanesthetic encephalopathy, 49/479
postencephalitic, *see* Postencephalitic
  parkinsonism
procyclidine, 6/219
profenamine, 6/219
progressive multifocal leukoencephalopathy,
  9/495
progressive supranuclear palsy, 6/557, 22/217,
  219-226, 49/239
pure autonomic failure, 2/120, 22/232, 236
reserpine intoxication, 37/438
resting tremor, 49/584
rigidity, 1/67, 266, 6/184
seleginine, 65/393
senile neuroaxonal dystrophy, 49/392
sleep disorder, 42/255
spastic paraplegia, 42/175
speech disorder, 42/255
sporadic amyotrophic lateral sclerosis, 59/233-235
stereoencephalotomy, 6/203
striatal necrosis, 49/502
striatonigral degeneration, 42/261
striatopallidodentate calcification, 6/707, 49/419,
  422
sweating crisis, 6/220
sympathetic nervous disorder, 2/120
taurine deficiency, 29/515, 42/255
tick bite intoxication, 37/111
tonic postural reflex (Foix-Thévenard), 6/141
trauma, 6/176, 228, 230-233
treatment, 6/220, 65/393
treatment assessment, 6/218-226
tremor, 49/597, 599, 603
tuberous sclerosis, 14/350
tyrosine hydroxylase, 42/256
valproic acid intoxication, 65/504
various causes, 6/227-247
vitamin B6, 28/123-129
Wilson disease, 49/225
Parolfactory gyrus (Broca), *see* Broca parolfactory
  gyrus
Parosmia
  acute amebic meningoencephalitis, 52/317
  anosmia, 2/54, 24/13
  olfactory illusion, 73/72
Parotid ephidrosis, *see* Auriculotemporal syndrome
Parotid tumor
  mixed, *see* Mixed parotid tumor
Parotitis
  mumps, 7/485, 34/411, 628, 56/113, 129, 429
Paroxetine
  selective serotonin reuptake inhibitor, 65/312
Paroxetine intoxication

headache, 65/320
insomnia, 65/320
tremor, 65/320
Paroxysmal aphasia
  ictal speech, 45/322
  transient ischemic attack, 45/508
Paroxysmal brain dysfunction
  episodic attack, 72/259
Paroxysmal choreoathetosis, *see* Hereditary
  paroxysmal kinesigenic choreoathetosis
Paroxysmal disorder, 45/507-511
  epilepsy, 45/507-509
  hypersomnia, 45/511
  migraine, 5/46, 45/509
  narcolepsy, 45/511
Paroxysmal dysarthria
  epilepsy, 45/508
  multiple sclerosis, 9/179
Paroxysmal dysphasia
  epilepsy, 45/508
Paroxysmal dystonia
  epilepsy, 72/255
  hypnogenic, *see* Hypnogenic paroxysmal dystonia
Paroxysmal dystonic choreoathetosis
  hereditary, *see* Hereditary paroxysmal dystonic
    choreoathetotis *and* Hereditary paroxysmal
    kinesigenic choreoathetosis
  hereditary exertional, *see* Hereditary exertional
    paroxysmal dystonic choreoathetosis
Paroxysmal hemicrania
  chronic, *see* Chronic paroxysmal hemicrania
Paroxysmal hypertension
  acute, *see* Acute paroxysmal hypertension
  headache, 26/494
  pheochromocytoma, 39/498, 42/764
  Sipple syndrome, 14/750
Paroxysmal kinesigenic choreoathetosis
  epilepsy, 72/255
  hereditary, *see* Hereditary paroxysmal kinesigenic
    choreoathetosis
Paroxysmal migrainous neuralgia
  ergotamine tartrate, 5/382
  methysergide, 5/382
Paroxysmal movement disorder
  epilepsy, 72/255
Paroxysmal nocturnal hemoglobinuria
  brain infarction, 53/167, 55/467
  brain ischemia, 53/167
  cavernous sinus thrombosis, 52/173
  cerebrovascular disease, 55/467
  juvenile cerebrovascular disease, 55/467
  venous sinus thrombosis, 55/467
Paroxysmal pain

epilepsy, 45/509
occipital neuralgia, 5/277, 369
Paroxysmal positional nystagmus
  benign, *see* Benign paroxysmal positional
    nystagmus
Paroxysmal speech arrest
  epilepsy, 45/508
Paroxysmal speech disorder
  pentetrazole, 45/508
Paroxysmal supraventricular tachycardia
  cardiac dysrhythmia, 39/264
Paroxysmal symptom
  multiple sclerosis, 47/170
  seizure, 47/170
Paroxysmal tachycardia
  cardiac dysrhythmia, 39/264-266
  migraine, 48/157
  progressive external ophthalmoplegia, 22/193
Paroxysmal tremor
  resting tremor, 49/585
Paroxysmal ventricular fibrillation
  cardiac dysrhythmia, 39/265
Paroxysmal ventricular tachycardia
  cardiac dysrhythmia, 39/264
Parrot node
  syphilitic osteoperiostitis, 33/370
Parrot syndrome
  Ehlers-Danlos syndrome, 14/111
Parry disease
  autosomal dominant inheritance, 42/467
  cerebellar ataxia, 42/467
  diagnostic problem, 42/467
  EEG, 42/467
  epilepsy, 42/467
  hypertension, 42/467, 60/667
  involuntary movement, 42/467
  myoclonus, 42/467
  rigidity, 42/467
Parry-Romberg syndrome, *see* Progressive
  hemifacial atrophy
Parsonage-Turner syndrome, *see* Neuralgic
  amyotrophy
Partial albinism
  Chédiak-Higashi syndrome, 60/647
  differential diagnosis, 14/106
  Klein-Waardenburg syndrome, 14/106, 115,
    30/408, 50/218
  Mende syndrome, 14/106, 119
  tuberous sclerosis, 14/496
Partial gigantism
  history, 14/390
  Klippel-Trénaunay syndrome, 14/522
  multiple dermal cylindroma, 14/589

neurofibromatosis type I, 14/397
neurogenic theory, 14/400
Rud syndrome, 38/9, 43/284
tuberous sclerosis, 14/102
Partial Horner syndrome
  cluster headache, 48/222
Partial seizure
  EEG, 72/299
  epilepsy, 72/182
  Takayasu disease, 63/53
Partial status epilepticus, *see* Status epilepticus
Parvovirus
  cerebellar agenesis, 50/178
  taxonomic place, 56/2
PAS stain
  mucoprotein, 9/24
  North American blastomycosis, 35/404, 52/390
  polysaccharide, 9/24
  skeletal muscle, 40/4
Pasini-Pierini syndrome
  Goltz-Gorlin syndrome, 14/113
Passow syndrome
  neurofibromatosis type I, 14/492
Passwell syndrome
  aminoaciduria, 21/216
  congenital ichthyosis, 21/216
  dwarfism, 21/216
  erythroderma, 21/216
  mental deficiency, 21/216
Pasteur rabies vaccine
  experimental allergic encephalomyelitis, 47/429,
    609
Pasteurella
  brain abscess, 52/149
Pasteurella multocida
  gram-negative bacillary meningitis, 33/103,
    52/104
Pastpointing
  neocerebellar syndrome, 2/420
Pastpointing test
  technique, 1/338
Patau syndrome
  aneuploidy, 31/504, 43/527
  anophthalmia, 14/120, 31/504, 513, 43/527
  apnea, 31/508, 50/557
  arhinencephaly, 14/120, 31/504, 508, 43/527,
    50/556, 559
  Arnold-Chiari malformation type I, 31/513,
    50/559
  arthrogryposis, 14/121
  birth incidence, 43/527
  brain atrophy, 43/527
  brain cortex disorganization, 50/249

brain stem hypoplasia, 43/527
cataract, 14/120, 31/513
cebocephaly, 31/510, 512, 50/559
cerebellar agenesis, 50/177, 183
cerebellar atrophy, 43/527
cerebellar heterotopia, 50/39, 251
cheilognathopalatoschisis, 14/120
cheilopalatoschisis, 14/120, 31/504, 506, 510
cheiloschisis, 31/504, 506, 43/527, 50/556
clinical symptomatology, 31/506, 43/527, 50/558
coloboma, 14/120, 31/504, 513
computer assisted tomography, 50/560
congenital deafness, 14/120
congenital heart malformation, 14/120, 31/504,
    43/528, 50/558
corpus callosum agenesis, 14/120, 43/527, 50/152
craniocephalic dysplasia, 14/120
cryptorchidism, 31/508, 43/528, 50/558
cutaneous scalp defect, 50/556
cyclopia, 31/504, 510, 513, 50/559
deafness, 14/120
definition, 31/504, 50/556
dermatoglyphics, 31/507, 43/528, 50/558
differential diagnosis, 14/121
dysplastic nail, 14/121
Edwards syndrome, 14/121
epicanthus, 14/120
epidemiology, 31/505, 50/556
ethmocephaly, 31/510, 512, 50/559
finger deformity, 14/121, 50/558
Gowers hemoglobin, 43/528
growth failure, 31/506, 43/527, 50/558
growth retardation, 31/506, 43/527, 50/558
heart disease, 43/528
hemoglobin, 31/514, 43/528
hexadactyly, 14/121, 31/504, 50/556
history, 31/504, 50/556
holoprosencephaly, 31/504, 507, 512, 43/527,
    50/235, 237, 557-559
Holt-Oram syndrome, 14/121
hydronephrosis, 31/508, 50/558
hydroureter, 31/508, 50/558
hypertelorism, 30/249, 31/504, 43/528
hypospadias, 31/508, 50/558
hypotelorism, 31/504, 512, 50/556
inguinal hernia, 31/507, 43/528
kidney malformation, 14/120, 31/508, 43/528,
    50/558
laboratory finding, 31/513, 50/559
low set ears, 31/507, 43/528
maternal age, 31/506, 43/527, 50/558
median facial cleft syndrome, 50/558
mental deficiency, 43/528, 46/69, 50/542, 556,

    558
microcephaly, 14/120, 31/507, 43/528, 50/556
micrognathia, 43/527
microphthalmia, 14/120, 31/504, 43/528
Mondini ear abnormality, 50/559
mosaicism, 31/506, 514, 50/558
multiple capillary hemangioma, 14/121, 31/504,
    507, 50/558
nail hypoplasia, 31/507, 43/528
nasal bone aplasia, 14/121
neurologic finding, 50/558
neuronal heterotopia, 43/527
nevus flammeus, 31/504, 507, 50/558
ocular abnormality, 31/504, 513, 50/556, 559
omphalocele, 31/504, 507, 50/558, 560
palatoschisis, 31/507, 43/528, 50/556
pathogenesis, 50/556
pes calcaneovalgus, 31/507, 50/558
pes equinovarus, 31/507, 50/558
polydactyly, 14/121, 31/507, 43/527, 50/558
prognosis, 31/514, 50/559
prune belly syndrome, 50/558
psychomotor retardation, 14/120, 31/508, 50/558
13q partial trisomy, 50/560
r(13) syndrome, 50/583, 585
Rossi syndrome, 14/121
Rubinstein-Taybi syndrome, 14/121
Russell syndrome, 14/121
scalp defect, 31/504, 50/558
Scheibe ear abnormality, 50/559
seizure, 31/508, 43/528
sensorineural deafness, 43/528
short neck, 31/507, 43/528, 50/558
sign, 50/557
symptom, 50/557
talipes, 43/528
thumb hyperflexia, 14/121
thumb hypoplasia, 14/121, 31/507
Ullrich-Feichtiger syndrome, 14/121
umbilical hernia, 31/507, 50/558
uterus bicornis, 31/508, 50/558
VACTERL syndrome, 50/514
Patchy anesthesia
    leprous neuritis, 51/219
Patchy blastomatous CSF infiltration, *see* Diffuse
    glioblastosis
Patchy pontine syndrome
    transient hemiparesis, 12/21
Patella absence
    trisomy 8, 50/563
Patellar hypoplasia
    Coffin-Siris syndrome, 43/240
    trisomy 8 mosaicism, 43/511

Patellar reflex
  supra, *see* Suprapatellar reflex
Patent central canal
  hydromyelia, 50/427
Pathergic granulomatosis
  Wegener granulomatosis, 8/116
Pathoclisis
  nature, 9/625
  neurotoxicity, 9/623
  summary, 9/654
Pathologic conduction
  axon, 51/51
Pathologic eyelid retraction, *see* Collier sign
Pathologic laughing
  epilepsy, 3/358, 14/347
  gelolepsy, 45/509
  tuberous sclerosis, 14/347
  vigabatrin intoxication, 65/511
Pathologic laughing and crying, 45/219-224
  affect lability, 45/220
  amyotrophic lateral sclerosis, 3/356, 22/115,
    45/219
  Behçet syndrome, 56/597
  brain infarction, 55/142
  brain lacunar infarction, 54/253
  brain peduncle, 3/362, 45/223
  capsula interna, 3/362, 45/223
  cerebrovascular disease, 55/142
  dacrystic epilepsy, 55/142
  dysphrenia hemicranica, 5/79
  emotional disorder, 3/356-363, 55/142
  epilepsy, 3/358, 45/221
  euphoria, 3/357, 45/220
  facial expression, 3/356, 45/221
  fou rire prodromique, 3/358, 45/221
  gelolepsy, 55/142
  hypothalamus, 3/362, 45/223
  left sided lesion, 3/351, 55/142
  olivopontocerebellar atrophy variant, 21/456
  operculum syndrome, 2/780
  pathogenesis, 3/361, 45/222
  pathology, 45/223
  pontine infarction, 2/241, 12/23
  pontine lesion, 2/241
  progressive bulbar palsy, 22/115
  progressive supranuclear palsy, 49/243
  pseudobulbar affect, 55/142
  pseudobulbar paralysis, 1/196, 2/241, 780, 3/358,
    40/307, 45/221, 55/142, 60/658
  pyramidal tract, 3/361, 45/223
  right sided lesion, 3/361, 55/142
  risus sardonicus, 3/357, 45/220
  subacute sclerosing panencephalitis, 56/419

substantia nigra, 3/362, 45/223
tetanus, 3/357, 45/220
thalamic lesion, 2/482
thalamus, 3/362, 45/223
Pathoneurophysiology
  multiple sclerosis, 47/131-143
Pathophysiology
  akinesia, 45/161-167, 49/65
  amyloidosis, 71/515
  aqueduct stenosis, 50/309
  athetosis, 49/77
  bacterial meningitis, 52/2
  ballismus, 6/484, 49/76, 371
  barotrauma, 61/215
  basal ganglion, 49/65
  Bassen-Kornzweig syndrome, 29/421-423
  benign essential tremor, 49/570
  brain arteriovenous malformation, 12/230, 31/176
  brain edema, 11/357
  brain embolism, 53/314
  brain infarction, 53/107, 314
  brain injury, 57/207
  brain ischemia, 11/268, 54/115, 57/210
  brain metastasis, 71/624
  brain tumor headache, 69/20
  candidiasis, 52/398
  carotid cavernous fistula, 24/411
  carotid system syndrome, 53/314
  childhood myoglobinuria, 62/565
  chorea, 49/73
  classic migraine, 48/69
  coma, 3/62-66
  compression neuropathy, 51/87
  congenital toxoplasmosis, 35/129
  consciousness focal disorder, 45/114-117
  cranial neuralgia, 5/270
  cyanide intoxication, 64/230
  decerebrate rigidity, 2/281
  decompression myelopathy, 61/222
  delirium, 46/548
  diastematomyelia, 32/241
  drug induced extrapyramidal syndrome,
    6/259-261
  dyspnea, 1/495
  dyssynergia cerebellaris myoclonica, 21/512,
    60/599
  dystonia, 6/137, 140-144, 49/71
  dystonia musculorum deformans, 6/536
  echinococcosis, 35/180
  electrical injury, 7/354, 23/689, 706
  emotional disorder, 3/343-363
  epilepsy, 72/39, 64
  ergotamine headache, 65/72

fat embolism, 23/632, 38/569-671
freezing phenomenon, 49/68
galactokinase deficiency, 29/260
galactosemia, 29/261
Gilles de la Tourette syndrome, 49/632
globoid cell leukodystrophy, 66/201
glossopharyngeal neuralgia, 5/275
glutamic acid antagonist, 57/210
hastening phenomenon, 49/68
headache, 5/2, 37, 48/406
hemiballismus, 49/371
hemi-inattention, 45/154
Hemophilus influenzae meningitis, 52/60
hepatic coma, 27/350
hereditary fructose intolerance, 29/257
hereditary paroxysmal kinesigenic
   choreoathetosis, 49/355
hereditary progressive diurnal fluctuating
   dystonia, 49/537
Huntington chorea, 6/139
hyperhidrosis, 75/70
hypernatremia, 28/448-450
hyperventilation syndrome, 63/433
hypomagnesemia, 28/555
hyponatremia, 28/496
hypotonia, 1/269-271, 49/71
infantile spasm, 73/204
infectious endocarditis, 63/114
intermittent spinal cord ischemia, 12/526
intravascular consumption coagulopathy, 55/494
iron deficiency anemia, 63/252
labyrinthine ischemia, 55/130
Landau-Kleffner syndrome, 73/286
levodopa induced dyskinesia, 49/197
lightning injury, 23/689
liver disease, 27/351-353
magnesium imbalance, 28/555
malabsorption syndrome, 7/599
malignant hyperthermia, 38/551-554, 62/571
metachromatic leukodystrophy, 66/172
migraine, 5/37, 48/59-79
mixed connective tissue disease, 71/127
motor function, 1/147-168
myoclonia, 6/765
myoclonic epilepsy, 6/775
myoclonus, 6/164-169, 38/585
myotonia, 40/549-557, 62/276
myotonia congenita, 40/549, 554
neurogenic bladder, 1/370
neuroleptic akathisia, 65/287
neuroleptic dyskinesia, 65/283
neuroleptic dystonia, 65/283
neuroleptic parkinsonism, 65/295
neuroleptic syndrome, 6/259
neuropathology, 6/259-261
obstructive sleep apnea, 63/452
ophthalmoplegic migraine, 5/73
organophosphate intoxication, 64/162
pain, 3/65
paracoccidioidomycosis, 52/457
Parkinson disease, 49/100
parkinsonism, 6/100-108
pellagra, 70/416
periodic paralysis, 41/167, 62/466
phosphofructokinase deficiency, 62/488
photogenic epilepsy, 2/618
pneumocephalus, 24/205
posttraumatic brain abscess, 57/321
posttraumatic intracranial hematoma, 57/259
posttraumatic meningitis, 57/318
postural reflex, 49/78
progressive myoclonus epilepsy, 6/166
pure autonomic failure, 22/233
rigidity, 6/101, 49/69
sarcoidosis, 71/468
Shy-Drager syndrome, 38/242-247
sickle cell anemia, 63/258
spasmodic torticollis, 6/585
spongy cerebral degeneration, 66/662
status epilepticus, 73/273
stress, 45/249-252
subacute combined spinal cord degeneration,
   28/143-147
subarachnoid hemorrhage, 55/7
Sydenham chorea, 49/361
symptomatic dystonia, 49/544
thalassemia, 63/257
thallium intoxication, 64/326
third ventricle colloid cyst, 68/327
transient ischemic attack, 53/267
transtentorial herniation, 54/48
tremor, 6/101, 49/73
trichopoliodystrophy, 29/294-300
trigeminal nerve, 2/54-59
trigeminal neuralgia, 5/272
uremic polyneuropathy, 51/360, 63/512
visuospatial agnosia, 45/168-172
vitamin B_{12} deficiency, 70/373
Patient-doctor relationship
   ethics, 24/800
Patient positioning
   paraplegic body, 26/218
   paraplegic limb, 26/220
   physical rehabilitation, 61/463
   spasticity, 26/478-480
   spinal cord injury, 26/218-220

Patient transport
   air, 26/298
   spinal injury, 26/295-298
Pattern recognition
   rehabilitation, 46/618
Pattern shift visual evoked response
   CSF oligoclonal IgG, 47/137
   multiple sclerosis, 47/134-138, 142
   optic neuritis, 47/135
Pattern visual evoked response
   multiple sclerosis, 47/133
Patterns
   amyloid disposition, *see* Amyloid disposition
     pattern
   bulbar cervical dissociation, *see* Bulbar cervical
     dissociation pattern
   cervical collar, *see* Cervical collar pattern
   onion skin, *see* Trigeminal descending spinal tract
     lesion
   sensory loss, *see* Sensory loss pattern
   suppresion burst, *see* Suppression burst pattern
Pavlov ratio
   definition, 61/10
Pavor nocturnus
   juvenile parasomnia, 45/140
Paxillus involutus intoxication
   arsenic intoxication, 36/214
   hemolytic anemia, 36/542
   renal effect, 36/542
PDMT, *see* Phenyldimethyltriazene
Peacock gait
   dystonia musculorum deformans, 49/521
Peapicker disease, *see* Leptospirosis
Pearly tumor
   suprasellar, *see* Suprasellar pearly tumor
Pearson syndrome
   erythropoietin, 66/427
   granulocyte colony stimulating factor, 66/427
   Kearns-Sayre-Daroff-Shy syndrome, 62/508,
     66/427
   liver, 66/397
   mitochondrial DNA, 66/397
   mitochondrial DNA defect, 62/508
   pancreatic dysfunction, 62/508, 66/427
   pancreatic exocrine function, 66/397, 427
   pancytopenia, 66/397, 427
   sideroblastic anemia, 62/508, 66/427
Pectoral nerve, *see* Anterior thoracic nerve
Pectoralis major muscle
   anencephaly, 50/84
Pectoralis major muscle reflex
   diagnostic value, 1/245
Pectoralis major muscle transfer

   brachial plexus, 51/152
Pectoralis minor syndrome
   subcoracoid, *see* Hyperabduction syndrome
Pectoralis muscle agenesis
   hemidysplasia, 43/416
   ichthyosis, 43/416
   Poland-Möbius syndrome, 43/461
Pectoralis muscle aplasia
   Poland-Möbius syndrome, 43/461, 50/215
Pectus carinatum
   Coffin-Lowry syndrome, 43/238
   congenital spondyloepiphyseal dysplasia, 43/478
   mental deficiency, 43/257
   myotonic myopathy, 43/165
   nemaline myopathy, 43/122
Pectus excavatum
   FG syndrome, 43/250
   glycosaminoglycanosis type VII, 66/310
   homocystinuria, 55/329
   LEOPARD syndrome, 43/28
   Marden-Walker syndrome, 43/457
   neurofibromatosis type I, 50/369
   otopalatodigital syndrome, 43/457
   r(13) syndrome, 50/585
Pedes plani, *see* Pes planus
Pediatric head injury, 23/445-456
   age, 23/450, 57/327
   battered child syndrome, 57/328
   bilateral subacute subdural hematoma, 57/339
   biomechanics, 23/448
   brain hemorrhage, 23/449
   brain stem death, 57/479
   cause, 57/328
   cause of death, 57/328
   cephalhematoma, 57/329
   Children Coma Score, 57/329
   chronic subdural hematoma, 24/306, 316, 318,
     57/336
   clinical symptom, 57/328
   computer assisted tomography, 57/329
   conscious child, 57/328
   convulsion, 57/329
   EEG, 23/453, 57/183
   epidemiology, 23/446, 57/327
   epidural hematoma, 57/335, 337
   epilepsy, 23/453, 57/341
   etiology, 57/328
   fall, 57/328
   Glasgow Coma Scale, 57/330
   head injury epidemiology, 23/446
   head injury outcome, 57/377
   incidence, 57/327
   intellectual impairment, 24/495

intelligence, 23/451
intracranial pressure monitoring, 57/329
motor deficit, 57/341
neuropsychologic deficit, 57/341
nuclear magnetic resonance, 57/329
persistent vegetative state, 57/341
ping-pong ball fracture, 57/334
prevention, 23/456
prognosis, 23/450, 24/833
radiodiagnosis, 23/449
recovery, 23/452
road accident, 57/328
scalp, 57/329
sex ratio, 57/328
skull fracture, 57/331
skull radiation, 57/331
statistic, 23/445
subcutaneous hematoma, 57/329
subdural hematoma, 57/49, 335
subgaleal hematoma, 57/329
transient cerebral blindness, 23/450
traumatic intellectual defect, 24/484
unconscious child, 57/329
vital sign, 57/329
vomiting, 57/329
Pediatric head injury outcome
mortality, 57/339
Pedicle
vertebral, *see* Vertebral pedicle
Pedicular arch hyperplasia
vertebral canal stenosis, 50/474
Peduncle
brain, *see* Brain peduncle
cerebellar, *see* Cerebellar peduncle
superior cerebellar, *see* Superior cerebellar
peduncle
ventral thalamic, *see* Ventral thalamic peduncle
Peduncle lesion
superior cerebellar, *see* Superior cerebellar
peduncle lesion
Peduncular artery
middle, *see* Middle peduncular artery
Peduncular hallucination
brain stem infarction, 53/392
definition, 45/353
encephalitis, 45/354
mesencephalic infarction, 53/392
psychosis, *see* Peduncular hallucinosis
thalamic infarction, 53/392
thalamic syndrome (Dejerine-Roussy), 45/354
toxicity, 45/354
vertebral artery angiography, 45/354
vertebrobasilar system syndrome, 53/392

Peduncular hallucinosis
alcoholic hallucinosis, 46/419
cause, 45/354
definition, 46/564
features, 55/146
visual hallucination, 2/617, 46/419
Pedunculopontine tegmental nucleus
enkephalin, 49/12
GABA, 49/12, 51
nucleus accumbens, 49/12
pallidotegmental fiber, 49/7
substance P, 49/12
tegmental subthalamic fiber, 49/11
Pedunculotomy
mesencephalic, *see* Mesencephalic
pedunculotomy
Peet, M.M., 1/5
Pefloxacin
iatrogenic neurological disease, 65/486
neurotoxin, 65/486
quinolone intoxication, 65/486
toxic encephalopathy, 65/486
Pefloxacin intoxication
epilepsy, 65/486
PEG
astrocytoma, 18/23
brain atrophy, 46/343
brain thromboangiitis obliterans, 39/203
brain tumor, 16/530
chordoma, 18/158
chronic subdural hematoma, 24/312
cluster headache, 48/237
coccidioidomycosis, 52/417
congenital cerebellar atrophy, 21/464
corpus callosum agenesis, 2/765, 50/153
dystonia musculorum deformans, 6/532
empty sella syndrome, 17/409, 431
glioblastoma multiforme, 18/60
happy puppet syndrome, 31/309
hereditary cerebello-olivary atrophy (Holmes),
21/410
hereditary olivopontocerebellar atrophy (Menzel),
21/445
Huntington chorea, 6/314, 49/288
hydrocephalic dementia, 46/328
intracerebral hematoma, 11/688
neurocysticercosis, 35/307
odontoid, *see* Odontoid peg
oligodendroglioma, 18/95
olivopontocerebellar atrophy (Dejerine-Thomas),
21/427
Parkinson disease, 6/195, 49/97
parkinsonism, 6/177

pinealoma, 42/767
pituitary adenoma, 17/401
porencephaly, 23/307
posttraumatic headache, 5/182
primary alcoholic dementia, 46/343
Refsum disease, 21/200
Rett syndrome, 29/317-319, 63/437
schizophrenia, 46/450, 481-487
subarachnoid hemorrhage, 12/151
traumatic hydrocephalus, 24/234, 249
tuberculoma, 18/418
tuberculous meningitis, 33/221
Peiffer classification
  leukodystrophy, 10/106
  orthochromatic leukodystrophy, 10/106
Pelamis platurus intoxication
  fatality, 37/91
Pelger-Huët abnormality
  epilepsy, 43/458
  genetic linkage, 43/108, 458
  late adult muscular dystrophy (Nevin), 62/184
  mutation rate, 43/458
  pain, 43/459
  prevalence, 43/458
Pelizaeus, F., 10/10
Pelizaeus-Merzbacher disease, 66/559-574
  adult Löwenberg-Hill type, 10/151, 176
  adult neuropathology, 10/177
  adult review, 10/176
  anoxic ischemic leukoencephalopathy, 47/529
  ataxia, 42/502, 505
  athetosis, 2/501, 6/457
  autosomal dominant optic atrophy, 13/38
  cerebroside, 10/33
  chemistry, 10/156
  chorioretinal degeneration, 13/39
  chromosome linkage, 66/562
  classic type, 10/152, 154, 156, 183, 42/502,
    66/566, 572
  classification, 10/189, 66/5, 572
  clinical features, 10/3, 186, 28/90-94
  Cockayne-Neill-Dingwall disease, 10/115, 151,
    182-184, 197, 13/436
  congenital, see Congenital Pelizaeus-Merzbacher
    disease
  connatal type, 66/567, 572
  Crome-Zapella disease, 10/197
  cysteine, 66/561
  dementia, 42/502-505
  demyelination, 42/503-505
  diffuse sclerosis, 9/472
  dystonia, 60/664
  dystonia musculorum deformans, 49/524

ENG, 66/569
epilepsy, 15/376, 42/503-505
etiology, 10/195
genetic basis, 66/573
genetic feature, 10/33
genetic study, 66/562
genetics, 10/154, 162, 176
glial dystrophy, 21/66
globoid cell leukodystrophy, 10/4
glycerophosphate dystrophy, 10/33
glycerophosphatide, 10/151
glycerophosphatide metabolism, 10/195
histochemistry, 10/192
history, 10/14, 33, 152
Huntington chorea, 49/293
hyperkinesia, 42/503
infantile neuroaxonal dystrophy, 10/151, 158, 160
inositol-CPM-phosphatidyltransferase, 10/193
kyphoscoliosis, 42/504
leukodystrophy, 10/19
long chain monounsaturated fatty aldehydes,
  10/33
Löwenberg-Hill type, 10/176, 42/505
lysocephalin, 10/195
mental deficiency, 42/503, 46/57, 59/357
microcephaly, 42/503
molecular biology, 66/82
molecular genetics, 66/559
mouse mutant, 66/560
multiple sclerosis, 47/73
myelin deficiency, 66/559
myelin gene, 66/560
myelin proteolipid protein, 66/561
neuroglia relationship, 28/412
neuropathology, 66/566
neurophysiology, 66/568
neuroradiology, 66/568
Norman type, 10/182
nosology, 10/196, 66/563
nystagmus, 42/502, 505, 59/357
oligodendrocyte, 66/560
optic atrophy, 13/39, 83
orthochromatic leukodystrophy, 10/106, 42/500
Parkinson disease, 49/96
pathogenesis, 10/191
pathology, 10/154, 162, 169
phosphatidylethanolamine, 10/195
proteolipid protein, 59/357, 66/561
proteolipid protein DM-20 gene, 66/571, 573
proteolipid protein gene, 66/559, 570, 573
proteolipid protein mutation, 66/563
proteolipid protein transgenic mouse, 66/563
psychiatric disorder, 42/505

psychomotor retardation, 42/503
pure type, 66/572
review, 66/559
secondary pigmentary retinal degeneration,
    13/331, 60/732
skeletal deformity, 42/503
spastic paraplegia, 42/503, 59/439, 66/572
spastic tetraparesis, 59/357
spasticity, 42/505
speech disorder, 42/502, 505
speech disorder in children, 42/502-505
sudanophilic leukodystrophy, 10/151, 13/436
sulfatide, 10/33
transitional type, 10/167-169, 180, 66/572
tremor, 42/502-505, 60/661
type, 66/14
variant, 10/179
X-linked spastic paraplegia, 66/571
Pellagra, 28/59-98
    see also Vitamin B3 deficiency
abdominal pain, 28/71
achlorhydria, 28/71
akinesia, 7/571
alcoholic dementia, 46/336
alcoholic encephalopathy, 28/85, 46/336, 399
alcoholism, 7/573, 28/85, 46/399
amblyopia, 7/571
amenorrhea, 28/71
amino acid, 7/574, 28/59, 46/336, 51/333
anorexia, 28/86, 38/648
antimetabolite, 9/627
antivitamin, 7/575
anxiety, 28/86, 38/648, 46/336
apathy, 28/71
ataxia, 28/90, 95, 46/399
autonomic nervous system, 75/492
Babinski sign, 46/336
beriberi, 46/399
biochemistry, 7/576, 28/76-80
biosynthesis, 28/70
burning feet syndrome, 7/573, 28/11, 71
cachexia, 7/574, 28/66, 92
carcinoid syndrome, 70/315
carcinoma, 38/648
cheilosis, 28/90
child, 28/94
classification, 28/67
clinical features, 7/573, 28/85-98, 38/648, 51/333
cogwheel rigidity, 28/85, 46/336
coma, 7/574, 38/648
confusion, 38/648, 46/336
consciousness, 7/573, 28/85, 46/336
course, 7/574

cramp, 7/573
cranial nerve involvement, 7/573
Creutzfeldt-Jakob disease, 6/748
death, 28/87, 46/336, 399
dehydration, 7/574
delirium, 6/748, 7/571, 573
dementia, 6/748, 7/571, 28/87, 46/399, 51/333,
    70/413
depression, 7/571, 28/71, 86, 46/336
dermatitis, 6/748, 28/59, 71, 86, 46/399, 51/333
diagnosis, 7/572, 28/93
diarrhea, 28/66, 71, 87, 38/648, 46/336, 399,
    51/333
diet, 28/66, 85, 46/336
disorientation, 7/573, 38/648
distractibility, 28/86
dizziness, 7/573
dysesthesia, 28/86
dysphagia, 28/71, 38/648
EMG, 7/573
encephalopathy, 28/59, 46/336, 70/410
endemic, *see* Endemic pellagra
epidemiology, 28/64-76
epilepsy, 6/748
ergotism, 28/93
erythema, 6/748, 7/572, 574, 28/87
etiology, 28/64-76
experimental deficiency, 28/66
extrapyramidal disorder, 28/90
fasciculation, 7/573
fatigue, 28/86, 46/336
gastrointestinal tract, 6/748, 7/572
geographic distribution, 28/73
glossitis, 7/574, 28/59
grasp reflex, 28/85, 46/337
hallucination, 7/571, 38/648, 46/336
headache, 7/573, 28/86
history, 7/572, 28/60-64
hydromania, 7/573
hypalgesia, 28/90
hyperesthesia, 28/90
hyperkeratosis, 28/89
hyperkinesia, 6/748, 7/571
hyporeflexia, 28/90
hypotonia, 6/748
insomnia, 28/86, 46/336
irritability, 7/573, 28/86, 38/648, 46/336
isoniazid induced, 28/97
kwashiorkor, 28/76
lethargy, 28/71
leucine, 28/72
Lieberkühn gland, 7/574
maize, 28/64

malabsorption syndrome, 7/572, 38/648
malignancy, 38/648
malnutrition, 28/18, 64, 46/336, 399
manic depression, 46/336
marasmus, 7/574
memory loss, 28/86, 38/648
metabolic disorder, 28/59
mortality rate, 28/74
mucous membrane, 7/572, 574, 46/337
myelopathy, 28/59
nerve conduction, 7/166
neurasthenia, 6/748, 28/86, 46/336
neuropathology, 7/575
neuropathy, 7/552, 70/414
nomenclature, 28/62
nystagmus, 28/95
optic nerve, 7/573
optic neuritis, 46/399
pain, 6/748
palpitation, 28/86
paranoid symptom, 46/336, 70/415
paresthesia, 6/748, 7/573
pathogenesis, 28/64-76
pathology, 7/575, 28/76-84, 38/648, 46/337
pathophysiology, 70/416
photophobia, 28/96, 46/336
polyneuropathy, 7/571-577, 28/71, 86, 46/337,
    51/333
psychic disorder, 6/748, 7/571, 573, 575
psychosis, 7/571, 573, 46/336
pupillary disorder, 28/90
respiratory chain defect, 28/77
rombergism, 28/90
seasonal incidence, 7/574, 28/66
seborrhea, 28/89
seizure, 7/571, 573
sensibility, 6/748, 7/573
skin manifestation, 6/748, 7/574, 28/88, 90,
    46/336
stage, 7/574, 28/87-93
stomatitis, 6/748, 28/59, 71, 87
subacute combined spinal cord degeneration,
    7/573, 575, 46/399
sucking reflex, 28/85
suicide, 28/66
survey, 70/423
symmetrical polyneuropathy, 7/572
symptom, 70/408
symptomatic, 70/410, 412
systemic lupus erythematosus, 28/93
treatment, 7/574
tremor, 7/571, 28/85, 90
tropical ataxic neuropathy, 7/641, 51/323

tropical myeloneuropathy, 56/526
tryptophan, 7/574, 28/59, 71, 38/648, 46/336
typhoid fever, 28/93
vertigo, 28/86
visual disorder, 7/571
vitamin A deficiency, 28/86
vitamin B1 deficiency, 28/85
vitamin B2 deficiency, 28/205
vitamin B3 deficiency, 7/572
vitamin B6, 28/71, 73
vitamin B6 deficiency, 46/336
vitamin PP, 7/572, 574, 28/59, 71, 46/336
vitamin PP deficiency, 7/576, 28/18, 59, 68,
    38/648, 46/336
vomiting, 6/748, 38/648
weakness, 7/573, 575, 28/90, 46/336
weight loss, 28/71
Wernicke encephalopathy, 38/648
Wernicke-Korsakoff syndrome, 6/748, 46/336,
    399
Pellagra like rash
    hereditary periodic ataxia, 21/564
Pellazzi syndrome
    pineal tumor, 2/444, 17/658
Pellet
    brain tumor carcinogenesis, 17/6, 9
Pelochelys bibroni intoxication
    ingestion, 37/89
Pelops of Smyrna
    history, 2/6
Pelvic abdominal reflex
    physiology, 1/391
Pelvic floor
    anatomy, 1/360
Pelvic fracture
    see also Spinal injury
    lumbosacral plexus injury, 25/475
    lumbosacral trunk injury, 25/475
    neurologic sign, 25/470
    spinal injury, 26/226
    traffic injury, 26/198
    type, 25/450, 470, 473
Pelvic injury
    spinal injury, 26/198
Pelvic migraine
    migraine equivalent, 48/160
Pelvic nerve
    action, 1/365
    anatomy, 1/361
    bladder neck, 1/370
Pelvic obliquity
    traumatic spine deformity, 26/168, 170
Pelvic pain

migraine, 5/46
neuroanatomy, 1/363
Pelvic plexus
anatomy, 1/361
bladder, 75/667
Pelvifemoral muscular dystrophy (Leyden-Möbius)
see also Limb girdle syndrome
features, 62/183
limb girdle syndrome, 62/179
Pemphigus
myasthenia gravis, 62/402
Pemphigus vulgaris
myasthenia gravis, 41/102, 240
Pena-Shokeir syndrome type I
anterior horn cell, 59/369
anterior horn cell degeneration, 59/71, 369
anterior horn syndrome, 43/438
arthrogryposis multiplex congenita, 43/437,
59/369
birth incidence, 43/438
brain atrophy, 43/438
camptodactyly, 43/437, 59/71
congenital amyelinating neuropathy, 59/369
cryptorchidism, 43/438
demyelination, 59/369
dermatoglyphics, 43/438
encephalomalacia, 43/438
gene frequency, 43/439
growth retardation, 43/437
Guillain-Barré syndrome, 59/369
hereditary motor and sensory neuropathy type II,
59/369
hereditary motor neuropathy, 59/369
hereditary sensory and autonomic neuropathy type
II, 59/369
heterozygote frequency, 43/439
high arched palate, 43/438
hydramnios, 59/71
hypertelorism, 43/438, 59/71
infantile spinal muscular atrophy, 59/71
low set ears, 43/438, 59/71
mandibular hypoplasia, 43/438
micrognathia, 43/438
motor end plate, 43/438
multiple ankylosis, 43/437
peripheral nerve, 43/438
pes equinovarus, 43/437
poliomyelitis, 59/369
polymicrogyria, 43/438
polyneuropathy, 59/369
pulmonary hypoplasia, 43/437, 59/71
rockerbottom foot, 43/437
sex ratio, 43/438

talipes, 43/437, 59/71
telecanthus, 43/438
ventricular dilatation, 43/438
Pena-Shokeir syndrome type II, see
Cerebro-oculofacioskeletal syndrome
Pencil injury
intraoral trauma, 12/348
Pencil tumor
spinal glioma, 20/327
Pendred syndrome
congenital deafness, 42/366
gene frequency, 42/369
goiter, 42/368-370
high frequency hearing loss, 42/368
hypothyroidism, 42/369
Mondini ear abnormality, 42/366
prevalence, 42/369
Pendular reflex
paleocerebellar syndrome, 2/419
Pendulous knee jerk
neocerebellar syndrome, 2/420
Penetrance
facioscapulohumeral muscular dystrophy, 62/165
gene, see Gene penetrance
incomplete, see Incomplete penetrance
Penetrating brain injury
brain scintigraphy, 23/288, 294
bullet wound, 23/100
glycogen, 27/206
missile, 23/100
spinal cord injury, 25/48
traumatic aphasia, 24/490
traumatic psychosis, 24/520
traumatic venous sinus thrombosis, 24/369, 372
Penetrating gunshot injury
brain abscess, 33/116
severe brain injury, 57/302
Penetrating head injury, 57/299-312
angiography, 57/308
antibiotic agent, 57/311
anticonvulsant, 57/311
anticonvulsant prophylaxis, 57/311
arterial vascular injury, 57/305
brain perfusion pressure, 57/300
carotid artery trauma, 57/309
carotid cavernous fistula, 24/405, 57/309
civilian gunshot injury, 57/299, 301
clinical presentation, 57/306
computer assisted tomography, 57/308
cortical blindness, 24/48, 50, 492
corticosteroid, 57/311
cranial infection, 57/310
cranial nerve lesion, 57/307

craniectomy, 57/304
craniofacial injury, 57/309
craniotomy, 57/304
CSF fistula, 57/311
diagnostic study, 57/302
echoEG, 23/279
EEG, 23/350, 57/185, 187
epidemiology, 57/299
face, 57/309
geniculocalcarine tract injury, 24/46
gunshot, 57/302
impact velocity, 57/299
incidence, 57/299
infection, 57/310
initial assessment, 57/301
intracranial pressure, 57/300
intubation, 57/302
knife, 57/308
knife blade, 57/308
management, 57/301, 308
mechanics, 23/100
mechanism, 57/300
meningitis, 57/310
missile injury, 57/299, 302
mortality rate, 57/303
neurologic deficit, 57/307
nonmissile, 57/306
operative management, 57/303
operative treatment, 57/304
orbitocranial injury, 57/310
pathology, 57/308
pneumocephalus, 57/312
reoperation, 57/305
repeat angiography, 57/308
rhinorrhea, 57/312
right sided subdural hematoma, 57/303
scalp, 57/309
seizure, 23/352
shock wave, 57/300
skull fracture, 57/301
slot fracture, 57/308
stab wound, 57/307, 309
supratentorial gunshot injury, 57/300
surgical management, 57/310
traumatic intellectual defect, 24/482, 492
traumatic intracerebral hematoma, 57/258
visual perception disorder, 24/491
Penfield, W., 1/12, 29, 31, 149-152
Penicillamine
    arsenic intoxication, 36/214
    arsenic polyneuropathy, 51/266, 268
    cystinosis, 29/179
    eosinophilic fasciitis, 63/385
    epilepsy, 28/116
    inorganic mercury intoxication, 36/183
    lead encephalopathy, 64/443
    lead intoxication, 36/27, 64/439
    mercury intoxication, 36/183
    Minamata disease, 36/76, 64/416
    myasthenia gravis, 41/126, 62/405
    neurologic complication, 38/499
    organic mercury intoxication, 36/76
    organolead intoxication, 64/134
    polymyositis, 62/372
    toxic myopathy, 62/597
    vitamin B6, 28/116, 51/297
    vitamin B6 deficiency, 51/297
    vitamin B6 metabolism, 70/430
    Wilson disease, 27/380, 402-404, 49/233
Penicillamine intoxication
    acute urticaria, 49/235
    agranulocytosis, 27/405
    cutis laxa, 49/235
    elastosis performans serpiginosa, 49/235
    Goodpasture syndrome, 49/235
    immune complex nephritis, 49/235
    leukopenia, 27/405
    nephrotic syndrome, 27/405
    skin fragility, 27/405
    systemic lupus erythematosus, 49/235
    thrombocytopenia, 27/405
    urticaria, 27/405, 49/235
    vitamin B6 deficiency, 27/405, 49/235
Penicillamine polyneuropathy
    chemistry, 51/297
Penicillin
    benign intracranial hypertension, 65/472
    brain abscess, 52/156
    brain gumma, 52/279, 283
    epilepsy, 65/472
    general paresis of the insane, 52/279
    gonococcal meningitis, 33/103
    gram-negative bacillary meningitis, 33/103,
        52/105, 108
    Lyme disease, 52/263
    meningococcal meningitis, 33/26, 52/30
    myoclonus, 38/582
    neuropathy, 7/515
    neurosyphilis, 33/373, 52/279, 282
    neurotoxicity, 46/602
    neurotoxin, 46/602, 65/472
    pneumococcal meningitis, 33/35, 42, 52/47
    staphylococcal meningitis, 33/72-74
    subarachnoid hemorrhage, 12/114
    superior longitudinal sinus thrombosis, 52/179
    tabes dorsalis, 52/280

Whipple disease, 52/140
Penicilliosis
brain abscess, 35/568, 52/490
Penicillium
fungal CNS disease, 35/567, 52/489
Penis
artificial, 26/460
electrovibrator, 26/458
Penis erection
see also Sexual function and Sexual impotence
Addison disease, 75/89
autonomic nervous system, 75/85
calcitonin gene related peptide, 75/86
Cushing syndrome, 75/89
dreaming, 3/89
drug, 75/90
EMG, 61/313
hyperthyroidism, 75/89
hypogastric plexus stimulation, 61/324
injection therapy, 61/320
innervation, 61/314
neurologic assessment, 75/89
neuropeptide Y, 61/314
neurotransmitter, 61/314
noradrenalin, 75/86
painful, 74/560
parasympathetic S3 centre, 61/314
penis prosthesis, 61/321
peptide histidine isoleucine, 75/86
physiologic aspect, 61/324
physiology, 61/313
pituitary tumor, 75/89
REM sleep, 74/560
renal insufficiency, 75/89
sacral parasympathetic efferent, 2/110
sexual dysfunction, 75/98
sleep, 3/82, 74/560
social factor, 61/324
spinal cord, 75/86
supraspinal pathway, 75/86
sympathetic L1 centre, 61/314
vacuum constrictor devices, 61/321
vas deferens aspiration, 61/324
vasoactive intestinal polypeptide, 61/314
Penis pain
sexual impotence, 75/92
Pennsylvania, USA
neurology, 1/5
Pentaborane intoxication
ataxia, 64/360
coma, 64/360
dizziness, 64/359
drowsiness, 64/359

epilepsy, 64/360
headache, 64/359
hiccup, 64/359
metal intoxication, 64/353
muscle spasm, 64/359
myalgia, 64/359
opisthotonos, 64/360
prevention, 64/360
rhabdomyolysis, 64/360
treatment, 64/360
tremor, 64/359
vertigo, 64/359
Pentaceropsis intoxication
family, 37/78
venomous spine, 37/78
Pentamethylenetetrazole, see Pentetrazole
Pentamidine
African trypanosomiasis, 52/343
toxic myopathy, 62/601
toxic neuropathy, 7/520
Pentasomy
definition, 50/540
Pentasomy X, see XXXXX syndrome
Pentavalent arsenic
arsenic polyneuropathy, 51/267
neurotoxicity, 51/265
Pentazocine
delirium, 46/539
opiate, 65/350
sternocleidomastoid tumor, 41/385
Pentazocine addiction
brain embolism, 55/518
Pentazocine intoxication
epilepsy, 65/357
hallucination, 65/357
headache, 65/357
symptom, 65/357
T and blue, 65/357
4-Pentenoic acid
gluconeogenesis inhibition, 65/99
hypoglycin intoxication, 65/85, 92
neurotoxin, 49/217, 65/126
Reye syndrome, 29/338, 49/217, 56/167
toxic encephalopathy, 49/217
valproic acid induced hepatic necrosis, 65/126
Pentetrazole
paroxysmal speech disorder, 45/508
puffer fish intoxication, 37/81
subarachnoid hemorrhage, 12/113
Penton protein
Reye syndrome, 56/165
Pentose phosphate pathway
myoglobinuria, 62/558

Pentose shunt
  oligodendrocyte, 9/14
Pentostatin
  neuropathy, 69/461
Pentylenetetrazol, *see* Pentetrazole
Peptic ulcer
  bismuth intoxication, 64/331
  hereditary amyloid polyneuropathy type 1, 42/522
  restless legs syndrome, 8/315, 51/546
Peptidase
  CSF, *see* CSF peptidase
  degradation, 66/61
  wallerian degeneration, 7/215
Peptide
  anorexigenic, *see* Anorexigenic peptide
  atrial natriuretic, *see* Atrial natriuretic peptide
  brain, 37/371
  delta sleep inducing, *see* Delta sleep inducing
    peptide
  encephalitogenic myelin basic protein, 9/509-511
  enteric nervous system, 75/617
  epilepsy, 72/96
  migraine, 48/61, 85, 99
  opiate, *see* Opiate peptide
  progressive dysautonomia, 59/153
  toxicity, 37/371
  tumor cell, 67/35
  vasoactive intestinal, *see* Vasoactive intestinal
    peptide
Peptide histidine isoleucine
  brain blood flow, 53/63
  penis erection, 75/86
Peptidoleukotriene
  smooth muscle function, 75/624
Peptiduria
  Wilson disease, 49/229
Peptostreptococcus
  brain abscess, 52/149
Perception
  aging, 46/491
  agnosia, 4/30
  audition, 45/521
  auditory, *see* Auditory perception
  brain injury, 23/11, 229
  brain tumor, 16/732
  calculation, 45/475
  cerebral dominance, 4/5
  color, *see* Color perception
  coma, 3/62, 71
  definition, 45/351
  delirium, 46/525-527
  depth, *see* Depth perception
  developmental dyslexia, 46/120
  dissociated, *see* Dissociated visual perception
  distance, *see* Distance perception
  eidetic imagery, 45/360
  epilepsy, 45/508
  hallucination, 45/351-369
  illusion, 45/361
  infantile autism, 46/190
  learning disability, 46/125
  metamorphopsia, 45/361
  migraine, 45/509, 48/162
  motion, *see* Motion perception
  motion sickness, 74/358
  musical, *see* Musical perception
  neuropsychologic defect, 45/521
  neuropsychology, 45/516
  pain, *see* Pain perception
  palinacousia, 45/367
  schizophrenia, 46/491
  shape, *see* Shape perception
  space, *see* Space perception
  tactile, *see* Tactile perception
  time sense, 2/712
  touch, *see* Tactile perception
  vision, 45/521
  visual, *see* Visual perception
  visual perseveration, 45/351-369
  Wernicke-Korsakoff syndrome, 45/199
Perception classification
  agnosia, 45/338-341
  apperceptive agnosia, 45/338-341
  object recognition, 45/339
  prosopagnosia, 45/341
Perchloric acid naphthoquinone method
  cholesterol, 9/26, 29
  histochemistry, 9/26, 29
  unsaturated sterol, 9/26
Perchloroethylene, *see* Tetrachloroethylene
Percussion myoedema
  nonspecific, 40/330
  rabies, 56/388
Percussion myotonia
  hereditary cylindric spirals myopathy, 62/348
  tetrodotoxin, 40/537
Percutaneous thermal rhizotomy
  spastic paraplegia, 61/370
  spinal cord injury, 61/370
  spinal spasticity, 61/370
  technique, 61/370
  transverse spinal cord lesion, 61/370
Percutaneous transluminal angiography
  brain angiography, 54/179
  subclavian steal syndrome, 54/179
Perforating foot ulcer

hereditary, *see* Hereditary perforating foot ulcer
hereditary sensory and autonomic neuropathy type
  II, 8/180
Perforating skin ulcer
  leprosy, 51/218
Performic acid alcian blue staining
  protein, 9/24
Perfusion pressure
  brain, *see* Brain perfusion pressure
Pergolide
  behavior disorder, 46/594
  pituitary tumor, 68/349
Perhexilene maleate, *see* Perhexiline maleate
Perhexiline
  chronic inflammatory demyelinating
    polyradiculoneuropathy, 51/530
  drug induced polyneuropathy, 51/294
  neurotoxicity, 51/294
  neurotoxin, 51/294, 479, 530, 62/606, 608, 64/10,
    14
  segmental demyelination, 64/14
  toxic myopathy, 62/606
  toxic neuropathy, 64/10, 14
Perhexiline intoxication
  action site, 64/10, 14
  autonomic polyneuropathy, 51/479
Perhexiline maleate
  cytoplasmic inclusion body, 65/484
  drug induced demyelinating neuropathy, 47/620
  drug induced polyneuropathy, 51/301
  toxic neuropathy, 51/301
Perhexiline maleate polyneuropathy
  amyotrophy, 51/302
  ataxia, 51/302
  CSF protein, 51/302
  deafness, 51/302
  debrisoquine, 51/302
  distal axonopathy, 51/302
  EMG, 51/301
  myopathy, 51/301
  papilledema, 51/302
  segmental demyelination, 51/302
  sensorimotor neuropathy, 51/302
Perhexiline myopathy
  sensorimotor polyneuropathy, 62/608
  ultrastructure, 62/608
Periaqueductal gray matter
  analgesia, 45/238-240
  autonomic nervous system, 74/61
  pain control, 45/238-240
  serotonin, 45/239
Periaqueductal syndrome
  aqueduct stenosis, 2/279, 50/311

Periarteritis
  multinodular stenosing amyloidosis, 11/613
  Wegener granulomatosis, 8/112
Periarteritis nodosa, *see* Polyarteritis nodosa
Periarthritis
  scapulohumeral, *see* Scapulohumeral periarthritis
Periaxonal space
  Schwann cell, 51/10
Periaxonal submyelinic space
  Schwann cell, 47/4
Pericallosal aneurysm
  brain aneurysm neurosurgery, 55/69
  surgery, 12/222
Pericallosal artery
  radioanatomy, 11/79
  topography, 11/79
  traumatic aneurysm, 24/390
Pericardial effusion
  congenital olivopontocerebellar atrophy, 60/668
  Jaeken syndrome, 60/668
Pericarotid sympathetic plexus
  headache, 48/336
Pericarotid syndrome, 48/329-339
  arteriography, 48/335
  carotid body tumor, 48/337
  cause, 48/337
  dissecting aneurysm, 48/337
  fibromuscular dysplasia, 48/337
  forehead anhidrosis, 48/336
  internal carotid artery, 48/335
  paratrigeminal syndrome, 48/329
Pericentral scotoma
  papillomacular lesion, 2/564
Pericentronuclear myopathy, *see* Myotubular
    myopathy
Perichiasmatic artery, 13/271
Peridural abscess
  spinal, *see* Spinal epidural empyema
Peridurography, *see* Canalography
Perikaryon volume ratio
  axon, *see* Axon-perikaryon volume ratio
Perilabyrinthitis
  vertigo, 2/365
Perimedullary artery
  spinal cord, 12/495
Perimeningitis, *see* Spinal epidural empyema
Perimetry
  visual field, 13/12
Perinatal asphyxia
  autonomic nervous system, 74/214
  epilepsy, 72/344
Perinatal brain injury
  dystonia, 6/545, 553

musicogenic epilepsy, 73/192
pallidal necrosis, 49/465
pallidonigral necrosis, 49/465
pallidostriatal necrosis, 49/465
subarachnoid hemorrhage, 12/158
symptomatic dystonia, 6/545
Perinatal encephalopathy
   striatal necrosis, 49/505
Perinatal hypoxia
   cystic leukoencephalopathy, 53/29
   double athetosis, 49/384
Perinatal hypoxic ischemia
   pallidal necrosis, 49/466
Perineum
   anatomy, 1/360
Perineural cyst
   spinal, *see* Spinal perineural cyst
Perineural fibroblast
   see also Neurofibroma
   neurofibromatosis, 14/20, 135
Perineural growth
   dendritic damage, 10/6
   glial cell, 10/6
Perineural lipomatosis
   hypercortisolism, 70/207
Perineural metastasis
   systemic tumor, 71/615
Perineurium
   nerve, 51/2
Perinodal area
   Ranvier node, 7/22, 51/13, 16
Periodic acid Schiff, *see* PAS stain
Periodic ammonia intoxication
   congenital lysine intolerance, 29/104
Periodic arthralgia
   genetics, 42/301
Periodic ataxia
   acute intermittent porphyria, 21/570
   argininosuccinic aciduria, 21/577
   basilar impression, 21/570
   Bassen-Kornzweig syndrome, 21/580
   calf hypertrophy, 60/659
   congenital lysine intolerance, 21/578
   downbeating nystagmus, 60/656
   encephalomyelitis, 21/570
   epilepsy, 21/577
   exacerbation, 60/650
   familial protein intolerance, 21/578
   Guillain-Barré syndrome, 21/570, 60/648
   hemiparesis, 21/571
   hereditary, *see* Hereditary periodic ataxia
   hereditary hyperammonemia, 21/573
   hereditary myokymia, 60/659

hyperornithinemia, 21/577
intention tremor, 21/579
Isaacs syndrome, 60/659
isovaleric acidemia, 60/647
labyrinthine disease, 21/570
Lonsdale-Blass syndrome, 21/578
migraine, 21/570
multiple neuronal system degeneration, 60/650
multiple sclerosis, 21/570, 60/648
myokymia, 60/659
myopathy, 60/660
neuronal intranuclear hyaline inclusion disease,
   60/648
nystagmus, 21/578
odontoid bone defect, 21/570
petit mal status, 60/648
pyruvate decarboxylase, 21/578
pyruvate decarboxylase deficiency, 60/647,
   62/502
pyruvate dehydrogenase complex deficiency,
   60/648
pyruvate dehydrogenase deficiency, 29/212
Refsum disease, 21/570, 581
Reye syndrome, 21/578
somnolence, 21/571
urea cycle enzyme, 60/647
vertebrobasilar insufficiency, 21/570
vomiting, 60/647
Wilson disease, 21/570
Periodic disease
   migraine, 5/46
Periodic dysarthria
   Guillain-Barré syndrome, 21/570
   multiple sclerosis, 21/570
Periodic dystonia, *see* Hereditary paroxysmal
      kinesigenic choreoathetosis
Periodic fever
   autonomic nervous system, 43/68
   carbon monoxide intoxication, 64/32
   etiocholanolone, 43/68
   Fabry disease, 60/172
   hereditary sensory and autonomic neuropathy type
      II, 60/11
   hereditary sensory and autonomic neuropathy type
      IV, 60/12
Periodic leg movement, *see* Nocturnal myoclonus
Periodic migrainous neuralgia, *see* Cluster headache
Periodic movement
   eye movement, 1/585
   sleep, *see* Sleep periodic movement
Periodic nystagmus
   hereditary periodic ataxia, 21/569
Periodic paralysis, 28/581-598, 41/147-170

acetazolamide, 40/285, 556, 41/156, 161, 167
ammonium chloride, 41/157
autophagic mechanism, 40/153
calcium deposition, 62/465
carbohydrate metabolism, 28/582, 597
causative mechanism, 41/169
classification, 28/581, 40/285
clinical features, 40/317, 323, 335, 41/288
concentric laminated body, 40/109
corticosteroid, 41/156
differential diagnosis, 28/592, 40/438, 41/163, 166
endemic, see Endemic periodic paralysis
familial, see Familial periodic paralysis
familial hypokalemic, see Familial hypokalemic
    periodic paralysis
genetics, 41/425
glycogen storage disease, 41/170
historic aspect, 28/581
history, 28/581, 41/147
hyperaldosteronism, 62/457
hyperkalemia, 63/558
hyperkalemic, see Hyperkalemic periodic
    paralysis
hypermagnesemia, 62/457
hyperthyroidism, 27/268, 28/597, 40/285, 41/165,
    241, 62/457, 468, 70/84
hypokalemic, see Hypokalemic periodic paralysis
hypothalamic tumor, 62/457
malignant hyperthermia, 62/570
migraine, 41/152
multiple mitochondrial DNA deletion, 66/429
muscle biopsy, 62/465
muscle cell membrane, 28/582, 41/150, 167, 170
muscle hypertrophy, 41/151, 160
myoglobinuria, 41/167
myopathy, 40/285
myotonia, 40/283, 544, 41/150, 158, 165
neuromuscular disease, 41/151, 425
normokalemic, see Normokalemic periodic
    paralysis
pathology, 62/464
pathophysiology, 41/167, 62/466
phosphorylation, 41/170
porphyria, 41/167
potassium, 28/597, 40/285, 544, 41/155
precipitating factor, 62/468
renal disease, 62/457
rimmed muscle fiber vacuole, 40/29, 106, 219,
    41/153, 160, 62/465
rimmed muscle fiber vacuole content, 62/18
rimmed muscle fiber vacuole type, 62/18
salbutamol, 41/161
sarcoplasmic reticulum, 41/154

sodium conductance, 62/467
sodium-potassium pump, 62/468
sodium transport channel, 41/167
spironolactone, 41/156
T tubule, 41/154
T tubule proliferation, 62/465
thyroid gland, 41/165, 241
thyrotoxic, see Thyrotoxic periodic paralysis
transverse tubule, 40/153
treatment, 62/469
tubular aggregate, 40/229, 41/154, 162, 62/465
ultrastructure, 28/582, 40/29, 106, 219
variant, 41/163
variant type, 62/463
Periodic paramyotonic paralysis, see Paralysis
    periodica paramyotonica
Periodic somnolence and overeating, see
    Kleine-Levin syndrome
Periodic syndrome
    migraine, 48/38
Periodic syndrome of children, see Abdominal
    migraine
Periodic system
    Mendelejev, see Mendelejev periodic system
Periodic vertigo
    White syndrome, 21/569
Periodic vestibulocerebellar ataxia
    diplopia, 42/116
    hereditary disease, 42/115
    nystagmus, 42/116
    vertigo, 42/116
Periodical psychosis
    epilepsy, 15/596
Periodontitis
    apical, see Apical periodontitis
Periorbital hematoma
    craniofacial injury, 23/372
Periorbital neurofibroma
    neurofibromatosis type I, 50/367
Periosteal chondroma
    benign, 19/303
Periostitis
    Bamberger-Marie syndrome, 1/488
    superior orbital fissure syndrome, 33/179
    syphilitic, see Syphilitic osteoperiostitis
    Tolosa-Hunt syndrome, 48/302
Peripachymeningitis, see Spinal epidural empyema
Peripapillary staphyloma
    basal encephalocele, 50/106
    basal meningoencephalocele, 50/106
Peripheral amyotrophy
    reflex atrophy, 1/231
Peripheral autonomic dysfunction, 75/113

Peripheral autonomic function
  survey, 75/105
Peripheral blood cell
  multiple sclerosis, 47/346-363
Peripheral circumscribed lipomatosis
  multiple, see Krabbe-Bartels disease
Peripheral demyelination
  uremia, 47/583
Peripheral facial paralysis
  auricular herpes, 7/483
  caudal tegmental syndrome, 12/20
  human immunodeficiency virus neuropathy,
    71/354
  lateral pontine syndrome, 12/17
  Sjögren syndrome, 71/74
  tetanus, 65/217, 219
Peripheral myoclonus
  critique, 49/614
Peripheral nerve
  ataxia telangiectasia, 60/365
  cerebrotendinous xanthomatosis, 66/605
  chronic relapsing neuropathy, 40/318
  experimental allergic neuritis, 9/531-536
  Friedreich ataxia, 60/303
  hereditary amyloid polyneuropathy, 60/102
  hexachlorophene intoxication, 37/479
  Hodgkin disease, 39/38
  ischemic neuropathy, 8/154-157
  leukemia, 39/13
  malabsorption syndrome, 70/229
  malignant lymphoma, 39/38
  metachromatic leukodystrophy, 66/166
  multiple neuronal system degeneration, 59/149
  multiple organ failure, 71/534
  myxedema neuropathy, 7/169, 8/57
  organotin intoxication, 64/137
  Pena-Shokeir syndrome type I, 43/438
  spongiform leukoencephalopathy, 37/479
  vasa nervorum disease, 12/644-661
  viral infection, 56/83
  vitamin E deficiency, 70/447
  X-linked adrenoleukodystrophy, 66/460
  zebra body, 22/212
Peripheral nerve lesion, see Nerve lesion
Peripheral nerve sheath tumor
  malignant, see Malignant peripheral nerve sheath
    tumor
Peripheral nervous system
  cancer, 69/71
  cerebrotendinous xanthomatosis, 66/603
  CNS myelin, 47/5-9
  definition, 51/1
  demyelination, 47/606

hexomonophosphate shunt, 27/79
isoniazid neuropathy, 7/582
miner experiment, 7/7
multiple organ failure, 71/533
myelin, 47/1-5
myelin basic protein P1, 47/5
myelin basic protein P2, 47/5
myelin lipid, 47/5
nerve, 51/1
organophosphorus compound intoxication, 9/652
pathology, 60/337
pernicious anemia, 7/592
pernicious psychosis, 7/592
polyarteritis nodosa, 11/139, 39/304, 51/452
Ranvier node, 7/22, 47/10
remyelination, 47/41
Sjögren syndrome, 71/83
stress, 74/313
xeroderma pigmentosum, 60/337
Peripheral nervous system compound axon
  CNS, see CNS-peripheral nervous system
    compound axon
Peripheral nervous system lesion
  body scheme disorder, 4/399
  trophic disorder, 1/485-487
Peripheral nervous system transitional zone
  nerve, 51/11
Peripheral neuritis
  ethchlorvynol intoxication, 37/354
  neurobrucellosis, 52/584, 589
  postinfectious complication, 47/481-489
  uremic encephalopathy, 27/338-340
Peripheral neurofibroma
  neurofibromatosis type I, 50/368
Peripheral neurofibromatosis, see
    Neurofibromatosis type I
Peripheral neurolymphomatosis, 56/179-190
  amyotrophy, 56/180
  cause, 56/181
  clinical features, 56/179, 181
  CNS reticulosis, 56/179
  cranial neuropathy, 56/181
  CSF, 56/180, 182
  diagnosis, 56/187
  duration, 56/181
  EMG, 56/182
  facial diplegia, 56/181
  Guillain-Barré syndrome, 56/179-181
  history, 56/179
  infiltrative type, 56/184
  leukemia, 56/180
  lymphoma, 56/185
  Marek disease, 56/179, 581

multiple cranial neuropathy, 56/181
neuropathology, 56/182
pain, 56/181
parenchymatous type, 56/184
remittent course, 56/180
stage, 56/182
Peripheral neuropathy, *see* Polyneuropathy
Peripheral reflex
autonomic nervous system, 74/41
Peripheral scotoma
retinal *vs* nerve lesion, 2/562
Peripheral trauma
segmental dystonia, 49/542
Periphlebitis retinae
multiple sclerosis, 9/115, 175
Peristalsis
hereditary amyloid polyneuropathy, 21/124
reflex, 75/618
Peristriate cortex, *see* Parastriate cortex
Perisylvian region
aprosody, 45/45
global aphasia, 45/20
right, *see* Right perisylvian region
Perithelial sarcoma
terminology, 18/234
Peritoneal dialysis
see also Renal dialysis
burning feet syndrome, 27/342
dialysis encephalopathy, 27/341
EEG, 27/341
headache, 27/323
neuropathy, 8/1
rubidium, 27/329
tin, 27/329
trace metal, 27/329
uremic polyneuropathy, 51/360
Periumbilical skin defect
Rieger syndrome, 43/470
Periungual fibroma
tuberous sclerosis, 50/371
Perivascular cuff
Guillain-Barré syndrome, 47/611
Perivenous encephalitis, *see* Acute disseminated
encephalomyelitis
Perivenous encephalomalacia
varicella zoster virus meningoencephalitis, 56/235
Periventricular calcification
achondroplasia, 50/258
encephalocele, 50/258
exencephaly, 50/258
microgyria, 50/258
schizencephaly, 50/258
Periventricular heterotopia

corpus callosum agenesis, 50/151
Periventricular infarction
brain matrix hemorrhage, 53/30
cerebrovascular disease, 53/28
cystic leukoencephalopathy, 53/29
hypoxia, 53/29
Periventricular leukomalacia
acute stage, 54/42
B-mode sonography, 54/40
cavity stage, 54/42
chronic stage, 54/42
infantile, *see* Infantile periventricular
leukomalacia
infantile cerebrovascular disease, 54/40
mental deficiency, 54/42
streptococcal meningitis, 52/81
subacute stage, 54/42
Periventricular necrosis
hydranencephaly, 50/347
Periventricular system
stress, 74/313
Periventricular tuber
tuberous sclerosis, 50/373
Periwinkle alkaloid
neuropathy, 7/517
Perlia nucleus
oculomotor nuclear complex, 2/301
Permethrin
distal axonopathy, 64/13
neurotoxin, 64/13, 212
pyrethroid, 64/212
pyrethroid intoxication, 64/212
toxic neuropathy, 64/13
Permethrin intoxication
symptom, 64/13
Perniciosa
endogenous, *see* Endogenous perniciosa
Pernicious anemia
see also Megaloblastic anemia *and* Vitamin B12
neuropathy
age, 63/253
astrocyte, 9/617
autoimmune disease, 41/73, 102, 240
behavior disorder, 42/609
classification, 38/18
clinical features, 38/18-24
dementia, 42/609
diphyllobothrium latum myelopathy, 7/631
dysesthesia, 70/376
Eaton-Lambert myasthenic syndrome, 62/421
endocrine candidiasis syndrome, 42/462
endocrine disorder, 42/609
hepatosplenomegaly, 42/609

history, 70/367
idiopathic, 70/372
intrinsic factor, 42/609
megaloblastic madness, 38/21, 42/609
myasthenia gravis, 41/240
neurologic aspect, *see* Subacute combined spinal cord degeneration
neuropathy, 42/609
optic atrophy, 42/609
optic neuritis, 38/22
paresthesia, 42/609
pathogenesis, 7/592
peripheral nervous system, 7/592
psychiatric disorder, 38/21
retrobulbar neuritis, 38/22
spastic paraplegia, 42/609
subacute combined spinal cord degeneration, 7/591, 28/159-164, 38/22, 63/254
tobacco, 65/32
vitamin B_{12}, 7/587, 592
vitamin B_{12}-cyanide relationship, 38/20
vitamin B_{12} deficiency, 38/19, 42/609, 51/338, 340, 70/371, 373
vitamin B_{12}-vitamin B_c relationship, 38/24
Pernicious psychosis
pathogenesis, 7/592
peripheral nervous system, 7/592
subacute combined spinal cord degeneration, 28/170-173
vitamin B_{12}, 7/587, 592
Pernio syndrome
cold injury, 8/341
vascular change, 8/341
Peroneal muscular atrophy, *see* Hereditary motor and sensory neuropathy type I *and* Hereditary motor and sensory neuropathy type II
Peroneal nerve
ankle fracture, 70/41
ankle injury, 70/40
common, *see* Common peroneal nerve
compression neuropathy, 51/104
femoral fracture, 70/38
knee fracture, 70/38
tumor, 70/46
Peroneal nerve compression
mononeuropathy, 7/315
rheumatoid arthritis, 38/484
Peroneal nerve conduction
nerve compression, 7/164
Peroneal nerve lesion
heart transplantation, 63/181
iatrogenic neurological disease, 63/181
Peroneal neuropathy

bacterial endocarditis, 52/300
rheumatoid arthritis, 71/18
Peroxidase deficiency, *see* Neuronal ceroid lipofuscinosis
Peroxisomal acetyl-coenzyme A acetyltransferase biochemistry, 66/519
Peroxisomal acetyl-coenzyme A acetyltransferase deficiency
peroxisomal disorder, 66/510
Peroxisomal acetyl-coenzyme A acyltransferase deficiency
facial dysmorphia, 66/518
hypotonia, 66/518
muscle weakness, 66/518
neuronal heterotopia, 66/518
renal cyst, 66/518
sudanophilic leukodystrophy, 66/518
very long chain fatty acid, 66/518
Peroxisomal chondrodysplasia
form, 66/531
Peroxisomal disease
acatalasia, 51/383
adrenoleukodystrophy, 51/383, 60/169
enzyme deficiency polyneuropathy, 51/367, 383
hyperpipecolatemia, 51/383
infantile Refsum disease, 51/383
rhizomelic chondrodysplasia punctata, 51/383
very long chain fatty acid, 60/674
Peroxisomal disorder
acatalasia, 66/510
acyl-coenzyme A oxidase deficiency, 66/510, 517
alanine glyoxylate aminotransferase deficiency, 66/510
alkylglycerone phosphate synthase deficiency, 66/510
bifunctional enzyme deficiency, 66/510, 517
biochemical classification, 66/510
cerebrohepatorenal syndrome, 66/17, 510
clinical classification, 66/510
dihydroxyacetone phosphate acyltransferase deficiency, 66/510
epilepsy, 72/222
form, 66/17
generalized, *see* Generalized peroxisomal disorder
glutaryl-coenzyme A oxidase deficiency, 66/510
hyperpipecolatemia, 66/510
infantile Refsum disease, 66/510
mevalonate kinase deficiency, 66/510
neonatal adrenoleukodystrophy, 66/510
peroxisomal acetyl-coenzyme A acetyltransferase deficiency, 66/510
Refsum disease, 66/494, 510
rhizomelic chondrodysplasia punctata, 66/510,

525
spinal muscular atrophy, 59/371
survey, 66/505
X-linked adrenoleukodystrophy, 66/510
Zellweger like syndrome, 66/510
Peroxisomal β-oxidation
biochemistry, 66/518-520
enzyme, 66/506
enzyme defect, 66/518
generalized peroxisomal disorder, 66/520
survey, 66/505
Peroxisomal 3-oxoacyl-coenzyme A thiolase
deficiency, see Peroxisomal acetyl-coenzyme A
acyltransferase deficiency
Peroxisomal thiolase, see Peroxisomal
acetyl-coenzyme A acetyltransferase
Peroxisomal thiolase deficiency, see Peroxisomal
acetyl-coenzyme A acetyltransferase deficiency
Peroxisome
adrenoleukodystrophy, 47/598
biochemistry, 66/519
biogenesis, 66/509
glyoxylic acid metabolism, 66/507
hypoglycin intoxication, 65/96
infantile Refsum disease, 66/513
Jamaican vomiting sickness, 65/96
Lafora progressive myoclonus epilepsy, 40/102
lysosome, 40/37, 99, 102
phytanic acid α-oxidation, 66/507
pipecolic acid metabolism, 66/508
Perphenazine
chemical formula, 65/275
neuroleptic agent, 65/275
phenothiazine, 65/275
Perseveration
acalculia, 4/187
aphasia, 4/91, 94
attention defect, 3/187
frontal lobe syndrome, 55/138
ideomotor apraxia, 4/50
migraine, 48/162
parietal lobe epilepsy, 73/101
Pick disease, 42/285
traumatic psychosyndrome, 24/550
visual, see Visual perseveration
Wernicke aphasia, 4/98
word pronunciation, 3/425
Perseverative agraphia
Alzheimer disease, 45/467
frontal lobe lesion, 45/467
multi-infarct dementia, 45/467
Pick disease, 45/467
Persistent crying

diphtheria pertussis tetanus vaccine, 52/245
pertussis vaccine, 52/242
Persistent hyaloid artery
connective tissue, 13/81
Persistent neurenteric canal, see Neurenteric cyst
Persistent vegetative state
brain death, 24/703
brain hypoxia, 57/473
brain injury prognosis, 24/672
brain stem death, 57/445
cardiac arrest, 38/157, 57/441
consciousness, 45/118
EEG, 57/189
Glasgow Outcome Scale, 57/115
head injury, 57/341
head injury outcome, 57/115, 368, 372
pediatric head injury, 57/341
persistent unconsciousness, 57/189
prediction, 24/678
prognosis, 24/678, 727
sign, 24/673, 703
Personality
anxiety, 45/266
chronic muscle contraction headache, 48/345
common migraine, 5/51
epilepsy, 73/472
frontal lobe lesion, 45/29
frontal lobe tumor, 45/29
head injury, 46/625-629
Huntington chorea, 46/306
hysteria, 46/578-580
migraine, see Migraine personality
multiple, see Multiple personality
multiple sclerosis, 47/59
neuropsychology, 45/4, 516
Parkinson disease, 6/194
parkinsonism, 6/194
premorbid, see Premorbid personality
rehabilitation, 46/624-630
sleep disorder, 3/92
temporal lobe epilepsy, 45/45
Personality change
Behçet syndrome, 56/597
carbamazepine intoxication, 65/507
carbon monoxide intoxication, 49/475
cardiac arrest, 63/217
chorea-acanthocytosis, 63/281
dentatorubropallidoluysian atrophy, 49/440
Friedreich ataxia, 21/338
frontal lobe syndrome, 45/29
head injury outcome, 57/407
herpes simplex virus encephalitis, 56/214
Huntington chorea, 49/282

hypocalcemia, 63/561
juvenile hereditary benign chorea, 49/346
Lafora progressive myoclonus epilepsy, 15/383, 27/172
myotonic dystrophy, 62/235
neurosyphilis, 52/277
Prader-Labhart-Willi syndrome, 31/324
progressive supranuclear palsy, 49/243
Personality disorder
    brain tumor, 16/735
    consciousness, 3/119
    epilepsy, 15/576
    muscle contraction headache, 48/356
    thalamic syndrome (Dejerine-Roussy), 2/483
    traumatic psychosyndrome, 24/554, 687, 693
Personality pattern
    abnormal, *see* Abnormal personality pattern
Personification
    paralysed limb, 45/380
Perspiration, *see* Sweating
Pertussis
    blindness, 52/234
    Bordetella, *see* Bordetella pertussis
    carrier state, 52/232
    clinical features, 52/233
    coma, 52/234
    CSF, 52/234
    100 days cough, 52/233
    deafness, 52/234
    EEG, 52/234
    epilepsy, 52/233
    history, 52/231
    hypocalcemia, 33/278
    immune palsy, 52/234
    immunity, 52/234
    immunization, 52/227, 235
    infection, 33/275-295
    meningoencephalitis, 52/234
    mental deficiency, 52/234
    mortality, 52/232, 237
    neurologic sequela, 52/227
    neurologic symptom, 52/233
    prevention, 52/237
    spastic paraplegia, 59/436
    spastic paresis, 52/234
Pertussis encephalopathy, 33/275-295
    acute, 33/288-290
    blindness, 33/289
    chronic, 33/290-292
    clinical features, 33/287-292
    coma, 9/551
    convulsion, 9/551
    cranial nerve, 33/279

death rate, 33/279
diabetes insipidus, 33/290
differential diagnosis, 33/292
edema, 9/551
EEG, 33/292
epidemiology, 33/279-282
etiology, 33/275
hippocampus, 9/551
history, 33/275
hypoglycemia, 33/277
incidence, 33/280-282
incubation period, 33/287
morbidity, 33/275
mortality, 9/551, 33/275, 280, 289
pathogenesis, 33/277-279, 282
pathology, 33/282-287
perivascular hemorrhage, 9/551
pertussis vaccine, 52/236
pulmonary disease, 9/551
Purkinje cell, 9/551
Reye syndrome, 33/292
spinal cord, 33/289
subdural hematoma, 33/289
symptom, 52/234
tetany, 33/278
treatment, 33/294
Pertussis immunization, *see* Pertussis vaccination
Pertussis vaccination
    CSF, 33/298
    neurologic sequela, 33/295-301
    neuropathy, 8/91
    poliomyelitis, 33/300
Pertussis vaccine
    acceptance rate, 52/237
    acellular, 52/238
    allutinogen 3, 52/237
    anaphylaxis, 52/236
    composition, 52/235
    contraindication, 52/241
    convulsion, 52/239
    EEG, 52/245
    epilepsy, 52/242, 245
    hypotonia, 52/242
    hypsarrhythmia, 52/242
    infantile spasm, 52/239
    long-term neurologic damage, 52/246
    mass vaccination, 52/238
    myoclonia, 52/242
    neurologic side effect, 52/236
    neurotoxicity, 52/243
    persistent crying, 52/242
    pertussis encephalopathy, 52/236
    protective effectiveness, 52/237

schema, 52/234
sialic acid receptor, 52/236
Pes arcuatus, *see* Pes cavus
Pes calcaneovalgus
    anencephaly, 50/84
    Patau syndrome, 31/507, 50/558
Pes cavovarus
    diastematomyelia, 50/436
    diplomyelia, 50/436
Pes cavus
    amyotrophy, 59/374
    Bassen-Kornzweig syndrome, 10/548, 13/418,
      51/394
    Behr disease, 13/88
    Charlevoix-Saguenay spastic ataxia, 60/669
    Ekbom syndrome type II, 60/669
    familial, *see* Familial pes cavus
    Friedreich ataxia, 21/338-341, 40/321, 42/143,
      60/315, 325, 669
    generalized myotonia (Becker), 43/163
    hereditary motor and sensory neuropathy type I,
      21/278, 42/339
    hereditary motor and sensory neuropathy type II,
      21/278, 42/339
    hereditary multiple recurrent mononeuropathy,
      60/62
    hereditary olivopontocerebellar atrophy (Menzel),
      21/444
    hereditary sensory and autonomic neuropathy type
      I, 21/77
    hypertrophic interstitial neuropathy, 21/152
    hypogonadal cerebellar ataxia, 21/468
    Marinesco-Sjögren syndrome, 21/556
    mechanism, 21/339
    myopathy, 40/321, 337, 427
    neurodegeneration, 21/3
    occurrence, 21/263
    olivopontocerebellar atrophy (Wadia-Swami),
      60/669
    Richards-Rundle syndrome, 43/264
    Roussy-Lévy syndrome, 22/425, 42/108
    scapuloperoneal myopathy, 43/130
    scapuloperoneal spinal muscular atrophy, 59/45,
      62/170
    spinocerebellar neuroaxonal dystrophy, 60/669
    treatment, 60/325
Pes equinovarus
    Bassen-Kornzweig syndrome, 51/394
    Brossard syndrome, 22/57
    Patau syndrome, 31/507, 50/558
    Pena-Shokeir syndrome type I, 43/437
    progressive pallidal atrophy (Hunt-Van Bogaert),
      6/457, 554, 21/531

Richards-Rundle syndrome, 22/516, 43/264
scapuloperoneal spinal muscular atrophy, 59/45
Pes excavatus, *see* Pes cavus
Pes planus
    early onset recessive stationary ataxia, 60/669
    Richards-Rundle syndrome, 43/264
Pesticide
    autonomic nervous system, 75/507
    epilepsy, 72/238
    high grade glioma, 68/89
    neurotoxicity, 51/265
    organophosphate intoxication, 64/151
    thallium intoxication, 64/323
    toxic neuropathy, 51/283
Pestivirus
    schizencephaly, 50/198
PET scanning, *see* Positron emission tomography
Peter abnormality, *see* Anophthalmia
Peter Pan syndrome
    craniopharyngioma, 18/545
    high pitched voice, 18/545
    Levi-Lorain dwarfism, 18/545
    skin, 18/545
Pethidine
    chemical structure, 65/354
    chemistry, 37/366
    delirium, 46/539
    designer drug, 65/349
    headache, 5/98
    opiate, 65/350
Pethidine addiction
    brain embolism, 55/518
Pethidine intoxication
    delirium, 65/357
    epilepsy, 65/357
    hallucination, 65/357
    myoclonus, 65/357
    norpethidine, 65/357
    symptom, 65/357
    tremor, 65/357
Petit mal epilepsy
    see also Absence *and* Akinetic petit mal
    anti, *see* Antipetit mal
    automatism, 42/680
    autonomic nervous system, 42/680
    classification, 15/3, 6, 22
    EEG, 15/470, 511, 42/681
    impulsive, *see* Juvenile myoclonic epilepsy
    juvenile, 15/473
    Lennox-Gastaut syndrome, 15/471
    microabsence, 15/130
    occipital delta rhythm, 42/672
    petit mal variant, 15/133

prognosis, 15/788
sleep, 15/470, 474
terminology, 15/3, 6, 22
triad, 15/8, 142
variant, 15/133, 135
Petit mal status
 see also Absence status
 confusion, 15/170
 lethargy, 15/170
 periodic ataxia, 60/648
 prognosis, 15/788
 sleep, 15/170
 status epilepticus, 15/169, 595
Petrobasilar suture syndrome
 abducent nerve paralysis, 2/313
 accessory nerve, 2/313
 cardiac dysrhythmia, 2/313
 dysphagia, 2/313
 glossopharyngeal nerve palsy, 2/313
 vagus nerve, 2/313
Petro-occipital sinus
 anatomy, 11/54
Petrosal apex syndrome, *see* Gradenigo syndrome
Petrosal nerve
 anatomy, 5/328, 48/489
 autonomic nervous system, 75/291
 cluster headache, 5/116
 deep, 5/335, 338, 351
 facial nerve, 48/488
 greater superficial, 5/338, 351
 large, 5/335, 338, 351
 small superficial, 5/338, 351
 sphenopalatine neuralgia, 5/328
Petrosal neuralgia
 cluster headache, 5/114
 differential diagnosis, 5/330
 sphenopalatine neuralgia, 5/330, 48/480
Petrosal sinus
 inferior, *see* Inferior petrosal sinus
 superior, *see* Superior petrosal sinus
 Trolard vein, 11/54
Petrosal sinus thrombophlebitis
 superior, *see* Superior petrosal sinus
  thrombophlebitis
Petrosal vein
 anatomy, 11/51
 embryology, 11/63
 radioanatomy, 11/110
 topography, 11/110
Petrositis
 acute, *see* Acute petrositis
 apical, *see* Apical petrositis
 paratrigeminal syndrome, 48/337

Petrosphenoid (crossway) syndrome, *see* Orbital
  apex syndrome
Petrosphenoid ligament
 Gruber, *see* Gruber petrosphenoid ligament
Petrosphenoid syndrome (Jaccodo)
 optic atrophy, 13/82
Petrosquamosal sinus
 anatomy, 11/58
Petrosum ganglion
 anatomy, 5/351
Petrous apex
 skull base tumor, 17/181
 trigeminal neuralgia, 5/273
Petrous pyramid
 Arnold-Chiari malformation type II, 50/408
Pette-Döring panencephalitis, *see* Subacute
  sclerosing panencephalitis
Pette, H., 1/2, 31
Petzetakis disease, *see* Cat scratch disease
Peutz-Jeghers syndrome
 acoustic neuroma, 14/602-604, 43/41
 anemia, 14/121, 43/41
 cachexia, 14/121
 café au lait spot, 14/121
 clinical features, 14/121
 coincidental lesion, 14/121, 603
 colonic polyposis, 14/121
 differential diagnosis, 14/121, 604
 differentiation, 14/604
 ectodermal dysplasia, 43/41
 ephelis, 14/121, 603
 gastrointestinal tract, 14/12, 121, 526, 43/41
 genetics, 14/526-528, 43/41
 histogenesis, 14/604
 history, 14/601
 intestinal cancer, 43/41
 intestinal neurinoma, 14/604, 791
 intestinal polyposis, 14/526, 601, 43/41
 mucocutaneous pigmentation, 14/526, 601, 43/41
 neurocutaneous syndrome, 14/101
 neurofibromatosis type I, 14/602, 604
 pathognomonic lesion, 14/602
 phakomatosis, 14/12, 526, 602
 pigmented nevus, 14/12, 121, 43/41
 polyposis, 14/12, 121, 526, 601, 43/41
 prognosis, 14/605, 43/42
Peutz syndrome, *see* Peutz-Jeghers syndrome
Peutz-Touraine-Jeghers disease, *see* Peutz-Jeghers
  syndrome
Peyer patch
 neurotropic virus, 56/29
Pfaundler-Hurler disease, *see* Hurler-Hunter disease
Pfeiffer syndrome, *see* Acrocephalosyndactyly

type V
Phagocytosis
  brain injury, 23/128
  multiple sclerosis, 47/218, 249
  muscle fiber, 62/6
  neuromuscular disease, 62/6
  subarachnoid hemorrhage, 55/7
Phakoblastoma
  malignant phakoma, 14/201, 562
Phakoblastomatosis
  pathogenesis, 14/5
Phakoma
  see also Hamartoma
  congenital spinal cord tumor, 32/355-386
  definition, 14/19, 201, 562
  hypothalamic, *see* Hypothalamic phakoma
  malignant, *see* Malignant phakoma
  retinal, *see* Retinal phakoma
  tuberous sclerosis, 14/343-345, 620
Phakomatosis, 31/1-30
  Albright syndrome, 31/27, 50/377
  associated type, 14/6-8
  ataxia telangiectasia, 50/377
  atypical form, 14/9
  Bonnet-Dechaume-Blanc syndrome, 14/13,
    50/377
  Brégeat syndrome, 14/14, 31/26
  chromosome, 14/531
  classification, 14/11-14
  clinical features, 14/254, 31/2-4
  clinical latent type, 14/707
  Cobb syndrome, 14/14, 72, 31/26
  definition, 14/668
  dermatoglyphics, 14/530
  dermatology, 14/562-618
  Ehlers-Danlos syndrome, 14/492
  embryology, 14/716
  endocrinology, 14/731-758
  epidemiology, 30/158-160, 31/1-3
  epilepsy, 15/325, 72/347
  general concept, 14/85-89
  general survey, 14/1-14, 50/365
  genetics, 14/6-8, 50, 482, 30/158-160, 31/30
  germ layer, 14/102
  glomus tumor, 14/10
  heterochromia iridis, 14/639
  hydrophthalmia, 14/643
  hypertrichosis, 14/10
  incontinentia pigmenti, 14/220, 31/26, 50/377
  intermediate form, 14/9
  intestinal neurinoma, 14/602
  intracerebral calcification, 49/417
  isolated lesion, 14/704

Jadassohn nevus, *see* Linear nevus sebaceous
  syndrome
Klippel-Trénaunay syndrome, 31/26, 50/377
Krabbe-Bartels disease, 14/12, 31/27
latent, 14/707
macrencephaly, 14/102
mental deficiency, 14/760, 46/42
mixed form, 14/532
muscular dystrophy, 14/492
myotonic dystrophy, 14/492
neurinoma, 14/7
neuroblast migration, 14/5
neurocutaneous angiomatosis, 68/296
neurocutaneous lipomatosis, 31/27
neurofibromatosis, 68/287
neurofibromatosis type I, 14/19, 25, 29-39
neurolipoma, 14/81-83
neuropathology, 14/19-100
nystagmus, 14/653
ophthalmology, 14/619-667
orthochromatic leukodystrophy, 10/113, 120-127
pathogenesis, 14/5, 31/30, 50/378
Peutz-Jeghers syndrome, 14/12, 526, 602
pheochromocytoma, 14/702, 750, 39/504
pigmentary, 14/103
psychiatric disorder, 14/759-771
rare syndrome, 31/26-30
spongioblast migration, 14/5
sporadic case, 14/6
striatopallidodentate calcification, 49/417
Sturge-Weber syndrome, 14/13, 19, 68/296
synonym, 14/4
systemic, *see* Ullmann syndrome
teratogenic agent, 14/5
transition form, 14/532
tuberous sclerosis, 14/19
Ullmann syndrome, 31/26
vascular, 14/103
visceral lesion, 14/668-730
Von Hippel-Lindau disease, 14/19
xeroderma pigmentosum, 14/13, 50/377
Phakos
  definition, 14/201, 562
  term derivation, 14/1, 485
Phakosis, *see* Phakomatosis
Phallotoxin
  Amanita phalloides intoxication, 36/538
  mushroom toxin, 65/37
Phanerotoxin intoxication
  cryptotoxin intoxication, 37/29
Phantom limb, 4/223-230, 45/395-401
  see also Phantom sensation
  amputation, 4/224, 400, 45/396

aneurysmal bone cyst, 3/29
body scheme, 4/210, 225, 227, 45/399
body scheme disorder, 2/478, 668
cerebral representation, 4/227
characteristics, 4/225
CNS lesion, 4/224, 226
congenital limb agenesis, 4/224, 399
denial of illness, 4/224
emotional trauma, 4/229
epilepsy, 4/224, 45/396
genesis, 4/227-229, 45/398
higher nervous activity disorder, 3/29
historic survey, 4/223
hypnosis, 4/227-230, 45/398
impulsive reaching, 4/226, 45/397
influencing factor, 4/227
macrosomatognosia, 4/226, 45/398
malignant, 69/93
movement, 4/225
mutilation, 4/226
nature, 4/227-229
nerve lesion, 4/224, 226, 45/397
neuroma, 4/227
occurrence, 4/224
orthopedic trauma, 70/45
paraplegia, 4/224, 226
paresthesia, 4/225, 45/398
phantom third limb, 4/226, 45/397
plexus lesion, 4/224
polyneuritis, 45/399
position, 4/225, 45/397
psychoanalysis, 45/400
psychotherapy, 45/398
recurrence after surgery, 4/228
sensory deprivation, 4/228, 45/399
sensory extinction, 4/228, 45/399
somatesthesia, 4/225, 45/397
somatesthetic deafferentation, 4/228, 45/399
spinal cord, 45/399
spinal cord lesion, 45/398
surgery, 1/110, 4/227-229, 45/398
tardive dyskinesia, 45/397
telescoping, 4/226, 45/397
thalamus, 45/398
topesthesia, 4/226, 45/397, 399
transverse spinal cord lesion, 4/224, 226
Phantom limb pain, 45/395-401
  amputation, 45/400
  carbamazepine, 45/401
  cordotomy, 45/400
  electroconvulsive treatment, 45/401
  gyrectomy, 1/110
  pathogenesis, 45/399

sciatica, 4/226, 229, 45/400
stump pain, 4/229, 45/400
surgery, 45/400
therapy, 45/400
topectomy, 1/110
treatment, 4/229
Phantom limb syndrome
  analgesic agent, 45/401
Phantom sensation
  see also Phantom limb
  amputation, 45/396
  brain stem, 45/400
  cause, 45/395
  class, 1/81
  classification, 26/490
  definition, 4/225
  description, 1/110
  incidence, 26/489
  parietal lobe, 45/400
  spinal cord injury, 26/489-499
  thalamic syndrome (Dejerine-Roussy), 2/478
  thalamus, 45/400
  transverse spinal cord syndrome, 2/187
  transverse syndrome, 2/187, 45/400
Phantom third limb
  epilepsy, 45/509
  phantom limb, 4/226, 45/397
Pharyngeal artery
  ascending, see Ascending pharyngeal artery
Pharyngeal hypoesthesia
  diphtheria, 51/183
Pharyngeal reflex
  amyotrophic lateral sclerosis, 22/293
  progressive bulbar palsy, 59/218
Pharyngeal system
  ascending, see Ascending pharyngeal system
Pharynx paresis
  diphtheria, 51/183
Phenacemide
  serum level, 15/692
Phenacemide intoxication
  aplastic anemia, 37/202
  behavior change, 37/202
  gastrointestinal disorder, 37/202
  hepatotoxicity, 37/202
  hydroxy metabolite, 37/200
  psychosis, 37/202
Phenacetin
  headache, 5/34
  migraine, 5/98
Phenazocine
  chemistry, 37/366
Phencyclidine

see also Hallucinogenic agent
acquired myoglobinuria, 62/573
analogue, 65/52
biokinetic, 65/53
brain hemorrhage, 55/522
brain metabolism, 57/81
chemical structure, 65/52
designer drug, 65/364
drug abuse, 55/522
drug dosage, 65/53
excitotoxin, 64/20
hallucinogenic agent, 65/41
pharmacology, 65/53
popular name, 65/52
psychosis, 46/594
sympathomimetic effect, 65/53
toxic myopathy, 62/601
Phencyclidine addiction
brain hemorrhage, 55/522
brain ischemia, 55/522
hypertension, 55/522
hypertensive encephalopathy, 55/522
subarachnoid hemorrhage, 55/522
toxic myopathy, 62/600
Phencyclidine intoxication
ataxia, 65/368
choreoathetosis, 49/385
coma, 65/54
epilepsy, 65/54, 368
fatality, 65/368
hallucination, 65/52, 367
hyperacusis, 65/368
myoglobinuria, 41/271
neonate, 65/54
nystagmus, 65/54, 367
symptom, 65/53, 368
treatment, 65/55
tremor, 65/54
Phenelzine
toxic myopathy, 62/601
vitamin B6 deficiency, 70/431
Phenethylamine
catechols, 37/232
derivative, 65/50
designer drug, 65/366
hallucinogenic agent, 65/41
headache, 48/97
metabolite, 65/50
methamphetamine, 65/366
3,4-methylenedioxyamphetamine, 65/366
3,4-methylenedioxymethamphetamine, 65/366
migraine, 48/93, 97
platelet MAO, 48/94

Phenethylamine intoxication
brain hemorrhage, 65/51
clinical effect, 65/51
epilepsy, 65/51
hallucination, 65/50
mydriasis, 65/51
pharmacology, 65/51
tremor, 65/51
Phenformin
myophosphorylase deficiency, 41/190
toxic myopathy, 62/601
Pheniprazine intoxication
optic atrophy, 13/63
Phenmetrazine
acquired myoglobinuria, 62/573
toxic myopathy, 62/601
Phenobarbital
antiepileptic agent, 65/497
barbiturate, 40/561
chloramphenicol, 40/561
epilepsy, 15/635, 670, 73/359
memory, 72/393
methylphenobarbital intoxication, 37/200
neonatal seizure, 73/257
primidone intoxication, 37/200
serum level, 15/678, 680, 683
side effect, 15/714
status epilepticus, 73/329
Phenobarbital intoxication
coma, 37/202
higher cortical function, 37/203
4-hydroxyphenobarbital, 37/200
lethargy, 37/202
metabolite, 37/200
reticular activating system, 37/201, 203
sedation, 37/199
somnolence, 37/202
stupor, 37/202
vestibular system, 37/203
Phenobarbitone, *see* Phenobarbital
Phenocopy
definition, 30/89
Phenol
intrathecal chemotherapy, 61/152
migraine, 48/94
spastic paraplegia, 61/369
spinal cord injury, 61/369
spinal spasticity, 61/369
transverse spinal cord lesion, 61/369
Phenol block
multiple sclerosis, 47/159, 166
Phenol injection
intrathecal, *see* Intrathecal phenol injection

Phenolsulfotransferase
  migraine, 48/94
  Parkinson disease, 49/122
Phenomenon
  Argyll Robertson, *see* Argyll Robertson pupil
  clasp knife, *see* Clasp knife phenomenon
  closing-in, *see* Closing-in phenomenon
  cogwheel, *see* Cogwheel phenomenon
  completion, *see* Completion phenomenon
  counter pressure, *see* Counter pressure
    phenomenon
  doll head eye, *see* Doll head eye phenomenon
  Flynn, *see* Flynn phenomenon
  freezing, *see* Freezing phenomenon
  Gordon, *see* Gordon phenomenon
  hastening, *see* Hastening phenomenon
  hemianopic pupillary, *see* Hemianopic pupillary
    phenomenon
  interossei, *see* Interossei phenomenon
  jitter, *see* Jitter phenomenon
  Koch, *see* Koch phenomenon
  look-through, *see* Look-through phenomenon
  Lucio, *see* Lucio phenomenon
  Marcus Gunn, *see* Marcus Gunn phenomenon
  Mizuo, *see* Mizuo phenomenon
  occupational Raynaud, *see* Occupational Raynaud
    phenomenon
  Raynaud, *see* Raynaud phenomenon
  reduplication, *see* Reduplication phenomenon
  second wind, *see* Second wind phenomenon
  signpost, *see* Signpost phenomenon
  tip of the tongue, *see* Tip of the tongue
    phenomenon
  Tournay, *see* Tournay phenomenon
  Westphal, *see* Paradoxical contraction (Westphal)
  zombi, *see* Zombi phenomenon
Phenoperidine
  severe head injury, 24/615
Phenothiazine
  antiphospholipid antibody, 63/330
  autonomic nervous system, 75/499
  behavior disorder, 46/597
  chlorpromazine, 65/275
  designer drug, 65/274
  dystonia, 6/560
  epilepsy, 15/653
  fluphenazine, 65/275
  mesoridazine besylate, 65/275
  migraine, 5/101, 48/197
  neuroleptic agent, 65/274
  neuroleptic parkinsonism, 42/651
  oxidative phosphorylation, 13/211
  Parkinson disease, 6/829, 49/108

  perphenazine, 65/275
  retinal pigmentation, 13/211
  senile dementia, 46/286
  side effect, 13/224
  sweating, 75/499
  thioridazine, 65/275
  tricyclic structure, 65/275
  trifluoperazine, 65/275
  triflupromazine, 65/275
  vitamin B_2, 28/206
  vitamin B_2 deficiency, 51/332
Phenothiazine intoxication
  Stargardt disease, 13/143
  tardive dyskinesia, 37/307-312
Phenotype
  adrenoleukodystrophy, 47/594
  Duchenne muscular dystrophy, 62/126
  Farber disease, 66/212
  glycosaminoglycan, 66/282
  glycosaminoglycanosis type IH-S, 66/287
  glycosaminoglycanosis type II, 66/293
  lipoprotein, *see* Lipoprotein phenotype
  lymphocyte, 56/67
  McLeod, *see* McLeod phenotype
  3-methylglutaconic aciduria, 66/652
  mitochondrial DNA depletion, 66/430
  Niemann-Pick disease type A, 66/135
  Niemann-Pick disease type B, 66/135
Phenotype-karyotype correlation
  holoprosencephaly, 50/240
Phensuximide
  serum level, 15/691
  side effect, 15/715
Phentaramine addiction
  carotid artery aneurysm, 55/521
Phentermine
  acquired myoglobinuria, 62/573
  amphetamine intoxication, 65/256
  drug addiction, 65/256
Phentolamine
  classification, 74/149
  hypertensive encephalopathy, 54/216
  primary pigmentary retinal degeneration, 13/197
  tetanus, 65/225
Phenurone, *see* Phenacemide
Phenylacetic acid
  headache, 48/174
  migraine, 48/174
Phenylalanine
  acquired hepatocerebral degeneration, 49/216
  dihydropteridine reductase deficiency, 29/38-42
  epilepsy, 72/222
  GM_2 gangliosidosis, 10/295

Phenylalanine hydroxylase
  phenylketonuria, 28/128, 29/31, 42/610
Phenylalanine transaminase deficiency
  high protein diet, 29/38-42
Phenylalaninemia
  differential diagnosis, 29/38-42
Phenylbutazone
  chronic paroxysmal hemicrania, 48/264
  headache, 48/174
  migraine, 48/174, 177
1-(1-Phenylcyclohexyl)-pyrrolidine, see
  Rolicyclidine
O-Phenyl-N-2-dimethylaminoethyl-4-phenyl-
  phosphonamidate methiodide intoxication
  chemical classification, 37/550
  organophosphorus compound, 37/550
Phenyldimethyltriazene
  brain tumor carcinogenesis, 17/20
Phenylenediamine
  acquired myoglobinuria, 62/573
  toxic myopathy, 62/601
Phenylephrine
  classification, 74/149
  dopamine β-mono-oxygenase deficiency, 59/163
Phenylethylamine, see Phenethylamine
Phenylketonuria, 29/29-46
    see also Hyperphenylalaninemia and
      Hyperphenylalaninemic embryopathy syndrome
  aminoaciduria, 10/138
  athetosis, 6/456
  autism, 29/30, 43/204
  autonomic nervous system, 75/234
  biochemistry, 28/128, 29/31-33, 42/610
  birth incidence, 42/611
  craniolacunia, 50/139
  dermatitis, 42/610
  diagnosis, 29/30, 38
  diet, 29/43, 42/611, 59/75
  differential diagnosis, 29/38-42
  dihydropteridine reductase deficiency, 29/38-42,
    42/611
  EEG, 29/30
  enamel hypoplasia, 42/611
  enzyme defect, 28/128, 29/31
  epilepsy, 72/222
  essential aminoacid metabolism, 29/33
  extensor plantar response, 29/30
  frequency, 29/29
  gait disturbance, 42/611, 59/75
  genetic linkage, 42/611
  genetics, 29/29
  history, 29/29
  hyperactivity, 42/611

  hyperphenylalaninemia, 29/44, 59/75
  hyperreflexia, 42/610
  hypsarrhythmia, 29/30
  infantile spinal muscular atrophy, 59/75
  irritability, 29/30
  locus ceruleus, 29/34
  maternal, see Maternal phenylketonuria
  mental deficiency, 29/30, 34-37, 42/610, 46/17,
    50, 59/75
  microcephaly, 29/30, 30/509, 42/610
  muscle stretch reflex, 42/610
  myelin, 10/139, 29/34, 42/610
  myelination, 29/33, 42/610
  myelinoclastic diffuse sclerosis, 9/481
  neuroglia relationship, 28/410
  newborn, 29/38-42
  odor, 29/30, 42/610
  osteoporosis, 29/43, 42/611
  oxitriptan, 29/25
  pathogenesis, 29/34-37
  pathologic anatomy, 29/33
  phenylalanine hydroxylase, 28/128, 29/31, 42/610
  pigmentary degeneration, 59/75
  polymicrogyria, 29/33
  pteridine coenzyme, 42/610
  pyridoxal, 28/128
  screening, 29/42
  seizure, 29/30, 42/610
  serotonin, 28/128, 29/35
  spinal muscular atrophy, 59/374
  substantia nigra, 29/34
  tetrahydrobiopterin metabolism, 75/234
  treatment, 29/42
  tremor, 42/611, 59/75
  tryptophan, 28/128, 29/31-33
  tyrosine, 28/129, 29/31-33, 42/610
  ulegyria, 29/33
  vitamin B6, 28/128
  vomiting, 29/30
  white matter change, 10/138
Phenylmercury derivative intoxication
  autonomic polyneuropathy, 51/478
Phenylmethane, see Toluene
Phenylphosphonothioic acid
    O-ethyl-O-(4-nitrophenyl)ester
  organophosphate intoxication, 64/155
Phenylpropanolamine
  headache, 55/520
  neurotoxin, 62/601, 65/259
  toxic myopathy, 62/601
Phenylpropanolamine addiction
  anorexigenic agent, 65/259
  subarachnoid hemorrhage, 55/520

Phenylpropanolamine intoxication
  agitation, 65/259
  alpha adrenergic receptor stimulating agent,
    65/259
  anxiety, 65/259
  bradycardia, 65/259
  brain hemorrhage, 65/259
  confusion, 65/259
  hypertension, 65/259
  hypertensive crisis, 65/259
  hypertensive encephalopathy, 65/259
  treatment, 65/260
2-Phenylpropionic acid, see Hydratropic acid
Phenylpyruvic oligophrenia, see Phenylketonuria
Phenylthiocarbamide
  multiple sclerosis, 9/99
Phenylthiocarbamide tasting
  prevalence, 42/423
  thyroid gland dysgenesis, 42/632
Phenytoin
  absorption, 15/627
  acetylcholinesterase, 40/558
  antiepileptic agent, 65/497
  antiphospholipid antibody, 63/330
  behavior disorder, 46/600
  brain infarction, 53/419
  cognitive impairment, 72/393
  demyelination, 47/38
  disulfiram, 40/561
  drug induced polyneuropathy, 51/307
  drug induced systemic lupus erythematosus,
    16/222
  epilepsy, 15/627, 670, 73/358
  head injury, 57/231
  hereditary paroxysmal kinesigenic
    choreoathetosis, 42/208
  hereditary periodic ataxia, 60/442
  metabolism, 15/628
  microcephaly, 30/511, 50/279
  motoneuron disease, 59/468
  motor unit hyperactivity, 40/286, 546, 41/296
  muscle fiber conductance, 40/560
  muscle hypertrophy, 40/558
  myopathy, 40/523, 560
  myotonia, 62/275
  myotonic dystrophy, 40/522, 62/241
  neonatal seizure, 73/257
  nerve conduction, 7/174
  neurotoxin, 7/519, 15/715, 37/200, 203, 40/286,
    42/641, 46/394, 600, 47/38, 49/365, 385, 51/68,
    507, 52/211, 56/234, 60/587, 656, 64/172,
    65/500
  organophosphate intoxication, 64/172
  pharmacologic action, 40/560
  postherpetic neuralgia, 56/234
  serum level, 15/684, 686
  side effect, 15/715-717
  sleep regulation, 15/486
  status epilepticus, 73/329
  steroid myopathy, 40/558, 41/252
  systemic lupus erythematosus, 55/370
  toxic encephalopathy, 65/500
  toxic polyneuropathy, 51/68
  tuberculous meningitis, 52/211
  uremic encephalopathy, 63/505
  vitamin Bc deficiency, 51/336
Phenytoin induced chorea
  predisposition, 49/365
Phenytoin intoxication
  apathy, 65/500
  areflexia, 42/641
  ataxia, 42/641, 65/500
  attention, 65/502
  cerebellar atrophy, 65/502
  choreoathetosis, 49/385
  cognition, 65/501
  coma, 65/501
  death, 65/501
  delirium, 65/500
  dementia, 65/500
  diplopia, 42/641, 65/500
  drug dosage, 40/561
  dysarthria, 65/500
  epilepsy, 65/500
  external ophthalmoplegia, 65/500
  extrapyramidal system, 37/203
  gingiva hyperplasia, 42/641
  hallucination, 65/500
  headache, 65/500
  hemiballismus, 49/374
  higher cortical function, 37/203
  5-(4-hydroxyphenyl)-5-phenylhydantoin, 37/200
  hyperreflexia, 42/641
  intelligence, 65/502
  isoniazid, 52/211
  isoniazid neuropathy, 42/641
  late cerebellar atrophy, 60/587
  memory, 65/501
  memory disorder, 65/500
  metabolite, 37/200
  myasthenic syndrome, 65/503
  myoclonus, 65/500
  myopathy, 65/503
  nystagmus, 42/641, 65/500
  ophthalmoplegia, 65/500
  plasma level, 65/500

plasma level symptom, 65/501
polyneuropathy, 65/503
seizure exacerbation, 65/502
supranuclear ophthalmoplegia, 60/656
tremor, 65/500
vertigo, 65/500
vestibular system, 37/203
vitamin B_c, 65/503
vitamin B_c deficiency, 42/641
Phenytoin polyneuropathy
  demyelination, 47/38
  toxic neuropathy, 7/519
  vitamin B_c deficiency, 51/308
Pheochromocytoma, 42/764-766
  abdominal pain, 39/498
  adrenal gland, 39/496-504, 42/765
  anorexia, 42/764
  anxiety, 39/498, 500, 42/764, 46/431, 48/427
  arteriography, 39/500, 42/765
  blood pressure, 75/476
  bradycardia, 39/498
  brain blood flow, 42/764
  brain embolism, 63/136
  brain ischemia, 39/501
  cardiomyopathy, 39/498, 42/764, 63/136
  catecholamine, 9/496, 14/252, 42/764
  catecholamine test, 8/481
  cerebellar hemangioblastoma, 18/285
  computer assisted tomography, 42/765
  congenital megacolon, 42/3
  constipation, 39/498
  contraindication, 37/436
  convulsion, 39/501, 42/764
  cortical blindness, 39/501
  diagnosis, 8/481, 42/765
  diaphoresis, 42/764
  differential diagnosis, 8/479, 39/502
  dizziness, 39/498
  dyspnea, 39/498
  embryology, 8/461
  exertional headache, 48/375
  fatigue, 39/501, 42/764
  flushing, 14/252, 39/498
  genetics, 14/492
  glomus jugulare tumor, 18/440
  guanethidine, 37/436
  headache, 5/230, 39/498, 500, 42/764, 48/6, 427
  heat intolerance, 39/498
  hereditary, 8/475, 20/358, 42/765
  hyperglycemia, 39/498
  hypertension, 11/458, 14/252, 702, 750, 39/498, 42/764, 48/427, 75/476
  hypertensive attack, 8/479
  hypertensive encephalopathy, 39/501, 54/212
  hyperventilation, 39/501, 42/764
  intestinal pseudo-obstruction, 51/493
  laboratory examination, 39/499
  multiple endocrine adenomatosis type III, 39/504, 42/752, 765
  mutation rate, 42/765
  nausea, 39/498, 500, 42/764, 48/427
  nerve tumor, 8/473-482
  neurofibromatosis, 8/475, 14/134, 252, 492, 702, 733, 750, 39/501, 503, 42/765, 50/366
  orthostatic hypotension, 14/252
  orthostatic intolerance, 74/326
  pallor, 39/498, 48/427
  palpitation, 39/498, 500, 48/427
  paresthesia, 39/498, 501
  paroxysmal hypertension, 39/498, 42/764
  phakomatosis, 14/702, 750, 39/504
  phlebography, 39/500
  pregnancy, 74/333
  prevalence, 8/475, 42/765
  psychiatric disorder, 42/764
  psychosis, 39/500, 42/764
  restlessness, 39/500
  scintillating scotoma, 39/501
  screening test, 8/481, 14/750, 42/765
  Sipple syndrome, 14/735, 738, 750, 39/503
  site, 39/496
  spinal ependymoma, 20/358
  sweating, 5/230, 14/252, 39/498, 500, 48/427
  symptomatology, 39/496
  synonym, 8/473
  systemic brain infarction, 11/458
  tachycardia, 14/252, 39/498, 48/427
  tissue culture, 17/69
  treatment, 39/500
  tremor, 39/498, 501, 42/764, 48/427
  tuberous sclerosis, 14/51
  vanilmandelic acid, 8/478, 481, 39/499, 42/765
  visual disorder, 39/498, 500
  vomiting, 39/500, 48/427
  Von Hippel-Lindau disease, 14/60, 249, 252, 747, 750, 39/503, 42/765, 43/21, 50/375
  weakness, 39/498, 501, 48/427
  weight loss, 39/498, 42/764
  Zuckerkandl organ, 8/473, 42/765
Pheochromocytoma myopathy
  creatine kinase, 62/537
  proximal muscle weakness, 62/537
Pheochromocytoma thyroid carcinoma multiple neurinoma syndrome
  neurofibromatosis type I, 14/134
Pheromone

autonomic nervous system, 74/373, 379
  behavior, 74/377
  classification, 74/374
  human, 74/373
  vomeronasal organ, 74/376
Philadelphia chromosome
  chromosome 22q deletion, 31/425
  myeloid leukemia, 31/435
Philadelphia, USA
  neurology, 1/5
Philippson reflex
  spinal cord lesion, 2/184
Philtrum absence
  fetal alcohol syndrome, 50/279
Phlebography
  ascending cerebral vein, 11/105
  orbital, *see* Orbital phlebography
  pheochromocytoma, 39/500
  spinal, *see* Spinal phlebography
  spinal angiography, 12/632
  spinal angioma, 12/632
  Tolosa-Hunt syndrome, 48/299
  vertebral, *see* Vertebral phlebography
Phlebostasis
  brain, 9/579
  multiple cystic encephalopathy, 9/579
Phlebothrombosis
  acute infantile hemiplegia, 53/32
  brain, *see* Brain phlebothrombosis
  CSF examination, 11/239
  spinal, *see* Spinal phlebothrombosis
Phlebotomus fever virus
  acute viral myositis, 56/195
Phlebovirus
  human pathogen, 56/15
Phlorizin
  retinal lesion, 13/208
Phobic anxiety
  free floating anxiety, 45/266
Phocca groenlandica intoxication
  vitamin A, 37/96
Phocomelia
  De Lange syndrome, 50/276
  Roberts syndrome, 30/247, 46/80
  thalidomide syndrome, 42/660
Phoenix, Arizona, USA
  neurology, 1/35
Phonemic jargon
  aphasia, 4/92
  Wernicke-Korsakoff syndrome, 45/315
Phonetic jargon
  Wernicke aphasia, 45/315
Phonology

aphasia, 45/307
  motor aphasia, 45/308
  Wernicke aphasia, 45/308
Phonophobia
  headache, 75/286
  migraine, 42/746, 48/4, 162
Phorate sulfone intoxication
  chemical classification, 37/547
  organophosphorus compound, 37/547
Phosphatase
  neurogenic osteoarthropathy, 26/385, 405, 517
Phosphate
  autonomic nervous system, 75/507
  brain scanning, 16/663
  CNS, 28/437
  CSF, 28/437
  plasma concentration, 63/564
  vitamin B_2 intoxication, 28/214
Phosphatidalcholine, *see* Plasmalogen
Phosphatidalethanolamine, *see* Plasmalogen
Phosphatidalserine, *see* Plasmalogen
Phosphatidic acid
  biochemistry, 9/2, 7
  chemistry, 10/237
Phosphatidopeptides
  biochemistry, 9/13
Phosphatidylcholine
  Alzheimer disease, 46/267
  biochemistry, 9/2, 8
  chemical structure, 3/311
  chemistry, 10/240, 266
  CSF, *see* CSF phosphatidylcholine
  neutral lipid, 9/30
Phosphatidylcholine sterol acyltransferase
  Bassen-Kornzweig syndrome, 63/279
  lipoprotein metabolism, 66/551
Phosphatidylcholine sterol acyltransferase gene
  lipoprotein metabolism, 66/551
Phosphatidylethanolamine
  see also Plasmalogen
  acidic lipid, 9/30
  biochemistry, 9/2, 8
  chemistry, 10/240
  CSF, *see* CSF phosphatidylethanolamine
  glia, 9/6
  Niemann-Pick disease, 10/488
  Pelizaeus-Merzbacher disease, 10/195
Phosphatidylethanolamine lipidosis
  inosamine phosphatide, 42/429
  mental deficiency, 10/552, 42/429
  mental deterioration, 10/504, 552
  Niemann-Pick disease, 10/504
  seizure, 42/429

splenomegaly, 10/504, 552
Phosphatidylglycerol
  chemistry, 10/237
Phosphatidylinositide
  acidic lipid, 9/30
Phosphatidylinositol
  biochemistry, 9/2, 8
  chemistry, 10/238
  glia, 9/6
Phosphatidylinositol kinase
  biochemistry, 9/16
Phosphatidylserine
  acidic lipid, 9/30
  biochemistry, 9/2, 8
  glia, 9/6
  multiple sclerosis, 3/313
Phosphene
  see also Elementary hallucination
  auditory hallucination, 45/368
  epilepsy, 45/367
  hallucination classification, 45/56
  induction, 46/567
  migraine, 45/365, 367
  nerve conduction, 47/32
  occipital lobe, 45/364
  sensory deprivation, 45/367
Phosphine 3 R
  myelin, 9/26
N-Phosphoacetyl aspartic acid, *see* Sparfosic acid
3'-Phosphoadenosine-5'-phosphosulfate
  biochemistry, 10/312
Phosphoamidase
  senile plaque, 46/255
Phosphodiesterase
  biochemistry, 10/245
  myopathy, 40/150
  nerve myelin, 51/25
  neuroleptic agent, 65/279
Phosphoethanolamine
  ataxia telangiectasia, 60/383
Phosphofructokinase
  biochemistry, 27/235
  brain glycolysis, 27/9
  Huntington chorea, 49/261
  myoglobinuria, 62/558
  phosphofructokinase deficiency, 27/222, 235
6-Phosphofructokinase
  Down syndrome, 50/531
Phosphofructokinase deficiency
  anemia, 43/183
  biochemical aspect, 27/235
  biochemistry, 40/190, 41/191, 427, 43/183, 62/488
  childhood myoglobinuria, 40/264, 62/564

chromosome 1, 62/488
chromosome 10, 62/488
chromosome 21, 62/488
clinical features, 27/235, 40/327, 339, 342, 41/190, 62/487
contracture, 40/327, 547, 41/297, 43/183
creatine kinase, 62/487
differential diagnosis, 40/547, 41/204
EMG, 40/327, 41/191, 62/487
enzyme defect, 41/191, 43/183
exercise intolerance, 27/235, 41/190, 43/183, 62/487
genetics, 41/191, 427, 62/488
glycogen storage disease, 27/235, 40/53, 190, 264, 327, 339, 342, 41/191, 62/487
hematology, 41/191
hemolysis, 41/191
hemolytic anemia, 41/191, 43/183, 62/487
histochemistry, 40/53, 41/427, 43/183
histopathology, 41/191, 62/488
inheritance, 62/488
ischemic exercise, 27/235, 40/327, 339, 547, 41/190, 62/487
laboratory examination, 27/235
metabolic myopathy, 41/190, 62/487
molecular genetics, 62/488
motor nerve conduction velocity, 41/191
muscle biopsy, 41/191, 62/488
muscle cramp, 62/487
muscle fiber feature, 62/17
muscle tissue culture, 27/236, 40/190
myoglobinuria, 27/235, 40/327, 339, 41/190, 265, 277, 43/183, 62/487
pathophysiology, 62/488
permanent weakness, 62/487
phosphofructokinase, 27/222, 235
phosphorus 31 nuclear magnetic resonance, 62/487
rhabdomyolysis, 40/264, 342
second wind phenomenon, 41/190, 43/183, 62/487
serum creatine kinase, 41/190, 62/487
subsarcolemmal myofibril, 40/53, 41/191, 427, 43/183, 62/488
treatment, 41/192
Phosphoglucoisomerase
  tibial nerve, 7/218
Phosphoglucomutase
  glycogen storage disease, 27/222
  white matter, 9/15
Phosphoglucomutase deficiency
  calf hypertrophy, 27/222, 236, 41/192
  clinical features, 27/222, 236, 41/192
  laboratory examination, 27/236, 41/192

myopathy, 27/236, 41/192
toe walker, 27/222, 41/192
weakness, 27/222, 41/192
Phosphoglycerate kinase
  brain glycolysis, 27/9
  galactosidase gene, 66/119
  lysosomal storage, 66/119
  myoglobinuria, 62/558
Phosphoglycerate kinase deficiency
  childhood myoglobinuria, 62/565
  epilepsy, 62/489
  exercise intolerance, 62/489
  glycogen storage disease, 62/489
  hemolytic anemia, 62/489
  inheritance, 62/489
  mental deficiency, 62/489
  metabolic encephalopathy, 62/489
  metabolic myopathy, 62/489
  muscle biopsy, 62/489
  muscle cramp, 62/489
  myoglobinuria, 62/489
  phosphorus 31 nuclear magnetic resonance,
    62/489
  X-linked recessive disorder, 62/489
Phosphoglycerate mutase
  brain glycolysis, 27/9
  muscle tissue culture, 62/94
  myoglobinuria, 62/558
Phosphoglycerate mutase deficiency
  childhood myoglobinuria, 62/565
  exercise intolerance, 62/489
  glycogen storage disease, 62/489
  inheritance, 62/489
  ischemic exercise, 62/489
  metabolic myopathy, 62/489
  muscle biopsy, 62/489
  muscle cramp, 62/489
  phosphorus 31 nuclear magnetic resonance,
    62/490
Phosphoglyceride
  ferric hydroxamate staining, 9/31
  hydroxylamine staining, 9/26
Phosphohexose isomerase
  CSF, see CSF phosphohexose isomerase
  glycogen storage disease, 27/222
Phosphohexose isomerase deficiency
  biochemical aspect, 27/236
  clinical features, 27/222, 236
  familial exercise intolerance, 27/236
  laboratory examination, 27/236
  myalgia, 41/192
  myopathy, 27/236
Phospholine intoxication, see Ecothiopate iodide

  intoxication
Phospholipase
  chemistry, 10/243
  experimental allergic encephalomyelitis, 9/38
  granulomatous amebic encephalitis, 52/325
  local anesthetic agent, 65/427
  neurotoxin, 65/185
  snake venom, 65/185
Phospholipase A
  chemistry, 10/244
  toxic myopathy, 62/610
Phospholipase A1
  amiodarone polyneuropathy, 51/303
Phospholipase A2
  amiodarone polyneuropathy, 51/303
  arthropod envenomation, 65/198
  local anesthetic agent, 65/427
  snake venom, 65/180
Phospholipase B
  chemistry, 10/244
Phospholipase C
  chemistry, 10/244
  experimental myopathy, 40/151
Phopspholipase C-β
  classification, 74/149
Phospholipase D
  chemistry, 10/244, 248
Phospholipid
  brown metachromasia, 10/25
  CSF, see CSF phospholipid
  degradation, 66/42
  globoid cell leukodystrophy, 10/85
  hereditary multiple recurrent mononeuropathy,
    42/316
  luxol fast blue staining, 9/25
  multiple sclerosis, 9/312
  myelin, 9/30
  nervous system tumor, 27/504
  nomenclature, 66/36
  serum, 9/321, 10/266
  total, 10/266
Phospholipidosis, see Mucolipidosis type IV
Phosphoribosylamine glycine ligase
  Down syndrome, 50/531
Phosphoribosyl pyrophosphate synthetase
  autism, 42/573
  gene localization, 42/573
  hyperuricemia, 42/572
  lacrimal gland, 42/573
  mental deficiency, 42/572
Phosphorolysis
  glycogen phosphorylase, 70/424
Phosphorotrithioic acid tributyl ester

organophosphate intoxication, 64/155
Phosphorus
  intracerebral calcification, 42/534
  leukoencephalopathy, 9/652
  myelopathy, 70/120
  spinal cord injury, 26/382, 387, 389, 401
  striatopallidodentate calcification, 6/717
  symptom, 70/115
Phosphorus intoxication
  pallidal necrosis, 6/652, 49/466
  tremor, 6/823
Phosphorylase
  antibody, 41/189
  liver, *see* Liver phosphorylase
  muscle tissue culture, 62/94, 109
  myophosphorylase deficiency, 41/185-190
Phosphorylase activation
  glycogen metabolism, 27/223
Phosphorylase isoenzyme
  chromosome 10, 62/486
  chromosome 11q13, 62/485
  chromosome 14, 62/486
  chromosome 20, 62/486
  metabolic myopathy, 41/188
  myophosphorylase deficiency, 62/486
Phosphorylase kinase
  glycogen metabolism, 27/222
  glycogen storage disease, 27/222, 234
  myoglobinuria, 62/558
Phosphorylase kinase deficiency
  autosomal recessive, 41/192
  biochemistry, 62/487
  cardiomyopathy, 62/487
  childhood myoglobinuria, 62/565
  chromosome 7, 62/487
  chromosome 11, 62/487
  chromosome 16q12-q13, 62/487
  chromosome Xq12-q13, 62/487
  clinical features, 27/222, 62/486
  creatine kinase, 62/487
  exercise intolerance, 62/487
  fasting hypoglycemia, 62/486
  glycogen storage disease, 62/486
  growth retardation, 41/192, 62/486
  hepatomegaly, 27/222, 235, 41/192
  hepatomyopathy, 62/487
  histopathology, 62/487
  ischemic exercise, 62/487
  liver disease, 62/486
  metabolic myopathy, 41/192, 62/486
  molecular genetics, 62/487
  muscle biopsy, 62/487
  myopathy, 62/487

  pigmenturia, 62/487
  sex linked, 41/192
Phosphorylation
  erythrocyte band 3 protein, 40/523
  marine toxin intoxication, 65/150
  mitochondrial myopathy, 41/214
  muscular dystrophy, 41/413, 489
  myotonic syndrome, 40/545
  oxidative, *see* Oxidative phosphorylation
  periodic paralysis, 41/170
Phosphorylcholine
  Duchenne muscular dystrophy, 40/380
Phosphorylcholine glycerine transferase, *see*
    Cholinephosphotransferase
Phosphorylethanolamine
  hypophosphatasia, 42/578
Phosphotransferase
  signal transduction, 67/24
Phosphotungstic acid
  metachromatic leukodystrophy, 10/47
Phosphotungstic acid hematoxylin
  hypertensive fibrinoid arteritis, 11/136
Photism
  sound induced, *see* Sound induced photism
Photogenic epilepsy, 2/618-622
  ataxia, 42/714
  centrencephalic EEG, 42/671
  clinical features, 2/621
  dementia, 42/714
  demyelination, 42/714
  diabetes mellitus, 42/714
  genetic factor, 42/709
  helicopter, 2/621
  management, 2/621
  mental deficiency, 42/715
  myoclonic epilepsy, 22/510
  natural stimulus, 2/620
  occipital delta rhythm, 42/672
  occurrence, 2/619
  pathophysiology, 2/618
  renal disease, 42/714
  self-induced seizure, 2/621
  sensorineural deafness, 42/714
  spastic diplegia, 42/715
  speech disorder, 42/714
  television, 2/620
  vertical stripe, 2/620
Photomyoclonus syndrome
  myoclonic epilepsy, 60/663
Photophobia
  basilar artery migraine, 48/138
  cerebral malaria, 35/146
  Chédiak-Higashi syndrome, 42/536

cinchonism, 65/452
cluster headache, 48/222
Cockayne-Neill-Dingwall disease, 43/351
congenital retinal blindness, 13/272, 22/530
distichiasis lymphedema syndrome, 43/377
far eastern tick-borne encephalitis, 56/139
headache, 75/286
hereditary progressive cone dystrophy, 13/134
Holmes-Adie syndrome, 74/411
hyperosmolar hyperglycemic nonketotic diabetic
  coma, 27/90
migraine, 42/746, 48/4, 7, 29, 70, 156, 162
oculocutaneous albinism, 42/404, 43/3-7
organotin intoxication, 64/138
pellagra, 28/96, 46/336
photosensitivity, 1/542, 22/530
quinidine intoxication, 37/448
Stargardt disease, 13/262
xeroderma pigmentosum, 43/12
Photopsia
  carotid artery stenosis, 55/111
  migraine, 48/128
Photosensitive dermatitis
  Cockayne-Neill-Dingwall disease, 51/399
  xeroderma pigmentosum, 51/399
Photosensitive epilepsy, 73/183
  EEG, 73/184
  incidence, 73/184
  pathogenic mechanism, 73/185
  treatment, 73/186
  visual stimulus, 73/185
Photosensitive porphyria
  congenital, see Congenital erythropoietic
    porphyria
Photosensitivity
  Bloom syndrome, 43/10
  Chédiak-Higashi syndrome, 14/783
  Cockayne-Neill-Dingwall disease, 13/432, 14/315
  congenital erythropoietic porphyria, 27/433
  EEG, 72/304
  Hartnup disease, 21/579, 42/146, 60/667
  hereditary acute porphyria, 51/391
  hereditary olivopontocerebellar atrophy, 60/667
  hereditary periodic ataxia, 21/564
  nosology, 13/436-438
  photophobia, 1/542, 22/530
  porphyria variegata, 42/620
  Rothmund-Thomson syndrome, 14/315, 777,
    39/405, 43/461
  tryptophanuria, 42/633
  xeroderma pigmentosum, 14/788, 60/666
Phrenic nerve
  lymphoma, 63/350

neuropathy, 8/388
Phrenic nerve lesion
  cervical plexus, 2/135
  heart transplantation, 63/181
  iatrogenic neurological disease, 63/181
Phrenic nerve palsy
  lymphoma, 63/350
Phrenic nerve surgery
  spinal cord injury, 61/270
Phrenitis
  delirium, 46/530
Phrenosin, see Cerebron
PHS syndrome, see Pseudoprogeria
  Hallermann-Streiff syndrome
Phthalate ester intoxication
  animal study, 36/444
Phthalazinol
  malignancy, 41/338
  motoneuron disease, 41/338
  muscle fiber fusion, 40/193
Phycomycosis, see Mucormycosis
Physalia intoxication
  autopsy, 37/54
  experimental data, 37/54
  rodenticide, 37/54
Physaliphorous cell
  chordoma, 18/160, 20/40
Physeter catodon intoxication
  ingestion, 37/95
  vitamin A intoxication, 37/95
Physical rehabilitation
  see also Physiotherapy
  active movement, 61/464
  acute transverse cord lesion, 61/458
  aim, 61/459
  ambulation, 61/471
  ankle foot orthosis, 61/472
  assisted coughing, 61/460
  communication, 61/470
  complete lesion, 61/474
  complication, 61/465
  contracture, 61/465
  deep vein thrombosis, 61/465
  duration, 61/476
  follow-up, 61/476
  functional electrical stimulation, 61/472
  functional potential, 61/473
  hip guidance orthosis, 61/472
  incomplete lesion, 61/473
  intermittent positive pressure breathing, 61/459
  knee ankle foot orthosis, 61/472
  leisure activity, 61/472
  lifting, 61/469

muscle strength, 61/468
muscle tone, 61/465
neurologic deficit assay, 61/458
orthosis, 61/463, 471
osteoporosis, 61/471
para-articular ossification, 61/465
passive movement, 61/460, 464
patient positioning, 61/463
postural drainage, 61/460
posture restraint, 61/466
pressure sore, 61/465
prosthesis, 61/489
reciprocal gait orthosis, 61/472
reflex sympathetic dystrophy, 61/465
resisted movement, 61/465
respiration, 61/459
respiratory exercise, 61/459
scope, 61/459
skill, 61/469
spinal cord injury, 61/457
splint, 61/462
sport, 61/472
tenodesis grip, 61/461
therapist role, 61/457
trunk balance, 61/468
vehicle driving, 61/470
verbal skill, 61/476
wheelchair ergonomics, 61/467
wheelchair phase, 61/465
wrist extension, 61/462
Physiognomy recognition, *see* Prosopagnosia
Physiologic jaundice
    newborn, *see* Newborn physiological jaundice
Physiologic striatopallidodentate calcification
    falx cerebri, 49/417
    tentorium, 49/417
Physiologic tremor
    accelerometry, 49/566, 584
    anxiety tremor, 6/818
    benign essential tremor, 49/565
    beta2 adrenergic receptor stimulation, 49/566
    EMG, 49/584
    enhancing factor, 49/576
    frequency analysis, 6/812
    list of causes, 49/586
    Parkinson disease, 49/100
    postural tremor, 6/818, 49/586
    root mean square amplitude, 49/584
    segmental stretch disease, 49/566
    type, 49/583
Physiotherapy, 41/457-497
    see also Physical rehabilitation
    brachial plexus, 51/150

brain infarction, 53/432
clumsiness, 46/165
dermatomyositis, 62/386
Duchenne muscular dystrophy, 62/132
inclusion body myositis, 62/386
multiple sclerosis, 47/174
neuropathy, 8/397
nonprogressive cerebellar ataxia, 42/125
pain, 26/496
polymyositis, 62/386
rehabilitation, 55/241
respiratory dysfunction, 26/344
shoulder hand syndrome, 8/330
spinal cord injury, 26/232, 344, 61/457
spinal pain, 26/496
Physostigmine
    alcohol withdrawal delirium, 46/541
    Alzheimer disease, 46/267
    cholinergic receptor blocking agent intoxixation,
        37/541, 46/538, 596
    cholinesterase inhibition, 37/541, 64/183
    organophosphate intoxication, 64/230
Phytanic acid
    alpha oxidation, 27/524-526
    beta oxidation, 27/525
    biochemistry, 66/519
    metabolism, 27/521, 524-527
    neonatal adrenoleukodystrophy, 60/674
    neurologic syndrome, 21/182
    omega oxidation, 27/525
    origin, 27/521
    Refsum disease, 8/23, 10/345, 13/314, 21/9, 186,
        192, 196, 202, 204, 210, 221, 27/520, 522-524,
        527-531, 28/410, 41/434, 42/148, 60/228, 232,
        674, 66/110, 485, 492-494, 498
Phytanic acid α-hydroxylase
    Refsum disease, 10/345, 41/434, 51/384, 60/728,
        66/495
Phytanic acid intoxication
    Refsum disease, 21/186, 36/347-349, 51/385,
        60/233, 237, 66/498
Phytanic acid α-oxidation
    peroxisome, 66/507
Phytanic acid storage disease, *see* Refsum disease
Phytohemagglutinin
    ataxia telangiectasia, 14/271
    experimental allergic encephalomyelitis, 9/531
    multiple sclerosis, 47/346, 354, 374
Phytol
    Refsum disease, 8/24, 21/187, 27/521, 528,
        41/434, 60/237, 728, 66/498
Pial blood flow
    brain vasospasm, 11/511-520

hypothermia, 11/515
Pial vessel
  adrenalin, 11/515
  topical agent, 11/515
Pian fungoide, *see* Mycosis fungoides
Pica syndrome
  iron deficiency anemia, 63/252
Pick body
  ultrastructure, 42/286
Pick disease, 46/233-244
  see also Presenile dementia
  acetylcholine receptor, 46/242
  achromatic cortical neuron, 21/63
  agraphia, 42/285
  Alzheimer disease, 42/285, 46/233
  amnesia, 42/285, 46/234, 240
  amyotrophic lateral sclerosis, 22/313, 59/410
  anatomoclinical correlation, 46/240
  aphasia, 42/285, 46/203, 234
  apragmatism, 46/233, 235, 240
  apraxia, 42/285, 46/234
  argyrophilic body, 21/54, 57, 27/489
  argyrophilic inclusion, 21/54, 46/233, 236, 243
  astrocytosis, 46/236, 238
  autonomic neuropathy, 46/235
  autosomal dominant inheritance, 46/243
  biochemistry, 27/489, 46/242
  brain atrophy, 27/489, 42/285, 46/236-239, 242
  brain biopsy, 10/685
  bulbar paralysis, 22/142
  bulimia, 46/236, 240
  cachexia, 46/236
  calcium edetic acid, 46/244
  catatonia, 46/424
  clinical features, 46/234-236
  compulsion, 46/234
  computer assisted tomography, 46/242
  confabulation, 46/234
  course, 46/235
  Creutzfeldt-Jakob disease, 6/746
  dementia, 46/200, 233, 59/410
  demyelination, 27/489, 46/236, 239
  depressed mood, 46/240
  detection, 46/200
  dysarthria, 42/285, 46/235, 240
  echolalia, 46/234, 240
  echopraxia, 46/234
  EEG, 46/241
  emotional disorder, 46/235
  epidemiology, 46/233
  experimental model, 46/243
  extrapyramidal syndrome, 6/668, 46/234, 240
  frontal dementia, 45/30

  frontal lobe atrophy, 27/489, 42/285, 46/239, 242, 424
  frontal lobe syndrome, 46/239, 432
  genetics, 46/243
  glutamate aspartate, 46/242
  granulovacuolar degeneration, 49/250
  grasp reflex, 46/234
  Hallervorden-Spatz syndrome, 6/614
  hippocampus, 46/238, 242
  histopathology, 27/489, 46/236-239
  Huntington chorea, 49/291, 293
  hydrocephalic dementia, 46/326
  ideational apraxia, 46/234
  inheritance, 42/285, 46/243
  instinct, 46/235
  iteration, 46/234
  Klüver-Bucy syndrome, 45/257, 260, 46/235
  late cortical cerebellar atrophy, 21/495
  lobar atrophy, 27/489, 46/236, 241
  logorrhea, 46/234, 240
  mastication, 46/235
  memory, 46/234
  micrographia, 46/234
  Morgagni-Stewart-Morel syndrome, 46/242
  moria, 46/240
  muscle rigidity, 46/234
  mutism, 46/233, 240
  neostriatal lesion, 6/64, 46/236
  neurochemistry, 27/489
  neurofibrillary tangle, 27/489, 42/286
  neuroleptic agent, 46/244
  neuronal swelling, 6/614, 42/286, 46/233, 236, 238, 243
  neuropathology, 21/64, 27/489
  neurotransmitter, 46/242
  occupational therapy, 46/244
  palilalia, 46/240
  pallidal atrophy, 6/645, 647, 46/236, 49/293
  pathogenesis, 46/242
  perseveration, 42/285
  perseverative agraphia, 45/467
  phonetic disintegration, 46/235
  physiopathology, 46/242
  pica, 46/235
  presenile astroglial dystrophy, 21/67, 46/236
  presenile dementia, 15/343, 42/285
  prevalence, 46/233
  pseudobulbar dysarthria, 46/235
  psychomotor syndrome, 46/234
  pyramidal syndrome, 46/234, 240
  sex ratio, 42/285, 46/233
  sexual deviation, 46/235
  spatial orientation disorder, 42/285, 46/234

speech disorder, 42/285, 46/234
sphincter, 46/235
sphincter control, 46/235
stereotypy, 46/240
steroid excretion, 46/242
striatal lesion, 6/668, 699
temporal disorientation, 46/234
temporal lobe atrophy, 27/489, 42/285, 45/260, 46/236, 239
thalamus, 2/491, 46/236
total frontal atrophy (Lüers-Spatz), 22/314
treatment, 46/244
zinc, 46/242-244
Pick, L., 10/227
Pick supermiliary aneurysm
brain hemorrhage, 11/582, 587
Pickwickian syndrome, 38/299-302
alveolar hypoventilation, 38/299
chronic respiratory insufficiency, 38/299, 63/413
clinical features, 38/300
cor pulmonale, 38/299, 63/455
criteria, 63/455
cyanosis, 3/96, 38/299
dyspnea, 38/300
edema, 38/300
hepatomegaly, 38/300
hypercapnia, 38/299, 63/455
hypersomnia, 38/299, 63/455
hypertrophy, 38/299
hyposensitivity, 38/301
hypoxemia, 38/300
hypoxia, 38/299
narcolepsy, 3/96
obesity, 3/96, 38/299, 63/455
obstructive sleep apnea, 63/455
periodic respiration, 3/96, 38/299
polycythemia, 3/96, 38/299, 63/455
respiratory disease, 3/96, 63/413
sleep disorder, 3/96, 38/302
somnolence, 38/299
tachycardia, 3/96, 38/300
Picolinic acid
brown glial, see Brown glial pigment
glial, see Glial pigment
Huntington chorea, 49/298
Picornavirus
see also Enterovirus infection
acute viral myositis, 56/193
aphthovirus, 56/309
aphthovirus genus, 56/11
cardiovirus, 56/309
cardiovirus genus, 56/11
cerebellar hypoplasia, 9/551

Coxsackie virus, 34/5, 133-139, 56/307
Coxsackie virus A, 56/195, 309
Coxsackie virus A1-A22, 56/11
Coxsackie virus A24, 56/11
Coxsackie virus B, 56/195, 309
Coxsackie virus B1-B6, 56/11
definition, 34/4
dermatomyositis, 41/71, 62/381
ECHO virus, 34/5, 133, 139, 56/195, 307, 309
ECHO virus 1-9, 56/11
ECHO virus 11-27, 56/11
ECHO virus 29-34, 56/11
encephalomyocarditis virus, 34/5, 140, 56/195, 62/381
enterovirus, 56/307, 309
enterovirus genus, 56/11
etymology, 56/309
experimental acute viral myositis, 56/195
features, 56/11
foot and mouth disease virus, 34/5, 140, 56/36, 309
genome, 56/2, 11, 310, 313
group, 56/307
hepatitis A virus, 56/11
human, 56/3
inclusion body myositis, 62/381
poliovirus, 34/133, 56/195, 307
poliovirus type 1, 34/5, 56/11, 309
poliovirus type 2, 34/5, 56/11, 309
poliovirus type 3, 34/5, 56/11, 309
polymyositis, 41/71, 62/381
property, 56/310
rhinovirus, 34/133, 56/309
rhinovirus genus, 56/11
RNA virus, 34/4, 133, 56/3, 11, 309, 62/381
taxonomic place, 34/133, 56/3
Theiler virus, 56/11
ultrastructure, 56/311
Vilyuisk virus, 34/140
viral infection, 56/11, 307
viral myositis, 41/71, 62/381
virus replication, 56/4, 11, 312-314
Picrotoxinin
chlorinated cyclodiene, 64/200
Piebald pigmentation
gene frequency, 43/43
prevalence, 43/43
sensorineural deafness, 43/42-44
Telfer syndrome, 22/513
white hair, 43/42
Pierre-Marie ataxia, see Hereditary cerebellar ataxia (Marie)
Pierre-Robin syndrome

chorioretinal degeneration, 13/41
optic atrophy, 13/41
Piesesthesia
terminology, 1/97
Pigmentary abnormality
hereditary vitreoretinal degeneration, 13/274
X-linked, *see* X-linked pigmentary abnormality
Pigmentary degeneration
lipofuscin, 22/45
phenylketonuria, 59/75
Pigmentary foveal dystrophy
butterfly shaped, 13/136
Pigmentary macular degeneration
familial striatopallidodentate calcification, 6/707,
49/420
onset in adult, 22/536
Pigmentary retinopathy
chloroquine, 13/148
complex I deficiency, 62/504, 66/422
complex III deficiency, 62/504, 66/423
congenital blindness, 22/528, 530
deficiency neuropathy, 51/326
definition, 60/717
diagnostic criteria, 60/718
experimental, *see* Experimental pigmentary
retinopathy
history, 60/717
Kearns-Sayre-Daroff-Shy syndrome, 62/308, 508
migraine, 48/27
primary pigmentary retinal degeneration, 62/308
progressive external ophthalmoplegia, 62/308
sign, 60/718
symptom, 60/718
Pigmentation
adrenoleukodystrophy, 60/667
cutaneous, *see* Cutaneous pigmentation
dyskeratosis congenita, 14/784, 43/392
fibrous dysplasia, 14/180-182
lipodystrophy, 8/346
mucocutaneous, *see* Mucocutaneous pigmentation
neurofibromatosis, 14/182
piebald, *see* Piebald pigmentation
reticulodentatocerebellar, *see*
Reticulodentatocerebellar pigmentation
retinal, *see* Retinal pigmentation
Rothmund-Thomson syndrome, 39/405, 43/460
skin, *see* Cutaneous pigmentation
Pigmented epulis
congenital, *see* Pigmented papillary
medulloblastoma
Pigmented glial and scavenger cells, *see*
Orthochromatic leukodystrophy
Pigmented nevus

diastematomyelia, 50/436
diplomyelia, 50/436
multiple, *see* Multiple pigmented nevi
neurocutaneous melanosis, 14/11, 119, 43/33
Peutz-Jeghers syndrome, 14/12, 121, 43/41
sweat gland agenesis, 14/597
Pigmented papillary medulloblastoma
melanin, 18/180
rare presentation, 18/180
terminology, 18/180
Pigmented paravenous retinochoroidal atrophy
case history, 13/268
chorioretinitis, 13/268
tuberculous spondylitis, 13/268
Pili torti
hereditary neuropathy, 60/263
sensorineural deafness, 42/387
trichopoliodystrophy, 14/784, 29/288, 42/584
Pill rolling movement
Parkinson disease, 6/183
Pillar fracture
cervical compression fracture, 61/30
vertebral column injury, 61/522
vertebral subluxation, 61/522
Pilocarpine
cluster headache, 48/236
epilepsy, 72/55
pure autonomic failure, 2/120, 22/232
tear, 74/427
Pilocytic astrocytoma, *see* Spongioblastoma
Piloerection
autonomic lesion, 1/69
autonomic nervous system, 1/467
gustatory, *see* Gustatory piloerection
spinal cord injury, 26/367, 61/280, 282
thermoregulation, 26/367
transverse spinal cord lesion, 61/280
Piloid astrocytoma, *see* Spongioblastoma
Piloid gliosis
tuberous sclerosis, 14/356
Pilomatricoma
myotonic dystrophy, 43/153, 62/239
Pilomotor neuron
autonomic nervous system, 74/32
Pilomotor response
spinal cord injury, 61/280
Pilonidal fistula, *see* Pilonidal sinus
Pilonidal sinus, 42/46-48
autosomal dominant inheritance, 50/65
birth incidence, 42/47
diastematomyelia, 50/436
diplomyelia, 50/436
epidemiology, 30/146, 150

incidence, 50/53, 65
lipomatosis, 42/47
meningitis, 42/47
sex difference, 50/56
spina bifida occulta, 42/47
spinal lipoma, 20/398
talipes, 42/47
vertebral canal dilation, 42/47
Piminodine
    chemistry, 37/366
Pimozide
    chemical formula, 65/276
    neuroleptic agent, 65/275
Pinacolyl methylphosphonofluoridate intoxication,
    see Soman intoxication
Pinacolylmethylphosphonofluoridate, *see* Soman
Pincers grip mechanism
    cervical spondylotic myelopathy, 19/116, 26/100
Pincers mechanism
    cervical spinal cord, 25/65, 260
Pindolol
    beta adrenergic receptor blocking agent, 65/434
    migraine, 48/180
    myotonia, 62/276
    toxic myopathy, 62/597
Pineal body
    blood-brain barrier, 53/84
    displacement, 1/561
Pineal cyst
    multilocular, *see* Multilocular pineal cyst
    pathology, 68/237
Pineal gland
    age, 74/503
    biological rhythm, 74/491, 499
    blood-brain barrier, 56/80
    brain tumor, *see* Pineal tumor
    light, 74/499, 502
    melatonin, 74/499
    sex, 74/503
    suprachiasmatic nucleus, 74/491
    teratoma, 18/361
    vampire hormone, 74/499
Pineal metastasis
    systemic tumor, 71/615
Pineal tumor
    see also Pinealoma
    akinetic mutism, 18/405
    Argyll Robertson pupil, 17/656
    Bell sign, 17/656
    Benedikt syndrome, 18/362
    brain biopsy, 16/723
    calcification, 17/659
    cerebellar disorder, 17/657

classification, 68/230
dermoid cyst, 17/654
diagnosis, 17/659, 68/239
embryology, 17/648
endocrine disorder, 17/658
epidemiology, 68/231
epidermoid cyst, 17/654
epilepsy, 72/181
eye disturbance, 17/655
eye movement, 17/656
gonadal dysfunction, 17/658
growth direction, 17/654
imaging, 67/177
intracranial hypertension, 17/655
location, 68/233
Nothnagel syndrome, 2/298, 18/362
nystagmus, 17/656
Parinaud syndrome, 17/656
pathology, 17/650
Pellazzi syndrome, 2/444, 17/658
pinealoma, 17/651, 42/766
posterior fossa tumor, 18/405
psychiatric disorder, 17/659
pupil, 17/656
staging, 68/243
supratentorial midline tumor, 18/361
surgery, 68/243
survey, 68/229
symptom, 17/654, 67/157
treatment, 17/661
ventriculography, 17/660
Pinealoblastoma
    chemotherapy, 68/248
    classic primitive neuroectodermal tumor, 68/222
    genetic factor, 42/766
    imaging, 67/179
    incidence, 68/232
    location, 68/232
    management, 68/252
    medulloblastoma resemblance, 42/742
    pathology, 68/234, 236
    radiotherapy, 68/248
    survival, 68/224
    tissue culture, 17/68
Pinealocytoma, *see* Pinealoma
Pinealoma
    see also Pineal tumor
    angiography, 18/365, 42/767
    brain tumor, 16/64, 18/508
    chemotherapy, 68/248
    CSF, 42/767
    CSF cytology, 16/399, 18/365, 42/767
    CSF flow, 18/362

definition, 16/23
distribution, 67/130
EEG, 18/362, 42/767
endocrine disorder, 16/352, 18/362, 42/767
epidemiology, 16/64
familial, 16/64, 42/767
hypertension, 42/767
hypothalamus tumor, 18/366
incidence, 18/361, 68/232
intracerebral calcification, 18/362, 42/767
intracranial hypertension, 42/767
intracranial pressure, 18/362
Japan, 16/65, 42/767
leptomeningeal metastasis, 71/662
location, 68/232
malignancy, 42/766
management, 68/252
metastasis, 71/662
Parinaud syndrome, 16/352, 18/362, 42/766
pathology, 18/361, 365, 68/236
PEG, 42/767
pineal tumor, 17/651, 42/766
posterior fossa tumor, 18/405
precocious puberty, 18/367
prognosis, 18/365
radiotherapy, 18/365, 508, 42/767, 68/248
sex ratio, 18/361, 42/767
sign, 18/362
symptom, 18/362
terminology, 42/766
tissue culture, 17/68, 82
treatment, 18/365, 42/767
type, 16/23, 42/766
ventriculoatrial shunt, 18/365, 42/767
ventriculography, 18/365, 42/767
Pineoblastoma, *see* Pinealoblastoma
Pineocytoma, *see* Pinealoma
Pink disease, *see* Acrodynia
Pinpoint pupil
    diamorphine intoxication, 65/352
    neurosyphilis, 52/280
    pontine infarction, 53/390
Pinsky-DiGeorge-Harley-Bird syndrome
    brain defect, 30/247
    hypertelorism, 30/247
    neurologic disorder, 30/247
Pipecolic acid
    biochemistry, 66/519
    CSF, *see* CSF pipecolic acid
L-Pipecolic acid
    cerebrohepatorenal syndrome, 66/515
Pipecolic acid metabolism
    peroxisome, 66/508

Piperazine
    designer drug, 65/274
    neuroleptic agent, 65/274
Piperazine derivative
    neuroleptic syndrome, 6/256
Piperidine
    designer drug, 65/274
    neuroleptic agent, 65/274
Piperidinedione intoxication
    thalidomide, 37/354
Piperidylchlorophenothiazine
    nyctalopia, 13/210
    secondary pigmentary retinal degeneration,
        13/210
Piracetam
    dyssynergia cerebellaris myoclonica, 60/603
Piriform sinus
    superior laryngeal nerve, 48/495
Piromen A
    multiple sclerosis, 9/388
Piromen B
    multiple sclerosis, 9/388
Piroxicam
    chronic paroxysmal hemicrania, 48/264
    headache, 48/174
    migraine, 48/174
Pisces intoxication, 37/69-87
    dogfish, 37/69
    stingray, 37/69, 71
    stonefish, 37/69, 75-77
    synanceja, 37/69
Pit
    coated, *see* Coated pit
    dental, *see* Dental pit
    optic disc, *see* Optic disc pit
Pitres sign
    terminology, 1/97
Pitressin, *see* Vasopressin
Pittsburgh pneumonia agent, *see* Legionellosis
Pituicyte
    terminology, 2/437
Pituitary abscess
    intraneoplastic abscess, 33/139
    meningism, 52/152
    Mollaret meningitis, 34/548, 56/630
    radiation, 52/153
    visual field defect, 52/152
Pituitary adenoma
    see also Chromophobe adenoma *and* Pituitary
        tumor
    acromegaly, 17/393-395, 422
    age distribution, 17/377
    angiography, 17/302

blindness, 17/385
brain scanning, 17/405
calcification, 17/397
carcinoma, 17/383
carotid cavernous fistula, 17/413
classification, 17/381, 383
clinical features, 17/384
corticotropin, 17/393, 403
cranial nerve, 17/390
craniopharyngioma, 17/397, 406
CSF, 17/405
Cushing syndrome, 2/454, 17/422
cyst, 17/427
development, 17/383
diabetes insipidus, 17/391
diagnosis, 17/396, 405
differential diagnosis, 17/406-408
distribution, 67/130
EEG, 17/405
endocrine test, 17/401, 403
epidemiology, 67/133
epilepsy, 17/390
galactorrhea, 17/423
giant cell tumor, 17/413
gonadotropin, 17/393, 68/358
headache, 17/390
hemianopia, 17/387
hyperpigmentation, 17/423
hyperpituitarism, 17/393, 423
hypoadrenocorticism, 17/393
hypopituitarism, 17/393, 421
imaging, 67/186
intracranial hypertension, 17/408
malignant adenoma, 17/383
massive adenoma, 17/383
metastatic carcinoma, 17/412
microscopy, 17/377
mortality, 17/415-417
nasopharyngeal carcinoma, 17/412
neurofibromatosis type I, 17/412
nonfunctioning, 68/358
optic chiasm, 17/385, 388
optic chiasm compression, 68/75
oxycephaly, 17/412
papilledema, 17/391
pathology, 17/377
PEG, 17/401
prognosis, 17/422
prolactin, 17/404
radiotherapy, 17/417
retrobulbar optic neuritis, 17/412
rhinorrhea, 17/391
sella turcica, 17/389

skull base, 68/466
skull base tumor, 68/482
statistic, 17/376
stereotaxic operation, 17/418
suprasellar pearly tumor, 17/412
surgery, 17/415, 418
teratoma, 17/412
thyrotropic hormone, 17/393, 404
Tolosa-Hunt syndrome, 48/302
treatment, 17/419
trigeminal neurilemoma, 17/412
visual disorder, 17/385-387
Pituitary adrenal axis disorder
    hypothalamic, see Hypothalamic pituitary adrenal
        axis disorder
Pituitary apoplexy
    headache, 5/141, 48/424
    mortality, 17/421
    nitrate intoxication, 65/444
    subarachnoid hemorrhage, 55/20
    symptom, 17/391
Pituitary carcinoma
    rare presentation, 68/358
Pituitary deficiency
    cerebellar ataxia, 42/129
Pituitary disorder
    adiposis dolorosa, 8/347
    headache, 48/432
    migraine, 48/432
Pituitary dwarfism type I
    hypoglycemia, 42/612
    obesity, 42/612
Pituitary dwarfism type II
    corticotropin, 42/613
    gonadotropin, 42/613
    growth hormone deficiency, 42/614
    sexual infantilism, 42/613
    thyrotropic hormone, 42/613
Pituitary dwarfism type III
    cherubism, 42/614
    growth hormone, 42/614
    obesity, 42/615
    somatomedin, 42/615
Pituitary dwarfism type IV
    hypoglycemia, 42/616
Pituitary dwarfism type V
    hypoglycemia, 42/616
    pituitary gland, 42/616
    sella turcica, 42/616
    sphenoid bone defect, 42/616
Pituitary fossa
    herniation, 1/571
Pituitary gigantism myopathy

acromegalic myopathy, 41/249, 62/537
denervation atrophy, 62/538
EMG, 41/249, 62/538
growth hormone excess, 41/248, 62/538
Pituitary gland
adrenal hypoplasia, 42/11
anencephaly, 30/180, 42/12
ataxia telangiectasia, 14/75, 293
basophil adenoma, 2/457
brain concussion, 23/465
brain injury, 23/60, 119
brain tumor, 16/342, 344
carcinoma, 17/384
chromophilic adenoma, 2/457
chromophobe adenoma, 2/456
β-endorphin, 45/239
epilepsy, 15/317, 42/11
glycosaminoglycanosis, 10/449
GM2 gangliosidosis, 10/406
headache, 5/141, 229
hemorrhage, 5/141
Levi-Lorain dwarfism, 2/455
mental deficiency, 43/279
pain, 45/239
pain control, 45/239
pathology, 23/59, 466
pituitary dwarfism type V, 42/616
radiation injury, 23/653
stress, 74/323
traumatic, 23/59, 466
Pituitary gland agenesis
hypoglycemia, 42/11
Pituitary gland aplasia
epidemiology, 30/150
holoprosencephaly, 30/470
Pituitary hemorrhage
headache, 48/279
snake envenomation, 65/180
Pituitary hemorrhagic necrosis
snake envenomation, 65/180
Pituitary hormone deficiency
multiple, see Pituitary dwarfism type II
Pituitary metastasis
systemic disease, 69/197
systemic tumor, 71/615
Pituitary microadenoma
imaging, 68/346
Pituitary tumor
see also Pituitary adenoma
acromegaly, 68/351
ACTH secreting tumor, 68/344
adenoma, see Pituitary adenoma
astrocytoma, 17/425

brain abscess, 17/412, 426
brain metastasis, 17/413, 421, 426
bromocriptine, 68/349
cabergoline, 68/349
carcinoma, 17/384
choristoma, 17/425
craniopharyngioma, 17/428
CSF fistula, 24/185
cyst, 17/427
epilepsy, 72/181
etiology, 68/344
follitropin/luteinizing hormone secreting tumor,
68/344
Fröhlich syndrome, 2/454
granuloma, 17/427
growth hormone secreting tumor, 68/344
growth hormone secretion, 68/351
headache, 48/422-424
hemorrhage, 68/345
history, 17/375
mucocele, 17/428
myoblastoma, 17/425
nonrecurrent nonhereditary multiple cranial
neuropathy, 51/571
optic chiasm, 68/344
penis erection, 75/89
pergolide, 68/349
prolactinoma, 68/344
quinagolide, 68/350
radiation injury, 67/330
Rathke cleft, 17/428
seesaw nystagmus, 16/323
sellar anatomy, 68/347
survey, 68/343
thyrotropic hormone secreting tumor, 68/344
tuberculoma, 17/427
type, 68/344
Pizotifen
chronic migrainous neuralgia, 48/251
cluster headache, 48/230
headache, 48/39
migraine, 48/39, 151, 179, 190-192, 201, 203
serotonin 1 receptor, 48/88
Placenta
Brucella suis, 52/582
teratogenesis, 30/115-117
Placental enlargement
triploidy, 43/564, 566
Placing reaction
cerebellar ablation, 2/403
component, 1/69
parietal lobe syndrome, 45/68, 77
Plagiocephaly

see also Craniosynostosis
  acrocephalosyndactyly type III, 50/123
  convulsion, 30/226
  craniosynostosis, 30/154, 222, 42/20, 50/113, 118
  incontinentia pigmenti, 14/216
  mental deficiency, 30/227
  Saethre-Chotzen syndrome, 43/322, 46/87
Planotopokinesia
  apraxia, 4/54
  constructional apraxia, 4/68, 45/492
Plant alkaloid
  adverse effect, 69/493
  chemotherapy, 39/106-114
Plantar pits
  palmar, *see* Palmar and plantar pit
Plantar reflex
  ataxia telangiectasia, 60/364
  cervical spinal cord injury, 61/56
  diagnostic value, 1/247
  extensor, *see* Extensor plantar response
  paraplegia, 26/248
  reflex loop, 1/253
Plaque
  amyloid, *see* Amyloid plaque
  arteriosclerosis, 11/331, 437
  astrocyte, 9/279-285
  astrocytoma, 9/294
  astrocytosis, 9/224
  atheromatous, *see* Atheromatous plaque
  axis cylinder, 9/264
  brain *vs* spinal cord, 9/250
  cerebroside sulfatide, 9/314
  character, 9/231
  complication, 9/291
  concentric plaque, 9/224
  CSF, 9/289
  Dawson finger, 9/237, 248, 251, 253, 258-260
  diameter, 9/228
  distribution, 9/226-228, 237-242
  edema, 9/220
  essential component, 9/229
  esterified cholesterol, 9/312
  fatty acid, 9/314
  fibrosis, 9/292
  Fog rule, 9/259
  fool, *see* Fool plaque
  forhead, *see* Forhead plaque
  form and size, 9/242-247
  growth, 9/242-252
  hemorrhage, 12/26
  Hollenhorst, *see* Hollenhorst plaque
  infantile neuroaxonal dystrophy, 9/265
  infection, 9/295

  Landkartenherde, 9/224
  linoleic acid, 9/315
  microglia, 9/272
  multiple sclerosis, *see* Multiple sclerosis plaque
  myelin sheath, 9/264
  natural history, 9/298
  oligodendrocyte, 9/274-279
  perivenular origin, 9/221, 257
  phosphatide, 9/313
  plasmalogen, 9/313
  primary infiltrate, 9/285
  prion amyloid, *see* Prion amyloid plaque
  senile, *see* Senile plaque
  shadow plaque, 9/296
  spinal meningioma, 20/186, 188, 206
  thrombosis, 9/294
  triglyceride, 9/313
  ultrastructure, 9/267-273, 287
  unicentric, *see* Unicentric plaque
  Wetterwinkel (Steiner), 9/252
Plaque forming cell
  multiple sclerosis plaque, 47/362
Plasma aminoacid
  Schut family ataxia, 60/485
Plasma catecholamine
  autonomic nervous system, 74/625
  orthostatic hypotension, 26/319, 63/159
  pure autonomic failure, 63/147
Plasma cell
  CSF, *see* CSF plasma cell
  experimental allergic encephalomyelitis, 47/442
  multiple myeloma, 20/10
  multiple sclerosis, 47/221, 228, 231
  multiple sclerosis plaque, 47/81-83
  progressive multifocal leukoencephalopathy,
      9/489
  serum IgA, 20/10
  serum IgG, 20/10
Plasma cell dyscrasia
  see also Monoclonal gammopathy *and*
      Paraproteinemia
  antibody molecule, 63/391
  B-lymphocyte, 63/391
  brain tumor, 71/438
  classification, 69/290
  cryoglobulinemia, 63/391
  heavy chain disease, 63/391
  IgM, 63/391
  multiple myeloma, 39/132, 63/391
  neurologic complication, 63/391
  primary amyloidosis, 63/391
  survey, 71/434
  symptom, 71/433

undetermined significant monoclonal
   gammopathy, 63/391
Waldenström macroglobulinemia, 63/391
Plasma cell leukemia
immunocompromised host, 56/470
Plasma exchange
demyelinating neuropathy, 47/618
multiple sclerosis, 47/200
vasculitis, 71/161
Plasma hydrocortisone
depression, 46/474
paraplegia, 26/397, 402
Plasma membrane
Duchenne muscular dystrophy, 40/48, 151, 213,
   252, 370, 43/106
muscle fiber, 62/32
Plasma reaction
histochemistry, 9/25
Plasma transfusion
multiple sclerosis, 9/385
Plasmacytoma
see also Multiple myeloma
incidence, 68/530
paraproteinemia, 69/306
solitary, see Solitary plasmacytoma
spinal cord, 67/198
spinal cord compression, 71/440
spinal epidural tumor, 20/103
spinal tumor, 68/518
terminology, 18/254
Plasmacytosis
African trypanosomiasis, 52/340
diffuse, see Diffuse plasmacytosis
Plasmalogen
see also Phosphatidylethanolamine
choline, 10/241
deficiency, 66/514
ethanolamine, 10/241
glia, 9/6
hexadecanol, 10/243
myelin, 9/2
myelinoclastic diffuse sclerosis, 9/482
plaque, 9/313
plasma staining, 9/25
Plasmalogen-hydrolase
chemistry, 10/245
Plasmapheresis
see also Antibody formation
acetylcholine receptor antibody, 41/136
Castleman disease, 63/395
cryoglobulinemia, 63/405
dermatomyositis, 62/386
Eaton-Lambert myasthenic syndrome, 62/426

Guillain-Barré syndrome, 51/255, 258
inclusion body myositis, 62/386
multiple myeloma, 18/256, 63/395
multiple sclerosis, 47/142, 367
myasthenia gravis, 41/96, 136, 62/420
myoglobinuria, 62/556
osteosclerotic myeloma, 63/395
POEMS syndrome, 51/432, 63/395
polymyositis, 62/386, 71/143
Refsum disease, 27/520, 36/348, 60/237, 728,
   66/498
undetermined significant monoclonal
   gammopathy, 63/403
Plasmarrhage, 9/648, 654
Plasmid DNA
gene therapy, 66/114
Plasminogen
wallerian degeneration, 51/33
Plasminogen activator, see Tissue plasminogen
   activator
Plasmocid
experimental chemical data, 40/157
I band, 40/157
mitochondrial abnormality, 40/160
myoglobinuria, 40/339, 41/271
toxic myopathy, 62/600
Z band, 40/157
Plasmocid intoxication
myoglobinuria, 40/339
rhabdomyolysis, 40/339
Plasmocytosis, see Plasmacytosis
Plasmodium falciparum
area preference, 52/367
brain metastasis, 69/116
fever persistence, 52/367
latent period, 52/366
life cycle, 52/366
malaria, 35/144, 52/365-367
malaria resistance, 52/365
maturation, 52/366
Plasmodium malariae
fever periodicity, 52/367
latent period, 52/366
life cycle, 52/366
long latency, 52/367
malaria, 35/144, 52/365-367
maturation, 52/366
pathogenic species, 52/365
Plasmodium ovale
latent period, 52/366
life cycle, 52/366
malaria, 35/144, 52/365-367
maturation, 52/366

pathogenic species, 52/365
Plasmodium vivax
  area preference, 52/367
  fever periodicity, 52/367
  latent period, 52/366
  life cycle, 52/366
  malaria, 35/144, 52/365-367
  maturation, 52/366
  pathogenic species, 52/365
Plaster bed
  pressure sore, 26/464
Plastic rigidity
  basal ganglion disease, 6/139-142
  distribution, 6/185
  hemirigor apoplecticus, 6/673
  Parkinson disease, 6/139
Plateau wave
  CSF, 55/206
  head injury, 57/218
  intracranial hypertension, 24/609
  intracranial pressure, 23/202, 55/206, 57/218
Platelet
  brain arteriosclerosis, 53/96
  brain infarction, 55/474
  cluster headache, 48/234
  count, *see* Platelet count
  headache, 48/281
  hemostasis, 63/301
  homocysteine, 55/326
  migraine, 48/89, 93
  serotonin, *see* Platelet serotonin
  trimethylaminuria, 29/231
Platelet activating factor
  eosinophil, 63/370
  hemostasis, 63/302
Platelet adhesiveness
  multiple sclerosis, 9/131-133
Platelet count
  multiple sclerosis, 47/340
Platelet dopamine uptake
  Huntington chorea, 49/290
Platelet 5-hydroxytryptamine uptake
  Huntington chorea, 49/290
Platelet MAO
  cluster headache, 48/94
  headache, 48/93
  migraine, 48/92-94
  phenethylamine, 48/94
  schizophrenia, 48/94
  tryptamine, 48/94
Platelet serotonin
  free fatty acid, 48/96
  function, 48/88

  migraine, 48/96
  wine, 48/96
Platinum
  chemotherapy, 67/284
  chronic axonal neuropathy, 51/531
  neurotoxicity, 51/265
  toxic polyneuropathy, 51/274
cis-Platinum, *see* Cisplatin
Platinum intoxication
  see also Cisplatin intoxication
  automobile catalyst, 64/357
  blood-brain barrier, 64/357
  blood level, 64/357
  concentration, 64/357
  excretion, 64/357
  hypersensitivity, 64/357
  metal intoxication, 64/353
  organ distribution, 64/357
  toxicokinetic, 64/357
  urine level, 64/357
Platinum polyneuropathy
  drug intoxication, 51/274
Platybasia
  see also Basilar impression
  anthropoid ape, 32/66
  Arnold-Chiari malformation type I, 42/17
  basal angle, 32/15
  basilar impression, 32/66, 42/17, 50/397
  cranial nerve, 42/17
  epidemiology, 30/150, 161
  genetics, 30/151
  hydrocephalus, 50/286
  malformation, 32/69, 71
  measurement, 32/15
  multiple nevoid basal cell carcinoma syndrome, 14/456
  occipital bone shortening, 32/14
  occipital dysplasia, 32/15
  osteitis deformans (Paget), 32/67, 38/365
  Prouzet line, 32/19
  sella turcica, 42/17
  short dura mater syndrome, 5/372
  syringomyelia, 50/446
  vertigo, 2/362
  Welcker angle, 32/15
Platypodia granulosa intoxication
  paralysis, 37/56
Platysma sign
  facial asymmetry, 1/183
  facial paralysis, 1/183
Platyspondyly
  glycosaminoglycanosis type IV, 66/304
Pleasure

brain substrate, 45/277
Plectonema tetebrans intoxication
  ciguatoxin, 37/97
Pleiotropism
  Lindenov-Hallgren syndrome, 13/454
  primary pigmentary retinal degeneration, 13/153,
    299
  tuberous sclerosis, 14/8, 351
  Usher syndrome, 13/447
Pleiotropy
  CNS malformation, 30/88
  definition, 30/6, 88
  environmental, 30/89
  genetics, 13/25, 30/89
  Laurence-Moon-Biedl syndrome, 30/6
Pleoconial myopathy, see Mitochondrial myopathy
Pleocytosis
  CSF, see CSF pleocytosis
  CSF Ig, 9/333
  CSF leukocyte, 9/330
  methotrexate intoxication, 65/529
  multiple sclerosis, 9/330, 333
Pleomorphic xanthoastrocytoma
  classification, 67/5
  imaging, 67/175
  low grade glioma, 68/55
  oligodendroglioma, 72/357
Plethysmography
  autonomic nervous system, 1/466
  blood flow, 1/463, 75/429
  cerebrovascular disease, 11/253
  ocular, see Ocular plethysmography
  spinal cord injury, 61/285
Pleura
  pain, 1/138
Pleurodynia, see Bornholm disease
Plexiform neurofibroma
  neurofibromatosis type I, 8/438, 440, 50/366
Plexiform neuroma
  definition, 14/32
  elongation, 14/25
  François syndrome, 14/627, 635, 638
  median nerve, 14/26
  nerve thickening, 14/25
  neurofibromatosis type I, 14/32
  tortuosity, 14/25
Plexopathy
  brachial, see Brachial plexopathy
  cancer, 69/79
  cervical, 69/79
  femoral neuropathy, 51/166
  hereditary neuropathy, 60/254
  lumbar, 69/87

lymphoma, 63/350
malignant, 69/71
malignant lymphoma, 69/278
mononeuritis multiplex, 51/166
pain, 69/72, 85
polyneuropathy, 60/254
survey, 69/71
traumatic, see Traumatic plexopathy
Plexus
  Auerbach myenteric, see Auerbach myenteric
    plexus
  basilar venous, see Basilar venous plexus
  brachial, see Brachial plexus
  celiac, see Celiac plexus
  cervical, see Cervical plexus
  choroid, see Choroid plexus
  dural venous, see Dural venous plexus
  epidural venous, see Epidural venous plexus
  Exner, see Exner plexus
  Hodgkin disease, 39/38
  hypogastric, see Hypogastric plexus
  lumbar, see Lumbar plexus
  lumbosacral, see Lumbosacral plexus
  malignant lymphoma, 39/38
  Meissner, see Submucosal plexus
  Meissner submucous, see Submucosal plexus
  myenteric, see Myenteric plexus
  pelvic, see Pelvic plexus
  pericarotid sympathetic, see Pericarotid
    sympathetic plexus
  pudendal, see Pudendal plexus
  sacral, see Sacral plexus
  sympathetic, see Sympathetic plexus
  tympanic, see Tympanic plexus
  vesical, see Vesical plexus
Plexus fibrosis
  choroid, see Choroid plexus fibrosis
Plexus hemisphericus
  cerebellar, see Cerebellar plexus hemisphericus
Plexus injury, 7/402-426
  see also Plexus lesion
  Bonney axon reflex test, 7/411
  brachial, see Brachial plexus
  cervical, 7/420
  lumbar, 7/422
  radiation, 7/425
  sacral, 7/424
Plexus lesion
  see also Plexus injury
  brachial, 7/402
  diaphragm, 7/409
  lumbar, see Lumbar plexus lesion
  phantom limb, 4/224

radiation injury, 17/425
supraclavicular nerve, 2/135
ulnar nerve, 2/146
Plexus metastasis
systemic tumor, 71/615
Plexus neuralgia
tympanic, *see* Tympanic plexus neuralgia
Plexus neuritis
brachial, *see* Brachial plexus neuritis
lumbosacral, *see* Lumbosacral plexus neuritis
Plexus papilloma
choroid, *see* Choroid plexus papilloma
malignant, 16/21
microscopy, 16/21
tissue culture, 17/82
Plexus paralysis
brachial, *see* Brachial plexus paralysis
lumbosacral, *see* Lumbosacral plexus palsy
Plexus pedunculi
radioanatomy, 11/93
topography, 11/93
Plexus syndrome
brachial, *see* Brachial plexus syndrome
cervical, *see* Cervical plexus syndrome
Pli courbe (Dilenge)
radioanatomy, 11/88
topography, 11/88
Plicature
pontine, *see* Pontine plicature
Plis capitonnés
trisomy 8, 50/563
trisomy 8 mosaicism, 43/510
trisomy 8 syndrome, 31/528
Plotosidae intoxication
family, 37/78
venomous spine, 37/78
Plotosus intoxication
family, 37/78
venomous spine, 37/78
Plumbism, *see* Lead intoxication
Pluriradicular syndrome
topographical anatomy, 20/579
Pneumatic mattress
spinal cord injury, 26/221
Pneumatic tool disease, *see* Vibration neuropathy
Pneumatosis intestinalis
dermatomyositis, 62/374
inclusion body myositis, 62/374
polymyositis, 62/374
Pneumobelt
respiratory dysfunction, 26/345
Pneumocephalus
antibiotic agent, 24/212

ball valve mechanism, 24/205
brain injury, 24/201
brain stab wound, 23/492
classification, 24/202
CSF fistula, 24/211
delayed, 24/208
depressed skull fracture, 24/211
diagnosis, 24/209, 221
epidural, 24/202, 211
epilepsy, 24/207
etiology, 24/201, 204
experimental, 24/206
extracranial, 24/201
gas forming organism, 24/206
gunshot injury, 57/312
head injury, 57/139, 312
iatrogenic factor, 61/164
iatrogenic neurological disease, 61/164
idiopathic, 24/208
immediate, 24/207
incidence, 24/201, 221
infection, 24/208
intracerebral, 24/204
inverted bottle mechanism, 24/205
lumbar puncture, 61/164
management, 24/211
meningitis, 24/212
missile injury, 24/211, 57/312
negative intracranial pressure, 24/205
neoplasm, 24/208
onset time, 24/207
pathophysiology, 24/205
penetrating head injury, 57/312
postoperative, 24/208
pressure wave, 24/206
radiology, 24/209
recurrent, 24/212
spinal anesthesia, 61/164
subarachnoid, 24/203, 211
subdural, 24/203, 211
symptom, 24/207
traumatic hydrocephalus, 24/234, 249
ventricular, 24/204
Pneumococcal meningitis, 33/35-49
acoustic nerve, 33/47, 52/51
antibiotic agent, 52/48
ataxia, 33/46, 52/44, 51
bacterial meningitis, 33/36, 47, 52/3, 52
brain abscess, 33/47, 52/43
brain edema, 33/38, 43, 52/49
Brudzinski sign, 33/38, 52/43
cefaloridine, 33/43
clinical features, 33/38, 52/43

computer assisted tomography, 52/51
counterimmunoelectrophoresis, 33/40
cranial bruit, 33/38, 52/44
cranial nerve palsy, 33/47, 52/43, 51
CSF, 33/39-42, 52/44-46
deafness, 33/47, 52/43, 51
diagnosis, 33/39-42, 52/44-46
gram stain, 33/40
hemorrhagic ependymitis, 33/37, 52/43
histology, 33/36, 52/43
history, 33/35, 52/41
hydrocephalus, 52/51
immunoelectrophoresis, 33/40
incidence, 52/42
Kernig sign, 33/38, 52/43
limulus lysate test, 33/41
meningococcal meningitis, 52/26
morbidity, 33/44, 52/49
mortality, 33/44-46, 52/49
natural course, 33/42, 52/46
neurologic sequela, 33/46-48, 52/50-52
neurosurgery, 33/43, 52/49
obstructive hydrocephalus, 52/43
parkinsonism, 33/38, 52/44
pathogenesis, 33/36-38, 52/42
pathology, 33/36-38, 52/42
penicillin, 33/35, 42, 52/47
petechial rash, 33/39, 52/44
predisposing factor, 33/36, 52/42
pregnancy, 33/36
recurrent meningitis, 33/48, 52/52
rhinorrhea, 33/48, 52/52
sickle cell anemia, 33/49, 52/53
skull fracture, 33/36, 52/42
splenectomy, 33/49, 52/53
sputum examination, 33/39, 52/44
status epilepticus, 33/47, 52/51
subdural effusion, 33/47, 52/51
tache cérébrale, 33/38, 52/44
tetracycline, 33/42
treatment, 33/42, 52/47
white blood cell count, 33/39, 52/44
Pneumococcus, see Streptococcus pneumoniae
Pneumocystis carinii pneumonia
acquired immune deficiency syndrome, 56/489
Pneumoencephalography, see PEG
Pneumonia
aspiration, see Aspiration pneumonia
delirium, 46/544
glycosaminoglycanosis type III, 66/299
head injury, 57/234
late infantile metachromatic leukodystrophy,
66/167

Legionella pneumophila, 52/253
meningococcal meningitis, 33/24, 52/27
multiple myeloma, 20/11
mycoplasma, see Mycoplasma pneumoniae
nocardiosis, 52/447
Pneumocystis carinii, see Pneumocystis carinii
pneumonia
primary meningococcal, see Primary
meningococcal pneumonia
relapsing, see Relapsing pneumonia
respiratory dysfunction, 26/345
spinal cord injury, 61/261
subarachnoid hemorrhage, 12/111
sulfa agent, 71/383
toxocariasis, 52/511
Pneumonitis
cat scratch disease, 52/128
chronic lymphoid interstitial, see Chronic
lymphoid interstitial pneumonitis
dermatomyositis, 62/374
Guillain-Barré syndrome, 51/255
inclusion body myositis, 62/374
lymphoid interstitial, see Lymphoid interstitial
pneumonitis
lymphoid interstitial pneumonitis, 39/530
polymyositis, 62/374
Pneumoplethysmography
thoracic outlet syndrome, 51/127
Pneumotaxic center
anatomy, 1/653
Hering-Breuer reflex, 63/479
Kölliker-Fuse nucleus, 63/479
nucleus parabrachialis medialis, 63/479
respiratory center, 63/432, 478
respiratory control, 63/432
site, 63/478
Pneumothorax
decompression illness, 63/419
thoracic spinal column injury, 26/224
Von Hippel-Lindau disease, 14/255
Podencephaly
encephalocele, 50/73
Podophyllin
drug induced polyneuropathy, 51/294
neuropathy, 69/471, 75/501
neurotoxicity, 51/294
Podophyllotoxin
chemotherapy, 39/114
neurologic toxicity, 39/114
POEMS syndrome
see also Castleman disease and Multiple myeloma
polyneuropathy
amenorrhea, 47/619, 63/394

anasarca, 63/394
ascites, 47/619, 63/394
axonal degeneration, 47/619, 51/431, 63/394
benign intracranial hypertension, 63/394
Castleman disease, 47/619, 63/394
chemotherapy, 63/396
confusional state, 63/395
cranial nerve palsy, 63/395
cranial neuropathy, 63/395
cyclophosphamide, 47/619, 63/396
demyelinating neuropathy, 47/619, 51/432, 63/395
EMG, 51/431, 63/394
finger clubbing, 47/619, 63/394
gynecomastia, 47/619, 51/430, 63/394
hepatomegaly, 47/619, 51/430, 63/394
hyperglycemia, 51/430
hyperhidrosis, 63/394
hyperpigmentation, 47/619, 51/430, 63/394
hypertrichosis, 47/619, 51/430, 63/394
hypogonadism, 51/430
hypothyroidism, 51/430
lymphadenopathy, 47/619, 51/430, 63/394
motor conduction velocity, 47/619, 51/431, 63/394
motor polyneuropathy, 63/394
multiple myeloma, 47/619, 51/430, 63/394
multiple myeloma polyneuropathy, 51/430
neuropathy, 51/430
osteosclerotic myeloma, 47/619, 51/430, 63/394
papilledema, 47/619, 51/430, 63/394
paraproteinemia, 47/619, 51/430, 432
plasmapheresis, 51/432, 63/395
polycythemia, 51/430
polyneuropathy pathogenesis, 51/431, 63/395
prednisolone, 47/619, 51/432, 63/396
radiotherapy, 51/432, 63/395
segmental demyelination, 47/619, 51/431, 63/394
sensorimotor polyneuropathy, 63/394
sexual impotence, 47/619, 63/394
splenomegaly, 47/619, 51/430, 63/394
symptom, 63/394
treatment, 51/432, 63/395
vasculopathy, 63/395
Poikiloderma
congenital, see Rothmund-Thomson syndrome
Goltz-Gorlin syndrome, 46/61
segmental, 14/108
Poikiloderma atrophicans vasculare
facioscapulohumeral muscular dystrophy, 40/420
Poikilodermatomyositis
mental deficiency, 43/258
Poikilomycosis
brain abscess, 35/567, 52/489
brain infarction, 35/567, 52/489

cerebral thromboembolism, 35/567, 52/489
endophthalmitis, 35/567, 52/489
Poikilothermia
globoid cell leukodystrophy, 10/303
holoprosencephaly, 50/239
tetraplegia, 2/199
transverse spinal cord lesion, 61/275
Poikilotonia
Huntington chorea, 49/280
Poison
exogenous, see Exogenous poison
Poker face
facial paralysis, 8/276
Pokeweed mitogen
multiple sclerosis, 47/344, 346, 354, 360-362, 374
Poland
neurology, 1/11
Poland-Möbius syndrome
arm hypoplasia, 43/461
birth incidence, 43/461
cranial nerve agenesis, 50/215
facial nerve agenesis, 50/215
Möbius syndrome, 50/215
monomelic spinal muscular atrophy, 59/376
pectoralis muscle agenesis, 43/461
pectoralis muscle aplasia, 43/461, 50/215
syndactyly, 43/461
Poland syndactyly, see Poland-Möbius syndrome
Poland syndrome, see Poland-Möbius syndrome
Polar bear intoxication
arsenic intoxication, 37/96
brain tumor, 37/96
diarrhea, 37/96
diplopia, 37/96
drowsiness, 37/96
headache, 37/96
ingestion, 37/95
irritability, 37/96
papilledema, 37/96
scurvy, 37/96
vitamin A intoxication, 37/96
vomiting, 37/96
Polar spongioblastoma
genetic factor, 42/730
neurofibromatosis, 14/33
neurofibromatosis type I, 14/33
Rosenthal fiber, 18/2
Policeman tip position
brachial plexus injury, 2/140
terminology, 2/140
Poliodystrophy
central visual disturbance, 60/655
CSF, 42/588

infantile progressive, *see* Alpers disease
intracerebral calcification, 42/588
lactic acidosis, 42/588
mitochondrial myopathy, 42/588
myopathy, 42/588
optic atrophy, 60/655
progressive, *see* Alpers disease
seizure, 42/588
Polioencephalitis
  see also Poliomyelitis
  chronic nodular, *see* Chronic nodular
    polioencephalitis
  poliomyelitis, 34/110
Polioencephalitis hemorrhagica superior, *see*
  Wernicke encephalopathy
Polioleukoencephalopathy
  hemorrhagic, *see* Hemorrhagic
    polioleukoencephalopathy
Poliomyelencephalic dysgenesia
  trisomy 13-15, 14/368
Poliomyelitis, 34/93-127
  see also Enterovirus infection, Polioencephalitis
    and Poliovirus
  acute, *see* Acute poliomyelitis
  acute cerebellar ataxia, 34/621, 56/329
  acute hemorrhagic conjunctivitis, 56/349
  age, 34/115, 56/314
  amyotrophic lateral sclerosis, 22/138, 34/114,
    42/65
  anhidrosis, 8/344
  anterior horn syndrome, 2/210
  antibody, 34/101, 118-120
  antibody response, 34/118, 56/323
  Aran-Duchenne disease, 42/86
  area postrema, 56/318
  autonomic nervous system, 42/652, 75/20
  axonal transport, 56/318
  biochemical effect, 34/104-106
  blood-brain barrier, 34/103, 42/652, 56/318
  bulbar, *see* Bulbar poliomyelitis
  bulbar syncope, 15/824
  calcium metabolism, 34/113
  carbon dioxide narcosis, 63/441
  cardiovascular disturbance, 34/107, 110, 113
  cellular immunity, 56/324
  central sleep apnea, 63/462
  cerebellar ataxia, 56/329
  circulating antibody, 34/118, 56/322
  clinical features, 34/108-117
  clinical type, 42/652, 56/317
  CNS, 34/103
  CNS infected site, 34/119, 56/318
  CNS invasion, 34/103, 115, 56/318

complication, 34/111-114
concentric laminated body, 40/109
cranial nerve, 42/652
cranial nerve palsy, 42/652, 56/328
CSF, 34/117, 56/333
deafness, 56/107
definition, 34/93
deformity, 40/337
differential diagnosis, 34/114, 41/166, 291, 336,
  56/349
disease syndrome, 56/325
duodenal antibody, 56/324
edrophonium test, 62/413
EMG, 1/638
encephalitis, 56/327
endemic, 34/93, 100-102, 42/653
endemic pattern, 34/101, 56/315
enterovirus, 34/95, 56/307
eosinophilic inclusion body, 56/319
epidemic pattern, 34/101, 292, 42/653, 56/315
epidemiology, 34/93, 99-102, 42/653, 56/314
extraneural infection, 34/102, 56/317
forgotten respiration, 1/654, 63/441, 488
genetic predisposition, 42/653, 56/315
genetic susceptibility, 34/116
geographic distribution, 9/123, 34/93, 56/314
guanidine, 22/24
hemiatrophy, 59/480
Hirayama disease, 59/114
history, 34/93-95, 56/308
HLA antigen, 42/653
human, 34/292
hypothalamic lesion, 2/444, 34/107
immune mechanism, 9/129
immunity, 56/322
immunity and prevention, 34/118-126
immunocompromised host, 56/476
immunodeficiency disease, 34/120
immunology, 56/323
immunopathology, 34/120
implantation, 34/102
incidence, 56/316
incubation period, 34/108, 56/316
inflammatory response, 34/107
injection, 56/315
laboratory diagnosis, 34/117
laboratory finding, 56/333
lead polyneuropathy, 64/436
lesion distribution, 56/319
limb girdle syndrome, 40/435
meningitis, 42/652
monomelic spinal muscular atrophy, 59/376
motoneuron disease, 22/138, 34/113

motor end plate, 56/318
multiple cranial neuropathy, 56/329
multiple sclerosis, 9/123
muscle, 56/318
musculoskeletal deformity, 34/113
neurogenic pulmonary edema, 63/496
neuronal change, 34/107
neuronophagia, 34/108, 56/321
neuropathology, 34/106, 56/318-320
neurovirulence, 34/101, 42/653, 56/321
oropharyngeal antibody, 34/119, 56/324
pallidal necrosis, 49/479
paralytic stage, 34/111, 42/653
Parkinson disease, 6/244
pathogenesis, 34/102-106, 56/316
pathology, 34/106-108, 56/316
Pena-Shokeir syndrome type I, 59/369
pertussis vaccination, 33/300
physical exertion, 56/315
polioencephalitis, 34/110
polyneuropathy, 62/413
postpoliomyelitic amyotrophy, 22/23, 138, 56/331
postvaccine era, 56/316
predisposing factor, 56/315
pregnancy, 31/217, 34/117, 42/563, 56/315
preparalytic stage, 34/111, 56/328
prevention and immunity, 34/118-126
prognosis, 34/115-117
progressive, see Progressive poliomyelitis
progressive bulbar palsy, 59/375
progressive muscular atrophy, 56/331
rabies, 56/391
respiratory involvement, 34/107, 109, 112, 42/652
scoliosis, 34/113, 40/337, 50/417
serology, 34/101, 117
sex ratio, 34/113, 56/315
spastic paraplegia, 59/436
spinal, see Spinal poliomyelitis
spinal deformity, 40/337
spinal muscular atrophy, 22/138, 34/109, 59/114,
    370, 373
spinal poliomyelitis, 34/109, 56/328
subacute anterior, see Subacute anterior
    poliomyelitis
symptom, 34/108, 56/328
syringomyelia, 32/285
terminology, 7/199, 34/93, 109
thoracic outlet syndrome, 51/124
tonsillectomy, 9/123, 34/116, 42/653, 56/315
transmission, 34/99, 42/652, 56/314
transverse myelitis, 56/330
trauma, 42/653, 56/315
treatment, 34/111-114

vaccine, 34/120-124
vascular response, 34/107
vertigo, 56/107
viral clearance, 56/324
viral infection, 42/653, 56/307
viral meningitis, 56/11, 128, 325
viremia, 34/94, 103, 108, 122
virus attenuation, 56/321
virus isolation, 34/117
Poliomyelitis like syndrome
    asthma, 63/420
Poliomyelomalacia
    aorta disease, 22/20
Poliopia
    temporal lobe epilepsy, 73/71
Poliosis
    lipofuscinosis, 60/647, 667
    secondary pigmentary retinal degeneration,
        13/186
    vitamin B12 deficiency, 60/667
    Vogt-Koyanagi-Harada syndrome, 56/622
Poliosis circumscripta
    tuberous sclerosis, 14/574
Poliovirus
    see also Poliomyelitis
    antigenicity, 34/118
    classification, 34/4, 95
    experimental acute viral myositis, 56/195
    gene localization, 42/653
    isolation, 34/117
    multiple sclerosis, 47/339
    neurovirulence, 56/38
    picornavirus, 34/133, 56/195, 307
    property, 34/95
    replication, 34/95-99
    Reye syndrome, 34/169, 49/217
    RNA virus, 34/95
    type, 56/310
    type 1, 56/321
    type 2, 56/321
    type 3, 56/321
    ultrastructure, 56/311
    vaccine strain, 56/321
    wild type strain, 56/321
Poliovirus infection, see Poliomyelitis
Poliovirus receptor
    gene localization, 42/653
Poliovirus vaccine
    Reye syndrome, 56/150
POLIP syndrome
    autonomic nervous system, 75/643
    complex IV deficiency, 62/506
    Kearns-Sayre-Daroff-Shy syndrome, 62/313

MNGIE syndrome, 62/505
progressive external ophthalmoplegia, 62/313
Pollen
headache, 5/237
Polyacrylamide gel electrophoresis
multiple sclerosis, 47/92, 111
Polyadenylation
mRNA, 66/73
terminal stretch, 66/73
Polyamine
uremic polyneuropathy, 51/361
Polyangiitis
microscopic, *see* Microscopic polyangiitis
Polyarteritis nodosa, 39/295-309, 55/353-365
    see also Allergic granulomatous angiitis *and*
        Kawasaki syndrome
    adrenocorticosteroids, 8/126, 11/451, 39/300,
        55/346, 357
    allergic angiopathy, 9/609-612
    allergic granulomatous angiitis, 51/453, 55/358,
        63/382
    amaurosis fugax, 55/356
    amphetamine addiction, 55/360, 520
    anemia, 8/125, 55/354
    angiopathic polyneuropathy, 51/446, 452
    autoimmune disease, 41/89, 385
    azathioprine, 39/308, 55/346, 358
    blindness, 51/452
    brain hemorrhage, 39/301, 48/284, 54/294
    brain infarction, 39/302, 48/284, 53/162, 55/356
    brain involvement, 11/139
    brain ischemia, 39/302, 53/162
    brain vasculitis, 39/302, 55/353, 396
    carotid dissecting aneurysm, 54/282
    carotid system syndrome, 53/308
    cerebrovascular disease, 11/139, 451, 39/301,
        48/285, 53/41
    chronic lymphocytic leukemia, 55/359
    classification, 71/4
    Cogan syndrome type II, 39/303, 51/446, 454
    collagen vascular disease, 8/124, 39/307, 51/446,
        452, 54/294
    confusional state, 55/356
    cranial nerve, 39/303, 51/452
    cranial neuropathy, 51/452, 55/355
    cryoglobulinemia, 39/181, 51/434, 55/359
    cryoglobulinemic polyneuropathy, 51/434
    cyclophosphamide, 39/308, 55/357
    dementia, 39/300, 55/356
    dermatomyositis, 39/307, 55/362
    diagnosis, 39/303, 55/357
    diagnostic guideline, 39/307
    differential diagnosis, 55/358

    drug abuse, 39/307, 55/359, 520
    EMG, 39/307
    encephalopathy, 39/301, 55/355
    eosinophilia, 8/125, 39/308, 63/371
    epidemiology, 55/353
    erythrocyte sedimentation rate, 55/354
    eye symptom, 55/363
    fever, 8/125, 39/295, 55/354, 396
    general features, 55/354
    general outline, 63/54
    granulomatous CNS vasculitis, 51/446, 55/396
    hallucinosis, 55/356
    headache, 8/125, 39/300, 48/284, 55/354
    hemiplegia, 39/301
    Henoch-Schönlein purpura, 55/359
    hepatitis B virus, 56/10
    hepatitis B virus infection, 39/296, 55/359, 56/302
    history, 39/295, 55/353
    hypertension, 8/125, 39/296, 308, 55/354, 356,
        396
    idiopathic vasculitis, 71/154
    immunopathology, 9/551
    infantile, *see* Kawasaki syndrome
    inflammatory myopathy, 55/355
    intracerebral hematoma, 11/672, 39/297
    ischemic myelopathy, 63/54
    ischemic neuropathy, 8/161
    ischemic optic neuropathy, 51/452
    juvenile, *see* Kawasaki syndrome
    Kawasaki syndrome, 48/285, 51/446, 453, 55/358
    leukoencephalopathy, 9/609
    livedo reticularis, 55/362
    lymphoma, 55/359
    lymphomatoid granulomatosis, 55/364
    meningeal sign, 39/300, 48/285
    meningitis serosa, 39/300, 55/356
    methamphetamine addiction, 39/296, 55/360
    mimick multiple sclerosis, 55/354
    mononeuritis multiplex, 7/200, 8/125, 9/551,
        12/659, 39/296, 305, 51/452, 55/355
    multiple cranial neuropathy, 51/452, 570
    multiple myeloma, 55/359
    myalgia, 8/125, 39/305, 55/354, 357
    myelomalacia, 55/102
    myocardial infarction, 11/451, 55/354
    myopathy, 39/306, 55/355, 357
    nephritis, 39/296, 55/354
    nerve ischemia, 12/659, 39/303
    neurobrucellosis, 52/588
    neurologic symptom, 39/296, 300, 55/354, 363
    neuropathy, 8/124, 161, 39/304-306
    ocular involvement, 39/302
    optic atrophy, 55/356

overlap syndrome, 55/358
pathogenesis, 8/659, 39/295
pathology, 39/296
peripheral nervous system, 11/139, 39/304, 51/452
polyneuropathy, 8/125, 9/609, 11/451, 39/304,
    51/452, 55/354, 396
prednisolone, 8/125, 39/308
prevalence, 55/353
prognosis, 39/308, 55/357
psychosis, 39/300, 55/356
radiculoneuropathy, 51/452
radiculopathy, 55/355
recurrent nonhereditary multiple cranial
    neuropathy, 51/570
relapsing course, 55/354
relapsing polychondritis, 55/362
retinal artery occlusion, 39/302, 55/355
retinal vasculitis, 55/356
retinopathy, 39/302, 55/355
rheumatoid arthritis, 8/126, 39/295, 308, 55/360
sarcoidosis, 55/364
scleroderma, 39/295, 55/361
segmental demyelination, 51/453
seizure, 39/302, 55/355, 357
sensory trigeminal neuropathy, 51/452
serum sickness, 39/295, 55/359
sex ratio, 8/124, 11/139, 39/295, 55/354
Sjögren syndrome, 55/361, 71/74
spinal cord, 39/297, 303
stroke, 39/297, 301, 55/354
subarachnoid hemorrhage, 39/297, 301, 48/284,
    55/102, 355
systemic brain infarction, 11/451
systemic lupus erythematosus, 39/295, 307,
    55/360
systemic vasculitis, 39/296, 55/353, 359, 396
Takayasu disease, 12/412, 46/359
temporal arteritis, 48/317, 55/346
terminology, 7/200, 39/295, 55/353
thrombocytosis, 55/354
transient ischemic attack, 55/356
treatment, 39/308, 55/357
trigeminal neuropathy, 51/452, 55/356
vasa nervorum disease, 8/125, 161, 12/659
vascular dementia, 46/359
vertebral dissecting aneurysm, 54/282
wallerian degeneration, 12/659, 51/453
Wegener granulomatosis, 9/610, 55/362
Polyarthritis
    hemochromatosis, 42/553
Polycationic liposome
    gene therapy, 66/114
Polychaeta intoxication

lumbriconereis heteropoda, 37/55
Polychlorinated biphenyl
    insulator fluid, 64/206
    neurotoxicity, 51/286
    neurotoxin, 51/286, 64/197, 206
    organochlorine, 64/206
    organochlorine insecticide intoxication, 64/206
    structural formula, 64/207
    toxic polyneuropathy, 64/197
Polychlorinated biphenyl intoxication
    dementia, 64/207
    epidemic, 64/207
    organochlorine insecticide intoxication, 64/207
    Yucheng disease, 64/207
    yusho disease, 64/207
Polychlorinated dibenzodioxin
    industry, 64/204
    neurotoxin, 64/204
Polychlorinated dibenzodioxin intoxication
    dementia, 64/205
    insomnia, 64/205
    LD50, 64/205
    myalgia, 64/205
    polyneuropathy, 64/205
    sensorimotor polyneuropathy, 64/205
    2,3,7,8-tetrachlorodibenzo-p-dioxin, 64/204
    toxic encephalopathy, 64/204
    toxic polyneuropathy, 64/204
    2,4,5-trichlorophenoxy acetic acid, 64/205
Polyclonal antibody
    multiple sclerosis, 47/232, 234
Polycystic kidney
    see also Polycystic renal disease
    axial mesodermal dysplasia spectrum, 50/516
    branchio-otorenal dysplasia, 42/361
    Cobb syndrome, 14/438
    congenital retinal blindness, 22/529
    Dandy-Walker syndrome, 50/332
    Gruber syndrome, 14/505, 31/243
    intracranial aneurysm, 14/519
    trisomy 9, 50/564
Polycystic osteodysplasia
    lipomembranous, see Lipomembranous polycystic
        osteodysplasia
Polycystic renal disease
    see also Polycystic kidney
    brain aneurysm, 55/81
    brain aneurysm neurosurgery, 55/81
    Ehlers-Danlos syndrome, 55/455
    subarachnoid hemorrhage, 55/2, 29, 81
Polycystic visceral disease
    brain aneurysm, 12/100, 102
Polycystoma, see Multiple nevoid basal cell

carcinoma syndrome
Polycythemia
  see also Polycythemia vera
  amaurosis fugax, 55/108
  brain abscess, 63/6
  brain hemorrhage, 63/251
  brain infarction, 53/165, 63/250
  brain ischemia, 53/165
  brain lacunar infarction, 63/251
  brain stem syndrome, 63/250
  brain venous infarction, 63/251
  carcinomatous encephalopathy, 9/596
  cavernous sinus thrombosis, 52/173
  cerebellar hemangioblastoma, 14/249, 18/285
  cerebrovascular disease, 53/31, 43
  chorea, 63/250
  cobalt intoxication, 36/330
  congenital heart disease, 63/3
  dementia, 46/398
  dizziness, 63/250
  extramedullary hematopoiesis, 63/250
  headache, 63/250
  hemoneurological disease, 63/249
  hereditary hemorrhagic telangiectasia (Osler),
    42/739, 55/449
  iatrogenic brain infarction, 53/165
  leukoencephalopathy, 9/594, 598
  multiple myeloma polyneuropathy, 51/430
  myelofibrosis, 63/250
  obstructive sleep apnea, 63/454
  pickwickian syndrome, 3/96, 38/299, 63/455
  POEMS syndrome, 51/430
  pruritus, 63/250
  respiratory encephalopathy, 63/415
  secondary, *see* Secondary polycythemia
  transient ischemic attack, 12/9, 53/266, 63/251
  venous sinus thrombosis, 54/435
  vertigo, 63/250
  visual impairment, 63/250
  Von Hippel-Lindau disease, 14/249, 43/21, 60/674
Polycythemia vera
  see also Erythrocytosis, Geisbock disease *and*
    Polycythemia
  brain, 16/235
  brain blood flow, 55/468
  brain infarction, 55/468, 63/250
  brain stem syndrome, 63/250
  cavernous sinus thrombosis, 54/410
  cerebellar hemangioblastoma, 14/505, 18/285
  chorea, 55/468, 63/250
  demyelination, 9/594
  diagnosis, 63/249
  differential diagnosis, 55/467

  dizziness, 63/250
  extramedullary hematopoiesis, 63/250
  features, 63/249
  headache, 55/469, 63/250
  Huntington chorea, 6/316
  hyperviscosity syndrome, 55/467, 484
  juvenile cerebrovascular disease, 55/467
  myelofibrosis, 63/250
  narcolepsy, 2/448
  neurologic symptom, 9/596
  neuropathy, 8/145
  peripheral sensory neuropathy, 55/468
  polyneuropathy, 55/468
  progressive multifocal leukoencephalopathy,
    9/596
  pruritus, 63/250
  secondary polycythemia, 55/469
  spastic paraplegia, 59/439
  spinal cord compression, 55/468
  systemic brain infarction, 11/460
  transient ischemic attack, 55/469
  vertigo, 55/469, 63/250
  visual impairment, 63/250
  Von Hippel-Lindau disease, 14/249, 701
Polydactyly
  acrocephalosyndactyly type I, 38/422
  acrocephalosyndactyly type II, 43/322-324,
    50/123
  acrocephalosyndactyly type V, 43/321
  Bardet-Biedl syndrome, 13/302, 385, 401, 22/508,
    536, 43/233
  Biemond syndrome type II, 43/334
  cheilopalatoschisis, 43/256
  Ellis-Van Creveld syndrome, 43/343
  Friedreich ataxia, 21/348
  Goltz-Gorlin syndrome, 14/113
  Gruber syndrome, 14/505, 43/391
  hemifacial hypertrophy, 22/550
  holoprosencephaly, 50/237
  Joubert syndrome, 63/437
  Klippel-Trénaunay syndrome, 14/524
  lissencephaly syndrome, 42/40
  mental deficiency, 43/256
  Minamata disease, 64/418
  Möbius syndrome, 42/324
  Mohr syndrome, 43/449
  multiple dermal cylindroma, 14/589
  Patau syndrome, 14/121, 31/507, 43/527, 50/558
  Saldino-Noonan syndrome, 31/243, 272
  Smith-Lemli-Opitz syndrome, 66/582
  trisomy 13-15, 14/774
  VACTERL syndrome, 50/513
  VATER syndrome, 50/513

Polydactyly syndrome
  short rib, *see* Saldino-Noonan syndrome
Polydipsia
  alcoholic, *see* Dipsomania
  Bardet-Biedl syndrome, 13/395
  oxidation-phosphorylation coupling defect,
    62/506
  water intoxication, 64/242
Polydysspondyly
  trisomy 8, 50/563
Polydystrophic dwarfism, *see*
    Glycosaminoglycanosis type VI
Polydystrophic oligophrenia, *see*
    Glycosaminoglycanosis type III
Polydystrophy
  pseudo-Hurler, *see* Mucolipidosis type III
Polygenic inheritance
  anencephaly, 50/79
  neural tube malformation, 30/127
Polyglot
  agraphia, 4/151, 169
  aphasia, 3/7, 45/305
  developmental dyslexia, 46/120
  verbal hallucination, 45/356
Polyglucosan
  Lafora progressive myoclonus epilepsy, 41/194
Polyglucosan body, *see* Bielschowsky body
Polyglucosan body disease
  adult, *see* Adult polyglucosan body disease
  choreoathetosis, 49/385
Polygyria
  Arnold-Chiari malformation type II, 50/407
  homocystinuria, 55/330
  Sturge-Weber syndrome, 14/66
Polyhydramnios
  anencephaly, 30/176, 178, 42/12
  Edwards syndrome, 43/535
  fumarate hydratase deficiency, 66/420
  thanatophoric dwarfism, 31/274
  triploidy, 43/564, 566
Polyisoprenol alcohol
  long chain, *see* Long chain polyisoprenol alcohol
Polymicrogyria
  see also Brain cortex heterotopia
  agyria, 50/257
  Arnold-Chiari malformation type II, 50/407
  Baller-Gerold syndrome, 43/371
  brain cortex dysplasia, 50/257
  brain cortex laminar necrosis, 50/257
  brain wart, 50/254
  carbon monoxide intoxication, 50/257
  cerebellar heterotopia, 50/251
  cerebrohepatorenal syndrome, 43/340

  congenital rubella, 50/278
  corpus callosum agenesis, 42/9, 50/165
  cytomegalovirus, 56/271
  cytomegalovirus infection, 50/257
  epilepsy, 72/109, 344
  Fukuyama syndrome, 41/38
  hemimegalencephaly, 50/263
  heterotopia, 50/40
  homocystinuria, 55/330
  iniencephaly, 50/131
  Lennox-Gastaut syndrome, 73/218
  macrencephaly, 50/257, 263
  meninges, 50/257
  mental deficiency, 46/11
  microcephaly, 50/262
  microgyria, 30/484, 50/254
  middle cerebral artery occlusion, 50/257
  neurofibromatosis type I, 14/37
  4p partial monosomy, 43/497
  4p syndrome, 31/557
  pachygyria, 50/257
  Pena-Shokeir syndrome type I, 43/438
  phenylketonuria, 29/33
  porencephaly, 50/257
  spina bifida, 50/486, 488
  status verrucosus, 30/489, 50/202, 253
  Sturge-Weber syndrome, 14/66, 31/21
  syphilis, 50/257
  thanatophoric dysplasia, 50/257
  toxoplasmosis, 50/257
Polymixin B, *see* Polymyxin B
Polymorphism
  oligodendroglioma, 16/14, 22
Polymyalgia
  temporal arteritis, 48/311, 326
Polymyalgia rheumatica
  see also Giant cell arteritis, Neuritis, Temporal
    arteritis *and* Temporal bone fracture
  angiopathic polyneuropathy, 51/446
  autoimmune disease, 41/60, 76, 83, 245
  classification, 71/4
  clinical features, 51/451
  diagnosis, 71/201
  differential diagnosis, 41/60, 245
  giant cell arteritis, 55/346, 71/193
  hypothyroidism, 70/101
  inflammatory myopathy, 41/83
  lymphocyte, 41/76, 55/346
  myalgia, 40/325
  prevalence, 39/327-330
  Sjögren syndrome, 71/74
  survey, 71/191
  symptom, 39/327-330

systemic brain infarction, 11/452
temporal arteritis, 5/119, 48/319, 55/341
Polymyoclonia
  myoclonus, 38/583
  striatonigral degeneration, 49/206
Polymyositis
  see also Dermatomyositis and Inflammatory
    myopathy
  acne fulminans, 62/373
  acquired immune deficiency syndrome, 56/202,
    499, 518, 62/372
  acquired myoglobinuria, 62/577
  acute, see Acute polymyositis
  adult, 71/130
  agammaglobulinemia, 56/203, 62/373
  age, 19/87, 62/372
  amyloid deposit, 62/377
  amyloid β-protein, 62/377
  ankylosing spondylitis, 62/373
  antibody test, 71/137
  anti-Jo-1 antibody, 62/374
  arthralgia, 62/374, 71/130
  aspiration pneumonia, 62/374
  autoimmune antibody, 62/380
  autoimmune disease, 41/54-83
  autoimmune mechanism, 40/264
  autoimmune thrombocytopenia, 62/373
  axonal neuropathy, 62/376
  azathioprine, 41/73, 62/385
  bacterial myositis, 62/372
  basophilic granular inclusion, 62/377
  Behçet syndrome, 62/373
  Bornholm disease, 41/54
  Borrelia burgdorferi, 62/373
  calcinosis, 41/57, 62/386
  calcinosis cutis universalis, 41/57
  cancer, 41/81
  cardiac disorder, 62/374
  cardiac dysrhythmia, 41/73
  cardiomyopathy, 62/374
  carnitine palmitoyltransferase deficiency, 41/56,
    60
  celiac disease, 62/373
  child, 41/56, 71/131
  chronic graft vs host disease, 62/373
  ciclosporin, 62/386, 71/142
  ciguatoxin intoxication, 65/163
  classification, 40/284, 41/55, 71/4
  clinical course, 62/372
  clinical features, 40/297, 324, 335, 41/56
  congestive heart failure, 62/374
  corticosteroid, 41/59, 62/167, 187, 383, 71/141
  corticotropin, 41/54, 74, 79

  course, 62/372
  Coxsackie virus, 56/202
  creatine kinase, 22/25, 38/485, 41/60, 62/375
  cricopharyngeal myotomy, 62/386
  Crohn disease, 62/373
  cyclophosphamide, 62/385, 71/142
  cysticercus, 62/373
  dermatitis herpetiformis, 62/373
  dermatomyositis, 71/129-131
  diagnosis, 62/375
  diagnostic criteria, 22/26, 62/378
  differential diagnosis, 41/59, 177, 202, 245
  Duchenne muscular dystrophy, 40/52, 253, 41/58
  dyspnea, 62/374, 71/130
  Eaton-Lambert myasthenic syndrome, 62/425
  electronmicroscopy, 41/69, 62/377
  EMG, 1/642, 19/87, 22/25, 38/485, 41/60, 62/187,
    375, 71/139
  eosinophilic, see Eosinophilic polymyositis
  eosinophilic cytoplasmic inclusion, 62/377
  epidemiology, 71/129
  erythematosus discoid lupus, 62/373
  erythrocyte sedimentation rate, 41/60
  exclusion criteria, 62/372
  experimental allergic myositis, 40/165
  experimental allergy, 40/164
  facioscapulohumeral muscular dystrophy, 41/59,
    43/99
  facioscapulohumeral syndrome, 62/167
  fever, 38/485, 62/373
  fructose-bisphosphate aldolase, 22/25
  gastrointestinal symptom, 62/374
  giant cell, see Giant cell polymyositis
  granulomatous disease, 62/373
  Hashimoto thyroiditis, 62/373
  heart disease, 41/73
  hematemesis, 62/374
  hepatitis B virus infection, 56/303
  hereditary complement deficiency, 62/373
  histochemistry, 40/52
  histopathology, 41/62
  Hodgkin disease, 39/56
  human foamy retrovirus, 62/382
  human immunodeficiency virus myopathy, 71/374
  human immunodeficiency virus type 1, 56/528,
    62/381
  human T-lymphotropic virus type I, 56/528,
    62/382
  hyperagammaglobulinemic purpura, 62/373
  hypereosinophilic syndrome, 41/65, 62/373
  hypocomplementemia, 41/78
  hypotonia, 40/335
  hypoxemia, 62/374

Ig, 71/142
IgA deficiency, 62/373
immunopathology, 62/379
immunosuppressive therapy, 62/385
inclusion body, 40/264
inflammatory myopathy, 40/284, 41/54-83, 274,
    62/369, 372
intravenous Ig, 62/386
irradiation, 71/143
joint contracture, 62/374
Kawasaki syndrome, 62/373
laboratory, 71/134
laboratory features, 41/60
lactate dehydrogenase, 22/25, 62/375
Legionella pneumophila, 62/373
legionellosis, 62/373
leukapheresis, 62/386, 71/143
limb girdle syndrome, 40/457, 62/187
Lyme disease, 62/373
lymphoid irradiation, 62/386
malaise, 62/373
malignancy, 41/56, 82, 62/374, 71/131
malignant lymphoma, 39/56
melena, 62/374
membranolytic attack complex, 62/379
mercaptopurine, 41/80
methotrexate, 41/80, 62/385, 71/142
mixed connective tissue disease, 62/374
monoclonal gammopathy, 62/373
muscle biopsy, 8/122, 19/87, 38/485, 41/60,
    62/375, 65/202
muscle ischemia, 41/57, 78
myalgia, 22/25, 38/485, 40/324, 41/56, 58, 62/187,
    370
myasthenia gravis, 41/73, 43/157, 62/373, 402
myocarditis, 41/73, 62/374, 402
myoglobinuria, 40/339, 41/58, 60, 274, 277,
    62/556, 577
myophosphorylase deficiency, 41/56, 60
nerve conduction, 71/137
neuropathy, 8/122
nuclear magnetic resonance, 62/379
overlap syndrome, 62/374, 71/131
paraneoplastic syndrome, 69/374, 384
parasitic polymyositis, 62/373
pathogenesis, 41/62, 74, 62/379
pathologic reaction, 40/264
penicillamine, 62/372
physiotherapy, 62/386
picornavirus, 41/71, 62/381
plasmapheresis, 62/386, 71/143
pneumatosis intestinalis, 62/374
pneumonitis, 62/374

practical guideline, 62/386
primary biliary cirrhosis, 62/373
primary inflammatory myositis, 62/369
prognosis, 41/81, 71/130
progressive spinal muscular atrophy, 22/25
psoriasis, 62/373
pulmonary dysfunction, 62/374
quadriceps myopathy, 40/440, 62/174
ragged red fiber, 40/220, 62/292
Raynaud phenomenon, 8/123, 40/438, 41/56, 58,
    62/187, 374
retrovirus, 56/202, 518, 62/381
rhabdomyolysis, 41/60
rheumatoid arthritis, 38/484, 41/56, 59, 77, 62/374
rimmed muscle fiber vacuole, 62/204, 376
sarcoidosis, 41/73, 62/373
schistosomiasis, 41/373
scleroderma, 62/374, 71/105
seasonal onset, 62/372
serum alanine aminotransferase, 62/375
serum aspartate aminotransferase, 62/375
serum creatine kinase, 41/80
serum muscle enzyme, 62/375, 71/135
sex ratio, 8/123, 41/55, 57
simian immunodeficiency virus, 62/382
simian retrovirus type I, 62/382
Sjögren syndrome, 8/17, 62/374, 71/74
skin, 71/130
steroid myopathy, 62/384
systemic lupus erythematosus, 8/122, 41/56, 59,
    62/374
systemic sclerosis, 41/56, 59, 62/374
tachyarrhythmia, 62/374
Toxoplasma gondii, 62/373
treatment, 41/62, 79, 62/383
Trichinella, 62/373
Trypanosoma, 62/373
tryptophan contaminant, 64/252
tubulofilament, 62/377
vasculitis, 41/56, 62/373
viral myositis, 62/381
virus, 40/264
virus isolation, 56/203
weight loss, 38/485, 62/374
Wohlfart-Kugelberg-Welander disease, 59/372
zidovudine, 62/372, 382
Polymyositis-dermatomyositis, see
    Dermatomyositis-polymyositis
Polymyxin
    encephalopathy, 46/602
Polymyxin B
    granulomatous amebic encephalitis, 52/330
    neurotoxin, 65/472

Polyneuritis
  see also Neuritis, Neuropathy *and* Polyneuropathy
  acromicria, 42/341
  acute, *see* Acute polyneuritis
  acute immune mediated, *see* Acute immune
    mediated polyneuritis
  alcoholism, 43/197
  arsphenamine, 7/529, 609, 9/648
  blennorrhagic, *see* Blennorrhagic polyneuritis
  chickenpox, 34/168
  chronic, *see* Chronic polyneuritis
  chronic progressive, *see* Chronic progressive
    polyneuritis
  chronic relapsing, *see* Chronic relapsing
    polyneuritis
  cranial, *see* Multiple cranial neuropathy
  definition, 7/473
  dengue fever, 56/12
  diphtheritic, *see* Diphtheritic neuropathy
  EMG, 1/639
  Epstein-Barr virus infection, 56/252
  Garcin syndrome, 2/313
  Guillain-Barré syndrome, 42/645
  herpes zoster, 34/173-175
  hyperlipoproteinemia type I, 42/443
  infectious mononucleosis, 34/188, 56/252
  Isaacs syndrome, 43/160
  lepromatous sensory, *see* Lepromatous sensory
    polyneuritis
  leprous neuritis, 51/218
  Marek disease, 56/180
  menopause, 42/341
  mental deficiency, 42/341
  mononeuritis multiplex, 7/473
  Morvan fibrillary chorea, 49/559
  motor polyneuropathy, 1/230
  myotonic dystrophy, 43/152
  neurobrucellosis, 33/316, 52/584
  opticoacoustic nerve degeneration, 42/418
  phantom limb, 45/399
  premature menopause, 42/341
  puerperal, *see* Puerperal polyneuritis
  recurrent relapsing, *see* Recurrent relapsing
    polyneuritis
  Refsum disease, 8/22, 10/345, 13/314, 21/199,
    202, 42/148, 66/497
  Rud syndrome, 38/9, 43/284
  subacute, *see* Subacute polyneuritis
  system degeneration, 42/97
  terminology, 7/199
  tetanus toxoid, 52/240
  thalidomide, 42/660
  typhoid fever, 51/187

  varicella zoster virus, 56/229
Polyneuropathy
  see also Neuritis, Neuropathy *and* Polyneuritis
  acetonylacetone intoxication, 64/81
  acquired immune deficiency syndrome, 56/493,
    517
  acquired toxoplasmosis, 35/126, 52/355
  acromegalic, *see* Acromegalic polyneuropathy
  acromegaly, 41/249
  acrylamide, *see* Acrylamide polyneuropathy
  acrylamide intoxication, 64/63
  acute intermittent porphyria, 42/617
  Addison disease, 42/512
  adrenoleukodystrophy, 60/168
  adrenomyeloneuropathy, 60/168, 170
  alcohol, *see* Alcoholic polyneuropathy
  alcoholism, 43/197, 51/68
  allergic granulomatous angiitis, 63/423
  almitrine dimesilate, *see* Almitrine dimesilate
    polyneuropathy
  almitrine intoxication, 63/421
  altretamine, 65/528
  altretamine intoxication, 65/534
  amiodarone, *see* Amiodarone polyneuropathy
  amiodarone intoxication, 65/458
  amyloid, *see* Amyloid polyneuropathy
  amyloid autonomic, *see* Amyloid autonomic
    polyneuropathy
  amyloidosis, 71/504, 507
  amyotrophic lateral sclerosis, 62/413
  angiopathic, *see* Angiopathic polyneuropathy
  anisocoria, 42/97
  antiepileptic agent, 65/499
  antihypertensive agent, 63/87
  areflexia, 60/256
  Argyll Robertson pupil, 42/314
  arsenic, *see* Arsenic polyneuropathy
  arsenic intoxication, 64/284, 75/504
  arthrogryposis multiplex congenita, 42/336
  associated, *see* Associated polyneuropathy
  ataxia, 42/117, 60/670
  ataxia telangiectasia, 60/365
  autonomic, *see* Autonomic polyneuropathy
  autonomic features, 60/256
  autonomic function measurement, 60/261
  autonomic nervous system, *see* Autonomic
    polyneuropathy
  axonal, *see* Axonal polyneuropathy
  axonal degeneration, 60/258
  azacitidine, 65/528
  Bassen-Kornzweig syndrome, 21/52, 22/195,
    60/165
  borderline, *see* Borderline polyneuropathy

botulism, 65/210
broxuridine, 65/529
broxuridine intoxication, 65/538
bulbar paralysis, 42/97
calcium carbimide, *see* Calcium carbimide
    polyneuropathy
cancer, 41/338
cancer patient, 71/674
captopril intoxication, 65/445
carbamate, 64/160
carbamate intoxication, 64/186, 191
carbamazepine, 47/38
carbon disulfide, *see* Carbon disulfide
    polyneuropathy
carbon disulfide intoxication, 6/658, 64/25
carboplatin, 65/529
carboplatin intoxication, 65/537
carcinoma, 38/679-688
cauda equina syndrome, 19/86
cerebellar ataxia, 60/653
cerebral malaria, 35/148, 52/370
cerebrotendinous xanthomatosis, 59/357, 60/166
Chagas disease, 52/346, 348
chloroquine, 65/484
chloroquine intoxication, 65/484
chronic demyelinating, *see* Chronic demyelinating
    polyneuropathy
chronic inflammatory demyelinating, *see* Chronic
    inflammatory demyelinating polyneuropathy
chronic obstructive pulmonary disease, 63/421
chronic progressive demyelinating, *see* Chronic
    progressive demyelinating polyneuropathy
chronic relapsing, *see* Chronic relapsing
    polyneuropathy
chronic superior mesenteric artery occlusion,
    7/631
ciclosporin intoxication, 65/552
ciguatoxin intoxication, 65/163
cisplatin, *see* Cisplatin polyneuropathy
cisplatin intoxication, 64/358, 65/536
clinical category, 60/254
clinical course, 60/258
clinical features, 41/340, 60/255
clinical work-up, 60/253-265
clioquinol, *see* Subacute myelo-optic neuropathy
CNS degeneration, 42/97
Cockayne-Neill-Dingwall disease, 43/351, 51/399
Cogan syndrome type II, 51/454
compression neuropathy, 51/86
concentric laminated body, 40/109
congenital analgesia, 42/296
congenital myasthenic syndrome, 62/413
cranial, *see* Multiple cranial neuropathy

critical illness, *see* Critical illness polyneuropathy
cryoglobulinemic, *see* Cryoglobulinemic
    polyneuropathy
CSF paraprotein, 42/604
cytarabine intoxication, 65/533
cytostatic agent, 69/459
dapsone intoxication, 65/485
deafness, 60/661
deficiency, *see* Deficiency neuropathy
definition, 60/253
deformity, 60/257
demyelinating, *see* Demyelinating polyneuropathy
demyelination, 41/339, 42/98, 47/38, 60/258
diabetes mellitus, 42/334, 70/133, 144
diabetic, *see* Diabetic polyneuropathy
diabetic autonomic, *see* Diabetic autonomic
    polyneuropathy
diamorphine, 37/389
diamorphine intoxication, 65/356
2,4-dichlorophenoxyacetic acid, *see*
    2,4-Dichlorophenoxyacetic acid polyneuropathy
differential diagnosis, 8/363, 41/291, 345
diketone, *see* Diketone polyneuropathy
dimethylaminopropionitrile, *see*
    Dimethylaminopropionitrile polyneuropathy
dioxin intoxication, 60/167
dipeptidyl carboxypeptidase I inhibitor, 65/445
diphenylhydantoin, *see* Phenytoin polyneuropathy
diphtheria, *see* Diphtheritic polyneuropathy
disability scale, 60/261
distal, *see* Distal polyneuropathy
distal axonal, *see* Distal axonal polyneuropathy
distal motor, *see* Distal motor polyneuropathy
distal symmetric, *see* Distal symmetric
    polyneuropathy
disulfiram, *see* Disulfiram polyneuropathy
dominant ataxia, 60/660
drug induced, *see* Drug induced polyneuropathy
drug related, 41/333
Eaton-Lambert myasthenic syndrome, 62/413,
    425
electrophysiologic study, 60/258
endocrine, *see* Endocrine polyneuropathy
enzyme deficiency, *see* Enzyme deficiency
    polyneuropathy
eosinophilia myalgia syndrome, 64/254
eosinophilic polymyositis, 63/385
ethambutol, *see* Ethambutol polyneuropathy
ethionamide, *see* Ethionamide polyneuropathy
ethionamide intoxication, 65/481
ethylene oxide, *see* Ethylene oxide
    polyneuropathy
etiology, 60/262

etoposide, 65/528
Fabry disease, 42/427, 60/171
facial spasm, 42/314-316
Farber disease, 60/166
flecainide intoxication, 65/456
force measurement, 60/259
Friedreich ataxia, 21/355
furaltadone, 7/537
gastrogenous, see Gastrogenous polyneuropathy
Gaucher disease, 21/65
giant follicular lymphadenopathy, 8/144
glyceryl trinitrate intoxication, 37/453
glycosaminoglycanosis, 60/175
gold, see Gold polyneuropathy
gold intoxication, 64/356
Guillain-Barré syndrome, 41/287-291, 346
hemophilia A, 42/738
hepatic, see Hepatic polyneuropathy
hepatic insufficiency, see Hepatic insufficiency
   polyneuropathy
hepatitis B virus, 56/10
hepatitis B virus infection, 56/297
hereditary, see Hereditary polyneuropathy
hereditary amyloid, see Hereditary amyloid
   polyneuropathy
hereditary cerebellar ataxia, 42/117
hereditary deafness, 22/504
hereditary motor and sensory neuropathy type I,
   41/432
hereditary motor and sensory neuropathy type II,
   41/432
hereditary recurrent, see Hereditary recurrent
   polyneuropathy
hereditary sensory and autonomic neuropathy type
   I, 41/434
hereditary tyrosinemia, 60/165
heredoataxia (Stephens), 22/207
heterocyclic antidepressant, 65/316
hexacarbon, see Hexacarbon polyneuropathy
hexane, see Hexane polyneuropathy
2-hexanone, see 2-Hexanone polyneuropathy
human immunodeficiency virus neuropathy,
   71/354
hydralazine, see Hydralazine polyneuropathy
hydralazine intoxication, 7/519, 37/432, 65/447
hypereosinophilic syndrome, 63/373
hyperlipidemia, 60/167
hyperthyroid, see Hyperthyroid polyneuropathy
hypertrophic, see Hypertrophic polyneuropathy
hypertrophic interstitial neuropathy, see
   Hypertrophic polyneuropathy
hypoesthesia, 60/661
hypothyroid, see Hypothyroid polyneuropathy

hypothyroidism, 41/244, 70/99
idiopathic hypereosinophilic syndrome, 63/138
IgM paraproteinemic, see IgM paraproteinemic
   polyneuropathy
inflammatory demyelinating, see Inflammatory
   demyelinating polyneuropathy
α-interferon intoxication, 65/529, 538
β-interferon intoxication, 65/529
involvement pattern, 60/257
isoniazid, see Isoniazid polyneuropathy
isoniazid intoxication, 65/480
Jamaican, see Jamaican polyneuropathy
juvenile metachromatic leukodystrophy, 66/168
Kearns-Sayre-Daroff-Shy syndrome, 62/309, 313
Kii Peninsula amyotrophic lateral sclerosis,
   22/371
Köhlmeier-Degos disease, 39/437, 51/454
late onset ataxia, 60/660, 670
lead, see Lead polyneuropathy
lead encephalopathy, 64/436
leukemia, 63/343
lipidosis, 60/165-177
lithium intoxication, 64/295
lymphocytic leukemia, 63/343
lymphoma, 69/278
macroglobulinemia, 8/146
malignant, 69/94
Marinesco-Sjögren syndrome, 60/343
mental deficiency, 42/335, 341
mercury, see Mercury polyneuropathy
metabolic, see Metabolic polyneuropathy
metachromatic leukodystrophy, 51/368, 59/356
methyl n-butyl ketone, see 2-Hexanone
   polyneuropathy
metronidazole, see Metronidazole polyneuropathy
metronidazole intoxication, 65/487
Minamata disease, 64/415
misonidazole, see Misonidazole polyneuropathy
misonidazole intoxication, 65/529, 538
MNGIE syndrome, 62/505
Möbius syndrome, 50/215, 59/375
monoclonal gammopathy, 63/398
mononeuropathy, 60/254
motor, see Motor polyneuropathy
motor sign, 60/255
motor symptom, 60/255
multifocal neuropathy, 60/254
multiple myeloma, see Multiple myeloma
   polyneuropathy
multiple organ failure, 71/534, 540
myasthenia gravis, 62/413
myeloid leukemia, 63/343
myeloma, 69/304

nerve biopsy, 60/259
nerve conduction velocity, 42/315, 60/670
nerve thickening, 60/257
neurofilament, *see* Neurofilament polyneuropathy
neurolipidosis, 60/165-177
neuropathy quantification, 60/259
Niemann-Pick disease type C, 21/66
nitrofural, 7/537
nitrofurantoin, *see* Nitrofurantoin polyneuropathy
nitrofurantoin intoxication, 65/487
nonsteroid anti-inflammatory agent intoxication,
    65/416
nutritional, *see* Beriberi
ophthalmoplegia, 22/194
opiate intoxication, 37/389, 65/356
optic atrophy, 60/661
opticoacoustic nerve degeneration, 42/418
organic acid metabolism, 66/641
organophosphate induced delayed, *see*
    Organophosphate induced delayed
    polyneuropathy
orthostatic hypotension, 22/262, 51/416, 63/146
pain, 69/72
paracarcinomatous, *see* Paraneoplastic
    polyneuropathy
paramalignancy, *see* Paraneoplastic
    polyneuropathy
paraneoplastic, *see* Paraneoplastic polyneuropathy
paraproteinemia, 63/398
paraproteinemic, *see* Paraproteinemic
    polyneuropathy
Parkinson disease, 49/108
pathogenesis, 41/346
pellagra, 7/571-577, 28/71, 86, 46/337, 51/333
Pena-Shokeir syndrome type I, 59/369
penicillamine, *see* Penicillamine polyneuropathy
perhexiline maleate, *see* Perhexiline maleate
    polyneuropathy
phenytoin, *see* Phenytoin polyneuropathy
phenytoin intoxication, 65/503
platinum, *see* Platinum polyneuropathy
plexopathy, 60/254
poliomyelitis, 62/413
polyarteritis nodosa, 8/125, 9/609, 11/451, 39/304,
    51/452, 55/354, 396
polychlorinated dibenzodioxin intoxication,
    64/205
polycythemia vera, 55/468
porphyria, 8/366, 41/434
postural tremor, 49/588
primary amyloidosis, 51/416
procarbazine, 65/528
procarbazine intoxication, 65/534

progressive external ophthalmoplegia, 22/194,
    207, 43/136, 62/309, 313
pyridoxine deficiency, *see* Vitamin B6 deficiency
    polyneuropathy
pyrithione zinc, 65/478
rapid ascending, *see* Rapid ascending
    polyneuropathy
recurrent nonhereditary, *see* Recurrent
    nonhereditary polyneuropathy
recurring, *see* Recurrent polyneuropathy
Refsum disease, 8/21, 21/199, 22/501, 27/523,
    36/348, 41/433, 42/148, 51/384, 60/165, 66/485,
    489
relapsing, *see* Relapsing peripheral neuropathy
remitting, *see* Remitting peripheral neuropathy
renal insufficiency, *see* Renal insufficiency
    polyneuropathy
rheumatoid arthritis, 8/127, 38/479-481, 51/446,
    71/16
rigidity, 60/660, 670
Rosenberg-Chutorian syndrome, 22/505
Rud syndrome, 51/398, 60/721
sarcoidosis, 51/196
scleroderma, 71/110
secondary pigmentary retinal degeneration,
    42/334, 60/653, 670
sensation measurement, 60/260
sensorimotor, *see* Sensorimotor polyneuropathy
sensorineural deafness, 42/373
sensory, *see* Sensory polyneuropathy
sensory loss, 42/97, 51/55, 60/256
sensory symptom, 60/255
serum paraprotein, 42/604
Sjögren syndrome, 71/83
sleep, 74/552
slow acetylator, 63/87
spinal cord electrotrauma, 61/192
streptomycin, *see* Streptomycin polyneuropathy
streptomycin intoxication, 65/483
subclinical, *see* Subclinical polyneuropathy
suramin, 65/529
suramin intoxication, 65/538
sweating measurement, 60/261
symmetrical, *see* Symmetrical polyneuropathy
Tangier disease, 8/20, 10/548, 21/52, 42/627,
    60/135, 165
taxol, 65/529
taxotere, 65/529
terminology, 7/199
tetanus, 65/218
thalidomide, *see* Thalidomide polyneuropathy
thalidomide intoxication, 65/559
thallium, *see* Thallium polyneuropathy

thallium intoxication, 75/503
toluene intoxication, 36/373
tongue paralysis, 42/97
toxic, *see* Toxic polyneuropathy
toxic oil syndrome, 63/380
toxoplasmosis, 52/355
treatment, 41/345
trichloroethylene, *see* Trichloroethylene
   polyneuropathy
tri-*o*-cresyl phosphate, *see* Tri-*o*-cresyl phosphate
   polyneuropathy
tri-*o*-cresyl phosphate intoxication, 64/160
tricyclic antidepressant, 65/316
trophic change, 60/257
tryptophan contaminant, 64/254
undetermined significant monoclonal
   gammopathy, 63/398, 402
undetermined significant monoclonal
   gammopathy IgM, 63/398, 402
upper limit, 42/337
uremic, *see* Uremic polyneuropathy
Valsalva maneuver, 60/261
vincristine, *see* Vincristine polyneuropathy
vincristine intoxication, 65/528, 534
vindesine intoxication, 65/535
vitamin B1 deficiency, 7/552
vitamin B3 deficiency, 51/333
vitamin B3 deficiency encephalopathy, 46/337
vitamin B5 deficiency, 51/341
vitamin B6 deficiency, *see* Vitamin B6 deficiency
   polyneuropathy
vitamin B6 intoxication, 65/480, 572
vitamin B12 deficiency, 7/553, 51/340
vitamin Bc deficiency, 51/336
Waldenström macroglobulinemia, *see*
   Waldenström macroglobulinemia
   polyneuropathy
Wegener granulomatosis, 8/128, 63/423
Whipple disease, 51/184, 52/138
Wolman disease, 60/165
xeroderma pigmentosum, 60/336
zimeldine, *see* Zimeldine polyneuropathy
Polyol pathway
   see also Sorbitol
   diabetic neuropathy, 27/93, 110
   diabetic polyneuropathy, 51/506
   hyperglycemia, 27/79-83, 86
   hyperosmolar hyperglycemic nonketotic diabetic
     coma, 27/91
   sorbitol, 27/80, 86
Polyoma
   progressive multifocal leukoencephalopathy,
     9/491

Polyomavirus
   acute viral myositis, 56/194
   neurotropic virus, 56/9
   progressive multifocal leukoencephalopathy,
     9/491, 47/504, 508, 511, 515, 519, 56/10, 71/417
Polyomavirus BK
   progressive multifocal leukoencephalopathy,
     47/504, 56/10
Polyomavirus JC
   progressive multifocal leukoencephalopathy,
     47/504, 56/10
Polyoncosis
   hereditary cutaneomandibular, *see* Multiple
     nevoid basal cell carcinoma syndrome
Polyopia
   hallucination classification, 45/56
   monocular, *see* Monocular polyopia
   occipital lobe tumor, 17/331
   parietal lobe, 45/363
   pseudofovea, 45/364
Polyostotic fibrocystic dysplasia
   Albright syndrome, 14/12, 163
Polyostotic fibrous dysplasia, *see* Albright
   syndrome
Polypeptide
   pancreatic, *see* Pancreatic polypeptide
   vasoactive intestinal, *see* Vasoactive intestinal
     polypeptide
Polyphagia, *see* Bulimia
Polypheny, *see* Pleiotropism
Polyphosphoinositide
   synthesis, 9/16
Polyploidy
   see also Chromosome number abnormality
   anophthalmia, 50/555
   arhinencephaly, 50/555
   Arnold-Chiari malformation type I, 50/555
   cerebellar hypoplasia, 50/555
   chromosomal aberration, 30/101
   chromosome number abnormality, 50/541
   corpus callosum agenesis, 50/555
   cyclopia, 50/555
   Dandy-Walker syndrome, 50/555
   definition, 50/540
   holoprosencephaly, 50/555
   hydranencephaly, 50/555
   hydrocephalus, 50/555
   meningomyelocele, 50/556
   microphthalmia, 50/555
   septum pellucidum agenesis, 50/555
   tetraploidy, 50/555
   triploidy, 50/555
   XXX syndrome, 50/553

XXXX syndrome, 50/553
XXXXX syndrome, 50/553
Polyposis
  colonic, *see* Colonic polyposis
  Peutz-Jeghers syndrome, 14/12, 121, 526, 601,
    43/41
Polyradiculitis
  see also Guillain-Barré syndrome
  brain, 16/223
  chickenpox, 34/168
  hepatitis, 7/486
  herpes simplex, 34/156
  herpes simplex virus, 56/221
  herpes zoster, 7/483
  influenza, 7/478
  leptospirosis, 7/489
  mumps, 7/485, 34/411
  nerve conduction, 7/175
  neurobrucellosis, 7/481, 52/594
  rubella, 7/486
  scarlet fever, 7/487
  smallpox, 7/487
Polyradiculomyelitis
  acquired immune deficiency syndrome, 56/513,
    71/264
Polyradiculomyelopathy
  tetanus toxoid, 52/240
Polyradiculoneuritis
  acute, *see* Guillain-Barré syndrome
  bulbar syncope, 15/824
  Landry paralysis, 7/487
Polyradiculoneuropathy
  acute inflammatory, *see* Acute inflammatory
    polyradiculoneuropathy
  arthropod envenomation, 65/198
  bee venom, 65/198
  chronic idiopathic demyelinating, *see* Chronic
    idiopathic demyelinating polyradiculo-
    neuropathy
  chronic inflammatory demyelinating, *see* Chronic
    inflammatory demyelinating polyradiculo-
    neuropathy
  eosinophilia myalgia syndrome, 64/254
  Guillain-Barré, *see* Guillain-Barré syndrome
  inflammatory, *see* Inflammatory polyradiculo-
    neuropathy
  paraneoplastic syndrome, 69/374, 379
  sensorimotor, *see* Guillain-Barré syndrome
  wasp venom, 65/198
Polyradiculopathy
  Burkitt lymphoma, 63/352
  chronic inflammatory, *see* Chronic inflammatory
    polyradiculopathy

cytomegalovirus, *see* Cytomegalovirus
    polyradiculopathy
  Köhlmeier-Degos disease, 51/454
  tetanus toxoid, 52/240
Polyradiculopathy syndrome
  dysarthria, 42/97
  dysphagia, 42/97
  familial, *see* Familial polyradiculopathy syndrome
Polysaccharide
  disorder, 27/143-163
  enzyme infusion, 27/163
  histochemistry, 9/24
  leukocyte infusion, 27/163
  myelin, 7/52
  PAS stain, 9/24
  plasma infusion, 27/163
  prenatal diagnosis, 27/162
  renal transplantation, 27/163
  senile plaque, 46/253
  treatment, 27/163
Polysaccharide antigen
  Escherichia coli, 33/61
  meningococcal meningitis, 33/26
Polysaccharide storage disease
  muscle fiber feature, 62/17
Polysialoganglioside
  sialic acid, 66/250
Polysomnography
  daytime hypersomnia, 45/135
  narcolepsy, 45/149
  obstructive sleep apnea, 63/457
  restless legs syndrome, 51/544
  sleep apnea syndrome, 45/132
Polysomy
  definition, 31/491
Polysyndactyly
  acrocephalosyndactyly type I, 50/122
  acrocephalosyndactyly type II, 38/422
  Saldino-Noonan syndrome, 31/243
Polytopic angioblastosis
  congenital, *see* Von Hippel-Lindau disease
Polyunsaturated fatty acid
  biochemistry, 9/5
  20 C, 9/6
  22 C, 9/6
  multiple sclerosis, 47/154, 187
  neuronal ceroid lipofuscinosis, 10/671
Polyuria
  Hand-Schüller-Christian disease, 38/95
  migraine, 48/156
  subarachnoid hemorrhage, 12/151
Pompe disease, *see* Acid maltase deficiency
Pons

apneustic center, 63/479
arterial hypertension, 63/74
autonomic nervous system, 74/152, 155
brain stem lacunar infarction, 53/394
hypertensive fibrinoid arteritis, 11/136
intrinsic arterial pattern, 11/30
mercury intoxication, 64/380
oligodendroglioma, 42/332
respiration, 1/652
thalamus, 74/155
Pons syndrome, *see* Pontine syndrome
Pontiac fever, *see* Legionellosis
Ponticulus
lateral, *see* Lateral ponticulus
posterior, *see* Foramen arcuale
Pontine artery
paramedian, *see* Paramedian pontine artery
Pontine degeneration
spinal, *see* Spinopontine degeneration
Pontine demyelination
concentric, 10/11
Pontine dystrophy
central, *see* Central pontine dystrophy
Pontine gaze center
topographical diagnosis, 2/335
Pontine gaze paralysis
bilateral, 2/336
Pontine glaucoma
neurofibromatosis, 18/397
Pontine glioma
see also Pontine tumor
Cheyne-Stokes respiration, 63/440
clinical features, 18/398-400
cyclic oculomotor paralysis, 42/332
episodic hyperpnea, 63/440
Horner syndrome, 18/398
hyperpnea, 63/440
hyperventilation, 63/440
posterior fossa tumor, 18/397
Pontine hematoma
brain embolism, 53/207
brain infarction, 53/207
cerebrovascular disease, 53/207
intracranial hemorrhage, 53/207
transient ischemic attack, 53/207
Pontine hemiplegia
clinical features, 1/187
transient ischemic attack, 12/23
unilateral paramedian pontine syndrome, 12/16
Pontine hemorrhage
age, 12/38
anatomic site, 12/40
angionecrosis, 12/49

ataxic hemiparesis, 54/313
basal, 54/313
basotegmental, 54/313
bilateral, *see* Bilateral pontine hemorrhage
cause, 12/38
cause hypertension, 12/39
cerebellar hemorrhage, 12/45
Charcot-Bouchard aneurysm, 12/48-50
Cheyne-Stokes respiration, 2/253
clinical features, 12/39
clinicopathologic correlation, 12/40, 50
coma, 12/44, 50
computer assisted tomography, 54/68
computer assisted tomography *vs* autopsy, 54/310
consciousness, 12/39
course, 12/44
cranial nerve, 12/22, 40
differential diagnosis, 12/43
duration of life, 12/38
early symptom, 12/39
epilepsy, 2/253, 12/39
gaze paralysis, 2/254
headache, 5/139, 48/279
hypertensive encephalopathy, 12/45, 49
hypothermia, 12/50
involuntary movement, 2/253
miosis, 2/253
mode of onset, 12/39
mydriasis, 2/253
nuclear magnetic resonance, 54/313
ocular bobbing, 12/43, 57
paramedian, 54/311
pathogenesis, 12/48
pathology, 12/46-48
posterior fossa tumor, 18/397
pre-existing condition, 12/38
prevalence, 12/38, 54/68
prevalence distribution, 12/38
primary, *see* Primary pontine hemorrhage
prodromic symptom, 12/39
pure motor hemiplegia, 54/313
sex ratio, 12/38
site, 12/40
subarachnoid hemorrhage, 12/45
symptom, 12/51
symptomatology, 2/252-254
syndrome, 2/251
tegmental, *see* Tegmental pontine hemorrhage
treatment, 12/51
trismus, 2/253
unilateral, *see* Unilateral pontine hemorrhage
vascular hypertension, 12/49
Pontine hyperpathia

156

lateral pontine syndrome, 12/17
Pontine infarction, 12/13-33
  see also Pontine syndrome
  abducent nerve, 12/16
  age, 12/14
  agitation, 12/15
  akinetic mutism, 12/15, 20
  anatomic aspect, 12/24
  anemic type, 12/26
  ataxia, 12/15, 53/390
  ataxic hemiparesis, 53/394
  basilar artery occlusion, 2/242, 12/13, 53/388
  basilar artery stenosis, 2/244, 12/13, 19-26
  bilateral site, 12/26
  brain blood flow, 53/61
  brain stem auditory evoked potential, 53/393
  cause, 2/242, 12/31
  Cestan-Chenais syndrome, 12/13
  Cheyne-Stokes respiration, 12/15
  clinical features, 12/14
  clinicopathologic correlation, 12/30
  clumsy hand dysarthria syndrome, 53/394
  coagulant treatment, 12/32
  coma, 2/249, 12/15, 20
  confusional state, 12/15, 20, 23, 30
  conjugate gaze, 12/16
  consciousness, 2/244, 12/15, 30
  convulsive seizure, 12/15, 30, 51
  cranial nerve, 12/16, 22
  decerebrate rigidity, 2/242, 12/15
  decerebration, 53/389
  diagnosis, 12/22
  differential diagnosis, 12/22
  diplegia, 53/389
  diplopia, 12/22
  divergent strabismus, 53/389
  dizziness, 12/15, 23
  dysarthria, 2/242, 261, 12/15, 15/23, 53/389, 394
  dysphagia, 12/15, 53/389
  early symptom, 12/15
  encephalosis, 12/28
  EOG, 54/163
  epidemiology, 12/14
  epilepsy, 12/20, 30
  facial nerve, 12/16
  facial paresthesia, 53/390
  fat embolism, 12/44
  fibrinolytic treatment, 12/33
  Foville syndrome, 2/242, 261, 12/13
  generalized weakness, 12/15
  headache, 2/241, 12/15
  hemiparesis, 12/15, 23
  hemiplegia, 12/15, 23

  hemorrhage, 2/250, 12/26
  history, 12/13
  hoarseness, 53/389
  horizontal gaze paralysis, 53/389
  hypertensive encephalopathy, 12/24
  internuclear ophthalmoplegia, 53/389, 394
  jaw lock, 53/389
  lacunar, 12/28
  locked-in syndrome, 53/389
  Marie-Foix-Alajouanine disease, 2/240, 242, 244, 264
  meiosis, 53/390
  Millard-Gubler syndrome, 2/242, 12/13
  multiple cranial nerve, 12/16
  myoclonus, 53/389
  nystagmus, 53/390, 394
  ocular bobbing, 53/390
  oculomotor nerve, 12/16, 22
  palatal myoclonus, 53/390
  paraplegia, 12/15, 23
  paresthesia, 2/241, 12/15
  pathologic laughing and crying, 2/241, 12/23
  pinpoint pupil, 53/390
  pontine syndrome, 12/16-23
  posterior inferior cerebellar artery, 2/244
  prodromal symptom, 12/15
  pseudobulbar paralysis, 2/241, 12/23, 31
  pseudobulbar symptom, 12/31
  ptosis, 53/390
  pure motor hemiplegia, 53/394
  respiratory rate, 12/15, 31
  scotoma, 12/22
  sex ratio, 12/14
  shivering, 53/389
  site, 2/243
  somnolence, 12/15
  stupor, 12/15
  symptom, 53/389
  table, 2/243
  tetraplegia, 12/15
  thermoregulation, 12/15, 31
  transient ischemic attack, 2/242, 12/14, 22
  treatment, 12/32, 51
  trigeminal nerve, 12/16
  type, 12/16
  vertebrobasilar system syndrome, 53/388
  vertigo, 2/241, 12/15, 22
  visual impairment, 12/15, 22
  Weber syndrome, 12/16
Pontine ischemia
  brain stem auditory evoked potential, 54/162
Pontine lacuna
  clumsy hand dysarthria syndrome, 54/249

one and one half syndrome, 54/245
pure motor hemiplegia, 54/240
Pontine lacunar infarction
  ataxic hemiparesis, 54/248
  sensorimotor stroke, 54/253
Pontine lesion
  decerebrate rigidity, 2/242
  dreaming, 3/99
  football injury, 23/574
  hereditary motor and sensory neuropathy type I,
    21/302
  hereditary motor and sensory neuropathy type II,
    21/302
  involuntary movement, 2/251
  multifocal, *see* Multifocal pontine lesion
  nystagmus, 2/249
  pathologic laughing and crying, 2/241
  trigeminal nerve agenesis, 50/214
Pontine myelinolysis
  central, *see* Central pontine myelinolysis
  extra, *see* Extra pontine myelinolysis
  leukemia, 39/12
Pontine paramedian reticular formation
  cerebrovascular disease, 55/124
Pontine paraplegia
  cause, 1/210
  symptom, 12/20
Pontine plicature
  Arnold-Chiari malformation type II, 50/405
Pontine syndrome, 2/238-265
  see also Foville syndrome *and* Pontine infarction
  anatomy, 12/24-29
  anterior, *see* Dejerine syndrome
  bilateral, *see* Bilateral pontine syndrome
  cavernous angioma, 2/259
  cerebrovascular disease, 55/124
  circumferential artery, 2/241, 244
  classification, 2/264
  crossed hemiplegia, 1/187, 2/240-261
  diagram, 2/264
  differential diagnosis, 12/22
  dorsocaudal type, *see* Foville syndrome
  dorsolaterocaudal type, *see* Gasperini syndrome
  Gellé syndrome, 2/241
  histopathology, 2/241-264
  infarction, 2/241
  ischemia, 2/241
  lateral, *see* Lateral pontine syndrome
  nystagmus, 55/125
  onset type, 2/241
  paramedian, *see* Paramedian pontine syndrome
  paraplegia, 1/210
  patchy, *see* Patchy pontine syndrome

pathology, 12/24-29
pontine infarction, 12/16-23
tegmental, *see* Tegmental pontine syndrome
telangiectasia, 2/259, 18/270
trauma, 2/262
tuberculoma, 2/259
tumor, 2/259
type, 2/263
unilateral paramedian, *see* Unilateral paramedian
  pontine syndrome
velopalatine myoclonus, 2/240
ventral, *see* Ventral pontine syndrome
ventrocaudal type, *see* Millard-Gubler syndrome
Pontine tegmentum syndrome, *see* Tegmental
  pontine syndrome
Pontine tumor
  see also Pontine glioma
  age at onset, 17/695
  astrocytoma, 17/700
  brain scanning, 17/699
  brain stem tumor, 68/369
  clinical syndrome, 17/695
  CSF, 17/697
  cytotoxic drug treatment, 17/704
  differential diagnosis, 17/696
  high voltage treatment, 17/703
  history, 17/693
  literature, 17/694
  pathology, 17/700
  radiology, 17/697
  symptom, 67/158
  treatment, 17/702
Pontine vein
  radioanatomy, 11/110
  topography, 11/110
  ventrolateral cerebellar, *see* Ventrolateral
    cerebellopontine vein
Pontocerebellar hypoplasia
  computer assisted tomography, 60/672
  infantile spinal muscular atrophy, *see* Norman
    disease
Pontomesencephalic vein
  anterior, *see* Anterior pontomesencephalic vein
Pontoneocerebellar atrophy (Brun)
  hereditary congenital cerebellar atrophy, 60/290
Pontoneocerebellar hypoplasia
  cerebellar agenesis, 50/192
Ponto-olivary atrophy
  olivopontocerebellar atrophy (Dejerine-Thomas),
    21/418
Popliteal nerve
  lateral, *see* Lateral popliteal nerve
Popliteal nerve palsy

lateral, *see* Lateral popliteal nerve palsy
Poppelreuter test
  agnosia, 4/16, 19
Porcine hemagglutinating encephalomyelitis virus
  animal model, 56/439
  coronavirus, 56/439
  pathogenesis, 56/441
  viral encephalitis, 56/441
  virus spread, 56/441
Porencephalic cyst, *see* Porencephaly
Porencephaly, 30/681-691
  amniocentesis, 50/359
  arachnoid cyst, 31/119
  birth incidence, 42/49
  blindness, 30/685, 50/359
  brain injury, 23/307
  brain scintigraphy, 23/307
  cerebellum, 30/683, 50/357
  clastic, 30/484
  computer assisted tomography, 42/49
  congenital hip dislocation, 50/360
  congenital tactile agnosia, 4/34
  corpus callosum agenesis, 2/764, 31/119, 42/7
  cryptorchidism, 50/360
  CSF, 42/48
  cyanotic heart disease, 50/359
  cystic brain degeneration, 30/682, 50/356
  cytomegalovirus, 56/271
  Dandy-Walker syndrome, 42/23
  decerebrate rigidity, 30/685, 50/359
  definition, 23/307, 30/681, 42/48
  diagnosis, 30/690, 50/360
  differential diagnosis, 30/691
  double, 30/484
  dysphagia, 30/685, 50/359
  EEG, 30/690
  encephalocele, 30/682, 50/356
  encephaloclastic, *see* Encephaloclastic
    porencephaly
  epidemiology, 30/158, 681, 50/355
  epilepsy, 30/686, 42/49, 50/359
  etiology, 30/685, 42/49, 50/358
  false, *see* False porencephaly
  familial incidence, 30/685, 42/49, 50/358
  genetics, 50/358
  gray matter heterotopia, 50/356
  head injury, 23/307, 42/48
  hemianopia, 42/49
  hemiplegia, 42/49
  hydranencephaly, 30/682, 684, 50/340, 346, 356,
    358
  hydrocephalus, 30/158, 684, 42/48
  incidence, 42/49, 50/63

  incontinentia pigmenti, 14/85
  intracranial pressure, 42/49
  ischemic infarction, 42/48
  Lowe syndrome, 42/607
  macrocephaly, 30/690, 50/360
  management, 30/691
  mental deficiency, 30/681, 685-687, 691, 46/9-11,
    50/359
  microcephaly, 30/685, 50/359
  Mohr syndrome, 43/450
  morbid anatomy, 1/199
  morphology, 50/356
  moyamoya disease, 50/359
  neuronal migration disorder, 42/48
  normal intelligence, 46/11
  optic atrophy, 30/685, 50/359
  optic nerve agenesis, 50/211
  pachygyria, 50/356
  pathogenesis, 30/683, 50/357
  pathology, 30/682, 686-690
  PEG, 23/307
  polymicrogyria, 50/257
  prenatal vascular disease, 53/27
  pseudo, *see* Pseudoporencephaly
  pseudobulbar paralysis, 1/199
  RIHSA, 23/307
  schizencephaly, 50/198, 357
  seizure, 42/48, 50/359
  spastic tetraplegia, 50/359
  spinal, 30/683
  spinal cord, 50/357
  synonym, 30/681
  thrombocytopenic purpura, 50/359
  toxoplasmosis, 50/359
  treatment, 42/49, 50/361
  true, *see* True porencephaly
  ventriculitis, 42/48
Porifera intoxication
  animal experiment, 37/36
  skin lesion, 37/35
Poriomania
  migraine, 48/162, 164
Porokeratosis (Mibelli), *see* Mibelli angiokeratoma
    syndrome
Porot-Filiu syndrome
  EEG, 42/688
  heterochromia iridis, 42/688
  mental deficiency, 42/688
Porphobilinogen
  acute intermittent porphyria, 8/9, 366, 41/334,
    42/618
  porphyria variegata, 8/9, 42/620
Porphobilinogen deaminase

acute intermittent porphyria, 51/390
Porphobilinogen synthase
  lead, 64/434
  lead polyneuropathy, 51/270
  organolead intoxication, 64/134
Porphyria, 27/429-443
  abdominal colic, 63/150
  acute intermittent, see Acute intermittent
    porphyria
  antiepileptic agent, 65/497
  autonomic dysfunction, 63/148, 75/695
  autonomic nervous system, 75/31
  barbiturate, 37/410
  biochemical features, 27/443
  bowel, 75/650
  brain infarction, 8/10
  bulbar syncope, 15/824
  chromosome 11, 60/119
  classification, 27/430
  clinical features, 60/117
  congenital erythropoietic, see Congenital
    erythropoietic porphyria
  copro, see Coproporphyria
  demyelination, 9/639
  differential diagnosis, 60/121, 70/308
  drug induced, 70/305
  electrophysiologic study, 60/118
  enzyme, 70/307
  enzyme deficiency polyneuropathy, 51/367
  enzyme involved, 60/121
  epidemiology, 60/119
  genetics, 60/119
  hepatic, see Acute intermittent porphyria,
    δ-Aminolevulinic acid dehydratase deficiency
    porphyria, Hereditary coproporphyria and
    Porphyria variegata
  hereditary acute, see Hereditary acute porphyria
  hereditary copro, see Hereditary coproporphyria
  history, 70/299
  inducing compound, 27/433
  intermittent, see Intermittent porphyria
  isolated paralysis, 60/117
  leukoencephalopathy, 9/639, 47/585
  metabolic defect, 60/119
  metabolism, see Porphyrin metabolism
  mixed hepatic, see Porphyria variegata
  nerve biopsy, 60/118
  neuropathology, 60/118
  neuropathy, 7/554, 8/9, 41/434, 60/117-121
  neurovisceral crisis, 60/117
  orthostatic hypotension, 22/263, 63/148
  pathogenesis, 27/442
  periodic paralysis, 41/167

polyneuropathy, 8/366, 41/434
porphyrin metabolism, 27/429
prognosis, 70/310
proto, see Protoporphyria
protocoproporphyria, see Porphyria variegata
rabies, 56/390
respiratory muscle paralysis, 60/118
sensation, 60/117
South African type, see Porphyria variegata
subacute polyneuritis, 60/117
survey, 70/299
Swedish type, see Acute intermittent porphyria
symptomatic, see Symptomatic porphyria
systemic neoplastic disease, 41/334
treatment, 60/119, 121, 70/309
tremor, 49/589
Porphyria cutanea tarda, see Symptomatic porphyria
Porphyria variegata
  aminolevulinic acid, 8/9, 42/620
  autonomic polyneuropathy, 51/479
  cutaneous manifestation, 27/437
  cutaneous pigmentation, 42/619
  features, 27/437
  ferrochelatase, 42/620
  hirsutism, 42/620
  incidence, 27/437
  metabolic neuropathy, 51/68
  neurologic manifestation, 27/437
  neuropathy, 60/120
  photosensitivity, 42/620
  porphobilinogen, 8/9, 42/620
  porphyrin metabolism, 27/438
  prevalence, 42/620
  protoporphyrinogen IX oxidase, 51/390
  protoporphyrinogen oxidase, 42/620
  skin fragility, 42/620
Porphyrin metabolism, 27/429-433
  acute intermittent porphyria, 27/435-437
  aminolevulinic acid synthetase, 27/429, 431, 433
  excretion, 27/440-442
  glia, 9/14
  heme regulation, 27/432
  lead polyneuropathy, 51/270
  porphyria, 27/429
  porphyria variegata, 27/438
  symptomatic porphyria, 27/437-440
  xeroderma pigmentosum, 14/13
Porpoise intoxication
  abdominal pain, 37/95
  cyanosis, 37/95
  hypersalivation, 37/95
  paralysis, 37/95
  tongue swelling, 37/95

Porropsia
  higher nervous activity disorder, 3/29
Port wine hemangioma, *see* Nevus flammeus
Port wine nevus, *see* Nevus flammeus
Portnoy valve
  hydrocephalus, 30/576
Portosystemic anastomosis encephalopathy
  Creutzfeldt-Jakob disease, 56/546
Portugal
  neurology, 1/11
Poser classification
  diffuse sclerosis, 9/472
Poser-Van Bogaert classification
  diffuse sclerosis, 9/472, 10/18
Position
  phantom limb, 4/225, 45/397
  sensation, 1/95
  syringobulbia, 50/457
Positional nystagmus
  benign paroxysmal type, *see* Benign paroxysmal
    positional nystagmus
  central type, 2/371
  differentiation, 2/372
  multiple sclerosis, 47/133
  otolith lesion, 2/371
  retinovestibular hereditary spastic ataxia, 21/387
  table, 2/372
  test, 2/370
  utriculitis, 2/371
Positive end expiratory pressure
  head injury, 57/229
  spinal cord injury, 61/265
Positive spike
  14 and 6 cps, 2/453
Positive wave
  EMG, 40/143
Positron emission tomography
  alcoholism, 60/673
  Alzheimer disease, 46/264
  aphasia, 45/299
  blood flow, 72/371, 375
  brain blood flow model, 54/109
  brain diaschisis, 54/125
  brain embolism, 53/232
  brain glucose metabolism, 49/36
  brain infarction, 53/232, 313, 54/65, 87, 120
  brain ischemia, 53/267
  brain lacunar infarction, 54/124, 256
  camera, 54/113
  cerebrovascular disease, 53/232, 54/105
  chorea, 49/42
  chorea-acanthocytosis, 63/281
  cocaine intoxication, 65/254

  computer assisted tomography, 54/65
  cyclotron, 54/107
  deoxyglucose C 11, 54/110
  deoxyglucose F 18, 54/110
  deoxyglucose model, 54/107
  depression, 46/428
  dystonia, 49/41
  epilepsy, 72/184, 371
  equipment, 54/112
  erythropoietin, 63/535
  external internal carotid bypass, 54/127
  extratemporal epilepsy, 72/375
  glucose, 54/112, 72/371, 380
  Huntington chorea, 49/42, 262, 288
  infarction stage, 54/121
  intracranial hemorrhage, 53/232
  lateral asymmetry, 46/486
  manganese intoxication, 64/310
  Marinesco-Sjögren syndrome, 60/673
  1-methyl-4-phenyl-1,2,3,6-tetrahydropyridine
    intoxication, 65/386
  moyamoya disease, 55/298
  multi-infarct dementia, 54/120
  neuroleptic parkinsonism, 65/295
  neuropsychology, 45/1, 7
  nuclear magnetic resonance, 54/86
  olivopontocerebellar atrophy (Schut-Haymaker),
    60/673
  oxygen, 54/110, 72/371
  Parkinson disease, 49/35, 98, 75/181
  penumbra, 54/125
  progressive dysautonomia, 59/139
  regional brain blood flow, 54/108
  schizophrenia, 46/486
  Schut family ataxia, 60/484
  sickle cell anemia, 55/511
  striatonigral degeneration, 60/543
  subarachnoid hemorrhage, 54/131
  superior to nuclear magnetic resonance, 54/87
  tetrahydrocannabinol intoxication, 65/47
  transient global amnesia, 45/213
  transient ischemic attack, 53/232, 313, 54/116
Postactivation exhaustion
  Eaton-Lambert myasthenic syndrome, 41/356
  myasthenia gravis, 41/356
Postactivation facilitation
  Eaton-Lambert myasthenic syndrome, 41/356
  myasthenia gravis, 41/356
Postalcohol hypotension
  postprandial, 75/169
Postanesthetic accident
  pallidonigral necrosis, 49/466
Postanesthetic complication

pallidostriatal necrosis, 49/466
Postanesthetic encephalopathy
  akinetic mutism, 49/479
  anoxic ischemic leukoencephalopathy, 9/577
  barbiturate, 9/577, 49/478
  choreoathetosis, 49/479
  clinical features, 49/479
  dementia, 49/479
  intermittent course, 49/479
  intermittent deterioration, 49/478
  nitrous oxide anesthesia, 9/577, 49/478
  pallidal necrosis, 49/466
  pallidonigral necrosis, 49/478
  parkinsonism, 49/479
  pathology, 49/478
  striatal necrosis, 49/500
Postanoxic encephalopathy
  diamorphine intoxication, 65/356
Postanoxic intention myoclonus
  brain injury, 38/580
  cardiac arrest, 38/580
Postanoxic myoclonic encephalopathy, *see*
  Posthypoxic myoclonus
Postanoxic myoclonus, *see* Posthypoxic myoclonus
Postcentral gyrus
  frontal lobe tumor, 17/272
  hemiballismus, 49/370
Postcentral gyrus resection
  sensory extinction, 3/193
Postconcussional syndrome
  accident neurosis, 57/422
  accommodation disorder, 57/424
  brain concussion, 23/424, 454
  characteristic features, 57/422
  child, 23/454, 57/425
  diplopia, 57/424
  dizziness, 23/454, 57/412, 424
  fatigability, 57/412
  head injury, 23/454, 57/375
  head injury outcome, 57/375
  headache, 23/454, 57/412, 424
  history, 57/421
  industrial injury, 57/424
  partner role, 57/426
  postconcussional encephalopathy, 57/421
  postconcussional neurosis, 57/421
  posttraumatic headache, 5/178
  posttraumatic neurosis, 57/412
  preaccident emotion instability, 57/422
  pretraumatic personality, 57/424
  psychiatric headache, 5/182
  railway spine, 57/421
  Renten-Neurose, 57/421

sociopsychologic predisposing factor, 57/424
  study, 57/425
  subjective symptom, 57/423
  symptom, 57/422
  symptom incidence, 57/423
Postencephalitic amyotrophy
  progressive muscular atrophy, 59/20
  review, 22/21-23
Postencephalitic hypersomnia
  secondary narcolepsy, 3/96
Postencephalitic parkinsonism
  Creutzfeldt-Jakob disease, 6/748
  differential diagnosis, 6/748
  dorsal spinal root section, 6/141
  encephalitis lethargica, 22/223, 34/455
  history, 6/174
  muscular atrophy, 22/22
  ocular spasm, 1/291, 6/748
  oculogyric crisis, 6/189, 748
  pallidal atrophy, 6/649
  Parkinson dementia, 46/377
  respiration, 63/481
Postencephalitic syndrome
  intestinal pseudo-obstruction, 51/493
Posterior alexia, 45/436-440
  clinical symptom, 45/436-438
  occipital lobe, 45/438
  pathology, 45/438-440
Posterior arch
  C1, *see* C1 posterior arch
Posterior arch abnormality
  diastematomyelia, 50/436
  diplomyelia, 50/436
Posterior arch hypertrophy
  vertebral canal stenosis, 50/474
Posterior atlantal ponticle, *see* Foramen arcuale
Posterior borderzone
  transcortical sensory aphasia, 45/20
Posterior cavernous sinus syndrome
  infraclinoid aneurysm, 12/89, 31/147
  paratrigeminal syndrome, 2/90
Posterior cerebellar vein
  inferior, *see* Inferior posterior cerebellar vein
  superior, *see* Superior posterior cerebellar vein
Posterior cerebral artery
  anatomy, 11/2, 4, 17-22, 53/409
  anomalous origin, 11/176
  asymmetry, 11/98
  bacterial endocarditis, 52/294
  brain aneurysm, 12/93
  brain embolism, 53/409
  brain infarction, 11/306, 53/409
  circular segment, 11/92

confusional state, 55/139
congenital aneurysm, 31/156
cortical segment, 11/97
Galen vein aneurysm, 18/279
headache, 5/130
middle temporal branch, 11/21
migraine, 48/165
migrainous transient global amnesia, 48/166
mycotic brain aneurysm, 52/294
ophthalmoplegic migraine, 2/295
pathology, 53/409
posterior temporal branch, 11/21
syndrome, 11/306
transcranial Doppler sonography, 54/12
transient global amnesia, 55/141
Willis circle, 11/176
Posterior cerebral artery infarction
computer assisted tomography, 54/54
EOG, 54/161
infarction site, 12/20, 53/412
Posterior cerebral artery syndrome, 53/409-414
agitated delirium, 53/413
alexia, 53/411
amnesia, 53/412
Anton syndrome, 53/413
asimultagnosia, 53/413
Balint syndrome, 53/413
bilateral, 53/413
color anomia, 53/411
color perception, 53/411
diagnosis, 53/414
frequency, 53/409
gaze apraxia, 53/413
left side, 53/412
memory disorder, 53/412
metamorphopsia, 53/411
oculomotor apraxia, 53/413
prosopagnosia, 53/413
pure alexia, 53/412
right side, 53/413
symptom, 53/410
transtentorial herniation, 53/410
treatment, 53/414
visual agnosia, 53/412
visual field defect, 11/306, 53/410
visual hallucination, 53/411
visual perseveration, 53/411
Posterior cerebral vein, see Dorsal callosal vein
Posterior cervical spinal cord stimulation
athetosis, 49/387
Posterior choroidal artery
anterior, see Anterior posterior choroidal artery
middle, see Middle posterior choroidal artery

Posterior column
Bassen-Kornzweig syndrome, 63/275, 278
dentatorubropallidoluysian atrophy, 49/442
familial amyotrophic lateral sclerosis, 21/30,
22/323
lamellar arrangement, 2/209
spina bifida, 32/544
spinal injury, 25/126
thalidomide polyneuropathy, 51/305
vitamin E deficiency, 70/448
Posterior column ataxia
areflexia, 42/164
cranial nerve, 42/164
hereditary, see Biemond syndrome
optic atrophy, 42/164
Purkinje cell loss, 42/164
scoliosis, 42/164
sensory loss, 42/164
Von Romberg sign, 42/165
Posterior column degeneration
amyotrophic lateral sclerosis, 22/325
Ferguson-Critchley syndrome, 22/437
hereditary dystonic paraplegia, 22/451
hereditary sensory and autonomic neuropathy type
I, 21/75, 60/9
hereditary spastic ataxia, 60/463
muscular atrophy, 42/80
spastic ataxia, 42/166
subacute necrotizing encephalomyelopathy,
42/625
Posterior column demyelination
Whipple disease, 51/184
Posterior column syndrome
lateral tract syndrome, 2/210
neuroanatomy, 2/222
transverse spinal cord lesion, 19/29
Posterior commissure
embryo, 50/11
respiration, 1/653
Posterior communicating artery
anatomy, 11/2, 17-19, 39, 53/296
brain aneurysm, 12/93
congenital aneurysm, 31/153
radioanatomy, 11/72
topography, 11/82
Posterior communicating artery aneurysm
brain aneurysm neurosurgery, 12/221, 55/67
Posterior condylar canal syndrome
multiple cranial neuropathy, 2/100
Posterior corpus callosum syndrome, see Corpus
callosum syndrome
Posterior cranial fossa
brain aneurysm, 12/95, 97

Towne projection, 32/10
Posterior cutaneous femoral nerve
  topographical diagnosis, 2/43
Posterior dysraphia
  intraspinal neurenteric cyst, 32/433
Posterior ethmoidal artery
  radioanatomy, 11/71
  topography, 11/71
Posterior fossa
  acoustic nerve tumor, 17/670
  arachnoid cyst, 31/107-127
  brain arteriovenous malformation, 54/369, 372
  cerebellopontine angle syndrome, 31/108
  nasopharyngeal tumor, 17/206
  skull fracture, 23/398
  spina bifida, 32/527
Posterior fossa astrocytoma
  neurofibromatosis type I, 50/368
Posterior fossa cyst
  Dandy-Walker syndrome, 32/59, 42/21
Posterior fossa disease
  ocular flutter, 60/657
Posterior fossa epidural hematoma
  occipital fracture, 54/350
Posterior fossa hematoma
  associated lesion, 24/347
  brain herniation, 24/346
  brain stem death, 57/461
  brain stem hematoma, 24/346
  cerebellar herniation, 24/346
  chronic subdural hematoma, 24/303
  computer assisted tomography, 57/157, 253
  cranial nerve, 24/347, 57/266
  diagnosis, 24/348
  differential diagnosis, 24/348
  epidural hematoma, 24/343, 57/159, 254
  head injury, 57/157
  hemorrhage source, 24/344
  history, 24/343
  hydrocephalus, 57/157, 254
  incidence, 24/343
  intracerebellar hematoma, 24/346, 57/253
  intracranial hypertension, 24/346
  lucid interval, 24/348
  management, 24/349
  mechanism, 24/344
  nystagmus, 57/266
  occipital bone, 57/157, 254
  pathogenesis, 24/344
  prepontine hemorrhage, 24/346
  radiodiagnosis, 24/349, 57/157
  respiratory disorder, 57/266
  result, 24/349

  sign, 24/347, 57/157
  skull fracture, 24/345
  subdural hematoma, 24/345, 57/253
  surgery, 24/349, 639
  traumatic intracranial hematoma, 57/157, 253, 266
  vertebral angiography, 24/344
Posterior fossa lesion
  olivopontocerebellar atrophy (Dejerine-Thomas),
    60/516
Posterior fossa tumor
  acoustic neuroma, 18/402
  Arnold-Chiari malformation type I, 18/407
  benign intracranial hypertension, 18/405-407
  brain stem, 63/230
  brain stem tumor, 18/397
  cerebellar astrocytoma, 18/388
  cerebellar ataxia, 18/407
  cerebellar medulloblastoma, 18/392
  child, 18/387-411
  chordoma, 18/404
  choroid plexus papilloma, 18/396
  cranial neuralgia, 5/312
  Dandy-Walker syndrome, 18/407
  dermoid cyst, 18/403
  differential diagnosis, 18/405-408
  ependymoma, 18/395
  glioblastoma, 18/397
  Gradenigo syndrome, 18/408
  hemangioblastoma, 18/401
  neuroblastoma, 18/407
  neurogenic arterial hypertension, 63/230
  obstructive hydrocephalus, 18/406
  opsoclonus, 18/407
  pineal tumor, 18/405
  pinealoma, 18/405
  pontine glioma, 18/397
  pontine hemorrhage, 18/397
  pseudopapilloma, 18/406
  sarcoma, 18/404
  trigeminal neuralgia, 5/312
Posterior frontal artery
  radioanatomy, 11/79
  topography, 11/79
Posterior groove occlusion
  hydrocephalus, 30/538
Posterior hypothalamic nucleus
  anatomy, 2/434-436
Posterior inferior cerebellar artery
  anatomy, 32/48
  branch group, 11/25
  headache, 5/137
  iron deficiency anemia, 55/464
  lateral branch, 11/91

low origin, 11/37
  medial branch, 11/91
  pontine infarction, 2/244
  progressive bulbar palsy, 2/257
  syndrome, 11/31
Posterior inferior cerebellar artery occlusion
  facial paralysis, 55/90
  hiccup, 55/90
  vertigo, 55/90
Posterior intercavernous sinus
  anatomy, 11/55
Posterior interosseous nerve syndrome
  anatomy, 51/100
  Frohse arcade, 51/100
  rheumatoid arthritis, 38/484, 71/18
Posterior lobe lesion
  musical alexia, 45/437
Posterior longitudinal ligament calcification
  cervical region, 19/154
  intervertebral disc prolapse, 19/154
  sex ratio, 19/154
  spondylosis, 19/154
Posterior migraine
  attack distribution, 5/371
  cervical region, 5/371
Posterior parietal vein
  anatomy, 11/55
Posterior pharyngeal space syndrome, see Villaret
  syndrome
Posterior ponticulus, see Foramen arcuale
Posterior spinal artery
  anatomy, 26/6, 64
  angiography, 32/48
  course, 12/495
  radioanatomy, 11/91
  spinal cord vascularization, 12/484
  topographical anatomy, 55/96
Posterior spinal artery aneurysm
  systemic lupus erythematosus, 63/54
Posterior spinal artery syndrome
  clinical features, 55/101, 63/42
  iatrogenic neurological disease, 55/101
  intrathecal phenol, 55/101
  spinal cord vascular disease, 12/499
Posterior spinal fusion
  cervical, 25/297
Posterior spinal nerve root, see Dorsal spinal nerve
  root
Posterior temporal artery
  radioanatomy, 11/86
  topography, 11/86
Posterior thalamic nuclear group
  pulvinaris intergeniculatis nucleus, 2/473

pulvinaris lateralis nucleus, 2/473
  pulvinaris medialis nucleus, 2/473
  pulvinaris superficialis nucleus, 2/473
Posterior tibial nerve
  entrapment, 7/316
Posterolateral thalamic syndrome
  analgic type, 2/485
  choreoathetosis, 2/485
  dissociated type, 2/485
  hemialgic type, 2/485
  hemiataxia, 2/485
  thalamopyramidal type, 2/485
  thalamostriate type, 2/485
Postexanthemata
  acute cerebellar ataxia, 34/626-628
Postganglionic nerve fiber
  nerve, 51/1
Postgastrectomy
  subacute combined spinal cord degeneration,
    28/164-169
Postgastrectomy syndrome
  see also Celiac disease and Malabsorption
    syndrome
  anemia, 28/234
  deficiency neuropathy, 51/327
  myelopathy, 28/233, 51/327
  neuropathy, 28/234, 51/327
  osteomalacia, 28/233, 51/327
  postvagotomy, 75/631
  vitamin D deficiency, 28/233, 51/327
Posthemiplegic athetosis
  focal athetosis, 49/383
  neuropathology, 49/386
  Vogt striatal athetotic syndrome, 49/386
Posthemiplegic chorea
  athetosis, 6/437
  hemiballismus, 6/437
  striatal necrosis, 49/501
Posthemiplegic dystonia
  neuropathology, 49/386
  symptomatic dystonia, 49/542
Postherpetic dysesthesia, see Postherpetic neuralgia
Postherpetic neuralgia, 5/323-325
  see also Herpes zoster
  aciclovir, 51/181, 56/234
  age, 5/324, 56/233
  amantadine, 51/181
  amitriptyline, 56/234
  carbamazepine, 51/181, 56/234
  chemotherapy, 5/324
  clinical features, 5/324
  corticosteroid, 51/181
  differential diagnosis, 5/316

drug treatment, 5/383
facial paralysis, 5/324
herpes zoster, 5/316, 56/230, 233
herpes zoster ganglionitis, 51/181
levodopa, 51/181
pain mechanism, 1/119
pain reaction, 5/324
phenytoin, 56/234
prednisolone, 56/234
prevalence, 56/233
prophylaxis, 5/324
scar, 5/316
sensory loss, 5/316, 323
skin lesion, 5/383, 51/180, 56/234
surgical treatment, 5/401
varicella zoster, 51/180
varicella zoster virus, 5/323, 56/229
vidarabine, 56/234
zoster localization, 56/234
Posthypoxic myoclonus
oxitriptan, 49/618
Postictal aggression
epilepsy, 73/480
Postictal blindness
epilepsy, 2/623
Postictal psychosis
epilepsy, 72/132, 73/481
Postictal symptom
grand mal epilepsy, 15/112
Posticteric encephalopathy
see also Acquired hepatocerebral degeneration,
Bilirubin encephalopathy, Hepatic
encephalopathy and Kernicterus
clinical features, 49/482
kernicterus, 6/452, 491
Postinfectious encephalomyelitis
see also Acute disseminated encephalomyelitis
and Disseminated encephalomyelitis
Bordetella pertussis, 47/479
cowpox virus, 47/326
diphtheria pertussis tetanus immunization, 47/479
influenza virus, 47/326
measles, 47/326, 476
multiple sclerosis, 9/130, 135, 138, 47/326
mumps, 34/412, 47/326, 56/129, 430
myelin basic protein, 47/326, 474
neuromyelitis optica, 9/427, 47/405
nuclear magnetic resonance, 56/430
varicella zoster virus, 47/326
Postinfectious encephalopathy
chickenpox, 47/479
influenza B, 47/479
myelinoclasis, 9/608

Postinfectious syndrome
survey, 71/547
Postlateral myelopathy
Von Hippel-Lindau disease, 42/156
Postlumbar puncture syndrome
age, 61/152
bed rest, 61/154
caffeine, 61/156
cranial nerve palsy, 61/163
CSF hypotension, 61/154
CSF leak, 61/154
desmopressin acetate, 61/156
differential diagnosis, 61/153
epidural blood patch, 61/156
epidural dextran 40, 61/156
epidural morphine, 61/156
epidural sodium chloride, 61/156
epilepsy, 61/166
fluid intake, 61/155
flunarizine, 61/156
headache, 61/152
incidence, 61/152
indometacin, 61/156
needle orientation, 61/157
needle size, 61/157
pathogenesis, 61/153
radicular syndrome, 61/158
sex ratio, 61/152
subcutaneous CSF, 61/158
symptom, 61/153
theophylline, 61/156
traumatic tap, 61/158
treatment, 61/154
unstyled needle, 61/158
vasopressin, 61/156
Postmastectomy syndrome
malignant, 69/93
Postmenopausal osteoporosis
radiologic finding, 20/12
senile osteoporosis, 20/12
vertebral myeloma, 20/12
Postnarcotic encephalopathy, see Postanesthetic
encephalopathy
Postoperative arachnoid cyst, see Postoperative
meningeal diverticulum
Postoperative bacterial endocarditis
systemic brain infarction, 11/445
Postoperative delirium
etiology, 46/547
Postoperative fungal endocarditis
cardiac surgery, 11/446
Postoperative meningeal diverticulum
diagnosis, 20/161

myelography, 20/162
Postoperative meningocele, *see* Postoperative
    meningeal diverticulum
Postoperative psychosis
    cardiac surgery, 11/446
    multiple embolism, 11/446
    potassium imbalance, 28/483
    sex ratio, 11/446
Postpneumonectomy ischemic myelopathy
    paraplegia, 61/142
Postpolio syndrome, *see* Postpoliomyelitic
    amyotrophy
Postpoliomyelitic amyotrophy, 59/35-39
    aging, 59/35
    amyotrophic lateral sclerosis, 22/23, 59/35
    anxiety, 59/35
    autonomic nervous system, 75/20, 416
    clinical features, 59/37
    cold induced pain, 51/482
    electromyographic abnormality, 59/37
    epidemiology, 59/36
    extensor plantar response, 59/37
    features, 56/331
    laboratory diagnosis, 59/37
    misdiagnosis, 59/35
    muscle biopsy, 59/38
    myogenic change, 59/38
    nerve conduction velocity, 59/37
    neuropathology, 59/38
    poliomyelitis, 22/23, 138, 56/331
    progressive muscular atrophy, 22/23, 59/19
    progressive spinal muscular atrophy, 22/23
    progressive weakness, 59/36
    secondary disease treatment, 59/38
    secondary illness, 59/36
    serum creatine kinase, 59/38
    treatment, 22/24, 59/38
Postpoliomyelitis
    autonomic polyneuropathy, 51/482
Postpoliomyelitis progressive muscular atrophy, *see*
    Postpoliomyelitic amyotrophy
Postradiation syndrome
    deficiency neuropathy, 51/326
Postrubella embryopathy
    optic atrophy, 13/82
Postsplenectomy state
    acanthocyte, 49/332
Postsynaptic inhibition
    strychnine intoxication, 41/305
    tetanus toxin, 41/304
Postsynaptic potential
    excitatory, *see* Excitatory postsynaptic potential
    inhibitory, *see* Inhibitory postsynaptic potential

Postthoracotomy syndrome
    malignant, 69/93
Postthyroidectomy
    striatopallidodentate calcification, 49/419
Posttraumatic amnesia
    see also Traumatic amnesia
    ageusia, 24/18
    anosmia, 24/6, 12, 18
    axonal disruption, 45/185
    brain concussion, 23/430, 433, 24/477, 479, 488
    brain contusion, 23/427, 430, 433, 45/185
    coma, 57/373, 382, 398
    commotio cerebri, 23/420
    computer assisted tomography, 45/187
    confusion, 24/479
    disorientation, 23/430, 45/188, 57/398
    duration, 24/489
    head injury, 23/430, 24/488, 684, 27/468,
        45/185-189, 57/373, 398
    head injury epidemiology, 57/4
    head injury outcome, 45/189, 57/373, 382, 398
    hematoma, 24/489
    hippocampus, 45/185
    hypnosis, 23/423
    intelligence, 45/188, 46/613
    learning, 27/468, 45/189
    memory, 3/293-295, 27/468, 45/187
    memory disorder, 3/293, 27/468, 57/398
    prognosis, 46/611
    recovery, 24/478
    social contact, 57/409
    test, 24/478, 488, 45/187, 46/611, 57/398
Posttraumatic brain abscess
    antibiotic treatment, 57/323
    death, 57/323
    depressed skull fracture, 57/321
    dural tears, 57/321
    epidemiology, 57/321
    head injury, 57/321
    incidence, 57/321
    morbidity, 57/323
    mortality, 57/323
    pathogenesis, 57/321
    pathophysiology, 57/321
    predisposing factor, 57/321
    prophylactic antibiotic agent, 57/323
Posttraumatic brain abscess outcome
    excision, 57/323
Posttraumatic coma
    EEG, 57/187
Posttraumatic dysautonomic cephalalgia
    cluster headache, 48/227
    migraine, 48/227

Posttraumatic encephalopathy
  classification, 23/9
  pallidostriatal necrosis, 49/466
Posttraumatic epilepsy
  acute, 72/195
  age, 24/446
  age influence, 72/191
  age of injury, 72/194
  basic mechanism, 15/287
  bone defect, 24/663
  brain injury, 24/834, 57/211
  definition, 24/446
  depressed skull fracture, 24/448, 451
  domestic, 72/193
  drug, 72/202
  early epilepsy, 57/192
  EEG, 57/192, 72/201
  epidemiology, 72/193
  epilepsy of late onset, 23/357, 57/193
  etiology, 15/274
  first seizure, 72/196
  follow-up, 72/199
  frequency, 24/448, 453
  head injury, 46/395, 57/192
  head injury outcome, 57/373, 375
  imaging, 72/202
  incidence, 24/445, 447, 450
  intracranial hematoma, 24/448, 451
  late, 24/446, 450-453
  late onset, 72/195
  location, 72/198
  long-term, 72/200
  missile injury, 24/448, 451, 474
  model, 72/198
  pathology, 72/197
  prevention, 24/453
  prophylaxis, 24/469
  recurrence rate, 24/447
  rehabilitation, 24/689
  risk, 72/200
  significance, 24/445, 449
  sport accident, 72/193
  traffic, 72/193
  traumatic amnesia, 24/451-453
  traumatic psychosyndrome, 24/566
  type, 24/447
  war study, 72/192
  workplace, 72/193
Posttraumatic headache, 5/178-184, 48/383-389
  arachnoid tear, 5/182
  cause, 24/504, 536, 580, 686
  cervical spine, 24/505, 507, 580
  cervical syndrome, 24/509

  child, 24/512
  classification, 48/5
  definition, 24/501, 48/10
  depression, 24/511
  differential diagnosis, 5/27
  etiology, 24/504, 48/383
  head injury, 24/501
  history, 24/502, 48/385
  incidence, 24/505
  injury severity, 24/510
  intracranial hematoma, 24/507
  intracranial hypertension, 5/182
  intracranial pressure, 24/507, 48/384
  literature, 24/502
  meningeal adhesion, 24/507
  migraine, 24/506
  neurosis, 24/511
  PEG, 5/182
  postconcussional syndrome, 5/178
  posttraumatic syndrome, 48/385
  psychiatric disorder, 5/182, 48/362
  psychiatric headache, 5/20, 182, 48/362-364
  psychological aspect, 24/509
  railway spine, 48/385
  rehabilitation, 24/687
  skull defect, 24/507
  subdural adhesion, 5/181
  traumatic psychosyndrome, 24/565
  treatment, 24/512, 581
  vascular factor, 24/505
  whiplash injury, 24/508
Posttraumatic intracranial hematoma
  alcohol, 57/259
  coagulopathy, 57/259
  pathophysiology, 57/259
Posttraumatic meningitis
  antibiotic regime, 57/320
  antimicrobial treatment, 57/319
  bacterial species, 57/319
  clinical presentation, 57/318
  CSF fistula, 57/318
  CSF glucose, 57/318
  diagnostic study, 57/318
  diagnostic test, 57/318
  epidemiology, 57/317
  head injury, 57/317
  incidence, 57/317, 320
  microbiology, 57/319
  missile injury, 57/322
  organism, 57/319
  otorrhea, 57/317
  pathogenesis, 57/318
  pathophysiology, 57/318

prophylactic antibiotic agent, 57/320
rhinorrhea, 57/317
treatment, 57/319
Posttraumatic neurosis
head injury outcome, 57/412
postconcussional syndrome, 57/412
Posttraumatic syndrome
armed forces, 24/460
axonal fracture, 48/386
brain injury, 24/838, 48/386
brain injury prognosis, 24/834
child, 24/494
consciousness, 48/386
constrained inactivity, 24/688
dizziness, 24/686
ENG, 24/537
head injury outcome, 57/376
headache, 24/501, 536, 686, 48/385, 387
history, 48/385
law, 24/838
management, 24/495
Maze test, 48/386
migraine, 48/388
missile injury, 24/460, 474, 48/386
motor impairment, 24/690
pathogenesis, 24/537
posttraumatic headache, 48/385
rehabilitation, 24/687, 46/629
review, 24/493, 686
sign, 24/836
treatment, 48/387
vertigo, 24/686
Posttraumatic syringomyelia, 26/113-155,
   61/375-396
arachnoiditis, 61/395
areflexia, 61/384
autonomic dysfunction, 61/383
brain stem, 26/126
case report, 26/117-122, 148-151
cavity extension, 26/131
cavity fluid, 26/135
circulation damage, 25/69
classification, 61/375
clinical features, 12/607, 26/124-128, 151
complete cord lesion surgery, 26/139, 145
computer assisted tomography, 50/454, 61/384
contrast study, 26/144
cough, 61/378
course, 26/138
CSF block, 61/376
CSF flow, 61/379, 396
cyst, 12/604, 26/144
cyst enlargement, 61/378

DAVID model, 61/386
dissociated sensory loss, 61/382
drainage tube selection, 61/390
endomyelography, 61/385
experimental, 26/154
extramedullary compression, 26/154
extrathecal drainage, 61/390
features, 32/309-312
hematomyelia, 61/378
hereditary sensory and autonomic neuropathy type
   II, 21/74
history, 26/114
hyperextension trauma, 61/526
incidence, 26/122, 61/377
initial symptom, 26/124, 151
latency period, 26/123, 151
liquefaction, 61/378
liquefaction necrosis, 25/60
mechanism, 61/378
meningeal fibrosis, 61/395
moderate trauma, 26/146
morphology, 61/386
mortality, 26/138
multiseptate syrinx, 61/385
muscle stretch reflex, 61/384
muscle weakness, 50/454
muscular atrophy, 26/127, 61/384
myelography, 12/607, 26/118, 120, 136, 142-146,
   152, 50/454, 61/384
myelomalacia, 12/599, 26/130, 154
neurogenic osteoarthropathy, 26/124
neuropathology, 61/385
neurosurgery, 61/388
nomenclature, 61/375
nuclear magnetic resonance, 61/55, 380, 384, 386,
   388, 392, 395
onset, 61/377
osteoarthrosis, 61/42
pain, 50/454, 61/381
paraplegia, 26/117, 50/454, 61/375, 393
partial cord lesion surgery, 26/141, 145
pathogenesis, 12/607, 26/116, 130, 134, 153
pathology, 26/116, 126, 128, 132, 136
pedicled omental grafting, 61/392
postoperative nuclear magnetic resonance, 61/395
progression, 61/533
progressive, 32/310
radiology, 26/142, 152, 61/384
segmental level, 26/123, 151
spinal arachnoiditis, 26/146, 148, 153
spinal cord division, 61/391
spinal cord injury, 12/592, 604, 50/454, 61/375
surgical outcome, 61/393

symptom, 50/454, 61/381
syringosubarachnoid shunt, 61/389, 393
syrinx tension, 61/386
terminal ventriculostomy, 61/392
terminology, 26/115
tetraplegia, 26/117, 50/454, 61/388
treatment, 12/608, 26/141, 152, 61/387
treatment decision tree, 61/390
trigeminal nerve, 26/125
Posttraumatic tremor
    features, 49/589
Postural apraxia
    developmental dyspraxia, 4/459
Postural deformity
    Parkinson dementia, 49/172
Postural hypotension, see Orthostatic hypotension
Postural myoclonus
    dyssynergia cerebellaris myoclonica, 6/767,
        21/513
Postural orthostatic tachycardia syndrome
    sleep, 74/555
Postural reduction
    hyperanteflexion sprain, 25/344
    result, 26/304
    spinal injury, 25/457, 460, 26/299
    thoracolumbar spine injury, 25/458, 460, 26/213,
        294
Postural reflex
    see also Posture
    autonomic nervous system, 1/441
    brain death, 24/717
    cerebellar lesion, 1/336
    1-methyl-4-phenyl-1,2,3,6-tetrahydropyridine
        intoxication, 65/372
    muscle tone, 1/264
    neuroleptic parkinsonism, 65/292
    nonconvulsive seizure, 15/137
    paleocerebellar syndrome, 2/418
    pathophysiology, 49/78
    physiology, 1/68
    synkinesis, 1/284, 411
    variant type, 1/255
Postural sensibility
    parietal lobe lesion, 2/683
    tuning fork test, 2/683
Postural tachycardia, see Idiopathic postural
    orthostatic tachycardia syndrome
Postural tone
    increased, 15/137
Postural tremor
    see also Static tremor
    adrenalin tremor, 6/818
    alcohol intoxication, 51/317

alcoholic polyneuropathy, 51/317
alcoholic tremor, 6/819, 49/590
anxiety tremor, 6/818
athetosis, 49/591
benign essential tremor, 6/819
cerebellar tremor, 6/820, 49/587
definition, 2/394
dystonia musculorum deformans, 49/522
essential tremor, 49/586
etiology, 45/465
fatigue, 6/818
head injury, 49/589
hereditary progressive diurnal fluctuating
    dystonia, 49/530
hyperkinetic agraphia, 45/465
Parkinson disease, 49/96
physiologic tremor, 6/818, 49/586
polyneuropathy, 49/588
stereotaxic surgery, 49/604
superior cerebellar peduncle, 49/588
thyrotoxic tremor, 6/818
undetermined significant monoclonal
    gammopathy, 63/399
Wilson disease, 49/589
Posture
    see also Astasia abasia, Equilibrium and Postural
        reflex
    asymmetry, 2/405
    ataxia, 1/311, 314, 322
    athetosis, 6/445
    autonomic nervous system, 75/436
    bajonet, see Bajonet posture
    basal ganglion, 49/78
    cerebellar component, 1/332
    coma, 3/65
    conjugate eye movement, 3/24
    control, 4/344, 347-349, 445
    coordination, 1/306
    decerebrate, see Decerebrate posture
    decorticate, see Decorticate posture
    disorder, 3/65
    dystonia, 6/520
    eye movement, 1/575, 598, 3/24
    frontal lobe tumor, 67/149
    hereditary motor and sensory neuropathy type I,
        21/281
    hereditary motor and sensory neuropathy type II,
        21/281
    intermittent spinal cord ischemia, 12/527
    lumbar intervertebral disc prolapse, 20/588
    movement, 1/298
    pallidal, see Pallidal posture
    Parkinson disease, 6/197

pseudo-Babinski, *see* Pseudo-Babinski posture
sensory afferent, 1/95
stooped, *see* Stooped posture
supporting reaction, 1/68
Sydenham chorea, 6/159
upright, 1/334
Postvaccinal encephalitis
see also Acute disseminated encephalomyelitis
epilepsy, 15/309
myelinoclasis, 9/651
rare type, 9/547-549
Postvaccinal measles encephalomyelitis
measles virus, 56/14
Postvaccinial encephalomyelitis
Bordetella pertussis, 47/479
diphtheria pertussis tetanus immunization, 47/479
multiple sclerosis, 9/135, 138
rabies, *see* Rabies postvaccinial encephalomyelitis
Postvaccinial transverse myelitis
clinical diagnosis, 47/485
Postviral encephalomyelitis
influenza virus, 56/168
Postviral fatigue syndrome
autonomic nervous system, 75/127
chronic mercury intoxication, 64/371
clinical features, 56/202
enterovirus infection, 56/332
features, 56/332
Sjögren syndrome, 71/74
sleep, 74/554
Potamotrygonidae intoxication
elasmobranche, 37/70
Potassium, 28/464-475
alteration, 28/430
brain, 9/16, 28/427, 464
brain infarction, 53/125
brain injury, 23/115
brain ischemia, 53/125
brain metabolism, 57/73
brain scanning, 16/664
cellular concentration, 63/556
CNS, 28/427-431
concentration gradient, 63/556
CSF, 28/427-429, 468-472
CSF to serum, 28/468
extracellular, *see* Extracellular potassium
extracellular fluid, 28/472
glia, 28/473
headache, 48/111
hyponatremia, 28/430
myasthenia gravis, 41/132
nerve cell, 72/50
neuromuscular excitability, 63/556

periodic paralysis, 28/597, 40/285, 544, 41/155
serum, *see* Serum potassium
spinal cord injury, 26/381, 387
spreading depression, 48/67, 113
thallium intoxication, 64/327
Potassium channel
arthropod envenomation, 65/194, 197
axolemma, 47/14-18
axon, 51/49
barium intoxication, 64/354
excitation contraction coupling, 40/550
muscle cell membrane, 40/128
myotonic dystrophy, 62/225
nerve-muscle unit, 40/126
nodal membrane, 51/49
voltage clamping, 51/49
Potassium chloride
memory disorder, 27/463
Potassium current
epilepsy, 72/46
nerve cell, 72/46
Potassium cyanide intoxication
concentric sclerosis, 9/440
Potassium imbalance
anorexia nervosa, 28/482
anticonvulsant, 28/480
clinical study, 28/479
depression, 28/480
epilepsy, 28/476-482
experimental study, 28/478
historic aspect, 28/463
hyponatremia, 28/482
lithium, 28/484
manic depressive psychosis, 28/483
mental illness, 28/482-485
postoperative psychosis, 28/483
Potentials
action, *see* Action potential
auditory evoked, *see* Auditory evoked potential
Bereitschafts, *see* Readiness potential
brain stem auditory evoked, *see* Brain stem auditory evoked potential
compound muscle action, *see* Compound muscle action potential
evoked, *see* Evoked potential
excitatory postsynaptic, *see* Excitatory postsynaptic potential
miniature motor end plate, *see* Miniature motor end plate potential
motor end plate, *see* Motor end plate potential
motor unit, *see* Motor unit potential
motor unit action, *see* Motor unit action potential
nerve action, *see* Nerve action potential

orthodromic sensory, *see* Orthodromic sensory
  potential
readiness, *see* Readiness potential
sensory action, *see* Sensory action potential
somatosensory evoked, *see* Somatosensory
  evoked potential
visual evoked, *see* Visual evoked response
Pott disease, *see* Tuberculous spondylitis
Potter face
  axial mesodermal dysplasia spectrum, 50/516
  caudal aplasia, 50/513
  prune belly syndrome, 50/515
Potter sequence, *see* Potter syndrome
Potter syndrome
  see also Caudal aplasia
  axial mesodermal dysplasia spectrum, 50/516
  birth incidence, 43/469
  caudal agenesis, 43/469
  heart disease, 43/469
  hydrocephalus, 43/469
  hypertelorism, 30/250
  renal agenesis syndrome, 43/469, 50/512
Potter syndrome type II
  multiple trichoepithelioma, 14/110
  Von Hippel-Lindau disease, 14/118
Powassan virus
  acute viral encephalitis, 34/78, 56/134
Poxvirus
  human, 56/2
  neurotropic virus, 56/5
  ortho, *see* Orthopoxvirus
  para, *see* Parapoxvirus
  Tana, *see* Tana poxvirus
  taxonomic place, 56/2
  Yaba, *see* Yaba poxvirus
PPS, *see* Pain producing substance
Practolol
  migraine, 48/181
Practosomatognosia
  body scheme disorder, 4/394-396
Prader-Labhart-Willi syndrome
  aneuploidy, 50/566
  appetite, 46/93
  chromosomal aberration, 43/464
  chromosome deletion syndrome, 50/590
  clinical features, 31/322, 50/590
  clinical presentation, 40/336
  CNS, 31/322-324
  dermatoglyphics, 43/463
  diabetes mellitus, 31/324, 43/464
  etiology, 31/321-323
  facial paralysis, 43/463
  feeding difficulty, 46/93, 50/591

hypogonadism, 31/322, 324, 40/336, 43/463, 464
hypotonia, 40/336, 46/93
mental deficiency, 31/322, 324, 43/463, 46/93,
  50/590
metabolic abnormality, 31/325
muscular hypotonia, 31/321
obesity, 31/322, 324, 40/336, 43/463, 46/93,
  50/590
pathology, 31/325
personality change, 31/324
prognosis, 31/325, 43/464
15q partial monosomy, 50/590
radiologic appearance, 31/324
sex ratio, 43/464
short stature, 31/322, 43/463
spinal muscular atrophy, 59/371
strabismus, 31/324, 43/463
talipes, 43/463
treatment, 31/325
Weiss-Alström syndrome, 13/463
Prader-Willi syndrome, *see* Prader-Labhart-Willi
  syndrome
Pragmatics
  aphasia, 45/309
Prague, Czechoslovakia
  neurology, 1/10
Praxagoras
  history, 2/5
Prazosin
  cataplexy, 65/448
  classification, 74/149
  iatrogenic neurological disease, 65/447
  narcolepsy, 65/448
  neurotoxin, 65/447
  urinary incontinence, 65/448
Prazosin intoxication
  cataplexy, 65/448
  cognitive impairment, 65/447
  dementia, 65/447
  depression, 65/447
  EEG, 65/447
  fatigue, 65/447
  hallucination, 65/447
  narcolepsy, 65/448
  neurologic adverse effect, 65/447
Prealbumin
  amyloid, 51/414
  CSF, 9/340
  hereditary amyloid polyneuropathy type 1,
    60/106-108, 71/512-514
  hereditary amyloid polyneuropathy type 2,
    60/106-108, 71/512-514
Prealbumin mutation

hereditary amyloid polyneuropathy, 60/91, 106-108
Preauricular tag
    trigeminal nerve agenesis, 50/214
Prebetalipoprotein
    hyperlipoproteinemia type IV, 10/275
    low density lipoprotein, 10/267
Precentral area
    frontal lobe tumor, 17/270, 272
Precentral artery, *see* Prerolandic artery
Precentral cerebellar vein
    radioanatomy, 11/110
    topography, 11/110
Precentral cortex
    anatomy, 45/23
    tremor, 49/597
Precentral gyrus
    hemiballismus, 49/370
Prechordal plate
    neural induction, 50/22
Preclinical detection
    angioneurotic edema, 43/61
    Becker muscular dystrophy, 43/87
    Duchenne muscular dystrophy, 43/106
    Emery-Dreifuss muscular dystrophy, 43/88
    progressive myositis ossificans, 43/190
Precocious puberty
    Albright syndrome, 14/12
    aqueduct stenosis, 50/311
    astrocytoma, 18/367
    diencephalic syndrome, 68/71
    fibrous dysplasia, 14/183
    hamartoma, 18/367
    hypothalamic hypophyseal syndrome, 2/444
    hypothalamus, 18/367
    neuroblastoma, 42/758
    neurofibromatosis type I, 14/156
    pathogenesis, 31/58
    pinealoma, 18/367
    teratoma, 18/367, 31/59
    tuberous sclerosis, 14/745
Precordial pain
    migraine, 5/47
    urticaria pigmentosa, 14/790
Prednisone, *see* Corticosteroid
Predominant axonal degeneration
    chronic inflammatory demyelinating
        polyradiculoneuropathy, 51/534
Prefrontal cortex
    anatomy, 45/23
Prefrontal leukotomy
    anxiety, 1/110
    frontal lobe syndrome, 45/25-27

medial thalamic nucleus, 2/491
    pain, 1/110
    schizophrenia, 45/26
Prefrontal region
    anatomy, 2/734
    behavior, 2/736
    function, 2/735
Prefrontal syndrome, 2/734-750
    see also Frontal lobe syndrome
    activity regulation, 2/740
    basic theory, 2/740
    choice reaction, 2/744
    conditioning, 2/744
    confabulation, 2/749
    conflicting reaction, 2/745
    emotion, 2/751
    impulsivity, 2/746
    intellectual impairment, 2/749
    learning, 2/748
    movement regulation, 2/743
    ocular search movement, 2/747
    orienting reaction, 2/742
    reproduction, 2/748
    self-criticism, 2/738
    successive action program, 2/745
    symptomatology, 2/738
    thematic apperception, 2/746
    tumor, 17/239-241
    variation, 2/751
Pregnancy
    anterior spinal artery syndrome, 63/42
    antiepileptic agent, 65/512
    arterial hypertension, 63/73
    astrocytoma, 18/19
    autonomic hyperreflexia, 74/332
    autonomic nervous system, 74/108, 328
    benzodiazepine intoxication, 65/338
    beriberi, 8/14
    brain aneurysm, 55/73
    brain aneurysm neurosurgery, 55/73
    brain infarction, 54/425
    brain metastasis, 16/73
    brain phlebothrombosis, 11/459
    brain tumor, 16/76
    brucellosis, 33/318
    carbon disulfide intoxication, 64/27
    carotid cavernous fistula, 24/408
    carpal tunnel syndrome, 7/292, 8/14
    cavernous sinus thrombosis, 52/173
    cerebrovascular disease, 53/28, 36
    Chagas disease, 52/345
    cluster headache, 48/219
    cocaine intoxication, 65/255

Coccidioides immitis, 52/410
congenital toxoplasmosis, 52/351
electrical injury, 61/194
epidural spinal lipoma, 20/407
epilepsy, 73/362, 372, 375
Guillain-Barré syndrome, 8/14
headache, 48/424, 433
hereditary brachial plexus neuropathy, 60/71
high altitude, 74/334
hypertensive encephalopathy, 63/73
lithium intoxication, 64/298
lumbar intervertebral disc prolapse, 8/14, 20/595
lumbosacral, 51/163
lumbosacral plexus, 51/163
meralgia paresthetica, 8/15
migraine, 5/54, 56, 48/18, 120, 433
Minamata disease, 64/419, 424
multiple sclerosis, 9/117, 418, 47/64, 152
myelomalacia, 55/98
myositis, 62/370
myotonia congenita, 62/264
myotonic dystrophy, 62/228
neuropathy, 8/13-15
optic atrophy, 13/64
orotic aciduria, 29/240
paradoxical embolism, 11/460
pheochromocytoma, 74/333
pneumococcal meningitis, 33/36
poliomyelitis, 31/217, 34/117, 42/563, 56/315
Refsum disease, 8/22, 21/191, 66/487, 496
restless legs syndrome, 8/316, 42/261, 51/546
seasonal rhythm, 74/490
sexual function, 26/452
spinal angioma, 20/461
spinal ependymoma, 20/366
stress, 74/328
subarachnoid hemorrhage, 12/115, 55/25, 73
systemic brain infarction, 11/459
Takayasu disease, 39/220
venous sinus occlusion, 54/425
venous sinus thrombosis, 54/425
vertebral hemangioma, 20/441
vitamin B1 deficiency, 51/331
Wilson disease, 27/407
Pregnancy toxemia
vitamin B6, 28/131
Prehemorrhagic softening
aneurysm formation, 12/49
angionecrosis, 12/49
apoplexy, 12/49
arteriolar fibrinohyalinoid degeneration, 12/49
Charcot-Bouchard aneurysm, 12/49
malignant hypertension, 12/49
Prehension
examination method, 4/447
finger to nose test, 4/447
motor activity, 4/447
Prehension test
dysmetria, 1/324
Preictal symptom
grand mal epilepsy, 15/108
Prekallikrein
hemostasis, 63/303
Prelipid
leukodystrophy, 10/12, 14
metachromatic leukodystrophy, 10/43, 52, 54
Prelipid substance (Alzheimer)
amaurotic idiocy, 10/20
Prelipoid
orthochromatic leukodystrophy, 10/115
Premature closure syndrome
classification, 38/415
clinical features, 38/415
diagnosis, 38/417
fontanelle, 38/417
history, 38/413
neurologic features, 38/416
pathogenesis, 38/413
radiologic features, 38/416
treatment, 38/417-419
trigonocephaly, 30/227
Premature graying, see Poliosis
Premenstrual headache
endocrine status, 5/230
Premenstrual syndrome
headache, 48/426
Premorbid personality
neuropsychologic syndrome, 3/23
Parkinson disease, 49/127
schizophrenia, 46/490
Premotor cortex, 2/725, 727
adversive seizure, 2/731
anatomy, 2/729, 45/23
function, 2/729
irritation, 2/731
motor aphasia, 2/734, 45/35
tapping test, 2/732
Premotor cortex syndrome
clinical aspect, 2/731-734
fist-ring test, 2/731
morphologic data, 2/729
physiologic aspect, 2/730
Prenatal anoxia
newborn epilepsy, 15/203
Prenatal diagnosis
aberration, see Chromosomal aberration

acid maltase deficiency, 43/178, 62/481
adrenoleukodystrophy, 47/598
amniocentesis, 32/564-568
amniotic fluid cell, 32/571
amniotic fluid protein, 32/570
anencephaly, 42/13, 50/85
anophthalmia, 42/14
aqueduct stenosis, 42/58
argininosuccinic aciduria, 42/526
Bardet-Biedl syndrome, 43/233
brancher deficiency, 43/180
Chédiak-Higashi syndrome, 42/537
corpus callosum agenesis, 42/9
debrancher deficiency, 43/179
Duchenne muscular dystrophy, 41/417
encephalomeningocele, 42/28
Fabry disease, 42/429
Fanconi syndrome, 43/18
Farber disease, 42/594
α-fetoprotein, 32/562, 564-572
fetoscopy, 32/570
galactosemia, 42/552
Gaucher disease, 42/438
globoid cell leukodystrophy, 42/491
glycosaminoglycanosis, 27/163
glycosaminoglycanosis type I, 27/162, 42/452
glycosaminoglycanosis type III, 42/456
glycosaminoglycanosis type VII, 66/312
GM1 gangliosidosis, 27/162
GM1 gangliosidosis type I, 42/432
GM1 gangliosidosis type II, 42/433
GM2 gangliosidosis, 27/162, 42/434
GM2 gangliosidosis type I, 42/434
GM2 gangliosidosis type II, 27/163, 42/436
GM2 gangliosidosis type III, 42/436
Gruber syndrome, 43/392
hemivertebra, 42/31
holoprosencephaly, 50/240
homocystinuria, 42/557
inborn error of carbohydrate metabolism, 27/163
iniencephaly, 30/264
Lesch-Nyhan syndrome, 42/154
mannosidosis, 27/157, 42/598
maple syrup urine disease, 29/70, 42/599
metachromatic leukodystrophy, 42/495, 60/127,
   66/176
mucolipidosis type I, 42/476
mucolipidosis type II, 27/161
myotonic dystrophy, 43/153, 62/215
polysaccharide, 27/162
Salla disease, 59/361
Smith-Lemli-Opitz syndrome, 66/590
spina bifida, 32/564-571, 42/56

xeroderma pigmentosum, 43/13
Prenatal infection
   hydrocephalus, 50/288
   macula dysplasia, 13/134
   mental deficiency, 46/19
Prenatal mercury syndrome, see Minamata disease
Prenatal radiation injury
   mental deficiency, 46/78
Prenatal rubella syndrome
   mental deficiency, 46/76
Prenatal toxoplasma syndrome, see Congenital
   toxoplasmosis
Preoptic hypothalamus
   anatomy, 2/434
Preoptic nucleus
   anatomy, 2/434, 437
Prepeduncular syndrome
   skull base tumor, 17/183
Prepontine hemorrhage
   posterior fossa hematoma, 24/346
Prerolandic artery
   radioanatomy, 11/86
   topography, 11/86
Presence response
   examination method, 4/448
   motor activity, 4/448
Presenile astroglial dystrophy
   myoclonia, 21/67
   Pick disease, 21/67, 46/236
Presenile dementia
   see also Pick disease
   Alzheimer disease, 15/343, 42/275-277, 46/248
   amnesia, 46/253
   anxiety, 42/286
   apathy, 42/287
   aphasia, 46/201, 253
   ataxia, 42/288
   behavior disorder, 42/286
   brain atrophy, 46/454
   catatonia, 42/286
   classification, 46/204
   clinical features, 46/253, 486
   CSF, 42/288
   demyelination, 42/288
   differential diagnosis, 15/343
   Ekbom other syndrome, 63/85
   epilepsy, 15/343, 21/67
   etiology, 42/287
   hyperreflexia, 42/288
   muscle stretch reflex, 42/288
   nystagmus, 42/288, 46/248
   Pick disease, 15/343, 42/285
   restlessness, 42/286

spastic paraplegia, 42/287
speech disorder, 42/287
ventricular dilatation, 42/288
Presenile dementia (Kraepelin)
  basal ganglion degeneration, 42/286
  brain atrophy, 42/286
Presenile glial dystrophy
  neuropathology, 21/59
Pressure
  arterial, *see* Arterial pressure
  blood, *see* Blood pressure
  brain perfusion, *see* Brain perfusion pressure
  colloid oncotic, *see* Colloid oncotic pressure
  CSF, *see* CSF pressure
  esophageal, *see* Esophageal pressure
  femoral neuropathy, 8/306
  increased intracranial, *see* Increased intracranial
    pressure
  intervertebral disc, 20/540
  intracranial, *see* Intracranial pressure
  intraocular, *see* Intraocular pressure
  intrathoracic, *see* Intrathoracic pressure
  mastoid tip, 48/503
  nerve conduction, 7/132
  nerve injury, 51/135
  neuropathy, *see* Compression neuropathy
  positive end expiratory, *see* Positive end
    expiratory pressure
  pressure sore, 61/348
  sense, 1/89, 96
Pressure cone
  cerebellar, *see* Cerebellar pressure cone
Pressure hydrocephalus
  normal, *see* Normal pressure hydrocephalus
Pressure palsy
  see also Hereditary multiple recurrent
    mononeuropathy
  hereditary multiple recurrent mononeuropathy,
    51/551
  hereditary recurrent neuropathy, 21/89
  nerve, *see* Nerve pressure palsy
Pressure sore
  aftercare, 26/475
  anemia, 26/471
  blister formation, 61/354
  blood flow, 26/463, 61/351, 355
  bone involvement, 26/276, 471
  bursitis, 61/354
  cancerous degeneration, 26/470
  classification, 26/468, 61/354
  complication, 26/471
  epidemiology, 61/347
  erythema, 61/354

  etiology, 61/348
  extrinsic factor, 26/464, 61/348
  hypoalbuminemia, 61/350
  identification, 61/351
  interface microenvironment, 61/349
  intrinsic factor, 26/463, 61/350
  juvenile spinal cord injury, 25/188, 61/245
  localization, 26/465
  mechanism, 26/463
  mobility, 61/350
  multiple sclerosis, 9/415
  muscular atrophy, 61/352
  nutrition, 61/350
  nutritional assessment, 26/471, 61/353
  paraplegia, 26/463, 61/347
  physical rehabilitation, 61/465
  plaster bed, 26/464
  predilection, 26/223
  pressure, 61/348
  pressure ischemia, 61/351
  prevention, 26/472, 61/347
  prophylaxis, 26/472, 61/358
  reactive hyperemia, 61/351
  risk assessment, 61/352
  sacral sore, 26/465
  sensation, 26/464, 61/350
  shear stress, 61/349
  sinus formation, 26/469
  soft tissue imaging, 61/357
  spasticity, 26/464, 480
  spinal cord injury, 25/188, 289, 26/223, 463-476,
    61/347
  stage, 26/468
  surgery, 26/474, 61/363
  technique advance, 61/363
  temperature, 61/349
  terminology, 26/463
  tetraplegia, 26/463, 61/347
  tissue breakdown, 61/354
  transverse spinal cord lesion, 61/347
  treatment, 26/472, 61/360
  vitamin C, 61/350
  Waterlow risk assessment, 61/352
  zinc, 61/350
Presynaptic neuromuscular blocking toxin
  acetylcholine release, 65/184
  neurotoxin, 65/183
  snake envenomation, 65/183
  snake venom, 65/183
Pretectal region
  anatomy, 2/532
  Edinger-Westphal nucleus, 2/533
Prevertebral ganglion

sympathetic neuron, 74/41
Prevertebral space
  cervical vertebral column injury, 25/252, 316
  widening, 25/316
Previous Sydenham chorea
  contraceptive drug induced chorea, 49/365
Priapism
  definition, 2/196
  misdiagnosis, 2/196
  spinal cord lesion, 19/34
  spinal shock, 2/196
Prichard classification
  epilepsy, 15/4
  uterine epilepsy, 15/4
Prichard, J.C., 15/4
Prilocaine intoxication
  local anesthetic agent, 37/405
  seizure, 37/405
Primaquine
  glucose-6-phosphate dehydrogenase deficiency,
    42/644
  malaria, 52/372
Primäre reizung, see Axonal reaction
Primary acalculia
  asymbolic acalculia, 4/185, 192
  asyntactic acalculia, 4/185, 192
Primary adrenal insufficiency
  classification, 70/211
Primary alcoholic dementia, 46/339-347
  amnesia, 46/342
  brain atrophy, 46/343-346
  clinical features, 46/339-342
  computer assisted tomography, 46/344-346
  confabulation, 46/342
  confusional state, 46/342
  CSF flow, 46/345
  neuropathology, 46/346
  neuroradiology, 46/343-346
  organic psychosis, 46/341
  PEG, 46/343
  psychological testing, 46/342
  theory, 46/339-342
  Wernicke-Korsakoff syndrome, 46/340-343
Primary alveolar hypoventilation
  carotid endarterectomy, 63/414
  congenital form, 63/415
  respiratory encephalopathy, 63/414
  sleep, 63/414
Primary amebic meningoencephalitis
  Acanthamoeba, 35/25, 52/314
  cerebral amebiasis, 35/34-37
  clinical features, 35/38-41, 52/316
  CSF, 52/319

differential diagnosis, 52/316
epidemiology, 35/34-37, 52/313
Hartmannella, 35/25
immunocompromised host, 52/314
Naegleria fowleri, 35/25
neuroamebiasis, 52/309
olfactory nerve, 52/316
pathogenesis, 35/48-52
pathology, 35/53-56
prognosis, 35/58-60
treatment, 35/58-60
Primary amyloidosis
  abdominal pain, 51/415
  age at onset, 51/415
  autonomic polyneuropathy, 51/416
  Bence Jones proteinuria, 51/416
  brain, 51/415
  chronic axonal neuropathy, 51/531
  CSF protein, 51/417
  diarrhea, 51/415
  erythrocyte sedimentation rate, 51/416
  familial brucellosis, 51/415
  fatigue, 51/415
  heart failure, 51/415
  hepatomegaly, 51/416
  hereditary, see Hereditary primary amyloidosis
  hereditary brain hemorrhage, 51/415
  hoarse voice, 51/416
  initial symptom, 51/415
  light chain Ig, 51/417
  macroglossia, 51/416
  monoclonal protein, 51/416
  multiple myeloma, 51/415, 419
  nausea, 51/415
  nephrotic syndrome, 51/415
  orthostatic hypotension, 22/253, 63/149
  paraproteinemic polyneuropathy, 51/429
  pathology, 51/417
  plasma cell dyscrasia, 63/391
  polyneuropathy, 51/416
  proteinuria, 51/415
  purpura, 51/416
  renal insufficiency, 51/415
  serum M component, 51/416
  Waldenström macroglobulinemia, 51/415
  weakness, 51/415
  weight loss, 51/415
Primary axonal degeneration
  critical illness polyneuropathy, 51/581
Primary biliary cirrhosis
  chronic axonal neuropathy, 51/531
  hepatic insufficiency polyneuropathy, 51/362
  Kayser-Fleischer ring, 49/228

central pontine myelinolysis, 47/588
cerebellar hemorrhage, 55/92
cerebellar medulloblastoma, 17/715, 18/188-190
cerebral amebiasis, 52/327
cerebral paragonimiasis, 35/259
cervical cord lesion, 35/281
Cheyne-Stokes respiration, 24/675
childhood absence epilepsy, 73/147
chlordane intoxication, 36/411
chondrodystrophic myotonia, 31/279
chordoma, 18/160, 19/291
chronic migrainous neuralgia, 48/250
chronic subdural hematoma, 24/313, 54/357
clumsiness, 46/165
coccidioidomycosis, 52/419
coma, 46/611
complete cord lesion, 25/282, 26/307
congenital brain tumor, 31/41
Conradi-Hünermann syndrome, 38/406
cough headache, 48/368
cranial epidural empyema, 33/156
cranial subdural empyema, 33/160
craniolacunia, 50/140
craniopharyngioma, 18/469
critical illness polyneuropathy, 51/575
CSF fistula, 24/191
CSF shunt, 46/330
Dandy-Walker syndrome, 30/644, 50/330
delirium, 46/532
dermal sinus and dermoid, 32/462
dermatomyositis, 71/130
desmoplastic cerebellar medulloblastoma, 18/189
developmental dyslexia, 4/389, 46/120
developmental dysphasia, 46/142
diabetic polyneuropathy, 27/112
dieldrin intoxication, 36/416
diffuse axonal injury, 57/380
dissecting aorta aneurysm, 39/249, 63/47
Down syndrome, 31/438
dyssynergia cerebellaris myoclonica, 21/515,
    60/602
eclampsia, 39/548
Edwards syndrome, 31/523
Emery-Dreifuss muscular dystrophy, 62/154
endrin intoxication, 36/416
ependymoma, 18/130
epidural hematoma, 54/349
epilepsy, 15/783
epilepsy of late onset, 15/269
epilepsy with myoclonic absence, 73/233
epilepsy with myoclonic astatic seizure, 73/227
ergot intoxication, 65/73
facial paralysis, 2/64, 8/289

familial pes cavus, 21/268
Friedreich ataxia, 21/357, 60/323
Galen vein, 31/196
Gaucher disease, 10/528
giant cell arteritis, 71/199
glioblastoma, 68/97
global aphasia, 45/323
globoid cell leukodystrophy, 10/306
glomus jugulare tumor, 18/448
glycosaminoglycanosis, 10/434
glycosaminoglycanosis type I, 10/435
glycosaminoglycanosis type II, 10/435, 66/295
glycosaminoglycanosis type III, 66/301
glycosaminoglycanosis type IV, 66/305
grand mal status, 15/151
granulomatous amebic encephalitis, 52/329
Guillain-Barré syndrome, 7/506
gunshot injury, 46/611
head injury, see Head injury prognosis
headache, 48/39
hemiballismus, 49/371
Hemophilus influenzae meningitis, 52/66
hepatic coma, 27/370
hereditary multiple recurrent mononeuropathy,
    51/554, 559
herpes simplex virus encephalitis, 34/154
histiocytosis X, 38/111
histologic anaplasia, 67/6
holoprosencephaly, 30/457, 50/239
hydrazine intoxication, 36/382
hydrocephalic dementia, 46/330
hydrocephalus, 32/552-554, 50/296
hypernatremia, 28/452
hypertensive encephalopathy, 54/217
hypokalemic periodic paralysis, 28/592
hypophosphatasia, 31/317
hypophyseal adenoma, 18/355
idiopathic facial paralysis, 8/289
incomplete cord lesion, 25/283, 26/308
infantile autism, 46/192
infantile spasm, 15/229, 788
infectious endocarditis, 63/118
intracerebral hematoma, 11/694-696, 24/364
intracranial pressure, 23/211, 217, 24/674, 57/379
isobenz intoxication, 36/416
juvenile absence epilepsy, 73/151
Klüver-Bucy syndrome, 24/704
Krabbe-Bartels disease, 14/412
kyphosis, 50/419
labyrinth concussion, 24/126
Landau-Kleffner syndrome, 73/287
lathyrism, 36/511, 65/7
Laurence-Moon-Bardet-Biedl syndrome, 31/328

Lowe syndrome, 31/304
lumbosacral nerve injury, 25/478
Lyme disease, 51/210
lymphocytic choriomeningitis, 56/130
malignant hyperthermia, 38/551
meningococcal meningitis, 52/29
metastatic cardiac tumor, 63/105
methanol intoxication, 36/356
midbrain tumor, 17/643
middle cerebral artery syndrome, 53/367
migraine, 48/39
motor aphasia, 45/323
moyamoya disease, 12/360
mucormycosis, 52/475
multiple myeloma, 18/256, 20/111
multiple nevoid basal cell carcinoma syndrome,
    14/585
multiple sclerosis, 9/199, 47/60
multiple trichoepithelioma, 14/589
mumps encephalitis, 34/628
narcolepsy, 3/94
neglect syndrome, 45/175-178
neuralgic amyotrophy, 51/173
neuroblastoma, 8/471
neuroborreliosis, 51/210
neurocutaneous melanosis, 14/423, 600
neurocysticercosis, 35/315
neurofibromatosis, 14/572
neurofibromatosis type I, 31/15
neurologic intensive care, 55/218
neuromyelitis optica, 9/430
neurosarcoidosis, 38/533
neurosyphilis, 33/382-387
newborn epilepsy, 15/214
Niemann-Pick disease, 10/505
nocardiosis, 52/451
North American blastomycosis, 35/401, 406,
    52/390
occipital encephalocele, 50/101
occipital lobe tumor, 17/314
oculomotor nerve injury, 24/83
oligodendroglioma, 68/128
optic nerve glioma, 18/346
optic nerve injury, 24/38
osteochondroma, 19/308
Paragonimus meningitis, 35/262
parietal encephalocele, 50/109
parietal lobe tumor, 17/306
Parkinson disease, 6/194, 49/107
Patau syndrome, 31/514, 50/559
pediatric head injury, 23/450, 24/833
persistent vegetative state, 24/678, 727
petit mal epilepsy, 15/788

petit mal status, 15/788
Peutz-Jeghers syndrome, 14/605, 43/42
pinealoma, 18/365
pituitary adenoma, 17/422
poliomyelitis, 34/115-117
polyarteritis nodosa, 39/308, 55/357
polymyositis, 41/81, 71/130
porphyria, 70/310
posttraumatic amnesia, 46/611
Prader-Labhart-Willi syndrome, 31/325, 43/464
primary amebic meningoencephalitis, 35/58-60
primary cardiac tumor, 63/105
primary CNS lymphoma, 68/265
primitive neuroectodermal tumor, 68/220
progressive bulbar palsy, 22/115
progressive multifocal leukoencephalopathy,
    9/493, 34/308, 38/656, 39/19, 47/518
progressive spinal muscular atrophy, 22/40
progressive supranuclear palsy, 46/302
psychosis, 15/796
rabies, 34/269
Refsum disease, 60/236, 66/497
rehabilitation, 26/310, 46/611, 55/239
respiration, 24/675
Reye syndrome, 29/341, 49/217
sacrococcygeal chordoma, 19/291
sarcoid neuropathy, 51/197
sarcoidosis, 38/533, 71/493
schistosomiasis, 52/540
schizencephaly, 50/201
sexual function, 26/311
Sneddon syndrome, 55/408
speech disorder in children, 4/412
sphenoid wing aplasia, 30/282
spina bifida, 32/553
spinal angioma, 20/467, 32/509
spinal cord abscess, 33/193, 52/191
spinal cord injury, see Spinal cord injury prognosis
spinal cord metastasis, 69/180
spinal cord missile injury, 25/219
spinal cord tumor, 19/10
spinal ependymoma, 20/363
spinal epidural empyema, 33/190, 52/188
spinal epidural hematoma, 12/565
spinal meningioma, 20/236
spinal nerve root injury, 25/423
spinal neurinoma, 20/285
spinal paragonimiasis, 35/264
spinal spongioblastoma, 19/360
spinal subdural empyema, 52/189
spinal tumor, 19/46
stroke, 46/611
subarachnoid hemorrhage, 12/179, 55/25, 45

subclavian steal syndrome, 53/374
subdural effusion, 24/339
subdural hematoma, 54/352
superior longitudinal sinus thrombosis, 33/169, 52/179
superior orbital fissure syndrome, 33/179
supratentorial brain tumor, 18/338
supratentorial midline tumor, 18/342
Sydenham chorea, 71/551
syphilis, 33/382-387
syringomyelia, 32/319
systemic lupus erythematosus, 55/381
tabes dorsalis, 33/385
Takayasu disease, 12/416
temporal arteritis, 48/324, 55/345
temporal lobe agenesis syndrome, 30/358
teratoma, 31/60
tetanus, 65/217, 71/560
thalamic tumor, 17/617
thallium intoxication, 36/272
third ventricle tumor, 17/472
Townes-Brocks syndrome, 42/380
transient global amnesia, 45/212
transient ischemic attack, 12/6
transverse sinus thrombophlebitis, 33/175
transverse sinus thrombosis, 52/177
transverse spinal cord lesion, 61/427
traumatic blindness, 24/38
traumatic carotid cavernous sinus fistula, 57/363
traumatic intracerebral hematoma, 57/379
traumatic intracranial hematoma outcome, 57/282
traumatic psychosyndrome, 24/569
traumatic venous sinus thrombosis, 24/376
trichinosis, 35/273-276
trichopoliodystrophy, 29/290
trochlear nerve injury, 24/84
tuberculosis, 71/283
tuberculous meningitis, 33/230-234, 52/214, 55/431, 71/283
tuberculous spondylitis, 33/250
tuberous sclerosis, 14/381, 580
uremic neuropathy, 8/3
venous sinus occlusion, 54/448
venous sinus thrombosis, 12/442, 54/414, 448
verruga peruana, 34/672
vertebral metastasis, 20/422
viral meningitis, 56/131
Wernicke-Korsakoff syndrome, 28/252
Wilson disease, 6/275
Wohlfart-Kugelberg-Welander disease, 22/77, 59/81
xeroderma pigmentosum, 31/291
Programmed cell death
  tumor cell, 67/29
Progressive amyotrophy
  facioscapulohumeral muscular atrophy, 21/29
  Möbius syndrome, 59/375
  paramyotonia congenita, 22/32
Progressive areolar choroidal sclerosis
  central, *see* Central choroidal atrophy
Progressive arthro-ophthalmopathy, *see* Stickler syndrome
Progressive ataxia
  organic acid metabolism, 66/641
Progressive bulbar palsy
  see also Motoneuron disease
  acute disseminated encephalomyelitis, 59/375
  adult, 22/111-145, 59/217-225
  amyotrophic lateral sclerosis, 22/112, 293, 42/65, 59/220
  aneurysm, 2/257
  basilar impression, 59/225
  botulism, 59/225, 375
  brain stem neoplasm, 59/375
  brain stem reflex, 22/113
  Brown-Vialetto-Van Laere syndrome, 59/374
  bulbar syndrome, 22/312
  bulbar type, 22/305, 308, 310, 312
  Bunina body, 59/224
  carcinomatous encephalopathy, 22/143
  cause, 22/125
  cause of death, 22/115
  cerebellar dysarthria, 59/409
  cerebellar symptom, 59/225
  child, *see* Fazio-Londe disease
  chorea, 59/409
  chorea-acanthocytosis, 59/225
  congenital abnormality, 59/375
  congenital myasthenic syndrome, 59/375
  congenital myopathy, 59/375
  congenital suprabulbar paresis, 59/375
  cutaneous abdominal reflex, 59/220
  cyst formation, 59/375
  deafness, 22/108, 114, 138, 516
  deglutition, 22/121, 59/218
  differential diagnosis, 59/224, 374, 383-401, 408
  diphtheria, 59/225
  diphtheritic polyneuropathy, 59/375
  dysarthria, 22/115-121, 59/218
  dysphonia, 22/116, 59/218
  dyspnea, 22/114, 59/219
  dystonia, 59/409
  EMG, 22/113
  eosinophilic inclusion body, 59/224
  epidemiology, 59/220
  external ophthalmoplegia, 22/140

facial muscle, 22/112, 59/218
forme fruste, 22/137
gastric dilatation, 22/114
globoid cell leukodystrophy, 59/375
gracile funiculus, 59/224
Guillain-Barré syndrome, 59/375
Guillain-Mollaret triangle, 22/140
Hallervorden-Spatz syndrome, 59/225
heart disease, 22/114
hereditary, 22/128
hereditary scapulohumeral type, 22/137
history, 22/111, 59/217
infantile, *see* Fazio-Londe disease
infantile Gaucher disease, 59/375
juvenile, *see* Fazio-Londe disease
limbic encephalitis, 22/143
locus ceruleus, 59/224
macroglobulinemia, 59/225
medial longitudinal fasciculus, 59/224
metachromatic leukodystrophy, 59/375
mitochondrial myopathy, 59/375
Möbius syndrome, 59/375
motoneuron ballooning, 59/224
muscular atrophy, 22/113
myasthenia gravis, 59/375, 408
myelinated fiber loss, 59/222
myotonic dystrophy, 59/375
neurofilament accumulation, 59/222
neuropathology, 22/126, 59/221-224
nosology, 22/128
nystagmus, 22/114
ocular myopathy, 22/140
oculomotor nucleus, 59/220
oculopharyngeal muscular dystrophy, 59/375
optic atrophy, 22/108, 138
oropharynx mechanic obstruction, 59/409
palatal myoclonus, 38/584, 59/225
paraproteinemia, 59/225
pathologic laughing and crying, 22/115
pharyngeal reflex, 59/218
pharyngeal retention, 59/219
poliomyelitis, 59/375
posterior inferior cerebellar artery, 2/257
presenting symptom, 22/112
prognosis, 22/115
progressive supranuclear palsy, 59/408
pyramidal tract, 59/224
respiration, 22/115, 59/218
retrograde axonal transport, 59/222
salivation, 59/219
sensory impairment, 59/220
sensory symptom, 22/114
Sjögren syndrome, 59/225

slow progression, 59/129
speech, 22/112, 116, 59/218
sphincter disturbance, 59/219
spinocerebellar syndrome, 59/129
spinocerebellar tract, 59/224
symptom, 22/113
symptomatic treatment, 59/459-471
symptomatology, 59/218-220
syringobulbia, 59/375
tachycardia, 22/114, 59/219
tetrodotoxin intoxication, 65/156
thyroid gland, 41/240
tick paralysis, 59/225, 375
treatment, 22/125
trichinosis, 59/225
vascular malformation, 59/375
visceromotoneuron, 59/219
vocal cord paralysis, 59/219
Wohlfart-Kugelberg-Welander disease, 22/129
X-linked variant, 22/137
Progressive bulbospinal muscular atrophy
hereditary spinal muscular atrophy, 59/23
Progressive cerebellar ataxia
ataxia telangiectasia, 60/360
celiac disease, 60/599
dyssynergia cerebellaris myoclonica, 60/599
hereditary periodic ataxia, 21/564
Progressive cerebellar dyssynergia
basal ganglion degeneration, 42/212
hydrocephalus, 42/211
mental deficiency, 42/213
Progressive cochleovestibular atrophy
hereditary, *see* Hereditary progressive
cochleovestibular atrophy
Van Buchem disease, 22/494
Progressive cone dystrophy
hereditary, *see* Hereditary progressive cone
dystrophy
Progressive cord lesion
cause, 25/276
cervical, 25/276
spinal cord injury, 25/276
spinal deformity, 25/277
Progressive degenerative subcortical
encephalopathy, *see* Spongy cerebral
degeneration
Progressive dementia
atrial myxoma, 63/98
lipomembranous polycystic osteodysplasia,
42/279-281
Progressive demyelinating polyneuropathy
chronic, *see* Chronic progressive demyelinating
polyneuropathy

Progressive diaphyseal dysplasia, 38/396-401
  course, 38/400
  features, 38/396-399
  genetics, 38/400
  history, 38/396
  laboratory finding, 38/399
  treatment, 38/400
Progressive distal muscular atrophy
  infantile spinal muscular atrophy, 59/70
Progressive diurnal fluctuating dystonia
  hereditary, *see* Hereditary progressive diurnal
    fluctuating dystonia
Progressive dysautonomia, 59/133-163
  acetylcholine, 59/147
  adrenalin, 59/152
  aged, 59/158
  Alzheimer disease, 59/136
  amino acid, 59/148
  amyotrophy, 59/138
  antecollis, 59/138
  apnea, 59/159
  arousal system, 59/160
  autoimmune disease, 59/162
  autonomic function, 59/158
  baroreflex arc reflex, 59/150
  bladder infection, 59/140
  brain stem auditory evoked potential, 59/139
  caffeine, 59/140, 158
  carbohydrate, 59/157
  cardiac pacemaker, 59/140
  cardiovascular test, 59/135
  catecholamine, 59/150
  choline acetyltransferase, 59/147
  cholinergic central pathway, 59/155
  cholinergic neuron, 59/147
  circadian rhythm, 59/155
  circopharyngeal myotomy, 59/140
  classification, 59/133
  CNS, 59/155
  computer assisted tomography, 59/139
  constipation, 59/140
  definition, 59/136
  diagnostic criteria, 59/137
  threo-3,4-dihydroxyphenylserine, 59/140
  dopamine, 59/153, 160
  dopamine β-mono-oxygenase deficiency, 59/162
  Eaton-Lambert myasthenic syndrome, 59/161
  electrophysiologic study, 59/139
  ergotamine, 59/140
  eye movement, 59/137
  failure, 59/135
  fasciculation, 59/138
  fluorocortisone, 59/140

  gastic emptying, 59/158
  gastrin, 59/154
  gastrostomy, 59/140
  glucagon, 59/154, 157
  glucose, 59/156
  growth hormone, 59/154
  gut peptide, 59/154
  hormone, 59/153
  hydrocortisone, 59/154
  hypercapnic ventilatory response, 59/160
  hypnotic agent, 59/160
  hypoglycemia, 59/155
  hypothalamic involvement, 59/160
  hypoxic ventilatory response, 59/160
  indometacin, 59/140, 158
  insulin, 59/157
  intermittent self-catherization, 59/140
  isoprenaline clearance, 59/150
  Joseph-Machado disease, 59/137
  Klein-Waardenburg syndrome, 59/162
  laboratory study, 59/139
  larynx stridor, 59/160
  levodopa, 59/136, 138, 140
  lipid, 59/157
  melatonin, 59/155
  metabolite excretion, 59/150
  methoxamine, 59/151
  multiple neuronal system degeneration,
    59/134-138, 141-143
  nerve lesion, 59/142
  neurochemistry, 59/145-149
  neuropathology, 59/140
  neuropeptide, 59/148
  neurotensin, 59/157
  neurotransmitter metabolism, 59/152-155
  noradrenalin, 59/139, 152
  noradrenergic nervous system, 59/149
  nuclear magnetic resonance, 59/139
  olivopontocerebellar atrophy, 59/137, 141
  olivopontocerebellar involvement, 59/136
  orthostatic hypotension, 59/136, 151
  pancreatic polypeptide, 59/157
  paraneoplastic syndrome, 59/161
  parasympathetic ganglion, 59/143
  Parkinson disease, 59/134
  peptide, 59/153
  peripheral autonomic ganglion, 59/143
  peripheral chemical pathology, 59/149-152
  pontomedullary region, 59/159
  positron emission tomography, 59/139
  postganglionic sympathetic neuron, 59/149
  postprandial hemodynamics, 59/156-158
  pressor drug response, 59/150

protein, 59/157
pure, 59/134
REM sleep, 59/158
respiration, 59/158
respiratory failure, 59/159
serotonin, 59/147, 152
Shy-Drager syndrome, 59/133-135
small cell carcinoma, 59/162
somatostatin, 59/151, 158
spinal cord, 59/148
sporadic type, 59/137
striatonigral degeneration, 59/141
striatonigral involvement, 59/136
substance P, 59/148
support stocking, 59/140
tracheostomy, 59/140
treatment, 59/140, 160
tyramine, 59/151
vagal afferent involvement, 59/160
vasopressin, 59/151
visceral nerve plexus, 59/143
wakefulness, 59/159
xylose, 59/158
Progressive encephalomyopathy
multiple mitochondrial DNA deletion, 66/429
Progressive external ophthalmoplegia, 13/309-314,
22/177-196, 62/287-315
see also Extraocular muscle weakness,
Kearns-Sayre-Daroff-Shy syndrome,
Mitochondrial myopathy, Ocular myopathy,
Oculomotor paralysis and Ophthalmoplegia
abiotrophic external ophthalmoplegia, 22/210
associated disease, 22/203-216
ataxia, 22/193, 43/136
Bassen-Kornzweig syndrome, 22/191, 204, 207,
60/656, 62/304
biochemistry, 43/137, 62/311
bulbar paralysis, 60/658
cardiac disorder, 43/137, 62/309
cardiac dysrhythmia, 41/215
cardiomyopathy, 22/192, 62/310
cataract, 62/306
celiac disease, 62/313
cerebellar ataxia, 43/137, 60/657
cerebellar syndrome, 62/308
Chédiak-Higashi syndrome, 60/657
cholinergic agent, 22/183
chorioretinal degeneration, 13/39
chronic, see Chronic progressive external
ophthalmoplegia
clinical features, 22/180, 209, 40/302, 342, 60/53,
62/292, 307
clinical syndrome, 41/214

complex IV, 62/293, 314
component, 41/214
computer assisted tomography, 62/295, 306, 311
concept history, 62/287
congenital fibrosis, 62/297
congenital myasthenia gravis, 62/297
congenital myopathy, 62/297, 333
congenital ocular muscle fibrosis, 62/297
congenital ptosis, 22/189, 62/297
cranial nerve, 43/136
cranial neuropathy, 62/313
creatine kinase, 22/183
criteria, 62/287
CSF, 62/311
CSF protein, 22/183, 43/137
curare, 22/183
curare sensitive ocular myopathy, 22/179, 188,
62/302
deafness, 22/193, 60/658, 62/313
definition, 60/47, 62/287, 307
degenerative type, 22/204
dementia, 62/308
dentatorubropallidoluysian atrophy, 60/656
diabetes mellitus, 62/309
diagnostic test, 62/294
differential diagnosis, 40/298, 60/48
diffuse Lewy body disease, 62/304
dysarthria, 22/187
dysphagia, 22/185, 187, 206
dysphonia, 22/206
dysthyroid ophthalmoplegia, 22/204
dystonia, 43/136, 62/304
ECG, 62/311
edrophonium test, 62/303
EMG, 22/181, 62/295
endocrine disorder, 62/309
epidemiology, 62/291
epilepsy, 62/308
etiology, 43/137, 60/55
Ferguson-Critchley syndrome, 62/304
Friedreich ataxia, 21/347, 22/191, 43/136, 60/310,
62/304
genetics, 22/185, 41/420, 43/137, 62/295, 312
glycogen storage disease, 22/189, 43/136, 62/298
glycosaminoglycanosis type I, 22/212
heart block, 13/311, 22/193, 204, 211
hereditary ataxia, 43/136, 62/304
hereditary cerebellar ataxia, 22/191, 208
hereditary motor and sensory neuropathy variant,
60/247
heredoataxia (Stephens), 22/204, 207
histochemistry, 40/47, 54
histopathology, 62/293, 310

history, 22/178, 62/288
hypoparathyroidism, 62/309
intestinal pseudo-obstruction, 62/313
Joseph-Machado disease, 60/656
Kearns-Sayre-Daroff-Shy syndrome, 22/192, 211, 40/302, 41/215, 43/137, 60/52-57, 62/287, 306, 508
Kiloh-Nevin syndrome, 43/136, 60/52
laboratory study, 60/54
limb weakness, 62/291
long acting thyroid stimulator, 22/184
macular degeneration, 22/194
maternal inherited, 66/428
MELAS syndrome, 60/56
meningitis, 22/194
mental deficiency, 62/308
MEPOP syndrome, 62/315
MERRF syndrome, 60/56
mitochondrial change, 40/220
mitochondrial disease, 22/189, 196, 60/52-57, 62/497
mitochondrial DNA, 62/294, 66/426
mitochondrial DNA deletion, 60/56
mitochondrial myopathy, 22/189, 60/657, 62/289, 313
MNGIE syndrome, 62/305, 313, 505
Möbius syndrome, 2/279, 43/136
molecular genetics, 62/311
motoneuron degeneration, 62/305
motor end plate, 43/136
multicore myopathy, 41/16
multiple mitochondrial DNA deletion, 62/510, 66/429
muscle biopsy, 60/54
muscle change, 22/182
muscle histology, 22/181
muscle histopathology, 62/310
muscle tissue culture, 40/186
muscular dystrophy, 2/279, 22/210, 43/136
myasthenia gravis, 22/181, 188, 204, 62/415
myasthenic ocular myopathy, 62/302
myopathic theory, 22/178
myopathy, 13/312, 22/182, 41/215, 43/136
myopia, 43/141
myotonia congenita, 62/305
myotonic dystrophy, 22/187, 62/302
myotubular myopathy, 22/179, 189, 41/11, 43/136, 62/297
neurogenic, 22/182
neurogenic ophthalmoplegia, 62/304
neuromuscular disease, 41/420
neuropathology, 22/182, 62/289
neuropathy, 43/136, 62/313

nigrospinodentate degeneration, 22/208
nondegenerative type, 22/204
ocular myasthenia gravis, 62/302
ocular myopathy, 22/186, 41/215, 62/288, 291
oculopharyngeal muscular dystrophy, 22/179, 187, 206, 40/302, 41/215, 43/136, 60/49-52, 62/295
olivopontocerebellar atrophy, 43/136, 62/304
olivopontocerebellar atrophy (Dejerine-Thomas), 60/656
ophthalmoplegic migraine, 2/324
optic atrophy, 13/37, 82, 62/305
orbital myositis, 22/190, 62/298
OSPOM syndrome, 62/313
oxidative phosphorylation, 41/215
paroxysmal tachycardia, 22/193
pathogenesis, 62/294
pathologic reaction, 40/91, 253
pigmentary retinopathy, 62/308
POLIP syndrome, 62/313
polyneuropathy, 22/194, 207, 43/136, 62/309, 313
primary pigmentary retinal degeneration, 13/304, 22/192, 43/137
primidone intoxication, 22/180
progressive supranuclear palsy, 22/219
pseudoprogressive external ophthalmoplegia, 22/204
ptosis, 22/179, 43/136, 62/297, 66/427
pure type, 22/185
ragged red fiber, 40/186, 60/54, 62/289, 291-293, 305, 314, 66/427
Refsum disease, 13/40, 22/204, 207
renal dysfunction, 62/309
retinal degeneration, 43/141
retinopathy, 62/305, 311
Schimke syndrome, 62/304
secondary pigmentary retinal degeneration, 13/310-314, 22/204, 60/654, 725
senile ptosis, 62/302
spastic ataxia, 62/304
spastic paraparesis, 62/305
spastic paraplegia, 13/307, 309, 43/136, 62/304
spongiform encephalopathy, 22/212
subacute necrotizing encephalomyelopathy, 60/56
T3, 22/184
T3 suppression test, 22/190, 204
thyroid disease, 41/240, 43/136
thyrotropic hormone, 22/184
treatment, 22/195, 60/57, 62/312
Wohlfart-Kugelberg-Welander disease, 22/210
zebra body, 22/194, 212
Progressive external ophthalmoplegia classification
amyotrophic lateral sclerosis, 62/290
Bassen-Kornzweig syndrome, 22/179, 62/290

congenital myopathic ptosis, 22/179, 62/290
dystonia musculorum deformans, 22/179, 62/290
familial spastic paraplegia, 22/179, 62/290
glycogen storage disease, 22/179, 62/290
hereditary ataxia, 22/179, 62/290
infantile spinal muscular atrophy, 22/179, 62/290
Joseph-Machado disease, 22/179, 62/290
Kearns-Sayre-Daroff-Shy syndrome, 22/179,
    40/281, 62/290
MELAS syndrome, 62/290
MERRF syndrome, 62/290
MNGIE syndrome, 62/290
Möbius syndrome, 22/179, 62/290
myasthenia gravis, 22/179, 62/290
myotonic dystrophy, 22/179, 62/290
myotubular myopathy, 22/179, 62/290
ocular myopathy, 22/179, 62/290
oculopharyngeal muscular dystrophy, 22/179,
    62/290
oculopharyngeal myopathy, 22/179, 62/290
ophthalmoplegia variant, 62/290
orbital myositis, 62/290
OSPOM syndrome, 62/290
practical guideline, 62/289
progressive supranuclear palsy, 22/179, 62/290
senile ptosis, 62/290
thyroid associated ophthalmopathy, 62/290
3 types, 60/47
Wohlfart-Kugelberg-Welander disease, 22/179,
    62/290
Progressive facial hemiatrophy, see Progressive
    hemifacial atrophy
Progressive familial choreoathetosis
neurocutaneous syndrome, 14/101
telangiectasia, 14/101
Progressive familial myoclonus epilepsy, see
    Progressive myoclonus epilepsy
Progressive glomerulopathy
choreoathetosis, 49/385
Progressive hemifacial atrophy
alopecia, 14/776, 22/548, 31/253, 43/408, 59/479
bulbar poliomyelitis, 22/549, 59/479
bulbar postpoliomyelitis amyotrophy, 59/479
cerebral hemiatrophy, 59/479
clinical features, 22/546, 59/477-479
CNS, 31/253, 43/408
congenital, see Congenital progressive hemifacial
    atrophy
coup de sabre deformity, 22/548, 31/253, 59/479
differential diagnosis, 31/254, 59/479
enophthalmos, 31/253, 43/409
epilepsy, 14/777, 31/253, 43/408, 59/479
etiology, 31/252

eye lesion, 31/253
face, 31/252
false type, 59/478
Goldenhar syndrome, 31/254, 59/479
heterochromia iridis, 31/253, 43/408
Horner syndrome, 8/347, 14/777, 31/253, 43/408
mental deficiency, 22/548, 59/479
paresthesia, 31/253, 43/408
ptosis, 31/253, 43/408
scleroderma, 8/347, 14/777, 22/548, 31/253,
    43/408, 59/479
seizure, 22/548
sex ratio, 8/347, 22/548, 43/408, 59/479
skin, 31/253
spread, 14/776, 22/547, 43/407, 59/479
Sturge-Weber syndrome, 14/512, 59/479
syndactyly, 22/550
syringomyelia, 14/777
teeth sparing, 59/479
trigeminal neuralgia, 14/777, 31/253, 43/408
true type, 59/478
Progressive hereditary craniodiaphyseal dysplasia,
    43/375-377
fibrous dysplasia, 14/198
Progressive hydromyelia
spina bifida, 50/494
Progressive lenticular degeneration, see Wilson
    disease
Progressive leukodystrophy
hereditary, see Hereditary progressive
    leukodystrophy
Progressive locomotor ataxia, see Tabes dorsalis
Progressive multifocal encephalitis
immunocompromise, 65/548
macrophage defense defect, 65/548
phagocyte defense defect, 65/548
T-lymphocyte, 65/548
Progressive multifocal leukoencephalopathy,
    34/307-326, 47/503-519
acquired immune deficiency syndrome, 47/503,
    506, 56/18, 495, 497, 514, 71/272
age, 9/492, 34/308, 47/507
animal model, 47/514
associated disease, 47/506, 71/400
astrocyte, 9/486, 39/524, 47/508, 56/514
ataxia, 9/494
autoimmune disease, 47/503
Binswanger disease, 54/231
BK virus, 34/313, 318-321, 47/504, 515, 518
brain biopsy, 34/310, 47/511, 515
brain scanning, 9/495
cancer, 69/449
carcinoma, 38/654-657

cause, 9/495
cerebellar syndrome, 9/494
cerebral lymphomatoid granulomatosis, 39/524
cerebral mass lesion, 47/504
child, 71/401
chorea, 9/494
clinical features, 9/492, 34/307-311, 38/656,
    47/504-507, 71/273
computer assisted tomography, 47/509, 511
corticosteroid, 47/506
course, 9/493, 34/308, 38/656, 39/19, 47/507
CSF, 9/495, 71/275
CSF examination, 47/510
cytarabine, 34/324, 47/518, 56/497
dementia, 38/656, 39/19, 46/388, 47/504, 54/231
demyelination, 9/486, 34/307, 46/388, 47/503,
    507-509, 511-513, 56/497, 514, 63/354
diagnosis, 34/321-323, 47/509-511, 71/275, 407
differential diagnosis, 71/276
diffuse sclerosis, 9/472
dystonia, 9/495
EEG, 9/495, 47/505, 510
electron microscopy, 34/307, 310, 47/508, 56/514
epidemiology, 34/313-321, 71/413
epilepsy, 9/495
etiology, 34/313-321, 47/515-518
granulomatous disease, 9/491, 34/307
headache, 9/495
histology, 71/272
histopathology, 47/511
history, 34/307, 71/399
Hodgkin lymphoma, 63/354
human immunodeficiency virus dementia, 71/277
idoxuridine, 47/519
immune response, 34/307, 47/507
immunocompromise, 56/480
immunocompromised host, 56/480
immunocytochemical study, 47/512
immunofluorescence test, 47/510
immunosuppression, 34/308, 47/503-506, 56/514,
    71/400
inflammatory response, 34/312, 47/509
JC virus, 34/315-317, 319-321, 38/656, 46/388,
    47/504-507, 514-518, 56/480, 497, 514, 63/354,
    71/412
laboratory, 71/410
lesion size, 47/513
leukemia, 39/19, 63/354, 71/400
lymphoma, 71/400
lymphoproliferative disease, 9/491, 34/307,
    47/503, 506
malignancy, 38/654-657, 47/503
malignant lymphoma, 39/45

multiple myeloma, 55/470
multiple sclerosis, 9/120, 126, 47/72, 327
myelin associated glycoprotein, 47/512
myelin basic protein, 47/512
myeloproliferative disease, 9/491, 34/307
neoplasm, 71/419
neuropathology, 71/402
non-Hodgkin lymphoma, 63/354
nuclear magnetic resonance, 47/509
oligodendrocyte, 9/486, 34/310, 39/524, 46/388,
    47/508, 511-513, 56/514
organ transplantation, 47/503-506
Papez circuit, 46/388
papilloma virus, 9/491
papovavirus, 9/327, 495, 34/307, 310, 313-321,
    39/19, 46/388, 47/503, 63/354
parkinsonism, 9/495
pathogenesis, 38/655, 47/511-515
pathology, 9/485, 34/311-313, 38/655,
    47/507-509, 71/273
plasma cell, 9/489
polycythemia vera, 9/596
polyoma, 9/491
polyomavirus, 9/491, 47/504, 508, 511, 515, 519,
    56/10, 71/417
polyomavirus BK, 47/504, 56/10
polyomavirus JC, 47/504, 56/10
prognosis, 9/493, 34/308, 38/656, 39/19, 47/518
psychiatric disorder, 9/494
Refsum disease, 21/215
renal transplantation, 34/308, 47/506
Russell body, 9/489
sarcoidosis, 71/400
senile plaque, 56/544
serologic test, 47/511
sign, 34/307, 47/504
simian virus 40, 34/313, 317-321, 38/656, 47/504,
    515, 56/514, 71/417
survey, 71/399
symptom, 34/307, 47/504
systemic disease, 9/491
toxoplasmosis, 63/354
treatment, 9/497, 34/323-326, 47/506, 518,
    71/277, 410
tuberculosis, 71/400
viral infection, 34/307, 47/503, 511-513
viral leukoencephalitis, 9/126
viral reactivation *vs* primary infection, 47/514
viral route, 47/514
virion, 9/489-492, 47/504, 508, 511
virology, 71/412
virus characteristics, 47/516
virus like structure, 47/215

Progressive muscular atrophy
  see also Amyotrophic lateral sclerosis and
    Wetherbee ail
  amyotrophic lateral sclerosis, 59/13, 18
  amyotrophic spondylotic myeloradiculopathy,
    59/17
  arteriosclerotic myelopathy, 59/19
  asthma, 59/22
  carcinomatous neuromyopathy, 59/21
  chronic inflammatory polyradiculopathy, 59/17
  cramp, 59/13
  differential diagnosis, 59/16-22
  electrical injury, 59/18
  EMG, 59/16
  endocrine disorder, 59/19
  ENG, 59/16
  fasciculation, 59/13
  hand, 59/13
  heavy metal intoxication, 59/18
  Ig dyscrasia, 59/16
  immunosuppression, 59/16
  incidence, 59/14
  infantile spinal muscular atrophy, 59/13
  Lyme disease, 59/21
  Madras disease, 59/22
  mental deficiency, 42/81
  metabolic disorder, 59/18
  motoneuron disease, 59/13
  multifocal nerve conduction block, 59/16
  muscle biopsy, 59/16
  myasthenia, 59/20
  nerve biopsy, 59/16
  nerve conduction velocity, 59/16
  neural conglomerate, 59/15
  neuromuscular transmission, 59/16
  onset, 59/13
  pathology, 59/14
  poliomyelitis, 56/331
  postencephalitic amyotrophy, 59/20
  postpoliomyelitic amyotrophy, 22/23, 59/19
  posttraumatic amyotrophy, 59/18
  proximal arm, 59/14
  respiratory muscle, 59/14
  risk factor, 59/14
  secondary pigmentary retinal degeneration,
    13/326
  sex ratio, 59/14
  Shy-Drager syndrome, 59/22
  subcutaneous fat, 59/13
  toxin, 59/18
  viral infection, 59/19
  Wohlfart-Kugelberg-Welander disease, 22/13
Progressive muscular dystrophy

  brain embolism, 55/163
  brain infarction, 55/163
  cerebrovascular disease, 55/163
  chorioretinal degeneration, 13/40
  history, 22/282
  Huntington chorea, 49/291
  intestinal pseudo-obstruction, 51/493
  limb girdle syndrome, 43/104
  optic atrophy, 13/40
Progressive myoclonus epilepsy
  see also Dentatorubropallidoluysian atrophy,
    Dyssynergia cerebellaris myoclonica, Glycogen,
    Juvenile neuroaxonal dystrophy, MELAS
    syndrome, MERRF syndrome, Myoclonic
    epilepsy, Neuronal ceroid lipofuscinosis and
    Ramsay Hunt syndrome
  ataxia, 42/114, 705
  atypical case, 27/187
  atypical inclusion body, 15/414, 49/609
  Baltic myoclonus epilepsy, 49/617
  behavior disorder in children, 42/705-708
  blindness, 42/702
  carbohydrate metabolism, 21/62
  cause, 49/616
  cerebellar ataxia, 42/698
  choreoathetosis, 42/699
  clinical features, 27/172, 73/295
  clinical laboratory study, 27/173
  consanguinity, 73/296
  Creutzfeldt-Jakob disease, 6/747
  dementia, 15/334, 42/114, 704, 49/617
  demographic data, 73/296
  dentatorubropallidoluysian atrophy, 21/528
  differential diagnosis, 27/188
  dyssynergia cerebellaris myoclonica, 42/211,
    60/599
  EEG, 27/174
  familial myoclonus, 6/768
  genetics, 27/173
  hallucination, 42/702
  hereditary myoclonus ataxia syndrome, 21/511
  Herrmann disease, 22/510
  heterogeneity, 1/282
  historic aspect, 27/171
  Lafora, see Lafora progressive myoclonus
    epilepsy
  Lafora body, 73/293
  Latham-Munro syndrome, 22/512
  marasmus, 15/334
  May-White syndrome, 22/512, 42/698
  mental deficiency, 42/704
  myocardial deposit, 27/182
  myoclonia, 6/765, 15/334

myoclonus, 27/172, 38/577
neuropathology, 27/175
optic atrophy, 13/39
orthochromatic leukodystrophy, 10/113
pathophysiology, 6/166
pharmacologic treatment, 73/301
protein electrophoresis, 27/174
secondary pigmentary retinal degeneration,
  42/704
sensorineural deafness, 42/698
status marmoratus, 42/700
survey, 73/293
treatment, 27/189, 73/300
tuberous sclerosis, 14/496
Unverricht-Lundborg, *see* Unverricht-Lundborg
  progressive myoclonus epilepsy
Progressive myositis ossificans, 43/190-192
collagen vascular disease, 43/191
congenital torticollis, 43/191
differential diagnosis, 40/324, 337
digital abnormality, 43/190
fever, 43/191
great toe shortening, 43/190
masseter muscle, 43/191
preclinical detection, 43/190
pseudohypoparathyroidism, 6/706, 49/420
severe trauma, 41/385
sex ratio, 43/191
spasmodic torticollis, 43/191
Progressive necrotic myelopathy, *see*
  Foix-Alajouanine disease
Progressive neuropathy
Chédiak-Higashi syndrome, 51/397
Progressive nuclear external ophthalmoplegia, *see*
  Progressive external ophthalmoplegia
Progressive oculocutaneous telangiectasia
ataxia telangiectasia, 60/360
Progressive ophthalmoplegia
myopathy, 40/281
Progressive pallidal atrophy (Hunt-Van Bogaert)
see also Pallidoluysian atrophy
athetosis, 6/455, 457, 554
basal ganglion degeneration, 6/164, 42/247
buccopharyngeal spasm, 21/531
choreoathetosis, 21/531, 42/248
clinical features, 6/640
course, 6/640, 21/531
critique, 2/500
dystonia, 21/531, 42/247
grimacing, 6/457, 21/531
hypokinesia, 42/247
levodopa, 42/248
locus ceruleus degeneration, 42/247

muscular hypotonia, 6/554
neurofibrillary tangle, 42/247
pes equinovarus, 6/457, 554, 21/531
rigidity, 6/457, 554, 21/531, 42/247
substantia nigra degeneration, 6/457, 42/247
thalamus, 2/491
torticollis, 6/457, 554
tremor, 42/247
Progressive pallidal degeneration
adult neuroaxonal dystrophy, 42/257
dystonia, 6/554
Hallervorden-Spatz syndrome, 42/257
symptomatic dystonia, 6/554
syndrome, 42/257
Progressive paraparesis
endemic paresis, 9/684
homocarnosinosis, 59/359
Sjögren-Larsson syndrome, 59/358
Progressive perceptive deafness
hereditary, *see* Dominant progressive
  sensorineural deafness
Progressive poliodystrophy, *see* Alpers disease
Progressive poliomyelitis
amyotrophic lateral sclerosis, 22/329
Progressive polyneuritis
chronic, *see* Chronic progressive polyneuritis
Progressive quadriparesis
achondroplasia, 50/395
Progressive rubella panencephalitis, 34/331-338,
  56/405-412
see also Rubella encephalitis
ataxia, 34/338, 56/409
brain atrophy, 56/410
brain biopsy, 34/338, 56/410
clinical features, 34/331-334, 56/408
Creutzfeldt-Jakob disease, 56/546
CSF, 34/334, 56/409
CSF oligoclonal band, 56/409
CSF rubella antigen, 34/334, 56/409
dementia, 34/338, 56/409
differential diagnosis, 34/338, 56/411
epidemiology, 56/405
epilepsy, 56/409
history, 34/331, 56/405
myoclonus, 34/338, 56/409
neuropathology, 56/410
ophthalmoplegia, 56/409
pathology, 34/336
rubella virus, 34/331, 56/405
slow virus disease, 56/405, 544
subacute sclerosing panencephalitis, 34/338,
  56/411
treatment, 56/411

ultrastructure, 56/410
viral infection, 34/56, 331, 56/406
Progressive senile vascular myelopathy, 12/497, 499
Progressive sensorineural deafness
  dominant, *see* Dominant progressive sensorineural
    deafness
Progressive sensorineural hearing loss
  recessive, *see* Recessive progressive sensorineural
    hearing loss
  sex linked, *see* Sex linked progressive
    sensorineural hearing loss
Progressive sensory neuropathy, *see* Hereditary
    sensory and autonomic neuropathy type II
Progressive spastic diplegia, *see* Progressive
    paraparesis
Progressive spinal muscular atrophy, 42/86-88
  see also Aran-Duchenne disease
  adult type, 22/1-56
  amyotrophic lateral sclerosis, 22/1, 42-44,
    310-312
  animal model, 22/32
  arteriosclerosis, 22/19
  astrocytosis, 22/17
  autoimmune disease, 22/24
  axonal neuropathy, 59/404
  axonal reaction, 22/14
  Betz cell degeneration, 22/18
  bulbar nerve symptom, 59/404
  bulbar paralysis, 22/134
  carcinoma, 22/31
  chronic inflammatory demyelinating
    polyradiculoneuropathy, 59/404-406
  combined peripheral and CNS inflammatory
    demyelinating syndrome, 59/406
  course, 22/2
  death rate, 22/11
  demyelinating neuropathy, 59/404
  diagnosis, 22/33
  differential diagnosis, 22/15, 59/383-401
  Eaton-Lambert myasthenic syndrome, 59/406
  electrical injury, 22/29
  EMG, 22/33-38
  etiology, 22/18-33
  familial cases, 22/12
  familial periodic paralysis, 22/32
  Friedreich ataxia, 22/13
  gastrectomy, 22/29
  glucose intolerance, 22/30
  Guillain-Barré syndrome, 59/406
  H-reflex, 22/33, 38
  hereditary, *see* Hereditary progressive spinal
    muscular atrophy
  hereditary motor and sensory neuropathy type I,

  21/29, 22/13
  hereditary motor and sensory neuropathy type II,
    21/29, 22/13, 59/406
  history, 22/2
  hypoglycemia, 22/29
  intermediolateral column, 22/17
  lateral column, 22/17
  lipofuscinosis, 22/17
  mental deficiency, 59/92
  metabolic disturbance, 22/31
  microcephaly, 59/92
  monomelic benign form, 59/403
  motoneuron degeneration, 22/9
  muscle cramp, 22/37
  muscle fiber, 22/18
  myasthenia gravis, 59/406
  myopathy, 59/407
  nephropathic cystinosis, 59/403
  nerve conduction velocity, 22/37
  nerve fiber, 22/17
  neurofibrillary tangle, 22/17
  neuron degeneration, 22/14
  neuronal conglomerate, 22/17
  neuropathology, 22/14-18
  neuropathy, 59/404
  neurosyphilis, 22/20
  normolipoproteinemic acanthocytosis, 59/402
  Norris series, 22/11
  oculopharyngeal amyotrophy, 22/12
  original case, 22/3-7
  parainfectious amyotrophy, 22/20
  polymyositis, 22/25
  postpoliomyelitic amyotrophy, 22/23
  posttraumatic amyotrophy, 22/28
  prognosis, 22/40
  protopathic motor system, 22/43
  proximal, *see* Wohlfart-Kugelberg-Welander
    disease
  proximal multifocal conduction block, 59/404
  pseudomyopathic, *see*
    Wohlfart-Kugelberg-Welander disease
  pseudomyotonic burst, 22/33
  pulmonary function, 22/39
  RNA virus type C, 22/33
  Ryukyan, *see* Ryukyan spinal muscular atrophy
  scapuloperoneal spinal muscular atrophy,
    22/57-68
  sex ratio, 22/12
  spinocerebellonigral amyotrophy, 22/14
  symptomatic treatment, 59/459-471
  toxic neuronopathy, 22/26
  treatment, 22/40
  ventral root fiber, 22/17

vestibular function, 22/40
viral infection, 22/21, 33
Wohlfart-Kugelberg-Welander disease, 22/12, 310, 59/92
X-linked bulbospinal form, 59/403
Progressive subacute encephalitis
  rubella, 46/389
Progressive subcortical encephalitis
  chronic, *see* Binswanger disease
Progressive subcortical encephalopathy, *see*
  Binswanger disease
Progressive subcortical leukoencephalopathy, *see*
  Binswanger disease
Progressive supranuclear palsy, 22/217-227,
  49/239-252
  Alzheimer neurofibrillary change, 22/222-224,
    49/249
  amnesia, 49/243
  amyotrophic lateral sclerosis, 22/226, 49/249, 64/5
  apathy, 49/243
  astrocytosis, 22/225, 46/302, 49/248, 250
  autonomic dysfunction, 75/189
  basal ganglion degeneration, 6/557, 42/259
  biochemical abnormality, 49/251
  blepharospasm, 22/219, 49/243
  bradykinesia, 22/219, 221, 46/301, 49/240
  bradyphrenia, 22/221, 49/243, 246
  bulbar paralysis, 22/140, 219
  captopril intoxication, 65/445
  choline acetyltransferase, 49/251
  clinical features, 22/218, 46/301
  cogwheel phenomenon, 22/221
  cogwheel saccade, 49/241
  coincident disorder, 49/246
  computer assisted tomography, 46/301, 49/245
  copper metabolism, 22/227
  course, 22/221, 42/259, 46/301, 49/239
  cranial nerve palsy, 6/557
  dementia, 6/557, 22/218, 220, 42/259, 46/301,
    49/239, 243
  depression, 49/239, 243
  differential diagnosis, 49/239, 246
  dipeptidyl carboxypeptidase I inhibitor, 65/445
  dipeptidyl carboxypeptidase I inhibitor
    intoxication, 65/445
  disease duration, 22/221, 49/239
  doll head eye phenomenon, 22/209, 219, 221,
    49/240, 243
  dopamine, 49/251
  drop attack, 49/239
  dysphasia, 49/243
  dystonia, 22/217, 219, 42/259, 46/301, 49/240,
    246
  dystonia musculorum deformans, 49/524
  EEG, 22/221, 46/301, 49/244
  encephalitis lethargica, 22/225
  euphoria, 49/243
  external ophthalmoplegia, 22/181, 221
  extrapyramidal syndrome, 22/219, 46/301, 64/5
  eye fixation, 49/241
  eye movement, 22/219, 49/240, 243
  eyelid apraxia, 49/243
  facial dystonia, 49/240
  gait apraxia, 22/219, 49/240
  gaze paralysis, 22/217, 219
  glutamate decarboxylase, 49/251
  glutamate dehydrogenase, 49/251
  granulovacuolar degeneration, 22/223, 49/248,
    250
  granulovacuolar neuronal degeneration, 22/223,
    49/250
  Guam motoneuron disease, 65/21
  history, 22/217
  homovanillic acid, 49/251
  hyperkalemic periodic paralysis, 49/246
  hypometric saccade, 49/241, 243
  incidence, 46/302
  insomnia, 46/301, 49/245
  internuclear ophthalmoplegia, 49/242
  lateral gaze palsy, 22/219
  lesion site, 22/225
  levodopa, 22/227, 49/247
  meningioma, 49/246
  mental change, 22/220, 49/240, 243
  mental manifestation, 46/302, 49/239
  multiple neuronal system degeneration, 49/239
  myoclonus, 49/618
  neurofibrillary tangle, 22/222-224, 42/259,
    46/274, 302, 49/248-250
  neuronal argyrophilic dystrophy, 21/63
  neuropathology, 22/221, 226, 46/302, 49/248
  neurotoxicology, 64/5
  normal pressure hydrocephalus, 49/247
  nuchal dystonia, 22/219, 49/247
  nuchal rigidity, 46/301, 49/240, 242
  nuclear magnetic resonance, 49/245
  nystagmus, 22/219, 49/242
  oculocephalic reflex, 22/219, 221, 49/241
  oculomotor apraxia, 49/243
  olivopontocerebellar atrophy (Dejerine-Thomas),
    49/246, 60/530
  ophthalmoplegia, 22/192, 219, 42/259, 46/301,
    49/240
  optokinetic nystagmus, 22/219, 221, 49/241
  orthostatic hypotension, 63/147
  paligraphia, 22/220

palilalia, 22/220
pallidoluysionigral degeneration, 49/456, 459
pallidoluysionigrothalamic degeneration, 49/456
Parinaud syndrome, 6/557
Parkinson dementia complex, 22/226
Parkinson disease, 49/94, 108, 240
parkinsonism, 6/557, 22/217, 219-226, 49/239
pathologic laughing and crying, 49/243
personality change, 49/243
prognosis, 46/302
progressive bulbar palsy, 59/408
progressive external ophthalmoplegia, 22/219
progressive external ophthalmoplegia
   classification, 22/179, 62/290
pseudobulbar paralysis, 6/557, 22/217, 219,
   42/259, 46/301
pyramidal syndrome, 46/301
REM, 22/219
REM sleep, 49/245
rigidity, 22/219, 46/302, 49/240
sex ratio, 22/218, 42/259
sleep, 74/548
sleep disorder, 49/244
splenomegaly, 22/227
striatonigral degeneration, 49/246, 60/540
subclavian steal syndrome, 49/247
subcortical dementia, 46/312, 378, 49/243
substantia nigra degeneration, 22/221, 42/259
supranuclear ophthalmoplegia, 42/259, 49/246,
   60/656
torticollis, 6/557
treatment, 49/247
trismus, 22/219
tyrosine hydroxylase, 46/302, 49/251
vertical gaze palsy, 22/219, 42/259, 49/239
vestibular nystagmus, 49/241
vestibulo-ocular response, 22/221, 49/241
visual impairment, 22/219, 49/240
Progressive systemic sclerosis
   see also Scleroderma
   angiopathic polyneuropathy, 51/446
   autoimmune disease, 41/55, 59
   bilateral sensory trigeminal neuropathy, 51/447
   brain infarction, 53/163
   brain ischemia, 53/163
   carpal tunnel syndrome, 51/448
   CREST syndrome, 51/448
   CSF protein, 51/448
   dissociated sensory loss, 51/448
   EMG, 51/448
   laboratory test, 51/448
   myopathy, 51/448
   neuropathic arthropathy, 51/448
   oligoclonal banding, 51/448
   Raynaud phenomenon, 43/71, 51/448
   sensorimotor polyneuropathy, 51/447
   sensory trigeminal neuropathy, 51/447
   Sjögren syndrome, 51/448, 71/74
Progressive torsion spasm in child (Flatau-Sterling),
   see Dystonia musculorum deformans
Progressive weakness
   familial hypobetalipoproteinemia, 13/427
   postpoliomyelitic amyotrophy, 59/36
Prolactin
   adenohypophysis, 2/442
   biological rhythm, 74/508
   brain injury, 23/123
   headache, 48/436
   Huntington chorea, 49/263
   mesolimbic dopamine system, 49/49
   migraine, 74/508
   pituitary adenoma, 17/404
   Reye syndrome, 56/152
   toluene intoxication, 64/50
Prolactin secretion
   depression, 46/474
   Huntington chorea, 49/290
   schizophrenia, 46/475
Prolactinoma
   pituitary tumor, 68/344
   radiotherapy, 68/348
   surgery, 68/348
   treatment, 68/348
Proliferation
   cell, see Cell proliferation
   retinal vasculoglial, see Retinal vasculoglial
      proliferation
   Schwann cell, see Schwann cell proliferation
Proliferation layer
   secondary, see Secondary proliferation layer
Proline
   iminoacidemia, 29/129-131
   metabolism, 29/129
   renal transport, 29/132
Proline oxidase
   hyperprolinemia type I, 42/569
   hyperprolinemia type II, 42/569
Prolinuria
   Goyer syndrome, 22/490
Prolonged transient ischemic attack
   acetylsalicylic acid, 53/274
   age, 53/259
   anticoagulant, 53/273
   arterial hypertension, 53/259
   asymptomatic carotid artery stenosis, 53/277
   blood viscosity, 53/272

carotid endarterectomy, 53/275, 277
clinical features, 53/261
coma, 53/262
computer assisted tomography, 53/258
congestive heart failure, 53/259
coronary heart disease, 53/259
definition, 53/257
diagnosis, 53/258
differential diagnosis, 53/262
dipyridamole, 53/275
epidemiology, 53/258
external internal carotid bypass, 53/278
incidence, 53/258
natural history, 53/259
pentoxifylline, 53/275
photon emission computerised tomography,
   53/272
prevalence, 53/258
risk factor, 53/258
sex ratio, 53/259
sulfinpyrazone, 53/274
suloctidil, 53/275
surgery, 53/275
surgical criteria, 53/277
ticlopidine, 53/275
treatment, 53/272
Prolotherapy
   muscle contraction headache, 5/165-168
Prolylhydroxyproline
   Duchenne muscular dystrophy, 40/380
Promedol intoxication, *see* Trimeperidine
Promyelocytic leukemia
   brain hemorrhage, 63/344
Pronator drift
   Sydenham chorea, 49/360
Pronator reflex
   diagnostic value, 1/245
Pronator sign
   hemiplegia, 1/184
   juvenile hereditary benign chorea, 49/345
   Sydenham chorea, 6/159
Pronator teres syndrome, *see* Anterior interosseous
   nerve syndrome
Pro-opiocortin, *see* Pro-opiomelanocortin
Pro-opiomelanocortin
   corticotropin, 65/350
   β-endorphin, 65/350
   migraine, 48/99
   protein processing, 66/79
   schematic structure, 65/351
Pro-otic sinus
   anatomy, 11/61
Propafenone intoxication

tremor, 65/457
vertigo, 65/457
visual impairment, 65/457
Propamidine
   toxic neuropathy, 7/520
Propanediol
   migraine, 5/101
Propanidid intoxication
   intravenous, 37/409
   motor dysfunction, 37/409
   rigidity, 37/409
   seizure, 37/409
2-Propanol
   toxic myopathy, 62/601
Propantheline bromide
   motoneuron disease, 59/467
Properdine deficiency
   meningococcal meningitis, 52/23
Proper-myl
   multiple sclerosis, 9/401
Prophylactic cranial irradiation
   brain metastasis, 69/222
Propionibacterium acnes
   cranial epidural empyema, 52/169
   Kawasaki syndrome, 56/638
Propionic acid
   autonomic nervous system, 74/367
   headache, 48/174
   migraine, 48/174
Propionic acidemia, 29/207-210
   see also Hyperglycinemia
   acid base metabolism, 28/520
   apathy, 27/72
   clinical features, 29/208
   clinical finding, 66/649
   coma, 27/72
   convulsion, 27/72
   diagnosis, 29/209
   enzyme defect, 29/208
   epilepsy, 72/222
   genetics, 29/209
   hyperammonemia, 29/105
   hyperglycinemia, 28/520, 29/204
   nuclear magnetic resonance, 66/644
   pathology, 29/208
   treatment, 27/73, 29/209
   vomiting, 27/72, 28/520, 29/208
Propionyl-coenzyme A carboxylase
   mitochondrial disease, 66/417
Propionyl-coenzyme A carboxylase deficiency, *see*
   Propionic acidemia
Propofol
   status epilepticus, 73/331

Propoxyphene, *see* Dextropropoxyphene
(+)-Propoxyphene, *see* Dextropropoxyphene
Propoxyphene hydrochloride, *see*
    Dextropropoxyphene
Propranolol
    Achilles reflex, 41/245
    action mechanism, 37/444
    aggression, 46/180
    antihypertensive agent, 37/444-447, 63/83
    antihypertensive agent intoxication, 37/447
    anxiety, 37/447
    ataxia telangiectasia, 60/390
    behavior disorder, 46/601
    benign essential tremor, 49/578
    beta adrenergic receptor blocking agent, 37/444,
        65/434
    brain infarction, 37/446
    cardiac dysrhythmia, 37/444
    carotidynia, 48/339
    carpal tunnel syndrome, 63/195
    chemical structure, 37/444
    classification, 74/149
    cough headache, 48/371
    delirium, 46/538, 63/195
    diplopia, 63/195
    headache, 48/39
    hypertension, 37/444
    hyperthyroidism, 37/446
    ischemia, 37/446
    migraine, 37/446, 48/39, 151, 178, 181, 203
    myasthenia, 63/195
    myopathy, 63/87
    myotonia, 40/547
    paresthesia, 63/195
    ptosis, 63/195
    schizophrenia, 37/447
    stupor, 63/195
    tachyarrhythmia, 37/445
    thyroid disorder, 37/445
    tremor, 37/446
Propranolol intoxication
    delirium, 37/447
    depression, 37/447
    gastrointestinal disorder, 37/447
    hallucination, 37/447
    sedation, 37/447
Proprioceptive impairment
    multiple sclerosis, 47/55
    thalamic lesion, 2/478
Proprioceptive sensation
    hereditary sensory and autonomic neuropathy type
        IV, 60/12
    Sherrington classification, 1/80

    type, 1/93
Proptosis, *see* Exophthalmos
N-Propylajmalin
    myotonia congenita, 40/557, 561
    myotonic dystrophy, 40/557, 561
2-Propyl-4-pentenoic acid
    neurotoxin, 65/126
    valproic acid induced hepatic necrosis, 65/126
Prosencephaly
    corpus callosum agenesis, 50/159
Prosody
    amusia, 4/198, 45/488
    aphasia, 45/309
    language, 4/198
Prosopagnosia
    agnosia, 4/16, 22, 31, 45/344
    bilateral occipital lobe infarction, 53/413
    bilateral occipital lobe syndrome, 45/53
    epilepsy, 45/509
    monochromatism, 45/335
    occipital lobe syndrome, 45/53
    occipital lobe tumor, 17/327
    perception classification, 45/341
    posterior cerebral artery syndrome, 53/413
    spatial orientation disorder, 24/491
    visual agnosia, 2/602, 655, 657, 4/17, 45/14
Prosopalgia, *see* Trigeminal neuralgia
Prosopitis granulomatosa
    Melkersson-Rosenthal syndrome, 8/213
Prosopoplegia
    terminology, 8/241
Prosoposchisis
    holoprosencephaly, 50/233
Prostacyclin
    brain infarction, 53/130, 429
    brain ischemia, 53/130
    brain microcirculation, 53/80
    cluster headache, 48/97
    headache, 48/97, 281
    hemostasis, 63/303
    local anesthetic agent, 65/428
    migraine, 48/97
    platelet aggregation, 48/96
Prostaglandin
    Bartter syndrome, 42/528
    chronic paroxysmal hemicrania, 48/263
    cluster headache, 48/97, 233
    cough headache, 48/371
    episodic edema, 43/63
    essential fatty acid, 47/199
    fever, 74/449
    headache, 48/97
    migraine, 48/92, 96, 194

multiple sclerosis, 47/199
nerve injury, 65/425
nociceptor, 45/229
Prostaglandin E
  CSF, *see* CSF prostaglandin E
Prostaglandin E1, *see* Alprostadil
Prostaglandin E2, *see* Dinoprostone
Prostaglandin I2, *see* Prostacyclin
Prostate cancer
  spinal cord metastasis, 69/178
Prostatic cancer
  enzyme, 66/56
Prostatic carcinoma
  vertebral column tumor, 20/42
  vertebral metastasis, 20/425
Prosthesis
  physical rehabilitation, 61/489
  spinal cord injury, 61/489
Prostigmine, *see* Neostigmine
Protagon
  metachromatic leukodystrophy, 10/47
Protagon hypothesis
  myelin sheath, 9/30
Protease
  brain metastasis, 69/109
  chemotherapy, 67/284
  CSF neutral, *see* CSF neutral protease
  metastasis, 71/645
Protease inhibitor
  metachromatic leukodystrophy, 47/593
  neuronal ceroid lipofuscinosis, 66/685
Protective protein deficiency
  genetics, 66/271
Protein
  adenovirus, 56/281
  alloxan Schiff staining, 9/23
  amido-black staining, 9/23
  amino acid composition, 9/32
  amniotic fluid, *see* Amniotic fluid protein
  amyloid A, *see* Amyloid A protein
  amyloid fibril, *see* Amyloid fibril protein
  antimyelin basic, *see* Antimyelin basic protein
  asparaginase linked, *see* Asparaginase linked
    protein
  Bence Jones, *see* Bence Jones protein
  biochemistry, 9/1
  brain injury, 23/125
  C-reactive, *see* C-reactive protein
  calcium binding, *see* Calcium binding protein
  carboxyl staining, 9/24
  cholesterol ester transfer, *see* Cholesterol ester
    transfer protein
  collagen like, *see* Collagen like protein

contractile, *see* Contractile protein
CSF, *see* CSF protein
CSF myelin basic, *see* CSF myelin basic protein
DDD technique, 9/24
disulfide group staining, 9/24
encephalitogenic myelin basic, *see*
  Encephalitogenic myelin basic protein
enzyme, 66/51
eosin staining, 9/23
erythrocyte membrane, *see* Erythrocyte membrane
  protein
F, *see* F protein
Folch-Lees proteolipid, *see* Folch-Lees proteolipid
  protein
free amino-group deamination, 9/23
galactosialidosis, 66/57
glial, *see* Glial protein
GM2 ganglioside activator, *see* GM2 ganglioside
  activator protein
guanine nucleoside binding, *see* Guanine
  nucleotide binding protein
H, *see* H protein
hair, *see* Hair protein
histochemistry, 9/23
hypochlorite Schiff staining, 9/23
immunofluorescence test, 9/24
L, *see* L protein
long chain acyl-coenzyme A dehydrogenase, *see*
  Long chain acyl-coenzyme A dehydrogenase
  protein
lysosomal, *see* Lysosomal protein
M, *see* Monoclonal protein
membrane, *see* Membrane protein
monoclonal, *see* Monoclonal protein
multiple sclerosis, 9/320
mumps virus, 56/431
myelin, *see* Myelin protein
myelin associated, *see* Myelin associated protein
myelin basic, *see* Myelin basic protein
myelin proteolipid, *see* Myelin proteolipid protein
myelin sheath, 7/56, 58
nerve myelin, 51/25
nervous system tumor, 27/505
neurofilament, *see* Neurofilament protein
ninhydrin Schiff staining, 9/23
nonencephalitogenic, *see* Nonencephalitogenic
  protein
NP, *see* NP protein
P, *see* P protein
P0, *see* P0 protein
P1, *see* P1 protein
P2, *see* P2 protein
penton, *see* Penton protein

performic acid alcian blue staining, 9/24
prion, 56/560, 585
progressive dysautonomia, 59/157
proteolipid, *see* Proteolipid protein
R, *see* Vitamin B12 binding protein
ribosomal S6, *see* Ribosomal S6 protein
ribosome, 66/80
S-100, *see* S-100 protein
Schiff leukofuchsin, 9/23
Schiff procedure, 9/23
sensitization, 9/517
sigma 1, *see* Sigma 1 protein
sulfatide activator, *see* Sulfatide activator protein
sulfhydryl staining, 9/24
synthesis, 66/75
synthesis blockade, 9/9
targeting, 66/80
γ-trace, *see* Cystatin C
turnover, 9/13
very long chain acyl-coenzyme A dehydrogenase,
    see Very long chain acyl-coenzyme A
    dehydrogenase protein
virology, 34/3
white matter, 9/1, 11, 13
Wolfgram, *see* Wolfgram protein
X-linked adrenoleukodystrophy, 66/464, 466
Protein C
hemostasis, 63/305
intravascular consumption coagulopathy, 55/494
Protein C deficiency
brain infarction, 53/167, 63/328
brain ischemia, 53/167
brain thrombosis, 63/328
brain venous infarction, 63/329
coagulopathy, 63/328
transient ischemic attack, 63/328
Protein cell dissociation, *see* Proteinocytologic
    dissociation
Protein deficiency
acute metabolic decompensation, 7/635
deficiency neuropathy, 7/633
GM2 ganglioside activator, *see* GM2 ganglioside
    activator protein deficiency
kwashiorkor, 7/635
protective, *see* Protective protein deficiency
starvation neuropathy, 7/633
Wernicke encephalopathy, 7/635
Protein energy malnutrition
deficiency neuropathy, 51/322
kwashiorkor, 51/322
marasmus, 51/322
Protein G, *see* Guanine nucleotide binding protein
Protein intolerance

familial, *see* Familial protein intolerance
hereditary, *see* Familial protein intolerance
hereditary periodic ataxia, 21/578
lysinuric, *see* Lysinuric protein intolerance
Protein kinase
nerve myelin, 51/25
Protein level
spinal cord injury, 26/382, 385
Protein metabolism
Down syndrome, 31/437
Duchenne muscular dystrophy, 40/379
Huntington chorea, 6/349
malnutrition, 29/9
Wilson disease, 27/386
Protein P1
myelin basic, *see* Myelin basic protein P1
Protein P2
amino acid, 47/453
characteristics, 47/492-494
disseminated encephalomyelitis, 47/467
experimental allergic neuritis, 47/453, 471
Guillain-Barré syndrome, 47/484
myelin basic, *see* Myelin basic protein P2
myelin basic protein, 47/492
Protein phosphorylation, *see* Phosphorylation
Protein processing
pro-opiomelanocortin, 66/79
regulation, 66/79
Protein S100
CSF, *see* CSF protein S100
Protein S deficiency
brain infarction, 63/328
brain phlebothrombosis, 54/410
brain thrombosis, 63/328
brain venous infarction, 63/329
coagulopathy, 63/328
transient ischemic attack, 63/328
Protein synthesis
brain infarction, 53/133
brain ischemia, 53/133
general step, 66/77
marine toxin intoxication, 65/144
mercury intoxication, 64/390
methionylalanine, 66/77
myofibrillar lysis myopathy, 62/339
translation, 30/95
Protein synthesis test
ribosomal, *see* Ribosomal protein synthesis test
Protein X
pyruvate dehydrogenase complex, 66/415
Proteinase
botulinum toxin, 65/245
CSF, *see* CSF proteinase

multiple sclerosis, 9/368, 47/214
neutral, *see* Neutral proteinase
wallerian degeneration, 9/36
Proteinocytologic dissociation
  CSF, 7/498
  cytomegalovirus, 56/270
  differential diagnosis, 19/134
  enterovirus 70, 56/351
  Epstein-Barr virus, 56/253
  Guillain-Barré syndrome, 7/495, 498
  Nonne-Froin syndrome, 7/498, 19/125
  spinal cord tumor, 20/217
  spinal tumor, 19/125
Proteinuria
  Behçet syndrome, 56/595
  Bence Jones, *see* Bence Jones proteinuria
  Fabry disease, 60/172
  facio-oculoacousticorenal syndrome, 43/399
  fructose intolerance, 42/548
  galactosemia, 42/552
  multiple myeloma, 20/11, 39/138
  primary amyloidosis, 51/415
  subarachnoid hemorrhage, 12/151
Proteolipid
  see also Trypsin resistant protein residue
  amino acid composition, 9/12
  CNS, 9/11
  CNS myelin, 9/40
  CSF, *see* CSF proteolipid
  encephalitogenic action, 9/11
  experimental allergic encephalomyelitis, 9/11
  myelin, 9/11, 10/135
  nerve *vs* white matter, 9/12
  protein moiety, 7/56, 58, 9/32
  Rosenthal fiber, 10/135
  trypsin resistant protein residue, 9/32
  trypsinization, 9/12
  white matter, 9/12
Proteolipid protein
  myelin, *see* Myelin proteolipid protein
  Pelizaeus-Merzbacher disease, 59/357, 66/561
Proteolipid protein DM-20 gene
  mutation, 66/565
  Pelizaeus-Merzbacher disease, 66/571, 573
Proteolipid protein gene
  dysmyelination, 66/564
  mutation, 66/570
  Pelizaeus-Merzbacher disease, 66/559, 570, 573
  X-linked spastic paraplegia, 66/571
Proteus
  adult meningitis, *see* Gram-negative bacillary
    meningitis
  gram-negative bacillary meningitis, 52/104, 117

infantile enteric bacillary meningitis, 33/62
Proteus mirabilis
  bacterial endocarditis, 52/301
  spinal cord abscess, 52/191
Proteus morganii intoxication
  bacterial infection, 37/87
  fish, 37/87
Proteus vulgaris
  brain abscess, 52/149
  spinal epidural empyema, 52/187
Prothrombin time
  Bassen-Kornzweig syndrome, 51/327
  hemostasis, 63/306
  Reye syndrome, 56/152, 65/118
Protirelin
  anorexia nervosa, 46/586
  basal ganglion, 49/33
  brain metabolism, 57/84
  depression, 46/449, 474
  Huntington chorea, 49/258, 263
  paraplegia, 61/403, 406
  Parkinson disease, 49/120
  spinal cord injury, 61/403, 406
  tetraplegia, 61/403, 406
  transverse spinal cord lesion, 61/403, 406
Protirelin test
  depression, 46/474
Protochordata intoxication, *see* Tunicata
    intoxication
Protocoproporphyria
  porphyria, *see* Porphyria variegata
Proton transporting adenosine triphosphate synthase
  Batten disease, 66/686
  complex V, 66/424
Protopathic motor system
  progressive spinal muscular atrophy, 22/43
Protoporphyria
  ataxia telangiectasia, 14/315
  erythrohepatic, *see* Erythrohepatic protoporphyria
  erythropoietic, *see* Erythrohepatic protoporphyria
Protoporphyrin
  chloroma, 18/252
  free erythrocyte, *see* Free erythrocyte
    protoporphyrin
Protoporphyrinogen IX oxidase
  porphyria variegata, 51/390
Protoporphyrinogen oxidase
  porphyria variegata, 42/620
Protozoa intoxication
  paralytic shellfish poisoning, 37/33
Protriptyline
  obstructive sleep apnea, 63/460
  tricyclic antidepressant, 65/312

Prouzet line
  basilar impression, 32/19
  platybasia, 32/19
Providencia
  gram-negative bacillary meningitis, 52/104
Provoked seizure, 72/16
Proximal muscle weakness
  acid maltase deficiency, 43/178
  areflexia, 43/107
  Becker muscular dystrophy, 43/86
  benign dominant myopathy, 43/111
  central core myopathy, 43/80
  Cushing syndrome, 40/317
  dementia, 43/132
  Eaton-Lambert myasthenic syndrome, 43/159
  Emery-Dreifuss muscular dystrophy, 43/88
  eosinophilic polymyositis, 63/385
  familial limb girdle myasthenia, 43/156
  female, 43/115
  hip dip, 40/314
  hyperparathyroidism, 40/317
  Isaacs syndrome, 43/160
  late onset muscular dystrophy, 43/107
  limb girdle syndrome, 43/104
  lipid storage myopathy, 43/184
  lower extremity, 40/314
  mitochondrial myopathy, 43/119, 188
  multicore myopathy, 43/120
  muscle shortening, 43/109
  muscular atrophy, 42/82
  muscular dystrophy, 43/107, 109
  myopathy, 43/115, 132
  myopathy hypogonadism, 43/123-125
  nemaline myopathy, 43/122
  osteomalacia, 40/317
  paramyotonia congenita, 43/168
  pheochromocytoma myopathy, 62/537
  Ryukyan spinal muscular atrophy, 42/93
  sarcotubular myopathy, 43/129
  spinal muscular atrophy type II, 42/90
  upper extremity, 40/309
  Wohlfart-Kugelberg-Welander disease, 42/91
Proximal myositis
  mixed connective tissue disease, 51/449
Proximal neuropathy
  diabetic, see Diabetic proximal neuropathy
  diabetic polyneuropathy, 51/501
Proximal pseudomyopathic spinal muscular atrophy,
    see Wohlfart-Kugelberg-Welander disease
Prune belly syndrome
  see also Caudal aplasia
  abdominal muscle absence, 43/466
  abdominal muscle hypoplasia, 43/466, 50/515

anencephaly, 50/84
birth incidence, 43/467
bladder wall hypertrophy, 43/467
congenital hip dislocation, 43/466
cryptorchidism, 43/466
oligohydramnios, 43/467, 50/515
Patau syndrome, 50/558
Potter face, 50/515
pulmonary hypoplasia, 50/515
renal abnormality, 43/466
retentio testis, 50/515
talipes, 43/466
urinary tract defect, 43/466, 50/515
wrinkled abdomen, 50/515
Pruritus
  African trypanosomiasis, 52/340
  brain embolism, 53/207
  brain infarction, 53/207
  cerebrovascular disease, 53/207
  eosinophilia myalgia syndrome, 63/375
  hereditary amyloid polyneuropathy type 3, 60/100
  intracranial hemorrhage, 53/207
  polycythemia, 63/250
  polycythemia vera, 63/250
  rabies, 34/253, 56/386
  transient ischemic attack, 53/207
  uremic polyneuropathy, 51/356
Pseudencephalus, see Anencephaly
Pseudoabducens nerve palsy
  brain embolism, 53/213
  brain infarction, 53/213
  cerebrovascular disease, 53/213
  intracranial hemorrhage, 53/213
  lesion site, 53/213
  transient ischemic attack, 53/213
Pseudoabducens nerve sign
  mesencephalic infarction, 53/391
  vertebrobasilar system syndrome, 53/391
Pseudoachondroplastic dysplasia
  atlantoaxial subluxation, 50/392
Pseudoagnosia
  shape discrimination, 45/334
  spatial orientation disorder, 3/225
  visual disorientation, 45/336
  visual sensory deficit, 4/25
Pseudoalexia
  psychogenic, 4/134, 45/445
  reading aloud, 4/133, 45/445
  unilateral paralexia, 4/133, 45/445
Pseudoallescheria
  amphotericin B, 52/492
  brain abscess, 52/492
  endophthalmitis, 52/491

fungal CNS disease, 52/490
  meningitis, 52/492
  miconazole, 52/492
  mycetoma, 52/491
Pseudoallescheria boydii
  brain abscess, 52/151
  chronic meningitis, 56/645
  fungal CNS disease, 52/479
Pseudoamyloid
  Alzheimer neurofibrillary change, 21/57
Pseudoamyotrophic lateral sclerosis
  human T-lymphotropic virus type I associated
    myelopathy, 59/451
Pseudoarthrosis
  neurofibromatosis type I, 50/369
Pseudoarylsulfatase A deficiency
  metachromatic leukodystrophy, 47/591
Pseudoarylsulfatase deficiency
  metachromatic leukodystrophy, 59/356, 60/124,
    66/176
Pseudoastrocytoma
  incontinentia pigmenti, 14/12, 43/23
Pseudoataxia
  differential diagnosis, 1/310
  ocular ataxia, 1/310
Pseudoathetosis
  anatomic classification, 1/279
  liver cirrhosis, 49/384
  tabes dorsalis, 52/280
Pseudoatrophic macula
  neurofibromatosis type I, 50/366
Pseudo-Babinski posture
  hereditary progressive diurnal fluctuating
    dystonia, 49/532
Pseudo-Babinski sign
  flaccid paralysis, 19/37
  lumbosacral lipoma, 19/37
  pallidal degeneration, 49/455
  pure pallidal degeneration, 6/641
  root lesion, 19/37
Pseudo-Bartter syndrome
  anorexia nervosa, 46/586
Pseudobasilar impression
  head movement, 19/336
  McRae line, 19/341
  rheumatoid arthritis, 19/326, 332
  spinal cord compression, 19/336
Pseudobulbar dysarthria
  Binswanger disease, 54/222
  Pick disease, 46/235
Pseudobulbar paralysis
  see also Operculum syndrome
  adrenoleukodystrophy, 60/658

adult, 22/114
adverse diuretic effect, 65/440
amyotrophic lateral sclerosis, 22/114, 294, 306
astasia abasia, 1/197
brain stem, 2/250
carbon monoxide intoxication, 1/198
central motor disorder, 1/196
central pontine myelinolysis, 2/263, 63/549
cerebrotendinous xanthomatosis, 10/532, 537,
  51/399, 60/167
clinical features, 1/196
diagnosis, 1/199
differentiation, 22/114, 40/307
diuretic agent intoxication, 65/440
Divry-Van Bogaert syndrome, 55/319
dysarthria, 2/241, 12/31, 40/307, 55/142
dysphagia, 2/241, 12/31, 40/307, 55/142
dysphonia, 12/31, 40/307
dyspneumia, 22/119
emotional disorder, 3/358
etiology, 1/198
extrapyramidal, 1/197, 6/138
facial paralysis, 2/241
Fazio-Londe disease, 22/107
gag reflex, 2/780
masseter muscle reflex, 2/780
methanol intoxication, 64/101
microgyria, 1/199
motoneuron disease, 40/307, 45/221
operculum syndrome, 2/777, 779
pallidal necrosis, 49/470
pathologic laughing and crying, 1/196, 2/241, 780,
  3/358, 40/307, 45/221, 55/142, 60/658
pathology, 1/199
pontine infarction, 2/241, 12/23, 31
porencephaly, 1/199
progressive supranuclear palsy, 6/557, 22/217,
  219, 42/259, 46/301
spastic paraplegia, 59/428
type, 1/197
Pseudobulbar syndrome
  bilateral lacuna, 54/253
  Binswanger disease, 46/359
  brain lacunar infarction, 46/355, 54/243, 253
Pseudocalcium
  see also Striatopallidodentate calcification
  hetereogeneous metal, 6/704
  hydroxyl apatite, 6/715
  iron, 6/714, 719
  magnesium, 6/704, 715, 718
  manganese, 6/716
  striatopallidodentate calcification, 49/417
  zinc, 6/715, 718

Pseudocamptodactyly
   inability to open mouth, 43/433
Pseudocholinesterase, *see* Butyrylcholinesterase
Pseudocholinesterase deficiency, *see*
   Butyrylcholinesterase deficiency
Pseudo-Crouzon syndrome, *see* Saethre-Chotzen
   syndrome
Pseudocyst
   intrasellar cyst, 17/430
Pseudodegeneration
   acquired exogenous origin, 21/48
   Creutzfeldt-Jakob disease, 21/59
   kuru, 21/59
Pseudodementia, 46/221-224
   depressive, *see* Depressive pseudodementia
   hysteria, 46/224, 575
   mania, 46/223
   schizophrenia, 46/223
Pseudodepression
   acquired immune deficiency syndrome dementia,
     56/491
Pseudoencephalitis hemorrhagica superior, *see*
   Wernicke encephalopathy
Pseudoepileptic attack
   psychogenic, 72/267
Pseudofovea
   monocular polyopia, 2/663, 17/331
   polyopia, 45/364
Pseudogeneral paresis
   alcoholic, *see* Alcoholic dementia
   thalamic dementia, 2/484
Pseudoglaucoma
   autosomal dominant, *see* Autosomal dominant
     pseudoglaucoma
Pseudoglioma, *see* Pseudoastrocytoma
Pseudo-Gradenigo syndrome, *see* Citelli syndrome
Pseudohallucination
   hallucination classification, 4/328
   hallucinosis, 46/563
Pseudohemianopia
   frontal lobe tumor, 17/260
Pseudohermaphroditism
   ARG syndrome, 50/582
   congenital adrenal hyperplasia, 42/514
   Gruber syndrome, 14/505
Pseudo-Hurler disease, *see* GM1 gangliosidosis
Pseudo-Hurler polydystrophy, *see* Mucolipidosis
   type III
Pseudohypertrophic muscular dystrophy, *see*
   Duchenne muscular dystrophy
Pseudohypertrophy
   see also Muscle hypertrophy
   Becker muscular dystrophy, 40/324, 388, 43/86

Duchenne like muscular dystrophy, 43/102
Duchenne muscular dystrophy, 40/250, 366,
   43/106
Fukuyama syndrome, 43/91
inferior olivary nucleus, 12/21
limb girdle syndrome, 40/436
muscle, *see* Muscle pseudohypertrophy
neuroaxonal dystrophy, 49/396
paralysis periodica paramyotonica, 43/167
paramyotonia congenita, 43/169
Pseudohypoparathyroidism
   Addison disease, 6/706, 49/420
   adenylate cyclase, 42/578
   aneurysmal bone cyst, 42/622
   brachymetacarpalism, 14/466
   cataract, 6/706, 42/621, 49/419
   Chvostek sign, 42/621, 63/560
   clinical features, 42/578
   convulsion, 46/85
   cyclic adenosine monophosphate, 42/622
   dementia, 46/396
   developmental abnormality, 42/621
   diabetes insipidus, 6/706, 49/420
   diabetes mellitus, 6/706, 49/420
   enamel defect, 42/621
   epileptic seizure, 42/621
   gonadal dysgenesis, 6/706, 49/420
   hereditary disorder, 42/621, 46/396
   hyperphosphatemia, 42/622, 46/85
   hypocalcemia, 42/621, 46/85, 63/558, 561
   intracerebral calcification, 42/534, 621
   lyonization, 42/622
   mental deficiency, 46/85, 396, 63/561
   moon face, 6/706, 42/621, 46/85, 396, 49/419
   obesity, 42/621
   parathyroid hormone metabolism, 42/578, 621
   progressive myositis ossificans, 6/706, 49/420
   pseudo, *see* Pseudopseudohypoparathyroidism
   seizure, 46/85, 63/560
   sex ratio, 6/706, 42/578, 622, 49/419
   short stature, 6/706, 42/578, 621, 46/85, 396,
     49/419
   skeletal deformity, 42/621
   striatopallidodentate calcification, 6/706, 42/621,
     49/418, 420, 63/561
   tetany, 42/621, 46/85, 63/558
   transport defect, 42/578
   treatment, 42/622, 46/85, 63/561
   Trousseau sign, 42/621, 63/560
   vitamin D2, 42/578
   X-linked dominant inheritance, 6/706, 42/622,
     46/85, 49/419
Pseudohypothyroidism

pseudo, *see* Pseudopseudohypothyroidism
Pseudoinflammatory foveal dystrophy (Sorsby)
  autosomal dominant inheritance, 13/138
  Bruch membrane, 13/138
  central scotoma, 13/138
  choroid atrophy, 13/138
  dyschromatopsia, 13/138
  edema, 13/138
  hemorrhage, 13/138
Pseudo-Kayser-Fleischer ring
  case report, 27/386
Pseudoleukemia, *see* Hodgkin disease
Pseudolipidosis
  Aberdeen calf, 21/52
  granular storage, 21/61
Pseudolipoma, *see* Hibernoma
Pseudolipomatosis
  symmetrical, *see* Symmetrical pseudolipomatosis
Pseudo-Meniere syndrome
  infectious disease, 7/485
  vertigo, 7/485
Pseudomeningitis, *see* Meningism
Pseudomeningocele
  spinal cord injury, 61/528
Pseudomental deficiency
  psychiatric disorder, 46/25
Pseudomeralgia paresthetica
  lumbar vertebral canal stenosis, 20/705
Pseudomicrogyria
  formalin fixation, 50/254
Pseudomonas
  adult meningitis, *see* Gram-negative bacillary
    meningitis
  bacterial meningitis, 52/3, 5
  infantile enteric bacillary meningitis, 33/62
Pseudomonas aeruginosa
  bacterial meningitis, 52/7
  brain abscess, 52/149
  gram-negative bacillary meningitis, 52/104
  infectious endocarditis, 52/301
  spinal epidural empyema, 52/187
Pseudomyasthenia, *see* Eaton-Lambert myasthenic
    syndrome
Pseudomycosis
  brain, 16/217
Pseudomyopathic familial spinal muscular atrophy,
    see Wohlfart-Kugelberg-Welander disease
Pseudomyotonia, *see* Isaacs syndrome
Pseudomyotonic burst
  infantile spinal muscular atrophy, 59/56
  progressive spinal muscular atrophy, 22/33
Pseudoneonatal adrenoleukodystrophy, *see*
    Acyl-coenzyme A oxidase deficiency

Pseudoneuralgic pain
  spastic paraplegia, 59/432
Pseudoneuralgic prephase
  intraspinal tumor, 19/24
  spinal meningioma, 20/204
  spinal neurinoma, 20/250, 254
Pseudoneuritis
  papilledema, 16/290-292, 294, 296
  terminology, 7/199
Pseudo-obstruction
  intestinal, *see* Intestinal pseudo-obstruction
  lead neuropathy, 64/437
Pseudo-ophthalmoplegia, *see* Oculomotor apraxia
Pseudopapilledema
  blind spot, 2/567
  congenital heart disease, 63/11
  macrocephaly, 14/480, 768
  Riley-Smith syndrome, 14/14, 76, 480, 486, 768,
    43/30
Pseudopapilloma
  differential diagnosis, 18/406
  posterior fossa tumor, 18/406
  pseudotumor, 67/116
Pseudoparesis
  inferior oblique muscle, *see* Brown syndrome
  ocular, *see* Ocular pseudoparesis
Pseudoporencephaly
  cause, 50/359
  clinical features, 30/691
  definition, 50/356
  epilepsy, 30/691
  frequent site, 42/48
  mental deficiency, 30/691
  pathogenesis, 30/684
  pathology, 30/682
Pseudoprogeria Hallermann-Streiff syndrome
  alopecia, 43/405
  bird like face, 43/405
  freckling, 43/405
  glaucoma, 43/405
  nanism, 43/405
  progressive spastic tetraplegia, 43/405
Pseudopseudohypoparathyroidism, 42/621-623
  see also Striatopallidodentate calcification
  age, 6/706
  convulsive seizure, 6/707
  forme fruste, 42/578
  hyperthyroidism, 14/185
  hypocalcemia, 63/561
  intracerebral calcification, 42/534
  mental symptom, 27/301
  striatopallidodentate calcification, 49/419
Pseudopseudohypothyroidism

mental deficiency, 46/85
Pseudopsoas shadow
 hourglass tumor, 20/297
 spinal neurinoma, 20/270
Pseudoptosis
 blepharospasm, 2/318
 unilateral elevator palsy, 2/338
Pseudoradicular pain
 lumbar intervertebral disc prolapse, 20/585
 lumbar stenosis, 20/757
Pseudoradicular syndrome
 see also Spinal nerve root syndrome
 low back pain, 20/578
Pseudo-Raynaud disease, see Vibration neuropathy
Pseudoretinitis pigmentosa, see Secondary
  pigmentary retinal degeneration
Pseudoretinoblastoma
 toxocariasis, 52/511
Pseudorosette, see Homer Wright rosette
Pseudosclerosis, see Rigid Huntington chorea
Pseudo-Stellwag sign
 bradykinesia, 2/344
Pseudosubluxation
 definition, 25/245
 spinal injury, 25/245
Pseudotabes
 Dejerine peripheral, 1/318
 diabetic, 27/121-123
Pseudothalamic pain
 cortical lesion, 2/478
 subcortical lesion, 2/478
Pseudothalamic syndrome
 parietal lobe syndrome, 45/65, 78
Pseudothalidomide syndrome, see Roberts
  syndrome
Pseudotumor
 Junius-Kuhnt disease, 13/265
 orbital, see Orbital pseudotumor
 pseudopapilloma, 67/116
Pseudotumor cerebri, see Benign intracranial
  hypertension
Pseudotumor symptom
 Arnold-Chiari malformation type I, 32/106
Pseudo-Von Gräfe sign
 oculomotor paralysis, 2/293, 319
Pseudoxanthoma cell
 hemangioblastoma, 14/244
 Von Hippel-Lindau disease, 14/244
Pseudoxanthoma elasticum, 43/45-47
 angioid streak, 11/465, 14/775, 43/45, 55/452
 autosomal dominant, 43/46, 55/451
 autosomal recessive, 11/465, 14/775, 43/46,
  55/451

brain aneurysm, 11/465, 43/45, 55/454
brain angiography, 55/453
brain atrophy, 43/45
brain hemorrhage, 55/453
brain infarction, 43/45, 53/165, 55/453
brain ischemia, 53/165
carotid artery occlusion, 43/45, 55/454
cerebrovascular disease, 55/451, 453
chorioretinal degeneration, 13/41, 43/45
choroid angioma, 43/45
epistaxis, 55/452
genetic counseling, 43/47, 55/454
Grönblad-Strandberg syndrome, 14/114, 55/452
heart disease, 43/45
hemoptysis, 55/452
hypertension, 11/465, 14/775, 43/45, 55/451-453
iatrogenic brain infarction, 43/45, 53/165
juvenile cerebrovascular disease, 55/453
myocardial infarction, 55/452
neurologic symptom, 55/453
optic atrophy, 13/41
prevalence, 43/45
psychiatric disorder, 14/776, 43/45
sex ratio, 11/465, 14/775, 43/46, 55/451
spinal cord intermittent claudication, 55/452
subarachnoid hemorrhage, 11/465, 43/45, 55/29
symptom, 14/775, 55/452
systemic brain infarction, 11/465
Takayasu disease, 12/413, 55/452
treatment, 43/47, 55/454
Pseudo-Zellweger syndrome, see Peroxisomal
  acetyl-coenzyme A acyltransferase deficiency
Psilocin
 see also Mushroom toxin
 hallucination, 65/40
 hallucinogenic agent, 65/42
 potency, 65/40
 psilocybe syndrome, 65/40
 psychosis, 46/594
 serotonin receptor, 65/40
Psilocin intoxication
 symptom, 65/40
Psilocybe syndrome
 Copelandia, 65/40
 Gymnopilus, 65/40
 Panaeolus, 65/40
 psilocin, 65/40
 Psilocybe, 65/40
 Stropharia, 65/40
Psilocybine
 see also Mushroom toxin
 hallucination, 4/333, 65/40
 hallucinogenic agent, 64/3, 65/42

neurotoxin, 4/333, 36/535, 46/594, 64/3, 65/40
  potency, 65/40
  psychosis, 46/594
  serotonin receptor, 65/40
  source, 64/3
  toxic encephalopathy, 64/3
Psilocybine intoxication
  symptom, 64/3, 65/40
Psittacosis
  infectious endocarditis, 52/289, 302
Psoas hematoma
  lumbosacral plexus, 51/165
Psoriasis
  acanthocyte, 49/332
  chorioretinal degeneration, 13/40
  β-naphthol, 13/224
  optic atrophy, 13/40
  polymyositis, 62/373
  unilateral hemimelic ichthyosiform erythromelia,
    43/416
Psoriatic arthropathy
  antiphospholipid antibody, 63/330
Psychiatric disorder
  see also Mental disorder *and* Psychosis
  adrenoleukodystrophy, 60/665
  alcoholism, 43/197
  amyotrophic lateral sclerosis, 22/296
  ataxia telangiectasia, 14/767
  centrofacial lentiginosis, 43/26
  cocaine intoxication, 65/254
  corpus callosum tumor, 17/504
  diabetes mellitus, 42/411
  eclampsia, 39/547
  ependymoma, 18/127
  epilepsy, 15/796, 804, 73/469
  Fabry disease, 42/428
  frontal lobe tumor, 17/240, 242, 244
  glioblastoma multiforme, 18/53
  Hartnup disease, 42/146
  heredopathia ophthalmo-otoencephalica, 42/150
  Huntington chorea, 42/225
  hypoparathyroidism, 42/577
  intracerebral calcification, 42/534
  mental deficiency, 46/32
  midbrain tumor, 17/634
  migraine, 5/78, 244, 247, 250, 253, 381
  multiple sclerosis, 9/204
  myelinoclastic diffuse sclerosis, 9/474
  neurocutaneous melanosis, 14/768
  neurocysticercosis, 35/302, 52/530
  neurofibromatosis, 14/762-765
  neurofibromatosis type I, 14/762-764
  neurology, 1/8, 19-21

osteitis deformans (Paget), 43/451
  parietal lobe tumor, 17/305
  Parkinson dementia complex, 42/250
  Pelizaeus-Merzbacher disease, 42/505
  pernicious anemia, 38/21
  phakomatosis, 14/759-771
  pheochromocytoma, 42/764
  pineal tumor, 17/659
  posttraumatic headache, 5/182, 48/362
  progressive multifocal leukoencephalopathy,
    9/494
  pseudomental deficiency, 46/25
  pseudoxanthoma elasticum, 14/776, 43/45
  Rett syndrome, 60/665
  spinal cord injury, 26/234
  Sturge-Weber syndrome, 14/765
  subarachnoid hemorrhage, 12/139
  systemic lupus erythematosus, 43/420
  temporal lobe epilepsy, 42/465
  temporal lobe tumor, 17/290
  thallium intoxication, 36/251
  thyroid gland, 27/272
  time sense, 3/244
  tuberous sclerosis, 14/760-762
  uremic encephalopathy, 63/506
  Usher syndrome, 60/665
  vitamin B_c, 28/202
  Von Hippel-Lindau disease, 14/766, 60/665
  Wilson disease, 27/385, 42/271, 49/223, 228
Psychiatric headache, 48/353-364
  anxiety, 5/20
  conversion headache, 5/20, 48/356
  delusion, 5/20, 249, 48/361
  depression, 5/20
  diagnostic guideline, 5/3
  differential diagnosis, 5/29
  emotion, 5/20
  muscle contraction headache, 48/355
  postconcussional syndrome, 5/182
  posttraumatic headache, 5/20, 182, 48/362-364
  stress, 5/20
  study, 48/358
  vascular headache, 48/355
Psychic blindness
  Klüver-Bucy syndrome, 2/602, 708
  temporal lobe syndrome, 45/43
Psychic deafness
  agnosia, 4/38, 428
Psychic gaze paralysis
  Balint syndrome, 3/214
  parietal lobe syndrome, 45/73
Psychogenic epilepsy, 72/267
  demography, 72/269

differential diagnosis, 72/268
EEG, 72/271
neurohormonal index, 72/271
psychological technique, 72/274
seizure provocation, 72/272
Psychogenic headache, 5/247-256
anxiety, 5/254
child, 48/34
classification, 48/6
cluster headache, 48/227
conversion, 5/249, 255
definition, 48/9
depression, 5/247, 253
disability, 48/40
psychotherapy, 5/256
repression, 5/255
review, 5/248, 251
terminology, 5/247
Psychogenic pain
clinical features, 1/111
hysteria, 43/207, 46/575
somatic signal processing, 46/581
temporomandibular cephalalgia, 48/394
thalamotomy, 46/575
Psychogenic polydipsia, see Water intoxication
Psychogenic water intoxication, see Water
intoxication
Psychometry
computerized test, 45/526
cutting score, 45/520
double dissociation, 45/520
frontal lobe lesion, 45/27-29
integrated interpretation, 45/521
learning disability, 46/124-126
neuropsychologic defect, 45/518, 520
pattern analysis, 45/520
test battery, 45/525
visuospatial inattention, 45/520
Psychomotor behavior
delirium, 46/530
hallucinosis, 46/562
mental deficiency, 46/26-28
Psychomotor development
Down syndrome, 50/527
Edwards syndrome, 50/560, 562
mental deficiency, 6/503
opticocochleodentate degeneration, 60/751
9p partial monosomy, 50/579
Psychomotor disorder, 4/443-462
developmental dysphasia, 4/401
etiology, 4/455-457
examination, 4/446-453
genetics, 4/445

mental deficiency, 46/26-28
Psychomotor epilepsy
EEG, 15/478, 513
emotional disorder, 3/344
radiology, 15/549-551
sleep, 15/476
Psychomotor hallucination
hallucinosis, 46/562
Psychomotor seizure
temporal lobe epilepsy, 73/72, 74
Psychomotor retardation
see also Minimal brain dysfunction
Aarskog syndrome, 43/314
achondroplasia, 43/315
Alpers disease, 42/486
argininosuccinic aciduria, 42/525
Batten disease, 42/461-463
Börjeson-Forssman-Lehmann syndrome, 42/530
cerebellar dysarthria, 42/137
childhood myoglobinuria, 62/564
citrullinuria, 42/538
CNS spongy degeneration, 42/506
complex I deficiency, 62/503, 66/422
dental abnormality, 42/691
dermatoleukodystrophy, 42/488
dihydrolipoamide dehydrogenase, 41/210
disequilibrium syndrome, 42/138-140
epilepsy, 42/691
epileptogenic encephalopathy, 42/693
Förster syndrome, 42/202
frontometaphyseal dysplasia, 38/410
fucosidosis, 42/550
glycosaminoglycanosis type I, 42/450
GM1 gangliosidosis type I, 42/431
GM1 gangliosidosis type II, 42/432
GM2 gangliosidosis type III, 42/436
Harvey syndrome, 42/223
hereditary striatal necrosis, 49/495
hyperargininemia, 29/102, 42/562
hypersarcosinemia, 29/228
infantile epileptogenic encephalopathy, 42/693
infantile neuroaxonal dystrophy, 6/627, 49/402
infantile neuronal ceroid lipofuscinosis, 66/689
isovaleric acidemia, 42/580
kernicterus, 6/503
leprechaunism, 42/590
Lesch-Nyhan syndrome, 42/153
macrencephaly, 42/41
metachromatic leukodystrophy, 42/491
microcephaly, 42/45
Miller-Dieker syndrome, 50/592
mucolipidosis type II, 42/448
myelin dysgenesis, 42/498

myotonic myopathy, 43/165
neonatal adrenoleukodystrophy, 47/595
neuronal ceroid lipofuscinosis, 42/461-463
Niemann-Pick disease type A, 42/469
orotic aciduria, 42/607, 63/256
Paine syndrome, 60/293
Patau syndrome, 14/120, 31/508, 50/558
Pelizaeus-Merzbacher disease, 42/503
pyruvate carboxylase deficiency, 41/211, 42/586
pyruvate decarboxylase deficiency, 62/502
pyruvate dehydrogenase deficiency, 29/212, 42/587
self-mutilation encephalopathy, 42/140
short chain acyl-coenzyme A dehydrogenase deficiency, 62/495
spastic paraplegia, 42/169, 180
subacute necrotizing encephalomyelopathy, 51/388
Troyer syndrome, 42/193
vitamin B6 deficiency, 42/716
yellow teeth, 42/691
Psychomotor syndrome
Pick disease, 46/234
Psychomotor type
motor performance, 4/457
Psychopathic behavior
Huntington chorea, 49/283
Psychosine
biochemistry, 9/9, 10/250
Psychosis
see also Psychiatric disorder
acrylamide polyneuropathy, 51/281
Addison disease, 46/462
adrenoleukodystrophy, 60/169
adrenomyeloneuropathy, 60/170
affective, see Affective psychosis
alcohol, 46/462
amnesia, see Wernicke-Korsakoff syndrome
arsenic polyneuropathy, 51/266
atypical, see Atypical psychosis
autoscopia, 4/232, 45/386
autosomal dominant nocturnal frontal lobe epilepsy, 72/132
body scheme disorder, 4/233, 241-247
brain arteriosclerosis, 11/126
calcium antagonist intoxication, 65/441
carbon disulfide polyneuropathy, 51/279
carbon monoxide intoxication, 49/475, 64/32
cerebrotendinous xanthomatosis, 59/357
childhood, see Childhood psychosis
chloroquine intoxication, 65/483
CNS stimulant, 46/591
congenital retinal blindness, 13/272
consciousness, 3/120
contingent negative variation, 46/502
cycloserine intoxication, 65/481
delirium, 46/552
dimethyltryptamine, 46/594
dipeptidyl carboxypeptidase I inhibitor intoxication, 65/445
disopyramide intoxication, 65/453
drug induction, 46/462
dysphrenia hemicranica, 5/79
EEG, 15/601, 46/495
endocrine disorder, 46/462
ephedrine intoxication, 65/260
epilepsy, 15/593, 46/461, 73/423, 475, 477
ethionamide intoxication, 65/481
ethosuximide, 73/477
ethosuximide intoxication, 65/507
experimental psychosis, 4/231
Fabry disease, 60/172
filariasis, 35/168
foramen magnum tumor, 20/199
functional, see Functional psychosis
glyceryl trinitrate intoxication, 37/453
hallucination, see Peduncular hallucinosis
hallucinogenic agent, 46/594
heart transplantation, 63/180
hereditary striatal necrosis, 49/495
hyperthyroidism, 46/462
hypnagogic imagery, 45/353
hypothyroidism, 27/259, 46/462
isoniazid intoxication, 65/480
Korsakoff, see Wernicke-Korsakoff syndrome
lead encephalopathy, 64/436
leptospirosis, 33/404
levodopa, 46/593
levodopa intoxication, 46/593
lidocaine intoxication, 37/450
Listeria monocytogenes meningitis, 33/88
lysergide, 46/594
macrosomatognosia, 4/231, 45/385
manic depressive, see Manic depressive psychosis
Melkersson-Rosenthal syndrome, 8/218, 227
mescaline, 46/594
metachromatic leukodystrophy, 59/356
microsomatognosia, 4/231
migraine, see Migraine psychosis
multiple myeloma, 20/11
nonsteroid anti-inflammatory agent intoxication, 65/416
oculomotor disorder, 46/498
OKT3, 63/536
organic, see Organic psychosis
organic brain syndrome, 46/460-462

organic illness, 46/460-462
organic mental syndrome, 46/460-462
pallidoluysian degeneration, 49/455
paraproteinemia, 20/11
parietal lobe syndrome, 45/76
peduncular hallucination, *see* Peduncular
  hallucinosis
pellagra, 7/571, 573, 46/336
periodical, *see* Periodical psychosis
pernicious, *see* Pernicious psychosis
phenacemide intoxication, 37/202
phencyclidine, 46/594
pheochromocytoma, 39/500, 42/764
polyarteritis nodosa, 39/300, 55/356
postictal, *see* Postictal psychosis
postoperative, *see* Postoperative psychosis
prognosis, 15/796
psilocin, 46/594
psilocybine, 46/594
puerperal, *see* Puerperal psychosis
reactive, *see* Reactive psychosis
Refsum disease, 21/209, 60/231
schizophrenia, 15/599
schizophreniform, *see* Schizophreniform
  psychosis
sleep-wake syndrome, 45/143
speech disorder in children, 4/424
spinal meningioma, 20/199
spinal neurinoma, 20/250
striatopallidodentate calcification, 49/419, 423
sulfonamide, 46/601
systemic lupus erythematosus, 39/277, 55/376
temporal lobe, *see* Temporal lobe psychosis
temporal lobe epilepsy, 45/46, 46/504, 73/85
tetrahydrocannabinol, 46/594
thalamic syndrome (Dejerine-Roussy), 2/484
thallium intoxication, 64/325
tocainide intoxication, 65/455
traumatic, *see* Traumatic psychosis
tuberous sclerosis, 14/349
vigabatrin intoxication, 65/511
Waldenström macroglobulinemia, 63/396
Psychosis polyneuritica, *see* Wernicke-Korsakoff
  syndrome
Psychosis-intermittent hyponatremia and polydipsia
  syndrome, *see* Water intoxication
Psychosurgery
  emotion, 45/276
  emotional disorder, 3/347
  mood change, 45/276
Psychosyndrome
  traumatic, *see* Traumatic psychosyndrome
Psychotherapy

anorexia nervosa, 46/589
chronic muscle contraction headache, 48/350
epilepsy, 73/441, 446
Friedreich ataxia, 60/324
headache, 5/256, 48/39, 360
migraine, 48/39
muscle contraction headache, 5/162
phantom limb, 45/398
psychogenic headache, 5/256
traumatic psychosyndrome, 24/571
Psychotic choreoathetosis
  chronic, *see* Chronic psychotic choreoathetosis
Psychotic migraine
  major migraine, 48/162
Psychotic polydipsia, *see* Water intoxication
Pteridine coenzyme
  phenylketonuria, 42/610
Pterois volitans intoxication
  bacterial infection, 37/75
  convulsion, 37/75
  cyanotic, 37/75
  death, 37/75
  delirium, 37/75
  diarrhea, 37/75
  gangrene, 37/75
  hypotension, 37/75
  hypothermia, 37/75
  local pain, 37/75
  nausea, 37/75
  pallor, 37/75
  picture, 37/73
  respiratory distress, 37/75
  swelling, 37/75
  vomiting, 37/75
Pteroylglutamic acid
  anencephaly, 30/184
Pterygium
  cervical, *see* Cervical pterygium
  Turner syndrome, 14/109
Pterygium colli
  distichiasis lymphedema syndrome, 43/377
  Edwards syndrome, 50/275
  Klippel-Feil syndrome, 42/37
  LEOPARD syndrome, 43/27
  9p partial monosomy, 50/579
  18p partial monosomy, 43/534
  9p partial trisomy, 43/516
  6q partial trisomy, 43/503
  r(13) syndrome, 50/585
  Turner syndrome, 30/249, 43/550, 50/543
Pterygium syndrome, *see* Turner syndrome
Pterygopalatine fossa syndrome
  orbital floor syndrome, 2/88

Pterygopalatine ganglion
  anatomy, 48/476
  sphenopalatine neuralgia, 48/475
Pterygopalatine neuralgia, *see* Sphenopalatine
  neuralgia
Ptosis
  Aarskog syndrome, 43/313
  acrocephalosyndactyly type I, 43/318
  acrocephalosyndactyly type II, 43/324
  acrocephalosyndactyly type III, 50/123
  Armendares syndrome, 43/366
  atenolol, 63/195
  autonomic dysfunction, 63/148
  Bassen-Kornzweig syndrome, 63/273
  Benedikt syndrome, 2/298, 6/162
  bilateral, *see* Bilateral ptosis
  Bitter syndrome, 14/107
  botulism, 65/210
  chronic bromide intoxication, 36/300
  chronic progressive external ophthalmoplegia,
    2/323
  congenital, *see* Congenital ptosis
  congenital acetylcholine receptor
    deficiency/synaptic cleft syndrome, 62/438
  congenital high conductance fast channel
    syndrome, 62/438
  congenital myopathy, 62/297, 333, 349
  congenital slow channel syndrome, 62/432
  cyclic oculomotor paralysis, 42/331
  dentatorubropallidoluysian atrophy, 60/655
  dilated pupil, 57/136
  Dubowitz syndrome, 43/247
  dwarfism, 43/382
  facioscapulohumeral muscular dystrophy, 43/98
  familial infantile myasthenia, 62/427
  fetal alcohol syndrome, 30/120, 50/279
  Friedreich ataxia, 21/347, 60/310
  headache, 48/330, 65/178, 75/282
  hereditary spastic ataxia, 21/366, 60/462, 655
  Horner syndrome, 43/64
  hypogonadism, 43/138
  hysteria, 1/617
  Joseph-Machado disease, 60/656
  juvenile progressive bulbar atrophy, 59/70
  Kearns-Sayre-Daroff-Shy syndrome, 38/224, 229,
    60/655
  LEOPARD syndrome, 43/27
  Marcus Gunn phenomenon, 2/326, 42/319
  Marinesco-Sjögren syndrome, 42/184
  multicore myopathy, 43/120
  multiple mitochondrial DNA deletion, 62/510
  muscle weakness, 40/298
  myogenic, 1/616
  myopathy, 40/356
  myotubular myopathy, 43/113, 62/297
  neonatal myasthenia gravis, 41/100
  neuronal intranuclear hyaline inclusion disease,
    60/655
  nuclear ophthalmoplegia, 2/279
  occipital lobe tumor, 17/336
  oculomotor nerve, 57/136
  oculomotor paralysis, 2/292
  oculopharyngeal muscular dystrophy, 43/100
  olivopontocerebellar atrophy with retinal
    degeneration, 60/507
  ophthalmoplegia, 22/185, 43/138, 145
  ophthalmoplegic migraine, 48/132
  orbital apex syndrome, 2/322
  orthostatic hypotension, 63/148
  otitis media, 48/332
  18p partial monosomy, 43/534
  paratrigeminal syndrome, 48/330, 334
  pontine infarction, 53/390
  progressive external ophthalmoplegia, 22/179,
    43/136, 62/297, 66/427
  progressive hemifacial atrophy, 31/253, 43/408
  propranolol, 63/195
  pseudo, *see* Pseudoptosis
  pseudoparalytic, 2/318
  pupil, 74/408
  pyruvate dehydrogenase complex deficiency,
    66/416
  18q partial monosomy, 43/532
  4q partial trisomy, 43/498
  19q partial trisomy, 43/538
  r(22) syndrome, 43/547
  Refsum disease, 8/22, 13/315, 21/211, 66/489
  Rubinstein-Taybi syndrome, 31/282, 43/234
  Saethre-Chotzen syndrome, 43/322
  scapuloperoneal spinal muscular atrophy, 59/46
  sea snake intoxication, 37/10, 92
  senile, *see* Senile ptosis
  Smith-Lemli-Opitz syndrome, 43/308, 50/276
  spastic, 2/318
  spastic ataxia, 62/304
  spinal muscular atrophy, 59/46
  sporadic progressive external ophthalmoplegia
    with ragged red fiber, 62/508
  sympathetic innervation, 1/617
  thallium intoxication, 64/325
  tick bite intoxication, 37/111
  trigeminal ganglion, 48/333
  Turner syndrome, 50/543, 545
  unilateral elevator palsy, 2/338
  vertebral fusion, 43/488
  vertebrobasilar system syndrome, 53/390

Ptosis epicanthus syndrome, *see* Klein-Waardenburg syndrome
Pubertas praecox, *see* Precocious puberty
Puberty
  precocious, *see* Precocious puberty
Pudendal nerve
  micturition, 1/366
  neuroanatomy, 1/362
  neurophysiology, 1/362
Pudendal plexus
  topography, 2/51
Pudendalgia
  lumbar vertebral canal stenosis, 20/689
Pudenz-Heyer drain
  hydrocephalus, 30/577
Puerperal polyneuritis
  infectious disease, 7/488
Puerperal psychosis
  mania, 46/450
  venous sinus occlusion, 54/429
  venous sinus thrombosis, 54/432
Puerperal thrombosis
  superior longitudinal sinus thrombosis, 12/428
PUFA, *see* Polyunsaturated fatty acid
Puffer fish intoxication
  see also Marine toxin intoxication, Tetraodontidae intoxication *and* Tetrodotoxin intoxication
  abdominal pain, 37/80
  acute symptom, 37/80
  apomorphine, 37/81
  artificial respiration, 37/81
  autonomic nervous system, 75/512
  bicarbonate, 37/81
  bulbar paralysis, 37/80
  classification, 37/78
  convulsion, 37/80
  death rate, 37/80
  dizziness, 37/80
  extraocular muscle paralysis, 37/80
  flaccid paralysis, 37/80
  fugu, 37/80, 65/155
  hypotension, 37/80
  nausea, 37/80
  pallor, 37/80
  paresthesia, 37/80
  pentetrazole, 37/81
  petechia, 37/80
  respiratory failure, 37/80
  salivation, 37/80
  sweating, 37/80
  tetrodotoxin, 37/78-80
  treatment, 37/81
  weakness, 37/80

Pulmonary arteriovenous fistula
  hereditary hemorrhagic telangiectasia (Osler), 42/739
  systemic brain infarction, 11/442
Pulmonary artery stenosis
  tuberous sclerosis, 14/51
Pulmonary complication
  incidence, 26/335
  rehabilitation, 41/493
  spinal cord injury, 26/335
Pulmonary cyst
  tuberous sclerosis, 14/51, 373
Pulmonary disease
  chronic obstructive, *see* Chronic obstructive pulmonary disease
  Duchenne muscular dystrophy, 40/360
  exertional headache, 48/375
  obstructive, *see* Obstructive pulmonary disease
  pertussis encephalopathy, 9/551
Pulmonary disorder
  Behçet syndrome, 34/483
  iniencephaly, 50/131
Pulmonary edema
  Anthopleura elegantissima intoxication, 37/54
  brain fat embolism, 55/184
  Brisket disease, 43/62
  cobalt intoxication, 36/330
  head injury, 57/235
  high altitude, 63/416
  intracranial hypertension, 16/139
  mountaineer, 43/62
  neurogenic, *see* Neurogenic pulmonary edema
  nucleus tractus solitarii, 63/230
  opiate intoxication, 65/353
  rhodactis howesi intoxication, 37/54
  scorpion sting intoxication, 37/112
Pulmonary embolism
  brain aneurysm neurosurgery, 55/54
  head injury, 57/236
  oral contraception, 55/524
  paraplegic deep vein thrombosis, 26/315, 345
  subarachnoid hemorrhage, 55/54
Pulmonary fibrosis
  aluminum intoxication, 64/274
Pulmonary function
  control mechanism, 1/494
  CSF ionic composition, 1/495
  dementia, 46/397
  Duchenne muscular dystrophy, 41/494, 496
  Friedreich ataxia, 60/318
  infantile spinal muscular atrophy, 59/56
  juvenile spinal cord injury, 61/247
  progressive spinal muscular atrophy, 22/39

test, 22/39, 41/493
Pulmonary hypertension
  arteriovenous malformation, 42/726
  central alveolar hypoventilation, 63/464
  diamorphine intoxication, 65/355
  eosinophilia myalgia syndrome, 63/376, 64/258
  hereditary hemorrhagic telangiectasia (Osler),
    42/739
  obstructive sleep apnea, 63/456
  opiate intoxication, 65/355
  respiratory encephalopathy, 63/415
  rheumatic heart disease, 63/17
  Rowley-Rosenberg syndrome, 43/473
  syncope, 15/823
  toxic oil syndrome, 63/380
Pulmonary hypoplasia
  anencephaly, 42/12
  Bowen syndrome, 43/335
  caudal aplasia, 50/511
  Pena-Shokeir syndrome type I, 43/437, 59/71
  prune belly syndrome, 50/515
Pulmonary hypoventilation
  secondary polycythemia, 63/250
Pulmonary interstitial fibrosis
  tuberous sclerosis, 50/373
Pulmonary lymphangiomyomatosis
  tuberous sclerosis complex, 68/292
Pulmonary myomatosis
  Sturge-Weber syndrome, 14/69
  tuberous sclerosis, 14/51
Pulmonary vascular reflex
  systemic manifestation, 1/494
Pulmonema cantonensis, *see* Angiostrongylus
  cantonensis
Pulpitis
  dental headache, 5/222, 48/391
  headache, 5/222, 48/391
  symptom, 5/223
Pulsating benign giant cell tumor, *see* Aneurysmal
  bone cyst
Pulsating exophthalmos
  meningocele, 14/629
  sphenoid wing aplasia, 30/278
Pulsation
  basal encephalocele, 50/106
  basal meningoencephalocele, 50/106
  carotid cavernous fistula, 24/414
  migraine, 48/4, 162
Pulse interval
  aging, 74/230
  autonomic nervous system, 74/230
Pulse rate
  glossopharyngeal neuralgia, 48/467

hyperventilation syndrome, 63/436
  intracranial hypertension, 16/137
  lag in sign, 2/150
  meningococcal meningitis, 52/24
  traumatic vegetative syndrome, 24/578
Pulsed wave Doppler sonography
  B scan, 54/12
  duplex sonography, 54/12
  velocity profile, 54/12
  volume flow, 54/12
Pulseless disease, *see* Takayasu disease
Pulvinar
  visual perception, 2/533
Pulvinaris intergeniculatis nucleus
  posterior thalamic nuclear group, 2/473
Pulvinaris lateralis nucleus
  posterior thalamic nuclear group, 2/473
Pulvinaris medialis nucleus
  posterior thalamic nuclear group, 2/473
Pulvinaris superficialis nucleus
  posterior thalamic nuclear group, 2/473
Punch drunk syndrome
  see also Boxing injury
  boxing injury, 23/540
  pathology, 23/62
Pupil
  see also Mydriasis
  acute subdural hematoma, 24/280, 290
  adrenergic receptor blocking agent, 74/419
  adrenergic receptor stimulating agent, 74/419
  age, 74/404
  anisocoria, 74/406
  Argyll Robertson, *see* Argyll Robertson pupil
  brain death, 24/719, 764
  brain injury, 57/128
  brain stem infarction, 55/125
  cerebral irritation, 1/620
  cholinergic receptor stimulating agent, 74/419
  chronic paroxysmal hemicrania, 75/292
  chronic subdural hematoma, 24/305
  clinical examination, 57/128
  diabetic, 27/114, 124
  disorder, 74/406
  dissociated reaction, 1/622
  drug, 74/417
  drug response, 2/107-111, 22/257
  epidural hematoma, 24/268
  epilepsy, 1/622, 15/140
  examination, 74/405
  geniculocalcarine tract, 1/621
  headache, 75/281
  hippus, 74/404
  Holmes-Adie syndrome, 74/411, 413

Horner syndrome, 74/408
intracerebral hematoma, 24/360
irregular, 1/623
light-near dissociation, 74/416
midbrain syndrome, 24/584, 74/416
migraine, 75/294
movement, 74/404
muscle involvement, 74/411
neurosis, 1/624
nonreactive, *see* Nonreactive pupil
oculomotor nerve, 1/623
oculomotor paralysis, 1/623, 74/411
pineal tumor, 17/656
pinpoint, *see* Pinpoint pupil
ptosis, 74/408
reaction, 1/620-622
reflex, 74/404
sea snake intoxication, 37/92
shape, 1/622
size, 1/619, 622
Tolosa-Hunt syndrome, 48/297
tonic, *see* Tonic pupil
traumatic, 74/414
traumatic intracranial hematoma, 57/265
Wernicke, *see* Wernicke pupil
Pupillary abnormality
autonomic polyneuropathy, 51/487
tabes dorsalis, 33/364
xeroderma pigmentosum, 60/336
Pupillary constriction
botulism, 51/189
Pupillary defect
afferent, 74/415
relative afferent, *see* Relative afferent pupillary
defect
Pupillary disorder
Gillespie syndrome, 60/655
hereditary motor and sensory neuropathy type I,
21/285
hereditary motor and sensory neuropathy type II,
21/285
neuronal intranuclear hyaline inclusion disease,
60/655
nuclear ophthalmoplegia, 2/279
pellagra, 28/90
Refsum disease, 8/22, 41/434, 42/148, 60/228, 655
Pupillary dysfunction
diabetes mellitus, 75/590
Pupillary function
diabetes mellitus, 70/140
metabolic encephalopathy, 45/122
Pupillary involvement
Miller Fisher syndrome, 51/246

Pupillary phenomenon
hemianopic, *see* Hemianopic pupillary
phenomenon
Pupillary reflex
brain stem death, 57/458, 473
head injury outcome, 57/383
hexachlorophene intoxication, 37/486
intracranial hypertension, 16/132
multiple sclerosis, 47/135
pathway, 74/403
reflex arc, 1/623
Pupillary reflex abnormality
amyloid polyneuropathy, 51/416
Anthopleura elegantissima intoxication, 37/54
rhodactis howesi intoxication, 37/54
Pupillometry
chronic paroxysmal hemicrania, 48/261
cluster headache, 48/235
Pupillomotor fiber
ciliary ganglion, 2/107
termination, 2/532
test, 74/596
Pupillomotor dysfunction
sympathetic tone, 75/117
Pupillomotor function
testing, 74/630
Pure akinesia
freezing phenomenon, 49/68
hastening phenomenon, 49/68
Pure autonomic failure
age at onset, 22/233, 264, 63/147
aldosterone, 22/234
anhidrosis, 2/120, 22/232
autonomic test, 75/119
baroreceptor, 22/231, 233, 63/147
cardiac arrest, 2/120, 63/215
cold pressor test, 22/233
cyproheptadine, 22/240
differential diagnosis, 75/118
eosinophilic body, 22/238
gasp reflex, 22/234
growth hormone, 75/173
heart rate, 22/232, 63/147
hereditary form, 22/233
Holmes-Adie syndrome, 22/236
levodopa, 22/236, 240
Lewy body, 2/120, 22/235, 239, 265
multiple neuronal system degeneration, 75/166,
172
Nageotte nodule, 22/238
neuropathology, 22/237
noradrenalin, 22/240, 63/147
nycturia, 22/232, 234

orthostatic hypotension, 22/231, 63/146
Parkinson disease, 49/94
parkinsonism, 2/120, 22/232, 236
pathophysiology, 22/233
physiology, 22/231, 233, 63/231
pilocarpine, 2/120, 22/232
plasma catecholamine, 63/147
primary familial dysautonomia, 22/236
related disorder, 22/236
sex ratio, 22/233, 63/147
slow progression, 22/264
sweating, 2/120, 22/232
syncope, 22/232
thermoregulation, 2/120
treatment, 22/239
tyramine, 22/240
Valsalva maneuver, 22/233
vascular hypotension, 22/231-242
Pure motor aphasia
    definition, 45/311
    diagnosis, 45/311
Pure motor hemiplegia
    abducent nerve paralysis, 54/244
    acalculia, 54/243
    aphasia, 54/243
    apraxia, 54/243
    brachiocrural hemiparesis, 54/243
    brachiofacial hemiparesis, 54/242
    brain hemorrhage, 54/242
    brain lacunar infarction, 53/364, 54/57, 240
    capsular claudication, 54/240
    capsular lacuna, 54/240
    carotid angiography, 53/364
    computer assisted tomography, 53/364, 54/57, 241
    cortical abscess, 54/242
    demyelinating disease, 54/242
    differentiation, 54/241
    dysarthria, 53/364
    dysphasia, 54/243
    infarction site, 54/240
    lacuna site, 54/241
    lesion site, 53/210
    middle cerebral artery, 53/364
    motor hemiparesis variant, 54/243
    Nocardia, 54/242
    nonproportional dificit, 54/242
    oculomotor paralysis, 54/244
    onset, 54/240
    perforating artery, 53/364
    pontine hemorrhage, 54/313
    pontine infarction, 53/394
    pontine lacuna, 54/240
    putaminal hemorrhage, 54/305

striatocapsular infarct, 53/364
    subdural hematoma, 54/242
    transient ischemic attack, 54/240
    vertebrobasilar system syndrome, 53/394
Pure motor radiculopathy
    systemic lupus erythematosus, 51/447
Pure pallidal degeneration
    choreoathetosis, 49/447
    dystonia, 49/447
    pseudo-Babinski sign, 6/641
    rigidity, 49/447
Pure pallidoluysian degeneration
    torticollis, 49/448
Pure sensory stroke
    ataxia, 54/250
    brain lacunar infarction, 53/395, 54/57, 250
    computer assisted tomography, 54/57, 251
    hypalgesia, 54/250
    paresthetic migraine, 54/250
    persistent sensory deficit, 54/250
    thalamic infarction, 53/395, 54/250
    thalamus, 54/250
    transient ischemic attack, 54/250
    vertebrobasilar system syndrome, 53/395
Pure stenosis
    aqueduct stenosis, 50/301, 305
Pure tone audiometry
    acoustic nerve tumor, 17/671
    lesion localization, 2/74
    testing method, 2/74
Pure word blindness
    alexia, 4/17, 100, 112, 115, 118-127
Pure word deafness
    alexia, 4/131
    aphasia, 4/100
    auditory agnosia, 4/100
    Wernicke aphasia, 45/315
Purified protein derivative, see Tuberculin
Purine
    autonomic nervous system, 74/93
    enteric nervous system, 75/617
    neurotransmitter, 74/93
Purine metabolism
    Lesch-Nyhan syndrome, 29/267
Purine nucleoside phosphorylase
    wallerian degeneration, 7/215, 9/36
Purine nucleoside phosphorylase deficiency
    spastic diplegia, 56/469
Purinethol
    myasthenia gravis, 41/133
Purkinje cell
    acrylamide intoxication, 64/74
    action, 2/416

arthropod envenomation, 65/200
boxing injury, 23/556
cerebellar cortex, 50/176
cerebellar stimulation, 2/409
chronic mercury intoxication, 64/372
epilepsy, 72/108
ethylene glycol intoxication, 64/124
Huntington chorea, 49/320
Lafora body, 15/394
LOC stimulus, 2/410, 416
mercury polyneuropathy, 51/272
off beam, 2/414
on beam, 2/414, 417
organotin intoxication, 64/142
paraquat intoxication, 65/398
pertussis encephalopathy, 9/551
physiology, 1/58
Purkinje cell heterotopia
cerebellar hypoplasia, 50/179
Purkinje cell loss
Batten disease, 42/461, 466, 468
Biemond syndrome, 42/164, 60/449
carnosinemia, 29/198, 42/532
cerebellar granular cell hypertrophy, 43/253
familial amyotrophic lateral sclerosis, 59/246
galactosemia, 42/551
late cerebellar atrophy, 42/136
neuronal ceroid lipofuscinosis, 42/461, 466, 468
olivopontocerebellar atrophy, 42/161
posterior column ataxia, 42/164
trichopoliodystrophy, 14/784, 42/584
Puromycin
ganglioside, 9/9
granulovacuolar degeneration, 46/259, 272
Purposive behavior
animal experiment, 2/737
frontal lobe lesion, 2/737
Purpura
allergic, see Henoch-Schönlein purpura
anaphylactoid, see Henoch-Schönlein purpura
brain microangiopathy, 9/598
cat scratch disease, 52/128
cryoglobulinemia, 63/403
cryoglobulinemic polyneuropathy, 51/434
Henoch-Schönlein, see Henoch-Schönlein purpura
hyperagammaglobulinemic, see
  Hyperagammaglobulinemic purpura
meningococcal meningitis, 33/23, 52/26
primary amyloidosis, 51/416
senile, see Senile purpura
splenogenic, see Splenogenic purpura
thrombocytopenic, see Thrombocytopenic purpura
thrombotic thrombocytopenic, see Thrombotic

thrombocytopenic purpura
toxi-infective, see Toxi-infective purpura
Werlhof essential thrombocytopenic, see Werlhof
  essential thrombocytopenic purpura
Purulent meningitis
noninfectious, see Noninfectious purulent
  meningitis
Purupuru
deficiency neuropathy, 7/642
Pushing test
ataxia, 1/334
Putamen
see also Caudate nucleus and Striatum
brain hemorrhage, 11/587
chorea-acanthocytosis, 63/283
function, 6/109
Huntington chorea, 49/318
rigidity, 49/70
senile chorea, 6/436
status marmoratus, 42/700
Putaminal degeneration
carbamate intoxication, 64/188
carbon disulfide intoxication, 64/188
disulfiram, 64/188
Putaminal hemorrhage
brain hemorrhage, 54/304
clumsy hand dysarthria syndrome, 54/305
headache, 5/138, 48/278
hemichorea, 54/305
pure motor hemiplegia, 54/305
symptom, 54/304
Putaminal lesion
symptomatic dystonia, 49/544
Putaminal necrosis
ethylene glycol intoxication, 64/124
hereditary, see Hereditary striatal necrosis
methanol intoxication, 64/99-101
Puusepp, L., 1/11
Pyelography
intravenous, see Intravenous pyelography
Pygmy dwarfism, see Pituitary dwarfism type IV
Pyknodysostosis
bone fragility, 43/468
mental deficiency, 46/49
short stature, 43/468
Pyknolepsy
myoclonic akinetic epilepsy, 42/671
myoclonic astatic epilepsy, 42/671
Pyle disease
blindness, 46/88
craniometaphyseal dysplasia, 13/83, 22/493,
  31/255, 60/773
deafness, 22/493, 46/88, 60/774

facial weakness, 22/493, 60/774
familial, *see* Familial metaphyseal dysplasia
fibrous dysplasia, 13/83, 14/198, 46/88
hearing loss, 60/773
hereditary progressive cochleovestibular atrophy, 60/773
hypertelorism, 13/83, 14/198
inheritance mode, 46/88
mental deficiency, 46/88
optic atrophy, 13/83, 22/493, 60/774
Schwartz-Lelek syndrome, 31/257
skull, 46/88
Pyomyositis, *see* Tropical polymyositis
Pyramid
petrous, *see* Petrous pyramid
Pyramidal cell
Meynert giant, *see* Meynert giant pyramidal cell
Pyramidal decussation
ethambutol polyneuropathy, 51/295
mirror movement, 42/234
Pyramidal hypoplasia
schizencephaly, 50/199
Pyramidal sign
carbon monoxide intoxication, 49/474
cat scratch disease, 52/130
corpus callosum tumor, 17/510
Gerstmann-Sträussler-Scheinker disease, 60/622
hereditary cerebello-olivary atrophy (Holmes), 21/410
hereditary motor and sensory neuropathy type I, 21/283
hereditary motor and sensory neuropathy type II, 21/283
hereditary olivopontocerebellar atrophy (Menzel), 21/443
hypertonia, *see* Spasticity
Mendel-Von Bechterew reflex, 1/185
Minamata disease, 64/414
multiple sclerosis, 9/163
olivopontocerebellar atrophy (Dejerine-Thomas), 21/426
optic atrophy, 42/412-414
organic acid metabolism, 66/641
spinopontine degeneration, 21/390
temporal lobe tumor, 17/292
Pyramidal syndrome
aqueduct stenosis, 50/311
barbiturate intoxication, 65/332
Behr disease, 13/88
complex III deficiency, 62/504, 66/423
dysarthria, 46/240
extra, *see* Extrapyramidal syndrome
hereditary sensory and autonomic neuropathy type

I, 21/77
Marinesco-Sjögren syndrome, 21/556
Pick disease, 46/234, 240
progressive supranuclear palsy, 46/301
spinal cord lesion, 2/210
superficial siderosis, 70/66
transcobalamin II deficiency, 63/255
transverse spinal cord lesion, 19/27
Pyramidal system dysfunction
Hallervorden-Spatz syndrome, 49/399
Pyramidal tract
see also Lateral tract syndrome
akinesia, 45/161
anatomy, 1/152-155
anencephaly, 50/83
apraxia, 45/431
Benedikt syndrome, 2/298
CNS, 30/30
embryo, 50/10
lamellar arrangement, 2/208
lathyrism, 7/644, 36/511, 65/2
motor activity, 1/298
opticocochleodentate degeneration, 21/545, 42/241
pathologic laughing and crying, 3/361, 45/223
progressive bulbar palsy, 59/224
synergy, 1/414
X-linked adrenoleukodystrophy, 66/455
Pyramidal tract absence
anencephaly, 50/83
carnosinemia, 42/532
corpus callosum agenesis, 42/7
Friedreich ataxia, 42/144
Pyramidal tract degeneration
carnosinemia, 42/532
Friedreich ataxia, 42/144
spastic paraplegia, 42/169, 177
spinocerebellar ataxia, 42/182
Pyramidal tract hypoplasia
corpus callosum agenesis, 50/150
Pyramidal tract lesion
Opalski syndrome, 1/189
synkinesis, 1/411, 414
Pyramidal tract sign
complex IV deficiency, 66/424
congenital ataxic diplegia, 42/121
hereditary paroxysmal kinesigenic choreoathetosis, 49/353
Lundberg syndrome, 13/91
striatonigral degeneration, 49/206
Pyramidal tract syndrome
hypereosinophilic syndrome, 63/372
Pyramidotomy

tremor, 49/597
Pyramis
    ablation, 2/402
    animal experiment, 2/402
Pyrazolones
    headache, 48/174
    migraine, 48/174
Pyrazoxon intoxication
    chemical classification, 37/545
    organophosphorus compound, 37/545
Pyrethroid
    alcohol moiety, 64/211
    allethrin, 64/212
    bioresmethrin, 64/213
    carboxylesterase, 64/211
    chrysanthemic acid, 64/211
    cipermethrin, 64/212
    2 classes, 64/214
    cyano pyrethroid, 64/211
    cyfluthrin, 64/212
    cyphenothrin, 64/215
    deltamethrin, 64/211
    fenpropathrin, 64/213
    fenvalerate, 64/212
    fluvalinate, 64/212
    insecticide, 64/211
    3-cis isomer, 64/211
    3-trans isomer, 64/211
    natural pyrethroid, 64/211
    noncyano pyrethroid, 64/211
    permethrin, 64/212
    tetramethrin, 64/213
    toxicity, 64/211
    wood preservative, 64/211
Pyrethroid intoxication
    action mechanism, 64/211
    acute intoxication, 64/212, 219
    allethrin, 64/212
    antidote, 64/215
    atropine, 64/216
    axonal damage, 64/215
    benzodiazepine, 64/216
    bioresmethrin, 64/213
    chlorine channel, 64/213
    chlorine conductance, 64/214
    choreoathetosis, 64/214
    cipermethrin, 64/212
    clomethiazole, 64/216
    clonic seizure, 64/214
    coma, 64/219
    CSF GABA, 64/219
    cyfluthrin, 64/212
    cyphenothrin, 64/215
    death, 64/214
    deltamethrin, 64/212
    diazepam, 64/216
    EEG, 64/219
    epilepsy, 64/219
    fasciculation, 64/219
    fenpropathrin, 64/213
    fenvalerate, 64/212
    fluvalinate, 64/212
    gastrointestinal symptom, 64/219
    headache, 64/219
    LD50, 64/216
    mephenesin, 64/217
    mephenesin carbamate, 64/217
    methocarbamol, 64/217
    occupational exposure, 64/218
    paresthesia, 64/218
    pediculosis, 64/217
    permethrin, 64/212
    repetitive nerve impulse, 64/211
    salivation, 64/214
    scabies, 64/217
    sodium current, 64/211
    sodium current prolongation, 64/215
    startle reaction, 64/214
    tetramethrin, 64/213
    tremor, 64/214
    voltage dependent sodium channel, 64/212
Pyrethrum intoxication
    allergic reaction, 36/435
    animal study, 36/436
    headache, 48/420
    mode of action, 36/436
Pyribenzamine addiction
    brain embolism, 55/518
Pyridostigmine
    myasthenia gravis, 41/131, 43/158, 62/417
    neuromuscular transmission, 7/112
    organophosphate intoxication, 64/174, 230
Pyridoxal
    metabolism, 28/105-109
    phenylketonuria, 28/128
    Schiff procedure, 28/112
Pyridoxal-5-phosphate
    acute intermittent porphyria, 28/130
    convulsion, 15/63
    cystathioninuria, 28/119, 29/118
    Down syndrome, 28/129
    epilepsy, 28/109, 112-117, 42/716
    homocystinuria, 28/120-122, 29/115
    metabolism, 28/105-109
    xanthurenic aciduria, 42/636
Pyridoxal phosphokinase

isoniazid, 51/297
Pyridoxamine
  metabolism, 28/105
Pyridoxic acid
  epilepsy, 28/109
  metabolism, 28/105
Pyridoxine, *see* Vitamin B6
Pyridoxine deficiency, *see* Vitamin B6 deficiency
Pyridoxine dependency
  enzyme defect, 72/134
  neonatal seizure, 73/258
  xanthurenic aciduria, *see* Vitamin B6 dependent
    xanthurenic aciduria
Pyridoxine dependent convulsion
  epilepsy, 72/222
Pyridoxine dependent xanthurenic aciduria, *see*
    Vitamin B6 dependent xanthurenic aciduria
Pyridoxine intoxication, *see* Vitamin B6 intoxication
N-3-Pyridylmethyl-N'-p-nitrophenyl urea
    intoxication, *see* Vacor intoxication
Pyrimidine
  deficiency neuropathy, 7/560
Pyrithione zinc
  neurotoxin, 65/478
  polyneuropathy, 65/478
Pyrogen
  bacterial, 74/438
  endogenous, 74/439
  fever, 74/438
  thermoregulation, 26/370
Pyrogenic cytokine
  fever, 74/446
Pyroglutamic aciduria
  acidosis, 29/230
  biochemistry, 29/230
  gait ataxia, 29/230
  mental deficiency, 29/230
  spastic tetraplegia, 29/230
Pyroracemic acid
  CSF, 16/371
1-Pyrroline-5-carboxylate dehydrogenase
  hyperprolinemia type II, 42/569
Pyrroloporphyria, *see* Acute intermittent porphyria
Pyruvate carboxylase
  Friedreich ataxia, 51/389
  gluconeogenesis, 66/418
  pyruvate metabolism, 66/414
  tricarboxylic acid cycle, 66/418
Pyruvate carboxylase deficiency
  acetoacetyl-coenzyme A carboxylase, 66/417
  acetyl-coenzyme A, 42/586
  acidemia, 42/516
  acyl-coenzyme A, 66/418

apocarboxylase, 66/418
beta oxidation defect, 66/396
biochemistry, 29/212
biocytin, 66/418
cerebral symptom, 41/211
convulsion, 29/212
developmental delay, 42/516
diagnosis, 66/419
group A, 66/418
group B, 66/418
group C, 66/418
hepatic, 29/212
hepatomegaly, 62/503
hyperalaninemia, 29/212, 42/516
hypoglycemia, 29/212, 42/516
infantile hypotonia, 41/211, 42/586
infantile lactic acidosis, 41/211
lactic acidosis, 41/211, 42/516, 586
mental deficiency, 29/212, 42/586
metabolic acidosis, 29/212, 62/503, 66/418
mitochondrial disease, 62/503
movement disorder, 41/211
psychomotor retardation, 41/211, 42/586
seizure, 41/211, 42/586
short stature, 42/586
subacute necrotizing encephalomyelopathy,
    29/212, 51/388
vitamin B1, 29/213
vitamin Bw, 42/586, 66/418
vitamin Bw metabolism, 66/396
Pyruvate decarboxylase
  Huntington chorea, 49/262
  periodic ataxia, 21/578
Pyruvate decarboxylase deficiency
  dexamethasone, 41/428
  epilepsy, 62/502
  episodic apnea, 62/502
  exercise intolerance, 62/502
  fetal alcohol syndrome, 62/502
  infantile hypotonia, 62/502
  infantile lactic acidosis, 62/502
  lethargy, 62/502
  movement disorder, 41/428
  ophthalmoplegia, 62/502
  optic atrophy, 62/502
  periodic ataxia, 60/647, 62/502
  psychomotor retardation, 62/502
  seizure, 62/502
  severe lactic acidosis, 62/502
  subacute necrotizing encephalomyelopathy,
    62/502
  vitamin B1, 41/428, 62/502
Pyruvate dehydrogenase

pyruvate dehydrogenase complex, 66/413, 415
pyruvate metabolism, 66/414
subunit, 66/415
tricarboxylic acid cycle, 27/14
Pyruvate dehydrogenase complex
  chromosome, 66/413
  dihydrolipoamide acetyltransferase, 66/413, 415
  dihydrolipoamide dehydrogenase, 66/413, 415
  genetics, 66/413
  organotin intoxication, 64/144
  protein X, 66/415
  pyruvate dehydrogenase, 66/413, 415
  pyruvate dehydrogenase kinase, 66/413, 415
  pyruvate dehydrogenase phosphatase, 66/413, 415
  subunit, 66/415
Pyruvate dehydrogenase complex deficiency
  ataxia, 66/416
  basal ganglion, 66/416
  benign phenotype, 66/416
  beta oxidation defect, 66/396
  chronic encephalopathy, 66/416
  complex IV deficiency, 66/425
  convulsion, 66/416
  dysphagia, 66/416
  enzyme, 66/415
  genotype, 66/416
  hydrocephalus, 66/416
  hypospadias, 66/416
  hypotonia, 66/416
  infantile type, 66/416
  lactic acidosis, 41/211
  lethargy, 66/416
  micrognathia, 66/416
  mitochondrial disease, 62/496, 502
  neuropathy, 66/416
  ophthalmoplegia, 66/416
  optic atrophy, 66/416
  periodic ataxia, 60/648
  ptosis, 66/416
  pyruvate dehydrogenase system, 41/210
  subacute necrotizing encephalomyelopathy,
    66/425
Pyruvate dehydrogenase deficiency
  acidemia, 42/516
  ataxia, 42/587, 63/439
  biochemistry, 29/212
  cerebellar ataxia, 29/212, 42/516
  choreoathetosis, 29/212, 42/516
  demyelination, 29/212
  dystonia, 60/664
  epilepsy, 72/222
  hyperalaninemia, 29/212, 42/516
  hyperpnea, 63/439

  hyperventilation, 60/667
  hypotonia, 29/212
  lactic acidosis, 42/587
  mental deficiency, 29/212, 42/587
  microcephaly, 29/212
  optic atrophy, 29/212
  periodic ataxia, 29/212
  psychomotor retardation, 29/212, 42/587
  seizure, 42/587
  short stature, 42/587
  treatment, 29/212
Pyruvate dehydrogenase kinase
  pyruvate dehydrogenase complex, 66/413, 415
Pyruvate dehydrogenase phosphatase
  pyruvate dehydrogenase complex, 66/413, 415
Pyruvate dehydrogenase phosphatase deficiency
  mitochondrial disease, 62/503
Pyruvate dehydrogenase system
  pyruvate dehydrogenase complex deficiency,
    41/210
Pyruvate kinase
  brain glycolysis, 27/9
  Duchenne muscular dystrophy, 40/357, 43/106
  limb girdle syndrome, 62/185
Pyruvate kinase deficiency
  acanthocyte, 49/332
  hyperviscosity syndrome, 55/484
Pyruvate metabolism
  acetyl-coenzyme A, 66/413
  mitochondrial disease, 66/398
  pyruvate carboxylase, 66/414
  pyruvate dehydrogenase, 66/414
  tricarboxylic acid cycle, 66/413
  X-linked recessive ataxia, 60/640
Pyruvate oxidation
  Friedreich ataxia, 42/144
Pyruvic acid
  arsenic polyneuropathy, 51/267
  ataxia, 21/578
  CSF, 6/349
  Friedreich ataxia, 51/388
  hepatic coma, 49/215
  hereditary periodic ataxia, 21/565, 569
  Huntington chorea, 6/349
  hypertrophic interstitial neuropathy, 21/161
  multiple sclerosis, 9/321
  oxidation, 27/14-19
  serum, 9/321
Pyruvoyltetrahydropterin synthase deficiency, see
    Dyspropterin synthase deficiency
Pythagoras, 2/5

Q band

chromosome, 31/343, 350
Quadrantanopia
  crossed type, 2/594-596
  homonymous, 2/587
  lower type, 2/587
  temporal lobe tumor, 17/290
  upper type, 2/587
  visual cortex, 2/558
  visual cortex injury, 24/47
Quadratus labii inferioris muscle
  cardiofacial syndrome, 42/307
Quadriceps myopathy
  see also Limb girdle syndrome
  autosomal dominant inheritance, 43/128, 62/185
  characteristics, 62/174, 184
  creatine kinase, 62/174
  EMG, 40/440, 43/128, 62/174
  features, 62/184
  heterogeneity, 62/174
  incomplete dystrophin deficiency, 62/135
  limb girdle syndrome, 40/438, 440, 43/128
  muscle weakness, 43/128, 62/184
  polymyositis, 40/440, 62/174
  sex ratio, 62/174
  spinal muscular atrophy, 40/440, 62/174
Quadriceps reflex, see Suprapatellar reflex
Quadrigeminal artery
  radioanatomy, 11/93
  topography, 11/93
Quadriparesis
  barium intoxication, 64/354
  cervical spondylotic myelopathy, 12/534
  hypokalemia, 63/557
  progressive, see Progressive quadriparesis
Quadriplegia, see Tetraplegia
Quadrupedal attitude
  hemiplegia flexor correction, 6/139
Quail eater disease
  acute tubular necrosis, 41/272
  myoglobinuria, 40/339, 41/272
  paralysis, 41/272
Quail myopathy
  toxic myopathy, 62/612
Quality of life
  anomia, 67/390
  bone marrow transplantation, 67/399
  brain tumor, 67/395, 397
  cancer, 67/390
  disability, 67/390
  epilepsy, 72/408, 421, 423, 426, 432
  handicap, 67/390
  immunotherapy, 67/399
  impairment, 67/390

instrument, 67/393
  intervention strategy, 67/401
  neurocognitive function, 67/389
  neuro-oncology, 67/389, 391
  performance status, 67/391
  rehabilitation, 67/401
  speech, 67/390
  stimulant therapy, 67/401
  test, 67/395
Quantal squander syndrome, see Isaacs syndrome
Quantitative sudomotor axon reflex test, 75/108
  thermoregulation, 75/65
Quantum theory
  myasthenia gravis, 41/109
Queckenstedt-Stookey test, see Queckenstedt test
Queckenstedt test
  abdominal compression, 19/97, 107, 20/273
  block accentuation, 19/102
  cervical tumor, 19/114
  cough headache, 48/370
  CSF pressure, 19/91, 97, 107, 110
  dural sinus, 33/165
  epidural hematoma, 26/21
  spinal cord injury, 26/186
  spinal ependymoma, 20/369
  spinal glioma, 20/283
  spinal lipoma, 20/401
  spinal meningioma, 20/231, 283
  spinal neurinoma, 20/231, 273, 283
  spinal subarachnoid block, 19/91
  spinal tumor, 19/61
  thrombophlebitis, 33/165
  valve phenomenon, 19/105
  venous sinus thrombophlebitis, 33/165
Queensland tick-borne typhus
  rickettsial infection, 34/657
Quinagolide
  pituitary tumor, 68/350
Quinazolinone
  hypnotic agent, 37/355
Quincke edema, see Angioneurotic edema
Quinidine
  Achilles reflex, 41/245
  antiarrhythmic agent, 37/447
  antiarrhythmic agent intoxication, 37/447-449
  antiphospholipid antibody, 63/191
  atrial fibrillation, 37/448
  brain hemorrhage, 63/192
  cardiac involvement, 40/509, 523
  cardiac pharmacotherapy, 63/190
  chemical structure, 37/448
  delirium, 46/538
  iatrogenic neurological disease, 63/190

malaria, 52/372
myasthenia gravis, 63/191
myotonia, 41/297
myotonic dystrophy, 40/509, 523
neurotoxin, 37/447, 46/538, 63/191, 65/451, 485
syncope, 65/452
systemic lupus erythematosus, 63/191
systemic lupus erythematosus like syndrome,
    63/191
thrombocytopenia, 63/192
Quinidine intoxication
  cinchonism, 37/448, 65/451
  convulsion, 37/449
  death, 37/449
  depression, 37/448
  diplopia, 37/448
  headache, 37/448
  hearing loss, 37/448
  mydriasis, 37/448
  nyctalopia, 37/448
  photophobia, 37/448
  respiratory arrest, 37/449
  tinnitus, 37/448
  vasospasm, 37/448
  vertigo, 37/448
  visual disorder, 37/448
Quiniform intoxication
  subacute myelo-optic neuropathy, 37/175-180
Quinine
  acoustic neuropathy, 64/9
  experimental animal, 40/548
  iatrogenic neurological disease, 65/485
  malaria, 52/371
  myotonia, 40/548, 558, 41/297, 62/275
  myotonic dystrophy, 40/513
  neurotoxin, 13/63, 224, 41/297, 559, 50/219,
      52/372, 64/8, 65/485
  nocturnal cramp, 41/310
  optic neuropathy, 64/8
  toxic neuropathy, 64/8
Quinine intoxication
  hearing loss, 13/224
  hereditary deafness, 13/63, 50/219
  optic atrophy, 13/63
  symptom, 64/9
  visual impairment, 13/224, 64/8
Quinine sulfate
  motoneuron disease, 59/468
  muscle relaxant, 41/310
Quinizarin
  sweat test, 2/17, 200
Quinolinate phosphoribosyltransferase
    (decarboxylating), see Nicotinate nucleotide

pyrophosphorylase (carboxylating)
Quinoline
  experimental pigmentary retinopathy, 13/150, 219
Quinolinic acid
  excitotoxin, 64/20
  Huntington chorea, 49/58, 261, 298
Quinolone intoxication
  benign intracranial hypertension, 65/486
  ciprofloxacin, 65/486
  CNS side effect, 65/486
  enoxacin, 65/486
  epilepsy, 65/486
  GABA inhibition, 65/486
  headache, 65/486
  nalidixic acid, 65/486
  norfloxacin, 65/486
  papilledema, 65/486
  pefloxacin, 65/486
Quisqualic acid
  striatal necrosis, 49/511

R band
  chromosome, 31/344, 351
R protein, see Vitamin B12 binding protein
Rabbit sign
  see also Parkinsonism
  neuroleptic agent intoxication, 49/187
Rabies, 34/235-270, 56/383-399
  allergic encephalitis, 56/390
  allergic neuritis, 9/501
  animal infection, 34/255-257
  areflexia, 56/388
  autonomic nervous system, 75/418
  Babe nodule, 56/395
  bulbar syncope, 15/824
  canine vaccination, 56/384
  Canis latrans, 56/384
  cat, 34/255, 56/388
  chemistry, 56/383
  clinical diagnosis, 34/257
  clinical features, 56/385
  coma, 34/254, 56/386, 388
  confusional state, 56/386
  corneal transplantation, 56/385
  course, 56/387
  CSF, 34/257, 56/389
  deglutition spasm, 1/285
  diagnosis, 56/389
  differential diagnosis, 56/390
  dog, 34/255, 56/384
  dysautonomia, 56/386
  ECG, 34/254, 56/389
  EEG, 34/254, 56/389

emotional disorder, 3/348
epidemiology, 34/250-252, 56/384
facial diplegia, 56/388
form, 56/386
fox, 34/257, 56/384
Guillain-Barré syndrome, 56/385, 388, 391
history, 34/235
human, 34/252-254, 56/385-389
hydrophobia, 34/235, 253, 56/387, 75/418
hyperimmune serum, 56/398
Ig, 34/266, 56/386
immune mediated encephalomyelitis, 34/414-416
immune response, 34/263, 56/393
immunity, 34/249
immunofluorescence test, 34/258-260, 56/396
incubation period, 34/253, 56/386
initial symptom, 75/418
inspiratory spasm, 56/387
interferon, 56/398
jackal, 34/257, 56/385
laboratory diagnosis, 34/257-261
laboratory finding, 56/389
Lyssa body, 56/395
lyssavirus, 56/13, 383
lyssavirus genus, 56/13
mongoose, 56/385
mortality, 56/385
motor unit hyperactivity, 59/377
multiple sclerosis, 9/138, 47/330
myalgia, 56/386
Negri inclusion body, 34/236, 56/383, 394
neurologic phase, 56/386
pathogenesis, 34/245-250, 56/391-394
pathology, 56/394-396
percussion myoedema, 56/388
poliomyelitis, 56/391
porphyria, 56/390
postexposure prophylaxis, 56/385, 397
postexposure treatment, 56/396
postexposure treatment schedule, 56/397
prevention, 34/258, 56/396
prodrome, 34/253, 56/386
prognosis, 34/269
prophylaxis, 34/260-263
pruritus, 34/253, 56/386
raccoon, 56/384
recovery, 56/389
replication, 56/383
resistance, 56/393
rhabdovirus family, 34/236, 56/13
segmental demyelination, 56/395
sinus tachycardia, 56/389
skeletal muscle, 56/32

skunk, 56/384
spheroid body, 6/626
spinal muscular atrophy, 59/370
stage, 56/386
terminal stage, 75/418
tetanus, 34/264, 56/390
treatment, 34/263-269, 56/398
vampire bat, 34/251, 254, 56/385
wallerian degeneration, 56/395
Rabies encephalitis
  clinical features, 56/386
  transmission, 56/385
Rabies paralysis
  Guillain-Barré syndrome, 56/387
  Landry paralysis, 34/254
  vampire bat, 34/254
Rabies postvaccinial encephalomyelitis
  IgG, 9/140
  immune mechanism, 9/547
  immune response, 56/394
  multiple sclerosis, 9/138
  survey, 34/414-416
Rabies vaccination
  neuropathy, 8/91
  schedule, 34/262
  treatment rationale, 56/396
  WHO schedule, 34/265-268
Rabies vaccine
  autoimmune disease, 9/538
  complication, 56/398
  Guillain-Barré syndrome, 9/544, 47/609
  immune mechanism, 9/547
  multiple sclerosis, 9/59
  Pasteur, see Pasteur rabies vaccine
Rabies virus
  antigenic property, 34/241, 244
  autonomic nervous system, 75/416
  biology, 34/236-245
  cell culture, 34/236-240
  dissemination, 56/392
  exoplasmic flow, 56/392
  experimental acute viral myositis, 56/195
  hemagglutinating property, 34/241
  immune mediated encephalomyelitis, 34/414-416
  limbic system, 56/392
  morphology, 34/236-245
  multiplication, 34/236-240
  neurotropism, 56/35
  neurovirulence, 56/38
  nucleic acid, 34/243
  nucleocapsid, 34/243
  property, 34/240-243
  rhabdovirus, 56/383

serotype, 34/248
soluble antigen, 34/245
structural protein, 34/243
variant, 34/248
viral genome, 56/392
Racemose angioma
Bonnet-Dechaume-Blanc syndrome, 14/109
spinal cord, 20/438, 444
type, 14/51
Racemose arteriovenous angioma, see Racemose
angioma
Rachischisis, 32/159-189
see also Craniorachischisis, Exencephaly,
Iniencephaly and Spina bifida
anencephaly, 32/179-187, 42/12
developmental stage, 32/160
dysraphia, 32/159
dysraphic stage, 32/165-171
embryology, 32/161-165
epidemiology, 30/150
exencephalic stage, 32/171-179
experimental, 32/159, 166-170, 172-179
history, 32/159
holoprosencephaly, 50/235
human, 32/165, 171, 179-187
iniencephaly, 50/129, 134
malformation, 32/161
mechanism, 32/165
neurulation, 32/162
pathology, 32/171, 174-176
primary, 32/161
renal malformation, 32/172
scoliosis, 32/177
secondary, 32/161
spina bifida, 50/480
spinal ganglion, 32/177
teratogenic agent, 32/169
terminology, 32/159
vertebral column, 32/177
viscerocranium, 32/180
Raclopride
classification, 74/149
Racoon sign
anterior cranial fossa fracture, 57/131
clinical examination, 57/131
skull fracture, 57/131
Rademaker, G.G.J., 1/69
Radial aplasia
axial mesodermal dysplasia spectrum, 50/516
Baller-Gerold syndrome, 43/370
craniosynostosis, 43/370
VATER syndrome, 50/513
Radial nerve

compression neuropathy, 51/99
conduction, 7/164
entrapment, 7/309
humeral fracture, 70/31
leprous neuritis, 51/219
lesion site scheme, 2/26
occupational lesion, 7/330
topographical diagnosis, 2/23-28
wrist fracture, 70/34
Radial nerve palsy
alcohol, 64/115
alcohol intoxication, 64/115
hemophilia, 63/311
rheumatoid arthritis, 38/484
Thomas extension, 8/405
Radial reflex
clinical value, 1/243
Radial tunnel syndrome
compression neuropathy, 51/101
Radiation
anencephaly, 30/182
apistus intoxication, 37/76
atom bomb, 50/276
blood-brain barrier, 28/392
brain arteriovenous malformation, 12/244, 252
brain tumor, 16/40, 76, 80
brain tumor carcinogenesis, 16/40
brain tumor origin, 16/40
cavernous sinus thrombosis, 52/174
cephalocele, 50/98
cerebellar agenesis, 50/180
cerebral fibrosarcoma, 42/731
chromosomal aberration, 30/103
coccidioidomycosis, 52/417
cognitive dysfunction, 67/371
cranial epidural empyema, 52/168
cranial subdural empyema, 52/170
cranial vault tumor, 17/118
dens, 50/391
experimental, 7/388
glycogen, 27/207
gymnapistes intoxication, 37/76
hearing loss, 7/397
immunocompromised host, 56/469, 471
injury, see Radiation injury
ionizing, see Ionizing radiation
leukemia, 39/14
lymphoid, see Lymphoid radiation
microcephaly, 50/276
multiple sclerosis, 47/303
neurotoxicity, 67/359
nothesthes intoxication, 37/76
occipital cephalocele, 50/100

ototoxicity, 67/359
pituitary abscess, 52/153
plexus injury, 7/425
retinoblastoma, 42/768
sarcoma, 17/118
scorpaena intoxication, 37/76
scorpaenodes intoxication, 37/76
scorpaenopsis intoxication, 37/76
sebastapistes intoxication, 37/76
subclavian artery stenosis, 53/372
syringomyelia, 32/279-283, 50/446
transverse sinus thrombosis, 52/176
traumatic blindness, 24/28
trichinosis, 52/568
Radiation exposure
  Down syndrome, 31/398
  hydrocephalus, 50/288
Radiation hypersensitivity
  Nijmegen breakage syndrome, 60/426
Radiation induced tumor
  nervous system, 68/380
  survey, 68/391
Radiation injury
  see also Radionecrosis
  acoustic nerve, 7/397
  acute effect, 67/325
  arteriosclerosis, 67/330
  astroglia, 23/646
  atom bomb, 6/625, 23/647
  blood-brain barrier, 23/645
  brain, 67/325
  brain edema, 23/646
  brain injury, 23/639
  brain stem injury, 23/656
  brain stem tumor, 23/657
  cerebral necrosis, 67/329
  cerebrovascular disease, 67/330
  child, 23/647
  CNS pathology, 67/383
  cochlear apparatus, 7/393
  cranial nerve, 67/342
  CSF, 23/657
  delayed clinical sign, 23/652
  delayed damage, 23/639, 648
  early clinical sign, 23/651
  electrolyte balance, 23/655
  empty sella, 23/654
  epilepsy, 23/652
  experimental injury, 23/639
  glycogen, 27/207
  hemianopia, 23/653
  hemisphere lesion, 23/652
  Hodgkin disease, 39/50-53

hypothalamus, 23/654, 67/330
inflammatory effect, 23/646
latency period, 23/640
liquefaction necrosis, 23/650
lumbosacral plexus, 51/162
malignancy, 67/331
malignant lymphoma, 39/50-53
nerve cell, 23/646
nerve lesion, 7/388
olfactory nerve, 7/397
oligodendrocyte, 23/646
optic, *see* Geniculocalcarine tract injury
pathomorphology, 23/642, 648
pituitary gland, 23/653
pituitary tumor, 67/330
plexus lesion, 17/425
prenatal, *see* Prenatal radiation injury
radiodiagnosis, 23/657
radiosurgery, 67/333
spinal cord, 7/395
survey, 67/325
thrombosis, 23/650
tolerance dose, 23/640
total necrosis, 23/647
treatment, 67/339
vestibular apparatus, 7/394, 398
Radiation myelopathy
  anterior spinal artery syndrome, 61/200
  astrocytosis, 26/90, 61/205
  Brown-Séquard syndrome, 26/85, 91, 61/200
  capillary change, 61/204
  capillary permeability, 61/203
  case report, 26/81-86
  central cord necrosis, 61/202
  cervical myelopathy, 12/501, 26/84, 91, 61/200
  chemo- and radiotherapy interaction, 61/208
  clinical features, 26/84, 91
  clinical form, 61/200
  demyelination, 61/201, 63/43
  differential diagnosis, 61/201
  dose-response relationship, 61/205, 208
  dose-time relation, 26/87
  experimental, 26/92
  experimental neuropathy, 61/203
  fractionation, 61/206
  free interval, 61/199, 209
  glucocorticoid, 61/210
  heparin, 61/210
  iatrogenic neurological disease, 61/199
  immunologic aspect, 26/93
  incidence, 26/81, 61/199
  intrathecal chemotherapy, 61/208
  ischemic myelopathy, 63/43

late injury, 63/43
Lhermitte sign, 26/85, 91, 61/199
lower motoneuron disease, 61/200
lumbosacral myelopathy, 61/200
muscular atrophy, 22/20
myelography, 26/86, 91
myelomalacia, 12/501, 61/202
nerve growth factor, 61/210
neuropathologic form, 61/203
neuropathology, 61/201
nuclear magnetic resonance, 61/201
oligodendrocyte, 26/93, 61/205
paraplegia, 12/501, 61/203
pathogenesis, 26/92
pathology, 26/86-90
progressive type, 12/501, 26/91, 61/199
radiation dose, 61/204
radiation features, 26/86, 92
radiculopathy, 61/204
radioprotective agent, 61/210
re-irradiation, 61/207
risk factor, 61/210
spinal cord injury, 26/81-95, 63/43
spinal cord volume, 61/207
Strandqvist plot, 61/206
Strandqvist slope, 61/209
symptom, 26/90, 61/199
therapy, 61/210
thoracic myelopathy, 26/84, 61/200
transient, 67/355
transient form, 26/91, 61/199
transverse spinal cord lesion, 61/200
warfarin, 61/210
Radiation neuropathy
   ionizing, *see* Ionizing radiation neuropathy
Radiation therapy, *see* Radiotherapy
Radicular artery
   anatomy, 55/96
   spinal cord, 12/481
   spinal cord vascularization, 12/481
Radicular neuropathy
   hereditary sensory, *see* Hereditary sensory and
      autonomic neuropathy type I
   sensory, *see* Sensory radicular neuropathy
Radicular pain
   extramedullary spinal tumor, 19/55
   hypertrophic pachymeningitis, 33/190
   pseudo, *see* Pseudoradicular pain
   spastic paraplegia, 59/431
   spinal epidural empyema, 52/187
   vertebral metastasis, 20/427
Radicular syndrome
   cervical, *see* Cervical radicular syndrome

cervical intervertebral disc prolapse, 20/555-559
cervical rib syndrome, 20/556
lumbar puncture, 61/158
Pancoast tumor, 20/556
postlumbar puncture syndrome, 61/158
pseudo, *see* Pseudoradicular syndrome
segment pointer muscle, 20/557
spinal neurinoma, 20/249
spinal tumor, 19/26
spondylolisthesis, 19/84
Radiculitis
   see also Radiculopathy
   acquired toxoplasmosis, 52/355
   African trypanosomiasis, 52/341
   angiostrongyliasis, 52/549, 554
   Brucella, 51/185
   herpes genitalis, 71/267
   herpes simplex, 34/156
   herpes simplex virus, *see* Herpes simplex virus
      radiculitis
   herpes zoster, 71/268
   motor, *see* Motor radiculitis
   neurobrucellosis, 33/311, 316, 52/585
   poly, *see* Polyradiculitis
   schistosomiasis, 52/538
   sensory, *see* Sensory radiculitis
   sporotrichosis, 52/496
   toxoplasmosis, 52/355
   varicella zoster, 71/268
Radiculography
   complication, 19/212
   diagnostic value, 19/208, 213
Radiculomedullary artery
   anatomy, 12/493
Radiculomyelitis
   angiostrongyliasis, 52/554, 556
   tuberculous meningitis, 52/202
Radiculomyeloencephalitis
   angiostrongyliasis, 52/551
Radiculoneuritis
   cat scratch disease, 52/130
Radiculoneuropathy
   African trypanosomiasis, 52/341
   cat scratch disease, 52/129
   definition, 7/474
   herpes zoster, 56/231
   influenza, 7/478
   lumbar vertebral canal stenosis, 20/683
   polyarteritis nodosa, 51/452
   spondylotic, *see* Spondylotic radiculopathy
   topographical diagnosis, 1/230
   undetermined significant monoclonal
      gammopathy, 63/399

Radiculopathy
  see also Radiculitis
  amyloidosis, 63/529
  ankylosing spondylitis, 38/506-508
  Bragard sign, 20/684
  Burkitt lymphoma, 39/75, 63/351
  cancer, 69/74
  diabetes mellitus, 27/113, 70/135
  diabetic neuropathy, 27/113
  dialysis, 63/529
  hypereosinophilic syndrome, 63/373
  leukemia, 39/13, 63/343
  lymphocytic leukemia, 63/343
  malignant, *see* Malignant radiculopathy
  meningeal lymphoma, 63/347
  methotrexate, 63/358
  β2-microglobulin, 63/529
  β2-microglobulin amyloidosis, 63/529
  multiple myeloma, 39/141, 55/470, 63/392
  myeloid leukemia, 63/343
  neurobrucellosis, 52/585, 590
  osteitis deformans (Paget), 38/366
  pain, 69/72
  polyarteritis nodosa, 55/355
  pure motor, *see* Pure motor radiculopathy
  radiation myelopathy, 61/204
  skeletal fluorosis, 36/476
  spinal epidural lymphoma, 63/349
  spinal pachymeningitis, 33/190
  spondylotic, *see* Spondylotic radiculopathy
  survey, 69/71
  thoracic, 70/137
  traumatic, *see* Traumatic radiculopathy
Radiculopial artery
  anatomy, 12/492
Radiodiagnosis
  bacterial endocarditis, 52/302
  bilateral cervical facet interlocking, 26/265
  bladder function, 26/271, 424
  blowout fracture, 24/100
  brain aneurysm, 12/96
  brain concussion, 23/433
  brain contusion, 57/149
  brain injury, 23/239, 57/143
  brain ischemia, 57/161
  brain swelling, 57/159
  cervical vertebral column injury, 26/263
  chronic subdural hematoma, 57/154
  computer assisted tomography, 57/149
  craniofacial injury, 23/374
  Creutzfeldt-Jakob disease, 46/292
  Cushing syndrome, 68/354
  depressed fracture, 57/331

  diffuse axonal injury, 57/151
  epidural hematoma, 24/271, 57/152
  intestinal function, 26/269
  intracerebral hematoma, 24/360, 57/156
  intracranial pressure, 23/209, 57/160
  intracranial vessel spasm, 57/166
  intraventricular hemorrhage, 57/158
  isotope, 57/165
  lateral skull projection, 57/145
  mucormycosis, 52/473
  neurogenic osteoarthropathy, 26/275, 277-281
  North American blastomycosis, 35/406, 52/389
  pediatric head injury, 23/449
  plain film, 57/143
  posterior fossa hematoma, 24/349, 57/157
  radiation injury, 23/657
  scout film, 57/144
  skull fracture, 23/390, 57/143, 331
  skull radiation, 57/143
  skull radiation criteria, 57/144
  spinal injury, 26/263-283
  subdural hematoma, 57/152
  suture diastasis, 57/331
  third ventricle tumor, 17/459, 461
  Towne projection, 57/145
  traumatic intracranial hematoma, 57/267
  ultrasound, 57/165
  urinary tract, 26/269-274
Radiography
  brain metastasis, 71/622
  chordoma, 19/290
  clinical cervical stability, 61/12
  clinical lumbar stability, 61/17
  compression neuropathy, 51/86
  CSF fistula, 24/190
  fossa cribrosa, 19/161
  lumbar intervertebral disc prolapse, 20/581
  occipitoatlantoaxial complex, 61/6
  paravertebral soft tissue, 19/163-166
  vertebral osteolysis, 19/166
  Villaret syndrome, 2/225
Radioisotope
  brain scanning, 11/249, 269, 16/663
  ionizing radiation neuropathy, 51/138
Radioisotope cisternography
  aqueduct stenosis, 50/313
Radioisotope diagnosis
  aphasia, 45/298
  astrocytoma, 18/29-33
  brain tumor, 45/298
  conduction aphasia, 45/298
  global aphasia, 45/298
  motor aphasia, 45/298, 310

stroke, 45/298
    Wernicke aphasia, 45/298
Radioisotope exchange
    multiple sclerosis, 47/90
Radioisotope scanning, *see* Radioisotope diagnosis
Radiolesion
    nerve injury, 51/134
Radionecrosis
    see also Radiation injury
    astrocytoma, 18/37
    delayed, *see* Delayed radionecrosis
    radiosurgery, 67/334
Radiosensitivity
    astrocytoma, 18/12
    ataxia telangiectasia, 60/352
    brain injury, 18/481
    brain tumor, 18/485
    oligodendroglioma, 18/100
Radiosensitizer
    brain tumor, 18/491
    bromouridine, 18/491
    fluorouridine, 18/491
    methotrexate, 18/491
Radiosensitizing agent
    neurotoxin, 67/364
Radiosurgery
    brain metastasis, 69/143
    craniopharyngioma, 68/322
    radiation injury, 67/333
    radionecrosis, 67/334
Radiotherapy
    see also Irradiation
    acquired immune deficiency syndrome, 71/387
    acromegaly, 68/352
    ankylosing spondylitis, 26/182
    astrocytoma, 18/36, 501, 67/269, 68/104
    ataxia telangiectasia, 60/390
    brain arteriovenous malformation, 54/384
    brain injury, 67/327
    brain metastasis, 18/229, 512, 69/140, 193,
        71/626, 629
    brain stem, 18/503
    brain stem tumor, 18/503
    brain tumor, 18/1, 39, 481-512, 67/273
    brain tumor headache, 69/23
    brain vascular tumor, 18/510
    Castleman disease, 63/395
    cellular response, 67/269
    cerebellar medulloblastoma, 18/189, 395, 492
    chordoma, 18/160
    choroid plexus papilloma, 18/512
    chronic subdural hematoma, 24/314
    cranial, 67/272

craniopharyngioma, 18/508, 68/322
Cushing syndrome, 68/355
disorder, 67/353
disseminated necrotizing leukoencephalopathy,
    63/355
encephalopathy, 39/51
ependymoma, 18/502, 68/170
epidural metastasis, 69/208
germinoma, 68/245-247
glioblastoma multiforme, 18/63, 489
glomus jugulare tumor, 18/451
hemangioblastoma, 18/510
Hodgkin disease, 39/50-53
Hodgkin lymphoma, 63/348
hypophyseal adenoma, 18/507
interstitial, 67/271
ionizing radiation neuropathy, 51/138
irradiation encephalopathy, 63/355
irradiation myelopathy, 63/356
leptomeningeal metastasis, 71/654
leukemia, 18/251, 63/354
malignant lymphoma, 39/50-53
meningeal leukemia, 69/247
meningioma, 18/506, 68/413
metastasis, 71/654
midbrain tumor, 17/642
mineralizing brain microangiopathy, 63/355
multiple myeloma, 18/256, 20/14, 111, 63/395
myelopathy, 39/51
neuropathy, 39/52
nongerminomatous germ cell tumor, 68/248
non-Hodgkin lymphoma, 63/348
oligodendroglioma, 18/38, 502, 68/128, 130
optic nerve glioma, 18/503
osteosclerotic myeloma, 63/395
pain, 69/35
pinealoblastoma, 68/248
pinealoma, 18/365, 508, 42/767, 68/248
pituitary adenoma, 17/417
POEMS syndrome, 51/432, 63/395
primary CNS lymphoma, 68/263
primitive neuroectodermal tumor, 68/216, 219
prolactinoma, 68/348
resistance, 67/269
sarcoma, 18/510
sensitization, 18/39
side effect, 67/272
skull base tumor, 68/472, 474
spinal cord metastasis, 69/182
spinal cord tumor, 68/504
spinal ependymoma, 19/359, 20/378
spinal hemangioblastoma, 20/451
spinal neurinoma, 20/285

stereotactic, 67/271
supratentorial brain tumor, 18/338
survey, 67/267
thymectomy, 41/135
thyrotropic hormone secreting adenoma, 68/357
total dose, 67/331
vasculopathy, 39/52
vertebral metastasis, 20/421
vestibular neurinoma, 68/452
whole brain, *see* Whole brain radiation therapy
Radiotrauma
acute form, 18/483
chronic type, 18/484
nerve, *see* Ionizing radiation neuropathy
subacute type, 18/483
Radioulnar dysostosis
Berant syndrome, 43/371
craniosynostosis, *see* Berant syndrome
Klinefelter variant XXXXY, 31/483
Klinefelter variant XXXY, 31/483, 43/559
Klinefelter variant XXYY, 31/484, 43/563
4p partial monosomy, 43/497
XXXX syndrome, 43/554
XXXXX syndrome, 43/554
Radioulnar synostosis
aneuploidy, 50/553
Saethre-Chotzen syndrome, 43/322
XXXX syndrome, 50/554
XXXXX syndrome, 50/555
XYY syndrome, 50/551
Radner vein, *see* Descending medial occipital vein
Raeder paratrigeminal syndrome, *see* Paratrigeminal
syndrome
Raeder syndrome, *see* Paratrigeminal syndrome
Rage
brain command system, 45/274
organolead intoxication, 64/131
organotin intoxication, 64/142
sham, *see* Sham rage
Ragged red fiber
see also Mitochondrial disease
combined complex I-V deficiency, 62/506
complex V deficiency, 62/506
exercise intolerance, 62/292
fatal infantile myopathy/cardiopathy, 62/510
histochemistry, 40/31-33
human menopausal gonadotropin myopathy,
62/292
infantile cytochrome-c oxidase deficiency, 62/292
Kearns-Sayre-Daroff-Shy syndrome, 40/220,
60/54, 62/292, 314, 508
MELAS syndrome, 62/292, 509
MERRF syndrome, 62/292

mitochondrial disease, 40/31-33, 187, 62/496
mitochondrial DNA depletion, 62/511
mitochondrial myopathy, 40/33, 220
MNGIE syndrome, 62/292, 505, 66/430
multiple mitochondrial DNA deletion, 62/510
muscle fiber, 62/24
muscle tissue culture, 40/186-188
polymyositis, 40/220, 62/292
progressive external ophthalmoplegia, 40/186,
60/54, 62/289, 291-293, 305, 314, 66/427
spinal muscular atrophy, 62/292
sporadic progressive external ophthalmoplegia,
see Sporadic progressive external
ophthalmoplegia with ragged red fiber
zidovudine myopathy, 62/383
Railway spine
ankylosing spondylitis, 26/176
headache, 48/385
postconcussional syndrome, 57/421
posttraumatic headache, 48/385
Raimiste sign
hemiplegia, 1/184
Ramsay Hunt syndrome
see also Dyssynergia cerebellaris myoclonica,
Dyssynergia cerebellaris progressiva, Herpes
zoster oticus *and* Progressive myoclonus
epilepsy
β-*N*-acetylhexosaminidase A deficiency, 59/402
β-*N*-acetylhexosaminidase B deficiency, 59/402
action myoclonus, 6/767
autosomal dominant transmission, 42/212
dyssynergia cerebellaris myoclonica, 21/509,
42/211, 60/593
Friedreich ataxia, 42/211, 49/616, 60/597
geniculate herpes, 5/209
herpes zoster, 34/171-173, 56/231
herpes zoster oticus, 2/360, 34/172
myoclonic epilepsy, 42/211, 49/616, 60/593, 663
myoclonus, 60/662
olivopontocerebellar atrophy (Dejerine-Thomas),
60/532
olivopontocerebellar atrophy (Wadia-Swami),
60/497
original case, 42/211
vertigo, 2/360
Ramses II
nose, 74/185
skull, 74/184
Ramus meningeus nervorum spinalium, *see*
Sinuvertebral nerve (Luschka)
Ranawat line
definition, 61/7
Rancho splint

median nerve palsy, 8/404
  upper extremity, 8/404
Random epilepsy
  generalized tonic clonic seizure, 73/177
Ranvier node, 47/9-12
  acetonylacetone intoxication, 64/89
  acetylcholinesterase, 9/27, 34
  action potential generation, 40/128
  astrocyte, 47/12, 51/16
  axon, 47/10, 51/5
  Cajal, S. Ramón y, 47/9
  CNS, 7/22, 47/9, 51/15
  conduction mechanism, 47/33
  diagram, 51/13
  electrical loading, 47/22
  electrogenesis, 51/41
  electrogenic mechanism, 51/16
  extracellular space, 7/22, 47/11
  freeze fracture, 47/9, 11, 51/14
  function, 51/16
  gap substance, 47/10, 51/14, 16
  internodal axolemma, 51/5, 15
  internode, 47/3
  internode distance, 47/19
  internode length, 51/11
  microvilli, 47/10, 51/12
  myelin chemistry, 9/41
  nerve, 7/22, 51/5, 8, 12
  nodal axolemma, 51/5, 14
  nodal axoplasm, 51/15
  nodal gap, 47/10
  paranodal axoglial junction, 47/11
  paranodal axolemma, 47/10, 51/14, 16
  perinodal area, 7/22, 51/13, 16
  peripheral nervous system, 7/22, 47/10
  remyelination, 47/42
  saltatory conduction, 47/13
  Schwann cell, 7/22, 9/41, 47/10, 51/9
  Schwann cell microvilli, 47/10, 51/14, 16
  Schwann cell paranode, 47/10, 51/13
  sodium channel, 51/15
  structure, 47/3-5
  terminal glial loop, 47/10
Raphania
  dancing mania, 6/300
Raphe magnus nucleus, see Nucleus raphe magnus
Rapid ascending polyneuropathy
  hexane polyneuropathy, 51/277
  2-hexanone polyneuropathy, 51/277
Rapid eye movement, see REM
Rapid reversible conduction blocking
  compression neuropathy, 51/88
Rapid reversible neuropathy

diabetic polyneuropathy, 51/498
Rarefactive necrosis
  spongy, see Spongy rarefactive necrosis
Rasmussen chronic encephalitis
  epilepsy, 72/350
  surgical procedure, 73/401
Rathke cleft
  intrasellar cyst, 17/428
  pituitary tumor, 17/428
Rathke cleft cyst
  classification, 68/310
  diagnosis, 68/326
  genesis, 68/325
  histology, 68/326
Rathke pouch
  anatomy, 2/433
  craniopharyngioma, 18/531
  median facial malformation, 30/435
Rathke pouch tumor, see Craniopharyngioma
Rauwolfia alkaloid intoxication
  antihypertensive agent, 37/436-438
  reserpine, 37/436-438
Raven Progressive Matrices
  constructional apraxia, 45/495
  head injury outcome, 57/403
Ray intoxication
  classification, 37/78
Raymond Cestan syndrome, see Cestan-Chenais
  syndrome
Raynaud disease
  acrocyanosis, 8/322, 22/261, 43/70
  autonomic nervous system, 43/70
  autosomal dominant transmission, 43/71
  blood vessel disorder, 8/322, 22/261
  cold, 8/321, 22/261, 43/70
  emotion, 8/321, 22/261, 43/70
  mixed connective tissue disease, 51/449
  sex ratio, 8/322, 22/261
  temporal arteritis, 48/316
  thoracic outlet syndrome, 51/123
  type, 8/321
  vasoconstriction, 22/261
  vasodilatation, 8/322, 22/261, 43/70
Raynaud phenomenon
  arsenic intoxication, 8/337
  belladonna alkaloid, 8/337
  beryllium intoxication, 8/337
  causative trauma, 8/326
  cervical rib, 43/71
  collagen disease, 43/71
  CRST syndrome, 43/71
  cryoglobulin, 43/71, 55/485
  cryoglobulinemia, 51/434, 63/403

cryoglobulinemic polyneuropathy, 51/434
dermatomyositis, 41/56, 62/374
diagnosis, 51/136
eosinophilic polymyositis, 63/385
ergotamine, 8/337
erythromelalgia, 8/339
exposure, 51/135
heavy metal, 8/337
hyperviscosity syndrome, 55/485
inclusion body myositis, 62/374
lead intoxication, 8/337
lepromatous leprosy, 51/218
limb girdle syndrome, 40/438
macroglobulinemia, 55/485
mercury intoxication, 8/337
methyllysergide, 8/338
migraine, 48/85
mixed connective tissue disease, 71/122
mushroom intoxication, 8/337
occupational, see Occupational Raynaud
  phenomenon
polymyositis, 8/123, 40/438, 41/56, 58, 62/187,
  374
progressive systemic sclerosis, 43/71, 51/448
reflex sympathetic dystrophy, 8/328, 51/481
rheumatoid arthritis, 43/71
scleroderma, 8/339, 43/71
smoking, 51/135
sweating, 8/344
syphilis, 43/71
systemic lupus erythematosus, 43/71, 420
thallium intoxication, 8/337
thromboangiitis obliterans, 43/71
treatment, 51/136
vascular hypertension, 8/345
vibration frequency, 51/134
Rb gene
  CNS tumor, 67/47
Reaching disorder
  ocular ataxia, 45/408
Reaction time
  attention, 3/143, 145
  mental deficiency, 46/27
  neglect syndrome, 45/173
  Parkinson disease, 49/103
  traumatic psychosyndrome, 24/553
  visuospatial agnosia, 45/169
Reactions
  adverse drug, see Adverse drug reaction
  alcohol, see Alcohol reaction
  alcohol flush, see Alcohol flush reaction
  allergic, see Allergic reaction
  allogeneic mixed lymphocyte, see Allogeneic

  mixed lymphocyte reaction
  arousal, see Arousal reaction
  autologous mixed lymphocyte, see Autologous
    mixed lymphocyte reaction
  enzyme catalyzed, see Enzyme catalyzed reaction
  ergotropic, see Ergotropic reaction
  Herxheimer, see Herxheimer reaction
  hopping, see Hopping reaction
  host, see Host reaction
  Koch, see Koch reaction
  leukemoid, see Leukemoid reaction
  magnet, see Magnet reaction
  mixed lymphocyte, see Mixed lymphocyte
    reaction
  myasthenic, see Myasthenic reaction
  orcinol, see Orcinol reaction
  orienting, see Orienting reaction
  pain, see Pain reaction
  paradaxical pupillary, see Paradoxical pupillary
    reaction
  placing, see Placing reaction
  plasma, see Plasma reaction
  resorcinol, see Resorcinol reaction
  retrograde cell, see Retrograde cell reaction
  startle, see Startle reaction
  supporting, see Supporting reaction
  tilting, see Tilting reaction
  tonic avoidin, see Tonic avoidin reaction
  transamination, see Transamination reaction
  transketolase, see Transketolase reaction
  trophotropic, see Trophotropic reaction
  Weil-Felix, see Weil-Felix reaction
  withdrawal, see Withdrawal reaction
Reactive astrocytosis
  glycogen, 27/205
Reactive hyperemia
  nerve injury, 7/271
  pressure sore, 61/351
Reactive meningitis, see Meningitis serosa
Reactive psychosis
  delirium, 46/552
Reactivity
  coma, 3/62, 71, 73
  EEG, 57/188
Readiness potential
  see also Event related potential
  akinesia, 49/66
  chorea-acanthocytosis, 49/74, 329, 63/282
  Gilles de la Tourette syndrome, 49/74, 632
  hereditary paroxysmal kinesigenic
    choreoathetosis, 49/354
  Huntington chorea, 49/74
  Parkinson disease, 49/66

Sydenham chorea, 49/74
Reading
  anterior alexia, 45/442
  Braille, *see* Braille reading
  central alexia, 45/440
  cerebral dominance, 45/448
  developmental dyslexia, 46/108-113
  frontal lobe tumor, 67/150
  hemialexia, 45/448
  hemisphere asymmetry, 45/448
  Kana, *see* Kana reading
  Kanji, *see* Kanji reading
  migraine, 48/130
  mirror, *see* Mirror reading
  neural substrate, 45/449-452
  normal process, 46/108
  number, *see* Number reading
  somatesthetic, *see* Somatesthetic reading
  speed, 4/136, 7/116, 125
Reading disability
  see also Developmental dyslexia
  developmental disorder, 4/435-439
  developmental dyslexia, 4/134
  handedness, 4/261, 320
  oculomotor apraxia, 42/329
  spatial orientation disorder, 3/219
Reading epilepsy
  clinical features, 73/188
  EEG, 2/621, 42/717, 73/188
  eye movement, 1/583
  grand mal epilepsy, 2/621
  jaw jerking, 2/621, 42/717
  pathogenic mechanism, 73/188
  primary, 73/187
  secondary, 73/187
Reading test
  alexia, 4/135-137
  developmental dyslexia, 46/109
Reagin
  multiple sclerosis, 9/501
Reasoning
  abstract, *see* Abstract reasoning
  intelligence, 3/306
  thinking, 45/523
Rebound insomnia
  benzodiazepine intoxication, 65/340
Rebound nystagmus
  Friedreich ataxia, 60/310
Rebound phenomenon
  cerebellar hypotonia, 1/329
Recall aphasia, *see* Nominal aphasia
Receptive amusia
  diagnosis, 45/487

  pure form, 4/202
  sensory, 4/202
  temporal lobe lesion, 45/488
Receptive aphasia, *see* Sensory aphasia
Receptive dysphasia
  developmental, *see* Developmental receptive
    dysphasia
Receptoma
  terminology, 18/438
Receptor
  acetylcholine, *see* Acetylcholine receptor
  aging, 74/226
  alveolar/arteriolar J, *see* Alveolar/arteriolar J
    receptor
  α-amino-3-hydroxy-5-methyl-4-isoxazoleprop-
    ionic acid, *see*
    α-Amino-3-hydroxy-5-methyl-4-isoxazoleprop-
    ionic acid receptor
  asialoglycoprotein, *see* Asialoglycoprotein
    receptor
  basal ganglion, 49/34
  benzodiazepine, *see* Benzodiazepine receptor
  C3, *see* C3 receptor
  cardiopulmonary, *see* Cardiopulmonary receptor
  dopamine, *see* Dopamine receptor
  epidermal growth factor, *see* Epidermal growth
    factor receptor
  Fc, *see* Fc receptor
  GABA, *see* GABA receptor
  glutamic acid, *see* Glutamic acid receptor
  growth factor, *see* Growth factor receptor
  growth hormone releasing factor, *see* Growth
    hormone releasing factor receptor
  ileal, *see* Ileal receptor
  integrin, *see* Integrin receptor
  J, *see* J receptor
  kainic acid, *see* Kainic acid receptor
  laminin, *see* Laminin receptor
  N-methyl-D-aspartic acid, *see*
    N-Methyl-D-aspartic acid receptor
  motility factor, *see* Motility factor receptor
  muscarinic, *see* Muscarinic receptor
  muscarinic acetylcholine, *see* Muscarinic
    acetylcholine receptor
  muscarinic cholinergic, *see* Muscarinic
    cholinergic receptor
  natural killer cell, *see* Natural killer cell receptor
  nicotinic, *see* Nicotinic receptor
  nigrostriatal dopamine, *see* Nigrostriatal dopamine
    receptor
  opiate, *see* Opiate receptor
  opiate intoxication, 37/369-371
  organophosphate intoxication, 64/226

pain, 1/114, 7/83
poliovirus, *see* Poliovirus receptor
ryanodine, *see* Ryanodine receptor
sensory, *see* Sensory receptor
serotonin, *see* Serotonin receptor
sympathetic beta, *see* Sympathetic beta receptor
transferrin, *see* Transferrin receptor
Receptor antibody
  acetylcholine, *see* Acetylcholine receptor antibody
  anticholinesterase, *see* Anticholinesterase receptor
    antibody
Recessive atopic dermatitis
  hereditary deafness, 50/218
Recessive generalized myotonia
  autosomal, *see* Autosomal recessive generalized
    myotonia
Recessive infantile optic atrophy
  CNS, 13/111
Recessive inheritance
  autosomal, *see* Autosomal recessive inheritance
  mental deficiency, 46/46-68
  Niemann-Pick disease, 10/485
  pattern, 30/95
  Stargardt disease, 13/266
  X-linked, *see* X-linked recessive inheritance
Recessive Meniere disease
  episodic vertigo, 22/486
  hearing loss, 22/486
Recessive muscular dystrophy
  Duchenne like, 43/102
Recessive olivopontocerebellar atrophy
  dystonia, 60/664
Recessive optic atrophy
  congenital, *see* Congenital recessive optic atrophy
  juvenile diabetes mellitus, 22/505
Recessive progressive sensorineural hearing loss
  hereditary progressive cochleovestibular atrophy,
    60/767
Recessive renal acidosis
  neural deafness, 22/488
Recessive tremor
  familial, *see* Familial recessive tremor
Reciprocal innervation
  ballismus, 49/76
Recklinghausen disease, *see* Neurofibromatosis
  type I
Recklinghausen neurofibromatosis, *see*
  Neurofibromatosis type I
Recognition
  calculation, 45/475
  letter, *see* Letter recognition
  memory, 3/264
  object, *see* Object recognition

pattern, *see* Pattern recognition
tactile, *see* Tactile recognition
tactile agnosia, 4/33, 279
time sense, 4/30
topographical memory, 2/669
visual, *see* Visual recognition
Wernicke-Korsakoff syndrome, 45/198
Recombinant insulin like growth factor I, *see*
  Recombinant somatomedin C
Recombinant somatomedin C
  benign intracranial hypertension, 67/111
Recoverin
  paraneoplastic syndrome, 69/330, 335
Recovery
  acetonylacetone intoxication, 64/87
  acrylamide intoxication, 64/69
  ageusia, 24/18
  anosmia, 24/11
  brain death, 63/218
  brain embolism, 53/203
  brain infarction, 53/203
  brain injury, 23/335
  brain injury prognosis, 24/673
  central pontine myelinolysis, 63/552
  cerebrovascular disease, 53/203
  cord lesion, 26/212
  cortical blindness, 45/51
  demyelination, 47/29-43
  extra pontine myelinolysis, 63/552
  Glasgow Coma Scale, 57/382
  head injury, 57/367, 436
  intracranial hemorrhage, 53/203
  myelomalacia, 61/113
  neglect syndrome, 45/175-178
  paraplegia, 61/399
  pediatric head injury, 23/452
  posttraumatic amnesia, 24/478
  rabies, 56/389
  Reye syndrome, 65/117
  spinal cord concussion, 2/180
  spinal cord injury, 61/46, 399, 421
  tetraplegia, 61/399
  transient ischemic attack, 53/203
  transverse spinal cord lesion, 61/399
  traumatic intellectual defect, 24/481
Recrudescent typhus
  rickettsial infection, 34/645
Rectal polyp
  tuberous sclerosis complex, 68/292
Rectum
  defecation, 1/508
  diabetes mellitus, 75/595
  motility dysfunction, 75/633

Rectus abdominis muscle syndrome
  compression neuropathy, 51/109
Recurrent abdominal pain
  headache, *see* Abdominal migraine
Recurrent acute demyelinating neuropathy
  systemic lupus erythematosus, 51/447
Recurrent aseptic meningitis, *see* Mollaret
  meningitis
Recurrent bronchial neuropathy, *see* Hereditary
  mononeuritis multiplex
Recurrent cranial nerve palsy
  hereditary, *see* Hereditary recurrent cranial nerve
    palsy
Recurrent fever, *see* Relapsing fever
Recurrent infantile hemiplegia
  alternate, 12/348
Recurrent laryngeal nerve
  lymphoma, 63/350
  tuberculosis, 51/186
Recurrent laryngeal nerve compression
  atrial myxoma, 63/99
  compression neuropathy, 63/99
Recurrent laryngeal nerve lesion
  heart transplantation, 63/181
  iatrogenic neurological disease, 63/181
Recurrent laryngeal nerve palsy
  aorta aneurysm, 63/49
  congenital, *see* Vagus nerve agenesis
  Gerhardt syndrome, 50/219
  lymphoma, 63/350
Recurrent meningitis
  see also Mollaret meningitis
  alcoholism, 52/53
  bacterial meningitis, 52/5
  complement deficiency, 52/53
  CSF fistula, 52/53
  diabetes insipidus, 52/52
  EEG, 52/52
  heal trauma, 52/53
  host factor, 52/52
  inappropriate antidiuretic hormone secretion,
    52/52
  intelligence quotient, 52/52
  lymphatic leukemia, 52/54
  Mollaret meningitis, 34/548, 56/631
  multiple myeloma, 52/54
  pathogenesis, 52/52
  pneumococcal meningitis, 33/48, 52/52
  predisposing factor, 52/53
  sickle cell anemia, 52/53
  splenectomy, 52/53
Recurrent mononeuropathy
  hereditary multiple, *see* Hereditary multiple

  recurrent mononeuropathy
Recurrent multifocal CNS disease
  Sjögren syndrome, 51/449
Recurrent multiple cranial neuropathy
  Tolosa-Hunt syndrome, 48/303
Recurrent neuropathy
  chronic, *see* Chronic recurring neuropathy
  hereditary, *see* Hereditary recurrent neuropathy
Recurrent nonhereditary multiple cranial neuropathy
  acute intermittent porphyria, 51/571
  arteritis, 51/571
  associated polyneuropathy, 51/570
  cause, 51/570
  collagen disease, 51/570
  cryoglobulinemia, 51/570
  diabetic neuropathy, 51/570
  Guillain-Barré syndrome, 51/570
  intracranial neoplasm, 51/570
  paraproteinemia, 51/570
  polyarteritis nodosa, 51/570
  sarcoidosis, 51/570
  Sjögren syndrome, 51/570
  Tolosa-Hunt syndrome, 51/571
  tropical disease, 51/570
  tuberculous meningitis, 51/570
Recurrent nonhereditary polyneuropathy
  classification, 21/88
Recurrent oculomotor paralysis
  familial type, 21/12
  ophthalmoplegic migraine, 1/611
  Sjögren syndrome, 51/449
Recurrent polyneuropathy
  Friedreich ataxia, 21/355
  hereditary, *see* Hereditary recurrent
    polyneuropathy
  hypertrophic interstitial neuropathy, 21/146
  immune mediated syndrome related to virus
    infection, 34/396
Recurrent relapsing polyneuritis
  immune mediated syndrome related to virus
    infection, 34/396
Recurrent transient paraplegia
  diver, 12/532
  spinal apoplexy, 12/498, 532
Red blood cell choline level, *see* Erythrocyte choline
  level
Red fire fish intoxication
  picture, 37/73
Red nucleus, *see* Nucleus ruber
Red reflex
  oculocutaneous albinism, 43/4
Red tide toxin
  marine toxin intoxication, 65/142

neurotoxic shellfish poisoning, 65/158
Red tides, *see* Red tide toxin
Reduced drive
  traumatic psychosyndrome, 24/554, 688
Reduced neostriatal GABA
  Huntington chorea, 49/297
Reduced neostriatal glutamic acid decarboxylase
  Huntington chorea, 49/297
Reduced nicotinamide adenine dinucleotide
  brain infarction, 53/127
  brain ischemia, 53/127
Reduced nicotinamide adenine dinucleotide
    dehydrogenase (ubiquinone), *see* Complex I
Reduced nicotinamide adenine dinucleotide
    phosphate dehydrogenase
  Huntington chorea, 49/260, 298, 319
  wallerian degeneration, 9/37
Reduced nicotinamide adenine dinucleotide
    tetrazolium reductase, *see* Tetrazolium reductase
Reducing body myopathy
  clinical course, 62/347
  clinical features, 40/306
  congenital myopathy, 41/17-20, 62/332
  Coxsackie virus B2, 62/348
  Coxsackie virus B4, 62/348
  Coxsackie virus B5, 62/348
  histochemistry, 40/38, 47
  muscle fiber, 40/38, 113
  muscle fiber type I predominance, 41/20
  pathologic reaction, 40/263
  rigid spine syndrome, 62/348
  ultrastructure, 40/113
Reduction
  cervical facet subluxation, 61/62
  cervical spine, 25/292
  cervical vertebral column injury, 25/290
  hyperanteflexion sprain, 25/345
  IgM, *see* IgM reduction
  nonsurgical, 25/292
  postural, *see* Postural reduction
  surgical, 25/292, 495
  thoracolumbar spine injury, 25/459
Redundant myelin loop
  hereditary multiple recurrent mononeuropathy,
    51/557
Reduplication
  dura mater, *see* Dura mater reduplication
  dura mater sheath, *see* Dura mater sheath
    reduplication
  filum, *see* Filum reduplication
  spatial orientation disorder, 3/251
Reduplication phenomenon
  autoscopia, 3/254

body part, 3/253
brain pathology, 3/255
meaning, 3/255
mental diplopia, 3/255
nonaphasic misnaming, 3/254
object, 3/254
orientation, 3/251-256
person, 3/252
self, 3/254
space, 3/251
time and event, 3/253
Reduplicative paramnesia
  memory disorder, 24/489
  senile dementia, 45/418
  topographical amnesia, 45/418
Reed cell, *see* Reed-Sternberg cell
Reed-Sternberg cell
  Hodgkin disease, 20/114
Referred pain
  autonomic nervous system, 22/253
  carotid artery, 48/336
  Head's zone, 2/174
  low back pain, 20/578
  lumbar intervertebral disc prolapse, 20/578
  meninx, 1/539
  nociceptive, 69/32
  sensation, 1/103
  type, 1/135, 138
  viscerogenic spinal reflex, 2/175
Reflex, 1/61-72, 237-255
  abdominal, *see* Abdominal reflex
  Achilles, *see* Achilles reflex
  anal, *see* Anal reflex
  anocutaneous, *see* Anocutaneous reflex
  autonomic, *see* Autonomic reflex
  autonomic nervous system, 74/19, 25
  axon, *see* Axon reflex
  baroreceptor-cardiac, *see* Baroreceptor-cardiac
    reflex
  biceps, *see* Biceps reflex
  biceps femoris, *see* Biceps femoris reflex
  bladder, 1/364, 375
  blink, *see* Blink reflex
  brachioradial, *see* Brachioradial reflex
  brain metastasis, 18/218
  brain stem, *see* Brain stem reflex
  bulbocavernosus, *see* Bulbocavernosus reflex
  C, *see* C-reflex
  cardiopulmonary, *see* Cardiopulmonary reflex
  cardiovascular, *see* Cardiovascular reflex
  carotid sinus, *see* Carotid sinus reflex
  cat eye, *see* Cat eye reflex
  cerebral diplegia, 4/349

chemoreceptor, *see* Chemoreceptor reflex
ciliospinal, *see* Ciliospinal reflex
clinical survey, 1/237
coma, 3/75
corneal, *see* Corneal reflex
corneomandibular, *see* Corneomandibular reflex
corneo-oculogyric, *see* Corneo-oculogyric reflex
coughing, *see* Coughing reflex
cremasteric, *see* Cremasteric reflex
crossed extensor, *see* Crossed extensor reflex
cuboid, *see* Mendel-Von Bechterew reflex
Cushing, *see* Cushing reflex
digestive neural, *see* Digestive neural reflex
dorsal cuboid, *see* Mendel-Von Bechterew reflex
enteroenteric inhibitory, *see* Enteroenteric
    inhibitory reflex
ephedrine, 1/62
epigastric, *see* Epigastric reflex
epipatellar, *see* Epipatellar reflex
extensor, *see* Extensor reflex
extensor plantar, *see* Extensor plantar response
extensor thrust, *see* Extensor thrust reflex
feeding, *see* Feeding reflex
finger flexor, *see* Finger flexor reflex
fixation, 2/287
flexor, *see* Flexor reflex
focal myoclonus, 49/612
foveal, *see* Foveal reflex
functional integration, 1/73
gag, *see* Gag reflex
gasp, *see* Gasp reflex
gastrointestinal, *see* Gastrointestinal reflex
gluteus maximus, *see* Gluteus maximus reflex
Gordon, *see* Gordon reflex
grasp, *see* Grasp reflex
grooming, *see* Grooming reflex
groping, *see* Groping reflex
H, *see* H-reflex
head paradoxical, *see* Head paradoxical reflex
head retraction, *see* Head retraction reflex
head turning, *see* Head turning reflex
headache, 48/6
hereditary amyloid polyneuropathy, 21/123
Hering-Breuer, *see* Hering-Breuer reflex
Hering-Breuer deflation, *see* Hering-Breuer
    deflation reflex
Hering-Breuer inflation, *see* Hering-Breuer
    inflation reflex
Hoffmann, *see* H-reflex
hyperactive carotid sinus, *see* Hyperactive carotid
    sinus reflex
hypertrophic interstitial neuropathy, 21/150
internal hamstring, *see* Internal hamstring reflex

inversion, 1/220
J, *see* J reflex
jaw, *see* Jaw reflex
Landau, *see* Landau reflex
lip, *see* Lip reflex
long latency, *see* Long latency reflex
long loop, *see* Long loop reflex
Marinesco-Sjögren syndrome, 21/556
mass, *see* Mass reflex
masseter muscle, *see* Masseter muscle reflex
mediating nerve, 2/16
Mendel-Von Bechterew, *see* Mendel-Von
    Bechterew reflex
micturition, *see* Micturition reflex
Moro, *see* Moro reflex
motor activity, 4/444
muscle, 2/16
muscle stretch, *see* Muscle stretch reflex
myotatic, *see* Myotatic reflex
nasofacial, *see* Nasofacial reflex
nasolacrimal, *see* Nasolacrimal reflex
nasopalpebral, *see* Nasopalpebral reflex
neck righting, *see* Neck righting reflex
newborn, 4/344-347
oculocephalic, *see* Oculocephalic reflex
oculovestibular, *see* Oculovestibular reflex
palatal, *see* Palatal reflex
palmomental, *see* Palmomental reflex
palpebral, *see* Palpebral reflex
parachute, *see* Parachute reflex
paradoxical, *see* Paradoxical reflex
parasympathetic system, 74/36
pattern, 26/248
pectoralis major muscle, *see* Pectoralis major
    muscle reflex
pelvic abdominal, *see* Pelvic abdominal reflex
pendular, *see* Pendular reflex
peripheral, *see* Peripheral reflex
peripheral motoneuron lesion, 1/220
peristalsis, 75/618
pharyngeal, *see* Pharyngeal reflex
phasic, 1/238
Philippson, *see* Philippson reflex
physiology, 1/61-63
plantar, *see* Plantar reflex
postural, *see* Postural reflex
pronator, *see* Pronator reflex
pulmonary vascular, *see* Pulmonary vascular
    reflex
pupil, 74/404
pupillary, *see* Pupillary reflex
radial, *see* Radial reflex
red, *see* Red reflex

reflex standing, 4/349
Refsum disease, 21/199, 202, 27/523, 60/228, 230, 66/489
regression, *see* Regression reflex
reinforcement, 1/238, 243
respiration, *see* Respiration reflex
righting, *see* Righting reflex
rooting, *see* Rooting reflex
scapulohumeral, *see* Scapulohumeral reflex
scratch, *see* Scratch reflex
semimembranous, *see* Semimembranous reflex
Sherrington study, 1/47, 49
spinal, *see* Spinal reflex
spinal cord function, 1/61-63
spinal cord segment, 2/16
stages in development, 4/349, 444
stapedius, *see* Stapedius reflex
startle, *see* Startle reflex
stepping, *see* Stepping reflex
stretch, *see* Stretch reflex
strychnine, 1/62
sucking, *see* Sucking reflex
sudomotor, *see* Sudomotor reflex
sudomotor axon, *see* Sudomotor axon reflex
superficial, *see* Cutaneous reflex
suprapatellar, *see* Suprapatellar reflex
swallowing, *see* Swallowing reflex
tapetal, *see* Tapetal reflex
tear, 74/427
tendon, *see* Muscle stretch reflex
tensor fasciae latae, *see* Tensor fasciae latae reflex
tibial adductor, *see* Tibial adductor reflex
tibialis posterior, *see* Tibialis posterior reflex
tonic, 1/238
tonic grasp, *see* Tonic grasp reflex
tonic neck, *see* Tonic neck reflex
tonic stretch, *see* Tonic stretch reflex
triceps brachii, *see* Triceps brachii reflex
trigeminal, *see* Trigeminal reflex
trigeminal palpebral, 1/614
trigeminopupillary, *see* Trigeminopupillary reflex
Trömner, *see* Trömner reflex
urinary sphincter, 1/254
vagovagal, *see* Vagovagal reflex
vestibulo-ocular, *see* Vestibulo-ocular reflex
viscerogenic spinal, *see* Viscerogenic spinal reflex
visual stimulation, 1/614
Wohlfart-Kugelberg-Welander disease, 59/86
Reflex abnormality
pupillary, *see* Pupillary reflex abnormality
Reflex atrophy
peripheral amyotrophy, 1/231
Reflex automatism

transverse spinal cord lesion, 2/181
Reflex behavior
corpus callosum tumor, 17/510
Reflex bradycardia
J reflex, 74/578
Reflex depression
spinal cord injury, 26/245
spinal shock, 2/180, 26/245
Reflex disorder
Creutzfeldt-Jakob disease, 46/291
frontal lobe tumor, 17/248
hydrocephalic dementia, 46/327
intramedullary spinal tumor, 19/32
Reflex dystrophy (Sudeck), *see* Reflex sympathetic dystrophy
Reflex epilepsy
chess and card epilepsy, 2/622
definition, 72/9
flashing light, 72/9
hereditary paroxysmal kinesigenic choreoathetosis, 42/205, 207, 49/356
kinesigenic type, 42/207
movement induced, 73/193
musicogenic epilepsy, 15/446
physiology, 15/451
sexual activity, 73/193
somatosensory precipitation, 73/193
Reflex myoclonus
definition, 49/612
EMG, 49/610
features, 49/615
hereditary paroxysmal kinesigenic choreoathetosis, 49/356
pathomechanism, 49/615
Reflex neurovascular syndrome
acrocyanosis, 8/341
Meniere disease, 8/349
scalenus anticus syndrome, 8/332
shoulder hand syndrome, 8/330
stage, 22/254
thoracic rib syndrome, 8/332
upper extremity, 8/321
Reflex return
spinal cord injury, 26/246-248, 251
spinal shock, 26/246-248
Reflex sneezing
trigeminal nerve lesion, 2/57
Reflex sympathetic dystrophy
adrenergic receptor blocking, 61/121
autoimmune mechanism, 61/126
autonomic nervous system, 75/35, 309
autonomic polyneuropathy, 51/479, 482
brachial plexus injury, 61/122

238

calcitonin gene related peptide, 61/126
caudate nucleus, 8/328
causalgia, 8/328, 61/124
cause, 61/121, 129
ciclosporin intoxication, 65/552
clinical stage, 61/121
component, 61/121
course, 61/122
dissociated symptom, 61/122
dopamine β-mono-oxygenase, 61/128
dynorphin, 61/124
dysautonomia, 22/254
dystonia, 61/122, 129
ectopic beat, 61/124
features, 1/513
hyperalgesia, 61/121
limb temperature, 61/127
livedo reticularis, 61/121
mental depression, 61/122
movement disorder, 61/128
muscle tone, 61/128
neurokinin, 61/126
neuroma model, 61/124
neuropeptide, 61/126
neuropeptide Y, 61/128
neurotrophic osteoarthropathy, 22/254, 38/472
nociception, 61/125
orthopedic trauma, 70/43
osteoporosis, 1/513, 8/328, 51/480
pain mechanism, 61/123
pathogenesis, 61/123
pathophysiologic model, 61/124
physical rehabilitation, 61/465
posttraumatic, 51/479
rat, 61/124
Raynaud phenomenon, 8/328, 51/481
receptor sensitization, 61/123
regional blood flow, 61/127
Rexed layer receptive field, 61/124
sex ratio, 51/480
shoulder hand syndrome, 1/513, 51/481
spinal mechanism, 61/124
spread, 61/121
stage I symptom, 61/121
stage II symptom, 61/122
stage III symptom, 61/122
substance P, 61/126
sympathetic nervous system, 48/112
tachykinin, 61/126
treatment, 61/129
tremor, 61/122, 128
Reflex syncope
  aged, 15/824
  coughing, 15/824
  exertion, 15/824
Reflex synergy
  transverse spinal cord lesion, 2/182
Reflex vasomotor response
  collateral circulation, 11/170
  nerve injury, 7/271
  Valsalva maneuver, 2/114
Reflexology
  Alzheimer disease, 46/252
Refractive disorder
  headache, 5/204
Refractory period
  saltatory conduction, 47/14
  Sherrington study, 1/50
Refsum disease, 21/231-257, 42/147-149,
    60/225-237, 66/485-498
  acute toxic symptom, 36/347
  adult, see Adult Refsum disease
  age at onset, 21/188, 60/227, 66/486
  alopecia, 60/231
  alpha oxidation, 27/519, 66/485, 492, 494, 497
  amyotrophy, 21/198
  angioneuroma, 21/246
  anosmia, 13/314, 21/197, 27/523, 42/148, 60/228,
    231, 652, 728, 66/485, 490
  arteriosclerosis, 21/246
  ataxia, 8/21, 21/9, 199, 36/348, 42/148, 51/384,
    60/228, 230, 728, 66/490
  ataxia telangiectasia, 60/235, 66/497
  autonomic nervous system, 21/244, 60/231
  autosomal recessive, 8/22, 10/345, 13/40, 242,
    314, 479, 21/9, 12, 185, 27/520, 528, 41/433,
    50/218, 51/384, 66/485, 494
  Bardet-Biedl syndrome, 21/215
  Bassen-Kornzweig syndrome, 13/316, 21/215,
    60/233, 235, 66/497
  biochemical aspect, 27/519-537
  biochemical finding, 60/232
  brachymetacarpalism, 21/206
  brachymetatarsalism, 21/206-208
  cardiac lesion, 21/200-202, 60/231
  cardiac manifestation, 66/485, 491
  cardiomyopathy, 8/22, 42/148, 60/228, 237,
    66/491
  cataract, 8/22, 13/314, 21/9, 193, 60/228, 653,
    728, 66/489
  cerebellar ataxia, 21/9, 41/434, 42/148, 60/228,
    231, 66/490
  cerebellar finding, 66/490
  cerebellar sign, 8/22, 10/345, 13/314, 21/199,
    27/523, 36/348, 42/148, 60/228, 230, 66/490
  cholesterol, 8/23, 21/202, 205, 27/521

chorioretinal degeneration, 13/40
choroid plexus, 8/23, 21/236
chromophobe adenoma, 21/246
chronic inflammatory demyelinating
    polyradiculoneuropathy, 51/530
chronic progressive demyelinating
    polyneuropathy, 51/385
chronic toxic symptom, 36/348
clinical features, 21/187, 51/384, 60/227,
    66/486-492
clinical sign, 66/491
Cockayne-Neill-Dingwall disease, 21/217,
    60/235, 66/497
course, 8/22, 21/190, 60/227
cranial nerve change, 60/231
CSF, 10/345, 42/148, 60/233, 66/491, 494
CSF protein, 8/21, 10/345, 13/314, 21/9, 202, 210,
    36/348, 41/434, 42/148, 51/384, 60/228, 233,
    728, 66/494
deafness, 8/22, 10/345, 13/242, 314, 479, 21/196,
    27/523, 60/228
dementia, 60/231
demyelinating neuropathy, 47/622, 66/490
demyelination, 21/238, 41/434
diagnosis, 27/522-524, 60/234, 66/496
diet, 8/24, 21/190, 218-221, 27/520, 530-537,
    36/347, 42/149, 60/229, 234, 236, 66/497
differential diagnosis, 21/211-216, 51/387,
    60/235, 66/496
ear, 21/244
ECG, 21/9, 200-202, 66/488, 491
ECG abnormality, 8/22, 13/314, 21/200, 60/670
EMG, 8/23, 21/198
enzyme deficiency polyneuropathy, 51/384
ependyma, 8/23, 21/234, 42/148
epidemiology, 21/12
epilepsy, 15/325
ERG, 8/23, 13/315, 21/193, 60/228, 66/488
erythroderma, 21/216
external ophthalmoplegia, 13/40, 22/212
eye, 21/244
Fabry disease, 21/216
familial spastic paraplegia, 59/327, 334
fat storage, 21/234
fatty acid, 8/23, 10/345, 13/314, 21/50, 186, 196,
    202, 27/519, 36/347, 40/514, 41/433, 66/485,
    491
features, 21/9
Flynn-Aird syndrome, 21/217, 22/520, 60/235,
    66/497
food phytanic acid content, 21/218, 60/237
forme fruste, 21/189
Friedreich ataxia, 8/23, 21/199, 60/235, 66/490,

497
genetics, 21/184-187, 41/434, 60/233, 66/487, 494
geography, 21/183, 66/484
Gerstmann-Sträussler-Scheinker disease, 60/661
grey matter, 21/236
Guillain-Barré syndrome, 60/235, 66/487, 497
hammer toe, 60/231
hearing loss, 8/22, 21/196, 22/501, 36/348,
    42/148, 66/490
heart, 8/22, 10/345, 21/246, 36/347, 41/434,
    60/231, 66/497
hemeralopia, 21/199, 27/522, 60/728
hereditary deafness, 50/218
hereditary hearing loss, 22/501
hereditary motor and sensory neuropathy type I,
    21/197, 60/235, 66/489, 497
heterozygote detection, 21/185, 189, 27/520,
    41/434, 42/149, 66/495
history, 21/181, 27/519, 60/225-227, 66/486
hypertrophic interstitial neuropathy, 7/181, 8/23,
    21/9, 146, 199, 41/434, 51/385, 60/235, 66/497
ichthyosis, 8/22, 13/479, 21/9, 202-205, 36/347,
    41/434, 42/148, 60/228, 232, 667, 728, 66/491
ichthyotic syndrome, 38/9
inclusion body, 51/385
infantile, see Infantile Refsum disease
inheritance, 10/345, 27/520
intention tremor, 21/199, 60/231, 66/490
internodal axolemma, 51/15
Jervell-Lange-Nielsen syndrome, 21/197
Joseph-Machado disease, 60/661
Kearns-Sayre-Daroff-Shy syndrome, 21/212
kidney, 10/345, 13/314, 21/238, 36/348, 42/148,
    66/497
laboratory data, 21/210
Laurence-Moon-Bardet-Biedl syndrome, 21/215
Laurence-Moon syndrome, 21/215
leptomeninx, 8/23, 21/234, 42/148
Lindenov-Hallgren syndrome, 21/216, 60/235,
    66/497
liver, 10/345, 13/314, 21/238, 36/348, 42/148,
    66/497
mental change, 21/209
metabolic ataxia, 21/575, 581
metabolic manipulation therapy, 66/110
metachromatic leukodystrophy, 21/216
miosis, 13/314, 21/193, 60/229, 66/489
mitochondrial alpha oxidation, 51/384
mitochondrial encephalomyopathy, 60/661
nerve biopsy, 21/240
nerve conduction, 7/181, 21/197, 42/149
nerve conduction velocity, 7/181, 8/23, 21/197,
    51/385, 60/230, 66/489

nerve hypertrophy, 21/199, 231
neuroacanthocytosis, 21/215
neuroglia relationship, 28/410
neurolipidosis, 8/23, 10/345, 13/314, 21/181, 231, 27/520
neurologic manifestation, 27/522
neuronal intranuclear hyaline inclusion disease, 60/661
neuropathy, 8/21-24, 21/197, 27/523, 40/303, 514, 60/228, 230
nyctalopia, 8/21, 13/314, 21/189, 192, 51/385, 60/228, 66/488
nystagmus, 8/22, 10/345, 21/193, 200, 41/434, 60/229, 231, 728, 66/489
ocular lesion, 21/192
ocular sign, 66/489
olivopontocerebellar atrophy (Dejerine-Thomas), 8/23, 60/661
omega oxidation, 13/314, 27/520
omega oxidation block, 51/387
onion bulb formation, 8/23, 21/9, 12, 233, 240, 248, 251, 254, 257, 51/385
ophthalmologic disturbance, 60/228
ophthalmoplegia, 21/211, 22/195, 60/228, 66/489
optic atrophy, 8/22, 13/40, 21/195, 51/385, 60/229, 66/489
paranoid psychosis, 60/231
paresis, 60/228
pathogenesis, 36/348, 349
PEG, 21/200
periodic ataxia, 21/570, 581
peroxisomal disorder, 66/494, 510
phytanic acid, 8/23, 10/345, 13/314, 21/9, 186, 192, 196, 202, 204, 210, 221, 27/520, 522-524, 527-531, 28/410, 41/434, 42/148, 60/228, 232, 674, 66/110, 485, 492-494, 498
phytanic acid α-hydroxylase, 10/345, 41/434, 51/384, 60/728, 66/495
phytanic acid intoxication, 21/186, 36/347-349, 51/385, 60/233, 237, 66/498
phytol, 8/24, 21/187, 27/521, 528, 41/434, 60/237, 728, 66/498
plasmapheresis, 27/520, 36/348, 60/237, 728, 66/498
polyneuritis, 8/22, 10/345, 13/314, 21/199, 202, 42/148, 66/497
polyneuropathy, 8/21, 21/199, 22/501, 27/523, 36/348, 41/433, 42/148, 51/384, 60/165, 66/485, 489
pregnancy, 8/22, 21/191, 66/487, 496
presenting sign, 60/227
prevalence, 21/183
primary pigmentary retinal degeneration, 66/488

prognosis, 60/236, 66/497
progressive external ophthalmoplegia, 13/40, 22/204, 207
progressive multifocal leukoencephalopathy, 21/215
psychosis, 21/209, 60/231
ptosis, 8/22, 13/315, 21/211, 66/489
pupillary disorder, 8/22, 41/434, 42/148, 60/228, 655
reflex, 21/199, 202, 27/523, 60/228, 230, 66/489
renal impairment, 42/148, 60/232, 66/491
retinal histology, 21/193, 195
4S protein, 21/210
Schwann cell, 42/148
secondary pigmentary retinal degeneration, 8/22, 10/345, 13/40, 242, 314, 479, 21/9, 182, 184, 189, 192, 211, 22/501, 27/522, 36/348, 40/514, 41/434, 42/148, 51/384, 60/228, 235, 653, 728, 66/485, 488
sensorineural deafness, 21/9, 196, 22/501, 41/434, 42/148, 60/228, 728, 66/485
sensory deficit, 8/22, 21/199, 27/523, 60/228, 230, 66/489
sensory deprivation, 60/232
sex ratio, 8/22, 21/9, 185, 60/234, 66/494
Sjögren-Larsson syndrome, 13/479
skeletal malformation, 8/22, 10/345, 13/314, 21/206-209, 42/148, 60/228, 231, 66/485, 491
skin, 21/246
skin change, 13/314, 21/204, 27/523, 42/148, 60/228, 232, 66/485, 491
skin lesion, 21/202-205
spinocerebellar degeneration, 60/235, 66/497
sudden death, 8/22, 21/191, 60/228, 66/488, 491, 498
survey, 21/232
symptom, 66/487
syndactyly, 21/208, 60/231, 66/491
tabes dorsalis, 60/235, 66/497
Tangier disease, 21/215
transiently raised phytanic acid level, 60/237
treatment, 8/24, 21/218-221, 51/387, 60/234, 236, 66/497
tremor, 21/200
ultrastructure, 21/252, 51/385
Usher syndrome, 21/215, 60/235, 66/497
Werner syndrome, 21/218, 60/235, 66/497
white matter, 21/238
Regeneration
axon, see Axon regeneration
brain, see Brain regeneration
Duchenne muscular dystrophy, 40/371
facial nerve, 8/254

facial paralysis, 8/270
  idiopathic facial paralysis, 8/254-258
  muscle, *see* Muscle regeneration
  muscle fiber, 40/20, 89, 62/9
  myopathy, 62/9
  nerve, *see* Nerve regeneration
  neuromuscular disease, 62/9
  olfaction, 24/11
  spinal cord, *see* Spinal cord regeneration
  spinal cord injury, 25/99
  spinal nerve root, *see* Spinal nerve root
    regeneration
Regional anesthesia
  complication rate, 65/422
Regional brain blood flow
  aphasia, 45/297
  brain arteriovenous malformation, 54/380
  brain infarction, 54/118
  brain injury, 57/69
  brain ischemia, 54/118
  cardiopulmonary bypass, 63/175
  classic migraine, 48/68-76
  EEG, 11/270
  heart transplantation, 63/175
  hebephrenia, 46/453
  language, 45/297
  mathematics, 54/140
  measurement, 48/63
  migraine, 48/69
  open heart surgery, 63/175
  papaverine, 37/455
  Parkinson disease, 49/97
  physiology, 11/122
  positron emission tomography, 54/108
  schizophrenia, 46/453
  serial measurement, 54/145
  severe head injury, 57/384
  single photon emission computer assisted
    tomography, 54/111
  spreading depression, 48/74
  transient ischemic attack, 54/117
  vascular dementia, 46/362-364
Regional brain metabolic rate for oxygen
  flow rate dependency, 54/116
  transient ischemic attack, 54/117
Regression
  caudal, *see* Caudal aplasia
Regression reflex
  kuru, 56/556
  spastic paraplegia, 59/428
  sucking reflex, 1/241
Regressive generalized chorea
  juvenile hereditary benign chorea, 49/346

Rehabilitation, 41/457-497
  ability return sequence, 55/237
  abnormal posture, 41/464
  adaptation, 55/240
  age, 46/612
  agraphia, 3/417-422, 4/102, 172-174
  aid in daily living, 41/490
  alexia, 3/422-426, 4/137
  amyotrophic lateral sclerosis, 59/197
  anoxia, 46/611
  aphasia, 3/373, 375, 378, 396-417, 4/101, 109,
    24/689, 45/324, 46/614-617, 55/238
  appliance, 55/245
  apraxia, 3/389, 46/620
  assessment, 41/483, 55/246
  astrocytoma, 18/36
  ataxia, 3/389
  ataxia telangiectasia, 60/391
  attention deficit, 46/624
  bed rest, 24/685
  behavior disorder, 46/624-630
  biofeedback, 55/242
  brachial plexus, 51/149
  bracing, 41/468
  brain death, 24/745
  brain hematoma, 55/233
  brain hemorrhage, 55/233
  brain infarction, 53/432, 55/233
  brain injury, 23/16, 24/683
  brain tumor, 18/523
  case history, 46/612
  center, 24/694
  cerebrovascular disease, 12/468-477, 46/610
  cervical spinal cord injury, 61/66
  cervical vertebral column injury, 61/65
  child, 25/186, 192
  classification, 24/683
  clinical neurology, 46/609-631
  community care, 46/630
  complication prevention, 55/241
  computer aid, 46/630
  constructional apraxia, 3/428, 46/620
  contracture, 41/466
  daily life activity, 55/237
  daily life activity scale, 55/239, 248
  definition, 24/683
  dementia, 3/426-430
  developmental dyslexia, 4/388, 438
  diffuse lesion, 12/469
  disability scoring, 55/247
  driving, 46/619
  dysarthria, 3/396-399, 4/101
  dysgraphia, 4/102, 172-174, 438

dyslexia, 46/617
dysphasia, 46/614-617
early mobilization, 24/684
emotional sequel, 55/246
encephalitis, 46/611
environment, 46/612-614
epilepsy, 15/805, 73/424, 453, 463
exercise, 41/463
familial sequel, 55/246
family adjustment, 55/240
female sexual function, 61/331
focal lesion, 12/468
fracture, 41/474
Friedreich ataxia, 60/325
future, 1/34
gamma system, 12/472
genetic counseling, 41/484-490
global amnesia, 46/622
H-reflex, 12/473
head injury, 46/610, 617-620, 624-629, 57/388, 436
head injury prognosis, 46/611
higher cortical function, 3/368-431
ideomotor apraxia, 46/621
inactivity, 24/688
inpatient therapy, 55/242
intellectual function, 41/479
juvenile spinal cord injury, 25/175, 192, 61/247
late recovery, 55/237
left handed children, 4/467
male sexual function, 61/313
medical management, 41/460
memory, 46/620-623
mental depression, 55/246
mental transformation, 46/619
minor head injury, 46/628
motor control, 46/620
multiple sclerosis, 9/416, 47/173-178
neglect syndrome, 45/178
neuronal recovery, 55/244
neuropsychologic assessment, 46/612
neuropsychology, 46/609-631
occupational therapy, 41/481, 55/241
organization, 24/693
orthopedic correction, 41/468-474
paraplegia, 20/830
pattern recognition, 46/618
personality, 46/624-630
physical, see Physical rehabilitation
physiotherapy, 55/241
planning, 46/613
poststroke state, 55/233
posttraumatic epilepsy, 24/689

posttraumatic headache, 24/687
posttraumatic syndrome, 24/687, 46/629
prerequisite, 46/612-614
prognosis, 26/310, 46/611, 55/239
prognosis accuracy, 55/240
prognostic factor, 24/692
proprioceptive neuromuscular facilitation, 55/243
pulmonary complication, 41/493
quality of life, 67/401
reading disorder, 3/422-426
recovery graph, 55/238
recovery limit, 55/237
recovery mechanism, 55/240
recovery percentage, 55/237
recovery prognosis, 55/236
recovery time, 55/237
school, 41/481
scoliosis, 41/475
sensorimotor impairment, 24/690
service organization, 55/236
social sequel, 55/246
spasticity, 3/389, 24/691
spatial cognition, 46/619
speech disorder, 3/396-409
speech disorder in children, 4/412, 421, 423, 426, 433
speech recovery, 55/238
speech therapy, 12/475, 55/243
sphincter control, 55/239
spinal cord injury, 25/186, 192, 26/310, 521-540, 61/457
spinal deformity, 25/192
splint, 55/245
strength preservation, 41/463
stroke, 46/629
stroke team, 55/235
stroke ward, 55/234
surgery, 55/245
technical aid, 24/692
thoracic spinal cord injury, 61/81
thoracolumbar spine injury, 25/463
thoracolumbar vertebral column injury, 61/100
traumatic intellectual defect, 24/691
traumatic psychosyndrome, 24/687
traumatic spine deformity, 26/163, 169
vertebral column tumor, 20/819-845
visual function, 55/245
visual processing, 46/618
vocational, 24/692
walking recovery, 55/238
weight control, 41/464
Rehabilitation of higher cortical function, see
Restoration of higher cortical function

Rehabilitative sport
  aim, 26/523
  archery, 26/525, 538
  basketball, 26/534
  bowling, 26/531
  cardiovascular system, 26/525
  competitive, 26/527
  disability classification, 26/537
  discus throwing, 26/530
  facility, 26/537
  fencing, 26/532
  history, 26/521
  Indian club throwing, 26/531
  javelin throwing, 26/527, 529
  Olympics for the Paralysed, 26/528
  physiologic aspect, 26/523
  postural control, 26/524
  psychological aspect, 26/526
  social reintegration, 26/526
  spinal cord injury, 26/521-540
  Stokes-Mandeville games, 26/528, 539
  swimming, 26/527, 533
  table tennis, 26/525
  weight lifting, 26/527, 532
  wheelchair dash, 26/534
Reiber formula
  multiple sclerosis, 47/94
Reiber graph
  multiple sclerosis, 47/88
Reich granule, *see* Schwann cell
Reich μ-granule
  Schwann cell, 51/9
Reich π-granule
  Schwann cell, 51/9
Reinnervation
  experimental myopathy, 40/139
  motor unit, 62/62
  muscle, *see* Muscle reinnervation
Reiter disease
  aortitis, 63/55
  carotid system syndrome, 53/308
  classification, 71/4
  multiple sclerosis, 47/329
  nonrecurrent nonhereditary multiple cranial
    neuropathy, 51/571
  Tolosa-Hunt syndrome, 48/303
Reiterative agraphia
  type, 45/467
Rejection encephalopathy
  confusional state, 63/530
  epilepsy, 63/530
  headache, 63/530
  kidney rejection, 63/530

rejection crisis, 63/530
renal transplantation, 63/530
Relapsing alternating ophthalmoplegia
  Bielschowsky, *see* Bielschowsky relapsing
    alternating ophthalmoplegia
Relapsing aphthous uveitis, *see* Behçet syndrome
Relapsing fever, 33/409-414
  CSF, 33/414
  multiple myeloma, 20/11
  neurologic manifestation, 33/411-413
Relapsing neuropathy
  chronic, *see* Chronic relapsing neuropathy
Relapsing peripheral neuropathy
  paraneoplastic polyneuropathy, 51/467
Relapsing pneumonia
  multiple myeloma, 20/11
Relapsing polychondritis
  aorta aneurysm, 63/48
  classification, 71/4
  polyarteritis nodosa, 55/362
Relapsing polyneuritis
  chronic, *see* Chronic relapsing polyneuritis
  recurrent, *see* Recurrent relapsing polyneuritis
Relative afferent pupillary defect
  autonomic nervous system, 74/413
Relative hemianopia, *see* Sensory extinction
Relaxation
  arousal, 74/365
  benign essential tremor, 49/577
  coumarin, 74/365
  epilepsy, 73/444
  esophageal sphincter, 75/629
  lavender, 74/365
  muscle, *see* Muscle relaxation
  odor, 74/365
  odorant, 74/365
  vanillin, 74/365
Religious state
  epilepsy, 15/95
REM
  dreaming, 1/489, 3/88
  EEG, 57/190
  limb girdle syndrome, 62/185
  progressive supranuclear palsy, 22/219
REM behavior disorder
  epilepsy, 72/254
REM sleep
  autonomic nervous system, 74/524
  behavior disorder, 74/559
  cardiac dysrhythmia, 63/239
  central alveolar hypoventilation, 63/464
  central sleep apnea, 63/462
  chronic paroxysmal hemicrania, 48/261

cluster headache, 48/239
congenital central alveolar hypoventilation, 63/464
daytime hypersomnia, 45/131, 134
delirium, 46/533
depression, 46/473
diamorphine intoxication, 65/352
dopamine β-mono-oxygenase deficiency, 59/163
dyssynergia cerebellaris myoclonica, 21/514
EEG, 46/495
epilepsy, 15/458, 476, 482
heart rate, 74/538
infantile autism, 46/191
insomnia, 45/138
irregular respiration, 1/490
manic depressive psychosis, 43/208
migraine, 48/76
model, 74/528
narcolepsy, 1/491, 2/448, 42/712, 45/136, 149
narcoleptic syndrome, 42/710-713
non, see Non-REM sleep
obstructive sleep apnea, 63/458
paradoxical sleep, 3/66, 81, 88
penis erection, 74/560
physiology, 74/539
progressive dysautonomia, 59/158
progressive supranuclear palsy, 49/245
restless legs syndrome, 51/544
Rett syndrome, 63/437
schizophrenia, 46/495
sleep apnea syndrome, 45/134
sleep phase, 1/489-491
snoring, 63/450
somnolence, 45/131
striatonigral degeneration, 49/208
Remak cell vacuole
a-alphalipoproteinemia, 51/397
Remak fiber, see Unmyelinated nerve fiber
Remak, R., 1/8
Remitting peripheral neuropathy
paraneoplastic polyneuropathy, 51/467
Remodeling
myelin, see Myelin remodeling
Remote effect of cancer, see Cancer
Remyelination
see also Demyelination
acute multiple sclerosis, 47/245
Bassen-Kornzweig syndrome, 51/394
cerebrotendinous xanthomatosis, 51/399
chronic inflammatory demyelinating polyradiculoneuropathy, 51/536
chronic multiple sclerosis, 47/241-244
CNS, 47/41

experimental allergic neuritis, 47/457
hypertrophic interstitial neuropathy, 51/72
mucolipidosis type I, 51/380
multiple sclerosis, 47/222, 227, 239-248, 250
nerve conduction velocity, 47/20
nerve myelin, 51/35
paraplegia, 61/402
peripheral nervous system, 47/41
phi node, 47/43
Ranvier node, 47/42
spinal cord injury, 61/402
tetraplegia, 61/402
transverse spinal cord lesion, 61/402
Renal abnormality
Bardet-Biedl syndrome, 43/233
Bowen-Conradi syndrome, 43/335
brain aneurysm, 42/723
branchio-otorenal dysplasia, 42/361
cerebrohepatorenal syndrome, 43/340
congenital torticollis, 43/146
cryptophthalmos syndrome, 43/374
Edwards syndrome, 43/536
EEC syndrome, 43/396
Fanconi syndrome, 43/16
Gruber syndrome, 43/391
keloid, 43/146
Lenz syndrome, 43/419
Marden-Walker syndrome, 43/424
mental deficiency, 43/257, 276
10p partial trisomy, 43/519
prune belly syndrome, 43/466
4q partial monosomy, 43/496
4q partial trisomy, 43/498
11q partial trisomy, 43/524
tetraploidy, 43/567
thalidomide syndrome, 42/660
triploidy, 43/566
trisomy 22, 43/548, 50/564
Turner syndrome, 43/550
Renal acidosis
hearing loss, 22/488
recessive, see Recessive renal acidosis
Renal agenesis syndrome
see also Caudal aplasia
anal atresia, 50/512
autosomal dominant inheritance, 50/513
autosomal recessive inheritance, 50/513
axial mesodermal dysplasia spectrum, 50/516
caudal aplasia, 50/511
diaphragm hernia, 50/512
esophagus atresia, 50/513
hydrocephalus, 50/513
hypospadias, 50/512

incidence, 50/513
microcephaly, 50/512
MURCS syndrome, 50/515
omphalocele, 50/512
Potter syndrome, 43/469, 50/512
urethra atresia, 50/512
X-linked recessive inheritance, 50/513
Renal angiolipoma
tuberous sclerosis, 50/373
Renal angiomyolipoma
tuberous sclerosis complex, 68/292
Renal artery inflammation
temporal arteritis, 48/317
Renal blood flow
aging, 74/232
Renal carcinoma
vertebral column tumor, 20/43
vertebral metastasis, 20/423
Von Hippel-Lindau disease, 43/21, 50/374
Renal cyst
peroxisomal acetyl-coenzyme A acyltransferase
deficiency, 66/518
tuberous sclerosis, 14/50
tuberous sclerosis complex, 68/292
Von Hippel-Lindau disease, 16/31
Renal dialysis
see also Peritoneal dialysis
myoglobinuria, 62/556
neurotoxicity, 51/294
renal insufficiency, 27/341-344
uremic neuropathy, 7/554, 8/1
Renal disease
brain edema, 16/195
cerebellar ataxia, 13/308
chorioretinal degeneration, 13/40
chronic, see Chronic renal disease
diamorphine addiction, 55/517
diamorphine intoxication, 65/355
Goyer syndrome, 22/490
headache, 5/28
hereditary amyloid polyneuropathy type 1,
42/522, 60/95
hereditary progressive cochleovestibular atrophy,
22/487
Herrmann disease, 22/510
hyperprolinemia type I, 29/133
Lemieux-Neemeh syndrome, 22/522
Muckle-Wells syndrome, 22/490
multiple myeloma, 20/11
opiate intoxication, 65/355
optic atrophy, 13/40
periodic paralysis, 62/457
photogenic epilepsy, 42/714

polycystic, see Polycystic kidney and Polycystic
renal disease
secondary pigmentary retinal degeneration,
13/308
syncope, 72/263
Renal dysfunction
arsenic intoxication, 41/288
Kearns-Sayre-Daroff-Shy syndrome, 62/309
mitochondrial DNA depletion, 62/511
progressive external ophthalmoplegia, 62/309
Wilson disease, 49/229
Renal dysplasia
carnitine cycle, 66/402
congenital retinal blindness, 22/529
hypoparathyroidism, 70/121
mitochondrial matrix, 66/410
Renal failure, see Renal insufficiency
Renal glycosuria, 29/182-185
familial glycinuria, 29/187
genetics, 29/184
inheritance, 29/184
physiology, 29/183
treatment, 29/185
Renal hypertension
hypertensive encephalopathy, 54/212
Renal insufficiency
see also Uremia
acute mercury intoxication, 64/375
amaurosis, 27/234
arterial hypertension, 63/71
aspergillosis, 27/343
asterixis, 27/324
autonomic nervous system, 27/338
benign intracranial hypertension, 67/111
blood urea nitrogen, 63/503
brain injury, 23/115
burning feet syndrome, 27/338, 342
carnitine palmitoyltransferase deficiency, 62/494
central pontine myelinolysis, 27/343
childhood myoglobinuria, 62/561
chorea, 27/336
choroid plexus, 27/340
chronic, see Chronic renal failure
Chvostek sign, 27/336
convulsion, 27/330-332
creatinine clearance, 63/503
cryptococcosis, 27/343
CSF, 27/340
cystinosis, 29/179
deafness, 27/335
dialysis risk, 63/526
drug, 27/323
EEG, 27/326, 332-334, 340

encephalopathy, 27/320-330, 341
epilepsy, 73/377
Fabry disease, 51/377
fasciculation, 27/324-326, 336
glomerular filtration rate, 63/503
headache, 27/323
heat stroke, 23/674, 679
hypertension, 75/475
hypertensive encephalopathy, 11/554
infection, 27/343
Listeria monocytogenes, 27/343
loxosceles intoxication, 37/112
mental change, 27/322-330
mental status abnormality, 27/322-330
mucormycosis, 27/343
multiple myeloma, 20/15, 63/392
muscle stretch reflex, 27/227
muscular disorder, 27/336-338
myoclonus, 27/324-326, 330
myophosphorylase deficiency, 27/233, 41/186, 62/485
nerve conduction, 27/340
neurologic complication, 63/503
neurologic manifestation, 27/321-344
neuropathy, 27/338-340
nitrofurantoin intoxication, 65/487
nocardiosis, 16/217, 27/343
orthostatic hypotension, 63/151
paralysis, 27/336
penis erection, 75/89
primary amyloidosis, 51/415
renal dialysis, 27/341-344
sea snake intoxication, 37/92
sensorineural deafness, 42/389-391
serum creatinine concentration, 63/503
sleep, 74/553
snake envenomation, 65/179
tetany, 27/336
toxoplasmosis, 27/343
trace metal, 27/329
Trousseau sign, 27/336
tuberous sclerosis, 50/371
Wernicke encephalopathy, 27/324, 327
Wernicke-Korsakoff syndrome, 27/324
Renal insufficiency polyneuropathy
see also Uremic polyneuropathy
axonal degeneration, 51/358
creatinine clearance, 51/358
EMG, 51/358
myelin constituent, 51/358
myelin degeneration, 51/358
nerve conduction, 51/358
nitrofurantoin, 51/358
paranodal demyelination, 51/358
pathology, 51/358
Renal malformation
rachischisis, 32/172
spina bifida, 32/541
VATER syndrome, 50/513
Renal osteodystrophy
Wilson disease, 49/229
Renal stones, see Nephrolithiasis
Renal toxicity
copper intoxication, 37/623
manganese intoxication, 37/623
Renal transplantation
Aspergillus fumigatus, 63/531
candidiasis, 35/419
ciclosporin, 63/534, 537
CNS infection, 63/530, 532
Cryptococcus neoformans, 63/531
cytomegalovirus, 63/531
deferoxamine, 63/534
encephalitis, 63/532
epilepsy, 63/535
Epstein-Barr virus, 63/533
erythropoietin, 63/534
femoral neuropathy, 63/530
fever, 63/531
headache, 63/531
immunosuppression, 63/530, 532
infection risk factor, 63/532
leukopenia, 63/532
Listeria monocytogenes, 63/531
listeriosis, 52/91
lymphoma, 63/533
meningitis, 63/531
neuroaxonal dystrophy, 6/626
neurologic complication, 63/523
OKT3, 63/534, 536
opportunistic infection, 63/533
polysaccharide, 27/163
primary CNS lymphoma, 63/533
progressive multifocal leukoencephalopathy, 34/308, 47/506
rejection encephalopathy, 63/530
survey, 70/262
Toxoplasma gondii, 63/533
treatment complication, 27/342-344
uremic neuropathy, 8/1
Renal tubular acidosis
bicarbonate, 29/181
classification, 29/181
hereditary, see Hereditary renal tubular acidosis
hypomagnesemia, 63/562
type I, 29/181

type II, 29/181
Renal tubular dysfunction
  Crome syndrome, 43/242
  Lowe syndrome, 42/606
  sensorineural deafness, 42/390
  tyrosyluria, 29/214-217
Renaut body
  function, 51/2
  leprosy, 51/231
  leprous neuritis, 51/231
  nerve, 51/2
Rendu-Osler-Weber disease, *see* Hereditary
    hemorrhagic telangiectasia (Osler)
Renin
  migraine, 48/98
  Parkinson disease, 49/122
  physiology, 22/234
  spinal cord injury, 26/320, 325
Renin-angiotensin system
  hypertension, 75/472
  orthostatic hypotension, 63/145
Renofacial dysplasia
  corpus callosum agenesis, 50/163
Renovascular hypertension
  neurofibromatosis type I, 50/369
Renshaw, B., 1/50-52
Renshaw cell
  inhibition, 1/52
  muscle tone, 1/258
  spastic paraplegia, 61/367
  spinal cord injury, 61/367
  transverse spinal cord lesion, 61/367
Reovirus
  acute viral encephalitis, 56/141
  Colorado tick fever, 56/11
  human, 56/3
  neurotropism, 56/35
  orbivirus genus, 56/11
  Reye syndrome, 9/549, 27/363, 29/332, 34/169,
    49/217, 56/150
  RNA virus, 56/10
  rotavirus genus, 56/11
  taxonomic place, 56/3
Repetition
  frontal lobe tumor, 67/150
Reproductive behavior
  autonomic nervous system, 74/1, 14
Reproductive dysfunction
  spinal cord injury, 75/576
RES brain tumor, 18/233-263
  see also Hodgkin disease
  histiocyte, 18/233
  Kiel classification, 18/235

RES tumor
  classification, 18/235
  Hodgkin disease, 18/235, 259-263
  leukemia, 18/247-254
  lymphosarcoma, 18/235
  microglia, 18/233
  microglioma, 18/236-247
  multiple myeloma, 18/254-258, 20/9
  myeloma, 18/235
  reticulosarcoma, 18/236-247
  Waldenström macroglobulinemia polyneuropathy,
    18/237
Reserpine
  Achilles reflex, 41/245
  action mechanism, 37/436
  antihypertensive agent, 37/436, 63/87
  antihypertensive agent intoxication, 37/437
  chemical structure, 37/436
  Chiari-Frommel syndrome, 2/444
  choreoathetosis, 37/437
  depression, 46/601
  drug induced parkinsonism, 6/234
  headache, 48/72, 78
  hypertensive encephalopathy, 54/217
  migraine, 5/40, 37/437, 48/88, 197
  monoamine, 74/145
  muscle rigidity, 1/67
  neuroleptic agent, 65/275
  neuroleptic akathisia, 65/287
  neuroleptic parkinsonism, 6/235
  neuroleptic syndrome, 6/256
  Parkinson disease, 49/108
  pharmacology, 74/145
  rauwolfia alkaloid intoxication, 37/436-438
Reserpine intoxication
  breast cancer, 37/438
  drowsiness, 37/438
  gastrointestinal disorder, 37/437
  headache, 37/438
  hypotension, 37/438
  lethargy, 37/438
  mental depression, 37/438
  parkinsonism, 37/438
  sedation, 37/438
  spasmodic torticollis, 6/169
  vertigo, 37/438
  weakness, 37/438
Resitox intoxication, *see* Coumafos intoxication
Resorcinol reaction
  glycolipid, 9/26
Respiration, 1/650-680
  see also Central neurogenic hyperventilation,
    Hyperpnea, Respiratory dysfunction,

Respiratory system *and* Ventilation
acute hypoxia, 75/260
amyotrophic lateral sclerosis, 22/120, 296
apneustic, *see* Apneusis
apraxia, 63/481
artificial, *see* Artificial respiration
ataxic, *see* Ataxic respiration
autonomic dysfunction, 63/482
autonomic nervous system, 74/72, 581, 75/430, 568
Biot, *see* Biot respiration
brain cortex, 1/654, 63/481
brain edema, 23/156
brain embolism, 53/208
brain infarction, 53/208
brain injury, 23/230, 24/602, 57/129
breath holding ability, 63/481
capacity, 8/387
capnography, 1/657
cardiovascular function, 74/597
center, 1/651-653
central disorder, 1/650
central neurogenic hyperventilation, 1/496, 675, 57/129
cerebellar hemorrhage, 12/57
cerebellum, 1/655
cerebrovascular disease, 53/208, 75/375
cervical vertebral column injury, 25/288
chemical control, 1/492-494
Cheyne-Stokes, *see* Cheyne-Stokes respiration
cingulate gyrus, 63/481
clinical examination, 57/128
complete cord lesion, 26/255-257
congenital myopathy, 62/334
couplet periodic, *see* Couplet periodic breathing
couplet periodic breathing, 1/674
CSF, 1/655
definition, 1/657
dementia, 63/481
diaphragmatic, *see* Diaphragmatic respiration
diencephalon, 1/653
epilepsy, 15/140, 63/481
exercise, 8/395
expiratory ventral group, 63/480
forgotten, *see* Forgotten respiration
functional neuroanatomy, 74/531
gasping, *see* Gasping respiration
glossopharyngeal, *see* Glossopharyngeal breathing
glossopharyngeal nerve, 1/655
head injury, 57/125
heart rate, 74/581, 603
heaving, 1/676
hedgehog, *see* Hedgehog breathing

hedgehog breathing, 1/674
Hering-Breuer deflation reflex, 74/572
Hering-Breuer inflation reflex, 74/572
hippocampus, 63/481
history, 63/477
hyperventilation, 1/659, 675
inspiratory dorsal group, 63/480
intracranial hemorrhage, 53/208
irregular, 1/650
larynx, 74/392
mechanism, 1/655
medulla oblongata, 1/651
mesencephalon, 1/653
metabolic encephalopathy, 45/121
motoneuron disease, 59/463-466
neuroanatomy, 63/430, 477
neurologic disorder, 74/581
nucleus ambiguus, 63/480
nucleus retroambigualis, 63/480
nucleus tractus solitarii, 63/477
oligodendrocyte, 9/624
orbital gyrus, 63/481
paradoxical, *see* Paradoxical respiration
periodicity, 1/650, 12/57
physical rehabilitation, 61/459
pons, 1/652
pontomedullary organization, 63/430
postencephalitic parkinsonism, 63/481
posterior commissure, 1/653
posthyperventilation apnea, 1/496, 659
prognosis, 24/675
progressive bulbar palsy, 22/115, 59/218
progressive dysautonomia, 59/158
rate, 12/15, 39
registration, 1/657
regular, 1/676
respiratory center, 63/430
respiratory failure, 63/477
rhythm, 1/657
rhythm generation, 63/479
sighing, *see* Sighing respiration
sleep, 3/82, 74/529, 539
sobbing, *see* Sobbing respiration
spinal cord injury, 26/204, 253
spinal shock, 26/253
spontaneous, *see* Spontaneous respiration
stroke, 75/375
superior colliculi, 1/653
superior temporal gyrus cortex, 63/481
swallowing, 74/393
tardive dyskinesia, 63/482
transient ischemic attack, 53/208
transverse spinal cord syndrome, 2/202

traumatic vegetative syndrome, 24/580, 586
tremor, 63/482
uncus, 63/481
vagus nerve, 1/655, 74/619
vasoconstriction, 1/440
volume, 1/657
Respiration reflex
  coughing, 63/491
  hiccup, 63/490
  sneezing, 63/492
  vasoconstriction, 1/440
Respiratory acidosis
  acid base balance, 28/511
  barium intoxication, 64/354
  CSF, 28/511
  space occupying lesion, 16/241
  systemic brain infarction, 11/463
Respiratory alkalosis
  acid base balance, 28/512
  central neurogenic hyperventilation, 63/484
  Cheyne-Stokes respiration, 63/483
  CSF, 28/512
  encephalopathy, 45/121
  hyperventilation syndrome, 38/315-317, 63/434
  Reye syndrome, 27/363, 29/333, 335, 56/152
  tetany, 63/559
Respiratory arrest
  brain injury, 23/156
  lidocaine intoxication, 37/451
  quinidine intoxication, 37/449
  striatal necrosis, 49/500
Respiratory center
  afferent, 63/430
  alveolar/arteriolar J receptor, 63/431
  apneustic center, 63/432, 478
  carotid sinus body chemoreceptor, 63/431
  cerebellar anterior lobe, 63/431
  gasping center, 63/478
  Hering-Breuer reflex, 63/431
  Hering nerve, 63/431
  Kölliker-Fuse nucleus, 63/432
  medullary chemosensitive area, 63/431
  nucleus ambiguus, 63/431
  nucleus parabrachialis medialis, 63/432
  nucleus tractus solitarii, 63/431
  pneumotaxic center, 63/432, 478
  respiration, 63/430
  site, 63/478
  vagus nerve, 63/431
Respiratory chain
  defect, 66/420
  defect classification, 66/396
  mitochondrial disease, 66/395

mitochondrial myopathy, 41/211
Respiratory chain complex
  complex II, 66/422
  mitochondrial DNA, 66/414
Respiratory chain defect
  mitochondria, 42/587
  mitochondrial DNA disease, 66/421
  oxidative phosphorylation disorder, 66/422
  pellagra, 28/77
Respiratory disease
  bacterial meningitis, 52/1
  chronic obstructive pulmonary disease, 63/413
  hypoventilation, 63/413
  hypoventilation syndrome, 63/413
  neurologic complication, 63/413
  obstructive apnea syndrome, 63/413
  pickwickian syndrome, 3/96, 63/413
  respiratory encephalopathy, 63/413
Respiratory disorder
  posterior fossa hematoma, 57/266
  sleep apnea syndrome, 45/131-135
  sleep paralysis, 42/712
Respiratory distress
  aorta body tumor, 8/494
  apistus intoxication, 37/76
  childhood myoglobinuria, 62/561
  congenital acetylcholine receptor
    deficiency/synaptic cleft syndrome, 62/438
  congenital megacolon, 42/3
  gymnapistes intoxication, 37/76
  hydrozoa intoxication, 37/40
  hyperammonemia, 42/560
  nothesthes intoxication, 37/76
  pterois volitans intoxication, 37/75
  scorpaena intoxication, 37/76
  scorpaenodes intoxication, 37/76
  scorpaenopsis intoxication, 37/76
  sebastapistes intoxication, 37/76
  spinal muscular atrophy, 59/369
  stingray intoxication, 37/72
  thyroid gland dysgenesis, 42/632
Respiratory dysfunction
  see also Lung volume, Respiration and
    Respiratory system
  adult acid maltase deficiency, 62/480
  artificial, see Artificial respiration
  atelectasis, 26/343
  basilar impression, 32/74
  cerebellar agenesis, 50/190
  chronic, see Chronic respiratory insufficiency
  congenital fiber type disproportion, 62/356
  congenital myopathy, 62/334, 349
  cord lesion, 26/337

250

corpus callosum agenesis, 50/167
craniosynostosis, 50/125
cuirass respirator, 26/345
Duchenne muscular dystrophy, 40/360
failure, 2/253
glossopharyngeal breathing, 26/346
intracranial hypertension, 16/136
iron lung, 26/345
laboratory investigation, 26/342
late infantile acid maltase deficiency, 27/195,
    62/480
limb girdle syndrome, 40/437
multiple neuronal system degeneration, 75/171
myotonic dystrophy, 40/516
physiotherapy, 26/344
pneumobelt, 26/345
pneumonia, 26/345
radiology, 26/343
sign, 26/342
sleep, 74/563
spinal cord injury, 26/337, 342
thyrotoxic myopathy, 62/528
tracheostomy, 26/344
treatment, 26/343, 41/496
Z band plaque myopathy, 62/344
Respiratory encephalopathy, 38/285-299
acidemia, 63/414
aspartic acid, 63/416
asterixis, 63/414
benign intracranial hypertension, 63/414
brain adenosine triphosphate, 63/416
brain dysfunction, 63/416
brain edema, 16/196
brain hypoxia, 63/416
central alveolar hypoventilation, 63/415
chronic respiratory insufficiency, 63/413
clinical features, 38/288-290
confusion, 63/414
confusional state, 63/414
consciousness, 63/414
cor pulmonale, 63/414
cough, 63/414
CSF, 38/290
CSF pH, 63/414
delirium, 63/414
diaphragmatic pacing, 63/415
disorientation, 63/414
diurnal hypersomnia, 63/414
dyspnea, 63/414
EEG, 63/414
EEG change, 38/290
exercise, 63/415
fluid retention, 63/414

free radical, 63/416
glutamic acid, 63/416
headache, 63/414
hyperreflexia, 63/414
hypersomnia, 63/414
hypoxia, 63/416
intracranial hypertension, 63/414
intracranial pressure, 38/290, 298
lactic acid, 63/416
mental change, 38/288-290
metabolic encephalopathy, 63/414
myoclonic jerk, 63/414
myoclonus, 63/414
ocular sign, 38/290
papilledema, 63/414
pathogenetic, 38/291-298
polycythemia, 63/415
positive pressure ventilation, 63/415
primary alveolar hypoventilation, 63/414
pulmonary hypertension, 63/415
reperfusion injury, 63/416
respiratory disease, 63/413
right ventricular failure, 63/415
serum bicarbonate, 63/415
sexual impotence, 63/415
sleep, 63/414
sleep apnea syndrome, 63/415
treatment, 38/298
tremor, 63/414
venous pressure, 63/414
Respiratory failure
calcium antagonist intoxication, 65/442
carbamate intoxication, 64/190
cardiac, see Cardiorespiratory failure
central, see Central respiratory disturbance
chlormethine intoxication, 64/233
congestin intoxication, 37/37
conidae intoxication, 37/60
delirium, 46/542
Duchenne muscular dystrophy, 62/122, 133
hepatic encephalopathy, 63/477
hydranencephaly, 30/675, 50/349
marine toxin intoxication, 37/33, 65/153
multiple neuronal system degeneration, 59/159
multiple sclerosis, 9/181
neurologic intensive care, 55/204
nifedipine intoxication, 65/442
nimodipine intoxication, 65/442
olivopontocerebellar atrophy, 49/208
organophosphate intoxication, 37/555, 557,
    64/169
pallidal necrosis, 49/466, 479
pallidostriatal necrosis, 49/466

palytoxin intoxication, 37/55
paralytic shellfish poisoning, 65/153
progressive dysautonomia, 59/159
puffer fish intoxication, 37/80
respiration, 63/477
salicylic acid intoxication, 63/477
sea snake intoxication, 37/92
Shy-Drager syndrome, 49/208
syringobulbia, 50/445
tetrodotoxin intoxication, 65/156
Respiratory frequency
    hyperventilation syndrome, 38/333, 63/433
    odor, 74/367
Respiratory function
    altitude, 74/586
    autonomic nervous system, 74/571, 75/425
    chemical transmission, 74/585
    heart disease, 75/425
    hemoglobin level, 74/589
    peripheral chemoreceptor, 74/587
    sleep disorder, 63/449
    Wohlfart-Kugelberg-Welander disease, 59/87
Respiratory infection
    glycosaminoglycanosis type III, 66/299
    neuromyelitis optica, 9/428
    r(14) syndrome, 50/590
Respiratory insufficiency
    complex II deficiency, 66/422
    complex IV deficiency, 62/505, 66/424
    critical illness polyneuropathy, 51/576, 581, 584
    Guillain-Barré syndrome, 7/499, 51/255
Respiratory muscle
    progressive muscular atrophy, 59/14
    spinal cord injury, 26/336
Respiratory muscle paralysis
    breathlessness, 1/495
    porphyria, 60/118
Respiratory syncytial virus
    animal model, 56/431
    animal viral disease, 56/431
    distemper virus, 34/344
    paramyxovirus, 56/431
    Reye syndrome, 49/217, 56/150
Respiratory system
    see also Lung volume, Respiration and
        Respiratory dysfunction
    abdominal pressure, 26/339
    assessment, 26/341
    chest wall, 26/340, 346
    chironex intoxication, 37/49
    chiropsalmus intoxication, 37/49
    chrysaora intoxication, 37/51
    diborane intoxication, 64/359

embolism, 26/315, 345, 394
glycosaminoglycanosis type IH, 66/294
glycosaminoglycanosis type VI, 66/307
hereditary amyloid polyneuropathy, 60/106
intrathoracic pressure, 26/339
lung capacity, 26/337
lung compliance, 26/341
lung volume, 26/337, 341
myotonic dystrophy, 62/238
respiratory force, 26/339
spinal cord injury, 26/335-350
Respiratory tract disorder
    bee sting intoxication, 37/107
    chronic obstructive, see Chronic obstructive
        respiratory tract disorder
Respiratory virus infection
    anencephaly, 31/219
    congenital malformation, 31/219
    influenza, 31/219
    influenza A, 31/219
    spina bifida, 31/219
Responsive extrapyramidal syndrome
    conditional, see Hereditary paroxysmal
        kinesigenic choreoathetosis
Resting eye potential
    synonym, 13/175
Resting tremor
    anterolateral thalamic syndrome, 2/485
    cerebellar lesion, 49/585
    cerebellar tremor, 49/585
    definition, 49/583
    frequency, 49/584
    hereditary chin tremor, 49/585
    Parkinson disease, 42/253, 49/73, 584
    parkinsonism, 49/584
    paroxysmal tremor, 49/585
    superior cerebellar peduncle lesion, 49/588
Restless legs syndrome, 8/311-319, 51/543-548
    acroparesthesia, 8/311
    action myoclonus, 51/356
    acute intermittent porphyria, 8/316, 51/546
    acute poliomyelitis, 8/316, 51/546
    age at onset, 8/313, 51/545
    akathisia, 1/290, 8/314, 51/546, 65/286
    amyloidosis, 8/316, 51/546
    anemia, 8/315, 42/261, 51/545, 63/252
    autosomal dominant, 42/260, 45/140, 51/546
    baclofen, 51/548
    benzodiazepine, 51/548
    bromocriptine, 51/548
    carcinoma, 8/315, 51/546
    chronic renal failure, 51/356
    clinical features, 8/312, 51/543

clonazepam, 51/356, 548
definition, 51/543
dextran, 8/318, 51/548
diabetes mellitus, 42/261, 51/546
diagnosis, 8/314, 51/545
disease course, 51/545
etiology, 8/314, 51/546
gastrectomy, 8/315, 51/545
glyceryl trinitrate, 8/318, 42/261
growing pain, 8/313, 51/545
hereditary, 8/315, 42/260, 51/546
history, 8/311, 51/543
inositol nicotinate, 8/318, 51/548
insomnia, 8/313, 51/544
iron deficiency, 8/311, 315, 51/546
iron deficiency anemia, 8/315, 63/252
monozygotic twins, 8/315, 42/260, 51/546
muscular pain fasciculation syndrome, 51/545
myoclonus, 49/621, 51/356
niceritrol, 51/548
nicotinyl alcohol, 8/318, 51/548
nocturnal myoclonus, 38/584, 42/260, 45/139,
    49/621, 51/544
oxapropanium iodide, 8/318, 51/548
pain, 8/313, 42/260
painful legs moving toes syndrome, 42/260,
    51/545
paresthesia, 8/311, 317, 27/338, 42/260, 49/621
pathogenesis, 8/317, 51/547
peptic ulcer, 8/315, 51/546
polysomnography, 51/544
pregnancy, 8/316, 42/261, 51/546
prevalence, 51/544
psychological factor, 8/316, 51/547
relative incidence, 51/547
REM sleep, 51/544
sleep disorder, 42/260, 45/140, 49/621
tolazoline, 8/318, 51/548
treatment, 8/318, 51/548
triazolam, 51/548
uremia, 8/316, 51/545
uremic neuropathy, 27/338
uremic polyneuropathy, 51/356
vitamin B$_C$ deficiency, 51/336, 546
vitamin C, 8/318
Restlessness
    see also Agitation
    ascariasis, 35/359
    atlas syndrome, 5/373
    chrysaora intoxication, 37/51
    cocaine intoxication, 65/253
    digoxin, 63/195
    dysphrenia hemicranica, 5/79

juvenile Gaucher disease, 10/307
migraine, 48/161
pheochromocytoma, 39/500
presenile dementia, 42/286
Restoration of higher cortical function
    acetylcholine, 3/370
    afferent paresis, 3/388
    agraphia, 3/382, 384, 417
    anticholinesterase agent, 3/371-377
    apraxia, 3/389
    ataxia, 3/389, 395
    blocking nerve cell function, 3/368
    brain cortex damaged area replacibility, 3/379
    cerebral dominance, 3/379
    cerebral localization, 3/381
    cholinesterase, 3/371
    constructional apraxia, 3/428
    control of action, 3/385
    deblocking, 3/368
    deinhibition, 3/368-378, 389-391
    diaschisis, 3/369
    differential restoration, 3/383
    disturbance type, 3/368
    disturbed function transfer, 3/379
    dynamic localization, 3/381
    dysarthria, 3/396-399
    feedback, 3/385, 394
    flaccid paralysis, 3/389, 395
    functional asynapsia, 3/371
    functional system reorganization, 3/380-387,
        391-396
    graded resistance method, 3/390
    handedness, 3/379
    head injury, 3/370, 373
    inactivation, 3/368
    inhibition nerve cell function, 3/368
    intellectual disability, 3/426-430
    intelligence test, 3/426
    mechanography, 3/376
    motor aphasia, 3/400-409
    motor function, 3/387-396
    motor plan, 3/381, 387, 393
    movement plan, 3/387, 393
    movement synchronization, 3/394
    movement visual control, 3/394
    muscle relaxation, 3/390
    nature of function, 3/380
    neuropsychologic syndrome, 3/368-431
    oral apraxia, 3/397
    pedagogic method, 3/378
    performance simplifying condition, 3/386
    pharmacologic agent, 3/368
    pharmacologic deblocking treatment, 3/372

pneumography, 3/376
postconcussional deafness, 3/370
programming motor activity, 3/394
programming restorative activity, 3/383-385, 392, 395
reading disorder, 3/422-426
reflex circuit, 3/388
retraining, 3/369
semantic aphasia, 3/409-417
sensorimotor area lesion, 3/373
sensory aphasia, 3/409-417
skill, 3/388, 394, 429
spastic hemiparesis, 3/389, 392
speech disorder, 3/373, 375, 378, 396-417
speech therapy, 3/398, 400-417
synaptic conduction, 3/370
synkinesis, 3/389
taking advance of the intact link, 3/384
tensometric dynamometer, 3/395
tensometric goniometer, 3/394
transfer to opposite hemisphere, 3/378
trigger movement, 3/390
unconditioned reflex connection, 3/389
Wernicke-Mann posture, 3/389, 392
writing disorder, 3/417-422
Restorative neurology
cervical vertebral fracture, 61/65
discomplete lesion, 61/45
spinal recovery, 61/47
Restrictive infiltrative cardiomyopathy
amyloidosis, 63/137
hypereosinophilic syndrome, 63/372
sarcoidosis, 63/137
symptom, 63/137
transient ischemic attack, 63/137
Resuscitation
cardiac, see Cardiac resuscitation
cardiac arrest, 55/209
cervical vertebral column injury, 61/61
neurologic intensive care, 55/209
Retardations
growth, see Growth retardation
mental, see Mental deficiency
psychomotor, see Psychomotor retardation
speech, see Speech retardation
Rete mirabile
brain, see Moyamoya disease
carotid, see Carotid rete mirabile
internal carotid artery, 53/294
Retentio testis
prune belly syndrome, 50/515
Reticular activating system
ascending, see Ascending reticular activating

system
consciousness, 3/49, 53
diazepam intoxication, 37/203
ethosuximide intoxication, 37/203
mesuximide intoxication, 37/203
metharbital intoxication, 37/201
phenobarbital intoxication, 37/201, 203
primidone intoxication, 37/203
valproic acid intoxication, 37/203
Reticular cell sarcoma, see Reticulosarcoma
Reticular formation
akinetic mutism, 2/281
anterolateral funiculus, 45/233
attention, 3/138, 160, 168, 171, 176-179
brain death, 24/739-741
brain stem, see Brain stem reticular formation
brain stem death, 57/459
consciousness, 3/49, 53, 55, 171
dentatorubropallidoluysian atrophy, 49/442
emotion, 3/322
encéphalé isolé, 2/280
ethambutol polyneuropathy, 51/295
facilitation, 2/533
function, 1/72
hereditary sensory and autonomic neuropathy type III, 8/345, 21/111, 60/30
hyperkinetic child syndrome, 3/197
infantile spinal muscular atrophy, 59/62
mesencephalic, see Mesencephalic reticular formation
midbrain, see Midbrain reticular formation
pontine paramedian, see Pontine paramedian reticular formation
sleep, 3/81
sneezing, 63/492
vision, 2/533
Reticular reflex myoclonus, see Reflex myoclonus
Reticularis polaris nucleus
extralamellar thalamic nucleus, 2/473
Reticulin fiber
hemangioblastoma, 14/55
hemangioendothelioma, 18/288
Reticuloblastoma
classification history, 18/234
Reticulocerebellar degeneration
thalamus degeneration, 21/597
Reticulocyte
spinal cord injury, 26/380, 398
Reticulodentatocerebellar pigmentation
Divry-Van Bogaert syndrome, 14/514
Reticuloendothelial brain tumor, see RES brain tumor
Reticuloendothelial disorder

familial pallidoluysionigral degeneration, 6/665, 49/483
familial pallidonigral degeneration, 6/665, 49/483
pallidoluysionigral degeneration, 6/665
pallidonigral degeneration, 42/242
Reticuloendothelial system
  bacterial meningitis, 55/418
  brain vasculitis, 55/418
  Japanese B encephalitis, 56/30
  lactate dehydrogenase elevating virus, 56/30
  neurotropic virus, 56/30
  pallidoluysionigral degeneration, 49/445
  St. Louis encephalitis, 56/30
Reticuloendothelial tumor, see RES brain tumor and RES tumor
Reticuloendotheliosis
  Cervós-Navarro encephalitis, 18/241
  inflammatory, 18/241
  proliferative, 18/241
Reticulosarcoma
  acanthocyte, 49/332
  acanthocytosis, 63/290
  acquired toxoplasmosis, 35/124, 52/353
  brain, 18/236-238, 246, 257, 321, 340
  cranial vault tumor, 17/121
  epilepsy, 72/180
  hyperviscosity syndrome, 55/487
  IgA, 18/237
  IgM, 18/237
  microglioma, 39/526-529
  paraproteinemia, 18/235
  primary cerebral, see Primary cerebral reticulosarcoma
  RES tumor, 18/236-247
  skull base tumor, 17/169
  spinal cord compression, 19/368
  spinal epidural tumor, 20/112
  spinal nerve root, 9/596
  swine fever, 18/246
  vertebral column, 20/26, 33
  Waldenström macroglobulinemia, 18/237
Reticulosarcomatosis
  granulomatous inflammation, 18/241
  tumor, 18/241
Reticulosis
  brain, 18/246
  brain malignant, see Primary CNS lymphoma
  chronic meningitis, 56/645
  CNS, see CNS reticulosis
  lymphatic leukemia, 18/247
  malignant, see Midline granuloma
  median neuropathy, 56/179
  monocytic leukemia, 18/247

  neuropathy, 8/143
Reticuloxanthoma
  vertebral column, 20/26
Reticulum
  endoplasmic, see Endoplasmic reticulum
Reticulum cell sarcoma, see Reticulosarcoma
Retina
  anatomy, 2/506-510
  brain vasculitis, 55/381
  carotid cavernous fistula, 24/415
  cirsoid aneurysm, 14/70, 650
  congenital blindness, see Congenital retinal blindness
  embryology, 13/1-3
  GM2 gangliosidosis, 10/405
  gyrate atrophy, 13/256-260
  Kearns-Sayre-Daroff-Shy syndrome, 38/224, 229
  monochromatic photography, 13/6
  neurofibromatosis type I, 14/638
  periphlebitis, see Periphlebitis retinae
  rodless, see Rodless retina
  spreading depression, 48/66
  visual fiber, 2/547
  vitamin E deficiency, 70/448
Retina abnormality
  r(14) syndrome, 50/588
Retinal amaurosis
  congenital, see Congenital retinal blindness
Retinal angioma
  angiodysgenetic necrotizing myelomalacia, 20/487
  Bonnet-Dechaume-Blanc syndrome, 14/70, 31/200
  Von Hippel-Lindau disease, 14/115, 16/31
Retinal angiomatosis, see Von Hippel-Lindau disease
Retinal anlage tumor, see Pigmented papillary medulloblastoma
Retinal aplasia
  Alström-Hallgren syndrome, 13/460
  congenital retinal blindness, 22/537
  hereditary, see Congenital retinal blindness
Retinal artery
  central, see Central retinal artery
Retinal artery embolism
  amaurosis fugax, 55/107
  aorta stenosis, 63/25
  brain embolism, 53/218, 319
  brain fat embolism, 55/178, 187
  brain infarction, 53/218
  cerebrovascular disease, 53/218
  coronary bypass surgery, 55/110
  fat embolism, 55/178

features, 55/110
intracranial hemorrhage, 53/218
transient ischemic attack, 53/218
Retinal artery occlusion
anterior ischemic optic neuropathy, 55/112
blood dyscrasia, 55/112
cardiac myxoma, 55/112
cardiac valvular disease, 55/112
carotid angiography, 55/112
carotid endarterectomy, 55/112
cause, 55/112
central, see Central retinal artery occlusion
cherry red spot, 55/112
collagen vascular disease, 55/112
diabetes mellitus, 55/112
giant cell arteritis, 55/112
hypertension, 55/112
migraine, 55/112
mitral valve prolapse, 63/21
polyarteritis nodosa, 39/302, 55/355
treatment, 55/112
typical history, 55/112
Retinal astrocytoma
tuberous sclerosis complex, 68/292
Retinal blindness
congenital, see Congenital retinal blindness
ERG, 13/16
Retinal blood vessel
abnormality, 2/517
Retinal degeneration
Barnard-Scholz syndrome, 22/471
Batten disease, 10/575, 42/461, 463
Cockayne-Neill-Dingwall disease, 13/433
congenital, see Congenital retinal blindness
congenital hyperphosphatasia, 31/258
diaminophenoxyalkane, 13/208
evoked cortical response, 13/22
experimental pigmentary, see Experimental
pigmentary retinopathy
glyceraldehyde-3-phosphate dehydrogenase,
13/202
glycosaminoglycanosis type I, 10/441
hereditary spastic ataxia, 60/463
homocarnosinosis, 59/359
hypogonadal cerebellar ataxia, 21/468
iodoacetate, 13/207
Kjellin syndrome, 22/470, 42/173-175
malate dehydrogenase, 13/202
Mulibrey dwarfism, 43/435
myotonic dystrophy, 40/513
neuronal ceroid lipofuscinosis, 42/461, 463
Niemann-Pick disease type A, 42/469
olivopontocerebellar atrophy, see

Olivopontocerebellar atrophy with retinal
degeneration
ophthalmoplegia, 43/141
primary pigmentary, see Primary pigmentary
retinal degeneration
progressive external ophthalmoplegia, 43/141
rodless retina, 13/200
secondary pigmentary, see Secondary pigmentary
retinal degeneration
sensorineural deafness, 42/391
spastic paraplegia, 42/169, 173-175, 177-179
spinocerebellar ataxia, 42/188
Retinal detachment
congenital spondyloepiphyseal dysplasia, 43/478
Ehlers-Danlos syndrome, 43/15
hereditary, see Hereditary retinal detachment
hereditary progressive arthro-ophthalmopathy,
13/41
Norrie disease, 43/441
secondary pigmentary retinal degeneration,
13/218
spondyloepiphyseal dysplasia, 43/478
Stickler syndrome, 43/484
Retinal dot
β-naphthol, 13/224
Sjögren-Larsson syndrome, 59/358
Retinal dysplasia
congenital spondyloepiphyseal dysplasia, 43/478
hereditary, see Congenital retinal blindness
verrucose, see Verrucose retinal dysplasia
Retinal fleck
Alport syndrome, 60/770
Retinal fluorescence angiography
cerebrovascular disease, 11/250-252
clinical application, 11/251
complication, 11/251
diagnostic value, 13/225-231
equipment, 11/250
interpretation, 11/251
theory, 11/250
Retinal fluorescence photography
retinal vessel, 13/7
retinopathy, 13/7
Retinal function
evaluation, 13/9-23
Goldmann perimeter, 13/12
Harrington-Flocks test, 13/11
Retinal ganglion cell
amacrine cell, 2/511
bipolar diffuse type, 2/511
bipolar midget type, 2/511
cone cell, 2/507
density, 2/513

function, 2/512
Lafora body, 15/394
methanol intoxication, 36/356, 64/101
nerve cell, 2/517
response type, 2/513
rod cell, 2/507
type, 2/510
Retinal hamartoma
tuberous sclerosis complex, 68/292
Von Hippel-Lindau disease, 50/374
Retinal hemangioblastoma
fluorescein angiography, 14/645
histology, 14/57, 648
Von Hippel-Lindau disease, 14/13, 245, 248, 645,
43/21
Retinal hemorrhage
bacterial endocarditis, 52/299
cryoglobulinemic polyneuropathy, 51/434
hemolytic uremic syndrome, 63/325
thrombotic thrombocytopenic purpura, 63/325
Waldenström macroglobulinemia, 63/396
Retinal hypoplasia
lymphocytic choriomeningitis virus, 56/358
Retinal infarction
brain microangiopathy, 55/381
cardiac catheterization, 63/176
systemic lupus erythematosus, 55/381
Retinal ischemia
temporal arteritis, 48/318
Retinal migraine
amaurosis fugax, 55/111
classification, 48/6
Retinal nerve fiber
anatomy, 2/514
myelinated, see Myelinated retinal nerve fiber
optic neuritis, 47/54
papillomacular bundle, 2/548
temporal raphe, 2/547
topography, 2/547
Retinal phakoma
tuberous sclerosis, 43/49
Retinal pigmentary dystrophy
Bassen-Kornzweig syndrome, 63/272
deferoxamine, 63/534
Joubert syndrome, 63/437
McLeod phenotype, 63/289
Retinal pigmentation
Cockayne-Neill-Dingwall disease, 51/399
familial spastic paraplegia, 59/312
homocarnosinase deficiency, 42/558
phenothiazine, 13/211
r(14) syndrome, 50/589
Retinal receptor field

ganglion cell, 2/513
size, 2/513
Retinal spot
colloid body, 14/620
Retinal telangiectasia
Alport syndrome, 60/770
Retinal tumor
colloid body, 14/620
retinoblastoma, 14/622
tuberous sclerosis, 14/47
Von Hippel-Lindau disease, 31/16
Retinal vascular tortuosity
Bonnet-Dechaume-Blanc syndrome, 14/261
Retinal vasculitis
polyarteritis nodosa, 55/356
Sjögren syndrome, 71/86
Retinal vasculoglial proliferation
Norrie disease, 22/508
Retinal venous thrombosis
central, see Central retinal venous thrombosis
Retinal vessel tortuosity
congenital heart disease, 63/11
Waldenström macroglobulinemia, 63/397
Retinitis pigmentosa, see Primary pigmentary retinal
degeneration and Secondary pigmentary retinal
degeneration
Retinitis punctata albescens
Bassen-Kornzweig syndrome, 13/317
Friedreich ataxia, 21/347
primary pigmentary retinal degeneration, 13/28,
31, 235-239, 247
Retinoblastoma
adenovirus, 56/291
bilateral, 30/101
birth incidence, 18/339, 42/768
cat eye reflex, 42/768
child, 18/339
chromosomal aberration, 42/768
esterase D locus, 50/587
gene localization, 42/468
genetics, 42/468, 50/586
intraocular calcification, 42/768
mutation rate, 42/768
pseudo, see Pseudoretinoblastoma
13q interstitial deletion, 50/586
13q partial monosomy, 43/526
r(13) syndrome, 50/586
radiation, 42/768
retinal tumor, 14/622
spinal, see Spinal retinoblastoma
spinal cord compression, 19/361
tissue culture, 17/68
treatment, 50/587

trilateral, *see* Trilateral retinoblastoma
Retinocerebellar angiomatosis, *see* Von
  Hippel-Lindau disease
Retinography
  central serous, *see* Central serous retinopathy
  myotonic dystrophy, 40/514
Retinohypothalamic tract
  biological rhythm, 74/476
  hypothalamus, 74/478
  suprachiasmatic nucleus, 74/476
Retinoic acid
  adverse effect, 69/499
  benign intracranial hypertension, 67/111
All-*trans*-Retinoic acid, *see* Retinoic acid
Retinol, *see* Vitamin A
Retinopathy
  acquired immune deficiency syndrome, 56/491,
    494
  cancer, 71/696
  carbon disulfide intoxication, 64/26
  carnitine deficiency, 66/406
  chloroquine intoxication, 13/212, 65/483
  chlorpromazine, 13/223
  complex I deficiency, 66/422
  complex V deficiency, 62/506
  diabetes mellitus, 27/119
  drug induced, 13/204
  facioscapulohumeral muscular dystrophy, 62/168
  glycosaminoglycanosis type VI, 66/308
  hexachlorophene intoxication, 37/479, 65/472
  hypertensive encephalopathy, 47/536, 54/212
  hyperviscosity syndrome, 55/484
  iron deficiency anemia, 55/463
  Kearns-Sayre-Daroff-Shy syndrome, 62/311
  macroglobulinemia, 39/190
  neonatal adrenoleukodystrophy, 47/595
  paraneoplastic syndrome, 71/696
  pigmentary, *see* Pigmentary retinopathy
  polyarteritis nodosa, 39/302, 55/355
  progressive external ophthalmoplegia, 62/305, 311
  retinal fluorescence photography, 13/7
  rubella syndrome, 31/214
  spongiform leukoencephalopathy, 37/479
  Waldenström macroglobulinemia polyneuropathy,
    39/190
Retinoschisis
  hyaloidotapetoretinal degeneration
    (Goldmann-Favre), 13/276
  sex linked idiopathic, *see* Sex linked idiopathic
    retinoschisis
  sex linked juvenile, *see* Sex linked juvenile
    retinoschisis
Retinotopic organization

axodendritic interaction, 2/546
convergence, 2/546
geniculocalcarine tract, 2/557
lateral geniculate nucleus, 2/554
optic chiasm, 2/551
optic nerve, 2/548
optic tract, 2/554
point to point concept, 2/547
receptor field relationship, 2/546
visual pathway, 2/545
Retinotoxic agent
  experimental pigmentary retinopathy, 13/204
  secondary pigmentary retinal degeneration,
    13/219-225
  septojod, 13/205
  sodium iodate, 13/205
Retinovestibular hereditary spastic ataxia,
    21/383-387
  age at onset, 21/384
  alcoholic positional nystagmus, 21/387
  autosomal dominant, 21/387
  caloric nystagmus, 21/387
  dysarthria, 21/384
  ENG, 21/385
  macular degeneration, 21/386
  mental state, 21/386
  ophthalmoscopy, 21/386
  positional nystagmus, 21/387
  spasticity, 21/384
  vestibular dysfunction, 21/387
  vision, 21/384, 386
Retotheliosarcoma
  chromatin, 18/234
Retractile tumor
  oligodendroglioma, 18/95
Retraction
  eyelid, *see* Eyelid retraction
  lip, *see* Lip retraction
  upper eyelid, *see* Upper eyelid retraction
Retraction nystagmus
  brain embolism, 53/212
  brain infarction, 53/212
  cerebrovascular disease, 53/212
  intracranial hemorrhage, 53/212
  localization value, 2/345
  transient ischemic attack, 53/212
Retraction syndrome, *see* Acquired retraction
    syndrome *and* Stilling-Türk-Duane syndrome
Retroauricular artery
  radioanatomy, 11/66
  topography, 11/66
Retrobulbar neuritis
  acquired immune deficiency syndrome, 56/18

alcoholism, 43/197
Alport syndrome, 60/770
arsacetin, 9/648
arsphenamine, 9/648
beriberi, 7/562
Brucella, 51/185
chloramphenicol, 51/294
cisplatin intoxication, 64/359
cluster headache, 48/270
Cogan syndrome type II, 51/454
ethambutol, 65/481
ethambutol intoxication, 52/212, 65/481
hepatitis B virus infection, 56/303
hereditary motor and sensory neuropathy type I,
  21/284
hereditary motor and sensory neuropathy type II,
  21/284
multiple sclerosis, 9/195-198, 47/52, 54
pernicious anemia, 38/22
sarcoid neuropathy, 51/195
sarcoidosis, 51/195
spastic paraplegia, 42/178
styrene intoxication, 64/56
tryparsamide, 9/648
tuberculous meningitis, 52/212
Whipple disease, 52/138
Retrobulbar neuropathy
  anterolateral myelopathy, 7/643
  cyanide intoxication, 65/32
  malnutrition, 28/19, 24
  nutritional cerebellar degeneration, 28/272
  tobacco, 65/32
  tropical spastic paraplegia, 56/536
  vitamin B_2, 28/20
  vitamin B_2 deficiency, 51/332
  vitamin B_3 deficiency, 28/20
  vitamin B_{12} deficiency, 28/24
Retrobulbar optic neuritis
  pituitary adenoma, 17/412
Retrocollis
  dystonia musculorum deformans, 49/521
  neuroleptic dystonia, 65/281
Retroflexion
  see also Cervical hyperdeflexion injury
  head, see Head retroflexion
  lower cervical spine, 25/234, 237
  occipitoaxial connection, 25/230
  spinal injury, 25/134
  vertebral column, see Spine retroflexion
Retrognathia
  Miller-Dieker syndrome, 50/592
  r(9) syndrome, 50/580
Retrograde amnesia

bilateral lobectomy, 2/712
brain concussion, 23/432
brain contusion, 23/432
commotio cerebri, 23/432
head injury, 45/185
head injury outcome, 57/397
memory disorder, 2/712, 3/269, 294
migraine, 48/164
transient global amnesia, 45/209
Wernicke-Korsakoff syndrome, 45/197
Retrograde axonal degeneration, see Axonal
  reaction
Retrograde axonal transport
  neuroaxonal dystrophy, 49/394
  progressive bulbar palsy, 59/222
Retrograde brachial angiography
  brain aneurysm neurosurgery, 12/207
Retrograde cell reaction
  features, 21/53
Retrograde degeneration, see Axonal reaction
Retrograde reaction, see Axonal reaction
Retrohyperflexion injury, see Cervical
  hyperdeflexion injury
Retrolental fibroplasia
  mental deficiency, 46/28
Retrolisthesis
  definition, 25/243
  spinal injury, 25/243
Retromicrognathia
  9p partial monosomy, 50/579
Retro-orbital tumor
  skull base, 67/146
Retroparotid space
  skull base tumor, 17/182
  syndrome, see Villaret syndrome
Retroperitoneal fibrosis
  ergot intoxication, 65/70
  ergotamine, 65/70
  methysergide, 65/70
  migraine, 5/56
Retroperitoneal hemorrhage
  lumbosacral plexus, 51/165
Retropharyngeal syndrome, see Villaret syndrome
Retrovirus
  acquired immune deficiency syndrome, 56/18
  acute viral myositis, 56/194
  amphotropic type, 56/17
  animal viral disease, 56/453
  autonomic nervous system, 75/411
  biology, 56/454
  disease duration, 56/459
  ecotropic type, 56/17
  experimental acute viral myositis, 56/196

features, 56/16
gene therapy, 66/114
human, 56/3
human immunodeficiency virus, 56/490
human immunodeficiency virus type 1, 56/528
human immunodeficiency virus type 2, 56/528
human T-lymphotropic virus type I, 56/18, 525, 528
human T-lymphotropic virus type Ib, 56/529
human T-lymphotropic virus type II, 56/18, 529
human T-lymphotropic virus type III, 56/18
human T-lymphotropic virus type V, 56/529
Kawasaki syndrome, 52/264
lentivirus, 56/460, 490, 529
lentivirus classification, 56/460
lentivirus subfamily, 56/16
leukemia, 56/16
lymphoma, 56/16
multiple sclerosis, 47/323, 56/525
murine leukemia virus, 56/455
murine neurotropic, see Murine neurotropic retrovirus
neurofibromatosis, 56/525
neurotropism, 56/525
OKT4, 56/529
oncogene, 56/17
oncovirus, 56/529
oncovirus subfamily, 56/16
polymyositis, 56/202, 518, 62/381
replication, 56/16
RNA virus, 56/16
sarcoma, 56/16
spumavirus, 56/529
spumavirus subfamily, 56/16
T-helper lymphocyte, 56/529
T-lymphocyte leukemia, 56/529
taxonomic place, 56/3
transcription, 56/6
visna, 56/16, 453, 459, 529
visna-maedi disease, 56/16, 529
xenotropic, 56/17
Retrovirus infection
  amyotrophic lateral sclerosis, 59/203
  macaques, 56/453
  murine disease, 56/453
  simian disease, 56/453
  type D, 56/453
  visna, 56/583
Retrovirus vector
  brain tumor, 67/300
Rett syndrome, 29/305-328
  amimia, 29/328, 63/437
  amino acid metabolism, 29/319, 323
angiography, 29/317-319
apnea, 60/636, 63/437
apraxia, 29/308, 60/636
ataxia, 60/636, 63/437
autism, 60/636, 63/437
blood ammonia level, 29/323
brain atrophy, 29/305, 63/437
breath holding, 60/636, 63/437
bruxism, 60/636
case report, 29/305-316
clinical features, 29/305-316
cognitive impairment, 63/437
computer assisted tomography, 63/437
CSF homovanillic acid, 63/437
CSF 3-methoxy-4-hydroxyphenylethylene glycol, 63/437
cytogenic study, 29/319
dementia, 60/636, 665
developmental history, 29/306-310
diagnostic criteria, 60/637
EEG, 29/316, 60/637, 63/437
EMG, 29/317
epilepsy, 60/666, 63/437
episodic apnea, 63/437
episodic hyperpnea, 63/437
familial history, 29/305
gait apraxia, 60/637
genetics, 63/437
growth retardation, 60/637
hand writing, 29/305, 312, 60/636, 664, 63/437
Huntington chorea, 49/293
hyperammonemia, 29/305, 325, 60/637
hyperpnea, 63/437
hyperventilation, 29/315, 60/636, 667, 63/437
hypotrophic foot, 60/669
involuntary movement, 63/437
laboratory evaluation, 29/319-323
liver biopsy, 29/319
mask like face, 29/311
mental deficiency, 29/305, 60/636, 63/437
neurologic features, 29/310-316
nuclear magnetic resonance, 63/437
ornithine carbamoyltransferase deficiency, 29/325, 63/437
pathology, 29/324
PEG, 29/317-319, 63/437
psychiatric disorder, 60/665
psychomotor response, 29/315
REM sleep, 63/437
scoliosis, 29/312, 60/637
spasticity, 60/636
stereotypic hand movement, 29/305, 312, 60/636, 664, 63/437

symptom, 29/328, 63/437
treatment, 29/319-324
weight loss, 60/637
X-linked recessive ataxia, 60/636
Retzius, G.M., 1/12
Retzius papillae, *see* Status verrucosus simplex
Reverse cholesterol transport
high density lipoprotein, 66/549
Reversible angiopathy
headache, 48/287
Reversible ischemic neurological deficit, *see*
Transient ischemic attack
Reversible neuropathy
rapid, *see* Rapid reversible neuropathy
Reye encephalopathy, *see* Reye syndrome
Reye syndrome, 29/331-342, 56/149-169,
70/267-288
see also Cowpox encephalopathy *and* Jamaican
vomiting sickness
acetylsalicylic acid, 56/163, 238, 287, 63/438,
65/116, 122, 70/281
acquired hepatocerebral degeneration, 49/217
acute hemorrhagic pancreatitis, 56/159
acute hepatitis, 56/159
acute pancreatitis, 56/159
acute viral encephalitis, 34/80
adenovirus, 9/549, 56/9, 150
adenovirus encephalitis, 56/287-289
adenovirus infection, 56/288
adenovirus type 3, 29/332, 56/287
adenovirus type 7, 56/287-289
aflatoxin B, 49/217
aflatoxin B1, 56/167
aflatoxin intoxication, 29/338, 56/159
age, 29/332, 336, 49/217, 56/149, 163, 238
alanine aminotransferase, 56/151
aminoacidemia, 29/335, 56/152
ammonia, 9/550, 29/335, 338, 34/80, 168, 56/151,
238, 63/438
animal model, 56/167
apnea, 29/333, 34/170, 56/151, 63/438
arginine deficiency, 56/167
aspartate aminotransferase, 9/549, 56/151
ataxia, 63/438
Babinski sign, 31/168, 56/151
bacterial meningitis, 56/159
Balb/cByJ mice, 56/168
biochemical abnormality, 27/363
brain, 70/271
brain biopsy, 70/272
brain edema, 9/549, 27/361, 364, 29/331, 337,
31/168, 33/292, 34/80, 169, 49/373, 56/15, 153,
238, 65/115

brain hemorrhage, 31/168
brain perfusion pressure, 56/161
carnitine deficiency, 49/217
chickenpox, 31/168, 34/168-170, 37/534, 56/149,
151
cholesterol, 56/238, 65/118
clinical characteristic, 65/116
clinical features, 27/363, 29/332-334
clinical stage, 65/117
coma, 27/363, 29/331, 333, 31/168, 34/169,
49/217, 56/150, 151, 238, 65/115
computer assisted tomography, 56/153
consciousness, 31/168, 49/218, 56/236
convulsion, 9/549, 27/363, 29/331, 333, 31/168,
34/169
cowpox vaccine, 34/169, 56/150
Coxsackie virus A, 29/332, 34/169, 49/217,
56/150
Coxsackie virus B, 29/332, 34/169, 49/217,
56/150
creatine kinase, 29/335, 56/152, 159, 65/118
CSF, 56/153
CSF 2-oxoglutaramic acid, 27/364
CSF pressure, 29/335, 31/168, 34/169, 49/218
cytomegalovirus, 56/150
death rate, 9/549, 27/364, 29/332, 31/168, 34/169,
56/149, 164
decerebrate posture, 27/363, 56/150
decerebration, 56/238, 65/116
decline varicella zoster virus incidence, 56/162
decorticate posturing, 27/363, 65/116
decortication, 65/116
delirium, 27/363, 29/333, 31/168, 34/80, 46/544,
56/151, 238, 63/438, 65/115
dengue virus, 56/150
diagnosis, 56/160, 65/122
diagnostic criteria, 29/335
dicarboxylic acid, 65/118
differential diagnosis, 56/158, 65/123, 70/276
disseminated encephalomyelitis, 47/479
dopamine, 56/152
ECHO virus, 9/549, 29/332, 49/217
EEG, 29/335, 31/168, 34/169, 56/153
electrolyte, 56/152
encephalitic stage, 29/332-334, 65/116
encephalitis, 55/215, 56/159
encephalomyocarditis virus, 56/239
encephalopathic stage, 56/151, 63/438, 65/116
endotoxemia, 56/165
endotoxin, 56/153
epidemiology, 29/336, 34/80, 168, 169, 56/161,
65/121
epilepsy, 65/116

episodic apnea, 56/151
Epstein-Barr virus, 29/332, 49/217, 56/150
etiology, 27/363, 29/338-340, 34/169, 56/164
febrile illness, 65/115
fibronectin, 56/152
free fatty acid, 49/218, 56/238, 65/118
genetic, 70/277
glucagon, 56/152
glutamine, 56/153
growth hormone, 56/152
hallucination, 29/333, 65/116
hemiballismus, 49/373
hepatic encephalopathy, 27/361-363, 363, 49/217
hepatomegaly, 29/334, 31/168, 56/150
herpes simplex virus, 29/332, 49/217, 56/150
herpes zoster virus, 34/169
histopathology, 65/120
history, 29/331, 65/115
homovanillic acid, 56/153
hydrocortisone, 56/152
hyperammonemia, 21/578, 27/361, 363, 29/331,
   335, 338, 34/169, 49/218
hyperpnea, 31/168, 63/438
hyperventilation, 29/333, 34/169, 49/218, 56/151,
   238, 63/438, 65/116
hypoglycemia, 9/549, 27/363, 29/335, 34/80, 169,
   56/152, 65/118
hypoglycin intoxication, 37/514, 65/80
immune mechanism, 9/549
incidence, 27/363, 56/161
influenza, 29/332, 34/80, 65/116, 70/288
influenza A subgroup, 56/162
influenza A virus, 34/169, 56/149, 161, 238
influenza B virus, 29/332, 336, 34/169, 56/15,
   149, 161, 238
influenza B virus infection, 33/293, 34/80, 169,
   49/217, 56/149, 63/438
influenza virus, 56/239
intracranial hypertension, 55/215, 56/161, 65/116
Jamaican vomiting sickness, 29/338, 37/533,
   65/80
kidney, 29/332, 34/168, 56/157
kidney function, 65/118
laboratory finding, 29/334
lactate dehydrogenase, 29/334, 56/152, 65/118
lactic acid, 56/152, 238
lactic acidosis, 49/218
liver, 9/549, 29/332, 31/168, 33/292, 49/217,
   70/274
liver biopsy, 33/293, 56/150
liver cell ultrastructure, 56/157
liver function, 29/334, 34/168, 49/217, 56/151,
   238, 65/116

liver function test, 9/550, 29/333, 34/80, 49/217,
   56/151
liver histopathology, 56/155
mannitol, 29/341
measles virus, 29/332, 34/169, 49/217
medium chain acyl-coenzyme A dehydrogenase
   deficiency, 62/495, 66/410
mengovirus, 56/239
metabolic acidosis, 27/363, 29/333, 335, 56/152,
   238
metabolic ataxia, 21/578
mitochondria, 29/337, 65/119
mitochondrial disease, 66/395
mitochondrial enzyme theory, 65/134
mitochondrial function, 56/165, 238
morbidity, 34/169, 56/164
mortality, 31/168, 34/80, 169, 56/164, 238
mumps virus, 56/150
muscle biopsy, 31/168, 56/158
myalgia, 65/116
myocardium, 56/157
neurologic intensive care, 55/215
neurologic symptom, 63/438
neuropathology, 27/351, 56/153, 65/120
nonsteroid anti-inflammatory agent intoxication,
   65/415
noradrenalin, 56/152
octanoic acid, 56/167
octopamine, 56/153
opisthotonos, 27/363, 65/116
organic acid, 56/152
organic acid metabolism, 66/641
ornithine carbamoyltransferase deficiency, 27/363,
   29/339, 56/160
outcome, 65/124
papilledema, 29/333, 31/168, 34/169, 56/151
parainfluenza virus, 27/363, 29/332, 34/169,
   56/150
pathogenesis, 29/338-340, 56/164, 239
pathology, 29/336-338, 56/153-159
4-pentenoic acid, 29/338, 49/217, 56/167
penton protein, 56/165
periodic ataxia, 21/578
pertussis encephalopathy, 33/292
plasma lactic acid, 65/118
poliovirus, 34/169, 49/217
poliovirus type 1, 29/332
poliovirus vaccine, 56/150
polyamine derivative, 56/166
prodromal illness, 29/332, 56/151, 363
prognosis, 29/341, 49/217
prolactin, 56/152
prothrombin time, 56/152, 65/118

262

recovery, 65/117
reovirus, 9/549, 27/363, 29/332, 34/169, 49/217, 56/150
respiratory alkalosis, 27/363, 29/333, 335, 56/152
respiratory syncytial virus, 49/217, 56/150
rheumatoid arthritis, 56/163
rhinovirus, 34/169
rotavirus, 56/150
rubella vaccine, 56/150
rubella virus, 29/332, 49/217
salicylic acid, 29/336, 49/217, 56/163
salicylic acid intoxication, 29/331, 56/159, 65/135
season, 56/162
seizure, 31/168, 56/151, 153, 238
septicemia, 56/159
serum alanine aminotransferase, 29/334, 56/151, 238, 65/118
serum aspartate aminotransferase, 29/334, 31/168, 33/293, 34/169, 49/217, 56/151, 238, 63/438, 65/118
serum ketone, 65/118
sex ratio, 29/336, 56/163
short chain fatty acid, 29/335, 339, 56/152, 166
sign, 70/268
stage, 56/149, 65/115
survey, 70/267
symptom, 65/115
systemic lupus erythematosus, 56/163
thermoregulation, 75/60
toxic encephalopathy, 27/363, 49/217, 65/115
toximul, 56/168
toxin, 70/276
treatment, 27/364, 29/340, 56/160, 65/123
ultrastructure, 29/337, 65/120
upper respiratory infection, 56/238, 65/116
valproic acid, 49/217, 56/159
valproic acid induced hepatic necrosis, 65/115
varicella zoster virus, 27/363, 34/168, 56/149, 161, 229, 238
varicella zoster virus infection, 9/549, 33/293, 34/169, 49/217, 56/163, 238, 63/438
viral infection, 9/549, 34/80, 49/217, 56/150, 65/116, 134
virus, 70/268
visceral pathology, 27/350
vomiting, 9/549, 27/363, 29/332, 31/168, 34/80, 169, 56/150, 238, 63/438, 65/116
Reynolds classification
epilepsy, 15/5
Reynolds number
brain arteriosclerosis, 63/80
brain blood flow, 63/81
carotid artery, 11/217

Reynolds, R.
epilepsy classification, 15/5
Rh incompatibility
kernicterus, 42/583
pallidal necrosis, 49/466
Rhabdoid tumor
nervous system, 68/380
Rhabdomyolysis
see also Meyer-Betz disease and Myoglobinuria
alcohol intoxication, 40/339
alcoholic polyneuropathy, 40/162, 339
aminocaproic acid intoxication, 40/339
amphotericin B intoxication, 40/339
artificial respiration, 63/420
carnitine palmitoyltransferase deficiency, 40/264
chloroquine intoxication, 40/339
cocaine intoxication, 65/252, 256
Coxsackie virus, 56/200
creatine kinase, 40/339
differential diagnosis, 40/338
hornet venom, 64/16
hypokalemia, 63/557
hypophosphatemia, 63/565
influenza virus, 56/199
intensive care complication, 63/420
3,4-methylenedioxyethamphetamine intoxication, 65/367
3,4-methylenedioxymethamphetamine intoxication, 65/367
Meyer-Betz disease, 40/263, 340, 41/60
myoglobinuria, 62/553
myophosphorylase deficiency, 40/264
myositis, 62/370
opiate intoxication, 37/389
pathologic reaction, 40/263
pentaborane intoxication, 64/360
phosphofructokinase deficiency, 40/264, 342
plasmocid intoxication, 40/339
polymyositis, 41/60
snake envenomation, 65/179
succinylcholine intoxication, 40/339
thyrotoxic myopathy, 62/529
toxic myopathy, 62/600
vincristine intoxication, 40/339
Rhabdomyoma
brain embolism, 55/163
brain infarction, 55/163
cardiac, see Heart rhabdomyoma
cerebrovascular disease, 55/163
congenital, see Congenital rhabdomyoma
heart, see Heart rhabdomyoma
muscle tumor, 41/381
primary cardiac tumor, 63/93

Sturge-Weber syndrome, 14/69
tuberous sclerosis, 14/51, 371, 687, 50/371
Rhabdomyosarcoma
  adriamycin, 41/382
  child, 18/340
  cobalt intoxication, 36/330
  multiple nevoid basal cell carcinoma syndrome,
    14/81
  muscle tumor, 41/381
  nervous system, 68/380
  neurofibromatosis type I, 50/369
  skull base, 68/466
  skull base tumor, 68/490
  survey, 68/385
Rhabdovirus
  Duvenhage virus, 56/384
  human, 56/3
  lyssavirus genus, 56/13
  lyssavirus group, 56/383
  mokola virus, 56/384
  rabies virus, 56/383
  replication, 56/13
  RNA virus, 56/13
  taxonomic place, 56/3
  vesiculovirus genus, 56/13
Rheobase
  myotonia, 40/553
Rheoencephalography
  brain blood flow, 11/270
  brain scanning, 11/252
  cerebrovascular disease, 11/252
Rheumatic disease
  discitis, 19/329
  headache, 5/28, 157
  heart, 11/435
  leukoencephalopathy, 9/609
  migraine, 5/56
  spondylosis, 11/427
Rheumatic disorder
  Sjögren syndrome, 71/60
Rheumatic fever
  bacterial endocarditis, 52/289
  brain embolism, 55/163
  brain infarction, 55/163
  cardiac valvular disease, 63/17
  cerebrovascular disease, 55/163
  endocarditis, 63/111
  multiple sclerosis, 9/80
  subarachnoid hemorrhage, 55/3
  Sydenham chorea, 49/359
  systemic brain infarction, 11/454
Rheumatic heart disease
  anticoagulant complication, 63/19

anticoagulant risk, 63/19
anticoagulation, 63/18
anticoagulation timing, 63/19
aorta stenosis, 63/24
atrial fibrillation, 63/18
brain embolism, 11/387, 53/155, 63/17
brain infarction, 53/155
brain ischemia, 53/155
cardiac valvular disease, 38/143
embolism incidence, 63/17
embolus size, 63/18
infectious endocarditis, 63/23
ischemic heart disease, 63/18
pulmonary hypertension, 63/17
recurrent embolism, 63/18
systemic brain infarction, 11/435
Rheumatic valvular disease, see Rheumatic heart
    disease
Rheumatoid arthritis, 38/479-499
  amodiaquine, 13/220
  angiopathic polyneuropathy, 51/446
  ankylosing spondylitis, 19/330
  anterior spinal artery syndrome, 55/101
  antiphospholipid antibody, 63/330
  atlantoaxial dislocation, 19/321, 323, 38/492-496
  atlantoaxial subluxation, 38/492-496, 50/392,
    71/24
  autoimmune disease, 41/59, 76, 84, 240
  autonomic dysfunction, 71/20
  autonomic nervous system, 38/487
  autonomic polyneuropathy, 51/446, 479
  axonal degeneration, 51/446
  brain, 38/492-496
  brain infarction, 53/163
  brain ischemia, 53/163
  Brown syndrome, 71/19
  carpal tunnel syndrome, 7/294, 38/482, 71/18
  cervical myelopathy, 71/21
  chloroquine, 13/220
  chronic axonal neuropathy, 51/531
  classification, 71/4, 15
  CNS, 38/487-496, 71/20
  CNS inflammation, 71/28
  collagen vascular disease, 8/126
  compression neuropathy, 38/481-484, 71/17
  cryoglobulinemia, 39/181
  cubital canal syndrome, 38/483
  dermatomyositis, 62/374
  EMG, 38/485-487, 51/447
  epidemiology, 71/21
  femoral neuropathy, 71/18
  Guyon canal, 38/483
  hydrocephalus, 71/29

hyperviscosity syndrome, 55/485
inclusion body myositis, 62/374
juvenile, *see* Juvenile rheumatoid arthritis
juvenile spinal cord injury, 61/233
juvenile vertebral column injury, 61/233
laryngeal symptom, 38/487
Lhermitte sign, 71/23
lymphocyte, 41/76
Marie-Strümpell disease, 19/86
mononeuropathy, 51/446
mummy, 74/187
myasthenia, 71/29
myasthenia gravis, 41/240, 62/402
myopathy, 71/29
myositis, 38/484
neuromuscular, 38/479-487
neuropathology, 51/447
neuropathy, 7/555, 8/126, 71/15
orthostatic hypotension, 63/152
pathology, 8/126
peroneal nerve compression, 38/484
peroneal neuropathy, 71/18
polyarteritis nodosa, 8/126, 39/295, 308, 55/360
polymyositis, 38/484, 41/56, 59, 77, 62/374
polyneuropathy, 8/127, 38/479-481, 51/446, 71/16
posterior interosseous nerve syndrome, 38/484,
    71/18
pseudobasilar impression, 19/326, 332
radial nerve palsy, 38/484
Raynaud phenomenon, 43/71
Reye syndrome, 56/163
segmental demyelination, 51/446
sensory polyneuropathy, 51/447
Sjögren syndrome, 8/17, 71/74
spastic paraplegia, 59/439
spinal, *see* Spinal rheumatoid arthritis
spinal tumor, 19/86
systemic brain infarction, 11/454
Takayasu disease, 63/55
tarsal tunnel syndrome, 38/484, 71/18
temporal arteritis, 48/320
tibial nerve compression, 38/484
treatment complication, 8/128, 38/496-499
ulnar nerve, 38/483
ulnar neuropathy, 71/18
Rheumatoid disease
aortitis, 63/55
vertebral canal stenosis, 50/474
Rheumatoid neuropathy
nerve conduction, 7/169
Rheumatoid vasculitis, 71/8
Rhinencephalic hypoplasia
influenza, 9/582

Rhinencephalon
limbic system, 73/57
Rhinitis
epilepsy, 5/235
Rhinocerebral infection
heart transplantation, 63/179
iatrogenic neurological disease, 63/179
Rhinocerebral syndrome
neutrophil defense defect, 65/549
Rhinophonia aperta
athetosis, 6/441
Rhinophyma
alcoholic polyneuropathy, 51/317
Rhinopteridae intoxication
elasmobranche, 37/70
Rhinorrhea
see also CSF fistula
anosmia, 24/5
antibiotic agent, 24/188, 223
carbamate intoxication, 64/190
cluster headache, 48/9, 222
CSF fistula, 23/310, 485, 24/211
ethmoid bone fracture, 24/184
meningitis, 24/188, 220
penetrating head injury, 57/312
pituitary adenoma, 17/391
pneumococcal meningitis, 33/48, 52/52
posttraumatic meningitis, 57/317
rhizostomeae intoxication, 37/53
sinus osteoma, 24/186
skull base fracture, 24/186
skull base tumor, 24/186
sphenoid bone defect, 24/187
spontaneous, 24/186
staphylococcal meningitis, 33/69
streptococcal meningitis, 33/69
treatment, 24/223
Rhinovirus
picornavirus, 34/133, 56/309
Reye syndrome, 34/169
Rhizomelic chondrodysplasia punctata
biochemistry, 66/528
cerebrohepatorenal syndrome, 66/525
clinical comparison, 66/526
Conradi-Hünermann syndrome, 66/525
diagnosis, 66/530
form, 66/531
peroxisomal disease, 51/383
peroxisomal disorder, 66/510, 525
radiology, 66/527, 529
ultrastructure, 66/528
Rhizomelic chondrodysplasia punctata complex
clinical classification, 66/510

Rhizopus microsporus, *see* Mucormycosis
Rhizopus oryzae, *see* Mucormycosis
Rhizopus rhizopodiformis, *see* Mucormycosis
Rhizostomeae intoxication
  anxiety, 37/53
  asthma, 37/53
  depression, 37/53
  dyspnea, 37/53
  muscle weakness, 37/53
  pain, 37/53
  respiratory symptom, 37/53
  rhinorrhea, 37/53
  skin, 37/53
  urticaria, 37/53
Rhizotomy
  dorsal, *see* Dorsal rhizotomy
  pain, 26/497
  percutaneous thermal, *see* Percutaneous thermal
    rhizotomy
  spastic paraplegia, 61/369
  spasticity treatment, 61/369
  spinal cord injury, 61/369
  spinal spasticity, 61/369
  subtemporal, *see* Subtemporal rhizotomy
  transoccipital, *see* Transoccipital rhizotomy
  transverse spinal cord lesion, 61/369
  trigeminal, *see* Trigeminal rhizotomy
  trigeminal root, 5/303-306
Rhodactis howesi intoxication
  alcohol, 37/54
  antitoxin, 37/54
  neurotoxic, 37/54
  pulmonary edema, 37/54
  pupillary reflex abnormality, 37/54
  suicide, 37/54
  treatment, 37/54
Rhodanese
  cyanide intoxication, 65/28
  cyanogenic glycoside intoxication, 65/28
  thiocyanate formation, 65/28
Rhodnius prolixus
  Chagas disease, 75/387
Rhodopsin
  jaundice, 13/193
  liver disease, 13/193
  nyctalopia, 13/193
  primary pigmentary retinal degeneration, 13/193
  rod cell, 2/508
Rhodotorula
  fungal CNS disease, 52/492
Rhodotorulosis
  amphotericin B, 52/493
  ataxia, 52/493

  blindness, 52/493
  epistaxis, 52/493
  headache, 52/493
  multiple cranial neuropathy, 52/493
  neomycin, 52/493
  nystatin, 52/493
Rhombencephalitis, *see* Brain stem encephalitis
Rhombencephalon
  CNS, 50/2
Rhombencephaloschisis
  cerebellar vermis, 50/188
Rhombencephalosynapsis
  cerebellar agenesis, 50/188
  Kundrat syndrome, 30/384
Rhomboschisis, *see* Cerebellar agenesis
Rhythm
  alpha, *see* Alpha rhythm
  annual, *see* Annual rhythm
  biological, *see* Biological rhythm
  circadian, *see* Circadian rhythm
  circannual, *see* Circannual rhythm
  diurnal, *see* Diurnal rhythm
  musical alexia, 45/486
  respiration, 1/657
  seasonal, *see* Seasonal rhythm
  theta, *see* Theta rhythm
Rhythmias
  anatomic classification, 1/279
  involuntary movement, 1/290
Rhythmic sense
  amusia, 4/196, 45/486
  language, 4/196
Rhythmic stabilization test
  Kabat, 2/421
  neocerebellar syndrome, 2/421
Rib
  bifid, *see* Bifid rib
  cervical, *see* Cervical rib
  supernumerary, *see* Supernumerary ribs
Rib abnormality
  anencephaly, 50/84
  cerebrocostomandibular syndrome, 43/237
  congenital torticollis, 43/146
  craniolacunia, 50/139
  Klippel-Feil syndrome, 42/37
  methotrexate syndrome, 30/122
  multiple nevoid basal cell carcinoma syndrome,
    14/114
  Sprengel deformity, 43/483
Rib cage
  lung volume, 26/340, 346
  spinal cord injury, 26/338, 346
Rib fracture

spinal injury, 26/190, 192
Rib notching
  neurofibromatosis type I, 14/158
Rib resection
  thoracic outlet syndrome, 51/128
Rib syndrome
  cervical, *see* Cervical rib syndrome
  first, *see* First rib syndrome
  thoracic, *see* Thoracic rib syndrome
Rib synostosis
  multiple nevoid basal cell carcinoma syndrome,
    14/459
Riboflavin, *see* Vitamin B2
Ribosomal protein synthesis test
  Becker muscular dystrophy, 41/411
  Duchenne muscular dystrophy, 41/411
Ribosomal S6 protein
  muscle tissue culture, 62/90
Ribosomal system
  Huntington chorea, 49/325
Ribosome
  muscle, 40/3, 67
  muscle fiber, 40/33
  muscle fiber type, 62/4
  protein, 66/80
Rice field fever, *see* Leptospirosis
Richards-Rundle-Sylvester syndrome, *see*
    Richards-Rundle syndrome
Richards-Rundle syndrome
  amenorrhea, 43/264
  3-amino-2-methylproprionic acid, 22/516
  amyotrophy, 22/516
  areflexia, 43/264
  ataxia, 22/516, 43/264
  autosomal recessive transmission, 22/516, 43/265
  claw hand, 43/264
  cochlear spiral ganglion degeneration, 21/359
  deafness, 22/516, 43/264, 60/657
  Friedreich ataxia, 21/359, 43/264
  hereditary hearing loss, 22/516
  hypogonadal cerebellar ataxia, 21/468, 42/129
  hypogonadism, 21/468, 22/516, 43/264, 60/668
  hyporeflexia, 43/264
  Jéquier-Deonna variant, 22/524
  kyphoscoliosis, 43/265
  mental deficiency, 22/516, 43/264
  muscle weakness, 22/516, 43/265
  muscular atrophy, 43/264
  neuropathology, 22/524
  nystagmus, 22/516, 43/264
  pes cavus, 43/264
  pes equinovarus, 22/516, 43/264
  pes planus, 43/264

Roussy-Lévy syndrome, 21/175, 22/524, 43/264
  sensorineural deafness, 43/264
  status epilepticus, 43/264
  talipes, 22/516, 43/264
Richardson-Steele-Olszewski syndrome, *see*
    Progressive supranuclear palsy
Richner-Hanhart syndrome
  autosomal recessive inheritance, 42/581
  Babinski sign, 42/581
  cataract, 42/581
  corneal dystrophy, 42/581
  cytosol tyrosine aminotransferase, 42/581
  diet, 42/582
  electron microscopy, 42/581
  histopathology, 42/581
  hyperkeratosis, 42/581
  mental deficiency, 42/581
  microcephaly, 42/581
  myopia, 42/581
  nystagmus, 42/581
  oculocutaneous tyrosinosis, 42/582
  palmoplantar keratosis, 42/581
  self-mutilation, 42/581
  talipes, 42/581
  tic, 42/581
  ultrastructure, 42/581
Rickets
  basilar impression, 50/399
  hyperparathyroid myopathy, 62/538
  hypocalcemia, 63/558
  Lowe syndrome, 42/606
Rickettsia, 34/641-643
  infectious endocarditis, 52/289, 302
  multiple sclerosis, 9/682
  Weil-Felix reaction, 34/643
Rickettsia encephalitis
  dementia, 46/389
Rickettsial infection, 34/641-649
  African tick-borne typhus, 34/657
  animal, 9/682
  Brill-Zinsser disease, 34/645
  classification, 34/642
  clinical features, 34/644
  eastern hemisphere tick-borne typhus, 34/657
  endemic typhus, 34/646
  epidemiology, 34/643
  flea-borne typhus, 34/646
  Indian tick-borne typhus, 34/657
  Kenyan tick-borne typhus, 34/657
  leukoencephalopathy, 9/609
  louse-borne typhus, 34/644-646
  mite-borne typhus, 34/647
  murine typhus, 34/646

neurologic manifestation, 34/648
North Asian tick-borne typhus, 34/657
Queensland tick-borne typhus, 34/657
recrudescent typhus, 34/645
Rocky Mountain spotted fever, 34/651-656
Russian tick-borne typhus, 34/657
scrub typhus, 34/647
tick bite intoxication, 37/111
tick-borne typhus, 34/651-658
Rickettsiosis, see Rickettsial infection
Ridge meningioma
sphenoidal, see Sphenoidal ridge meningioma
Ridley-Jopling classification
leprosy, 51/217
Rieger syndrome
autosomal dominant inheritance, 43/471
cataract, 43/470
dental abnormality, 43/470
glaucoma, 43/470
goniodysgenesis, 43/470
hypertelorism, 30/247
iris dysplasia, 30/247, 43/471
ocular defect, 30/247
periumbilical skin defect, 43/470
Riese, W., 1/3
Rifampicin
acute amebic meningoencephalitis, 52/329
bacterial meningitis, 52/13
encephalopathy, 46/602
gram-negative bacillary meningitis, 52/121
leprosy, 51/217
leprous neuritis, 51/232
Listeria monocytogenes meningitis, 52/95
meningococcal meningitis, 33/29, 52/32
neurobrucellosis, 52/596
toxic myopathy, 62/597, 609
tuberculous meningitis, 33/228, 52/211
Rifampin, see Rifampicin
Rifamycin
leprous neuritis, 51/232
Rifle sling palsy
compression neuropathy, 51/92
neuralgic amyotrophy, 51/174
Rift Valley fever
hydranencephaly, 50/340
Right hemisphere
arousal, 45/173
intention, 45/173
language, 45/304
motor sequence stimulation, 15/82
neglect syndrome, 45/173
Right left discrimination
body scheme, 4/221

body scheme disorder, 4/219-222, 395
Right left disorientation
autotopagnosia, 45/382
developmental dyslexia, 4/401
Gerstmann syndrome, 1/107, 2/687, 45/70, 383
spatial agnosia, 2/604
testing method, 2/688
Wernicke aphasia, 4/98
Right perisylvian region
aprosody, 45/45
Right sided aortic arch
mental deficiency, 43/296
Right temporal lobe
affective psychosis, 46/504
Right thalamus
minor hemisphere syndrome, 45/88-90
Righting reflex
component, 1/255
description, 1/264
neck, see Neck righting reflex
type, 1/68
Rigid Huntington chorea
age distribution, 6/391
clinical picture, 49/287
homovanillic acid, 49/285, 295
juvenile case, 6/321
muscle tone, 49/69
neuropathology, 49/285, 321
spastic, see Creutzfeldt-Jakob disease
Westphal-Strümpell, see Creutzfeldt-Jakob disease
Westphal type, 6/323-329
Rigid spine syndrome
congenital fiber type disproportion, 62/355
Emery-Dreifuss muscular dystrophy, 40/390,
62/155
EMG, 62/173
family, 62/173
heredofamilial myosclerosis, 62/173
muscle fibrosis, 62/173
myosclerosis, 62/173
reducing body myopathy, 62/348
Rigid syndrome
akinetic, see Akinetic rigid syndrome
Rigidity
see also Striatal syndrome
adult neuronal ceroid lipofuscinosis, 42/466
alpha, see Alpha rigidity
amyotrophic lateral sclerosis, 59/235
benzodiazepine intoxication, 65/339
Carter-Sukavajana syndrome, 21/412
cataract, 60/653
central motor disorder, 1/170
central pontine myelinolysis, 2/263

cerebellar catalepsy (Babinski), 21/521
chronic bromide intoxication, 36/300
chronic mercury intoxication, 64/370
cogwheel, *see* Cogwheel rigidity
congenital myopathy, 62/335
Creutzfeldt-Jakob disease, 46/291
cyanide intoxication, 64/232
decerebrate, *see* Decerebrate rigidity
decortication, 1/268
dentatorubropallidoluysian atrophy, 21/520
diazoxide intoxication, 37/433
Divry-Van Bogaert syndrome, 55/319, 321
dominant olivopontocerebellar atrophy, 60/661
dopamine, 49/70
EMG, 1/646
fasciculation, 60/659
gamma, *see* Gamma rigidity
globus pallidus, 49/71
globus pallidus internus, 21/520, 526
Hallervorden-Spatz syndrome, 49/399, 66/712, 714
hereditary spastic ataxia, 60/661
hereditary striatal necrosis, 49/495
homovanillic acid, 49/70
Huntington chorea, 49/69
hyperactive alpha system, 6/180
hyperactive gamma system, 6/180
hypercalcemia, 63/561
hyperkinesia, 6/851
hypernatremia, 63/554
internal pallidal segment, 49/457
Joseph-Machado disease, 42/261
juvenile parkinsonism, 49/155
kwashiorkor, 7/635, 51/322
Lafora progressive myoclonus epilepsy, 15/386, 27/173
late cerebellar atrophy, 60/661
late onset ataxia, 60/659-661, 670
lissencephaly, 30/486
long latency reflex, 49/70
manganese intoxication, 64/310
Minamata disease, 36/123, 64/414
multiple neuronal system degeneration, 75/162
muscle, *see* Muscle rigidity
neck, *see* Nuchal rigidity
nerve conduction velocity, 60/670
neuroleptic parkinsonism, 65/292
neuropathy, 60/659
neurosurgery, 6/851
nuchal, *see* Nuchal rigidity
olivopontocerebellar atrophy, 49/69
pallidal necrosis, 49/465
pallidoluysian degeneration, 49/455

pallidoluysionigral degeneration, 49/446, 456
pallidonigral necrosis, 49/465
pallidostriatal necrosis, 6/657, 49/465
Parkinson dementia, 49/171
Parkinson dementia complex, 49/69
Parkinson disease, 49/54, 69, 91, 102
parkinsonism, 1/67, 266, 6/184
Parry disease, 42/467
pathophysiology, 6/101, 49/69
plastic, *see* Plastic rigidity
polyneuropathy, 60/660, 670
progressive pallidal atrophy (Hunt-Van Bogaert), 6/457, 554, 21/531, 42/247
progressive supranuclear palsy, 22/219, 46/302, 49/240
propanidid intoxication, 37/409
pure pallidal degeneration, 49/447
putamen, 49/70
spasticity, 6/851
striatal necrosis, 49/502
striatonigral degeneration, 49/69, 206
supranuclear palsy, 49/69
tetanus, 65/217
thalamic ventral oral posterior nucleus, 49/600
thalamotomy, 6/185
type, 1/170
water intoxication, 64/241
Wilson disease, 49/69
Woods-Schaumburg variant, 22/170
RIHSA
  brain scanning, 16/664, 669
  brain tumor, 18/29
  myeloscintigraphy, 19/247-255
  porencephaly, 23/307
RIHSA cisternography
  brain injury, 23/301, 304
  brain scintigraphy, 23/301, 304
RIHSA content
  brain edema, 16/173
Riley-Day syndrome, *see* Hereditary sensory and autonomic neuropathy type III
Riley, H.A., 1/14, 35
Riley-Smith syndrome, 14/480-486
  clinical picture, 14/480-482
  diagnosis, 14/486
  genetics, 14/76, 483
  Krabbe-Bartels disease, 14/83
  macrocephaly, 14/14, 76, 83, 480, 768, 43/30
  multiple cutaneous hemangioma, 14/14, 76, 480, 768, 43/30
  pathogenesis, 14/482
  pseudopapilledema, 14/14, 76, 480, 486, 768, 43/30

Rilmenidine
  blood pressure, 74/171
Rimmed muscle fiber vacuole
  acid maltase deficiency, 41/178
  autophagic vacuole, 40/231
  carnitine deficiency, 41/197, 199
  chloroquine, 40/151, 339, 438
  debrancher deficiency, 41/182
  dermatomyositis, 62/376
  distal myopathy, 40/475, 62/204
  glycogen accumulation, 40/36, 98, 240
  inclusion body myositis, 62/376
  lipid accumulation, 40/95, 242
  multiple hydropic vacuole, 40/230
  myopathy, 40/153, 229
  oculopharyngeal muscular dystrophy, 40/253,
    62/297
  periodic paralysis, 40/29, 106, 219, 41/153, 160,
    62/465
  polymyositis, 62/204, 376
  red rimmed vacuole, 40/253
  subsarcolemmal, 41/187, 191
  Wohlfart-Kugelberg-Welander disease, 62/200,
    204
Rimmed vacuole distal myopathy
  acid maltase deficiency, 62/481
  chloroquine, 40/151
  clinical course, 62/352
  congenital myopathy, 62/332
  course, 62/202
  creatine kinase, 62/204
  debrancher deficiency, 62/483
  distal myopathy, 62/199
  EMG, 62/204
  glycogen storage disease, 62/490
  histochemistry, 62/352
  histopathology, 62/204
  myophosphorylase deficiency, 62/485
  symptom, 62/202
Ring fiber
  muscle fiber, 40/27, 74, 141
  myotonic dystrophy, 40/497
Ring scotoma
  glaucoma, 2/565
  hyaloidoretinal degeneration (Wagner), 13/274
  hyaloidotapetoretinal degeneration
    (Goldmann-Favre), 13/276
Ringbinden, see Subsarcolemmal myofibril
Rinne test
  deafness, 2/74
  examination method, 2/74
  mastoid process, 2/74
  tuning fork test, 2/74

Rio de Janeiro, Brazil
  Neurological Institute, 1/12, 34
Riolan nosegay
  anatomy, 48/505
Riolan nosegay syndrome
  stylohyoid syndrome, 48/504
Risus sardonicus
  emotional disorder, 3/357
  pathologic laughing and crying, 3/357, 45/220
  strychnine intoxication, 41/305
  tetanus, 41/304, 52/230, 65/217
River ray intoxication
  elasmobranche, 37/70
RNA
  amyotrophic lateral sclerosis, 59/177
  DNA, 30/95
  memory, 3/262
  messenger, see mRNA
  synthesis, 9/9
RNA splicing
  mRNA, 66/73
RNA virus
  bunyavirus, 34/5, 56/15
  coronavirus, 56/12
  human, 56/3
  orbivirus, 56/10
  orthomyxovirus, 56/15
  paramyxovirus, 56/13
  picornavirus, 34/4, 133, 56/3, 11, 309, 62/381
  poliovirus, 34/95
  reovirus, 56/10
  retrovirus, 56/16
  rhabdovirus, 56/13
  rotavirus, 56/10
  Togavirus, 56/11
  transcription, 56/6
  transcription strategy, 56/5
  virology, 34/4-7
  virus taxonomy, 56/2
Ro 3-0658 intoxication, see Diethyl
    trichloromethylphosphonate intoxication
Roberts syndrome
  cheilopalatoschisis, 30/247
  EEG, 46/80
  hydrocephalus, 30/100
  hypertelorism, 30/247
  mental deficiency, 46/79
  phocomelia, 30/247, 46/80
Robin syndrome
  cheilopalatoschisis, 43/471
  feeding disorder, 43/471
  glaucoma, 43/472
  glossoptosis, 43/471

hydrocephalus, 43/472
mental deficiency, 43/472
microcephaly, 43/472
micrognathia, 43/471
Möbius syndrome, 43/472
11q partial trisomy, 43/524
Stickler syndrome, 43/484
Robinov-Silverman-Smith syndrome, *see* Fetal face
syndrome
Robinow syndrome, *see* Fetal face syndrome
Robotic
surgery, 67/260
Robustotoxin
arthropod envenomation, 65/195
arthropod venom, 65/195
neurotoxin, 65/194
Rochon-Duvigneau syndrome, *see* Orbital apex
syndrome
Rocio virus
flavivirus, 56/12
Rockerbottom foot
Edwards syndrome, 50/275, 561
hereditary motor and sensory neuropathy type I,
21/277
hereditary motor and sensory neuropathy type II,
21/277
9p partial monosomy, 50/579
Pena-Shokeir syndrome type I, 43/437
Rocky Mountain spotted fever
meningitis rash, 52/25
rickettsial infection, 34/651-656
tick-borne typhus, 34/651-657
Rocuronium
intensive care, 71/536
multiple organ failure, 71/536
Rod body
see also Nemaline myopathy
electron micrography, 40/232
electron microscopy, 40/237
histochemistry, 40/24, 237
myopathy, *see* Nemaline myopathy
nemaline myopathy, 40/259, 287
tenotomy, 40/144, 240, 287
ultrastructure, 40/87
Z band, 40/240
Rod myopathy, *see* Nemaline myopathy
Rodless retina
inherited, 13/200
mouse, 13/200
retinal degeneration, 13/200
Rodriguez-Arias, B., 1/35
Roentgen radiation, *see* Radiation
Rohypnol, *see* Flunitrazepam

Rokitanski-Van Bogaert syndrome, *see*
Neurocutaneous melanosis
Rolandic spike
benign partial epilepsy with centrotemporal spike,
73/14
brain mapping, 73/17
epilepsy, 73/15
Rolando artery
radioanatomy, 11/86
spinal cord vascularization, 12/486
topography, 11/86
Rolando vein
brain phlebothrombosis, 54/398
radioanatomy, 11/106
superior longitudinal sinus occlusion, 54/397
Rolicyclidine
designer drug, 65/364
Rollet syndrome, *see* Orbital apex syndrome
Romana sign
Chagas disease, 75/390
Romano-Ward syndrome, *see* Hereditary long QT
syndrome
Romberg, M.H., *see* Von Romberg, M.H.
Romberg sign, *see* Von Romberg sign
Rome, Italy
neurology, 1/10
Romilar, *see* Dextromethorphan
Ron-Pearce syndrome
deafness, 60/658
Ronnel, *see* Fenclofos
Root
brachial plexus, 2/135, 137
cervical, *see* Cervical root
cervical spinal, *see* Cervical spinal root
pain, 26/492
posterior spinal, *see* Dorsal spinal nerve root
spinal, *see* Spinal nerve root
trigeminal, *see* Trigeminal root
Root avulsion, *see* Spinal nerve root avulsion
Root compression
spinal, *see* Spinal nerve root compression
Root entrapment
spinal, *see* Spinal nerve root entrapment
Root entry zone
dorsal spinal, *see* Dorsal spinal root entry zone
Root ganglion
dorsal spinal, *see* Dorsal spinal nerve root
ganglion
Root injury
spinal, *see* Spinal nerve root injury
Rooting reflex
brain death, 24/718
feeding reflex, 4/345

newborn behavior, 4/345
Rosenberg-Bergström disease
  amyotrophy, 22/514
  choreoathetosis, 22/514
  hereditary hearing loss, 22/514
Rosenberg-Chutorian syndrome
  autosomal recessive inheritance, 22/507
  deafness, 60/658
  hereditary hearing loss, 22/505, 42/418
  hyporeflexia, 22/506
  Jéquier-Deonna variant, 22/524
  muscular atrophy, 22/506, 42/418
  nerve conduction velocity, 22/506
  nyctalopia, 22/506
  optic atrophy, 22/505, 42/418, 60/654
  opticocochleodentate degeneration, 60/758
  polyneuropathy, 22/505
  sensory loss, 22/506, 42/418
Rosenberg Self-Esteem Scale
  epilepsy, 72/422
Rosenthal fiber
  Alexander disease, 10/94, 100, 42/485, 66/701
  Binswanger disease, 54/227
  cerebellar astrocytoma, 18/2
  glial dystrophy, 21/67
  leukodystrophy, 42/484-486
  neurofibromatosis type I, 14/33, 37
  oligodendroglioma, 20/335
  polar spongioblastoma, 18/2
  proteolipid, 10/135
  spinal glioma, 20/331
  spongioblastoma, 16/15
  syringomyelia, 32/292, 297
Rosenthal vein
  absence, 30/422
  anatomy, 11/48, 51, 63
  radioanatomy, 11/108
  topography, 11/108
Roseola infantum
  benign intracranial hypertension, 16/160
Ross river virus
  acute viral myositis, 56/194
  alphavirus, 56/12
  experimental acute viral myositis, 56/195
Ross syndrome
  anhidrosis, 75/125
  Holmes-Adie syndrome, 75/125
  ophthalmic finding, 74/430
Rossi syndrome
  Patau syndrome, 14/121
  Turner syndrome, 14/14, 109
Rossolimo sign
  foot, 1/248

hand, 1/244
  technique, 1/185
Rostral neuropore
  anencephaly, 50/3
  neural tube, 50/4
Rostral paravermal zone
  cerebellar dysarthria, 60/643
Rostral tegmental lesion
  Cestan-Chenais syndrome, 2/265
Rostral tegmental syndrome
  conjugate gaze, 12/20
  extremity involvement, 12/20
  sensory trigeminal neuropathy, 12/20
  tegmental pontine syndrome, 12/20
Rotation injury
  boxing injury, 23/564, 566
  cervical spine, 25/67, 125
  spinal column, 25/6
  spinal cord, 25/62, 67
  thoracolumbar spine, 25/443
  vertebral artery injury, 26/58-60
Rotation nystagmus
  localization value, 2/384-387
  vestibular system, 16/322
Rotation vestibular nystagmus
  after-nystagmus, 2/333
Rotavirus
  Reye syndrome, 56/150
  RNA virus, 56/10
Rotenone intoxication
  clinical aspect, 36/445
  laboratory, 36/445
  pathology, 36/445
  treatment, 36/457
Roth spot
  bacterial endocarditis, 52/299
  septic embolism, 11/214
Rothmann, M., 1/8
Rothmund-Petges-Clejat syndrome, see
  Rothmund-Thomson syndrome
Rothmund syndrome, see Rothmund-Thomson
  syndrome
Rothmund-Thomson syndrome
  age at onset, 39/405, 43/460, 46/61
  alopecia, 43/460, 46/61
  ataxia telangiectasia, 14/315
  bird like face, 14/777
  bone abnormality, 14/777, 39/405, 43/460
  cataract, 14/777, 39/405, 43/460, 46/61
  Cockayne-Neill-Dingwall disease, 13/436, 31/238
  cutis marmorata, 43/460
  dermatitis, 14/777, 43/460
  dwarfism, 13/436

ectodermal dysplasia, 14/112
EEG, 14/777
frontal bossing, 43/460
Goltz-Gorlin syndrome, 14/112, 31/288
growth retardation, 14/315, 46/61
hereditary connective tissue disease, 39/405
hypogonadism, 14/777, 39/405, 43/460, 46/61
hypotrichosis, 39/405
mental deficiency, 14/777, 43/461, 46/61
microcephaly, 14/777, 43/460
nail hypoplasia, 14/777, 43/460, 46/61
photosensitivity, 14/315, 777, 39/405, 43/461
pigmentation, 39/405, 43/460
progeria, 14/112
sex ratio, 14/777, 43/460
short stature, 14/777, 39/405, 43/460
telangiectasia, 14/777, 43/460
Rottnest quokka
  myotonic disorder, 40/533
  vitamin E deficiency, 40/171
Round sting ray intoxication
  elasmobranche, 37/70
Rous sarcoma virus
  acute viral myositis, 56/194
  virolysis, 56/60
Roussy-Lévy syndrome, 21/171-177
  amyotrophy, 21/176, 42/109
  areflexia, 21/175, 42/108
  ataxia, 42/108
  autonomy critique, 21/173
  autosomal dominant, 21/10, 42/108
  Babinski sign, 42/108
  benign essential tremor, 42/108, 49/569, 587
  cardiomyopathy, 21/175
  cause, 21/171
  clinical features, 21/175
  club claw feet, 21/175
  clumsiness, 42/108
  cochlear spiral ganglion degeneration, 21/359
  course, 21/176
  deaf mutism, 21/175
  distal muscle weakness, 42/108
  EMG, 21/176, 42/108
  epilepsy, 21/175
  essential tremor, 21/175
  Friedreich ataxia, 21/12, 172-174, 351, 42/108
  gait, 21/175, 42/108
  hereditary motor and sensory neuropathy type I,
    21/172-174, 307, 42/108
  hereditary motor and sensory neuropathy type II,
    21/172-174, 307, 42/108
  hereditary sensory and autonomic neuropathy type
    I, 21/77

history, 21/172
hypertrophic interstitial neuropathy, 21/148, 177,
  60/220
hypogonadal cerebellar ataxia, 21/476
kyphoscoliosis, 21/176, 42/108
lightning pain, 21/173
mental deficiency, 21/174
migraine, 21/174
muscular atrophy, 42/108
nerve conduction velocity, 21/177, 42/108
neuropathy, 60/659
old taxonomy, 21/3
onion bulb formation, 21/177
pes cavus, 22/425, 42/108
Pierre Marie-Boveri type, 21/177
prevalence, 21/22
Richards-Rundle syndrome, 21/175, 22/524,
  43/264
scapulohumeral muscular atrophy, 21/174
sensory impairment, 21/176, 42/108
talipes, 42/108
tremor, 21/171, 175, 42/108, 49/588
Rowley-Rosenberg syndrome
  aminoaciduria, 43/473
  growth retardation, 43/473
  muscular dystrophy, 43/473
  pulmonary hypertension, 43/473
Royal Free Hospital disease, see Postviral fatigue
  syndrome
Rubber
  brain tumor carcinogenesis, 17/28
Rubbing nose sign
  frontal lobe tumor, 17/259
Rubella
  acute cerebellar ataxia, 34/627
  aqueduct stenosis, 50/308
  arthralgia, 56/406
  arthritis, 56/406
  brachial neuritis, 7/486
  clinical features, 56/108
  CNS complication, 56/407
  congenital, see Congenital rubella
  diagnosis, 56/108
  familial spastic paraplegia, 59/327
  management, 56/108
  mental deficiency, 46/19
  multiple sclerosis, 34/438, 47/329
  nerve lesion, 7/486
  polyradiculitis, 7/486
  progressive subacute encephalitis, 46/389
  season, 56/406
  secondary pigmentary retinal degeneration,
    13/214

spina bifida, 50/480
Stargardt disease, 13/142
temporal bone pathology, 56/109
teratogenic agent, 34/372-375
toxic encephalopathy, 56/407
vestibular hypofunction, 56/108
Rubella encephalitis
see also Progressive rubella panencephalitis
acute cerebellar ataxia, 34/627
CSF, 56/407
immune mediated encephalomyelitis, 34/419-421
multiple sclerosis, 34/438
noncongenital, see Noncongenital rubella
encephalitis
Rubella encephalomyelitis
CSF, 56/407
Rubella encephalopathy
CSF, 56/407
Rubella panencephalitis
progressive, see Progressive rubella
panencephalitis
spastic paraplegia, 59/436
Rubella syndrome
autism, 43/204
clinical features, 31/213
deafness, 31/212
epidemiology, 31/213
hydrocephalus, 31/212
immune response, 31/213
malformation, 31/212
mechanism, 31/214
mental deficiency, 31/212
microcephaly, 30/508, 510, 31/212
prenatal, see Prenatal rubella syndrome
prevention, 31/214
retinopathy, 31/214
Rubella vaccine
Reye syndrome, 56/150
Rubella virus
acute postinfectious disseminated
encephalomyelitis, 56/407
biology, 56/405
cataract, 56/408
chronic progressive panencephalitis, 56/405
congenital deafness, 56/107
congenital rubella syndrome, 56/405
deafness, 56/12, 107, 408
epilepsy, 56/407
fetal malformation, 56/12
German measles, 56/12
leukoencephalitis, 56/410
mental deficiency, 56/408
microcephaly, 56/408

molecular structure, 56/405
neurotropic virus, 56/405
noncongenital rubella encephalitis, 56/405
optic atrophy, 56/408
optic neuritis, 56/407
panencephalitis, 56/12
persistence, 56/406, 408
progressive rubella panencephalitis, 34/331,
56/405
Reye syndrome, 29/332, 49/217
Togaviridae family, 56/405
Togavirus, 56/12
vertigo, 56/107
Rübenzieherlähmung
nerve stretch, 7/334
sciatic nerve, 7/334
Rubeola, see Measles and Mumps
Rubeosis
iris, see Iris rubeosis
Ruber
nucleus, see Nucleus ruber
Rubidium
peritoneal dialysis, 27/329
Rubinstein-Taybi syndrome
arhinencephaly, 30/472, 50/242
broad thumb, 13/83, 31/282, 43/234, 46/91,
50/276
cataract, 13/83, 46/92
cerebrohepatorenal syndrome, 30/472
chromosomal aberration, 30/472, 31/284
clinodactyly, 31/283, 43/234
CNS, 31/283
corpus callosum agenesis, 31/284, 50/163, 276
craniofacial abnormality, 43/234
cryptorchidism, 31/282, 284, 43/234, 50/276
dermatoglyphics, 31/284
Down syndrome, 31/427
dysmorphia, 13/83
EEG, 31/284
epicanthus, 43/234
epidemiology, 31/282, 43/235
etiology, 31/284, 43/235
face, 31/282, 43/234, 50/276
frontal bossing, 43/234
grimacing, 31/282, 43/234
growth retardation, 31/282, 284, 43/234
heart disease, 43/235, 46/92
hemangioma, 31/284
high arched palate, 31/282, 43/234
hirsutism, 31/284, 43/234
hypertelorism, 31/282, 43/234, 46/91
hypotonia, 31/283, 43/234
iris coloboma, 13/83, 31/232

kyphosis, 31/284, 43/234
lordosis, 31/284, 43/234
mental deficiency, 13/83, 31/282, 43/234, 46/91,
  50/276
microcephaly, 30/508, 31/282, 43/234, 46/91,
  50/276
microcrania, 43/234
micrognathia, 43/234
motor retardation, 31/282, 43/234, 60/92
ocular abnormality, 31/282, 43/234
olfactory bulb, 30/472
optic atrophy, 13/83, 31/282
Patau syndrome, 14/121
prevalence, 43/235
ptosis, 31/282, 43/234
radiologic appearance, 31/284
recurrence risk, 43/235
scoliosis, 31/284, 43/234
seizure, 31/284, 50/276
short stature, 43/234, 50/276
skeletal abnormality, 31/283
spastic gait, 31/283, 43/234, 50/276
spastic paraplegia, 50/276
spina bifida, 31/284, 43/234
strabismus, 31/282, 43/234, 46/91, 50/276
Rubivirus
  neurotropic virus, 56/12
Rubral tremor
  see also Cerebellar tremor
  absorption, 6/162-164
  Benedikt syndrome, 21/529
  critique, 6/437
  dentatorubrothalamic pathway, 2/481
  EMG, 49/587
  involuntary movement, 2/481
  Parkinson disease, 49/96
  striatonigral degeneration, 49/206
Rubrofugal connection
  anatomy, 6/40
Rubropetal fiber
  anatomy, 6/41
Rubrospinal tract
  anatomy, 1/300
Rubrothalamic syndrome, see Anterolateral
  thalamic syndrome
Rucksack paralysis
  thoracic outlet syndrome, 51/121
Rud syndrome
  deafness, 51/398, 60/721
  diabetes mellitus, 13/321, 38/9, 43/284
  epilepsy, 13/321, 479, 14/13, 38/9, 43/284, 46/59,
    51/398, 60/721
  hypogonadism, 13/321, 43/284, 51/398, 60/721

ichthyotic syndrome, 13/321, 479, 38/9, 43/284,
    46/59, 51/398, 60/721
  infantilism, 13/321, 479, 14/13, 38/9, 43/284
  iron deficiency anemia, 38/9, 43/284
  mental deficiency, 13/321, 479, 14/13, 38/9,
    43/284, 46/59, 51/398, 60/721
  motor conduction velocity, 51/398
  motor nerve conduction velocity, 51/398
  onion bulb formation, 51/398
  partial gigantism, 38/9, 43/284
  polyneuritis, 38/9, 43/284
  polyneuropathy, 51/398, 60/721
  secondary pigmentary retinal degeneration,
    13/321, 479, 51/398, 60/721
  short stature, 14/13, 43/284, 51/398
  Sjögren-Larsson syndrome, 13/479, 22/478, 38/10
  spastic paraplegia, 14/13
  tetany, 38/9
Rudderfish intoxication
  CNS, 37/86
  hallucination, 37/86
Ruffini ending
  sensation, 7/82
Rufous oculocutaneous albinism
  classification, 43/6
Rugby football injury
  brain concussion, 23/570
  headache, 23/570
Rumania
  neurology, 1/11
Rupp, C., 1/20
Rupture
  aorta, see Aorta rupture
  aorta aneurysm, 63/49
  bladder, 25/470
  brain aneurysm, 12/83
  brain arteriovenous malformation, 54/374
  Charcot-Bouchard aneurysm, 63/74
  diaphragm, 26/192, 195-197, 226
  dura mater, see Dura mater rupture
  ependymal, see Ependymal rupture
  spinal nerve root, see Spinal nerve root rupture
  tympanic membrane, see Tympanic membrane
    rupture
Russell brain syndrome, see Exophthalmos
Russell-Silver dwarf, see Russell-Silver syndrome
Russell-Silver syndrome
  see also Edwards syndrome
  asymmetry, 43/476
  blue sclerae, 43/477
  café au lait spot, 43/476, 46/21, 82
  chromosomal aberration, 46/21
  clinodactyly, 43/476

craniofacial abnormality, 46/21, 82
dermatoglyphics, 43/477
dwarfism, 30/248
Edwards syndrome, 46/21
etiology, 43/477
feeding disorder, 43/477, 46/21
growth retardation, 43/476, 46/21
hemihypertrophy, 22/550, 59/483
hypertelorism, 30/248
intrauterine growth retardation, 46/82
mental deficiency, 46/21, 82
micrognathia, 43/477, 46/21
mosaic trisomy 18, 46/21
short stature, 22/550, 43/476, 46/21, 82
skeletal malformation, 30/248
syndactyly, 43/476
Russell syndrome
cachexia, 16/349
euphoria, 16/349
growth hormone secretion, 16/349
hyperkinesia, 16/349
hypothalamus, 16/349
optic nerve glioma, 16/349
Patau syndrome, 14/121
spontaneous nystagmus, 16/349
Russian spring-summer encephalitis, see Central
    European tick-borne encephalitis
Russian tick-borne typhus
rickettsial infection, 34/657
Ruvalcaba syndrome
microcephaly, 43/290
Ryania intoxication
animal study, 36/447
Ryanodine receptor
malignant hyperthermia, 62/567, 569
Ryukyan spinal muscular atrophy
areflexia, 42/93
creatine kinase, 42/93
epidemiology, 21/25
fasciculation, 42/93
hyporeflexia, 42/93
kyphoscoliosis, 42/93
prevalence, 21/28, 42/94
proximal muscle weakness, 42/93
stationary end state, 59/43
talipes, 42/93

S-100 protein
allergic encephalomyelitis, 9/514, 520
physicochemical property, 9/514
Sabin-Feldman dye test
toxoplasmosis, 35/120
Saccade

cerebrovascular disease, 55/121
cogwheel, see Cogwheel saccade
hypometric, see Hypometric saccade
Saccadic eye movement, see Opsoclonus
Saccadomania, see Opsoclonus
Saccharomyces, see Cryptococcus neoformans
Saccharopine dehydrogenase
hyperlysinemia, 42/567
Saccharopine oxidoreductase
hyperlysinemia, 42/567
Saccharopinuria
amino acid, 29/222
CSF, 29/222
diet, 29/222
hyperlysinemia, 29/221, 42/567
mental deficiency, 29/222
Saccoradiculography, see Radiculography
Sachs, B., 1/5, 14, 10/226
Sachsalber syndrome
see also Hemifacial hypertrophy
buphthalmos, 14/398
facial hemihypertrophy, 14/398
neurofibromatosis, 14/398
Sacral agenesis, see Caudal aplasia
Sacral arachnoid cyst, 32/408-418
clinical features, 32/414
epidemiology, 32/410
hydrostatic pressure, 32/413
intrasacral meningocele, 31/93, 95
localization, 32/412
malformation, 32/415
meningeal diverticulum type, 32/412
pathogenesis, 32/409, 413
perineurial, 32/409
radiologic diagnosis, 32/415-418
Schreiber type, 32/409, 411
size, 32/412
spinal perineural cyst, 32/409
Sacral meningocele
anterior, see Anterior sacral meningocele
neurofibromatosis type I, 14/158
Sacral nerve
anatomy, 1/361
injury, 25/476
Sacral parasympathetic efferent
penis erection, 2/110
Sacral plexus
anatomy, 7/423
injury, 7/424
lumbosacral plexus, 51/159
Sacrococcygeal chordoma
constipation, 19/289
origin, 18/151

pain, 19/289
prognosis, 19/291
symptom duration, 19/289
urinary incontinence, 19/289
Sacrococcygeal teratoma
    congenital spinal cord tumor, 32/355-386
Sacroiliac fracture dislocation
    obturator nerve injury, 25/473
    superior gluteal nerve injury, 25/475
Saddle nose, see Swayback nose
Saethre-Chotzen syndrome
    acrobrachycephaly, 43/322
    acrocephalosyndactyly type III, 43/326
    acrocephalosyndactyly type V, 43/322
    autosomal dominant transmission, 43/322
    coloboma, 43/322
    craniofacial dysostosis (Crouzon), 31/234
    craniosynostosis, 43/322
    dermatoglyphics, 43/322
    dystopia canthorum, 43/322
    facial asymmetry, 43/322
    fontanelle, 30/100
    foramina parietalia permagna, 50/142
    high arched palate, 43/322
    hypertelorism, 30/246, 43/322
    low set ears, 43/322
    mental deficiency, 43/322, 46/87
    plagiocephaly, 43/322, 46/87
    ptosis, 43/322
    radioulnar synostosis, 43/322
    seizure, 43/322
    skull dysplasia, 30/246
    strabismus, 43/322
    syndactyly, 43/322
Sagittal sinus
    epidural hematoma, 57/252, 256
    inferior, see Inferior sagittal sinus
Sagittal sinus thrombosis
    see also Venous sinus thrombosis
    asparaginase, 69/495
    headache, 5/155
Sagittal suture
    see also Craniosynostosis and Scaphocephaly
    clinical features, 30/224
    digital marking, 30/224
    epidemiology, 30/224
    external appearance, 30/224
    malformation, 30/225
    mental deficiency, 30/225
    premature fusion, 30/224
Sagittal synostosis
    craniosynostosis, 50/116
Saimiri-ateles virus

gammaherpes virus, 56/8
Sakati-Nyhan-Tisdale syndrome
    acrocephalopolysyndactyly, 43/326
    acrocephaly, 43/325
    craniosynostosis, 43/326
Salaam spasm, see Infantile spasm
Salam-Idriss syndrome
    glycosaminoglycanosis type III, 10/343
Salazosulfapyridine
    drug induced polyneuropathy, 51/309
    ototoxicity, 51/309
Salbutamol
    hyperkalemia, 41/161
    hyperkalemic periodic paralysis, 62/472
    periodic paralysis, 41/161
    toxic myopathy, 62/597
Saldino-Noonan syndrome
    bone abnormality, 31/243
    cheiloschisis, 31/243
    genital abnormality, 31/243
    Gruber syndrome, 31/243
    hemopoiesis, 31/243
    hepatomegaly, 31/243
    hypoplasia, 31/243
    low set ears, 31/243
    polydactyly, 31/243, 272
    polysyndactyly, 31/243
    short rib, 31/243, 272
Salicylic acid
    acoustic neuropathy, 64/9
    delirium, 46/540
    headache, 48/174
    metabolic encephalopathy, 46/602
    migraine, 5/98, 378, 48/174
    neurologic complication, 38/497
    neurotoxin, 50/219
    nonsteroid anti-inflammatory agent, 65/416
    Reye syndrome, 29/336, 49/217, 56/163
    toxic myopathy, 62/601
    toxic neuropathy, 64/9
Salicylic acid derivative
    anencephaly, 50/80
    hereditary deafness, 50/219
Salicylic acid intoxication, 37/417-421
    acid base balance, 37/418-420
    carbohydrate metabolism, 37/418
    clinical picture, 37/417-420
    complication, 37/420
    diarrhea, 37/417
    drowsiness, 37/417
    fever, 37/417
    headache, 37/417
    hearing loss, 37/417

mental confusion, 37/417
mental disorder, 37/417
methyl, *see* Methyl salicylate intoxication
nausea, 37/417
neuropathology, 37/420
respiratory failure, 63/477
Reye syndrome, 29/331, 56/159, 65/135
sweating, 37/417
symptom, 64/9
tinnitus, 37/417
treatment, 37/420
vertigo, 37/417
vomiting, 37/417
Salicylic acid methyl ester intoxication, *see* Methyl
    salicylate intoxication
Saline, *see* Sodium chloride
Saliva
    larynx, 74/392
Salivary flow, *see* Salivation
Salivation
    carbamate intoxication, 64/190
    chronic paroxysmal hemicrania, 48/262, 75/293
    cluster headache, 48/222, 237
    facial paralysis, 8/265
    glossopharyngeal neuralgia, 48/466
    headache, 75/284
    hereditary sensory and autonomic neuropathy type
      III, 21/109, 60/26
    ichthyohaemotix fish intoxication, 37/87
    Melkersson-Rosenthal syndrome, 8/220
    paralytic shellfish poisoning, 37/33
    parasympathetic insufficiency, 75/117
    progressive bulbar palsy, 59/219
    puffer fish intoxication, 37/80
    pyrethroid intoxication, 64/214
    scorpion sting intoxication, 37/112
    thermal, *see* Thermal salivation
    Wilson disease, 49/227
Salla disease
    ataxia, 43/306, 59/360
    athetosis, 43/306, 59/360
    Babinski sign, 43/306
    choreoathetosis, 49/385, 59/360
    coarsened facies, 59/360
    computer assisted tomography, 59/361
    cytoplasmic inclusion body, 43/306
    defect, 66/362
    dyspraxia, 51/382
    EEG, 43/306
    enzyme deficiency polyneuropathy, 51/370
    facial dysmorphism, 59/360
    genetics, 66/364
    hepatosplenomegaly, 59/360

hyperreflexia, 43/306
infantile sialic acid storage disease, 59/361
mental deficiency, 43/306, 51/382, 59/360
neuropathology, 59/361
prenatal diagnosis, 59/361
Schwann cell vacuole, 51/382
seizure, 59/360
sialic acid, 66/354
sialic acid transport, 51/382
spastic tetraparesis, 59/360
urinary sialic acid, 51/382, 59/360
vacuolated lymphocyte, 43/306, 51/382, 59/360
Salleras-Ortiz syndrome
    cranial nerve defect, 30/100
Salmonella
    acquired myoglobinuria, 62/577
    brain abscess, 52/149
    gram-negative bacillary meningitis, 52/104, 117,
      121
    mackeral intoxication, 37/87
    multiple sclerosis, 47/339
    scombroid intoxication, 37/85
    spinal epidural empyema, 52/187
Salmonella Citrobacter
    adult meningitis, *see* Gram-negative bacillary
      meningitis
    infantile enteric bacillary meningitis, 33/62
Salmonella typhi
    cranial subdural empyema, 52/171
Salmonella typhimurium
    cranial subdural empyema, 52/171
Salmonella typhosa
    neuritis, *see* Typhoid neuritis
Salt appetite
    aldosterone, 1/501
    homeostasis, 28/497-499
    mitochondrial myopathy, 43/120, 127
Salt regulation
    aging, 74/235
Salt and water balance
    anatomy, 1/499-501
    headache, 5/231
    hypothalamus, 2/438
Salt and water imbalance
    myoglobinuria, 41/273
Saltatory conduction
    myelinated nerve fiber, 47/12-14
    nonmyelinated nerve fiber, 47/12-14
    Ranvier node, 47/13
    refractory period, 47/14
Salvarsan, *see* Arsphenamine
Salvarsan encephalopathy, *see* Arsphenamine
    encephalopathy

Sandhoff disease, *see* GM2 gangliosidosis type II
Sandhoff variant, *see* GM2 gangliosidosis type II
Sanfilippo disease
  type IIIA, *see* Glycosaminoglycanosis type IIIA
  type IIIB, *see* Glycosaminoglycanosis type IIIB
  type IIIC, *see* Glycosaminoglycanosis type IIIC
  type IIID, *see* Glycosaminoglycanosis type IIID
Sanger Brown ataxia, *see* Hereditary spastic ataxia
Santiago de Chile
  Neurological Institute, 1/34
Santorini vein, *see* Posterior parietal vein
Santos disease
  congenital megacolon, 75/635
Sao Paulo, Brazil
  neurology, 1/12
Saphenous nerve
  compression neuropathy, 51/107
Saphenous neuralgia
  aorta surgery, 63/58
Saposin b deficiency, *see* Sulfatase activator
    deficiency
Saramycetin
  North American blastomycosis, 35/407, 52/392
Sarcoid
  hypothalamic tumor, 68/71
  optic chiasm compression, 68/75
  optic pathway tumor, 68/74
Sarcoid neuropathy
  acute polyneuritis, 51/197
  Argyll Robertson pupil, 51/196
  ataxia, 51/196
  corticosteroid, 71/489
  cranial mononeuritis multiplex, 51/195
  cranial nerve, 38/525-529
  CSF, 51/196
  demyelinating neuropathy, 51/197
  epilepsy, 51/196
  facial paralysis, 51/195
  Guillain-Barré syndrome, 51/196
  hyperacusis, 51/195
  incidence, 51/195
  Kveim test, 51/196
  multiple cranial neuropathy, 51/196
  pathogenesis, 51/197
  prognosis, 51/197
  retrobulbar neuritis, 51/195
  sensorimotor neuropathy, 51/196
  spinal nerve, 38/525-529
  steroid, 51/195
  treatment, 51/195, 197
Sarcoidosis, 38/521-539
  see also Granulomatous myositis *and*
    Neurosarcoidosis
  ataxia, 51/196
  azathioprine, 71/493
  benign intracranial hypertension, 16/160
  brain, 16/229
  brain biopsy, 10/685
  brain embolism, 55/163, 63/137
  brain infarction, 55/163
  carpal tunnel syndrome, 7/295
  cerebral, 38/530
  cerebrovascular disease, 53/41, 55/163
  chronic axonal neuropathy, 51/531
  chronic meningitis, 56/644-647
  ciclosporin, 71/493
  clinical features, 40/317
  clinical form, 71/486
  CNS, 38/529-531, 71/472, 477, 489
  corticosteroid, 71/492
  cranial nerve, 38/525-529
  cranial neuropathy, 71/475
  CSF, 38/533, 71/483
  dementia, 46/390
  diagnosis, 38/531-533, 71/486
  differential diagnosis, 71/481
  encephalopathy, 71/478
  endocrine disorder, 71/477
  eosinophilia, 63/371
  epidemiology, 71/470
  facial paralysis, 71/485
  granulomatous CNS vasculitis, 55/396
  hereditary cranial nerve palsy, 60/42
  histology, 40/52
  history, 71/463
  hydrocephalus, 38/530, 71/476, 488, 490
  hypothalamic, 38/530
  immunopathogenic, 71/469
  incidence, 38/521
  inflammatory myopathy, 40/52, 317
  intractable hiccup, 63/490
  listeriosis, 52/91
  lung damage, 71/465
  manifestation, 38/524-531
  meningitis, 71/488
  meningoencephalitis, 38/531
  methotrexate, 71/494
  multiple cranial neuropathy, 51/195
  mycobacterial infection, 71/472
  myelopathy, 71/480
  myopathy, 38/535-539, 71/481, 489
  nerve lesion, 7/479
  neurologic symptom, 55/365
  neuropathology, 71/472
  neuropathy, *see* Sarcoid neuropathy
  organ involvement, 71/466

pathology, 38/522-524
pathophysiology, 71/468
polyarteritis nodosa, 55/364
polymyositis, 41/73, 62/373
polyneuropathy, 51/196
prognosis, 38/533, 71/493
progressive multifocal leukoencephalopathy, 71/400
recurrent nonhereditary multiple cranial neuropathy, 51/570
restrictive infiltrative cardiomyopathy, 63/137
retrobulbar neuritis, 51/195
scanning, 71/494
seizure, 71/479, 489
serum dipeptidyl carboxypeptidase level, 71/467
spastic paraplegia, 59/439
spinal cord, 38/529
survey, 71/463
treatment, 38/533-535, 71/485, 491
tuberculosis, 71/471
Vogt-Koyanagi-Harada syndrome, 56/623
Sarcolemma
  see also Delta lesion and Muscle cell
  Clostridium perfringens toxin type A, 40/151
  conductance, 40/549
  Duchenne muscular dystrophy, 40/74, 252, 370, 384
  muscle fiber, 62/31
  muscle fiber type, 62/3
  myotonia congenita, 40/128
  myotonic dystrophy, 40/74, 213, 494, 497
  myotonic syndrome, 62/261
  pathologic change, 40/74, 151, 212
  ultrastructure, 40/67
Sarcoma
  ataxia telangiectasia, 46/58
  brain, see Brain sarcoma
  brain metastasis, 18/205
  cerebellar arachnoid, see Cerebellar arachnoid sarcoma
  congenital spinal cord tumor, 32/355-386
  cranial vault tumor, 17/117
  Ewing, see Ewing sarcoma
  histiocytic, see Histiocytic sarcoma
  idiopathic multiple hemorrhagic, see Idiopathic multiple hemorrhagic sarcoma
  intraorbital, 18/340
  Kaposi, see Kaposi sarcoma
  lateral ventricle tumor, 17/606
  metastatic cardiac tumor, 63/93
  middle ear, see Middle ear sarcoma
  monstrocellular, see Monstrocellular sarcoma
  nerve tumor, 8/446-453

neurogenic, see Neurogenic sarcoma
  neuropathy, 8/144
  orbital tumor, 17/177
  osteitis deformans (Paget), 17/118, 170
  osteogenic, see Osteosarcoma
  perithelial, see Perithelial sarcoma
  posterior fossa tumor, 18/404
  primary, see Primary sarcoma
  primary cerebral reticulum, see Primary cerebral reticulosarcoma
  radiation, 17/118
  radiation induced, 68/393
  radiotherapy, 18/510
  retrovirus, 56/16
  SBLA syndrome, 42/769
  scalp tumor, 17/132
  skull base tumor, 17/167, 18/404
  soft tissue, see Soft tissue sarcoma
  spinal cord metastasis, 20/425
  spinal epidural tumor, 20/103, 113
  supratentorial brain tumor, 18/321
  systemic neoplastic disease, 41/384
  vertebral metastasis, 20/415, 425
  Werner syndrome, 43/489
Sarcomatosis
  primary meningeal, see Primary meningeal sarcomatosis
Sarcoplasmic body myopathy
  congenital myopathy, 62/332
  desmin intermediate filament, 62/345
Sarcoplasmic mass
  cap disease, 62/341
  muscle fiber, 40/27, 244, 497, 62/44
  muscle filamentous body, 40/249
  myotonic dystrophy, 40/27, 244, 497
Sarcoplasmic reticulum
  see also Transverse tubule
  chloroquine, 40/151
  colchicine, 40/154
  histochemistry, 40/2
  malignant hyperthermia, 38/553, 41/269
  muscle fiber, 62/37
  muscle fiber type, 62/3
  pathologic change, 40/33, 103
  periodic paralysis, 41/154
  structure and function, 40/66, 102, 224, 549
  troponin, 40/553
  vincristine, 40/154
Sarcoplasmic reticulum adenosine triphosphate deficiency syndrome
  malignant hyperthermia, 62/570
Sarcosinemia, see Hypersarcosinemia
Sarcotubular myopathy

classification, 40/288
clinical features, 40/306
congenital myopathy, 41/21, 62/332
creatine kinase, 43/129
facial paralysis, 43/129
histochemistry, 40/47
hyporeflexia, 43/129
muscle fiber type II, 43/129
pathologic reaction, 40/263
proximal muscle weakness, 43/129
rimmed muscle fiber vacuole content, 62/18
rimmed muscle fiber vacuole type, 62/18
Sarin
    chemical structure, 64/225
    neurotoxin, 37/543, 64/152, 223
    organophosphate, 64/152
    organophosphate intoxication, 37/556, 64/152, 223
Sarin intoxication
    chemical classification, 37/543
    organophosphorus compound, 37/543
Satellite cell
    culture, 40/184
    Duchenne muscular dystrophy, 40/371
    infantile spinal muscular atrophy, 59/59
    muscle regeneration, 40/22
    myelination, 9/40
    ultrastructure, 40/89
Satellitosis
    neoplastic, see Neoplastic satellitosis
    neuronal loss, 22/14
    oligodendroglioma, 16/12
    wallerian degeneration, 9/37
Satoyoshi disease, see Phosphohexose isomerase
    deficiency
Satoyoshi syndrome
    alopecia universalis, 64/256
    diarrhea, 64/256
    eosinophilia myalgia syndrome, 64/256
    motor unit hyperactivity, 41/307, 59/377
    muscle spasm, 64/256
Saturday night palsy
    alcohol intoxication, 64/115
    compression neuropathy, 51/100
    neurapraxia, 51/77
Saturnism, see Lead intoxication
Satyr ear
    ectodermal dysplasia, 14/112
Saxitoxin
    Alexandrium genus, 37/33
    axon, 51/47
    emerita analoga intoxication, 37/56
    lethal dose, 65/154

marine toxin intoxication, 37/33, 65/146, 152
muscle action potential, 65/152
myasthenic syndrome, 64/15
nerve conduction block, 65/152
neurotoxin, 37/1, 33, 56, 41/291, 64/14, 65/142,
    146, 152-154
paralytic shellfish poisoning, 37/33, 65/152
Protogonyaulax, 64/14
sodium channel, 65/146
structural formula, 37/33
structure, 65/146
toxic dose, 65/152
toxic myopathy, 64/15
toxic neuromuscular junction disorder, 64/14
Saxitoxin intoxication
    see also Dinoflagellates intoxication
    action site, 64/15
    acute muscle weakness, 41/291
    pathomechanism, 64/15
    symptom, 64/14
    venom, 37/56
Say-Gerald syndrome, see VATER syndrome
SBLA syndrome, 42/769-771
    adrenocortical carcinoma, 42/769
    brain tumor, 42/769
    genetic counseling, 42/771
    intracranial tumor, 42/769
    laryngeal cancer, 42/769
    leukemia, 42/769
    lung cancer, 42/769
    mammary carcinoma, 42/769
    sarcoma, 42/769
Scale beam squint, see Skew deviation
Scalene triangle, see Scalenus anticus syndrome
Scalenus anticus syndrome
    see also Thoracic outlet syndrome
    cases, 7/434
    critique, 2/146
    history, 7/432
    Horner syndrome, 2/150
    neurovascular compression, 2/150, 7/430
    reflex neurovascular syndrome, 8/332
    thoracic outlet syndrome, 7/443, 51/121, 124, 129
Scalloping
    vertebral, see Vertebral scalloping
Scalp
    brain arteriovenous malformation, 12/229
    brain injury mechanism, 57/20
    bruising, 57/131
    clinical examination, 57/131
    craniofacial injury, 57/309
    cyclopia, 30/442
    head injury, 57/131

laceration, 57/131
missile injury, 23/511, 519
pediatric head injury, 57/329
penetrating head injury, 57/309
swelling, 57/131
tenderness, 48/4
traumatic aneurysm, 24/381
venous malformation, 12/236
Scalp defect
Adams-Oliver syndrome, 30/100
amniotic band syndrome, 30/116
cutaneous, see Cutaneous scalp defect
Patau syndrome, 31/504, 50/558
Scalp injury
see also Head injury
headache, 5/180
hematoma, 24/655
intracranial infection, 24/215
laceration, 24/654
wound, 24/215
Scalp muscle contraction
headache, 5/181
Scalp tumor
hemangioma, 17/124
malignant melanoma, 17/132
neurofibromatosis type I, 17/124
sarcoma, 17/132
squamous carcinoma, 17/126
Scandinavia
neurology, 1/11
Washington Psychosocial Seizure Inventory,
72/412
Scanning electron microscopy
Duchenne erythrocyte, 40/386
striatopallidodentate calcification, 49/426
Scaphocephaly
see also Craniosynostosis and Sagittal suture
craniosynostosis, 30/154, 221, 42/20, 50/113, 116
mental deficiency, 30/225
Smith-Lemli-Opitz syndrome, 50/276
Summit syndrome, 43/325
Scapula
elevation, see Elevated scapula
Scapula alata, see Winged scapula
Scapular nerve
dorsal, see Dorsal scapular nerve
supra, see Suprascapular nerve
Scapular winging, see Winged scapula
Scapulohumeral distal muscular dystrophy
cardiac dysrhythmia, 43/130
contracture, 43/130
Scapulohumeral muscular atrophy
hereditary spinal muscular atrophy, 59/24

Roussy-Lévy syndrome, 21/174
Scapulohumeral muscular atrophy (Vulpian-
Bernhardt), see Scapulohumeral spinal muscular
atrophy
Scapulohumeral muscular dystrophy
facioscapulohumeral muscular dystrophy type,
62/170
neurogenic muscular atrophy, 21/29
winged scapula, 62/170
Scapulohumeral muscular dystrophy (Erb)
see also Limb girdle syndrome
features, 62/183
limb girdle syndrome, 62/179
Scapulohumeral myopathy
X-linked, 43/130
Scapulohumeral periarthritis
β2-microglobulin amyloidosis, 63/529
Scapulohumeral reflex
diagnostic value, 1/244
Scapulohumeral spinal muscular atrophy
Aran-Duchenne disease, 22/75
bulbar involvement, 22/134
clinical type, 22/138
description, 22/310
EMG, 62/171
hereditary motor and sensory neuropathy type I,
21/304, 42/108
hereditary motor and sensory neuropathy type II,
21/304
muscle biopsy, 62/171
neurogenic type, 21/29
separate entity, 22/136
Scapulohumeral syndrome, see
Facioscapulohumeral syndrome
Scapulohumeroperoneal muscular atrophy
areflexia, 42/343
heterozygote detection, 42/343
hyporeflexia, 42/343
Scapuloilioperoneal muscular atrophy
areflexia, 42/344
creatine kinase, 42/344
hyporeflexia, 42/344
Scapulolaryngeal hemiplegia, see Gignoux
syndrome
Scapuloperoneal amyotrophy, see Scapuloperoneal
spinal muscular atrophy
Scapuloperoneal muscular dystrophy
age at onset, 62/169
autosomal dominant, 62/168
clinical features, 40/309, 319, 322, 423-428
differential diagnosis, 40/481
EMG, 42/99, 62/169
facioscapulohumeral muscular dystrophy, 62/169

282

hereditary motor and sensory neuropathy type I,
    40/425
hereditary motor and sensory neuropathy type II,
    40/425
inflammatory, *see* Inflammatory scapuloperoneal
    muscular dystrophy
muscle computerized assisted tomography, 62/169
myotonic dystrophy, 22/59
neuromuscular disease, 41/411
pathology, 40/425
scapuloperoneal syndrome, 40/425, 41/411, 42/99
serum creatine kinase, 62/169
shoulder girdle, 42/99
sporadic, 62/168
winged scapula, 42/99
X-linked muscular dystrophy, 22/59, 40/426
Scapuloperoneal myopathy
    areflexia, 43/131
    cardiomyopathy, 43/131
    contracture, 43/130
    creatine kinase, 43/131
    genetic linkage, 43/131
    hyporeflexia, 43/131
    mental deficiency, 43/131
    pes cavus, 43/130
Scapuloperoneal neuropathy, *see* Scapuloperoneal
    spinal muscular atrophy
Scapuloperoneal spinal muscular atrophy
    adult, 22/58, 60
    adult onset, 59/45
    age at onset, 59/44
    areflexia, 62/170
    autosomal dominant, 62/169
    bulbar lesion, 22/137
    cardiomyopathy, 59/45
    demyelination, 62/170
    descending course, 59/45
    differential diagnosis, 59/375
    dominant, 22/58, 60
    EMG, 59/45, 62/170
    foot deformity, 59/45
    genetics, 22/75
    hereditary motor and sensory neuropathy type II,
        21/304, 22/60, 59/44, 62/170
    hereditary motor and sensory neuropathy variant,
        60/247
    hypertrophic interstitial neuropathy, 62/170
    hypertrophic nerve, 62/170
    juvenile onset, 59/45
    motor conduction velocity, 62/170
    muscle biopsy, 59/46
    nerve biopsy, 62/170
    neuropathology, 59/45

pes cavus, 59/45, 62/170
pes equinovarus, 59/45
progressive spinal muscular atrophy, 22/57-68
ptosis, 59/46
recessive infantile, 22/58, 62
recessive subtype, 62/170
sporadic case, 59/46
symptom, 62/170
Wohlfart-Kugelberg-Welander disease, 59/92
Scapuloperoneal syndrome, 22/57-63, 40/423-428
    clinical features, 40/309, 319, 322
    differential diagnosis, 40/473, 41/202
    differentiation, 62/161
    Emery-Dreifuss muscular dystrophy, 40/392, 426,
        62/171
    facioscapulohumeral syndrome, 40/416, 418, 423
    fasciculation, 42/99
    history, 40/434
    nemaline myopathy, 40/426
    neurogenic, 40/427
    nosology, 62/168, 170
    scapuloperoneal muscular dystrophy, 40/425,
        41/411, 42/99
    sensory loss, 42/99
Scapulothoracic fixation
    facioscapulohumeral muscular dystrophy, 40/423
Scar formation
    spinal cord repair, 61/485
Scaritoxin
    ciguatoxin, 65/159
    marine toxin intoxication, 65/159
    neurotoxin, 65/142
Scarlet fever
    facial paralysis, 7/487
    nerve lesion, 7/487
    oculomotor paralysis, 7/487
    polyradiculitis, 7/487
    streptococcal meningitis, 52/77
Scarpa ganglion
    acoustic neuroma, 2/361
Scatophagidae intoxication
    family, 37/78
    venomous spine, 37/78
SCH 33390
    classification, 74/149
Schachter syndrome
    mental deficiency, 14/792
Schäfer sign
    Babinski sign, 1/184, 250
Schaffer, K., 10/227
Schaffer-Spielmeyer cell change
    adult neuronal ceroid lipofuscinosis, 10/218
    definition, 21/43

familial amaurotic idiocy, 10/213
GM2 gangliosidosis, 10/216, 386
neuronal ceroid lipofuscinosis, 10/652
neuronal storage dystrophy, 21/61
Schaltenbrand, G., 1/35
Scheibe ear abnormality
  Patau syndrome, 50/559
Scheidt melanism
  lentiginosis profusa (Darier), 14/600
Scheie disease, *see* Glycosaminoglycanosis type IS
Scheie syndrome, *see* Glycosaminoglycanosis
    type IS
Scherer, H.J., 10/6, 7
Scheuermann disease
  autosomal dominant transmission, 50/419
  lumbar lordosis, 32/147
  spinal epidural cyst, 20/137, 146
  syringomyelia, 32/283
Scheuthauer-Marie-Saiton syndrome, *see*
    Cleidocranial dysplasia
Schiff leukofuchsin
  histochemistry, 9/23
  protein, 9/23
Schilder disease
  diffuse cerebral sclerosis, *see* Diffuse sclerosis
  type (1912), *see* Myelinoclastic diffuse sclerosis
  type (1913), *see* Adrenoleukodystrophy
  type (1924), *see* Subacute sclerosing
    panencephalitis
Schilder, P., 10/1-3
Schilling test
  subacute combined spinal cord degeneration,
    28/150
  vitamin B12 deficiency, 7/553
Schimke syndrome
  chorea, 62/304
  hearing loss, 62/304
  mental deficiency, 62/304
  progressive external ophthalmoplegia, 62/304
Schindler disease, *see* α-*N*-Acetylgalactosaminidase
    deficiency
Schistosoma
  helminthiasis, 52/511, 535-541
  nematode, 52/535
Schistosoma haematobium
  helminthiasis, 52/506
  schistosomiasis, 52/535
  transverse myelitis, 52/511
Schistosoma japonicum
  brain abscess, 52/151
  helminthiasis, 52/506
  lumbosacral plexus neuritis, 51/167
  schistosomiasis, 52/535, 537

Schistosomiasis, 35/231-240
  acute, *see* Acute schistosomiasis
  anterior spinal artery syndrome, 52/538
  antimony, 52/540
  brain, 16/228
  brain granuloma, 52/537
  brain injury, 52/537
  clinical features, 52/536
  clinical stage, 52/536
  CNS, 52/537
  computer assisted tomography, 52/540
  corticosteroid, 52/541
  CSF, 52/540
  diagnosis, 52/540
  ectopic lesion, 52/537
  encephalitis, 52/538
  encephalomyelitis, 52/538
  eosinophilia, 52/540, 63/371
  epilepsy, 52/511, 538, 72/169
  geography, 52/535
  granulomatous CNS vasculitis, 55/388
  headache, 52/538
  helminthiasis, 35/216
  histopathology, 52/538
  immune response, 52/539
  intracranial hypertension, 52/538
  multiple cranial neuropathy, 52/538
  myelomalacia, 63/42
  neurosurgery, 52/541
  organophosphate, 64/155
  paraplegia, 52/538
  parasitic disease, 35/14
  pathogenesis, 52/539
  polymyositis, 41/373
  praziquantel, 52/540
  prognosis, 52/540
  psychiatric symptom, 52/538
  radiculitis, 52/538
  Schistosoma haematobium, 52/535
  Schistosoma japonicum, 52/535, 537
  Schistosoma mansoni, 52/535, 537
  spastic paraplegia, 59/436
  spinal cord compression, 52/538
  spinal cord embolism, 63/42
  spinal cord granuloma, 52/537
  spinal muscular atrophy, 59/370
  systemic neoplasm, 41/373
  transverse myelitis, 52/511, 538
  treatment, 52/540
  tropical myeloneuropathy, 56/526
Schizencephaly, 30/338-347
  anencephaly, 50/72
  brain cortex heterotopia, 50/199

callosal commissure, 50/176
cerebral mantle defect, 30/341
clinical features, 30/346, 50/201
CNS, 30/346
commissura anterior, 50/176
computer assisted tomography, 50/201
definition, 50/197
diagnosis, 30/346
EEG, 50/201
epilepsy, 30/346, 72/109, 342
etiology, 30/341-343, 346
fused cleft, 50/198
history, 30/338-341, 50/198
hydranencephaly, 30/665-667, 50/198, 342, 347
hydrocephalus, 30/339, 345, 662
lamb, 50/198
management, 30/347
mental deficiency, 30/346, 46/9
microcephaly, 30/346, 50/198
microgyria, 30/341, 345, 483, 50/199
pathology, 30/343-346, 50/198
periventricular calcification, 50/258
pestivirus, 50/198
porencephaly, 50/198, 357
prenatal vascular disease, 53/27
prognosis, 50/201
pyramidal hypoplasia, 50/199
septum pellucidum agenesis, 50/200
ultrasound, 50/201
unfused cleft, 50/198
Schizocyte
  thrombotic thrombocytopenic purpura, 9/607
Schizoidism
  affective disorder, 43/209
  anhidrotic ectodermal dysplasia, 14/788
  genetics, 43/209
Schizophrenia, 46/481-510
  acetylcholine, 29/451
  aging, 46/491
  agraphesthesia, 46/490
  alcoholic hallucinosis, 46/419
  amphetamine intoxication, 65/257
  amusia, 45/484
  amygdalectomy, 46/497
  amygdaloid nucleus, 46/507
  anorexia nervosa, 46/588
  anterior posterior asymmetry, 46/485
  aphasia, 45/313
  attention, 46/501
  attention disorder, 3/140, 188
  auditory evoked potential, 46/502
  auditory hallucination, 45/357, 46/562
  auditory threshold, 46/505

autointoxication, 9/620
autonomic nervous system, 46/496-498
B-mitten pattern, 46/496
backward masking, 46/502
basal ganglion, 46/486
behavior, 46/507-510
behavior disorder, 46/490
brain asymmetry, 46/504-507
brain atrophy, 46/421, 450-452, 454
brain blood flow, 46/486, 73/477
brain cortex atrophy, 46/481
brain stem, 46/509
brain tumor, 46/461
brain ventricle enlargement, 46/481
Broca area, 46/486
callosectomy, 46/504
catatonia, 46/422-424
celiac disease, 28/230
cerebellar atrophy, 46/483
cerebral dominance, 46/456-460
childhood, see Childhood schizophrenia
childhood psychosis, 46/192
chronicity, 46/488
classification, 46/444
clinicopathologic correlation, 46/421
cognition, 46/482, 488, 490-492, 501
computer assisted tomography, 46/451, 481-487
congenital pain insensitivity, 51/566
consciousness, 3/120
contingent negative variation, 46/499, 502
corpus callosum, 46/454, 458
corpus callosum disconnection syndrome, 46/501,
  504, 508
corpus callosum transmission, 46/500, 506
delirium, 46/419, 552
delusion, 43/212
dementia, 46/123, 419
developmental dysphasia, 46/141
dichotic listening, 46/503, 505
diffuse focal sign, 46/484
dopamine, 46/422, 446
dopamine β-mono-oxygenase, 43/214
dopamine receptor, 46/422
EEG, 46/421, 460, 492-496, 505
electrodermal asymmetry, 46/504
electrodermal habituation, 46/497
encephalitis lethargica, 46/455
epilepsy, 15/599, 46/461, 493, 507, 73/477
Epstein-Barr virus, 56/257
evoked brain stem response, 46/500
evoked cortical response, 46/499-504
evoked potential, 46/503-505
finger agnosia, 46/490

focal abnormality, 46/486
focal impairment, 46/457
frontal lobe, 46/458, 509
functional psychosis, 46/444
genetics, 46/498
global impairment, 46/457
growth hormone secretion, 46/474
H-reflex, 46/505
hallucination, 43/212, 45/355, 46/419
hallucinogenic agent, 46/448
handedness, 46/456, 505
head injury, 46/461
headache, 48/361
hippocampectomy, 46/497
hippocampus, 46/506-508
Huntington chorea, 46/306, 49/282
hyperactivity, 46/485
hypergraphia, 45/469
hypofrontality, 46/485, 487
hypokinesia, 73/477
hysteria, 46/576, 581
intelligence quotient, 46/457
Klinefelter syndrome, 43/557
lateral asymmetry, 46/486
lateralization, 46/494, 503
lateralized dysfunction, 46/489
left hemisphere, 46/489
left temporal lobe, 46/504
lilliputian hallucination, 45/358
limbic system, 46/454, 497
Lindenov-Hallgren syndrome, 13/455, 458
macrosomatognosia, 4/231
MAO, 43/214
metachromatic leukodystrophy, 66/168
microsomatognosia, 4/231
migraine, 48/361
minimal brain damage, 46/456
motor disorder, 46/455
multiple personality, 46/576
negative symptom syndrome, 46/501
neurochemistry, 46/446-448
neuroleptic agent, 46/446
neurologic soft sign, 46/489-491
neurologic substrate, 46/509
neuronal inhibitory deficit, 46/502
neuropathology, 46/453-455
neuropsychologic assessment, 46/457-460, 487-489
neuropsychology, 46/570
neuropsychophysiology, 46/505
neurosyphilis, 46/455
neurotransmitter, 43/214
noradrenalin, 46/448

obstetric complication, 46/456
occipital reversal, 46/484
oculomotor disorder, 46/498, 505
organic aspect, 46/419-422
organolead intoxication, 64/133
oxitriptan, 46/496
9p partial monosomy, 43/512
paranoid, see Paranoid schizophrenia
parietal lobe, 46/458, 489
Parkinson disease, 46/447
PEG, 46/450, 481-487
perception, 46/491
platelet MAO, 48/94
positron emission tomography, 46/486
power spectrum analysis, 46/492
prefrontal leukotomy, 45/26
premorbid personality, 46/490
prevalence, 43/212
prolactin secretion, 46/475
propranolol, 37/447
pseudodementia, 46/223
psychological task, 46/486
psychopharmacology, 46/446-448
psychosis, 15/599
regional brain blood flow, 46/453
REM sleep, 46/495
secondary pigmentary retinal degeneration, 13/321
sleep, 46/495
somatosensory evoked potential, 46/500
somatosensory extinction, 46/505
speech disorder, 46/490
stereognosis, 46/490
striatopallidodentate calcification, 49/423
symptom correlation, 46/487
temporal lobe, 46/458, 493, 508
temporal lobe epilepsy, 46/419, 460
thalamus, 2/431
three syndrome model, 46/510
traumatic psychosis, 24/527
vestibular system, 46/491
visual hallucination, 46/562
Wernicke aphasia, 45/313
whole brain estimate, 46/485
XXX syndrome, 43/552
Schizophrenia type I syndrome
  features, 46/510
Schizophrenia type II syndrome
  features, 46/510
Schizophrenic syndrome
  adrenomyeloneuropathy, 60/170
  childhood, see Childhood schizophrenic syndrome
Schizophreniform psychosis

amphetamine intoxication, 65/257
brain tumor, 46/420
encephalitis lethargica, 46/420
head injury, 46/420
organic delusional syndrome, 46/420
systemic lupus erythematosus, 46/420
temporal lobe epilepsy, 46/420
Schluckatmung, see Swallow breath
Schmid-Fraccaro syndrome, see Cat eye syndrome
Schmidt-Lantermann incisure
    diagram, 51/12
    dystrophic mouse, 51/12
    function, 51/11
    infantile spinal muscular atrophy, 51/12
    myelin, 9/27
    nerve, 51/9, 12
    segmental demyelination, 9/34, 51/69
    wobbler mouse, 51/12
Schmidt syndrome
    definition, 2/100
    features, 1/189
    Gignoux syndrome, 2/100
    Jackson syndrome, 1/189
    medulla oblongata syndrome, 2/224
    nonrecurrent nonhereditary multiple cranial
        neuropathy, 51/571
Schmincke tumor
    chordoma, 18/159
Schmorl school
    history, 7/447
    thoracic spine, 7/447
Schnabel cavern
    optic atrophy, 13/52
    optic nerve, 13/78
Schollen, see Spheroid body
Scholz-Bielschowsky-Henneberg diffuse sclerosis,
    see Metachromatic leukodystrophy
Scholz diffuse sclerosis, see Metachromatic
    leukodystrophy
Scholz drüsige angiopathy, see Brain amyloid
    angiopathy
Scholz, W., 10/11
Schönlein-Henoch purpura, see Henoch-Schönlein
    purpura
Schönlein purpura, see Henoch-Schönlein purpura
School
    migraine, 5/244
    occupational therapy, 41/481
    rehabilitation, 41/481
    Schmorl, see Schmorl school
Schradan, see Octamethyl pyrophosphoramide
Schröder, A.H., 1/17
Schuller formula

multiple sclerosis, 47/94
Schultz, F., 1/3, 8, 17, 22
Schultz procedure
    cholesterol, 9/26, 29
Schunkelreaktion
    cerebellar ablation, 2/397
Schut family ataxia, 60/481-488
    biochemical study, 60/485
    brain amino acid, 60/485
    bulbar paralysis, 60/658
    chromosome 6, 60/488
    clinical features, 60/481
    computer assisted tomography, 60/484
    enzyme study, 60/486
    excitatory aminoacid receptor, 60/486
    genetic linkage, 60/487
    laboratory study, 60/483
    motor nerve conduction velocity, 60/483
    neuropathology, 60/483
    nuclear magnetic resonance, 60/484
    oculomotor study, 60/484
    plasma aminoacid, 60/485
    positron emission tomography, 60/484
    psychometric study, 60/484
Schützenberger-Barré-Liéou syndrome
    cervical headache, 5/192
Schut-Haymaker olivopontocerebellar atrophy, see
    Olivopontocerebellar atrophy
        (Schut-Haymaker)
Schwabach test
    deafness, 2/73
    examination method, 2/73
    mastoid process, 2/73
    tuning fork test, 2/73
Schwann cell
    acid phosphatase, 51/10
    adrenoleukodystrophy, 47/593, 51/384
    autoantigen, 9/505
    axon diameter, 51/18
    blue nevus, 14/599
    cytoplasm, 47/10
    dermatology, 14/599
    Elzholz body, 51/10
    experimental allergic neuritis, 9/531-536
    experimental demyelination, 47/7
    function, 51/16
    histogenesis, 8/426
    inner mesaxon, 51/9
    lamina basalis, 47/5
    Lantermann-Schmidt incisure, 47/3, 7
    leprous neuritis, 51/226
    metachromatic leukodystrophy, 8/16, 10/372,
        42/494, 51/368

multiple sclerosis, 47/244
Mycobacterium leprae, 51/223
myelin formation, 47/1
myelin remodeling, 51/18
myelinated nerve fiber, 7/10
myelination, 7/24, 51/17
nerve, 51/1, 9, 11
nerve myelin, 51/23
neurinoma, 8/461, 14/24
neurinoma *vs* neurofibrome, 14/30
neurogenic sarcoma, 8/446
neuroma, 8/416
organochlorine insecticide intoxication, 64/204
organophosphate intoxication, 64/229
origin, 51/17
outer mesaxon, 51/9
periaxonal space, 51/10
periaxonal submyelinic space, 47/4
Ranvier node, 7/22, 9/41, 47/10, 51/9
Refsum disease, 42/148
Reich μ-granule, 51/9
Reich π-granule, 51/9
unmyelinated nerve fiber, 7/10, 12, 24, 73
Schwann cell-axon interaction
  myelination, 51/23
Schwann cell development
  nerve growth factor, 51/24
Schwann cell disorder
  hereditary multiple recurrent mononeuropathy,
    51/558
Schwann cell inclusion
  Cockayne-Neill-Dingwall disease, 51/399
Schwann cell microvilli
  Ranvier node, 47/10, 51/14, 16
Schwann cell paranode
  Ranvier node, 47/10, 51/13
Schwann cell process
  unmyelinated nerve fiber, 51/17
Schwann cell proliferation
  hypertrophic interstitial neuropathy, 51/72
Schwann cell vacuole
  a-alphalipoproteinemia, 51/397
  Farber disease, 51/379
  fucosidosis, 51/381
  galactosialidase deficiency, 51/380
  mucolipidosis type II, 51/382
  Salla disease, 51/382
Schwannoglioma, *see* Neurinoma
Schwannoid oligodendroglioma
  optic nerve, 18/83
Schwannoma, *see* Neurinoma
Schwannomatosis, 68/564
Schwannosis

intramedullary, *see* Intramedullary schwannosis
Schwartz-Jampel syndrome, *see* Chondrodystrophic
  myotonia
Schwartz-Lelek syndrome
  alkaline phosphatase, 31/257
  frontal bossing, 31/257
  genu varum, 31/257
  hyperostosis, 31/257
  Pyle disease, 31/257
Sciatic nerve
  compression neuropathy, 51/102
  entrapment, 7/315
  femoral fracture, 70/38
  Guillain-Barré syndrome, 51/254
  hip fracture, 70/37
  lesion site diagram, 2/49
  lipocalcinogranulomatosis, 43/185
  neurinoma, 8/428
  occupational lesion, 7/33
  Rübenzieherlähmung, 7/334
  stretching, 7/34
  topographical diagnosis, 2/43
Sciatic nerve palsy
  aorta aneurysm, 63/49
Sciatic neuropathy
  aorta aneurysm, 63/49
  iatrogenic neurological disease, 63/177
  intra-aortic balloon, 63/178
Sciatica
  aorta aneurysm, 63/49
  Fincher syndrome, 20/367
  melitococcal, *see* Melitococcal sciatica
  paralysing, *see* Paralysing sciatica
  phantom limb pain, 4/226, 229, 45/400
  proctalgia, 20/689
  spinal tumor, 19/71
Sciatique paralysante, *see* Paralysing sciatica
Scimitar sacrum
  anterior sacral meningocele, 32/194
Scintigraphy
  brain, *see* Brain scintigraphy
  spinal angiography, 12/633
  spinal angioma, 12/633
  subarachnoid hemorrhage, 12/157
Scintillating scotoma
  headache, 48/442
  migraine, 48/4, 65, 155
  pheochromocytoma, 39/501
Scintillation
  brain scanning, 11/247-249
  hallucination classification, 45/56
Sclerodactyly
  carpal tunnel syndrome, 7/295

Scleroderma, 39/355-372
  see also Progressive systemic sclerosis
  autonomic dysfunction, 71/113
  cardiovascular, 39/363
  cerebrovascular disease, 71/114
  circumscribed, *see* Morphea
  classification, 71/4
  clinical features, 39/358-360, 362
  CNS, 39/365-367
  congenital hemifacial atrophy, 22/548
  corticosteroid myopathy, 71/110
  dermatomyositis, 62/374, 71/106
  diffuse cutaneous systemic sclerosis, 71/101
  distal axonal polyneuropathy, 71/111
  eosinophilia myalgia syndrome, 63/375, 71/115
  eosinophilic fasciitis, 71/115
  gastrointestinal, 39/363
  generalized, 71/104
  hemiballismus, 49/373
  Huntington chorea, 49/291
  inclusion body myositis, 62/374, 71/109
  inflammatory cell, 71/107
  kidney, 39/363
  linear, 71/104
  liver, 39/363
  localized, 71/103
  lung, 39/362
  mental change, 39/367
  morphea, 71/104
  musculoskeletal, 39/364
  myasthenia gravis, 71/114
  myelopathy, 71/114
  myopathy, 71/105, 108
  myositis, 71/105
  neuropathy, 8/121, 39/365, 71/111, 113
  ocular, 39/364
  pathology, 39/361
  polyarteritis nodosa, 39/295, 55/361
  polymyositis, 62/374, 71/105
  polyneuropathy, 71/110
  primary, 71/99
  progressive hemifacial atrophy, 8/347, 14/777,
    22/548, 31/253, 43/408, 59/479
  Raynaud phenomenon, 8/339, 43/71
  silicone breast implant, 71/115
  skin, 39/359-361
  spinal cord, 39/366
  survey, 71/99
  systemic, *see* Progressive systemic sclerosis
  systemic brain infarction, 11/453
  systemic sclerosis, 71/100, 102
  toxic oil syndrome, 63/380, 71/115
  treatment, 39/371
  trigeminal sensory neuropathy, 71/112
  vasculitic neuropathy, 71/112
  vasculopathy, 71/105, 108
Sclerohyalinosis, *see* Arteriolar fibrinohyalinoid
    degeneration
Sclérose centrolobaire symmétrique, *see*
    Pelizaeus-Merzbacher disease
Sclérose en îles, *see* Multiple sclerosis
Sclérose en plaques, *see* Multiple sclerosis
Sclérose en taches, *see* Multiple sclerosis
Sclerosing leukoencephalitis
  subacute, *see* Subacute sclerosing panencephalitis
Sclerosing panencephalitis
  subacute, *see* Subacute sclerosing panencephalitis
Sclerosis
  acute multiple, *see* Acute multiple sclerosis
  Ammon horn, *see* Ammon horn sclerosis
  amyotrophic lateral, *see* Amyotrophic lateral
    sclerosis
  cerebral, *see* Diffuse sclerosis
  choroid, *see* Choroid sclerosis
  chronic multiple, *see* Chronic multiple sclerosis
  concentric, *see* Concentric sclerosis
  cortical laminar, *see* Cortical laminar sclerosis
  diffuse, *see* Diffuse sclerosis
  diffuse cutaneous systemic, *see* Diffuse cutaneous
    systemic sclerosis
  disseminated, *see* Multiple sclerosis
  familial amyotrophic lateral, *see* Familial
    amyotrophic lateral sclerosis
  familial infantile cerebral, *see* Familial infantile
    cerebral sclerosis
  hereditary multiple, *see* Hereditary multiple
    sclerosis
  hippocampal, *see* Hippocampal sclerosis
  inflammatory diffuse, *see* Inflammatory diffuse
    sclerosis
  lobular, *see* Lobular sclerosis
  mesial temporal, *see* Mesial temporal sclerosis
  Mönckeberg, *see* Mönckeberg sclerosis
  multiple, *see* Multiple sclerosis
  myelinoclastic diffuse, *see* Myelinoclastic diffuse
    sclerosis
  myocardial, *see* Myocardial sclerosis
  neurogenic acro-osteolysis, 42/294
  progressive systemic, *see* Progressive systemic
    sclerosis
  pseudo, *see* Rigid Huntington chorea
  systemic, *see* Systemic sclerosis
  tuberous, *see* Tuberous sclerosis
  Westphal-Strümpell pseudo, *see* Creutzfeldt-Jakob
    disease
Sclerosteosis

alkaline phosphatase, 31/258, 43/475
calvaria, 31/258
clavicle, 31/258
compression neuropathy, 43/475
facial paralysis, 31/258, 43/475
frontal bossing, 43/474
gene frequency, 43/475
headache, 31/258, 38/412, 43/475
hereditary cranial nerve palsy, 60/43
hyperostosis, 31/258
hypertelorism, 31/258
mandible, 31/258
optic atrophy, 31/258, 43/475
osteosclerosis, 31/258
pain, 43/475
papilledema, 43/475
sensorineural deafness, 31/258, 43/474
syndactyly, 43/474
Van Buchem disease, 31/258
Sclerotic cerebellar atrophy, see Focal cerebellar
    cortical panatrophy
Sclerotic myeloma
    multiple myeloma, 47/618
Sclerotome
    cervical, see Cervical sclerotome
Scoliosis
    see also Kyphoscoliosis and Lordosis
    achondroplasia, 50/417
    adolescent idiopathic type, 50/418
    Aicardi syndrome, 42/696
    angiodysgenetic necrotizing myelomalacia,
        20/487
    anterior longitudinal ligament, 50/415
    Arnold-Chiari malformation type I, 40/337
    arthrogryposis, 50/417
    ataxia, 42/348
    autosomal dominant, 50/418
    Beals syndrome, 30/101
    Biemond syndrome, 42/164
    bracing, 41/477
    cleidocranial dysostosis, 50/418
    Coffin-Lowry syndrome, 30/101
    congenital, 32/148-156
    congenital fiber type disproportion, 41/15
    congenital spondyloepiphyseal dysplasia, 43/478
    diastematomyelia, 32/253, 42/24, 50/416, 435
    Duchenne muscular dystrophy, 40/362, 62/119
    duplication, 50/416
    dysautonomia, 50/418
    Ehlers-Danlos syndrome, 43/14, 50/418
    embryology, 50/415
    epidemiology, 30/162, 50/417
    etiology, 50/417

facioscapulohumeral muscular dystrophy, 40/422
Friedreich ataxia, 51/388, 60/313, 325
genetics, 30/162
glycosaminoglycanosis type VII, 66/310
Goltz-Gorlin syndrome, 14/788
hemivertebra, 42/31
hereditary factor, 50/418
hereditary sensory and autonomic neuropathy type
    I, 21/77
history, 50/413
homocystinuria, 50/418
horizontal gaze paralysis, 42/332
idiopathic, see Idiopathic scoliosis
incidence, 50/54
intermediate spinal muscular atrophy, 42/90
intervertebral disc, 50/413
juvenile spinal cord injury, 25/183
Klippel-Feil syndrome, 40/337, 42/38
kyphosis, 50/418
late onset, 32/151
lordosis, 50/414, 416
lumbar intervertebral disc prolapse, 20/588
Marfan syndrome, 50/417
mechanics, 50/414
mental deficiency, 43/124
multicore myopathy, 43/120
muscle stretch, 50/414
muscular dystrophy, 41/478, 50/417
myopathy, 40/337, 362, 442, 43/124
neurodegeneration, 21/3
neurofibromatosis, 50/417
neurofibromatosis type I, 50/369
neuroleptic dystonia, 65/281
nevus unius lateris, 43/40
operation, 50/421
osteogenesis imperfecta, 50/418
osteoporosis, 50/418
poliomyelitis, 34/113, 40/337, 50/417
posterior column ataxia, 42/164
rachischisis, 32/177
radiology, 50/420
rehabilitation, 41/475
Rett syndrome, 29/312, 60/637
Rubinstein-Taybi syndrome, 31/284, 43/234
segmentation failure, 50/416
self-mutilation encephalopathy, 42/141
sensory neuropathy, 42/348
sex ratio, 50/64
spasmodic torticollis, 42/264
spina bifida, 32/545
spinocerebellar ataxia, 50/418
spinocerebellar degeneration, 42/190
Sprengel deformity, 1/483

syringomyelia, 32/258, 271, 42/59, 50/418, 445
treatment, 41/475, 60/325
vertebral column, 32/132, 148, 151, 177
vertebral torsion, 50/417
X-linked dominant inheritance, 50/64
xeroderma pigmentosum, 60/336
Scombroid intoxication
  classification, 37/78
  Clostridium, 37/85
  Escherichia coli, 37/85
  histamine, 37/85
  Proteus morganii, 37/85
  Salmonella, 37/85
Scopolamine
  see also Hallucinogenic agent
  anticholinergic activity, 65/48
  behavior disorder, 46/596
  belladonna alkaloid, 65/48
  biokinetic, 65/48
  Datura stramonium, 65/48
  delirium, 46/537
  hallucinogenic agent, 65/42
  memory, 46/596
Scopolamine intoxication
  hallucination, 65/48
  mydriasis, 65/48
  symptom, 65/48
  treatment, 65/49
Scorpaena intoxication
  convulsion, 37/76
  death, 37/76
  delirium, 37/76
  diarrhea, 37/76
  hypotension, 37/76
  hypothermia, 37/76
  ischemia, 37/76
  local pain, 37/76
  nausea, 37/76
  pallor, 37/76
  radiation, 37/76
  respiratory distress, 37/76
  vomiting, 37/76
Scorpaenidae intoxication
  picture, 37/73
Scorpaenodes intoxication
  convulsion, 37/76
  death, 37/76
  delirium, 37/76
  diarrhea, 37/76
  hypotension, 37/76
  hypothermia, 37/76
  ischemia, 37/76
  local pain, 37/76

  nausea, 37/76
  pallor, 37/76
  radiation, 37/76
  respiratory distress, 37/76
  vomiting, 37/76
Scorpaenopsis intoxication
  convulsion, 37/76
  death, 37/76
  delirium, 37/76
  diarrhea, 37/76
  hypotension, 37/76
  hypothermia, 37/76
  ischemia, 37/76
  local pain, 37/76
  nausea, 37/76
  pallor, 37/76
  radiation, 37/76
  respiratory distress, 37/76
  vomiting, 37/76
Scorpion sting intoxication
  anticonvulsant, 37/112
  convulsion, 37/112
  heart failure, 37/112
  hypertension, 37/112
  muscle rigidity, 37/112
  mydriasis, 37/112
  myocardial infarction, 37/112
  pain, 37/112
  pulmonary edema, 37/112
  salivation, 37/112
  skeletal muscle, 37/112
  sweating, 37/112
  swelling, 37/112
Scorpion venom
  arthropod envenomation, 65/193, 196
  autonomic nervous system, 75/513
  dysphagia, 65/196
  epilepsy, 65/196
  neurotoxin, 37/112, 65/193, 196
  nystagmus, 65/196
  toxic encephalopathy, 65/193
Scotoma
  see also Visual field defect
  altitudinal, see Altitudinal scotoma
  amyl nitrate, 48/65
  asthenopic, see Asthenopic scotoma
  Bassen-Kornzweig syndrome, 10/549
  cardiovascular agent intoxication, 37/426
  cecocentral, see Centrocecal scotoma
  central, see Central scotoma
  central serous retinopathy, 13/142
  cervical spondylosis, 7/450
  chloroquine intoxication, 65/484

classic migraine, 48/156
fortification spectra, 48/129
glycoside intoxication, 37/426, 429
hallucination classification, 45/56
homonymous paracentral, *see* Homonymous
    paracentral scotoma
junction, *see* Junction scotoma
metamorphopsia, 48/129
migraine, 5/50, 45/510, 48/3, 64, 126, 135
negative, 48/126
pericentral, *see* Pericentral scotoma
peripheral, *see* Peripheral scotoma
pontine infarction, 12/22
progressive central, 48/126
ring, *see* Ring scotoma
scintillating, *see* Scintillating scotoma
Seidel, *see* Seidel scotoma
sieve, *see* Sieve scotoma
spectral appearance, 48/128
spreading depression, 48/65
visual disorder, 48/127
visual hallucination, 45/356
visual loss, 48/129
Scott syndrome, *see* Tropical ataxic neuropathy
Scrapie
    Alzheimer disease, 46/270
    animal viral disease, 9/679-681, 34/275, 287, 302,
        56/585
    ataxia, 42/656, 56/585
    autonomic nervous system, 42/656
    Creutzfeldt-Jakob disease, 46/289
    maternal effect, 42/656
    metachromatic leukodystrophy of mink, 9/679
    mink encephalopathy, 34/287
    multiple sclerosis, 9/78, 120, 126, 128, 47/330
    prion, 56/566
    prion disease, 56/543
    senile plaque, 56/555
    slow virus disease, 34/275, 287, 302, 56/585
    subacute spongiform encephalopathy, 42/655
    transmissible demyelination, 9/679-681
    transmission, 46/295
    viral mechanism, 56/20
Scrapie virus, *see* Prion
Scratch reflex
    spinal dog, 1/49
Screening
    maple syrup urine disease, 29/66-68, 71
    phenylketonuria, 29/42
    traumatic intracranial hematoma, 57/271
    valproic acid induced hepatic necrosis, 65/128
    vestibular neurinoma, 68/432
Scrivener palsy, *see* Writers cramp

Scrotum
    shawl, *see* Shawl scrotum
Scrotum hyperplasia
    Klinefelter syndrome, 50/550
Scrub typhus
    rickettsial infection, 34/647
Scurvy
    fox intoxication, 37/96
    hemorrhage, 38/71
    leukoencephalopathy, 9/607
    polar bear intoxication, 37/96
    vitamin C deficiency, 28/200, 208, 38/71
    wolf intoxication, 37/96
Scyphozoa intoxication
    actiniaria, 37/53
    anthozoa, 37/53
    boxjelly, 37/41
    carybdeidae, 37/43
    chirodropus, 37/45
    chironex, 45/50
    chrysaora, 37/51-53
    jellyfish, 37/40
    Palythoa vestitus, 37/55
    physalia, 37/54
    rhizostomeae, 37/52
    rhodactis howesi, 37/54
    sea wasp, 37/41
    semaeostomeae, 37/50
SDH, *see* Succinate dehydrogenase
Sea anemone intoxication
    anthopleura elegantissima, 37/54
    equinatoxin, 37/54
    local pain, 37/53
    Palythoa tuberculosa, 37/55
    Palythoa vestitus, 37/55
    physalia, 37/54
    rhodactis howesi, 37/54
    skin, 37/53
    ulceration, 37/53
Sea bass intoxication
    vitamin A intoxication, 37/87
Sea blue histiocyte
    albinism, 60/674
    Batten disease, 60/674
    bone marrow, 60/674
    epilepsy, 60/674
    glycosaminoglycanosis type IH, 60/674
    Hallervorden-Spatz syndrome, 49/401
    juvenile dystonic lipidosis, 42/447
    juvenile neuronal ceroid lipofuscinosis, 60/674
    Niemann-Pick disease type B, 60/674
    Niemann-Pick disease type C, 60/674
    spinal muscular atrophy, 59/103, 372

vertical supranuclear ophthalmoplegia, 42/605
Sea blue histiocyte disease
  Biemond syndrome, 60/449
  bleeding tendency, 42/473
  hepatosplenomegaly, 42/472
  mental deficiency, 42/473
  prevalence, 42/474
  seizure, 42/473
  sphingomyelinase, 42/474
  thrombocytopenia, 42/472
Sea cucumber intoxication
  holothurioidea, 37/66
Sea lillie intoxication
  Echinodermata, 37/66
Sea slug intoxication
  holothurioidea, 37/66
Sea snake intoxication, 37/91-95
  antivenene, 37/11, 93
  anxiety, 37/92
  biochemistry, 37/92
  bulbar paralysis, 37/92
  cyanosis, 37/92
  euphoria, 37/92
  facial paralysis, 37/92
  hemoglobinuria, 37/92
  hyperkalemia, 37/92
  hypoxia, 37/92
  local pain, 37/10
  malaise, 37/92
  myalgia, 37/10, 92
  myoglobin, 37/10
  myoglobinuria, 37/92
  nausea, 37/92
  paralysis, 37/92
  parotid swelling, 37/92
  pathology, 37/92
  pharmacology, 37/92
  ptosis, 37/10, 92
  pupil, 37/92
  renal insufficiency, 37/92
  respiratory failure, 37/92
  respiratory paralysis, 37/92
  speech reduction, 37/92
  swallowing, 37/92
  sweating, 37/92
  symptomatology, 37/92
  teeth, 37/92
  tongue, 37/92
  toxicology, 37/93
  treatment, 37/11, 93
  trismus, 37/10, 92
  venom, 37/56
  vomiting, 37/92

Sea squirt intoxication
  chordata, 37/68
Sea star intoxication
  asteroidea, 37/66
Sea urchin intoxication
  echinoidea, 37/66-68
Sea wasp intoxication
  fatty degeneration, 37/41
  necrosis, 37/41
  skin, 37/41
Seasonal rhythm
  see also Biological rhythm
  disease, 74/490
  light, 74/490
  ovulation, 74/490
  pregnancy, 74/490
  suprachiasmatic nucleus, 74/490
Seat belt injury
  axial compression burst, 61/34
  chance fracture, 61/90
  columnar mechanics, 61/36
  thoracolumbar spine injury, 25/448
Sebaceous adenoma
  Balzer-Menetrier type, 14/48, 574, 578
  Balzer type, 14/342
  Caspary type, 14/342
  Darier type, 14/342
  Hallopeau-Leredde type, 14/48, 342, 574, 578
  multiple trichoepithelioma, 14/579
  neurocutaneous syndrome, 14/101
  Pringle type, 14/8, 48, 342, 574, 578
  tuberous sclerosis, 14/11, 48, 341, 572, 31/6, 43/9, 49, 50/371
Sebaceous gland agenesis
  neurocutaneous melanosis, 14/597
Sebaceous secretion
  behavior, 74/377
Sebaceous syndrome
  linear nevus, see Linear nevus sebaceous syndrome
Sebastapistes intoxication
  convulsion, 37/76
  death, 37/76
  delirium, 37/76
  diarrhea, 37/76
  hypotension, 37/76
  hypothermia, 37/76
  ischemia, 37/76
  local pain, 37/76
  nausea, 37/76
  pallor, 37/76
  picture, 37/73
  radiation, 37/76

respiratory distress, 37/76
vomiting, 37/76
Sebastapistes mcadamsi intoxication
marine toxin, 37/73
picture, 37/73
Seborrhea
ataxia telangiectasia, 60/370
blepharitis, 60/370
dermatitis, 60/370
Parkinson disease, 6/191, 75/179
pellagra, 28/89
Seckel dwarf
classification, 30/508
hypertelorism, 30/248
microcephaly, 30/100, 508
skeletal malformation, 30/248
Seckel syndrome
birth incidence, 43/379
cheilopalatoschisis, 43/379
craniosynostosis, 43/379
cryptorchidism, 43/379
ear abnormality, 43/379
hypertelorism, 43/379
mental deficiency, 31/285, 43/378, 46/47
microcephaly, 31/285, 43/378
micrognathia, 43/379
multiple joint dislocation, 43/378
talipes, 43/379
Second pain
anatomy, 1/85
Second wind phenomenon
myophosphorylase deficiency, 41/186, 189,
62/486
phosphofructokinase deficiency, 41/190, 43/183,
62/487
Secondary acalculia, 4/186-192
constructive acalculia, 4/190
frontal acalculia, 4/189
occipital acalculia, 4/188
parietal acalculia, 4/189
Secondary adrenal insufficiency
classification, 70/211
Secondary agraphia, 4/141-144
classification, 4/144
Secondary amyloidosis
hereditary sensory and autonomic neuropathy type
II, 60/17
orthostatic hypotension, 22/253, 63/149
Secondary autonomic failure
sleep, 74/547
Secondary dystonia
symptomatic dystonia, 49/541
Secondary narcolepsy

postencephalitic hypersomnia, 3/96
Secondary neuroaxonal dystrophy
alcoholic encephalopathy, 49/408
alcoholism, 49/398
biliary atresia, 49/408
glycogen storage disease, 49/408
GM2 gangliosidosis, 49/408
mucoviscidosis, 49/408
neurolipidosis, 49/398
Parkinson disease, 49/408
trichopoliodystrophy, 49/408
vitamin E deficiency, 49/408
Wilson disease, 49/408
Secondary pigmentary retinal degeneration,
13/213-225
see also Primary pigmentary retinal degeneration
acute sensory neuropathy, 60/736
adenosine triphosphatase 6 mutation syndrome,
62/510
Alezzandrini syndrome, 13/186
Alström-Hallgren syndrome, 13/451, 460-463,
60/720
angioid streak, 13/237
apoceruloplasmin deficiency, 60/730
Armendares syndrome, 43/366
arteriosclerosis, 13/218
associated ocular disease, 13/178, 249-252
associated otological disease, 13/239-242
ataxia, 60/654, 660, 670, 674, 722
ataxia telangiectasia, 60/654
atypical form, 13/179-191
Bardet-Biedl syndrome, 13/242, 288-297, 337,
43/233, 60/719
Bassen-Kornzweig syndrome, 10/279,
13/316-318, 413-416, 21/580, 42/511, 51/326,
393-395, 60/653, 727
Batten disease, 10/575, 13/327-330, 42/464
benign essential tremor, 60/731
Biemond syndrome type II, 60/720
biochemistry, 13/295
blood pressure, 13/197
Bonnet-Dechaume-Blanc syndrome, 13/335,
60/736
brain amyloid angiopathy, 54/337
CAMAK syndrome, 60/730
CAMFAK syndrome, 60/730
central choroidal atrophy, 13/279, 281
cerebellar ataxia, 13/304-307, 60/653, 735
Charlevoix-Saguenay spastic ataxia, 60/654
chickenpox, 13/214
chloroquine intoxication, 60/736
choreoathetosis, 49/385
choroid sclerosis, 13/278-288

chronic progressive external ophthalmoplegia, 60/725
Cockayne-Neill-Dingwall disease, 13/322-324, 43/350-352, 60/654, 729
congenital retinal blindness, 22/536, 60/724
congenital seesaw nystagmus, 60/737
deaf mutism, 13/150, 239-242, 453
deafness, 60/722, 724
deferoxamine intoxication, 60/736
dementia, 60/654, 722
diabetes mellitus, 60/660, 674
dominant olivopontocerebellar atrophy, 60/654
Down syndrome, 13/338
Duchenne muscular dystrophy, 60/723
dwarfism, 13/152, 324-326
dystonia musculorum deformans, 13/335, 60/730
epilepsy, 13/337-340
ethambutol intoxication, 60/736
external ophthalmoplegia, 13/179
familial hemiplegic migraine, 48/27, 60/737
familial hypobetalipoproteinemia, 60/728
familial spastic paraplegia, 22/467-473, 60/733
Flynn-Aird syndrome, 14/112, 22/519, 42/327, 60/730
Friedreich ataxia, 13/297-304, 21/347, 60/310, 654, 734
α-L-fucosidase deficiency, 60/726
Gardner disease, 60/737
glioma polyposis syndrome, 60/737
glycosaminoglycanosis, 13/332
glycosaminoglycanosis type I, 10/436, 42/450-452
glycosaminoglycanosis type IH, 60/176
glycosaminoglycanosis type IS, 42/452-455
glycosaminoglycanosis type II, 10/436, 42/454
glycosaminoglycanosis type IIA, 60/176
glycosaminoglycanosis type IIIA, 60/176
Goltz-Gorlin syndrome, 14/113
Gordon-Capute-Konigsmark syndrome, 60/721
Hallervorden-Spatz syndrome, 6/606, 610, 13/336, 60/731
hereditary ataxia, 13/297-307
hereditary hearing loss, 22/499-503
hereditary hemorrhagic telangiectasia (Osler), 13/251
hereditary motor and sensory neuropathy variant, 60/246
hereditary neuropathy, 60/265, 733
hereditary sensory and autonomic neuropathy type II, 60/11
hereditary spastic ataxia, 21/383-389, 60/734
homocarnosinosis, 42/557-559, 60/727
human T-lymphotropic virus type I associated myelopathy, 60/736

hyperpipecolatemia, 13/326
hypertrophic interstitial neuropathy, 13/316, 21/151
hypertrophic polyneuropathy, 13/316
hypobetalipoproteinemia, 60/653
hypogonadism, 60/722
infantile neuronal ceroid lipofuscinosis, 60/654
infantile Refsum disease, 60/654
influenza, 13/214
inverse type, see Inverse secondary pigmentary retinal degeneration
Jaeken syndrome, 60/726
Joubert syndrome, 60/734
Kearns-Sayre-Daroff-Shy syndrome, 22/211, 43/142, 60/654, 725
Kjellin syndrome, 60/733
late infantile ceroid lipofuscinosis, 60/654
Laurence-Moon-Bardet-Biedl syndrome, 22/536
Laurence-Moon syndrome, 43/253-255, 60/719, 733
Lindenov-Hallgren syndrome, 13/451-460
Loken-Senior syndrome, 60/735
Lowry syndrome, 60/720
Marinesco-Sjögren syndrome, 13/320, 21/560, 60/654
Matthews-Rundle syndrome, 60/654
McLeod phenotype, 63/289
MELAS syndrome, 60/725
mental deficiency, 13/337-340, 43/293-296, 60/654, 722
mental deficiency syndrome, 60/719
MERRF syndrome, 60/725
metachromatic leukodystrophy, 60/732
Mirhosseini-Holmes-Walton syndrome, 60/720
mitochondrial encephalomyopathy, 60/725
mitochondrial genomic disease, 60/725
mitochondrial myopathy, 60/725
mucolipidosis type IV, 60/727
muscular atrophy, 13/306, 308
muscular dystrophy, 13/319
myoclonus, 13/339
myophosphorylase deficiency, 60/727
myopia, 13/218
myotonic dystrophy, 13/318, 60/723
neonatal adrenoleukodystrophy, 60/654, 729
nerve conduction velocity, 60/670
neuroacanthocytosis, 13/427
neurogenic muscular atrophy, 13/308, 60/674
neuronal ceroid lipofuscinosis, 10/299, 42/464, 60/732
neuropathy, 42/334, 60/660
Niemann-Pick disease, 13/330
nystagmus, 13/296, 321

295

oculo-otocerebrorenal syndrome, 60/721
oculorenocerebellar syndrome, 60/721
Oliver-McFarlane syndrome, 60/721
olivopontocerebellar atrophy, 13/305, 60/735
olivopontocerebellar atrophy variant, 21/452
olivopontocerebellar atrophy with retinal
  degeneration, 60/653
onchocerciasis, 13/218
ophthalmoplegia, 22/191-193, 310-314
orthochromatic leukodystrophy, 10/113
Parinaud syndrome, 13/304
Pelizaeus-Merzbacher disease, 13/331, 60/732
piperidylchlorophenothiazine, 13/210
poliosis, 13/186
polyneuropathy, 42/334, 60/653, 670
progressive external ophthalmoplegia,
  13/310-314, 22/204, 60/654, 725
progressive muscular atrophy, 13/326
progressive myoclonus epilepsy, 42/704
Refsum disease, 8/22, 10/345, 13/40, 242, 314,
  479, 21/9, 182, 184, 189, 192, 211, 22/501,
  27/522, 36/348, 40/514, 41/434, 42/148, 51/384,
  60/228, 235, 653, 728, 66/485, 488
renal disease, 13/308
retinal detachment, 13/218
retinotoxic agent, 13/219-225
rubella, 13/214
Rud syndrome, 13/321, 479, 51/398, 60/721
schizophrenia, 13/321
sensorineural deafness, 42/392
Sjögren-Larsson syndrome, 13/319, 22/477,
  43/307, 60/733
skeletal abnormality, 60/735
spastic paraplegia, 13/307, 42/177, 179
sphingolipidosis, 13/326-330
spinal muscular atrophy, 59/374
spinocerebellar ataxia, 13/304-307
Stargardt disease, 13/28
striatopallidodentate calcification, 13/336, 49/424,
  60/730
Sturge-Weber syndrome, 13/334, 14/512, 60/735
synonym, 60/718
syphilis, 13/216
syphilitic neuroretinitis, 60/736
tetraplegia, 13/307
trauma, 13/216
Usher syndrome, 13/253, 443, 60/654, 722
vascular occlusion, 13/216
virus, 13/216
vitamin A deficiency, 60/736
vitamin E, 13/194
vitiligo, 13/186
Von Hippel-Lindau disease, 60/735

Secondary polycythemia
  carbon monoxide encephalopathy, 63/250
  cause, 63/250
  cerebellar hemangioblastoma, 63/250
  cystic kidney disease, 63/250
  2,3-diphosphoglyceric acid deficiency, 63/250
  hepatocellular hemangioblastoma, 63/250
  high altitude, 63/250
  hypernephroma, 63/250
  nocturnal oxygen desaturation, 63/250
  polycythemia vera, 55/469
  pulmonary hypoventilation, 63/250
  renal artery stenosis, 63/250
Secondary proliferation layer
  CNS, 30/65-70, 50/35
Secondary subcortical epilepsy
  classification, 15/11
Secondary syphilis
  condyloma lata, 52/273
  nervous system, 33/349
  temporal arteritis, 55/347
Secretion
  calcium, 28/531
  CSF production, 32/530
  growth hormone, see Growth hormone secretion
  inappropriate antidiuretic hormone, see
    Inappropriate antidiuretic hormone secretion
  nasal, see Nasal secretion
  prolactin, see Prolactin secretion
  sebaceous, see Sebaceous secretion
  stomach, 1/506
  sweat, see Sweat secretion
  vasopressin, see Inappropriate antidiuretic
    hormone secretion
Secretor gene
  ABH, see ABH secretor gene
Sedation
  brain infarction, 53/432
  clonidine intoxication, 37/443
  delirium, 46/553
  methyldopa intoxication, 37/441
  migraine, 5/381, 48/39
  phenobarbital intoxication, 37/199
  propranolol intoxication, 37/447
  reserpine intoxication, 37/438
Sedative agent, see Hypnotic agent
Sedimentation rate
  erythrocyte, see Erythrocyte sedimentation rate
Seesaw nystagmus
  astrocytoma, 16/323
  congenital, see Congenital seesaw nystagmus
  craniopharyngioma, 16/323
  optic chiasm injury, 1/605, 16/323, 24/44

pituitary tumor, 16/323
vestibular system, 16/323
Segawa dystonia
biological rhythm, 74/473
Segment
cervical sclerotome, 50/384
internal carotid artery, 53/293
medial pallidal, *see* Medial pallidal segment
vertebral artery, 53/374
Segment pointer muscle
definition, 2/161
radicular syndrome, 20/557
ventral spinal nerve root, 2/160-162
Segmental athetosis
definition, 49/382
dystonia, 49/382
Segmental demyelination
alcoholic polyneuropathy, 51/318
buckthorn toxin, 64/14
carbon disulfide intoxication, 64/26
Chagas disease, 52/346, 348
chloroquine, 65/484
chloroquine intoxication, 65/484
chronic inflammatory demyelinating
polyradiculoneuropathy, 51/532
diphtheria toxin, 64/14
diphtheritic neuropathy, 51/70
eosinophilia myalgia syndrome, 64/255
ethambutol polyneuropathy, 51/295
experimental allergic neuritis, 51/70
globoid cell leukodystrophy, 51/71, 373
Guillain-Barré syndrome, 51/70, 243
hereditary motor sensory neuropathy type I,
51/264
hereditary multiple recurrent mononeuropathy,
51/551
hexachlorophene, 64/14
human T-lymphotropic virus type I associated
myelopathy, 56/531
hypertrophic interstitial neuropathy, 21/162, 51/72
lead, 51/263
lead intoxication, 9/642, 51/71
lead polyneuropathy, 51/271, 64/436
leprous neuritis, 51/224, 231
macrophage, 51/68
metachromatic leukodystrophy, 51/71
monoclonal gammopathy, 51/71
multiple myeloma polyneuropathy, 51/431, 63/393
muscarine, 64/14
nerve conduction, 7/155
neuropathologic features, 7/222-225
neuropathology, 51/63
neuropathy target esterase inhibition, 64/160

neurotoxicity, 51/294
organophosphate, 64/160
paraneoplastic polyneuropathy, 51/467
perhexiline, 64/14
perhexiline maleate polyneuropathy, 51/302
POEMS syndrome, 47/619, 51/431, 63/394
polyarteritis nodosa, 51/453
rabies, 56/395
rheumatoid arthritis, 51/446
Schmidt-Lantermann incisure, 9/34, 51/69
thallium, 64/14
uremic neuropathy, 7/166, 8/2, 27/339
uremic polyneuropathy, 63/511
Waldenström macroglobulinemia, 51/71
Segmental demyelination neuropathy
hereditary recurrent neuropathy, 21/90-98
Segmental dystonia
classification, 49/521
encephalitis, 49/541
forel field lesion, 6/151
peripheral trauma, 49/542
tardive dyskinesia, 49/186
Segmental fibrinoid necrosis
Henoch-Schönlein purpura, 9/607
Segmental innervation, 2/157-175
see also Dermatome
histamine skin test, 2/173
history, 2/157-159
lower urinary tract, 61/292
methodology, 2/159
phylogeny, 2/130
terminology, 2/159
topographical diagnosis, 2/170, 172
Segmental intubation
cerebrovascular disease, 11/581
Segmental stretch disease
physiologic tremor, 49/566
Segmentation failure
Klippel-Feil syndrome, 32/143
neural arch, 32/133
scoliosis, 50/416
vertebral column, 32/133-137
Segregation
neurofilament, *see* Neurofilament segregation
Sei whale intoxication, *see* Balaenoptera borealis
intoxication
Seidel scotoma
glaucoma, 2/565
Seitelberger disease, *see* Infantile neuroaxonal
dystrophy
Seitelberger, F., 10/32
Seiwal intoxication, *see* Balaenoptera borealis
intoxication

Seizure
  see also Convulsion, Epilepsy *and* Epileptiform
    discharge
  abdominal, *see* Abdominal seizure
  absence, *see* Absence seizure
  acquired toxoplasmosis, 35/125, 52/354
  acrodysostosis, 43/291
  Addison disease, 42/512
  adrenoleukodystrophy, 10/132
  adult neuronal ceroid lipofuscinosis, 42/466
  adversive, *see* Adversive seizure
  afebrile, *see* Afebrile seizure
  Aicardi syndrome, 31/226, 50/165
  amygdala, 73/77
  anencephaly, 30/176
  angiostrongyliasis, 52/554
  anoxic, *see* Syncope
  anterior frontal, 73/43
  anxiety, 72/420
  argininosuccinic aciduria, 42/524
  aspergillosis, 52/378
  athetotic hemiplegia, 42/200
  atonia, 73/224
  atonia astasia, 73/224
  atypical benign partial epilepsy, 73/224
  autism, 43/204
  Bean syndrome, 43/11
  Beckwith-Wiedemann syndrome, 31/330
  bee sting intoxication, 37/107
  Behçet syndrome, 71/225
  benign partial epilepsy with centrotemporal spike,
    73/10
  benign partial epilepsy with occipital paroxysm,
    73/20
  benign partial epilepsy with somatosensory
    evoked potential, 73/26
  Binswanger disease, 46/318
  blood-brain barrier, 28/391
  Bonnet-Dechaume-Blanc syndrome, 43/36
  brain arteriovenous malformation, 53/35
  brain embolism, 53/206
  brain glioma, 72/179
  brain infarction, 53/206
  brain tumor, 67/143, 69/10
  bupivacaine intoxication, 37/405
  cardiovascular agent intoxication, 37/426
  cause, 69/12
  cavernous angioma, 72/209
  cavernous hemangioma, 73/400
  cerebellar hypoplasia, 42/18
  cerebellar vermis agenesis, 42/4
  cerebrohepatorenal syndrome, 31/305
  cerebrovascular disease, 53/36, 206

  childhood absence epilepsy, 73/146
  cingulate, 73/43
  citrullinuria, 42/538
  classification, 72/1-12, 28
  clonic, *see* Clonic seizure
  CNS spongy degeneration, 15/426, 42/506
  Cockayne-Neill-Dingwall disease, 43/350
  Coffin-Lowry syndrome, 31/245
  complex III deficiency, 62/504, 66/423
  concentric sclerosis, 47/414
  congenital Pelizaeus-Merzbacher disease, 15/426
  convulsive, *see* Convulsive seizure
  corpus callosum agenesis, 2/765, 30/287, 42/7, 9,
    101, 50/161
  cri du chat syndrome, 43/501
  CSF shunt, 46/330
  cytomegalovirus infection, 31/212, 215, 56/268
  De Lange syndrome, 43/246, 50/276
  definition, 73/143
  diabetes mellitus, 42/411
  dorsolateral, 73/42
  Down syndrome, 31/406-408, 50/529
  drug, 72/263
  drug induced, *see* Drug induced seizure
  dwarfism, 43/385
  Edwards syndrome, 31/517
  endocrine candidiasis syndrome, 42/642
  enflurane intoxication, 37/412
  epileptic, *see* Epileptic seizure
  Fabry disease, 42/427, 55/457, 60/172
  febrile, *see* Febrile seizure
  fluroxene intoxication, 37/412
  foramina parietalia permagna, 30/275, 278, 42/29
  Friedman-Roy syndrome, 43/251
  frontal lobe epilepsy, 73/39
  frontopolar, 73/43
  fructose-1,6-diphosphatase deficiency, 42/586
  Fukuyama syndrome, 43/92
  generalized minor seizure, 73/224
  glycogen storage disease type I, 42/586
  glycosaminoglycanosis type III, 66/301
  glycoside intoxication, 37/426
  GM$_1$ gangliosidosis type II, 42/433
  GM$_2$ gangliosidosis type II, 42/435
  GM$_2$ gangliosidosis type III, 42/436
  Goldberg syndrome, 42/439
  happy puppet syndrome, 31/309
  hemophilia A, 42/738
  hereditary acute porphyria, 51/391
  herpes virus hominis syndrome, 31/212, 216
  heterochromia iridis, 42/688
  holoprosencephaly, 30/452, 456, 459, 42/32,
    50/238, 240

Huntington chorea, 42/226
hydranencephaly, 30/675, 50/349
hydrocephalus, 30/555, 42/691, 50/291
hyperargininemia, 29/102, 42/562
hyperglycinemia, 42/565
hyperosmolar hyperglycemic nonketotic diabetic
    coma, 27/89
hyperphosphatasia, 42/568
hyperprolinemia type I, 42/569
hyperprolinemia type II, 29/138, 42/569
hypoglycemia, 72/261
hypoparathyroidism, 27/300, 303-305, 42/577
hypothalamic phakoma, 42/737
hypothyroidism, 70/93
hysterical, see Hysterical seizure
iatrogenic agent, 72/227
idiopathic generalized myoclonic astatic epilepsy,
    73/224
idiopathic hypoglycemia, 42/576
intracranial hemorrhage, 53/206
juvenile absence epilepsy, 73/150
juvenile dystonic lipidosis, 42/446
juvenile Gaucher disease, 10/307, 42/438
juvenile Gaucher disease type A, 66/125
kernicterus, 42/583
ketamine intoxication, 37/408
Klinefelter syndrome, 43/558
Köhlmeier-Degos disease, 39/437
late infantile metachromatic leukodystrophy,
    66/167
Lennox-Gastaut syndrome, 73/224
linear nevus sebaceous syndrome, 31/289, 43/38
lissencephaly, 30/486
Listeria monocytogenes meningitis, 63/531
low grade glioma, 68/51
Lowe syndrome, 31/302
macrencephaly, 30/651, 42/41
mannosidosis, 42/597
Marcus Gunn phenomenon, 30/403
Marinesco-Sjögren syndrome, 60/343
megalocornea, 43/270
mental deficiency, 42/498, 43/270, 278, 282, 285
metabolic, 72/221
metabolic cause, 72/261
metabolic encephalopathy, 69/399
methanol intoxication, 36/353, 356
microcephaly, 42/45, 43/429, 50/269
migraine, 48/19
movement induced, see Hereditary paroxysmal
    kinesigenic choreoathetosis
mucolipidosis type I, 42/475
multiple neurofibromatosis, 43/35
multiple sclerosis, 47/56-58, 170

mumps encephalitis, 34/628, 56/430
myelin dysgenesis, 42/498-500
myoclonia, 73/224
myoclonic astatic, 73/224
myoclonic epilepsy, 73/224
myophosphorylase deficiency, 41/186, 189,
    43/181
neonatal, see Neonatal seizure
neuroacanthocytosis, 42/209
neurocutaneous melanosis, 43/33
neurocysticercosis, 46/389, 52/530
neurofibromatosis type I, 31/3
neuron death, 72/108
neuronal ceroid lipofuscinosis, 42/461, 463, 466
nevus unius lateris, 43/40
Niemann-Pick disease type A, 42/468
Niemann-Pick disease type C, 42/471
occipital lobe tumor, 67/154
opercular, 73/42
opticocochleodentate degeneration, 21/541
orbital frontal, 73/43
orthochromatic leukodystrophy, 10/123, 42/500
4p partial monosomy, 43/497
18p partial monosomy, 43/534
4p partial trisomy, 43/500
Paine syndrome, 43/433
palilalia, 43/226
parietal lobe tumor, 67/152
paroxysmal symptom, 47/170
partial, see Partial seizure
Patau syndrome, 31/508, 43/528
pellagra, 7/571, 573
penetrating head injury, 23/352
phenylketonuria, 29/30, 42/610
phosphatidylethanolamine lipidosis, 42/429
poliodystrophy, 42/588
polyarteritis nodosa, 39/302, 55/355, 357
porencephaly, 42/48, 50/359
prilocaine intoxication, 37/405
progressive hemifacial atrophy, 22/548
propanidid intoxication, 37/409
provocation, 72/272
provoked, see Provoked seizure
pseudohypoparathyroidism, 46/85, 63/560
psychogenic, 72/270
psychomotor, see Psychomotor seizure
psychosocial inventory, 72/408
pyruvate carboxylase deficiency, 41/211, 42/586
pyruvate decarboxylase deficiency, 62/502
pyruvate dehydrogenase deficiency, 42/587
18q partial monosomy, 43/533
4q partial trisomy, 43/499
19q partial trisomy, 43/538

r(21) syndrome, 43/543
r(22) syndrome, 43/547
recombinant chromosome 3 syndrome, 43/495
Reye syndrome, 31/168, 56/151, 153, 238
Rubinstein-Taybi syndrome, 31/284, 50/276
Saethre-Chotzen syndrome, 43/322
Salla disease, 59/360
sarcoidosis, 71/479, 489
sea blue histiocyte disease, 42/473
self-induced, *see* Self-induced seizure
sensorineural deafness, 42/385, 688
sickle cell anemia, 42/623, 63/259
Sjögren-Larsson syndrome, 43/307, 59/358
spastic paraplegia, 42/178
Sturge-Weber syndrome, 15/303, 43/48, 50/376
subacute necrotizing encephalomyelopathy,
    51/388
subarachnoid hemorrhage, 55/14, 21
supplementary motor area, 73/41
systemic lupus erythematosus, 39/278, 43/420,
    55/374-376, 71/43
Taybi-Linder syndrome, 43/386
temporal lobe epilepsy, 73/67, 70
temporal lobe tumor, 67/151
tetracaine intoxication, 37/405
tetraploidy, 43/567
thalassemia major, 42/629
thyroid disease, 70/91
tonic, *see* Tonic seizure
tonic clonic, *see* Tonic clonic seizure
Toxoplasma gondii syndrome, 31/212, 219
transient ischemic attack, 53/206
traumatic intracranial hematoma, 57/266, 281
Treponema pallidum syndrome, 31/212, 218
trichopoliodystrophy, 29/286, 42/584, 46/61
trisomy 8 mosaicism, 43/511
tuberous sclerosis, 50/371-373
ubidecarenone deficiency, 66/423
Urbach-Wiethe disease, 42/595
vascular, 72/209
verbal auditory agnosia, 43/229
vertical supranuclear ophthalmoplegia, 42/605
vinyl ether intoxication, 37/412
visual, *see* Visual seizure
vitamin B6, 70/431
vitamin B6 deficiency, 28/109, 42/716, 51/334
vitamin Bc transport defect, 43/278
W syndrome, 43/271
Wegener granulomatosis, 71/182
xeroderma pigmentosum, 43/13
Seizure threshold
dynamic system theory, 72/40
Seldinger technique

brain angiography, 54/169
Selective IgA deficiency
glycosaminoglycanosis type II, 66/294
Selective nucleoperikaryon necrobiosis
cerebrovascular disease, 11/152
Selective serotonin reuptake inhibitor
adrenergic receptor affinity, 65/315
ataxia, 65/316
chemical structure, 65/314
delirium, 65/316
dopamine reuptake, 65/315
dopaminergic receptor affinity, 65/315
elimination half-life, 65/314
epilepsy, 65/316
extrapyramidal effect, 65/316
fluoxetine, 65/312
fluvoxamine, 65/312
headache, 65/316, 320
histaminic receptor affinity, 65/315
insomnia, 65/320
muscarinic receptor affinity, 65/315
neuropathy, 65/316
noradrenalin reuptake, 65/315
orthostatic hypotension, 65/316
paroxetine, 65/312
receptor affinity, 65/315
serotonin reuptake, 65/315
sertraline, 65/312
steady state concentration, 65/314
tremor, 65/316, 320
visual accommodation, 65/316
Selective serotonin reuptake inhibitor intoxication
tremor, 65/323
Selenium
mercury intoxication, 64/385
Selenotoca intoxication
family, 37/78
venomous spine, 37/78
Self-criticism
prefrontal syndrome, 2/738
Self-induced epilepsy
rare presentation, 15/449
visual stimulus, 73/190
Self-induced seizure
photogenic epilepsy, 2/621
Self-induced water intoxication, *see* Water
    intoxication
Self-induced water intoxication and schizophrenic
    disorder, *see* Water intoxication
Self-mutilation
childhood psychosis, 46/193
chorea-acanthocytosis, 49/328, 63/280-282
dystonia deafness syndrome, 42/372

Gilles de la Tourette syndrome, 42/221
hereditary sensory and autonomic neuropathy type
    IV, 60/11, 12, 17
indifference to pain, 4/222
Klinefelter syndrome, 43/560
Lesch-Nyhan syndrome, 6/456, 22/514, 29/264,
    42/153, 46/66, 59/360, 60/19, 63/256
organotin intoxication, 64/139
Richner-Hanhart syndrome, 42/581
Self-mutilation encephalopathy
    brain atrophy, 42/141
    craniosynostosis, 42/141
    exophthalmos, 42/141
    hypertelorism, 42/141
    hypoxanthine phosphoribosyltransferase, 42/140
    mental deficiency, 42/140
    psychomotor retardation, 42/140
    scoliosis, 42/141
Self-negligence
    Huntington chorea, 49/283
Sella
    empty, see Empty sella
    J-shaped, see J-shaped sella
Sella turcica
    arachnoid cyst, 31/98
    Biemond syndrome type II, 43/334
    centrofacial lentiginosis, 43/26
    chromophobe adenoma, 17/389
    cutis verticis gyrata, 43/244
    facio-oculoacousticorenal syndrome, 43/400
    glycosaminoglycanosis type I, 42/451
    glycosaminoglycanosis type VII, 42/459
    GM1 gangliosidosis type I, 42/431
    Goldberg syndrome, 42/439
    hemangioendothelioma, 18/288
    hypophyseal adenoma, 18/352
    imaging, 68/345
    multiple nevoid basal cell carcinoma syndrome,
        43/31
    myotonic dystrophy, 43/152
    pituitary adenoma, 17/389
    pituitary agenesis, 42/11
    pituitary dwarfism, 42/616
    pituitary dwarfism type V, 42/616
    platybasia, 42/17
    radiology, 68/345
    skull base tumor, 68/468
    Stargardt disease, 42/420
    tumor, 16/690, 23/178
Sellar meningioma
    brain vasospasm, 23/178
Sellar region
    anatomy, 68/347

Sellar tumor
    brain scanning, 16/690
    craniotomy, 23/176
Semaeostomeae intoxication
    chrysaora, 37/50
    cyanea, 37/50
    marine toxin, 37/50-53
Semantic alexia
    classification, 4/116
    definition, 45/440
    symptom, 45/440
Semantic anomia
    anomic aphasia, 45/319
Semantic jargon
    Alzheimer disease, 45/315
    aphasia, 4/92
    Wernicke aphasia, 45/315
Semantic paragraphia
    deep agraphia, 45/463
Semantic paralexia
    deep dyslexia, 45/463
    diagnosis, 45/448
Semantics
    acephaly, 50/71
    acrania, 50/71
    agnosia, 3/36
    aphasia, 45/308
    barotrauma, 61/215
    congenital pain insensitivity, 51/563
    episodic headache, 48/3
    manic depressive psychosis, 46/444
    muscle contraction headache, 5/157
    sphenopalatine neurosis, 48/478
    Unverricht-Lundborg progressive myoclonus
        epilepsy, 49/616
Semilobar holoprosencephaly
    anatomy, 30/438, 50/227
    classification, 30/438
    definition, 50/230
    nomenclature, 30/438
Semimembranous reflex
    lower extremity, 1/247
Semiochemical
    classification, 74/374
Semliki forest virus
    acute viral myositis, 56/194
    cell surface receptor, 56/34
    experimental acute viral myositis, 56/195
    natural killer cell, 56/73
Senescence
    attention, 3/145-147
    progeric, see Progeric senescence
    striatopallidodentate calcification, 49/424

Senile chorea
  critique, 49/558
  dementia, 6/436
  Huntington chorea, 6/159
  late onset, 6/436
  putamen, 6/436
Senile cutaneous elastosis
  angioid streak, 13/141
Senile dementia, 42/275-277
  Alzheimer disease, 46/248, 287
  amyloid, 46/284
  amyloidosis, 46/285
  attention disorder, 3/145-147
  brain biopsy, 10/685
  classification, 46/204
  clinical features, 46/253, 286
  complement, 46/284
  computer assisted tomography, 46/129, 286
  differential diagnosis, 46/129, 286
  Down syndrome, 31/405
  EEG, 46/286
  historic note, 46/283
  Ig, 46/284
  insomnia, 46/286
  intelligence test, 3/311
  memory disorder, 3/265
  neuropathology, 21/64
  phenothiazine, 46/286
  reduplicative paramnesia, 45/418
  senile plaque, 46/285
  treatment, 46/286
Senile dystonia
  brain arteriosclerosis, 6/556
Senile glial dystrophy
  morphology, 21/54
Senile neuroaxonal dystrophy
  parkinsonism, 49/392
Senile osteoporosis
  postmenopausal osteoporosis, 20/12
  vertebral myeloma, 20/12
Senile plaque
  acetylcholine, 46/268
  acetylcholinesterase, 46/255, 269
  acid phosphatase, 46/255
  aging, 42/288, 46/210
  Alzheimer disease, 27/486, 42/275, 46/247,
    253-258, 56/555, 569
  amyloid, 46/253-258, 56/569
  biochemical aspect, 27/486
  brain amyloid angiopathy, 54/334
  cerebellar cortex, 21/64
  clinicopathologic correlation, 46/260-264
  complex IV, 46/255

Creutzfeldt-Jakob disease, 56/551, 569
  dehydrogenase, 46/255
  dementia, 27/486
  Down syndrome, 50/530
  dyshoric angiopathy, 42/728, 46/257
  electron microscopy, 46/255
  form difference, 56/569
  formation, 21/57
  Gerstmann-Sträussler-Scheinker disease, 21/553,
    56/554, 569
  glycoprotein, 46/253
  incidence, 46/274
  kuru, 46/274, 56/551, 555
  kuru plaque, 56/551
  localization, 46/257
  microscopy, 46/253-258
  multicentric plaque, 56/555
  neuronal argyrophilic dystrophy, 21/64
  neurotransmitter change, 46/268
  Parkinson dementia, 49/178
  Parkinson disease, 42/246
  phosphoamidase, 46/255
  polysaccharide, 46/253
  progressive multifocal leukoencephalopathy,
    56/544
  scrapie, 56/555
  senile dementia, 46/285
  spinocerebellar ataxia, 21/553, 42/182
  synaeresis, 21/44
  unicentric plaque, 56/555
Senile ptosis
  progressive external ophthalmoplegia, 62/302
  progressive external ophthalmoplegia
    classification, 62/290
Senile purpura
  aged, 38/71
  bleeding, 38/71
  skin, 38/71
Senile vascular myelopathy
  progressive, see Progressive senile vascular
    myelopathy
Senium, see Aged
Sensation, 1/80-111
  ataxia telangiectasia, 60/364
  bladder, 1/378, 75/672
  combined, see Combined sensation
  electricity, 1/105
  epicritic, 1/80
  exteroceptive, 1/80, 89
  general, 1/80
  gnostic, 1/80
  Golgi-Mazzoni corpuscle, 7/82
  hypoesthesia, 1/89

interoceptive, 1/80, 99
J receptor, 74/576
Krause end bulb, 7/82
Meissner corpuscle, 7/82
Merkel disc, 7/82
micturition, 1/378
Minamata disease, 36/74
motion, 1/95
nerve ending, 7/83
nociceptive, 1/80
organic mercury intoxication, 36/74
pacinian corpuscles, 7/82
phantom, *see* Phantom sensation
porphyria, 60/117
position, 1/95
pressure sore, 26/464, 61/350
proprioceptive, *see* Proprioceptive sensation
protopathic, 1/80
referred pain, 1/103
Ruffini ending, 7/82
somatic, 1/80
special, 1/80
spectral, *see* Spectral sensation
tactile, *see* Tactile sensation
temperature, *see* Temperature sensation
type, 1/80
vibration, 1/95
visceral, *see* Visceral sensation
vital, 1/80
Sense of balance
anatomy, 2/358
Sensibility
deep, *see* Deep sensibility
discriminative, *see* Discriminative sensibility
dissociated, *see* Dissociated sensibility
pellagra, 6/748, 7/573
postural, *see* Postural sensibility
Sensitivity
baroreceptor, *see* Baroreceptor sensitivity
baroreflex, 74/623
cold, 8/341
corneal, *see* Corneal sensitivity
CSF pressure, 19/94
Doppler sonography, 54/7
leucine, *see* Leucine sensitivity
meninx, 1/539, 541
succinylcholine, 42/654
suxamethonium, *see* Suxamethonium sensitivity
Sensitization
CNS, 9/517
glioblastoma multiforme, 18/64
mechanism, 74/150
pharmacology, 74/150

protein, 9/517
radiotherapy, 18/39
Sensorimotor adaptability
traumatic psychosyndrome, 24/553, 690
Sensorimotor nerve
neurotransmission, 74/102
Sensorimotor neuropathy
acrylamide polyneuropathy, 51/281
acute, *see* Acute sensorimotor neuropathy
alcoholic polyneuropathy, 51/316
amiodarone intoxication, 65/458
Andermann syndrome, 50/166
arsenic intoxication, 41/288
ascending, *see* Ascending sensorimotor
neuropathy
birth incidence, 42/103
carcinoma, 38/685
chronic, *see* Chronic sensorimotor neuropathy
corpus callosum agenesis, 42/100-103
diabetic polyneuropathy, 51/500
dipeptidyl carboxypeptidase I inhibitor
intoxication, 65/445
eosinophilia myalgia syndrome, 63/376
ethylene oxide polyneuropathy, 51/275
gold intoxication, 64/356
heterozygote frequency, 42/103
Kawasaki syndrome, 56/639
lathyrism, 65/9
multiple myeloma, 38/687
neurosarcoidosis, 63/422
nitrofurantoin polyneuropathy, 51/296
paraneoplastic polyneuropathy, 51/466
perhexiline maleate polyneuropathy, 51/302
sarcoid neuropathy, 51/196
toxic oil syndrome, 63/380
Waldenström macroglobulinemia, 63/397
Waldenström macroglobulinemia polyneuropathy,
51/432
Sensorimotor organization
supraspinal level, 1/72
Sensorimotor paraneoplastic neuropathy
paraneoplastic syndrome, 71/675
Sensorimotor polyneuropathy
a-alphalipoproteinemia, 51/397
acetonylacetone, 64/86
acquired immune deficiency syndrome, 56/493,
517
acrylamide intoxication, 64/63, 67
allergic granulomatous angiitis, 51/453, 63/383
amiodarone intoxication, 65/458
amiodarone myopathy, 62/608
amyloid polyneuropathy, 51/416
angiopathic polyneuropathy, 51/445

arsenic polyneuropathy, 51/266
asparaginase, 63/359
Bassen-Kornzweig syndrome, 51/394
captopril intoxication, 65/445
carbamate intoxication, 64/186
carbon disulfide intoxication, 64/25
chronic mercury intoxication, 64/371
cisplatin intoxication, 64/358
dipeptidyl carboxypeptidase I inhibitor, 65/445
diphtheria, 51/183
eosinophilia myalgia syndrome, 63/376, 64/254
gold intoxication, 64/356
hereditary amyloid polyneuropathy type 1, 51/424
hexane polyneuropathy, 51/276
2-hexanone polyneuropathy, 51/276
hyperthyroid polyneuropathy, 51/518
α-interferon, 63/359
methanol intoxication, 64/99, 101
multiple myeloma, 18/255, 51/419, 63/393
neurosarcoidosis, 63/422
perhexiline myopathy, 62/608
POEMS syndrome, 63/394
polychlorinated dibenzodioxin intoxication,
    64/205
procarbazine, 63/359
progressive systemic sclerosis, 51/447
Sjögren syndrome, 51/449
systemic lupus erythematosus, 51/447
2,3,7,8-tetrachlorodibenzo-p-dioxin, 64/205
toxic oil syndrome, 63/380
undetermined significant monoclonal
    gammopathy, 63/399
undetermined significant monoclonal
    gammopathy IgM, 63/402
uremia, 63/510
uremic polyneuropathy, 51/357, 63/510
vinblastine intoxication, 65/535
vincristine, 63/358
vincristine intoxication, 65/535
vindesine intoxication, 65/535
Waldenström macroglobulinemia, 63/397
Wegener granulomatosis, 51/451
xeroderma pigmentosum, 51/399
Sensorimotor polyradiculoneuropathy, see
    Guillain-Barré syndrome
Sensorimotor stroke
  anterior choroidal artery, 54/252
  blepharospasm, 54/253
  brain lacunar infarction, 54/251
  eye movement, 54/253
  lacuna site, 54/252
  pontine lacunar infarction, 54/253
  prevalence, 54/251

thalamic infarction, 54/252
  variant, 54/252
Sensorineural deafness
  see also Deafness
  achondroplasia, 43/316
  acoustic nerve agenesis, 50/217
  acoustic neuroma, 42/756
  albinoidism, 43/3
  alopecia, 42/374, 395
  Alpers disease, 42/486
  aminoaciduria, 43/126
  amyloidosis, 42/396
  anhidrotic ectodermal dysplasia, 42/363
  arthrogrypotic like hand malformation, 43/333
  ataxia, 42/112, 386
  blue sclerae, 42/381
  branchio-otodysplasia, 42/359
  branchio-otorenal dysplasia, 42/361
  bulbopontine paralysis, 42/95
  camptobrachydactyly, 43/236
  cataract, 42/393
  cerebellar ataxia, 42/131, 698
  cervico-oculoacusticus syndrome, 32/123, 125
  Cockayne-Neill-Dingwall disease, 43/350
  complex III deficiency, 62/504, 66/423
  congenital analgesia, 42/299
  congenital megacolon, 42/3
  convulsion, 42/383
  corneal dystrophy, 42/367
  craniodiaphyseal dysplasia, 43/355
  craniometaphyseal dysplasia, 31/255, 43/362
  cutaneous pigmentation, see Margolis-Ziprkowski
    syndrome
  cytomegalic inclusion body disease, 56/109
  cytomegalovirus infection, 56/266, 270
  deferoxamine, 63/534
  dental abnormality, 42/419
  diabetes mellitus, 42/391
  diaphyseal dysplasia, 43/376
  digital abnormality, 42/378
  dominant ataxia, 60/660
  dominant progressive, see Dominant progressive
    sensorineural deafness
  dyssynergia cerebellaris myoclonica, 60/599
  dystonia, 42/372
  ECG abnormality, 42/388
  EEC syndrome, 43/395
  erythrokeratodermia, 42/373
  facio-oculoacousticorenal syndrome, 43/399
  Flynn-Aird syndrome, 42/327
  frontometaphyseal dysplasia, 43/401
  gastric motility, 42/394
  gene frequency, 43/43

glycosaminoglycanosis type II, 42/454
glycosaminoglycanosis type III, 42/455
goiter, 42/375
Goldberg syndrome, 42/439
Harboyan syndrome, 42/367
hemorrhagic nephritis, 42/376
hereditary diabetes mellitus, 60/775
hereditary hypophosphatemic osteomalacia,
    60/771
hereditary motor and sensory neuropathy type I,
    42/370
hereditary motor and sensory neuropathy variant,
    60/246
hereditary progressive cochleovestibular atrophy,
    60/770, 775
hereditary renal tubular acidosis, 60/770
heredopathia ophthalmo-otoencephalica,
    42/150-152
heterochromia iridis, 42/688
hyperkeratosis, 42/395
hyperostosis corticalis deformans, 43/409
hypertelorism, 43/417
hypertension, 42/389
hypertrophic interstitial neuropathy, 21/152
hyperuricemia, 42/571
hypogonadism, 42/374, 377, 389, 43/272
hypoparathyroidism, 42/383, 70/123
imperforate anus, 42/379
infantile neuroaxonal dystrophy, 42/229
inverse secondary pigmentary retinal
    degeneration, 42/377
Kearns-Sayre-Daroff-Shy syndrome, 43/142,
    62/508
keratopachydermia, 42/378
kernicterus, 42/583
Klein-Waardenburg syndrome, 30/408, 43/51,
    50/218
Klippel-Feil syndrome, 42/37
Lemieux-Neemeh syndrome, 42/370
LEOPARD syndrome, 43/27
lop ear, 42/379
macrothrombocytopathia, 42/380
Marfan like syndrome, 42/381
marfanoid features, 42/381
Margolis-Ziprkowski syndrome, 43/44
mental deficiency, 42/131, 385, 389, 43/236, 272,
    295, 46/28
mitochondrial myopathy, 43/188
mucolipidosis type I, 42/475
myocardial sclerosis, 42/131
myoclonic epilepsy, 42/698
myopathy, 43/126
myopia, 42/381, 389, 393

myotonic dystrophy, 43/152
N syndrome, 43/283
nail hypoplasia, 42/384-386
nephritis, 42/370, 380, 383, 396
neurogenic acro-osteolysis, 42/293
neuropathy, 42/373, 60/660
Norrie disease, 43/441
obesity, 42/391
oculocutaneous albinism, 42/404, 43/3, 6
onychodystrophy, 42/384-386
ophthalmoplegia, 42/605, 43/138
optic atrophy, 42/386, 411, 414
opticocochleodentate degeneration, 42/241
orbital hypertelorism, 43/417
osteogenesis imperfecta, 43/453
osteopetrosis, 43/456
ota nevus, 43/37
otodental dysplasia, 42/419
Patau syndrome, 43/528
photogenic epilepsy, 42/714
piebald pigmentation, 43/42-44
pili torti, 42/387
polyneuropathy, 42/373
prevalence, 43/43
progressive myoclonus epilepsy, 42/698
Refsum disease, 21/9, 196, 22/501, 41/434,
    42/148, 60/228, 728, 66/485
renal insufficiency, 42/389-391
renal tubular dysfunction, 42/390
reproductive fitness, 43/45
retinal degeneration, 42/391
Richards-Rundle syndrome, 43/264
sclerosteosis, 31/258, 43/474
secondary pigmentary retinal degeneration,
    42/392
seizure, 42/385, 688
sensory neuropathy, 42/394
sensory radicular neuropathy, 42/346, 351, 394
skeletal abnormality, 42/379
small intestine diverticulitis, 42/394
spastic paraplegia, 42/178
spinal muscular atrophy, 59/373
spinocerebellar ataxia, 42/181
spiny hyperkeratosis, 42/395
sporadic goiter, 42/368-370
Stilling-Türk-Duane syndrome, 50/213
stippled epiphysis, 42/375
sudden death, 42/388
swayback nose, 42/393
thalidomide syndrome, 42/660
thrombopathia, 42/380
thyroid gland, 42/366
urticaria, 42/396

Van Buchem disease, 31/258, 38/408, 43/410
viral infection, 56/105
white hair, 43/42
Wildervanck syndrome, 43/343
xeroderma pigmentosum, 43/13
Sensory action potential
brachial plexus, 51/148
chorea-acanthocytosis, 63/282
multiple myeloma, 63/393
undetermined significant monoclonal
gammopathy, 63/399
Sensory aphasia
apraxia, 45/429
auditory agnosia, 4/37
bilinguist, 2/599
constructional apraxia, 45/501
developmental dysphasia, 46/140
parietal lobe syndrome, 45/75
restoration of higher cortical function, 3/409-417
transcortical, see Transcortical sensory aphasia
Wernicke aphasia, 4/98, 45/313
Sensory ataxia
vitamin B6 intoxication, 51/298
Sensory conduction velocity
see also Motor conduction velocity and Nerve
conduction velocity
leprous neuritis, 51/221
thalidomide polyneuropathy, 51/305
Sensory crest
neurofibromatosis type I, 14/86
Sensory deficit
amyotrophic lateral sclerosis, 59/412
Biemond syndrome, 21/377
body scheme disorder, 4/400
dementia, 59/412
foramen magnum tumor, 19/58
hereditary olivopontocerebellar atrophy (Menzel),
21/443
hereditary sensory and autonomic neuropathy type
III, 75/145
intramedullary spinal tumor, 19/58
Refsum disease, 8/22, 21/199, 27/523, 60/228,
230, 66/489
spinal tumor, 19/58
trigeminal, see Trigeminal sensory deficit
visual, see Visual sensory deficit
Sensory deprivation
consciousness, 3/49, 126
delirium, 46/534
EEG, 46/534
emotion, 3/334
hallucination, 4/329
intelligence, 3/305

phantom limb, 4/228, 45/399
phosphene, 45/367
Refsum disease, 60/232
sleep, 3/85, 90
Sensory disorder
attention disorder, 3/187, 195
congenital spinal cord tumor, 32/366
frontal lobe tumor, 17/259
mental deficiency, 46/28
parietal lobe tumor, 17/300
somatic, see Somatosensory disorder
spinal shock, 26/252
spinal spongioblastoma, 19/360
syringomyelia, 32/262-266
thallium intoxication, 36/245, 248
Sensory eclipse, see Sensory extinction
Sensory extinction
see also Tactile inattention
age, 3/190-193
agnosia, 4/30
allesthesia, 3/191
amorphosynthesis, 3/193
amytal test, 3/193
aphasia, 3/193
attention, 3/160-167
attention disorder, 3/189-195, 4/30
attention theory, 3/194
audition, 3/191
barognosis, 3/191
body scheme, 3/195
body scheme disorder, 4/213
body scheme structure, 4/210
brain cortex, 45/160
clinical significance, 3/192, 195
corpus callosum, 45/159
displacement, 3/190-192
dominance pattern, 3/190, 195
electroshock, 3/193
exosomatesthesia, 3/191
face-hand test, 3/190
general anesthesia, 3/193
hemidepersonalization, 3/56
hemispherectomy, 3/193
lesion site, 1/105
limited attention theory, 45/160
local adaptation, 3/194
neglect syndrome, 45/158-161
obscuration, 3/190, 45/159
organic mental syndrome, 3/193
parietal lobe lesion, 3/193
parietal lobe syndrome, 45/65
pathophysiologic theory, 45/159-161
phantom limb, 4/228, 45/399

postcentral gyrus resection, 3/193
reciprocal inhibition theory, 45/161
simultaneous double stimulation, 3/190-192, 45/159
simultaneous stimulation, 45/158-161
somatesthesia, 3/190
spinal cord injury, 3/193
stereognosis, 3/191
stimulation threshold, 3/194
synesthesia, 3/191
tactile agnosia, 4/34
tactile inattention, 45/65
taste, 3/191
testing, 45/159
thalamic lesion, 3/193
thalamus, 45/160
theory, 3/193-195
vision, 3/191
Sensory feedback
  cerebral dominance, 4/286
Sensory impairment
  agnosia, 4/24-27
  chronic mercury intoxication, 64/370
  familial spastic paraplegia, 22/434, 59/305
  leprous neuritis, 51/230
  progressive bulbar palsy, 59/220
  Roussy-Lévy syndrome, 21/176, 42/108
Sensory inattention
  lesion site, 1/105
Sensory induced seizure, see Reflex epilepsy
Sensory integration
  depersonalization, 2/711
  derealization, 2/711
  Sylvian fissure, 2/700, 711
Sensory loss
  see also Sensory neuropathy
  adult polyglucosan body disease, 62/484
  Biemond syndrome, 42/164
  brachial neuritis, 42/303
  carpal tunnel syndrome, 42/309
  caudal agenesis, 42/50
  cavernous angioma, 42/725
  cerebellar atrophy, 42/135
  Chédiak-Higashi syndrome, 42/536
  chorea-acanthocytosis, 63/281
  CNS degeneration, 42/97
  craniometaphyseal dysplasia, 43/362
  CSF paraprotein, 42/603
  dissociated, see Dissociated sensory loss
  facial paralysis, 42/313
  familial spastic paraplegia, 60/244
  Friedreich ataxia, 21/320, 330, 42/143
  hemophilia A, 42/738

hereditary amyloid polyneuropathy, 42/518, 521
hereditary motor and sensory neuropathy type I, 21/282, 42/339
hereditary motor and sensory neuropathy type II, 21/282, 42/339
hereditary motor and sensory neuropathy variant, 60/244
hereditary multiple recurrent mononeuropathy, 42/316
hereditary neuropathy, 60/256
hypertrophic interstitial neuropathy, 42/317
intramedullary spinal tumor, 19/32
isoniazid neuropathy, 42/648
Klinefelter variant XXYY, 43/563
lumbar intervertebral disc prolapse, 20/585, 591
Machado disease, 42/155
meralgia paresthetica, 42/323
migrant sensory neuritis, 51/461
Muckle-Wells syndrome, 42/396
muscular atrophy, 42/80
neurogenic acro-osteolysis, 42/293-295
neuromyelitis optica, 9/429, 47/401
olivopontocerebellar atrophy, 42/161
pain, see Pain sensory loss
polyneuropathy, 42/97, 51/55, 60/256
posterior column ataxia, 42/164
postherpetic neuralgia, 5/316, 323
Rosenberg-Chutorian syndrome, 22/506, 42/418
scapuloperoneal syndrome, 42/99
sensory radicular neuropathy, 42/351
serum paraprotein, 42/603
spastic paraplegia, 42/169, 177
spina bifida, 42/54
spinal arachnoiditis, 42/107
spinal ependymoma, 19/360, 20/368
syringobulbia, 42/59
syringomyelia, 42/59
Tangier disease, 10/548, 42/627
temperature, see Temperature sensory loss
thalamic syndrome (Dejerine-Roussy), 42/771
thalidomide, 42/660
Sensory loss pattern
  alcoholic polyneuropathy, 51/56
  diabetic polyneuropathy, 51/57
Sensory neglect
  akinesia, 45/162
Sensory nerve
  conduction velocity, 7/143
Sensory nerve conduction
  hereditary multiple recurrent mononeuropathy, 51/555
  myxedema neuropathy, 8/58
  technique, 7/143

uremic neuropathy, 7/166, 8/1, 27/340
Sensory neuritis
  migrant, *see* Migrant sensory neuritis
Sensory neuropathy
  see also Sensory loss *and* Sensory radicular
    neuropathy
  acute, *see* Acute sensory neuropathy
  adenosine triphosphatase 6 mutation syndrome,
    62/510
  antibody, 71/443
  antichondroitin sulfate, 71/444
  areflexia, 42/348
  ataxia, 42/348
  barium intoxication, 64/354
  behavior disorder in children, 42/348
  carcinoma, 8/369, 38/684-686
  chronic, *see* Chronic sensory neuropathy
  chronic obstructive pulmonary disease, 63/421
  cisplatin polyneuropathy, 51/300
  complex I deficiency, 62/504, 66/422
  complex III deficiency, 62/504
  congenital, *see* Hereditary sensory and autonomic
    neuropathy type I
  deafness, 8/181
  diabetic polyneuropathy, 8/30, 367, 51/500
  differential diagnosis, 8/366
  dimethylaminopropionitrile polyneuropathy,
    51/282
  dipeptidyl carboxypeptidase I inhibitor
    intoxication, 65/445
  disialosyl ganglioside, 71/443
  distal, *see* Distal sensory neuropathy
  dorsal spinal nerve root ganglion, 8/201
  dorsal spinal root ganglion degeneration, 42/346
  ethionamide polyneuropathy, 51/298
  familial spastic paraplegia, 21/16
  Friedreich ataxia, 8/184
  hereditary, *see* Hereditary sensory and autonomic
    neuropathy type II
  hereditary amyloid polyneuropathy, 21/121
  hereditary motor, *see* Hereditary motor sensory
    neuropathy
  Hirschowitz type, 22/524
  hypereosinophilic syndrome, 63/373
  hypertrophic, *see* Hypertrophic sensory
    neuropathy
  hypothermia, 42/346
  ischemic neuropathy, 8/367
  isoniazid polyneuropathy, 51/297
  Marinesco-Sjögren syndrome, 51/398
  mental deficiency, 42/346
  metronidazole intoxication, 65/487
  metronidazole polyneuropathy, 51/301

Minamata disease, 64/415
misonidazole polyneuropathy, 51/301
monoclonal antibody, 71/444
multiple myeloma, 47/618, 63/393
nerve conduction, 7/167
paraneoplastic, *see* Paraneoplastic sensory
  neuropathy
paraneoplastic polyneuropathy, 51/466
paraneoplastic syndrome, 69/354, 364
progressive, *see* Hereditary sensory and
  autonomic neuropathy type II
scoliosis, 42/348
sensorineural deafness, 42/394
Sjögren syndrome, 71/83
spastic paraplegia, 42/179
subacute, *see* Subacute sensory neuropathy
toxic oil syndrome, 63/380
trigeminal, *see* Trigeminal sensory neuropathy
vitamin B6 deficiency polyneuropathy, 51/298
vitamin B6 intoxication, 65/572
Von Romberg test, 42/348
Waldenström macroglobulinemia, 63/397
Sensory overload
  delirium, 46/535
Sensory paralysis
  crossed, *see* Grenet syndrome
  nerve injury, 7/273
Sensory polyneuritis
  lepromatous, *see* Lepromatous sensory
    polyneuritis
Sensory polyneuropathy
  acrylamide polyneuropathy, 51/281
  almitrine dimesilate polyneuropathy, 51/308
  cisplatin intoxication, 65/536
  diabetic neuropathy, 8/31
  flecainide intoxication, 65/456
  Friedreich ataxia, 51/388
  hepatic insufficiency polyneuropathy, 51/362
  isoniazid, 41/333, 52/211
  metronidazole intoxication, 65/487
  pain, 8/31
  rheumatoid arthritis, 51/447
  taxol intoxication, 65/537
  taxotere intoxication, 65/537
  thalidomide intoxication, 65/559
  vitamin B6 intoxication, 65/572
  Waldenström macroglobulinemia, 63/397
Sensory radicular neuropathy
  see also Sensory neuropathy
  areflexia, 42/351
  dorsal spinal root ganglion degeneration, 42/351
  gastric aperistalsis, 21/82
  hereditary, *see* Hereditary sensory and autonomic

neuropathy type I
  hereditary deafness, 22/524
  hereditary hearing loss, 22/518
  hereditary perforating foot ulcer, 8/201, 21/73, 76
  hereditary sensory and autonomic neuropathy type
    II, 21/74
  hyporeflexia, 42/351
  jejunoileal diverticulitis, 21/82
  mental deficiency, 42/351
  nerve conduction velocity, 42/351
  neuropathy, 42/351
  pain, 42/351
  sensorineural deafness, 42/346, 351, 394
  sensory loss, 42/351
  sensory syndrome, 8/199, 201
  serosal fibrosis, 22/524
  talipes, 42/352
  trophic ulcer, 42/351
Sensory radiculitis
  varicella zoster, 51/180
Sensory receptor
  autonomic nervous system, 74/33
  physiologic property, 7/82-85
Sensory representation
  brain cortex, 1/85
  cerebral dominance, 4/303, 305, 307, 324
Sensory sequence stimulation
  brain hemisphere, 15/78
Sensory suppression, see Sensory extinction
Sensory symptom
  amyotrophic lateral sclerosis, 22/114, 294, 59/195
  Bannwarth syndrome, 51/201
  foramen magnum tumor, 17/724
  hereditary neuropathy, 60/255
  hydralazine neuropathy, 7/532
  hypertrophic interstitial neuropathy, 21/151
  lathyrism, 65/3
  migraine, 48/130
  multiple sclerosis, 9/169
  neuroborreliosis, 51/201
  paracentral lobule syndrome, 53/345
  polyneuropathy, 60/255
  progressive bulbar palsy, 22/114
  spinal meningioma, 20/197, 201
  visual, see Visual sensory system
Sensory syndrome
  hereditary sensory and autonomic neuropathy type
    III, 8/198
  indifference to pain, 8/199
  sensory radicular neuropathy, 8/199, 201
Sensory trigeminal neuropathy
  bilateral, see Bilateral sensory trigeminal
    neuropathy

mixed connective tissue disease, 51/450
  polyarteritis nodosa, 51/452
  progressive systemic sclerosis, 51/447
  rostral tegmental syndrome, 12/20
  Sjögren syndrome, 51/449
Sepedonium
  fungal CNS disease, 52/493
Sepedoniumosis
  CSF, 52/493
  ventriculitis, 52/493
Sepsis
  cervical spinal cord injury, 61/58
  critical illness polyneuropathy, 51/575, 584,
    63/420
  headache, 48/284
  intravascular consumption coagulopathy, 55/493
  subarachnoid hemorrhage, 55/3
  transverse sinus thrombosis, 54/397
Septal area
  anterior critical point, 2/771
  behavior, 2/771
  brain infarction, 55/140
  emotion, 2/770
  posterior critical point, 2/771
  sham rage, 2/770
  startle reaction, 2/770
Septal defect
  atrial, see Atrial septal defect
  ventricular, see Cardiac ventricular septum defect
Septal vein
  radioanatomy, 11/107
  topography, 11/107
Septic encephalopathy
  metabolic encephalopathy, 69/405
  multiple organ failure, 71/540
Septojod
  retinotoxic agent, 13/205
Septo-optic dysplasia
  clinical features, 30/320, 471, 50/211
  congenital retinal blindness, 42/407
  diagnosis, 30/321
  epidemiology, 30/321
  epilepsy, 30/320
  genetics, 30/321
  Kallmann syndrome, 30/471
  mental deficiency, 30/320
  pathogenesis, 30/321
  septum pellucidum agenesis, 30/314-320, 471
Septum pellucidum
  acrocephalosyndactyly type I, 43/320
  agenesis, see Septum pellucidum agenesis
  anatomoclinical relationship, 30/329
  brain tumor, 17/539

cavum, *see* Cavum septi pellucidi
cavum vergae, 30/302-304, 321-324
corpus callosum agenesis, 50/150
cyst, 30/322, 326
embryology, 30/299-302
ependymal cyst, 17/538
epilepsy, 72/109
mental symptom, 2/770
microcephaly, 30/314-321, 325
septum gangliosum, 30/304, 310
septum gliosum, 30/304
tumor, *see* Septum pellucidum tumor
Septum pellucidum agenesis, 30/311-321
acrocephalosyndactyly type I, 30/315, 326, 43/320
associated defect, 2/767, 17/538
congenital rubella, 50/278
craniolacunia, 50/139
epilepsy, 30/314-321
hydrocephalus, 30/314-321, 326, 328
limbic system, 30/329
mental deficiency, 30/312, 314-321, 46/13
polyploidy, 50/555
schizencephaly, 50/200
septo-optic dysplasia, 30/314-320, 471
spina bifida, 30/314-321
status verrucosus, 50/203
Septum pellucidum fenestration
hydrocephalus, 50/290
Septum pellucidum tumor
anatomy, 17/474
astrocytoma, 17/539
cavum vergae, 17/537
clinical features, 17/476, 539
contrast study, 2/769
corpus callosum agenesis, 50/157
cyst, 2/767, 17/537, 18/134
EEG, 17/478
histology, 17/475
intracranial pressure, 2/769
radiology, 17/479
sex ratio, 17/539
supratentorial midline tumor, 18/357
treatment, 17/481
Sepulveda Epilepsy Battery
epilepsy, 72/431
Sequestration epidermoid, *see* Acquired epidermoid
Serax intoxication, *see* Oxazepam intoxication
Serial dysmetria
intention tremor, 49/590
Serialization
developmental dyslexia, 46/113
Serine
chondroitin, 10/431

GM2 gangliosidosis, 10/295
homocystinuria, 55/325
L-Serine
biochemistry, 9/8
Serine palmitoyltransferase
vitamin B6 deficiency, 51/297
Serine protease
Duchenne muscular dystrophy, 40/380
Serine proteinase inhibitor
neurotoxin, 65/186
snake venom, 65/186
Serogenetic neuropathy
clinical picture, 8/99
CSF, 8/102
differential diagnosis, 8/108
etiology, 8/104
Guillain-Barré syndrome, 8/100
immune serum, 8/86, 95
lesion site, 8/102
neuralgic amyotrophy, 8/99, 51/172
pathology, 8/107
terminology, 8/98
treatment, 8/107
Serologic test
acute amebic meningoencephalitis, 52/320
African trypanosomiasis, 52/342
Eaton-Lambert myasthenic syndrome, 62/425
granulomatous amebic encephalitis, 52/320
paracoccidioidomycosis, 35/534
progressive multifocal leukoencephalopathy,
47/511
syphilis, 33/341
Serosal fibrosis
sensory radicular neuropathy, 22/524
Serotonin
affective disorder, 46/475
alcohol intoxication, 64/114
Alzheimer disease, 46/266
basal ganglion, 49/28, 33
biochemical anatomy, 29/466
biological inactivation, 29/462
biosynthesis, 29/460
blood-brain barrier, 48/64
brain microcirculation, 53/79
catabolism, 29/460
cephalopoda intoxication, 37/62
cough headache, 48/371
Down syndrome, 31/433-435
Duchenne muscular dystrophy, 40/168, 383
dyssynergia cerebellaris myoclonica, 60/603
dystonia musculorum deformans, 49/526
eosinophilia myalgia syndrome, 63/378
headache, 48/88, 90, 97, 446

hyperalgesia, 48/194
intravenous injection, 48/88
ischemic myopathy, 40/167
malnutrition, 29/5
manganese intoxication, 64/306
micturition, 75/670
migraine, 5/18, 39, 100, 42/747, 48/38, 86, 88, 90, 97, 113, 182, 188, 192-196
multiple neuronal system degeneration, 75/192
mummy, 74/192
nervous system tumor, 27/510
nociceptor, 45/229
octopoda intoxication, 37/62
pain, 1/142
Parkinson disease, 29/478, 49/119
periaqueductal gray matter, 45/239
phenylketonuria, 28/128, 29/35
physiologic effect, 48/87
plasma, 48/88
platelet, see Platelet serotonin
progressive dysautonomia, 59/147, 152
regional distribution, 29/466
release, 29/462
sleep, 3/104
smooth muscle function, 75/624
storage, 29/462
substantia nigra, 49/8
suprachiasmatic nucleus, 74/489
temporal artery, 48/87
turning behavior, 45/167
turnover, 29/463
urinary excretion, 48/90
Serotonin antagonism
methysergide, 48/88
Serotonin creatinine sulfate
injection, 48/88
Serotonin inhibitor
chronic migrainous neuralgia, 48/251
Serotonin receptor
dihydroergotamine, 65/66
Huntington chorea, 49/295
migraine, 48/87
psilocin, 65/40
psilocybine, 65/40
Serotonin 1 receptor
cyproheptadine, 48/88
pizotifen, 48/88
Serotonin 1A receptor
buspirone intoxication, 65/343
Serotonin 2 receptor
catamenial migraine, 48/92
ketanserin, 48/88
lysergide, 65/44

spiperone, 48/88, 92
Serotonin releasing factor
headache, 48/89
migraine, 42/747, 48/89
Serotonin syndrome
autonomic nervous system, 75/13, 499
Serotoninergic agent
behavior disorder, 46/596
Serous encephalitis
acute, see Reye syndrome
Serous inflammation
neuropathy, 9/608
Serous meningitis, see Meningitis serosa
Serous retinopathy
central, see Central serous retinopathy
Serranid intoxication
vitamin A intoxication, 37/87
Serrated ligament
Arnold-Chiari malformation type II, 50/406
Serratia
adult meningitis, see Gram-negative bacillary meningitis
infantile enteric bacillary meningitis, 33/62
Serratia marcescens
bacterial endocarditis, 52/298, 301
brain abscess, 52/150, 298
brain microabscess, 52/298
gram-negative bacillary meningitis, 52/104
multiple brain abscess, 52/298
Serratus anterior paralysis
neuralgic amyotrophy, 8/77, 51/171
terminology, 8/77, 51/177
Serrulatotoxin
arthropod envenomation, 65/198
neurotoxin, 65/194, 198
Sertraline
selective serotonin reuptake inhibitor, 65/312
Sertraline intoxication
headache, 65/320
insomnia, 65/320
tremor, 65/320
Serum alanine aminotransferase
chorea-acanthocytosis, 49/330
dermatomyositis, 62/375
facioscapulohumeral muscular dystrophy, 62/166
inclusion body myositis, 62/375
McLeod phenotype, 63/287
metastatic cardiac tumor, 63/96
polymyositis, 62/375
primary cardiac tumor, 63/96
Reye syndrome, 29/334, 56/151, 238, 65/118
Serum angiotensin converting enzyme level, see
Serum dipeptidyl carboxypeptidase level

311

Serum antibody
demyelination, 47/446
Serum aspartate aminotransferase
chorea-acanthocytosis, 49/330
dermatomyositis, 62/375
inclusion body myositis, 62/375
metastatic cardiac tumor, 63/96
polymyositis, 62/375
primary cardiac tumor, 63/96
Reye syndrome, 29/334, 31/168, 33/293, 34/169,
49/217, 56/151, 238, 63/438, 65/118
Serum copper
multiple sclerosis, 9/322
Wilson disease, 9/322, 27/399, 49/231
Serum creatine kinase
chorea-acanthocytosis, 49/329
corticosteroid, 40/343
Duchenne muscular dystrophy, 41/80
electrolyte disorder, 63/557
lithium, 40/344
Marinesco-Sjögren syndrome, 60/674
methanol intoxication, 64/98
mevalonate kinase deficiency, 60/674
myophosphorylase deficiency, 27/233, 41/185,
187, 62/485
phosphofructokinase deficiency, 41/190, 62/487
polymyositis, 41/80
postpoliomyelitic amyotrophy, 59/38
scapuloperoneal muscular dystrophy, 62/169
spinal muscular atrophy, 59/373
ubiquinone, 40/343
Wohlfart-Kugelberg-Welander disease, 59/88
Serum creatine phosphokinase, *see* Serum creatine
kinase
Serum dipeptidyl carboxypeptidase level
sarcoidosis, 71/467
Serum esterone
Wohlfart-Kugelberg-Welander disease, 59/88
Serum α-fetoprotein
ataxia telangiectasia, 60/375
Serum hepatitis
hepatitis B virus, 56/295
Serum human T-lymphotropic virus type I antibody
human T-lymphotropic virus type I associated
myelopathy, 59/453
Serum IgA
plasma cell, 20/10
Serum IgG
human T-lymphotropic virus type I associated
myelopathy, 59/450
multiple sclerosis, 9/326
plasma cell, 20/10
Serum lactate dehydrogenase isoenzyme 3

chorea-acanthocytosis, 49/330
Serum lipoprotein
chorea-acanthocytosis, 60/144
ciclosporin intoxication, 65/553
Serum M component
multiple myeloma, 20/14, 39/133, 139, 63/394
primary amyloidosis, 51/416
Serum nerve antibody
Guillain-Barré syndrome, 51/253
Serum paralysis
immunization, 7/555
Serum paraprotein
areflexia, 42/604
hyporeflexia, 42/604
muscular atrophy, 42/603
paralysis, 42/603
polyneuropathy, 42/604
sensory loss, 42/603
Serum potassium
blood biochemistry, 26/401
hereditary hyperkalemic periodic paralysis,
43/151
hyperkalemic periodic paralysis, 43/151
hypokalemic periodic paralysis, 43/170
myotonic hyperkalemic periodic paralysis,
43/152, 62/267
normokalemic periodic paralysis, 28/597, 41/162,
425, 43/171, 62/463
paramyotonia congenita, 43/169
spinal cord injury, 26/401
urine, 26/401
Serum sickness
antigen-antibody reaction, 9/537
hydralazine intoxication, 37/431
injection, 9/551
leukoencephalopathy, 9/608
lumbosacral plexus neuritis, 51/167
neuropathy, 7/555, 8/82, 96, 9/608
polyarteritis nodosa, 39/295, 55/359
tetanus antitoxin, 52/240
Setting sun sign
hydrocephalus, 30/546
kernicterus, 6/494, 509, 511
Seven day fever, *see* Leptospirosis
Severe trauma
critical illness polyneuropathy, 51/581
progressive myositis ossificans, 41/385
Severinghaus, A.E., 1/25
Severity index
limb girdle syndrome, 40/447
Sex
biological rhythm, 74/502
hypothalamus, 74/470

melatonin, 74/502
  pineal gland, 74/503
  suprachiasmatic nucleus, 74/497
  thermoregulation, 75/58
  vasoactive intestinal polypeptide neuron, 74/495
Sex chromatin
  chromosome, 31/345-347
  genetics, 13/24
  intraspinal neurenteric cyst, 20/70
  Lyon principle, 50/542
  sex chromosome relation formula, 31/475-478
Sex chromosome fragility
  computer assisted tomography, 60/672
  megalotestes, 43/287
  mental deficiency, 43/287, 303
  nuclear magnetic resonance, 60/673
  Renpenning type, 43/303
  X-linked recessive ataxia, 60/635
Sex chromosomes
  Aicardi syndrome, 50/165
  aneuploidy, 50/540-542
  chromosome number abnormality, 50/541
  hereditary motor and sensory neuropathy type I,
    60/202
  hereditary motor and sensory neuropathy type II,
    60/202
  hereditary neuropathy, 60/265
  intraspinal neurenteric cyst, 32/423
  late cerebellar atrophy, 60/586
  MAO, 48/94
  McLeod phenotype, 63/289
  mental deficiency, 46/73-76
  pentasomy X, see XXXXX syndrome
  tetrasomy X, see XXXX syndrome
  trisomy, 31/491-495
  trisomy X, see XXX syndrome
  Turner syndrome, 43/550-552, 46/74, 50/543
  XX male, 43/555-557, 46/75
  XXX syndrome, 46/74
  XXXX syndrome, 46/74
  XXXXX syndrome, 46/74
  49,XXXXY variant, see Klinefelter syndrome
  XXXY variant, see Klinefelter variant XXXY
  48,XXXY variant, see Klinefelter syndrome
  XXXYY variant, see Klinefelter variant XXXYY
  47,XXY, see Klinefelter syndrome
  XXY male, see Klinefelter syndrome
  XXYY variant, see Klinefelter variant XXYY
  48,XXYY variant, see Klinefelter syndrome
  46,XY variant, see Klinefelter syndrome
  47,XYY variant, see Klinefelter syndrome
Sex hormone
  annual rhythm, 74/489
  chemotherapy, 39/121
  epilepsy, 15/320
  headache, 48/425
  migraine, 48/425, 434
  teratogenic agent, 30/125
Sex linked cerebellar ataxia
  infantile, see Infantile sex linked cerebellar ataxia
Sex linked early onset neural deafness
  clinical features, 22/485
Sex linked idiopathic retinoschisis
  vitreoretinal degeneration, 13/37
Sex linked juvenile retinoschisis
  Stargardt disease, 13/126
  vitreotapetoretinal dystrophy, 13/139
Sex linked neurodegenerative disease
  monilethrix, see Trichopoliodystrophy
Sex linked progressive sensorineural hearing loss
  hereditary progressive cochleovestibular atrophy,
    60/767
Sex linked recessive aqueduct stenosis
  hydrocephalus, 50/62
  mental deficiency, 46/66
Sex predominance
  mental deficiency, 46/66-68
  spinal cord tumor, 19/7
Sex ratio
  abnormal theta wave, 42/667
  acoustic neuroma, 42/756
  acute amebic meningoencephalitis, 52/316
  acute hemorrhagic conjunctivitis, 38/608, 56/350
  acute infantile hemiplegia, 12/341
  acute intermittent porphyria, 42/618
  Aicardi syndrome, 42/697
  alcoholic polyneuropathy, 51/315
  alcoholism, 43/197
  allergic granulomatous angiitis, 51/453, 63/383
  Alpers disease, 42/487
  amyloid polyneuropathy, 51/416
  amyotrophic lateral sclerosis, 22/288, 341, 42/65,
    59/179
  aneurysmal bone cyst, 19/300, 20/2, 4
  angiodysgenetic necrotizing myelomalacia,
    20/482
  anterior cerebral artery syndrome, 53/342
  anterior spinal meningocele, 42/43
  anxiety neurosis, 43/199
  aorta aneurysm, 63/48
  aphasia, 45/304
  aqueduct stenosis, 42/57
  astrocytoma, 18/4
  Behçet syndrome, 56/596
  benign intracranial hypertension, 42/740
  benign osteoblastoma, 19/299

benign partial epilepsy with centrotemporal spike, 73/8

Bloom syndrome, 43/10

brain aneurysm, 12/85, 42/723

brain arteriovenous malformation, 12/239

brain embolism, 53/200

brain fat embolism, 55/182

brain infarction, 11/297, 53/200

brain metastasis, 18/203

brain thromboangiitis obliterans, 12/389, 39/203, 55/308, 312

brain tumor, 67/131

Brown hyperparathyroid tumor, 19/301

cardiofacial syndrome, 42/308

carnitine palmitoyltransferase deficiency, 62/494

carpal tunnel syndrome, 42/309

cat scratch disease, 52/126

cataplexy, 42/713

caudal aplasia, 50/65

cebocephaly, 50/61

central sleep apnea, 63/460

cephalocele, 50/97

cerebellar granular cell hypertrophy, 43/253

cerebellar medulloblastoma, 18/392

cerebrovascular disease, 53/14, 200

cervical spinal canal stenosis, 42/58

cervico-oculoacusticus syndrome, 30/162, 32/123, 125

childhood myoglobinuria, 62/561

chondroma, 19/303

chordoma, 19/288

choroid plexus papilloma, 42/761

chronic migrainous neuralgia, 48/249

chronic muscle contraction headache, 48/344

chronic subdural hematoma, 24/303

citrullinuria, 42/538

cluttering, 43/221

CNS malformation, 30/151, 50/56

Cobb syndrome, 14/438

Cockayne-Neill-Dingwall disease, 43/352

concentric sclerosis, 47/414

congenital deafness, 42/366

congenital megacolon, 42/3

congenital retinal blindness, 13/65, 42/399

congenital spondyloepiphyseal dysplasia, 43/479

coproporphyria, 42/539

corpus callosum agenesis, 50/61

corpus callosum tumor, 17/497

cranial neuralgia, 42/353

craniolacunia, 50/139

craniopharyngioma, 18/540

craniosynostosis, 42/21

cri du chat syndrome, 43/502

criminality, 43/205

cutis verticis gyrata, 43/244

cyclopia, 50/61

Dandy-Walker syndrome, 50/63

depressed skull fracture, 23/404

dermoid cyst, 20/72, 78, 68/330

developmental articulatory defect, 43/219

developmental dyslexia, 43/222

diastematomyelia, 50/64, 440

diplomyelia, 50/440

distal muscular dystrophy, 43/96

drug induced extrapyramidal syndrome, 6/255

dysosteosclerosis, 43/394

Edwards syndrome, 43/537

eosinophilia myalgia syndrome, 64/251

eosinophilic fasciitis, 63/384

eosinophilic granuloma, 63/422

ependymoma, 18/112, 19/12

epidermoid cyst, 20/78

epidural hematoma, 24/262, 26/14

epidural spinal lipoma, 20/403, 405

epilepsy, 72/22

familial amyotrophic lateral sclerosis, 59/180, 243

febrile convulsion, 15/251

fibromuscular dysplasia, 11/368, 55/283, 288

fingerprint body myopathy, 43/82

Foix-Alajouanine disease, 9/454

Friedreich ataxia, 21/327

Fukuyama syndrome, 43/92

glioblastoma multiforme, 18/50, 53

globoid cell leukodystrophy, 10/70

glossopharyngeal neuralgia, 48/460

Goltz-Gorlin syndrome, 14/786, 43/20

gram-negative bacillary meningitis, 52/117

granulomatous amebic encephalitis, 52/316

granulomatous CNS vasculitis, 55/389

head injury, 23/26

hemochromatosis, 42/553

hereditary amyloid polyneuropathy type 2, 51/423

hereditary cerebello-olivary atrophy (Holmes), 21/408

hereditary exertional paroxysmal dystonic choreoathetosis, 49/350

hereditary olivopontocerebellar atrophy (Menzel), 21/440

hereditary paroxysmal dystonic choreoathetotis, 49/350

hereditary paroxysmal kinesigenic choreoathetosis, 42/206, 208, 49/350

Hirayama disease, 59/107

histiocytosis X, 63/422

holoprosencephaly, 50/239

homocystinuria, 55/325

homosexuality, 43/215
human T-lymphotropic virus type I associated
    myelopathy, 56/531, 59/452
hydrocephalus, 50/62
hyperaldosteronism, 39/493
hypertrophic cardiomyopathy, 63/136
hyperventilation syndrome, 38/327, 63/433
hypophyseal diabetes insipidus, 42/543
idiopathic scoliosis, 32/132, 42/52
IgM paraproteinemic polyneuropathy, 51/437
inclusion body myositis, 62/204
incontinentia pigmenti, 43/23
iniencephaly, 50/60
intermittent spinal cord ischemia, 12/515
internal carotid artery syndrome, 53/305
intracerebral hematoma, 11/679
intracranial hemorrhage, 53/200
intractable hiccup, 63/491
intraspinal neurenteric cyst, 20/71
juvenile hereditary benign chorea, 49/345
Kawasaki syndrome, 56/637
Kii Peninsula amyotrophic lateral sclerosis,
    22/359, 367
kinesigenic choreoathetosis, 42/208
Köhlmeier-Degos disease, 39/435, 51/454, 55/275
kuru, 42/657, 56/556
lathyrism, 28/37, 36/506, 65/3, 5
Leber hereditary optic neuropathy, 62/509
Leber optic atrophy, 42/399
Lesch-Nyhan syndrome, 42/154
lumbar intervertebral disc prolapse, 20/582
lumbar vertebral canal stenosis, 20/678
lymphomatoid granulomatosis, 51/451
macrencephaly, 42/42
malignant hyperthermia, 38/550, 43/117
manic depressive psychosis, 43/208
missile injury, 61/253
mitral valve prolapse, 55/165, 63/20
monostotic fibrous dysplasia, 19/301
Morgagni-Stewart-Morel syndrome, 43/412
motoneuron disease, 21/31
moyamoya disease, 12/360, 42/751, 55/293
multi-infarct dementia, 53/14
multiple myeloma, 18/254, 39/137
multiple sclerosis, 42/496, 47/268, 302
mumps, 56/113, 429
mumps meningitis, 9/545
myasthenia gravis, 41/101, 43/157, 62/399
myoclonic jerk, 73/159
myoglobinuria, 62/559
myophosphorylase deficiency, 41/189, 43/181
myositis, 62/370
myotonia congenita, 62/266

nemaline myopathy, 43/122
neuralgic amyotrophy, 51/173
neurofibromatosis type I, 14/147
neuroleptic akathisia, 65/287
neuroleptic parkinsonism, 42/651, 65/294
neuromyelitis optica, 9/428, 47/400
neuropsychology, 45/3
North American blastomycosis, 35/402, 52/386
occipital cephalocele, 50/99
organophosphate intoxication, 64/169
orthochromatic leukodystrophy, 42/501
osteitis deformans (Paget), 17/118, 170, 43/451
osteochondroma, 19/307
osteoid osteoma, 19/297, 300
osteoma, 19/295
paraganglioma, 42/763
Parkinson dementia complex, 42/250
Parkinson disease, 42/246, 49/125
pediatric head injury, 57/328
Pena-Shokeir syndrome type I, 43/438
Pick disease, 42/285, 46/233
pinealoma, 18/361, 42/767
poliomyelitis, 34/113, 56/315
polyarteritis nodosa, 8/124, 11/139, 39/295,
    55/354
polymyositis, 8/123, 41/55, 57
pontine hemorrhage, 12/38
pontine infarction, 12/14
posterior longitudinal ligament calcification,
    19/154
postlumbar puncture syndrome, 61/152
postoperative psychosis, 11/446
Prader-Labhart-Willi syndrome, 43/464
primary cerebral reticulosarcoma, 18/236
progressive hemifacial atrophy, 8/347, 22/548,
    43/408, 59/479
progressive muscular atrophy, 59/14
progressive myositis ossificans, 43/191
progressive spinal muscular atrophy, 22/12
progressive supranuclear palsy, 22/218, 42/259
prolonged transient ischemic attack, 53/259
pseudohypoparathyroidism, 6/706, 42/578, 622,
    49/419
pseudoxanthoma elasticum, 11/465, 14/775,
    43/46, 55/451
pure autonomic failure, 22/233, 63/147
18q partial monosomy, 43/533
10q partial trisomy, 43/518
15q partial trisomy, 43/531
quadriceps myopathy, 62/174
Raynaud disease, 8/322, 22/261
reflex sympathetic dystrophy, 51/480
Refsum disease, 8/22, 21/9, 185, 60/234, 66/494

Reye syndrome, 29/336, 56/163
Rothmund-Thomson syndrome, 14/777, 43/460
scoliosis, 50/64
septum pellucidum tumor, 17/539
Shy-Drager syndrome, 38/234
Sjögren syndrome, 51/449
slow eye movement, 42/187
Sneddon syndrome, 55/402
snoring, 63/450
spasmodic torticollis, 6/568
specific developmental dyslexia, 43/222
sphenoid wing aplasia, 50/143
spina bifida, 32/548, 50/56
spinal angioma, 20/438
spinal arachnoid cyst, 20/80-82
spinal cord injury, 25/142
spinal cord stab wound, 25/198
spinal cord tumor, 19/7
spinal ependymoma, 20/354
spinal epidural empyema, 52/185
spinal lipoma, 20/393
spinal meningioma, 20/190, 226
spinal neurinoma, 19/358, 20/245, 279
spinal perineural cyst, 20/152
spinal tumor, 19/53
spinocerebellar ataxia, 42/187
steroid myopathy, 62/535
Stilling-Türk-Duane syndrome, 42/306, 50/213
striatopallidodentate calcification, 49/422
stuttering, 43/227
subacute necrotizing encephalomyelopathy,
    28/351
subclavian artery atherosclerosis, 53/371
subclavian artery stenosis, 53/371
subdural effusion, 24/334
Sydenham chorea, 49/360
temporal arteritis, 48/325, 55/341
thoracic intervertebral disc prolapse, 20/565
thoracic outlet syndrome, 51/122
Tolosa-Hunt syndrome, 48/292-296
transient ischemic attack, 53/200, 259
trigeminal neuralgia, 5/311, 42/353
trisomy 8 mosaicism, 43/510
tropical spastic paraplegia, 56/531
undetermined significant monoclonal
    gammopathy, 63/399
uremic polyneuropathy, 63/510
verbal auditory agnosia, 43/229
vertebral dissecting aneurysm, 54/277
vertebral metastasis, 20/415
Wegener granulomatosis, 39/343, 51/450
Wildervanck syndrome, 43/344
Wohlfart-Kugelberg-Welander disease, 21/28,

    22/69, 59/87
X-linked ichthyosis, 13/477
Sexual activity
  body odor, 74/377
  dreaming, 3/89
  exertional headache, 5/148, 48/379
  olfaction, 74/377
  reflex epilepsy, 73/193
  subarachnoid hemorrhage, 55/2
Sexual ateleiotic dwarf, see Pituitary dwarfism
    type I
Sexual behavior
  Klüver-Bucy syndrome, 3/352, 355, 45/259, 262
  suprachiasmatic nucleus, 74/467, 498
  vasoactive intestinal polypeptide, 74/498
  vasopressin, 74/498
Sexual deviation
  Pick disease, 46/235
  twin study, 43/215
Sexual dimorphic nucleus of preoptic area
  homosexual man, 74/500
Sexual disturbance
  mass reflex, 2/196
  multiple sclerosis, 47/168
Sexual dysfunction
  aorta aneurysm, 63/50
  aorta surgery, 63/59
  category, 75/96
  epilepsy, 75/98, 352
  female, 75/95
  head injury, 75/100
  multiple sclerosis, 75/97
  Parkinson disease, 75/99
  penis erection, 75/98
  sexual impotence, 75/98
  stroke, 75/99
  transverse spinal cord syndrome, 2/196-198
Sexual function
  see also Penis erection and Sexual impotence
  adoption, 26/460
  arousal, 26/455
  artificial insemination, 26/459
  artificial penis, 26/460
  autonomic nervous system, 74/540, 75/85
  basilar impression, 32/72
  coitus, 26/448
  complete cord lesion, 26/448
  counseling, 26/458
  delivery, 26/452
  drug, 75/90
  ejaculation, 26/440, 442, 448
  electrovibrator, 26/458
  epileptiform discharge, 75/352

erection, 26/437, 442, 448
family life, 26/453
female, *see* Female sexual function
genital reflex, 26/439, 443-447
gonadal function, 26/451
hormone, 26/451
hormone treatment, 26/460
incomplete cord lesion, 26/449
libido, 26/454
male, *see* Male sexual function
management, 26/457
marriage, 26/452
masturbation, 26/448, 457
nerve supply, 26/443-447
orgasm, 26/440, 442, 448
paraplegia, 26/454-457
physiology, 26/437, 443
pleasing, 26/455
pregnancy, 26/452
prognosis, 26/311
psychosexual aspect, 26/454-457
reflex arc, 26/440
role change, 26/457
sleep, 74/540
spermatogenesis, 26/451
spinal cord center, 26/440
spinal cord injury, 26/437-461
spinal cord injury prognosis, 25/281, 26/311
spinal shock, 26/258, 448
Sexual impotence
   see also Penis erection *and* Sexual function
   adrenomyeloneuropathy, 60/170
   amyloid polyneuropathy, 51/416
   anatomy, 26/442
   autonomic nervous system, 74/118
   autonomic polyneuropathy, 51/487
   brain embolism, 53/207
   brain infarction, 53/207
   carbon disulfide intoxication, 64/25
   cerebrovascular disease, 53/207
   chordoma, 18/155
   clonidine intoxication, 37/443
   constriction device, 75/92
   diabetes mellitus, 27/115, 70/141, 74/118
   diabetic autonomic visceral neuropathy, 8/38, 27/115
   diabetic polyneuropathy, 51/500
   disulfiram intoxication, 37/321
   drug, 75/90
   hereditary amyloid polyneuropathy type 1, 8/9, 368
   hereditary sensory and autonomic neuropathy type II, 60/11

   hexane polyneuropathy, 51/277
   2-hexanone polyneuropathy, 51/277
   hormone, 75/94
   hypopituitarism, 18/155
   injection therapy, 75/91
   intracranial hemorrhage, 53/207
   lathyrism, 36/511, 65/3
   methyldopa intoxication, 37/440
   multiple myeloma polyneuropathy, 63/393
   multiple neuronal system degeneration, 75/168
   multiple sclerosis, 9/171
   obstructive sleep apnea, 63/454
   oral agent, 75/93
   orthostatic hypotension, 63/146
   penis pain, 75/92
   POEMS syndrome, 47/619, 63/394
   psychology, 26/457
   respiratory encephalopathy, 63/415
   sexual dysfunction, 75/98
   Shy-Drager syndrome, 1/474
   surgical treatment, 75/93
   transient ischemic attack, 53/207
   uremic polyneuropathy, 63/511
   vincristine polyneuropathy, 51/299
   vitamin B12 deficiency, 51/341
Sexual infantilism
   pituitary dwarfism type II, 42/613
   Turner syndrome, 50/544
Sexual intercourse
   headache, 48/10, 373-381
   paraplegia, 61/313
   spinal cord injury, 61/313
   tetraplegia, 61/313
   transient global amnesia, 45/208
Sexual interest
   vaginal secretion, 74/378
Sexual organ
   autonomic nervous system, 74/14
Sexuality
   brain substrate, 45/277
   hypothalamus, 45/277
   spinal cord injury, 61/545
Sézary syndrome
   see also Mycosis fungoides
   adult T-cell leukemia, 18/235, 56/527
SGOT, *see* Serum aspartate aminotransferase
SGPT, *see* Serum alanine aminotransferase
Shagreen skin
   tuberous sclerosis, 14/102, 342, 576, 43/49, 50/371
Sham rage
   emotional disorder, 3/345
   posterior hypothalamus lesion, 2/440

septal area, 2/770
Shape perception
  agnosia, 45/334
  constructional apraxia, 45/413
  space perception, 45/413
  visual sensory deficit, 45/334
Shapiro syndrome
  asterixis, 50/167
  ataxia, 50/167
  autonomic nervous system, 75/10
  confusion, 50/167
  corpus callosum agenesis, 50/163, 167
  diaphoresis, 50/167
  epilepsy, 50/167
  episodic hypothermia, 50/167
  fatigue, 50/167
  hyperhidrosis, 50/167
  hypothermia, 50/167
  late onset, 50/167
  lethargy, 50/167
  mutism, 50/167
  survey, 75/74
  thermoregulation, 75/60, 74
  tremor, 50/167
  withdrawal, 50/167
Shark intoxication
  classification, 37/78
Sharp disease, see Mixed connective tissue disease
Sharp-Purser test
  spinal rheumatoid arthritis, 19/337
Shaver disease, see Pulmonary fibrosis
Shawl scrotum
  Aarskog syndrome, 43/313
Shawl sign
  dermatomyositis, 62/371
Shear strain
  cervical vertebral column injury, 25/254
Shearing injury, see Diffuse axonal injury
Sheath
  carotid, see Carotid sheath
  myelin, see Myelin sheath
  nerve, see Nerve sheath
  optic nerve, 2/521
  spinal nerve root, see Spinal nerve root sheath
Sheehan syndrome
  adenohypophyseal postpartum necrosis, 2/455
  adenohypophyseal syndrome, 2/455
  anorexia nervosa, 46/585, 588
  chromophobe adenoma, 18/354
  craniopharyngioma, 2/460
  dementia, 46/397
  hemorrhage, 2/455
  panhypopituitarism, 46/397

Sheep erythrocyte rosetting technique
  multiple sclerosis, 47/341, 344, 356, 359
Sheldon syndrome
  antiphospholipid antibody, 63/122
Shellfish poisoning
  amnesic, see Amnesic shellfish poisoning
  diarrheic, see Diarrheic shellfish poisoning
  neurotoxic, see Neurotoxic shellfish poisoning
  paralytic, see Paralytic shellfish poisoning
Shepherd crook
  fibrous dysplasia, 14/169
Sherrington, C.S., 1/7
Sherrington classification
  proprioceptive sensation, 1/80
Sherrington study
  cell conductivity, 1/47, 49
  central excitatory state, 1/50
  nerve cell, 1/47
  reflex, 1/47, 49
  refractory period, 1/50
Shift worker
  delayed sleep phase syndrome, 45/142
  sleep-wake cycle, 45/142
Shigella
  acquired myoglobinuria, 62/577
  childhood myoglobinuria, 62/561
  gram-negative bacillary meningitis, 52/121
Shigella sonnei
  infantile enteric bacillary meningitis, 33/63
Shingles, see Herpes zoster
Shivering
  autonomic nervous system, 1/451
  migraine, 48/156
  pontine infarction, 53/389
  spinal cord injury, 61/279
  thermoregulation, 1/451, 26/362, 75/55
  transverse spinal cord lesion, 61/279
Shock
  brain concussion, 23/437
  brain death, 55/263
  brain fat embolism, 55/179
  brain injury outcome, 57/66
  brain metabolism, 57/66
  brain stem death, 57/471
  difference with spinal shock, 26/187
  electric, see Electric shock
  head injury, 57/66
  head injury outcome, 57/66
  hemorrhagic, see Hemorrhagic shock
  hydrozoa intoxication, 37/38
  hypoglycemic, see Hypoglycemic shock
  pallidal necrosis, 49/466
  spinal, see Spinal shock

318

spinal cord injury, 26/187
  traumatic, *see* Traumatic shock
Shock wave
  brain concussion, 23/98
  brain injury, 23/98
  intracranial pressure, 57/300
  penetrating head injury, 57/300
Shooting pain
  Fabry disease, 60/172
  hereditary amyloid polyneuropathy type 1, 51/424
  hereditary perforating foot ulcer, 8/201
  hereditary sensory and autonomic neuropathy type
    I, 60/8
Short bowel syndrome
  deficiency neuropathy, 51/325, 328
Short chain acyl-coenzyme A dehydrogenase
  metabolic myopathy, 62/492
  mitochondrial matrix system, 66/409
  muscle carnitine deficiency, 66/404
Short chain acyl-coenzyme A dehydrogenase
    deficiency
  lipid metabolic disorder, 62/495
  mitochondrial matrix system, 66/409
  psychomotor retardation, 62/495
Short chain alkylmercury intoxication
  mercury polyneuropathy, 51/271, 273
Short chain fatty acid
  hepatic coma, 27/361
  hypoglycin intoxication, 65/131
  Reye syndrome, 29/335, 339, 56/152, 166
Short chain 3-hydroxyacyl-coenzyme A
    dehydrogenase
  childhood myoglobinuria, 62/563
  myoglobinuria, 62/558
Short chain 3-hydroxyacyl-coenzyme A
    dehydrogenase deficiency
  cardiomyopathy, 62/495
  lipid metabolic disorder, 62/495
  metabolic encephalopathy, 62/495
  metabolic myopathy, 62/495
  mitochondrial matrix system, 66/412
  myoglobinuria, 62/495
  Reye like episode, 62/495
Short dura mater syndrome
  Arnold-Chiari malformation type I, 5/372
  atlanto-occipital abnormality, 5/372
  craniectomy, 5/372
  platybasia, 5/372
Short increment sensitivity index
  acoustic nerve tumor, 17/676
Short latency evoked potential
  brain stem death, 57/469
  syringomyelia, 50/446

Short metacarpal
  Turner syndrome, 50/543
Short metatarsal
  Turner syndrome, 50/544
Short neck
  congenital hyperphosphatasia, 31/258
  congenital spondyloepiphyseal dysplasia, 43/478
  Down syndrome, 50/526
  Edwards syndrome, 43/535
  Hurler-Hunter disease, 10/436
  Miller-Dieker syndrome, 50/592
  Patau syndrome, 31/507, 43/528, 50/558
  r(14) syndrome, 50/590
Short rib polydactyly syndrome, *see*
    Saldino-Noonan syndrome
Short stature
  see also Dwarfism *and* Growth retardation
  acrodysostosis, 43/291
  Bartter syndrome, 42/528
  Biemond syndrome type II, 43/334
  brachial neuritis, 42/304
  complex III deficiency, 62/504, 66/423
  Down syndrome, 43/544
  dyschondrosteosis, 43/389-391
  dysosteosclerosis, 43/393
  epiphyseal dysplasia, 43/396
  fetal face syndrome, 43/400
  formiminotransferase deficiency, 42/546
  Freeman-Sheldon syndrome, 38/413
  hair brain syndrome, 43/297-299
  Hallermann-Streiff syndrome, 43/403
  hyperargininemia, 42/562
  hyperdibasic aminoaciduria, 42/563
  infantile optic glioma, 42/733
  Kearns-Sayre-Daroff-Shy syndrome, 43/142,
    62/508
  Laurence-Moon syndrome, 43/253
  Marden-Walker syndrome, 43/423
  megalocornea, 43/270
  mental deficiency, 43/270, 273-275
  microcephaly, 43/429
  18p partial monosomy, 43/534
  Prader-Labhart-Willi syndrome, 31/322, 43/463
  pseudohypoparathyroidism, 6/706, 42/578, 621,
    46/85, 396, 49/419
  pyknodysostosis, 43/468
  pyruvate carboxylase deficiency, 42/586
  pyruvate dehydrogenase deficiency, 42/587
  r(15) syndrome, 43/530
  Rothmund-Thomson syndrome, 14/777, 39/405,
    43/460
  Rubinstein-Taybi syndrome, 43/234, 50/276
  Rud syndrome, 14/13, 43/284, 51/398

Russell-Silver syndrome, 22/550, 43/476, 46/21, 82
sickle cell anemia, 42/623
Sjögren-Larsson syndrome, 59/358
spondyloepiphyseal dysplasia, 43/481
spondylometaphyseal dysplasia, 43/482
Turner syndrome, 30/249, 31/499, 43/550, 50/544
Urbach-Wiethe disease, 42/595
XXXX syndrome, 43/554
XXXXX syndrome, 43/554
Shortening
    leg, *see* Leg shortening
    occipital bone, *see* Occipital bone shortening
Shot injury, *see* Gunshot injury
Shoulder
    congenital elevation, 1/483
Shoulder arthroplasty
    axillary nerve, 70/31
Shoulder girdle
    Brossard syndrome, 22/57
    compression syndrome, 7/430
    scapuloperoneal muscular dystrophy, 42/99
Shoulder hand syndrome
    compression syndrome, 7/430
    dysautonomia, 22/254
    migraine, 48/144
    myocardial infarction, 1/513, 8/330
    physiotherapy, 8/330
    reflex neurovascular syndrome, 8/330
    reflex sympathetic dystrophy, 1/513, 51/481
    sympathectomy, 8/330
Shoulder neuritis, *see* Brachial neuralgia
Shoulder pain
    neuralgic amyotrophy, 8/78
Shulman syndrome
    eosinophilia myalgia syndrome, 64/261
Shunt
    CSF, *see* CSF shunt
    hexomonophosphate, *see* Hexomonophosphate shunt
    lumboperitoneal, *see* Lumboperitoneal shunt
    pentose, *see* Pentose shunt
    syringoperitoneal, *see* Syringoperitoneal shunt
    syringosubarachnoid, *see* Syringosubarachnoid shunt
    ventriculoatrial, *see* Ventriculoatrial shunt
    ventriculoperitoneal, *see* Ventriculoperitoneal shunt
Shunt procedure
    hydromyelia, 50/431
Shy-Drager syndrome, 38/233-252
    see also Orthostatic hypotension
    age, 38/234

age at onset, 22/264
anhidrosis, 1/474
atonic bladder, 1/474
autonomic dysfunction, 63/146
autonomic manifestation, 38/234
autonomic polyneuropathy, 51/479
brain blood flow, 38/244, 247, 53/61
bulbar paralysis, 22/139
cause of death, 38/237
central sleep apnea, 63/462
clinical features, 38/234-238
clock function, 74/468
course, 38/235-237
diagnosis, 38/238
dysautonomia, 22/253, 264
etiology, 38/252
external ophthalmoplegia, 1/474
fecal incontinence, 1/474
Ferguson-Critchley syndrome, 59/322
genetics, 38/234, 75/245
history, 38/233
intestinal pseudo-obstruction, 51/493
laboratory finding, 38/238
muscle fasciculation, 1/474
muscular atrophy, 1/474
neuropathology, 22/265-272, 381, 390, 59/140
oculomotor disorder, 22/141
olivopontocerebellar atrophy, 49/211
olivopontocerebellar atrophy (Dejerine-Thomas), 60/530
olivopontocerebellar atrophy (Wadia-Swami), 60/497
onset, 38/234
ophthalmic finding, 74/430
orthostatic hypotension, 1/474, 2/121, 43/66, 63/145
Parkinson disease, 49/95, 108, 59/134
pathology, 38/239-242
pathophysiology, 38/242-247
pharmacology, 38/247-249
progressive dysautonomia, 59/133-135
progressive muscular atrophy, 59/22
respiratory failure, 49/208
sex ratio, 38/234
sexual impotence, 1/474
sleep, 74/543
striatonigral degeneration, 49/206, 210, 60/540
tachyphylaxis, 22/268
treatment, 38/249-252
tremor, 1/474
urinary incontinence, 1/474
weakness, 1/474
widespread lesion, 22/232

SIADH, *see* Inappropriate antidiuretic hormone
 secretion
Sialic acid
 biochemistry, 66/353
 chemistry, 10/252
 enzyme, 66/355
 galactosialidosis, 66/354
 gangliosidosis, 66/250
 GM2 gangliosidosis, 10/292
 infantile free sialicylic acid storage disease,
  66/354
 metabolism, 66/354
 polysialoganglioside, 66/250
 Salla disease, 66/354
 sialidase deficiency, 66/354
 sialuria, 66/354
Sialic acid transport
 Salla disease, 51/382
Sialidase
 enzyme deficiency polyneuropathy, 51/370
Sialidase deficiency
 mucolipidosis, 66/356
 mucolipidosis type I, 66/353
 sialic acid, 66/354
 spinal muscular atrophy, 59/102
Sialidasis type I
 mucolipidosis type I, 51/380
Sialidosis, *see* Mucolipidosis type I
Sialolipidosis, *see* Mucolipidosis type IV
Sialorrhea
 cause, 6/187
Sialuria
 sialic acid, 66/354
 survey, 66/366
Sialyloligosaccharide
 mucolipidosis type II, 66/380
Sibling
 juvenile hereditary benign chorea, 49/345
 multiple sclerosis, 47/290-292
 syringomyelia, 50/451
Sicard-Collet syndrome
 history, 24/179
 management, 24/180
 multiple cranial neuropathy, 2/100, 24/179
 nonmedullary cause, 2/225
 nonrecurrent nonhereditary multiple cranial
  neuropathy, 51/571
 skull base tumor, 17/182
 symptom, 2/100, 24/180
 vagus nerve injury, 24/179
 Villaret syndrome, 2/100
Sicard syndrome, *see* Sicard-Collet syndrome
Sicca symptom

Sjögren syndrome, 71/60
Sicca syndrome, *see* Sjögren syndrome
Sick motoneuron
 Duchenne muscular dystrophy, 40/383
 muscular dystrophy, 41/409
Sick sinus syndrome
 brain embolism, 55/153
 brain hemorrhage, 55/152
 brain infarction, 55/152
Sickle cell anemia, 38/33-44, 55/503-513
 see also Hemoglobinopathy
 acquired myoglobinuria, 62/572
 anesthesia, 55/466
 angioid streak, 13/141, 55/452
 bacterial meningitis, 63/258
 birth incidence, 42/624
 borderzone ischemia, 55/508
 brain angiography, 55/505-507
 brain blood flow, 55/511
 brain embolism, 55/507
 brain hemorrhage, 38/37-40, 53/34, 55/465, 504,
  510, 63/262
 brain infarction, 53/33, 166, 55/464, 504, 508,
  63/259
 brain infarction mechanism, 55/505
 brain infection, 38/34
 brain ischemia, 53/166, 55/464
 brain thrombosis, 38/34
 brain watershed ischemia, 63/259
 cavernous sinus thrombosis, 52/173
 cerebral vessel occlusion, 38/36
 cerebrovascular disease, 53/28, 31, 33, 43, 55/464
 chorioretinal degeneration, 13/41
 classification, 63/258
 coma, 55/510
 computer assisted tomography, 55/506
 constitutional symptom, 55/503
 convulsion, 55/510
 fat embolism, 38/42-44
 headache, 63/259
 hematomyelia, 38/60
 hemoglobin C, 42/624
 hemoglobin S, 42/624, 55/503
 hemoglobin SC disease, 55/465
 hemolysis, 55/503
 heterozygote frequency, 42/624
 heterozygous hemoglobin SC disease, 55/465
 history, 38/33
 hyperviscosity syndrome, 55/484
 immunocompromised host, 56/469, 473
 infection, 38/40-42
 intracranial hemorrhage, 63/259
 juvenile brain infarction, 55/464

juvenile cerebrovascular disease, 55/504
large vessel vasculopathy, 55/507
leukoencephalopathy, 9/614
meningitis, 38/40-42
mental deficiency, 42/623
moyamoya disease, 53/164, 55/505, 511
multiple white matter lesion, 55/507
neuropathology, 55/508
nuclear magnetic resonance, 55/506, 512
obliterative endarterectomy, 55/504
occlusive vasculopathy, 55/509
optic atrophy, 13/41
pain, 42/623
pain attack, 63/258
pallidoluysionigral degeneration, 49/483
pallidonigral degeneration, 49/450
pathophysiology, 63/258
pneumococcal meningitis, 33/49, 52/53
positron emission tomography, 55/511
progressive intimal proliferation, 63/261
race, 55/503
recurrent brain infarction, 55/464
recurrent meningitis, 52/53
retinal sea fan lesion, 63/262
sea-fan retinopathy, 55/465
seizure, 42/623, 63/259
short stature, 42/623
skeletal deformity, 42/624
stroke, 42/623
subarachnoid hemorrhage, 55/465, 510, 63/262
subdural hematoma, 55/510
symptom, 55/503
systemic brain infarction, 11/460
treatment, 55/512, 63/263
vascular lesion, 55/508
vascular lesion pathogenesis, 55/509
vascular occlusion, 55/503
vertebral artery occlusion, 55/506
vertebrobasilar system, 55/509
Willis circle lesion, 63/259
Sickle cell trait
see also Hemoglobinopathy
brain infarction, 55/513
brain phlebothrombosis, 55/466
cerebrovascular disease, 55/466
hemoglobin AS, 55/512
hemoglobin SC, 55/513
juvenile cerebrovascular disease, 55/466
migraine, 55/466
myelomalacia, 55/466
oral contraceptive agent, 55/466
pallidoluysionigral degeneration, 6/665
venous sinus thrombosis, 55/513

Sideroblastic anemia
Pearson syndrome, 62/508, 66/427
X-linked recessive ataxia, 60/638
Sideropenia, see Iron deficiency
Siderosis
marginal zone, see Marginal zone siderosis
superficial, see Superficial siderosis
Siebenmann syndrome, see Jugular foramen
syndrome
Siemens-Bloch disease, see Incontinentia pigmenti
Siemens syndrome, see Incontinentia pigmenti
Siemerling-Creutzfeldt disease, see
Adrenoleukodystrophy
Sieve scotoma
dominant infantile optic atrophy, 13/114
visual field, 13/114
Siganidae intoxication
family, 37/78
venomous spine, 37/78
Siganus intoxication
family, 37/78
venomous spine, 37/78
Sighing respiration
capnography, 1/668
childhood, 1/667
tuberculous meningitis, 1/667
Sigma 1 protein
neurotropism, 56/35
Sigmoid sinus, see Transverse sinus
Sigmoid sinus thrombophlebitis, see Transverse
sinus thrombophlebitis
Sign language
aphasia, 46/616
Sign of the setting sun, see Setting sun sign
Signe de la chiquenaude, see Flip sign
Signe du journal, see Froment sign
Signe du rideau de Vernet
jugular foramen syndrome, 2/97
Signpost phenomenon
Parkinson disease, 6/184
Signs
Abadie, see Abadie sign
Alfödi, see Alföldi sign
autonomic nervous, see Autonomic nervous sign
Babin cloth peg, see Babin cloth peg sign
Babinski, see Babinski sign
Barré, see Barré sign
Bell, see Bell sign
Biernacki, see Biernacki sign
Bing, see Bing sign
Bragard, see Bragard sign
Brudzinski, see Brudzinski sign
Chaddock, see Chaddock sign

Chvostek, *see* Chvostek sign
collapsing cord, *see* Collapsing cord sign
Colle, *see* Colle sign
Collier, *see* Collier sign
cord, *see* Cord sign
cracked pot, *see* Cracked pot sign
creased hand, *see* Creased hand sign
Crowe, *see* Crowe sign
Dalrymple, *see* Dalrymple sign
Dauer-Babinski, *see* Pseudo-Babinski sign
dense delta, *see* Dense delta sign
digito-ocular, *see* Digito-ocular sign
direct tumor, *see* Direct tumor sign
doughnut, *see* Doughnut sign
Duchenne, *see* Duchenne sign
dysmorphic, *see* Dysmorphic sign
Elsberg, *see* Elsberg sign
empty delta, *see* Empty delta sign
Erb, *see* Erb sign
extrapyramidal, *see* Extrapyramidal sign
false localization, *see* False localization sign
flexion adduction, *see* Flexion adduction sign
flick, *see* Flick sign
flip, *see* Flip sign
Froment, *see* Froment sign
Fuchs, *see* Fuchs sign
Gonda, *see* Gonda sign
Gordon-Hay, *see* Gordon-Hay sign
Gowers, *see* Gowers sign
Griesinger, *see* Griesinger sign
groove, *see* Groove sign
Guttmann, *see* Guttmann sign
Hoffmann, *see* Hoffmann sign
hollow hand, *see* Hollow hand sign
Howship-Romberg, *see* Howship-Romberg sign
Hyndham, *see* Hyndham sign
Jeanné, *see* Jeanné sign
Kemp, *see* Kemp sign
Kernig, *see* Kernig sign
lag, *see* Lag in sign
Lasègue, *see* Lasègue sign
Léri, *see* Léri sign
Lhermitte, *see* Lhermitte sign
lid twitch, *see* Lid twitch sign
lower motoneuron, *see* Lower motoneuron sign
MacCarthy, *see* MacCarthy sign
Mayer, *see* Mayer sign
meningeal, *see* Meningeal sign
Myerson, *see* Myerson sign
negative delta, *see* Negative delta sign
neuroma, *see* Neuroma sign
ocular, *see* Ocular sign
Oppenheim, *see* Oppenheim sign

Pitres, *see* Pitres sign
platysma, *see* Platysma sign
pronator, *see* Pronator sign
pseudoabducens nerve, *see* Pseudoabducens nerve
    sign
pseudo-Babinski, *see* Pseudo-Babinski sign
pseudo-Von Gräfe, *see* Pseudo-Von Gräfe sign
pyramidal, *see* Pyramidal sign
pyramidal tract, *see* Pyramidal tract sign
rabbit, *see* Rabbit sign
racoon, *see* Racoon sign
Raimiste, *see* Raimiste sign
Romana, *see* Romana sign
Rossolimo, *see* Rossolimo sign
rubbing nose, *see* Rubbing nose sign
Schäfer, *see* Schäfer sign
shawl, *see* Shawl sign
Souques, *see* Souques sign
Stewart-Holmes, *see* Stewart-Holmes sign
string, *see* String sign
Strümpell, *see* Strümpell sign
Sugiura, *see* Sugiura sign
sunset, *see* Sunset sign
Terson, *see* Terson sign
Tinel, *see* Tinel sign
Trendelenburg, *see* Trendelenburg sign
Trousseau, *see* Trousseau sign
tumor, *see* Tumor sign
upper motoneuron, *see* Upper motoneuron sign
Von Gräfe, *see* Von Gräfe sign
Von Romberg, *see* Von Romberg sign
Wadia, *see* Wadia sign
Silastic imprint method
    skin, 75/108
    thermoregulation, 75/65
Silent lesion
    central pontine myelinolysis, 63/549
    Köhlmeier-Degos disease, 55/276
Silicone breast implant
    scleroderma, 71/115
Siluridae intoxication
    family, 37/78
    venomous spine, 37/78
Silurus intoxication
    family, 37/78
    venomous spine, 37/78
Silver
    neurotoxicity, 51/307
    striatopallidodentate calcification, 6/716, 49/425
Silver disease
    spinal muscular atrophy, 59/374
Silver mononeuropathy
    case history, 51/307

Silver-Russell syndrome, *see* Russell-Silver
  syndrome
Simian acquired immune deficiency virus type D
  acute viral myositis, 56/194
  experimental acute viral myositis, 56/195
Simian disease
  retrovirus infection, 56/453
Simian hand
  Aran-Duchenne amyotrophy, 22/291
  hereditary motor and sensory neuropathy type I,
    21/279
  hereditary motor and sensory neuropathy type II,
    21/279
  muscular atrophy, 22/291
  thumb, 21/279, 22/291
Simian T-cell tropic retrovirus
  lentivirus, 56/460
  meningoencephalitis, 56/454
Simian vacuolating virus, *see* Simian virus 40
Simian virus 5
  multiple sclerosis, 47/330
Simian virus 40
  acute viral myositis, 56/193
  enhancer, 56/37
  neural tumor, 71/418
  oncogenic virus, 18/132
  papovavirus, 56/9
  progressive multifocal leukoencephalopathy,
    34/313, 317-321, 38/656, 47/504, 515, 56/514,
    71/417
Simmonds cachexia, *see* Sheehan syndrome
Simola test
  Huntington chorea, 6/348
Simultanagnosia
  alexia, 4/123, 45/439
  Balint syndrome, 2/657, 45/15, 409, 55/144
  cardiac arrest, 63/216
  features, 3/222
  image, 4/16-18
  left sided lesion, 3/226
  nature, 3/215
  occipital lesion, 2/655-657
  occipital lobe tumor, 17/327
  situation, 4/16-18
  spelling dyslexia, 2/657, 661
  visual agnosia, 45/55
  Wolpert interpretation, 3/215, 4/17
Sincipital cephalocele
  foramen cecum, 50/102
  frontoethmoid, 50/102
  Malaysian race, 50/97
  metopic suture, 50/102
  morphologic classification, 50/102

  nasofrontal type, 50/102
Sincipital encephalocele
  agyria, 50/103
  brain stem, 50/103
  computer assisted tomography, 50/104
  holoprosencephaly, 50/103
  hydrocephalus, 50/103
  hypertelorism, 50/104
  microgyria, 50/103
  nasoethmoidal type, 50/103
  naso-orbital type, 50/103
  optic nerve, 50/103
  paternal age, 50/103
Sindbis virus
  blood-brain barrier, 56/84
  experimental acute viral myositis, 56/195
  natural killer cell, 56/73
Singer, Ch., 2/4, 8
Singing
  aphasia, 45/305
Single fiber electromyography
  see also Jitter phenomenon
  Duchenne muscular dystrophy, 40/374
  Eaton-Lambert myasthenic syndrome, 62/79
  Emery-Dreifuss muscular dystrophy, 62/150
  familial infantile myasthenia, 62/427
  limb girdle syndrome, 62/186
  motor unit, 62/49, 53
  myasthenia gravis, 41/128, 62/77, 79, 414
  myopathy, 62/49, 53
Single photon emission computer assisted
    tomography
  blood flow, 72/375
  brain blood flow index, 54/119
  brain blood volume index, 54/119
  brain infarction, 53/313, 54/65
  brain lacunar infarction, 54/256
  brain stem death, 57/464
  carbon monoxide intoxication, 64/33
  cerebrovascular disease, 54/105
  2-2.5 cm, 54/106
  cocaine intoxication, 65/254
  epilepsy, 72/184, 372
  epileptic focus, 72/377
  equipment, 54/112
  extratemporal epilepsy, 72/375
  gamma radiation, 54/106
  glucose metabolism, 72/373
  iodine 123, 54/107
  kreon 81m, 54/107
  limited angle, 54/106
  model, 54/111
  multi-infarct dementia, 54/123

324

neuropsychology, 45/7
neuroreceptor, 72/377
oxygen extraction rate, 54/119
Parkinson disease, 75/184
partial epilepsy, 72/373
radioisotope cohort, 54/107
radiopharmaceuticals, 54/113
regional brain blood flow, 54/111
rotating gamma camara, 54/114
system, 54/114
technetium 99m, 54/107
transient ischemic attack, 53/313, 54/119, 131
wide angle, 54/106
Singultus, *see* Hiccup
Sinoatrial disorder
chronic, *see* Chronic sinoatrial disorder
Sinopulmonary infection
ataxia telangiectasia, 14/268, 286, 60/360, 371,
647, 667
Sinus
anterior intercavernous, *see* Anterior
intercavernous sinus
Breschet, *see* Breschet sinus
carotid, *see* Carotid sinus
cavernous, *see* Cavernous sinus
dermal, *see* Dermal sinus
dura mater, 30/416-418
dural venous, *see* Venous sinus
endoneural, *see* Endoneural sinus
headache, 48/6
inferior petrosal, *see* Inferior petrosal sinus
inferior sagittal, *see* Inferior sagittal sinus
marginal, *see* Marginal sinus
maxillary, *see* Maxillary sinus
occipital, *see* Occipital sinus
paranasal, *see* Paranasal sinus
petro-occipital, *see* Petro-occipital sinus
petrosal, *see* Petrosal sinus
petrosquamosal, *see* Petrosquamosal sinus
pilonidal, *see* Pilonidal sinus
posterior intercavernous, *see* Posterior
intercavernous sinus
pro-otic, *see* Pro-otic sinus
sagittal, *see* Sagittal sinus
sphenoid, *see* Sphenoid sinus
straight, *see* Straight sinus
superior longitudinal, *see* Superior longitudinal
sinus
superior petrosal, *see* Superior petrosal sinus
tentorial, *see* Tentorial sinus
transverse, *see* Transverse sinus
Sinus artery
inferior cavernous, *see* Inferior cavernous sinus

artery
Sinus osteoma
rhinorrhea, 24/186
Sinus pain
infection, 5/330
trigeminal neuralgia, 5/330
tumor, 5/330
Sinus pyriformis, *see* Piriform sinus
Sinus rectus, *see* Straight sinus
Sinus syndrome
anterior cavernous, *see* Anterior cavernous sinus
syndrome
carotid, *see* Carotid sinus syndrome
cavernous, *see* Cavernous sinus syndrome
middle cavernous, *see* Middle cavernous sinus
syndrome
posterior cavernous, *see* Posterior cavernous sinus
syndrome
sphenoidal, *see* Sphenoidal sinus syndrome
Sinus thrombophlebitis
cavernous, *see* Cavernous sinus thrombophlebitis
dural, *see* Venous sinus thrombophlebitis
superior longitudinal, *see* Superior longitudinal
sinus thrombophlebitis
superior petrosal, *see* Superior petrosal sinus
thrombophlebitis
transverse, *see* Transverse sinus thrombophlebitis
Sinus thrombosis
aseptic cavernous, *see* Aseptic cavernous sinus
thrombosis
brain arteriovenous malformation, 54/364
carcinoid syndrome, 70/316
cavernous, *see* Cavernous sinus thrombosis
dural, *see* Venous sinus thrombosis
inferior petrosal, *see* Inferior petrosal sinus
thrombosis
lateral, *see* Transverse sinus thrombosis
sagittal, *see* Sagittal sinus thrombosis
straight, *see* Straight sinus
superior longitudinal, *see* Superior longitudinal
sinus thrombosis
superior petrosal, *see* Superior petrosal sinus
thrombosis
transverse, *see* Transverse sinus thrombosis
traumatic venous, *see* Traumatic venous sinus
thrombosis
venous, *see* Venous sinus thrombosis
Sinus tract
dermal, *see* Dermal sinus tract
Sinus venous thrombosis, *see* Venous sinus
thrombosis
Sinusitis
actinomycosis, 35/386

aspergillosis, 52/379
brain abscess, 52/144
cavernous sinus thrombosis, 52/172
cranial epidural empyema, 52/168
cranial subdural empyema, 52/170
ear nose and throat disease, 5/218
headache, 5/207, 218, 330, 48/33
maxillary, *see* Maxillary sinusitis
migraine, 48/227
orbital cellulitis, 33/179
pain, 5/330
paranasal, 5/213
sphenoid, *see* Sphenoid sinusitis
sphenopalatine neuralgia, 5/115
subarachnoid hemorrhage, 55/20
subdural empyema, 52/73
superior longitudinal sinus thrombosis, 52/178
Wegener granulomatosis, 8/114, 39/343
Sinuvertebral nerve (Luschka)
course, 26/7
Siphon incisivum line
radioanatomy, 11/86
topography, 11/86
Sipple syndrome
C cell hyperplasia, 42/752
hypercalcemia, 42/752
hypopituitarism, 42/752
medullary thyroid carcinoma, 14/750
neurofibromatosis, 14/736, 738, 750
paroxysmal hypertension, 14/750
pheochromocytoma, 14/735, 738, 750, 39/503
Sirenomelia, *see* Caudal aplasia
Situation agnosia, *see* Simultanagnosia
Situs inversus
iniencephaly, 50/131
myoclonus, 42/240
nystagmus, 42/240
Sixth nerve palsy, *see* Abducent nerve paralysis
Sjögren-Larsson-Richards syndrome, *see*
    Sjögren-Larsson syndrome
Sjögren-Larsson syndrome, 22/475-478, 66/615-619
autosomal recessive, 13/141
Babinski sign, 43/307
biochemical defect, 66/617
biochemical study, 66/618
biochemistry, 13/475, 22/476
brain atrophy, 43/307
chorioretinal degeneration, 13/41
choroid atrophy, 13/473
clinical features, 13/471, 22/476, 59/358
congenital ichthyosis, 13/141, 66/615
convulsion, 22/477
defective enamel, 13/474

dental dysplasia, 22/477
dermatoglyphics, 13/475, 22/477
diagnosis, 66/619
differential diagnosis, 13/134, 476
EEG, 22/477
enzyme defect, 66/618
epilepsy, 13/471
epileptic focus, 22/477
eye fundus white dot, 66/616
familial spastic paraplegia, 59/326-328
fatty alcohol, 66/617
fatty alcohol oxidoreductase, 66/617, 619
fatty aldehyde dehydrogenase, 66/617
gene frequency, 43/307
genetics, 13/470, 22/475
geography, 13/470
hereditary spinal muscular atrophy, 59/26
heterozygote frequency, 43/307
histopathology, 66/618
history, 13/468
hyperreflexia, 43/307
hypertelorism, 13/474, 30/248, 66/615
ichthyosiform erythroderma, 13/477, 43/307
ichthyosis, 13/468, 472, 476, 22/476, 59/358,
    66/616
ichthyotic syndrome, 38/9
incontinentia pigmenti, 13/474
kyphoscoliosis, 66/615
long chain alcohols, 59/358
macular degeneration, 66/615
mental deficiency, 13/141, 468, 472, 22/477,
    30/248, 43/307, 46/60, 59/358, 66/615
Miescher syndrome type II, 14/120
neurocutaneous syndrome, 14/101
neurologic symptom, 22/476
neuropathology, 22/477, 66/618
nosology, 22/477
optic atrophy, 13/41
pathology, 13/479
prevalence, 13/470
primary pigmentary epithelium degeneration,
    13/141
progressive paraparesis, 59/358
race, 13/470
Refsum disease, 13/479
retinal dot, 59/358
Rud syndrome, 13/479, 22/478, 38/10
secondary pigmentary retinal degeneration,
    13/319, 22/477, 43/307, 60/733
seizure, 43/307, 59/358
short stature, 59/358
skin manifestation, 30/248
spastic diplegia, 13/141, 66/615

spastic paraplegia, 13/468, 471, 478, 59/439
speech disorder, 13/471, 66/615
treatment, 13/480, 59/358, 66/619
Usher syndrome, 13/479
ventricular dilatation, 43/307
xeroderma pigmentosum, 22/478
Sjögren syndrome
age distribution, 71/66
Alzheimer disease, 71/74
angiography, 71/67
anosmia, 71/74
antibody, 71/69-73
antiphospholipid antibody, 63/330
autoantibody, 71/59, 80
autoimmune disease, 41/102, 240
autoimmune disorder, 71/60
autonomic involvement, 75/699
autonomic neuropathy, 71/84
axonal degeneration, 51/449
bilateral facial paresis, 51/449
Binswanger disease, 71/74
brain infarction, 53/163
brain ischemia, 53/163
Brown-Séquard syndrome, 71/75
bulbar paralysis, 71/74
carotid system syndrome, 53/308
cerebellar speech, 71/75
chronic meningitis, 56/644
classification, 71/4
clinical description, 71/62
CNS, 39/429
compression neuropathy, 71/85
cranial nerve, 39/426-429
cryoglobulinemia, 39/181
CSF, 71/65, 69
CSF IgG, 51/449
dementia, 71/74
dermatomyositis, 62/374, 71/74
diagnosis, 39/419, 424, 71/61
differential diagnosis, 71/73
dorsal spinal nerve root ganglionitis, 51/449
dysarthria, 71/75
dysgeusia, 71/74
Eaton-Lambert myasthenic syndrome, 62/421
EEG, 71/65
EMG, 71/65
eosinophilia myalgia syndrome, 64/260
epidemiology, 39/420
epilepsy, 71/74
focal, 71/62
glossopharyngeal neuralgia, 71/74
hereditary amyloid polyneuropathy, 21/125
hereditary cranial nerve palsy, 60/42

histopathology, 71/80
Holmes-Adie syndrome, 51/449
human immunodeficiency virus, 71/74
human T-lymphotropic virus type I associated
    myelopathy, 59/450
hyperreflexia, 71/75
hypertrophic sensory neuropathy, 51/449
hyperviscosity syndrome, 55/485
hypogeusia, 71/74
hypokalemic periodic paralysis, 71/74, 89
immunogenetic, 71/75
inclusion body myositis, 62/374, 71/74
iritis, 71/86
ischemic vasculitis, 51/449
laboratory test, 51/449
Lhermitte sign, 71/75
Lyme disease, 71/74
lymphoid interstitial pneumonitis, 39/530
Meniere disease, 71/74
meningoencephalitis, 71/74
Mollaret meningitis, 56/630
mononeuritis multiplex, 51/449
mononeuropathy, 51/449
motor neuropathy, 71/85
multiple cranial neuropathy, 51/449
multiple sclerosis, 51/449, 71/74, 75
myasthenia gravis, 39/430, 41/240, 62/402, 71/74,
    89
myelopathy, 63/55, 71/74
myopathy, 39/430, 71/88
myositis, 71/74
neurodiagnostic test, 71/65
neuroimaging, 71/87
neurologic syndrome, 71/62
neuromuscular, 39/422-429
neuronopathy, 71/83
neuropathology, 71/77
neuropathy, 8/17, 39/423-426
nonfocal, 71/63
nuclear magnetic resonance, 71/65
nystagmus, 71/75
optic neuritis, 71/74
optic neuropathy, 71/85
oral, 71/61
paraneoplastic, 71/64
Parkinson disease, 71/74
peripheral facial paralysis, 71/74
peripheral nervous system, 71/83
polyarteritis nodosa, 55/361, 71/74
polymyalgia rheumatica, 71/74
polymyositis, 8/17, 62/374, 71/74
polyneuropathy, 71/83
postviral fatigue syndrome, 71/74

prevalence, 71/61
procainamide intoxication, 37/452
progressive bulbar palsy, 59/225
progressive systemic sclerosis, 51/448, 71/74
recurrent multifocal CNS disease, 51/449
recurrent nonhereditary multiple cranial
    neuropathy, 51/570
recurrent oculomotor paralysis, 51/449
retinal vasculitis, 71/86
rheumatic disorder, 71/60
rheumatoid arthritis, 8/17, 71/74
sensorimotor polyneuropathy, 51/449
sensory neuropathy, 71/83
sensory trigeminal neuropathy, 51/449
sex ratio, 51/449
sicca symptom, 71/60
spasticity, 71/75
subarachnoid hemorrhage, 63/55
survey, 71/59
symptom, 71/75
syphilis, 71/74
systemic lupus erythematosus, 71/74
T-lymphocyte, 51/449
Takayasu disease, 71/74
temporal arteritis, 71/74
therapy, 71/81
toxic oil syndrome, 63/380
transverse myelitis, 63/55
trigeminal neuropathy, 39/427, 71/74, 85
vascular disease, 71/79
Sjögren, T., 10/227
Sjöqvist operation
    trigeminal neuralgia, 5/397
Skeletal abnormality
    acrocephalosyndactyly type I, 31/231
    acrocephalosyndactyly type II, 31/235
    acrodysostosis, 31/229
    Albright syndrome, 31/312
    ataxia telangiectasia, 60/373
    Bardet-Biedl syndrome, 13/384
    chondrodystrophic myotonia, 31/278, 40/328, 546
    Coffin-Lowry syndrome, 31/244
    De Lange syndrome, 31/280
    diabetes insipidus, 60/668
    Down syndrome, 31/400
    Goldenhar syndrome, 31/247
    Goltz-Gorlin syndrome, 14/787, 31/287
    Gruber syndrome, 31/243
    hypophosphatasia, 31/314
    incontinentia pigmenti, 31/242
    Klinefelter variant XXXXY, 31/483
    Klinefelter variant XXYY, 31/484
    Loken-Senior syndrome, 60/735

Marinesco-Sjögren syndrome, 21/556, 60/343
    orofaciodigital syndrome type I, 31/251
    Rubinstein-Taybi syndrome, 31/283
    secondary pigmentary retinal degeneration,
        60/735
    sensorineural deafness, 42/379
    Townes-Brocks syndrome, 42/379
    Wolf-Hirschhorn syndrome, 31/558, 43/497
    xeroderma pigmentosum, 60/336
    XYY syndrome, 31/490
Skeletal deformity
    anencephaly, 30/181
    aspartylglycosaminuria, 27/159, 42/527
    congenital myopathy, 62/334
    Fabry disease, 42/428
    facial paralysis, 42/313
    Friedreich ataxia, 60/313
    glycosaminoglycanosis type I, 10/435, 42/450
    glycosaminoglycanosis type IS, 42/453
    glycosaminoglycanosis type II, 10/435
    glycosaminoglycanosis type IVB, 66/265
    glycosaminoglycanosis type VI, 42/458
    glycosaminoglycanosis type VIII, 42/460
    GM1 gangliosidosis type I, 42/431
    hereditary neuropathy, 60/263
    homocystinuria, 55/329
    hypophosphatasia, 42/578
    mucolipidosis type II, 42/448
    mucolipidosis type III, 42/449
    multiple sulfatase deficiency, 42/493
    myopathy, 40/337
    orbital hypertelorism, 30/248
    Pelizaeus-Merzbacher disease, 42/503
    pseudohypoparathyroidism, 42/621
    sickle cell anemia, 42/624
    spastic paraplegia, 42/172
    Stilling-Türk-Duane syndrome, 42/306
    unilateral facial paresis, 42/313
Skeletal dysplasia, see Bone dysplasia
Skeletal fluorosis
    clinical features, 36/471-477
    cranial nerve lesion, 36/476
    differential diagnosis, 36/492
    EMG, 36/476
    epidemiology, 36/466-468
    fluoride absorption, 36/468-470
    fluoride distribution, 36/468-470
    fluoride excretion, 36/470
    fluoride source, 36/466-468
    history, 36/465, 475
    laboratory finding, 36/477-479
    myelopathy, 36/476
    neurologic manifestation, 36/475-477

neuropathy, 36/476
pathogenesis, 36/484-490
pathology, 36/484-490
prevention, 36/493
radiculopathy, 36/476
radiologic finding, 36/479-484
treatment, 36/492-495
Skeletal lesion
  ataxia telangiectasia, 14/291
  Flynn-Aird syndrome, 22/519
  fucosidosis, 10/466
  multiple nevoid basal cell carcinoma syndrome,
    14/458
  neurofibromatosis type I, 14/157
  tuberous sclerosis, 14/374-376
Skeletal malformation
  achondroplasia, 30/248
  cleidocranial dysostosis, 30/248, 32/41
  Coffin-Lowry syndrome, 30/248
  Conradi-Hünermann syndrome, 30/248
  craniodiaphyseal dysplasia, 30/248
  Dandy-Walker syndrome, 32/59, 50/332
  De Lange syndrome, 30/248
  F syndrome, 30/249
  Fanconi syndrome, 30/249
  fetal face syndrome, 30/248
  Klippel-Feil syndrome, 30/249
  Larsen syndrome, 30/249
  Melnick-Needles syndrome, 30/249
  neurofibromatosis type I, 50/369
  Nielsen syndrome, 30/249
  Noonan syndrome, 30/249
  otopalatodigital syndrome, 30/248
  r(9) syndrome, 50/581
  Refsum disease, 8/22, 10/345, 13/314,
    21/206-209, 42/148, 60/228, 231, 66/485, 491
  Russell-Silver syndrome, 30/248
  Seckel dwarf, 30/248
  Smith-Lemli-Opitz syndrome, 66/587
  Sprengel deformity, 30/248, 32/41
  trisomy 22, 50/564
Skeletal muscle pumping
  autonomic failure, 75/721
Skew deviation
  benign intracranial hypertension, 55/125
  bilateral, 2/309
  brain embolism, 53/213
  brain infarction, 53/213
  cerebellar hemorrhage, 12/56
  cerebrovascular disease, 53/213
  eye movement, 1/605
  features, 2/343, 55/125
  intracranial hemorrhage, 53/213

mesencephalic infarction, 53/391
Nothnagel syndrome, 2/343
nystagmus, 2/343
ocular imbalance, 24/75
superior cerebellar artery syndrome, 53/397
transient ischemic attack, 53/213
trochlear nerve paralysis, 2/309
vertebrobasilar ischemia, 55/125
vertebrobasilar system syndrome, 53/391, 397
SK&F 39162, see Auranofin
Skiing injury
  carotid artery thrombosis, 23/579
  facial injury, 23/581
  prevention, 23/581
  skull fracture, 23/578
  statistic, 23/577
  traumatic carotid thrombosis, 23/579
Skill
  intelligence, 3/313, 4/6
  physical rehabilitation, 61/469
  restoration of higher cortical function, 3/388, 394,
    429
  verbal, see Verbal skill
Skin
  actiniaria intoxication, 37/53
  angioma, 14/392
  atrophy, see Cutaneous atrophy
  blood flow, see Cutaneous blood flow
  boxjelly intoxication, 37/41
  cluster headache, 48/219
  color, see Cutaneous pigmentation
  cutaneous vasomotor activity, 26/365
  denervated, 7/267
  dermatomyositis, 71/130
  Doppler flowmetry, 75/109
  dysplasia, see Cutaneous dysplasia
  glycosaminoglycanosis, 10/454
  hereditary amyloid polyneuropathy, 60/100, 106
  hereditary amyloid polyneuropathy type 1, 60/95
  hypereosinophilic syndrome, 63/372
  hyperextensible, see Hyperextensible skin
  hypopigmentation, see Cutaneous
    hypopigmentation
  hypoplasia, see Cutaneous hypoplasia
  infantile spinal muscular atrophy, 59/56
  Köhlmeier-Degos disease, 39/436
  leiomyoma, see Cutaneous leiomyoma
  lesion, see Skin lesion
  mummy, 74/191
  nerve, see Cutaneous nerve
  Peter Pan syndrome, 18/545
  pigmentation, see Cutaneous pigmentation
  polymyositis, 71/130

progressive hemifacial atrophy, 31/253
Refsum disease, 21/246
rhizostomeae intoxication, 37/53
scleroderma, 39/359-361
sea anemone intoxication, 37/53
sea wasp intoxication, 37/41
senile purpura, 38/71
shagreen, see Shagreen skin
silastic imprint method, 75/108
sympathetic activity, 75/106
sympathetic nerve, 74/649
temperature, see Thermoregulation
Skin abnormality
   diastematomyelia, 50/436
   diplomyelia, 50/436
   thallium intoxication, 36/247
Skin blood flow, see Cutaneous blood flow
Skin flare
   autonomic nervous system, 74/607
   response, 74/607
Skin fragility
   penicillamine intoxication, 27/405
   porphyria variegata, 42/620
Skin hypopigmentation, see Cutaneous
      hypopigmentation
Skin innervation
   diagram, 2/18
   scheme, 2/130
   sympathetic innervation, 26/355-358, 61/277
Skin lesion
   acrodysostosis, 31/230
   acrylamide intoxication, 64/67
   Aicardi syndrome, 31/227
   Albright syndrome, 31/312
   amyotrophic lateral sclerosis, 22/296, 341
   arsenic intoxication, 64/286
   ataxia telangiectasia, 14/75, 290
   Behçet syndrome, 34/480, 71/213
   Cobb syndrome, 14/441
   Cockayne-Neill-Dingwall disease, 14/315, 31/237
   Coffin-Lowry syndrome, 31/245
   Coffin-Siris syndrome, 31/239
   Fabry disease, 55/456
   Flynn-Aird syndrome, 22/519
   Hallermann-Streiff syndrome, 31/249
   Hartnup disease, 14/315, 29/154
   incontinentia pigmenti, 14/85, 214
   lyngbya majuscula intoxication, 37/97
   multiple nevoid basal cell carcinoma syndrome,
      14/80, 457
   neurocutaneous melanosis, 14/416, 43/33
   neurofibromatosis type I, 14/148
   orofaciodigital syndrome type I, 31/250

porifera intoxication, 37/35
postherpetic neuralgia, 5/383, 51/180, 56/234
Refsum disease, 21/202-205
striatopallidodentate calcification, 6/710
tuberous sclerosis, 14/48, 341, 573
Skin potential
   odor, 74/367
   thermoregulation, 75/65
Skin resistance
   odor, 74/367
Skin response
   autonomic nervous system, 74/629
   galvanic, see Galvanic skin response
   hereditary sensory and autonomic neuropathy type
      III, 75/151
Skin temperature
   autonomic nervous system, 75/430
   blood flow, 1/463
   foot sweat test, 75/57
   migraine, 48/77
   odor, 74/367
   sweat rate, 75/57
   test, 74/602
   thermoregulation, 1/463, 7/252, 266, 26/361, 365,
      369
Skin test
   cat scratch disease, 34/461
   coccidioidin, see Coccidioidin skin test
   coccidioidomycosis, 52/416
   N,N-diethyl-m-toluamide intoxication, 36/443
   histamine, see Histamine skin test
   lepromin, see Lepromin skin test
   mumps, see Mumps skin test
   neurobrucellosis, 52/585
   paracoccidioidomycosis, 35/535
   trichinosis, 35/280, 52/570
Skull
   acrocephalosyndactyly type II, 31/235
   anencephaly, 30/180, 50/83
   aneurysmal bone cyst, 20/3
   brain injury mechanism, 57/20
   caliper, see Skull traction
   cloverleaf, see Cloverleaf skull
   dolichocephalic, see Dolichocephalic skull
   electrical injury, 7/374, 376, 23/713
   fibrous dysplasia, 14/193, 46/88
   frontal projection, 57/145
   headache, 48/50
   hydrocephalus, 30/600
   intracranial infection, 24/215
   lacunar, see Craniolacunia
   lateral skull projection, 57/145
   obstructive hydrocephalus, 16/94

osteitis deformans (Paget), 17/118, 38/362-365, 43/450
Pyle disease, 46/88
Ramses II, 74/184
suture, 57/144
Towne view, 57/144
traumatic aneurysm, 24/383
vascular marking, 57/144
Skull base
  adenoid cystic carcinoma, 68/466
  brain metastasis, 69/123, 71/612
  chondrosarcoma, 68/466
  chordoma, 68/466
  esthesioneuroblastoma, 68/466
  Ewing sarcoma, 68/466
  facial neurinoma, 68/466
  fibrosarcoma, 68/34
  Foix-Jefferson syndrome, 67/146
  Foster Kennedy syndrome, 67/146
  gasserian ganglion syndrome, 67/146
  glomus jugulare tumor, 68/482
  Gradenigo syndrome, 67/146
  Hand-Schüller-Christian disease, 18/322
  histiocytosis X, 18/322
  intracranial hypertension, 67/146
  jugular foramen neurinoma, 68/466
  jugular foramen syndrome, 67/147
  juvenile angiofibroma, 68/466
  meningioma, 68/411, 466
  nasopharyngeal carcinoma, 68/466
  paraganglioma, 68/466
  pituitary adenoma, 68/466
  primary sarcoma, 68/466
  retro-orbital tumor, 67/146
  rhabdomyosarcoma, 68/466
  spina bifida, 32/541
  superior orbital fissure syndrome, 67/146
  trigeminal neurinoma, 68/466
  vestibular neurinoma, 68/466
  Villaret syndrome, 67/147
Skull base fracture
  see also Skull fracture
  abducent nerve injury, 24/68, 82, 84
  anterior fossa, 23/393
  carotid artery thrombosis, 26/43, 57/345
  carotid cavernous fistula, 24/406, 57/345
  cranial nerve palsy, 57/345
  dissecting pseudoaneurysm, 57/346
  Garcin syndrome, 2/313
  intracranial infection, 24/220
  rhinorrhea, 24/186
  skull fracture, 57/146
  traumatic carotid cavernous sinus fistula, 57/345

  trochlear nerve injury, 24/65
  vault, 23/387
Skull base metastasis, 69/92
  jugular foramen, 69/128
  middle fossa, 69/128
  occipital condylus, 69/128
  orbital, 69/128
  parasellar, 69/128
  survey, 69/128
Skull base synchondrosis
  closure, 50/113
Skull base syndrome
  half, see Garcin syndrome
Skull base tumor, 67/146
  see also Skull tumor
  adenoid cystic carcinoma, 68/489
  aneurysmal bone cyst, 17/163
  angiography, 68/470
  anterior occipital condyle, 17/182
  auditory canal, 68/469
  benign, 17/137, 68/467
  bone marrow, 17/172
  bone tumor, 68/490
  cavernous sinus, 68/468, 474
  cavernous sinus syndrome, 17/181
  cerebellopontine angle syndrome, 17/182
  ceruminal gland adenoma, 17/190
  chemotherapy, 68/473
  chloroma, 17/200, 18/252
  cholesteatoma, 17/196, 206
  chondroma, 17/147, 150, 154, 156, 158, 194
  chondrosarcoma, 17/168, 68/484
  chordoma, 18/404, 68/484
  classification, 17/137, 68/465, 467
  clinical syndrome, 17/181
  craniopharyngioma, 17/194
  cylindroma, 17/186
  diagnosis, 17/137, 68/469
  differential diagnosis, 17/192, 68/471
  ear tumor, 17/190
  eosinophilic granuloma, 17/175
  esthesioneuroblastoma, 68/486
  Ewing sarcoma, 17/169
  fibroma, 17/147, 177
  fibrosarcoma, 17/168
  fibrous dysplasia, 17/173
  Foix-Jefferson syndrome, 17/181
  foramen magnum, 68/469
  frontal base, 68/468
  Garcin syndrome, 2/103, 17/182
  giant cell tumor, 17/160
  Gradenigo syndrome, 17/181, 68/468
  hemangioma, 17/163

histogenesis, 17/139
Hodgkin disease, 17/200
hormone treatment, 68/473
incidence, 68/465
infraclinoid syndrome, 17/181
jugular foramen, 68/469
jugular foramen syndrome, 17/182
juvenile angiofibroma, 68/488
leukemia, 17/174
lipoma, 17/159
literature, 17/139
location, 68/466-468
malignancy, 68/466
malignant, 17/137, 68/467
meningioma, 17/176, 68/477
metastasis, 17/191
middle ear carcinoma, 17/190
middle ear sarcoma, 17/190
mixed parotid tumor, 17/191
nasopharyngeal carcinoma, 68/488
nasopharyngeal fibroma, 17/178
neurofibroma, 17/176
optic nerve glioma, 17/196
orbit, 17/176
orbital apex syndrome, 17/181, 68/468
orbital tumor, 17/201
origin, 17/139
ossifying fibroma, 17/146
osteitis deformans (Paget), 17/170, 172
osteoid osteoma, 17/147
osteoma, 17/138, 140, 143
osteosarcoma, 17/167
paraganglioma, 68/482
parasellar cholesteatoma, 17/197
petroclival, 68/477
petrous apex, 17/181
pituitary adenoma, 68/482
prepeduncular syndrome, 17/183
radiotherapy, 68/472, 474
reticulosarcoma, 17/169
retroparotid space, 17/182
rhabdomyosarcoma, 68/490
rhinorrhea, 24/186
sarcoma, 17/167, 18/404
sella turcica, 68/468
Sicard-Collet syndrome, 17/182
sphenoid bone, 17/181
sphenoid sinus, 68/468
sphenoid wing, 68/468
superior orbital fissure syndrome, 17/181
surgery, 68/472
survey, 68/465
tomography, 17/137

treatment, 68/470
tumor type, 68/467
type, 68/466
Vernet syndrome, 17/182, 68/469
Villaret syndrome, 17/182
Skull defect
    amniotic band syndrome, 30/116
    cosmetics, 24/664
    epilepsy, 24/663
    meroanencephaly, 42/12
    microcephaly, 30/277
    posttraumatic headache, 24/507
    site, 24/663
    size, 24/663
    surgery, 24/663
    traumatic headache, 24/507
Skull fracture
    see also Skull base fracture and Skull injury
    anosmia, 24/5
    anterior fossa, 23/393
    arterial lesion, 23/400
    bacterial meningitis, 33/3, 52/6
    bilateral periorbital hematoma, 57/131
    biomechanics, 23/69, 389
    brain injury, 24/39
    brain stem concussion, 23/459
    comminuted depressed fracture, 57/148
    comminuted fracture, 57/102
    compound, see Compound skull fracture
    cranial nerve, 23/398
    cranial subdural empyema, 52/170
    CSF fistula, 24/185, 57/135, 139
    depressed, see Depressed skull fracture
    double density shadow, 57/148
    EEG, 23/318, 327, 350, 57/186
    epidural hematoma, 57/255
    epilepsy, 15/545
    football injury, 23/573
    fracture healing, 57/147
    growing fracture, 57/147, 334
    head injury, 57/102, 147
    head injury outcome, 57/376
    headache, 5/180
    hematoma risk, 57/435
    immediate, 23/37
    Kohlu injury, 23/597
    linear, 57/102, 331
    meningitis, 23/387, 24/218
    middle fossa, 23/395
    missile injury, 23/511, 519
    occipital bone, 57/254
    oculomotor nerve injury, 24/61, 81
    optic canal fracture, 24/44

orbital roof fracture, 57/146
panda eyes, 57/131
parietotemporal fracture, 57/147
pediatric head injury, 57/331
penetrating head injury, 57/301
pneumococcal meningitis, 33/36, 52/42
posterior fossa, 23/398
posterior fossa hematoma, 24/345
racoon sign, 57/131
radiodiagnosis, 23/390, 57/143, 331
skiing injury, 23/578
skull base fracture, 57/146
spinal injury, 26/190
subdural hematoma, 57/337
temporal bone, *see* Temporal bone fracture
traumatic aneurysm, 24/392
traumatic intracranial hematoma, 57/269
traumatic venous sinus thrombosis, 24/370
trochlear nerve injury, 24/65
type, 23/391
vault, 23/387, 57/146
Skull injury
  see also Brain concussion, Brain injury,
    Commotio cerebri, Craniofacial injury, Head
    injury *and* Skull fracture
  craniospinal, *see* Craniospinal injury
  electrical, 7/374, 376, 23/713
  stab wound, 23/477
  thermal injury, 23/714
Skull osteomyelitis
  actinomycosis, 35/386
  headache, 5/219
  surgery, 24/216, 665
Skull syndrome
  cloverleaf, *see* Cloverleaf skull syndrome
Skull traction
  cervical vertebral column injury, 25/277, 286, 291,
    26/215-219, 301, 303
  complication, 25/331
  excessive, 25/277
Skull tumor
  see also Skull base tumor
  paranasal sinus, 17/204
  vault, *see* Cranial vault tumor
Slant
  antimongoloid, *see* Antimongoloid eyes
  mongoloid, *see* Mongoloid eyes
SLDH3, *see* Serum lactate dehydrogenase
  isoenzyme 3
SLE, *see* Systemic lupus erythematosus
Sleep, 1/488-491, 3/80-105
  see also Hypersomnia, Lethargy *and* Sleep rhythm
  acquired immune deficiency syndrome, 74/554
  aging, 74/542
  Alzheimer disease, 74/496, 549
  apnea, *see* Sleep apnea syndrome
  asthma, 74/553
  attention, 3/168
  autonomic nervous system, 74/14, 523, 537,
    75/437
  awakening, 3/84
  barometric pressure, 3/85
  basilar artery migraine, 48/137
  benign neonatal, *see* Benign neonatal sleep
  benign partial epilepsy with centrotemporal spike,
    73/12
  biological rhythm, 74/495
  bladder control, 3/100
  blood pressure, 3/82
  body jerks, 3/86
  body movement, 3/84, 101
  body twitch, 3/82
  brain blood flow, 3/82
  brain death, 24/729, 733
  brain stem, 3/81
  central neurogenic hyperventilation, 63/484
  cerebellar degeneration, 74/547
  cervicogenic headache, 48/408
  chemical basis, 3/104
  Cheyne-Stokes respiration, 63/482, 484
  chronic paroxysmal hemicrania, 48/261
  circadian rhythm, 74/536
  cluster headache, 48/239
  congenital central hypoventilation syndrome,
    74/560
  conjugate gaze, 1/604
  cycle, 15/457
  cytokine, 74/534
  deep sleep, 3/81, 85
  depth of sleep, 3/83-85
  diplopia, 3/90
  diuresis, 3/82
  diurnal rhythm, 3/85
  drowsiness, 3/86, 90, 114
  drug, 3/103, 15/484
  ECG, 3/82
  EEG, 3/80, 82, 46/495, 72/286, 294
  endocrinology, 74/535
  epilepsy, 3/86, 15/457, 770, 74/547, 75/353
  eye movement, 1/579, 3/82
  factor promoting sleep, 3/85
  familial insomnia, 74/549
  frontal lobe tumor, 17/259
  gastric secretion, 3/82
  gastrointestinal tract, 74/540
  glycosaminoglycanosis type IH, 66/297

grand mal epilepsy, 15/463
headache, 74/550
heart death, 74/552
heart rate, 3/82
hereditary sensory and autonomic neuropathy type III, 74/546
hygiene measure, 74/564
hypnagogic hallucination, 3/86, 90
hypnagogic phenomena, 3/86
hypnagogic system, 3/67
hypothalamus, 74/548
hypsarrhythmia, 15/471
immune system, 74/534
infant, 3/85
infantile spasm, 15/125
juvenile myoclonic epilepsy, 73/164
learning, 3/87
Lennox-Gastaut syndrome, 73/216
light sleep, 3/81, 85
low voltage fast sleep, 3/67, 81
melatonin, 74/501
mental life, 3/103
mental stress, 74/651
migraine, 48/76, 121, 156
monotony, 3/85
motion sickness, 74/357
multiple neuronal system degeneration, 75/171
multiple sclerosis, 74/549
muscle tone, 3/84
myocardial infarction, 74/553
myoclonus, see Nocturnal myoclonus
myotonic dystrophy, 62/235
narcolepsy, 74/558
need, 74/526
neonate, 3/85
nerve cell enzyme activity, 3/105
neuroanatomy, 3/66, 80-82, 74/534
neuromuscular disorder, 74/550
neurophysiology, 3/66, 80-82, 74/526
nocturnal myoclonus, 49/622
non-REM, see Non-REM sleep
ontogeny, 74/525
optokinetic nystagmus, 2/533
orthodox phase, 3/81
paradoxical phase, 3/67, 81
parasomnia, 74/558
Parkinson disease, 74/547
penis erection, 3/82, 74/560
petit mal epilepsy, 15/470, 474
petit mal status, 15/170
pharmacotherapy, 74/563
polygraphic recording, 3/68
polyneuropathy, 74/552

postural orthostatic tachycardia syndrome, 74/555
postviral fatigue syndrome, 74/554
premature baby, 3/85
primary alveolar hypoventilation, 63/414
progressive supranuclear palsy, 74/548
psychomotor epilepsy, 15/476
rapid eye movement, see REM sleep
regulation, 2/447
regulation by drug, 15/484
REM, see REM sleep
renal insufficiency, 74/553
respiration, 3/82, 74/529, 539
respiratory dysfunction, 74/563
respiratory encephalopathy, 63/414
reticular formation, 3/81
schizophrenia, 46/495
secondary autonomic failure, 74/547
sensory deprivation, 3/85, 90
serotonin, 3/104
sexual function, 74/540
Shy-Drager syndrome, 74/543
slow, see Slow sleep
slow wave sleep, 3/67, 81
stage, 3/82, 74/523
status epilepticus, 73/267
steroid, 3/105
stroke, 74/549
suppression burst pattern, 73/236
suprachiasmatic nucleus, 74/497
tachypnea, 63/484
thalamic reticular system, 2/484
thermoregulation, 74/541
tonic seizure, 15/120
tooth grinding, 74/559
tricyclic antidepressant, 65/318
type, 3/80
ventilation, 74/533
vigilance, 3/86, 90
Sleep apnea
  carbamate intoxication, 64/186
  glycosaminoglycanosis, 66/293
  hypertension, 75/477
  obesity, 75/477
  treatment, 66/294
Sleep apnea syndrome, 74/546
  acromegaly, 45/132
  Biot respiration, 63/487
  brain stem disorder, 45/133
  central alveolar hypoventilation, 63/464
  central sleep apnea, 63/461
  chronic alveolar hypoventilation, 45/133
  chronic obstructive pulmonary disease, 45/133
  cluster headache, 48/239

craniosynostosis, 50/125
etiology, 45/133
forgotten respiration, 63/488
hypothyroidism, 45/132
menopause, 45/132
micrognathia, 45/132
multiple neuronal system degeneration, 63/146
myotonic dystrophy, 62/217, 235
obstructive, 74/555
orthostatic hypotension, 63/146
polysomnography, 45/132
REM sleep, 45/134
respiratory disorder, 45/131-135
respiratory encephalopathy, 63/415
snoring, 63/450
striatonigral degeneration, 49/208
sudden infant death, 45/132
symptom, 74/556
syringobulbia, 50/445
treatment, 45/133
Sleep cycle, *see* Sleep rhythm
Sleep deprivation, 3/89-91
circadian rhythm, 1/492
daytime hypersomnia, 45/135
deep sleep, 3/84
delirium, 46/533
epilepsy, 15/480
hallucination, 4/329
Sleep disorder, 3/80-105, 45/129-143
age, 3/92
autonomic deficit, 74/555
autonomic failure, 74/543
autoscopia, 4/232
autosomal dominant nocturnal frontal lobe
epilepsy, 72/132
barotrauma, 63/416
brain hypoxia, 63/416
brain injury, 24/592
bruxism, 3/101
central alveolar hypoventilation, 63/449
central sleep apnea, 63/449
classification, 45/129
CNS stimulant, 46/592
congenital central alveolar hypoventilation,
63/449
delayed sleep, *see* Delayed sleep phase syndrome
delirium tremens, 3/99
drug, 3/103
drug induced extrapyramidal syndrome, 6/254
dysbarism, 63/416
encephalitis lethargica, 3/80
enuresis, 3/100
epilepsy, 72/253

glycosaminoglycanosis type III, 42/456
head injury outcome, 57/413
hyperparathyroidism, 27/288
hypersomnia, 3/67, 96
hypnotic agent, 3/91
hypothalamus disorder, 46/418
hypothyroidism, 70/95
insomnia, 3/67, 81, 91-93
jactatio capitis nocturna, 3/101-103
Kleine-Levin syndrome, 3/97
levodopa, 46/593
manic depressive psychosis, 3/93, 43/208
Marinesco-Sjögren syndrome, 60/344
melatonin, 74/501
mental depression, 42/255
mountain sickness, 63/416
neuroticism, 43/210
nightmare, 3/98
nocturnal myoclonus, 45/139
obstructive sleep apnea, 63/449
parasomnia, 45/140-142
Parkinson dementia complex, 42/249
Parkinson disease, 75/180
parkinsonism, 42/255
personality, 3/92
physiologic study, 3/92
pickwickian syndrome, 3/96, 38/302
progressive supranuclear palsy, 49/244
respiratory function, 63/449
restless legs syndrome, 42/260, 45/140, 49/621
sleep paralysis, 3/94
sleep talking, 3/99, 103
snoring, 3/101, 63/449
somnambulism, 3/99
sudden infant death, 63/449
thallium intoxication, 36/249
yawning, 3/101
Sleep disturbance, *see* Dyssomnia
Sleep drunkenness, *see* Awakening disorder
Sleep myoclonus, *see* Nocturnal myoclonus
Sleep paralysis
anxiety, 42/712
hypnagogic hallucination, 42/712
idiopathic narcolepsy, 3/94
narcolepsy, 2/448, 3/94, 15/844, 45/148
narcoleptic syndrome, 42/712
respiratory disorder, 42/712
sleep disorder, 3/94
Sleep periodic movement
electrophysiology, 49/621
Sleep related myoclonus, *see* Nocturnal myoclonus
Sleep rhythm
see also Sleep

thalamic syndrome (Dejerine-Roussy), 2/482
uremia, 63/504
Sleep-wake cycle
  CSF homovanillic acid, 45/136
  delirium, 46/529, 533
  disorder, 74/470
  epilepsy, 15/457
  generalized tonic clonic seizure, 73/177
  multi-infarct dementia, 46/354
  shift worker, 45/142
Sleep-wake pattern
  biological rhythm, 74/495
Sleep-wake syndrome
  clock function, 74/468
  psychosis, 45/143
Sleep walking, *see* Somnambulism
Sleeping sickness, *see* African trypanosomiasis
Slice fracture
  vertebral discoligament lesion, 61/90
Slow acetylation
  isoniazid, 51/297
Slow acetylator
  hydralazine, 63/87
  polyneuropathy, 63/87
Slow axon transport block
  acrylamide polyneuropathy, 51/282
Slow channel syndrome
  congenital, *see* Congenital slow channel syndrome
  congenital myopathy, 62/332
Slow eye movement
  dementia, 42/187
  dentatorubropallidoluysian atrophy, 49/440
  hereditary spinocerebellar degeneration, 21/443
  neuropathy, 42/187
  olivopontocerebellar atrophy, 42/187
  organophosphate, 64/156
  sex ratio, 42/187
  spinocerebellar ataxia, 42/187
Slow sleep, 3/66
  epilepsy with continuous spike wave, 72/9
  Landau-Kleffner syndrome, 73/274
  Lennox-Gastaut syndrome, 73/274
Slow virus
  epilepsy, 72/145
  neurotropic, *see* Neurotropic slow virus
Slow virus disease, 42/655-659
  acquired immune deficiency syndrome, 56/544
  animal model, 56/439, 583
  canine distemper, 56/544
  carcinomatous cerebellar degeneration, 21/495
  CNS, 34/275-288
  coronavirus, 56/439
  Creutzfeldt-Jakob disease, 34/285

  definition, 56/543
  degenerative process, 21/60
  host immune response, 56/19
  host tolerance, 56/20
  immune response, 56/20
  kuru, 9/551, 34/275-287
  late cortical cerebellar atrophy, 21/495
  latency, 56/19
  measles virus, 56/417
  motoneuron disease, 22/21
  murine leukemia virus, 56/455
  murine neurotropic retrovirus, 56/586
  nonpermissive cell infection, 56/20
  persistence, 56/20
  progressive rubella panencephalitis, 56/405, 544
  scrapie, 34/275, 287, 302, 56/585
  subacute sclerosing panencephalitis, 56/544
  virus nonimmunogenicity, 56/20
  virus sequestration, 56/20
  visna, 9/681, 34/294, 56/459, 544
  visna-maedi disease, 9/681, 56/583
Sluder neuralgia, *see* Sphenopalatine neuralgia
Sluder sphenopalatine neuralgia, *see* Sphenopalatine
  neuralgia
Sluder syndrome, *see* Sphenopalatine neuralgia
Slurred speech
  ethylene glycol intoxication, 64/123
  hereditary paroxysmal dystonic choreoathetosis,
    49/349
  hyperserotonemia, 29/225
  Mast syndrome, 42/282
  Minamata disease, 36/105
Sly syndrome, *see* Glycosaminoglycanosis type VII
Small cell lung cancer
  brain metastasis, 69/221
  prophylaxis, 69/219
Small disease
  amyotrophy, 22/509
  Coats disease, 22/509
  hereditary hearing loss, 22/509
Small fiber damage
  amyloid polyneuropathy, 51/419
Small fiber neuropathy
  diabetic, 70/135
  idiopathic, *see* Idiopathic small fiber neuropathy
Small intestine diverticulitis
  sensorineural deafness, 42/394
Small intestine diverticulosis
  malabsorption syndrome, 28/235
  neurologic manifestation, 28/235
Smallpox
  Guillain-Barré syndrome, 7/487
  polyradiculitis, 7/487

teratogenic agent, 34/378
Smallpox vaccine
  neuropathy, 7/487, 8/88
Smell
  see also Anosmia *and* Olfaction
  indifference to pain, 8/195
  multiple sclerosis, 9/180
  radiation effect, 7/397
  taste, 24/20
  testing method, 2/53
Smelter chill, *see* Zinc intoxication
Smiling response in children
  age, 4/357
  social behavior, 4/357
Smith-Lemli-Opitz syndrome, 66/581-595
  abnormality, 66/593
  apnea, 60/667
  autosomal recessive inheritance, 50/276
  biochemistry, 66/588, 591
  birth incidence, 43/308
  blepharoptosis, 66/582
  cataract, 66/582
  cheilopalatoschisis, 43/308
  cholesterol biosynthesis, 66/587
  clinical characteristic, 66/582, 585
  CNS, 66/585
  cryptorchidism, 43/308, 50/276
  dermatoglyphics, 43/308
  differential diagnosis, 66/593
  ear abnormality, 43/308
  EEG, 43/308
  epilepsy, 31/254
  gastrointestinal tract, 66/586
  genetics, 66/592
  genital malformation, 66/586
  genital system, 66/584
  Gruber syndrome, 31/243
  heart, 66/585
  heterozygote frequency, 43/308
  hyperpnea, 60/667, 63/438
  hyperventilation syndrome, 63/438
  hypospadias, 43/308, 50/276
  kidney, 66/586
  mental deficiency, 31/254, 43/308, 46/66
  microcephaly, 30/508, 31/254, 50/276, 66/582
  micrognathia, 50/276
  oral malformation, 66/587
  palatoschisis, 66/582
  pathology, 66/584
  polydactyly, 66/582
  prenatal diagnosis, 66/590
  ptosis, 43/308, 50/276
  pulmonary malformation, 66/586

scaphocephaly, 50/276
  skeletal malformation, 66/587
  sterol metabolism, 66/589
  survey, 66/581
  syndactyly, 43/308
  teratology, 66/591
  treatment, 66/594
  Ullrich-Feichtiger syndrome, 31/244
Smith-McCort syndrome
  Dyggve-Melchior-Clausen syndrome, 43/266
Smoking
  arterial hypertension, 54/207
  brain hemorrhage, 55/524
  brain infarction, 55/524
  brain thromboangiitis obliterans, 39/201, 55/308
  head injury outcome, 57/378
  oral contraception, 55/525
  oral contraceptive agent, 53/36
  Parkinson disease, 49/126
  Raynaud phenomenon, 51/135
  subarachnoid hemorrhage, 55/524
  vibration neuropathy, 51/135
SMON, *see* Subacute myelo-optic neuropathy
Smooth muscle function
  adenosine, 75/624
  histamine, 75/624
  humoral regulation, 75/624
  peptidoleukotriene, 75/624
  serotonin, 75/624
Smooth pursuit eye movement
  multiple sclerosis, 47/133
Snack food
  deficiency neuropathy, 51/322
Snake
  venomous, *see* Venomous snake
Snake envenomation
  acetylcholine release, 65/187
  Agkistrodon halys, 65/182
  anticholinesterase treatment, 65/181
  antivenom, 65/180
  apnea, 65/178
  Astrotia stokesii, 65/180
  banded krait, 65/182
  Bitis caudalis, 65/182
  blood incoagulability, 65/179
  bulbar paralysis, 65/178
  α-bungarotoxin, 65/182
  α-bungarotoxin receptor, 65/182
  β-bungarotoxin, 65/182
  κ-bungarotoxin, 65/182
  Bungarus fasciatus, 65/182
  Bungarus flaviceps, 65/183
  Bungarus multicinctus, 65/181-183

cardiotoxin, 65/183
caudoxin, 65/182
ceruleotoxin, 65/182, 184
clinical effect, 65/178
cobra snake venom, 65/181
coma, 65/178
consumption coagulopathy, 65/179
coral snake, 65/179
Crotalus durissus terrificus, 65/182, 185
Crotalus scutullatus scutullatus, 65/182
Crotalus viridis viridis, 65/185
crotamine, 65/185
crotoxin, 65/182
crotoxin A, 65/184
CSF, 65/179
curare like mechanism, 65/181
cytotoxin, 65/183
Dendroapsis angusticeps, 65/185
Dendroaspis polylepsis, 65/185
Dendroaspis viridis, 65/185
diagnosis, 65/177
dysarthria, 65/178
dysphagia, 65/178
Echis carinatus, 65/179
elapid, 65/182
EMG, 65/179
erabutoxin b, 65/182
first aid, 65/181
headache, 65/178
hemoglobinuria, 65/179
immunologic diagnosis, 65/178
intracranial hemorrhage, 65/179
krait, 65/179
Laticauda semifasciata, 65/182
LD50 value, 65/182
local pain, 65/178
local swelling, 65/178
mamba, 65/182
membrane toxin, 65/183
Mojave toxin, 65/182
muscarinic toxin, 65/187
muscle spasm, 65/180
myasthenic syndrome, 65/179, 182
myoglobinuria, 65/179
myopathy, 65/179
myotoxin, 65/185
Naja naja philippinensis, 65/182
Naja naja siamensis, 65/181
Naja nigricollis, 65/182
neuromuscular blocking neurotoxin, 65/181
neuromuscular junction block, 65/181
α-neurotoxin family, 65/182
notechis II-5, 65/182
Notechis scutatus scutatus, 65/179, 182
notexin, 37/16, 65/182
ophthalmoplegia, 65/178
opisthotonos, 65/180
Oxyuranus scutellatus, 65/179, 182
paradoxin, 65/182
pit viper, 65/179, 182
pituitary hemorrhage, 65/180
pituitary hemorrhagic necrosis, 65/180
preparalytic symptom, 65/178
presynaptic neuromuscular blocking toxin, 65/183
Pseudechis australis, 65/179, 182
Pseudocerastes fieldi, 65/182
Pseudonaja textilis, 65/182
renal insufficiency, 65/179
respiratory paralysis, 65/178
rhabdomyolysis, 65/179
sea snake, 65/179
spitting cobra, 65/179
startle reflex, 65/180
striatum, 65/187
subarachnoid hemorrhage, 65/179
taipoxin, 65/182
textilotoxin, 65/182
toxic myopathy, 65/179
treatment, 65/180
Trimeresurus mucrosquamatus, 65/182
Tropidechis carinatus, 65/179
type, 65/177
ultrastructure, 65/184
viper, 65/179, 182
Vipera ammodytes, 65/182
Vipera russelli, 65/179
Snake venom
  absorption, 65/177
  acetylcholine release, 65/187
  acquired myoglobinuria, 62/577
  α-bungarotoxin, 65/181
  β-bungarotoxin, 65/181
  κ-bungarotoxin, 65/183
  composition, 65/181
  crotamine, 65/185
  crotoxin, 65/183
  crotoxin A, 65/184
  demyelination, 9/39
  dendrotoxin, 65/186
  fasciculin, 65/186
  fasciculin 1, 65/186
  fasciculin 2, 65/186
  intravascular consumption coagulopathy, 63/315
  LD50 value, 65/182
  mamba venom, 65/185
  muscarinic toxin, 65/187

myotoxin, 65/185
neurotoxin, 9/39, 37/1-18, 62/601, 610,
    65/177-185
α-neurotoxin family, 65/182
phospholipase, 65/185
phospholipase A2, 65/180
postsynaptic type, 65/181
presynaptic neuromuscular blocking toxin, 65/183
presynaptic type, 65/181
serine proteinase inhibitor, 65/186
striatum, 65/187
toxic dose, 65/177
toxic encephalopathy, 65/177
toxic myopathy, 62/601, 610
SNAP-25, *see* Synaptosomal associated protein 25
Snapper intoxication
    vitamin A intoxication, 37/87
Sneddon syndrome, 55/401-408
    see also Divry-Van Bogaert syndrome
    age at onset, 55/402
    allergic granulomatous angiitis, 55/408
    angiopathic polyneuropathy, 51/446
    anticardiolipin antibody, 55/403
    antiphospholipid antibody, 55/403, 63/330
    autosomal dominant transmission, 55/403
    brain angiography, 55/402
    brain infarction, 55/401
    brain ischemia, 55/408
    brain thromboangiitis obliterans, 55/403
    carotid system syndrome, 53/308
    cause, 55/403
    chronic ischemic leukoencephalopathy, 55/408
    clinical features, 55/401
    computer assisted tomography, 54/64
    cutis marmorata, 55/401
    dementia, 55/402, 406
    diagnosis, 55/402
    differential diagnosis, 55/403
    disease duration, 55/404
    Divry-Van Bogaert syndrome, 55/403, 408
    epilepsy, 55/402, 406
    exclusion criteria, 55/408
    frequency, 55/401
    granulomatous CNS vasculitis, 55/408
    histology, 55/403
    hypertension, 55/404
    immune induced angiopathy, 55/408
    laboratory diagnosis, 55/402
    lateral medullary infarction, 55/407
    livedo racemosa, 55/401
    livedo reticularis, 55/317, 319, 401
    middle cerebral artery, 55/402
    moyamoya phenomenon, 55/407

    multifocal brain infarction, 55/402
    multiple brain infarction, 55/402
    neurologic deficit, 55/406
    oral contraceptive agent, 55/404
    prognosis, 55/408
    reported cases, 55/404
    sex ratio, 55/402
    skin biopsy, 55/403
    transient ischemic attack, 55/402
    treatment, 55/408
Sneezing
    autosomal dominant compelling helio-ophthalmic
        outburst, 74/431
    bright light exposure, 63/492
    carotid dissecting aneurysm, 54/271
    descending trigeminal nucleus, 63/492
    ethmoidal nerve, 63/492
    headache, 48/367
    indifference to pain, 8/195
    reflex, *see* Reflex sneezing
    respiration reflex, 63/492
    reticular formation, 63/492
    sphenopalatine neuralgia, 48/477
    trigeminal reflex, 2/57
    yawning, 63/493
Snellen chart
    visual acuity, 13/9
Snoring
    acromegaly, 63/451
    airflow resistance, 63/450
    alcohol, 63/451
    alpha arousal, 63/450
    alveolar ventilation, 63/451
    apnea, 63/450
    arterial hypertension, 63/451
    barbiturate, 63/451
    benign snoring, 63/450
    benzodiazepine, 63/451
    body position, 63/451
    central sleep apnea, 63/461
    chronic obstructive pulmonary disease, 63/451
    continuous snoring, 63/449
    cyclic snoring, 63/450
    daytime fatigue, 63/450
    definition, 63/449
    epidemiology, 63/450
    familial history, 63/451
    gasp, 63/450
    hyoid bone, 63/450
    hypothyroidism, 63/451
    ischemic heart disease, 63/451
    macroglossia, 63/450
    non-REM sleep, 63/450

obesity, 63/450
oropharyngeal muscle hypotonia, 63/450
oropharyngeal stenosis, 63/450
prevalence, 63/450
REM sleep, 63/450
sex ratio, 63/450
sleep apnea syndrome, 63/450
sleep disorder, 3/101, 63/449
symptom, 63/450
treatment, 63/451
Snout reflex, *see* Lip reflex
Snowshoe hare virus
California encephalitis, 56/140
Snub-nosed dwarfism
cryptorchidism, 43/387
swayback nose, 43/387
Soamin, *see* Arsanilic acid
Sobbing respiration
subacute necrotizing encephalomyelopathy,
60/667
Sobbing syncope
bradypnea, 15/827
gasping respiration, 15/827
Soccer injury
carotid artery thrombosis, 23/569
chronic subdural hematoma, 23/568
sport injury, 23/567
traumatic carotid thrombosis, 23/569
Social behavior
behavior development, 4/356-360
infantile autism, 46/190
smiling response in children, 4/357
Social reintegration
head injury outcome, 57/410
rehabilitative sport, 26/526
Société de Neurologie de Paris, France, 1/7
Socioeconomic class
anencephaly, 30/195, 42/13, 50/80
developmental articulatory defect, 43/220
multiple sclerosis, 9/100, 114
Sodium
autonomic nervous system, 74/76
brain, 9/16
brain infarction, 53/125
brain injury, 23/109
brain ischemia, 53/125
brain metabolism, 57/73
CNS, 28/431-433
CSF, 28/431, 445
headache, 48/436
migraine, 5/231
serum concentration, 63/545
spinal cord injury, 26/381, 387

Sodium arsanilate, *see* Arsanilic acid
Sodium arsenate
cephalocele, 50/98
Sodium azide intoxication
striatal necrosis, 49/500
Sodium bicarbonate, *see* Bicarbonate
Sodium channel
arthropod envenomation, 65/194
axolemma, 47/14-18, 33-37, 51/47
axon, 51/47
ciguatoxin, 65/160
depolarization, 47/18
internode membrane, 51/47
local anesthetic agent, 65/419, 426
marine toxin intoxication, 65/146
muscle cell membrane, 40/126, 558
myotonia, 62/278
myotonic dystrophy, 62/224
myotonic hyperkalemic periodic paralysis, 62/224
paralytic shellfish poisoning, 65/152
Ranvier node, 51/15
saxitoxin, 65/146
structure, 65/146
tetrodotoxin, 28/447, 51/263, 65/147
Sodium chloride
intrathecal chemotherapy, 61/152
normokalemic periodic paralysis, 28/592, 597,
41/162
Sodium cromoglycate, *see* Cromoglicate disodium
Sodium current
epilepsy, 72/44
nerve cell, 72/44
pyrethroid intoxication, 64/211
Sodium deficiency
inappropriate antidiuretic hormone secretion,
1/502
Sodium glutamate
autonomic nervous system, 75/502
Chinese restaurant syndrome, 75/502
headache, 48/6, 420, 444
Sodium hypothesis
membrane potential, 7/63-70
Sodium iodate
retinotoxic agent, 13/205
Sodium iodoacetate, *see* Iodoacetate sodium
Sodium isocyanate
neurotoxin, 51/68
toxic polyneuropathy, 51/68
Sodium metabolism, 28/443-447, 495-499
thirst regulation, 28/497
vasopressin, 28/498
Sodium nitrate intoxication
pallidal necrosis, 49/479

Sodium nitrite intoxication
  pallidal necrosis, 6/652, 49/466
Sodium nitroprusside
  action mechanism, 37/434
  adverse effect, 37/435
  antihypertensive agent, 37/433
  antihypertensive agent intoxication, 37/435
  delirium, 46/538
  hypertensive encephalopathy, 54/216
  mode of action, 37/434
  therapeutic use, 37/434
Sodium nitroprusside intoxication
  acidosis, 37/435
  hypotension, 37/435
  hypothyroidism, 37/435
Sodium-potassium adenosine triphosphatase
  diabetic polyneuropathy, 51/508
  mercury intoxication, 64/389
  uremic polyneuropathy, 51/361
Sodium salicylate intoxication
  diarrhea, 37/417
  drowsiness, 37/417
  fever, 37/417
  headache, 37/417
  hearing loss, 37/417
  mental confusion, 37/417
  nausea, 37/417
  sweating, 37/417
  tinnitus, 37/417
  vertigo, 37/417
  vomiting, 37/417
Sodium transport channel
  periodic paralysis, 41/167
Sodium valproate, see Valproic acid
Soft tissue
  cranial vault tumor, 17/123, 125
  Hand-Schüller-Christian disease, 38/95
  paravertebral, see Paravertebral soft tissue
  turban tumor, 17/124
Soft tissue sarcoma
  neurofibromatosis type I, 50/369
Solaster papposus intoxication
  asteroidea, 37/66
  cat, 37/66
Soldier heart, see Hyperventilation syndrome
Solitary myeloma
  criteria, 20/13, 15
  life expectancy, 20/13
  multiple myeloma polyneuropathy, 51/430
  vertebral column, 20/9, 35-39
Solitary plasmacytoma, 71/435
  multiple myeloma, 18/256
  multiple myeloma polyneuropathy, 63/393

  paraproteinemia, 69/306
Solitary tract
  baroreceptor-cardiac reflex, 74/153
  central neural site, 74/153
  hypothalamus, 74/153
  medulla, 74/153
  nucleus, 74/153
  projection, 74/153
Soluble N-ethylmaleimide sensitive fusion
    attachment protein 25, see Synaptosomal
    associated protein 25
Solvent
  industrial, see Industrial solvent
  organic, see Organic solvent
  stoddard, see Stoddard solvent
Solvent abuse
  delirium, 46/546
Solvent intoxication
  organic, see Organic solvent intoxication
Solvent mixture
  organic solvent intoxication, 36/374, 64/42
Soman
  chemical structure, 64/225
  neurotoxin, 37/543, 64/223
  organophosphate, 64/152
  organophosphate intoxication, 37/556, 64/152,
    223
Soman intoxication
  chemical classification, 37/543
  organophosphorus compound, 37/543
Somatesthesia
  disconnection syndrome, 45/102
  phantom limb, 4/225, 45/397
  sensory extinction, 3/190
Somatesthetic deafferentation
  body scheme disorder, 4/218
  phantom limb, 4/228, 45/399
Somatesthetic illusion
  temporal lobe epilepsy, 73/72
Somatesthetic reading
  alexia, 4/121, 45/437
Somatic cell therapy
  lysosomal storage disease, 66/113
Somatoform disorder, see Hysteria
Somatognosia
  body scheme, 4/208
  body scheme disorder, 2/668
  Wernicke aphasia, 4/98
Somatomedin
  pituitary dwarfism type III, 42/615
Somatomotor inhibition
  autonomic nervous system, 74/259
Somatoparaphrenia

body scheme disorder, 4/219, 45/380
  hemiplegia, 4/219
  organic delusional syndrome, 46/420
  right parietal lobe lesion, 46/420
Somatopsyche
  body scheme, 4/207
Somatosensory disorder
  cerebral dominance, 4/303, 305
  parietal lobe syndrome, 45/64
Somatosensory evoked potential
  acetonylacetone intoxication, 64/87
  aorta surgery, 63/57
  ataxia telangiectasia, 60/366
  Bassen-Kornzweig syndrome, 51/327, 63/275
  brain stem death, 57/469
  Charlevoix-Saguenay spastic ataxia, 60/455
  chorea-acanthocytosis, 63/282
  Down syndrome, 50/529
  dyssynergia cerebellaris myoclonica, 60/600
  EEG, 54/160
  Friedreich ataxia, 60/321
  head injury, 23/80, 82, 57/196
  hereditary multiple recurrent mononeuropathy,
    51/556
  hereditary paroxysmal kinesigenic
    choreoathetosis, 49/354
  Huntington chorea, 49/289
  lead polyneuropathy, 64/437
  locked-in syndrome, 54/163
  multiple sclerosis, 47/140-142
  neurologic intensive care, 55/218
  schizophrenia, 46/500
  thoracic outlet syndrome, 51/126
  vitamin E deficiency, 70/445
Somatosensory extinction
  schizophrenia, 46/505
Somatostatin
  Alzheimer disease, 46/268
  basal ganglion, 49/28, 33
  gastrointestinal disease, 39/451
  headache, 48/109
  Huntington chorea, 49/75, 256, 260
  multiple neuronal system degeneration, 75/193
  Parkinson disease, 49/112, 120, 123
  progressive dysautonomia, 59/151, 158
Somatostatin like immunoreactivity
  Huntington chorea, 49/258
Somatostatin neuron
  Huntington chorea, 49/298
Somatotopagnosia
  terminology, 1/107
Somersault posture, see Pallidal posture
Somite dysplasia

cervicothoracic, see Cervicothoracic somite
    dysplasia
Somnambulism
  consciousness, 3/117
  epilepsy, 72/254
  Gilles de la Tourette syndrome, 49/631
  hysteria, 43/207
  juvenile parasomnia, 45/140
  migraine, 48/38, 161
  sleep disorder, 3/99
Somniosus microcephalus
  elasmobranche intoxication, 37/85
Somnolence
  acquired toxoplasmosis, 35/125, 52/354
  chelonia intoxication, 37/91
  frontal lobe tumor, 17/259
  hexachlorophene intoxication, 37/486, 65/476
  Listeria monocytogenes encephalitis, 33/87
  meningeal leukemia, 63/341
  mephenytoin intoxication, 37/202
  mesencephalic infarction, 53/391
  mesuximide intoxication, 37/202
  metharbital intoxication, 37/202
  obstructive sleep apnea, 63/453
  periodic ataxia, 21/571
  phenobarbital intoxication, 37/202
  pickwickian syndrome, 38/299
  pontine infarction, 12/15
  REM sleep, 45/131
  vertebrobasilar system syndrome, 53/391
Sonography
  occipital encephalocele, 50/100
  transcranial Doppler, see Transcranial Doppler
    sonography
Sorbitol
  see also Polyol pathway
  brain infarction, 53/427
  CSF pressure, 16/361
  diabetic neuropathy, 27/93, 110
  hyperglycemia, 27/79-83
  hyperosmolar hyperglycemic nonketotic diabetic
    coma, 27/91
  polyol pathway, 27/80, 86
Sorghum vulgare, see Millet jowar
Soroche, see Altitude illness
Sorsby pseudoinflammatory foveal dystrophy, see
    Pseudoinflammatory foveal dystrophy (Sorsby)
Sotos syndrome, see Cerebral gigantism
Sound induced photism
  hallucinosis, 46/562
Souques sign
  synkinesis, 1/184
South African porphyria, see Porphyria variegata

South African tick-bite fever
  tick-borne typhus, 34/657
South American blastomycosis, *see*
  Paracoccidioidomycosis
Space
  see also Gravity
  aging, 74/297
  animal research, 74/287
  arterial baroreflex, 74/283
  autonomic nervous system, 74/273
  bed rest, 74/288
  blood pressure, 74/286
  body fluid, 74/293
  cardiopulmonary baroreflex, 74/283
  cardiovascular function, 74/277
  confinement, 74/296
  CSF, *see* CSF space
  epidural, *see* Epidural space
  exercise, 74/279
  experience, 3/27
  extracellular, *see* Extracellular space
  fitness, 74/295
  gender effect, 74/295
  heart rate, 74/296
  periaxonal, *see* Periaxonal space
  periaxonal submyelinic, *see* Periaxonal
    submyelinic space
  prevertebral, *see* Prevertebral space
  reduplication phenomenon, 3/251
  retroparotid, *see* Retroparotid space
  spinal epidural, *see* Spinal epidural space
  spinal subdural, *see* Spinal subdural space
  subarachnoid, *see* Subarachnoid space
  vasoconstrictor response, 74/285
  vestibular interaction, 74/292
  Virchow-Robin perivascular, *see* Virchow-Robin
    perivascular space
  visual, *see* Visual space
Space exploration
  spatial orientation disorder, 45/405-411
Space occupying lesion
  abacterial encephalitis, 16/218
  Abt-Letterer-Siwe disease, 16/231
  brain phlebothrombosis, 16/235
  CSF fistula, 16/246
  epidural hematoma, 16/240
  Foix-Alajouanine disease, 19/378
  galactosemia, 16/242
  granulomatous encephalitis, 16/231
  Hand-Schüller-Christian disease, 16/231
  hypertensive encephalopathy, 16/239
  meningeal leukemia, 16/234
  meningitis sarcomatosa, 19/354

  multiple myeloma, 18/255, 39/135
  osteitis deformans (Paget), 17/170, 38/361
  parasellar, *see* Parasellar space occupying lesion
  Parkinson disease, 49/108
  respiratory acidosis, 16/241
  spastic paraplegia, 59/435
  spinal glioma, 20/278
  spinal sympathicoblastoma, 19/363
  stroke, 16/239
  subacute necrotizing encephalomyelopathy,
    16/241
  subarachnoid abscess, 16/211
  subarachnoid hemorrhage, 16/241
  syringomyelia, 32/285, 293
  venous sinus thrombosis, 16/235-237
  Wilson disease, 16/242
Space perception
  anatomy, 45/411-414
  depth perception, 45/412
  line orientation, 45/412
  object localization, 45/411
  parietal lobe syndrome, 45/80
  shape perception, 45/413
  spatial orientation disorder, 45/411-414
  spatial thinking, 45/413
  visuospatial agnosia, 45/411
Spain
  neurology, 1/11
Sparfosic acid
  adverse effect, 69/501
  antineoplastic agent, 65/529
  chemotherapy, 39/120
  encephalopathy, 65/529
  epilepsy, 65/529
  neurologic toxicity, 39/120
  neurotoxin, 39/95, 113, 119, 65/529, 538
Sparfosic acid intoxication
  encephalopathy, 65/538
Sparganosis
  epilepsy, 72/169
Sparid intoxication
  vitamin A intoxication, 37/87
Sparing
  macula, *see* Macular sparing
  vertical axis, *see* Vertical axis sparing
Spasm
  after, *see* After-spasm
  arterial, *see* Arterial spasm
  blepharo, *see* Blepharospasm
  brain vascular, *see* Brain vasospasm
  buccopharyngeal, *see* Buccopharyngeal spasm
  carbamate intoxication, 64/190
  carotid injury, 11/426

carpopedal, *see* Carpopedal spasm
cerebrotendinous xanthomatosis, 29/374
cervical arterial, *see* Cervical arterial spasm
clonic facial, *see* Facial clonus
conjugate lateral, *see* Conjugate lateral spasm
definition, 1/285
deglutition, *see* Deglutition spasm
esophageal, *see* Esophageal spasm
esophageal motility disorder, 75/630
extensor, *see* Extensor spasm
eye convergence, 1/607, 2/341
eye fixation, 1/583, 599
eyelid, *see* Blepharospasm
facial, *see* Facial spasm
facial clonus, 1/285
flexor, *see* Flexor spasm
gaze, *see* Gaze spasm
glossal, *see* Glossal spasm
hemifacial, *see* Hemifacial spasm
infantile, *see* Infantile spasm
infantile myoclonic, *see* Infantile myoclonic
    spasm
laryngeal, *see* Laryngeal spasm
late infantile metachromatic leukodystrophy,
    10/346, 29/357, 66/167
massive, *see* Massive spasm
muscle, *see* Muscle spasm
ocular, *see* Ocular spasm
origin, 1/279
torsion, *see* Dystonia musculorum deformans
upper eyelid, *see* Upper eyelid spasm
vascular, *see* Vascular spasm
Spasmodic torticollis, 6/567-596, 42/263-265
  age at onset, 6/568
  agitation, 6/799
  anatomic classification, 1/279
  anxiety, 6/799
  archicerebellar syndrome, 2/418
  behavior disorder, 42/264
  caloric test, 6/575
  cause, 6/577, 582
  concept, 6/567
  counter pressure phenomenon, 6/574, 42/264
  course, 49/382
  cryptorchidism, 43/146
  description, 1/288
  drug treatment, 6/589
  dystonia, 42/263
  dystonia musculorum deformans, 6/529,
    42/215-218, 49/521-523
  EEG, 6/592-596
  EMG, 1/648, 6/569-577
  focal athetosis, 49/382

  Gegendruckphänomen, 6/573
  hereditary olivopontocerebellar atrophy (Menzel),
    21/441
  hyperthyroidism, 42/264
  hypoparathyroidism, 27/308
  keloid, 43/146
  migraine, 48/38
  motor function, 6/569
  neuroanatomy, 6/586-589
  neuropathology, 6/582-585
  neuropharmacology, 49/57
  pain, 42/264
  pallidal atrophy, 6/638
  pallidoluysian degeneration, 49/455
  pallidoluysionigral degeneration, 49/455
  pathophysiology, 6/585
  progressive myositis ossificans, 43/191
  reserpine intoxication, 6/169
  scoliosis, 42/264
  sex ratio, 6/568
  spinal epidural tumor, 20/105
  sternocleidomastoid muscle hypertrophy, 42/763
  striatal change, 6/584
  surgical treatment, 6/590-592
  symptomatic type, 6/568, 584
  symptomatology, 6/568
  syringomyelia, 50/445
  tegmental lesion, 6/106
  thoracic outlet syndrome, 51/124
  tonic neck reflex, 6/576
  tremor, 42/264
  Wildervanck syndrome, 43/343
Spasmus nutans
  congenital nystagmus, 1/597, 16/323
  oscillation, 2/367
Spastic amaurotic axial idiocy (Seitelberger), *see*
    Infantile neuroaxonal dystrophy
Spastic ataxia
  Bedouin, *see* Bedouin spastic ataxia
  beriberi, 7/644
  Charlevoix-Saguenay, *see* Charlevoix-Saguenay
    spastic ataxia
  chorea, 42/165
  dementia, 42/166
  dysphagia, 42/165
  hereditary, *see* Hereditary spastic ataxia
  hyperreflexia, 42/165
  optic atrophy, 42/165
  posterior column degeneration, 42/166
  progressive external ophthalmoplegia, 62/304
  ptosis, 62/304
  retinovestibular hereditary, *see* Retinovestibular
    hereditary spastic ataxia

spasticity, 60/661
spinocerebellar degeneration, 42/166
striatonigral degeneration, 42/261
Spastic ataxia (Marie), *see* Hereditary cerebellar
    ataxia (Marie)
Spastic cerebral palsy
  birth incidence, 42/168
  contracture, 42/167
  epilepsy, 42/167
  mental deficiency, 42/167
  microcephaly, 42/167
  myelin dysgenesis, 42/498
  paraplegia, 42/167
  tetraplegia, 42/167
Spastic diplegia
  Bardet-Biedl syndrome, 43/233
  cerebral palsy, 46/14
  hyperammonemia, 42/560
  lymphopenia, 56/469
  mental deficiency, 42/715, 46/14
  microcephaly, 50/269
  New Caledonia, 21/34
  optic atrophy, 42/415
  Paine syndrome, 30/514, 43/432
  photogenic epilepsy, 42/715
  progressive, *see* Progressive paraparesis
  purine nucleoside phosphorylase deficiency,
    56/469
  Sjögren-Larsson syndrome, 13/141, 66/615
  thyroid gland dysgenesis, 42/632
Spastic dysphonia
  athetosis, 49/382
  olivopontocerebellar atrophy (Dejerine-Thomas),
    21/425
Spastic dystonia
  definition, 6/140
  hemiplegic dystonia, 6/146, 149
Spastic gait
  Rubinstein-Taybi syndrome, 31/283, 43/234,
    50/276
Spastic hemiplegia
  brain arteriovenous malformation, 54/378
  cerebral palsy, 46/14
  mental deficiency, 46/14
  symptom, 1/180
Spastic myalgia
  terminology, 5/157
Spastic paraparesis
  acetonylacetone intoxication, 64/87
  adrenomyeloneuropathy, 60/170
  carbon disulfide intoxication, 64/25
  cerebrotendinous xanthomatosis, 60/167
  cytarabine intoxication, 65/533

dementia, 46/401
diuretic agent intoxication, 65/440
familial dementia, 46/401
hereditary striatal necrosis, 22/455
neurobrucellosis, 33/314, 52/584
primary lateral sclerosis, 59/172
progressive external ophthalmoplegia, 62/305
tetanus toxin, 65/245
tropical, *see* Tropical spastic paraplegia
Waldenström macroglobulinemia, 63/396
Spastic paraplegia, 42/168-170
  acquired bone disease, 59/433
  Addison disease, 42/512
  adrenoleukodystrophy, 60/169
  adrenomyeloneuropathy, 59/439
  adult leukodystrophy, 59/439
  adverse diuretic effect, 65/440
  alcohol, 61/369
  alcohol intoxication, 59/438
  alpha motor system, 61/367
  amyotrophy, 59/374
  angiostrongyliasis, 59/436
  ankylosing spondylitis, 59/439
  anterior horn syndrome, 42/169, 177
  anterolateral myelopathy, 7/643
  aorta thrombosis, 59/437
  argininemia, 59/439
  aspartic acid, 61/367
  ataxia, 42/288
  autosomal dominant, *see* Autosomal dominant
    spastic paraplegia
  Babinski sign, 42/180
  baclofen, 61/368
  Bardet-Biedl syndrome, 13/307
  Bartonella infection, 59/436
  bartonellosis, 59/436
  basal ganglion degeneration, 42/175, 177
  Behçet syndrome, 59/436
  bilateral paracentral angioma, 59/429
  bilateral peduncular lesion, 59/430
  bilateral subdural hematoma, 59/429
  bismuth, 64/10
  brain atrophy, 42/176
  celiac plexus block, 59/437
  cerebellar ataxia, 13/307
  cerebellar atrophy, 42/177
  cerebral palsy, 42/167
  cerebrotendinous xanthomatosis, 59/439
  choreoathetosis, 42/175
  chorioretinal degeneration, 13/39
  citrullinuria, 42/538
  Clarke column, 42/169
  contracture, 42/169

cordotomy, 61/369
corticospinal tract degeneration, 42/180
Creutzfeldt-Jakob disease, 59/436
Cryptococcus neoformans, 59/436
cyst, 59/434
cytomegalovirus infection, 50/277
dantrolene, 61/368
deafferentation, 61/370
decompression myelopathy, 61/220
dementia, 42/176
demyelination, 59/434
dermatomyositis, 59/439
developmental abnormality, 59/434
diamorphine, 59/438
diazepam, 61/368
differential diagnosis, 59/425-439
disseminated choroiditis, 42/178
dorsal spinal root entry zone microcoagulation,
    61/371
Down syndrome, 50/529
dysphagia, 42/175
dystonia, 42/175
Eales disease, 59/439
electrical injury, 61/195
epidural hematoma, 59/438
etiology, 59/435-439
exercise effort, 59/428
extrapyramidal sign, 42/175
falcine meningioma, 19/86, 59/428
familial, *see* Familial spastic paraplegia
familial spinal arachnoiditis, 42/107
Fanconi syndrome, 59/439
fasciculation, 42/172
fetal Minamata disease, 64/418
filariasis, 59/436
fluoride intoxication, 59/438
Foix-Alajouanine disease, 59/437
foramen magnum tumor, 59/431
Friedreich ataxia, 13/307, 21/19
frontal ataxia, 59/429
GABA, 61/367
gamma motor system, 61/367
Gegenhalten, 59/429
gene frequency, 42/170
Gerstmann-Sträussler-Scheinker disease, 59/436
glutamic acid, 61/367
glycine, 61/367
glycosaminoglycanosis type I, 60/175
glycosaminoglycanosis type IH, 60/175
glycosaminoglycanosis type IS, 60/175
glycosaminoglycanosis type II, 60/175
glycosaminoglycanosis type III, 60/175
glycosaminoglycanosis type VII, 60/175

Hallervorden-Spatz syndrome, 59/439
Hartnup disease, 21/564
heart disease, 42/169
hereditary ataxia, 22/434
hereditary cerebral amyloid angiopathy, 59/439
hereditary sensory and autonomic neuropathy type
    I, 21/77
hereditary sensory and autonomic neuropathy type
    II, 60/17
homocarnosinase deficiency, 42/557
homocarnosinase metabolism, 42/179
homocarnosinosis, 42/557
hot bath test, 59/428
human T-lymphotropic virus type I, 59/436
human T-lymphotropic virus type I associated
    myelopathy, 56/525
human T-lymphotropic virus type III, 59/436
Huntington chorea, 49/291
Hutchinson-Laurence-Moon syndrome, 13/381,
    59/439
hyperglycinemia, 42/170
hyperreflexia, 42/172, 180
iatrogenic paraplegia, 59/437
incontinentia pigmenti, 43/23
infection, 59/435
intrathecal agent, 61/368
ischemic myelopathy, 59/437
Kjellin syndrome, 22/469-471, 42/173-175
Klippel-Feil syndrome, 59/431
lathyrism, 7/520, 643, 28/37, 36/505, 59/438,
    64/9, 65/2
Laurence-Moon syndrome, 13/307, 43/254
Lhermitte sign, 59/432
lysosomal storage disease, 59/439
malignant hypertension, 59/438
malnutrition, 28/31-37
mannosidosis, 59/439
measles virus, 59/436
mental deficiency, 42/169, 176, 43/277
MERRF syndrome, 59/429
metabolic and deficiency disease, 59/439
methotrexate intoxication, 65/530
microcephaly, 50/277
morphine, 61/368
mucormycosis, 59/436
multiple sclerosis, 42/496, 59/431
muscular atrophy, 42/171-175
myalgia, 42/180
myelomalacia, 59/437
myelotomy, 61/369
Neisseria meningitidis, 59/436
neurocysticercosis, 59/436
neurolipidosis, 59/439

neuropathy, 42/171
neurostimulation, 61/372
nocardiosis, 59/436
obturator neurotomy, 61/369
ophthalmoplegia, 22/191, 471
optic atrophy, 13/39, 22/434, 42/177
organophosphate induced delayed polyneuropathy, 64/176
orthochromatic leukodystrophy, 42/500
paracoccidioidomycosis, 59/436
paramyxovirus, 59/436
parkinsonism, 42/175
Pelizaeus-Merzbacher disease, 42/503, 59/439, 66/572
percutaneous thermal rhizotomy, 61/370
pernicious anemia, 42/609
pertussis, 59/436
pharmacotherapy, 61/368
phenol, 61/369
poliomyelitis, 59/436
polycythemia vera, 59/439
posttraumatic type, 61/367
presenile dementia, 42/287
prevalence, 42/170
progressive external ophthalmoplegia, 13/307, 309, 43/136, 62/304
pseudobulbar paralysis, 59/428
pseudoneuralgic pain, 59/432
psychomotor retardation, 42/169, 180
pyramidal tract degeneration, 42/169, 177
radicular pain, 59/431
regression reflex, 59/428
Renshaw cell, 61/367
retinal degeneration, 42/169, 173-175, 177-179
retrobulbar neuritis, 42/178
rheumatoid arthritis, 59/439
rhizotomy, 61/369
rubella panencephalitis, 59/436
Rubinstein-Taybi syndrome, 50/276
Rud syndrome, 14/13
sarcoidosis, 59/439
schistosomiasis, 59/436
secondary pigmentary retinal degeneration, 13/307, 42/177, 179
seizure, 42/178
sensorineural deafness, 42/178
sensory loss, 42/169, 177
sensory microrootlet section, 61/371
sensory neuropathy, 42/179
Sjögren-Larsson syndrome, 13/468, 471, 478, 59/439
skeletal deformity, 42/172
space occupying lesion, 59/435

space occupying process, 59/434
spasticity, 61/367
speech disorder in children, 42/175
spinal angioma, 59/438
spinal arachnoiditis, 42/107, 59/437
spinal cord diaschisis, 61/367
Sprengel deformity, 59/431
Staphylococcus, 59/436
striatopallidodentate calcification, 59/439
subacute combined spinal cord degeneration, 59/439
subacute necrotizing encephalomyelopathy, 59/439
superior sagittal sinus, 59/428
surgical treatment, 61/369
syphilis, 59/435
syringomyelia, 59/434
systemic bone disease, 59/432
talipes, 42/169, 171
tethered cord, 59/434
tizanidine, 61/368
total frontal atrophy (Lüers-Spatz), 22/314
trauma, 59/434
trichinosis, 59/436
trophic ulcer, 42/180
tropical, see Tropical spastic paraplegia
Troyer syndrome, 42/193
tuberculosis, 59/435
type, 1/206
varicella zoster, 59/436
varicella zoster virus myelitis, 56/236
vascular lesion, 59/437
verruga peruana, 59/436
vertebral hemangioma, 59/438
vertebral metastasis, 59/437
Vogt-Koyanagi-Harada syndrome, 59/436
Von Hippel-Lindau disease, 42/156, 43/21
X-linked, see X-linked spastic paraplegia
X-linked recessive ataxia, 60/639
xeroderma pigmentosum, 60/336
Spastic paresis
  Parkinson disease, 49/96
  pertussis, 52/234
Spastic pseudosclerosis, see Creutzfeldt-Jakob disease
Spastic quadriplegia, see Spastic tetraplegia
Spastic tabes dorsalis, see Mills syndrome
Spastic tetraparesis
  Pelizaeus-Merzbacher disease, 59/357
  Salla disease, 59/360
Spastic tetraplegia
  adrenoleukodystrophy, 43/483
  Barnard-Scholz syndrome, 22/471

carnosinemia, 42/532
cerebral palsy, 42/167, 46/14
corpus callosum agenesis, 42/8
cytoside lipidosis, 10/545
familial spastic paraplegia, 42/168-170
fetal Minamata disease, 64/418, 421
late infantile metachromatic leukodystrophy, 66/167
mental deficiency, 43/295, 46/14
Minamata disease, 64/418, 420
N syndrome, 43/283
opticocochleodentate degeneration, 21/540
11p partial trisomy, 43/525
porencephaly, 50/359
pyroglutamic aciduria, 29/230
6q partial trisomy, 43/503
syringomyelia, 32/268
Von Hippel-Lindau disease, 43/21
Spasticity
  acetonylacetone intoxication, 64/87
  adrenomyeloneuropathy, 60/661
  adult GM1 gangliosidosis, 60/661
  alcohol, 26/483
  baclofen, 26/482
  basis spasticity, 26/477
  carbon monoxide intoxication, 64/31
  central motor disorder, 1/169
  cerebral palsy, 60/661
  cerebrotendinous xanthomatosis, 51/399, 60/661
  Charlevoix-Saguenay spastic ataxia, 60/453
  Chédiak-Higashi syndrome, 60/661
  chemical interruption, 26/483
  clinical features, 1/169, 264
  coma, 3/65
  concentric sclerosis, 47/414
  constipation, 26/480
  contracture, 26/478-480, 60/461
  Creutzfeldt-Jakob disease, 46/291
  diazepam, 26/482
  early onset recessive stationary ataxia, 60/661
  Edwards syndrome, 50/561
  EMG, 1/646
  emotional aspect, 26/481
  excess spasticity, 26/478
  frontometaphyseal dysplasia, 38/410
  gamma system, 1/67
  general, 26/233
  Gerstmann-Sträussler-Scheinker disease, 60/661
  globoid cell leukodystrophy, 10/303
  globoid cell leukodystrophy cerebellar variant, 60/661
  GM2 gangliosidosis type III, 60/661
  Hallervorden-Spatz syndrome, 66/712

hemiplegia, 1/180
hereditary cerebello-olivary atrophy (Holmes), 21/410
hereditary sensory and autonomic neuropathy type I, 60/9
hereditary spastic ataxia, 60/461
hereditary striatal necrosis, 49/495
history, 1/261
Hooft-Bruens syndrome, 42/508
hyperargininemia, 42/562
hypertonia, 6/851
infantile Gaucher disease, 42/438, 66/124
infantile neuroaxonal dystrophy, 49/402
intrathecal phenol injection, 9/413, 26/483
Joseph-Machado disease, 42/261, 60/661
juvenile Gaucher disease type A, 66/125
late cerebellar atrophy, 60/661
mercury polyneuropathy, 51/272
MERRF syndrome, 62/509
microcephaly, 43/427
multiple sclerosis, 9/412, 47/156-158
muscle contracture, 3/389, 26/478-480
myelotomy, 26/486
myotomy, 26/485
neurectomy, 26/485
neuronal intranuclear hyaline inclusion disease, 60/661
pallidoluysionigral degeneration, 6/665
pallidonigral degeneration, 49/456
paraplegia, 1/206
patient positioning, 26/478-480
Pelizaeus-Merzbacher disease, 42/505
pharmacotherapy, 26/481
physiology, 1/66
pressure sore, 26/464, 480
prevention, 26/477
rehabilitation, 3/389, 24/691
retinovestibular hereditary spastic ataxia, 21/384
Rett syndrome, 60/636
rigidity, 6/851
Sjögren syndrome, 71/75
spastic ataxia, 60/661
spastic paraplegia, 61/367
spinal, see Spinal spasticity
spinal cord injury, 26/233, 323-333, 477-488, 61/367
spinal type, 1/268
spino-olivocerebellar degeneration, 60/661
supraspinal lesion, 1/73
surgery, 26/484
syringomyelia, 42/59
tenotomy, 26/484
transverse spinal cord lesion, 61/367

348

treatment, 3/389, 26/481
tri-*o*-cresyl phosphate polyneuropathy, 51/285
X-linked ataxia, 60/661
xeroderma pigmentosum, 43/13, 51/399, 60/336
Spät Apoplexie
  brain hemorrhage, 24/354, 357, 365
  traumatic intracerebral hematoma, 24/354, 365
  traumatic intracranial hematoma, 57/254
Spatial acalculia
  definition, 45/474
  diagnosis, 45/479
Spatial agnosia
  constructional apraxia, 45/493
  hemiasomatognosia, 45/375
  oculomotor apraxia, 2/606
  parietal lobe syndrome, 45/78, 80
  right left disorientation, 2/604
  spatial orientation disorder, 3/222-224
  unilateral, *see* Visuospatial agnosia
  visual, *see* Visuospatial agnosia
Spatial agraphia
  occipital lobe lesion, 4/147
Spatial cognition
  rehabilitation, 46/619
Spatial discrimination
  terminology, 1/105
Spatial disorder
  head injury, 46/617-620
Spatial disorientation, *see* Spatial orientation
    disorder
Spatial dyskinesia
  developmental dyspraxia, 4/458
  ideomotor apraxia, 4/51, 60
Spatial experience
  development, 3/27
Spatial localization
  occipital lobe tumor, 17/329
  orientation, 3/216
  spatial orientation disorder, 3/216
Spatial memory
  agnosia, 45/416
  spatial orientation disorder, 45/414-418
  topographical memory, 45/416
Spatial memory disorder
  pathology, 45/418
  topographical amnesia, 45/416
Spatial neglect
  unilateral, *see* Visuospatial agnosia
Spatial orientation disorder, 3/212-227, 45/405-418
  alexia, 3/219, 4/124, 45/438
  autoscopia, 24/491
  Balint syndrome, 45/406-411
  body scheme, 3/23

body scheme disorder, 3/223
cerebral dominance, 3/27
classification, 3/212
constructional apraxia, 4/52, 45/498
counting disability, 3/219
diffuse brain lesion, 2/697
dyscalculia, 24/491
dyslexia, 24/491
estimating distance, 3/217
false memory, 3/276
frontal lobe tumor, 17/259
historic note, 3/212-216, 45/405
memory disorder, 3/276
occipital lobe lesion, 2/665
occipital lobe tumor, 17/329
parietal lobe syndrome, 2/697, 45/70-74
Parkinson dementia, 46/374
Pick disease, 42/285, 46/234
prosopagnosia, 24/491
pseudoagnosia, 3/225
reading disability, 3/219
reduplication, 3/251
route finding, 3/218
short-term memory for spatial location, 3/217
space exploration, 45/405-411
space perception, 45/411-414
spatial agnosia, 3/222-224
spatial localization, 3/216
spatial memory, 45/414-418
spatial thinking, 45/413
three dimensional block construction, 3/221
topographical memory, 3/219-221
visual attention, 3/214, 226
visual fixation, 3/214, 226
visuoconstructive disability, 3/221
visuospatial agnosia, 3/30, 216, 218, 224, 45/411
Spatial thinking
  callosotomy, 45/414
  space perception, 45/413
  spatial orientation disorder, 45/413
Spatz system atrophy, *see* System atrophy (Spatz)
Specific anosmia
  prevalence, 42/358
SPECT, *see* Single photon emission computer
    assisted tomography
Spectral sensation
  terminology, 1/110
Spectrin
  chorea-acanthocytosis, 63/286
  dystrophin, 62/126
  muscle fiber, 62/35
Spectrometry
  atomic absorption, *see* Atomic absorption

spectrometry
electrothermal atomic absorption, *see*
   Electrothermal atomic absorption spectrometry
Minamata disease, 64/420
Spectrophotometry
atomic absorption, *see* Atomic absorption
   spectrophotometry
Speech
amusia, 45/488
aphasia, 4/84, 45/306
athetosis, 6/441
Binswanger disease, 46/319
brain injury, 23/233
cancer, 67/390
cerebellar, *see* Cerebellar speech
cerebral dominance, 4/280, 292, 322, 43/223
Cockayne-Neill-Dingwall disease, 13/433
dentatorubropallidoluysian atrophy, 21/519
developmental dyslexia, 46/114
Down syndrome, 31/404
expression, 45/523
Friedreich ataxia, 21/329, 337
frontal lobe lesion, 45/32-36
glycosaminoglycanosis type III, 66/300
handedness, 43/223
hereditary sensory and autonomic neuropathy type
   III, 21/110, 60/27
hyperactivity, 46/181
ictal, *see* Ictal speech
infantile autism, 46/190
median facial cleft syndrome, 46/30
mental deficiency, 46/30
middle cerebral artery syndrome, 53/362
otopalatodigital syndrome, 46/30
palatoschisis, 46/30
progressive bulbar palsy, 22/112, 116, 59/218
quality of life, 67/390
slurred, *see* Slurred speech
temporal lobe epilepsy, 2/719
Speech arrest
paralytic shellfish poisoning, 37/33
paroxysmal, *see* Paroxysmal speech arrest
Speech audiometry
acoustic nerve tumor, 17/671
technique, 16/312
Speech development
course, 4/350-352
disorder, 4/361
Speech discrimination
hearing loss, 24/134, 137
Speech disorder
see also Dysarthria *and* Stuttering
amyotrophic dystonic paraplegia, 42/199

basilar artery migraine, 48/136
basilar impression, 42/17
Bean syndrome, 43/11
benign choreoathetosis, 42/205
benign essential tremor, 42/267
brain aneurysm, 42/723
brain embolism, 53/206
brain infarction, 53/206
brain injury, 4/361, 23/233
cerebellar atrophy, 42/135
cerebral dominance, 4/5
cerebrovascular disease, 53/206
citrullinuria, 42/538
corpus callosum tumor, 17/509
craniosynostosis, 50/125
disconnection syndrome, 4/17, 21, 28, 37
Dumon-Radermecker syndrome, 42/477
dyslalia, 4/428
dyssynergia cerebellaris progressiva, 42/212
echinoidea intoxication, 37/68
Ferguson-Critchley syndrome, 42/142
Friedreich ataxia, 60/309
frontal lobe tumor, 17/253, 271, 67/150
headache, 48/118
hemiballismus, 42/224
heredopathia ophthalmo-otoencephalica, 42/150
intracranial hemorrhage, 53/206
kuru, 42/657
late cerebellar atrophy, 42/135
late infantile metachromatic leukodystrophy,
   66/94, 167
maxillofacial dysostosis, 43/425
mental deficiency, 43/257
mental depression, 42/255
methyldopa intoxication, 37/441
migraine, 42/746, 48/143
multiple sclerosis, 42/496
myopathy, 43/124
neurofibromatosis type I, 50/368
neuroticism, 43/210
ophthalmoplegia, 43/139
20p partial trisomy, 43/539
paramyotonia congenita, 43/168
parietal lobe syndrome, 45/74
Parkinson dementia complex, 42/249
Parkinson disease, 6/188, 42/245
parkinsonism, 42/255
paroxysmal, *see* Paroxysmal speech disorder
Pelizaeus-Merzbacher disease, 42/502, 505
photogenic epilepsy, 42/714
Pick disease, 42/285, 46/234
presenile dementia, 42/287
rehabilitation, 3/396-409

restoration of higher cortical function, 3/373, 375, 378, 396-417
schizophrenia, 46/490
Sjögren-Larsson syndrome, 13/471, 66/615
spinocerebellar ataxia, 42/182
spinopontine degeneration, 42/192
striatonigral degeneration, 42/261
supplementary motor area, 45/34
Toxopneustes pileolus intoxication, 37/68
transient ischemic attack, 53/206
Willvonseder syndrome, 42/194
Wilson disease, 42/270
XXX syndrome, 43/552
Speech disorder in children, 4/407-439
  see also Stuttering
  acquired dysphasia, 4/425
  age of speech aquisition, 4/422
  angioneurotic edema, 43/61
  aspartylglucosaminuria, 29/225
  aspartylglycosaminuria, 27/159
  athetotic hemiplegia, 42/200
  autism, 43/202
  behavior disorder in children, 4/423-425
  benign choreoathetosis, 42/205
  benign essential tremor, 42/267
  cerebral dominance, 4/465-469
  cheilopalatoschisis, 4/415
  classification, 4/408, 428
  congenital retinal blindness, 13/272, 42/401
  corpus callosum agenesis, 42/100
  craniodiaphyseal dysplasia, 43/355
  developmental articulatory defect, 43/219
  developmental expressive dysphasia, 4/428
  developmental receptive dysphasia, 4/428
  disequilibrium syndrome, 42/139
  dysphasia, 4/419
  dysphonia, 4/409
  dystonia, 42/219
  dystonia deafness syndrome, 42/372
  environmental factor, 4/426
  examination, 4/410-414, 432
  Förster syndrome, 42/202
  Friedreich ataxia, 42/143
  glycosaminoglycanosis type III, 42/455
  hearing loss, 4/420
  hereditary benign chorea, 42/203
  hereditary paroxysmal kinesigenic choreoathetosis, 42/205
  hereditary sensory and autonomic neuropathy type III, 1/477, 43/59, 60/27
  histidinemia, 42/554
  Huntington chorea, 42/225
  kinesigenic choreoathetosis, 42/207

kuru, 42/657
  Lafora progressive myoclonus epilepsy, 42/207
  Leber optic atrophy, 42/401
  Marinesco-Sjögren syndrome, 42/184
  maxilla abnormality, 4/418
  maxillofacial dysostosis, 43/425
  mental deficiency, 4/421-425, 43/257
  metachromatic leukodystrophy, 42/493
  mutism, 4/424
  nasal obstruction, 4/418
  neuroticism, 43/210
  nonprogressive cerebellar ataxia, 42/124
  20p partial trisomy, 43/539
  palatal dysproportion syndrome, 4/415
  pallidopyramidal degeneration, 42/244
  paramyotonia congenita, 43/168
  Pelizaeus-Merzbacher disease, 42/502-505
  prevalence, 43/219
  prognosis, 4/412
  psychosis, 4/424
  rehabilitation, 4/412, 421, 423, 426, 433
  secondary speech disorder, 4/419-427
  spastic paraplegia, 42/175
  specific retardation of speech development, 4/427-434
  speech dysarthria, 4/409-413
  speech dysrhythmia, 4/409-413
  speech sonogram, 4/414
  structural abnormality, 4/414-419
  subsarcolemmal myofibril, 42/219
  tongue agenesis, 4/416
  tongue-tie, 4/416
  treatment, 4/412, 421, 423, 426, 433
  Troyer syndrome, 42/193
  verbal apraxia, 4/459
  verbal auditory agnosia, 4/428, 43/228
  Willvonseder syndrome, 42/194
  Wilson disease, 42/270
  word deafness, 4/428
  XXX syndrome, 43/552
Speech isolation syndrome, *see* Mixed transcortical aphasia
Speech retardation
  organic acid metabolism, 66/641
Speech sonogram
  speech disorder in children, 4/414
Speech test
  intracranial pressure, 16/336
  statoacoustic nerve, 2/73
Speech therapy
  aphasia, 46/152-155, 616
  clumsiness, 46/164
  herpetic meningoencephalitis, 46/616

infantile autism, 46/192
motoneuron disease, 59/461
multiple sclerosis, 47/175
nonprogressive cerebellar ataxia, 42/125
rehabilitation, 12/475, 55/243
restoration of higher cortical function, 3/398,
    400-417
stuttering, 46/171
Speed
  illusory visual, *see* Illusory visual speed
  reading, 4/136, 7/116, 125
Spelling disorder
  agraphia, 4/147
Spelling dyslexia
  developmental dyslexia, 46/111-113
  occipital lobe tumor, 17/328
  simultanagnosia, 2/657, 661
Spencer, H., 1/72
Spermatogenesis
  sexual function, 26/451
Sphaeroides hamiltoni intoxication, *see* Puffer fish
    intoxication
Sphenocavernous angle syndrome
  superior orbital fissure syndrome, 2/321
Sphenoid body aplasia
  brain herniation, 30/278
Sphenoid bone
  embryology, 30/280
  skull base tumor, 17/181
  tumor, 13/82
Sphenoid bone defect
  anencephaly, 42/12, 50/84
  pituitary dwarfism type V, 42/616
  rhinorrhea, 24/187
Sphenoid sinus
  skull base tumor, 68/468
Sphenoid sinusitis
  cavernous sinus thrombosis, 52/172
  chronic meningitis, 56/644
Sphenoid syndrome
  optic atrophy, 13/82
Sphenoid wing
  herniation, 1/571
  meningioma, 68/410
  skull base tumor, 68/468
Sphenoid wing aplasia, 30/278-282
  clinical features, 30/279, 50/144
  computer assisted tomography, 50/143
  craniorachischisis, 50/144
  cranioschisis, 50/144
  differential diagnosis, 30/280
  embryopathogenesis, 30/280, 50/137
  embryopathology, 50/144

epidemiology, 30/278, 50/143
etiology, 50/144
exophthalmos, 50/143
neurofibromatosis, 30/278, 281, 50/144
orbital encephalocele, 50/144
prognosis, 30/282
pulsating exophthalmos, 30/278
radiology, 30/280, 50/144
sex ratio, 50/143
treatment, 30/282, 50/144
vascular abnormality, 30/281
Sphenoid wing dysplasia
  see also Neurofibromatosis type I
  neurofibromatosis type I, 50/367, 369
Sphenoidal fissure syndrome, *see* Superior orbital
    fissure syndrome
Sphenoidal ridge meningioma
  frontal lobe tumor, 17/238
Sphenoidal sinus syndrome
  anterior cranial fossa, 2/296
  facial nerve, 2/296
  features, 2/296
  meningioma, 2/296
  oculomotor nerve, 2/296
Sphenoiditis
  sphenopalatine neuralgia, 48/479
Sphenopalatine ganglion
  anatomy, 5/351
Sphenopalatine neuralgia, 48/475-481
  anatomy, 5/326
  carotidynia, 48/226, 338
  Charlin neuralgia, 48/478, 485
  classification, 48/6
  clinical features, 48/477
  cluster headache, 5/330, 48/217
  coryza, 48/477
  cranial neuritis, 5/22
  description, 48/225
  diagnosis, 5/329
  differential diagnosis, 5/330
  drug treatment, 5/383
  glossopharyngeal neuralgia, 48/468, 478
  maxillary artery, 5/328
  Melkersson-Rosenthal syndrome, 8/217
  menopause, 5/115
  Mondrum-Benisty syndrome, 5/330
  nasal pain, 5/329
  neurotic, 48/478
  orbital pain, 48/480
  pain, 5/115, 48/477
  paratrigeminal syndrome, 5/331
  pathogenesis, 5/328
  petrosal nerve, 5/328

352

petrosal neuralgia, 5/330, 48/480
pterygopalatine ganglion, 48/475
sinusitis, 5/115
sneezing, 48/477
sphenoiditis, 48/479
surgery, 5/402
treatment, 5/330, 48/480
uvula, 48/477
Vidian nerve, 48/479
Sphenopalatine neurosis
    semantics, 48/478
Sphenoparietal sinus, *see* Breschet sinus
Sphenopetrous fissure syndrome, *see* Orbital apex
    syndrome
Spherocytosis
    familial, *see* Familial spherocytosis
    hyperviscosity syndrome, 55/484
Spheroid body
    see also Neuroaxonal dystrophy
    aging, 6/626
    amyotrophic lateral sclerosis, 6/626, 22/17, 59/291
    Bassen-Kornzweig syndrome, 63/278
    biliary atresia, 6/626
    brain arteriosclerosis, 6/626
    brain contusion, 6/626
    brain tumor, 6/626
    carbon monoxide intoxication, 6/626
    cardiac insufficincy, 6/626
    diabetes mellitus, 6/626
    encephalitis, 6/625
    encephalomalacia, 6/626
    fibrocystic pancreatic disease, 6/626
    foamy, *see* Foamy spheroid body
    Guam amyotrophic lateral sclerosis, 59/287-291
    Hallervorden-Spatz syndrome, 6/605, 22/403
    hereditary ataxia, 6/626
    kuru, 6/626
    lepra, 6/626
    liver cirrhosis, 6/626
    Macaca irus, 59/291
    malabsorption syndrome, 6/628
    multiple sclerosis, 6/626
    neuroaxonal dystrophy, 49/391
    neurosyphilis, 6/626
    origin, 6/607
    posterior pituitary lobe, 6/626
    rabies, 6/626
    sheep, 6/626
    spongy cerebral degeneration, 6/626
    substance, 6/610
    substantia nigra, 59/287-291
    tabes dorsalis, 6/626
    ultrastructure, 21/55

Spheroid body myopathy
    congenital myopathy, 62/332
    histopathology, 62/345
Spherophakia
    Alport syndrome, 42/376
    microcoria, 42/403
Spherula insularis
    multiple sclerosis, 9/59, 109, 47/319
Sphincter
    anal, *see* Anal sphincter
    esophageal, *see* Esophageal sphincter
    iris, 74/400
    Oddi, *see* Oddi sphincter
    Pick disease, 46/235
    urinary, *see* Urinary sphincter
Sphincter disturbance
    angiodysgenetic necrotizing myelomalacia,
        20/487
    basilar impression, 42/17
    blennorrhagic polyneuritis, 7/478
    cat scratch disease, 52/130
    congenital spinal cord tumor, 32/366
    diastematomyelia, 50/436
    diplomyelia, 50/436
    familial spastic paraplegia, 59/305
    hereditary olivopontocerebellar atrophy (Menzel),
        21/444
    hyperkinetic child syndrome, 43/201
    multiple sclerosis, 47/58
    olivopontocerebellar atrophy (Dejerine-Thomas),
        21/427
    progressive bulbar palsy, 59/219
    spinal ependymoma, 20/367-369
    spinal lipoma, 20/400
    spinal meningioma, 20/200
    vertebral metastasis, 20/418
Sphincter reflex
    anal, *see* Anal sphincter reflex
Sphincter resection
    urethral, *see* Urethral sphincter resection
Sphingolipid
    N-acetylthreoceramide, 9/9
    catabolism, 66/149
    ceramide lactoside lipidosis, 29/350
    cerebroside sulfatide, 10/327
    chemistry, 10/246, 266, 66/36
    erythroconfiguration, 9/9
    fucosidosis, 29/350
    ganglioside, 10/327
    globoid cell leukodystrophy, 10/89, 351, 29/350
    glycerophosphatide, 10/266
    GM3 hematoside lipodystrophy, 29/350
    metabolism, 66/43

metachromatic leukodystrophy, 10/25
myelin, 7/46, 9/2, 10/136
neuronal ceroid lipofuscinosis, 10/589
synthesis, 9/8
Sphingolipid activator protein deficiency
neurolipidosis, 66/61
Sphingolipidosis
enzymopathy, 21/50
epidemiology, 10/556-587
genetics, 10/556-587
glycosaminoglycanosis, 29/372
macrencephaly, 50/263
major form, 66/45
myelin storage dystrophy, 21/51
neurolipidosis, 66/20
secondary pigmentary retinal degeneration,
13/326-330
Sphingomyelin
catabolism, 10/284
chemistry, 9/311, 10/266
CSF, see CSF sphingomyelin
fatty acid, 10/284
lipidosis, see Niemann-Pick disease
membranous cytoplasmic inclusion body, 10/287
metabolism, 66/139
metachromatic leukodystrophy, 10/49
multiple sclerosis, 9/311, 368
myelin, 9/31
nerve myelin, 9/4
neutral lipid, 9/30
Niemann-Pick disease, 10/286, 327, 487, 501,
29/350
synthesis, 9/9, 10/284
Sphingomyelin lipidosis, see Niemann-Pick disease
Sphingomyelin phosphodiesterase
metabolism, 66/139
Sphingomyelin phosphodiesterase deficiency, see
Niemann-Pick disease
Sphingomyelin synthesis
vitamin B6 deficiency, 51/297
Sphingomyelinase
enzyme deficiency polyneuropathy, 51/369, 373
Niemann-Pick disease, 10/281, 332, 29/350,
42/469-471, 46/57
sea blue histiocyte disease, 42/474
vertical supranuclear ophthalmoplegia, 42/605
Sphingomyelinosis, see Niemann-Pick disease
Sphingophospholipid
chemistry, 10/248
Sphingosine
biochemistry, 9/8
chemistry, 10/246
erythro, see Erythrosphingosine

Farber disease, 66/216
structure, 66/250
threo, see Threosphingosine
Spider bite intoxication, 37/111-113
autonomic nervous system, 75/512
symptom, 37/112
treatment, 37/112
Spider cell
astrocytoma, 18/2
Spider envenomation
Androctonus, 65/194
Araneus, 65/194
Buthus, 65/194
Centruroides, 65/194
Clubonia, 65/194
Latrodectus, 65/104
Leiurus, 65/194
Loxosceles, 65/194
Lycosa, 65/194
Phoneutria, 65/194
scorpion genus, 65/194
spider species, 65/194
tick species, 65/194
Tityus, 65/194
Spiegler tumor, see Multiple dermal cylindroma
Spielmeyer-Vogt-Sjögren disease, see Neuronal
ceroid lipofuscinosis
Spielmeyer, W., 1/8, 27, 10/10, 213-217
Spiller-Frazier operation
history, 5/303
surgical procedure, 5/392
trigeminal neuralgia, 5/397
Spiller, W.G., 1/5
Spina bifida
see also Arnold-Chiari malformation type I,
Arnold-Chiari malformation type II,
Arnold-Chiari malformation type III,
Arnold-Chiari malformation type IV,
Craniorachischisis, Dysraphia, Hydrocephalus,
Lumbosacral lipoma, Meningocele,
Myelomeningocele and Rachischisis
abortion, 30/148
Aicardi syndrome, 42/696
anencephaly, 30/163, 176, 188, 42/55, 50/75, 77,
79
anterior sacral meningocele, 50/500
antiepileptic agent, 65/516
aqueduct stenosis, 32/532, 42/55, 50/309
area medullovasculosa, 50/485
areflexia, 42/54
Arnold-Chiari malformation type I, 30/162,
32/521, 523-526, 42/54
Arnold-Chiari malformation type II, 32/523-526,

50/482, 490
Arnold-Chiari malformation type III, 32/523
Arnold-Chiari malformation type IV, 32/523
basilar impression, 32/73
biochemical lesion, 50/490
birth incidence, 42/55
bladder function, 32/550, 42/54
blood vessel, 32/541
brain stem valve, 32/531
brain stem valvular action, 32/531
cauda equina tumor, 19/87
cavum septi pellucidi, 30/325, 50/487
cavum vergae, 30/325
centrofacial lentiginosis, 14/782, 43/26
clinical features, 32/548-552
CNS malformation, 50/57
complication, 32/555
congenital hip dislocation, 42/54
congenital spinal cord tumor, 32/357, 362
corpus callosum agenesis, 50/150, 487
cranial lesion, 32/533, 535
craniolacunia, 32/521, 541, 50/487, 489
cranium bifidum, 50/477
CSF, 32/528, 532, 42/55
Dandy-Walker syndrome, 42/23, 50/487
dermal sinus tract, 50/496
dermoid cyst, 50/494, 496
diagnostic procedure, 32/548-552
diastematomyelia, 50/482, 497
diet, 50/479
differential diagnosis, 19/87
early closure, 32/557, 559
Edwards syndrome, 43/537
embryogenesis, 50/488
embryology, 32/527, 533
EMG, 32/548
encephalocranial disproportion, 32/527
environmental factor, 32/547
epidemiology, 30/149-151, 163, 32/546-548,
    50/477
epidural spinal lipoma, 20/406
epilepsy, 50/494
etiologic heterogeneity, 50/479
etiology, 50/479
experimental, 32/527
experimental animal, 50/28
familial hydrocephalus, 50/289
α-fetoprotein, 30/179, 42/56
filum terminale, 50/496, 501
fontanelle, 32/541
foot deformity, 32/522, 545, 552
forebrain malformation, 32/540
fracture, 50/495

Friedreich ataxia, 21/348
gait, 50/493
genetics, 30/163, 32/533, 547, 548
geography, 50/478
Goltz-Gorlin syndrome, 14/788, 43/19
Gruber syndrome, 14/115, 505, 43/391
Hallermann-Streiff syndrome, 43/403
hemivertebra, 32/545, 42/30, 55
hereditary motor and sensory neuropathy type I,
    21/286
hereditary motor and sensory neuropathy type II,
    21/286
hereditary sensory and autonomic neuropathy type
    I, 21/77, 60/9
heterotopia, 32/525
hindbrain malformation, 32/539
history, 32/519-528, 50/477
holoprosencephaly, 50/237
hydrocephalus, 30/157, 535, 32/528-533, 535,
    42/35, 50/287, 482-485, 492
hydromyelia, 50/486
incidence, 50/58, 477
incontinentia pigmenti, 43/23
intelligence, 32/555-557, 50/494
intestinal function, 50/494
Klippel-Trénaunay syndrome, 14/395, 43/25
Krabbe-Bartels disease, 14/407
kyphosis, 32/545
laryngeal abductor paralysis, 42/319
late complication, 50/494
level, 50/481
lipomeningocele, 50/499
lissencephaly, 30/487
lordosis, 32/545
lumbosacral lipoma, 50/499
malformation, 32/533-535
malignancy, 50/494
massa intermedia, 50/487
maternal diabetes mellitus, 30/126
maternal factor, 50/480
meningocele, 19/187, 50/477, 482
meningomyelocele, 20/58
mental deficiency, 32/553, 555-557
mesencephalic tectum beaking, 50/487
metabolic factor, 50/479
microgyria, 30/484, 42/55, 50/486
midbrain malformation, 32/540
mobility, 50/493
myelocystocele, 50/482
myelodysplasia, 14/523
myelography, 19/87, 187
myelomeningocele, 50/480, 487, 490
myeloschisis, 32/527, 536

neural tissue volume, 32/527
neurenteric canal, 50/501
neurenteric cyst, 50/501
neurologic finding, 32/548
neuron population, 32/543
neuronal heterotopia, 42/55
neuropathic ulcer, 50/495
neurotrophic osteoarthropathy, 38/459
obesity, 50/495
occipital cephalocele, 50/101
occipital encephalocele, 42/26
occulta, see Spina bifida occulta
paralysis, 42/54
pathogenesis, 32/519-533
pathology, 32/526, 534-544
polymicrogyria, 50/486, 488
posterior column, 32/544
posterior fossa, 32/527
postnatal development, 32/545
potato blight, 30/124
prenatal diagnosis, 32/564-571, 42/56
principal spinal cord lesion, 50/480
prognosis, 32/553
progressive hydromyelia, 50/494
11q partial trisomy, 43/524
race, 50/478
rachischisis, 50/480
radiologic appearance, 32/550-552
recurrence risk, 42/55, 50/478
renal malformation, 32/541
respiratory virus infection, 31/219
retained conceptus, 50/480
rubella, 50/480
Rubinstein-Taybi syndrome, 31/284, 43/234
ruptured neural tube hypothesis, 32/531
scoliosis, 32/545
seasonal variation, 32/546
sensory loss, 42/54
septum pellucidum agenesis, 30/314-321
sex ratio, 32/548, 50/56
skull base, 32/541
spina bifida aperta, 32/542
spina bifida occulta, 30/163, 50/495
spinal cord lesion, 50/485, 489
spinal cord tumor, 32/357, 359
spinal lesion, 32/535
spinal lipoma, 19/362, 20/110, 397
spinal muscular atrophy, 59/370
spinal nerve, 32/543
spinal nerve root, 32/543
surgical treatment, 32/554
survival, 32/552, 555, 50/493
suture, 32/541

syringomyelia, 32/305, 61/376
syringomyelocele, 50/482
talipes, 42/54
teratogenic agent, 32/528, 531, 50/479
teratologic process, 32/534
terminology, 32/534-536
tethered spinal cord, 50/494
19th century, 32/519-521
tissue increase, 32/543
treatment, 32/554, 50/491
triploidy, 43/564-566
trisomy 8, 50/563
trypan blue, 32/528, 531
tuberous sclerosis, 14/368, 398, 31/6
urinary bladder, 50/493
ventricular dilatation, 32/532
vertebral abnormality, 32/540
vitamin A, 32/528
white matter, 32/544
X-linked transmission, 50/479
Spina bifida aperta
  definition, 32/537
  juvenile spinal cord injury, 25/176
  spina bifida, 32/542
Spina bifida complex
  iniencephaly, 50/134
Spina bifida cystica
  hydromyelia, 50/425
Spina bifida occulta
  Coffin-Lowry syndrome, 43/238
  cyst, 20/76
  definition, 32/537
  history, 32/522
  hydromyelia, 50/425
  multiple nevoid basal cell carcinoma syndrome,
    14/456, 459
  pilonidal sinus, 42/47
  plain X-ray, 19/159
  r(13) syndrome, 50/585
  spina bifida, 30/163, 50/495
  spinal cord compression, 19/159
  spinal subdural cyst, 20/93
Spinal accessory nerve
  anatomy, 2/79
  subarachnoid hemorrhage, 55/18
Spinal anesthesia
  back pain, 61/150
  brain hemorrhage, 61/164
  cardiac arrest, 61/165
  cauda equina syndrome, 61/159
  complication, 61/147
  complication frequency, 65/420
  complication rate, 65/422

cranial nerve palsy, 61/164
gram-negative bacillary meningitis, 52/103
hearing loss, 61/165
intracranial subdural hematoma, 61/164
local anesthetic agent, 65/420
multiple sclerosis, 47/151
neurotoxicity, 61/150
pneumocephalus, 61/164
puncture-tied complication, 61/149
subarachnoid hemorrhage, 61/164
systemic hypotension, 61/150
Spinal angiography, 12/631-642, 19/229-241
   angiodysgenetic necrotizing myelomalacia,
     20/490-497
   arterial vascular injury, 19/232
   complication, 26/65
   iatrogenic neurological disease, 63/44
   method, 19/229
   normal angiogram, 19/230
   phlebography, 12/632
   scintigraphy, 12/633
   spinal tumor, 19/232
   vascular malformation, 19/231
   vertebral tumor, 19/232
Spinal angioma, 20/435-468
   see also Spinal arteriovenous malformation *and*
     Spinal cord vascular malformation
   Adamkiewicz artery, 32/474
   age, 20/438
   aneurysm, 20/448
   angiography, 20/455-458, 32/470-490
   autopsy prevalence, 20/26
   Brown-Séquard syndrome, 32/466
   cervical site, 12/637
   clinical features, 32/466
   CNS, 32/466
   Cobb syndrome, 14/429, 442
   CSF, 20/459
   cutaneous manifestation, 32/469
   diagnosis, 20/451
   differential diagnosis, 20/28
   draining vein, 32/484-486
   embolization, 32/487
   15% epidural, 20/436
   extramedullary, 32/481
   feeding vessel, 32/476-482
   gas myelography, 32/470
   hemangioblastoma, 20/448
   hereditary hemorrhagic telangiectasia (Osler),
     32/467, 474, 482, 484
   histologic classification, 20/438
   history, 20/436, 32/465
   intermittent course, 20/461

   intramedullary, 32/478-481
   intramedullary hematoma, 32/489
   level, 32/470
   localization, 32/484
   malformation, 32/469
   mixed, 32/481
   myelography, 12/632, 19/190, 20/110, 454
   myelomalacia, 55/103
   myeloscintigraphy, 20/455
   osseous manifestation, 20/26
   phlebography, 12/632
   plain X-ray, 32/470
   pregnancy, 20/461
   prevalence, 20/435
   prognosis, 20/467, 32/509
   pseudoaneurysm, 32/482-484
   radiologic appearance, 32/470
   radiology, 12/631, 20/453
   scintigraphy, 12/633
   segmental lesion distribution, 20/437
   sex ratio, 20/438
   site, 20/436
   spastic paraplegia, 59/438
   spinal bruit, 20/451
   spinal cord compression, 32/487
   spinal nerve root syndrome, 32/466
   spinal phlebography, 20/455
   spinal subarachnoid hemorrhage, 12/134
   spread, 20/436
   steal syndrome, 32/487
   surgical strategy, 20/462
   surgical technique, 20/463
   thoracic site, 12/639
   tomography, 32/470
   trauma, 20/461
   treatment, 20/461, 32/495-508
Spinal angioma racemosum venosum, *see*
   Foix-Alajouanine disease
Spinal apoplexy
   arteriosclerosis, 12/532
   Brown-Séquard syndrome, 19/41
   decompression illness, 12/532
   differential diagnosis, 20/445
   malaria, 12/532, 35/148
   recurrent transient paraplegia, 12/498, 532
   syphilis, 12/532
   total transverse spinal cord syndrome, 12/498
   transverse spinal cord lesion, 19/41
Spinal arachnoid cyst, 32/393-420
   age, 20/80-82
   case series, 20/83
   clinical features, 20/84-86, 32/398, 405
   communicating, 32/406

course, 32/398
developmental, 20/80-86
diagnosis, 32/407
differential diagnosis, 32/406
differentiation from meningocele, 32/393
epidemiology, 32/398, 404
epidural, 32/394-400, 407
familial incidence, 32/395
hydrostatic pressure, 32/396
hypertrichosis, 20/82
inflammatory, 20/86-89
intradural, 32/400-408
intraspinal, 31/82-96
localization, 32/405
location, 20/80-82
malformation, 32/399
meningocele, 32/394
origin, 32/401
pathogenesis, 32/396
radiologic finding, 32/399, 407
secluded, 32/407
sex ratio, 20/80-82
size, 32/395, 401
surgical treatment, 32/397
syringomyelia, 20/82
treatment, 32/418-420
volume, 32/395
Spinal arachnoid diverticulum, *see* Spinal arachnoid
  cyst
Spinal arachnoiditis
  basilar impression, 32/70, 81
  bladder function, 42/107
  chronic meningitis, 56/644
  CSF, 42/107
  familial, *see* Familial spinal arachnoiditis
  myelography, 19/85
  pain, 42/107
  posttraumatic syringomyelia, 26/146, 148, 153
  review, 33/263-272
  sensory loss, 42/107
  spastic paraplegia, 42/107, 59/437
  spinal cord intermittent claudication, 12/535
  syringomyelia, 26/146, 148, 153, 32/312, 50/455
  vascular dementia, 46/360
Spinal arteriovenous malformation, 20/481-515
  see also Spinal angioma *and* Spinal cord vascular
    malformation
  angiography, 20/490-499
  characteristics, 68/570
  hereditary hemorrhagic telangiectasia (Osler),
    38/275-277
  history, 20/481
  treatment, 20/504

varicosis medullae spinalis, 20/481
vertebral, 12/289
Spinal artery
  anastomotic ansa of the conus, 12/494
  anterior, *see* Anterior spinal artery
  posterior, *see* Posterior spinal artery
Spinal artery injury
  see also Spinal cord blood flow, Spinal cord
    vascular disease *and* Spinal cord vascularization
  abdominal aorta aneurysm, 26/74
  angiography, 26/65
  anticoagulant, 26/74
  aorta coarctation, 26/72
  aorta rupture, 26/72
  aorta surgery, 26/63, 69
  aortography, 26/65
  arteria radicularis magna, 26/65, 67, 69
  arteriosclerosis, 26/66
  dissecting aneurysm, 26/71
  experimental, 26/73
  malignant tumor, 26/69
  pathogenesis, 26/66
  physiology, 26/66
  thoracic aortic aneurysm, 26/68, 70
  thoracotomy, 26/69
  vasculitis, 26/67
Spinal astrocytoma, *see* Spinal glioma
Spinal ataxia
  malnutrition, 28/22
  syringomyelia, 32/265
  tropical ataxic neuropathy, 7/640
Spinal automatism
  extensor reflex, 1/251
  flexor reflex, 1/251
  Foix-Alajouanine disease, 19/38
Spinal autonomous efferent
  dorsal spinal nerve root, 2/167
  spinal nerve root syndrome, 2/162
Spinal bruit
  angiodysgenetic necrotizing myelomalacia,
    20/489
  spinal angioma, 20/451
Spinal canal
  cervical, *see* Cervical spinal canal
  stenosis, *see* Vertebral canal stenosis
Spinal cholesterosis
  xanthoma, 27/250
  xanthomatosis, 14/779
Spinal collapse
  traumatic spine deformity, 26/169
Spinal column injury
  lumbar, *see* Lumbar spinal column injury
Spinal column tumor

358

atrial myxoma metastasis, 63/99
Spinal compression syndrome, *see* Spinal cord
   compression
Spinal cord
  Adamkiewicz artery, 12/478
  aneurysmal bone cyst, 67/194
  angiography, 26/65
  arterial anatomy, 12/479, 483, 486, 493
  arterial supply, 26/6, 64, 55/95
  arteriovenous lesion, 19/15
  astrocytoma, 19/13
  ataxia telangiectasia, 14/74
  autonomic innervation, 22/246
  autonomic nervous system, 74/65, 70, 201, 75/567
  bacterial encephalitis, 33/187-193
  blood supply, 12/478
  bowel, 75/626
  brain death, 24/763, 817
  Burkitt lymphoma, 39/75
  capillary angioma, 20/444
  carcinoma effect, 38/669-676
  cerebellar hemangioblastoma, 18/286
  cervical, *see* Cervical spinal cord
  chondrosarcoma, 67/197
  chordoma, 67/197
  choriocarcinoma, 32/361
  CNS, 30/33
  complete lesion, *see* Complete cord lesion
  compression injury, 25/72, 260
  craniospinal injury, 25/43, 66, 93
  critical illness polyneuropathy, 51/581
  dermoid cyst, 19/16
  diaschisis, 2/181
  echinococcosis, 52/524
  embryo, 50/10
  embryology, 32/450-452
  embryonal artery, 55/95
  eosinophilic granuloma, 67/196
  epidermoid cyst, 19/16
  epidural tumor, 67/196
  ethambutol polyneuropathy, 51/295
  evoked potential, 1/51
  Ewing sarcoma, 67/197
  exercise, 74/257
  flavivirus, 56/135
  Friedreich ataxia, 60/301
  functional anatomy, 25/235
  glycine, 29/508
  glycosaminoglycanosis, 10/445
  GM2 gangliosidosis, 10/400
  gunshot, 25/49
  gut, 75/626
  hemangioendothelioma, 18/288

hemophilia, 38/60
hereditary cerebello-olivary atrophy (Holmes),
   21/408
hereditary olivopontocerebellar atrophy (Menzel),
   21/439
heterotopic hemopoietic tissue, 20/443
hourglass tumor, 8/429
Huntington chorea, 6/340-342, 49/320
hydranencephaly, 30/671, 50/347
hydrocephalus, 30/671-673
hypertonia type, 1/2, 68
incomplete lesion, *see* Incomplete cord lesion
interneuron, 1/57
intramedullary arterial system, 12/496
intrauterine injury, 25/96
Langerhans cell histiocytosis, 70/60
late lesion, *see* Late cord lesion
leukemia, 39/13, 67/198
lymphoma, 19/14, 67/198
meningeal leukemia, 16/234, 63/342
motor lesion topography, 1/228
motor unit hyperactivity, 41/303-308
multiple myeloma, 67/198
multiple organ failure, 71/532
multiple sclerosis, 47/245
muscle metaboreflex, 74/257
myxedema neuropathy, 8/59
necrosis, *see* Myelomalacia
neurinoma, 67/201
neurofibroma, 67/201
neuroleptic akathisia, 65/288
neurosarcoidosis, 38/529
oligodendroglioma, 18/83
olivopontocerebellar atrophy (Dejerine-Thomas),
   21/424
organophosphate induced delayed polyneuropathy,
   64/175
osteitis deformans (Paget), 38/361, 366, 70/14
osteoblastoma, 67/195
osteoid osteoma, 67/194
osteoporosis, 70/20
osteosarcoma, 67/197
oxygen requirement, 63/40
penis erection, 75/86
perimedullary artery, 12/495
periodic ischemia, 12/508
pertussis encephalopathy, 33/289
phantom limb, 45/399
pial plexus, 12/484
plasmacytoma, 67/198
polyarteritis nodosa, 39/297, 303
porencephaly, 50/357
progressive dysautonomia, 59/148

racemose angioma, 20/438, 444
radiation injury, 7/395
radicular artery, 12/481
rotation injury, 25/62, 67
sarcoidosis, 38/529
scleroderma, 39/366
spinal ependymoma, 19/12
spinal epidermoid, 19/16
spinal nerve root artery, 26/65
spinopontine degeneration, 21/391
stress, 74/317
subacute combined degeneration, *see* Subacute
    combined spinal cord degeneration
substance P, 45/230
Takayasu disease, 63/53
telangiectasia, 14/520, 19/15
tethered, *see* Tethered spinal cord
transection, 1/62
transient ischemic attack, *see* Intermittent spinal
    cord ischemia
transverse lesion, *see* Transverse spinal cord lesion
transverse section, *see* Transverse spinal cord
    lesion
traumatic vascular disease, 12/556
tri-*o*-cresyl phosphate intoxication, 64/160
trophic function, 1/59
varicocele, 14/431
vascular disorder, 55/95
vascular supply, 63/38
vascular tumor, *see* Spinal vascular tumor
vascularization, 61/112
venous system, 12/487
vitamin B12 deficiency, 70/384
watershed, 12/588, 591
Spinal cord abscess
    antibiotic agent, 52/192
    back pain, 52/191
    bacterial endocarditis, 52/300
    cause, 52/190
    computer assisted tomography, 52/191
    dermal sinus, 52/191
    diagnosis, 33/193, 52/191
    differential diagnosis, 52/191
    Hemophilus parainfluenzae, 52/191
    iatrogenic neurological disease, 52/191
    Listeria monocytogenes encephalomyelitis,
        52/191
    lumbar puncture, 52/191, 61/162
    meningitis, 52/190
    mortality, 52/192
    myelography, 52/191
    neurosurgery, 52/191
    nuclear magnetic resonance, 52/191

paraplegia, 61/161
pathology, 33/191, 52/190
prognosis, 33/193, 52/191
Proteus mirabilis, 52/191
recurrence, 52/192
retroperitoneal abscess, 52/190
spinal phlebothrombosis, 52/191
Staphylococcus aureus, 52/191
Streptococcus pneumoniae, 52/191
Streptococcus viridans, 52/191
torulopsiasis, 35/571, 52/496
transverse myelitis, 52/191
treatment, 33/193, 52/191
vertebral osteomyelitis, 52/190
Spinal cord angiography, *see* Spinal angiography
Spinal cord angioma, *see* Spinal angioma
Spinal cord aplasia
    spinal muscular atrophy, 59/370
Spinal cord arterial territory
    anatomy, 12/492
    cervicothoracic, 12/493
    middle thoracic, 12/493
    thoracolumbar, 12/494
Spinal cord arteriovenous malformation
    aneurysm, 32/469
    brain arteriovenous malformation, 54/369
    hereditary hemorrhagic telangiectasia (Osler),
        55/450
    pseudoaneurysm, 32/482-484
Spinal cord atherosclerosis
    autopsy prevalence, 63/38
    myelopathy, 63/41
Spinal cord atrophy
    electrical injury, 23/711
    gas myelography, 19/181
    spinal cord injury, 25/40
Spinal cord autoregulation
    spinal cord injury, 25/14
Spinal cord blood flow
    see also Spinal artery injury
    arteritis, 12/498
    autoregulation, 63/39
    collateral circulation, 12/492, 495
    laminectomy, 12/489
    posttraumatic change, 25/83
    regulation, 12/576
    spinal cord experimental injury, 25/14, 19
    spinal cord injury, 26/313-333
Spinal cord compression
    see also Vertebral compression
    Aarskog syndrome, 43/314
    achondroplasia, 32/329-331
    aneurysmal bone cyst, 19/368

angiosarcoma, 19/368
ankylosing spondylitis, 19/375, 70/4
Arnold-Chiari malformation type I, 42/15
aspergillosis, 35/398, 52/380
astrocytoma, 19/359
benign chondroblastoma, 19/367
blood dyscrasia, 19/366
brucellous spondylitis, 19/370
calcium, 69/170
Candida granuloma, 52/401
candidiasis, 52/401
cause, 19/357
cervical, see Cervical spinal cord compression
cervical spondylosis, 7/456
chloroma, 63/343
chondroblastoma, 19/367
chondrodystrophia fetalis, 19/365
chondromyxofibroma, 19/367
chondrosarcoma, 19/368
chordoma, 19/369
clinical features, 19/72
coccidioidomycosis, 52/414
course, 19/348
critique of term, 19/347
demyelination, 69/170
dermoid, 19/363
diastrophic dwarfism, 43/384
differential diagnosis, 19/347-386
dinoprostone, 69/170
dracunculiasis, 52/514
echinococcosis, 35/193-202, 349
ependymoma, 19/359
epidermoid, 19/363
epidural, 69/74
epidural abscess, 19/370
epidural lymphoma, 63/349
epidural tumor, 19/356
Ewing sarcoma, 19/368
extradural Nocardia granuloma, 52/449
extramedullary tumor, 19/352
false diagnosis, 19/375-378
familial spastic paraplegia, 59/306
fibrosarcoma, 19/368
filariasis, 35/163, 167, 170, 52/514, 516
Foix-Alajouanine disease, 19/378
glioblastoma, 19/361
glutamic acid, 69/170
glycosaminoglycanosis type IV, 66/304
glycosaminoglycanosis type IVB, 66/306
hemangioblastoma, 19/362
hemorrhage, 19/372, 69/170
histiocytosis X, 19/365
Hodgkin disease, 19/364, 366

Hodgkin lymphoma, 63/346, 349
hourglass tumor, 19/161, 20/295
hypertrophic pachymeningitis, 33/190
intervertebral disc calcification, 19/153
intramedullary spinal tumor, 19/351
ischemia, 69/170
Landry paralysis, 19/377
leukemia, 63/343
lipoma, 19/361
localization, 19/350
lymphoma, 63/346, 69/277
malignant angiosarcoma, 19/369
malignant hemangioendothelioma, 19/369
meningeal leukemia, 63/343
meningioma, 19/358
meningomyelitis, 19/377
metastasis, 69/167
motor symptom, 19/56
multiple myeloma, 18/255, 19/366, 20/13, 111,
    39/135, 141, 63/392, 71/440
multiple sclerosis, 19/376
myelitis necroticans, 19/378
neuralgic stage, 19/348
neurinoma, 19/358
neuroastrocytoma, 19/363
neurocysticercosis, 35/303, 52/530
neurogenic sarcoma, 19/369
nocardiosis, 35/519, 52/449
non-Hodgkin lymphoma, 63/349
oligodendroglioma, 19/361
osteitis deformans (Paget), 19/172, 20/782,
    38/366, 46/400
osteoblastoma, 19/367
osteochondroma, 19/174
osteochondrosis, 19/372
osteoid osteoma, 19/173, 367
osteoporosis, 19/365
osteosarcoma, 19/368
pachymeningitis, 52/188
parasite, 19/371
paravertebral soft tissue, 19/163-166
plasmacytoma, 71/440
polycythemia vera, 55/468
pseudobasilar impression, 19/336
reticulosarcoma, 19/368
retinoblastoma, 19/361
schistosomiasis, 52/538
spina bifida occulta, 19/159
spinal angioma, 32/487
spinal cord trauma, 12/557
spinal cord vascular disease, 12/570
spinal ependymoma, 19/360
spinal epidural empyema, 19/370

spinal ganglioneuroma, 19/363
spinal glioblastoma, 19/361
spinal lymphogranulomatosis, 19/366
spinal medulloblastoma, 19/361
spinal sympathicoblastoma, 20/112
spongioblastoma, 19/360
stage, 19/347-349, 20/205
subacute combined spinal cord degeneration,
   19/378
subdural lymphoma, 63/349
sympathicoblastoma, 19/363
symptom, 19/24
syringomyelia, 19/378
tabes dorsalis, 19/377
teratoma, 19/363
thalassemia, 63/257
thalassemia major, 42/629
thoracolumbar vertebral column injury, 61/92
topical diagnosis, 19/350
transection syndrome, 20/205
transverse myelitis, 19/376
tuberculous spondylitis, 19/370
tumor site, 19/351
tumor type, 19/52, 351
vascular tumor, 19/362
venous pressure, 69/170
vertebral compression, 19/365
vertebral degenerative process, 19/372
vertebral enchondroma, 19/367
vertebral hemangioma, 19/368
vertebral joint ganglion, 20/128
vertebral lesion, 19/364
vertebral metastasis, 19/369
vertebral osteolysis, 19/166
vertebral osteoma, 19/367
vertebral osteomyelitis, 19/369
vertebral tumor, 19/367
Waldenström macroglobulinemia, 63/396
Spinal cord concussion
   definition, 25/47, 61/37
   function recovery, 25/275
   history, 23/3, 25/71
   missile injury, 25/218
   pathogenesis, 25/71
   princeps record, 25/44
   recovery, 2/180
Spinal cord conduction
   experimental injury, 25/11
Spinal cord contusion
   acute stage, 25/76
   cervical, 25/273
   definition, 61/37
   edema, 25/47, 76

hemorrhagic necrosis, 25/47
petechial hemorrhage, 25/47
spinal cord hemorrhage, 25/76
Spinal cord cyst
   see also Syringomyelia
   classification, 25/90
   sympathetic sign, 25/386
   traumatic cervical, 25/86, 90, 278
Spinal cord degeneration
   subacute combined, see Subacute combined spinal
      cord degeneration
Spinal cord disease
   hyponatremia, 75/581
   osmoregulation, 75/581
Spinal cord edema
   blood flow disorder, 12/557
   corticosteroid, 24/631
   organotin intoxication, 64/136
   spinal cord vascular disease, 12/574
   spinal shock, 12/575
   swelling, 12/557
Spinal cord electrotrauma
   amyotrophic lateral sclerosis, 61/194
   brain hemorrhage, 61/192
   cardiac arrest, 61/192
   coma, 61/192
   concept, 61/191
   epidemiology, 61/193
   keraunomyelopathy, 61/195
   keraunoneurosis, 61/195
   keraunoparalysis, 61/192
   lightning parameter, 61/191
   motoneuron disease, 61/194
   natural electricity, 61/192
   paraplegia, 61/195
   physical parameter, 61/191
   polyneuropathy, 61/192
   technical electricity, 61/192
   tetraplegia, 61/195
   transverse myelopathy, 61/195
Spinal cord embolism
   aorta arteriosclerosis, 63/41
   atrial myxoma, 63/41, 98
   diabetes mellitus, 63/42
   fibrocartilaginous disc embolus, 12/601, 63/42
   malaria, 63/42
   prevalence, 63/41
   schistosomiasis, 63/42
Spinal cord experimental injury
   see also Spinal cord injury
   biochemical effect, 25/13
   biophysics, 25/45
   Brown-Séquard syndrome, 25/22

conduction, 25/11
edema, 25/22
electrical event, 25/10
evoked potential, 25/10
H-reflex, 25/13
hemorrhagic necrosis, 25/21
history, 25/45
lesion sequence, 25/9
microangiography, 25/17
microcirculation, 25/20
model, 25/76, 82
myelotomy, 25/22
neuropathologic history, 25/44
noradrenalin, 25/13
pathology, 25/18
spinal cord blood flow, 25/14, 19
tissue carbon dioxide, 25/17
tissue oxygen, 25/17, 19
tissue perfusion, 25/17
vessel, 25/20
Spinal cord granuloma
schistosomiasis, 52/537
Spinal cord gumma
see also Neurosyphilis
symptom, 52/278
Spinal cord hemangioblastoma
Von Hippel-Lindau disease, 14/59, 244, 248,
50/375, 68/498, 507, 575
Spinal cord hemangioma
imaging, 67/194
spinal epidural space, 19/15
Spinal cord hemorrhage
aorta aneurysm, 63/37
decompression myelopathy, 61/222
hypoprothrombinemia, 38/77
lumbar puncture, 61/160
paraplegia, 61/160
spinal cord contusion, 25/76
spinal cord vascular disease, 12/577
Spinal cord hyperthermia
multiple sclerosis, 9/413
Spinal cord infarction, see Myelomalacia
Spinal cord infection
bacterial, 33/187-194
traumatic, 25/60
Spinal cord injury
see also Paraplegia, Spinal cord experimental
injury, Spinal cord lesion, Spinal injury and
Spinal shock
abdominal aorta aneurysm, 26/74
abscess, 25/61
acute abdomen, 26/211
acute stage, 61/39

acute stage urinary bladder, 61/299
adjustment vs age, 61/548
adjustment vs education, 61/548
adjustment vs injury severity, 61/548
adjustment vs patient personality, 61/548
adjustment stage, 61/537
age, 25/141, 184
age vs mortality, 61/504
alcohol, 61/369
aldosterone, 26/320, 325
alpha motor system, 61/367
alternating synergics, 26/250
American Spinal Injury Association Motor Scale,
61/424
4-aminopyridine, 61/401
anal reflex, 61/295
analgesic agent, 26/233
anemia, 26/380, 398
angiography, 61/515
angiotensin II, 61/444
annual incidence, 61/22
anocutaneous reflex, 61/295
anterior cervical cord syndrome, 25/270, 283
anterior cord syndrome, 25/270, 283, 335, 359
anticoagulant, 26/316
aorta aneurysm, 26/70, 74
aorta coarctation, 26/72
aorta rupture, 26/72
apnea, 61/262
arterial lesion, 61/111
arterial vascular injury, 26/63-79, 61/527
artificial respiration, 26/343
artificial sphincter, 61/496
artificial ventilation, 61/265
aspartic acid, 61/367
associated injury, 26/188, 224
atelectasis, 26/343, 61/264
autonomic dysfunction, 26/253, 313-333
autonomic dysreflexia, 61/447
autonomic nervous system, 75/569
axon regeneration, 61/407
axon survival, 61/400
axonal dysfunction, 61/401
baclofen, 61/368
baclofen pump, 61/489
baroreflex, 61/444
basal cardiovascular state, 61/440
biochemistry, 26/381-408
birth control, 61/342
birth incidence, 25/44, 94-96, 155-170
bladder function, 26/412-426, 61/291
bladder neck resection, 61/304
bladder reflex activity, 61/301

blast injury, 25/221
blood cell count, 26/379
blood gas, 26/385
blood pressure, 26/205, 313-333, 61/439
body position, 61/263
bowel dysfunction, 75/646
brachial monoplegia, 25/274
bradycardia, 61/436
brain blood flow, 26/328
Brown-Séquard syndrome, 25/273
bulbar cervical dissociation pattern, 25/275
bulbocavernosus reflex, 61/295
calcium, 26/382, 387, 390, 402
cardiac arrest, 61/439, 446
cardiovascular drug response, 61/438
cardiovascular posture response, 61/441
cardiovascular system, 26/313-333, 61/435
catecholamine, 26/317
catheter, 26/208-211
catheterization, 25/290, 61/299
cauda equina injury, 61/47
cause, 25/142, 176, 259, 61/21
cause of death, 61/500
central cervical cord syndrome, 25/65, 271, 283,
    335, 359
central cord necrosis, 61/40
central hemorrhage, 61/39
central venous pressure, 61/452
cervical, 25/145, 269, 282, 287, 335, 359, 26/337,
    342
cervical discography, 26/187
cervical extension, 61/37
cervical flexion, 61/37
cervical spinal cord compression, 61/37
cervical spondylotic myelopathy, 26/97-112
change segmental level, 26/308
child, see Juvenile spinal cord injury
chloride, 26/381, 387
cholesterol, 26/382
chronic posttraumatic change, 61/528
chronic stage, 61/41
ciliospinal reflex, 25/386
circulatory disturbance, 25/262
classification, 25/46, 145, 147, 151
clinical trial, 61/399
clinicopathologic correlation, 61/47
closed, 25/61
coagulopathy, 26/391
coitus, 61/315
colliquation necrosis, 25/58
complete, see Complete cord lesion
complication, 25/182, 61/42
compression, 25/72, 260

computer assisted tomography, 61/515
concussion, 61/37
conservative management, 26/285-306
contact pressure, 25/261
contracture, 25/289
contusio cervicalis posterior, 25/273, 284
contusion, 61/37
coping success, 61/543
cordotomy, 61/369
coronary blood flow, 26/206
corticosteroid, 26/222, 61/399
coughing, 61/262
creatine kinase, 26/384, 403
creatinine, 26/389, 402
Crede expression, 61/300
cruciate paralysis, 25/275, 391-393
cure research, 61/479
cutaneous urinary diversion, 61/307
cutaneous vasomotor activity, 61/280
cystic necrosis, 25/86, 90
cystometry, 61/296
dantrolene, 61/368
data base, 61/21
deafferentation, 61/370
degenerating sequence, 61/480
delayed cord lesion, 25/93, 276
demyelination, 61/43
dentate ligament, 25/271
dermatographism, 26/259
detrusor areflexia, 75/576
detrusor dyssynergia, 61/296
diagnosis, 25/287, 26/185
diaphragm function, 26/335, 337
diaphragm paralysis, 61/266
diaphragmatic pacing, 61/261
diazepam, 61/368
dioxide consumption, 61/279
direct, 25/61
disability acceptance, 61/546
discomplete, see Discomplete spinal cord injury
discomplete lesion, 61/43
dissecting aneurysm, 26/71, 74
distress, 61/537
dopamine β-mono-oxygenase, 61/448
dorsal spinal root entry zone microcoagulation,
    61/371
drug induced azoospermia, 61/317
drug induced hypothermia, 61/283
dysbarism, 23/611
early care, 25/284
ECG, 61/286, 439
edema, 25/54, 79, 262
ejaculation, 61/315, 322

electrical injury, 23/711
electroejaculation, 61/323
electrophrenic respiration, 61/267
electrophrenic respiration candidate selection, 61/268
electrophrenic respiration results, 61/271
electrostimulation, 61/491
embolism, 26/68, 207
employment, 61/550
enzyme level, 26/383-385, 400, 405
epidemiology, 25/141-143, 61/499
epidural empyema, 25/60
epidural hemorrhage, 61/38
erection, 61/313
erection dysfunction treatment, 61/320
erroneous depression assessment, 61/540
erythrocyte sedimentation rate, 26/380
exercise induced hyperthermia, 61/284
experimental, *see* Spinal cord experimental injury
experimental therapy, 61/399
extension, 26/252
external sphincterotomy, 61/304
eye movement disorder, 25/387
female sexual adjustment, 61/334
female sexual function, 61/331
fertility, 61/313, 316
fibrinolysis, 26/392
first aid, 25/285
flexion, 26/252
Flying Squad, 26/297
food ingestion, 61/446
Frankel Scale, 61/422
functional motor level, 61/262
functional recovery, 61/421
functional restoration, 61/489
GABA, 61/367, 401
gamma motor system, 61/367
gastrointestinal dysfunction, 75/577
genitourinary examination, 61/294
glutamic acid, 61/367
glycine, 61/367
gunshot injury, 25/49
haptoglobin, 26/380, 398
head injury, 61/545
heart rate, 26/313-333, 61/439
heat loss, 61/280
heat production, 61/278
hematocrit, 26/379, 398
hematology, 26/377-381, 397-399
hematomyelia, 61/523
hemoglobin, 26/379, 397
hemorrhage, 25/266
hemorrhagic necrosis, 25/30-42, 80, 84, 266

history, 25/44, 61/506
hormone change, 61/316
Horner syndrome, 25/383, 415, 417
hyperacute stage, 61/39
hyperkalemia, 61/286
hyperthermia, 26/369, 372, 61/283
hyperventilation, 61/263, 445
hypogastric plexus stimulator, 61/493
hypoglycemia, 61/446
hypothermia, 26/368, 372, 61/282
impotentia coeundi, 61/315
incidence, 25/41, 43, 141, 175
incomplete lesion, *see* Incomplete cord lesion
indirect, 25/61
infection, 25/61
initial clinical appraisal, 26/185
intelligence, 61/543
intestinal function, 26/258
intestinal tract, 26/211, 269
intramural ureter obstruction, 61/308
intrathecal agent, 61/368
intravascular consumption coagulopathy, 26/391
intubation response, 61/437
joint contracture, 25/184
K opiate receptor blocker, 61/399, 403, 406
17-ketosteroid, 26/397, 402, 451
kidney function, 26/390, 402
laceration, 61/37
laminectomy, 25/304, 420, 459
late sequela, 25/69, 278
leptomeningitis, 25/60
life expectancy, 25/193, 26/310, 61/499
life quality, 61/541
life satisfaction, 61/547
life table method, 61/507
lightning injury, 23/703, 721
limb stimulator, 61/493
lipid peroxidation inhibitor, 61/399
liquefaction necrosis, 25/80, 84
long-term recovery, 61/430
long-term urinary management, 61/302
lower urinary tract, 61/291
lumbar puncture, 25/179, 303, 26/186
lumbodorsal, 25/453
lung capacity, 26/337
lung volume, 26/337, 341
macroscopic pathology, 25/27-42
male sexual function, 61/313
marital status, 61/543
mass reflex, 61/447
maturity stage, 61/537
mechanism, 25/62, 181
meningeal thickening, 25/40, 59, 89, 94, 281

menstruation, 61/342

mental depression, 61/537

meta studies, 61/424

metabolism, 25/288

N-methylaspartic acid receptor blocker, 61/399, 406

methylprednisolone, 61/399, 402

microcirculation, 25/57, 83

micturition, 26/414

micturition mechanism, 61/294

miscellaneous pattern, 25/276

missile, *see* Spinal cord missile injury

missile associated lesion, 61/256

missile ballistics, 61/254

missile composition, 61/254

missile injury incidence, 61/253

missile injury sex ratio, 61/253

missile injury treatment, 61/257

monosialoganglioside, 61/399, 406

morphine, 61/368

mortality, 25/142, 281, 26/310, 61/291, 500

mortality cause, 61/261

myelography, 25/179, 303, 395, 418, 26/186, 268, 61/515

myelomalacia, 25/80, 84

myelotomy, 25/305, 61/369

naloxone, 61/399

National Acute Spinal Cord Injury Study Scale, 61/423

nerve graft, 61/480

nerve regeneration, 61/481

neurogenic osteoarthropathy, 26/277, 501-519

neurologic examination, 26/185

neurologic recovery, 61/421

neuropathic limb fracture, 26/276

neuropathology, 25/44-104, 61/37

neuropeptide Y, 61/450

neurostimulation, 61/372, 491

noradrenalin, 26/319, 325

nuclear magnetic resonance, 61/515

nuclear magnetic resonance based prognosis, 61/427

nystagmus, 25/385

obturator neurotomy, 61/369

occupational therapy, 26/233, 61/457

ocular dysfunction, 25/381

orthostatic hypotension, 22/262, 26/318, 61/442

outcome, 61/457

overstretching, 25/261, 268

pain, 26/489-499, 75/579

panting, 26/367, 61/282

paradoxical reflex, 26/250

paraplegia, 26/252

parathyroid function, 26/397

pathogenesis, 25/45, 180

pathology, 25/41, 180, 61/21

patient adjustment, 61/537

patient personality characteristics, 61/542

patient positioning, 26/218-220

penetrating brain injury, 25/48

penetrating injury pathology, 61/254

penis prosthesis, 61/321, 496

percutaneous thermal rhizotomy, 61/370

perinatal, 25/94

phantom sensation, 26/489-499

pharmacologic management, 61/430

pharmacotherapy, 26/222, 61/368

phenol, 61/369

phosphorus, 26/382, 387, 389, 401

phrenic nerve stimulator, 61/491

phrenic nerve surgery, 61/270

phrenic nerve viability, 61/269

physical rehabilitation, 61/457

physiotherapy, 26/232, 344, 61/457

piloerection, 26/367, 61/280, 282

pilomotion, 61/276

pilomotor response, 61/280

pinching, 25/260

plasma adrenalin, 61/436

plasma aldosterone, 61/443

plasma cortisol, 26/397, 402

plasma noradrenalin, 61/436

plasma renin, 61/443

plethysmography, 61/285

pneumatic mattress, 26/221

pneumonia, 61/261

positive end expiratory pressure, 61/265

posttraumatic myelopathy, 25/89

posttraumatic progression, 61/45

posttraumatic syringomyelia, 12/592, 604, 50/454, 61/375

postural control, 26/524

potassium, 26/381, 387

pressure sore, 25/188, 289, 26/223, 463-476, 61/347

principle, 61/457

progressive cord lesion, 25/276

proliferating cell, 61/480

prosthesis, 61/489

protein level, 26/382, 385

protirelin, 61/403, 406

pseudomeningocele, 61/528

psychiatric disorder, 26/234

psychology, 61/537

psychosocial effect, 61/334

pulmonary complication, 26/335

pulmonary infection control, 61/264
Queckenstedt test, 26/186
radiation myelopathy, 26/81-95, 63/43
radiology, 26/186, 61/515
rebellion stage, 61/537
recovery, 61/46, 399, 421
recovery determining factor, 61/422
recovery vs lesion type, 61/428
recovery scoring, 61/422
recovery vs spinal syndrome, 61/428
recovery substrate, 61/400
recovery vs treatment type, 61/428
reflex depression, 26/245
reflex pattern, 26/248
reflex return, 26/246-248, 251
reflex vasodilatation, 61/283
regeneration, 25/99
regenerative therapy, 61/406
rehabilitation, 25/186, 192, 26/310, 521-540,
    61/457
rehabilitative sport, 26/521-540
remyelination, 61/402
renal function, 61/291
renin, 26/320, 325
renin release, 61/443
renin response, 61/443
Renshaw cell, 61/367
repair stage, 25/59, 86
reproductive dysfunction, 75/576
residual urine volume, 61/296
resorption, 25/58
respiration, 26/204, 253
respiration pathophysiology, 61/262
respiratory dysfunction, 26/337, 342
respiratory force, 26/339
respiratory management, 61/261
respiratory muscle, 26/336
respiratory pressure, 61/263
respiratory system, 26/335-350
respiratory test, 61/264
reticulocyte, 26/380, 398
rhizotomy, 61/369
rib cage, 26/338, 346
rib cage motion, 61/263
sacral anterior root stimulator, 61/492
secondary degeneration, 25/60
segmental frequency distribution, 61/22
segmental pathology difference, 61/22
self-concept, 61/547
sensory disturbance, 26/252
sensory extinction, 3/193
sensory microrootlet section, 61/371
serum level, 26/377-408

serum potassium, 26/401
sex difference, 61/546
sex ratio, 25/142
sexual capability, 61/315
sexual counseling, 61/319
sexual function, 26/437-461
sexual function test, 61/319
sexual intercourse, 61/313
sexuality, 61/545
shivering, 61/279
shock, 26/187
skin care, 26/223
sociology, 61/537
sodium, 26/381, 387
spasticity, 26/233, 323-333, 477-488, 61/367
sperm quality, 61/316
spinal cord atrophy, 25/40
spinal cord autoregulation, 25/14
spinal cord blood flow, 26/313-333
spinal cord diaschisis, 61/367
spinal deformity, 26/159-174
spinal injury, 25/177, 184
spinal muscular atrophy, 59/370
spinal nerve root, 61/41
spinal nerve root avulsion, 61/526
spinal nerve root compression, 61/525
spinal nerve root lesion, 61/42
spinal nerve root neuroma, 61/41
spinal shock, 26/203, 243-263, 61/439
spinal subarachnoid hemorrhage, 25/70, 267
stab injury incidence, 61/253
stab injury sex ratio, 61/253
stab injury treatment, 61/257
stab wound associated lesion, 61/255
stab wound pathology, 61/255
steroid antioxidant, 61/405
stress ulcer, 26/351-353
subacute stage, 61/40
subdural empyema, 25/60
subtotal, 25/270
Sunnybrook Cord Injury Scale, 61/423
surgical treatment, 61/369
survival, 25/41, 61/272
survival study, 61/508
sweating, 26/357, 366, 369, 61/276, 281
sympathetic nervous system, 61/275
sympathoneural activation, 61/437
systemic hypotension, 61/436
tachycardia, 26/320, 61/436
terminology, 25/145, 147
testis histology, 61/317
tethered cord, 61/529
thermoregulation, 26/211, 260, 355-376,

61/275-287
thermoregulation management, 61/286
thermoregulatory dysfunction, 75/578
thoracic, *see* Thoracic spinal cord injury
thoracic aortic aneurysm, 26/68, 70
thoracolumbar, *see* Thoracolumbar spinal cord
   injury
thrombosis, 26/207, 315, 394
thyroid function, 26/395, 400
time factor, 61/46
tirilazad, 61/405
tizanidine, 61/368
Toronto data, 61/508
total lung capacity, 61/263
tracheal stimulus response, 61/437
tracheostomy, 26/204, 344, 61/267, 437
transport, 25/285
trauma category, 61/499
traumatic arachnoiditis, 61/41
tubular necrosis, 25/80, 84
turning frame, 26/221
upper urinary tract, 61/291
urea, 26/382, 387, 401
urethral pressure, 61/297
urethral resistance restoration, 61/306
urinary calculi, 61/308
urinary complication, 25/289
urinary incontinence, 61/291
urinary retention, 75/576
urinary tract, 26/208, 269-274, 409-436
urine chemistry, 26/387
urodynamics, 61/296
uroflow measurement, 61/297
urology, 25/190
US survival data, 61/511
vacuum constrictor devices, 61/321
Valsalva maneuver, 61/445
vascular tone, 61/436
vasculitis, 26/67
vasodepressor agent, 61/454
vasomotor control, 26/259
vasopressor agent, 61/451
ventilator care, 61/261
vesicoureteric reflux, 61/308
vesicourethral neuropathy forms, 61/295
vesicourethral reflex, 61/291
visceral stimulation, 61/451
vital capacity, 26/337, 61/262, 266
wallerian degeneration, 61/480
Yale Scale, 61/423
Spinal cord injury prognosis, 26/307-311
   complete cord lesion, 26/307
   conservative management, 26/304

general, 26/307
imaging data, 61/427
incomplete lesion, 26/308
life expectancy, 26/310
mental function, 25/281
sexual function, 25/281, 26/311
Spinal cord intermittent claudication, 12/507-547,
   20/611-801
   see also Cauda equina intermittent claudication
      and Lumbar vertebral canal stenosis
   aorta disease, 20/796, 32/338, 63/41
   aorta thrombosis, 63/41
   aortoiliac atherosclerosis, 63/41
   history, 20/611-613
   ischemic myelopathy, 63/40
   lumbar vertebral canal stenosis, 20/749, 760,
      32/337-339
   myelomalacia, 63/40
   nature, 20/620-624
   pain, 20/756, 32/337-339
   pathogenesis, 20/749
   pseudoxanthoma elasticum, 55/452
   spinal arachnoiditis, 12/535
   spinal cord lesion, 20/613-620
   spinal nerve root compression, 20/749
   spinal nerve root ischemia, 20/749
   steal syndrome, 32/487
   total transverse spinal cord syndrome, 12/498
Spinal cord ischemia, *see* Ischemic myelopathy
Spinal cord laceration
   definition, 61/37
Spinal cord lesion
   see also Spinal cord injury
   cervical vertebral column injury, 61/27, 30
   clinical symptomatology, 2/178-213
   complete, *see* Complete cord lesion
   cystometry, 26/424
   dentatorubropallidoluysian atrophy, 60/615
   Friedreich ataxia, 21/322-324
   Guttmann sign, 2/190, 202
   Horner syndrome, 2/209
   kit test, 2/205
   late, *see* Late cord lesion
   mass reflex, 2/196
   orthostatic hypotension, 63/153
   paralysis, 2/190, 202
   phantom limb, 45/398
   Philippson reflex, 2/184
   priapism, 19/34
   principal, *see* Principal spinal cord lesion
   pyramidal syndrome, 2/210
   spina bifida, 50/485, 489
   spinal cord intermittent claudication, 20/613-620

symptomatology, 2/178-212
trichinosis, 52/567
tuberous sclerosis, 14/47
Spinal cord malformation
  Kii Peninsula amyotrophic lateral sclerosis,
    22/371
  neurofibromatosis type I, 14/38
Spinal cord metastasis, 20/415-430
  bone scan, 69/178
  brain metastasis, 69/191
  breast cancer, 69/205
  cauda equina, 69/176
  chemotherapy, 69/184
  clinical, 69/173
  CNS involvement, 68/507
  CSF, 20/419
  differential diagnosis, 69/177
  erythrocyte sedimentation rate, 20/419
  glucocorticoid, 69/183
  imaging, 67/204
  incidence, 69/172
  intradural extramedullary, 69/172
  intramedullary, 69/198
  leptomeningeal, 69/198
  location, 69/171
  lung carcinoma, 69/169
  melanoma, 20/425
  myelography, 69/179
  nuclear magnetic resonance, 69/179
  pain, 69/174
  pathogenesis, 69/168
  prognosis, 69/180
  prostate cancer, 69/178
  radiotherapy, 69/182
  sarcoma, 20/425
  steroid, 69/183
  surgery, 69/184
  surgical treatment, 20/429
  survey, 69/167, 71/630
  systemic tumor, 71/615
  transection syndrome, 20/418
  treatment, 69/181
Spinal cord missile injury
  associated injury, 25/212
  ballistics, 25/209
  classification, 25/49
  clinical features, 25/211, 218
  neurology, 25/211, 218
  pathogenesis, 25/210
  pathology, 25/212
  prognosis, 25/219
  radiology, 25/212, 214
  surgery, 25/214, 217

treatment, 25/213, 217
Spinal cord necrosis, see Myelomalacia
Spinal cord regeneration
  animal, 25/98, 103
  axon, 25/100, 103
  fiber origin, 25/102
  functional recovery, 25/103
  mammal, 25/99, 103
Spinal cord repair
  action, 61/483
  basic data, 61/479
  embryonic nerve tissue, 61/485
  growth inhibitor antibody, 61/485
  myelin associated axon growth protein, 61/484
  myelin basic protein antibody, 61/481
  nerve growth factor, 61/484
  nerve growth inhibitory protein, 61/482
  scar formation, 61/485
  tubulin, 61/483
Spinal cord stab wound
  age, 25/198
  associated injury, 25/201, 204
  bone fragment, 25/52
  clinical features, 25/200
  cord lesion, 25/199, 201
  incidence, 25/197
  knife, 25/52
  mechanism, 25/199
  neck blood vessel, 25/203
  pathology, 25/198
  result, 25/206
  sex ratio, 25/198
  surgery, 25/206
  thoracic injury, 25/204
  treatment, 25/206
  weapon, 25/52, 198, 202
Spinal cord stimulation
  posterior cervical, see Posterior cervical spinal
    cord stimulation
Spinal cord teratoma
  statistic, 19/16
  vertebral column, 19/363
  yolk sac tumor, 32/361
Spinal cord transection, see Transverse spinal cord
  lesion
Spinal cord transient ischemic attack, see
  Intermittent spinal cord ischemia
Spinal cord transverse lesion, see Transverse spinal
  cord lesion
Spinal cord transverse syndrome, see Transverse
  spinal cord syndrome
Spinal cord trauma
  aorta dissection, 61/76

myelomalacia, 12/557
neurologic intensive care, 55/204
spinal cord compression, 12/557
Spinal cord tumor
    see also Spinal tumor
    age, 19/6
    amyotrophy, 19/69
    angiography, 19/229-243
    annual incidence, 19/2
    arteriovenous malformation, 12/248
    astrocytoma, 68/498
    Brown-Séquard syndrome, 19/24, 30
    cavernous malformation, 68/574
    child, 19/10
    chondrosarcoma, 68/524
    congenital, *see* Congenital spinal cord tumor
    diagnosis, 67/161
    EMG localization, 19/280
    ependymoma, 68/507
    epidemiology, 68/497
    ethnic group, 19/8
    Ewing sarcoma, 68/525
    extramedullary, 67/161
    frequency, 19/1
    gas myelography, 19/181
    geographic distribution, 19/8
    giant cell tumor, 68/516
    gliofibroma, 68/383
    hemangiopericytoma, 68/576
    Hirayama disease, 59/114
    histologic type, 20/179-182
    Horner syndrome, 19/59
    imaging, 68/500
    initial symptom, 19/9
    intensity duration curve, 19/279
    intramedullary, 67/161
    lymphomatoid granulomatosis, 63/424
    mortality, 19/2
    muscular atrophy, 19/26
    neurinoma, 68/543
    neurofibromatosis type I, 14/136
    neurogenic bladder, 19/34
    nystagmus, 19/40, 59
    osteoblastoma, 68/513
    osteoid osteoma, 68/513
    osteosarcoma, 68/523
    papilledema, 19/38, 88
    pathology, 68/498
    plain X-ray, 19/139-177
    postlaminectomy deformity, 68/505
    prognosis, 19/10
    proteinocytologic dissociation, 20/217
    radiotherapy, 68/504
    relative frequency, 20/181
    segmental lesion distribution, 20/182
    sex predominance, 19/7
    sex ratio, 19/7
    site frequency, 19/5
    spina bifida, 32/357, 359
    spinal monitoring, 68/504
    spinal muscular atrophy, 59/114
    statistic, 19/1-22
    surgery, 68/506
    survey, 68/497
    symptom, 19/9, 68/499
    symptomatology, 19/23-49
    syringomyelia, 19/9, 61/377
    treatment, 68/502, 506
    tuberculous meningitis, 32/362
    type frequency, 19/3
    Von Hippel-Lindau disease, 16/31, 43/21
    xanthochromia, 19/62
Spinal cord vascular disease
    see also Spinal artery injury
    anterior spinal artery, 12/499
    aorta lesion, 12/592
    arterial obstruction, 12/497
    blood flow reduction, 12/498
    cause, 12/500-503
    chemical trauma, 12/613
    classification, 12/497
    clinical syndrome, 12/498
    cystic degeneration, 12/604
    epidural hematoma, 12/558
    hematomyelia, 12/557, 580
    ischemic myelopathy, 12/592
    late sequela, 12/602
    myelomalacia, 12/582
    pathology, 12/603
    posterior spinal artery syndrome, 12/499
    spinal cord compression, 12/570
    spinal cord edema, 12/574
    spinal cord hemorrhage, 12/577
    spinal epidural hematoma, 12/558
    spinal subarachnoid hemorrhage, 12/567
    spinal subdural hematoma, 12/565
    total transverse syndrome, 12/498
Spinal cord vascular malformation, 12/548-554
    see also Spinal angioma *and* Spinal arteriovenous
        malformation
    cause, 12/550
    classification, 12/548, 68/569
    clinical features, 12/551
    congenital, 32/355-386
    differential diagnosis, 12/551
    laboratory finding, 12/552

roentgenology, 12/552
spinal subarachnoid hemorrhage, 19/15
treatment, 12/553
Spinal cord vascularization, 12/478-490
see also Spinal artery injury
anterior radicular artery, 12/481
anterior spinal artery, 12/483
capillary, 12/487
development, 12/478
diagram, 25/456
extraspinal artery, 12/479
pial arterial plexus, 12/484
posterior radicular artery, 12/482
posterior spinal artery, 12/484
radicular artery, 12/481
Rolando artery, 12/486
spinal artery system, 20/501-503
venous system, 12/487, 20/503
Spinal cord vein
anatomy, 55/97
Spinal cordectomy
syringomyelia, 50/460
Spinal cyst
Brown-Séquard syndrome, 32/437
neurocysticercosis, 20/92, 52/530
Spinal deformity
arthrogryposis multiplex congenita, 32/517
bony damage, 26/169
cervical spinal canal, 25/278-281
child, 25/183, 185, 192
complete cord lesion, 26/166, 172
congenital, 19/159-163
hereditary sensory and autonomic neuropathy type
IV, 60/12
juvenile spinal cord injury, 25/183
laminectomy, 25/462
Milwaukee bracing, 41/478
muscle tone, 26/165
muscular imbalance, 25/183
neurofibromatosis type I, 32/141-143
poliomyelitis, 40/337
posttraumatic, 25/183
progressive cord lesion, 25/277
rehabilitation, 25/192
respiratory complication, 25/187
spinal cord injury, 26/159-174
spinal injury, 26/159-174
surgery, 25/186
traumatic, *see* Traumatic spine deformity
treatment, 25/185
vertebral fusion, 26/171
Spinal dermoid
chemical meningitis, 19/364

meningitis, 19/364
myelography, 19/364
Spinal dura mater
anatomy, 26/5
Spinal dysraphism, *see* Spina bifida
Spinal eosinophilic granuloma
biopsy, 20/111
differential diagnosis, 20/111
treatment, 20/111
Spinal ependymal cyst
age, 20/89
case series, 20/89-90
cervical spinal cord compression, 32/440
clinical features, 20/90
location, 20/90
terminal ventricle cyst, 20/91
Spinal ependymoma, 20/353-384
age, 19/360, 20/354
angiography, 20/374
astrocytoma grade II, 20/362
benign intracranial hypertension, 20/367
blepharoplast, 20/362
Brown-Séquard syndrome, 19/360, 20/368
calcification, 19/360, 20/371
cauda equina, 19/359, 20/354-356
cause, 20/354
chemotherapy, 20/382
child, 20/366
clinical features, 20/364
coccygodynia, 20/365
CSF, 20/369
CSF protein, 19/360, 20/369
dense variety, 20/359
differential diagnosis, 19/359, 20/375
dissociated sensory loss, 20/368
Elsberg sign, 20/370
entire spinal cord, 20/378
enuresis, 20/367
extrusion method, 20/378
factor deciding mortality, 20/381
fasciculated variety, 20/360
filum terminale, 19/359, 20/356, 363, 375
giant tumor (Elsberg), 20/376
histogenesis, 20/363
histologic classification, 20/353
histologic diagnosis, 20/362
histologic distinction, 20/361
histology, 20/359, 361
Homer Wright rosette, 20/359
intramedullary type, 20/355
kyphoscoliosis, 20/366
lesion level, 20/278
metastasis, 20/357

microsurgery, 20/384
mortality, 20/379
motor deficit, 20/367
motor disorder, 19/360
multiple, *see* Multiple spinal ependymoma
myelography, 19/360, 20/372
myxopapillary variety, 20/361
Nonne-Froin syndrome, 20/369
oligodendroglioma, 20/362
onset, 20/364
osteolysis, 20/370
pain, 19/360, 20/364-366
papillary placenta like variety, 20/360
paresis, 19/360, 20/365
pheochromocytoma, 20/358
pregnancy, 20/366
prevalence, 20/354
prognosis, 20/363
Queckenstedt test, 20/369
radiology, 20/370
radiotherapy, 19/359, 20/378
relapse, 20/384
rosette, 20/357
segmental lesion distribution, 20/182
sensory loss, 19/360, 20/368
sex ratio, 20/354
sphincter disturbance, 20/367-369
spinal cord, 19/12
spinal cord compression, 19/360
spine rigidity, 20/366
Stookey test, 20/369
subarachnoid hemorrhage, 20/367
surgical indication, 20/376
syringomyelia, 20/357
terminal ventricle (Krause), 20/363
trauma, 20/365
treatment, 20/375
virus like inclusion, 20/360, 364
Spinal epidermoid
    anterospinal, 20/79
    chemical meningitis, 19/364
    meningitis, 19/364
    myelography, 19/364
    retrospinal, 20/79
    spinal cord, 19/16
Spinal epidural angioma
    surgery, 20/463
    vascular malformation, 20/441
Spinal epidural cyst, 20/137-168
    age, 20/143
    ankylosing spondylitis, 20/163
    associated disease, 20/138-141, 143
    case series, 20/138-142

clinical features, 20/146
congenital type, 20/137
development, 20/144
distichia, 20/141
distichiasis lymphedema syndrome, 43/377
echinococcosis, 20/164
ependymal cyst, 20/167
hydatid cyst, 20/164
location, 20/138
microscopy, 20/144
multiple cyst, 20/144
race, 20/137
radiology, 20/146
schema, 20/167
Scheuermann disease, 20/137, 146
spinal perineural cyst, 20/146
twin study, 20/143
vertebral joint ganglion, 20/163
Spinal epidural empyema
    Actinomyces israelii, 52/187
    alcoholism, 52/73
    antibiotic agent, 52/188
    aspergillosis, 52/187, 380
    back pain, 52/187
    cause, 52/186
    classification, 33/187
    clinical features, 33/189, 52/187
    clinicopathologic features, 33/155-157
    Coccidioides immitis, 52/187
    coccidioidomycosis, 52/414
    computer assisted tomography, 52/188
    CSF, 52/73, 187
    diagnosis, 33/189, 52/187
    Echinococcus granulosus, 52/187
    epidural anesthesia, 52/186
    Escherichia coli, 52/187
    etiology, 33/188
    Fusobacterium necrophorum, 52/187
    iatrogenic neurological disease, 52/186
    incidence, 52/185
    injury, 52/186
    lumbar puncture, 52/186
    metastasis, 52/186
    mucormycosis, 52/187
    Mycobacterium tuberculosis, 52/187
    myelography, 52/188
    neurosurgery, 52/188
    nocardiosis, 52/187
    nuclear magnetic resonance, 52/188
    prognosis, 33/190, 52/188
    Proteus vulgaris, 52/187
    Pseudomonas aeruginosa, 52/187
    radicular pain, 52/187

Salmonella, 52/187
sex ratio, 52/185
site, 52/186
spinal cord compression, 19/370
Staphylococcus, 33/190
Staphylococcus aureus, 52/186
Streptococcus pneumoniae, 52/187
synonym, 19/370
traumatic, 25/60
treatment, 33/190, 52/188
type, 52/186
vertebral osteomyelitis, 52/73, 186
Spinal epidural hematoma, 26/1-26
age distribution, 61/139
anesthesia complication, 61/137
anticoagulation, 20/448, 61/140
birth, 25/95
Brown-Séquard syndrome, 61/138
catheter introduction, 61/140
cause, 12/559-562, 20/448
clinical features, 12/562
clinical type, 61/138
complete cord lesion, 26/21
course, 61/138
differential diagnosis, 12/564
frequency, 19/15
general, 25/29, 47, 266
hemostasis material, 61/142
iatrogenic neurological disease, 61/137
incidence, 25/69
literature review, 61/138
lumbar puncture, 61/159
lumbar puncture level, 61/140
myelography, 26/22, 25
paraplegia, 61/137
pathology, 12/564
prognosis, 12/565
radiology, 12/564
segmental lesion distribution, 61/139
spinal cord vascular disease, 12/558
spinal site, 12/502
symptom, 61/138
treatment, 12/565
Spinal epidural lesion
cauda equina, 19/86
Spinal epidural lipoma
corticosteroid intoxication, 65/555
Spinal epidural lymphoma
back pain, 63/349
Burkitt lymphoma, 63/351
paraplegia, 63/349
radiculopathy, 63/349
Spinal epidural space, 19/113

anatomy, 26/5, 34
blood supply, 20/118
innervation, 26/7
physiology, 26/11
spinal cord hemangioma, 19/15
Spinal epidural tumor, 20/103-130
anatomy, 20/104
angioma, 20/110
benign type, 20/103
Burkitt lymphoma, 20/128
cerebellar medulloblastoma, 20/129
chordoma, 20/129
clinical features, 20/104
CSF, 20/106
differential diagnosis, 20/106
Elsberg sign, 20/107
eosinophilic granuloma, 20/111
ganglion vertebral joint, 20/128
giant cell tumor, 20/103, 128
hemangiopericytoma, 20/103
hemopoiesis, 20/103
Hodgkin disease, 20/113-128
leukemia, 20/103, 112
lipoma, 20/109
lymphogranuloma, 20/103
lymphosarcoma, 20/112
malignancy, 19/63, 67
meningioma, 20/104
metastasis, 20/129
multiple myeloma, 20/111
mycosis fungoides, 20/128
myelography, 19/66, 195, 20/107
myeloscintigraphy, 20/108
neuroblastoma, 20/103
neurofibroma, 20/104
origin, 20/104
osteosarcoma, 20/129
plasmacytoma, 20/103
radiology, 20/106
reticulosarcoma, 20/112
sarcoma, 20/103, 113
spasmodic torticollis, 20/105
spinal medulloblastoma, 20/129
symptom mechanism, 20/105
treatment, 20/108
tumor variety, 20/130
type frequency, 20/104
Spinal extradural space, see Spinal epidural space
Spinal fusion
Duchenne muscular dystrophy, 41/478
posterior, see Posterior spinal fusion
Spinal ganglion
hereditary motor and sensory neuropathy type I,

21/299

hereditary motor and sensory neuropathy type II,
21/299

hereditary sensory and autonomic neuropathy type
IV, 60/13

rachischisis, 32/177

Spinal ganglioneuroma

benign, 19/363

spinal cord compression, 19/363

Spinal glioblastoma

brain tumor, 19/361

spinal cord compression, 19/361

Spinal glioma, 20/323-340

animal, 20/340

astrocytoma, 19/359, 20/329-331, 333, 336

clinical features, 20/336

CSF, 20/231

differential diagnosis, 19/359, 20/338

ependymoma, 19/359

frequency, 20/278

glioblastoma, 19/359, 361, 20/335

histologic type, 20/327

history, 20/323

multiple sclerosis, 20/338

oligodendroglioma, 19/359, 361, 20/332

oncogenesis, 20/340

pachymeningitis cervicalis hypertrophicans,
20/339

pencil tumor, 20/327

prevalence, 20/181, 324

Queckenstedt test, 20/283

radiology, 20/338

relative prevalence, 20/325

Rosenthal fiber, 20/331

segmental lesion distribution, 20/224

space occupying lesion, 20/278

spinal medulloblastoma, 19/359, 361

spongioblastoma, 19/359, 20/327, 329

syringomyelia, 20/327, 339

treatment, 20/339

type, 20/335

Spinal granuloma

cryptococcosis, 52/432

North American blastomycosis, 35/403, 52/389

paracoccidioidomycosis, 52/457

Spinal hemangioblastoma

angiography, 20/449

classification, 19/362

radiotherapy, 20/451

Spinal histiocytosis X

vertebral compression, 19/365

Spinal hyperflexion injury

cervical spine, 25/125, 259

Spinal injury, 12/556-613

see also Cervical vertebral column injury,
Craniospinal injury, Pelvic fracture, Spinal cord
injury, Thoracic spinal cord injury,
Thoracolumbar spine injury, Transverse spinal
cord lesion and Unilateral cervical facet
interlocking

abdominal injury, 26/198, 226

angulation, 25/243

ankylosing spondylitis, 26/175, 38/515-517

anterior column, 25/127

anterolisthesis, 25/243

araeosoma intoxication, 37/67

associated injury, 26/188, 224

asthenosoma intoxication, 37/67

atlas fracture, 25/127, 314

biomechanics, 25/123-140

brain contusion, 26/188

centrostephanus intoxication, 37/67

cervical hyperanteflexion injury, 25/63

cervical hyperdeflexion injury, 25/63, 26/285

cervical spine, 25/227-379

cervical subtotal cord lesion, 25/270

chest injury, 26/190, 204

child, 25/177, 184, 187, 191

chordata intoxication, 37/69

classification, 25/127, 145, 148, 26/289

combined mechanism, 25/67

compression, 25/67

compression fracture, 25/128, 345

conservative management, 26/285-306

craniospinal, see Craniospinal injury

diadematidae intoxication, 37/67

diaphragm rupture, 26/192, 195-197, 226

disruption, 25/31, 340

echinoidea intoxication, 37/67

echinothrix intoxication, 37/67

echinothuridae intoxication, 37/67

edema, 12/557

epidural abscess, 26/39-43

epidural hematoma, 12/558, 25/29, 47, 26/1-30

experimental, see Spine experimental injury

eye injury, 26/190

facet interlocking, 25/136, 329-340

features, 25/42

first aid, 26/295

fracture, 26/200

fracture cervical spine, 25/27-40

fracture management, 26/495

general, 26/212

hangman fracture, 25/32, 134, 326-329

head injury, 26/188, 224

hematomyelia, 12/557, 580

hematothorax, 26/194, 225
heterodontus francisci intoxication, 37/70
incidence, 25/41, 43
instability, 25/245, 280, 293, 340, 447
intestinal atony, 26/199
late sequela, 12/602
lateral flexion, 25/125, 135, 259
limb fracture, 26/200, 226-232
macroscopic pathology, 25/27-42
mechanism, 26/285
motor end plate bulging, 25/125
multiple sclerosis, 9/206
occult fracture, 25/245
painful, 37/67
pathology, 25/42
patient transport, 26/295-298
pelvic fracture, 26/226
pelvic injury, 26/198
posterior column, 25/126
posttraumatic edema, 12/575
postural reduction, 25/457, 460, 26/299
pseudosubluxation, 25/245
radiodiagnosis, 26/263-283
retroflexion, 25/134
retrolisthesis, 25/243
rib fracture, 26/190, 192
rotation, 25/67
skull fracture, 26/190
spinal canal deformity, 25/27-42, 183, 185, 277
spinal cord injury, 25/177, 184
spinal deformity, 26/159-174
spinal shock, 26/203
stab wound, 25/52, 198-206
stable injury, 26/287
sternum fracture, 26/193
stingray intoxication, 37/71
Stokes-Mandeville-Egerton bed, 26/300, 302
subdural hematoma, 26/31-38
surgical treatment, 25/187
terminology, 25/145, 148, 26/285
thoracolumbar, see Thoracolumbar spine injury
traction, 26/301
transport, 26/295-298
treatment, 25/191, 26/202
turning, 26/299
unstable injury, 26/287, 294
vertebral artery, 26/55-62
vertebral fusion, 26/213
wound, 37/71
Spinal intradural cyst, 20/55-94
arachnoid cyst, 20/72, 80, 82-87
Arnold-Chiari malformation type I, 20/82
embryology, 20/55-57

hypertrichosis, 20/78
iniencephaly, 20/58
meningitis serosa, 12/425
type, 20/72
Spinal ischemia
intermittent, see Intermittent spinal cord ischemia
Spinal lesion classification
Brown-Séquard syndrome, 25/269
Spinal leukemia
hemorrhage, 20/112
thrombocytopenia, 20/112
Spinal lipoma, 20/389-409
age, 20/393
angiolipoma, 20/109
animal, 20/393
associated lesion, 20/396
ataxia, 20/400
case series, 20/390-392
cause, 20/398
cervical rib, 20/397
clinical features, 20/400
congenital lipoma, 20/394
craniopharyngioma, 20/397
CSF, 20/401
dermoid cyst, 20/397
diagnosis, 20/402
dissociated sensory loss, 20/400
duration, 20/394
epidural, see Epidural spinal lipoma
epidural vs intradural, 20/408
flaccid paralysis, 20/400
hamartoma, 20/397
histology, 20/396
hydrocephalus, 20/398
Klippel-Feil syndrome, 20/397
Krabbe-Bartels disease, 14/411
lacuna sacralis, 20/397
maldevelopment, 20/389, 397
meningocele, 20/110
multiple epidural, see Multiple epidural spinal
   lipoma
myelography, 20/402
pathology, 20/395
pilonidal sinus, 20/398
precipitating factor, 20/394
prevalence, 20/389
primitive perivascular mesenchyma, 20/399
Queckenstedt test, 20/401
radiology, 19/362, 20/401
segmental lesion distribution, 20/395
sex ratio, 20/393
site, 20/395
sphincter disturbance, 20/400

spina bifida, 19/362, 20/110, 397
  surgical treatment, 20/402
  syringomyelia, 20/398
Spinal lipomatosis
  hypercortisolism, 70/207
Spinal lymphogranulomatosis, 20/113-128
  spinal cord compression, 19/366
Spinal medulloblastoma
  glioma polyposis syndrome, 42/735
  ivory vertebra, 19/171
  spinal cord compression, 19/361
  spinal epidural tumor, 20/129
  spinal glioma, 19/359, 361
Spinal meningioma, 20/177-237, 67/201
  age, 19/358, 20/190, 192, 226
  anesthesia dolorosa, 20/198
  angioblastic type, 20/188, 192
  autonomic nervous system, 20/198
  case series, 20/179-181
  cauda equina, 20/233
  chemotherapy, 20/233
  child, 20/191-193
  clinical features, 20/190
  cord compression, 20/205
  CSF, 20/216-219, 231
  CSF protein, 19/68, 358, 20/216
  dermatographism, 20/199
  diagnosis, 20/208
  differential diagnosis, 19/68, 358, 20/189, 223
  endotheliomatous type, 20/188
  epidural, 20/183
  epilepsy, 20/199
  etiology, 20/177
  female, 19/358
  fibroblastic type, 20/192, 244
  foramen magnum tumor, 19/68, 20/184
  funicular pain, 20/198, 201
  globular type, 20/188
  histogenesis, 20/177
  histologic classification, 20/187
  history, 20/177
  history length, 20/193, 227
  hormone treatment, 20/236
  hourglass tumor, 19/358, 20/184
  kyphosis, 20/214
  lesion level, 20/184-186, 278
  malignant degeneration, 20/189
  medulla oblongata, 20/206, 278
  meningiomatosis, 20/189
  metastasis, 20/189
  mortality, 20/237
  motor disorder, 19/68, 358
  motor symptom, 20/195, 201
  multiple tumors, 20/189
  myelography, 19/68, 358, 20/219
  myxomatous type, 20/188
  neurinoma, 20/244
  neurofibroma, 20/184
  neurofibromatosis, 20/240
  ovoid type, 20/188
  pain, 19/68, 358, 20/200
  paraplegia dolorosa, 20/198
  plaque, 20/186, 188, 206
  prevalence, 20/178
  prognosis, 20/236
  psammomatous type, 20/187, 192
  pseudoneuralgic prephase, 20/204
  psychosis, 20/199
  Queckenstedt test, 20/231, 283
  radicular symptom, 20/198
  radiology, 20/209
  recurrence, 20/189
  scintimyelography, 20/221
  segmental dermographism, 20/199
  segmental lesion distribution, 20/182, 224
  sensory symptom, 20/197, 201
  sex ratio, 20/190, 226
  site, 20/183, 186, 278
  size, 20/187
  sphincter disturbance, 20/200
  spinal nerve root syndrome, 20/198, 201
  surgical treatment, 20/232
  symptom duration, 20/202
  symptom order, 20/202, 228
  symptom prevalence, 20/199, 228
  transitional type, 20/188
  trauma, 20/178
  vertebral column syndrome, 20/207
  xanthomatous type, 20/188
Spinal meningitis
  Escherichia coli, 52/190
  Hemophilus influenzae, 52/190
  meningococcal meningitis, 52/190
  obsolete term, 52/189
  Staphylococcus aureus, 52/190
  Streptococcus pneumoniae, 52/190
  Streptococcus pyogenes, 52/190
Spinal meningocele
  anterior, *see* Anterior spinal meningocele
Spinal muscular atrophy
  β-*N*-acetylhexosaminidase A deficiency,
    59/99-101, 401
  β-*N*-acetylhexosaminidase B deficiency, 59/99
  adolescent, *see* Adolescent spinal muscular
    atrophy
  adult onset, 42/84, 59/98

aminoaciduria, 59/371
amyotrophic lateral sclerosis, 59/15, 113
amyotrophy, 59/368, 374
arsenic, 59/370
arthrogryposis multiplex congenita, 59/373
associated condition, 59/20-22
ataxia, 59/374
ataxia telangiectasia, 14/316
autosomal dominant, 59/372
autosomal dominant spastic paraplegia, 59/374
black widow spider venom intoxication, 59/370
bulbar, 42/82, 59/44, 99
cardiomyopathy, 59/45
cataract, 59/373
central motor defect, 59/372
cervical myelopathy, 59/113
chorea, 59/373
chromosome 4 linkage, 59/26
chromosome 5 linkage, 59/26
chronic progressive peripheral neuropathy, 59/373
classification, 59/367
clinical features, 40/279, 324, 337
Coenurus cerebralis, 59/370
cold paresis, 59/107
congenital cervical spinal atrophy, 59/370
congenital retardation, 59/372
Coxsackie virus, 59/370
creatine kinase, 22/25
CSF pressure, 59/110
defined metabolic derangement, 59/99-103
diabetes mellitus, 59/374
differential diagnosis, 40/438, 41/59, 59/20-22,
    367-373, 373
distal myopathy, 40/478, 59/373
Down syndrome, 59/371
dyssynergia cerebellaris myoclonica, 60/599
Echinococcus granulosus, 59/370
electrical injury, 51/139
enzyme deficiency polyneuropathy, 51/375
external ophthalmoparesis, 59/373
facioscapulohumeral, see Facioscapulohumeral
    spinal muscular atrophy
facioscapulohumeral syndrome, 40/423
Finkel, see Finkel spinal muscular atrophy
forearm, 59/108
galactosialidase deficiency, 51/380
genetics, 41/440
glycogen storage disease, 59/371
GM2 gangliosidosis, 51/375
Guillain-Barré syndrome, 59/370
gynecomastia, 59/44
hereditary, see Hereditary spinal muscular atrophy
hereditary motor and sensory neuropathy type II,

59/373, 60/200
hereditary progressive, see Hereditary progressive
    spinal muscular atrophy
hereditary spinal muscular atrophy, 59/22
history, 40/434
Huntington chorea, 49/291
3-hydroxy-3-methylbutyric aciduria, 59/370
hyperlipidemia, 59/103
hypogonadism, 59/44
incontinentia pigmenti, 59/368, 370
infantile, see Infantile spinal muscular atrophy
infantile neuroaxonal dystrophy, 59/368, 370
infantile neuronal degeneration, 59/368
infantile spinal muscular atrophy, 59/368
intermediate, see Intermediate spinal muscular
    atrophy
intermittent porphyria, 59/370
Joseph-Machado disease, 59/26
juvenile, see Wohlfart-Kugelberg-Welander
    disease
juvenile proximal, see
    Wohlfart-Kugelberg-Welander disease
Kuskokwim arthrogryposis, 59/373
late onset, 42/82
leukodystrophy, 59/371
lipofuscin, 59/45
live oral poliovirus vaccine, 59/370
lysosomal enzyme deficiency, 59/374
meningomyelocele, 59/370
mental deficiency, 59/373
mental subnormality, 59/372
methylcrotonoyl-coenzyme A carboxylase, 59/368
3-methylcrotonyl-coenzyme A carboxylase
    deficiency, 59/100
β-methylcrotonylglycinuria, 59/370
β-methylcrotonylglycinuria type I, 59/100
microcephaly, 59/372
Möbius syndrome, 59/368, 370
monomelic, see Monomelic spinal muscular
    atrophy
multiple lysosomal enzyme deficiency, 59/100
mumps, 59/370
myopathic characteristic, 59/46
myotonic dystrophy, 59/373
neural deafness, 59/93
neuroacanthocytosis, 59/374
neuroaxonal dystrophy, 59/374
neuromuscular disease, 41/435
neuropathology, 59/114
nonpolio enterovirus infection, 59/370
nonprogressive, see Nonprogressive spinal
    muscular atrophy
nonprogressive autosomal recessive spinal

muscular atrophy, 59/372
nonprogressive juvenile, *see* Hirayama disease
nonprogressive mental deficiency, 59/372
nosology, 59/46
oblique amyotrophy, 59/108
optic atrophy, 59/93, 373
organic aciduria, 59/103, 371
pallidoluysionigral degeneration, 49/454, 457
peroxisomal disorder, 59/371
phenylketonuria, 59/374
poliomyelitis, 22/138, 34/109, 59/114, 370, 373
Prader-Labhart-Willi syndrome, 59/371
progressive, *see* Progressive spinal muscular
    atrophy
ptosis, 59/46
quadriceps myopathy, 40/440, 62/174
rabies, 59/370
ragged red fiber, 62/292
recessive scapuloperoneal, 59/46
respiratory distress, 59/369
risk factor, 59/15
Ryukyan, *see* Ryukyan spinal muscular atrophy
scapuloperoneal, 59/44
schistosomiasis, 59/370
scorpion sting, 59/370
sea blue histiocyte, 59/103, 372
secondary pigmentary retinal degeneration,
    59/374
sensorineural deafness, 59/373
serum creatine kinase, 59/373
sialidase deficiency, 59/102
Silver disease, 59/374
snake bite, 59/370
special form, 59/41-46
spina bifida, 59/370
spinal cord aplasia, 59/370
spinal cord injury, 59/370
spinal cord tumor, 59/114
syringomyelia, 32/285, 50/452, 59/113, 373
Taenia solium, 59/370
tick paralysis, 59/370
tremulous movement, 59/108
Troyer syndrome, 59/374
umbilical artery catheterization, 59/370
unilateral diagram, 59/373
unilateral paralysis, 59/373
vocal cord paralysis, 59/373
Waldenström macroglobulinemia, 63/396
WFN classification, 59/1-7
wobbler mouse, 22/137
Wohlfart-Kugelberg-Welander disease, 22/134,
    59/92, 371
xeroderma pigmentosum, 59/368, 370, 372

Spinal muscular atrophy type II
    proximal muscle weakness, 42/90
Spinal muscular atrophy type III, *see*
        Wohlfart-Kugelberg-Welander disease
Spinal myoclonus
    clinical features, 49/614
    clonazepam, 41/308
    motor unit hyperactivity, 41/308
    tetrabenazine, 41/308
Spinal nerve, 25/403
    osteitis deformans (Paget), 70/15
    sarcoid neuropathy, 38/525-529
    spina bifida, 32/543
Spinal nerve root
    angulation, 25/399
    anterior, *see* Ventral spinal nerve root
    cervical, *see* Cervical spinal root
    chronic inflammatory demyelinating
        polyradiculoneuropathy, 51/536
    critical illness polyneuropathy, 51/581
    dentate ligament, 25/400
    droplet metastasis, 20/285
    elasticity, 25/398
    experimental injury, 25/394
    Friedreich ataxia, 60/303
    hereditary motor and sensory neuropathy type I,
        21/299
    hereditary motor and sensory neuropathy type II,
        21/299
    intervertebral foramen, 25/400
    ionizing radiation neuropathy, 51/137
    length, 25/398
    leukemia, 63/343
    lymphocytic leukemia, 63/343
    meningeal diverticulum, 20/152
    meningeal leukemia, 18/252
    metastasis, 69/76
    mummy, 74/190
    myeloid leukemia, 63/343
    neurinoma, 8/417, 68/546
    posterior, *see* Dorsal spinal nerve root
    reticulosarcoma, 9/596
    spina bifida, 32/543
    spinal cord injury, 61/41
    thickness, 25/398
    topography, 1/228
    traction injury, 25/393-435
Spinal nerve root artery
    cervical, 25/266
    lesion, 12/598
    spinal cord, 26/65
Spinal nerve root avulsion, 2/157-175
    see also Ventral spinal nerve root avulsion

378

anesthesia, 1/89
brachial plexus, 2/135, 137, 7/407, 51/145-148
differentiation nerve root, 2/171
history, 25/393
meningeal diverticulum, 20/157, 160
plexus traction, 7/407
spinal cord injury, 61/526
symptom, 51/147
synopsis, 2/172
Spinal nerve root compression
   Hodgkin lymphoma, 63/346
   lymphoma, 63/346
   β2-microglobulin, 63/529
   motor unit hyperactivity, 41/303
   multiple myeloma, 63/392
   spinal cord injury, 61/525
   spinal cord intermittent claudication, 20/749
Spinal nerve root entrapment, 20/611-801
Spinal nerve root entry zone lesion
   dorsal, see Dorsal spinal root entry zone lesion
Spinal nerve root ganglion
   dorsal, see Dorsal spinal root ganglion
Spinal nerve root ganglionitis
   dorsal, see Dorsal spinal nerve root ganglionitis
Spinal nerve root herniation
   stylet injury syndrome, 61/158
Spinal nerve root injury
   anatomy, 25/396
   arachnoid, 25/397
   avulsion, see Spinal nerve root avulsion
   biopsy, 25/422
   cauda equina, 25/450, 454, 470
   cause, 25/259, 411
   cervical, 25/274
   clinical features, 25/413-417
   cold vasodilatation test, 25/421
   CSF, 25/417
   diverticulum, 25/419
   dura mater, 25/397
   EMG, 25/417
   evoked potential, 25/422
   exponential injury, 25/394
   histamine flare response, 25/420
   Horner syndrome, 25/415, 417
   laminectomy, 25/420, 424
   lumbar, 25/450
   lumbar puncture, 25/417
   lumbosacral, 25/471
   management, 25/423-427
   mechanism, 25/97, 407
   myelography, 25/395, 418
   pathology, 25/97
   prognosis, 25/423

   radiology, 25/418
   sacral root, 25/468
   spondylotic radiculopathy, 26/97-112
   surgery, 25/425
Spinal nerve root ischemia
   spinal cord intermittent claudication, 20/749
Spinal nerve root lesion
   epidural anesthesia, 61/137
   spinal cord injury, 61/42
Spinal nerve root neuropathy
   hereditary sensory, see Hereditary sensory and
     autonomic neuropathy type I
Spinal nerve root pain
   cervical intervertebral disc prolapse, 20/548
   clinical picture, 20/552
   myalgic component, 20/554
   paraplegia, 26/492
   pseudo, 20/585, 757
   Wolfson syndrome, 20/428
Spinal nerve root regeneration
   posttraumatic, 61/42
   significance, 61/48
   transection, 25/100
Spinal nerve root rupture
   brachial plexus, 51/146
Spinal nerve root sheath
   nerve, 51/2
Spinal nerve root swelling
   Guillain-Barré syndrome, 51/243
Spinal nerve root syndrome
   see also Pseudoradicular syndrome
   C4 syndrome, 20/555
   C5 syndrome, 20/557
   C6 syndrome, 20/557
   C7 syndrome, 20/556-558
   C8 syndrome, 20/556
   cauda equina, 20/594
   cervical, see Cervical radicular syndrome
   differential diagnosis, 20/556
   epidural ganglion cyst, 20/605
   high lumbar syndrome, 20/595
   L4 syndrome, 20/579, 583
   L5 syndrome, 20/579, 583
   lumbar, 20/573-604
   meningeal leukemia, 63/341
   muscle stretch reflex, 2/167-170
   paralysing sciatica, 20/592
   spinal angioma, 32/466
   spinal autonomous efferent, 2/162
   spinal meningioma, 20/198, 201
   spinal neurinoma, 20/294
   thoracic level, 20/566
Spinal neurinoma, 20/238-312

age, 20/245, 279
autonomic nervous system, 20/252
Brown-Séquard syndrome, 20/256
cause, 20/238
chemotherapy, 20/285
child, 20/246
clinical features, 20/245
clinical syndrome, 20/254, 256
CSF, 20/231, 270-274, 283
CSF protein, 20/271
diagnosis, 20/257
differential diagnosis, 19/358, 20/244, 277
direct tumor sign, 20/258
Elsberg sign, 20/260
epilepsy, 20/250
fibroblastic meningioma, 20/244
foramen magnum tumor, 20/253
funicular paresis, 20/255-257
histogenesis, 20/238
histology, 20/243
history, 20/238
history length, 20/227, 248, 279
hormone treatment, 20/285
hourglass tumor, 19/358, 20/242
kyphosis, 20/262-266
lesion level, 20/242
lesion site, 20/240
malignant degeneration, 20/244
metastasis, 20/244
multiple tumors, 20/244
myelography, 20/274
neurofibromatosis, 20/244
pain, 20/250
paraplegia, 20/251
presenting symptom, 20/253
prevalence, 20/182, 239
prognosis, 20/285
pseudoneuralgic prephase, 20/250, 254
pseudopsoas shadow, 20/270
psychosis, 20/250
Queckenstedt test, 20/231, 273, 283
radicular syndrome, 20/249
radiology, 20/258
radiotherapy, 20/285
recurrence, 20/244
scintimyelography, 20/277
segmental lesion distribution, 20/182, 224, 243, 253, 279
sex ratio, 19/358, 20/245, 279
site, 20/247
spinal nerve root syndrome, 20/294
surgical treatment, 20/284
symptom duration, 20/253

symptom order, 20/228, 253, 281
symptom prevalence, 20/228, 251, 280
tumor sign, 20/258
vertebral column syndrome, 20/257
vertebral phlebography, 20/277
Spinal neurofibroma, 20/238-312
    cause, 20/238
    histogenesis, 20/238
    history, 20/238
    prevalence, 20/239
    solitary, 42/759
Spinal neuronitis
    subacute myoclonic, see Subacute myoclonic
        spinal neuronitis
Spinal oligodendroglioma
    curiosity, 19/361
Spinal ossovenography, see Spinal phlebography
Spinal osteogenic tumor
    benign, see Benign spinal osteogenic tumor
Spinal pachymeningitis
    compressive myelopathy, 33/190
    radiculopathy, 33/190
Spinal pain, 26/489-497
    see also Dorsal rhizotomy
    acupuncture, 26/496
    cordotomy, 26/497
    dorsal column stimulation, 26/497
    myelotomy, 26/497
    nerve block, 26/496
    pathway, 45/234
    physiotherapy, 26/496
    sympathectomy, 26/497
Spinal paragonimiasis
    clinical features, 35/263
    CSF, 35/263
    differential diagnosis, 35/264
    prognosis, 35/264
    treatment, 35/264
Spinal parasympathetic efferent
    segmental distribution, 2/164
Spinal perineural cyst
    clinical features, 20/149
    features, 31/86
    histology, 20/147
    myelography, 20/149
    sacral arachnoid cyst, 32/409
    sex ratio, 20/152
    spinal epidural cyst, 20/146
    symptomatic series, 20/150
Spinal phlebography
    abnormal phlebogram, 19/237
    intervertebral disc prolapse, 19/239
    normal phlebogram, 19/236

osseous, 19/66
spinal angioma, 20/455
spondylosis, 19/240
trauma, 19/240
vascular malformation, 19/239
vertebral tumor, 19/237
Spinal phlebothrombosis
spinal cord abscess, 52/191
subacute, *see* Foix-Alajouanine disease
Spinal poliomyelitis
distribution, 34/109
poliomyelitis, 34/109, 56/328
Spinal pseudotumor, *see* Spinal arachnoiditis
Spinal reflex
tetany, 28/535
viscerogenic, *see* Viscerogenic spinal reflex
Spinal retinoblastoma
curiosity, 19/361
Spinal rheumatoid arthritis, 38/487-492
Bland criteria, 19/327
cervical site, 19/326-331
histology, 71/23
neuropathic effect, 19/339
odontoid process, 19/331-334
palate sign (Mathews), 19/337
pathology, 19/328
prevalence, 19/324
radiology, 19/331
Sharp-Purser test, 19/337
symptomatology, 19/336
treatment, 19/339
Spinal shock
see also Diaschisis *and* Spinal cord injury
autonomic nervous system, 75/569
birth incidence, 25/157
bladder function, 2/188, 26/254
child, 25/179
clinical features, 26/245
degree, 26/245
dermatographism, 26/259
diaphragm, 26/337
duration, 1/61, 2/181, 26/246
experimental, 26/244
histology, 26/243
history, 26/243
intestinal function, 26/258
ischuria paradoxa, 2/188
physiology, 26/243
priapism, 2/196
reflex depression, 2/180, 26/245
reflex return, 26/246-248
respiration, 26/253
sensory disorder, 26/252

sexual function, 26/258, 448
spinal cord edema, 12/575
spinal cord injury, 26/203, 243-263, 61/439
spinal injury, 26/203
tetraplegia, 61/439
thermoregulation, 26/260
thoracolumbar spine injury, 25/441
transverse spinal cord lesion, 61/439
traumatic shock, 26/246
urinary retention, 2/188
vasomotor control, 26/259
Spinal spasticity
alcohol, 61/369
alpha motor system, 61/367
baclofen, 61/368
cordotomy, 61/369
dantrolene, 61/368
deafferentation, 61/370
diazepam, 61/368
dorsal spinal root entry zone microcoagulation,
61/371
gamma motor system, 61/367
intrathecal agent, 61/368
morphine, 61/368
myelotomy, 61/369
neurostimulation, 61/372
obturator neurotomy, 61/369
percutaneous thermal rhizotomy, 61/370
pharmacotherapy, 61/368
phenol, 61/369
rhizotomy, 61/369
sensory microrootlet section, 61/371
surgical treatment, 61/369
tizanidine, 61/368
Spinal spongioblastoma
age, 19/360
ataxia, 19/360
motor disorder, 19/360
prognosis, 19/360
sensory disorder, 19/360
Spinal subarachnoid block
CSF, 19/125-138
Queckenstedt test, 19/91
Spinal subarachnoid hemorrhage
attitude en chien de fusil, 12/133
cause, 12/567
clinical features, 12/568, 55/32
congenital heart disease, 63/4
differential diagnosis, 12/173, 569
frequency, 55/32
irritation, 12/173
leukemia, 63/344
nerve root pain, 55/32

pain, 12/133, 173
review, 12/132-135
spinal angioma, 12/134
spinal cord injury, 25/70, 267
spinal cord vascular disease, 12/567
spinal cord vascular malformation, 19/15
systemic lupus erythematosus, 63/54
thermoregulation, 12/173
treatment, 12/570
Spinal subdural cyst
  spina bifida occulta, 20/93
  undetermined, 20/93
Spinal subdural empyema
  aspergillosis, 52/189
  clinical features, 52/189
  CSF, 52/189
  dermal sinus, 52/189
  myelography, 52/189
  neurosurgery, 52/189
  prognosis, 52/189
  Staphylococcus aureus, 33/190, 52/189
  Streptococcus pneumoniae, 52/189
  Streptococcus viridans, 52/189
  transverse myelitis, 52/189
  traumatic, 25/60
  treatment, 52/189
  vertebral osteomyelitis, 52/189
Spinal subdural hematoma
  anticoagulant, 12/566
  birth lesion, 25/70
  blood dyscrasia, 12/566
  cause, 61/160
  cyst, 20/92
  hemorrhagic pachymeningitis, 12/567
  spinal cord vascular disease, 12/565
  symptom, 61/160
  trauma, 12/566
Spinal subdural hemorrhage
  hemophilia, 55/475
  leukemia, 63/344
Spinal subdural space
  anatomy, 26/34, 35
Spinal subdural tumor, 19/65
Spinal subluxation
  ankylosing spondylitis, 26/181
  cervical spine, 25/131, 341, 26/181, 289
  definition, 25/242
  thoracolumbar spine, 26/293
Spinal sulcal artery
  topography, 55/97
Spinal surgery
  orthopedic, 70/28
Spinal sympathetic efferent

distribution, 2/163
Spinal sympathicoblastoma
  Horner syndrome, 20/112
  papilledema, 20/112
  paraplegia, 20/112
  space occupying lesion, 19/363
  spinal cord compression, 20/112
Spinal sympathoblastoma, see Spinal
  sympathicoblastoma
Spinal sympathogonioma, see Spinal
  sympathicoblastoma
Spinal syndrome
  central, see Central spinal syndrome
Spinal syphilis
  syringomyelia, 32/293, 50/453
Spinal trauma
  cervical spondylosis, 25/252
  orthopedic trauma, 70/27
Spinal tumor
  see also Spinal cord tumor and Vertebral column
    tumor
  age, 19/53
  amyotrophic lateral sclerosis, 19/87
  aneurysmal bone cyst, 68/515
  angiography, 19/66, 229-243, 67/211
  autonomic symptom, 19/59
  benign, 68/513
  biopsy, 68/513
  blind vertebra, 20/22
  calcification, 19/166
  cervical site, 19/67
  child, 19/53
  chordoma, 68/520
  clinical stage, 19/25
  computer assisted tomography, 67/206
  conus, see Conus medullaris tumor
  course, 19/40
  cranial nerve sign, 19/39
  cranial vault tumor, 68/515
  CSF, 19/61, 125-138
  CSF cytology, 19/61
  CSF pressure, 19/114
  CSF protein, 19/61
  diagnosis, 19/45, 61, 67/206
  differential diagnosis, 19/33, 51-75
  dissociated sensory loss, 19/59
  Elsberg sign, 19/59
  EMG, 19/66, 267-286
  eosinophilic granuloma, 20/111, 68/517
  epidemiology, 68/511
  epidural, 67/193
  extramedullary, see Extramedullary spinal tumor
  funicular pain, 19/55

granulomatous CNS vasculitis, 55/390
hemangioblastoma, 19/70
hemangioma, 68/515
hemorrhage, 19/56
Horner syndrome, 19/35
hypertrophic polyneuropathy, 19/87
incidence, 68/530
intradural, 67/201, 68/527
intramedullary, *see* Intramedullary spinal tumor
Lasègue sign, 19/72
level, 19/34, 36
Lindau, 19/362
lumbosacral site, 19/70
malignant, 67/203
mammary carcinoma, 20/43
meningioma, 19/64
multiple, *see* Multiple spinal tumor
multiple myeloma, 68/518
muscular atrophy, 19/26
myelography, 19/65, 179-203, 67/206
myeloscintigraphy, 19/67, 245-265
neurinoma, 19/64, 68
neurogenic bladder, 19/59
neuroimaging, 67/193-211
nystagmus, 19/40
one eyed vertebra, 20/22
pain, 19/54
papilledema, 19/60, 88
paresthesia, 19/57
plain indication, 19/145
plain X-ray, 19/62, 139-177
plasmacytoma, 68/518
prevalence, 19/51
primary, 68/513
primary lymphoma, 68/519
prognosis, 19/46
proteinocytologic dissociation, 19/125
Queckenstedt test, 19/61
radicular syndrome, 19/26
radiologic features, 67/208-210
radiology, 68/512
remittent course, 19/42
rheumatoid arthritis, 19/86
sciatica, 19/71
sensory deficit, 19/58
sex ratio, 19/53
spinal angiography, 19/232
subarachnoid hemorrhage, 19/60, 88
surgery, 68/529
survey, 68/511
symptom, 19/54
symptomatology, 19/23
thoracic site, 19/70

tomography, 19/137
vascular, 68/515
vertebral scalloping, 19/63
X-ray finding, 19/64
Spinal unit
functional, *see* Functional spinal unit
spine experimental injury, 25/2
transport, 26/297
vertebral experimental injury, 25/2
Spinal vascular disease, *see* Vascular myelopathy
Spinal vascular tumor
arteriovenous malformation type I, 68/570
arteriovenous malformation type II, 68/572
arteriovenous malformation type III, 68/573
arteriovenous malformation type IV, 68/573
cavernous malformation, 68/574
classification, 68/569
Spinal vein
anterior, *see* Anterior spinal vein
Spine experimental injury
cervical spine, 25/3
intervertebral disc, 25/1
ligament, 25/2
model, 25/1
spinal unit, 25/2
whiplash injury, 25/4
Spine malformation
caudal aplasia, 50/512
Spine manipulation
cervical spondylotic myelopathy, 7/464
cervical vertebral column injury, 25/292, 26/302
vertebral artery injury, 26/61
Spine metastasis
cancer, 67/200
Spine retroflexion
iniencephaly, 50/129
Spine syndrome
rigid, *see* Rigid spine syndrome
Spine traction
cervical spondylotic myelopathy, 7/464
Spinnenzellgliom, *see* Spider cell
Spinocerebellar ataxia
anterior horn syndrome, 42/182
areflexia, 42/182
blindness, 42/181
brain atrophy, 42/182, 190
cerebellar atrophy, 42/182, 190
deafness, 42/181
dementia, 42/182
dysmorphism, 60/664
genetic linkage, 42/128
hereditary, *see* Hereditary spinocerebellar ataxia
HLA antigen, 42/128

hyporeflexia, 42/182
inherited, *see* Inherited spinocerebellar ataxia
Marinesco-Sjögren syndrome, 42/184-186
mental deficiency, 42/181, 60/664
muscular atrophy, 42/182
mydriasis, 42/186
optic atrophy, 42/181
prevalence, 42/128
pyramidal tract degeneration, 42/182
retinal degeneration, 42/188
scoliosis, 50/418
secondary pigmentary retinal degeneration,
    13/304-307
senile plaque, 21/553, 42/182
sensorineural deafness, 42/181
sex ratio, 42/187
slow eye movement, 42/187
speech disorder, 42/182
spinocerebellar degeneration, 42/182
vertical gaze palsy, 42/188
X-linked spinocerebellar degeneration,
    42/189-192
Spinocerebellar degeneration
  age at onset, 60/647, 649
  Alpers disease, 10/210
  Alzheimer disease, 21/15
  amyotrophic lateral sclerosis, 21/15, 22/371
  amyotrophy, 21/15
  Bassen-Kornzweig syndrome, 8/19, 10/279
  biochemistry, 60/673
  brain biopsy, 60/678
  carnosinemia, 42/532
  catalogeu, 60/644-647
  cerebellum dissection, 60/286
  cerebellum inspection, 60/286
  classification suggestion, 60/275
  classification variable use, 60/272
  conjunctiva biopsy, 60/677
  continuum, 21/416
  corneal dystrophy, 60/652
  Creutzfeldt-Jakob disease, 21/15
  CSF, 60/676
  cylindrical tubule, 40/39
  dementia, 46/130
  diabetes insipidus, 60/650
  diabetes mellitus, 60/674
  endocrine change, 60/675
  enzyme study, 60/676
  epidemiology, 21/14
  familial spastic paraplegia, 21/15
  Fazio-Londe disease, 22/104
  first symptom, 60/647
  Friedreich ataxia, 42/144

giant axonal neuropathy, 60/82
hemolytic anemia, 60/674
hereditary, *see* Hereditary spinocerebellar
    degeneration
hereditary motor and sensory neuropathy type I,
    21/306, 60/245
hereditary motor and sensory neuropathy type II,
    21/306, 60/245
hereditary motor and sensory neuropathy variant,
    60/245
hereditary neuropathy, 60/733
hyperreflexia, 42/190
hypoplastic cerebellar artery, 60/651
ICD classification, 60/272
incidence, 60/647
intestinal gut biopsy, 60/678
Kii Peninsula amyotrophic lateral sclerosis,
    22/371
molecular biology, 60/678-682
multiple sclerosis, 9/205
muscle biopsy, 60/677
nerve biopsy, 60/678
nerve conduction velocity, 42/190
nosology, 21/434
number of cases, 60/648
nystagmus, 42/190
optic atrophy, 13/97
ota nevus, 43/37
Ota syndrome, 14/791
Parkinson dementia complex, 21/15
prevalence, 60/647
Refsum disease, 60/235, 66/497
scoliosis, 42/190
skin biopsy, 60/677
spastic ataxia, 42/166
spinocerebellar ataxia, 42/182
stabilometry, 60/651
Stargardt disease, 13/132
talipes, 42/190
tremor, 42/190
urine, 60/676
WFN classification proposal, 60/273
Woods-Schaumburg variant, 22/157
X-linked, *see* X-linked spinocerebellar
    degeneration
Spinocerebellar degeneration classification
  common dominant ataxia type, 60/279
  congenital ataxia, 60/274
  familial spastic paraplegia, 60/281
  Friedreich ataxia, 60/275, 277
  other dominant ataxia type, 60/279
  recessive ataxia onset after age 20, 60/278
Spinocerebellar neuroaxonal dystrophy

myoclonic epilepsy, 60/663
pes cavus, 60/669
Spinocerebellar syndrome
  globoid cell leukodystrophy, 51/373
  progressive bulbar palsy, 59/129
Spinocerebellar tract
  anterior, *see* Ventral spinocerebellar tract
  Bassen-Kornzweig syndrome, 8/19, 63/278
  dentatorubropallidoluysian atrophy, 49/442
  familial amyotrophic lateral sclerosis, 21/30
  lathyrism, 65/9
  progressive bulbar palsy, 59/224
  spinopontine degeneration, 21/391
Spinocerebellofoveal dystrophy
  Stargardt disease, 13/132
Spinocerebellonigral amyotrophy
  progressive spinal muscular atrophy, 22/14
Spinocerebellonigral degeneration
  Parkinson disease, 49/94
Spinocortical tract
  degeneration, *see* Corticospinal tract degeneration
Spinocranial tumor, *see* Foramen magnum tumor
Spinodentatonigro-oculomotor degeneration
  hereditary, *see* Hereditary
    spinodentatonigro-oculomotor degeneration
Spino-olivocerebellar degeneration
  dementia, 60/665
  distal amyotrophy, 60/659
  optic atrophy, 60/655
  spasticity, 60/661
  supranuclear ophthalmoplegia, 60/656
  upper eyelid retraction, 60/655
  vestibular function, 60/658
Spinopontine atrophy, *see* Spinopontine
    degeneration
Spinopontine degeneration, 21/389-401
  amyotrophy, 21/446
  ataxia, 42/192
  autosomal dominant, 21/371, 389
  brain atrophy, 42/192
  buccopharyngeal spasm, 21/390
  cerebellar atrophy, 42/192
  Clarke column, 21/391, 42/192
  clinical features, 21/389
  corpus restiforme, 21/391
  differential diagnosis, 21/397
  diplopia, 42/192
  dysarthria, 21/390
  Friedreich ataxia, 21/397, 400
  gaze paralysis, 21/371, 390
  hereditary cerebellar ataxia (Marie), 21/397, 400
  hereditary olivopontocerebellar atrophy (Menzel),
    21/446

hereditary spastic ataxia, 21/399, 60/463
hyperreflexia, 42/192
intention tremor, 21/390
Klippel-Feil syndrome, 21/390
late cortical cerebellar atrophy, 21/397
mental state, 21/390
neuropathology, 21/390-398
nucleus ruber, 21/391
nystagmus, 21/371, 390, 42/192
oculomotor paralysis, 21/390
olivopontocerebellar atrophy, 21/397
olivopontocerebellar atrophy variant, 22/172
pontine atrophy, 21/391
pontocerebellar tract, 21/392
pyramidal sign, 21/390
speech disorder, 42/192
spinal cord, 21/391
spinocerebellar tract, 21/391
striatonigral degeneration, 60/544
tremor, 21/390
Woods-Schaumburg syndrome, 21/401
Spinopontocerebellar degeneration
  autosomal dominant optic atrophy, 13/38
  pallidoluysionigral degeneration, 49/457
Spinoreticulodiencephalic fiber
  pain, 45/235
Spinothalamic tract
  anterolateral funiculus, 45/233
  Cestan-Chenais syndrome, 2/239
  lateral, *see* Lateral spinothalamic tract
Spinothalamic tract degeneration
  Biemond syndrome, 42/299
  congenital analgesia, 42/299
Spinous process
  cervical, 25/366
  fracture, 25/366
Spiny neuron type I
  neostriatum, 49/3
Spiny neuron type II
  neostriatum, 49/3
Spiperone
  catamenial migraine, 48/92
  serotonin 2 receptor, 48/88, 92
Spirochaeta argentinensis
  multiple sclerosis, 9/59, 108
Spirochaeta myelophthora
  multiple sclerosis, 9/59, 127, 47/319
Spirogermanium intoxication
  ataxia, 65/538
Spirolactone
  brain edema, 16/204
  hyperaldosteronism, 39/494
Spiromustine intoxication

hallucination, 65/534
myoclonus, 65/534
Spironolactone
diuretic agent, 65/438
periodic paralysis, 41/156
Spitz-Holter valve
component, 30/576
Splanchnocystic dysencephaly, *see* Gruber
syndrome
Splanchnocystic dysencephaly (Gruber), *see* Gruber
syndrome
Spleen
accessory, *see* Accessory spleen
glycosaminoglycanosis, 10/454
glycosaminoglycanosis type VII, 66/311
hereditary amyloid polyneuropathy, 60/106
Splenectomy
Hodgkin disease, 39/50
immunocompromised host, 56/469
malignant lymphoma, 39/50
pneumococcal meningitis, 33/49, 52/53
recurrent meningitis, 52/53
Tangier disease, 8/21
Splenio-occipital alexia
brain tumor, 45/444
corpus callosum infarction, 45/444
diagnosis, 45/443
occipital infarction, 45/444
Splenogenic purpura
leukoencephalopathy, 9/598
Splenomegaly
Abt-Letterer-Siwe disease, 38/97, 42/441
cytomegalic inclusion body disease, 56/267
cytoside lipidosis, 10/545
glycosaminoglycanosis type III, 66/300
human T-lymphotropic virus type I associated
myelopathy, 56/527
infantile Gaucher disease, 10/565
juvenile Gaucher disease, 10/522, 565
liver cirrhosis, 27/351
phosphatidylethanolamine lipidosis, 10/504, 552
POEMS syndrome, 47/619, 51/430, 63/394
progressive supranuclear palsy, 22/227
Tangier disease, 10/547
trimethylaminuria, 29/231
tropical, *see* Visceral leishmaniasis
Split brain
animal study, 45/100
aphasia, 46/153
neuropsychology, 45/2, 10
Split brain syndrome, 45/99-104
anterior cerebral artery syndrome, 53/347
apraxia, 45/99

brain commissurotomy, 45/99
brain infarction, 53/347
callosotomy, 45/99
corpus callosum syndrome, 45/100, 53/347
disconnection syndrome, 45/101-104
dominant hemisphere, 45/103
epilepsy, 45/101
Split notochord syndrome
classification, 20/62
diastematomyelia, 20/62, 32/241
summary, 20/60
yolk sac fistula, 20/60
Splitting of the macula, *see* Macular splitting
Spondylarthropathy
β2-microglobulin, 63/529
Spondylarthrosis
cervical, *see* Cervical spondylarthrosis
Spondylarthrotic myelopathy
cervical, *see* Cervical spondylarthrotic
myelopathy
Spondylitis
ankylosing, *see* Ankylosing spondylitis
Brucella, 51/185, 52/585
brucellosis, 19/368, 52/592
neurobrucellosis, 33/310, 52/585, 590, 595
tuberculous, *see* Tuberculous spondylitis
Spondylodesis
thoracic spinal cord injury, 61/83
Spondyloepiphyseal dysplasia
atlantoaxial subluxation, 50/392
cheilopalatoschisis, 43/478
compression neuropathy, 43/481
congenital, *see* Congenital spondyloepiphyseal
dysplasia
cytoplasmic inclusion body, 43/480
dwarfism, 43/478
glycosaminoglycanosis type IV, 42/457
kyphoscoliosis, 43/478
retinal detachment, 43/478
short stature, 43/481
talipes, 43/478
vertebral abnormality, 43/478-482
vertebral canal stenosis, 43/480
Spondyloepiphyseal dysplasia tarda, 43/480-482
osteoarthrosis, 43/481
X-linked recessive, 43/482
Spondylolisthesis
cervical, 25/251
lumbar vertebral canal stenosis, 20/781
mechanical component, 19/84
radicular syndrome, 19/84
spondylolysis, 19/85
traumatic, *see* Hangman fracture

Spondylolysis
  spondylolisthesis, 19/85
Spondylometaphyseal dysplasia
  cytoplasmic inclusion body, 43/483
  kyphoscoliosis, 43/482
  short stature, 43/482
  vertebral abnormality, 43/482
Spondylosis
  anterior spinal artery syndrome, 55/101
  cervical, see Cervical spondylosis
  cervical myelopathy, 7/456
  Klippel-Feil syndrome, 42/37
  ligamentum flavum hypertrophy, 20/812
  lumbar, see Lumbar spondylosis
  myelography, 19/200
  occupational, 7/333
  posterior longitudinal ligament calcification,
    19/154
  rheumatic disease, 11/427
  spinal phlebography, 19/240
  spondylotic ridge, 20/669, 676
Spondylotic kyphosis
  ankylosing spondylitis, 50/419
Spondylotic myelopathy
  cervical, see Cervical spondylotic myelopathy
Spondylotic radiculopathy
  amyotrophic, see Amyotrophic spondylotic
    myeloradiculopathy
  cervical spinal root, 26/97-112
  clinical features, 26/102
  differential diagnosis, 26/103
  foraminotomy, 26/107
  history, 26/97
  laminectomy, 26/107
  myelography, 26/104
  pathogenesis, 26/101
  pathology, 26/98
  radiology, 26/103
  spinal nerve root injury, 26/97-112
  surgery, 26/107
  treatment, 26/107
  tumor, 19/85
  vertebral fusion, 26/107
Sponge intoxication, see Porifera intoxication
Spongiform cortical encephalopathy
  uremia, 9/639
Spongiform encephalitis
  murine neurotropic retrovirus, 56/586
Spongiform encephalomyelopathy
  mouse, 56/455
  murine leukemia virus, 56/455
Spongiform encephalopathy
  consciousness, 45/119, 121

  diamorphine intoxication, 65/357
  Gerstmann-Sträussler-Scheinker disease, 56/555
  infantile type, 6/671
  lead encephalopathy, 64/437
  neuropathology, 21/60, 68
  opiate intoxication, 65/357
  organic acid metabolism, 66/641
  organotin intoxication, 64/136
  pallidal degeneration, 6/651
  presenile, 10/209
  progressive external ophthalmoplegia, 22/212
  status spongiosus, 11/129, 159-162
  striatal lesion, 6/671
  subacute, see Subacute spongiform
    encephalopathy
Spongiform leukoencephalopathy
  antineoplastic agent, 47/568
  cycloleucine, 47/569
  dactinomycin, 47/568
  diamorphine, 47/566-568
  ethylnitrosourea, 47/570
  experimental, see Experimental spongiform
    leukoencephalopathy
  hexachlorophene, 65/471
  hexachlorophene intoxication, 37/479, 488,
    47/565, 65/471
  homidium bromide, 47/569
  peripheral nerve, 37/479
  retinopathy, 37/479
  toxic, 47/562
  uremic leukoencephalopathy, 9/639
Spongiform myelopathy
  animal model, 56/455, 587
  hexachlorophene intoxication, 37/499
  organotin intoxication, 64/136
Spongioblast migration
  phakomatosis, 14/5
Spongioblastoma
  biopsy, 67/226
  calcospherite, 16/15
  chemotherapy, 68/49
  congenital, 31/47
  CSF cytology, 16/394
  diagnosis, 68/42
  differentiation, 67/4
  granulated body, 16/15
  histologic differentiation, 18/124
  microscopy, 16/21
  neurofibromatosis type I, 14/33, 102, 142, 50/367
  optic chiasm, see Oligodendroglioma
  optic chiasm compression, 68/75
  optic pathway tumor, 68/77
  polar, see Polar spongioblastoma

Rosenthal fiber, 16/15
spinal, see Spinal spongioblastoma
spinal cord compression, 19/360
spinal glioma, 19/359, 20/327, 329
supratentorial brain tumor, 18/311
terminology, 18/1, 67/225
thalamic tumor, 68/65
third ventricle tumor, 17/444
tissue culture, 16/38
tuberous sclerosis, 14/46, 102, 350, 365, 380,
    18/300
Spongioblastoma cerebelli, see Cerebellar
    medulloblastoma
Spongioblastoma indifferentiate, see Cerebellar
    medulloblastoma
Spongioblastosis centralis circumscripta
    (Bielschowsky), see Tuberous sclerosis
Spongiosis
    Creutzfeldt-Jakob disease, 6/751
    status, see Status spongiosus
Spongy cerebral degeneration, 10/203-210,
    66/661-666
    N-acetylaspartic acid, 66/662
    Alexander disease, 10/100, 209
    Alpers disease, 10/210
    Alzheimer type I astrocyte, 10/207
    biochemical study, 10/352
    brain scanning, 66/665
    brain stem, 22/195
    carrier frequency, 66/665
    clinical features, 10/204, 572
    congenital type, 66/662
    cortical, 6/731, 734
    Creutzfeldt-Jakob disease, 6/744, 46/293
    diagnosis, 66/665
    differential diagnosis, 10/208
    DNA mutation, 66/664
    epidemiology, 10/575
    genetics, 10/574, 66/664
    glial dystrophy, 21/68
    glycogen, 27/203
    macrencephaly, 10/206, 209
    neuropathology, 10/206, 66/662
    nosology, 10/28, 114, 203
    organic acid metabolism, 66/639
    pallidal degeneration, 6/651
    pathophysiology, 66/662
    spheroid body, 6/626
    spongy dystrophy, 21/68
    survey, 10/573
    terminology, 10/352
Spongy degeneration
    brain biopsy, 10/685

CNS, see CNS spongy degeneration
familial amaurotic idiocy, 10/203
leukodystrophy, 10/19
orthochromatic leukodystrophy, 10/114
Spongy dystrophy
    Creutzfeldt-Jakob disease, 21/68
    infantile, see Infantile spongy dystrophy
    kuru, 21/68
    spongy cerebral degeneration, 21/68
Spongy edema
    hydrocephalus, 50/290
Spongy encephalopathy, see Spongiform
    encephalopathy
Spongy glial dystrophy
    globoid cell leukodystrophy, 21/66
    triethyltin intoxication, 21/66
Spongy rarefactive necrosis
    neuroaxonal dystrophy, 49/396
Spongy state
    neuroaxonal, see Neuroaxonal spongy state
Spontaneous abortion
    chromosome number abnormality, 50/541
    CNS malformation, 50/55
Spontaneous activity
    amyotrophic lateral sclerosis, 59/185
    magnetoencephalography, 72/324
    motor unit, 62/50, 67
    myopathy, 62/50, 67
    nerve cell, 1/60
Spontaneous cerebellar hemorrhage
    age, 12/54
    blood dyscrasia, 12/54
    Charcot-Bouchard aneurysm, 12/54
    hypertension, 12/54
    pathology, 12/54
    thermoregulation, 12/56
    vascular malformation, 12/54
Spontaneous CSF fistula
    aqueduct stenosis, 50/310
Spontaneous electrical muscle activity
    critical illness polyneuropathy, 51/576, 578, 584
Spontaneous lactic acidosis, 28/514-516
    acid base metabolism, 28/515
Spontaneous myoclonus
    definition, 49/612
Spontaneous nystagmus
    acoustic neuroma, 2/361, 16/317
    Alexander law, 2/366, 378
    Russell syndrome, 16/349
    variety, 2/367
Spontaneous respiration
    critical illness polyneuropathy, 51/581
Spontaneous striatal degeneration

striatal lesion, 6/679
Sporadic amyotrophic lateral sclerosis
    clinical features, 59/233-235
    dementia, 59/233
    neuropathology, 59/233, 235
    Parkinson dementia, 59/235
    parkinsonism, 59/233-235
Sporadic progressive external ophthalmoplegia with
        ragged red fiber
    age at onset, 62/508
    clinical features, 62/508
    cytochrome c oxidase negative, 62/508
    differential diagnosis, 62/508
    mitochondrial DNA defect, 62/508
    muscle biopsy, 62/508
    muscle weakness, 62/508
    oculopharyngeal muscular dystrophy, 62/508
    ophthalmoplegia, 62/508
    ptosis, 62/508
Sporadic visceral myopathy
    type, 51/493, 70/328
Sporothrix schenkii
    chronic meningitis, 56/645
Sporotrichosis
    acquired immune deficiency syndrome, 71/300
    amphotericin B, 52/495
    ataxia, 52/495
    brain abscess, 52/494
    brain granuloma, 16/230
    brain infarction, 35/570, 52/496
    chronic meningitis, 56/645
    CSF, 35/569, 52/495
    flucytosine, 52/495
    headache, 35/569, 52/495
    lymphocytic meningitis, 35/569, 52/495
    meningitis, 35/569, 52/494
    radiculitis, 52/496
Sporotrichum
    fungal CNS disease, 35/568-570, 52/493
Sport
    head injury epidemiology, 23/26, 57/11
    headache, 48/124
    physical rehabilitation, 61/472
    rehabilitative, see Rehabilitative sport
Sport injury
    athletics, 23/576
    baseball, 23/575
    basketball, 23/575
    boxing, see Boxing injury
    brain injury, 23/527
    dementia, 46/395
    epilepsy, 23/530
    football, see Football injury

golf, 23/575
gymnastics, 23/577
handball, 23/575
head injury, 23/27
horse riding, 23/577
ice hockey, see Ice hockey injury
jiujitsu, 23/567
judo, 23/567
jukado, 23/567
juvenile spinal cord injury, 61/237
karate, 23/567
Kung Fu, 23/567
miscellaneous sport, 23/572, 582, 585
prophylaxis, 23/528
rugby football, see Rugby football injury
sailing, 23/585
skiing, see Skiing injury
soccer injury, 23/567
statistic, 23/528
surfing, 23/585
swimming, 23/585
traumatic epilepsy, 23/530
Spot
    ash leaf, see Ash leaf spot
    blind, see Blind spot
    Brushfield, see Brushfield spot
    café au lait, see Café au lait spot
    cherry red, see Cherry red spot
    retinal, see Retinal spot
    Roth, see Roth spot
Spotted fever, see Meningococcal meningitis
Sprain
    cervical hyperanteflexion, see Cervical
        hyperanteflexion sprain
    cervical hyperdeflexion, see Cervical
        hyperdeflexion sprain
    cervical vertebral column injury, 26/289
    definition, 25/240, 243
    hyperanteflexion, see Cervical hyperanteflexion
        sprain
    hyperdeflexion, see Cervical hyperdeflexion
        sprain
    wry neck, 25/311
Spreading depression
    aura, 48/59, 68, 113
    behavior disorder, 48/67
    brain blood flow, 48/68, 73
    cortical blood flow, 48/67
    headache, 48/72
    hemiparesis, 48/68
    hippocampus, 48/65, 68
    hypermetabolism, 53/125
    metabolic inhibition, 48/66

migraine, 5/41, 48/65, 68, 72-74
migrainous acute confusional state, 48/166
migrainous transient global amnesia, 48/166
piribedil, 48/68
potassium, 48/67, 113
regional brain blood flow, 48/74
regional cerebral blood flow, 48/68
retina, 48/66
scotoma, 48/65
stimuli, 48/66
thalamus, 48/65
trigeminal nerve fiber, 48/113
133xenon, 48/68
Sprengel deformity
  birth incidence, 43/483
  cervico-oculoacusticus syndrome, 32/123-130
  elevated scapula, 43/483
  hemivertebra, 42/30, 43/483
  hypertelorism, 30/248
  Klippel-Feil syndrome, 1/483, 42/37, 43/483
  multiple nevoid basal cell carcinoma syndrome,
    14/459
  rib abnormality, 43/483
  scoliosis, 1/483
  skeletal malformation, 30/248, 32/41
  spastic paraplegia, 59/431
  vertebral abnormality, 43/483
Spring-Summer encephalitis
  Russian, see Central European tick-borne
    encephalitis
Sprue
  deficiency neuropathy, 7/618
  neuropathy, 28/25
  nontropical, 28/25
  tropical, see Tropical sprue
Spumavirus
  retrovirus, 56/529
Spur
  vertebral bony, see Vertebral bony spur
Spurling test
  meningeal irritation, 1/543
Squaliformes intoxication
  elasmobranche, 37/70
Squalus acanthias intoxication
  elasmobranche, 37/70
Squamata intoxication
  fatality, 37/91
Squamous carcinoma
  Hermansky-Pudlak syndrome, 43/6
  oculocutaneous albinism, 43/5
  scalp tumor, 17/126
  xeroderma pigmentosum, 43/12
Squatting

autonomic nervous system, 74/619
  blood pressure, 74/619
  heart rate, 74/619
  hereditary multiple recurrent mononeuropathy,
    51/554
  orthostatic hypotension, 75/719
Squatting test
  autonomic nervous system, 74/619
Squill intoxication
  cardiac glycoside, 37/425
  historic aspect, 37/425
Squint
  headache, 5/205
SSPE, see Subacute sclerosing panencephalitis
St. Anthony fire, see Ergot intoxication
St. John disease, see Sydenham chorea
St. Louis encephalitis
  acute viral encephalitis, 34/73-75, 56/134, 137
  animal viral disease, 34/292
  brain blood flow, 56/87
  cerebellum, 56/138
  Culex nigripalpus, 56/137
  Culex pipiens, 56/137
  Culex tarsalis, 56/137
  epilepsy, 56/138
  fatality rate, 56/138
  flavivirus, 56/12
  high rainfall, 56/137
  reticuloendothelial system, 56/30
  sewage, 56/137
  substantia nigra, 56/138
  thalamus, 2/486, 56/138
  Togavirus, 56/12
  tremor, 56/138
St. Louis encephalitis virus
  deafness, 56/107
  experimental acute viral myositis, 56/195
  vertigo, 56/107
St. Modestus dance, see Sydenham chorea
St. Vitus dance, see Sydenham chorea
Stab wound
  abdominal cavity, 25/205
  brachial plexus, 25/203
  brain, see Brain stab wound
  Brown-Séquard syndrome, 25/199
  congenital arteriovenous fistula, 31/200
  cord transection, 25/52
  cranial nerve, 23/485
  diaphragm, 25/205
  hemorrhagic necrosis, 25/52
  lumbar puncture, 25/201
  myelography, 25/201
  penetrating head injury, 57/307, 309

skull injury, 23/477
spinal cord, *see* Spinal cord stab wound
spinal cord injury, *see* Spinal cord stab wound
spinal injury, 25/52, 198-206
vertebral column, 25/52
Stabilization
  cervical spine, 25/293-302
  cervical vertebral column injury, 25/293
  nonsurgical, 25/293
  surgical, 25/293-302
  thoracic vertebral column injury, 61/81
Stagnant anoxia
  brain anoxia, 27/44, 46/358
Stagnation thrombosis
  thromboangiitis obliterans, 11/133
Stammering, *see* Stuttering
Standing
  autonomic nervous system, 74/609, 611
  blood pressure, 74/611
  cardiovascular function, 74/597
  cardiovascular response, 74/610
  gravity, 74/609
Standing disturbance, *see* Astasia abasia
Stapedial artery
  anastomosis, 12/309
Stapedial reflex testing
  vestibular neurinoma, 68/427
Stapedius muscle
  facial nerve injury, 24/111
  stuttering, 46/170
Stapedius reflex
  acoustic neuroma, 16/315
  facial paralysis, 8/263
  testing method, 2/63
Staphylococcal meningitis
  aminoglycoside, 33/72-74
  antibiotic agent, 52/72
  bacteremia, 52/71
  bacterial meningitis, 52/4
  chloramphenicol, 33/72-74
  clinical features, 33/70
  craniotomy, 33/69
  CSF, 33/71
  CSF shunt, 52/73
  endocarditis, 52/71
  epidural empyema, 52/72
  gentamicin, 33/72-74
  head trauma, 33/69
  hypoglycorrhachia, 33/71
  incidence, 33/69, 52/73
  laboratory data, 33/71
  mortality rate, 52/71
  neurosurgical intervention, 52/71

Ommaya reservoir, 52/73
pathogenesis, 33/69
penicillin, 33/72-74
rhinorrhea, 33/69
risk factor, 52/73
subdural empyema, 52/73
tobramycin, 33/72-74
treatment, 33/72-74, 52/72
vancomycin, 33/72-74
Staphylococcus
  acquired myoglobinuria, 62/577
  bacterial food poisoning, 37/87
  brain embolism, 55/166
  cerebrovascular disease, 55/166
  food intoxication, 37/87
  spastic paraplegia, 59/436
  spinal epidural empyema, 33/190
Staphylococcus aureus
  bacterial endocarditis, 52/292, 295, 298, 301
  bacterial meningitis, 52/5
  brain abscess, 52/145, 149
  cavernous sinus thrombosis, 52/173
  cranial epidural empyema, 52/169
  drug abuse, 55/517
  infectious endocarditis, 63/114
  mycotic brain aneurysm, 52/295
  spinal cord abscess, 52/191
  spinal epidural empyema, 52/186
  spinal meningitis, 52/190
  spinal subdural empyema, 33/190, 52/189
  subdural empyema, 52/73
  superior longitudinal sinus thrombosis, 52/179
  toxic shock syndrome, 52/257
  transverse sinus thrombosis, 52/177
  tropical polymyositis, 62/373
  venous sinus thrombosis, 54/431
Staphylococcus epidermidis
  bacterial meningitis, 52/5
  brain abscess, 52/150
  CSF shunt infection, 52/73
  gram-negative bacillary meningitis, 52/110
Staphyloma
  peripapillary, *see* Peripapillary staphyloma
Staple food
  livona pica intoxication, 37/82
  vitamin B2 deficiency, 51/332
Starfish intoxication
  asteroidea, 37/66
Stargardt disease, 13/124-144
  associated disease, 13/132
  Bruch membrane, 13/126
  central chorioretinitis, 13/142
  central choroidal atrophy, 13/138

centroperipheral tapetoretinal dystrophy, 13/127
color vision, 13/129
deafness, 13/133
dermatitis herpetiformis, 13/133
differential diagnosis, 13/133
dominant type, 13/131
elliptocytosis, 13/133
EOG, 13/130
ERG, 13/128
Friedreich ataxia, 13/132
fundus flavimaculatus, 13/28, 129, 133, 136,
    42/420
Grönblad-Strandberg syndrome, 13/134
hemeralopia, 13/262
hereditary, 13/130
heterochromia iridis, 13/133
histology, 13/131
Hutchinson-Tay choroidopathy, 13/133
hyperkinesia, 13/132, 42/420
hyperreflexia, 42/420
inverse primary pigmentary retinal degeneration,
    13/28, 133
keratoconus, 13/133
Laurence-Moon-Bardet-Biedl syndrome, 13/133
microphthalmia, 13/133
myopia, 13/140
myotonic dystrophy, 13/140
neurolipidosis, 13/132, 134
optic atrophy, 13/83
paraplegia, 13/132
peripheral involvement, *see* Centroperipheral
    tapetoretinal dystrophy
phenothiazine intoxication, 13/143
photically-evoked response, 13/130
photophobia, 13/262
primary pigmentary retinal degeneration, 13/190,
    252
recessive inheritance, 13/266
retinal vein occlusion, 13/133
rubella, 13/142
secondary pigmentary retinal degeneration, 13/28
sella turcica, 42/420
sex linked juvenile retinoschisis, 13/126
spinocerebellar degeneration, 13/132
spinocerebellofoveal dystrophy, 13/132
symptomatology, 13/125
transition, 13/128
tremor, 42/420
type, 13/125
vision loss, 13/262
visual acuity, 13/129, 270
visual adaptation, 13/129
visual field, 13/129, 270

vitreoretinal traction disease, 13/134
Stargazer intoxication, 37/83
Stark-Kaeser syndrome, *see* Scapuloperoneal spinal
    muscular atrophy
Starr, M.A., 1/5
Startle
    autosomal recessive generalized myotonia, 62/266
    fall, 62/266
Startle epilepsy
    see also Hyperexplexia *and* Syncinésie sursaut
    auditory stimulation, 15/441
    clinical features, 49/620
    involuntary movement, 6/301
    musicogenic epilepsy, 73/192
    organolead intoxication, 64/131
    organotin intoxication, 64/139
    stereotaxic exploration, 15/771
    visual stimulation, 15/441
Startle myoclonus
    Creutzfeldt-Jakob disease, 56/545
    Gilles de la Tourette syndrome, 49/636
Startle reaction
    Creutzfeldt-Jakob disease, 46/291
    globoid cell leukodystrophy, 10/72, 303, 51/373
    GM_2 gangliosidosis, 10/288, 391, 42/433
    GM_2 gangliosidosis type I, 42/433
    GM_2 gangliosidosis type II, 42/435
    jumpers of the Maine, 42/231
    pyrethroid intoxication, 64/214
    septal area, 2/770
    striatopallidodentate calcification, 49/422
Startle reflex
    snake envenomation, 65/180
Startle syndrome
    audiogenic, *see* Audiogenic startle syndrome
    Creutzfeldt-Jakob disease, 49/620
    essential, *see* Essential startle disease
    hyperexplexia, 49/620
    myoclonus, 49/619
Starvation neuropathy
    cachexia, 8/15
    deficiency neuropathy, 7/644
    protein deficiency, 7/633
Static tremor, 6/814-816
    see also Postural tremor
    clinical features, 6/814
    definition, 2/394, 49/583
    mechanism, 6/815
    neocerebellar syndrome, 2/420
    olivopontocerebellar atrophy (Dejerine-Thomas),
        21/425
    treatment, 6/816
Stationary generalized chorea

juvenile hereditary benign chorea, 49/346
Statoacoustic nerve, 2/72-77
  see also Acoustic nerve
  abnormality, 30/407-410
  acoustic neuroma, 16/307
  benign paroxysmal positional nystagmus, 2/377
  cerebellopontine angle syndrome, 2/318
  cerebellopontine angle tumor, 2/377
  embryology, 30/398
  Gellé syndrome, 2/241
  Meniere disease, 2/377
  speech test, 2/73
Statognosis
  terminology, 1/95
Statokinesimetry
  Huntington chorea, 49/289
Stature
  Klinefelter syndrome, 50/550
  short, *see* Short stature
  XYY syndrome, 31/487
Status asthmaticus
  acquired myoglobinuria, 62/573
  critical illness polyneuropathy, 51/575
Status Bonnevie-Ullrich, *see* Turner syndrome
Status cavernosus
  features, 6/674
Status corticis verrucosus deformans, *see* Status
  verrucosus deformis
Status cribrosus
  brain embolism, 53/316
  brain infarction, 53/316
  carotid system syndrome, 53/316
  neuropathology, 6/673
  striatal necrosis, 49/502
  thalamus, 2/486
  torticollis, 6/582
  transient ischemic attack, 53/316
Status dysmyelinisatus, *see* Hallervorden-Spatz
  syndrome
Status dysmyelinisatus pallidi, *see* Infantile
  neuroaxonal dystrophy
Status dysraphicus, *see* Dysraphia
Status epilepticus
  see also Epilepsy *and* Koshevnikoff epilepsy
  absence, 73/323
  absence status, 73/474
  acquired myoglobinuria, 62/573
  aminophylline, 46/539
  barbiturate, 73/329
  benzodiazepine, 73/328
  brain injury, 15/180
  brain tumor, 16/261
  cardiac arrest, 63/217

  cavernous angioma, 42/724
  child, 73/326
  chloral hydrate, 73/331
  ciclosporin intoxication, 65/552
  classification, 73/318
  clomethiazole, 73/330
  clonic, 73/323
  complex partial, 73/324
  corpus callosum agenesis, 42/7
  definition, 72/16
  diazepam, 15/184
  EEG, 15/513, 72/308, 73/270
  encephalopathy, 73/267
  epilepsy, 72/61
  Epstein-Barr virus, 56/252
  etomidate, 73/331
  febrile, *see* Febrile status epilepticus
  generalized, 15/2, 107, 169
  generalized convulsive, 73/318
  generalized myoclonic status, 15/160
  generalized tonic, 15/153
  generalized tonic clonic, 15/181
  grand mal status, 15/146, 150, 462
  head injury, 57/192
  hemiclonic, 15/182
  hemigrand mal status, 15/168
  Huntington chorea, 15/339
  hypernatremia, 63/554
  hyponatremia, 63/545
  hysteria, 46/576
  infantile, 15/178
  infection, 72/145
  Koshevnikoff epilepsy, 15/176
  Landau-Kleffner syndrome, 73/274
  legionellosis, 52/254
  Lennox-Gastaut syndrome, 73/274
  lissencephaly, 42/40
  lorazepam, 73/328
  lysergide, 65/45
  midazolam, 73/328
  multiple myeloma, 18/255
  myoclonic, 73/323
  myoclonic epilepsy, 15/160
  myoclonus, 15/160
  neurochemistry, 73/318
  neurogenic pulmonary edema, 63/495
  neurologic intensive care, 55/204
  neuropathology, 73/318-320
  neurophysiology, 15/54, 73/318
  opticocochleodentate degeneration, 42/241
  partial, 15/175, 181, 73/327
  pathophysiology, 73/273
  petit mal status, 15/169, 595

phenobarbital, 73/329
phenytoin, 73/329
pneumococcal meningitis, 33/47, 52/51
poststatus, 72/61
prevalence, 73/317
propofol, 73/331
Richards-Rundle syndrome, 43/264
simple partial, 73/325
sleep, 73/267
striatal convulsion, 6/169
survey, 15/145
Talairach AC-PC quideline, 15/760
thiopental, 73/329
tonic, 73/322
tonic clonic, 15/146, 73/320
treatment, 15/183, 73/327, 332
treatment protocol, 73/331
unilateral, 15/160
vitamin B6 deficiency, 42/716
water intoxication, 63/547
Status lacunaris, *see* Brain lacunar infarction
Status marmoratus
   athetosis, 2/502, 6/16, 60
   brain infarction, 53/190
   cause, 6/448
   cerebral palsy, 6/678, 46/16
   cerebrovascular disease, 53/28
   clinical features, 6/449
   double athetosis, 6/155, 447, 534, 49/384, 386,
      507
   dystonia, 6/533, 546-551
   dystonia musculorum deformans, 49/524
   hypermyelination, 6/676
   mental deficiency, 46/16
   perinatal hypoxemia, 6/678
   prenatal carbon monoxide, 6/584
   progressive myoclonus epilepsy, 42/700
   putamen, 42/700
   striatal necrosis, 49/500, 505
   symptomatic dystonia, 6/546, 49/541
   thalamus, 2/486
Status marmoratus dysmyelinisatus
   epilepsy, 72/109
Status marmoratus (Vogt), *see* Status marmoratus
Status migrainosus
   duration, 48/125
   migraine, 5/380
Status porosus
   infantile neuroaxonal dystrophy, 42/230
Status spongiosus, 27/202-205
   cerebral hemiatrophy, 50/206
   Creutzfeldt-Jakob disease, 46/293
   glycogen, 27/203

spongiform encephalopathy, 11/129, 159-162
Status subchoreaticus
   evidence, 6/329
   Huntington chorea, 49/287
   rigid type, 6/307
Status verrucosus
   brain cortex heterotopia, 50/203
   brain stem, 50/202
   cauliflower brain, 50/202
   cell migration, 30/489
   cerebellum, 50/202
   clinical features, 30/352, 50/204
   corpus callosum agenesis, 50/203
   definition, 30/347, 50/201
   epilepsy, 50/204
   etiology, 30/351
   Exner plexus, 30/349
   histologic features, 30/488-491
   history, 30/347, 50/202
   localization, 30/488
   management, 30/352
   mental deficiency, 30/352
   pathogenesis, 30/489, 492
   pathology, 30/347-351, 488, 50/202
   polymicrogyria, 30/489, 50/202, 253
   septum pellucidum agenesis, 50/203
   verrucose retinal dysplasia, 50/203
Status verrucosus deformis
   brain cortex dysplasia, 50/253
   brain wart, 50/253
   microgyria, 30/479, 50/202
   nomenclature, 50/201
   pathogenesis, 30/489
   status verrucosus simplex, 50/251
Status verrucosus simplex
   brain cortex embryology, 30/489
   brain cortex heterotopia, 50/253
   status verrucosus deformis, 50/251
Stavudine
   neurotoxicity, 71/368
   toxicity, 71/372
Steal syndrome
   Adamkiewicz artery, 32/487
   angiography, 32/487
   ischemic myelopathy, 32/487
   paraplegia, 32/487
   spinal angioma, 32/487
   spinal cord intermittent claudication, 32/487
   subclavian, *see* Subclavian steal syndrome
Stearic acid
   biochemistry, 66/35
   chorea-acanthocytosis, 63/286
   migraine, 48/95

plasma, 48/96
N-Stearylglucocerebrosidase
  Gaucher disease, 10/565
Steatorrhea
  Bassen-Kornzweig syndrome, 13/421, 51/327
  deficiency neuropathy, 51/324, 327
  idiopathic, *see* Celiac disease
  vitamin E deficiency, 60/668
  Wolman disease, 51/379
Steele-Richardson-Olszewski syndrome, *see*
    Progressive supranuclear palsy
Steely hair, *see* Trichopoliodystrophy
Steindler procedure
  brachial plexus, 51/152
Steiner Wetterwinkel, *see* Wetterwinkel (Steiner)
Steinert disease, *see* Myotonic dystrophy
Stelazine, *see* Trifluoperazine
Stellate ganglion
  cardiac dysrhythmia, 63/238
  ECG change, 63/235
Stellio
  history, 6/300
Stellwag sign
  pseudo, *see* Pseudo-Stellwag sign
Stenogyria
  Arnold-Chiari malformation type II, 50/407
Stenosing amyloidosis
  multinodular, *see* Multinodular stenosing
    amyloidosis
Stenosis
  acquired, 50/466, 52/358
  aorta, *see* Aorta stenosis
  aqueduct, *see* Aqueduct stenosis
  artery, *see* Artery stenosis
  basilar artery, *see* Basilar artery stenosis
  brain arteriosclerosis, 53/94
  carotid artery, *see* Carotid artery stenosis
  cervical, *see* Cervical stenosis
  cervical spinal canal, *see* Cervical spinal canal
    stenosis
  combined, *see* Combined stenosis
  craniovertebral, *see* Craniovertebral stenosis
  dysostotic, *see* Dysostotic stenosis
  dysplastic, *see* Dysplastic stenosis
  foramen magnum, *see* Foramen magnum stenosis
  innominate artery, *see* Innominate artery stenosis
  interpedicular, *see* Interpedicular stenosis
  intervertebral foramen, *see* Intervertebral foramen
    stenosis
  lumbar, *see* Lumbar stenosis
  lumbar vertebral canal, *see* Lumbar vertebral canal
    stenosis
  lumbar vertebral canal stenosis, 20/780

  mitral, *see* Mitral stenosis
  pulmonary artery, *see* Pulmonary artery stenosis
  pure, *see* Pure stenosis
  sex linked recessive aqueduct, *see* Sex linked
    recessive aqueduct stenosis
  spinal canal, *see* Vertebral canal stenosis
  subclavian artery, *see* Subclavian artery stenosis
  thoracic, *see* Thoracic stenosis
  toxoplasmosis, 52/358
  vertebral artery, *see* Vertebral artery stenosis
  vertebral canal, *see* Vertebral canal stenosis
  vertebrobasilar, *see* Vertebrobasilar stenosis
Stensen, N., 2/10
Stenvers projection
  acoustic nerve tumor, 17/680
  craniocervical bone abnormality, 32/10
Step voltage
  electrical injury, 61/194
Steppage gait
  hereditary amyloid polyneuropathy type 1, 51/420
  organophosphate induced delayed neuropathy,
    64/229
Stepping reflex
  newborn behavior, 4/346
Stereoanesthesia
  agnosia, 4/34
  terminology, 1/106
Stereoencephalotomy, 6/856-864
  brain arteriovenous malformation, 12/254
  clinical work-up, 6/863
  complication, 6/867
  contraindication, 6/863
  dystonia, 6/562
  epilepsy, 15/758
  Huntington chorea, 6/360
  indication, 6/863
  Parkinson disease, 6/203
  parkinsonism, 6/203
  postoperative care, 6/874
  procedure type, 6/864
  psychological preparation, 6/865
  stereotaxic exploration, 15/758, 760
  Talairach AC-PC quideline, 15/760
  target, 6/861
  thalamus, 2/491
Stereognosis
  cerebral dominance, 4/278
  disorder, 2/667
  labiobuccal, *see* Labiobuccal stereognosis
  schizophrenia, 46/490
  sensory extinction, 3/191
  terminology, 1/104
  testing method, 2/683

Stereopsis, *see* Depth perception
Stereoscopy
  parietal lobe, 45/362
Stereotactic lesion
  epilepsy, 73/407
Stereotaxic thalamotomy
  benign essential tremor, 49/577
  tremor, 49/597
Stereotaxic ventral intermedius thalamotomy
  tremor, 49/603
Stereotaxy
  frameless system, 67/243
  localizer, 67/247
Stereotyped behavior
  frontal lobe lesion, 2/737
Stereotypy
  acute neuroleptic effect, 65/278
  akathisia, 65/286
  Alzheimer disease, 46/252
  amphetamine, 46/592
  aphasia, 4/89-91
  behavior disorder, 46/592
  Huntington chorea, 49/277
  infantile autism, 46/190
  motor, *see* Motor stereotypy
  neuroleptic akathisia, 65/286
  neuroleptic difference, 65/280
  neuroleptic dyskinesia, 65/278
  Pick disease, 46/240
  synkinesis, 1/410
  traumatic psychosyndrome, 24/551
Sterile meningitis, *see* Meningitis serosa
Stern-Garcin syndrome, *see* Thalamic dementia
Sternberg giant cell, *see* Reed-Sternberg cell
Sternberg-Reed cell, *see* Reed-Sternberg cell
Sternocleidomastoid muscle hypertrophy
  spasmodic torticollis, 42/763
Sternocleidomastoid tumor
  pentazocine, 41/385
Sternum fracture
  spinal injury, 26/193
Sternum malformation
  glycosaminoglycanosis type VI, 10/437
  Sturge-Weber syndrome, 14/512
Steroid
  see also Corticosteroid *and* Glucocorticoid
  adrenal, *see* Adrenal steroid
  brain abscess, 33/137
  brain edema, 16/176, 203, 57/214, 67/73
  brain metastasis, 69/194
  cardiotonic, *see* Cardiotonic steroid agent
  carotidynia, 48/339
  chronic idiopathic demyelinating

  polyradiculoneuropathy, 47/614
  chronic migrainous neuralgia, 48/252
  chronic subdural hematoma, 57/288
  fever inducing, 1/448
  Guillain-Barré syndrome, 51/257
  headache, 48/434, 437
  intracranial pressure, 23/212
  migraine, 5/56
  olfaction, 74/380
  sarcoid neuropathy, 51/195
  sleep, 3/105
  spinal cord metastasis, 69/183
  temporal arteritis, 48/313, 323, 325
  thromboxane, 48/202
Steroid excretion
  Pick disease, 46/242
Steroid hormone
  signal transduction, 67/24
Steroid myopathy
  atrophy fiber type II, 40/161, 62/384
  cardiac pharmacotherapy, 63/189
  corticosteroid, 41/250, 62/384
  creatine kinase, 62/535, 605
  creatinuria, 62/535
  Cushing syndrome, 40/162, 41/250-253, 62/535
  dermatomyositis, 62/384
  dexamethasone, 41/252
  diagnosis, 62/535
  dose-response relationship, 62/605
  EMG, 62/535
  endocrine myopathy, 41/250-253
  exogenous steroids, 40/162
  experimental, 40/160-162
  inclusion body myositis, 62/384
  muscle fiber type II, 40/161
  muscle fiber type II atrophy, 62/605
  muscle fiber type IIb atrophy, 62/535
  muscle ultrastructure, 62/535
  muscular atrophy, 62/535
  pathogenesis, 62/536
  phenytoin, 40/558, 41/252
  polymyositis, 62/384
  sex ratio, 62/535
  treatment, 62/536
  triamcinolone, 41/252
Steroid responsive element
  gene transcription, 66/72
Steroid sulfatase deficiency, *see* X-linked ichthyosis
Steroid therapy
  benign intracranial hypertension, 67/111
Steryl sulfatase deficiency, *see* X-linked ichthyosis
Stewart-Holmes sign
  ataxia, 1/333

neocerebellar syndrome, 2/420
technique, 1/329
Stewart principle
brain blood flow, 11/120
Stewart syndrome, *see* Midline granuloma
Stibler-Jaeken syndrome, *see* Jaeken syndrome
Stickler syndrome
cataract, 43/485
chorioretinal degeneration, 13/41
hereditary, *see* Hereditary progressive
arthro-ophthalmopathy
hyperextensible joint, 43/484
hypertelorism, 30/247
kyphoscoliosis, 43/484
myopia, 43/484
ocular defect, 30/247
optic atrophy, 13/41
retinal detachment, 43/484
Robin syndrome, 43/484
vertebral abnormality, 43/484
Stiff-man syndrome
autonomic nervous system, 75/18
diazepam, 40/546, 41/306
differential diagnosis, 41/348
familial form, 41/306
hypertonia, 1/268
motor unit hyperactivity, 41/305, 348, 59/377
muscle cramp, 41/306
myotonia, 62/274
myotonic syndrome, 62/274
neuromuscular block, 41/306
paraneoplastic syndrome, 69/363, 374, 380,
71/682
symptom, 62/274
synonym, 40/330
therapy, 71/682
Stiff-man syndrome like disorder
hereditary type, 43/145
Stilbamidine
toxic neuropathy, 7/520
Stilling-Türk-Duane syndrome, 42/305-307
see also Abducent nerve abnormality, Acquired
retraction syndrome, Oculomotor nerve,
Oculomotor nerve abnormality *and* Trochlear
nerve abnormality
abducent nerve paralysis, 42/306
cervico-oculoacusticus syndrome, 32/123, 127
crocodile tear syndrome, 50/213
genetics, 50/213
Klippel-Feil syndrome, 50/213
Marcus Gunn phenomenon, 30/403, 50/213
neuropathy, 42/307
oculomotor disorder, 2/326

oculomotor nerve abnormality, 30/403
optic atrophy, 13/83
orbital injury, 24/89
sensorineural deafness, 50/213
sex ratio, 42/306, 50/213
skeletal deformity, 42/306
strabismus, 42/306
thalidomide, 42/307
trochlear nerve abnormality, 30/403
Wildervanck syndrome, 43/343, 50/213
Stimulant
excretion, 37/237-241
history, 37/223-232
toxicity, 37/223-297
Stimulus
LOC, *see* LOC stimulus
sensory, 7/83
stress, 45/247
Stimulus isolation unit
motor conduction velocity, 7/120
technique, 7/120
Stingray intoxication
anxiety, 37/72
arrhythmia, 37/72
cramp, 37/72
death, 37/70
diarrhea, 37/72
elasmobranche, 37/70
hemorrhage, 37/70
muscle fasciculation, 37/72
nausea, 37/72
pain, 37/72
respiratory distress, 37/72
spinal injury, 37/71
sweating, 37/72
swelling, 37/72
syncope, 37/72
treatment, 37/72
venom, 37/72
vomiting, 37/72
weakness, 37/72
wound, 37/71
Stippled epiphysis
congenital, *see* Conradi-Hünermann syndrome
fetal warfarin syndrome, 46/50
sensorineural deafness, 42/375
Stoddard solvent
organic solvent intoxication, 64/40
Stokes-Adams syndrome, *see* Adams-Stokes
syndrome
Stokes-Mandeville-Egerton bed
paraplegia, 26/221, 301, 472
spinal injury, 26/300, 302

Stokes-Mandeville games
  rehabilitative sport, 26/528, 539
Stomach
  Chagas disease, 75/398
  diabetes mellitus, 75/592
  dilatation, 1/506
  function, 23/126
  hypothalamus, 1/506
  motility, 1/509
  neural regulation, 1/505
  secretion, 1/506
Stomach ulcer
  see also Stress ulcer
  brain injury, 23/126
  hypothalamic lesion, 1/510
Stomatitis
  acute mercury intoxication, 64/375
  bismuth intoxication, 64/331
  chelonia intoxication, 37/91
  dental headache, 5/226
  herpetic, see Herpetic stomatitis
  pellagra, 6/748, 28/59, 71, 87
Stonefish intoxication
  antivenene, 37/77
  venom, 37/69, 76
Stooped posture
  Parkinson disease, 6/189
Storage disease
  cholesterol ester, see Cholesterol ester storage
    disease
  glycogen, see Glycogen storage disease
  glycoprotein, see Glycoprotein storage disease
  histochemistry, 40/52-54
  lipid, see Lipid storage disease
  lysosome, see Lysosomal storage disease
  muscle biopsy, 62/17
  myopathy, 62/16
  neurovisceral, see Neurovisceral storage disease
  polysaccharide, see Polysaccharide storage
    disease
  triglyceride, see Triglyceride storage disease
Storage dystrophy
  myelin, see Myelin storage dystrophy
  neuronal, see Neuronal storage dystrophy
Storage myopathy
  lipid, see Lipid storage myopathy
Störring thalamic blindness, see Thalamic blindness
    (Störring)
Strabismus
  acrocephalosyndactyly type I, 43/318
  acrocephalosyndactyly type V, 43/320
  amblyopia, 42/421
  ARG syndrome, 50/581

ataxic diplegia, 42/121
Behr disease, 42/416
Chédiak-Higashi syndrome, 42/536
Coffin-Lowry syndrome, 43/238
concomitant, see Concomitant strabismus
congenital cerebellar ataxia, 42/124
congenital heart disease, 63/11
congenital nystagmus, 42/159
convergent, see Convergent strabismus
corpus callosum agenesis, 42/101
craniofacial dysostosis (Crouzon), 43/359
cutis verticis gyrata, 43/244
cyclic oculomotor paralysis, 42/331
De Lange syndrome, 43/246
divergent, see Divergent strabismus
Down syndrome, 50/526
dysosteosclerosis, 43/393
Ehlers-Danlos syndrome, 43/14
elfin face syndrome, 31/318
Fanconi syndrome, 43/16
Förster syndrome, 42/202
Friedman-Roy syndrome, 43/251
Gaucher disease, 42/438
Goltz-Gorlin syndrome, 43/20
hereditary vitreoretinal degeneration, 13/274
incontinentia pigmenti, 43/23
infantile Gaucher disease, 10/307, 565
juvenile Gaucher disease, 29/352, 42/438
Marinesco-Sjögren syndrome, 21/556, 42/184
microcephaly, 43/429, 50/269
Mietens-Weber syndrome, 43/304
Mulibrey dwarfism, 43/435
oculocutaneous albinism, 43/4, 7
ophthalmoplegic migraine, 48/132
orthochromatic leukodystrophy, 42/500
9p partial monosomy, 50/578
18p partial monosomy, 43/534
11p partial trisomy, 43/525
20p partial trisomy, 43/539
paralytic, see Paralytic strabismus
Prader-Labhart-Willi syndrome, 31/324, 43/463
18q partial monosomy, 43/532
Rubinstein-Taybi syndrome, 31/282, 43/234,
    46/91, 50/276
Saethre-Chotzen syndrome, 43/322
Stilling-Türk-Duane syndrome, 42/306
Summitt syndrome, 43/372
XXXX syndrome, 43/554
XXXXX syndrome, 43/554
Strabismus fixus
  medial rectus fibrosis, 2/327
Strachan-Scott-Stannus syndrome, see Tropical
    ataxic neuropathy

Strachan-Scott syndrome, *see* Tropical ataxic
   neuropathy
Strachan syndrome, *see* Tropical ataxic neuropathy
Straight leg raising test, *see* Lasègue sign
Straight sinus
   thrombophlebitis, 33/172
Straight sinus thrombosis
   anatomy, 33/172
   pathology, 32/172
   symptom, 33/172, 54/398
   thalamic infarction, 54/398
Strains, *see* Stress
Strambio ventricle, *see* Cavum vergae
Strandin
   GM$_2$ gangliosidosis, 10/390
Strangulation
   pallidal necrosis, 49/466
   pallidostriatal necrosis, 49/466
   striatal necrosis, 49/500
Strangulation encephalopathy
   pallidostriatal necrosis, 49/480
Strasbourg, France
   neurology, 1/8
Sträussler disease, *see*
   Gerstmann-Sträussler-Scheinker disease
Streeter band, *see* Amniotic band syndrome
Strength-duration curve
   Hodgkin-Huxley model, 7/67
Strephosymbolia
   developmental dyslexia, 46/107
Streptococcal meningitis, 33/69-74
   adult type, 52/83
   antibiotic agent, 52/84
   bacteremia, 52/78
   blindness, 52/81
   cell mediated immunity, 52/80
   chemoprophylaxis, 52/82
   clinical features, 52/80
   co-trimoxazole, 52/84
   CSF, 33/71, 52/81
   deafness, 52/81
   diabetes mellitus, 52/79
   early onset infection, 52/81
   epilepsy, 52/81
   group A streptoccoci, 52/77
   group B streptococci, 52/77
   hydrocephalus, 52/81
   hypoglycorrhachia, 33/71
   immunization, 52/82
   inappropriate antidiuretic hormone secretion,
     52/82
   late onset infection, 52/81
   lipoteichoic acid, 52/79

   mental deficiency, 52/81
   mortality, 52/81
   patient, 33/69
   periventricular leukomalacia, 52/81
   predisposing factor, 52/79
   prevention, 52/82
   purpura fulminans, 52/82
   rapid test, 52/82
   rhinorrhea, 33/69
   scarlet fever, 52/77
   sequela, 52/81
   Streptococcus agalactiae, 52/78
   Streptococcus milleri, 52/80
   treatment, 52/84
Streptococcus
   bacterial endocarditis, 52/292, 298
   classification, 52/78
   fibronectin, 52/79
   group B serotype, 52/79
   group D, 52/77
   group N, 52/77
   hemolysis, 52/78
   α-hemolysis, 52/77
   β-hemolysis, 52/77
   Lancefield group, 52/78
   lipoteichoic acid, 52/79
   nerve lesion, 7/487
   tropical polymyositis, 62/373
Streptococcus agalactiae
   cranial subdural empyema, 52/171
   streptococcal meningitis, 52/78
Streptococcus faecalis
   bacterial endocarditis, 52/301
Streptococcus milleri
   brain abscess, 52/149
   streptococcal meningitis, 52/80
Streptococcus pneumoniae
   acquired myoglobinuria, 62/577
   bacterial endocarditis, 52/301
   bacterial meningitis, 52/3, 55/416
   brain metastasis, 69/116
   carrier state, 52/41
   cavernous sinus thrombosis, 52/173
   gram-negative bacillary meningitis, 52/110
   meningitis, *see* Pneumococcal meningitis
   serotyping, 52/41
   spinal cord abscess, 52/191
   spinal epidural empyema, 52/187
   spinal meningitis, 52/190
   spinal subdural empyema, 52/189
   superior longitudinal sinus thrombosis, 52/179
Streptococcus pyogenes
   brain abscess, 52/150

spinal meningitis, 52/190
Streptococcus viridans
  bacterial endocarditis, 52/300
  brain abscess, 52/149
  brain embolism, 55/166
  cerebrovascular disease, 55/166
  leukoencephalopathy, 9/609
  mononeuritis multiplex, 52/300
  spinal cord abscess, 52/191
  spinal subdural empyema, 52/189
Streptokinase
  anticoagulant, 63/308
  basilar artery thrombosis, 54/180
  bleeding risk, 63/320
  brain aneurysm neurosurgery, 55/59
  brain hemorrhage, 65/461
  iatrogenic neurological disease, 65/461
  middle cerebral artery syndrome, 53/368
  subarachnoid hemorrhage, 55/59
  vertebrobasilar artery thrombosis, 54/181
  vertebrobasilar insufficiency, 53/385
  vertebrobasilar system syndrome, 53/385
Streptomyces
  actinomycosis, 35/383
  fungal CNS disease, 35/570, 52/479
Streptomycin
  acoustic neuropathy, 64/9
  deafness, 51/295
  drug induced polyneuropathy, 51/295
  Hemophilus influenzae meningitis, 33/56
  neurobrucellosis, 33/317, 52/596
  neuropathy, 7/515
  neurotoxin, 7/515, 50/219, 51/295, 64/9, 65/483
  optic neuropathy, 65/483
  toxic neuropathy, 64/9
  tuberculous meningitis, 33/226-229, 52/211-213
Streptomycin intoxication
  optic neuropathy, 65/483
  polyneuropathy, 65/483
  symptom, 64/9
Streptomycin polyneuropathy
  acoustic nerve, 51/295
  calcium, 51/296
  sensorimotor type, 51/295
Streptomycosis
  brain infection, 35/570
  mycetoma, 35/570
Streptonigrin
  iniencephaly, 50/135
Streptozocin
  diabetic polyneuropathy, 51/510, 512
Streptozotocin, see Streptozocin
Stress, 45/245-254

acquired immune deficiency syndrome, 74/335
adaptive response, 74/307
adrenalin, 45/248, 74/313
aging, 74/232
altitude, see Altitude stress
anger, 45/251
antidepressant, 45/254
anxiety, 45/250, 253
anxiety tremor, 6/818
arcuate nucleus, 74/312
autonomic nervous system, 74/79-81, 232, 307,
  75/439
behavior therapy, 45/253
benzodiazepine, 45/254
blood pressure, 74/309
cardiac dysrhythmia, 63/239
catecholamine, 74/310
chemodectoma, 74/315
chronic muscle contraction headache, 48/345
cognition, 45/246
cold, 74/318
conditioning, 45/248
coping, 45/247
corticotropin, 45/248, 250, 74/308
corticotropin releasing factor, 74/307
delirium, 46/533
depression, 74/504
descending spinal fiber, 74/313
electricity, 74/320
emotional, see Emotional stress
endurance exercise training, 74/325
environmental, see Environmental stress
epilepsy, 73/454
fatigue, 74/325
feeling, 45/245
fever, 74/455
free floating anxiety, 45/249-251
general adaptation syndrome, 45/247
glucocorticoid, 45/248
gravitational, see Gravitational stress
head trauma, 74/318
heart rate, 74/309
heat, 74/320
hippocampus, 45/251
histopathology, 45/252
Horner syndrome, 74/317
human immunodeficiency virus, 74/335
hyperthermia, 74/320, 455
hypothalamus, 45/251, 74/504
hypothalamus-pituitary-adrenal axis, 74/308
hypothermia, 74/319
incertohypothalamic system, 74/313
interaction model, 45/247

intervertebral disc, 20/528
limbic system, 45/249-251
locus ceruleus, 74/307, 310
malaise, 45/248
mental, *see* Mental stress
migraine, 5/52, 45/253, 48/36, 121
multiple sclerosis, 47/179
nerve, 74/317
neuralgic amyotrophy, 51/174
neuroblastoma, 74/315
neurophysiology, 74/308
neurotransmission, 45/251
noradrenalin, 45/248, 74/310
noradrenergic nervous system, 74/309
olfactory system, 74/313
orthostatic, *see* Orthostatic stress
orthostatic hypotension, 74/325
orthostatic intolerance, 74/325
orthostatic intolerance syndrome, 74/324
panic disorder, 74/325
Parkinson disease, 49/124
pathophysiology, 45/249-252
peripheral nervous system, 74/313
periventricular system, 74/313
pituitary gland, 74/323
pregnancy, 74/328
psychiatric headache, 5/20
spinal cord, 74/317
stimulus, 45/247
temperature, 74/319
testing, 45/248
tetrahydrobiopterin deficiency, 74/316
thermal, *see* Thermal stress
thermoregulation, 45/249
trauma, 74/317
tuberoinfundibular system, 74/312
vegetative dystonia syndrome, 74/322
vibration neuropathy, 74/317
Stress incontinence
bladder, 75/670
Kegel exercise, 75/677
Stress ulcer
see also Stomach ulcer
etiology, 26/351
incidence, 26/351
radiology, 26/352
spinal cord injury, 26/351-353
treatment, 26/352
Stress wave
biomechanics, 23/94
brain concussion, 23/94, 98
brain injury, 23/94
brain injury mechanism, 57/21

head injury, 23/94, 98
Stretch
muscle, *see* Muscle stretch
nerve, *see* Nerve stretch
syncope, 75/218
Stretch disease
segmental, *see* Segmental stretch disease
Stretch reflex, 1/63-68
Brondgeest, 1/63
muscle, *see* Muscle stretch reflex
muscle spindle, 1/64, 6/95
tonic, *see* Tonic stretch reflex
Striatal afferent
diagram, 2/498
neostriatum, 49/5
Striatal convulsion
familial paroxysmal choreoathetosis, 6/169
hereditary paroxysmal kinesigenic
choreoathetosis, 6/169, 556
status epilepticus, 6/169
Striatal degeneration
classification, 6/632
dystonia musculorum deformans, 2/502, 6/666
manganese encephalopathy, 49/477
spontaneous, *see* Spontaneous striatal
degeneration
Sydenham chorea, 6/59
Striatal fiber
Huntington chorea, 6/399
Striatal hemiplegia
dystonia, 2/504
evidence, 2/503
Striatal involvement
pallidal degeneration, 49/450, 454
pallidoluysian degeneration, 49/454
pallidopyramidal degeneration, 49/446
Striatal lesion
Creutzfeldt-Jakob disease, 6/671, 699
electricity, 7/363
epilepsy, 6/676
hemorrhage, 6/673
hypoglycemic encephalopathy, 6/675
infarction, 6/673
olivocerebellar atrophy, 6/669
olivopontocerebellar atrophy, 6/669
orthostatic hypotension, 6/670
Parkinson disease, 6/670
Pick disease, 6/668, 699
spongiform encephalopathy, 6/671
spontaneous striatal degeneration, 6/679
trauma, 6/672
Wilson disease, 2/499, 6/671
Striatal necrosis

alcoholism, 49/500, 504
athetosis, 49/499, 501
brain anoxia, 6/674
brain lacunar infarction, 49/500, 502
carbon monoxide intoxication, 49/500, 504
cardiac arrest, 49/500
cardiorespiratory failure, 49/500
cavernous state, 49/502
central pontine myelinolysis, 49/507
cerebral palsy, 49/506
cerebral vein, 49/500
chorea, 49/499-501
comparative neurology, 49/511
corticostriate encephalitis, 6/679, 49/500, 512
cyanide intoxication, 49/500, 504
death, 6/676
decorticate posture, 49/499
double athetosis, 49/506
dystonia, 49/499, 501
eclampsia, 49/500
epilepsy, 6/676, 49/500
ethylene glycol intoxication, 64/124
excitotoxic transmitter, 49/500
experiment, 49/510
familial spastic paraplegia, 22/455
Förster syndrome, 49/509
hemiplegia, 49/501
hereditary, see Hereditary striatal necrosis
hereditary dystonic paraplegia, 22/455
Huntington chorea, 49/509, 511
hypoglycemia, 49/500
hypoglycemic encephalopathy, 49/503
hypoxic hypotensive state, 49/502
ibotenic acid, 49/511
insect bite, 6/676
insect bite encephalopathy, 49/503
kainic acid, 49/511
kainic acid intoxication, 49/500
lenticulostriate artery, 49/501
Macaca irus, 49/511
methanol intoxication, 6/676, 49/500, 503, 64/101
middle cerebral artery occlusion, 49/500
neuroleptic induced, 49/505
newborn asphyxia, 49/500, 506
parkinsonism, 49/502
pathology, 49/505
perinatal encephalopathy, 49/505
postanesthetic encephalopathy, 49/500
posthemiplegic chorea, 49/501
quisqualic acid, 49/511
respiratory arrest, 49/500
rigidity, 49/502
sodium azide intoxication, 49/500

status cribrosus, 49/502
status marmoratus, 49/500, 505
strangulation, 49/500
subacute necrotizing encephalomyelopathy,
    49/500, 507, 510
thrombosis, 49/500
vascular lesion, 49/500
Striatal syndrome, 2/497-504
    see also Rigidity
    characteristics, 6/633
    dystonia, 2/499
    Hallervorden-Spatz syndrome, 2/500
    Huntington chorea, 6/322
    involuntary movement, 2/499
Striate artery
    medial, see Heubner artery
    thalamo, see Thalamostriate artery
Striate cortex, see Visual cortex
Striate vein
    thalamo, see Thalamostriate vein
Striatonigral degeneration, 6/694-701, 60/537-546
    age at onset, 49/206
    Alzheimer disease, 49/209
    ataxia, 49/206
    athetosis, 42/262
    autonomic dysfunction, 49/206, 63/146
    Babinski sign, 42/262, 49/206
    basic material, 60/539
    cause, 6/700
    central respiratory disturbance, 49/208
    Cheyne-Stokes respiration, 49/208
    classification, 22/172
    clinical anatomical correlation, 49/209
    clinical criteria, 60/539
    clinical features, 6/695, 49/206
    Creutzfeldt-Jakob disease, 6/699
    CSF, 42/262
    dentate nucleus, 42/262
    diagnosis, 60/543
    diagnostic criteria, 60/539
    differential diagnosis, 49/209
    disease duration, 49/206
    dystonia, 42/262
    fasciculation, 42/262
    features, 6/667
    Ferguson-Critchley syndrome, 59/322
    hastening phenomenon, 49/68
    history, 60/537
    homovanillic acid, 42/262
    Huntington chorea, 6/355, 697, 699, 49/206, 293
    Joseph-Machado disease, 42/262, 60/468, 543
    levodopa, 42/263, 60/541
    Lewy body, 6/667, 698

material since 1964, 60/539
multiple neuronal system degeneration, 22/265,
    49/211, 60/540
muscular atrophy, 49/206
myokymia, 42/262
neuropathology, 6/695-697
neurophysiologic criteria, 60/539
nigrospinodentate degeneration, 60/543
normal pressure hydrocephalus, 60/540
nosology, 6/692-700
nuclear magnetic resonance, 60/543
olivocerebellar atrophy, 6/699
olivopontocerebellar atrophy, 49/205, 208-211,
    60/540
olivopontocerebellar atrophy (Dejerine-Thomas),
    21/417, 60/528
Onuf nucleus, 49/209
orthostatic hypotension, 63/146
pallidal atrophy, 6/649
pallidoluysionigral degeneration, 49/459
Parinaud syndrome, 49/206
Parkinson dementia complex, 60/540
Parkinson disease, 49/94, 206, 211, 60/540
parkinsonism, 42/261
pathogenesis, 6/700
pathologic finding, 49/208
polymyoclonia, 49/206
positron emission tomography, 60/543
progressive dysautonomia, 59/141
progressive supranuclear palsy, 49/246, 60/540
pure case, 60/544
pyramidal tract sign, 49/206
recent pure case, 60/544
REM sleep, 49/208
rigidity, 49/69, 206
rubral tremor, 49/206
Shy-Drager syndrome, 49/206, 210, 60/540
sleep apnea syndrome, 49/208
spastic ataxia, 42/261
speech disorder, 42/261
spinopontine degeneration, 60/544
substantia nigra degeneration, 42/262
supranuclear ophthalmoplegia, 49/209
treatment, 49/210
Striatonigral fiber
    dopamine, 29/464, 49/4
    enkephalin, 49/8
    GABA, 49/8
    substance P, 49/8
    substantia nigra, 49/8
Striatopallidal fiber
    enkephalin, 49/6
    GABA, 49/6
substance P, 49/6
Striatopallidodentate calcification, 6/703-720
    see also Intracerebral calcification, Pseudocalcium
        and Pseudopseudohypoparathyroidism
    acquired immune deficiency syndrome, 49/424
    alopecia, 49/423
    aluminum, 6/716, 49/425
    animal, 49/430
    ataxia, 49/420
    athetosis, 49/423
    biochemistry, 6/716-718, 49/425
    brain biopsy, 10/685
    brain tumor, 49/423
    brittle nail, 49/423
    calcium, 6/715-719, 49/425
    carbon monoxide intoxication, 49/417
    cataract, 49/419, 423
    cerebellar hypoplasia, 49/421
    chorea, 49/422
    clinical features, 6/707-712, 49/421
    cobalt, 6/716, 49/425
    Cockayne-Neill-Dingwall disease, 13/435
    computer assisted tomography, 49/428, 60/672
    copper, 6/716, 49/425
    course, 49/423
    cranial radiotherapy, 49/428
    CSF pressure, 49/422
    cystic cerebellar astrocytoma, 49/424
    cystine, 6/715
    deafness, 49/423
    delusion, 49/423
    dementia, 46/130, 49/419, 422
    dentate nucleus calcification, 49/426
    dentatorubropallidoluysian atrophy, 21/520
    development rate, 49/423
    differential diagnosis, 6/719
    dilution, 49/423
    Down syndrome, 49/424, 50/531
    dry skin, 49/423
    dysarthria, 6/709, 49/419, 423
    dystonia, 6/557, 60/664
    dystonia musculorum deformans, 49/524
    EEG, 6/708, 49/422
    epilepsy, 6/707, 49/419, 422
    familial, see Familial striatopallidodentate
        calcification
    fifth ventricle, 6/712
    fourth ventricle, 6/712
    frontal cortex nodular dysgenesia, 49/424
    genetics, 42/534
    Geyelin-Penfield calcification, 6/720
    glycosaminoglycan, 6/715
    Hallervorden-Spatz syndrome, 49/417

hallucination, 49/423
hereditary osteopetrosis, 49/424
hidrotic ectodermal dysplasia, 49/424
histochemistry, 6/714-716
history, 6/704
Huntington chorea, 49/293
hydroxyapatite, 49/426
hyperparathyroidism, 27/292
hypocalcemia, 63/561
hypoparathyroidism, 27/305-310, 49/418
idiopathic hypoparathyroidism, 6/705, 49/418
infectious, *see* Infectious striatopallidodentate
   calcification
intracerebral calcification, 6/703, 42/534
intracranial hypertension, 49/422
iron, 6/716, 49/425
Kearns-Sayre-Daroff-Shy syndrome, 49/424
late onset ataxia, 60/671
lead encephalopathy, 64/436
lead intoxication, 49/417
lipofuscin, 6/715
lipomembranous polycystic osteodysplasia,
   49/424
macroscopic calcification, 49/426
macular degeneration, 6/720
magnesium, 6/716, 49/425
manganese, 49/425
MELAS syndrome, 49/424, 62/510
mental depression, 49/423
MERRF syndrome, 49/424
micrencephaly, 49/421
microscopic calcification, 49/425
microscopy, 49/426
mitochondrial encephalomyopathy, 60/671
mitochondrial myopathy, 49/421, 424
molybdenum, 6/716, 49/425
multiple sclerosis, 49/423
myoclonia, 49/422
natural history, 49/422
neurofibromatosis type I, 14/152
neuropathology, 6/712-716
neuroradiology, 6/710, 49/426
nodular dysgenesis, 49/424
nosology, 6/704, 717
pallidal degeneration, 6/650
papilledema, 6/707-709, 49/422
paraplegia, 49/423
parkinsonism, 6/707, 49/419, 422
phakomatosis, 49/417
phosphorus, 6/717
physiologic, *see* Physiologic striatopallidodentate
   calcification
physiopathology, 6/718, 49/429

postthyroidectomy, 49/419
prevalence, 49/428
pseudocalcium, 49/417
pseudohypoparathyroidism, 6/706, 42/621,
   49/418, 420, 63/561
pseudopseudohypoparathyroidism, 49/419
psychiatric symptom, 6/709
psychosis, 49/419, 423
regional pathology, 21/45
scanning electron microscopy, 49/426
schizophrenia, 49/423
secondary pigmentary retinal degeneration,
   13/336, 49/424, 60/730
senescence, 49/424
sex ratio, 49/422
silver, 6/716, 49/425
skin lesion, 6/710
spastic paraplegia, 59/439
startle reaction, 49/422
stupor, 49/423
sulcus calcification, 49/426
symptomatic dystonia, 6/557, 49/543
teeth loss, 49/423
tremor, 6/707, 49/422
Urbach-Wiethe disease, 49/424
vascular, *see* Vascular striatopallidodentate
   calcification
vascular mineralization, 11/141
white matter vein calcification, 49/427
zinc, 6/704, 49/425
Striatum
   see also Caudate nucleus, Neostriatum *and*
      Putamen
   akinesia, 45/165
   anatomy, 6/7-19
   ballismus, 6/481
   brain lacunar infarction, 2/499, 6/673
   cell, 6/13-16
   corpus, *see* Corpus striatum
   cytology, 49/2
   dorsal, *see* Dorsal neostriatum
   embryology, 6/10-13
   function, 6/634
   histochemistry, 6/15
   kernicterus, 6/502
   lesion, *see* Striatal necrosis
   Parkinson disease, 6/212
   snake envenomation, 65/187
   snake venom, 65/187
   survey, 6/632
   ventral, *see* Ventral neostriatum
Striatum apoplexy
   brain embolism, 11/585

brain hemorrhage, 11/585
embolic, *see* Embolic striatum apoplexy
Stride potential
  electrical injury, 7/345
Stridor
  Arnold-Chiari malformation type II, 50/410
  cause, 50/219
  cerebellar hypoplasia, 60/667
  congenital, *see* Vagus nerve agenesis
  congenital laryngeal abductor paralysis, 42/318
  Gerhardt syndrome, 50/219
  glossopharyngeal nerve agenesis, 50/219
  laryngeal abductor paralysis, 42/318
  larynx, *see* Larynx stridor
  multiple neuronal system degeneration, 63/146
  neuroleptic dystonia, 65/282
  orthostatic hypotension, 63/146
  tetany, 63/560
String sign
  carotid dissecting aneurysm, 54/272
Striofugal fiber
  anatomy, 6/16
Striopetal fiber
  anatomy, 6/17-19
Striosome
  Huntington chorea, 49/257
Striothalamic connection
  anatomy, 6/17
Stroban
  organochlorine insecticide intoxication, 36/423
  toxaphene intoxication, 36/423
Stroke
  see also Brain hemorrhage, Brain infarction, Brain
    thrombosis *and* Carotid artery thrombosis
  acute, 11/299-301
  age, *see* Age at stroke
  amphetamine intoxication, 65/258
  amusia, 45/487
  anticoagulant, 12/459
  aphasia, 45/296, 46/150, 614, 629
  apraxia, 45/430
  basal forebrain, *see* Basal forebrain stroke
  Binswanger disease, 46/318
  blood pressure, 53/459-466, 75/373
  brain arteriovenous malformation, 54/372
  brain blood flow, 54/144
  brain edema, 16/194
  cardiac dysrhythmia, 63/238, 75/370
  cardiac papillary fibroelastoma, 63/103
  cerebellar infarction, 63/8
  cerebrogenic, 75/371
  chemotherapy, 67/355
  Cheyne-Stokes respiration, 75/375

chronic, 11/299-301
clinical trial, 53/459-462
clinical trial methodology, 53/462-466
completed, 11/299-301, 339
concentric sclerosis, 9/449
congenital heart disease, 38/123-127, 63/3, 5
congestive heart failure, 63/134
depression, 46/426
diabetes mellitus, 75/366
driving, 46/619
ECG abnormality, 75/370
eclampsia, 39/545
epilepsy, 72/116
evolution, 11/339
evolving, 11/299
general management, 12/445
Glasgow Coma Scale, 53/210
heat, *see* Heat stroke
hereditary hemorrhagic telangiectasia (Osler),
  42/739
homocystinuria, 42/555
hyperlipoproteinemia type II, 42/445
hypertension, 75/368
juvenile, 12/158
lightning injury, 7/41, 51, 362, 23/691, 719
macroglobulinemia, 39/191
mania, 46/429
memory disorder, 46/623
migraine, 5/80-82
mitral valve prolapse, 63/21
myocardial infarction, 55/161
neurogenic pulmonary edema, 55/157
neurologic intensive care, 55/204
open heart surgery, 63/8
polyarteritis nodosa, 39/297, 301, 55/354
progeria, 43/465
prognosis, 46/611
pure sensory, *see* Pure sensory stroke
radioisotope diagnosis, 45/298
recurrent, 11/299-303
rehabilitation, 46/629
respiration, 75/375
risk assessment, 54/144
sensorimotor, *see* Sensorimotor stroke
sexual dysfunction, 75/99
sickle cell anemia, 42/623
sleep, 74/549
space occupying lesion, 16/239
sudomotor dysfunction, 75/376
systemic lupus erythematosus, 71/44
Takayasu disease, 63/53
thalamic syndrome (Dejerine-Roussy), 42/771
threatening, 11/299

transient global amnesia, 45/210
vasculitis, 46/359
verbal auditory agnosia, 43/229
Waldenström macroglobulinemia polyneuropathy,
    39/191
Strongyloides stercoralis
    acquired immune deficiency syndrome, 71/307
    brain abscess, 52/151
    CNS infection, 69/446
    helminthiasis, 52/506
Stroop test
    Huntington chorea, 46/309
Strümpell, A., 1/8, 16
Strümpell-Lorrain disease, see Familial spastic
    paraplegia
Strümpell sign
    anterior tibial, 1/185
Strychnine
    acquired myoglobinuria, 62/573
    inhibition, 1/51
    myoglobinuria, 41/271
    reflex, 1/62
    toxic myopathy, 62/601
    trigeminal neuralgia, 5/299
Strychnine intoxication
    motor unit hyperactivity, 41/305, 59/377
    myopathy, 40/330
    opisthotonos, 41/305
    postsynaptic inhibition, 41/305
    risus sardonicus, 41/305
Stryker frame
    cervical spine, 25/293
    decubitus, 20/820
    thoracolumbar spine, 25/458
Stump pain
    malignant, 69/93
    phantom limb pain, 4/229, 45/400
Stupor
    arsenic polyneuropathy, 51/266
    ascariasis, 35/359
    atenolol, 63/195
    brain death, 24/702
    brain edema, 63/416
    cerebellar infarction, 53/383
    consciousness, 3/113
    digoxin, 63/195
    ethylene glycol intoxication, 64/123
    hepatic, see Hepatic stupor
    hypernatremia, 63/554
    hypertensive encephalopathy, 54/212
    hypomagnesemia, 63/562
    hyponatremia, 63/545
    isoniazid intoxication, 65/480

lead encephalopathy, 64/435
Marchiafava-Bignami disease, 17/542, 28/324
mephenytoin intoxication, 37/202
mercury polyneuropathy, 51/272
methanol intoxication, 36/353, 64/99
metharbital intoxication, 37/202
migraine, 5/52, 48/6, 161
mountain sickness, 63/416
nifedipine, 63/194
phenobarbital intoxication, 37/202
pontine infarction, 12/15
primary pontine hemorrhage, 12/39
propranolol, 63/195
striatopallidodentate calcification, 49/423
toxic oil syndrome, 63/380
uremic encephalopathy, 63/504
valproic acid intoxication, 65/504
Sturge-Weber-Dimitri syndrome, see Sturge-Weber
    syndrome
Sturge-Weber syndrome
    acromegaly, 14/512
    adiposogenital dystrophy, 14/512
    agyria, 14/66
    Arieti-Gray syndrome, 14/227
    associated disease, 14/512
    associated malformation, 14/512
    biochemistry, 14/229
    birth incidence, 43/49
    brain, 12/234
    brain angiography, 55/446
    brain arteriovenous malformation, 31/196-199
    brain atrophy, 14/65, 43/48, 55/445
    brain cortex atrophy, 14/63, 65, 43/48
    brain cortex calcification, 14/63-65
    brain hemorrhage, 55/444
    brain infarction, 55/444, 446
    brief sketch, 18/275
    buphthalmos, 14/224, 234
    cerebral malformation, 31/21
    cerebrovascular disease, 55/443
    choroid angioma, 14/68, 639-643, 31/18, 23
    choroid plexus papilloma, 42/761
    clinical features, 31/22, 50/376
    clonic seizure, 31/22
    computer assisted tomography, 55/445
    cortical heterotopia, 31/21
    course, 14/226
    Crouzon-Christophe-Gaucher type, 14/227
    Cushing-Bailey-Cobb type, 14/227
    diagnosis, 31/23, 55/444
    Divry-Van Bogaert syndrome, 14/227
    endocrinology, 14/747
    epidemiology, 30/159, 31/1

epilepsy, 14/227, 232-234, 31/3, 46/95, 55/443, 72/210, 347
facial angioma, 14/66
facial hemihypertrophy, 22/549
fibrous dysplasia, 14/199, 201
gene, 72/137
genetics, 14/238, 507, 30/159, 31/19, 50/377
glaucoma, 14/223, 227, 234, 644, 43/48, 50/376
hamartoma, 18/276
hemicerebral atrophy, 18/276
hemifacial hypertrophy, 14/67
hemiplegia, 43/48
heterochromia iridis, 14/512
history, 14/223
homonymous hemianopia, 50/376
inheritance mode, 14/509-512
intracerebral calcification, 31/18-20, 43/47
intracranial calcification, 14/227
iris coloboma, 14/512
Jahnki type, 14/227
Klippel-Trénaunay syndrome, 14/7, 228, 397, 532, 43/49
Krabbe-Bartels disease, 14/82, 227
Lawford type, 14/227
lens luxation, 14/512
leptomeningeal angiomatosis, 14/228, 232, 43/47, 50/376
leptomeningocortical angioma, 55/447
macrocephaly, 50/377
melanin, 14/102
meningeal angioma, 14/62
meningioma, 50/376
mental deficiency, 31/3, 22, 43/48, 46/95, 50/376, 55/443
microgyria, 14/66
microphthalmia, 14/512
Milles type, 14/227
neurocutaneous melanosis, 14/8, 424, 532
neurocutaneous syndrome, 14/101
neurofibromatosis, 14/6, 532
neurofibromatosis type I, 14/568
neuropathology, 14/61-68, 228, 55/447
neuroradiology, 55/445
nevus distribution, 14/231
nevus flammeus, 14/2-4, 31/18, 43/47, 55/443
ocular lesion, 14/643
ophthalmology, 14/234, 639-644
ota nevus, 43/37
oxycephaly, 14/512
pathology, 31/19-21
phakomatosis, 14/13, 19, 68/296
polygyria, 14/66
polymicrogyria, 14/66, 31/21

prevalence, 14/507, 43/49
progressive hemifacial atrophy, 14/512, 59/479
psychiatric disorder, 14/765
pulmonary myomatosis, 14/69
railway calcification, 55/445
rhabdomyoma, 14/69
sagittal sinus nonfilling, 54/412
Schirmer type, 14/227
secondary pigmentary retinal degeneration, 13/334, 14/512, 60/735
seizure, 15/303, 43/48, 50/376
sternum malformation, 14/512
subarachnoid hemorrhage, 12/105
superior longitudinal sinus thrombosis, 55/446
syndactyly, 14/512
thyroid adenoma, 14/69
transient ischemic attack, 55/444
treatment, 14/235-237, 31/23, 55/448
trigeminal nerve, 43/47
tuberous sclerosis, 14/47, 532, 578
tumor biology, 16/31
Ullmann syndrome, 14/448
urogenital agenesis, 14/512
Usher syndrome, 14/228
variant, 14/227
venous malformation, 12/234
venous sinus thrombosis, 55/446
visceral lesion, 14/696-699
Von Hippel-Lindau disease, 14/7, 13, 251-253, 532
xanthogranuloma, 50/377
XYY syndrome, 14/235
Sturgeon fish intoxication
  bacterial infection, 37/86
  fish roe, 37/68
Sturgeon roe intoxication
  bacterial infection, 37/86
Stuttering, 4/465-468, 46/169-172
  see also Dysarthria, Speech disorder and Speech disorder in children
  aluminum neurotoxicity, 63/525
  aphasia, 46/171
  auditory system, 46/170
  cerebral dominance, 46/170
  cluttering, 43/221, 46/171
  developmental dysarthria, 4/409-413
  developmental dysphasia, 46/140
  dialysis, 63/525
  dichotic listening, 46/171
  differential diagnosis, 46/171
  etiology, 46/171
  handedness, 4/261
  historic note, 46/169

incidence, 46/169
inheritance, 46/169
larynx, 46/170
palilalia, 46/171
prevalence, 43/226
sex ratio, 43/227
speech therapy, 46/171
stapedius muscle, 46/170
Stylalgia, *see* Stylohyoid syndrome
Stylocarotid syndrome, *see* Stylohyoid syndrome
Stylohyoid apparatus
  embryology, 48/505
Stylohyoid syndrome, 48/501-511
  see also Elongated styloid process syndrome
  anatomy, 48/504
  clinical type, 48/503
  diagnosis, 48/502
  globus hystericus, 48/502
  otalgia, 48/510
  pathogenesis, 48/509
  Riolan nosegay syndrome, 48/504
  swallowing, 48/503
  tonsillitis, 48/503
Styloid chain
  anatomy, 48/507
Styloid chain bone
  ossification, 48/506
Styloid process
  anatomy, 48/505
  elongation, *see* Elongated styloid process
    syndrome
  fracture, 48/502
Styloid syndrome, *see* Stylohyoid syndrome
Stylomandibular ligament
  ossification, 48/507
Stylomastoid artery
  radioanatomy, 11/66
  topography, 11/66
Styrene
  neurotoxin, 64/39, 53, 56
  organic solvent intoxication, 64/39
Styrene intoxication
  see also Organic solvent intoxication
  acroparesthesia, 64/54
  attention, 64/54
  autonomic nervous system, 75/506
  chromatic discrimination, 64/56
  concentration, 64/54
  EEG, 64/54
  mandelic acid, 64/53
  phenylglyoxylic acid, 64/53
  retrobulbar neuritis, 64/56
  symptom, 64/54

Suarez classification
  epilepsy, 15/10
Subacute anterior poliomyelitis
  amyotrophic lateral sclerosis, 22/311
Subacute bacterial endocarditis, 33/469-476
  brain embolism, 11/393
  congenital heart disease, 38/136
  hemiplegia, 63/111
  meningitis, 33/473
  neurologic deficit, 11/394
Subacute cerebellar degeneration
  Eaton-Lambert myasthenic syndrome, 62/421
  Hodgkin disease, 39/54
  malignant lymphoma, 39/54
Subacute combined degeneration
  autonomic nervous system, 75/30
  gliofibrillogenesis, 9/617, 625
Subacute combined spinal cord degeneration,
    70/430
  acquired immune deficiency syndrome, 56/493,
    512
  amblyopia, 28/173-178
  amino acid, 59/359
  anatomy, 2/210
  astrocyte, 9/614
  biochemical aspect, 28/148-150
  clinical features, 28/148, 159-170, 59/359
  demyelination, 9/614-617
  diagnosis, 28/150-154, 184
  diphyllobothrium latum myelopathy, 7/631
  EEG, 28/163
  etiology, 28/143-147
  experimental aspect, 28/158
  familial spastic paraplegia, 59/306
  history, 28/141-143
  homocystinuria, 63/255
  intrinsic factor, 28/146
  lesion, 70/391
  leukoencephalopathy, 9/614, 47/584
  malnutrition, 28/21
  mental change, 63/254
  methylmalonic aciduria, 63/255
  multiple sclerosis, 47/72
  murine leukemia virus, 56/455
  myelin sheath, 47/584
  myelopathy, 28/178
  neurologic manifestation, 28/159-170, 178-184
  neuropathology, 28/154-158, 56/531, 63/254
  neuropathy, 28/179-184
  nitrous oxide, 63/256
  optic nerve, 28/173-178
  pathophysiology, 28/143-147
  pellagra, 7/573, 575, 46/399

408

pernicious anemia, 7/591, 28/159-164, 38/22,
    63/254
pernicious psychosis, 28/170-173
postgastrectomy, 28/164-169
Schilling test, 28/150
spastic paraplegia, 59/439
spinal cord compression, 19/378
symptom, 63/254
treatment, 28/184-187, 63/254
vitamin B$_{12}$, 7/587
vitamin B$_{12}$ deficiency, 46/399, 51/339, 63/254,
    70/375
vitamin B$_{12}$ deficiency anemia, 63/254
vitamin B$_{12}$ metabolism, 59/359
vitamin B$_{12}$ neuropathy, 7/587, 592
vitamin B$_c$ deficiency, 63/255
vitamin B$_c$ deficiency anemia, 63/255
Subacute disabling neuropathy
    paraneoplastic polyneuropathy, 51/467
Subacute encephalitis
    acquired immune deficiency syndrome, 56/507
    leukodystrophy classification, 10/20
    progressive, *see* Progressive subacute encephalitis
Subacute malignant cerebellar degeneration
    Hodgkin disease, 39/54
Subacute motor neuronopathy
    paraneoplastic syndrome, 69/374, 377
Subacute myelo-optic neuropathy
    abdominal pain, 51/296
    ascending sensorimotor neuropathy, 51/296
    ataxia, 51/296
    clioquinol, 64/8
    clioquinol intoxication, 37/115-194, 47/405
    diarrhea, 51/296
    encephalopathy, 51/296
    hyperbaric oxygenation, 37/135
    myelopathy, 51/296
    neuromyelitis optica, 9/435
    optic atrophy, 51/296
    quiniform intoxication, 37/175-180
    symptom, 64/8
Subacute myelo-optic neuropathy like virus
    multiple sclerosis, 47/330
Subacute myoclonic spinal neuronitis
    painful, 6/775
Subacute necrotic myelitis, *see* Angiodysgenetic
    necrotizing myelomalacia
Subacute necrotizing encephalomyelopathy,
    28/349-360
    acid base metabolism, 28/517
    adenosine triphosphatase 6 mutation syndrome,
        62/510
    adenosine triphosphate, 42/588

age at onset, 28/350
ataxia, 42/588, 625, 51/387
athetosis, 51/387
autosomal recessive, 51/387
basal ganglion degeneration, 42/588, 625
blindness, 42/625
bulbar paralysis, 60/658
carbohydrate metabolism, 42/626
cardiomyopathy, 51/388
case report, 28/349
central pontine myelinolysis, 28/293, 308
childhood myoglobinuria, 62/563
chorea, 51/387
choreoathetosis, 49/385
clinical features, 10/209, 28/350-353
cocarboxylase, 28/355, 42/588
complex IV deficiency, 41/212, 62/505, 66/424
convulsion, 42/588
cranial nerve, 42/625
CSF, 28/352, 42/625
dementia, 46/402
dentate nucleus, 42/588
dihydrolipoamide dehydrogenase, 66/419
duration of illness, 28/351
dystonia, 60/664
EEG, 28/353
epilepsy, 42/625
eye sign, 28/352
fever, 42/625
form, 51/388
hereditary periodic ataxia, 21/564
hereditary secondary dystonia, 59/343
history, 28/349
hyperpyruvic acidemia, 29/211-213
hypotonia, 10/209, 28/352
lactic acidosis, 41/212, 42/588, 625
maternal inherited, 66/425
metabolic aspect, 28/355
metabolic ataxia, 21/578
mitochondrial abnormality, 41/212
mitochondrial disease, 41/212, 51/388, 62/497
mitochondrial protein, 66/17
NARP syndrome, 66/425
nerve conduction velocity, 51/388, 60/670
neuropathy, 42/588, 60/660
nystagmus, 42/625
ophthalmoplegia, 42/588
optic atrophy, 10/209, 51/388, 60/654
orthochromatic leukodystrophy, 10/114
pathology, 28/356-360
posterior column degeneration, 42/625
progressive external ophthalmoplegia, 60/56
psychomotor retardation, 51/388

pyruvate carboxylase deficiency, 29/212, 51/388

pyruvate decarboxylase deficiency, 62/502

pyruvate dehydrogenase complex deficiency, 66/425

pyruvic acid metabolism, 28/354, 29/213

seizure, 51/388

sex ratio, 28/351

sobbing respiration, 60/667

space occupying lesion, 16/241

spastic paraplegia, 59/439

striatal necrosis, 49/500, 507, 510

substantia nigra degeneration, 42/625

symptom, 28/351

thalamus, 42/488, 625

thiamine triphosphate, 42/625

vitamin B$_1$, 28/353-356

vitamin B$_1$ transferase inhibitor, 42/626

Subacute polyneuritis
  nerve conduction, 7/178
  porphyria, 60/117

Subacute sclerosing leukoencephalitis, *see* Subacute sclerosing panencephalitis

Subacute sclerosing panencephalitis, 34/343-364, 56/417-431
  animal model, 56/587
  ataxia, 56/419
  athetosis, 56/419
  autonomic dysfunction, 56/419
  ballismus, 56/419
  brain biopsy, 10/685, 56/427
  canine distemper, 9/673
  cellular immunity, 56/425
  chorea, 56/419
  chorioretinitis, 56/419
  clinical course, 56/418
  clinical diagnosis, 56/418
  clinical features, 34/349-354
  clinical stage, 56/419
  consciousness, 45/120
  Creutzfeldt-Jakob disease, 56/546
  CSF, 9/327, 349, 352, 34/354, 56/419
  CSF IgG, 56/419
  CSF measles antibody, 56/420
  CSF oligoclonal band, 56/419
  decerebrate rigidity, 56/419
  decorticate posture, 56/419
  definition, 34/345
  dementia, 46/389, 56/418
  demyelination, 56/417, 422
  diffuse sclerosis, 9/470, 472, 10/20, 47/421
  EEG, 34/355, 56/419
  eosinophilic inclusion body, 56/422
  epidemiology, 34/356-358, 56/418
  epilepsy, 15/351, 56/419
  Epstein-Barr virus, 56/256
  Epstein-Barr virus infection, 56/252
  etiology, 9/126, 34/346-349
  F protein antibody, 56/427
  ganglioside, 9/4
  geography, 56/418
  giant plasmatic glial cell, 10/7
  H protein antibody, 56/427
  herpes simplex titer, 9/327
  history, 34/345, 56/417
  humoral immune response, 56/425
  IgG, 9/140, 349, 352
  immune mechanism, 9/543
  immunocytochemical study, 47/236
  immunologic aspect, 34/55
  infectious mononucleosis, 56/252
  α-interferon, 56/421
  laboratory diagnosis, 56/419
  leukodystrophy, 10/2
  measles virus, 34/344, 346-349, 362-364, 56/14
  methisoprinol, 56/421
  monoclonal protein antibody, 56/426
  multiple sclerosis, 9/120, 126, 47/86, 104, 326
  myoclonia, 56/419
  myoclonus, 38/584, 49/618
  myxovirus, 9/551
  neurofibrillary tangle, 46/274
  neuropathology, 56/421
  nosology, 10/8
  NP protein antibody, 56/427
  nuclear magnetic resonance, 56/420
  optic atrophy, 56/419
  paramyxovirus, 56/14
  pathogenesis, 34/362-364, 56/428
  pathologic laughing and crying, 56/419
  pathology, 34/358-362
  progressive rubella panencephalitis, 34/338, 56/411
  rubella titer, 9/327
  slow virus disease, 56/544
  thalamus, 2/487
  tissue protein, 56/427
  treatment, 34/364, 56/421
  viral infection, 34/55, 47/215
  viral persistence, 56/428
  virus recovery, 56/427

Subacute sclerosing panencephalitis virus
  measles virus, 56/422

Subacute sensory neuronopathy
  autonomic nervous system, 75/534

Subacute sensory neuropathy
  autonomic nervous system, 75/15

paraneoplastic syndrome, 71/674
Subacute spinal phlebothrombosis, *see*
    Foix-Alajouanine disease
Subacute spongiform encephalopathy
  amyloid plaque, 42/655
  brain biopsy, 10/685
  Canavan type, 11/159
  cerebrovascular disease, 11/129
  Creutzfeldt-Jakob disease, 6/735, 11/159,
    42/655-659, 46/289, 291
  differential diagnosis, 10/209
  kuru, 42/655
  scrapie, 42/655
  Van Bogaert type, 11/159
Subangular alexia
  brain infarction, 45/444
  brain tumor, 45/444
  clinical features, 45/443
Subarachnoid abscess
  meningitis, 16/211
  space occupying lesion, 16/211
Subarachnoid block
  spinal, *see* Spinal subarachnoid block
Subarachnoid hemorrhage, 12/68-185, 55/1-33
  see also Cerebrovascular disease
  abducent nerve, 55/18
  acoustic nerve, 55/18
  activity at onset, 55/12
  acute papilledema, 53/212
  adolescent, 12/158
  adrenalin, 12/112
  adrenergic receptor stimulating agent, 12/112
  aftereffect, 12/183
  age, 12/127, 55/44
  alcohol, 12/113, 55/523
  allergic granulomatous angiitis, 63/383
  allergic reaction, 12/114
  allergy, 12/114
  ε-aminocaproic acid, 12/177, 55/52
  p-aminomethylbenzoic acid, 12/178
  amphetamine addiction, 55/360, 520
  amphetamine intoxication, 65/258
  aneurysm, 12/77, 55/2
  aneurysm distribution, 55/46
  aneurysm site, 55/46
  aneurysm size, 55/46
  aneurysma bilobata, 12/81
  angiodysgenetic necrotizing myelomalacia,
    20/488
  angiopathy, 12/110
  angiostrongyliasis, 52/549
  annual incidence, 55/2, 43
  anterior cerebral artery syndrome, 53/341

anterior communicating artery aneurysm, 12/92
anthrax, 12/112
anticoagulant, 12/113, 55/3
antifibrinolysis, 55/52
aorta coarctation, 55/2, 29, 63/4
aphasia, 12/132
apnea, 57/473
apoplexy, 12/165
arachnoid cyst, 12/184
arrhythmia, 55/9
arterial hypertension, 12/128
arterial spasm, 55/5
arteriovenous angioma, 12/106
arteriovenous malformation, 55/2
aspergillosis, 35/397, 52/379, 55/435
autonomic nervous system, 12/150, 55/14, 75/39
autopsy finding, 12/75
autopsy incidence, 55/44
B-mode sonography, 54/38
bacterial endocarditis, 52/291, 293, 295, 55/3
basilar artery vasoconstriction, 53/388
basilar dissecting aneurysm, 54/280
biological rhythm, 74/472
birth incidence, 12/158
blood, 12/150
blood clotting factor XII deficiency, 63/313
blood disease, 12/109
blood dyscrasia, 12/109, 55/2
blood residual N, 12/151
boxer encephalopathy, 12/121
boxing injury, 23/548
brain aneurysm, 12/77, 104, 42/723
brain aneurysm neurosurgery, 12/215
brain angiography, 12/124, 154, 55/24, 48
brain arteriovenous malformation, 12/241, 247,
  53/35, 54/372, 374, 376, 55/73
brain blood flow, 54/147
brain cavernous angioma, 12/104
brain edema, 55/6
brain embolism, 12/170, 55/167
brain hemorrhage, 12/165, 31/160, 55/20, 167
brain herniation, 55/5
brain infarction, 53/163, 55/5, 9, 20, 167
brain ischemia, 53/163, 55/57
brain metastasis, 12/120, 18/216
brain phlebothrombosis, 12/170, 55/3, 20
brain scanning, 12/157
brain stem death, 57/473
brain tumor, 12/118, 169
brain vasospasm, 11/520, 12/155, 23/168, 174,
  187, 55/10, 55-58
brain ventricle hemorrhage, 12/167
brucellosis, 12/111

calcium antagonist intoxication, 65/441
calcium channel blocking agent, 55/59
candidiasis, 52/401
capillary telangiectasia, 55/2
carbon monoxide intoxication, 12/113
cardiac dysrhythmia, 55/10, 15, 54, 154, 63/236
cardiac effect, 55/15
cardiovascular complication, 55/9
carotid artery stenosis, 55/73
carotid cavernous fistula, 12/102
cause, 12/67, 55/2, 28
cause prevalence, 55/4
cavernous angioma, 12/104, 42/724, 55/2
cavernous sinus thrombosis, 55/20
cerebellar hemorrhage, 12/167
cerebrovascular disease, 55/167
cervicogenic headache, 55/19
cheese reaction, 12/112
chronic paroxysmal hemicrania, 55/20
clinical features, 12/126, 160, 55/11
clinical grading, 55/50
clock function, 74/468
cluster headache, 55/20
coagulopathy, 55/3
Cobb syndrome, 14/439
cocaine addiction, 55/521
cocaine intoxication, 65/50
coitus, 48/377
collagen vascular disease, 55/2
coma, 12/172, 55/21
computer assisted tomography, 54/60, 70, 55/48,
    57/157
conjugate deviation, 12/130
consciousness, 55/14
controlled hypotension, 12/212
convulsive seizure, 12/132
course, 12/142
cranial computer assisted tomography, 55/21
cranial nerve deficit, 55/9, 16
cranial nerve palsy, 12/129
critical size, 55/47
CSF, 12/144, 54/195, 55/23
CSF examination, 11/238
CSF glucose, 54/196
CSF hypertension, 12/177
CSF hypotension, 12/172
CSF pressure, 12/150, 172
CSF protein, 12/149
deafness, 55/134
deep vein thrombosis, 55/54
definition, 55/1
dehydrating agent, 12/215
diabetes insipidus, 55/9, 53

diabetes mellitus, 55/9
diagnosis, 12/123, 144-158, 55/48
diagnosis confirmation, 55/48
diamorphine intoxication, 65/355
diapedesis, 12/73, 55/6
diapedetic hemorrhage, 12/73
differential diagnosis, 12/162, 55/19
dissecting aneurysm, 55/79
drug abuse, 55/3
dysphasia, 55/18
dysproteinemia, 12/113
early surgery, 55/59
ECG, 55/10, 15, 24, 54
ECG change, 12/129, 63/233
ECG pathogenesis, 63/234
ECG pathology, 63/233
ECG pathophysiology, 63/234
ECG QT interval, 63/233
ECG ST depression, 63/233
ECG T wave inversion, 63/233
echoEG, 12/157
eclampsia, 55/3
EEG, 11/280, 12/157
Ehlers-Danlos syndrome, 55/29, 454
electrical injury, 12/121, 23/692
electrolyte disorder, 55/53
embolism, 12/170
emotional excitement, 55/2
encephalitis, 12/170
endocrine complication, 55/9
ephedrine, 55/521
epidemiology, 12/70, 55/1, 43
epidural hematoma, 12/162
epilepsy, 55/9, 53, 72/215
erythropoietic disease, 12/110
etiology, 12/76
examination protocol, 55/27
exertional headache, 48/374
facial nerve, 55/18
familial aneurysm, 55/81
fever, 55/15
fibrinolytic agent overdose, 65/462
fibromuscular dysplasia, 11/368, 371, 55/2, 29,
    284
fluid derangement, 55/53
focal symptom, 55/15
form, 12/143
frequency, 12/160
frontal lobe syndrome, 46/432
fungal origin, 52/295
furosemide, 55/53
fusiform aneurysm, 55/78
gastrointestinal bleeding, 55/54

gastrointestinal complication, 55/9
general management, 12/175
general medical management, 55/50
giant aneurysm, 55/47
glioblastoma multiforme, 12/119
glossopharyngeal nerve, 55/18
hamartoma, 12/104
head injury, 12/74, 120, 57/157
headache, 12/183, 48/6, 33, 273, 275, 277, 286,
    53/205, 240, 55/11, 19
headache frequency, 55/13
headache type, 55/13
heat stroke, 12/122, 23/673
hemangioblastoma, 12/119
hematocephalus externus, 12/159
hematologic complication, 55/9
hematoma, 12/71
hematomyelia, 12/173
hematuria, 12/151
hemiplegia, 12/131
hemosiderin deposit, 55/6
hemostasis, 12/176
Henoch-Schönlein purpura, 55/3
hereditary hemorrhagic telangiectasia (Osler),
    12/104, 42/739, 55/450
history, 12/68
Hunt-Hess Scale, 53/209
hydrocephalic dementia, 46/330
hydrocephalus, 12/152, 179, 55/5, 8, 55
hyperbilirubinemia, 12/113
hypernatremia, 12/113, 63/554
hypertension, 55/2
hypertensive encephalopathy, 55/20
hypoglossal nerve, 55/18
hypophyseal adenoma, 12/119
hypothermia, 12/211
iatrogenic neurological disease, 61/164
inappropriate antidiuretic hormone secretion,
    55/9, 53
incidence, 55/1
infant, 12/158, 161, 348, 55/28
infection, 55/32
infectious disease, 12/111
infectious endocarditis, 63/23
inflammatory aneurysm, 55/78
influenza, 12/111
initial symptom, 55/12
insomnia, 46/418
insulin, 12/113
intoxication, 55/3
intracerebral hematoma, 12/70, 167, 55/54
intracranial hematoma, 55/54
intracranial hypertension, 12/177, 55/55, 59

intracranial pressure, 55/8
intravascular consumption coagulopathy, 55/3,
    497
intraventricular hemorrhage, 12/167, 55/54
juvenile, 12/161, 55/28
ketonuria, 12/151
laboratory finding, 55/21
leptospirosis, 12/111
leukemia, 12/110, 55/3, 63/344
Listeria monocytogenes, 12/111
Listeria monocytogenes meningitis, 33/85
listeriosis, 33/86
lumbar puncture, 12/178, 61/164
lymphoma, 63/350
macroglobulinemia, 39/191, 55/485
malaria, 12/111
malignant astrocytoma, 55/2
management, 12/175, 55/27
mania, 46/429
Marfan syndrome, 39/390, 55/29
marginal zone siderosis, 12/72
mass lesion, 55/6
medical management, 55/22
megadolichobasilar artery, 12/81
megadolichocarotid artery, 12/81
melanoma, 12/119
melanoma brain metastasis, 18/27
meningeal bleeding, 12/163
meningeal fibrosis, 55/7
meningeal irritation, 12/128, 55/14
meningeal reaction, 12/74
meningioma, 12/119
meningitis, 12/112, 169, 55/20
meningitis serosa, 55/7
meningococcal meningitis, 33/23, 52/26
meningoencephalitis, 55/3
mental confusion, 55/19
mental disorder, 12/139
methamphetamine addiction, 55/360
methamphetamine intoxication, 65/366
methanol intoxication, 36/356, 64/102
α-methylcyclohexane carboxylic acid, 55/52
3,4-methylenedioxymethamphetamine
    intoxication, 65/367
microangioma, 12/106
middle cavernous sinus syndrome, 12/89, 31/147
middle meningeal artery, 12/162
migraine, 12/110, 112, 55/19
Millard-Gubler syndrome, 12/95
misdiagnosis, 53/240
moria, 12/143
morphine, 12/113
mortality, 55/44

mortality rate, 12/179, 55/1
moyamoya disease, 12/360, 374, 42/749, 53/33, 55/29, 293
mucormycosis, 52/471
multiple aneurysm, 55/47
muscle contraction headache, 55/19
muscle stretch reflex, 12/131
mycotic aneurysm, 12/80, 63/23
mycotic brain aneurysm, 12/112, 52/295
myelography, 12/157, 61/164
myocardial infarction, 55/9
natal injury, 12/158
negative angiography, 55/49
negative angiography cause, 55/49
neoplasm, 55/30
neoplastic aneurysm, 55/80
neurobrucellosis, 33/311, 314, 52/584, 588
neurocutaneous melanosis, 43/33
neurogenic pulmonary edema, 55/53
neurologic complication, 55/9
neurologic intensive care, 55/204
neuropathology, 55/4
newborn, 12/158, 160, 63/326
nicotine intoxication, 65/263
nimodipine, 55/59
noradrenalin, 63/234
normal pressure hydrocephalus, 46/325
nuchal rigidity, 12/128
nuclear magnetic resonance, 55/21
occipital headache, 48/137
ocular symptom, 12/135
oculomotor nerve, 55/17
ophthalmoscopy, 11/214
opiate intoxication, 65/355
optic atrophy, 12/139
oral contraception, 55/524
organic psychosyndrome, 12/141
papilledema, 12/135, 137, 55/16
paraglioma, 12/119
paranoid psychosis, 12/140
paratrigeminal syndrome, 55/20
pathologic anatomy, 12/70
pathology, 12/70, 75
pathophysiology, 55/7
PEG, 12/151
penicillin, 12/114
pentetrazole, 12/113
perforation, 12/72
perinatal brain injury, 12/158
phagocytosis, 55/7
phencyclidine addiction, 55/522
phenylpropanolamine addiction, 55/520
physical exertion, 12/125, 55/2

pituitary apoplexy, 55/20
pneumonia, 12/111
polyarteritis nodosa, 39/297, 301, 48/284, 55/102, 355
polycystic renal disease, 55/2, 29, 81
polyuria, 12/151
pontine hemorrhage, 12/45
positron emission tomography, 54/131
posttraumatic type, 55/2
pregnancy, 12/115, 55/25, 73
premature infant, 12/158
preoperative management, 55/47
primary type, 55/1
primitive trigeminal artery, 12/101
prognosis, 12/179, 55/25, 45
proteinuria, 12/151
pseudoxanthoma elasticum, 11/465, 43/45, 55/29
psychiatric disorder, 12/139
pulmonary embolism, 55/54
rebleed, 55/52
rebleed prevention strategy, 55/51
recurrence, 12/180
respiratory complication, 55/9
rheumatic fever, 55/3
saccular aneurysm anatomy, 55/45
scintigraphy, 12/157
season, 55/1
seizure, 55/14, 21
sepsis, 55/3
sequela, 55/5
sex incidence, 55/44
sexual activity, 55/2
sickle cell anemia, 55/465, 510, 63/262
sinusitis, 55/20
site, 55/44
Sjögren syndrome, 63/55
skull radiation, 55/22
smoking, 55/524
snake bite, 12/113
snake envenomation, 65/179
space occupying lesion, 16/241
spinal, see Spinal subarachnoid hemorrhage
spinal accessory nerve, 55/18
spinal anesthesia, 61/164
spinal ependymoma, 20/367
spinal tumor, 19/60, 88
streptokinase, 55/59
Sturge-Weber syndrome, 12/105
subdural blood, 55/6
subdural hematoma, 12/163, 55/54
subdural hemorrhage, 12/109
subhyaloid hemorrhage, 53/212
suboccipital puncture, 12/123

surgery timing, 55/59
surgical complication, 55/54
surgical technique, 55/60
surgical treatment, 12/205
survival, 12/182
syphilis, 12/111, 55/29
systemic lupus erythematosus, 12/111, 63/54
systemic manifestation, 55/9
Terson sign, 55/16
Terson syndrome, 12/139
thermal injury, 12/121
tissue plasminogen activator, 55/59
tonsillitis, 12/111
transcranial Doppler sonography, 54/12, 55/22
trauma, 12/120
traumatic aneurysm, 55/78
traumatic carotid cavernous sinus fistula, 57/347
treatment, 12/173
trigeminal nerve, 55/18
trochlear nerve, 55/17
tuberous sclerosis, 55/29
tumor, 12/118
type, 12/143
unexplained cause, 55/31
uremia, 12/114
urine, 12/150
vagus nerve, 55/18
Valsalva maneuver, 12/125
vascular dementia, 46/360
vascular malformation, 12/104, 110
vascular spasm, 53/83
vasopressin, 28/499
vasospasm cause, 55/55
venous angioma, 12/104, 55/2
ventricular enlargement, 12/152
ventricular hemorrhage, 12/167
vertebral dissecting aneurysm, 54/279
vertigo, 55/134
viremia, 55/20
visual field defect, 12/136
visual phenomenon, 55/13
vitamin K_1 deficiency, 12/126
vitamin K_2 deficiency, 12/126
vitreous hemorrhage, 12/138
vomiting, 55/14
Waldenström macroglobulinemia, 12/110, 55/485,
    63/396
Waldenström macroglobulinemia polyneuropathy,
    12/110, 39/191, 532, 51/432
warning leak, 55/48
Wernicke-Korsakoff syndrome, 12/140
Subarachnoid space
  desmoplastic cerebellar medulloblastoma, 18/184

echoEG, 23/268
  metastasis, 71/660
  ventriculostium, 31/126
Subcallosal gyrus
  anatomy, 2/770
Subclavian angiography
  brain tumor, 16/633
Subclavian artery
  anatomy, 53/371
  arteritis, 48/316
  Doppler sonography, 54/9
  lesion, 12/592
  pathology, 53/371
  radioanatomy, 11/67
  thrombosis, see Subclavian artery occlusion
Subclavian artery atherosclerosis
  sex ratio, 53/371
  side, 53/371
Subclavian artery occlusion
  temporal arteritis, 48/317
  thoracic outlet syndrome, 51/124
Subclavian artery stenosis
  atresia, 53/372
  baseball pitcher, 53/372
  Blalock-Taussig procedure, 53/372
  cause, 53/372
  cervical rib, 53/372
  cricket bowler, 53/372
  fibromuscular dysplasia, 53/372
  kinking, 53/372
  mediastinal tumor, 53/372
  oarsmen, 53/372
  radiation, 53/372
  sex ratio, 53/371
  side, 53/371
  Takayasu disease, 53/372
  thoracic outlet syndrome, 7/434
  trauma, 53/372
Subclavian bruit
  brain embolism, 53/216
  brain infarction, 53/216
  cerebrovascular disease, 53/216
  intracranial hemorrhage, 53/216
  thoracic outlet syndrome, 51/124
  transient ischemic attack, 53/216
Subclavian steal syndrome
  angiography, 53/374
  ataxia, 53/373
  Blalock-Taussig anastomosis, 63/5
  blood pressure, 53/374
  brain infarction, 53/159
  brain ischemia, 53/159
  brain perfusion pressure, 53/159

bruit, 53/374
compression syndrome, 7/435
diagnosis, 53/373
diplopia, 53/373
duplex sonography, 53/226
headache, 5/149, 53/373
hemiballismus, 49/373
ischemic arm exercise, 53/373
percutaneous transluminal angiography, 54/179
perioral paresthesia, 53/373
prognosis, 53/374
progressive supranuclear palsy, 49/247
radial pulse, 53/374
surgery, 53/278, 374
symptom, 53/373
Takayasu disease, 39/218, 55/338, 71/202
temporal arteritis, 55/343
transient ischemic attack, 12/8, 53/265
treatment, 53/374
vertebrobasilar transient ischemic attack, 12/11
vertigo, 53/373
visual impairment, 53/373
Subclavius nerve
topographical diagnosis, 2/20
Subclinical polyneuropathy
hyperthyroid polyneuropathy, 51/516
Subcoracoid pectoralis minor syndrome, *see*
Hyperabduction syndrome
Subcortical argyrophilic dystrophy, *see* Progressive
supranuclear palsy
Subcortical arteriosclerotic encephalopathy, *see*
Binswanger disease
Subcortical dementia, 46/311-314
Alzheimer disease, 46/312-314
classification, 46/204
clinical features, 46/312
depressive pseudodementia, 46/312
frontal lobe, 46/313
historic note, 46/311
Huntington chorea, 46/312-314, 378
language, 46/312
Parkinson disease, 46/312-314, 378
pathology, 46/313
progressive supranuclear palsy, 46/312, 378,
49/243
Subcortical encephalitis
chronic progressive, *see* Binswanger disease
Subcortical heterotopia
hydranencephaly, 50/337, 342
Subcortical lesion
aphasia, 45/320
attention disorder, 3/150, 189
pseudothalamic pain, 2/478

Subcutaneous fat absence
Berardinelli-Seip syndrome, 42/591
infantile optic glioma, 42/733
progeria, 43/465
Subcutaneous lipoma
adiposis dolorosa, 8/347, 43/57
Subcutaneous thrombophlebitis
Behçet syndrome, 56/594
Subdural CSF collection
headache, 5/182
Subdural cyst
spinal, *see* Spinal subdural cyst
Subdural effusion
see also Subdural hematoma
accidental injury, 24/331
age, 24/334
air study, 24/337
angiography, 24/337
bacterial meningitis, 33/15, 52/14
battered child syndrome, 24/331
birth incidence, 24/330
bleeding tendency, 24/333
brain scintigraphy, 24/336
brain thrombosis, 24/332
burr hole, 24/337
child, 24/329
clinical features, 33/162
craniectomy, 24/338
definition, 24/329
differential diagnosis, 33/162
epilepsy, 24/339
etiology, 24/330
fluid composition, 24/335
Hemophilus influenzae meningitis, 33/58
history, 24/329
incidence, 24/334
infant, 24/329
infantile enteric bacillary meningitis, 33/67
lumbar puncture, 24/338
meningitis, 24/332
mortality, 24/339
pathogenesis, 24/333
pneumococcal meningitis, 33/47, 52/51
prognosis, 24/339
radiology, 24/337
result, 24/339
sex ratio, 24/334
sign, 24/334
surgery, 24/337, 642, 644, 656
surgical complication, 24/33
symptom, 24/334
transillumination, 24/335
treatment, 24/337, 33/162

tuberculous meningitis, 52/216
venous sinus thrombosis, 24/333
Subdural empyema
   see also Intracranial infection
   actinomycosis, 35/385
   angiography, 57/323
   antibiotic agent, 24/220
   antibiotic treatment, 57/323
   brain angiography, 52/73
   burr hole, 57/324
   cerebrovascular disease, 55/418
   clinical presentation, 57/323
   computer assisted tomography, 52/73, 57/323
   cranial, *see* Cranial subdural empyema
   craniotomy, 57/324
   diagnostic study, 57/323
   EEG, 23/252
   epidemiology, 57/323
   epilepsy, 52/73
   head injury, 52/73, 57/237, 323
   infection, 57/237
   intracranial infection, 24/219
   irrigation, 57/324
   lumbar puncture, 57/323
   mortality, 57/324
   organism, 57/323
   paranasal sinus, 24/219
   sinusitis, 52/73
   spinal, *see* Spinal subdural empyema
   spinal cord injury, 25/60
   staphylococcal meningitis, 52/73
   Staphylococcus aureus, 52/73
   surgical drainage, 57/324
   symptom, 16/210, 24/219
   traumatic, 24/219
   treatment, 24/220, 57/323
Subdural fluid collection, *see* Subdural hygroma
Subdural granuloma
   candidiasis, 52/401
Subdural hematoma
   see also Intracranial hematoma *and* Subdural
     effusion
   acute, *see* Acute subdural hematoma
   afibrinogenemia, 38/72
   alcoholism, 63/313
   anticoagulant, 26/32
   arousal disorder, 45/112
   arterial origin, 57/257
   autopsy, 26/34
   autoradiography, 57/261
   battered child syndrome, 23/605
   bilateral, 54/351
   blood clotting factor X deficiency, 63/312

boxing injury, 23/547
brain aneurysm neurosurgery, 55/54
brain infarction, 11/312
brain injury mechanism, 57/34
brain scintigraphy, 23/291, 297, 299
brain tumor, 69/424
burr hole, 57/338
cause, 54/350
chronic, *see* Chronic subdural hematoma
clinical features, 26/31
clinical status, 57/337
computer assisted tomography, 23/260, 262,
   54/350, 57/152, 274
congenital heart disease, 63/7
convulsion, 57/337
craniotomy, 57/338
CSF, 26/34, 37
CSF examination, 11/239
CSF shunt, 46/329
dementia, 46/395
development, 57/256
diagnosis, 54/350
dialysis, 63/523, 525
differential diagnosis, 16/240, 53/237
EEG, 11/280, 23/324, 333, 344, 57/186
epidural hematoma, 24/279
epilepsy, 72/216
experimental study, 57/59
external subdural drainage, 57/338
fire-hose rupture, 57/257
football injury, 23/572, 574
gunshot injury, 57/303
head injury, 57/34, 48, 103, 335
head injury outcome, 57/378
head injury prognosis, 57/379
hemispheric swelling, 57/279
hemophilia, 26/33, 63/310
hydrocephalus, 30/583
hypernatremia, 63/554
hypoprothrombinemia, 38/76
incidence, 57/251, 337
infant, 57/49
intracerebral calcification, 49/417
intracranial, *see* Intracranial subdural hematoma
layering, 57/158
leukemia, 39/8, 63/344
location, 57/252
lucid interval, 55/182
lumbar puncture, 26/31, 35, 37, 61/160
lymphoma, 63/350
macrocrania, 57/337
management, 57/335
mechanism, 57/337

morbidity, 57/338
mortality, 57/103, 338
myelography, 26/34
newborn, 63/326
nuclear magnetic resonance, 57/174
outcome, 57/282
pathology, 23/50, 57/48
pediatric head injury, 57/49, 335
pertussis encephalopathy, 33/289
posterior fossa hematoma, 24/345, 57/253
postoperative management, 57/278
prevention, 26/36
primary hemorrhagic thrombocythemia, 38/72
prognosis, 54/352
pure motor hemiplegia, 54/242
radiodiagnosis, 57/152
right sided, 57/303
segmental lesion distribution, 26/34
sickle cell anemia, 55/510
sign, 24/661
skull fracture, 57/337
spinal, see Spinal subdural hematoma
spinal injury, 26/31-38
spontaneous, 26/33
subarachnoid hemorrhage, 12/163, 55/54
subdural hygroma, 57/286
subdural tap, 57/338
superior longitudinal sinus occlusion, 54/397
supratentorial, 57/337
surgery indication, 57/273
surgery timing, 57/379
surgical technique, 57/277
symptomatology, 57/337
thrombocytopenia, 26/31
traumatic, 23/605, 26/33, 54/345
traumatic CSF hypotension, 24/256
traumatic intracerebral hematoma, 24/352
traumatic intracranial hematoma, 57/152, 251
traumatic intracranial hematoma outcome, 57/284
treatment, 54/351
trivial trauma, 26/32
uremic polyneuropathy, 63/513
vascular striatopallidodentate calcification, 49/417
venous origin, 57/257
vomiting, 57/337
Subdural hygroma
arachnoid cyst, 31/106
bacterial meningitis, 52/8
child, 57/287
drainage, 57/287
echoEG, 23/278
EEG, 23/352
increased vessel permeability, 57/286

intracranial pressure, 57/286
mortality, 57/287
pathogenesis, 57/286
posttraumatic, 57/286
subdural hematoma, 57/286
temporal lobe agenesis syndrome, 31/105
Subdural lymphoma
epilepsy, 63/349
headache, 63/349
Hodgkin lymphoma, 63/349
non-Hodgkin lymphoma, 63/349
spinal cord compression, 63/349
Subdural tumor
spinal, see Spinal subdural tumor
Subependymal astrocytoma
see also Cranial vault tumor
cervical spinal cord, 20/362
classification, 18/3, 110
epilepsy, 72/181
infratentorial, 68/169
intraventricular, 18/14
supratentorial, 68/169
third ventricle tumor, 17/443
Subependymal cell layer
cerebrum, 30/65-68
CNS, 50/36
Subependymal giant cell astrocytoma
classification, 67/5
differentiation, 67/6
localization, 67/6
low grade glioma, 68/55
Subependymal giant cell tumor
thalamic tumor, 68/65
Subependymal glioma, see Subependymal
astrocytoma
Subependymal glomerular astrocytoma
tuberous sclerosis, 18/14
Subependymal granuloma
Whipple disease, 70/243
Subependymal hemorrhage
B-mode sonography, 54/36
brain hemorrhage, 54/36
infantile cerebrovascular disease, 54/36
Subependymal heterotopia
Aicardi syndrome, 50/168
corpus callosum agenesis, 50/156
orofaciodigital syndrome type I, 50/168
Subependymal nodule
calcification, 14/364
tuberous sclerosis, 14/349, 363, 379
Subependymoma, see Subependymal astrocytoma
Subgaleal hematoma
EEG, 23/318

pediatric head injury, 57/329
Subiculum
    cardiac dysrhythmia, 63/239
Subirana, A., 1/14, 31
Subluxation
    atlantoaxial, see Atlantoaxial subluxation
    atlantoaxial dislocation, 25/131
    C1-C2, see Atlantoaxial subluxation
    cervical, see Cervical subluxation
    cervical facet, see Cervical facet subluxation
    cervical vertebral column injury, 61/25
    facet, see Facet subluxation
    pseudo, see Pseudosubluxation
    spinal, see Spinal subluxation
    vertebral, see Vertebral subluxation
Submento-occipital projection
    craniocervical bone abnormality, 32/10
    maxillary sinus, 32/10
Submucosal plexus
    anatomy, 75/615
    Wolman disease, 51/379
Submyelinic space
    periaxonal, see Periaxonal submyelinic space
Suboccipital dysplasia, 32/28-41
    atlas, 32/26
    atlas dysplasia, 32/36-40
    axis dysplasia, 32/40
    basilar process, 32/30
    basilar transverse fissure, 32/29
    bony masses on foramen magnum, 32/31
    epitransverse process, 32/32, 34
    infratransverse process, 32/33-35
    occipitomastoid suture, 32/34
    occipitoparacondylar suture, 32/33
    paracondylar process, 32/32
    paracondylar suture, 32/33
    paramastoid suture, 32/34
    persistent terminal ossicle, 32/32
    third condyle, 32/30
    transverse process, 32/35
Suboccipital intersegmental artery
    anatomy, 11/36
Suboccipital puncture
    complication, 12/123, 61/147
    subarachnoid hemorrhage, 12/123
Subordinate hemisphere, see Cerebral dominance
Subperiosteal giant cell tumor, see Aneurysmal bone
    cyst
Subperiosteal ossifying hematoma, see Aneurysmal
    bone cyst
Subpial granular layer
    CNS, 50/37
Subsarcolemmal myofibril

benign autosomal dominant myopathy, 43/112
    chorea, 42/219
    dystonia, 42/219
    EEG, 42/219
    hyperreflexia, 42/219
    isoniazid neuropathy, 7/584
    lordosis, 42/253
    phosphofructokinase deficiency, 40/53, 41/191,
        427, 43/183, 62/488
    speech disorder in children, 42/219
Subscapular nerve
    lesion localization, 2/137
    topographical diagnosis, 2/21
Substance K
    Huntington chorea, 49/296, 298
Substance P
    Alzheimer disease, 46/268
    basal ganglion, 49/33
    basal ganglion tier II, 49/21
    basal ganglion tier III, 49/22
    brain blood flow, 53/63
    chorea-acanthocytosis, 63/283
    chronic paroxysmal hemicrania, 48/263
    cluster headache, 75/291
    gastrointestinal disease, 39/453
    headache, 48/111, 274, 75/291
    hereditary sensory and autonomic neuropathy type
        III, 45/230, 60/29
    histamine, 48/98
    Huntington chorea, 49/75, 258, 260, 296
    intracranial artery, 48/109
    migraine, 48/98, 113
    multiple neuronal system degeneration, 75/193
    neostriatum, 49/47
    pain perception, 45/228
    pallidonigral fiber, 49/8
    Parkinson disease, 49/120
    pedunculopontine tegmental nucleus, 49/12
    progressive dysautonomia, 59/148
    reflex sympathetic dystrophy, 61/126
    spinal cord, 45/230
    striatonigral fiber, 49/8
    striatopallidal fiber, 49/6
    substantia nigra, 49/8
    trigeminal ganglion, 48/110
    vasoactive intestinal polypeptide, 48/99
    vasodilatation, 48/98
Substance X, see Ganglioside
Substances
    gap, see Gap substance
Substantia gelatinosa
    enkephalin, 45/230
    nociceptor afferent, 45/230

opiate receptor, 45/230
Substantia innominata
  basal ganglion tier II, 49/21
  mesolimbic dopamine system, 49/50
  terminological history, 6/30
Substantia nigra
  alcohol intoxication, 64/114
  Benedikt syndrome, 2/298
  boxing injury, 23/557
  cell, 6/42-44
  chorea-acanthocytosis, 63/283
  cytology, 49/7
  dorsal nucleus raphe, 49/9
  enkephalin, 49/8
  function, 6/106
  GABA, 49/8
  glutamate decarboxylase, 49/8
  Guam amyotrophic lateral sclerosis, 22/287,
    59/287-291
  Hallervorden-Spatz syndrome, 22/403, 49/395
  histochemistry, 6/45
  Huntington chorea, 6/407, 49/320
  infantile neuroaxonal dystrophy, 49/396
  1-methyl-4-phenyl-1,2,3,6-tetrahydropyridine
    intoxication, 65/371
  neuroanatomy, 6/42-46
  nigral afferent, 49/8
  nigrotectal fiber, 49/10
  nigrotegmental fiber, 49/10
  nigrothalamic fiber, 49/10
  nondopaminergic, see Nondopaminergic
    substantia nigra
  pallidonigral fiber, 49/8
  paraquat intoxication, 65/398
  Parkinson disease, 6/214
  pars compacta, 6/19, 44
  pars reticulata, 6/19, 44, 407
  pathologic laughing and crying, 3/362, 45/223
  phenylketonuria, 29/34
  serotonin, 49/8
  spheroid body, 59/287-291
  St. Louis encephalitis, 56/138
  striatonigral fiber, 49/8
  substance P, 49/8
  subthalamonigral fiber, 49/8
  zona reticulata, see Zona reticulata substantia
    nigra
Substantia nigra degeneration
  Batten disease, 42/468
  cerebello-olivary atrophy, 42/136
  Hallervorden-Spatz syndrome, 42/257
  neuronal ceroid lipofuscinosis, 42/468
  Parkinson dementia complex, 6/176, 42/250

Parkinson disease, 42/246
  progressive pallidal atrophy (Hunt-Van Bogaert),
    6/457, 42/247
  progressive supranuclear palsy, 22/221, 42/259
  striatonigral degeneration, 42/262
  subacute necrotizing encephalomyelopathy,
    42/625
  xeroderma pigmentosum, 60/337
Substantia nigra depigmentation
  familial amyotrophic lateral sclerosis, 59/246
Subtemporal rhizotomy
  history, 5/303, 305
  trigeminal neuralgia, 5/303-305
Subthalamic fiber
  tegmental, see Tegmental subthalamic fiber
Subthalamic nucleus, 6/33-38
  afferent, 49/10
  ballismus, 6/59, 62
  cell, 6/33
  connection, 6/482
  corticosubthalamic fiber, 49/11
  cytology, 49/10
  dentatorubropallidoluysian atrophy, 49/440
  dystonia, 49/457
  efferent, 49/12
  hemiballismus, 42/224, 49/369
  histochemistry, 6/33
  Huntington chorea, 49/320
  manganese intoxication, 64/307
  mercury intoxication, 64/381
  pallidosubthalamic fiber, 49/11
  pallidosubthalamic projection, 49/11
  tegmental subthalamic fiber, 49/11
  thalamosubthalamic fiber, 49/11
Subthalamofugal fiber
  anatomy, 6/33
Subthalamonigral fiber
  substantia nigra, 49/8
Subthalamopallidal fiber
  GABA, 49/6
Subthalamopetal fiber
  anatomy, 6/38
Subthalamus, 6/33-38
  anatomy, 6/5
Subtle vitamin B12 deficiency
  vitamin B12 deficiency, 70/394
Subtraction method
  brain artery, 11/71
Subungual fibroma
  tuberous sclerosis, 50/372
Subunit strand
  neurofilament, 7/17
Succimer

arsenic intoxication, 64/289
bismuth intoxication, 64/344
lead encephalopathy, 64/443
lead intoxication, 64/439
mercury intoxication, 64/394
Succinate cytochrome-c reductase
  complex II deficiency, 62/504
Succinate dehydrogenase
  defect, 66/396
  Levine experimental anoxia ischemia, 9/583
  mitochondrial disease, 40/31, 62/497
  multiple sclerosis plaque, 9/37
  skeletal muscle, 40/4
  tricarboxylic acid cycle, 27/16
  wallerian degeneration, 9/37
Succinate dehydrogenase (ubiquinone), *see*
  Complex II
Succinate semialdehyde dehydrogenase deficiency
  computer assisted tomography, 60/671
  GABA metabolism, 66/654
  mental deficiency, 60/664
  organic acid metabolism, 66/641
Succinic acid
  CSF, 16/371
  multiple sclerosis, 9/385
Succinimide
  epilepsy, 15/643, 670
Succinimide intoxication
  anticonvulsant, 37/200
Succinylcholine
  acute muscle weakness, 41/291
  chondrodystrophic myotonia, 41/299
  halothane, 41/268, 270
  Isaacs syndrome, 40/546
  malignant hyperthermia, 38/550, 552, 40/546,
    41/268, 270, 62/568, 63/439
  metabolic encephalopathy, 69/400
  prolonged paralysis, 41/291
  sensitivity, 42/654
Succinylcholine intoxication
  myoglobinuria, 40/339
  rhabdomyolysis, 40/339
Sucking reflex
  brain death, 24/718
  pellagra, 28/85
  regression reflex, 1/241
Sucrose
  intracranial pressure, 23/212
Suction ophthalmodynamometry
  carotid artery stenosis, 11/228
Sudan black B staining
  histochemistry, 9/25, 28
  skeletal muscle, 40/4

Sudan dye
  multiple sclerosis plaque, 9/37
Sudan III staining
  lipid, 9/25, 28
Sudan IV staining
  lipid, 9/25, 28
Sudanophilia
  histochemistry, 10/30, 34
  triglyceride, 10/28
Sudanophilic leukodystrophy
  see also Orthochromatic leukodystrophy
  classification, 66/5
  Cockayne-Neill-Dingwall disease, 13/436
  Crome-Zapella disease, 10/197
  cyanide intoxication, 9/634
  definition, 10/32, 34, 37
  form, 66/13
  illness duration effect, 10/35
  lead intoxication, 9/642
  lipid hexosamine, 10/34
  meningeal angiomatosis, 42/508
  multiple sclerosis, 10/34
  neuropathology, 21/51
  nosology, 10/106
  Pelizaeus-Merzbacher disease, 10/151, 13/436
  peroxisomal acetyl-coenzyme A acyltransferase
    deficiency, 66/518
  primary pigmentary retinal degeneration, 13/340
  X-linked type, 42/500
Sudanophilic leukodystrophy with meningeal
    angiomatosis, *see* Divry-Van Bogaert syndrome
Sudden death
  amphetamine intoxication, 65/258
  autonomic nervous system, 74/171
  cardiac dysrhythmia, 63/238
  cocaine intoxication, 65/256
  congenital central alveolar hypoventilation,
    63/415
  congestive heart failure, 63/131
  ependymal cyst, 18/135
  epilepsy, 63/496
  Friedreich ataxia, 21/346
  metastatic cardiac tumor, 63/94
  neurogenic pulmonary edema, 63/496
  oxoglutarate dehydrogenase deficiency, 66/419
  primary cardiac tumor, 63/94
  Refsum disease, 8/22, 21/191, 60/228, 66/488,
    491, 498
  sensorineural deafness, 42/388
  unexplained, 18/135
Sudden drop
  epilepsy, 73/213
  Lennox-Gastaut syndrome, 73/213

Sudden infant death
  adenoidectomy, 63/466
  botulism, 51/190
  congenital central alveolar hypoventilation,
    63/415
  congenital disorder, 63/465
  epilepsy, 63/466
  infection, 63/465
  medium chain acyl-coenzyme A dehydrogenase
    deficiency, 66/410
  periodic respiration, 63/465
  preterm infant, 63/464
  prolonged apnea, 63/465
  reflux apnea, 63/466
  respiratory pattern, 63/465
  sleep apnea syndrome, 45/132
  sleep disorder, 63/449
  tonsillectomy, 63/466
  trauma, 63/466
  treatment, 63/466
Sudeck dystrophy, see Reflex sympathetic
    dystrophy
Sudeck reflex dystrophy, see Reflex sympathetic
    dystrophy
Sudeck reflex sympathetic dystrophy, see Reflex
    sympathetic dystrophy
Sudhoff, K., 1/3
Sudomotor abnormality
  localization value, 2/116
  sweat test, 2/116
Sudomotor axon reflex
  orthostatic hypotension, 63/160
  test, 74/169
Sudomotor burst
  sweat rate, 74/652
  sympathetic nerve, 74/652
Sudomotor fiber
  cervical cord, 2/209
  parasympathetic, 2/164
  test, 74/596
Sudomotor function
  aging, 74/235
  test, 75/65
  transverse spinal cord lesion, 61/275
Sudomotor neuron
  autonomic nervous system, 74/31
Sudomotor reflex
  autonomic nervous system, 74/629
Sudomotor test
  nerve injury, 7/251
Suffocation
  Finkel spinal muscular atrophy, 59/43
  pallidonigral necrosis, 49/466

Sugiura sign
  Vogt-Koyanagi-Harada syndrome, 56/614
Suicide
  acute mercury intoxication, 64/375
  angatea intoxication, 37/56
  Anthopleura elegantissima intoxication, 37/54
  chrysaora intoxication, 37/51
  ethylene glycol intoxication, 64/122
  head injury epidemiology, 57/12
  Huntington chorea, 6/313
  manic depressive psychosis, 43/208
  migraine, 48/162
  neuroleptic akathisia, 65/285
  organolead intoxication, 64/133
  pellagra, 28/66
  rhodactis howesi intoxication, 37/54
Suicide gene
  brain tumor, 67/301
Sulcus calcification
  striatopallidodentate calcification, 49/426
Sulfa agent
  acquired immune deficiency syndrome, 71/383
  pneumonia, 71/383
Sulfamethoxazole
  ataxia, 65/472
  headache, 65/472
  neurotoxin, 65/472
  paracoccidioidomycosis, 35/537
Sulfanilamide
  encephalomyelitis, 63/55
  spinal cord vasculopathy, 63/55
Sulfanilamide elixir
  ethylene glycol intoxication, 64/122
Sulfasalazine, see Salazosulfapyridine
Sulfatase activator deficiency
  metachromatic leukodystrophy, 66/170
Sulfatase deficiency
  glycosaminoglycanosis type II, 66/314
Sulfate sodium
  barium intoxication, 64/354
Sulfated glucuronic acid lactosaminyl
    paragloboside, see Glucuronylparagloboside
    sulfate
Sulfated glucuronic acid paragloboside, see
    Glucuronylparagloboside sulfate
Sulfatide
  biochemistry, 9/3
  biosynthesis, 66/165
  body fluid, 66/174
  brain development, 9/18
  brown metachromasia, 10/25
  chemistry, 10/25-27
  degradation, 66/165

function, 66/164
globoid cell leukodystrophy, 10/21, 304
2-hydroxytetracosanoic acid, 9/6
lipidosis, *see* Metachromatic leukodystrophy
metachromatic leukodystrophy, 10/311-313, 348, 29/357, 42/491
myelin, 9/10
myelin synthesis, 10/138
nerve myelin, 51/29
neuropathy, *see* Metachromatic leukodystrophy
Pelizaeus-Merzbacher disease, 10/33
15-tetracosanoic acid, 9/6
Sulfatide activator protein
metachromatic leukodystrophy, 47/589, 60/126
Sulfatide lipidosis, *see* Metachromatic leukodystrophy
Sulfatide neuropathy, *see* Metachromatic leukodystrophy
Sulfatidosis, *see* Metachromatic leukodystrophy
Sulfhydryl compound, *see* Thiol compound
Sulfide oxidation
organophosphate, 64/156
Sulfite oxidase deficiency
amino acid, 29/121
biochemistry, 29/121
cerebrovascular disease, 53/43
decerebration, 29/121
epilepsy, 72/222
hyperacusis, 29/121
muscular hypotonia, 29/121
opisthotonos, 29/121
retardation, 29/121
Sulfogalactoglycerolipid
enzyme deficiency polyneuropathy, 51/367
Sulfolipid
biochemistry, 9/10
chemical structure, 66/164
metachromatic leukodystrophy, 10/141
structure, 66/163
Sulfonamide
bacterial meningitis, 52/9
Hemophilus influenzae meningitis, 33/56
meningococcal meningitis, 52/30
neuropathy, 7/538
nocardiosis, 16/218, 35/526, 52/450
paracoccidioidomycosis, 35/537
psychosis, 46/601
Sulfur containing amino acid
disorder, 29/111-124
history, 29/111
metabolism, 29/112
vitamin B_6, 28/117, 122
Sulfurtransferase

cyanide intoxication, 64/231
Sulindac
chemical meningitis, 56/130
headache, 48/174
migraine, 48/174
Sulphatide, *see* Sulfatide
Sulpiride
anorexia nervosa, 46/588
antiepileptic agent, 65/497
chemical formula, 65/276
migraine, 48/92
neuroleptic agent, 65/275
specific action, 65/280
Sulthiame, *see* Sultiame
Sultiame
antiepileptic agent, 65/497
epilepsy, 15/651, 670
serum level, 15/691
side effect, 15/718
Summation
definition, 1/49
Summitt syndrome
acrocephaly, 43/325
craniosynostosis, 43/372-374
epicanthus, 43/325, 372
obesity, 43/325, 372-374
scaphocephaly, 43/325
strabismus, 43/372
syndactyly, 43/325, 372-374
SUNCT syndrome
autonomic nervous system, 75/299
blood pressure, 75/299
cluster headache, 75/299
heart rate, 75/299
hemicrania continua, 75/299
Sundowning
biological rhythm, 74/495
Sunfish intoxication
classification, 37/78
Sunflower cataract
Wilson disease, 27/386, 49/228
Sunnybrook Cord Injury Scale
spinal cord injury, 61/423
spinal cord injury recovery score, 61/423
transverse spinal cord lesion, 61/423
Sunset sign
aqueduct stenosis syndrome, 42/57
Sunstroke
heat stroke, 23/669
Superficial brain venous system
anatomy, 11/45
radioanatomy, 11/105
topography, 11/105

Superficial middle cerebral vein
  anatomy, 11/48, 60
Superficial reflex, *see* Cutaneous reflex
Superficial siderosis
  chemistry, 70/72
  clinical features, 70/67-71
  etiology, 70/72
  features, 70/65
  history, 70/73-75
  meta-analysis, 70/66
  nerve lesion, 70/66
  nervous system, 70/65
  neuropathology, 70/76
  neuroradiology, 70/72
  pyramidal syndrome, 70/66
  survey, 70/65
  therapy, 70/77
Superficial temporal artery
  anatomy, 11/65
  chronic migrainous neuralgia, 48/254
  Doppler sonography, 54/9
Superior anastomotic vein, *see* Trolard vein
Superior artery to dentate nucleus
  anatomy, 11/26
Superior cerebellar artery
  see also Marginal artery
  basilar artery stenosis, 53/386
  chorea, 55/90
  cortical branch, 11/27
  lateral branch, 11/93
  marginal ramus, 11/27
  medial branch, 11/93
  superior paravermal branch, 11/27
  vertebrobasilar system syndrome, 53/397
Superior cerebellar artery aneurysm
  aqueduct stenosis, 50/308
Superior cerebellar artery syndrome
  abducent nerve, 12/20
  artery occlusion, 12/20
  ataxia, 53/397
  cerebellar sign, 12/20
  chorea, 53/397
  conjugate gaze, 12/20
  dysarthria, 12/20
  Horner syndrome, 2/240, 53/397
  internuclear ophthalmoplegia, 53/397
  lower extremity, 12/20
  pontine, 2/240
  skew deviation, 53/397
  superior cerebellar peduncle, 12/20
  tegmental pontine syndrome, 12/20
  terminology, 11/31
  trigeminal nerve, 12/20

  vertigo, 12/20
  vomiting, 12/20
Superior cerebellar peduncle
  cerebellar tremor, 49/587
  Cestan-Chenais syndrome, 2/239
  postural tremor, 49/588
  superior cerebellar artery syndrome, 12/20
Superior cerebellar peduncle lesion
  resting tremor, 49/588
Superior cerebellar peduncle syndrome, *see*
    Cestan-Chenais syndrome
Superior cerebellar vein
  medial, *see* Precentral cerebellar vein
Superior cerebellar vermis hypoplasia
  iniencephaly, 50/131
Superior colliculi
  dentatorubropallidoluysian atrophy, 49/442
  Parinaud syndrome, 2/338
  physiologic study, 2/532
  respiration, 1/653
Superior colliculi syndrome, *see* Parinaud syndrome
Superior frontal gyrus
  hemiballismus, 49/370
Superior frontal vein
  anatomy, 11/47
Superior gluteal nerve injury
  sacroiliac fracture dislocation, 25/475
Superior laryngeal nerve
  anatomy, 48/496
  piriform sinus, 48/495
Superior laryngeal neuralgia, 48/495-499
  alcohol injection, 48/498
  carbamazepine, 48/498
  glossopharyngeal neuralgia, 48/468
  laryngeal itch, 48/496
  laryngeal nerve block, 48/495
  laryngeal pain, 48/496
  otalgia, 48/495
  swallowing, 48/497
  toux des aboyeurs, 48/496
  virus, 48/498
Superior longitudinal sinus
  anatomy, 11/52
  brain phlebothrombosis, 12/427
  dura mater, 30/424
  hydrocephalus, 12/427
  radioanatomy, 11/111
  traumatic venous sinus thrombosis, 24/370, 372
  venous sinus thrombosis, 12/427
Superior longitudinal sinus occlusion
  brain infarction, 54/397, 438
  Rolando vein, 54/397
  subdural hematoma, 54/397

Trolard vein, 54/397
Superior longitudinal sinus thrombophlebitis,
    33/167-172
  benign intracranial hypertension, 33/170-172
Superior longitudinal sinus thrombosis
  see also Venous sinus thrombosis
  antecedent factor, 52/178
  antibiotic agent, 52/179
  benign intracranial hypertension, 33/170
  brain angiography, 52/179
  ceftazidime, 52/179
  cilastatin, 52/179
  clinical features, 33/168, 171, 52/178
  coma, 52/178
  computer assisted tomography, 52/178
  confusion, 52/178
  cord sign, 52/178
  co-trimoxazole, 52/179
  CSF, 11/52, 52/178
  dental infection, 52/178
  differential diagnosis, 16/237, 33/169
  EEG, 52/178
  empty delta sign, 52/179
  epilepsy, 52/178
  functional anatomy, 52/177
  headache, 48/283, 52/178
  Hemophilus influenzae, 52/179
  imipenem, 52/179
  Köhlmeier-Degos disease, 55/277
  malignant lymphoma, 55/489
  meningitis, 52/178
  metronidazole, 52/179
  mortality, 52/179
  multiple cranial neuropathy, 55/277
  otitic hydrocephalus, 33/171
  paralysis, 54/396
  paraplegia, 54/396
  penicillin, 52/179
  prognosis, 33/169, 52/179
  puerperal thrombosis, 12/428
  sequela, 33/169, 52/179
  sinusitis, 52/178
  Staphylococcus aureus, 52/179
  Streptococcus pneumoniae, 52/179
  Sturge-Weber syndrome, 55/446
  treatment, 33/172, 52/179
Superior medialis paralysis
  differential diagnosis, 2/329
Superior mesenteric artery occlusion
  chronic, see Chronic superior mesenteric artery
    occlusion
Superior oblique tendon sheath syndrome, see
    Brown syndrome

Superior ophthalmic vein
  anatomy, 11/54
  radioanatomy, 11/114
  topography, 11/114
  traumatic carotid cavernous sinus fistula, 57/361
Superior orbital fissure syndrome, 2/86-88
  abducent nerve paralysis, 33/179
  blindness, 33/180
  cavernous sinus syndrome, 48/301
  course, 33/179
  exophthalmos, 33/179
  external ophthalmoplegia, 33/179
  lesion site, 2/320
  localization, 2/320
  megadolichocarotid artery, 12/81
  miosis, 2/321
  mydriasis, 2/321
  nonrecurrent nonhereditary multiple cranial
    neuropathy, 51/571
  oculomotor paralysis, 33/180
  optic nerve injury, 2/569
  orbital apex syndrome, 2/296
  periostitis, 33/179
  prognosis, 33/179
  skull base, 67/146
  skull base tumor, 17/181
  sphenocavernous angle syndrome, 2/321
  symptom, 33/179
  Tolosa-Hunt syndrome, 48/291, 296
Superior orbital fracture
  symptom, 24/98
  treatment, 24/99
Superior paravermal branch
  superior cerebellar artery, 11/27
Superior petrosal sinus
  anatomy, 11/49, 54, 62, 33/180
  radioanatomy, 11/113
  topography, 11/113
Superior petrosal sinus thrombophlebitis
  Crowe test, 33/180
  facial pain, 33/180
  symptom, 33/180
  Tobey-Ayer test, 33/180
Superior petrosal sinus thrombosis
  trigeminal neuropathy, 54/398
Superior petrosal vein
  anatomy, 11/49
Superior posterior cerebellar vein
  anatomy, 11/51
  embryology, 11/63
  radioanatomy, 11/110
  topography, 11/110
Superior red nucleus syndrome, see Nucleus ruber

superior syndrome
Superior sagittal sinus occlusion, *see* Superior longitudinal sinus occlusion
Superior sagittal sinus thrombosis, *see* Superior longitudinal sinus thrombosis
Superior temporal gyrus
    Down syndrome, 50/274
Superior temporal lobe hypoplasia
    Down syndrome, 50/531
Superior temporal sulcus
    hemi-inattention, 45/157
Superior vena cava obstruction
    dissecting aorta aneurysm, 63/47
Superior vena cava syndrome
    metastatic cardiac tumor, 63/94
    primary cardiac tumor, 63/94
Supernumerary ribs
    r(13) syndrome, 50/585
Superoxide dismutase
    free radical scavengers, 64/19
    Wilson disease, 49/229
Superoxide dismutase 1
    Down syndrome, 50/521, 530
Supinator syndrome
    Frohse arcade, 51/100
Supplementary motor area
    anatomy, 1/151
    seizure, 73/41
    speech disorder, 45/34
Supplementary motor area syndrome
    akinesia, 53/346
    alien hand sign, 53/346
    anterior cerebral artery, 53/345
    anterior cerebral artery syndrome, 53/345
    bilateral lesion, 53/346
    brain infarction, 53/345
    dysphasia, 53/346
    dyspraxia, 53/346
    ideomotor apraxia, 53/346
    mutism, 53/346
    palilalia, 53/346
Supporting reaction
    cerebellar ablation, 2/396, 402
    cerebellar lesion, 1/264
    posture, 1/68
Suppression burst pattern
    early infantile epileptic encephalopathy, 73/240
    EEG, 73/235
    sleep, 73/236
Suppressor gene
    brain tumor, 67/44
    tumor cell, 67/19
Suppressor T-lymphocyte, *see* T-suppressor

lymphocyte
Suppurative myositis, *see* Inflammatory myopathy
Suprabulbar paresis
    congenital, *see* Congenital suprabulbar paresis
Suprachiasmatic nucleus
    Alzheimer disease, 74/496
    annual rhythm, 74/489
    biological rhythm, 74/476, 495
    chemical anatomy, 74/477
    chemoarchitecture, 74/474
    circadian rhythm, 74/487, 492
    circannual rhythm, 74/488
    cluster headache, 74/507
    CSF, 74/487
    depression, 74/506
    development, 74/493
    dopamine, 74/489
    fetal, 74/492
    homosexual man, 74/500
    hormone, 74/487
    hypnic headache syndrome, 74/507
    hypothalamus, 74/487, 489
    maturity at birth, 74/492
    menstrual cycle, 74/491
    migraine, 74/507
    neurotensin, 74/481
    neurotransmitter, 74/489
    optic nerve, 74/480
    optic tract, 74/481
    paraventricular nucleus, 74/479
    pineal gland, 74/491
    reproduction, 74/498
    retinohypothalamic tract, 74/476
    seasonal rhythm, 74/490
    serotonin, 74/489
    sex, 74/497
    sex difference, 74/474, 497
    sexual behavior, 74/467, 498
    sleep, 74/497
    third ventricle, 74/480
    vasoactive intestinal polypeptide, 74/475, 477
    vasopressin, 74/477, 488, 495
    vasopressinergic fiber, 74/487
Suprachiasmatic recess
    gross anatomy, 2/525
Supraclavicular nerve
    cervical spinal cord injury, 61/56
    plexus lesion, 2/135
Supracondylar ligament
    compression neuropathy, 51/92
Supranuclear ophthalmoplegia
    congenital heart disease, 63/7
    dementia, 42/605

426

Gaucher disease, 60/656
Huntington chorea, 60/656
Niemann-Pick disease type C, 60/656
olivopontocerebellar atrophy, 60/657
phenytoin intoxication, 60/656
postopen heart surgery, 63/7
progressive supranuclear palsy, 42/259, 49/246,
    60/656
spino-olivocerebellar degeneration, 60/656
striatonigral degeneration, 49/209
Wilson disease, 60/656
Supranuclear palsy
    glutamate dehydrogenase deficiency, 60/553
    progressive, see Progressive supranuclear palsy
    rigidity, 49/69
Supraoptic commissure, see Ganser dorsal
    commissure and Gudden-Meynert ventral
    commissure
Supraoptic decussation
    anatomy, 6/34-36
Supraoptic hypothalamus
    anatomy, 2/434-436
Supraoptic nucleus
    anatomy, 2/434, 437
    Broca diagonal band, 2/434
Supraoptic nucleus degeneration
    hypophyseal diabetes insipidus, 42/543
Supraopticohypophyseal tract
    anatomy, 2/437
Supraorbital nerve
    paratrigeminal syndrome, 48/333
Supraorbital neuralgia
    avulsion, 5/392
    cluster headache, 5/115
    injection, 5/388
Suprapatellar reflex
    loop anatomy, 1/246
Suprapubic cystotomy
    paraplegia, 26/409
Suprascapular nerve
    compression neuropathy, 51/102
    occupational lesion, 7/332
    topographical diagnosis, 2/20
Suprasellar pearly tumor
    pituitary adenoma, 17/412
Supratentorial brain abscess, 33/116-118, 122-128
    brain scanning, 33/124
    carotid angiography, 33/123
    computer assisted tomography, 33/126
    CSF, 33/118
    differential diagnosis, 33/116
    EEG, 33/122
    intracavity contrast media, 33/127

intracranial pressure, 33/117
meningeal irritation, 33/118
radionuclide scan, 33/124
ventriculography, 33/124
Supratentorial brain tumor
    astrocytoma grade I, 18/310
    astrocytoma grade II, 18/310
    astrocytoma grade III, 18/311
    astrocytoma grade IV, 18/311
    brain angiography, 18/336
    brain scanning, 18/334
    central neuroblastoma, 18/310
    chemotherapy, 18/338
    child, 18/305-384
    choroid plexus papilloma, 18/314
    clinical features, 18/323
    craniopharyngioma, 18/347-352
    CSF, 18/337
    desmoid cyst, 18/317
    differential diagnosis, 18/327
    echoEG, 18/333
    EEG, 18/333
    ependymoma, 18/313
    epidermoid cyst, 18/317
    fibroma, 18/321
    focal sign, 18/326
    intracranial hypertension, 18/321
    less than 2 yr old, 18/327
    meningioma, 18/318
    metastasis, 18/340, 368
    misdiagnosis, 18/332
    multiplicity, 18/309
    neuroastrocytoma, 18/309
    neuroblastoma, 18/340
    neuroradiology, 18/335
    oligodendroglioma, 18/317
    optic nerve glioma, 18/343-347
    pathologic anatomy, 18/307
    plain X-ray, 18/334
    prevalence, 18/307
    prognosis, 18/338
    radiotherapy, 18/338
    sarcoma, 18/321
    skull transillumination, 18/336
    spongioblastoma, 18/311
    teratoma, 18/317
    third ventricle tumor, 18/358
    treatment, 18/337
    tuberous sclerosis, 18/322
    type, 18/309
    ventriculography, 18/336
Supratentorial midline tumor
    clinical features, 18/341

corpus callosum tumor, 18/355
diagnostic test, 18/342
exophthalmos, 18/339
hydrocephalus, 16/112
pineal tumor, 18/361
prognosis, 18/342
septum pellucidum tumor, 18/357
treatment, 18/342
Supratrochlear artery
carotid artery stenosis, 54/2, 4
Doppler sonography, 54/3
flow pattern, 54/2-4
headache, 75/285
Supraventricular tachycardia
paroxysmal, *see* Paroxysmal supraventricular
tachycardia
Sural nerve
amyloid polyneuropathy, 51/418
ankle fracture, 70/41
chronic idiopathic demyelinating
polyradiculoneuropathy, 47/614
compression neuropathy, 51/106
eosinophilia myalgia syndrome, 64/254
histology, 60/343
leprous neuritis, 51/224
Marinesco-Sjögren syndrome, 60/343
metachromatic leukodystrophy, 47/593
Sural nerve biopsy
acrylamide intoxication, 64/70
ataxia telangiectasia, 60/376
cerebrotendinous xanthomatosis, 60/167
chorea-acanthocytosis, 60/141
chronic inflammatory demyelinating
polyradiculoneuropathy, 51/536
congenital pain insensitivity, 51/564
critical illness polyneuropathy, 51/578, 581
diabetic polyneuropathy, 51/510
Fabry disease, 51/378
hereditary sensory and autonomic neuropathy type
I, 60/7
hereditary sensory and autonomic neuropathy type
II, 60/7, 11
hereditary sensory and autonomic neuropathy type
III, 60/7, 31
hereditary sensory and autonomic neuropathy type
IV, 60/7, 14
hereditary sensory and autonomic neuropathy type
V, 60/7, 15
Klein-Waardenburg syndrome, 59/162
metachromatic leukodystrophy, 51/372
Sural nerve stimulation
cluster headache, 48/235
Suramin

antineoplastic agent, 65/529
neuropathy, 69/461, 472
neurotoxin, 65/529, 538
polyneuropathy, 65/529
Suramin intoxication
polyneuropathy, 65/538
Surface antigen
hepatitis B virus, 56/295
Surface dyslexia
alexia, 45/445
Surgical decompression
hydromyelia, 50/431
Survival
anencephaly, 30/176
aqueduct stenosis, 30/619
astrocytoma, 18/4, 33, 38, 68/5, 10-12
brain arteriovenous malformation, 54/374
brain injury, 23/16
brain metastasis, 18/227
brain stem death, 57/466
cardiac arrest, 63/206
cerebellar hemorrhage, 12/66
cerebellar medulloblastoma, 18/189
dissecting aorta aneurysm, 61/114
electrocerebral silence, 57/466
ependymoma, 18/121, 131
glioblastoma, 68/5, 102, 112
glioblastoma multiforme, 18/62, 66, 68/113
hydranencephaly, 50/349
oligodendroglioma, 18/101
pinealoblastoma, 68/224
primary CNS lymphoma, 68/265
spina bifida, 32/552, 555, 50/493
spinal cord injury, 25/41, 61/272
subarachnoid hemorrhage, 12/182
temporal arteritis, 48/325
Suspected cerebrovascular disease, 53/199-241
Sustained handgrip, *see* Isometric exercise
Suture
coronal, *see* Coronal suture
hydrocephalus, 32/541
lambdoid, *see* Lambdoid suture
metopic, *see* Metopic suture
nerve repair, 8/513
skull, 57/144
spina bifida, 32/541
Suture bulge
metopic, *see* Metopic suture bulge
Suxamethonium
neuromuscular block, 40/129
toxic myopathy, 62/597, 601
Suxamethonium induced muscle rigidity
malignant hyperthermia, 62/569, 63/439

Suxamethonium sensitivity
  heterozygote frequency, 42/655
  malignant hyperthermia, 43/117
  prevalence, 42/654
SV5, *see* Simian virus 5
SV40, *see* Simian virus 40
Swallow breath
  larynx, 74/393
Swallow syncope
  see also Vasovagal syndrome
  aorta aneurysm, 63/58
  aorta surgery, 63/58
  dysautonomia, 22/264
Swallowing
  autonomic nervous system, 74/631
  esophageal pressure, 74/634
  pain, 48/502
  pathology, 1/509
  physiology, 1/504
  respiration, 74/393
  sea snake intoxication, 37/92
  stylohyoid syndrome, 48/503
  superior laryngeal neuralgia, 48/497
Swallowing reflex
  anencephaly, 30/177
  newborn behavior, 4/345
  physiology, 22/122-125
Swamp fever, *see* Leptospirosis
Swan neck deformity
  myotonic dystrophy, 40/308
Swayback
  copper, 27/389
  lamb, 9/112, 634, 645, 671, 675-679, 27/389
  multiple sclerosis, 9/78, 112, 678
Swayback nose
  Aarskog syndrome, 43/313
  achondroplasia, 43/315
  cleidocranial dysplasia, 43/348
  Coffin-Siris syndrome, 43/240
  Conradi-Hünermann syndrome, 43/345
  De Lange syndrome, 43/246
  Dubowitz syndrome, 43/248
  facio-oculoacousticorenal syndrome, 43/399
  hypertelorism, 43/439
  mental deficiency, 43/279, 435
  Mulibrey dwarfism, 43/435
  9p partial monosomy, 43/512
  4p partial trisomy, 43/500
  21q partial monosomy, 43/541
  2q partial trisomy, 43/493
  6q partial trisomy, 43/503
  10q partial trisomy, 43/517
  sensorineural deafness, 42/393

  snub-nosed dwarfism, 43/387
  thanatophoric dwarfism, 43/388
  Wegener granulomatosis, 51/451
  XXXX syndrome, 43/554
  XXXXX syndrome, 43/554
Sweat
  body odor, 74/375
  odor, 74/375
Sweat gland
  autonomic nervous system, 74/407
  human, 74/376
  odor, 74/375
Sweat gland agenesis
  pigmented nevus, 14/597
Sweat production
  Guillain-Barré syndrome, 75/686
Sweat rate
  skin temperature, 75/57
  sudomotor burst, 74/652
  sympathetic nerve, 74/652
Sweat secretion
  thermoregulation, 75/58
Sweat test
  amyloid polyneuropathy, 51/416
  anhidrosis, 75/66
  autonomic nervous system, 74/628
  cervical cord lesion, 2/200
  foot, *see* Foot sweat test
  hereditary sensory and autonomic neuropathy type
    IV, 60/13
  iodine starch test, 2/17, 200
  Manuila-Jadassohn, 14/106
  Minor iodine starch method, 2/17
  peripheral lesion, 2/173
  quinizarin, 2/17, 200
  sudomotor abnormality, 2/116
  sympathetic nervous disorder, 2/116
  testing method, 2/17, 173
  thermoregulatory, *see* Thermoregulatory sweat
    test
  topographical diagnosis, 2/15
Sweating
  absence, *see* Anhidrosis
  alcohol, 75/495
  auriculotemporal syndrome, 2/103
  autonomic nervous system, 1/452-459, 74/627
  cervical cord lesion, 2/200
  chelonia intoxication, 37/89
  chronic paroxysmal hemicrania, 48/263, 75/292
  cyanea intoxication, 37/52
  defect, 75/116
  disturbance, 1/452-457
  electrolyte balance, 1/501

excessive, *see* Excessive sweating
facial, *see* Facial sweating
foot, 28/517
forehead, *see* Forehead sweating
frontal lobe tumor, 17/259
glutaryl-coenzyme A dehydrogenase deficiency, 66/647
Guillain-Barré syndrome, 7/499, 51/245
gustatory, *see* Gustatory sweating
headache, 75/282, 284
heart disease, 1/458
Horner syndrome, 1/455
hypothalamic syndrome, 2/450
innervation, 2/163
Melkersson-Rosenthal syndrome, 8/221
mental stress, 1/455
methyl salicylate intoxication, 37/417
migraine, 48/8, 75/295
multiple sclerosis, 9/180
muscle contraction headache, 75/295
neonatal, 1/455
organophosphate intoxication, 37/554, 64/170
oxidation-phosphorylation coupling defect, 62/506
paratrigeminal syndrome, 75/293
phenothiazine, 75/499
pheochromocytoma, 5/230, 14/252, 39/498, 500, 48/427
physiologic factor, 75/70
puffer fish intoxication, 37/80
pure autonomic failure, 2/120, 22/232
Raynaud phenomenon, 8/344
salicylic acid intoxication, 37/417
scorpion sting intoxication, 37/112
sea snake intoxication, 37/92
sodium salicylate intoxication, 37/417
spinal cord injury, 26/357, 366, 369, 61/276, 281
stingray intoxication, 37/72
thalamic syndrome (Dejerine-Roussy), 2/482
thermoregulation, 26/357, 366, 369
traumatic vegetative syndrome, 24/579
uremic polyneuropathy, 51/357
zinc intoxication, 64/362
Sweating disorder
  multiple neuronal system degeneration, 75/171
Sweating excess
  hereditary sensory and autonomic neuropathy type III, 43/58, 60/28
Sweaty foot syndrome
  acid base metabolism, 28/517
  isovaleric acidemia, 42/579
Sweden
  training, 1/24

Swedenborg, E., 2/12
Swedish porphyria, *see* Acute intermittent porphyria
Swellfish intoxication, *see* Puffer fish intoxication
Swelling
  anterior tibial syndrome, 12/538, 20/798
  astrocytic, *see* Astrocytic swelling
  axonal, *see* Axonal swelling
  bee sting intoxication, 37/107
  brain, *see* Brain swelling
  brain edema, 16/169
  carotid artery bifurcation, 48/338
  central serous retinopathy, 13/142
  cyanea intoxication, 37/52
  echinoidea intoxication, 37/67
  giant axonal, *see* Giant axonal swelling
  Gubler, *see* Gubler swelling
  head injury, 57/131
  hereditary angioedema, 43/60
  lip, 8/205
  mackeral intoxication, 37/87
  neuronal, *see* Neuronal swelling
  pterois volitans intoxication, 37/75
  scalp, 57/131
  scorpion sting intoxication, 37/112
  spinal cord edema, 12/557
  spinal nerve root, *see* Spinal nerve root swelling
  stingray intoxication, 37/72
Swimming
  rehabilitative sport, 26/527, 533
  sport injury, 23/585
Swine fever
  animal study, 18/246
  reticulosarcoma, 18/246
Swine influenza vaccination
  complication, 47/478, 480, 483-485
  Guillain-Barré syndrome, 47/483
  neuromyelitis optica, 47/478, 480, 485
Swineherd disease, *see* Leptospirosis
Switzerland
  neurology, 1/10
Sydenham chorea, 6/409-432
  action tremor, 49/363
  acute disseminated encephalomyelitis, 71/550
  age, 6/409, 49/360
  agraphia, 45/466
  antecedent history, 49/361
  antistreptolysin titer, 6/426
  benign intracranial hypertension, 16/160
  cardiac valvular disease, 49/363
  cardiopathy, 6/416-418
  cause, 6/411
  chorea mollis, 6/427, 49/360
  chorea paralytica, 6/427, 49/359

chronic meningitis, 56/645
clinical features, 6/414-416, 49/360
clinical variant, 6/427
course, 6/426
diagnosis, 6/430, 71/551
diagnostic criteria, 6/411
differential diagnosis, 71/551
disease duration, 49/361
dysarthria, 49/360
dyskinesia, 49/363
ECG, 49/361
EEG, 6/421-426, 49/361
EMG, 6/418-421, 49/73
erythema marginatum, 6/411
erythrocyte sedimentation rate, 6/426
estrogen, 49/364
familial history, 49/360
forme fruste, 6/427
α-2-globulin, 6/426
H-reflex, 6/421
HLA antigen, 49/362
Huntington chorea, 6/354
hypotonia, 6/415, 49/71, 360
IgG antibody, 49/362
immune complex, 49/362
incidence, 6/409
incoordination, 49/363
innervation inhibition, 49/74
involuntary movement, 2/501
long-term sequel, 49/363
milking grip, 49/360
mortality, 49/364
neuropathology, 6/429, 49/364
pathogenesis, 6/430, 71/551
pathophysiology, 49/361
posture, 6/159
previous, see Previous Sydenham chorea
prior history, 49/364
progesterone, 49/364
prognosis, 71/551
pronator drift, 49/360
pronator sign, 6/159
psychiatric residue, 49/364
readiness potential, 49/74
relapse, 6/429
rheumatic fever, 49/359
season, 6/409, 49/360
serum fibrinogen, 6/411
serum mucopolysaccharide, 6/412
sex ratio, 49/360
streptococcal sequel, 6/412
striatal degeneration, 6/59
survey, 71/550

systemic lupus erythematosus, 49/364
treatment, 6/431, 49/362
tremor, 49/363
Syllable mutilation
migraine, 48/162
Sylvester disease
amyotrophy, 22/505
distal amyotrophy, 60/659
Friedreich ataxia, 22/505
hereditary hearing loss, 22/505
nystagmus, 22/505
Sylvian aqueduct stenosis, see Aqueduct stenosis
Sylvian aqueduct syndrome, see Aqueduct stenosis
Sylvian fissure
echoEG, 23/268
sensory integration, 2/700, 711
Sylvian group, see Ansa insulare
Sylvian triangle
radioanatomy, 11/86
topography, 11/86
Sylvian vein group
radioanatomy, 11/105
topography, 11/105
Symbolic agnosia
definition, 2/607
picture, 4/16, 300
Symmelia
caudal aplasia, 42/50
Symmetric lipomatosis
multiple, see Multiple symmetric lipomatosis
Symmetrical cerebellar hypogenesis syndrome
cerebellar hypoplasia, 30/388
Symmetrical encephalomalacia
lamb, 49/484
pallidonigral necrosis, 49/484
Symmetrical neuropathy
ascending, see Ascending symmetrical neuropathy
Symmetrical polyneuropathy
diabetic polyneuropathy, 51/499
distal, see Distal symmetric polyneuropathy
hereditary neuropathy, 60/254
hypothyroid polyneuropathy, 51/515
pellagra, 7/572
Symmetrical pseudolipomatosis
Krabbe-Bartels disease, 14/406
Symmetrical spastic cerebral palsy
contracture, 42/167
epilepsy, 42/167
mental deficiency, 42/167
microcephaly, 42/167
recurrence risk, 42/168
Symonds, C. Sir, 15/11
Symonds classification

epilepsy, 15/11
Symonds headache
  cluster headache, 5/114
Sympathectomy
  autonomic nervous system, 74/110, 75/17
  headache, 5/105
  migraine, 5/104
  myelomalacia, 55/100
  pain, 26/497
  pain relief, 1/102
  shoulder hand syndrome, 8/330
  spinal pain, 26/497
Sympathetic activity
  muscle, 74/250, 75/109
  regulation, 74/259
  skin, 75/106
Sympathetic apraxia
  definition, 45/428
Sympathetic beta receptor
  aging, 74/226
Sympathetic dyspraxia
  agraphia, 45/463
Sympathetic dystonia
  balance, 22/251
Sympathetic dystrophy
  reflex, see Reflex sympathetic dystrophy
Sympathetic ganglion
  aging, 74/225
  autonomic nervous system, 74/225
  development, 7/3, 8
  function, 74/39
  nerve tumor, 8/459-472
  normal, 75/2
  Parkinson disease, 49/110
  tumor, 8/459
Sympathetic meningitis, see Meningitis serosa
Sympathetic nerve
  aberrant regeneration, 2/121
  aging, 74/226, 657
  autonomic nervous system, 74/658
  awake state, 74/652
  catecholamine, 74/162
  hypotension, 74/659
  muscle, 74/649
  neurotransmission, 74/100
  noradrenalin, 74/162
  parasympathetic nerve, 74/104
  skin, 74/649
  spontaneous discharge, 74/649
  sudomotor burst, 74/652
  sweat rate, 74/652
  yawning, 74/652
Sympathetic nerve fiber

orbit, 48/333
  paratrigeminal syndrome, 48/332
  tetrodotoxin intoxication, 65/156
Sympathetic nerve terminal
  calcium, 74/143
  exocytosis, 74/143
Sympathetic nervous disorder
  acute polyneuropathy, 2/119
  cerebrovascular disease, 2/121
  diabetic polyneuropathy, 2/119
  drug, 2/121
  examination method, 2/114-117
  localization, 2/117
  ocular sign, 25/382
  old age, 2/121
  parkinsonism, 2/120
  spinal transverse lesion, 2/118
  sweat test, 2/116
  tabes dorsalis, 2/120
  vasomotor change, 2/117
Sympathetic nervous system
  see also Autonomic nervous system
  adrenomedullary hormonal system, 74/138
  aged, 75/5
  altitude stress, 74/654
  anatomy, 1/429, 431, 74/6
  catecholamine, 75/458
  causalgia, 48/112
  environmental stress, 74/653
  exercise, 74/246
  gravitational stress, 74/653
  headache, 48/112
  hereditary sensory and autonomic neuropathy type
    III, 21/113, 42/300, 60/23, 25, 29, 31
  history, 1/427
  immune system, 75/557
  noradrenalin, 74/138
  pain, 75/314
  physiology, 22/247
  reflex sympathetic dystrophy, 48/112
  spinal cord injury, 61/275
  spinal efferent, see Spinal sympathetic efferent
  thermal stress, 74/655
  tone, 22/251
Sympathetic nervous system ablation
  immune response, 75/555
Sympathetic neuron
  autonomic nervous system, 75/620
  classification, 74/24
  prevertebral ganglion, 74/41
  type, 74/36
  viscus, 74/34
Sympathetic ophthalmia

Vogt-Koyanagi-Harada syndrome, 56/612
Sympathetic pathway
  enteroenteric inhibitory reflex, 75/625
  headache, 48/331
  paratrigeminal syndrome, 48/331
  testing, 74/624
Sympathetic plexus
  internal carotid artery, 48/335
  paratrigeminal syndrome, 48/334
  pericarotid, *see* Pericarotid sympathetic plexus
Sympathicoblastoma
  Homer Wright rosette, 18/177
  spinal, *see* Spinal sympathicoblastoma
  spinal cord compression, 19/363
Sympathoblast
  neuroblastoma, 8/461
Sympathocyte
  neuroastrocytoma, 8/461
Sympathogonia
  neuroblastoma, 8/461
Sympathogonioma, *see* Sympathicoblastoma
Sympathomimetic agent, *see* Adrenergic receptor
    stimulating agent
Sympathotonia
  body type, 22/252
Symphalangism
  Möbius syndrome, 50/215
Symptomatic cutaneous hepatic porphyria, *see*
    Symptomatic porphyria
Symptomatic dystonia, 6/544-563
  see also Hemidystonia
  adult onset lipidosis, 49/543
  anoxic encephalopathy, 49/542
  ataxia telangiectasia, 49/543
  basal ganglion, 6/553
  Batten disease, 6/556
  bilirubin encephalopathy, 49/541
  brain arteriosclerosis, 6/556
  brain arteriovenous malformation, 49/543
  brain tumor, 49/543
  carbon dioxide intoxication, 49/543
  classification, 49/71
  copper intoxication, 49/543
  craniocerebral trauma, 6/559
  demyelinating disorder, 49/542
  drug induced dystonia, 49/543
  dystonia musculorum deformans, 49/523
  encephalitis, 49/541
  encephalitis lethargica, 6/558
  familial amyotrophic dystonic paraplegia, 6/556
  familial paroxysmal choreoathetosis, 6/556,
    42/205
  Friedreich ataxia, 6/555

glutaconic aciduria, 49/543
glutamic aciduria, 49/543
glutaryl-coenzyme A dehydrogenase deficiency,
    49/543
glycosaminoglycanosis, 49/543
GM1 gangliosidosis, 49/543
GM2 gangliosidosis, 49/543
Hallervorden-Spatz syndrome, 6/551, 554
Hartnup disease, 49/543
head injury, 49/542
hereditary myoclonic dystonia, 6/556
homocystinuria, 49/543
Huntington chorea, 6/555
hysteria, 6/561
ischemic encephalopathy, 49/542
juvenile dystonic lipidosis, 49/543
keratansulfaturia, 49/543
lenticulostriate artery, 49/542
Lesch-Nyhan syndrome, 49/543
mercury intoxication, 49/543
neoplasm, 6/560
neuroleptic agent, 6/560, 49/543
neuronal ceroid lipofuscinosis, 49/543
Niemann-Pick disease type C, 49/543
Parkinson disease, 6/555
pathophysiology, 49/544
perinatal brain injury, 6/545
posthemiplegic dystonia, 49/542
primary disorder, 6/545
primary dystonia, 49/541
progressive pallidal degeneration, 6/554
putaminal lesion, 49/544
secondary dystonia, 49/541
status marmoratus, 6/546, 49/541
striatopallidodentate calcification, 6/557, 49/543
syphilis, 6/559
syringomyelia, 49/543
treatment, 6/561-563, 49/544
tuberculous meningitis, 6/559
Wilson disease, 6/553, 49/543
Symptomatic paroxysmal kinesigenic
    choreoathetosis
  see also Hereditary paroxysmal kinesigenic
    choreoathetosis
  hereditary paroxysmal kinesigenic
    choreoathetosis, 49/355
  thalamic infarction, 49/355
Symptomatic porphyria
  cutaneous manifestation, 27/439
  differential diagnosis, 27/439
  liver, 27/439
  porphyrin metabolism, 27/437-440
Symptoms

autonomic, *see* Autonomic symptom
paroxysmal, *see* Paroxysmal symptom
postictal, *see* Postictal symptom
preictal, *see* Preictal symptom
sensory, *see* Sensory symptom
sicca, *see* Sicca symptom
Synaeresis
  senile plaque, 21/44
Synanceja intoxication
  antivenom, 37/69
  death, 37/69
  pisces, 37/69
  stonefish, 37/69
Synanceja trachynis Richardson intoxication
  picture, 37/74
  stonefish, 37/74
Synapse
  acetylcholine, 1/48, 53, 56
  atropine, 1/48
  biochemical study, 29/5
  contact theory, 1/47
  curare, 1/48
  electrophysiology, 1/55
  ganglioside, 29/6
  memory, 3/262
  microelectrode, 1/52, 55-57
  mitochondria, 1/53
  morphologic study, 29/4
  morphology, 1/54
  muscarine, 1/48
  nicotine, 1/48
  visual cortex, 2/538
Synapse number
  neuron death, 51/17
Synaptic connectivity
  nerve cell, 72/43
Synaptic transmission
  adrenalin, 1/48
  amino acid, 1/48, 53, 29/22
  brain amino acid metabolism, 29/22
Synaptic vesical paucity syndrome
  congenital, *see* Congenital synaptic vesical
    paucity syndrome
Synaptic vesicle
  acetylcholinesterase, 41/105
  botulism, 41/290
  Eaton-Lambert myasthenic syndrome, 41/355
  experimental autoimmune myasthenia gravis,
    41/115
  hemicholinium, 41/105, 111
  lanthanum, 41/105
  myasthenia gravis, 41/97-105, 109
  normal, 41/104

Synaptobrevin
  botulinum toxin, 65/248
  isoform, 65/248
  peptide bond, 65/248
  tetanus toxin, 65/248
Synaptosomal associated protein 25
  botulinum toxin, 65/248
  isoform, 65/248
  site, 65/248
Synaptotagmin
  paraneoplastic syndrome, 69/330, 337
Synchondrosis
  cranial base, *see* Skull base synchondrosis
  dens, 25/250
  skull base, *see* Skull base synchondrosis
Syncinésie sursaut
  see also Hyperexplexia *and* Startle epilepsy
  involuntary movement, 6/301
Syncopal attack
  epilepsy, 72/250
Syncope, 11/532-547
  see also Vasovagal syndrome
  Addison disease, 39/479
  aorta stenosis, 15/823, 38/148, 63/24
  Arnold-Chiari malformation type I, 75/218
  asphyxic, 15/826
  astasia abasia, 15/824
  autonomic, *see* Autonomic syncope
  autonomic dysfunction, 22/262
  baroreflex, 75/207
  bee sting intoxication, 37/107
  blood pressure, 75/715
  brain anoxia, 15/815-834
  brain blood flow, 75/209
  breath holding, *see* Breath holding syncope
  bulbar, *see* Bulbar syncope
  cardiac, 15/822
  cardiac cause, 72/260
  cardiac valvular disease, 38/148
  cardiogenic, 63/236
  cardiovascular agent intoxication, 37/426
  cardiovascular function, 75/208
  carotid compression, 15/825
  carotid sinus hypersensitivity, 75/214
  carotid sinus syndrome, 7/452
  cervical plexus syndrome, 7/452
  cervical spondylosis, 7/452
  chronic sinoatrial disorder, 39/262
  cinchonism, 65/485
  clinical features, 75/72
  congenital heart disease, 38/119-123
  congestive heart failure, 63/132
  convulsive, *see* Convulsive syncope

coronary atherosclerosis, 15/823
cough, *see* Cough syncope
differential diagnosis, 15/830, 53/236
disorder, 75/204
dissecting aorta aneurysm, 63/45
diving, 75/217
drug, 72/263
dysautonomia, 22/262, 264
EEG, 75/209
Emery-Dreifuss muscular dystrophy, 62/153
emotion, 15/824, 22/263, 75/210
epidemiology, 75/209
ergot intoxication, 65/67
etiology, 15/821
exercise, 75/217
fish roe intoxication, 37/86
G force, 75/217
gastrointestinal, 75/212
glycoside intoxication, 37/426
Hartnup disease, 21/564
head hyperextension, 75/218
headache, 48/443
heart block, 22/193
heart embolus, 63/96
heart tumor, 63/94
heat, *see* Heat syncope
hepatic encephalopathy, 72/263
hereditary long QT syndrome, 42/718, 63/238
high altitude, 75/216
hyperosmolar hyperglycemic nonketotic
    syndrome, 72/262
hypertrophic cardiomyopathy, 63/136
hyperventilation syndrome, 63/441
hypocalcemia, 72/262
hypoglycemia, 72/261
hypomagnesemia, 72/262
hyponatremia, 72/261
hypo-osmolar hyponatremia, 72/261
ictal sign, 15/818
ischemia, 15/817
Klippel-Feil syndrome, 7/453
mechanism, 75/210
metastatic cardiac tumor, 63/94, 96
micturition, *see* Micturition syncope
migraine, 48/156
mitral valve prolapse, 63/21
neural mediated, 75/205, 210
neurogenic, 63/236, 72/260
neurohumoral factor, 75/208
nonepileptic, 72/259
ocular, 75/215
orthostatic hypotension, 63/143, 75/715
pain, *see* Pain syncope

pericarditis, 63/94
physiology, 75/204
primary cardiac tumor, 63/94, 96
psychogenic, 75/211
pulmonary hypertension, 15/823
pure autonomic failure, 22/232
quinidine, 65/452
reflex, *see* Reflex syncope
reflex exertion, 15/825
renal disease, 72/263
sobbing, *see* Sobbing syncope
stingray intoxication, 37/72
stretch, 75/218
survey, 75/203
swallow, *see* Swallow syncope
Takayasu disease, 7/453, 55/337, 71/202
temporal arteritis, 55/344
third ventricle tumor, 75/218
transient ischemic attack, 53/234
treatment, 15/830, 75/219
urogenital, 75/214
urticaria pigmentosa, 14/790
vasodilatation, 75/207
vasovagal, *see* Vasovagal syncope
vomiting, *see* Vomiting syncope
white and blue, 15/829
Syncopic cervical syndrome
  Wallenberg artery, 7/452
Syncytial virus
  respiratory, *see* Respiratory syncytial virus
Syndactyly
  acrocephalosyndactyly type I, 38/422, 43/319
  acrocephalosyndactyly type II, 38/422,
      43/322-324, 50/123
  acrocephalosyndactyly type V, 43/320
  Bardet-Biedl syndrome, 13/302, 385
  Biemond syndrome type II, 43/334
  cerebral gigantism, 43/336
  corpus callosum agenesis, 42/101
  craniodigital syndrome, 43/243
  cryptophthalmos syndrome, 43/374
  De Lange syndrome, 50/276
  Ehlers-Danlos syndrome, 14/111
  familial spastic paraplegia, 22/468
  FG syndrome, 43/250
  Goltz-Gorlin syndrome, 43/19
  Gruber syndrome, 14/13, 114
  hemifacial hypertrophy, 22/550
  Herrmann-Opitz syndrome, 43/325
  holoprosencephaly, 50/237
  hypertelorism microtia clefting syndrome, 43/415
  Klippel-Trénaunay syndrome, 14/395, 523
  lissencephaly, 42/40

Miller-Dieker syndrome, 50/592
Möbius syndrome, 42/324, 50/215
multiple nevoid basal cell carcinoma syndrome,
    14/456, 462
neuralgic amyotrophy, 51/175
Poland-Möbius syndrome, 43/461
progressive hemifacial atrophy, 22/550
Refsum disease, 21/208, 60/231, 66/491
Russell-Silver syndrome, 43/476
Saethre-Chotzen syndrome, 43/322
sclerosteosis, 43/474
Smith-Lemli-Opitz syndrome, 43/308
Sturge-Weber syndrome, 14/512
Summitt syndrome, 43/325, 372-374
triploidy, 43/564-566
Syndrome de vasodilatation hemicéphalique
    d'origine sympathique
cluster headache, 48/217
Syndromes
Aarskog, *see* Aarskog syndrome
Aarskog-Scott, *see* Aarskog syndrome
Aberfeld, *see* Myotonic myopathy
abnormal acetylcholine-acetylcholine receptor
    interaction, *see* Abnormal
    acetylcholine-acetylcholine receptor interaction
    syndrome
acquired immune deficiency, *see* Acquired
    immune deficiency syndrome
acquired retraction, *see* Acquired retraction
    syndrome
acrocallosal, *see* Acrocallosal syndrome
Adamkiewicz artery, *see* Adamkiewicz artery
    syndrome
Adams-Oliver, *see* Adams-Oliver syndrome
Adams-Stokes, *see* Adams-Stokes syndrome
ADD, *see* Attention deficit disorder syndrome
adducted thumb, *see* Adducted thumb syndrome
adenohypophyseal, *see* Adenohypophyseal
    syndrome
adenosine triphosphatase 6 mutation, *see*
    Adenosine triphosphatase 6 mutation syndrome
Adie, *see* Holmes-Adie syndrome
adiposogenital, *see* Adiposogenital syndrome
ADR, *see* ADR syndrome
adrenogenital, *see* Adrenogenital syndrome
Aicardi, *see* Aicardi syndrome
akinetic rigid, *see* Akinetic rigid syndrome
Alajouanine, *see* Alajouanine syndrome
Albright, *see* Albright syndrome
alcohol withdrawal, *see* Alcohol withdrawal
    syndrome
Alezzandrini, *see* Alezzandrini syndrome
Alice in Wonderland, *see* Alice in Wonderland
    syndrome
Ålien hand, *see* Ålien hand syndrome
Allgrove, *see* Allgrove syndrome
Alport, *see* Alport syndrome
Ålport, *see* Forsius-Eriksson syndrome
Alström, *see* Alström-Hallgren syndrome
Alström-Hallgren, *see* Alström-Hallgren
    syndrome
Amalric, *see* Amalric syndrome
aminopterin, *see* Aminopterin syndrome
amnesic, *see* Amnesic syndrome
amniotic band, *see* Amniotic band syndrome
Andermann, *see* Andermann syndrome
anetic, *see* Anetic syndrome
Angelman, *see* Happy puppet syndrome
anoetic, *see* Anetic syndrome
antebrachial pain, *see* Antebrachial pain syndrome
anterior cavernous sinus, *see* Anterior cavernous
    sinus syndrome
anterior cerebellar lobe, *see* Anterior cerebellar
    lobe syndrome
anterior cerebral artery, *see* Anterior cerebral
    artery syndrome
anterior cervical cord, *see* Anterior cervical cord
    syndrome
anterior cingulate, *see* Anterior cingulate
    syndrome
anterior condylar canal, *see* Anterior condylar
    canal syndrome
anterior corpus callosum, *see* Corpus callosum
    syndrome
anterior ethmoidal nerve, *see* Anterior ethmoidal
    nerve syndrome
anterior horn, *see* Anterior horn syndrome
anterior inferior cerebellar artery, *see* Anterior
    inferior cerebellar artery syndrome
anterior interosseous nerve, *see* Anterior
    interosseous nerve syndrome
anterior spinal artery, *see* Anterior spinal artery
    syndrome
anterior tarsal tunnel, *see* Anterior tarsal tunnel
    syndrome
anterior tibial, *see* Anterior tibial syndrome
anterolateral thalamic, *see* Anterolateral thalamic
    syndrome
anterolateral tract, *see* Anterolateral tract
    syndrome
antiphospholipid antibody, *see* Antiphospholipid
    antibody syndrome
Antley-Bixler, *see* Multisynostotic
    osteodysgenesis
Anton, *see* Anton syndrome
Anton-Babinski, *see* Anton-Babinski syndrome

436

aorta arch, *see* Takayasu disease
apallic, *see* Akinetic mutism *and* Persistent
  vegetative state
Apert, *see* Acrocephalosyndactyly type I
Apert-Crouzon, *see* Acrocephalosyndactyly
  type II
aqueduct, *see* Aqueduct stenosis
archicerebellar, *see* Archicerebellar syndrome
ARG, *see* ARG syndrome
Arieti-Gray, *see* Arieti-Gray syndrome
Armadillo, *see* Isaacs syndrome
Armendares, *see* Armendares syndrome
atlas, *see* Atlas syndrome
atonia astasia, *see* Förster syndrome
attention deficit disorder, *see* Attention deficit
  disorder syndrome
audiogenic startle, *see* Audiogenic startle
  syndrome
auriculotemporal, *see* Auriculotemporal syndrome
autonomic failure, *see* Autonomic failure
  syndrome
Avellis, *see* Avellis syndrome
Babinski-Fröhlich, *see* Babinski-Fröhlich
  syndrome
Babinski-Nageotte, *see* Babinski-Nageotte
  syndrome
Balint, *see* Balint syndrome
Balint-Holmes, *see* Balint syndrome
Baller-Gerold, *see* Baller-Gerold syndrome
Bamberger-Marie, *see* Bamberger-Marie
  syndrome
Bandler, *see* Bandler syndrome
Bannayan-Zonana, *see* Bannayan-Zonana
  syndrome
Bannwarth, *see* Bannwarth syndrome
Bardet-Biedl, *see* Bardet-Biedl syndrome
Barnard-Scholz, *see* Barnard-Scholz syndrome
Barraquer-Simons, *see* Lipodystrophia
  progressiva (Barraquer-Simons)
Barré, *see* Barré syndrome
Barré-Liéou, *see* Barré-Liéou syndrome
Barth type I, *see* X-linked myotubular myopathy
Barth type II, *see* X-linked neutropenic
  cardioskeletal myopathy
Bartter, *see* Bartter syndrome
basal cell nevus, *see* Multiple nevoid basal cell
  carcinoma syndrome
basal ganglion, *see* Basal ganglion syndrome
basilar artery, *see* Basilar artery syndrome
basofrontal, *see* Basofrontal syndrome
Bassen-Kornzweig, *see* Bassen-Kornzweig
  syndrome
battered child, *see* Battered child syndrome

Beals, *see* Beals syndrome
Bean, *see* Bean syndrome
Beckwith-Wiedemann, *see* Beckwith-Wiedemann
  syndrome
Behçet, *see* Behçet syndrome
Bencze, *see* Bencze syndrome
Benedikt, *see* Benedikt syndrome
Berant, *see* Berant syndrome
Berardinelli-Seip, *see* Berardinelli-Seip syndrome
Berenbruch-Cushing-Cobb, *see* Cobb syndrome
Berlin, *see* Berlin syndrome
Bernard-Horner, *see* Horner syndrome
Bernard-Soulier, *see* Bernard-Soulier syndrome
Bernfeld, *see* Stylohyoid syndrome
Bertolotti-Garcin, *see* Garcin syndrome
Bethlem-Van Wijngaarden, *see* Bethlem-Van
  Wijngaarden syndrome
Bickers-Adams, *see* Bickers-Adams syndrome
BIDS, *see* Hair brain syndrome
Biemond, *see* Biemond syndrome
Biemond type II, *see* Biemond syndrome type II
Biermer, *see* Addison disease
bilateral pontine, *see* Bilateral pontine syndrome
Bing-Neel, *see* Waldenström macroglobulinemia
  polyneuropathy
Bitter, *see* Bitter syndrome
blind loop, *see* Blind loop syndrome
Bloch-Sulzberger, *see* Incontinentia pigmenti
Bloom, *see* Bloom syndrome
blue diaper, *see* Blue diaper syndrome
blue rubber bleb nevus, *see* Bean syndrome
BO, *see* Branchio-otodysplasia
bobble head doll, *see* Bobble head doll syndrome
Bonham-Carter, *see* Bonham-Carter syndrome
Bonnet-Bonnet, *see* Orbital apex syndrome
Bonnet-Dechaume-Blanc, *see*
  Bonnet-Dechaume-Blanc syndrome
Bonnet, *see* Bonnet syndrome
Bonnevie-Ullrich, *see* Turner syndrome
Bonnier, *see* Bonnier syndrome
BOR, *see* Branchio-otorenal dysplasia
borderline, *see* Borderline syndrome
Börjeson-Forssman-Lehmann, *see*
  Börjeson-Forssman-Lehmann syndrome
Boucher-Neuhäuser, *see* Boucher-Neuhäuser
  syndrome
Bowen, *see* Bowen syndrome
Bowen-Conradi, *see* Bowen-Conradi syndrome
brachial plexus, *see* Brachial plexus syndrome
Bradbury-Eggleston, *see* Pure autonomic failure
brain stem dysfunction, *see* Brain stem
  dysfunction syndrome
Brandt, *see* Brandt syndrome

Brégeat, *see* Brégeat syndrome
Breughel, *see* Dystonia musculorum deformans
Briquet, *see* Briquet syndrome
Brissaud-Sicard, *see* Brissaud-Sicard syndrome
Brooke, *see* Multiple trichoepithelioma
Brossard, *see* Brossard syndrome
Brown's, *see* Brown syndrome
Brown-Séquard, *see* Brown-Séquard syndrome
Brown-Vialetto-Van Laere, *see*
    Brown-Vialetto-Van Laere syndrome
Bruns-Garland, *see* Diabetic proximal neuropathy
bulbar, *see* Bulbar syndrome
bulbar brain, *see* Bulbar brain syndrome
bulbar jugular, *see* Bulbar jugular syndrome
bulbar nerve, *see* Bulbar nerve syndrome
bulbus retraction, *see* Stilling-Türk-Duane
    syndrome
C, *see* C syndrome
CAMAK, *see* CAMAK syndrome
CAMFAK, *see* CAMFAK syndrome
camptomelic, *see* Camptomelic syndrome
canal of Guyon, *see* Guyon canal
cancer family, *see* Cancer family syndrome
Capute-Rimoin-Konigsmark, *see*
    Capute-Rimoin-Konigsmark syndrome
carbohydrate deficient glycoprotein, *see*
    Carbohydrate deficient glycoprotein syndrome
carcinoid, *see* Carcinoid syndrome
cardiofacial, *see* Cardiofacial syndrome
Carol-Godfried-Prakken-Prick, *see*
    Carol-Godfried-Prakken-Prick syndrome
carotid sinus, *see* Carotid sinus syndrome
carotid system, *see* Carotid system syndrome
carpal tunnel, *see* Carpal tunnel syndrome
Carpenter, *see* Acrocephalosyndactyly type II
Carter-Sukavajana, *see* Carter-Sukavajana
    syndrome
cat eye, *see* Cat eye syndrome
cauda equina, *see* Cauda equina syndrome
central cervical cord, *see* Central cervical cord
    syndrome
central spinal, *see* Central spinal syndrome
cerebellar, *see* Cerebellar syndrome
cerebellar artery, *see* Cerebellar artery syndrome
cerebellopontine angle, *see* Cerebellopontine
    angle syndrome
cerebroasthenic, *see* Cerebroasthenic syndrome
cerebrocostomandibular, *see*
    Cerebrocostomandibular syndrome
cerebrohepatorenal, *see* Cerebrohepatorenal
    syndrome
cerebro-oculofacioskeletal, *see*
    Cerebro-oculofacioskeletal syndrome

cervical, *see* Cervical syndrome
cervical plexus, *see* Cervical plexus syndrome
cervical rib, *see* Cervical rib syndrome
cervicobrachial outlet, *see* Cervical rib syndrome
cervicocephalic, *see* Barré-Liéou syndrome
cervico-oculoacusticus, *see*
    Cervico-oculoacusticus syndrome
cervicothoracic, *see* Cervicothoracic syndrome
Cestan-Chenais, *see* Cestan-Chenais syndrome
Changi Camp, *see* Changi Camp syndrome
Charcot, *see* Charcot syndrome
Charcot-Wilbrand, *see* Charcot-Wilbrand
    syndrome
Charles-Bonnet, *see* Charles-Bonnet syndrome
Charles Symonds, *see* Multiple cranial neuropathy
Charlin, *see* Charlin neuralgia
Chédiak-Higashi, *see* Chédiak-Higashi syndrome
cheiro-oral, *see* Cheiro-oral syndrome
cherry red spot myoclonus, *see* Mucolipidosis
    type I
Chester-Erdheim, *see* Chester-Erdheim syndrome
Chiari-Frommel, *see* Chiari-Frommel syndrome
childhood schizophrenic, *see* Childhood
    schizophrenic syndrome
Chinese restaurant, *see* Chinese restaurant
    syndrome
chondrodysplasia hemangioma, *see* Maffucci-Kast
    syndrome
chorda tympani, *see* Intermedius neuralgia
choreiform, *see* Choreiform syndrome
choroidal artery, *see* Choroidal artery syndrome
Christ-Siemens-Touraine, *see*
    Christ-Siemens-Touraine syndrome
chronic brain, *see* Dementia *and* Organic
    psychosis
chronic fatigue, *see* Postviral fatigue syndrome
chronic regional pain, *see* Chronic regional pain
    syndrome
Churg-Strauss, *see* Allergic granulomatous
    angiitis
chylomicronemia, *see* Chylomicronemia
    syndrome
CINCA, *see* CINCA syndrome
Citelli, *see* Citelli syndrome
Claude, *see* Claude syndrome
Claude Bernard-Horner, *see* Horner syndrome
cloverleaf skull, *see* Cloverleaf skull syndrome
clumsy hand dysarthria, *see* Clumsy hand
    dysarthria syndrome
cluster headache, *see* Cluster headache syndrome
cluster migraine, *see* Cluster migraine syndrome
cluster tic, *see* Cluster tic syndrome
cluster variant, *see* Cluster variant syndrome

COACH, *see* COACH syndrome
Cobb, *see* Cobb syndrome
Coffin-Lowry, *see* Coffin-Lowry syndrome
Coffin-Siris, *see* Coffin-Siris syndrome
Coffin-Siris-Wegienka, *see* Coffin-Lowry syndrome
Cogan, *see* Cogan syndrome
Cogan type I, *see* Cogan syndrome type I
Cogan type II, *see* Cogan syndrome type II
Collet-Bonnet, *see* Collet-Bonnet syndrome
Collet-Sicard, *see* Sicard-Collet syndrome
combined peripheral and CNS inflammatory demyelinating, *see* Combined peripheral and CNS inflammatory demyelinating syndrome
compartment, *see* Compartment syndrome
compression, *see* Compression syndrome
conditional responsive extrapyramidal, *see* Hereditary paroxysmal kinesigenic choreoathetosis
congenital acetylcholine receptor deficiency/short channel opentime, *see* Congenital acetylcholine receptor deficiency/short channel opentime syndrome
congenital acetylcholine receptor deficiency/synaptic cleft, *see* Congenital acetylcholine receptor deficiency/synaptic cleft syndrome
congenital ε-acetylcholine receptor subunit mutation, *see* Congenital ε-acetylcholine receptor subunit mutation syndrome
congenital bilateral perisylvian, *see* Congenital bilateral perisylvian syndrome
congenital central hypoventilation, *see* Congenital central hypoventilation syndrome
congenital chemosensitive hypoventilation, *see* Congenital chemosensitive hypoventilation syndrome
congenital high conductance fast channel, *see* Congenital high conductance fast channel syndrome
congenital myasthenic, *see* Congenital myasthenic syndrome
congenital rubella, *see* Congenital rubella syndrome
congenital slow channel, *see* Congenital slow channel syndrome
congenital synaptic vesical paucity, *see* Congenital synaptic vesical paucity syndrome
Conradi-Hünermann, *see* Conradi-Hünermann syndrome
continuous muscle fiber activity, *see* Isaacs syndrome
conus medullaris, *see* Conus medullaris syndrome

Cornelia De Lange, *see* De Lange syndrome
corpus callosum, *see* Corpus callosum syndrome
corpus callosum disconnection, *see* Corpus callosum disconnection syndrome
corticopontocerebellar, *see* Neocerebellar syndrome
corticostriatocerebellar, *see* Gökay-Tükel syndrome
Costen, *see* Costen syndrome
costoclavicular, *see* Costoclavicular syndrome
couvade, *see* Couvade syndrome
Cowden, *see* Cowden syndrome
cranial half-base, *see* Garcin syndrome
craniocarpotarsal, *see* Freeman-Sheldon syndrome
craniodigital, *see* Craniodigital syndrome
CREST, *see* CREST syndrome
cri du chat, *see* Cri du chat syndrome
Crigler-Najjar, *see* Crigler-Najjar syndrome
Critchley, *see* Critchley syndrome
crocodile tear, *see* Crocodile tear syndrome
Crome, *see* Crome syndrome
Cronckhite-Canada, *see* Cronckhite-Canada syndrome
crossed nucleus rubber, *see* Crossed nucleus ruber syndrome
Crouzon, *see* Craniofacial dysostosis (Crouzon)
Crow-Fukase, *see* POEMS syndrome
CRSM, *see* Mucolipidosis type I
CRST, *see* CRST syndrome
crush, *see* Crush syndrome
cryptophthalmos, *see* Cryptophthalmos syndrome
cubital canal, *see* Cubital canal syndrome
Cushing, *see* Cushing syndrome
Cutius type V, *see* Curtius syndrome type V
Dandy, *see* Infraclinoid syndrome
Dandy-Walker, *see* Dandy-Walker syndrome
Darier-Grönblad-Strandberg, *see* Grönblad-Strandberg syndrome
Davidenkow, *see* Scapuloperoneal spinal muscular atrophy
Dawidenkow, *see* Scapuloperoneal spinal muscular atrophy
De Lange, *see* De Lange syndrome
De Sanctis-Cacchione, *see* Xeroderma pigmentosum
decerebration, *see* Decerebration syndrome
Dechaume, *see* Dechaume syndrome
decortication, *see* Decortication syndrome
defibrination, *see* Intravascular consumption coagulopathy
deiterospinal, *see* Barré syndrome
Dejean, *see* Orbital floor syndrome
Dejerine, *see* Dejerine syndrome

Dejerine-Klumpke, *see* Dejerine-Klumpke
  syndrome
Dejerine-Roussy, *see* Thalamic syndrome
  (Dejerine-Roussy)
dementia, *see* Dementia syndrome
Denny Brown, *see* Sensory radicular neuropathy
DeToni-Fanconi-Debré, *see*
  DeToni-Fanconi-Debré syndrome
Di Ferrante, *see* Glycosaminoglycanosis type VIII
dialysis disequilibrium, *see* Dialysis
  disequilibrium syndrome
diamorphine withdrawal, *see* Diamorphine
  withdrawal syndrome
DIDMOAD, *see* Wolfram syndrome
diencephalic, *see* Diencephalic syndrome
DiGeorge, *see* DiGeorge syndrome
disconnection, *see* Disconnection syndrome
disequilibrium, *see* Disequilibrium syndrome
distichiasis lymphedema, *see* Distichiasis
  lymphedema syndrome
Divry-Van Bogaert, *see* Divry-Van Bogaert
  syndrome
DOOR, *see* DOOR syndrome
dorsal medullary, *see* Dorsal medullary syndrome
dorsal midbrain, *see* Dorsal midbrain syndrome
double Y, *see* XYY syndrome
Down, *see* Down syndrome
Doyne-Grönblad-Strandberg, *see*
  Grönblad-Strandberg syndrome
drug induced extrapyramidal, *see* Drug induced
  extrapyramidal syndrome
drug induced myotonic, *see* Drug induced
  myotonic syndrome
Duane, *see* Stilling-Türk-Duane syndrome
Duane retraction, *see* Acquired retraction
  syndrome *and* Stilling-Türk-Duane syndrome
Dubowitz, *see* Dubowitz syndrome
Duchenne-Erb, *see* Duchenne-Erb syndrome
Dumon-Radermecker, *see* Dumon-Radermecker
  syndrome
Dyggve-Melchior-Clausen, *see*
  Dyggve-Melchior-Clausen syndrome
Dyke-Davidoff, *see* Dyke-Davidoff syndrome
dyschondroplasia hemangioma, *see* Maffucci
  syndrome
dysequilibrium, *see* Disequilibrium syndrome
dyspallic, *see* Akinetic mutism
dystonia deafness, *see* Dystonia deafness
  syndrome
dystrophia retinae pigmentosa dysacusis, *see*
  Lindenov-Hallgren syndrome
Eaton-Lambert, *see* Eaton-Lambert myasthenic
  syndrome

Eaton-Lambert myasthenic, *see* Eaton-Lambert
  myasthenic syndrome
Edwards, *see* Edwards syndrome
EEC, *see* EEC syndrome
Ehlers-Danlos, *see* Ehlers-Danlos syndrome
Ekbom, *see* Restless legs syndrome
Ekbom other, *see* Ekbom other syndrome
Eldridge-Berlin-Money-McKusick, *see*
  Eldridge-Berlin-Money-McKusick syndrome
elfin face, *see* Elfin face syndrome
Ellis-Van Creveld, *see* Ellis-Van Creveld
  syndrome
elongated styloid process, *see* Elongated styloid
  process syndrome
embryopathy, *see* Embryopathy syndrome
Emery-Dreifuss, *see* Emery-Dreifuss muscular
  dystrophy
empty sella, *see* Empty sella syndrome
endocrine candidiasis, *see* Endocrine candidiasis
  syndrome
Engel, *see* Engel syndrome
eosinophilia myalgia, *see* Eosinophilia myalgia
  syndrome
eosinophilic granuloma, *see* Eosinophilic
  granuloma syndrome
epiconus medullaris, *see* Epiconus medullaris
  syndrome
epidermal nevus, *see* Linear nevus sebaceous
  syndrome *and* Nevus unius lateris
erythrokeratodermia ataxia, *see*
  Erythrokeratodermia ataxia syndrome
excessive CSF protein, *see* Excessive CSF protein
  syndrome
extrapyramidal, *see* Extrapyramidal syndrome
F, *see* F syndrome
facilitating myasthenic, *see* Eaton-Lambert
  myasthenic syndrome
faciodigital genital, *see* Aarskog syndrome
facio-oculoacousticorenal, *see*
  Facio-oculoacousticorenal syndrome
facioscapulohumeral, *see* Facioscapulohumeral
  syndrome
familial akinetorigid, *see* Pure akinesia
familial polyradiculopathy, *see* Familial
  polyradiculopathy syndrome
familial tremor, *see* Familial tremor syndrome
fascia lata, *see* Fascia lata syndrome
fasciculation, *see* Fasciculation syndrome
Faulk-Epstein-Jones, *see* Faulk-Epstein-Jones
  syndrome
Ferguson-Critchley, *see* Ferguson-Critchley
  syndrome
Ferguson-Critchley like, *see* Ferguson-Critchley

like syndrome
fetal alcohol, *see* Fetal alcohol syndrome
fetal hydantoin, *see* Fetal hydantoin syndrome
fetal trimethadione, *see* Fetal trimethadione
  syndrome
FG, *see* FG syndrome
Fincher, *see* Fincher syndrome
first and second brachial arch, *see* Goldenhar
  syndrome
Fisher, *see* Fisher syndrome
fissure, *see* Fissure syndrome
flecked retina, *see* Fundus flavimaculatus
FLM, *see* Medial longitudinal fasciculus
floating-harbor, *see* Floating-Harbor syndrome
flocculonodular lobe, *see* Flocculonodular lobe
  syndrome
floppy baby, *see* Floppy baby syndrome
Flynn-Aird, *see* Flynn-Aird syndrome
focal dermal hypoplasia, *see* Goltz-Gorlin
  syndrome
focal neuropsychological, *see* Focal
  neuropsychological syndrome
Foix, *see* Foix syndrome
Foix-Chavany-Marie, *see* Operculum syndrome
Foix-Jefferson, *see* Foix-Jefferson syndrome
foramen lacerum (Jefferson), *see* Foramen
  lacerum syndrome (Jefferson)
Forney, *see* Forney syndrome
Forney-Robinson-Pascoe, *see* Forney syndrome
Forsius-Erikson, *see* Forsius-Eriksson syndrome
Forssmann carotid, *see* Forssmann-Skoog
  syndrome
Forssmann-Skoog, *see* Forssmann-Skoog
  syndrome
Förster, *see* Förster syndrome
Foster Kennedy, *see* Foster Kennedy syndrome
Foville, *see* Foville syndrome
Foville-Millard-Gubler, *see* Foville syndrome
Foville pontine, *see* Foville syndrome
François, *see* François syndrome
Fraser, *see* Fraser syndrome
Freeman-Sheldon, *see* Freeman-Sheldon
  syndrome
Fried-Emery, *see* Intermediate spinal muscular
  atrophy
Friedman-Roy, *see* Friedman-Roy syndrome
Friedreich ataxia like Silvester, *see* Friedreich
  ataxia like Sylvester syndrome
Fröhlich, *see* Fröhlich syndrome
Froin, *see* Nonne-Froin syndrome
frontal lobe, *see* Frontal lobe syndrome
Fuchs, *see* Gyrate atrophy (Fuchs)
Fukuyama, *see* Fukuyama syndrome

G, *see* G syndrome
Galloway-Mowat, *see* Galloway-Mowat
  syndrome
Gamstorp-Wohlfart, *see* Isaacs syndrome
Ganser, *see* Ganser syndrome
Garcin, *see* Garcin syndrome
Garel, *see* Stylohyoid syndrome
Garel-Bernfeld, *see* Stylohyoid syndrome
Garin-Bujadoux-Bannwarth, *see* Bannwarth
  syndrome
Gaskoyen, *see* Bean syndrome
Gasperini, *see* Gasperini syndrome
gasserian ganglion, *see* Gasserian ganglion
  syndrome
Gellé, *see* Gellé syndrome
Gerhardt, *see* Gerhardt syndrome
Gerstmann, *see* Gerstmann syndrome
Geschwind, *see* Geschwind syndrome
Gignoux, *see* Gignoux syndrome
Gilles de la Tourette, *see* Gilles de la Tourette
  syndrome
Gillespie, *see* Gillespie syndrome
Giroux-Barbeau, *see* Erythrokeratodermia ataxia
  syndrome
glioma polyposis, *see* Glioma polyposis syndrome
Godfried-Prick-Carol-Prakken, *see*
  Godfried-Prick-Carol-Prakken syndrome
Godin, *see* Godin syndrome
Gökay-Tükel, *see* Gökay-Tükel syndrome
Goldberg, *see* Goldberg syndrome
Goldenhar, *see* Goldenhar syndrome
Goldenhar-Gorlin, *see* Goldenhar syndrome
Goltz, *see* Goltz-Gorlin syndrome
Goltz-Gorlin, *see* Goltz-Gorlin syndrome
Goltz-Peterson-Gorlin-Ravits, *see* Goltz-Gorlin
  syndrome
Goodpasture, *see* Goodpasture syndrome
Gordon-Capute-Konigsmark, *see*
  Gordon-Capute-Konigsmark syndrome
Gorlin, *see* Multiple nevoid basal cell carcinoma
  syndrome
Gorlin-Chaudry-Moss, *see* Gorlin-Chaudry-Moss
  syndrome
Gorlin-Goltz-Ward, *see* Multiple nevoid basal cell
  carcinoma syndrome
Gorlin LEOPARD, *see* LEOPARD syndrome
Gougerot-Sjögren, *see* Sjögren syndrome
Goyer, *see* Goyer syndrome
Gradenigo, *see* Gradenigo syndrome
Gradenigo-Lannois, *see* Gradenigo syndrome
Grebe-Myle-Löwenthal, *see*
  Grebe-Myle-Löwenthal syndrome
Gregg, *see* Postrubella embryopathy

Greig, *see* Hypertelorism

Grenet, *see* Grenet syndrome

Grisel, *see* Grisel syndrome

Groll-Hirscshowitz, *see* Groll-Hirschowitz syndrome

Grönblad-Strandberg, *see* Grönblad-Strandberg syndrome

Grönblad-Strandberg-Touraine, *see* Grönblad-Strandberg syndrome

growth deficiency, *see* Growth deficiency syndromes

Gruber, *see* Gruber syndrome

Guérin-Stern, *see* Guérin-Stern syndrome

Guillain-Barré, *see* Guillain-Barré syndrome

Guillain-Garcin, *see* Garcin syndrome

Haferkamp, *see* Haferkamp syndrome

hair brain, *see* Hair brain syndrome

Hajdu-Cheney, *see* Arthrodento-osteodysplasia

half base, *see* Garcin syndrome

half skull base, *see* Garcin syndrome

Hallermann-Streiff, *see* Hallermann-Streiff syndrome

Hallervorden-Spatz, *see* Hallervorden-Spatz syndrome

Hallgren, *see* Hallgren syndrome

Haltia-Santavuori, *see* Infantile neuronal ceroid lipofuscinosis

happy puppet, *see* Happy puppet syndrome

Harboyan, *see* Harboyan syndrome

HARD, *see* HARD syndrome

HARDE, *see* HARDE syndrome

harlequin, *see* Harlequin syndrome

HARP, *see* HARP syndrome

Harvey, *see* Harvey syndrome

Hashimoto-Pritzker, *see* Hashimoto-Pritzker syndrome

Haw River, *see* Haw River syndrome

heat, *see* Heat syndrome

Heerfordt, *see* Heerfordt syndrome

Helweg-Larsen-Ludvigsen, *see* Helweg-Larsen-Ludvigsen syndrome

hemi-3, *see* Hemi-3 syndrome

hemibase, *see* Garcin syndrome

hemicerebellopyramidal, *see* Hemicerebellopyramidal syndrome

hemineglect, *see* Hemineglect syndrome

hemiskull base, *see* Garcin syndrome

hemolytic uremic, *see* Hemolytic uremic syndrome

Hennebert, *see* Hennebert syndrome

hereditary long QT, *see* Hereditary long QT syndrome

hereditary motor and sensory neuropathy plus, *see* Hereditary motor and sensory neuropathy plus syndrome

hereditary myoclonus ataxia, *see* Hereditary myoclonus ataxia syndrome

hereditary myotonic, *see* Hereditary myotonic syndrome

Herman, *see* Herman syndrome

Hermansky-Pudlak, *see* Hermansky-Pudlak syndrome

herpes virus hominis, *see* Herpes virus hominis syndrome

Herrmann-Opitz, *see* Herrmann-Opitz syndrome

HHE, *see* HHE syndrome

high-pressure nervous, *see* High-pressure nervous syndrome

Hill-Sherman, *see* Hill-Sherman syndrome

HMC, *see* Hypertelorism microtia clefting syndrome

Hoffmann, *see* Hoffmann syndrome

Hoffmann-Zurhelle, *see* Hoffmann-Zurhelle syndrome

Holmes-Adie, *see* Holmes-Adie syndrome

Holt-Oram, *see* Holt-Oram syndrome

homolateral ataxia crural paresis, *see* Homolateral ataxia crural paresis syndrome

Hooft-Bruens, *see* Hooft-Bruens syndrome

Horner, *see* Horner syndrome

Horton, *see* Chronic migrainous neuralgia *and* Cluster headache

Horton-Magath-Brown, *see* Temporal arteritis

Hunter-Russell, *see* Hunter-Russell syndrome

Hutchinson-Gilford, *see* Progeria

Hutchinson-Laurence-Moon, *see* Hutchinson-Laurence-Moon syndrome

hydrocephalus, *see* Hydrencephaly syndrome

hypereosinophilic, *see* Hypereosinophilic syndrome

hyperkinetic child, *see* Hyperkinetic child syndrome

hyperlexia, *see* Hyperlexia syndrome

hyperosmolar hyperglycemic nonketotic, *see* Hyperosmolar hyperglycemic nonketotic syndrome

hyperosmolarity, *see* Hyperosmolarity syndrome

hyperphenylalaninemic embryopathy, *see* Hyperphenylalaninemic embryopathy syndrome

hypertelorism hypospadias, *see* Hypertelorism hypospadias syndrome

hypertelorism microtia clefting, *see* Hypertelorism microtia clefting syndrome

hyperventilation, *see* Hyperventilation syndrome

hyperviscosity, *see* Hyperviscosity syndrome

hypnic headache, *see* Hypnic headache syndrome

442

hypolipidemia, *see* Hooft disease
hypophyseal, *see* Hypophyseal syndrome
hypothalamic, *see* Hypothalamic syndrome
hypothalamic hypophyseal, *see* Hypothalamic
  hypophyseal syndrome
hypoventilation, *see* Hypoventilation syndrome
ichthyotic, *see* Ichthyotic syndrome
idiopathic postural orthostatic tachycardia, *see*
  Idiopathic postural orthostatic tachycardia
  syndrome
inactivation, *see* Motor ritardando
inferior vena cava, *see* Inferior vena cava
  syndrome
infraclinoid, *see* Infraclinoid syndrome
infratentorial, *see* Infratentorial syndrome
intensive care, *see* Intensive care syndrome
internal carotid artery, *see* Internal carotid artery
  syndrome
interosseous nerve, *see* Interosseous nerve
  syndrome
interparietal, *see* Interparietal syndrome
interpeduncular, *see* Interpeduncular syndrome
irregular sleep-wake, *see* Irregular sleep-wake
  syndrome
Isaacs, *see* Isaacs syndrome
Issacs-Mertens, *see* Isaacs syndrome
jabs and jolts, *see* Jabs and jolts syndrome
Jaccodo, *see* Petrosphenoid syndrome (Jaccodo)
Jackson, *see* Jackson syndrome
Jacod, *see* Orbital apex syndrome
Jaeken, *see* Jaeken syndrome
Jansky-Bielschowsky, *see* Neuronal ceroid
  lipofuscinosis
Jefferson, *see* Foramen lacerum syndrome
  (Jefferson)
Jervell-Lange-Nielsen, *see* Jervell-Lange-Nielsen
  syndrome
Johanson-Blizzard, *see* Johanson-Blizzard
  syndrome
Joseph, *see* Joseph syndrome
Joubert, *see* Joubert syndrome
Joubert-Boltshauser, *see* Joubert syndrome
Juberg-Hayward, *see* Juberg-Hayward syndrome
jugular foramen, *see* Jugular foramen syndrome
Kallman, *see* Kallmann syndrome
Kasabach-Merritt, *see* Kasabach-Merritt
  syndrome
Kast, *see* Kast syndrome
Kawasaki, *see* Kawasaki syndrome
Kearns-Sayre-Daroff, *see*
  Kearns-Sayre-Daroff-Shy syndrome
Kearns-Sayre-Daroff-Shy, *see*
  Kearns-Sayre-Daroff-Shy syndrome

Kearns-Shy, *see* Kearns-Sayre-Daroff-Shy
  syndrome
Kennedy, *see* Foster Kennedy syndrome
Kestenbaum, *see* Aqueduct stenosis
Khalifeh-Zellweger-Plancherel, *see*
  Khalifeh-Zellweger-Plancherel syndrome
Kiloh-Nevin, *see* Kiloh-Nevin syndrome
King-Denborough, *see* King-Denborough
  syndrome
kinky hair, *see* Trichopoliodystrophy
Kjellin, *see* Kjellin syndrome
Kleeblattschädel, *see* Cloverleaf skull syndrome
Kleine-Levin, *see* Kleine-Levin syndrome
Klinefelter, *see* Klinefelter syndrome
Klintworth, *see* Acute myoclonic ataxia
Klippel-Feil, *see* Klippel-Feil syndrome
Klippel-Trénaunay, *see* Klippel-Trénaunay
  syndrome
Klippel-Trénaunay-Weber, *see* Klippel-Trénaunay
  syndrome
Klüver-Bucy, *see* Klüver-Bucy syndrome
Kocher-Debré-Sémélainge, *see*
  Kocher-Debré-Sémélaigne syndrome
Könnecke-Richards-Rundle, *see* Richards-Rundle
  syndrome
Korsakoff, *see* Wernicke-Korsakoff syndrome
Kugelberg-Welander, *see*
  Wohlfart-Kugelberg-Welander disease
Kuhlendahl, *see* Kuhlendahl syndrome
Kundrat, *see* Kundrat syndrome
Kvistad-Dahl-Skre, *see* Kvistad-Dahl-Skre
  syndrome
Lambert-Brody, *see* Lambert-Brody syndrome
Lambert-Eaton, *see* Eaton-Lambert myasthenic
  syndrome
Lance-Adams, *see* Posthypoxic myoclonus
Landau-Kleffner, *see* Landau-Kleffner syndrome
Landouzy-Dejerine, *see* Facioscapulohumeral
  muscular dystrophy
Landry-Guillain-Barré, *see* Guillain-Barré
  syndrome
Langer-Giedion, *see* Langer-Giedion syndrome
Larsen, *see* Larsen syndrome
lateral hypothalamic, *see* Lateral hypothalamic
  syndrome
lateral medullary, *see* Lateral medullary syndrome
lateral pontine, *see* Lateral pontine syndrome
lateral tract, *see* Lateral tract syndrome
Latham-Munro, *see* Latham-Munro syndrome
Laurence-Moon, *see* Laurence-Moon syndrome
Laurence-Moon-Bardet-Biedl, *see*
  Laurence-Moon-Bardet-Biedl syndrome
Laurence-Moon-Biedl, *see* Laurence-Moon-Biedl

Lawrence, *see* Lawrence syndrome
Lemieux-Neemeh, *see* Lemieux-Neemeh
  syndrome
Lennox-Gastaut, *see* Lennox-Gastaut syndrome
lenticulo-optic, *see* Lenticulo-optic syndrome
lentiginosis, *see* Lentiginosis syndrome
Lenz, *see* Lenz syndrome
leonine-mouse, *see* Leonine-mouse syndrome
LEOPARD, *see* LEOPARD syndrome
Leriche, *see* Leriche syndrome
Lesch-Nyhan, *see* Lesch-Nyhan syndrome
Leschke-Ullman, *see* Leschke-Ullmann syndrome
lethal multiple pterygia, *see* Lethal multiple
  pterygia syndrome
Levine-Critchley, *see* Chorea-acanthocytosis
Lewandowsky-Lutz, *see* Lewandowsky-Lutz
  syndrome
LGB, *see* Guillain-Barré syndrome
Li-Fraumeni, *see* Li-Fraumeni syndrome
limb girdle, *see* Limb girdle syndrome
Lindenov-Hallgren, *see* Lindenov-Hallgren
  syndrome
linear nevus sebaceous, *see* Linear nevus
  sebaceous syndrome
lissencephaly, *see* Lissencephaly syndrome
lobster eye, *see* Gerhardt syndrome
locked-in, *see* Locked-in syndrome
Loken-Senior, *see* Loken-Senior syndrome
long QT interval, *see* Long QT syndrome
Lonsdale-Blass, *see* Lonsdale-Blass syndrome
Louis-Bar, *see* Ataxia telangiectasia
Lowry, *see* Lowry syndrome
Lowry-Wood, *see* Lowry syndrome
Luft, *see* Luft syndrome
Lundberg, *see* Lundberg syndrome
luxury perfusion, *see* Luxury perfusion syndrome
Lyle, *see* Lyle syndrome
Mabry, *see* Mabry syndrome
Maffucci, *see* Maffucci syndrome
Maffucci-Kast, *see* Maffucci-Kast syndrome
Magendie, *see* Skew deviation
Magendie-Hertwig, *see* Skew deviation
Majewski, *see* Saldino-Noonan syndrome
malabsorption, *see* Malabsorption syndrome
Malamud-Cohen, *see* Infantile sex linked
  cerebellar ataxia
Maley, *see* Maley syndrome
malignant neuroleptic, *see* Malignant neuroleptic
  syndrome
malocclusion, *see* Malocclusion syndrome
Marchesani-Wirz, *see* Grönblad-Strandberg
  syndrome

Marden-Walker, *see* Marden-Walker syndrome
Marfan, *see* Marfan syndrome
marfan like, *see* Marfan like syndrome
Marfan-Weve, *see* Marfan syndrome
Margolis-Ziprkowski, *see* Margolis-Ziprkowski
  syndrome
Marin Amat, *see* Marin Amat syndrome
Marinesco-Sjögren, *see* Marinesco-Sjögren
  syndrome
Maroteaux-Lamy, *see* Glycosaminoglycanosis
  type VI
Marshall, *see* Marshall syndrome
Martorell-Fabré, *see* Takayasu disease
MASA, *see* MASA syndrome
Mast's, *see* Mast syndrome
Matthews-Rundle, *see* Matthews-Rundle
  syndrome
May-White, *see* May-White syndrome
McCormick-Lemmi, *see* Familial
  pallidoluysionigral degeneration *and*
  Pallidonigral degeneration
McCune-Albright, *see* Albright syndrome
McNicholl, *see* Cerebrocostomandibular
  syndrome
Means, *see* Oligosymptomatic hyperthyroidism
Meckel-Gruber, *see* Gruber syndrome
medial medullary, *see* Medial medullary syndrome
medial thalamic, *see* Medial thalamic syndrome
median cleft lip hypotelorism, *see* Median cleft lip
  hypotelorism syndrome
median facial cleft, *see* Median facial cleft
  syndrome
median longitudinal fasciculus, *see* Median
  longitudinal fasciculus syndrome
medulla oblongata, *see* Medulla oblongata
  syndrome
megalocorneal mental deficiency, *see*
  Megalocorneal mental deficiency syndrome
MELAS, *see* MELAS syndrome
Melkersson-Rosenthal, *see* Melkersson-Rosenthal
  syndrome
Melnick-Needles, *see* Melnick-Needles syndrome
Mende, *see* Mende syndrome
meningeal, *see* Meningeal syndrome
Menkes, *see* Trichopoliodystrophy
menstrual associated, *see* Menstrual associated
  syndrome
MEPOP, *see* MEPOP syndrome
MERRF, *see* MERRF syndrome
mesencephalic, *see* Mesencephalic syndrome
mesencephalic diencephalic, *see* Mesencephalic
  diencephalic syndrome
mesencephalic tegmentum, *see* Mesencephalic

tegmentum syndrome

mesial occipitotemporal, *see* Mesial occipitotemporal syndrome

methadone withdrawal, *see* Methadone withdrawal syndrome

methionine malabsorption, *see* Methionine malabsorption syndrome

methotrexate, *see* Methotrexate syndrome

Mibelli angiokeratoma, *see* Mibelli angiokeratoma syndrome

midbrain, *see* Midbrain syndrome

middle cavernous sinus, *see* Middle cavernous sinus syndrome

middle cerebral artery, *see* Middle cerebral artery syndrome

Miescher type II, *see* Miescher syndrome type II

Mietens-Weber, *see* Mietens-Weber syndrome

Millard-Gubler, *see* Millard-Gubler syndrome

Miller, *see* Miller syndrome

Miller-Dieker, *see* Miller-Dieker syndrome

Miller Fisher, *see* Miller Fisher syndrome

Mills, *see* Mills syndrome

minor hemisphere, *see* Minor hemisphere syndrome

Mirhosseini-Holmes-Walton, *see* Mirhosseini-Holmes-Walton syndrome

MNGIE, *see* MNGIE syndrome

Möbius, *see* Möbius syndrome

Moersch-Woltman, *see* Stiff-man syndrome

Mogan-Baumgartner-Cogan, *see* Cogan syndrome type II

Mohr, *see* Mohr syndrome

Mondrum-Benisty, *see* Mondrum-Benisty syndrome

Morgagni, *see* Morgagni-Stewart-Morel syndrome

Morgagni-Stewart-Morel, *see* Morgagni-Stewart-Morel syndrome

Morquio, *see* Glycosaminoglycanosis type IV

Morton, *see* Morton neuralgia

Morvan, *see* Hereditary sensory and autonomic neuropathy type II

motor, *see* Motor syndrome

Muckle-Wells, *see* Muckle-Wells syndrome

mucocutaneous lymph node, *see* Kawasaki syndrome

mucosal neuroma, *see* Multiple endocrine adenomatosis type III

multiple deficiency, *see* Multiple deficiency syndrome

multiple endocrine adenomatosis, *see* Multiple endocrine adenomatosis syndrome

multiple endocrine neoplasia, *see* Multiple endocrine neoplasia syndrome

multiple hamartoma, *see* Multiple hamartoma syndrome

multiple lentigines, *see* LEOPARD syndrome

multiple nevoid basal cell carcinoma, *see* Multiple nevoid basal cell carcinoma syndrome

MURCS, *see* MURCS syndrome

muscular pain fasciculation, *see* Muscular pain fasciculation syndrome

myasthenic, *see* Myasthenic syndrome

myoclonic epilepsy, *see* Myoclonic epilepsy syndrome

myofascial pain dysfunction, *see* Myofascial pain dysfunction syndrome

myotonic, *see* Myotonic syndrome

Naegeli, *see* Naegeli syndrome

Nageotte, *see* Babinski-Nageotte syndrome

narcoleptic, *see* Narcoleptic syndrome

NARP, *see* NARP syndrome

nasal nerve, *see* Charlin neuralgia

nasocilliary, *see* Charlin neuralgia

negative symptom, *see* Negative symptom syndrome

neglect, *see* Neglect syndrome

Nelson, *see* Nelson syndrome

neocerebellar, *see* Neocerebellar syndrome

nephrotic, *see* Nephrotic syndrome

Netherton, *see* Netherton syndrome

Nettleship-Falls, *see* Nettleship-Falls syndrome

Neu, *see* Neu syndrome

Neu-Laxova-COFS, *see* Neu-Laxova-COFS syndrome

neural crest, *see* Hereditary sensory and autonomic neuropathy type II

neuro-Behçet, *see* Behçet syndrome

neurocirculatory, *see* Neurocirculatory syndrome

neurocutaneous, *see* Neurocutaneous syndrome

neuroleptic, *see* Neuroleptic syndrome

neuroleptic malignant, *see* Neuroleptic malignant syndrome

neurologic, *see* Neurologic syndrome

neuropsychologic, *see* Neuropsychologic syndrome

neurovascular compression, *see* Neurovascular compression syndrome

nevoid basal cell carcinoma, *see* Multiple nevoid basal cell carcinoma syndrome

nevus flammeus glaucoma, *see* Sturge-Weber syndrome

Nielsen, *see* Nielsen syndrome

Noack, *see* Noack syndrome

Nonne-Froin, *see* Nonne-Froin syndrome

Noonan, *see* Noonan syndrome

Noonan-Ehmke, *see* Noonan-Ehmke syndrome

Norman-Roberts, *see* Norman-Roberts syndrome

Northern epilepsy, *see* Northern epilepsy syndrome

Nothnagel, *see* Nothnagel syndrome

nucleus ruber superior, *see* Nucleus ruber superior syndrome

Nijmegen breakage, *see* Nijmegen breakage syndrome

Nyssen-Van Bogaert, *see* Opticocochleodentate degeneration

obesity cardiorespiratory, *see* Pickwickian syndrome

obstructive apnea, *see* Obstructive apnea syndrome

occipital condyle, *see* Occipital condyle syndrome

occipital lobe, *see* Occipital lobe syndrome

oculoauriculovertebral, *see* Goldenhar syndrome

oculocerebrocutaneous, *see* Oculocerebrocutaneous syndrome

oculocerebrofacial, *see* Oculocerebrofacial syndrome

oculocerebrorenal, *see* Lowe syndrome

oculomotor, *see* Oculomotor syndrome

oculonasal, *see* Charlin neuralgia

oculonuclear artery, *see* Oculonuclear artery syndrome

oculo-otocerebrorenal, *see* Oculo-otocerebrorenal syndrome

oculorenocerebellar, *see* Oculorenocerebellar syndrome

oculovertebral, *see* Oculovertebral syndrome

OFD I, *see* Orofaciodigital syndrome type I

OFD II, *see* Mohr syndrome

Ogilvie, *see* Ogilvie syndrome

olfactory fissure, *see* Olfactory fissure syndrome

Oliver-McFarlane, *see* Oliver-McFarlane syndrome

Ollier-Klippel-Trénaunay, *see* Klippel-Trénaunay syndrome

one and one half, *see* One and one half syndrome

Opalski, *see* Opalski syndrome

operculum, *see* Operculum syndrome

ophthalmoplegia, *see* Progressive external ophthalmoplegia

opiate withdrawal, *see* Opiate withdrawal syndrome

Opitz-Kaveggia, *see* FG syndrome

Oppenheim-Scholz-Morel, *see* Dyshoric angiopathy

opsoclonus myoclonus, *see* Opsoclonus myoclonus syndrome

opsoclonus myoclonus ataxia, *see* Opsoclonus myoclonus ataxia syndrome

optic chiasm, *see* Optic chiasm syndrome

orbital, *see* Orbital syndrome

orbital apex, *see* Orbital apex syndrome

orbital floor, *see* Orbital floor syndrome

organic affective, *see* Organic affective syndrome

organic brain, *see* Organic brain syndrome

organic delusional, *see* Organic delusional syndrome

organic mental, *see* Organic mental syndrome

organic personality, *see* Organic personality syndrome

orofaciodigital, *see* Orofaciodigital syndrome

orofaciodigital type I, *see* Orofaciodigital syndrome type I

orofaciodigital type II, *see* Mohr syndrome

orthopedic, *see* Orthopedic syndrome

orthostatic intolerance, *see* Orthostatic intolerance syndrome

Osler, *see* Hereditary hemorrhagic telangiectasia (Osler)

OSPOM, *see* OSPOM syndrome

Ostrum-Furst, *see* Ostrum-Furst syndrome

Osuntokun, *see* Osuntokun syndrome

Ota, *see* Ota syndrome

otopalatodigital, *see* Otopalatodigital syndrome

overlap, *see* Overlap syndrome

pain, *see* Pain syndrome

Paine, *see* Paine syndrome

painful feet, *see* Burning feet syndrome

painful legs moving toes, *see* Painful legs moving toes syndrome

paleocerebellar, *see* Paleocerebellar syndrome

pallidal, *see* Pallidal syndrome

pallidopyramidal, *see* Pallidopyramidal degeneration

Papillon-Léage, *see* Orofaciodigital syndrome type I

Papillon-Léage-Psaume, *see* Orofaciodigital syndrome type I

paracentral lobule, *see* Paracentral lobule syndrome

paramedian medullary, *see* Paramedian medullary syndrome

paramedian pontine, *see* Paramedian pontine syndrome

paraneoplastic, *see* Paraneoplastic syndrome

paranoid, *see* Paranoid syndrome

parapharyngeal space, *see* Parapharyngeal space syndrome

parasellar, *see* Parasellar syndrome

paratrigeminal, *see* Paratrigeminal syndrome

paresthetic causalgic, *see* Paresthetic causalgic syndrome

parietal aphasia, *see* Parietal aphasia syndrome
parietal lobe, *see* Parietal lobe syndrome
Parinaud, *see* Parinaud syndrome
Parinaud oculoglandular, *see* Parinaud syndrome
Parkes Weber, *see* Parkes Weber syndrome
Parkes Weber-Klippel, *see* Klippel-Trénaunay syndrome
Parkinson, *see* Parkinsonism
Parkinson plus, *see* Multiple neuronal system degeneration
Parrot, *see* Parrot syndrome
Parry-Romberg, *see* Progressive hemifacial atrophy
Parsonage-Turner, *see* Neuralgic amyotrophy
partial Horner, *see* Partial Horner syndrome
Pasini-Pierini, *see* Pasini-Pierini syndrome
Passow, *see* Passow syndrome
Passwell, *see* Passwell syndrome
Patau, *see* Patau syndrome
patchy pontine, *see* Patchy pontine syndrome
Pearson, *see* Pearson syndrome
Pellazzi, *see* Pellazzi syndrome
Pena-Shokeir type I, *see* Pena-Shokeir syndrome type I
Pena-Shokeir type II, *see* Cerebro-oculofacioskeletal syndrome
Pendred, *see* Pendred syndrome
periaqueductal, *see* Periaqueductal syndrome
pericarotid, *see* Pericarotid syndrome
periodic, *see* Periodic syndrome
Pernio, *see* Pernio syndrome
Peter Pan, *see* Peter Pan syndrome
petrobasilar suture, *see* Petrobasilar suture syndrome
petrophenoid (crossway), *see* Orbital apex syndrome
petrosal apex, *see* Gradenigo syndrome
petrosphenoid (Jaccodo), *see* Petrosphenoid syndrome (Jaccodo)
Peutz, *see* Peutz-Jeghers syndrome
Peutz-Jeghers, *see* Peutz-Jeghers syndrome
Pfeiffer, *see* Acrocephalosyndactyly type V
phantom limb, *see* Phantom limb syndrome
pheochromocytoma thyroid carcinoma multiple neurinoma, *see* Pheochromocytoma thyroid carcinoma multiple neurinoma syndrome
photomyoclonus, *see* Photomyoclonus syndrome
pica, *see* Pica syndrome
pickwickian, *see* Pickwickian syndrome
Pierre-Robin, *see* Pierre-Robin syndrome
Pinsky-DiGeorge-Harley-Bird, *see* Pinsky-DiGeorge-Harley-Bird syndrome
pluriradicular, *see* Pluriradicular syndrome

Poland, *see* Poland-Möbius syndrome
Poland-Möbius, *see* Poland-Möbius syndrome
poliomyelitis like, *see* Poliomyelitis like syndrome
POLIP, *see* POLIP syndrome
polyradiculopathy, *see* Polyradiculopathy syndrome
pons, *see* Pontine syndrome
pontine, *see* Pontine syndrome
pontine tegmentum, *see* Tegmental pontine syndrome
Porot-Filiu, *see* Porot-Filiu syndrome
postconcussional, *see* Postconcussional syndrome
postencephalitic, *see* Postencephalitic syndrome
posterior cavernous sinus, *see* Posterior cavernous sinus syndrome
posterior cerebral artery, *see* Posterior cerebral artery syndrome
posterior column, *see* Posterior column syndrome
posterior condylar canal, *see* Posterior condylar canal syndrome
posterior corpus callosum, *see* Corpus callosum syndrome
posterior interosseous nerve, *see* Posterior interosseous nerve syndrome
posterior pharyngeal space, *see* Villaret syndrome
posterior spinal artery, *see* Posterior spinal artery syndrome
posterolateral thalamic, *see* Posterolateral thalamic syndrome
postgastrectomy, *see* Postgastrectomy syndrome
postinfectious, *see* Postinfectious syndrome
postlumbar puncture, *see* Postlumbar puncture syndrome
postmastectomy, *see* Postmastectomy syndrome
postpolio, *see* Postpoliomyelitic amyotrophy
postradiation, *see* Postradiation syndrome
postthoracotomy, *see* Postthoracotomy syndrome
posttraumatic, *see* Posttraumatic syndrome
postural orthostatic tachycardia, *see* Postural orthostatic tachycardia syndrome
postviral fatigue, *see* Postviral fatigue syndrome
Potter, *see* Potter syndrome
Potter type II, *see* Potter syndrome type II
Prader-Labhart-Willi, *see* Prader-Labhart-Willi syndrome
Prader-Willi, *see* Prader-Labhart-Willi syndrome
prefrontal, *see* Prefrontal syndrome
premature closure, *see* Premature closure syndrome
premenstrual, *see* Premenstrual syndrome
premotor cortex, *see* Premotor cortex syndrome
prenatal mercury, *see* Minamata disease
prenatal rubella, *see* Prenatal rubella syndrome

prenatal toxoplasma, *see* Congenital
toxoplasmosis
prepeduncular, *see* Prepeduncular syndrome
pronator teres, *see* Anterior interosseous nerve
syndrome
prune belly, *see* Prune belly syndrome
pseudo-Bartter, *see* Pseudo-Bartter syndrome
pseudobulbar, *see* Pseudobulbar syndrome
pseudo-Crouzon, *see* Saethre-Chotzen syndrome
pseudo-Gradenigo, *see* Citelli syndrome
pseudo-Meniere, *see* Pseudo-Meniere syndrome
pseudoprogeria Hallermann-Streiff, *see*
Pseudoprogeria Hallermann-Streiff syndrome
pseudoradicular, *see* Pseudoradicular syndrome
pseudothalamic, *see* Pseudothalamic syndrome
pseudothalidomide, *see* Roberts syndrome
Psilocybe, *see* Psilocybe syndrome
psychomotor, *see* Psychomotor syndrome
pterygium, *see* Turner syndrome
pterygopalatine fossa, *see* Pterygopalatine fossa
syndrome
ptosis epicanthus, *see* Klein-Waardenburg
syndrome
punch drunk, *see* Punch drunk syndrome
pyramidal, *see* Pyramidal syndrome
pyramidal tract, *see* Pyramidal tract syndrome
quantal squander, *see* Isaacs syndrome
radial tunnel, *see* Radial tunnel syndrome
radicular, *see* Radicular syndrome
Raeder, *see* Paratrigeminal syndrome
Raeder paratrigeminal, *see* Paratrigeminal
syndrome
Ramsay Hunt, *see* Ramsay Hunt syndrome
Raymond Cestan, *see* Cestan-Chenais syndrome
rectus abdominis muscle, *see* Rectus abdominis
muscle syndrome
reflex neurovascular, *see* Reflex neurovascular
syndrome
renal agenesis, *see* Renal agenesis syndrome
restless legs, *see* Restless legs syndrome
retraction, *see* Acquired retraction syndrome *and*
Stilling-Türk-Duane syndrome
retropharyngeal, *see* Villaret syndrome
Rett, *see* Rett syndrome
Reye, *see* Reye syndrome
rhinocerebral, *see* Rhinocerebral syndrome
Richards-Rundle, *see* Richards-Rundle syndrome
Richards-Rundle-Sylvester, *see* Richards-Rundle
syndrome
Richardson-Steele-Olszewski, *see* Progressive
supranuclear palsy
Richner-Hanhart, *see* Richner-Hanhart syndrome
Rieger, *see* Rieger syndrome

rigid spine, *see* Rigid spine syndrome
Riley-Day, *see* Hereditary sensory and autonomic
neuropathy type III
Riley-Smith, *see* Riley-Smith syndrome
Riolan nosegay, *see* Riolan nosegay syndrome
Roberts, *see* Roberts syndrome
Robin, *see* Robin syndrome
Robinow, *see* Fetal face syndrome
Robinow-Silverman-Smith, *see* Fetal face
syndrome
Rochon-Duvigneau, *see* Orbital apex syndrome
Rokitanski-Van Bogaert, *see* Neurocutaneous
melanosis
Rollet, *see* Orbital apex syndrome
Romano-Ward, *see* Hereditary long QT syndrome
Ron-Pearce, *see* Ron-Pearce syndrome
Rosenberg-Chutorian, *see* Rosenberg-Chutorian
syndrome
Ross, *see* Ross syndrome
Rossi, *see* Rossi syndrome
rostral tegmental, *see* Rostral tegmental syndrome
Rothmund, *see* Rothmund-Thomson syndrome
Rothmund-Petges-Clejat, *see* Rothmund-Thomson
syndrome
Rothmund-Thomson, *see* Rothmund-Thomson
syndrome
Roussy-Lévy, *see* Roussy-Lévy syndrome
Rowley-Rosenberg, *see* Rowley-Rosenberg
syndrome
rubella, *see* Rubella syndrome
Rubinstein-Taybi, *see* Rubinstein-Taybi syndrome
rubrothalamic, *see* Anterolateral thalamic
syndrome
Rud, *see* Rud syndrome
Russell, *see* Russell syndrome
Russell brain, *see* Exophthalmos
Russell-Silver, *see* Russell-Silver syndrome
Ruvalcaba, *see* Ruvalcaba syndrome
Sachsalber, *see* Sachsalber syndrome
Saethre-Chotzen, *see* Saethre-Chotzen syndrome
Sakati-Nyhan-Tisdale, *see* Sakati-Nyhan-Tisdale
syndrome
Salam-Idriss, *see* Salam-Idriss syndrome
Saldino-Noonan, *see* Saldino-Noonan syndrome
Salleras-Ortiz, *see* Salleras-Ortiz syndrome
Sanfilippo, *see* Glycosaminoglycanosis type III
sarcoplasmic reticulum adenosine triphosphate
deficiency, *see* Sarcoplasmic reticulum
adenosine triphosphate deficiency syndrome
Satoyoshi, *see* Satoyoshi syndrome
Say-Gerald, *see* VATER syndrome
SBLA, *see* SBLA syndrome
scalenus anticus, *see* Scalenus anticus syndrome

scapulohumeral, *see* Facioscapulohumeral syndrome

scapuloperoneal, *see* Scapuloperoneal syndrome

Schachter, *see* Schachter syndrome

Scheie, *see* Glycosaminoglycanosis type IS

Scheuthauer-Marie-Saiton, *see* Cleidocranial dysplasia

Schimke, *see* Schimke syndrome

schizophrenia type I, *see* Schizophrenia type I syndrome

schizophrenia type II, *see* Schizophrenia type II syndrome

schizophrenic, *see* Schizophrenic syndrome

Schmid-Fraccaro, *see* Cat eye syndrome

Schmidt, *see* Schmidt syndrome

Schützenberger-Barré-Liéou, *see* Schützenberger-Barré-Liéou syndrome

Schwartz-Jampel, *see* Chondrodystrophic myotonia

Schwartz-Lelek, *see* Schwartz-Lelek syndrome

Scott, *see* Tropical ataxic neuropathy

Seckel, *see* Seckel syndrome

sensory, *see* Sensory syndrome

serotonin, *see* Serotonin syndrome

Sézary, *see* Sézary syndrome

Shapiro, *see* Shapiro syndrome

Sheehan, *see* Sheehan syndrome

Sheldon, *see* Sheldon syndrome

short bowel, *see* Short bowel syndrome

short dura mater, *see* Short dura mater syndrome

short rib polydactyly, *see* Saldino-Noonan syndrome

shoulder hand, *see* Shoulder hand syndrome

Shulman, *see* Shulman syndrome

Shy-Drager, *see* Shy-Drager syndrome

Sicard, *see* Sicard-Collet syndrome

Sicard-Collet, *see* Sicard-Collet syndrome

sicca, *see* Sjögren syndrome

sick sinus, *see* Sick sinus syndrome

Siebenmann, *see* Jugular foramen syndrome

Siemens, *see* Incontinentia pigmenti

Silver-Russell, *see* Russell-Silver syndrome

Sipple, *see* Sipple syndrome

Sjögren, *see* Sjögren syndrome

Sjögren-Larsson, *see* Sjögren-Larsson syndrome

Sjögren-Larsson-Richards, *see* Sjögren-Larsson syndrome

sleep apnea, *see* Sleep apnea syndrome

sleep-wake, *see* Sleep-wake syndrome

slow channel, *see* Slow channel syndrome

Sluder, *see* Sphenopalatine neuralgia

Sly, *see* Glycosaminoglycanosis type VII

Smith-Lemli-Opitz, *see* Smith-Lemli-Opitz syndrome

Smith-McCort, *see* Smith-McCort syndrome

Sneddon, *see* Sneddon syndrome

Sotos, *see* Cerebral gigantism

speech isolation, *see* Mixed transcortical aphasia

sphenocavernous angel, *see* Sphenocavernous angle syndrome

sphenoid, *see* Sphenoid syndrome

sphenoidal fissure, *see* Superior orbital fissure syndrome

sphenoidal sinus, *see* Sphenoidal sinus syndrome

sphenopetrous fissure, *see* Orbital apex syndrome

spinal cord transverse, *see* Transverse spinal cord syndrome

spinal nerve root, *see* Spinal nerve root syndrome

spinocerebellar, *see* Spinocerebellar syndrome

split brain, *see* Split brain syndrome

split notochord, *see* Split notochord syndrome

Stark-Kaeser, *see* Scapuloperoneal spinal muscular atrophy

startle, *see* Startle syndrome

steal, *see* Steal syndrome

Steele-Richardson-Olszewski, *see* Progressive supranuclear palsy

Stern-Garcin, *see* Thalamic dementia

Stewart, *see* Midline granuloma

Stibler-Jaeken, *see* Jaeken syndrome

Stickler, *see* Stickler syndrome

stiff-man, *see* Stiff-man syndrome

Stilling-Türk-Duane, *see* Stilling-Türk-Duane syndrome

Stokes-Adams, *see* Adams-Stokes syndrome

Strachan, *see* Tropical ataxic neuropathy

Strachan-Scott, *see* Tropical ataxic neuropathy

Strachan-Scott-Stannus, *see* Tropical ataxic neuropathy

striatal, *see* Striatal syndrome

Sturge-Weber, *see* Sturge-Weber syndrome

Sturge-Weber-Dimitri, *see* Sturge-Weber syndrome

stylocarotid, *see* Stylohyoid syndrome

stylohyoid, *see* Stylohyoid syndrome

styloid, *see* Stylohyoid syndrome

subclavian steal, *see* Subclavian steal syndrome

subcoracoid pectoralis minor, *see* Hyperabduction syndrome

Summitt, *see* Summitt syndrome

SUNCT, *see* SUNCT syndrome

superior cerebellar artery, *see* Superior cerebellar artery syndrome

superior cerebellar peduncle, *see* Cestan-Chenais syndrome

superior colliculi, *see* Parinaud syndrome

superior oblique tendon sheath, *see* Brown syndrome

superior red nucleus, *see* Nucleus ruber superior syndrome

superior vena cava, *see* Superior vena cava syndrome

supinator, *see* Supinator syndrome

supplementary motor area, *see* Supplementary motor area syndrome

sweaty foot, *see* Sweaty foot syndrome

Sylvian aqueduct, *see* Aqueduct stenosis

symmetrical cerebellar hypogenesis, *see* Symmetrical cerebellar hypogenesis syndrome

syncopic cervical, *see* Syncopic cervical syndrome

syringomyelia like, *see* Syringomyelia like syndrome

Takatsuki, *see* POEMS syndrome

Tapia, *see* Tapia syndrome

tarsal tunnel, *see* Tarsal tunnel syndrome

Taybi-Linder, *see* Taybi-Linder syndrome

tegmental pontine, *see* Tegmental pontine syndrome

Telfer, *see* Telfer syndrome

temporal lobe, *see* Temporal lobe syndrome

temporal lobe agenesis, *see* Temporal lobe agenesis syndrome

Terson, *see* Terson syndrome

tethered cord, *see* Tethered cord syndrome

thalamoperforate, *see* Anterolateral thalamic syndrome

thalidomide, *see* Thalidomide syndrome

Thomson, *see* Rothmund-Thomson syndrome

thoracic outlet, *see* Thoracic outlet syndrome

thoracic rib, *see* Thoracic rib syndrome

Tiling-Wernicke, *see* Operculum syndrome

Tolosa-Hunt, *see* Tolosa-Hunt syndrome

total transverse spinal cord, *see* Total transverse spinal cord syndrome

Touraine-Solente-Golé, *see* Neurocutaneous melanosis

Tourette, *see* Gilles de la Tourette syndrome

Townes-Brocks, *see* Townes-Brocks syndrome

toxic oil, *see* Toxic oil syndrome

toxic shock, *see* Toxic shock syndrome

toxin mediated, *see* Toxin mediated syndrome

Toxoplasma gondii, *see* Toxoplasma gondii syndrome

transection, *see* Transection syndrome

transverse, *see* Transverse syndrome

transverse spinal cord, *see* Transverse spinal cord syndrome

traumatic diencephalic, *see* Traumatic diencephalic syndrome

traumatic frontobasal, *see* Traumatic frontobasal syndrome

traumatic parietal, *see* Traumatic parietal syndrome

traumatic psycho, *see* Traumatic psychosyndrome

traumatic temporobasal, *see* Traumatic temporobasal syndrome

traumatic vegetative, *see* Traumatic vegetative syndrome

Treacher-Collins, *see* Treacher-Collins syndrome

tremor dysarthria ataxia, *see* Tremor dysarthria ataxia syndrome

Treponema pallidum, *see* Treponema pallidum syndrome

triad, *see* Prune belly syndrome

triple Y, *see* Sex chromosomes

triploidy, *see* Triploidy syndrome

trisymptomatic, *see* Behçet syndrome

Troyer, *see* Troyer syndrome

Turcot, *see* Glioma polyposis syndrome

Türk-Stilling-Duane, *see* Stilling-Türk-Duane syndrome

Turner, *see* Turner syndrome

Ullmann, *see* Ullmann syndrome

Ullrich, *see* Ullrich syndrome

Ullrich-Feichtiger, *see* Ullrich-Feichtiger syndrome

Ullrich-Fremerey-Dohna, *see* Ullrich-Fremerey-Dohna syndrome

Ullrich-Turner, *see* Ullrich-Turner syndrome

upper airway sleep apnea, *see* Sleep apnea syndrome

upper pontine tegmentum, *see* Cestan-Chenais syndrome *and* Foville syndrome

Usher, *see* Usher syndrome

uveoencephalitic, *see* Vogt-Koyanagi-Harada syndrome

uveomeningoencephalitic, *see* Vogt-Koyanagi-Harada syndrome

uveo-otocutaneous, *see* Vogt-Koyanagi-Harada syndrome

VACTERL, *see* VACTERL syndrome

Van den Bosch, *see* Van den Bosch syndrome

Van der Hoeve-Halbertsma-Waardenburg, *see* Klein-Waardenburg syndrome

Van der Hoeve-Halbertsma-Waardenburg-Gauldi, *see* Klein-Waardenburg syndrome

Van der Hoeve-Waardenburg-Klein, *see* Klein-Waardenburg syndrome

Van Laere, *see* Brown-Vialetto-Van Laere syndrome

Van Meekeren-Danlos-Ehlers, *see* Ehlers-Danlos syndrome

vasovagal, *see* Vasovagal syndrome
VATER, *see* VATER syndrome
vegetative dystonia, *see* Vegetative dystonia syndrome
ventral pontine, *see* Ventral pontine syndrome
Verger-Dejerine, *see* Verger-Dejerine syndrome
Vernet, *see* Vernet syndrome
Vernet-Sargnon, *see* Sicard-Collet syndrome
vertebral column, *see* Vertebral column syndrome
vertebrobasilar system, *see* Vertebrobasilar system syndrome
vertebrogenic, *see* Vertebrogenic syndrome
vertical one and half, *see* Vertical one and half syndrome
vestibular, *see* Vestibular syndrome
vestibular complex, *see* Vestibular complex syndrome
vestibulocerebellar, *see* Archicerebellar syndrome
Vialetto-Van Laere, *see* Brown-Vialetto-Van Laere syndrome
Villaret, *see* Villaret syndrome
Vogt-Koyanagi, *see* Vogt-Koyanagi-Harada syndrome
Vogt-Koyanagi-Harada, *see* Vogt-Koyanagi-Harada syndrome
Vogt striatal athetotic, *see* Vogt striatal athetotic syndrome
Vogt-Waardenburg, *see* Vogt-Waardenburg syndrome
Von Gräfe, *see* Progressive external ophthalmoplegia
Von Gräfe-Liebreich-Hallgren, *see* Usher syndrome
Von Gräfe-Lindenov, *see* Lindenov-Hallgren syndrome
Von Gräfe-Sjögren, *see* Von Gräfe-Sjögren syndrome
W, *see* W syndrome
Waardenburg, *see* Klein-Waardenburg syndrome
Waardenburg-Shah, *see* Waardenburg-Shah syndrome
Wadia-Swami, *see* Olivopontocerebellar atrophy (Wadia-Swami)
Walker, *see* HARD syndrome *and* HARDE syndrome
Walker-Warburg, *see* HARD syndrome *and* HARDE syndrome
Wallenberg, *see* Lateral medullary infarction
Wallgren, *see* Wallgren syndrome
Walsh, *see* Walsh syndrome
Warburg, *see* HARD syndrome *and* HARDE syndrome
Ward, *see* Multiple nevoid basal cell carcinoma syndrome
Ward-Gorlin-Goltz, *see* Multiple nevoid basal cell carcinoma syndrome
WARG, *see* WARG syndrome
Waterhouse-Friderichsen, *see* Waterhouse-Friderichsen syndrome
Weaver, *see* Weaver syndrome
Weber, *see* Weber syndrome
Weber-Von Leyden, *see* Weber syndrome
Weiss-Alström, *see* Weiss-Alström syndrome
Wermer, *see* Wermer syndrome
Werner, *see* Werner syndrome
Wernicke-Korsakoff, *see* Wernicke-Korsakoff syndrome
Wernicke-Mann, *see* Wernicke-Mann syndrome
West, *see* Infantile spasm
White's, *see* White syndrome
Wieacker-Wolff, *see* Wieacker-Wolff syndrome
Wildervanck, *see* Wildervanck syndrome
Willi-Prader, *see* Prader-Labhart-Willi syndrome
Williams, *see* Williams syndrome
Willvonseder, *see* Willvonseder syndrome
Wiskott-Aldrich, *see* Wiskott-Aldrich syndrome
withdrawal, *see* Withdrawal syndrome
Wolf-Hirschhorn, *see* Wolf-Hirschhorn syndrome
Wolff-Parkinson-White, *see* Wolff-Parkinson-White syndrome
Wolfram, *see* Wolfram syndrome
Wolfslast-Blumel-Subrahmanyam, *see* Wolfslast-Blumel-Subrahmanyam syndrome
Wolfson, *see* Wolfson syndrome
Woods-Pendleton, *see* Familial pallidonigral degeneration
Woods-Schaumburg, *see* Woods-Schaumburg syndrome
Wyburn-Mason, *see* Bonnet-Dechaume-Blanc syndrome
X-linked lymphoproliferative, *see* X-linked lymphoproliferative syndrome
X-linked recessive spastic ataxic, *see* X-linked recessive spastic ataxic syndrome
X-linked scapuloperoneal, *see* Emery-Dreifuss muscular dystrophy
XXX, *see* XXX syndrome
XXXX, *see* XXXX syndrome
XXXXX, *see* XXXXX syndrome
XXXXY, *see* Klinefelter syndrome *and* Klinefelter variant XXXXY
XXXY, *see* Klinefelter syndrome *and* Klinefelter variant XXXY
XXY, *see* Klinefelter syndrome
XXYY, *see* Klinefelter syndrome
XYY, *see* XYY syndrome

Zackai, *see* Zackai syndrome
Zellweger, *see* Cerebrohepatorenal syndrome
Zellweger like, *see* Zellweger like syndrome
Ziehen-Oppenheim, *see* Dystonia musculorum
   deformans
Zinsser-Engman-Cole, *see* Congenital
   dyskeratosis
Zollinger-Ellison, *see* Zollinger-Ellison syndrome
Syneresis, *see* Synaeresis
Synergomone
   classification, 74/374
Synergy
   alternating, *see* Alternating synergics
   arm, 1/414
   oculomandibular visual direction, *see*
     Oculomandibular visual direction synergy
   pyramidal tract, 1/414
   reflex, *see* Reflex synergy
   synkinesis, 1/409, 412
Synesthesia
   sensory extinction, 3/191
   terminology, 1/89
Synkinesia, *see* Synkinesis
Synkinesis, 1/404-422
   aberrant nerve regeneration, 1/417
   Alzheimer disease, 46/252
   anatomic classification, 1/279
   anterior tibial phenomenon, 1/185
   basal ganglion lesion, 1/175
   central motor disorder, 1/182
   classification, 1/407, 410
   clinical form, 1/418
   coordinated, 1/266, 408, 411, 418, 421
   cortical cerebellar atrophy, 60/644
   De Kleijn type, 1/411
   developmental dyslexia, 46/114
   eye movement, 1/610
   facial nerve, 1/416
   global, 1/266, 408, 410, 413, 416, 419
   handedness, 4/255
   head position, 1/408
   hemiplegia, 1/182
   history, 1/405
   identical, 1/411, 419
   imitative, 1/408, 411
   interossei phenomenon, 1/184
   Magnus type, 1/411
   Marie-Foix classification, 1/408
   mass, 1/410
   mirror movement, 1/412
   oculomandibular visual direction synergy, 1/410
   ophthalmoplegia, 2/294
   pathologic, 1/410

   postural reflex, 1/284, 411
   pyramidal tract lesion, 1/411, 414
   restoration of higher cortical function, 3/389
   Souques sign, 1/184
   spasmodic, 1/266, 408, 411
   stereotypy, 1/410
   synergy, 1/409, 412
   therapy, 1/421
   trigemino-oculomotor, *see* Trigemino-oculomotor
     synkinesis
   yawning, 63/493
Synophrys
   De Lange syndrome, 43/246, 50/276
   Klein-Waardenburg syndrome, 14/116, 30/408,
     43/51, 50/218
Synophthalmia cyclopia
   congenital heart disease, 63/11
Synostosis
   atlanto-occipital, *see* Atlanto-occipital synostosis
   cloverleaf skull syndrome, 30/245
   coronal, *see* Coronal synostosis
   lambdoid, *see* Lambdoid synostosis
   metopic, *see* Metopic synostosis
   radioulnar, *see* Radioulnar synostosis
   sagittal, *see* Sagittal synostosis
Synovial cyst
   femoral nerve, 71/19
Synovial fluid antibody
   multiple sclerosis, 47/112
Syntactic alexia
   diagnosis, 45/441
Syntaxin
   botulinum toxin, 65/248
   isoform, 65/248
   tetanus toxin, 65/248
Syntaxis
   aphasia, 45/306
   motor aphasia, 45/307
   Wernicke aphasia, 45/307
Syphilis, 33/337-390
   see also Tabes dorsalis *and* Treponema pallidum
   acquired, 33/339
   acquired immune deficiency syndrome, 56/494,
     71/286
   acquired immunity, 33/347
   acute cerebellar ataxia, 34/625
   adiposis dolorosa, 8/347
   anterior horn syndrome, 2/210
   anterior spinal artery syndrome, 55/101
   antibody test, 52/274
   antiphospholipid antibody, 63/330
   aortitis, 63/52
   arhinencephaly, 9/582

atypical, 52/274
autonomic nervous system, 75/20
bismuth intoxication, 64/331
blindness, 52/281
bone involvement, 33/352
brain angiography, 55/413
brain infarction, 55/412
brain lacunar infarction, 54/240
cardiovascular, 33/353
cerebrovascular complication, 55/411
cerebrovascular disease, 53/34
chancre immunity, 33/348
chorea, 6/438
chronic meningitis, 56/645
classification, 33/339
clinical features, 33/348-353
clinical type, 52/273
Columbus, Christopher, 33/338
computer assisted tomography, 55/413
congenital, see Congenital syphilis
congenital brain granuloma, 42/736
cross immunity, 33/348
CSF, 33/345-347, 52/274, 276
CSF examination, 11/239
deafness, 52/275
diagnosis, 71/288
downbeating nystagmus, 60/656
dystonia, 6/559
early acquired, 33/339
early congenital, 33/340
early latent stage, 33/350
epidemiology, 33/376-382
etiology, 33/338
experimental animal, 33/339
false positive reaction, 33/342, 52/276
fluorescent treponemal antibody absorption test,
    33/343, 52/276
HATTS, 52/276
headache, 5/28, 52/274
history, 33/337
Huntington chorea, 6/315
hyperviscosity syndrome, 55/485
immunology, 33/347
incidence, 33/377-379
infection immunity, 33/348
infectious striatopallidodentate calcification,
    49/417
laboratory test, 52/275
late acquired, 33/340
late congenital, 33/340
late latent stage, 33/350
latent stage, 33/350, 52/274
lymphocytic meningitis, 52/275

meningeal gumma, 55/412
meningovascular, see Neurosyphilis
MHA-TP, 52/276
microscopic diagnosis, 52/275
migraine, 5/56
mouth, 33/351
multiple sclerosis, 9/58
natural course, 52/281
natural immunity, 33/347
nerve lesion, 7/479
neuro, see Neurosyphilis
nonrecurrent nonhereditary multiple cranial
    neuropathy, 51/571
nuclear magnetic resonance, 71/287
optic atrophy, 33/366
optic neuritis, 52/281
origin, 33/338
Oslo study, 52/281
osteitis, 52/274
pachymeningitis, 33/161
Parkinson disease, 6/243
pathology, 33/340, 351-353
polymicrogyria, 50/257
pre-Columbian, 33/377-379
prevalence, 33/377-379
primary stage, 33/348
prodrome, 55/412
prognosis, 33/382-387
Raynaud phenomenon, 43/71
reagin test, 33/341
relapse, 33/350
secondary, see Secondary syphilis
secondary pigmentary retinal degeneration,
    13/216
serodiagnosis, 33/340-345
serologic test, 33/341
serology, 52/275
seronegative, 33/341
Sjögren syndrome, 71/74
skin manifestation, 33/351
spastic paraplegia, 59/435
specific test, 33/342-345
spinal, see Spinal syphilis
spinal apoplexy, 12/532
stigmata, 33/340
subarachnoid hemorrhage, 12/111, 55/29
symptomatic dystonia, 6/559
systemic brain infarction, 11/430
Takayasu disease, 12/413, 63/52
tertiary, see Tertiary syphilis
tongue, 33/351
transmission, 33/339
treatment, 52/282, 55/413, 71/290

Treponema pallidum immobilization test, 33/343, 52/276

treponemal test, 33/342-345, 52/276

Tuskegee study, 52/281

uveitis, 52/274

Venereal Disease Research Laboratory test, 33/341, 52/276

visceral, 33/353

Syphilitic neuroretinitis
secondary pigmentary retinal degeneration, 60/736

Syphilitic osteoperiostitis
Parrot node, 33/370
Tolosa-Hunt syndrome, 48/291

Syringobulbia, 32/273-276
see also Hydromyelia *and* Syringomyelia
anhidrosis, 8/344
autonomic nervous system, 75/10
bilateral facial paresis, 8/258
carbon dioxide narcosis, 63/441
cardiovascular disorder, 32/276
clinical features, 32/274-276
course, 32/276
cranial nerve impairment, 32/274
differential diagnosis, 32/294, 50/451
dysphagia, 50/445
ependymal rupture, 50/427
esophagus dysfunction, 50/445
first symptom, 32/273
Foix-Alajouanine disease, 9/462
forgotten respiration, 63/441, 490
general, 50/444
hemiatrophy, 59/480
Horner syndrome, 50/445
isolated, 32/273
muscle weakness, 32/275
nystagmus, 42/59, 50/445
orthostatic hypotension, 50/445, 63/153
oscillopsia, 50/445
palatal myoclonus, 38/584
pathology, 32/298-300
position, 50/457
progressive bulbar palsy, 59/375
respiratory failure, 50/445
sensory loss, 42/59
sleep apnea syndrome, 50/445
trigeminal descending spinal tract lesion, 50/444
trophic skin change, 32/275
true, 32/273
vertigo, 2/361
vocal cord paralysis, 50/445
Wolff-Parkinson-White syndrome, 32/276

Syringoma, 14/591

Syringomyelia, 32/255-319, 42/59-61, 50/443-461
see also Hydromyelia, Spinal cord cyst *and* Syringobulbia
age at onset, 32/257
agraphia, 45/464
amyotrophy, 32/266
angiodysgenetic necrotizing myelomalacia, 20/446
anhidrosis, 8/344
aqueduct stenosis, 50/302
Arnold-Chiari malformation type I, 32/104, 107, 300, 304, 42/16, 59, 50/444, 453, 458
Arnold-Chiari malformation type II, 19/188
atlanto-occipital abnormality, 32/280
autonomic dysfunction, 32/268-270, 50/445
autonomic nervous system, 75/10
basal arachnoiditis, 32/281, 42/60, 50/455
basilar impression, 42/17, 59, 50/400, 446
blood supply abnormality, 32/269
bone change, 32/270
Bonnet-Dechaume-Blanc syndrome, 31/200
brain arteriovenous malformation, 14/520
brain stem glioma, 50/453
carpal tunnel syndrome, 32/284
central sleep apnea, 63/463
central spinal syndrome, 2/211
cervical arthropathy, 32/271
cervical collar pattern, 50/444
cervical root lesion, 32/283
cervical spondylarthrotic myelopathy, 50/452
child, 32/257
chromosomal aberration, 43/115
chronic poliomyelitis, 22/282
classification, 50/443
clinical features, 32/259-273, 50/444
CNS tumor, 32/286-293
communicating, *see* Communicating syringomyelia
complicated hydromyelia, 50/427
computer assisted tomography, 50/447
cough, 61/378
course, 32/261
CSF, 32/277, 42/60, 50/446
CSF pressure, 19/116
Dandy-Walker syndrome, 30/632, 32/59, 42/23
DAVID model, 61/386
definition, 50/443
developmental, 61/376
diagnosis, 32/293
differential diagnosis, 50/451
dissociated sensory loss, 50/444
dissociation anesthesia, 32/263-265
EEG, 32/278

embryology, 32/302-308
EMG, 1/639, 32/277, 50/445
endomyelography, 50/447
ependymal rupture, 50/427
epidemiology, 30/164, 32/256-258, 50/451
etiology, 32/313-315, 50/454
experimental, 32/315
extramedullary spinal tumor, 50/456
familial, *see* Familial syringomyelia
familial spastic paraplegia, 59/306
fetal loss, 32/316
first symptom, 32/260, 262
Foix-Alajouanine disease, 9/462
foramen magnum decompression, 50/459
gas myelography, 19/181
general, 50/443
genetics, 30/164, 32/258, 50/427, 451
group, 61/376
gustatory sweating, 50/445
hemangioblastoma, 14/60
hemi-3 syndrome, 50/453
hereditary motor and sensory neuropathy type I, 32/294
hereditary motor and sensory neuropathy type II, 32/294
hereditary sensory and autonomic neuropathy type I, 21/76, 32/259, 294
hereditary sensory and autonomic neuropathy type II, 50/452, 60/6
Hirayama disease, 59/113
history, 32/255, 50/426
Horner syndrome, 32/261, 268, 50/445
Huntington chorea, 6/315, 49/291
hydrocephalus, 50/446
hydromyelia, 32/235, 256, 42/60, 50/425, 457
hyperhidrosis, 50/445
idiopathic, 32/300, 318
inflammatory, 61/376
infratentorial tumor, 50/456
intramedullary spinal tumor, 50/455
Klippel-Trénaunay syndrome, 14/116
Krabbe-Bartels disease, 14/12, 529
leprosy, 32/293
lipoma, 61/377
lower motoneuron, 32/266
lumbosacral, *see* Hereditary sensory and autonomic neuropathy type I *and* Lumbosacral syringomyelia
malformation, 32/261
management, 32/318
manometry, 19/92
Melkersson-Rosenthal syndrome, 14/529
metrizamide computer assisted tomography, 50/447
monomelic spinal muscular atrophy, 59/376
motor sign, 50/444
multiple sclerosis, 32/284, 50/452
muscular disorder paresis, 32/294
myelography, 50/446
myelotomy, 50/459
myotonic dystrophy, 43/155, 50/454
neoplastic, 61/376
nerve conduction, 50/445
neurofibromatosis, 32/287
neurofibromatosis type I, 14/37, 133, 146, 31/12, 14, 32/287
neurogenic osteoarthropathy, 26/124, 32/270
neurosurgical procedure, 50/459
neurotrophic osteoarthropathy, 38/456-459
noncommunicating, 32/256, 281, 287, 308, 313, 317
noncommunicating type, 50/458
Noonan syndrome, 50/454
nuclear magnetic resonance, 50/447
nystagmus, 42/59
obex plugging, 50/458
occupation, 32/252-258
old taxonomy, 22/282
orthostatic hypotension, 63/153
pachymeningitis, 61/376
pain, 32/263
pathogenesis, 32/259, 300-313, 50/457
pathology, 32/295-298, 50/456
platybasia, 50/446
poliomyelitis, 32/285
posttraumatic, *see* Posttraumatic syringomyelia
prevalence, 42/60, 50/65
prognosis, 32/319
progressive hemifacial atrophy, 14/777
radiation, 32/279-283, 50/446
respiratory pattern, 63/463
Rosenthal fiber, 32/292, 297
sagittal diameter, 19/156
Scheuermann disease, 32/283
scoliosis, 32/258, 271, 42/59, 50/418, 445
sensory disorder, 32/262-266
sensory loss, 42/59
short latency evoked potential, 50/446
sibling, 50/451
space occupying lesion, 32/285, 293
spasmodic torticollis, 50/445
spastic paraplegia, 59/434
spastic tetraplegia, 32/268
spasticity, 42/59
spina bifida, 32/305, 61/376
spinal arachnoid cyst, 20/82

spinal arachnoiditis, 26/146, 148, 153, 32/312, 50/455
spinal ataxia, 32/265
spinal canal dilation, 32/271-273
spinal cord compression, 19/378
spinal cord tumor, 19/9, 61/377
spinal cordectomy, 50/460
spinal ependymoma, 20/357
spinal glioma, 20/327, 339
spinal lipoma, 20/398
spinal muscular atrophy, 32/285, 50/452, 59/113, 373
spinal syphilis, 32/293, 50/453
symptom, 50/444
symptomatic dystonia, 49/543
syringoperitoneal shunt, 50/460
syringostomy, 50/459
syringosubarachnoid shunt, 50/459
Tangier disease, 50/452
terminal ventriculostomy, 50/460
thoracic outlet syndrome, 50/453
treatment, 32/315-318
twins, 50/451
ultrasound, 50/450
Von Hippel-Lindau disease, 14/13, 31/17, 32/287, 43/21
Syringomyelia like syndrome
a-alphalipoproteinemia, 51/397
Syringomyelocele
hydromyelia, 50/427
spina bifida, 50/482
Syringoperitoneal shunt
syringomyelia, 50/460
Syringostomy
syringomyelia, 50/459
Syringosubarachnoid shunt
posttraumatic syringomyelia, 61/389, 393
syringomyelia, 50/459
System atrophy (Spatz)
concept, 21/43
System degeneration
areflexia, 42/97
Divry-Van Bogaert syndrome, 10/124
heterogeneous, see Progressive supranuclear palsy
multiple, see Multiple neuronal system degeneration
multiple neuronal, see Multiple neuronal system degeneration
orthochromatic leukodystrophy, 10/114
polyneuritis, 42/97
Systemic amyloidosis
multiple myeloma polyneuropathy, 51/430
primary, see Primary systemic amyloidosis

Systemic angiomatosis, see Ullmann syndrome
Systemic angiopathy, see Polymyositis
Systemic brain infarction, 53/200-202
acquired heart disease, 11/449
air embolism, 11/437
aldosteronism, 11/458
allergic granulomatous angiitis, 11/453
arterial hypertension, 11/432
arterial hypotension, 11/433
atheromatosis, 11/437
brain embolism, 11/434
brain thromboangiitis obliterans, 11/454
cardiac catheter, 11/443
cardiac surgery, 11/442, 444
cause, 11/421-510
cervical rib, 11/429
collagen disease, 11/442
congenital heart disease, 11/447
cryoglobulinemia, 11/461
diabetes mellitus, 11/457
eclampsia, 11/459
electrical injury, 11/464
endocardial fibroelastosis, 11/441
endocarditis, 11/433
extracorporeal circulation, 11/443
Fabry disease, 11/464
fat embolism, 11/438
fibromuscular dysplasia, 11/454
fungal infection, 11/431
heart tumor, 11/441
Henoch-Schönlein purpura, 11/453
hereditary hemorrhagic telangiectasia (Osler), 11/465
homocystinuria, 11/462
hypercoagulation, 11/456
hyperlipidemia, 11/462
hypernatremia, 11/463
hyperosmolar coma, 11/462
hyperparathyroidism, 11/458
hypoglycemia, 11/457
hypothyroidism, 11/458
ionizing radiation, 11/463
ketotic coma, 11/463
Köhlmeier-Degos disease, 11/464
leukemia, 11/461
macroglobulinemia, 11/461
malaria, 11/431
migraine, 11/465
myeloma, 11/461
myocardial infarction, 11/436
myocardiopathy, 11/441
oral contraception, 11/457
paradoxical embolism, 11/441

pheochromocytoma, 11/458
polyarteritis nodosa, 11/451
polycythemia vera, 11/460
polymyalgia rheumatica, 11/452
postoperative bacterial endocarditis, 11/445
pregnancy, 11/459
pseudoxanthoma elasticum, 11/465
puerperium, 11/460
pulmonary arteriovenous fistula, 11/442
respiratory acidosis, 11/463
rheumatic fever, 11/454
rheumatic heart disease, 11/435
rheumatoid arthritis, 11/454
scleroderma, 11/453
sickle cell anemia, 11/460
syphilis, 11/430
systemic lupus erythematosus, 11/450
Takayasu disease, 11/448, 451
temporal arteritis, 11/452
thrombocytosis, 11/461
thrombotic thrombocytopenic purpura, 11/450
tumor embolism, 11/442
water intoxication, 11/463
Wegener granulomatosis, 11/453
Systemic carnitine deficiency
  carnitine cycle, 66/401
  carnitine deficiency, 62/493
  hepatic encephalopathy, 62/493
  metabolic encephalopathy, 62/493
  metabolic myopathy, 62/493
Systemic hamartosis, see Ullmann syndrome
Systemic hypertension, see Arterial hypertension
Systemic hypotension
  anoxic ischemic leukoencephalopathy, 9/576
  brain vasospasm, 11/124
  central cord necrosis, 61/114
  cervical spinal cord injury, 61/58
  myelomalacia, 61/112
  paraplegia, 61/114
  spinal anesthesia, 61/150
  spinal cord injury, 61/436
  tetraplegia, 61/436
  transverse spinal cord lesion, 61/436
Systemic infection
  aspergillosis, 52/377
  delirium, 46/544
  headache, 48/5
  intracranial infection, 24/228
Systemic late infantile lipidosis, see GM1
  gangliosidosis
Systemic lupus erythematosus, 55/369-382
  acute dysautonomia, 51/447
  akinetic mutism, 55/376

alopecia, 43/420
angiopathic polyneuropathy, 51/446
anterior ischemic optic neuropathy, 55/377
anticardiolipin antibody, 55/372
anticonvulsant, 55/381
antineuron antibody, 55/372
antiphospholipid antibody, 55/372, 63/122, 330,
  71/40
ataxia, 55/377
athetosis, 55/377
autoimmune disease, 22/24, 41/55, 59, 102, 275,
  55/370
autonomic polyneuropathy, 51/479
azathioprine, 71/49
benign intracranial hypertension, 39/286
brain angiography, 55/380
brain embolism, 53/156
brain hemorrhage, 39/280, 53/34, 54/294, 55/371
brain infarction, 53/156, 55/371, 374
brain ischemia, 53/156
brain stem, 39/281, 71/44
brain vasculitis, 55/371
carpal tunnel syndrome, 7/296, 51/447
catatonia, 55/376
cerebellar, 39/281
cerebral disorder, 39/279-281
cerebrovascular disease, 53/34
chorea, 55/374, 377
choreoathetosis, 49/385
chronic axonal neuropathy, 51/531
chronic demyelinating polyneuropathy, 51/447
chronic discoid, 8/213
chronic idiopathic demyelinating
  polyradiculoneuropathy, 47/614
chronic meningitis, 56/645
cinchonism, 65/452
classification, 71/4
clinical features, 51/447
coma, 55/376
computer assisted tomography, 54/63, 55/380,
  71/47
confusional state, 55/370
corticosteroid, 71/48
cranial neuropathy, 51/447, 55/374, 376
criteria, 71/36
cryoglobulinemia, 39/181
cryptococcosis, 52/433
CSF, 39/287, 43/420, 54/198, 55/381, 71/46
CSF examination, 11/239
cyclophosphamide, 55/381, 71/49
cytokine, 71/41
deafness, 55/377
definition, 39/274

delirium, 55/376
demyelination, 55/373
depression, 46/427
dermatomyositis, 62/374
diagnostic criteria, 55/379
drug induced, *see* Drug induced systemic lupus
   erythematosus
EEG, 39/287, 55/380, 71/46
epidemiology, 55/369, 71/35
epilepsy, 55/373, 73/378
epileptic seizure, 55/375
erythrocyte sedimentation rate, 55/379
ethosuximide, 55/370
ethosuximide intoxication, 65/508
etiology, 71/38
eye symptom, 55/377
fibrinoid necrosis, 71/38
focal cerebral lesion, 39/279-281
general features, 55/369
general outline, 63/54
general symptom, 55/370
Guillain-Barré syndrome, 51/447
hallucinosis, 55/376
headache, 39/286, 48/284, 55/374, 378, 71/44
hematology, 55/379
hematoxylin body, 11/139
hemiballismus, 49/373, 55/377
hemiplegia, 39/279, 43/420
history, 39/273
hydralazine neuropathy, 42/647
hyperviscosity syndrome, 55/485
immune complex vasculitis, 71/39
immunosuppression, 55/381
incidence, 71/36
inclusion body myositis, 62/374
internuclear ophthalmoplegia, 55/377
intestinal pseudo-obstruction, 51/493
intracranial hypertension, 16/220
intracranial pressure, 39/286
ischemic optic neuropathy, 51/447
LE cell, 43/420
Libman-Sacks endocarditis, 63/29
lupoid sclerosis, 55/378
lupus, 71/46
meningitis serosa, 39/285, 55/378
meningococcal meningitis, 52/23
mental function, 71/42
Mollaret meningitis, 56/629
mononeuritis multiplex, 51/447
movement disorder, 39/281, 71/45
multiple brain microinfarction, 55/381
multiple sclerosis, 47/71, 109, 329, 338, 55/373
myasthenia gravis, 62/402

myelomalacia, 55/102
myelopathy, 39/282, 55/378, 63/54, 71/45
myoglobinuria, 40/339, 41/275, 277
myopathy, 39/284
nephritis, 43/420
neurologic complication, 63/54
neurologic symptom percentage, 55/374
neuro-ophthalmology, 71/44
neuropathy, 7/555, 8/119, 11/139, 215, 450,
   39/283, 43/420, 71/44
nuclear magnetic resonance, 71/47
ophthalmoscopy, 11/215
optic atrophy, 55/377
orthostatic hypotension, 63/152
painful ophthalmoplegia, 55/377
pathogenesis, 39/274-276, 71/38
pathology, 39/275, 55/370
pellagra, 28/93
penicillamine intoxication, 49/235
phenytoin, 55/370
polyarteritis nodosa, 39/295, 307, 55/360
polymyositis, 8/122, 41/56, 59, 62/374
posterior spinal artery aneurysm, 63/54
procainamide, 63/191
procainamide intoxication, 65/452
prognosis, 55/381
psychiatric disorder, 43/420
psychosis, 39/277, 55/376
pure motor radiculopathy, 51/447
quinidine, 63/191
radiographic features, 39/288
Raynaud phenomenon, 43/71, 420
recurrent acute demyelinating neuropathy, 51/447
relapsing course, 55/373
retinal infarction, 55/381
Reye syndrome, 56/163
RNA antibody, 43/421
scanning, 71/47
schizophreniform psychosis, 46/420
seizure, 39/278, 43/420, 55/374-376, 71/43
sensorimotor polyneuropathy, 51/447
Sjögren syndrome, 71/74
spinal subarachnoid hemorrhage, 63/54
stroke, 71/44
subarachnoid hemorrhage, 12/111, 63/54
Sydenham chorea, 49/364
symptom, 71/42
systemic brain infarction, 11/450
thermoregulation, 75/60
thrombocytopenia, 55/379
transverse myelitis, 63/54
treatment, 39/289, 55/381, 71/48
tremor, 55/377

458

trigeminal neuropathy, 55/377
valproic acid intoxication, 65/505
vascular dementia, 46/359
vasculitis, 71/38
vasculopathy, 55/371
Systemic lupus erythematosus endocarditis
  see also Libman-Sacks endocarditis and
    Nonbacterial thrombotic endocarditis
  brain embolism, 63/122
  brain infarction, 63/122
  brain injury, 55/372
  neurologic complication, 63/121
  pathology, 63/121
Systemic lymphoma
  classification, 69/262
  human immunodeficiency virus infection, 71/341
  methotrexate, 69/273
  neurologic complication, 69/261
  primary CNS lymphoma, 68/265
  treatment, 69/269
Systemic malignancy
  epidemiology, 71/611
  nervous system, 71/611
Systemic sclerosis
  antiphospholipid antibody, 63/330
  autonomic nervous system, 74/120
  dermatomyositis, 62/371, 374
  diffuse cutaneous, see Diffuse cutaneous systemic
    sclerosis
  inclusion body myositis, 62/374
  intestinal pseudo-obstruction, 51/493
  polymyositis, 41/56, 59, 62/374
  progressive, see Progressive systemic sclerosis
  scleroderma, 71/100, 102
Systemic vasculitis
  aortitis, 63/51
  cause, 56/297
  hepatitis B virus, 56/301
  hepatitis B virus infection, 56/297
  immune complex, 56/297
  necrosis, 56/301
  polyarteritis nodosa, 39/296, 55/353, 359, 396
  type, 56/297
Systems
  acoustic, see Acoustic system
  adrenomedullary hormonal, see Adrenomedullary
    hormonal system
  alpha, see Alpha system
  arousal, see Arousal system
  arterial, see Arterial system
  ascending pharyngeal, see Ascending pharyngeal
    system
  ascending reticular activating, see Ascending

  reticular activating system
  auditory, see Auditory system
  autonomic, see Autonomic system
  autonomic nervous, see Autonomic nervous
    system
  axial mesodermal, see Axial mesodermal system
  basilar artery, see Basilar artery system
  biliary, see Biliary system
  brain command, see Brain command system
  cardiovascular, see Cardiovascular system
  central nervous, see CNS
  centrencephalic, see Centrencephalic system
  circadian, see Circadian system
  comb, see Comb system
  complement, see Complement system
  dentatorubral, see Dentatorubral system
  digestive, see Digestive system
  Edinger Comb, see Edinger comb system
  enteric nervous, see Enteric nervous system
  extrapyramidal, see Extrapyramidal system
  fastigovestibular, see Fastigovestibular system
  fibrinolytic, see Fibrinolytic system
  frameless, see Frameless system
  gamma, see Gamma system
  gamma efferent, see Gamma efferent system
  gate control, see Gate control system
  glycine cleavage enzyme, see Glycine cleavage
    enzyme system
  Golgi, see Golgi system
  HLA, see HLA system
  HLA histocompatibility antigen, see HLA
    histocompatibility antigen system
  hypothalamic hypophyseal, see Hypothalamic
    hypophyseal system
  hypothalamic hypophyseal adrenal, see
    Hypothalamic hypophyseal adrenal system
  immune, see Immune system
  incertohypothalamic, see Incertohypothalamic
    system
  inner mitochondrial membrane, see Inner
    mitochondrial membrane system
  kallikrein-kinin, see Kallikrein-kinin system
  lacrimal, see Lacrimal system
  limbic, see Limbic system
  Mendelejev periodic, see Mendelejev periodic
    system
  mesolimbic dopamine, see Mesolimbic dopamine
    system
  mesolimbic noradrenalin, see Mesolimbic
    noradrenalin system
  middle meningeal, see Middle meningeal system
  mitochondrial matrix, see Mitochondrial matrix
    system

motor, *see* Motor system

nervous, *see* Nervous system

nigrostriatal dopamine, *see* Nigrostriatal dopamine system

noradrenergic nervous, *see* Noradrenergic nervous system

NOSPECS, *see* NOSPECS system

oculomotor, *see* Oculomotor system

olfactory, *see* Olfactory system

pallidoluysian, *see* Pallidoluysian system

parasympathetic, *see* Parasympathetic system

parasympathetic nervous, *see* Parasympathetic nervous system

peripheral nervous, *see* Peripheral nervous system

periventricular, *see* Periventricular system

protopathic motor, *see* Protopathic motor system

pyruvate dehydrogenase, *see* Pyruvate dehydrogenase system

renin-angiotensin, *see* Renin-angiotensin system

reticular activating, *see* Reticular activating system

reticuloendothelial, *see* Reticuloendothelial system

ribosomal, *see* Ribosomal system

superficial brain venous, *see* Superficial brain venous system

sympathetic nervous, *see* Sympathetic nervous system

T tubule, *see* T tubule system

tectoreticular, *see* Tectoreticular system

trigeminovascular, *see* Trigeminovascular system

trophotropic, *see* Trophotropic system

tuberoinfundibular, *see* Tuberoinfundibular system

ventricular, *see* Ventricular system

vertebrobasilar, *see* Vertebrobasilar system

vestibular, *see* Vestibular system

visual sensory, *see* Visual sensory system

Systox, *see* Demeton

T-cell, *see* T-lymphocyte

T-helper cell
  multiple sclerosis, 47/189

T-helper lymphocyte
  human immunodeficiency virus, 56/490
  retrovirus, 56/529

T-helper/T-suppressor cell
  CSF, *see* CSF T-helper/T-suppressor cell
  multiple sclerosis, 47/85

T-lymphocyte
  brain abscess, 65/548
  CD3+, 56/67
  CD4+, 56/67

CD8+, 56/67

CD8+ cytotoxic, *see* CD8+ cytotoxic T-lymphocyte

CSF, *see* CSF T-lymphocyte

cytotoxic, *see* Cytotoxic T-lymphocyte

encephalitis, 65/548

experimental allergic encephalomyelitis, 47/234, 434, 442, 449

Guillain-Barré syndrome, 47/611, 51/248

helper, *see* T-helper cell

Listeria monocytogenes, 52/90

lymphocytic choriomeningitis virus, 56/358

malignant lymphoma, 18/325

meningitis, 65/548

meningoencephalitis, 65/548

multiple sclerosis, 47/68, 81, 232, 234, 341-345

Mycobacterium leprae, 51/215

neutropenia, 63/340

progressive multifocal encephalitis, 65/548

Sjögren syndrome, 51/449

suppressor, *see* T-suppressor lymphocyte

viral infection, 56/66

T-lymphocyte leukemia
  adult, *see* Adult T-cell leukemia
  retrovirus, 56/529

T-lymphocyte lymphoma
  acquired immune deficiency syndrome, 56/489
  human immunodeficiency virus, 56/489
  non-Hodgkin lymphoma, 69/267

T. pallidum, *see* Treponema pallidum

T-suppressor lymphocyte
  multiple sclerosis, 47/189, 342, 347
  viral infection, 56/71

T tubule
  hereditary cylindric spirals myopathy, 62/348
  muscle fiber, 62/23, 37
  muscle fiber type, 62/3
  periodic paralysis, 41/154

T tubule system
  myotonia, 62/277

T3 suppression test
  progressive external ophthalmoplegia, 22/190, 204
  thyroid associated ophthalmopathy, 62/531

T4, *see* Thyrotropic hormone

TAB vaccination
  neuropathy, 8/88

Tabernathe iboga intoxication
  confusion, 37/329
  dysphoria, 37/329
  hallucination, 37/329

Tabes dorsalis
  see also Neurosyphilis *and* Syphilis

Abadie sign, 52/280
anticonvulsant, 45/230
Argyll Robertson pupil, 19/86, 33/364, 52/280
ataxia, 33/363, 52/280
autonomic polyneuropathy, 51/486
Biernacki sign, 52/280
clinical features, 33/363-365, 52/279
deafness, 52/280
Herxheimer reaction, 52/280
lightning pain, 8/201, 33/363, 52/280
lymphocytic meningitis, 52/280
malperforans ulcer, 52/280
neurogenic osteoarthropathy, 19/86, 52/280
neurosyphilis, 33/362-365, 52/277
neurotrophic osteoarthropathy, 38/453-456
nociceptor afferent, 45/230
optic atrophy, 13/53-56, 19/86, 52/280
orthostatic hypotension, 1/472, 2/120, 22/262
pain, 19/86, 45/230
pathology, 33/363
penicillin, 52/280
prognosis, 33/385
pseudoathetosis, 52/280
pupillary abnormality, 33/364
Refsum disease, 60/235, 66/497
spheroid body, 6/626
spinal cord compression, 19/377
sympathetic nervous disorder, 2/120
tabetic face, 52/280
trophic disorder, 33/365
urinary incontinence, 33/363, 52/280
Von Romberg sign, 52/280
Tabetic ataxia
   clinical features, 1/311-313
Tabetic glossodynia
   glossopharyngeal neuralgia, 48/468
Table tennis
   rehabilitative sport, 26/525
Tabun
   chemical structure, 64/225
   neurotoxin, 37/544, 64/152, 223
   organophosphate, 64/152
   organophosphate intoxication, 64/152, 223
Tabun intoxication
   chemical classification, 37/544
   organophosphorus compound, 37/544
Tacaribe virus
   murine encephalopathy, 56/358
Tachycardia
   amphetamine intoxication, 65/257
   autonomic nervous system, 74/614
   brain air embolism, 55/193
   brain fat embolism, 55/186

brain hemorrhage, 55/152
brain infarction, 55/152
breath holding syncope, 15/828
caffeine intoxication, 65/260
carbon monoxide intoxication, 64/31
chelonia intoxication, 37/89
cingulate gyrus, 2/767
critical illness polyneuropathy, 51/576, 581
diabetes mellitus, 75/597
diphtheria, 51/183, 65/241
ephedrine intoxication, 65/260
Guillain-Barré syndrome, 7/499, 51/245, 257
hereditary acute porphyria, 51/391
hyperventilation syndrome, 38/319, 63/436
insula, 63/205
mackeral intoxication, 37/87
malignant hyperthermia, 38/551, 41/268, 43/117,
   62/567, 63/439
mercury polyneuropathy, 51/273
migraine, 48/156
nereistoxin intoxication, 37/55
nicotine intoxication, 36/425, 429, 65/262
organophosphate intoxication, 64/228
oxidation-phosphorylation coupling defect,
   62/506
paroxysmal, see Paroxysmal tachycardia
paroxysmal supraventricular, see Paroxysmal
   supraventricular tachycardia
paroxysmal ventricular, see Paroxysmal
   ventricular tachycardia
pheochromocytoma, 14/252, 39/498, 48/427
pickwickian syndrome, 3/96, 38/300
progressive bulbar palsy, 22/114, 59/219
spinal cord injury, 26/320, 61/436
test, 1/467
tetraplegia, 61/436
transverse spinal cord lesion, 61/436
Tachykinin
   reflex sympathetic dystrophy, 61/126
Tachyphemia, see Tachyphrasia
Tachyphrasia
   cluttering, 43/220
Tachyphylaxis
   Shy-Drager syndrome, 22/268
Tachypleus gigas intoxication
   horseshoe crab, 37/58
Tachypleus tridentatus intoxication
   horseshoe crab, 37/58
Tachypnea
   see also Central neurogenic hyperventilation
   brain fat embolism, 55/183
   capnography, 1/665
   carbon monoxide intoxication, 64/31

classification, 1/657
congenital hyperammonemia, 42/560
hypopnea, 1/665
methanol intoxication, 64/98
sleep, 63/484
Tacrolimus, *see* Tsukubaenolide
Tactile agnosia, 4/32-37
  ahylognosia, 45/66
  amorphognosia, 45/66
  amytal test, 4/36
  anatomic localization, 4/36
  aphasia, 4/34
  apraxia, 4/34
  asymbolia, 4/34
  bedside examination, 45/66
  bilateral, 4/35
  cerebral palsy, 4/34
  characteristics, 4/22
  clinical features, 4/33-35
  congenital, *see* Congenital tactile agnosia
  corpus callosum, 4/35
  examination, 4/35
  graphesthesia, 45/66
  hand, 4/35
  higher nervous activity, 4/14
  parietal lobe lesion, 4/36
  parietal lobe syndrome, 45/66
  parietal lobe tumor, 17/300
  pathogenesis, 4/36
  primary, *see* Primary tactile agnosia
  recognition, 4/33, 279
  sensory extinction, 4/34
  somatosensory, 4/33
  tactile amytal test, 4/36
  terminology, 4/14
  test technique, 1/104
  thalamic syndrome (Dejerine-Roussy), 2/485
Tactile amytal test
  tactile agnosia, 4/36
Tactile anomia
  anterior cerebral artery syndrome, 53/347
  brain infarction, 53/347
  corpus callosum syndrome, 53/347
  unilateral, *see* Unilateral tactile anomia
Tactile aphasia
  touch, 4/34
Tactile apraxia
  developmental disorder, 4/34
  parietal lobe lesion, 4/34
Tactile asymbolia
  parietal lobe syndrome, 45/66
Tactile discrimination
  technique, 1/88

Tactile extinction, *see* Sensory extinction
Tactile hallucinosis
  chronic, *see* Chronic tactile hallucinosis
Tactile inattention
  see also Sensory extinction
  parietal lesion, 2/683
  parietotemporo-occipital area, 3/189
  sensory extinction, 45/65
Tactile pathway
  anatomy, 1/88
Tactile perception
  neuropsychology, 45/521
  parietal lobe syndrome, 45/79
Tactile recognition
  terminology, 4/33
  test, 4/279
Tactile recognition disorder, *see* Tactile agnosia
Tactile sensation
  class, 1/87
  hereditary sensory and autonomic neuropathy type
    II, 60/11
Taenia solium
  see also Neurocysticercosis
  acquired immune deficiency syndrome, 71/307
  cestode, 52/529
  epilepsy, 72/147, 167
  helminthiasis, 52/506, 511
  neurocysticercosis, 16/225, 35/296-298, 46/389,
    52/529
  spinal muscular atrophy, 59/370
Taeniasis
  chronic meningitis, 56/645
Tahyna virus
  California encephalitis, 56/140
Taille de guèpe
  congenital atonic sclerotic muscular dystrophy,
    41/44
Taipoxin
  autonomic nervous system, 75/512
  chemistry, 65/184
  neurotoxin, 37/17, 62/610, 65/182, 184
  snake envenomation, 65/182
  toxic myopathy, 62/610
Takatsuki syndrome, *see* POEMS syndrome
Takayasu aortic arch arteritis, *see* Takayasu disease
Takayasu arteritis, *see* Takayasu disease
Takayasu disease
  see also Temporal arteritis
  amyloidosis, 12/413
  aneurysm, 12/414
  angiography, 12/406, 71/202
  angiopathic polyneuropathy, 51/446
  aorta disease, 63/37

aortitis, 63/52
arm symptom absence, 63/52
arteriocirrhosis, 12/410
arteriosclerosis, 12/414
arthralgia, 55/337
blunt trauma, 12/415
brain arteritis, 12/411
brain infarction, 53/162
brain ischemia, 53/162
cardiac symptom, 63/52
cardiovascular manifestation, 39/218
carotid sinus hypersensitivity, 39/213-232, 217
carotid system syndrome, 53/308
cerebrovascular disease, 53/41
classification, 55/335, 71/4
clinical features, 12/400, 405, 39/214-220, 55/337
collateral circulation, 12/407, 39/221-223
congenital abnormality, 12/415
corticosteroid, 55/339
cranial neuropathy, 51/452, 63/53
cyclophosphamide, 55/339
diagnosis, 39/230
differential diagnosis, 55/335
embolism, 12/415
epidemiology, 55/335
epilepsy, 55/337
erythrocyte sedimentation rate, 55/338
etiology, 12/409, 46/359, 55/336
fever, 55/337
granulomatous CNS vasculitis, 55/395
headache, 5/149, 48/284, 55/337, 71/202
history, 12/398, 39/213
hypertensive encephalopathy, 55/337
infantile hypertension, 63/53
laboratory finding, 39/221
Leriche syndrome, 63/53
migraine, 5/56
morbid anatomy, 12/409
myalgia, 55/337
myelopathy, 63/53
neurologic features, 12/404, 39/216
ophthalmologic features, 12/404, 39/215, 46/359
optic atrophy, 13/60, 83
optic disc, 55/335
partial seizure, 63/53
pathogenesis, 12/409, 39/229, 55/336
pathology, 12/409, 39/223-229, 55/336
polyarteritis nodosa, 12/412, 46/359
pregnancy, 39/220
prevalence, 55/335
prodromal state, 12/400
prognosis, 12/416
pseudoxanthoma elasticum, 12/413, 55/452

radiologic features, 39/220
retinal symptom, 63/53
rheumatoid arthritis, 63/55
rheumatoid symptom, 63/53
Sjögren syndrome, 71/74
spinal cord, 63/53
stroke, 63/53
subclavian artery stenosis, 53/372
subclavian steal syndrome, 39/218, 55/338,
    71/202
survey, 71/191
symptom, 71/202
syncope, 7/453, 55/337, 71/202
syphilis, 12/413, 63/52
systemic brain infarction, 11/448, 451
temporal arteritis, 12/412
thromboangiitis obliterans, 12/411, 46/359
total aortography, 63/53
transient ischemic attack, 71/202
treatment, 12/416, 39/231, 55/338
trophic disorder, 12/405
tuberculosis, 12/414
vascular compression, 12/416
vascular dementia, 46/357-359, 358
vascular surgery, 55/339
vasculitis, 71/8
weight loss, 55/337
Talairach AC-PC quideline
    status epilepticus, 15/760
    stereoencephalotomy, 15/760
Talipes
    adducted thumb syndrome, 43/331
    arthrogryposis multiplex congenita, 43/78
    Bowen syndrome, 43/335
    central core myopathy, 43/80
    cerebrohepatorenal syndrome, 43/339
    congenital spondyloepiphyseal dysplasia, 43/478
    corpus callosum agenesis, 42/7
    craniolacunia, 50/139
    cri du chat syndrome, 43/501
    diastematomyelia, 42/24
    diastrophic dwarfism, 43/382-384
    distal myopathy, 43/114
    flexion contracture, 20/827
    Freeman-Sheldon syndrome, 38/413, 43/353
    Friedman-Roy syndrome, 43/251
    Friedreich ataxia, 21/320
    Harvey syndrome, 42/223
    hereditary motor and sensory neuropathy type I,
        21/278, 42/339
    hereditary motor and sensory neuropathy type II,
        21/278, 42/339
    holoprosencephaly, 50/237

inability to open mouth, 43/434
incontinentia pigmenti, 43/23
Lowry syndrome, 43/368
Möbius syndrome, 42/324
monosomy 21, 43/540
Muckle-Wells syndrome, 42/396
myopathic arthrogryposis multiplex, 43/78
myopathy, 40/337
nemaline myopathy, 43/122
nystagmus myoclonus, 42/240
optic atrophy, 42/406, 412-414, 417
4p partial trisomy, 43/500
10p partial trisomy, 43/519
Patau syndrome, 43/528
Pena-Shokeir syndrome type I, 43/437, 59/71
pilonidal sinus, 42/47
Prader-Labhart-Willi syndrome, 43/463
prune belly syndrome, 43/466
18q partial monosomy, 43/533
6q partial trisomy, 43/503
11q partial trisomy, 43/524
r(22) syndrome, 43/547
Richards-Rundle syndrome, 22/516, 43/264
Richner-Hanhart syndrome, 42/581
Roussy-Lévy syndrome, 42/108
Ryukyan spinal muscular atrophy, 42/93
Seckel syndrome, 43/379
sensory radicular neuropathy, 42/352
spastic paraplegia, 42/169, 171
spina bifida, 42/54
spinocerebellar degeneration, 42/190
spondyloepiphyseal dysplasia, 43/478
Troyer syndrome, 42/193
Talipes cavus, see Pes cavus
Talipes equinovarus
anencephaly, 50/84
myopathy, 40/337, 427, 480
osteotomy, 40/365
Tall stature
Klinefelter variant XXXY, 31/483, 43/559
Klinefelter variant XXYY, 43/562
Tamiami virus
cerebellar heterotopia, 56/358
Tamoxifen
adverse effect, 69/499
antineoplastic agent, 65/529
encephalopathy, 65/529
neurotoxin, 39/95, 65/529, 538, 67/363
optic neuropathy, 65/529
Tamoxifen intoxication
optic neuropathy, 65/538
Tamoya intoxication
carybdeidae, 37/43

local pain, 37/43
Tamsulosin
classification, 74/149
Tana poxvirus
neurotropic virus, 56/5
Tanderil, see Oxyphenbutazone
Tangier disease
apolipoprotein AI, 42/628
autonomic nervous system, 42/627, 75/32
autosomal recessive, 10/548, 21/8
cholesterol ester, 66/16
cholesterol metabolism, 10/278, 42/627
clinical features, 60/133
differential diagnosis, 60/19
electrophysiologic study, 60/135
foam cell, 10/548, 42/627
gene frequency, 42/627
hepatosplenomegaly, 10/278, 42/627
hereditary neuropathy, 21/8
high density lipoprotein, 10/269, 278, 547, 42/627,
66/549
lipid metabolic disorder, 10/278, 21/12
multiple mononeuropathy, 60/135
neuropathology, 60/135
neuropathy, 8/20, 10/547, 21/52, 42/627
orthostatic hypotension, 63/156
pathology, 66/550
polyneuropathy, 8/20, 10/548, 21/52, 42/627,
60/135, 165
prevalence, 42/627
pseudosyringomyelic type, 60/134
Refsum disease, 21/215
sensory loss, 10/548, 42/627
splenectomy, 8/21
splenomegaly, 10/547
syringomyelia, 50/452
tonsil coloration, 42/627
transport defect, 10/136
Whipple disease, 8/21
yellow orange tonsil, 10/547, 60/133
Tapetal reflex
chorioretinal degeneration, 13/27
female, 13/27, 155
ophthalmoscopy, 13/155
Tapetoretinal amaurosis
congenital, see Congenital retinal blindness
Tapetoretinal degeneration, see Primary pigmentary
retinal degeneration and Secondary pigmentary
retinal degeneration
Tapetoretinal degeneration with marginal corneal
dystrophy (Bietti)
misdiagnosis, 13/278
Tapetoretinal dystrophy

central, *see* Stargardt disease
diffuse, *see* Centroperipheral tapetoretinal
  dystrophy
Tapia syndrome
  medulla oblongata syndrome, 2/224
  multiple cranial neuropathy, 2/99
  nonrecurrent nonhereditary multiple cranial
    neuropathy, 51/571
  tongue paralysis, 2/102
  vocal cord paralysis, 2/102
Tapioca
  cyanogenic glycoside intoxication, 65/27
Tapping test
  premotor cortex, 2/732
Tarantulism
  history, 6/300
Tardive amaurotic idiocy
  epilepsy, *see* Dumon-Radermecker syndrome
Tardive dyskinesia
  agraphia, 45/466
  akathisia, 49/186
  amoxapine intoxication, 65/319
  animal model, 49/190
  antipsychotic agent intoxication, 37/307-312
  axial dystonia, 49/186
  biogenic amine, 29/476-478
  chlorpromazine intoxication, 37/307-312
  chorea, 49/186
  classification, 49/186
  clinical features, 49/185
  differential diagnosis, 49/187
  dopamine, 29/476-478
  dopamine decreased turnover, 49/189
  dopamine hypersensitivity, 49/191
  drug induced movement disorder, 49/185
  dyspnea, 63/482
  dystonia, 49/186
  epidemiology, 49/188
  focal dystonia, 49/186
  heterocyclic antidepressant, 65/319
  methyldopa, 37/440
  milking grip, 49/187
  muscarinic cholinergic nerve cell, 49/189
  myoclonus, 49/186
  neuroleptic agent, 49/186
  neuropathology, 49/190
  papaverine, 37/456
  phantom limb, 45/397
  pharmacology, 49/189
  phenothiazine intoxication, 37/307-312
  prevention, 49/192
  respiration, 63/482
  segmental dystonia, 49/186

  spontaneous, 49/187
  tic, 49/186
  treatment, 49/188, 192
Tardive myopathy
  adult acid maltase deficiency, 62/480
Tardy neuritis
  carpal tunnel syndrome, 7/260
  ulnar, *see* Tardy ulnar paralysis
Tardy ulnar paralysis
  ulnar nerve, 7/434, 440
Target cell
  JHM virus, 56/445
  multiple sclerosis, 47/249
Target fiber
  central core, 40/244, 41/5
  denervation, 40/31, 75, 142, 243
  description, 40/75
  mitochondria, 40/91
  muscle fiber type I, 40/13
  muscle regeneration, 40/13
  muscle reinnervation, 40/145
  tenotomy, 40/144
  Z band, 40/215
Targetoid fiber
  organophosphate intoxication, 40/80
Taricha egg
  tetrodotoxin, 37/80
Tarichotoxin, *see* Tetrodotoxin
Tarlov cyst, *see* Spinal perineural cyst
Tarsal tunnel syndrome
  anatomy, 51/105
  anterior, *see* Anterior tarsal tunnel syndrome
  compression neuropathy, 51/105
  foot injury, 70/42
  rheumatoid arthritis, 38/484, 71/18
  toxic shock syndrome, 52/259
Tarui disease, *see* Phosphofructokinase deficiency
Tasikinesia
  neuroleptic symptom, 6/253
Task specific tremor
  orthostatic tremor, 49/592
  primary writing tremor, 49/591
  vocal tremor, 49/592
Taste
  see also Ageusia
  disulfiram intoxication, 37/321
  facial nerve injury, 24/111
  indifference to pain, 8/195
  radiation effect, 7/397
  sensory extinction, 3/191
  smell, 24/20
  testing method, 2/63
Taste blindness

mendelian threshold detection, 42/422
Taurine
  amino acid neurotransmitter, 29/514-516
  biochemical consideration, 29/514
  brain, 9/16
  clinical consideration, 29/515
  CSF, 42/255
  epilepsy, 29/515
  Friedreich ataxia, 42/144
  mental depression, 42/255
  myotonia, 62/276
  physiologic consideration, 29/514
  toluene intoxication, 64/50
Taurine deficiency
  depression, 29/515
  levodopa, 42/256
  Parkinson disease, 29/515
  parkinsonism, 29/515, 42/255
  tyrosine hydroxylase, 42/256
Taxiphyllin
  cyanogenic glycoside intoxication, 65/26
Taxol
  adverse effect, 69/499
  antineoplastic agent, 65/529
  autonomic nervous system, 75/38, 497
  chemotherapy, 67/284
  epilepsy, 65/529
  neuropathy, 69/460, 467, 469
  neurotoxin, 65/529, 537
  polyneuropathy, 65/529
Taxol intoxication
  autonomic polyneuropathy, 65/537
  axonal polyneuropathy, 65/537
  sensory polyneuropathy, 65/537
  treatment, 65/537
Taxotere
  adverse effect, 69/499
  antineoplastic agent, 65/529
  chemotherapy, 67/284
  neuropathy, 69/460, 467, 470
  neurotoxin, 65/529, 537
  polyneuropathy, 65/529
Taxotere intoxication
  autonomic polyneuropathy, 65/537
  axonal polyneuropathy, 65/537
  sensory polyneuropathy, 65/537
  treatment, 65/537
Tay-Sachs disease, see GM2 gangliosidosis
Tay-Sachs disease and visceral involvement, see
    GM1 gangliosidosis
Tay, W., 10/226
Taybi-Linder syndrome
  convulsion, 43/386

corpus callosum agenesis, 43/386
  dwarfism, 43/385
  microcephaly, 43/385
  neuronal heterotopia, 43/386
  seizure, 43/386
  vertebral abnormality, 43/385
TCE, see Trichloroethylene
Tea and toast diet
  vitamin Bc deficiency, 51/336
Tear
  see also Lacrimal function
  arachnoid, see Arachnoid tear
  disorder, 74/428
  methacholine, 74/427
  pilocarpine, 74/427
  psychogenic, 74/427
  reflex, 74/427
  Schirmer tear test, 74/429
  test, 74/428
  trigeminal nerve, 74/427
Tectocerebellar dysraphia
  cerebellar agenesis, 50/177
Tectoreticular system
  conjugate eye movement, 3/24
  higher nervous activity, 3/24
Tectum beaking
  mesencephalic, see Mesencephalic tectum beaking
Teenage disc prolapse
  lumbar intervertebral disc prolapse, 20/594
Tegmental area
  ventral, see Ventral tegmental area
Tegmental atrophy
  dentatorubropallidoluysian atrophy, 49/442
Tegmental pontine hemorrhage
  frequency, 12/43
  symptom, 54/313
Tegmental pontine syndrome
  caudal tegmental syndrome, 12/20
  classification, 12/22
  convulsive seizure, 12/20
  dorsocaudal type, see Foville syndrome
  dorsolaterocaudal type, see Gasperini syndrome
  Foville syndrome, 12/20
  rostral tegmental syndrome, 12/20
  superior cerebellar artery syndrome, 12/20
  symptom, 2/309, 317
  ventrocaudal type, see Millard-Gubler syndrome
Tegmental subthalamic fiber
  nucleus accumbens, 49/11
  pedunculopontine tegmental nucleus, 49/11
  subthalamic nucleus, 49/11
Tegmental tract
  central, see Central tegmental tract

Tegmentum
    astrocytoma, 17/633
    brain stem, *see* Brain stem tegmentum
    tuberculoma, 17/633
Tegmentum syndrome
    caudal, *see* Caudal tegmental syndrome
    mesencephalic, *see* Mesencephalic tegmentum
        syndrome
    pontine, *see* Tegmental pontine syndrome
    rostral, *see* Rostral tegmental syndrome
Tegretol, *see* Carbamazepine
Teichopsia
    hallucination classification, 45/56
Telangiectasia
    angiomatous malformation, 14/52
    ataxia, *see* Ataxia telangiectasia
    Bloom syndrome, 14/775, 43/10
    brain, 12/104, 232, 14/520, 54/365
    brain arteriovenous malformation, 12/232, 54/364,
        375
    brain vascular tumor, 18/269
    cerebellar hemorrhage, 12/55
    corticomedullary junction, 18/270
    cryptic hemangioma, 12/104
    cryptic type, 54/375
    dyskeratosis congenita, 14/784, 43/392
    genetics, 14/520
    Goltz-Gorlin syndrome, 43/198
    hereditary, *see* Hereditary telangiectasia
    hereditary hemorrhagic, *see* Hereditary
        hemorrhagic telangiectasia (Osler)
    histopathology, 14/72
    moyamoya disease, 53/32, 55/293
    multiple interstitial, *see* Multiple interstitial
        telangiectasia
    neurocutaneous melanosis, 14/424, 597
    neuronal intranuclear hyaline inclusion disease,
        60/666
    neuropathology, 14/51-53
    pia, 14/424
    pontine syndrome, 2/259, 18/270
    progressive familial choreoathetosis, 14/101
    progressive oculocutaneous, *see* Progressive
        oculocutaneous telangiectasia
    retinal, *see* Retinal telangiectasia
    Rothmund-Thomson syndrome, 14/777, 43/460
    spinal cord, 14/520, 19/15
    twin study, 14/520
    Ullmann syndrome, 14/450
Telangiectatic angiomatosis
    hereditary, *see* Hereditary hemorrhagic
        telangiectasia (Osler)
Telecanthus

adducted thumb syndrome, 43/331
    hypertelorism hypospadias syndrome, 43/413
    Pena-Shokeir syndrome type I, 43/438
Telegram style
    aphasia, 4/90
Telencephaloschisis, *see* Schizencephaly
Telencephalosynapsis, *see* Holoprosencephaly
Teleopsia
    hallucination classification, 45/56
    localization value, 2/614
    migraine, 48/162
Telescoping
    phantom limb, 4/226, 45/397
Television hallucination
    hallucinosis, 46/567
Telfer syndrome
    ataxia, 22/513
    deafness, 60/658
    hereditary hearing loss, 22/513
    Klein-Waardenburg syndrome, 22/513
    mental deficiency, 22/514
    piebald pigmentation, 22/513
Tellurium intoxication
    demyelination, 9/645
    leukoencephalopathy, 9/645
    liver glia, 9/646
Telodendrion, *see* Motor end plate
Temazepam
    benzodiazepine intoxication, 65/335
Temefos
    organophosphate intoxication, 64/155
Temephos, *see* Temefos
Temperature
    ambient, *see* Ambient temperature
    biological rhythm, 74/492
    body, *see* Thermoregulation
    corneal, *see* Corneal temperature
    glycogen, 27/210
    migraine, 48/162
    multiple sclerosis, 47/151
    nerve conduction, 7/132
    olfaction, 74/363
    pressure sore, 61/349
    skin, *see* Skin temperature
    stress, 74/319
Temperature regulation, *see* Thermoregulation
Temperature reversal
    benzodiazepine withdrawal, 65/340
    ciguatoxin intoxication, 65/162
    marine toxin intoxication, 65/158
    neurotoxic shellfish poisoning, 65/158
    paralytic shellfish poisoning, 65/158
Temperature sensation

type, 1/86
Temperature sensitivity
  defect, 8/200
Temperature sensory loss
  amyloid polyneuropathy, 51/416
Temporal agnosia
  definition, 2/606
Temporal arteriography
  temporal arteritis, 48/324
Temporal arteritis, 39/313-339, 55/341-347
  see also Giant cell arteritis, Polymyalgia
    rheumatica *and* Takayasu disease
  abducent nerve paralysis, 48/313
  acquired immune deficiency syndrome, 56/499
  acute amaurosis, 55/342
  acute ischemic papillopathy, 48/312
  age, 48/325, 55/341
  age distribution, 39/315
  albumin, 48/321
  allergic granulomatous angiitis, 55/346
  amaurosis, 55/341
  anemia, 55/342
  angiopathic polyneuropathy, 51/446
  anorexia, 48/309
  anterior ischemic optic neuropathy, 55/114
  antiphospholipid antibody, 63/330
  Anton syndrome, 48/316
  aortic involvement, 63/54
  aortitis, 11/452, 48/316, 63/54
  arteriosclerosis, 48/322
  autoimmune disease, 55/341
  azathioprine, 48/324
  blindness, 11/133, 51/452
  brain, 5/121
  brain angiography, 55/344
  brain arteritis, 55/343
  brain infarction, 48/314, 319, 323, 53/162, 55/343
  brain ischemia, 48/314, 53/162
  brain stem infarction, 48/314
  carotid infarction, 48/315
  carotid sinus hypersensitivity, 55/344
  carotid system syndrome, 53/308
  carotidynia, 48/338
  cataract, 48/319
  central retinal artery, 48/318
  central retinal artery occlusion, 48/313
  cerebellar infarction, 55/344
  cerebrovascular disease, 53/41
  chronic migrainous neuralgia, 48/250
  ciliary artery, 48/318
  clinical features, 39/313-326, 41/48, 55/342
  confusional state, 55/344
  cortical blindness, 55/344
  corticosteroid, 48/323
  CSF, 55/344
  cytotoxic T-suppressor cell, 55/346
  death, 48/326
  definition, 55/341
  dementia, 55/344
  diabetes insipidus, 55/344
  diamorphine addiction, 55/518
  differential diagnosis, 55/346
  diplopia, 48/313
  Doppler ultrasonography, 55/344
  epidemiology, 55/341
  epilepsy, 55/344
  erythrocyte sedimentation rate, 39/336, 48/10,
    51/452, 55/344
  external carotid artery, 48/322
  eye muscle, 48/314
  facial nerve, 55/342
  familial, 48/309, 55/341
  fatigue, 55/342
  features, 51/451
  femoral artery, 48/316
  fever, 55/342
  fibrinogen, 48/321
  gaze paralysis, 55/344
  general outline, 63/54
  girdle pain, 55/342
  glaucoma, 5/120
  α-2-globulin, 55/342
  glycoregulation, 55/343
  granulomatous arteritis, 48/320
  granulomatous CNS vasculitis, 55/346
  hallucination, 55/344
  headache, 5/149, 39/316-319, 48/6, 77, 284, 310,
    316, 55/341, 342
  hemoglobin, 48/320
  hepatic arteritis, 48/317
  herpes zoster, 55/346
  histology, 11/133, 55/345
  history, 39/313
  HLA-DR4 antigen, 55/346
  immune disease, 55/345
  immunosuppressive agent, 48/324
  increased erythrocyte sedimentation rate, 55/342
  internal carotid artery stenosis, 48/314
  iris rubeosis, 48/319
  ischemic keratopathy, 48/319
  ischemic myelopathy, 55/342, 344
  ischemic optic neuropathy, 51/452
  ischemic papillopathy, 48/315
  ischemic syndrome, 39/319-326
  ischemic visual loss, 39/326
  juvenile hypereosinophilic temporal arteritis,

55/345, 347
juvenile temporal arteritis, 55/341
laboratory investigation, 48/320
lateral medullary infarction, 55/344
Leptospira interrogans, 55/415
leukoencephalopathy, 9/609
liver function test, 48/321
localization, 55/345
mandibular intermittent claudication, 48/310
mesenteric artery, 48/317
migraine, 48/77
mononeuritis multiplex, 51/452
mononeuropathy, 51/452
multiple cranial neuropathy, 51/452
multiple myeloma, 48/320
myelopathy, 63/54
myocardial infarction, 48/316
normal pressure hydrocephalus, 55/344
occipital infarction, 48/315
ocular pain, 48/312
oculomotor nerve, 55/342
oculomotor paralysis, 48/313
ophthalmic artery, 11/133
ophthalmoplegia, 5/120
optic atrophy, 13/60, 82
optic disc, 48/312
optic nerve, 48/312, 319, 326
optic nerve infarction, 48/317
orbital arteritis, 48/318
orbital cellulitis, 48/312
orbital pain, 48/314
pain, 39/316-319, 48/312
paratrigeminal syndrome, 48/337
pathology, 39/330-333, 48/322
plasma protein, 48/321
polyarteritis nodosa, 48/317, 55/346
polymyalgia, 48/311, 326
polymyalgia rheumatica, 5/119, 48/319, 55/341
prevalence, 48/313
prognosis, 48/324, 55/345
Raynaud disease, 48/316
renal artery inflammation, 48/317
retinal ischemia, 48/318
rheumatic polymyalgia, 39/327-330
rheumatoid arthritis, 48/320
scalp artery, 48/310
scalp ischemic necrosis, 48/311
scalp necrosis, 55/342
scalp tenderness, 48/310
secondary syphilis, 55/347
sex ratio, 48/325, 55/341
Sjögren syndrome, 71/74
steroid, 48/313, 323, 325

steroid administration, 48/10
subclavian artery occlusion, 48/317
subclavian steal syndrome, 55/343
survival, 48/325
syncope, 55/344
systemic brain infarction, 11/452
Takayasu disease, 12/412
temporal arteriography, 48/324
temporal artery biopsy, 39/333-336, 48/10, 320,
    55/344, 346
temporomandibular joint, 39/318
terminology, 5/119
Tolosa-Hunt syndrome, 48/302
tongue necrosis, 55/342
tongue swelling, 48/311
treatment, 48/313, 323, 55/347
vasa nervorum, 48/314
vascular dementia, 46/357, 359
vascular lesion, 48/314
vertebral arteritis, 48/322
vertebral artery, 48/323, 53/379
vertebral artery thrombosis, 48/315
vertebral infarction, 48/315
vertebrobasilar insufficiency, 55/344
vertebrobasilar system syndrome, 53/379
visual loss, 48/312, 314
Wegener granulomatosis, 55/346, 71/182
weight loss, 55/342
Temporal artery
    anterior, see Anterior temporal artery
    cluster headache, 48/78
    granulomatous arteritis, 48/309
    middle, see Middle temporal artery
    noradrenalin, 48/87
    posterior, see Posterior temporal artery
    serotonin, 48/87
    superficial, see Superficial temporal artery
Temporal artery biopsy
    temporal arteritis, 39/333-336, 48/10, 320, 55/344,
        346
Temporal bone fracture
    see also Neuro-otology and Polymyalgia
        rheumatica
    Battle sign, 57/131
    clinical features, 24/120
    CSF fistula, 24/121, 125, 185, 57/139
    diagnosis, 24/125
    dizziness, 24/121
    facial nerve injury, 24/105, 124
    hearing loss, 24/123, 125, 185
    hemotympanum, 24/120
    incidence, 24/119
    labyrinth concussion, 24/122, 125, 185

liquor tympanum, 24/121
longitudinal, 24/106, 120, 125
mastoid ecchymosis, 24/121
mastoid process, 24/107
middle fossa fracture, 23/395
nystagmus, 24/122, 125, 127
oblique, 24/120
ossicular chain, 24/121, 125, 185
otorrhagia, 24/120, 125, 185
otorrhea, 24/120, 125, 185
pathology, 24/119
tinnitus, 24/123, 125
transverse, 24/105, 120, 125
vertigo, 24/121, 125
vestibular finding, 24/121
Temporal bone tumor
herpes simplex virus encephalitis, 56/214
Temporal cortex
Alzheimer disease, 46/249
Temporal crescent
cerebrovascular disease, 55/118
migraine, 55/118
occipital lobe, 2/597
visual field defect, 2/596
Temporal diploic vein
anterior, see Anterior temporal diploic vein
Temporal disc pallor
multiple sclerosis, 9/173
Temporal disorientation
Pick disease, 46/234
Temporal horn
epilepsy, 15/539
Temporal lobe
acetylcholine, 46/266
aggression, 2/718
Alzheimer disease, 46/266
anxiety, 45/275
ataxia, 1/343
auditory hallucination, 45/356
autonomic nervous system, 2/720
autoscopia, 2/714, 45/357
basal ganglion, 73/64
birth lesion, 15/101
brain abscess, 52/144
cerebral dominance, 4/3, 301
cerebrovascular disease, 55/117
congenital heart disease, 63/2
EEG, 72/300
function, 17/281, 283
functional anatomy, 15/41
functional division, 2/701
glioblastomatosis, 14/496
hallucination, 45/355-358

hallucinosis, 46/566
homonymous hemianopia, 2/708
hyperactivity, 46/182
Klüver-Bucy syndrome, 2/602, 24/724
language, 46/148
left, see Left temporal lobe
lilliputian hallucination, 45/358
limbic system, 2/719
magnetoencephalography, 72/330
medial, see Medial temporal lobe
memory function, 15/99
mesial vs lateral, 15/41
morphology, 2/700
nonverbal memory, 45/44
right, see Right temporal lobe
schizophrenia, 46/458, 493, 508
seizure pattern, 15/95
symptom pattern, 2/702
syndrome, 2/700-724
traumatic intellectual defect, 24/532
traumatic intracerebral hematoma, 24/356, 358
traumatic psychosis, 24/532
traumatic psychosyndrome, 24/558
tumor, see Temporal lobe tumor
vestibular system, 16/328
visual field defect, 2/587
visual hallucination, 2/709, 45/356
Wernicke aphasia, 45/18
Temporal lobe abscess
intractable hiccup, 63/490
memory disorder, 46/623
Temporal lobe agenesis syndrome, 30/356-359
arachnoid cyst, 31/105
clinical features, 30/357
course, 30/358
CSF, 31/105
diagnosis, 30/359
EEG, 30/358
history, 30/356
hydrocephalus, 31/105, 50/289
mental deficiency, 46/13
middle fossa arachnoid cyst, 30/357-359
pathology, 30/357
prognosis, 30/358
radiologic appearance, 30/358
sex distribution, 30/357
subdural hygroma, 31/105
Temporal lobe atrophy
Pick disease, 27/489, 42/285, 45/260, 46/236, 239
Temporal lobe epilepsy
affect, 45/45
aggression, 46/433
Ammon horn sclerosis, 2/718, 73/69

amnesia, 2/713
amygdala, 73/61
amygdaloid nucleus, 73/55
amygdalotomy, 2/719
amytal test, 73/83
auditory association area, 2/705
auditory hallucination, 2/706, 711
auditory illusion, 73/72
automatism, 15/95, 42/685, 73/71
autonomic seizure, 73/72
autoscopia, 2/711
behavior, 45/45
birth lesion, 15/101
carbamazepine, 73/78
cardiovascular symptom, 73/72
cingulate cortex, 73/59
classification, 15/89, 72/10
clinical characteristic, 73/70
concept history, 15/88
corpus amygdaloideum, 73/58
cryptogenic, 73/81
déjà vécu, 2/706
déjà vu, 2/706, 46/433
dentate gyrus, 73/80
depersonalization, 2/711
derealization, 2/711
differential diagnosis, 73/78
EEG, 73/73
emotion, 2/706, 45/274
emotional aura, 73/72
emotional disorder, 2/716
epidemiology, 73/68
epileptic mastication, 15/97
erythropsia, 73/71
features, 42/685
focal, 42/685
functional nuclear magnetic resonance, 73/82
genetics, 73/68
Geschwind syndrome, 73/84
hallucination, 2/706, 42/685
hallucinosis, 46/566
hippocampal sclerosis, 73/69
hippocampus, 72/86, 116, 73/59
hyperactivity, 46/182
hypersexuality, 46/433
intractable hiccup, 63/490
Klüver-Bucy syndrome, 45/259
lesion, 73/80
limbic system, 73/56
literature, 15/87
memory disorder, 2/713
memory function, 15/99
mental deficiency, 42/685

mesial, 73/393
mesial temporal sclerosis, 73/69
monochromatism, 73/71
multiple personality, 46/577
musicogenic epilepsy, 15/446, 45/484
mydriasis, 73/73
narcolepsy, 45/148
nonverbal memory, 45/44
olfactory illusion, 73/72
oral grasping, 45/259
organic personality syndrome, 46/432-434
personality, 45/45
physiologic property, 72/68
poliopia, 73/71
positron emission tomography experiment, 73/82
psychiatric aspect, 73/84
psychiatric disorder, 42/465
psychomotor seizure, 73/72, 74
psychomotor symptom, 73/70
psychosis, 45/46, 46/504, 73/85
radiology, 15/539
recall, 2/714
schizophrenia, 46/419, 460
schizophreniform psychosis, 46/420
seizure, 73/67, 70
seizure distribution, 73/73
seizure pattern, 15/95
somatesthetic illusion, 73/72
speech, 2/719
surgery, 73/392
surgical treatment, 15/98, 100
survey, 73/53, 394
temporal lobe tumor, 17/285
temporal lobectomy, 45/44
twilight state, 2/714
vertiginous illusion, 73/72
vestibular function, 2/706
visual perseveration, 73/71
yawning, 63/493
Temporal lobe hypoplasia
superior, see Superior temporal lobe hypoplasia
Temporal lobe infarction
Klüver-Bucy syndrome, 45/260
transient global amnesia, 45/215
Temporal lobe lesion
acalculia, 4/189
agraphia, 4/3, 153
alexia, 4/3, 45/13
amusia, 4/3, 204
apraxia, 4/54-57
attention disorder, 3/147, 352
auditory agnosia, 4/38
behavior, 2/717

cerebral dominance, 4/301
déjà vu, 4/3
depersonalization, 2/709, 711
derealization, 2/709, 711
ectopallium, 2/720
EEG, 2/718
emotional disorder, 3/346
entopallium, 2/720
hallucination, 4/331
homonymous hemianopia, 2/586
Klüver-Bucy syndrome, 2/708, 23/11
memory disorder, 2/713, 3/147
mesopallium, 2/720
musical agnosia, 4/203
musical hallucination, 45/484
optokinetic nystagmus, 2/707
receptive amusia, 45/488
temporal lobectomy, 4/302
topographical memory disorder, 4/16, 299
uncinate fit, 2/586, 616
verbal auditory agnosia, 43/229
visual agnosia, 4/20-22
Wernicke aphasia, 4/98
Temporal lobe psychosis
depersonalization, 24/532
Temporal lobe syndrome, 2/700-720, 45/43-47
affect, 45/44
dementia, 46/124
indifference, 45/44
Klüver-Bucy syndrome, 45/43
localization, 45/43
neglect, 45/44
psychic blindness, 45/43
time sense, 45/47
Temporal lobe triad
classification, 15/8
Temporal lobe tumor
aphasia, 17/291, 67/152
ataxia, 17/292
auditory hallucination, 46/461, 568
déjà vu, 17/284
depersonalization, 17/284
EEG, 17/287-289
emotional disorder, 3/344, 348
gustatory hallucination, 17/286
hallucination, 17/286, 46/568
headache, 17/291, 67/152
hearing loss, 17/289
hemianopia, 17/290
illusion, 17/289
literature, 17/281
mania, 46/429
memory, 67/152

mental disorder, 17/290
miosis, 17/292
mydriasis, 17/292
nominal aphasia, 4/98
olfactory hallucination, 17/286
papilledema, 17/292
psychiatric disorder, 17/290
pyramidal sign, 17/292
quadrantanopia, 17/290
seizure, 67/151
symptom, 17/282, 293
temporal lobe epilepsy, 17/285
tinnitus, 17/289
tremor, 17/292
vertigo, 17/293
vestibular disorder, 17/289
visual deficit, 67/151
visual disorder, 17/290
visual hallucination, 17/290
Temporal lobectomy
agnosia, 45/336
amusia, 45/46, 484
animal experiment, 2/708
apperceptive agnosia, 45/336
behavior, 2/715
emotion, 45/276
emotional disorder, 3/344, 346
epilepsy, 15/750
geniculocalcarine tract, 2/558
Klüver-Bucy syndrome, 2/708
Labbé vein, 2/708
lateralization, 46/503
nonverbal memory, 45/44
sorrow, 45/276
temporal lobe epilepsy, 45/44
temporal lobe lesion, 4/302
verbal memory, 45/44
Temporal occipital artery
anatomy, 11/21
Temporal pressure cone
brain herniation, 16/119
Temporal sulcus
superior, see Superior temporal sulcus
Temporary blindness
occipital lobe tumor, 17/326
Temporobasal syndrome
traumatic, see Traumatic temporobasal syndrome
Temporomandibular ankylosis
congenital myopathy, 62/335
nemaline myopathy, 62/342
Temporomandibular arthrosis
see also Costen syndrome
differential diagnosis, 5/316

Temporomandibular cephalalgia, 48/413-415
 headache, 48/413
 migraine, 48/415
 osteoarthrosis, 48/413
 pain, 48/413
 psychogenic pain, 48/394
 temporomandibular joint, 48/413
 trigeminal neuralgia, 48/415
Temporomandibular joint
 clicking, 48/414
 pain, 5/316, 345
 radiographic examination, 5/346
 temporal arteritis, 39/318
 temporomandibular cephalalgia, 48/413
Temporomandibular joint dysfunction
 Dechaume syndrome, 5/345
 headache, 48/34, 47
 Hennebert syndrome, 5/345
Tenascin
 undetermined significant monoclonal
  gammopathy, 63/400
Tendon hypertrophy
 glutamic aciduria, 60/669
Tendon reflex, see Muscle stretch reflex
Tendon transfer
 brachial plexus, 51/149
Tendon xanthoma
 cerebral cholesterosis, 14/779
 cerebrotendinous xanthomatosis, 10/532, 547,
  27/250, 29/375, 42/430, 59/357, 60/669
Tenocyclidine
 designer drug, 65/364
Tenodesis grip
 physical rehabilitation, 61/461
 technique, 61/461
Tenotomy
 central core, 40/45
 central core myopathy, 62/334
 cytoarchitectonic change, 40/143-145
 Duchenne muscular dystrophy, 40/364
 experimental, 41/5, 10
 experimental myopathy, 40/143
 muscular atrophy, 40/143
 muscular dystrophy, 41/471
 rod body, 40/144, 240, 287
 spasticity, 26/484
 target fiber, 40/144
 Z band, 40/144
Tensilon test, see Edrophonium test
Tension headache, see Muscle contraction headache
Tension vascular headache
 classification, 48/6
 definition, 48/8

Tensor fasciae latae reflex
 diagnostic value, 1/247
Tentorial artery
 normal anatomy, 12/271
 radioanatomy, 11/69
Tentorial hypoplasia
 Arnold-Chiari malformation type II, 50/405, 408
Tentorial incisure
 anatomy, 1/553
 Arnold-Chiari malformation type II, 50/406
 brain herniation, 1/565
Tentorial sinus
 anatomy, 11/61
Tentorium
 innervation, 1/538
 notch, 1/550-554
 physiologic striatopallidodentate calcification,
  49/417
 venous malformation, 12/236
Tephromalacia
 anterior, see Anterior tephromalacia
TEPP intoxication, see Tetraethyl pyrophosphate
 intoxication
Teratogenesis
 antiepileptic agent, 65/515
 cephalocele, 50/98
 CNS, 30/34, 50/41
 placenta, 30/115-117
Teratogenic agent
 alcohol, 30/119-121
 aminopterin, 30/122
 animal viral disease, 34/379-384
 anticonvulsant, 30/121
 chickenpox, 34/378
 Coxsackie virus infection, 34/378
 criteria, 30/112
 cytochalasin B, 30/124
 cytomegalovirus infection, 34/374-377
 fetal infection, 34/369-372
 herpes simplex virus, 34/377
 herpes zoster, 34/378
 hydrocephalus, 34/383
 infectious agent, 30/118
 influenza, 34/379
 inhalation anesthetics, 30/122
 ionizing radiation, 30/118
 maternal diabetes mellitus, 30/126
 maternal hyperthyroidism, 30/126
 measles, 34/378
 methotrexate, 30/122
 pathogenesis, 34/369-372
 phakomatosis, 14/5
 potato blight, 30/124

rachischisis, 32/169
rubella, 34/372-375
sensitive period, 30/113
sex hormone, 30/125
smallpox, 34/378
spina bifida, 32/528, 531, 50/479
thalidomide syndrome, 30/123
tranquilizing agent, 30/125
trypan blue, 32/169
D-tubocurarine, 30/123
vaccinia, 34/378
Venezuelan equine encephalitis, 34/378
vitamin A, 32/169, 172
warfarin, 30/123, 31/77
Teratoid cyst, *see* Epidermoid cyst
Teratoid tumor
  nervous system, 68/380
Teratology
  congenital brain tumor, 31/37
  definition, 30/1
  Smith-Lemli-Opitz syndrome, 66/591
Teratoma
  chiasmal, 31/57
  clinical features, 31/53-55
  congenital, 32/355-386
  congenital brain tumor, 31/41-43, 50-61
  congenital spinal cord tumor, 32/358-386
  cystic, *see* Intraspinal neurenteric cyst
  definition, 31/50
  EEG, 31/57
  embryology, 31/52
  epidemiology, 31/42, 51
  α-fetoprotein, 31/60
  hypothalamic tumor, 68/71
  imaging, 67/179
  intracranial, 18/165
  intraorbital, 18/341
  lateral ventricle tumor, 17/606
  location, 31/42, 52
  malignant, *see* Malignant teratoma
  management, 68/252
  mode of growth, 31/52
  pathology, 31/53, 68/236
  pineal gland, 18/361
  pituitary adenoma, 17/412
  precocious puberty, 18/367, 31/59
  prognosis, 31/60
  radiologic appearance, 31/55-59
  sacrococcygeal, *see* Sacrococcygeal teratoma
  spinal cord, *see* Spinal cord teratoma
  spinal cord compression, 19/363
  suprasellar, 31/58
  supratentorial brain tumor, 18/317

tissue culture, 17/77
  treatment, 31/60
Teratomatous cyst, *see* Epidermoid cyst
Terbutaline
  acquired myoglobinuria, 62/577
  classification, 74/149
Terminal axonic neoformation
  nerve regeneration, 7/226
  nonmyelinated nerve fiber, 7/226
Terminal innervation ratio
  Duchenne muscular dystrophy, 40/384
  myotonic dystrophy, 40/519
Terminal ventricle
  diastematomyelia, 50/438
  diplomyelia, 50/438
Terminal ventricle cyst
  spinal ependymal cyst, 20/91
Terminal ventricle (Krause)
  spinal ependymoma, 20/363
Terminal ventriculostomy
  posttraumatic syringomyelia, 61/392
  syringomyelia, 50/460
Terrapin intoxication
  ingestion, 37/89
Terror
  dysphrenia hemicranica, 5/78
  migraine, 48/162
Terson sign
  subarachnoid hemorrhage, 55/16
Terson syndrome
  subarachnoid hemorrhage, 12/139
Tertiary Lyme disease, *see* Lyme disease
Tertiary myelopathy
  carcinoma, 38/675
  malignancy, 38/675
Tertiary syphilis, 33/351-353
  familial spastic paraplegia, 59/327
  manifestation, 52/274
  treponemal test, 52/276
Testicular atrophy
  cauda equina syndrome, 20/199
  dyskeratosis congenita, 43/392
Testicular heterotopia
  r(9) syndrome, 50/580
  r(14) syndrome, 50/590
  tuberous sclerosis, 14/51
Testicular tumor
  Leydig cell, *see* Leydig cell testicular tumor
Testis
  adrenoleukodystrophy, 47/595
  hereditary amyloid polyneuropathy, 60/106
  Klinefelter syndrome, 50/550
  retentio, *see* Retentio testis

undescended, *see* Cryptorchidism
X-linked adrenoleukodystrophy, 66/460
Testosterone
  annual rhythm, 74/489
  biological rhythm, 74/508
  brain injury, 23/123
  cluster headache, 48/239, 74/508
  headache, 48/427, 433
  migraine, 48/433
  myotonic dystrophy, 62/248
Tests
  amytal, *see* Amytal test
  André-Thomas extensibility, *see* André-Thomas extensibility test
  Anton, *see* Anton test
  auditory recognition, *see* Auditory recognition test
  Austin urinary, *see* Austin urinary test
  autonomic, *see* Autonomic test
  Babinski-Weil, *see* Babinski-Weil test
  Berry spot, *see* Berry spot test
  Bikeles, *see* Bikeles test
  Bonney, *see* Bonney test
  Bonney axon reflex, *see* Bonney axon reflex test
  caloric, *see* Caloric test
  carotid compression, *see* Carotid compression test
  carotid sinus stimulation, *see* Carotid sinus stimulation test
  catecholamine, *see* Catecholamine test
  cetylpyridiniumchloride-carbazole, *see* Cetylpyridiniumchloride-carbazole test
  cocarboxylase loading, *see* Cocarboxylase loading test
  coccidioidin skin, *see* Coccidioidin skin test
  cold pressor, *see* Cold pressor test
  cold pressure, *see* Cold pressor test
  cold vasodilatation, *see* Cold vasodilatation test
  corticotropin, *see* ACTH test
  crossed Lasègue, *see* Crossed Lasègue test
  Crowe, *see* Crowe test
  dazzling, *see* Dazzling test
  dexamethasone suppression, *see* Dexamethasone suppression test
  edrophonium, *see* Edrophonium test
  Ellsworth-Howard, *see* Ellsworth-Howard test
  equal loudness balance, *see* Equal loudness balance test
  extended arms, *see* Extended arms test
  face-hand, *see* Face-hand test
  finger to ear, *see* Finger to ear test
  fist-ring, *see* Fist-ring test
  fluorescent treponemal antibody absorption, *see* Fluorescent treponemal antibody absorption test
  foot sweat, *see* Foot sweat test

Galveston orientation and amnesia, *see* Galveston orientation and amnesia test
genetic hypothesis, *see* Genetic hypothesis test
glucose tolerance, *see* Glucose tolerance test
glyceryl trinitrate, *see* Glyceryl trinitrate test
Gordon, *see* Gordon test
Hallipre, *see* Hallipre test
Halstead Category, *see* Halstead Category Test
hand turning, *see* Hand turning test
head up tilt, *see* Head up tilt test
Head's hand-ear-eye, *see* Head's hand-ear-eye test
hippuran excretion, *see* Hippuran excretion test
Hirch-Peiffer, *see* Hirch-Peiffer test
histamine skin, *see* Histamine skin test
immunofluorescence, *see* Immunofluorescence test
intelligence, *see* Intelligence test
iodine starch, *see* Iodine starch test
ischemic exercise, *see* Ischemic exercise test
kit, *see* Kit test
Kveim, *see* Kveim test
lepromin, *see* Lepromin skin test
levodopa provocation, *see* Levodopa provocation test
limulus lysate, *see* Limulus lysate test
loudness balance, *see* Loudness balance test
loudness discomfort level, *see* Loudness discomfort level test
lymphocyte transformation, *see* Lymphocyte transformation test
Matas, *see* Matas test
mental stress, *see* Mental stress test
migration inhibition, *see* Migration inhibition test
multiple sleep latency, *see* Multiple sleep latency test
mumps skin, *see* Mumps skin test
muscle strength, *see* Muscle strength test
neurophysiologic, *see* Neurophysiologic test
neuropsychologic, *see* Neuropsychologic test
nitroblue tetrazolium, *see* Nitroblue tetrazolium test
Ozeretski, *see* Ozeretski test
pastpointing, *see* Pastpointing test
Poppelreuter, *see* Poppelreuter test
prehension, *see* Prehension test
prothrombin, *see* Prothrombin time
protirelin, *see* Protirelin test
pushing, *see* Pushing test
quantitative sudomotor axon reflex, *see* Quantitative sudomotor axon reflex test
Queckenstedt, *see* Queckenstedt test
reading, *see* Reading test
rhythmic stabilization, *see* Rhythmic stabilization

test
ribosomal protein synthesis, *see* Ribosomal
   protein synthesis test
Rinne, *see* Rinne test
Sabin-Feldman dye, *see* Sabin-Feldman dye test
Schilling, *see* Schilling test
Schwabach, *see* Schwabach test
serologic, *see* Serologic test
Sharp-Purser, *see* Sharp-Purser test
Simola, *see* Simola test
skin, *see* Skin test
speech, *see* Speech test
Spurling, *see* Spurling test
squatting, *see* Squatting test
Stroop, *see* Stroop test
sudomotor, *see* Sudomotor test
sweat, *see* Sweat test
T3 suppression, *see* T3 suppression test
tactile amytal, *see* Tactile amytal test
tapping, *see* Tapping test
thyrotropin releasing hormone, *see* Protirelin test
tilting chair, *see* Tilting chair test
Tobey-Ayer, *see* Tobey-Ayer test
Token, *see* Token test
tone decay, *see* Tone decay test
TPI, *see* Treponema pallidum immobilization test
Treponema pallidum immobilization, *see*
   Treponema pallidum immobilization test
tryptophan load, *see* Tryptophan load test
tuberculin, *see* Tuberculin test
tuning fork, *see* Tuning fork test
Venereal Disease Research Laboratory, *see*
   Venereal Disease Research Laboratory test
Von Romberg, *see* Von Romberg test
Von Stein, *see* Von Stein test
Wada, *see* Wada test
Wadia, *see* Wadia test
Weber, *see* Weber test
Wechsler, *see* Wechsler test
Tetanoid chorea
   history, 6/517
   Wilson disease, 6/553
Tetanospasmin
   Clostridium tetani, 52/229, 64/10, 16
   myasthenic syndrome, 64/16
   myoglobinuria, 64/16
   neurotoxin, 52/229, 231, 64/10, 16
   tetanus, 52/231
   toxic myelopathy, 64/10
Tetanospasmin intoxication
   action site, 64/10
Tetanus, 33/491-538
   see also Clostridium tetani

active immunization, 33/517-522
age, 52/229
antitoxin treatment, 65/220
asphyxia, 52/230
autonomic dysfunction, 65/217, 225
autonomic nervous system, 52/230
baclofen, 65/224
beta adrenergic receptor blocking agent, 65/225
betanidine, 65/225
blood pressure, 65/218
cardiac arrest, 52/230
cardiac dysrhythmia, 52/230
cephalic, *see* Cephalic tetanus
chemoprophylaxis, 33/514
classification, 33/523
clinical features, 33/520, 52/230
complication, 33/526-528
corticosteroid, 41/305
course, 33/520-529
cranial neuropathy, 65/218
deep vein thrombosis, 65/218
definition, 33/491
deglutition spasm, 1/285
diamorphine, 37/393
diazepam, 41/305, 65/223
differential diagnosis, 65/220, 71/560
drug addict, 52/229
dysautonomia, 65/217
dysphagia, 52/230
EMG, 1/643, 65/219
epidemiology, 33/495-500, 52/229, 65/216,
   71/559
epilepsy, 52/230
etiology, 71/559
GABA, 29/506
general, 52/230
generalized form, 65/217
geographic distribution, 33/497
glycine, 29/511
history, 33/491
hypoglossal paralysis, 7/476
Ig, 65/222
immunization, 33/515-522, 52/227, 235, 65/220
incidence, 52/229
intensive care unit, 65/225
local form, 65/217
magnesium sulfate, 65/225
mortality rate, 33/496-498
motoneuron, 52/231
motor end plate, 52/231
motor unit hyperactivity, 40/330, 41/303, 59/377
muscle cramp, 41/306
muscle spasm, 52/230

myotonia, 40/547
neonatal tetanus, 33/528, 52/229, 65/217, 219
nerve lesion, 7/476
neurologic sequela, 52/227
neuromuscular block, 41/305
nuchal rigidity, 65/217
ophthalmoplegia, 7/476
opiate intoxication, 37/393, 65/355
opisthotonos, 52/230, 65/219
paralytic ileus, 65/218
pathogenesis, 33/500-513, 52/231
pathologic laughing and crying, 3/357, 45/220
peripheral facial paralysis, 65/217, 219
phentolamine, 65/225
polyneuropathy, 65/218
population susceptibility, 33/498
prevention, 33/513-520, 71/560
prognosis, 65/217, 71/560
rabies, 34/264, 56/390
reflex spasm, 65/216
rigidity, 65/217
risus sardonicus, 41/304, 52/230, 65/217
survey, 71/558
sympathetic overactivity, 52/231
symptom, 65/216, 71/559
tetanic attack, 65/217
tetanospasmin, 52/231
tetanus antiserum, 65/222
tetanus toxoid, 65/221
tetraplegia, 65/218
treatment, 33/529-538, 65/220
trismus, 41/304, 52/230, 65/217
vaccine, 52/235
vocal cord paralysis, 65/218
Waldenström macroglobulinemia, 33/532
wound care, 65/220
Tetanus antitoxin
multiple sclerosis, 9/60
serum sickness, 52/240
Tetanus toxin
binding, 65/245
biochemistry, 33/500-502
cell internalization, 65/247
central core myopathy, 62/338
central effect, 33/509-513
chain, 65/245
chemistry, 65/226
CNS interaction, 33/507
epilepsy, 72/54
inhibition, 1/51
intoxication mechanism, 65/246
membrane translocation, 65/247
molecular features, 65/226

neuroexocytosis block, 65/247
neurotoxin, 1/51, 7/476, 33/491, 37/393, 41/303,
    52/229, 62/338, 65/215, 226, 245
peripheral effect, 33/509-513
pharmacodynamics, 33/508-513
pharmacokinetics, 33/504-508
postsynaptic inhibition, 41/304
receptor binding, 65/245
retroaxonal flow, 65/245
spastic paraparesis, 65/245
synaptobrevin, 65/248
syntaxin, 65/248
zinc binding, 65/245
zinc endopeptidase, 65/245
Tetanus toxoid, 33/502-504
allergic neuritis, 52/240
anaphylactoid reaction, 52/240
brachial plexus neuritis, 2/139
brachial plexus neuropathy, 52/240
composition, 52/235
cranial nerve palsy, 52/240
delayed hypersensitivity, 52/240
demyelinating neuropathy, 52/240
efficacy, 52/236
mononeuritis, 52/240
neurologic complication, 52/240
neuropathy, 7/555, 8/91
polyneuritis, 52/240
polyradiculomyelopathy, 52/240
polyradiculopathy, 52/240
side effect, 52/240
tetanus, 65/221
Tetany
    see also Calcium and Calcium imbalance
brain injury, 24/594
Burkitt lymphoma, 63/352
carpopedal spasm, 41/301
central manifestation, 28/534-536
Chvostek sign, 40/330, 41/302, 63/560
clinical features, 63/559
cramp, 40/330
diuretic agent intoxication, 65/440
EEG, 28/534
electrolyte disorder, 63/558
EMG, 1/643, 646, 63/560
epilepsy, 63/560
ethylene glycol intoxication, 64/122
hyperventilation syndrome, 38/315, 318, 63/435
hypocalcemia, 27/273, 28/532-536, 33/278,
    63/558
hypokalemia, 63/558
hypokalemic alkalosis, 63/559
hypomagnesemia, 63/559, 562

hypoparathyroid myopathy, 62/541
hypoparathyroidism, 27/298-300, 70/123
Karpfenmaul, 63/559
malabsorption syndrome, 70/226
mental change, 28/536
motor unit hyperactivity, 40/330, 41/301
multiple sclerosis, 47/57
muscle cramp, 40/330
myotonia, 40/547
nerve ischemia, 63/560
opisthotonos, 41/301, 63/559
paresthesia, 63/560
peripheral manifestation, 28/532-534
pertussis encephalopathy, 33/278
Pfötchenstellung, 63/559
pseudohypoparathyroidism, 42/621, 46/85, 63/558
renal insufficiency, 27/336
respiratory alkalosis, 63/559
Rud syndrome, 38/9
spinal reflex, 28/535
stridor, 63/560
Trousseau sign, 41/302, 63/560
uremic encephalopathy, 63/504
Tethered cord
    aqueduct stenosis, 50/309
    diastematomyelia, 50/438
    diplomyelia, 50/438
    posttraumatic, 61/530
    spastic paraplegia, 59/434
    spinal cord injury, 61/529
    vertebral column injury, 61/529
Tethered cord syndrome
    cauda equina tumor, 19/87
    diastematomyelia, 50/435
    myelography, 19/189
    symptom, see Diastematomyelia and Diplomyelia
Tethered spinal cord
    spina bifida, 50/494
(2,3,4,6-Tetra-O-acetyl-1-thio-β-D-glucopyrano-
    sato-S)(triethylphosphine)-gold, see Auranofin
Tetra-alkyllead
    antiknock agent, 64/129
    organolead, 64/129
Tetrabenazine
    hemiballismus, 49/375
    neuroleptic agent, 65/275
    spinal myoclonus, 41/308
Tetracaine intoxication
    seizure, 37/405
2,3,7,8-Tetrachlorodibenzo-p-dioxin
    dementia, 64/205
    insomnia, 64/205
    myalgia, 64/205

polychlorinated dibenzodioxin intoxication,
    64/204
sensorimotor polyneuropathy, 64/205
Tetrachloroethylene
    organic solvent intoxication, 64/40, 42
15-Tetracosanoic acid
    biochemistry, 10/249
    cerebroside, 9/6
    sulfatide, 9/6
Tetracyclic antidepressant
    behavior disorder, 46/597
Tetracycline
    bacterial meningitis, 52/9
    benign intracranial hypertension, 65/472, 67/111
    headache, 48/34
    Hemophilus influenzae meningitis, 33/56
    iniencephaly, 50/135
    malaria, 52/373
    meningococcal meningitis, 33/29
    neurobrucellosis, 33/318, 52/596
    neurosyphilis, 52/282
    neurotoxin, 65/472, 568
    pneumococcal meningitis, 33/42
    toxic myopathy, 62/609
    vitamin A intoxication, 65/568
    Whipple disease, 52/140
Tetraethylammonium chloride
    multiple sclerosis, 9/387
Tetraethyllead
    encephalopathy, 51/265
    neurotoxicity, 51/265
    organolead intoxication, 64/130
Tetraethyllead intoxication
    symptom, 64/443
Tetraethylphosphorodiamidic fluoride intoxication
    chemical classification, 37/544
    organophosphorus compound, 37/544
Tetraethyl pyrophosphate
    chemical structure, 64/225
    neurotoxin, 37/541, 549, 64/151, 223
    organophosphate, 64/152
    organophosphate intoxication, 64/151, 155, 223
Tetraethyl pyrophosphate intoxication
    chemical classification, 37/548
    fatal, 37/541
    organophosphorus compound, 37/548
Tetraethylthiuram disulfide, see Disulfiram
Tetrahydrobiopterin
    aromatic amino acid, 75/235
    dystonia musculorum deformans, 49/524
    Parkinson disease, 49/123
Tetrahydrobiopterin deficiency
    stress, 74/316

Tetrahydrobiopterin metabolism, 75/235
  autonomic nervous system, 75/234
  phenylketonuria, 75/234
Tetrahydrocannabinol
  see also Cannabis and Hallucinogenic agent
  active component, 65/46
  active metabolite, 65/46
  arachidonylethanolamide, 65/47
  biokinetic, 65/46
  cannabinoid receptor, 65/47
  cannabis, 65/46
  Cannabis sativa, 65/46
  chemical structure, 65/47
  dronabinol, 65/46
  hallucinogenic agent, 65/42
  11-hydroxydronabinol, 65/46
  multiple sclerosis, 47/169
  psychosis, 46/594
  symptom, 65/47
  tetrahydrocannabinolic acid, 65/46
Δ9-Tetrahydrocannabinol, see Dronabinol
Tetrahydrocannabinol intoxication
  ataxia, 65/47
  dysarthria, 65/47
  hallucination, 65/47
  positron emission tomography, 65/47
Tetrahydrocannabinolic acid
  cannabis, 65/46
  tetrahydrocannabinol, 65/46
Tetrahydro-β-carboline, see Tryptoline
Tetrahydrofolic acid, see Vitamin Bc
Tetrahydroisoquinoline
  cytochrome P450, 65/400
  designer drug, 65/399
  1-methyl-4-phenyl-1,2,3,6-tetrahydropyridine
    intoxication, 65/399
  pharmacology, 65/399
1,2,3,6-Tetrahydro-1-methyl-4-phenylpyridine, see
    1-Methyl-4-phenyl-1,2,3,6-tetrahydropyridine
1,2,3,6-Tetrahydroxy-1-methyl-4-phenylpyridine
    intoxication
  symptom, 64/6
Tetraisopropyl dithionopyrophosphate intoxication
  chemical classification, 37/549
  organophosphorus compound, 37/549
Tetraisopropyl pyrophosphate intoxication
  chemical classification, 37/549
  organophosphorus compound, 37/549
Tetralogy
  Fallot, see Fallot tetralogy
Tetramethrin
  neurotoxin, 64/213
  pyrethroid, 64/213

pyrethroid intoxication, 64/213
Tetramethylammonium hydroxide
  neptunea intoxication, 37/64
3,7,11,15-Tetramethylhexadecanoic acid, see
    Phytanic acid
Tetramethyllead
  organolead intoxication, 64/130
Tetramethyllead intoxication
  organolead, 64/133
Tetramethylphosphorodiamidic fluoride
    intoxication, see Dimefox intoxication
Tetramethyl pyrophosphate intoxication
  chemical classification, 37/549
  organophosphorus compound, 37/549
Tetramine, see Tetramethylammonium hydroxide
Tetramonoisopropyl pyrophosphortetramide
    intoxication, see Iso-Ompa intoxication
Tetraodontidae intoxication
  see also Puffer fish intoxication and
    Tetraodontiformes intoxication
  geographic distribution, 37/27, 79
Tetraodontiformes intoxication
  see also Tetraodontidae intoxication
  geographic distribution, 37/79
Tetraparesis
  Behçet syndrome, 56/598
  ciclosporin, 63/10
  hypotonic, see Hypotonic tetraparesis
  Lesch-Nyhan syndrome, 59/360
  spastic, see Spastic tetraparesis
Tetraplegia
  adjustment vs age, 61/548
  adjustment vs education, 61/548
  adjustment vs injury severity, 61/548
  adjustment vs patient personality, 61/548
  adjustment stage, 61/537
  4-aminopyridine, 61/401
  amyloidosis, 63/529
  angiotensin II, 61/444
  ankylosing spondylitis, 26/175-184
  autonomic dysreflexia, 61/447
  axon regeneration, 61/407
  axon survival, 61/400
  axonal dysfunction, 61/401
  baroreflex, 61/444
  basal cardiovascular state, 61/440
  basilar impression, 42/17
  blood pressure, 61/439
  bradycardia, 61/436
  cardiac arrest, 61/264, 439, 446, 63/215
  cardiovascular drug response, 61/438
  cardiovascular posture response, 61/441
  cardiovascular system, 61/435

central pontine myelinolysis, 2/263, 63/549

central venous pressure, 61/452

chorioretinal degeneration, 13/39

ciclosporin, 63/537

ciclosporin intoxication, 63/537

clinical trial, 61/399

CNS spongy degeneration, 42/506

coitus, 61/315

corpus callosum agenesis, 42/100, 50/152

corticosteroid, 61/399

decompression myelopathy, 61/216

dialysis, 63/529

diaphragmatic pacing, 61/261

diphtheria, 65/240

disability acceptance, 61/546

dissecting aorta aneurysm, 63/37, 44

distress, 61/537

dopamine β-mono-oxygenase, 61/448

ECG, 61/439

ejaculation, 61/315, 322

electrical injury, 61/195

electroejaculation, 61/323

employment, 61/550

erroneous depression assessment, 61/540

experimental therapy, 61/399

familial hyperargininemia, 59/358

female sexual function, 61/331

fertility, 61/313

food ingestion, 61/446

GABA, 61/401

glutaryl-coenzyme A dehydrogenase deficiency, 66/647

headache, 26/494

heart rate, 61/439

Hooft-Bruens syndrome, 42/508

hyperkalemia, 63/558

hypermagnesemia, 63/563

hyperthermia, 61/275

hyperventilation, 61/445

hypoglycemia, 61/446

hypokalemia, 63/557

hypothermia, 61/275

impotentia coeundi, 61/315

intubation response, 61/437

K opiate receptor blocker, 61/399, 403, 406

Laurence-Moon syndrome, 13/307

life quality, 61/541

life satisfaction, 61/547

lipid peroxidation inhibitor, 61/399

male sexual function, 61/313

mass reflex, 61/447

maturity stage, 61/537

mental depression, 61/537

N-methylaspartic acid receptor blocker, 61/399, 406

methylprednisolone, 61/399, 402

β2-microglobulin, 63/529

β2-microglobulin amyloidosis, 63/529

mitochondrial DNA depletion, 62/511

monosialoganglioside, 61/399, 406

naloxone, 61/399

neuropeptide Y, 61/450

nevus unius lateris, 43/40

optic atrophy, 13/40

orthostatic hypotension, 26/318, 61/442

osteitis deformans (Paget), 38/367

paradoxical respiration, 61/262

Parkinson dementia complex, 42/250

pathology specimen, 25/27-42

patient adjustment, 61/537

patient personality characteristics, 61/542

penis prosthesis, 61/321

plasma adrenalin, 61/436

plasma aldosterone, 61/443

plasma noradrenalin, 61/436

plasma renin, 61/443

poikilothermia, 2/199

pontine infarction, 12/15

posttraumatic syringomyelia, 26/117, 50/454, 61/388

pressure sore, 26/463, 61/347

protirelin, 61/403, 406

psychology, 61/537

rebellion stage, 61/537

recovery, 61/399

recovery substrate, 61/400

regenerative therapy, 61/406

remyelination, 61/402

renin release, 61/443

renin response, 61/443

respiratory management, 61/261

secondary pigmentary retinal degeneration, 13/307

self-concept, 61/547

sex difference, 61/546

sexual capability, 61/315

sexual intercourse, 61/313

sociology, 61/537

spastic, see Spastic tetraplegia

spastic cerebral palsy, 42/167

spinal cord electrotrauma, 61/195

spinal shock, 61/439

steroid antioxidant, 61/405

sympathoneural activation, 61/437

systemic hypotension, 61/436

tachycardia, 61/436

tetanus, 65/218
tirilazad, 61/405
tracheal stimulus response, 61/437
tracheostomy, 61/437
traffic injury, 25/142
uremic polyneuropathy, 51/360
vacuum constrictor devices, 61/321
Valsalva maneuver, 61/445
vascular tone, 61/436
vasodepressor agent, 61/454
vasopressor agent, 61/451
visceral stimulation, 61/451
Waldenström macroglobulinemia, 63/396
Tetraploidy
    anophthalmia, 43/567
    carp mouth, 43/567
    definition, 50/540
    dermatoglyphics, 43/567
    digital abnormality, 43/467
    epicanthus, 43/568
    mental deficiency, 43/567
    pachygyria, 43/567
    pathogenesis, 50/555
    polyploidy, 50/555
    renal abnormality, 43/567
    seizure, 43/567
Tetrasomy
    chromosome number abnormality, 50/539
    definition, 50/540
Tetrasomy X, see XXXX syndrome
Tetrathiomolybdic acid
    Wilson disease, 49/235
Tetrazolium reductase
    mitochondrial disease, 62/497
    muscle biopsy, 40/3
    muscle histochemistry, 40/134
Tetrodotoxin
    Atelopus, 64/15
    chordata intoxication, 37/79-82
    Echinodermata, 64/15
    fish intoxication, 37/79-82
    Gastropoda, 64/15
    Gobius, 64/15
    Hapalochlaena, 64/15
    leatherjacket intoxication, 37/79
    lethal dose, 65/154
    marine toxin intoxication, 37/32, 65/146, 152, 155
    motor end plate block, 40/143
    muscle fiber conductance, 40/552
    myasthenic syndrome, 64/15
    nerve axon, 65/155
    neurotoxin, 28/447, 37/1, 68, 79, 40/143, 41/292,
        51/263, 64/15, 65/141, 155

paralytic shellfish poisoning, 65/152
percussion myotonia, 40/537
puffer fish intoxication, 37/78-80
sodium channel, 28/447, 51/263, 65/147
sodium channel unavailability, 40/558
structural formula, 37/81
structure, 65/147
Tarich, 64/15
taricha egg, 37/80
Tetraodon, 64/15
toxic dose, 65/155
toxic myopathy, 64/15
toxic neuromuscular junction disorder, 64/15
Tetrodotoxin intoxication
    see also Marine toxin intoxication and Puffer fish
        intoxication
    action site, 64/15
    acute muscle weakness, 41/291
    area postrema, 65/155
    arterial dilatation, 65/155
    arterial hypotension, 65/155
    ataxia, 65/156
    autonomic nervous system, 75/511
    bulbar paralysis, 65/156
    diagnosis, 65/156
    emetic chemoreceptor, 65/155
    EMG, 65/156
    epidemiology, 65/155
    extraocular muscle paralysis, 65/156
    flaccid paralysis, 65/156
    fugu, 37/80, 65/155
    history, 65/155
    Japanese puffer fish, 37/80, 41/292
    Lagocephalus lunaris lunaris, 65/155
    management, 65/157
    marine toxin intoxication, 65/143
    nerve action potential, 65/156
    nerve conduction, 65/156
    nerve conduction block, 65/155
    paresthesia, 65/156
    pathomechanism, 64/15
    progressive bulbar palsy, 65/156
    puffer fish, 65/155
    respiratory failure, 65/156
    sodium channel block, 65/155
    sympathetic nerve fiber, 65/156
    symptom, 64/15, 65/156
    tetraodontiformes, 65/155
    treatment, 65/157
    vertigo, 65/156
    zombi phenomenon, 65/155
Thalamencephalic angioma, see Brégeat syndrome
Thalamic aphasia

anomia, 54/309
mutism, 54/309
paraphasia, 54/309
Thalamic artery
  choroidal group, 2/476
  pulvinar group, 2/474, 476
  radioanatomy, 11/93
  thalamogeniculate group, 2/474
  thalamoperforating group, 2/473
  thalamotuberal group, 2/473
  topography, 11/93
Thalamic asymmetry
  language, 45/78
Thalamic blindness (Störring)
  critique, 2/480
Thalamic centromedian parafascicular nuclear
    complex
  pedunculopontine fiber, 49/9
Thalamic cyst
  congenital pain insensitivity, 51/565
Thalamic degeneration
  association with other system atrophies, 60/712
  clinical study, 60/703
  clinicopathologic correlation, 60/712
  dementia, 21/590
  history, 60/701
  literature review, 60/704-712
  neuroanatomy, 60/701
  neurophysiologic correlate, 60/713
  neurophysiologic features, 60/703
  nosology, 60/712
  olivocerebellar atrophy, 2/483
Thalamic dementia
  akinetic mutism, 21/594
  apathy, 21/596, 46/357
  brain metastasis, 18/218
  cause, 45/91
  chorea, 21/590
  choreoathetosis, 6/355, 21/591
  confabulation, 21/595
  Creutzfeldt-Jakob disease, 2/483, 6/355, 21/590
  criticism, 21/589
  diagnosis, 45/91
  inertia, 21/595
  medial thalamic syndrome, 2/486
  myoclonus, 21/591
  nucleus reuniens, 2/483
  pseudogeneral paresis, 2/484
  rejected case, 21/595
  review, 21/589
  thalamic ischemia, 45/92
  thalamic syndrome (Dejerine-Roussy), 45/95
  thalamoperforating artery, 2/483

tremor, 21/590, 592
tumor, 2/484
Thalamic hand
  anterolateral thalamic syndrome, 2/485
  definition, 2/481
Thalamic hemorrhage
  aphasia, 45/320
  brain hemorrhage, 54/307
  central neurogenic hyperventilation, 63/441
  computer assisted tomography, 54/310, 312
  headache, 5/139, 48/278
  symptom, 54/307, 312
  vertebrobasilar insufficiency, 54/309
Thalamic infarction
  abulia, 53/395
  akinetic mutism, 53/209, 55/138
  amnesia, 53/392, 395
  aphasia, 45/91
  ataxia, 53/396
  ataxic hemiparesis, 54/248
  basilar apex syndrome, 53/392
  behavior, 53/395
  brain lacunar infarction, 54/239
  chorea, 53/396
  cognition, 53/395
  computer assisted tomography, 53/392
  dystonia, 53/396
  facial paresthesia, 53/395
  Galen vein thrombosis, 54/399
  hereditary paroxysmal kinesigenic
    choreoathetosis, 49/355
  hypersomnia, 53/395
  nuclear magnetic resonance, 53/392
  peduncular hallucination, 53/392
  pure sensory stroke, 53/395, 54/250
  sensorimotor stroke, 54/252
  straight sinus thrombosis, 54/398
  symptom, 53/396
  symptomatic paroxysmal kinesigenic
    choreoathetosis, 49/355
  verbal perseveration, 53/395
  vertical gaze palsy, 53/395
  visual evoked response, 54/160
Thalamic ischemia
  thalamic dementia, 45/92
  thalamic syndrome (Dejerine-Roussy), 45/91
Thalamic lacunar infarction
  ataxic hemiparesis, 54/248
Thalamic lesion
  akinetic mutism, 2/483
  cerebellar atrophy, 2/489
  chorea, 6/59
  dysacusis, 2/480

482

frontal lobe syndrome, 21/590
hemiplegia, 2/486
hypersomnia, 2/482
hyposmia, 2/480
macrostereognosia, 2/479
metamorphopsia, 2/479
narcolepsy, 2/482
olivocerebellar atrophy, 2/489
olivopontocerebellar atrophy, 2/489
painful anosognosia, 2/479
pathologic laughing and crying, 2/482
proprioceptive impairment, 2/478
sensory extinction, 3/193
Thalamic microneuron
  Huntington chorea, 49/319
Thalamic neglect
  thalamic syndrome (Dejerine-Roussy), 45/92-94
Thalamic nuclear group
  intralamellar, *see* Intralamellar thalamic nuclear
    group
  lateral, *see* Lateral thalamic nuclear group
  medial, *see* Medial thalamic nuclear group
  paraventricular, *see* Paraventricular thalamic
    nuclear group
  posterior, *see* Posterior thalamic nuclear group
Thalamic nucleus
  anterior, *see* Anterior thalamic nucleus
  centromedian, *see* Centromedian thalamic nucleus
  classification, 21/587
  connection, 21/589
  dorsomedial, *see* Dorsomedial thalamic nucleus
  extralamellar, *see* Extralamellar thalamic nucleus
  function, 21/587
  lateral, *see* Lateral thalamic nuclear group
  medial, *see* Medial thalamic nuclear group
  organophosphate intoxication, 64/228
  sub, *see* Subthalamic nucleus
  type, 2/468
Thalamic nucleus malformation
  aqueduct stenosis, 50/309
Thalamic pain
  African trypanosomiasis, 52/341
  characteristics, 1/106
  medial nucleus, 2/484
  pseudo, 2/478
  syndrome, 42/771
  thalamic syndrome (Dejerine-Roussy), 2/476
Thalamic peduncle
  ventral, *see* Ventral thalamic peduncle
Thalamic syndrome (Dejerine-Roussy), 45/87-95
  amnesia, 45/92
  anesthesia dolorosa, 1/106
  anterolateral, *see* Anterolateral thalamic syndrome

apathy, 2/484
aphasia, 45/90, 94
ataxia, 1/319, 42/771
brain lacunar infarction, 54/250
causalgia, 2/482
cause, 2/486-491
character change, 2/484
chorea, 6/437
choreoathetosis, 42/771
chronotaraxis, 2/486
classic type, *see* Posterolateral thalamic syndrome
clinical features, 2/476-479
cyanosis, 2/482
dementia, 45/91
dysarthria, 2/482
early bibliography, 6/67
epilepsy, 2/485
euphoria, 2/484
frontal lobe syndrome, 2/484
gaze paralysis, 2/485
headache, 42/771
hemiplegia, 42/771
hyperesthesia, 42/771
hyperthermia, 2/482
hypotonia, 2/481
involuntary movement, 2/481
language, 45/90, 94
medial, *see* Medial thalamic syndrome
memory, 45/91
mental depression, 2/484
mental disorder, 2/482-486
neuropathology, 2/486-491
pain, 45/235
peduncular hallucination, 45/354
personality disorder, 2/483
phantom sensation, 2/478
posterolateral, *see* Posterolateral thalamic
  syndrome
psychosis, 2/484
sensory loss, 42/771
sleep rhythm, 2/482
stroke, 42/771
sweating, 2/482
symptom, 1/106, 53/396
tactile agnosia, 2/485
thalamic dementia, 45/95
thalamic ischemia, 45/91
thalamic neglect, 45/92-94
thalamic pain, 2/476
thalamic tumor, 45/91
thalamus ischemia, 45/91
trigeminal neuralgia, 5/275
vasovegetative lability, 2/482

vegetative crisis, 2/482, 486
vigilance, 2/482
Villaret syndrome, 2/482
Thalamic tumor
  anaplastic astrocytoma, 68/65
  aphasia, 45/320
  astrocytoma, 18/8, 11, 408, 68/65
  basal ganglion tumor, 18/329
  cavernous angioma, 68/65
  diagnosis, 17/611, 68/64
  EEG, 17/617
  epidemiology, 68/63
  germinoma, 68/65
  glioblastoma multiforme, 68/65
  granuloma, 68/65
  hamartoma, 68/65
  heterotopia, 68/65
  incidence, 17/610
  metastasis, 68/65
  oligodendroglioma, 68/65
  pathology, 17/618, 68/64
  primary CNS lymphoma, 68/65
  primitive neuroectodermal tumor, 68/65
  prognosis, 17/617
  radiology, 17/611, 613, 615
  spongioblastoma, 68/65
  subependymal giant cell tumor, 68/65
  symptom, 68/63
  thalamic syndrome (Dejerine-Roussy), 45/91
  treatment, 68/67
Thalamic ventral anterior nucleus
  pallidothalamic projection, 49/6
Thalamic ventral lateral nucleus
  hypotonia, 49/71
  pallidothalamic projection, 49/6
Thalamic ventral lateral nucleus-anterior ventral
    nucleus complex
  pallidal afferent, 49/600
  ventral globus pallidus, 49/26
Thalamic ventral oral posterior nucleus
  pallidal afferent, 49/600
  rigidity, 49/600
Thalamic ventralis intermedius nucleus
  kinesthetic neuron, 49/600
  Parkinson disease, 49/599
  proprioceptive afferent, 49/600
  spinothalamic afferent, 49/600
  tremor, 49/599
  tremorogenesis, 6/108
Thalamocortical connection
  anatomy, 2/498
Thalamogeniculate artery
  anatomy, 11/21

brain lacunar infarction, 54/236
migraine, 5/85
tremor, 2/481
Thalamolenticular artery
  radioanatomy, 11/82
  topography, 11/82
Thalamoneocortical fiber
  pain, 45/236-238
Thalamoperforate syndrome, see Anterolateral
    thalamic syndrome
Thalamoperforating artery
  anatomy, 11/20
  brain lacunar infarction, 54/236
  thalamic dementia, 2/483
Thalamostriate artery
  radioanatomy, 11/82
  topography, 11/82
Thalamostriate fiber
  glutamic acid, 49/4
  terminal mosaic pattern, 49/4
Thalamostriate vein
  radioanatomy, 11/107
  topography, 11/107
Thalamosubthalamic fiber
  subthalamic nucleus, 49/11
Thalamotomy
  aphasia, 45/320
  athetosis, 49/387
  benign essential tremor, 49/570
  cerebellar tremor, 49/588
  dyssynergia cerebellaris myoclonica, 60/603
  hemiballismus, 1/288
  multiple sclerosis, 47/169
  pain, 2/491
  Parkinson disease, 2/491, 49/599
  psychogenic pain, 46/575
  rigidity, 6/185
  stereotaxic, see Stereotaxic thalamotomy
  stereotaxic ventral intermedius, see Stereotaxic
    ventral intermedius thalamotomy
  tremor, 74/473
  ventrolateral, see Ventral lateral thalamotomy
Thalamotuberal artery
  anatomy, 11/20
Thalamus, 45/87-95
  akinesia, 45/163
  Alpers disease, 42/486
  anatomy, 2/468-476, 21/587, 588
  anterolateral funiculus, 45/232
  arousal disorder, 45/109
  arterial hypertension, 63/74
  autonomic nervous system, 74/61, 155
  ballismus, 6/482

blood supply, 2/473-476

brain diaschisis, 54/126

brain hemorrhage, 11/591, 54/68, 309

brain lacunar infarction, 2/486, 6/673, 54/58

central European tick-borne encephalitis, 2/486

CNS, 30/31

Creutzfeldt-Jakob disease, 2/487, 489, 6/733, 741-745, 21/590-593

eastern equine encephalitis, 2/486

EEG, 54/157, 72/290, 293

embryo, 50/11

emotion, 3/319

encephalitis lethargica, 2/486

epilepsy, 15/615

extralamellar nucleus, *see* Extralamellar thalamic nucleus

flavivirus, 56/135

function, 1/73, 6/107-109

GM2 gangliosidosis, 2/487

hemiballismus, 49/369, 373

hereditary olivopontocerebellar atrophy (Menzel), 21/438

hereditary sensory and autonomic neuropathy type I, 60/9

history of pathology, 6/66-68

Huntington chorea, 2/491, 6/338, 49/319

hydranencephaly, 30/670, 50/346

hydrocephalus, 30/670

intention tremor, 6/108

Japanese B encephalitis, 2/486

kernicterus, 6/502

left, *see* Left thalamus

lissencephaly, 42/40

locus ceruleus, 74/155

migraine, 5/85, 48/166

motor syndrome, 1/186

neglect syndrome, 45/156

neuropsychology, 45/2

nuclear group, 2/472

olivopontocerebellar atrophy (Dejerine-Thomas), 21/425

opiate receptor, 45/236

opticocochleodentate degeneration, 21/547, 42/241

pain, 45/234-236

paresthesia, 1/109

pathologic laughing and crying, 3/362, 45/223

phantom limb, 45/398

phantom sensation, 45/400

Pick disease, 2/491, 46/236

pons, 74/155

progressive pallidal atrophy (Hunt-Van Bogaert), 2/491

pure sensory stroke, 54/250

right, *see* Right thalamus

schizophrenia, 2/431

sensory extinction, 45/160

spreading depression, 48/65

St. Louis encephalitis, 2/486, 56/138

status cribrosus, 2/486

status marmoratus, 2/486

stereoencephalotomy, 2/491

sub, *see* Subthalamus

subacute necrotizing encephalomyelopathy, 42/488, 625

subacute sclerosing panencephalitis, 2/487

symptomatology, 2/476

tick-borne encephalitis, 2/486

tremor, 2/485, 49/597

trigeminal neuralgia, 5/275

vascular supply, 5/86

venous drainage, 2/476

ventral oral anterior nucleus, 6/108

ventrolateral nucleus, 6/107-109

visuospatial agnosia, 45/170

Wernicke encephalopathy, 2/487

Wernicke-Korsakoff syndrome, 45/197

Thalamus degeneration, 21/587-604

  Alzheimer disease, 2/491

  amyotrophic lateral sclerosis, 22/141

  association nucleus, 21/598

  clinicopathologic correlation, 21/600

  congenital retinal blindness, 21/598

  degeneration type, 21/598

  Friedreich ataxia, 21/598, 600

  Huntington chorea, 21/598

  infantile spinal muscular atrophy, 21/598

  multiple neuronal system degeneration, 21/596

  neurologic disease, 21/596

  nonspecific sensory relay nucleus, 21/597

  nosology, 21/598

  reticulocerebellar degeneration, 21/597

  specific sensory relay nucleus, 21/598

  Wernicke encephalopathy, 21/593

Thalamus fusion

  Arnold-Chiari malformation type II, 50/405

Thalassemia

  see also Hemoglobinopathy

  acoustic nerve compression, 63/257

  alpha, *see* Thalassemia major

  beta, *see* Thalassemia major

  brain hemorrhage, 55/466, 513

  brain phlebothrombosis, 55/466

  cerebrovascular disease, 55/466

  classification, 38/44, 63/257

  erythropoietin, 63/257

headache, 55/466
nerve root compression, 63/257
neurologic manifestation, 38/45
neurologic symptom, 55/466
neurologic symptom category, 38/45
pain attack, 63/257
pathophysiology, 63/257
review, 38/44-46
spinal cord compression, 63/257
thalassemic face, 63/257
treatment, 63/258
Thalassemia intermedia
hemoglobin H, 42/630
Thalassemia major
absent beta chain, 38/44
Bart hemoglobin, 42/630
birth incidence, 42/624
cerebrovascular disease, 55/466
cranial bone hypertrophy, 42/629
EEG, 42/629
fetal hemoglobin, 42/629
hemiplegia, 42/629
hepatosplenomegaly, 42/629
heterozygote frequency, 42/629
muscular atrophy, 42/629
seizure, 42/629
spinal cord compression, 42/629
transfusion dependence, 63/257
visual disorder, 42/629
Thalidomide
areflexia, 42/660
Babinski sign, 42/660
Bitter syndrome, 14/107
chronic axonal neuropathy, 51/531
dermatitis, 42/660
drug induced polyneuropathy, 51/304
human immunodeficiency virus infection, 71/388
hyporeflexia, 42/660
myalgia, 42/660
nerve conduction, 7/174
neuropathy, 7/540, 71/388
neurotoxicity, 51/304, 71/368
neurotoxin, 7/174, 540, 42/660, 46/79, 50/219,
    51/68, 304, 531, 65/559
paresthesia, 42/660
piperidinedione intoxication, 37/354
polyneuritis, 42/660
sensory loss, 42/660
Stilling-Türk-Duane syndrome, 42/307
susceptibility, 42/660
toxic polyneuropathy, 51/68
Thalidomide intoxication
hereditary deafness, 50/219

polyneuropathy, 65/559
sensory polyneuropathy, 65/559
teratogenic effect, 37/355
Thalidomide polyneuropathy
incidence, 51/305
posterior column, 51/305
sensory conduction velocity, 51/305
Thalidomide syndrome
facial paralysis, 42/660
gastrointestinal disorder, 42/660
heart disease, 42/660
iris coloboma, 42/660
mental deficiency, 46/79
microphthalmia, 42/660
phocomelia, 42/660
pseudo, see Roberts syndrome
renal abnormality, 42/660
sensorineural deafness, 42/660
teratogenic agent, 30/123
Thallium
autonomic nervous system, 75/38
brain edema, 64/4
chronic axonal neuropathy, 51/531
distal axonopathy, 64/12
element data, 64/323
neuronal mitochondrial vacuole, 51/266
neuropathy, see Thallium polyneuropathy
neurotoxicity, 51/265
neurotoxin, 7/515, 543, 8/337, 9/646, 30/123,
    36/239-273, 46/392, 51/68, 265, 274, 332, 479,
    531, 63/152, 64/4, 8, 10, 12, 14, 323, 325
optic neuropathy, 64/8
segmental demyelination, 64/14
speciation, 64/323
toxic encephalopathy, 64/4, 12
toxic neuropathy, 64/8, 10, 14
toxic polyneuropathy, 51/68, 274
toxicity, 64/325
Thallium intoxication
action site, 64/10
activated carbon, 64/326
alopecia, 64/325
ataxia, 64/325
autonomic nervous system, 75/503
autonomic polyneuropathy, 51/479
biochemical aspect, 36/255-257
biokinetic, 64/323
bulbar paralysis, 64/325
cardiac, 36/249
cardiovascular symptom, 64/325
central compartment, 64/323
chorea, 49/556, 64/325
circulatory, 36/247, 249

clinical features, 36/245-253, 64/325
CNS, 64/323
coma, 64/325
constipation, 64/324
cutaneous symptom, 64/325
definition, 36/241
dementia, 46/392
demyelination, 9/646
diethyldithiocarbamic acid, 64/326
differential diagnosis, 36/253-255
dimercaprol, 36/272
distribution volume, 64/324
dysarthria, 64/325
dysphagia, 64/325
edetic acid, 36/272
EEG, 36/253, 64/325
elimination half-life, 64/324
EMG, 36/253
encephalopathy, 36/252
endocrinic, 36/249
enteroenteral cycle, 64/323
epilepsy, 64/325
etiology, 36/241-245
excretion, 64/324
fast-exchange compartment, 64/323
fecal, 64/324
ferric ferrocyanide, 64/326
forced diuresis, 64/327
furosemide, 64/327
gastric lavage, 64/328
gastrointestinal, 64/323
gastrointestinal symptom, 64/325
hallucination, 64/325
hemoperfusion, 64/327
history, 36/239-241
industry, 64/323
insecticide, 64/323
kidney, 36/249
lethal dose, 64/325
leukoencephalopathy, 9/646
Mees line, 64/326
meningeal irritation, 36/248
motor disturbance, 36/246
motor weakness, 64/325
myelopathy, 64/8
neuropathy, 7/515, 543, 64/8
orthostatic hypotension, 63/152
pain, 36/245, 64/325
paresthesia, 64/325
pathology, 36/257-271
pathophysiology, 64/326
pesticide, 64/323
plasma level, 64/323, 325

polyneuropathy, 75/503
potassium, 64/327
prognosis, 36/272
psychiatric disorder, 36/251
psychosis, 64/325
ptosis, 64/325
Raynaud phenomenon, 8/337
renal symptom, 64/326
respiratory involvement, 36/247
rodenticide, 64/323
sensory disorder, 36/245, 248
skin abnormality, 36/247
sleep disorder, 36/249
slow exchange, 64/323
steady state, 64/324
symptom, 64/4, 12
therapy, 64/326
treatment, 36/272
tremor, 64/325
urinary excretion, 64/324
visual impairment, 64/8
Thallium polyneuropathy
    alopecia, 51/274
    painful, 51/274
    treatment, 51/274
    vitamin B2 deficiency, 51/332
Thanatophoric dwarfism
    achondrogenesis (Fraccaro-Parenti), 31/272
    cloverleaf skull syndrome, 31/272, 43/388
    CNS, 31/274
    corpus callosum agenesis, 43/388
    encephalomeningomyelocele, 43/388
    genetics, 31/274
    hydrocephalus, 31/274, 43/388
    mental deficiency, 46/88
    microgyria, 43/388
    neuronal heterotopia, 43/388
    polyhydramnios, 31/274
    radiologic appearance, 31/273
    swayback nose, 43/388
    vertebral abnormality, 43/388
Thanatophoric dysplasia
    polymicrogyria, 50/257
Tharlarctos maritimus intoxication
    vitamin A, 37/96
Thassalophryne intoxication
    family, 37/78
    venomous spine, 37/78
The Netherlands
    neurology, 1/10, 24
    paraneoplastic syndrome, 69/351
Thebaine
    chemistry, 37/365

toxic, 37/365
Thebaine intoxication, 37/365
Theiler murine encephalomyelitis virus, *see* Theiler
   virus
Theiler virus
   animal viral disease, 34/295, 302
   demyelination, 34/295
   experimental acute viral myositis, 56/195
   multiple sclerosis, 47/324
   picornavirus, 56/11
Theiler virus encephalomyelitis
   multiple sclerosis, 47/431
Theophylline
   acquired myoglobinuria, 62/573, 577
   Cheyne-Stokes respiration, 63/484
   postlumbar puncture syndrome, 61/156
   toxic myopathy, 62/601
Thephorin
   extrapyramidal disorder, 6/828
Thermal discrimination threshold
   diabetic polyneuropathy, 51/504
Thermal dolorimeter
   precise measurement, 1/86
Thermal injury
   see also Heat injury
   skull injury, 23/714
   subarachnoid hemorrhage, 12/121
Thermal salivation
   gustatory sweating, 2/122
Thermal stress
   sympathetic nervous system, 74/655
Thermanesthesia
   terminology, 1/86
Thermography
   cerebrovascular disease, 11/253
   cluster headache, 48/235
   cranial vault tumor, 17/106
   leprous neuritis, 51/218
   migraine, 48/77, 235
   thermistor recording, 11/254
Thermohyperesthesia
   terminology, 1/86
Thermolesion
   nerve injury, 51/134
Thermometry
   development, 74/437
Thermoreceptor
   thermoregulation, 75/54
Thermoregulation
   see also Heat conservation, Heat injury, Heat loss,
     Heat stroke *and* Heat syncope
   active rewarming, 75/61
   age, 75/58

aging, 74/233
amphetamine, 75/64
anhidrosis, 75/64
atropine, 75/61
autonomic, 75/54
autonomic disorder, 75/68
autonomic nervous system, 1/437-451, 74/1, 14,
   76, 207
behavior, 75/54
blood flow, 1/463, 26/371
brain embolism, 53/207
brain infarction, 53/207
brain injury, 23/231
cardiac arrest, 75/61
central temperature, 26/361, 368
cerebrovascular disease, 53/207
core rewarming, 75/61
core temperature, 74/443, 75/61
cortical control, 75/54
cutaneous vasomotor activity, 26/365
degenerative disorder, 75/72
dehydration, 75/58
demyelination, 47/38
electrical injury, 74/321
engineering model, 74/440
epilepsy, 15/140
exercise, 74/263
failure, 75/59, 72
fever, 1/447, 74/439
Gilles de la Tourette syndrome, 49/631
heat gain, 26/361, 363
heat loss, 26/359, 364
history, 1/437, 26/355
homeostasis, 61/275
hyperthermia, 26/369, 372, 75/62
hypohidrosis, 75/64
hypothalamus, 1/443, 2/438, 61/277, 71/570,
   75/53, 59
hypothermia, 2/447, 26/368, 372, 71/569, 75/59
impaired, 75/116
intracranial hemorrhage, 53/207
intracranial hypertension, 16/135
malignant hyperthermia, 75/73
management, 26/371
mechanism, 26/358
memory disorder, 27/463
mental activity, 75/58
multiple sclerosis, 9/182, 47/136, 142, 150
myoglobinuria, 62/557
neonatal, 74/208
nerve conduction velocity, 47/20
nerve lesion, 7/252, 266
neuroleptic malignant syndrome, 75/74

outline, 74/439
panting, 26/367
paradoxical sleep, 1/491
Parkinson disease, 75/179
passive rewarming, 75/61
pathway, 75/53
piloerection, 26/367
pontine infarction, 12/15, 31
primary pontine hemorrhage, 12/39
pure autonomic failure, 2/120
pyrogen, 26/370
quantitative sudomotor axon reflex test, 75/65
Reye syndrome, 75/60
segmental variability, 61/276
sensor, 74/441
sex, 75/58
Shapiro syndrome, 75/60, 74
shivering, 1/451, 26/362, 75/55
silastic imprint method, 75/65
skin potential, 75/65
skin temperature, 1/463, 7/252, 266, 26/361, 365,
    369
sleep, 74/541
spinal cord injury, 26/211, 260, 355-376,
    61/275-287
spinal shock, 26/260
spinal subarachnoid hemorrhage, 12/173
spontaneous cerebellar hemorrhage, 12/56
stress, 45/249
sudomotor failure, 75/68
survey, 74/437, 75/53
sweat secretion, 75/58
sweating, 26/357, 366, 369
sympathetic influence, 26/356
systemic lupus erythematosus, 75/60
test, 26/370
testing, 1/466
thermoreceptor, 75/54
thermoregulatory sweat test, 75/65, 108
transient ischemic attack, 53/207
transverse spinal cord lesion, 61/275
transverse spinal cord syndrome, 2/199
traumatic vegetative syndrome, 24/579
vagal afferent, 74/447
vasodilator agent, 75/61
vasomotoricity, 1/440
Wernicke encephalopathy, 75/60
Thermoregulatory dysfunction
    spinal cord injury, 75/578
Thermoregulatory sweat test
    anhidrosis, 75/108
    thermoregulation, 75/65, 108
Theta rhythm

fakirism, 45/240
    grand mal epilepsy, 42/667
Theta rhythm abnormality
    myoclonic akinetic epilepsy, 42/667
Thévenard disease, see Hereditary sensory and
    autonomic neuropathy type I
Thiamine, see Vitamin B1
Thiamine pyrophosphate, see Cocarboxylase
Thiamine triphosphate
    subacute necrotizing encephalomyelopathy,
    42/625
Thiazide derivative
    central pontine myelinolysis, 47/587
Thick muscle filament
    skeletal muscle, 40/63
Thickness
    myelin, see Myelin thickness
    spinal nerve root, 25/398
1-[1-(2-Thienyl)cyclohexyl]-piperidine, see
    Tenocyclidine
Thigmanesthesia
    terminology, 1/89
Thigmesthesia
    general touch sense, 1/87
Thimet sulfone intoxication, see Phorate sulfone
    intoxication
Thin muscle filament
    skeletal muscle, 40/63
Thinking
    abstraction, 45/523
    delirium, 46/528
    neurophysiologic aspect, 3/37
    neuropsychologic defect, 45/522
    neuropsychology, 45/516
    reasoning, 45/523
    spatial, see Spatial thinking
Thioctic acid
    amanitin, 65/37
    arsenic polyneuropathy, 51/267
Thiocyanate
    cyanide intoxication, 65/28, 30
    cyanogenic glycoside intoxication, 65/28, 30
    excretion, 7/642
Thiocyanate intoxication
    organic, see Organic thiocyanate intoxication
Thiodan intoxication, see Endosulfan intoxication
Thioether oxidation, see Sulfide oxidation
6-Thioguanine, see Tioguanine
Thiol compound
    Minamata disease, 64/416
Thiolase deficiency, see Acetyl-coenzyme A
    acetyltransferase deficiency
Thiopental

neurologic intensive care, 55/209
status epilepticus, 73/329
Thiopental intoxication
barbiturate, 37/410
Thiophene intoxication
late cortical cerebellar atrophy, 21/497
Thioridazine
chemical formula, 65/275
neuroleptic agent, 65/275
phenothiazine, 65/275
primary pigmentary retinal degeneration, 13/225
Thioridazine intoxication
familial spastic paraplegia, 59/327
Thiosulfate
metachromatic leukodystrophy, 47/593
Thiotepa
adverse effect, 69/492
alkylating agent, 39/97
antineoplastic agent, 65/528
chemotherapy, 39/97
CNS leukemia, 69/251
encephalopathy, 63/359
leukemia, 63/359
meningeal leukemia, 63/359
myelopathy, 63/359, 65/528
neurologic toxicity, 39/97
neurotoxicity, 63/359
neurotoxin, 39/94, 97, 119, 63/359, 65/528, 534,
67/357
Thiotepa intoxication
headache, 65/534
myelopathy, 65/534
Thiothixene, see Tiotixene
Thioxanthene
chemical structure, 65/276
designer drug, 65/274
neuroleptic agent, 65/274
Third occipital condylus
bone abnormality, 50/388
features, 50/389
Third ventricle colloid cyst
aqueduct stenosis, 50/309
clinical presentation, 68/327
diagnosis, 68/328
histology, 68/327
pathophysiology, 68/327
treatment, 68/328
Third ventricle cyst, see Ependymal cyst
Third ventricle tumor
see also Ependymal cyst
actinomycosis, 17/445
anatomy, 16/113
angiography, 17/462

brain scanning, 17/459
choroid plexus papilloma, 17/442
classification, 17/441
colloid cyst, 17/440, 446-448, 450, 18/359
craniopharyngioma, 17/450
CSF, 17/456
cyst, see Ependymal cyst
dysembryoma, 17/450
echoEG, 17/459
EEG, 17/457
ependymoma, 17/442, 18/127
hemangioma, 17/444
histology, 17/442, 18/359
history, 17/440
incidence, 17/441
meningioma, 17/444
neuroastrocytoma, 17/444
prognosis, 17/472
radiodiagnosis, 17/459, 461
spongioblastoma, 17/444
subependymal astrocytoma, 17/443
supratentorial brain tumor, 18/358
surgery, 17/470
symptom, 17/452, 454
syncope, 75/218
treatment, 17/470
ventriculography, 17/465, 467, 469
Thirst
hypernatremia, 63/554
Thirst regulation
sodium metabolism, 28/497
Thomas, A., 1/7
Thomas extension
radial nerve palsy, 8/405
upper extremity, 8/405
Thomsen disease, see Myotonia congenita and
Phosphoglucomutase deficiency
Thomson syndrome, see Rothmund-Thomson
syndrome
Thoracic angioma
neuroradiology, 12/639
Thoracic duct obstruction
thoracic outlet syndrome, 51/124
Thoracic injury
spinal cord stab wound, 25/204
Thoracic intervertebral disc prolapse, 20/565-570
age, 20/565
angiography, 20/568
case series, 20/569
cause, 20/565
classification, 20/567
clinical features, 20/566
CSF, 20/567

myelography, 20/568
prevalence, 20/565
radiology, 20/567
sex ratio, 20/565
site, 20/565
surgical treatment, 20/568
table, 20/569
Thoracic nerve
anterior, *see* Anterior thoracic nerve
long, *see* Long thoracic nerve
Thoracic neurenteric cyst, *see* Intraspinal
neurenteric cyst
Thoracic outlet syndrome
see also Scalenus anticus syndrome
Adson-Coffey maneuver, 51/124
anatomy, 51/119
arterial compression, 51/123
cause, 51/119
cervical fibrous band, 2/146
cervical rib, 51/120
cervical rib syndrome, 51/121
classification, 51/124
compression neuropathy, 51/87
compression syndrome, 7/430, 443
costoclavicular maneuver, 51/124
costoclavicular syndrome, 51/121
diagnosis, 51/125
differential diagnosis, 51/128
Doppler flowmetry, 51/127
EMG, 51/126
F wave latency, 51/126
forearm intermittent claudication, 51/124
hyperabduction maneuver, 51/124
incidence, 51/121
muscular atrophy, 51/122
nerve conduction, 51/125
occupational group, 51/125
pain, 51/122
paresthesia, 51/122
pneumoplethysmography, 51/127
poliomyelitis, 51/124
Raynaud disease, 51/123
rib resection, 51/128
rucksack paralysis, 51/121
scalenus anticus syndrome, 7/443, 51/121, 124,
129
selective arteriography, 51/127
sex ratio, 51/122
somatosensory evoked potential, 51/126
spasmodic torticollis, 51/124
subclavian artery occlusion, 51/124
subclavian artery stenosis, 7/434
subclavian bruit, 51/124

surgery complication, 51/129
symptom group, 51/125
syringomyelia, 50/453
thoracic duct obstruction, 51/124
trauma, 51/124
treatment, 51/128
Thoracic rib syndrome
cervical rib, 8/332
reflex neurovascular syndrome, 8/332
Thoracic spinal column injury
hematothorax, 26/224
pneumothorax, 26/224
Thoracic spinal cord injury
see also Spinal injury
aorta dissection, 61/76
arthrodesis, 61/83
canal encroachment, 61/76
case report, 25/31
compression injury, 25/137
decompression, 61/83
diagnosis, 61/76
final outcome, 61/83
fracture, 26/265
general, 26/292
prevalence, 25/442
rehabilitation, 61/81
spondylodesis, 61/83
surgery, 61/81
surgical complication, 61/81
treatment, 61/81
Thoracic stenosis
vertebral canal stenosis, 50/471
Thoracic vertebral column
facet joint plane, 61/8
Thoracic vertebral column biomechanics
anatomy, 61/12
Thoracic vertebral column injury
anatomy, 61/75
anterior *vs* posterior column, 61/77
aorta dissection, 61/76
burst fracture, 61/78
compression fracture, 61/77
computer assisted tomography, 61/76
fracture dislocation, 61/79
incidence, 61/75
intervertebral disc prolapse, 61/77
lateral dislocation, 61/79
management, 61/81
multilevel injury, 61/76
nuclear magnetic resonance, 61/77
planar tomography, 61/76
sagittal dislocation, 61/79
secondary cord compression, 61/77

stabilization, 61/81
surgery, 61/81
Thoracic vertebral column stability
  checklist, 61/13
  coupled motion, 61/13
  scoring system, 61/13
Thoracodorsal nerve
  topographical diagnosis, 2/20
Thoracolumbar spinal cord injury
  conus medullaris central necrosis, 25/443
  course, 61/96
  outcome, 61/103
  surgical criteria, 61/96
Thoracolumbar spine injury, 25/437-464
  see also Spinal injury
  anal reflex, 25/455
  anatomy, 25/439
  autonomic dysfunction, 25/454
  bony fusion, 25/462
  bulbocavernosus reflex, 25/455
  classification, 25/440
  diagnosis, 25/453
  facet interlocking, 26/296
  fracture dislocation, 26/267, 294
  fusion, 25/462
  general, 26/292
  history, 25/437
  instability, 25/447
  intestinal function, 25/455
  laminectomy, 25/459, 461
  management, 25/451, 456
  mechanism, 25/137
  missile injury, 25/440
  muscle stretch reflex, 25/441, 454
  pain, 25/463
  pathology, 25/450
  postural reduction, 25/458, 460, 26/213, 294
  radiology, 25/455
  reduction, 25/459
  rehabilitation, 25/463
  seat belt injury, 25/448
  spinal shock, 25/441
  surgical, 25/459
  type, 25/447
Thoracolumbar vertebral column injury
  anatomy, 61/34
  anterior element, 61/34
  biomechanics, 61/34
  burst injury, 61/35
  canal encroachment, 61/97
  classification, 61/34, 91
  compression fracture, 61/90
  computer assisted tomography, 61/91

conservative treatment, 61/93
conservative treatment criteria, 61/95
corrective surgery history, 61/97
dural tears, 61/96
empirical axiom, 61/104
facet fracture, 61/36
facet interlocking, 61/93
facet necrosis, 61/38
facet orientation, 61/34
fixateur externe, 61/99
fixateur interne, 61/99
flexion injury, 61/35
flexion rotation injury, 61/35
frequency, 61/89
Guttmann principle, 61/93
injury type, 61/90
instability classification, 61/95
instability definition, 61/90
instrumentation, 61/97
interventionist view, 61/89
intervertebral disc hemorrhage, 61/36
ligamentotaxis, 61/97
Magnus method, 61/93
neuroanatomy, 61/92
neurologic deficit, 61/92
neurologic deterioration, 61/102
neurologic outcome, 61/92
noninterventionist view, 61/89
nuclear magnetic resonance, 61/92
occult injury, 61/36
open injury, 61/96
open reduction, 61/95
osteoporosis, 61/95
outcome, 61/103
polytrauma, 61/96
posterior element, 61/34
Rauchfusz suspension, 61/93
reangulation, 61/95
recent knowledge, 61/89
redislocation, 61/95
rehabilitation, 61/100
segmental lesion distribution, 61/34
spinal cord compression, 61/92
stability dilemma, 61/89
surgery timing, 61/96
surgical complication, 61/102
surgical criteria, 61/96
three column concept, 61/90
treatment choice, 61/92
vertebral end plate fracture, 61/36
vertebral recollapse, 61/95
vertebral wedge compression fracture, 61/35
Threoceramide

biochemistry, 9/9
Threonine
  CSF, 6/348
  familial spastic paraplegia, 59/313
  Huntington chorea, 6/348
Threosphingosine
  biochemistry, 9/9
Throat
  foreign body, 48/502
  pain, 48/502
  raw, 48/502
Thrombasthenia
  Glanzmann, see Glanzmann thrombasthenia
Thromboangiitis obliterans, 55/307-314
  brain, see Brain thromboangiitis obliterans
  brain infarction, 53/179
  carotid system syndrome, 53/308
  cerebral, see Brain thromboangiitis obliterans
  deafness, 55/132
  histology, 11/131
  intracerebral hematoma, 11/672
  leukodystrophy, 10/127
  nerve ischemia, 12/655
  Raynaud phenomenon, 43/71
  stagnation thrombosis, 11/133
  Takayasu disease, 12/411, 46/359
  vasa nervorum disease, 12/655
Thromboangiitis obliterans cerebri, see Brain
    thromboangiitis obliterans
Thrombocyte, see Platelet
Thrombocythemia
  amaurosis fugax, 55/108, 473
  brain infarction, 55/473
  cerebrovascular disease, 55/473
  epilepsy, 55/473
  essential, see Essential thrombocythemia
  headache, 55/473
  hereditary hemorrhagic telangiectasia (Osler),
    55/449
  inherited platelet disease, 55/472
  neurologic symptom, 55/473
  primary hemorrhagic, see Primary hemorrhagic
    thrombocythemia
  primary type, 55/473
  transient ischemic attack, 55/473
Thrombocytopenia
  Abt-Letterer-Siwe disease, 42/441
  adult Gaucher disease, 10/564, 66/123
  alcohol intoxication, 64/113
  antiphospholipid antibody, 63/122, 330
  Argentinian hemorrhagic fever, 56/364
  blood transfusion, 63/321
  brain fat embolism, 55/186

brain hemorrhage, 54/71, 55/472, 63/322
cerebrovascular disease, 55/472
coagulopathy, 63/309, 321
cytomegalic inclusion body disease, 56/267
cytotoxic chemotherapy, 63/321
etiology, 38/64
Fanconi syndrome, 43/16
fat embolism, 23/635
giant hemangioma, 14/773
headache, 55/472, 63/322
hematomyelia, 38/69
hereditary (Jürgens-Von Willebrand), see
    Jürgens-Von Willebrand hereditary
    thrombocytopenia
histiocytosis X, 42/441
hyperglycinemia, 42/565
immunothrombocytopenic purpura, 63/321
intracranial hemorrhage, 38/65-68
intravascular consumption coagulopathy, 55/496,
    63/321
juvenile Gaucher disease type A, 66/125
juvenile Gaucher disease type B, 66/125
leukemia, 63/321, 343
listeriosis, 52/91
lymphocytic leukemia, 63/343
lysinuric protein intolerance, 29/210
megaloblastic anemia, 63/322
mephenytoin intoxication, 37/202
methylmalonic aciduria, 42/601
myelodysplasia, 63/322
myelofibrosis, 63/322
myeloid leukemia, 63/343
nerve, 38/69
organic acid metabolism, 66/641
penicillamine intoxication, 27/405
postneurosurgery intracranial bleeding, 63/322
quinidine, 63/192
sea blue histiocyte disease, 42/472
spinal leukemia, 20/112
subdural hematoma, 26/31
systemic lupus erythematosus, 55/379
transient ischemic attack, 53/266
treatment, 38/69
Thrombocytopenic purpura
  see also Thrombotic thrombocytopenic purpura
  brain hemorrhage, 53/34, 63/322
  brain infarction, 63/323
  cat scratch disease, 52/128
  immune type, 63/323
  leukoencephalopathy, 9/598
  porencephaly, 50/359
  thrombotic, see Thrombotic thrombocytopenic
    purpura

treatment, 63/323
Werlhof essential, *see* Werlhof essential
thrombocytopenic purpura
Thrombocytosis
brain infarction, 53/166
brain ischemia, 53/166
iron deficiency anemia, 55/464
Kawasaki syndrome, 52/266
metastatic cardiac tumor, 63/96
polyarteritis nodosa, 55/354
primary cardiac tumor, 63/96
systemic brain infarction, 11/461
transient ischemic attack, 53/266
Thromboembolic disease
Klinefelter syndrome, 50/549
Thrombolytic agent, *see* Fibrinolytic agent
Thrombopathia
sensorineural deafness, 42/380
Thrombophlebitis
Behçet syndrome, 71/217
brain, *see* Brain thrombophlebitis
brain abscess, 52/144
cavernous sinus, *see* Cavernous sinus
thrombophlebitis
cerebrovascular disease, 53/32
diamorphine intoxication, 65/355
dural sinus, *see* Venous sinus thrombophlebitis
Galen vein, 33/172
inferior sagittal sinus, 33/172
intracranial, *see* Intracranial thrombophlebitis
intracranial infection, 24/218
intravascular consumption coagulopathy, 55/496
nonbacterial thrombotic endocarditis, 11/415, 417,
63/120
opiate intoxication, 65/355
Queckenstedt test, 33/165
sign, 11/417
straight sinus, 33/172
subcutaneous, *see* Subcutaneous thrombophlebitis
superior longitudinal sinus, *see* Superior
longitudinal sinus thrombophlebitis
superior petrosal sinus, *see* Superior petrosal sinus
thrombophlebitis
transverse sinus, *see* Transverse sinus
thrombophlebitis
Thromboplastic activity
CSF, *see* CSF thromboplastic activity
Thromboplastin time
a-alphalipoproteinemia, 51/396
Thrombosis
anterior cerebral artery, 53/339
anterior spinal artery, 61/111
anterolateral tract syndrome, 2/210

aorta, *see* Aorta thrombosis
arterial hypertension, 11/130
aseptic cavernous sinus, *see* Aseptic cavernous
sinus thrombosis
basilar artery, *see* Basilar artery thrombosis
basilar artery aneurysm, 12/22
brain, *see* Brain thrombosis
brain aneurysm, 12/84
brain arteriosclerosis, 12/25, 53/95
brain venous, *see* Brain phlebothrombosis
brain venous infarction, 54/438, 63/3
carotid artery, *see* Carotid artery thrombosis
carotid cavernous, *see* Carotid cavernous
thrombosis
cavernous sinus, *see* Cavernous sinus thrombosis
central retinal venous, *see* Central retinal venous
thrombosis
deep vein, *see* Deep vein thrombosis
dural sinus, *see* Venous sinus thrombosis
endothelium, 11/130
Galen vein, *see* Galen vein thrombosis
human immunodeficiency virus infection, 71/337
inferior petrosal sinus, *see* Inferior petrosal sinus
thrombosis
middle cerebral artery, 53/355
plaque, 9/294
puerperal, *see* Puerperal thrombosis
radiation injury, 23/650
sagittal sinus, *see* Sagittal sinus thrombosis
sinus, *see* Sinus thrombosis
spinal cord injury, 26/207, 315, 394
stagnation, *see* Stagnation thrombosis
striatal necrosis, 49/500
subclavian artery, *see* Subclavian artery occlusion
superior longitudinal sinus, *see* Superior
longitudinal sinus thrombosis
superior petrosal sinus, *see* Superior petrosal sinus
thrombosis
transverse sinus, *see* Transverse sinus thrombosis
traumatic, 24/159
traumatic carotid, *see* Traumatic carotid
thrombosis
traumatic venous sinus, *see* Traumatic venous
sinus thrombosis
tuberculous meningitis, 52/199
venous, *see* Venous thrombosis
venous sinus, *see* Venous sinus thrombosis
vertebral artery, *see* Vertebral artery thrombosis
vertebral artery injury, 26/57-62
vertebrobasilar artery, *see* Vertebrobasilar artery
thrombosis
vertebrobasilar insufficiency, 2/245, 251
Thrombotic endocarditis

brain embolism, 55/489
nonbacterial, *see* Nonbacterial thrombotic
   endocarditis
nonseptic, *see* Nonseptic thrombotic endocarditis
Thrombotic thrombocytopenic purpura
   see also Thrombocytopenic purpura
   anoxic ischemic leukoencephalopathy, 47/541
   ataxia, 55/472, 63/324
   brain infarction, 53/34, 166, 63/324
   brain ischemia, 53/166
   burr erythrocyte, 9/600, 607
   cerebrovascular disease, 53/34, 55/472
   coma, 55/472, 63/324
   cranial nerve palsy, 55/472
   cranial neuropathy, 63/324
   demyelination, 9/599-607
   EEG, 55/473
   encephalopathy, 45/119
   epilepsy, 55/472, 63/325
   headache, 55/472, 63/324
   helmet erythrocyte, 9/600, 607
   hemiparesis, 63/324
   histology, 11/139
   leukoencephalopathy, 9/599
   mortality, 55/473
   multiple cranial neuropathy, 55/472
   neurologic symptom, 38/70
   papilledema, 63/324
   retinal hemorrhage, 63/325
   schizocyte, 9/607
   symptom, 63/324
   systemic brain infarction, 11/450
   triangular erythrocyte, 9/600, 607
Thromboxane
   headache, 48/281
   migraine, 48/202
   steroid, 48/202
Thrombus
   arteriolar fibrinohyalinoid degeneration, 11/627
   development, 11/334
   headache, 48/281
   organization, 11/335
   resolution, 11/336
Thrush, *see* Candidiasis
Thujone, *see* Absinthol
Thumb
   adducted, *see* Adducted thumb syndrome
   broad, *see* Broad thumb
   drummer palsy, 7/332
   simian hand, 21/279, 22/291
Thumb agenesis
   holoprosencephaly, 50/237
Thumb hyperflexia

Patau syndrome, 14/121
Thumb hypoplasia
   Patau syndrome, 14/121, 31/507
Thumb sucking
   cerebral dominance, 4/468
   lateralization pattern, 4/468
Thunderclap headache
   carotid dissecting aneurysm, 54/275
   intracranial, 54/275
   vertebral dissecting aneurysm, 54/277
Thuyon, *see* Absinthol
Thymectomy
   multiple sclerosis, 47/201
   myasthenia gravis, 41/96, 134, 43/158, 62/419
   radiotherapy, 41/135
   transcervical, 41/135
   transthoracic, 41/135
Thymicolymphatic involution
   general adaptation syndrome, 45/252
Thymidine
   chemotherapy, 39/120
   neurologic toxicity, 39/120
Thymidine kinase
   brain tumor, 67/301
Thymocyte antibody
   multiple sclerosis, 47/197-199
Thymoma
   endocrine candidiasis syndrome, 42/642
   myasthenia gravis, 41/361, 43/157, 62/402
Thymopoietin
   myasthenia gravis, 62/406
   thymus, 41/115
Thymus
   anencephaly, 30/181
   ataxia telangiectasia, 14/76, 31/24
   germinal center, 41/114
   thymopoietin, 41/115
Thymus aplasia
   ataxia telangiectasia, 60/361
Thymus gland
   myasthenia gravis, 43/157, 62/405
Thymus hyperplasia
   anencephaly, 42/12
   myasthenia gravis, 41/96, 102, 114, 135, 361
   myositis, 41/85
Thymus hypoplasia
   congenital ichthyosis, 43/406
   immunocompromised host, 56/468
Thyroid adenoma
   Sturge-Weber syndrome, 14/69
Thyroid associated ophthalmopathy
   classification, 62/530
   clinical features, 62/530

computer assisted tomography, 62/531
diagnostic triad, 62/531
diplopia, 62/531
endocrine myopathy, 62/528
Hashimoto thyroiditis, 62/530
hereditary progressive arthro, *see* Hereditary
   progressive arthro-ophthalmopathy
hyperthyroidism, 41/240
immunosuppression, 62/532
lid lag, 62/530
myotonic dystrophy, 22/179, 204
NOSPECS system, 62/530
nuclear magnetic resonance, 62/531
orbital myositis, 62/298
pathogenesis, 62/531
progressive external ophthalmoplegia
   classification, 62/290
protirelin stimulation test, 62/531
T3 suppression test, 62/531
thyroid disease, 70/88
treatment, 62/531
Thyroid autoimmune antibody
   Down syndrome, 50/532
Thyroid cancer
   opsoclonus, 60/657
   Parkinson disease, 27/272, 49/126
Thyroid carcinoma
   medullary, *see* Medullary thyroid carcinoma
   multiple endocrine adenomatosis type III, 42/752
   neurofibromatosis type I, 14/134
   vertebral column tumor, 20/43
   vertebral metastasis, 20/424
Thyroid cyst
   B-mode sonography, 54/21, 25
Thyroid disease
   see also Hyperthyroidism *and* Hypothyroidism
   cornea, 70/89
   encephalopathy, 70/91
   endocrine myopathy, 41/240
   mental change, 70/91
   optic nerve, 70/90
   progressive external ophthalmoplegia, 41/240,
     43/136
   seizure, 70/91
   survey, 70/81
   thyroid associated ophthalmopathy, 70/88
Thyroid dysfunction
   eye muscle, 1/612
   Parkinson disease, 49/126
Thyroid gland
   autonomic nervous system, 27/261-263
   brain development role, 27/355
   congenital deafness, 42/366

Down syndrome, 50/532
epilepsy, 15/318
exophthalmos, 41/240
function, 26/395, 400
headache, 5/230
myasthenia gravis, 27/263-267, 41/240, 43/157
neurologic relationship, 27/255-274
periodic paralysis, 41/165, 241
progressive bulbar palsy, 41/240
psychiatric disorder, 27/272
sensorineural deafness, 42/366
Thyroid gland dysgenesis
   anencephaly, 42/12
   ataxia, 42/632
   feeding disorder, 42/632
   hypothermia, 42/632
   mental deficiency, 42/631
   myxedema, 42/632
   phenylthiocarbamide tasting, 42/632
   postmaturity, 42/631
   respiratory distress, 42/632
   spastic diplegia, 42/632
   terminology, 42/631
Thyroid hormone
   anorexia nervosa, 46/586
Thyroid releasing factor, *see* Protirelin
Thyroid stimulating hormone, *see* Thyrotropic
   hormone
Thyroidectomy
   post, *see* Postthyroidectomy
Thyroiditis
   autoimmune, *see* Hashimoto thyroiditis
   Freund adjuvant, 9/506
   Hashimoto, *see* Hashimoto thyroiditis
Thyrotoxic myopathy, 70/82
   acute, 27/268
   amyotrophic lateral sclerosis, 62/529
   aspartate aminotransferase, 62/529
   biochemistry, 62/530
   chronic, 27/267
   creatine kinase, 62/529
   differential diagnosis, 27/264
   dysphagia, 62/528
   dysphonia, 62/528
   EMG, 62/529
   endocrine myopathy, 41/235-240, 62/528
   fructose-bisphosphate aldolase, 62/529
   muscle biopsy, 62/529
   muscle cramp, 62/528
   muscle stretch reflex, 62/528
   muscle weakness, 62/528
   myalgia, 62/528
   myasthenia gravis, 27/263, 41/240, 43/157

myoglobinuria, 62/529
myokymia, 62/528
neuropathy, 62/528
oxidative phosphorylation, 41/239
respiratory dysfunction, 62/528
rhabdomyolysis, 62/529
weakness distribution, 62/528
Thyrotoxic ocular myopathy, *see* Thyroid associated
    ophthalmopathy
Thyrotoxic periodic paralysis
  Asian race, 62/532
  attack, 62/461
  clinical features, 62/532
  endocrine myopathy, 41/165, 241-243, 62/528
  9α-fluorohydrocortisone, 41/242
  genetics, 62/461
  HLA haplotype, 62/532
  inheritance, 62/532
  oxidative phosphorylation, 41/242
  precipitating factor, 62/532
  prevalence, 62/461
  treatment, 62/532
Thyrotoxic tremor
  hyperthyroidism, 6/818
  postural tremor, 6/818
Thyrotoxicosis, *see* Hyperthyroidism
Thyrotropic hormone
  adenohypophysis, 2/442
  depression, 46/474
  Down syndrome, 50/532
  Levi-Lorain dwarfism, 2/454
  orbital myositis, 62/300
  pituitary adenoma, 17/393, 404
  pituitary dwarfism type II, 42/613
  progressive external ophthalmoplegia, 22/184
Thyrotropic hormone secreting adenoma
  dopaminergic agent, 68/357
  octreotide, 68/357
  radiotherapy, 68/357
  surgery, 68/357
Thyrotropic hormone secreting tumor
  pituitary tumor, 68/344
Thyrotropin, *see* Thyrotropic hormone
Thyrotropin releasing hormone, *see* Protirelin
Thyroxine
  benign intracranial hypertension, 67/111
  brain injury, 23/123
  epilepsy, 15/731
  myotonia congenita, 27/272
Tiagabine
  epilepsy, 73/348, 361
Tibial adductor reflex
  diagnostic value, 1/247

Tibial nerve
  compression neuropathy, 51/105
  phosphoglucoisomerase, 7/218
  posterior, *see* Posterior tibial nerve
  topographical diagnosis, 2/45
Tibial nerve compression
  rheumatoid arthritis, 38/484
Tibial syndrome
  anterior, *see* Anterior tibial syndrome
Tibialis posterior reflex
  lower extremity, 1/247
Tic, 6/782-803
  see also Gilles de la Tourette syndrome
  anatomic classification, 1/279
  childhood chronic multiple, *see* Childhood chronic
    multiple tic
  chorea, 6/789
  classification, 6/785
  distribution, 49/628, 630
  drug induced, 6/792-794
  facial, *see* Facial tic
  facial spasm, 6/801
  features, 1/291
  Gilles de la Tourette syndrome, 42/221
  Guillain-Mollaret triangle, 6/798
  hyperactivity, 46/181
  motor, *see* Motor tic
  multiple, *see* Multiple tics
  neuroacanthocytosis, 42/209
  psychogenic tic, 6/785-787
  Richner-Hanhart syndrome, 42/581
  striatal, 6/794-798
  tardive dyskinesia, 49/186
  Uhrmacher, *see* Uhrmacher tic
Tic disease
  chorée variable des dégénérés, 6/301
  striatal type, 6/668
Tic douloureux, *see* Cranial neuralgia
Tic douloureux of the chorda tympani, *see*
    Intermedius neuralgia
Tick bite intoxication
  arbovirus, 37/111
  ataxia, 37/111
  atrophy, 37/111
  CNS, 37/111
  CSF pleocytosis, 37/111
  diplopia, 37/111
  facial paralysis, 37/111
  fasciculation, 37/111
  fatigue, 37/111
  flaccid paralysis, 37/111
  irritability, 37/111
  nystagmus, 37/111

parkinsonism, 37/111
ptosis, 37/111
rickettsial infection, 37/111
tick paralysis, 37/111
viral infection, 37/111
weakness, 37/111
Tick-borne disease
  Ixodes dammini, 51/199
  Ixodes ricinus, 51/199
Tick-borne encephalitis
  acute viral encephalitis, 34/76
  amyotrophy, 22/22
  central eastern, *see* Central European tick-borne
    encephalitis
  central European, *see* Central European tick-borne
    encephalitis
  clinical features, 56/139
  far eastern, *see* Far eastern tick-borne encephalitis
  flavivirus, 56/12
  Ixodes persulcatus, 56/139
  Ixodes ricinus, 56/139
  thalamus, 2/486
  Togavirus, 56/12
Tick-borne encephalitis flavivirus
  multiple sclerosis, 47/330
Tick-borne encephalitis virus
  deafness, 56/107
  experimental acute viral myositis, 56/195
  vertigo, 56/107
Tick-borne typhus
  African, *see* African tick-borne typhus
  brucellosis, 34/657
  eastern hemisphere, *see* Eastern hemisphere
    tick-borne typhus
  fièvre boutonneuse, 34/657
  Indian, *see* Indian tick-borne typhus
  Kenyan, *see* Kenyan tick-borne typhus
  North Asian, *see* North Asian tick-borne typhus
  Queensland, *see* Queensland tick-borne typhus
  rickettsial infection, 34/651-658
  Rocky Mountain spotted fever, 34/651-657
  Russian, *see* Russian tick-borne typhus
  South African tick-bite fever, 34/657
Tick paralysis
  see also Acute muscle weakness
  acute cerebellar ataxia, 34/627
  acute muscle weakness, 41/290
  arthropod envenomation, 65/194
  botulism, 40/308
  Dermacentor andersoni, 41/290, 65/195
  evoked muscle action potential, 41/291
  Ixodus holocyclus, 65/195
  Ixodus matus, 65/195

progressive bulbar palsy, 59/225, 375
spinal muscular atrophy, 59/370
tick bite intoxication, 37/111
weakness, 40/297
Tick venom
  arthropod envenomation, 65/193
  ataxia, 65/195
  facial paralysis, 65/195
  myasthenic syndrome, 64/15
  neurotoxin, 37/111, 64/15, 65/194
  toxic encephalopathy, 65/193
  toxic myopathy, 64/15
  toxic neuromuscular junction disorder, 64/15
Tick venom intoxication
  action site, 64/15
  pathomechanism, 64/15
  symptom, 64/15
Tight junction
  blood-brain barrier, 53/84, 56/79
  nerve, 51/2
Tigretier
  involuntary movement, 6/301
Tigrolysis, *see* Axonal reaction
Tiling-Wernicke syndrome, *see* Operculum
    syndrome
Tilney, F., 1/5
Tilted optic disc
  Bedouin spastic ataxia, 60/655
  hereditary spastic ataxia, 60/464
Tilting chair test
  cerebellar lesion, 1/336
Tilting reaction
  basal ganglion, 49/78
Tilting and turning bed, *see*
    Stokes-Mandeville-Egerton bed
Time-dose effect plot
  brain tumor, 18/483
Time sense
  age, 3/242
  biological rhythm, 3/231-234
  biological time, 3/231
  consciousness, 3/118
  disorder, 3/32, 236-249
  drug, 3/248
  emotion, 3/330
  emotional distress, 3/331
  epilepsy, 45/509
  gnosis, 3/229
  government time, 3/233
  occipital lobe tumor, 17/331
  orientation, 3/229-249
  overactivity state, 3/236
  perception, 2/712

personal time, 3/233
physiologic change, 3/242
physiologic clock, 3/231-234
psychiatric disorder, 3/244
recognition, 4/30
rhythmic activity nervous system, 3/235
rhythmicity, 3/231
temporal lobe syndrome, 45/47
underactivity state, 3/237
Timolol
migraine, 48/180
Tin
chronic dialysis, 27/329
organic compound intoxication, 9/647
peritoneal dialysis, 27/329
Tinel sign
brachial plexus, 51/149
carpal tunnel syndrome, 51/94
migrant sensory neuritis, 51/462
nerve regeneration, 1/97
Tinnitus
acetylsalicylic acid intoxication, 65/460
acoustic nerve tumor, 17/669
basilar artery migraine, 48/136
brain embolism, 53/207
brain infarction, 53/207
carotid dissecting aneurysm, 54/272
cerebellopontine angle syndrome, 2/93
cerebellopontine angle tumor, 2/377
cerebrovascular disease, 53/207
cinchonism, 65/452, 485
cisplatin intoxication, 64/358
clinical examination, 57/138
diuretic agent intoxication, 65/440
fish roe intoxication, 37/86
furosemide, 65/440
Gellé syndrome, 2/241
head injury, 57/138
intracranial hemorrhage, 53/207
isoniazid intoxication, 65/480
lidocaine intoxication, 65/454
local anesthetic agent, 65/420
mefloquine intoxication, 65/485
Meniere disease, 2/377
methyl salicylate intoxication, 37/417
mexiletine intoxication, 65/455
nonsteroid anti-inflammatory agent intoxication,
    65/416, 563
osteitis deformans (Paget), 38/362
pulsatile, 54/272
quinidine intoxication, 37/448
salicylic acid intoxication, 37/417
sodium salicylate intoxication, 37/417

temporal bone fracture, 24/123, 125
temporal lobe tumor, 17/289
transient ischemic attack, 53/207
unilateral pulsating, see Unilateral pulsating
    tinnitus
vestibular neurinoma, 68/424, 444
Vogt-Koyanagi-Harada syndrome, 56/622
Tioguanine
antimetabolite, 39/106
chemotherapy, 39/106
neurologic toxicity, 39/106
Tiopronin
toxic myopathy, 62/597, 609
Tiotixene
chemical formula, 65/276
neuroleptic agent, 65/276
Tip of the tongue phenomenon
anomic aphasia, 45/319
TIPP intoxication, see Tetraisopropyl pyrophosphate
    intoxication
Tirilazad
paraplegia, 61/405
spinal cord injury, 61/405
tetraplegia, 61/405
transverse spinal cord lesion, 61/405
Tissot classification
epilepsy, 15/2
Tissue culture
acoustic neuroma, 17/82
astrocytoma, 16/38, 17/48, 50, 53, 60, 18/13
brain metastasis, 17/85
brain tumor, 16/37, 17/42, 45-48
carotid body tumor, 17/69
cell division, 17/87, 90, 92
cell migration, 17/86
cerebellar medulloblastoma, 17/71, 73-79
chemodectoma, 17/69
chromaffinoma, 17/69
chromosome, 17/90
contact inhibition, 17/86
covering, 17/48
craniopharyngioma, 17/82
CSF cytology, 17/43
dura mater, 17/48
ependymoma, 16/38, 17/65, 68, 18/110
fibrillary astrocytoma, 18/13
gemistocytic astrocytoma, 17/56
giant cell glioblastoma, 17/63
glioblastoma multiforme, 17/63
glomangioma, 17/70
history, 17/42
invasiveness, 17/86
juxtacarotid chemodectoma, 17/69

lemmocytoma, 17/83
medulloepithelioma, 17/69
meningioma, 17/83
mercury polyneuropathy, 51/273
muscle regeneration, 40/184
myelination, 9/40
neurinoma, 16/38, 17/82
neuroastrocytoma, 17/68, 77
neuroblastoma, 17/67
neurocytoma, 17/67
neuroepithelium, 17/44
neurofibrosarcoma, 17/83
neuroglia, 17/54
neuroma, 17/82
oligodendroglioma, 16/38, 17/65, 18/84
pheochromocytoma, 17/69
pinealoblastoma, 17/68
pinealoma, 17/68, 82
plexus papilloma, 17/82
ploidy, 17/90
retinoblastoma, 17/68
simplification, 17/86
spongioblastoma, 16/38
teratoma, 17/77
tuberous sclerosis, 17/77
vasoformative tumor, 17/85
Tissue level
neurotoxicity, 51/264
Tissue plasminogen activator
anticoagulant, 63/308
brain aneurysm neurosurgery, 55/59
brain infarction, 63/321
hemostasis, 63/305
middle cerebral artery syndrome, 53/368
risk, 63/321
subarachnoid hemorrhage, 55/59
vertebrobasilar artery thrombosis, 54/181
Titeca-Van Bogaert disease, see
Dentatorubropallidoluysian atrophy
Titin
limb girdle syndrome, 62/187
muscle fiber, 62/26
Tityustoxin
myasthenic syndrome, 64/15
neurotoxin, 64/15, 65/194
sodium permeability, 64/15
toxic myopathy, 64/15
toxic neuromuscular junction disorder, 64/15
Tityustoxin intoxication
action site, 64/15
pathomechanism, 64/15
symptom, 64/15
TMNU, see Trimethylnitrosourea

Toadfish intoxication, see Puffer fish intoxication
Toados intoxication, see Puffer fish intoxication
Tobacco addiction
brain blood flow, 55/524
brain hemorrhage, 55/524
brain infarction, 55/524
Tobacco alcoholic amblyopia, 28/331-340
clinical features, 28/332
etiology, 28/338-340
history, 28/332
Leber optic atrophy, 13/67
nicotine intoxication, 13/62, 65/264
pathology, 28/333-338
Tobacco amblyopia
cyanide intoxication, 7/643, 13/62, 65/30
cyanogenic glycoside intoxication, 36/521, 65/31
plasma thiocyanate, 7/643
vitamin B$_{12}$, 28/174
vitamin B$_{12a}$, 28/175
Tobey-Ayer test
superior petrosal sinus thrombophlebitis, 33/180
venous sinus thrombophlebitis, 33/165
Tocainide
asterixis, 63/192
cardiac pharmacotherapy, 63/192
encephalopathy, 63/192
epilepsy, 63/192
iatrogenic neurological disease, 63/192, 65/455
myotonia, 62/275
tremor, 63/192
Tocainide intoxication
delirium, 65/455
epilepsy, 65/455
hallucination, 65/455
paresthesia, 65/455
psychosis, 65/455
tremor, 65/455
vertigo, 65/455
visual impairment, 65/455
α-Tocopherol, see Vitamin E
α-Tocopherol intoxication, see Vitamin E
intoxication
TOCP, see Tri-o-cresyl phosphate
Toe
hammer, see Hammer toe
Toe walker
congenital ataxic diplegia, 42/121
Duchenne muscular dystrophy, 40/321, 361
late infantile acid maltase deficiency, 27/228,
62/480
phosphoglucomutase deficiency, 27/222, 41/192
Togavirus
see also Arbovirus, Arenavirus, Arthropod-borne

virus *and* Bunyamwera virus
  acute viral encephalitis, 34/68-78
  acute viral myositis, 56/193
  alphavirus genus, 56/12
  biology, 34/68-78
  cerebellar hypoplasia, 50/179
  eastern encephalitis, 56/136
  epidemic encephalitis, 56/30
  experimental acute viral myositis, 56/196
  features, 34/68-78, 56/11
  flavivirus genus, 56/12
  human, 56/3
  hydranencephaly, 50/340
  Japanese B encephalitis, 56/12
  lactate dehydrogenase elevating virus, 56/28
  multiple sclerosis, 47/325
  neurotropic virus, 56/12
  replication, 56/12
  RNA virus, 56/11
  rubella virus, 56/12
  rubivirus genus, 56/12
  St. Louis encephalitis, 56/12
  taxonomic place, 56/3
  tick-borne encephalitis, 56/12
  viral class, 34/5
Token test
  head injury outcome, 57/405
Tolazoline
  aura, 5/379
  restless legs syndrome, 8/318, 51/548
Tolbutamide
  multiple sclerosis, 9/398
Tolfenamic acid
  chronic paroxysmal hemicrania, 48/264
  headache, 48/174
  migraine, 48/174, 203
Tolmetin
  chemical meningitis, 56/130
  headache, 48/174
  migraine, 48/174
Tolosa-Hunt syndrome, 48/291-304
  age, 48/292, 296
  aneurysm, 31/147, 48/302
  angiography, 48/298
  aseptic cavernous sinus thrombosis, 48/303
  Bielschowsky relapsing alternating
    ophthalmoplegia, 48/303
  child, 48/37
  cluster headache, 48/226
  corticosteroid, 48/291, 299, 304
  cranial nerve, 48/297
  craniotomy, 48/300
  differential diagnosis, 48/300

  EMG, 48/299
  etiology, 48/300
  etiopathogenesis, 48/300
  granulomatous basal meningitis, 48/303
  headache, 48/6
  hereditary cranial nerve palsy, 60/42
  hereditary recurrent cranial nerve palsy, 48/303
  inflammation, 48/291
  laboratory finding, 48/298, 301
  literature, 48/292, 296
  myositis, 48/302
  natural history, 48/299
  neurologic finding, 48/297
  nonrecurrent nonhereditary multiple cranial
    neuropathy, 51/571
  oculomotor paralysis, 31/147
  oligosymptomatic hyperthyroidism, 48/303
  onset of attack, 48/296
  ophthalmoplegia, 48/296, 299, 302
  ophthalmoplegic migraine, 42/748, 48/302
  orbital apex syndrome, 48/291, 300
  orbital myositis, 62/301
  orbital pain, 48/291
  orbital phlebography, 48/299
  orbital pseudotumor, 48/302
  parasellar space occupying lesion, 48/302
  paratrigeminal syndrome, 48/301, 303, 337
  pathogenesis, 48/300
  pathologic anatomy, 48/300
  periostitis, 48/302
  phlebography, 48/299
  pituitary adenoma, 48/302
  pupil, 48/297
  recurrent multiple cranial neuropathy, 48/303
  recurrent nonhereditary multiple cranial
    neuropathy, 51/571
  Reiter disease, 48/303
  sex ratio, 48/292-296
  skull radiography, 48/299
  superior orbital fissure syndrome, 48/291, 296
  superior orbital vein, 48/298
  syphilitic osteoperiostitis, 48/291
  temporal arteritis, 48/302
  treatment, 48/291, 304
Toluene
  ataxia, 64/3
  cerebellar encephalopathy, 64/5
  chemical formula, 64/82
  leukoencephalopathy, 66/721
  neuropathology, 66/726
  neurotoxin, 36/372, 46/393, 62/601, 613, 64/3, 5,
    42, 52, 82
  organic solvent, 64/82

organic solvent intoxication, 64/39, 42
toxic encephalopathy, 64/3, 5
toxic myopathy, 62/601
Toluene intoxication
  see also Organic solvent intoxication
  acetylcholinesterase inhibition, 64/50
  acute effect, 64/51
  adenosine triphosphatase inhibition, 64/50
  autonomic nervous system, 75/506
  benzoic acid, 64/50
  blood level, 64/51
  body burden, 64/50
  bone marrow depression, 36/372
  brain level, 64/51
  chronic effect, 64/52
  CNS lesion, 36/373
  CSF, 64/50
  dementia, 64/42
  2-hexanone, 64/50
  hippuric acid, 64/50
  hormone, 64/50
  hypothalamus, 64/50
  long-term effect, 64/52
  memory, 64/52
  neuropathy, 36/372
  polyneuropathy, 36/373
  prolactin, 64/50
  symptom, 64/3, 5
  taurine, 64/50
  time-weighted mean exposure, 64/52
  toluene abuse, 64/53
  toluene addiction, 64/53
  toxic myopathy, 62/613
Toluene sniffing
  brain infarction, 55/523
Tolypocladium inflatum
  adverse effect, 69/503
Tomacula
  hereditary multiple recurrent mononeuropathy,
    51/551, 555, 558, 60/64
  neuralgic amyotrophy, 51/559
  paranodal, *see* Paranodal tomacula
Tomacula absence
  hereditary brachial plexus neuropathy, 60/72
Tomacular neuropathy, *see* Hereditary multiple
    recurrent mononeuropathy
Tomaculous neuropathy, *see* Hereditary multiple
    recurrent mononeuropathy
Tomé body, *see* Tomé-Fardeau body
Tomé-Fardeau body
  inclusion body, 62/103
  muscle tissue culture, 62/103
  oculopharyngeal muscular dystrophy, 62/103

Tomography
  basophil adenoma, 2/457
  computer assisted, *see* Computer assisted
    tomography
  conventional, *see* Conventional tomography
  craniocervical region, 32/10-14
  intermittent spinal cord ischemia, 12/523
  metrizamide computer assisted, *see* Metrizamide
    computer assisted tomography
  odontoid fracture, 25/323
  osteoid osteoma, 19/173
  positron emission, *see* Positron emission
    tomography
  residual image, 19/140
  single photon emission computer assisted, *see*
    Single photon emission computer assisted
    tomography
  skull base tumor, 17/137
  spinal angioma, 32/470
  spinal tumor, 19/137
  type, 19/140
  xenon computer assisted, *see* Xenon computer
    assisted tomography
Tone
  bladder, 1/381
  definition, 2/394
  muscle, *see* Muscle tone
  postural, *see* Postural tone
  sympathetic nervous system, 22/251
  venous, *see* Venous tone
  vestibular function, 16/302
  vestibular system, 16/302
Tone decay, 16/312
  brain stem tumor, 16/321, 17/672, 675
Tone decay test, 2/94
  acoustic nerve tumor, 17/672, 675
Tongue
  fissured, *see* Lingua plicata
  furrowed, 8/205, 208, 214
  hemiplegia, 1/180
  hereditary sensory and autonomic neuropathy type
    III, 75/149
  hyperserotonemia, 29/225
  multilobulated, *see* Multilobulated tongue
  pain, 48/502
  sea snake intoxication, 37/92
  syphilis, 33/351
Tongue agenesis
  speech disorder in children, 4/416
Tongue atrophy
  Bassen-Kornzweig syndrome, 60/132
  bulbopontine paralysis, 42/96
  infantile spinal muscular atrophy, 22/85

Wohlfart-Kugelberg-Welander disease, 22/70
Tongue fibrillation
  Fazio-Londe disease, 42/95
  Northern Swedish recessive ataxia, 60/659
Tongue hamartoma
  Mohr syndrome, 43/450
  orofaciodigital syndrome type I, 43/447, 50/167
Tongue hyperplasia
  maxillofacial dysostosis, 43/425
Tongue paralysis
  atrophy, 30/411
  fasciculation, 30/411
  hypoglossal nerve, 30/411
  Jackson syndrome, 1/189, 2/223
  polyneuropathy, 42/97
  Tapia syndrome, 2/102
Tongue protrusion
  Down syndrome, 50/526
  neuroleptic dystonia, 65/281
Tongue-tie
  speech disorder in children, 4/416
Tonic avoidin reaction
  athetosis, 6/155
Tonic clonic epilepsy
  bilateral spike-wave, 42/670
  classification, 15/107
  generalized epilepsy, 15/108-116
  primary generalised epilepsy, 42/682
Tonic clonic seizure
  dopamine β-mono-oxygenase deficiency, 59/163
  EEG, 15/115
  generalized, see Generalized tonic clonic seizure
  ketotic hyperammonemia, 29/106
  pallidonigral degeneration, 42/242, 49/456
Tonic control center
  basal ganglion, 1/161
Tonic fit, see Decerebrate rigidity
Tonic grasp reflex
  athetosis, 6/155
Tonic motoneuron
  alpha motoneuron type, 1/67
Tonic neck reflex
  anatomy, 1/264
  behavior development, 4/346
  cerebral lesion, 4/346
  developmental disorder, 4/346
  features, 1/255
  higher nervous activity disorder, 4/346
  newborn behavior, 4/346
  spasmodic torticollis, 6/576
Tonic postural reflex (Foix-Thévenard)
  double athetosis, 6/141
  parkinsonism, 6/141

Wilson disease, 6/141
Tonic pupil
  cause, 74/412
  ciliary ganglion, 74/410
  features, 1/621
  hereditary sensory and autonomic neuropathy type
    II, 60/18
  Holmes-Adie syndrome, 2/109, 22/259
  hyporeflexia, 42/72
Tonic seizure
  classification, 15/107
  dermatoleukodystrophy, 42/488
  generalized epilepsy, 15/116-120
  hereditary long QT syndrome, 42/718
  Lennox-Gastaut syndrome, 15/116
  multiple sclerosis, 9/178, 46/170, 47/57, 71
  sleep, 15/120
Tonic stretch reflex
  physiology, 1/64
Tonsil
  yellow orange, see Yellow orange tonsil
Tonsil coloration
  Tangier disease, 42/627
Tonsillar fossa
  glossopharyngeal neuralgia, 48/461
  pain, 48/503
Tonsillar herniation
  see also Brain herniation
  brain injury, 57/134
  cerebellar infarction, 16/122, 53/383
  cerebellum, 1/565-567
  clinical examination, 57/134
  craniometaphyseal dysplasia, 43/365
  foramen magnum, 1/565-567, 16/120
  head injury, 57/50
  intracranial pressure, 57/50, 134
  pathology, 23/56
  Von Hippel-Lindau disease, 14/115
Tonsillectomy
  multiple sclerosis, 9/123, 47/151
  poliomyelitis, 9/123, 34/116, 42/653, 56/315
  sudden infant death, 63/466
Tonsillitis
  moyamoya disease, 55/295, 302
  stylohyoid syndrome, 48/503
  subarachnoid hemorrhage, 12/111
Tonus, see Tone
Tooth grinding
  sleep, 74/559
Toothache
  altitude illness, 48/399
Topagnosia
  somato, see Somatotopagnosia

terminology, 1/104
Topectomy
  causalgia, 1/110
  phantom limb pain, 1/110
Topesthesia
  body scheme, 4/279
  body scheme disorder, 4/394-396
  phantom limb, 4/226, 45/397, 399
  terminology, 1/87, 104
Topiramate
  epilepsy, 73/361
Topoanesthesia
  terminology, 1/89, 104
Topognosis, *see* Topesthesia
Topographical amnesia
  pathology, 45/418
  reduplicative paramnesia, 45/418
  spatial memory disorder, 45/416
Topographical memory
  agnosia, 4/16
  cerebral dominance, 4/299
  memory disorder, 3/219-221, 4/16, 229, 300
  occipital lobe, 2/613
  parietal lobe syndrome, 45/73
  recognition, 2/669
  spatial memory, 45/416
  spatial orientation disorder, 3/219-221
Topographies
  nerve lesion, *see* Nerve lesion topography
Torcular Herophili, *see* Confluens sinuum
Torkildsen procedure
  Dandy-Walker syndrome, 30/643
Torpedo
  ataxia telangiectasia, 14/75
  glycosaminoglycanosis type IH, 10/444
  GM2 gangliosidosis, 10/397, 410
Torsion
  dystonia, *see* Dystonia musculorum deformans
  spasm, *see* Dystonia musculorum deformans
  vertebral, *see* Vertebral torsion
  visual field, 17/321
Torsion dystonia, *see* Dystonia musculorum
  deformans
Torsion dystonia (Mendel), *see* Dystonia
  musculorum deformans
Torsion neurosis
  history, 6/518
  terminology, 6/518
Torsion spasm, *see* Dystonia musculorum deformans
Torticollis
  amyotrophic lateral sclerosis, 59/235
  benign paroxysmal, *see* Benign paroxysmal
    torticollis

congenital, *see* Congenital torticollis
  juvenile spinal cord injury, 61/233
  juvenile vertebral column injury, 61/233
  neuroleptic dystonia, 65/281
  progressive pallidal atrophy (Hunt-Van Bogaert),
    6/457, 554
  progressive supranuclear palsy, 6/557
  pure pallidoluysian degeneration, 49/448
  spasmodic, *see* Spasmodic torticollis
  status cribrosus, 6/582
Tortipelvis, *see* Dystonia musculorum deformans
Torula histolytica, *see* Cryptococcus neoformans
Torulopsiasis
  brain abscess, 35/571, 52/496
  spinal cord abscess, 35/571, 52/496
Torulopsis, *see* Cryptococcus neoformans
Torulosis, *see* Cryptococcosis
Tosylchloramide sodium
  Schiff staining, 9/23
Total frontal atrophy (Lüers-Spatz)
  Pick disease, 22/314
  spastic paraplegia, 22/314
Total transverse spinal cord syndrome
  flaccid paralysis, 12/498
  spinal apoplexy, 12/498
  spinal cord intermittent claudication, 12/498
Touraine disease, *see* Neurocutaneous melanosis
Touraine-Solente-Golé syndrome, *see*
    Neurocutaneous melanosis
Tourette syndrome, *see* Gilles de la Tourette
  syndrome
Tournay phenomenon
  anisocoria, 74/413
Tourniquet
  ischemia, 70/40
  nerve impairment, 70/40
Tourtellotte formula
  multiple sclerosis, 47/88-91
Toux des aboyeurs
  superior laryngeal neuralgia, 48/496
Towne projection
  acoustic nerve tumor, 17/680
  basilar impression, 32/51
  posterior cranial fossa, 32/10
  radiodiagnosis, 57/145
Townes-Brocks syndrome
  autosomal dominant, 42/379
  deafness, 42/379
  imperforate anus, 42/379
  lop ear, 42/379
  prognosis, 42/380
  skeletal abnormality, 42/379
Toxaphene intoxication

convulsion, 36/419
gastrointestinal complication, 36/419
laboratory finding, 36/420
pathology, 36/420
stroban, 36/423
treatment, 36/420
Toxic agent
acquired myoglobinuria, 62/575
arsanilic acid, 9/648, 13/204
cerebellar agenesis, 50/180
experimental pigmentary retinopathy, 13/204-213
Toxic encephalitis
gold intoxication, 64/356
Toxic encephalopathy
see also Morvan fibrillary chorea
acetonylacetone, 64/12
acrylamide, 64/12
acute, *see* Reye syndrome
aflatoxin B, 49/217
alcohol, 64/2, 5
aldrin, 64/201
aluminum, 64/4, 274
2-amino-3-methylaminopropionic acid, 64/5
amphetamine, 65/251
arsenic, 64/12
arthropod venom, 65/193
bacterial endocarditis, 52/291, 296
barbiturate, 65/508
bismuth, 64/4
brain microinfarction, 52/297
α-bungarotoxin, 65/177
cancer, 69/396
carbamazepine, 65/505
carbon monoxide, 64/7
carnitine deficiency, 49/217
chloramphenicol, 65/486
chlordane, 64/202
chorea, 49/556
ciprofloxacin, 65/486
cocaine, 65/251
copper, 49/230
cyanogenic glycoside, 64/12, 65/25
cytotoxic brain edema, 57/213
dendrotoxin, 65/177
dieldrin, 64/201
domoic acid, 64/3
endosulfan, 64/201
enoxacin, 65/486
eosinophilia myalgia syndrome, 64/257
ergot alkaloid, 65/61
fasciculin, 65/177
fluorouracil, 64/5
gold, 64/4

gyromitrine, 64/3
headache, 52/296
Hemophilus influenzae meningitis, 52/61
heptachlor, 64/201
hexachlorophene, 65/471
hexane, 64/12
2-hexanone, 64/12
hydrazine, 64/3
hymenoptera venom, 65/193
hypoglycin, 49/217
hypoglycin intoxication, 65/79
immune complex, 52/297
infectious endocarditis, 63/112, 117
isoniazid, 65/480
Jamaican vomiting sickness, 65/79, 115
lamotrigine, 65/512
manganese, 64/6
marine toxin, 65/141
mefloquine, 65/485
mercury, 64/4
methanol, 64/2, 7
methylazoxymethanol glucoside, 64/5
methyl chloride, 64/3
methylmercury, 64/5
metronidazole, 65/487
middle cerebral artery syndrome, 53/363
multiple brain abscess, 52/297
myoclonus, 49/618
nalidixic acid, 65/486
nicotine, 65/251
nonbacterial thrombotic endocarditis, 63/120
norfloxacin, 65/486
organotin, 64/4
pallidal necrosis, 49/465
pallidoluysionigral degeneration, 49/460
pallidonigral necrosis, 49/465
pallidostriatal necrosis, 49/465
paralytic shellfish poisoning, 65/141
pefloxacin, 65/486
4-pentenoic acid, 49/217
phenytoin, 65/500
polychlorinated dibenzodioxin intoxication, 64/204
psilocybine, 64/3
psychiatric symptom, 52/296
Reye syndrome, 27/363, 49/217, 65/115
rubella, 56/407
scorpion venom, 65/193
snake venom, 65/177
spider venom, 65/193
thallium, 64/4, 12
tick venom, 65/193
toluene, 64/3, 5

valproic acid, 49/217
vasogenic brain edema, 57/213
vigabatrin, 65/511
Toxic hydrocephalus
  CSF, 16/236
  nasal sinus infection, 12/426, 16/236
  terminology, 12/426
Toxic meningitis, *see* Meningitis serosa
Toxic motoneuron disorder
  WFN classification, 59/5
Toxic myelopathy
  bismuth, 64/10
  doxorubicin, 64/10
  epidural anesthesia, 61/137
  lathyrism, 64/9
  latrodectus venom, 64/10
  3-oxalylaminoalanine, 64/10
  tetanospasmin, 64/10
Toxic myopathy
  alcohol, 62/601
  ε-aminocaproic acid, 62/598, 601
  amiodarone, 62/606
  amoxapine, 62/601
  amphetamine, 62/601
  amphotericin B, 62/601
  antihistaminic agent, 62/601
  autophagia, 62/606
  barbiturate, 62/601
  basic mechanism, 62/595
  beta adrenergic receptor blocking agent, 62/604
  bezafibrate, 62/601
  black widow spider venom, 62/611
  botulinum toxin, 64/15
  botulism, 62/612
  brown spider venom, 62/601, 611
  captopril, 62/597
  carbenoxolone, 62/601
  carbon monoxide, 62/601
  cardiac glycoside, 62/598
  chloralose, 62/601
  chloroquine, 62/606
  chlorpromazine, 62/601
  cimetidine, 62/597, 609
  clofibrate, 62/597, 601
  clofibride, 62/597
  Clostridium welchii, 62/612
  cocaine, 62/601
  colchicine, 62/597, 600, 606
  copper sulfate, 62/601
  corticosteroid, 62/605
  creatine kinase, 62/596
  crotamine, 62/610
  crotoxin, 62/610
  cytotoxic agent, 62/597
  danazol, 62/597
  diamorphine, 62/601
  diamorphine addiction, 62/600
  diazepam, 62/601
  dihydrocodeine, 62/601
  diuretic agent, 62/597, 601
  emetine, 62/600
  enalapril, 62/597
  eosinophilia myalgia syndrome, 62/604
  ethchlorvynol, 62/597, 609
  ethylene glycol, 62/601
  experimental myotonia, 62/597
  fasciculation, 62/596
  fenfluramine, 62/601
  fenofibrate, 62/601
  flunitrazepam, 62/601
  fluphenazine, 62/601
  focal myopathy, 62/611
  gasoline, 62/601
  gasoline intoxication, 62/613
  gemfibrozil, 62/601
  glutethimide, 62/601
  glycyrrhiza, 62/601, 603, 64/17
  gold, 62/597
  Haff disease, 62/601, 612
  haloperidol, 62/601
  hornet venom, 62/601, 612, 64/16
  Hydrophiidae, 64/16
  hypocholesterolemic agent, 62/598
  hypokalemia, 62/603
  inflammatory myopathy, 62/604
  isoetarine, 62/597
  isoniazid, 62/601
  labetalol, 62/597
  α-latrotoxin, 64/16
  lindane, 62/601
  lithium, 62/597, 601
  loxapine, 62/601
  lysergide, 62/601
  malignant hyperthermia, 38/553, 62/601
  meprobamate, 62/601
  mercuric chloride, 62/601
  metal fume intoxication, 62/613
  metaldehyde, 62/601
  methadone, 62/601
  metolazone, 62/597
  mevinolin, 62/601
  monensin intoxication, 62/612
  morphine, 62/601
  mulgatoxin a, 62/610
  muscle cramp, 62/596
  myalgia, 62/596

myokymia, 62/596
myotonia, 62/597
myotoxin a, 62/610
necrotizing myopathy, 62/598
needle, 62/611
nifedipine, 62/597
notechis II-5, 62/610
notexin, 62/610
opiate addiction, 62/598
organophosphate intoxication, 62/612
oxprenolol, 62/601
paracetamol, 62/601
penicillamine, 62/597
pentamidine, 62/601
perhexiline, 62/606
phencyclidine, 62/601
phencyclidine addiction, 62/600
phenelzine, 62/601
phenformin, 62/601
phenmetrazine, 62/601
phenylenediamine, 62/601
phenylpropanolamine, 62/601
phospholipase A, 62/610
pindolol, 62/597
plasmocid, 62/600
procainamide, 62/597
2-propanol, 62/601
quail ingestion, 62/601
quail myopathy, 62/612
rhabdomyolysis, 62/600
rifampicin, 62/597, 609
salbutamol, 62/597
salicylic acid, 62/601
saxitoxin, 64/15
sea snake venom, 64/16
snake envenomation, 65/179
snake venom, 62/601, 610
strychnine, 62/601
suxamethonium, 62/597, 601
taipoxin, 62/610
tarantula spider venom, 62/611
tetracycline, 62/609
tetrodotoxin, 64/15
theophylline, 62/601
tick venom, 64/15
tiopronin, 62/597, 609
tityustoxin, 64/15
toluene, 62/601
toluene intoxication, 62/613
trifluoperazine, 62/601
tryptophan, 62/597
vasopressin, 62/601
vidarabine, 62/609

vincristine, 62/600, 606
wasp venom, 62/601, 612
zidovudine, 62/597, 600, 602
zimeldine, 62/597
zinc phosphide, 62/601
Toxic neuromuscular junction disorder
batrachotoxin, 64/15
botulinum toxin, 64/15
α-bungarotoxin, 64/15
β-bungarotoxin, 64/15
carbamate, 64/15
ciguatoxin, 64/15
curare, 64/15
α-latrotoxin, 64/15
organophosphate, 64/15
saxitoxin, 64/14
tetrodotoxin, 64/15
tick venom, 64/15
tityustoxin, 64/15
Toxic neuropathy, 7/510-523
acromegalic neuropathy, 7/556
acrylamide, 64/10
aluminum, 64/10
amiodarone, 51/302
arsenic, 64/10
buckthorn toxin, 64/10, 14
cadmium, 64/10
carbon disulfide, 64/8, 10
carbon disulfide intoxication, 64/25
chlordane, 64/13
chloroquine, 51/302
cisplatin, 51/299
clioquinol, 64/8, 10, 13
colchicine, 64/13
cuprizone, 64/10
cyanogenic glycoside, 64/8, 10
dactinomycin, 64/10
dapsone, 65/485
debrisoquine, 51/302
2,4-dichlorophenoxyacetic acid, 7/523, 51/286
differential diagnosis, 7/552
diphtheria toxin, 64/10, 14
disulfiram, 64/8
doxorubicin, 51/300, 64/10
drug, 7/510
ethambutol, 64/8, 65/481
freon, 64/13
gentamicin, 64/9
gold, 64/10, 13
herbicide, 51/283
hexacarbon, 64/10
hexachlorophene, 64/10, 14, 65/471
homidium bromide, 64/10

507

3,3'-iminodipropionitrile, 64/10
isoniazid, 65/480
isoniazid neuropathy, 7/516
kanamycin, 64/9
lead, 64/10, 13
lupin seed, 64/13
mercury, 64/10
metal, 7/510
methanol, 64/8
metronidazole, 51/301
misonidazole, 51/299, 301
muscarine, 64/14
nitrofurantoin, 65/487
organophosphate, 51/283, 64/11, 151
pentamidine, 7/520
perhexiline, 64/10, 14
perhexiline maleate, 51/301
permethrin, 64/13
pesticide, 51/283
phenytoin polyneuropathy, 7/519
propamidine, 7/520
quinine, 64/8
salicylic acid, 64/9
stilbamidine, 7/520
streptomycin, 64/9
thallium, 64/8, 10, 14
trichloroethylene, 51/279, 64/9
tri-o-cresyl phosphate, 51/285, 64/10, 13
triethyltin, 64/10
vinblastine, 64/13
vincristine, 51/299, 64/10, 13
Toxic oil syndrome
    acute respiratory insufficiency, 63/380
    alopecia, 63/380
    amnesia, 63/381
    anorexia, 63/380
    axonal degeneration, 63/381
    brain infarction, 63/381
    carpal tunnel syndrome, 63/380
    confusion, 63/381
    confusional state, 63/381
    cough, 63/380
    dyspnea, 63/380
    EMG, 63/381
    encephalopathy, 63/381
    endomysial fibrosis, 63/381
    eosinophilia, 63/371
    eosinophilia myalgia syndrome, 63/379
    fever, 63/380
    gasping respiration, 63/480
    headache, 63/380
    hyporeflexia, 63/381
    irritability, 63/380

muscle biopsy, 63/381
muscle cramp, 63/381
muscle weakness, 63/380
muscular atrophy, 63/380
myalgia, 63/380
myoclonus, 63/381
myopathy, 63/381
nerve biopsy, 63/381
paresthesia, 63/381
pathogenesis, 63/381
polyneuropathy, 63/380
pulmonary hypertension, 63/380
rapeseed oil, 63/380
scleroderma, 63/380, 71/115
segmental endovasculitis, 63/380
sensorimotor neuropathy, 63/380
sensorimotor polyneuropathy, 63/380
sensory neuropathy, 63/380
Sjögren syndrome, 63/380
stupor, 63/380
symptom, 63/380
treatment, 63/382
weight loss, 63/380
Toxic polyneuropathy
    acetonylacetone, 51/68, 64/81, 86
    acrylamide, 51/68, 280
    amitriptyline, 51/68
    arsenic polyneuropathy, 51/68, 266
    carbamate, 64/160
    carbon disulfide, 51/68, 64/23
    carbon disulfide intoxication, 64/25
    chloramphenicol, 51/68
    clioquinol, 51/68
    dapsone, 51/68
    dibenzo-1,4-dioxin, 64/197
    diphtheria toxin, 51/68
    disulfiram, 51/68
    drug induced, see Drug induced polyneuropathy
    ethylene oxide, 51/275
    giant axonal neuropathy, 60/82
    gold, 51/274
    hexachlorophene, 51/68
    hexane, 51/68
    2-hexanone, 51/68
    hydralazine, 51/68
    isoniazid, 51/68
    ketone, 51/68
    lead, 51/68, 268
    mercury, 51/68, 271
    mipafox, 51/68
    nitrofurantoin, 51/68
    organochlorine, 64/197
    organophosphate, 64/151

phenytoin, 51/68
platinum, 51/274
polychlorinated biphenyl, 64/197
polychlorinated dibenzodioxin intoxication, 64/204
sodium isocyanate, 51/68
thalidomide, 51/68
thallium, 51/68, 274
trichloroethylene, 51/68
tri-*o*-cresyl phosphate, 51/68
triethyltin, 51/68
vincristine, 51/68
Toxic shock syndrome
acquired myoglobinuria, 62/574, 577
adult respiratory distress syndrome, 52/258
brain edema, 52/260
brain hemorrhage, 52/260
carpal tunnel syndrome, 52/259
clinical features, 52/258
CNS, 52/253
coma, 52/258
creatine kinase, 52/259
CSF, 52/259
delirium, 46/544, 52/258
diarrhea, 52/258
differential diagnosis, 52/260
dysgenesia, 52/259
EEG, 52/259
encephalopathy, 52/259
epidemiology, 52/258
epilepsy, 52/259
headache, 52/259
histopathology, 52/260
laboratory finding, 52/259
malignant hyperthermia, 62/574
menstruation, 52/258
mental symptom, 52/258
metabolic encephalopathy, 62/574
metallic taste, 52/259
multisystem organ failure, 52/258
muscle tenderness, 52/258
muscle weakness, 52/258
myalgia, 52/259
myoglobinuria, 62/574
myopathy, 52/259
myositis, 52/259
neurologic symptom, 52/258
postural syncope, 52/258
Staphylococcus aureus, 52/257
surgical wound, 52/258
symptom, 62/574
tarsal tunnel syndrome, 52/259
toxin, 52/257

treatment, 52/260
vagina tampon, 52/257
vascular leakage syndrome, 52/258
Toxi-infective purpura
leukoencephalopathy, 9/598
Toxin
acquired immune deficiency syndrome, 75/414
acquired myoglobinuria, 62/576
amyotrophic lateral sclerosis, 59/176
aspergillosis, 35/397, 52/377
Bordetella pertussis, 52/232
botulism, 40/143, 65/209
brain embolism, 55/163
brain infarction, 55/163
Buckthorn, *see* Buckthorn toxin
cerebrovascular disease, 55/163
Clostridium tetani, 65/215
Clostridium welchii, 40/151
Corynebacterium diphtheriae, 65/236
Dinophysis, *see* Dinophysiotoxin
diphtheria, *see* Diphtheria toxin
diphtheritic neuritis, 51/182
industrial, *see* Industrial toxin
marine, *see* Marine toxin
myoclonus, 38/581
myoglobinuria, 41/271, 62/557
Parkinson dementia, 49/178
progressive muscular atrophy, 59/18
Reye syndrome, 70/276
toxic shock syndrome, 52/257
vasculitis, 71/158
Toxin fragment A
Corynebacterium diphtheriae, 51/182
diphtheria, 65/238
Toxin fragment B
Corynebacterium diphtheriae, 51/182
diphtheria, 65/238
Toxin mediated syndrome
botulism, 71/555
survey, 71/555
symptom, 71/555
Toxocariasis
eye, 52/511
helminthiasis, 35/212, 52/510
pathology, 52/510
pneumonia, 52/511
pseudoretinoblastoma, 52/511
Toxoid
diphtheria, *see* Diphtheria toxoid
neuropathy, 8/87
tetanus, *see* Tetanus toxoid
Toxoplasma encephalitis
acquired immune deficiency syndrome, 71/303

Toxoplasma gondii
  see also Toxoplasmosis
  acquired immune deficiency syndrome, 56/495, 514
  aqueduct stenosis, 50/308
  brain abscess, 52/148, 151
  CNS infection, 69/444
  inflammatory myopathy, 62/373
  Melkersson-Rosenthal syndrome, 8/226
  multiple sclerosis, 47/320
  parasitology, 52/351
  polymyositis, 62/373
  renal transplantation, 63/533
Toxoplasma gondii infection
  immunocompromised host, 65/549
Toxoplasma gondii syndrome
  clinical features, 31/218
  hydrocephalus, 31/212
  laboratory finding, 31/219
  mental deficiency, 31/212, 219
  microcephaly, 30/510, 31/212
  prevention, 31/219
  seizure, 31/212, 219
  treatment, 31/219
Toxoplasmosis, 35/115-138
  see also Toxoplasma gondii
  accidental transmission, 35/116
  acquired, see Acquired toxoplasmosis
  acquired immune deficiency syndrome, 52/356, 71/300
  acute cerebellar ataxia, 34/626, 52/355
  aqueduct stenosis, 50/306
  ataxia, 52/355
  brain biopsy, 52/356
  brain cyst, 52/352
  brain embolism, 55/163
  brain infarction, 55/163
  cavum septi pellucidi, 30/326
  cerebral, see Cerebral toxoplasmosis
  cerebrovascular disease, 55/163
  chorioretinitis, 50/277
  chronic meningitis, 56/645
  clindamycin, 52/352
  computer assisted tomography, 52/356
  congenital, see Congenital toxoplasmosis
  congenital brain granuloma, 42/736
  CSF, 52/356
  dementia, 46/389
  EEG, 35/133
  encephalitis, 52/352
  encephalopathy, 42/662
  epidemiology, 35/117
  epilepsy, 72/164, 351

heart cyst, 52/352
histologic diagnosis, 35/120
Hodgkin disease, 39/43
host defence, 52/356
hydranencephaly, 50/337
hydrocephalus, 30/545, 35/130, 50/277, 52/358
hypothalamic tumor, 68/71
immune response, 42/662
immunity, 35/118
immunocompromised patient, 52/351
infectious striatopallidodentate calcification, 49/417
intracerebral calcification, 42/662, 50/277, 52/359
intracranial calcification, 35/131-134
laboratory diagnosis, 35/120
leukemia, 63/354
life cycle, 35/115
lymphocytic meningitis, 52/354, 359
main cerebral form, 52/354
malignant lymphoma, 39/43
meningitis, 42/662
meningoencephalitis, 52/354, 358
mental deficiency, 35/130, 46/19, 52/358
microcephaly, 35/130, 50/277
microgyria, 30/483
muscle cyst, 52/352
neonatal manifestation, 35/134
nerve lesion, 7/488
nuclear magnetic resonance, 52/356
ocular involvement, 35/132
pathogenesis, 35/119
pathology, 35/127-129
pigmented retinal lesion, 13/141
polymicrogyria, 50/257
polyneuropathy, 52/355
porencephaly, 50/359
prenatal, see Congenital toxoplasmosis
prevalence, 42/661, 52/351
progressive multifocal leukoencephalopathy, 63/354
pyrimethamine + sulfadiazine, 52/351
radiculitis, 52/355
renal insufficiency, 27/343
retinochoroiditis, 52/355
Sabin-Feldman dye test, 35/120
serologic finding, 35/121
serology, 52/356
spiramycin, 52/352
stenosis, 52/358
syndrome, 30/508
systemic neoplasm, 41/373
transmission, 35/116
treatment, 35/128, 137, 52/351

Toxopneustes pileolus intoxication
 dizziness, 37/68
 dyspnea, 37/68
 facial paralysis, 37/68
 speech disorder, 37/68
TP53 gene
 tumor cell, 67/20
TPI test, *see* Treponema pallidum immobilization
 test
TPN diaphorase, *see* Reduced nicotinamide adenine
 dinucleotide phosphate dehydrogenase
Trace metal
 CSF, *see* CSF trace metal
 multiple sclerosis, 9/113
 peritoneal dialysis, 27/329
 renal insufficiency, 27/329
Trace minerals
 Parkinson dementia, 49/178
γ-Trace protein, *see* Cystatin C
Trachea
 larynx, 74/392
Tracheal malformation
 associated malformation, 50/514
Tracheoesophageal fistula
 Down syndrome, 50/527
 VATER syndrome, 50/513
Tracheostomy
 brain arteriovenous malformation, 12/263
 brain injury, 24/611, 614
 glycosaminoglycanosis type IH, 66/297
 glycosaminoglycanosis type VI, 66/307
 obstructive sleep apnea, 63/457
 progressive dysautonomia, 59/140
 respiratory dysfunction, 26/344
 spinal cord injury, 26/204, 344, 61/267, 437
 tetraplegia, 61/437
 transverse spinal cord lesion, 61/437
Trachinidae intoxication
 family, 37/78
 venomous spine, 37/78
Trachinus intoxication
 family, 37/78
 venomous spine, 37/78
Traction
 cervical facet subluxation, 61/62
 halter, *see* Halter traction
 headache, 5/21, 48/5
 nerve injury, 7/245
 skull, *see* Skull traction
 spinal injury, 26/301
 vertebral column, *see* Spine traction
Traction injury
 birth, 25/162

 brachial plexus, 51/145
 brachial plexus injury, 7/405-407, 25/395,
  409-435
 brachial plexus paralysis, 2/137, 141
 lumbosacral plexus, 51/160
 spinal nerve root, 25/393-435
 type, 51/146
Traction response
 automatism development, 1/71
Tractotomy
 medullary, *see* Medullary tractotomy
Tracts
 biliary, *see* Biliary tract
 dermal sinus, *see* Dermal sinus tract
 gastrointestinal, *see* Gastrointestinal tract
 geniculocalcarine, *see* Geniculocalcarine tract
 genital, *see* Genital tract
 hypothalamic hypophyseal, *see* Hypothalamic
  hypophyseal tract
 Lissauer, *see* Lissauer tract
 mamillotegmental, *see* Mamillotegmental tract
 mamillothalamic, *see* Mamillothalamic tract
 neospinothalamic, *see* Neospinothalamic tract
 olivocochlear, *see* Olivocochlear tract
 optic, *see* Optic tract
 paleospinothalamic, *see* Paleospinothalamic tract
 pyramidal, *see* Pyramidal tract
 retinohypothalamic, *see* Retinohypothalamic tract
 rubrospinal, *see* Rubrospinal tract
 solitary, *see* Solitary tract
 spinocerebellar, *see* Spinocerebellar tract
 spinothalamic, *see* Spinothalamic tract
 tuberohypophyseal, *see* Tuberohypophyseal tract
 urinary, *see* Urinary tract
 urogenital, *see* Urogenital tract
Traffic injury
 cause, 25/142
 head injury, 23/27, 32
 juvenile spinal cord injury, 25/176, 61/237
 mortality, 25/142
 paraplegia, 25/142
 pelvic fracture, 26/198
 statistic, 25/142
 tetraplegia, 25/142
Tranexamic acid, *see* α-Methylcyclohexane
 carboxylic acid
Tranquilizing agent
 hyperkinetic child syndrome, 3/197
 ingestion, 30/125
 medazepam, 37/355
 mental deficiency, 30/125
 oxazepam, 37/355, 359
 teratogenic agent, 30/125

Transamination reaction
  brain, 27/17
Transcobalamin
  vitamin B12 deficiency, 51/338
Transcobalamin II deficiency
  ataxia, 63/255
  epilepsy, 63/255
  megaloblastic anemia, 63/255
  mental deficiency, 63/255
  pyramidal syndrome, 63/255
Transcortical aphasia, 45/316-318
  alexia, 4/132
  brain embolism, 53/215
  brain infarction, 53/215
  cerebrovascular disease, 53/215
  intracranial hemorrhage, 53/215
  mixed, *see* Mixed transcortical aphasia
  transient ischemic attack, 53/215
Transcortical motor aphasia
  definition, 45/317
  frontal lobe, 45/19
  mesial cortex, 45/20
  nonfluent agraphia, 45/459
Transcortical sensory aphasia
  alexia, 45/447
  definition, 45/317
  fluent agraphia, 45/460
  parietal lobe syndrome, 45/75
  posterior borderzone, 45/20
Transcranial Doppler sonography
  anterior communicating artery, 54/12
  autoregulation, 54/12
  basilar artery, 54/12
  brain arteriovenous malformation, 53/226
  brain collateral blood flow, 54/12
  carotid siphon, 54/12
  middle cerebral artery, 54/12
  posterior cerebral artery, 54/12
  subarachnoid hemorrhage, 54/12, 55/22
  tilt table orthostasis, 54/12
  vasospasm, 54/12
Transcription factor
  brain metastasis, 69/109
Transcutaneous nerve stimulation
  brachial plexus, 51/154
Transection syndrome
  spinal cord compression, 20/205
  spinal cord metastasis, 20/418
  stage 1, 20/205
  stage 2, 20/205
  stage 3, 20/205
Transfer factor
  ataxia telangiectasia, 60/390

  coccidioidomycosis, 52/421
  multiple sclerosis, 47/203
Transferrin
  carbohydrate deficient glycoprotein syndrome,
    66/633
  free radical scavengers, 64/19
  neuroleptic akathisia, 65/287
Transferrin receptor
  astrocytoma, 67/297
Transient cerebral blindness
  see also Transient cortical blindness
  pediatric head injury, 23/450
Transient cerebral ischemia, *see* Transient ischemic
    attack
Transient cortical blindness
  see also Transient cerebral blindness
  child, 24/51
  EEG, 24/49
Transient global amnesia, 45/205-216
  behavior, 45/209
  Binswanger disease, 54/222
  brain angiography, 55/141
  brain infarction, 55/141
  brain tumor, 45/211
  cardiovascular agent intoxication, 37/426
  cause, 53/235
  cerebrovascular disease, 45/206, 216, 55/141
  clioquinol, 45/207
  cold water immersion, 45/208
  computer assisted tomography, 45/213
  demography, 45/207
  digoxin intoxication, 65/450
  dissecting aorta aneurysm, 63/47
  driving, 45/209
  EEG, 45/206
  emotional stress, 45/209, 245
  epilepsy, 45/206, 55/141
  etiology, 45/216
  exercise, 45/209
  glycoside intoxication, 37/426
  head injury, 45/206, 55/141
  headache, 45/209
  hot water immersion, 45/209
  hypertension, 45/211
  intelligence, 45/212
  ischemic heart disease, 45/211
  memory deficit, 45/209
  migraine, *see* Migrainous transient global amnesia
  positron emission tomography, 45/213
  posterior cerebral artery, 55/141
  prognosis, 45/212
  retrograde amnesia, 45/209
  risk factor, 45/210

sexual intercourse, 45/208
stroke, 45/210
temporal lobe infarction, 45/215
transient ischemic attack, 45/210
vertebral artery angiography, 45/208, 214
Transient global ischemia, *see* Transient ischemic attack
Transient infantile hemiplegia
angioma, 12/348
carotid artery, 12/348
hemiplegic migraine, 12/348
Transient ischemic attack, 53/257-278
acetylsalicylic acid, 53/274, 324, 55/474
action myoclonus, 53/211
adriamycin, 63/136
age, 53/200, 259
akinetic mutism, 53/209
amaurosis fugax, 53/265, 309, 315, 318
amnesia, 12/3, 53/206
amyloidosis, 63/137
anemia, 55/463
angioscopy, 53/315
annual mortality, 53/260
anticoagulant, 12/11, 458, 53/273, 325
antiphospholipid antibody, 63/330
aphasia, 53/214
arterial compression, 12/8
arterial hypertension, 53/259
arterial hypotension, 12/9
arteriosclerosis, 53/200
artery stenosis, 53/265
asterixis, 53/211
asymptomatic carotid artery stenosis, 53/277, 291, 293, 320
ataxia, 12/5, 53/214
ataxic hemiparesis, 54/247
atrial myxoma, 63/97
auditory symptom, 53/207
autonomic dysfunction, 53/207
B-mode echoCG, 53/270
bacterial endocarditis, 52/291
basilar artery aneurysm, 53/387
basilar artery syndrome, 12/13
bee venom, 65/198
Behr disease, 53/213
bilateral, 53/318
bilateral carotid lesion, 53/320
Binswanger disease, 54/222
blepharospasm, 53/211
blindness, 53/207
blood pressure, 53/207
blood viscosity, 53/266, 272
brain amyloid angiopathy, 54/337

brain angiography, 45/212, 53/228, 270
brain arteriovenous malformation, 54/375
brain blood flow, 54/146, 148
brain damage discrimination, 53/291
brain infarction, 53/307
brain lacunar infarction, 53/264, 54/241, 253
brain scintigraphy, 53/232
brain stem auditory evoked potential, 53/233
Canadian Neurologic Scale, 53/209
cardiac disorder, 53/200
cardiac dysrhythmia, 12/10
cardiac papillary fibroelastoma, 63/103
carotid angiography, 11/340
carotid artery occlusion, 11/340
carotid bruit, 53/216, 321
carotid dissecting aneurysm, 54/272
carotid endarterectomy, 53/275-277, 292, 322
carotid restenosis, 53/276
carotid sinus massage, 53/219
carotid siphon stenosis, 53/307
carotid system, 53/261, 291
cause, 12/7-10, 53/201
cause of death, 53/260
cerebrovascular disease, 53/17
cervical spondylosis, 12/7
cheiro-oral syndrome, 12/4
chronic vegetative state, 53/209
circulating water distribution, 54/117
climate, 53/200
clinical course, 53/202
clinical evaluation, 53/199
clinical features, 53/261
cocaine intoxication, 65/255
coma, 53/208, 262
computer assisted tomography, 53/220, 258, 268, 313, 315, 54/53
conduction aphasia, 53/215
confusion, 53/206
confusional state, 12/3
congestive heart failure, 53/259
conjugate gaze deviation, 53/212
consciousness loss, 53/206
coronary bypass patient, 63/185
coronary bypass surgery, 63/185
coronary heart disease, 53/259
correct diagnosis, 53/199
course, 53/312
cryoglobulinemia, 53/266, 63/403
CSF, 53/224
CSF spectrophotometry, 53/224
deafness, 55/357
decerebrate rigidity, 53/211
decortication, 53/211

definition, 53/257
diagnosis, 53/258, 267, 312
diagnostic error, 53/220
diagnostic speed, 54/85
differential diagnosis, 53/233, 262
digital subtraction angiography, 53/270
diplopia, 12/4, 53/207
dipyridamole, 53/275
directional Doppler velocimetry, 53/225
Doppler signal analysis, 53/268
Doppler sonography, 53/268
drop attack, 12/5, 53/262
duplex sonography, 53/225
duration, 53/203, 257
dysarthria, 12/5
dysphagia, 12/5, 53/207
dysphonia, 53/207
dystonia, 53/211
ECG, 53/230
echoCG, 53/230
echoEG, 53/233
EEG, 45/212, 53/232, 55/299
Ekbom other syndrome, 63/85
embolism, 53/26
Emery-Dreifuss muscular dystrophy, 62/154
emotional disorder, 53/206
epidemiology, 53/258
epileptic focus, 53/234
episodic limb shaking, 53/318
essential thrombocythemia, 63/331
etiology, 12/7
external internal carotid bypass, 53/278, 316, 323,
    54/130
eyelid nystagmus, 53/213
Fabry disease, 55/457
fibromuscular dysplasia, 11/369, 371, 53/163, 266
genetic predisposition, 53/199
Glasgow Coma Scale, 53/209
global aphasia, 53/215
hallucination, 53/206
headache, 12/6, 53/263
hematologic abnormality, 53/266
hematotachygraphy, 53/225
hemianesthesia, 12/5
hemianopia, 53/207
hemiballismus, 53/211
hemichorea, 53/211
hemiparesis, 53/206
hemiplegia, 12/5
hemodynamic mechanism, 53/264, 316
heparin, 53/325
history, 53/199, 257
Holter monitoring, 53/230, 270

Horner syndrome, 53/213
hyperaldosteronism, 39/496
hypercalcemia, 63/561
hypereosinophilic syndrome, 63/372
hypertensive crisis, 12/9
hyperthermia, 53/207
hypertrophic cardiomyopathy, 63/137
hyperviscosity syndrome, 55/485
incidence, 53/258
infarction risk, 53/321
infectious endocarditis, 63/116
internal carotid artery syndrome, 53/307
internuclear ophthalmoplegia, 53/212
intravenous digital subtraction angiography,
    53/227
ischemic heart disease, 53/291
lacuna, 53/264
libido, 53/207
macroglobulinemia, 53/266
management, 12/9-11
mental function, 53/214
metabolic water distribution, 54/117
methyldopa intoxication, 37/441
micturition, 53/207
migraine, 48/19, 156, 53/234, 263
mitral valve prolapse, 55/165, 63/21
monocular vision loss, 53/207
morbidity, 53/321
moyamoya disease, 53/33, 55/294
muscle tone, 53/211
mycotic brain aneurysm, 52/295
natural history, 53/259, 319
neurobrucellosis, 33/314, 52/588
neurosyphilis, 12/9
nomenclature, 53/257
nuchal rigidity, 53/207
nuclear magnetic resonance, 53/231, 313, 54/85
nystagmus, 53/213
ocular bobbing, 53/213
oculoplethysmography, 53/225, 268
oculopneumoplethysmography, 53/225
onset, 53/202
ophthalmodynamometry, 53/219, 268
ophthalmoscopy, 53/218
oscillopsia, 53/207
oxygen extraction rate, 54/117
palatal myoclonus, 53/211
Parinaud syndrome, 53/212
paroxysmal aphasia, 45/508
pathogenesis, 53/264
pathologic bruit, 53/217
pathophysiology, 53/267
pentoxifylline, 53/275

514

phonoangiography, 53/218
photon emission computerised tomography, 53/272
physiologic bruit, 53/216
polyarteritis nodosa, 55/356
polycythemia, 12/9, 53/266, 63/251
polycythemia vera, 55/469
pontine hematoma, 53/207
pontine hemiplegia, 12/23
pontine infarction, 2/242, 12/14, 22
positron emission tomography, 53/232, 313, 54/116
precipitating factor, 53/202
premonitory, 54/247
prevalence, 53/258
prognosis, 12/6
prolonged, see Prolonged transient ischemic attack
prosthetic cardiac valve, 55/165
protein C deficiency, 63/328
protein S deficiency, 63/328
pruritus, 53/207
pseudoabducens nerve palsy, 53/213
pure motor hemiplegia, 54/240
pure sensory stroke, 54/250
race, 53/200
radionuclide angiography, 53/270
recovery, 53/203
regional brain blood flow, 54/117
regional brain metabolic rate for oxygen, 54/117
respiration, 53/208
restrictive infiltrative cardiomyopathy, 63/137
retinal artery embolism, 53/218
retraction nystagmus, 53/212
risk factor, 53/200, 258, 291
seizure, 53/206
sex ratio, 53/200, 259
sexual impotence, 53/207
single photon emission computer assisted tomography, 53/313, 54/119, 131
skew deviation, 53/213
Sneddon syndrome, 55/402
speech disorder, 53/206
spinal cord, see Intermittent spinal cord ischemia
status cribrosus, 53/316
Sturge-Weber syndrome, 55/444
subclavian bruit, 53/216
subclavian steal syndrome, 12/8, 53/265
sulfinpyrazone, 53/274
suloctidil, 53/275
surgery, 12/11, 53/275
surgical criteria, 53/277
surgical result, 53/292
surgical treatment, 53/322

symptom, 53/204
syncope, 53/234
systemic disease, 53/200
Takayasu disease, 71/202
thermoregulation, 53/207
thrombocythemia, 55/473
thrombocytopenia, 53/266
thrombocytosis, 53/266
thromboembolism, 53/265
thrombophilia, 63/328
ticlopidine, 53/275
tinnitus, 53/207
transcortical aphasia, 53/215
transient global amnesia, 45/210
treatment, 53/272, 322
tremor, 53/263
unilateral pulsating tinnitus, 53/207
vascular pulsation, 53/215
vascular spasm, 11/522, 53/264
vasculitis, 53/266
vertebral artery, 12/9-11
vertebrobasilar, see Vertebrobasilar transient ischemic attack
vertebrobasilar arterial system, 12/2
vertebrobasilar ischemia, 2/249
vertebrobasilar system, 53/261
vertical gaze palsy, 53/212
vertical nystagmus, 53/213
vertical one and half syndrome, 53/212
vertigo, 12/4, 53/207, 263
visual evoked response, 53/232
visual field defect, 12/3
visual hallucination, 53/207
visual manifestation, 53/207
vomiting, 53/206
wasp venom, 65/198
xanthochromia, 53/224
Transient neonatal hypoglycemia
asymptomatic, 27/56
convulsion, 27/56
neurologic sequela, 27/55
symptomatic, 27/56
Transient neonatal myasthenia gravis
disease duration, 62/401
symptom, 62/401
Transient paraplegia
recurrent, see Recurrent transient paraplegia
Transillumination
hydrocephalus, 30/548, 50/292
subdural effusion, 24/335
Transitional sclerosis, 9/469-482
age at onset, 9/474
classification, 10/36

multiple sclerosis, 9/473
Transitional zone
  CNS, *see* CNS transitional zone
  nerve, *see* Nerve transitional zone
  peripheral nervous system, *see* Peripheral nervous
    system transitional zone
Transketolase
  alcoholic polyneuropathy, 51/315
  dialysis encephalopathy, 63/528
  hypertrophic interstitial neuropathy, 21/161
  uremic polyneuropathy, 51/361
  vitamin B1 deficiency, 51/332
Transketolase reaction
  deficiency neuropathy, 51/330
Transmission blockade
  neuromuscular, *see* Neuromuscular transmission
    blocking
Transmission computer assisted tomography
  brain infarction, 54/117
Transmissions
  X-linked, *see* X-linked transmission
Transmitter substance
  autonomic nervous system, 74/38
Transoccipital rhizotomy
  trigeminal neuralgia, 5/305
Transorbital leukotomy
  anxiety, 1/110
  pain, 1/110
Transplantation
  bone marrow, *see* Bone marrow transplantation
  brain tumor, 17/2
  corneal, *see* Corneal transplantation
  fetal thymus, 60/389
  heart, *see* Heart transplantation
  hepatobiliary, *see* Hepatobiliary transplantation
  organ, *see* Organ transplantation
  pancreas, *see* Pancreas transplantation
  renal, *see* Renal transplantation
  uremic polyneuropathy, 51/360
Transport
  axonal, *see* Axonal transport
  bilirubin, 27/416
  brain injury, 24/610
  cervical vertebral column injury, 25/277
  inexpert, 25/277
  nerve, 51/6
  patient, *see* Patient transport
  reverse cholesterol, *see* Reverse cholesterol
    transport
  sialic acid, *see* Sialic acid transport
  spinal cord injury, 25/285
  spinal injury, 26/295-298
  spinal unit, 26/297

Transport block
  slow axon, *see* Slow axon transport block
Transsexualism
  unknown origin, 43/216
Transtentorial herniation
  see also Brain herniation *and* Clivus ridge
    syndrome
  arrhythmia pathogenesis, 63/237
  brain injury, 23/55, 57/133
  cardiac dysrhythmia, 63/237
  catecholamine, 63/237
  cerebellar pressure cone, 16/119
  cervicomedullary injury, 24/98
  dehydrating agent, 1/564
  diagnosis, 1/561-563
  EEG, 1/561
  head injury, 57/50
  hemiplegia, 1/560
  homonymous hemianopia, 24/64
  hypothermia, 1/564
  intracranial pressure, 55/206, 57/50, 133
  mechanics, 55/206
  nuchal rigidity, 1/560
  occipital lobe tumor, 17/333
  oculomotor nerve injury, 24/61, 64
  oculomotor paralysis, 1/559
  pallidal necrosis, 49/469
  pathogenesis, 1/554
  pathology, 1/555-557, 23/55
  pathophysiology, 54/48
  posterior cerebral artery syndrome, 53/410
  prevalence, 16/117
  symptomatology, 54/48
  traumatic intracranial hematoma, 57/263
  treatment, 1/563
Transthyretin, *see* Prealbumin
Transverse atlantal ligament laxity
  Down syndrome, 50/392
Transverse basilar fissure
  bone abnormality, 50/388
  features, 50/389
Transverse cord lesion, *see* Transverse spinal cord
    lesion
Transverse myelitis
  bacterial meningitis, 52/14
  cat scratch disease, 52/130
  cerebral malaria, 35/149, 52/370
  coccidioidomycosis, 52/414, 420
  Coxsackie virus, 56/330
  diamorphine addiction, 37/387
  diphtheria toxoid, 52/241
  ECHO virus, 56/330
  enterovirus 72, 56/330

Epstein-Barr virus, 56/253
Epstein-Barr virus infection, 56/252
Epstein-Barr virus myelitis, 56/8, 253
forgotten respiration, 63/489
immunocompromise, 56/477
immunocompromised host, 56/477
infectious mononucleosis, 56/252
lymphocytic choriomeningitis, 56/360
malaria, 52/370
multiple sclerosis, 47/72
neuromyelitis optica, 47/401
poliomyelitis, 56/330
postinfectious complication, 47/484
postvaccinial, see Postvaccinial transverse
   myelitis
Schistosoma haematobium, 52/511
schistosomiasis, 52/511, 538
Sjögren syndrome, 63/55
spinal cord abscess, 52/191
spinal cord compression, 19/376
spinal subdural empyema, 52/189
systemic lupus erythematosus, 63/54
tuberculous meningitis, 52/202
Vogt-Koyanagi-Harada syndrome, 56/620
Transverse myelopathy
anterior spinal artery syndrome, 63/42
ascending, see Foix-Alajouanine disease
atrial myxoma, 63/98
intrathecal methotrexate, 63/358
methanol intoxication, 64/101
methotrexate, 63/358
methotrexate intoxication, 65/557
spinal cord electrotrauma, 61/195
Transverse sinus
caudal swing, 11/63
diploic vein, 11/55
dura mater, 30/424, 426
embryology, 11/60, 62
epidural hematoma, 57/252
radioanatomy, 11/112
variation, 11/53
venous sinus thrombosis, 12/425
Transverse sinus thrombophlebitis
anatomy, 33/173
clinical features, 33/174
differential diagnosis, 33/175
Griesinger sign, 33/174
prognosis, 33/175
Transverse sinus thrombosis
see also Venous sinus thrombosis
anatomy, 52/176
anemia, 54/397
antecedent factor, 52/176

Bacteroides fragilis, 52/177
brain angiography, 52/177
brain phlebothrombosis, 12/425
ceftazidime, 52/177
cerebellar abscess, 54/397
cholesteatoma, 52/176, 54/397
cilastatin, 52/177
clinical features, 52/176
computer assisted tomography, 52/176
Crowe sign, 54/397
CSF, 52/176
ear ache, 52/176
Escherichia coli, 52/177
fever, 54/397
Gradenigo syndrome, 52/176
Griesinger sign, 52/176, 54/397
headache, 5/154, 48/283, 52/176
hydrocephalus, 52/176
imipenem, 52/177
intracranial hypertension, 52/177
metronidazole, 52/177
nafcillin, 52/177
nystagmus, 52/176
otitis media, 52/176, 54/397
otogenic hydrocephalus, 54/397
prognosis, 52/177
radiation, 52/176
sepsis, 54/397
Staphylococcus aureus, 52/177
treatment, 52/177
vertigo, 52/176
weight loss, 54/397
Transverse spinal cord lesion
see also Spinal injury
adjustment vs age, 61/548
adjustment vs education, 61/548
adjustment vs injury severity, 61/548
adjustment vs patient personality, 61/548
adjustment stage, 61/537
alcohol, 61/369
alpha motor system, 61/367
American Spinal Injury Association Motor Scale,
   61/424
4-aminopyridine, 61/401
anal reflex, 61/295
angiotensin II, 61/444
anocutaneous reflex, 61/295
antithrombin III, 26/394
apoplectic type, 19/37, 56
aspartic acid, 61/367
autonomic dysreflexia, 61/447
axon regeneration, 61/407
axon survival, 61/400

axonal dysfunction, 61/401
baclofen, 61/368
baroreflex, 61/444
basal cardiovascular state, 61/440
Bastian rule, 2/180
blood pressure, 61/439
bradycardia, 61/436
bulbocavernosus reflex, 61/295
cardiac arrest, 61/439, 446
cardiovascular posture response, 61/441
cardiovascular system, 61/435
child, 19/41
clinical trial, 61/399
coitus, 61/315
complete, 19/29
cordotomy, 61/369
corticosteroid, 61/399
dantrolene, 61/368
deafferentation, 61/370
definition, 61/37
diazepam, 61/368
dioxide consumption, 61/279
disability acceptance, 61/546
distress, 61/537
dorsal spinal root entry zone microcoagulation, 61/371
ECG, 61/286, 439
ejaculation, 61/315, 322
electrical injury, 61/195
electroejaculation, 61/323
experimental therapy, 61/399
female sexual function, 61/331
food ingestion, 61/446
Frankel Scale, 61/422
GABA, 61/367, 401
gamma motor system, 61/367
glutamic acid, 61/367
glycine, 61/367
heart rate, 61/439
heat gain, 26/358, 361
hyperkalemia, 61/286
hyperventilation, 61/445
hypoglycemia, 61/446
impotentia coeundi, 61/315
intrathecal agent, 61/368
intubation response, 61/437
K opiate receptor blocker, 61/399, 403, 406
leukemia, 63/343
level, 19/29
life quality, 61/541
life satisfaction, 61/547
lipid peroxidation inhibitor, 61/399
long-term recovery, 61/430

male sexual function, 61/313
mass reflex, 61/447
maturity stage, 61/537
mental depression, 61/537
meta studies, 61/424
metabolism, 61/278
N-methylaspartic acid receptor blocker, 61/399, 406
methylprednisolone, 61/399, 402
monosialoganglioside, 61/399, 406
morphine, 61/368
motor symptom, 19/56
myelotomy, 61/369
naloxone, 61/399
National Acute Spinal Cord Injury Study Scale, 61/423
neurostimulation, 61/372
nuclear magnetic resonance based prognosis, 61/427
obturator neurotomy, 61/369
orthostatic hypotension, 22/263, 61/442, 63/153, 232
pathology specimen, 25/27-42
patient adjustment, 61/537
patient personality characteristics, 61/542
penis prosthesis, 61/321
percutaneous thermal rhizotomy, 61/370
phantom limb, 4/224, 226
pharmacologic management, 61/430
pharmacotherapy, 61/368
phenol, 61/369
piloerection, 61/280
plasma adrenalin, 61/436
plasma aldosterone, 61/443
plasma noradrenalin, 61/436
plasma renin, 61/443
poikilothermia, 61/275
posterior column syndrome, 19/29
pressure sore, 61/347
prognosis, 61/427
protirelin, 61/403, 406
psychology, 61/537
pyramidal syndrome, 19/27
radiation myelopathy, 61/200
rapid, 19/41
rebellion stage, 61/537
recovery, 61/399
recovery determining factor, 61/422
recovery vs lesion type, 61/428
recovery scoring, 61/422
recovery vs spinal syndrome, 61/428
recovery substrate, 61/400
recovery vs treatment type, 61/428

reflex automatism, 2/181
reflex synergy, 2/182
regenerative therapy, 61/406
remyelination, 61/402
renin release, 61/443
renin response, 61/443
Renshaw cell, 61/367
rhizotomy, 61/369
self-concept, 61/547
sensory microrootlet section, 61/371
sex difference, 61/546
sexual capability, 61/315
shivering, 61/279
sociology, 61/537
spasticity, 61/367
spinal apoplexy, 19/41
spinal cord diaschisis, 61/367
spinal shock, 61/439
steroid antioxidant, 61/405
sudomotor activity, 61/278
sudomotor function, 61/275
Sunnybrook Cord Injury Scale, 61/423
surgical treatment, 61/369
sympathoneural activation, 61/437
systemic hypotension, 61/436
tachycardia, 61/436
thermoregulation, 61/275
thermoregulation management, 61/286
tirilazad, 61/405
tizanidine, 61/368
tracheal stimulus response, 61/437
tracheostomy, 61/437
vacuum constrictor devices, 61/321
Valsalva maneuver, 61/445
vascular tone, 61/436
vasopressor agent, 61/451
visceral stimulation, 61/451
Yale Scale, 61/423
Transverse spinal cord syndrome
  bladder function, 2/188-193
  complete, 2/178-200
  incomplete, 2/207-212
  intestinal function, 2/192-195
  level, 2/201-207
  loss of sensation, 2/184
  motor disorder, 2/180-187
  nociceptive border zone reaction, 2/187
  orthostatic hypotension, 2/198
  phantom sensation, 2/187
  reflex pattern, 2/183
  respiration, 2/202
  rhythmic alternating contraction, 2/184
  sexual dysfunction, 2/196-198
  thermoregulation, 2/199
  total, *see* Total transverse spinal cord syndrome
Transverse syndrome
  orthostatic hypotension, 2/198
  phantom sensation, 2/187, 45/400
  total spinal, *see* Total transverse spinal cord syndrome
  vascular myelopathy, 12/498
Transverse tubule
  see also Sarcoplasmic reticulum
  chloroquine, 40/151
  colchicine, 40/154
  histochemistry, 40/2
  honeycomb structure, 40/106, 224
  normal structure, 40/66, 224, 549
  periodic paralysis, 40/153
  tubular aggregate, 40/33, 112, 226, 229
  vincristine, 40/154
Tranylcypromine
  multiple sclerosis, 9/397
Trapezoidocephaly, *see* Multisynostotic osteodysgenesis
Traube-Hering wave
  CSF pressure, 19/113
Trauma
  acquired myoglobinuria, 62/572
  amyotrophic lateral sclerosis, 22/309, 328, 59/178
  anoxic ischemic leukoencephalopathy, 9/587
  aorta rupture, 61/112, 116
  atlantoaxial dislocation, 32/83
  autonomic nervous system, 75/39
  basilar impression, 32/68, 50/399
  bitemporal hemianopia, 2/575
  blood-brain barrier, 28/371-374
  brain, *see* Brain injury
  brain abscess, 33/116
  brain hemorrhage, 11/608
  brain infarction, 53/165
  brain ischemia, 53/165
  carotid artery thrombosis, 12/348, 23/539, 550, 26/43, 45
  carotid dissecting aneurysm, 53/39, 54/271
  carpal tunnel syndrome, 7/292
  cavum septi pellucidi, 30/326
  chorea, 6/438, 49/550
  corpus callosum, 17/542
  dissecting aorta aneurysm, 63/45
  elbow, *see* Elbow trauma
  epidural hematoma, 54/345
  epidural spinal lipoma, 20/407
  femoral neuropathy, 8/306
  glossopharyngeal neuralgia, 48/464
  head, *see* Head trauma

immunocompromised host, 56/474
intravascular consumption coagulopathy, 55/493
local anesthetic agent, 65/421
lumbar intervertebral disc prolapse, 20/584, 594
migraine, 48/73, 85
multiple, *see* Multiple trauma
multiple sclerosis, 9/115-117, 47/64, 152
myoclonus, 49/618
nerve, *see* Nerve injury
orthopedic, *see* Orthopedic trauma
palatal myoclonus, 38/584
Parkinson disease, 6/176, 228, 230-233, 49/108
parkinsonism, 6/176, 228, 230-233
peripheral, *see* Peripheral trauma
poliomyelitis, 42/653, 56/315
pontine syndrome, 2/262
secondary pigmentary retinal degeneration,
   13/216
severe, *see* Severe trauma
spastic paraplegia, 59/434
spinal, *see* Spinal trauma
spinal angioma, 20/461
spinal cord, *see* Spinal cord trauma
spinal ependymoma, 20/365
spinal meningioma, 20/178
spinal phlebography, 19/240
spinal subdural hematoma, 12/566
stress, 74/317
striatal lesion, 6/672
subarachnoid hemorrhage, 12/120
subclavian artery stenosis, 53/372
sudden infant death, 63/466
thoracic outlet syndrome, 51/124
vertebral dissecting aneurysm, 53/39
vibration neuropathy, 74/317
Traumatic amnesia, 45/185-190
   see also Memory disorder *and* Posttraumatic
      amnesia
   brain concussion, 23/427, 24/562
   control function, 24/548
   definition, 45/185
   immediate memory, 24/546, 562
   memory disorder, 3/293-295
   memory recall, 24/548, 564
   memory scale, 24/548
   memory storage, 24/547
   occupational competence, 24/484
   posttraumatic epilepsy, 24/451-453
   recent memory, 24/546, 563
   test, 24/488, 546
   traumatic psychosis, 24/526
   Wernicke-Korsakoff syndrome, 24/478
Traumatic aneurysm

brain aneurysm, 55/78
brain aneurysm neurosurgery, 55/78
brain artery, 24/389
brain stab wound, 23/488
cavernous sinus, 24/386
classification, 24/381
cranial nerve, 24/386
dissecting aneurysm, 24/389
epidural hematoma, 24/392
epistaxis, 24/383
head injury, 24/381
internal carotid artery, 24/384
intracerebral hematoma, 24/393
meningeal artery, 24/392-394
mortality, 24/391
pericallosal artery, 24/390
scalp, 24/381
skull, 24/383
skull fracture, 24/392
subarachnoid hemorrhage, 55/78
treatment, 24/396, 664
Traumatic aphasia
   brain injury, 24/490
   child, 24/485
   general, 24/483
   penetrating brain injury, 24/490
Traumatic blindness
   see also Cortical blindness *and* Optic nerve injury
   brain injury, 24/27
   child, 24/51
   clinical examination, 24/28
   cortical blindness, 24/48-52
   delayed, 24/37
   history, 24/27
   immediate, 24/37
   incidence, 24/31, 52
   optic chiasm injury, 24/28, 44
   optic nerve injury, 24/27
   prognosis, 24/38
   radiation, 24/28
   transient, 24/51
   treatment, 24/32, 34, 42
   unilateral, 24/37
   visual cortex, 24/28, 48-52
Traumatic carotid cavernous sinus fistula,
      57/345-366
   see also Carotid cavernous fistula
   abducent nerve paralysis, 57/346, 350, 362
   angiographic technique, 57/353
   angiography, 57/347, 353
   arterial steal, 57/355
   balloon occlusion, 57/351, 354
   balloon overinflation, 57/360

balloon rupture, 57/360
bruit, 57/352
calibrated leak balloon catheter, 57/363
cause, 57/345
cavernous sinus surgical exposure, 57/364
cavernous sinus varix, 57/347
chemosis, 57/350
classification, 57/345
collateral circulation, 57/364
common carotid artery ligation, 57/355
complication, 57/360
computer assisted tomography, 57/353
cranial nerve palsy, 57/346
death, 57/347
detachable balloon catheter, 57/359
direct repair, 57/355
electrothrombosis, 57/355
embolization, 57/347
endovascular technique, 57/357
endovascular therapy, 57/357
etiology, 57/345
exophthalmos, 57/352
external carotid artery, 57/353
Fogarty catheter, 57/357
head injury, 57/345
hemodynamic disturbance, 57/355
internal maxillary artery, 57/353
intracranial bruit, 57/350
intracranial hemorrhage, 57/347
nasal hemorrhage, 57/347
natural history, 57/347
neuro-ophthalmology, 57/347
oculomotor paralysis, 57/352, 361
ophthalmoplegia, 57/352
palpebral chemosis, 57/346
prognosis, 57/363
progressive ligation, 57/355
pseudoaneurysm, 57/361
pulsatile exophthalmos, 57/354
radiologic evaluation, 57/353
retrogasserian procedure, 57/346
retrograde packing, 57/355
secondary glaucoma, 57/353
site, 57/353
size, 57/353
skull base fracture, 57/345
spontaneous closure, 57/347
spontaneous regression, 57/347
spontaneous thrombosis, 57/363
subarachnoid hemorrhage, 57/347
subtraction angiography, 57/362
superior ophthalmic vein, 57/361
surgical treatment, 57/355

technical problem, 57/360
transvascular embolization, 57/361
treatment indication, 57/363
treatment result, 57/360
trigeminal nerve, 57/353
trochlear nerve paralysis, 57/346
venous approach, 57/361, 364
venous drainage, 57/354
visual loss, 57/352
Traumatic carotid thrombosis
    boxing injury, 23/539, 550
    football injury, 23/574
    skiing injury, 23/579
    soccer injury, 23/569
Traumatic CSF hypotension
    mental deterioration, 24/255-257
    subdural hematoma, 24/256
Traumatic delayed apoplexy
    brain infarction, 53/190
Traumatic dementia
    see also Traumatic intellectual defect
    clinical features, 24/233
    pathology, 23/62, 24/233
    sign, 24/552
    test, 24/552
    traumatic psychosyndrome, 24/552
Traumatic diencephalic syndrome
    traumatic psychosyndrome, 24/582, 704
Traumatic encephalopathy
    occipital lobe, 3/69
    white matter demyelination, 3/69
Traumatic epidural arachnoid cyst, see Postoperative
    meningeal diverticulum
Traumatic epilepsy, 72/116, 191, 360
    armed forces, 24/469
    boxing injury, 23/530
    classification, 15/281
    depressed skull fracture, 23/410
    EEG, 23/357
    missile injury, 23/522, 24/468
    prevalence, 15/280
    seizure pattern, 15/276
    sport injury, 23/530
Traumatic exophthalmos
    orbital injury, 24/89
Traumatic frontobasal syndrome
    traumatic psychosyndrome, 24/582
Traumatic headache
    battered child syndrome, 48/32
    migraine, 24/506
    post, see Posttraumatic headache
    skull defect, 24/507
    traumatic psychosyndrome, 24/565

Traumatic hydrocephalus
  brain cortex, 24/242
  brain scintigraphy, 24/239, 248
  clinical features, 24/232
  contraindication, 24/246
  CSF, 24/237
  differential diagnosis, 24/234
  echoEG, 24/236
  EEG, 24/234
  head injury, 24/231
  incidence, 24/231
  intracranial hypertension, 24/244, 248
  intracranial pressure, 24/239, 244, 248
  isotope cisternography, 24/236, 238, 248
  leptomeningeal fibrosis, 24/242
  morbidity, 24/247
  mortality, 24/247
  neuropathology, 24/240
  pathogenesis, 24/233, 238, 243-246
  PEG, 24/234, 249
  pneumocephalus, 24/234, 249
  surgery, 24/247
  treatment, 24/246
  treatment result, 24/247, 250
Traumatic hypotension
  CSF, 24/255
Traumatic intellectual defect
  see also Traumatic dementia
  age, 24/482
  incidence, 24/481, 492
  occupational competence, 24/484
  pediatric head injury, 24/484
  penetrating head injury, 24/482, 492
  persistent dementia, 24/481, 702
  premorbid intelligence, 24/482
  recovery, 24/481
  rehabilitation, 24/691
  temporal lobe, 24/532
  test, 24/481, 483, 488
Traumatic intracerebral hematoma
  acute, 24/353, 365
  basal ganglion, 24/356, 359
  bifrontal contusion, 57/150
  bilateral, 24/355, 361
  brain contusion, 24/353, 57/257
  brain swelling, 57/159
  bullet wound, 57/258
  burst lobe, 57/253
  compound polar injury, 57/253
  computer assisted tomography, 57/149, 156
  depressed fracture, 57/258
  development, 57/256
  diffuse axonal injury, 57/156

distribution, 57/253
  EEG, 57/186
  epidural hematoma, 24/352
  head injury, 57/104
  head injury outcome, 57/379
  history, 24/351
  incidence, 24/352
  intrafrontal hematoma, 57/158
  intravascular consumption coagulopathy, 57/258
  ischemic neuronal necrosis, 57/261
  local effect, 57/260
  mechanism, 24/354
  pathology, 24/353
  penetrating head injury, 57/258
  prognosis, 57/379
  Spät Apoplexie, 24/354, 365
  subdural hematoma, 24/352
  temporal lobe, 24/356, 358
  traumatic intracranial hematoma, 57/252
Traumatic intracranial hematoma
  absolute risk, 57/270
  admission policy, 57/271
  adult, 57/269
  annual occurrence rate, 57/250
  asymptomatic, 57/254
  basal ganglion hematoma, 57/152, 162
  brain angiography, 57/267
  brain contusion, 57/252, 280
  brain hypoxia, 57/262
  brain stem hematoma, 57/152
  burr hole, 57/277
  changing nuclear magnetic resonance pattern,
    57/171
  chest complication, 57/281
  child, 57/255, 269
  chronic subdural collection, 57/155
  chronic subdural hematoma, 57/154, 251
  classification, 57/251
  clinical deterioration, 57/266
  clinical features, 57/264
  computer assisted tomography, 57/152, 267,
    272-274
  conscious level, 57/264, 269
  craniectomy, 57/277
  craniotomy, 57/277
  delayed, 57/156, 254
  detection, 57/268
  deterioration, 57/281
  development, 57/254
  diagnosis, 57/268
  diffuse brain swelling, 57/280
  epidemiology, 57/250
  epidural hematoma, 57/152

failure to improve, 57/281
frequency, 57/255
general anesthesia, 57/275
head injury, 57/249
hematoma volume, 57/263
hyponatremia, 57/281
incidence, 57/250
intracerebral hematoma, 57/156, 253
intracranial pressure, 57/258, 275, 279
intraoperative bleeding, 57/278
intraoperative complication, 57/278
intraventricular hemorrhage, 57/148, 158
layering, 57/158
lucid interval, 57/264
mechanism, 57/254
meningitis, 57/281
motor weakness, 57/265
neck pain, 57/265
neurologic symptom, 57/264
new hematoma, 57/280
nuchal rigidity, 57/265
nuclear magnetic resonance, 57/171, 267
observation policy, 57/271
occult intracranial hematoma, 57/273
onset of symptom, 57/266
operative treatment, 57/272
plain X-ray, 57/267
posterior fossa hematoma, 57/157, 253, 266
postoperative complication, 57/279
postoperative hematoma, 57/280
postoperative management, 57/278
posttraumatic subdural hygroma, 57/286
preoperative management, 57/275
pupil, 57/265
radiodiagnosis, 57/267
relative risk, 57/268
risk, 57/143
screening, 57/271
seizure, 57/266, 281
skull fracture, 57/269
skull radiation, 57/267, 271
Spät Apoplexie, 57/254
spectrum, 57/253
subdural hematoma, 57/152, 251
surgery complication, 57/278
surgery indication, 57/273
surgical evaluation, 57/275
surgical technique, 57/277
transtentorial herniation, 57/263
traumatic intracerebral hematoma, 57/252
vasopressin, 57/281
Traumatic intracranial hematoma outcome
  age, 57/282
conscious level, 57/282
delay in diagnosis, 57/285
delay in surgery, 57/285
Glasgow Coma Scale, 57/284
head injury, 57/282
hematoma size, 57/285
injury-diagnosis interval, 57/286
intradural lesion, 57/282
prognosis, 57/282
result, 57/282
subdural hematoma, 57/284
surgery timing, 57/285
type, 57/282
Traumatic memory deficit, see Traumatic amnesia
Traumatic meningioma
  history, 24/441
  incidence, 24/441
  mechanism, 24/444
Traumatic motoneuron disorder
  WFN classification, 59/5
Traumatic myelopathy
  Hirayama disease, 59/114
  late, 25/89
  monomelic spinal muscular atrophy, 59/376
  morphologic data, 25/90
Traumatic neuroma
  amputation, 70/44
Traumatic neurosis
  brain concussion, 23/425
Traumatic parietal syndrome
  vegetative disorder, 24/582
Traumatic plexopathy
  monomelic spinal muscular atrophy, 59/376
Traumatic psychosis
  see also Traumatic psychosyndrome
  aggression, 24/527
  alcoholism, 24/529
  anoxia, 24/519
  arteriosclerosis, 24/529
  bilateral brain injury, 24/533
  brain concussion, 24/521
  brain edema, 24/519
  brain injury, 24/520
  clinical features, 24/524
  confabulation, 24/525
  consciousness, 24/527
  diagnosis, 24/521
  differential diagnosis, 24/530
  disorientation, 24/525
  dysphasia, 24/531
  EEG, 24/523
  focal sign, 24/529
  frontal lobe, 24/531

frontobasal brain injury, 24/531
hippocampal gyri, 24/533
history, 24/517
intracerebral hematoma, 24/532
morbid state, 24/526, 529
neurosis, 24/526
pathogenesis, 24/518
penetrating brain injury, 24/520
schizophrenia, 24/527
temporal lobe, 24/532
traumatic amnesia, 24/526
treatment, 24/533
Traumatic psychosyndrome
    see also Traumatic psychosis
    affect regulation, 24/555, 567
    air study, 24/539
    alcoholism, 24/568, 570
    behavior therapy, 24/556
    brain atrophy, 24/539
    brain injury, 24/838
    bulbar brain syndrome, 24/583, 705, 727
    compensation, 24/569
    comprehension disturbance, 24/551
    concentration disturbance, 24/553
    criminality, 24/567
    depression, 24/838
    diagnosis, 24/545
    differential diagnosis, 24/555
    dysarthria, 24/547
    dysphasia, 24/546
    EEG, 24/540
    fatigue, 24/552, 686
    frontal lobe, 24/558, 560
    law, 24/838
    memory disorder, 24/546
    memory storage, 24/547
    memory training, 24/556
    midbrain syndrome, 24/583, 705
    neurosis, 24/565, 587
    nomenclature, 24/535
    parietal syndrome, 24/582, 704
    pathology, 24/557
    perseveration, 24/550
    personality disorder, 24/554, 687, 693
    posttraumatic epilepsy, 24/566
    posttraumatic headache, 24/565
    premorbid organic brain lesion, 24/565, 687
    premorbid state, 24/564
    prognosis, 24/569
    psychiatric examination, 24/543
    psychotherapy, 24/571
    reaction time, 24/553
    reduced drive, 24/554, 688

rehabilitation, 24/687
sensorimotor adaptability, 24/553, 690
sign, 24/538, 558, 836
stereotypy, 24/551
temporal lobe, 24/558
test, 24/544, 553
thought disturbance, 24/550, 552
traumatic dementia, 24/552
traumatic diencephalic syndrome, 24/582, 704
traumatic frontobasal syndrome, 24/582
traumatic headache, 24/565
traumatic temporobasal syndrome, 24/582
traumatic vegetative syndrome, 24/575
treatment, 24/555, 561, 570, 693
visual perception disorder, 24/549
Traumatic radiculopathy
    monomelic spinal muscular atrophy, 59/376
Traumatic shock
    spinal shock, 26/246
Traumatic spine deformity
    age, 26/167
    complete cord lesion, 26/166, 172
    cord damage degree, 26/165
    emotional state, 26/165
    fusion, 26/171
    general, 26/161
    Harrington rod, 25/187, 26/168
    history, 26/159
    instability, 26/162
    limb deformity, 26/167
    mechanics, 26/164
    muscle tone, 26/165
    neurologic disorder, 26/164
    neuromuscular, 26/164
    pelvic obliquity, 26/168, 170
    pharmacologic control, 26/165
    prevention, 26/163
    rehabilitation, 26/163, 169
    segmental level, 26/167
    skeletal, 26/164
    spinal collapse, 26/169
    surgery, 26/170
    treatment, 26/169, 171
Traumatic spondylolisthesis, *see* Hangman fracture
Traumatic syndrome
    post, *see* Posttraumatic syndrome
Traumatic syringomyelia
    post, *see* Posttraumatic syringomyelia
Traumatic temporobasal syndrome
    traumatic psychosyndrome, 24/582
Traumatic vasospastic disease, *see* Vibration
        neuropathy
Traumatic vegetative syndrome

attack, 24/593
blood pressure, 24/578
cardiovascular system, 24/579
definition, 24/575
dizziness, 24/577
EEG, 24/579
gasping respiration, 24/586
headache, 24/580
metabolism, 24/581
nausea, 24/577
pulse rate, 24/578
respiration, 24/580, 586
sign, 24/576
sweating, 24/579
thermoregulation, 24/579
traumatic psychosyndrome, 24/575
vasomotor disorder, 24/578
vertigo, 24/577
vomiting, 24/577
Traumatic venous sinus thrombosis
angiography, 24/378
anticoagulant, 24/379
birth incidence, 24/371
closed head injury, 24/370
corticosteroid, 24/379
history, 24/369
incidence, 24/370
infant, 24/371
mortality, 24/376
papilledema, 24/376
penetrating brain injury, 24/369, 372
prognosis, 24/376
sign, 24/372, 378
sinography, 24/371
skull fracture, 24/370
superior longitudinal sinus, 24/370, 372
surgery, 24/373-375
treatment, 24/373, 378
Trazodone
heterocyclic antidepressant, 65/312
Trazodone intoxication
delirium, 65/320
headache, 65/320
vertigo, 65/320
Treacher-Collins syndrome
annual incidence, 43/422
antimongoloid eyes, 43/421
cheilopalatoschisis, 43/422
dermoid, 32/127
eyelid coloboma, 43/421
mental deficiency, 46/46
micrognathia, 43/421
microtia, 43/421

zygomatic arch absence, 43/421
Treisman theory
motion sickness, 74/354
Trematode
helminthiasis, 52/505, 511
Tremor, 6/809-824
see also Asterixis
accelerometry, 49/574
acquired immune deficiency syndrome dementia,
56/491
acrodynia, 64/374
acrylamide intoxication, 64/63, 68
action, see Action tremor
Addison disease, 42/512
adrenalin, see Adrenalin tremor
akathisia, 65/286
alcohol intoxication, 64/115
alcoholic, see Alcoholic tremor
Alzheimer disease, 46/252
amiodarone, 63/190
amiodarone intoxication, 65/458
amiodarone myopathy, 63/190
amiodarone polyneuropathy, 51/302, 63/190
amphetamine intoxication, 65/51, 257
anatomic classification, 1/279
anatomic site, 49/597
ant venom, 65/199
anxiety, see Anxiety tremor
aprindine intoxication, 65/455
areflexia, 60/660
Argentinian hemorrhagic fever, 56/365
arsenic intoxication, 6/823
arthropod envenomation, 65/199
ataxia telangiectasia, 21/581
barbiturate intoxication, 65/332
batwing, see Asterixis
Benedikt syndrome, 1/187, 6/162, 49/597
benign essential, see Benign essential tremor
benign hereditary, see Benign essential tremor
beta adrenergic receptor blocking agent
intoxication, 65/437
bismuth intoxication, 64/338
Bolivian hemorrhagic fever, 56/369
brachium conjunctivum lesion, 21/514
brain arteriovenous malformation, 53/36
brain infarction, 63/345
bromide intoxication, 36/299
caffeine intoxication, 65/260
carbamate intoxication, 64/191
carbon disulfide intoxication, 64/25
carbon monoxide intoxication, 6/823, 64/32
central pontine myelinolysis, 2/263
cerebellar, see Cerebellar tremor

cerebellar dysarthria, 42/137
cerebello-olivary atrophy, 60/661
cerebral palsy, 49/604
Chédiak-Higashi syndrome, 14/783, 60/661
chlordecone intoxication, 64/204
chlorphenotane, 64/199
chronic bromide intoxication, 36/299
chronic enterovirus infection, 56/330
chronic hepatocerebral degeneration, 49/219
chronic mercury intoxication, 64/370, 394
chronic relapsing neuritis, 49/588
ciclosporin, 63/10, 189, 537
ciclosporin intoxication, 63/537, 65/551
circadian fluctuation, 74/472
citrullinuria, 42/538
classification, 6/811
cocaine, 65/251
cocaine intoxication, 65/251, 253
congenital, *see* Congenital tremor
congenital myopathy, 62/335
corpus callosum agenesis, 42/100
corticosteroid, 63/360
cycloserine, 65/481
cycloserine intoxication, 65/481
definition, 1/286, 49/583
dentate nucleus, 21/514
dentatorubral, *see* Dentatorubral tremor
diabetes mellitus, 49/589
diazoxide intoxication, 37/433, 65/446
diborane intoxication, 64/359
differential diagnosis, 49/583, 592
disequilibrium syndrome, 42/139
dominant cerebello-olivary atrophy, 42/136
drug induced extrapyramidal syndrome, 6/262
drug induced tremor, 6/263
Dumon-Radermecker syndrome, 42/477
dyspnea, 63/482
dyssynergia cerebellaris myoclonica, 42/211
dyssynergia cerebellaris progressiva, 42/212
dystonia, 42/253, 49/586
dystonia musculorum deformans, 42/218
EMG, 1/647, 49/571
EMG differentiation, 49/568
ergot intoxication, 65/67
essential, *see* Essential tremor
essential familial, *see* Essential tremor
ethionamide, 65/481
ethionamide intoxication, 65/481
exanthema, 6/823
facial spasm, 42/315
familial benign, *see* Benign essential tremor
familial recessive, *see* Familial recessive tremor
fatigue, 6/818, 62/52

flapping, *see* Asterixis
flecainide, 63/192
flecainide intoxication, 65/456
fluoxetine intoxication, 65/320
fluvoxamine intoxication, 65/320
frequency, 49/574
Friedreich ataxia, 21/330, 60/661
frontal lobe tumor, 17/257
gabapentin intoxication, 65/512
general paresis of the insane, 52/279
generator, 49/572, 597
Gerstmann-Sträussler-Scheinker disease, 60/661
Guillain-Barré syndrome, 49/589
Hallervorden-Spatz syndrome, 6/606, 626, 66/712
head, *see* Head tremor
head injury, 49/604
Henneman size principle, 49/573
hereditary, *see* Essential tremor
hereditary ataxia, 60/660
hereditary cerebello-olivary atrophy (Holmes),
    21/403, 409
hereditary chin, *see* Hereditary chin tremor
hereditary dystonic paraplegia, 22/458
hereditary essential, *see* Hereditary essential
    tremor
hereditary motor and sensory neuropathy type II,
    60/661
hereditary motor and sensory neuropathy variant,
    60/245
hereditary multiple recurrent mononeuropathy,
    51/552
hereditary polyneuropathy, 49/588
hereditary striatal necrosis, 49/495
heredopathia ophthalmo-otoencephalica, 42/150
heterocyclic antidepressant, 65/316, 320
high-pressure nervous syndrome, 48/402
homovanillic acid, 49/73
horizontal gaze paralysis, 42/333
Huntington chorea, 6/328, 42/226, 49/293
hyperkinetic agraphia, 45/465-467
hypernatremia, 63/554
hyperthyroidism, 70/87
hypertrophic interstitial neuropathy, 42/317
hyperventilation syndrome, 38/319, 63/435
hypomagnesemia, 63/562
hyponatremia, 63/545
hypophosphatemia, 63/564
hysterical, *see* Hysterical tremor
IgM paraproteinemic polyneuropathy, 51/437
inertial load, 49/572
infant, 6/824
intention, *see* Intention tremor
intermediate spinal muscular atrophy, 42/90

involuntary movement, 42/238
Japanese B encephalitis, 56/138
jaw, *see* Jaw tremor
Joseph-Machado disease, 42/262, 60/661
juvenile Gaucher disease, 10/307
juvenile hereditary benign chorea, 49/345
juvenile parkinsonism, 49/155
Klinefelter syndrome, 21/467, 43/558, 50/549
kuru, 42/657
kwashiorkor, 7/635, 51/322
Lafora progressive myoclonus epilepsy, 27/173
Lassa fever, 56/370
lead intoxication, 64/435
leukemia, 63/345
lidocaine, 63/192
lidocaine intoxication, 65/454
lithium, 51/304
lithium intoxication, 64/295
lysergide, 65/45
lysergide intoxication, 65/44
manganese intoxication, 6/823, 64/308
Matthews-Rundle syndrome, 60/661
mechanical resetting, 49/571
mercury intoxication, 6/823, 51/272, 64/394
mercury polyneuropathy, 51/272
mescaline intoxication, 65/46
1-methyl-4-phenyl-1,2,3,6-tetrahydropyridine,
   49/73, 65/369
mexiletine, 63/192
mexiletine intoxication, 65/455
mianserin intoxication, 65/320
micrographia, 45/464
migraine, 48/147
Minamata disease, 36/105, 64/418
Minor benign essential, *see* Benign essential
   tremor
modulation, 49/600
motor unit, 62/52
motor unit behavior, 49/572
multiple mitochondrial DNA deletion, 62/510
multiple neuronal system degeneration, 75/162
multiple sclerosis, 9/181, 42/496, 49/604
muscimol intoxication, 65/40
muscular atrophy, 42/78
myoclonia, 6/772-774
myopathy, 43/132
neocerebellar syndrome, 2/420, 423
neuroleptic parkinsonism, 65/292
neuropharmacology, 49/57
neurophysiology, 6/180
nicotine intoxication, 36/425, 65/261
nodding, *see* Nodding tremor
normal, *see* Physiologic tremor

norpethidine intoxication, 65/357
nystagmus myoclonus, 42/240
olivopontocerebellar atrophy, 21/397, 455
olivopontocerebellar atrophy (Dejerine-Thomas),
   60/661
olivopontocerebellar atrophy with retinal
   degeneration, 60/661
organochlorine insecticide intoxication, 64/199,
   202, 204
organolead intoxication, 64/131, 134
organophosphate intoxication, 37/555, 64/167,
   226
organotin intoxication, 64/139
origin, 6/179
orthochromatic leukodystrophy, 42/500
orthostatic, *see* Orthostatic tremor
pallidoluysionigral degeneration, 49/446, 455
pallidonigral degeneration, 42/242
pallidopyramidal degeneration, 42/243
paraproteinemic neuropathy, 49/589
paraquat, 51/287
Parkinson, *see* Parkinson tremor
Parkinson disease, 6/161, 178-180, 196, 42/245,
   49/90
parkinsonism, 49/597, 599, 603
paroxetine intoxication, 65/320
paroxysmal, *see* Paroxysmal tremor
pathophysiology, 6/101, 49/73
Pelizaeus-Merzbacher disease, 42/502-505,
   60/661
pellagra, 7/571, 28/85, 90
pentaborane intoxication, 64/359
pethidine intoxication, 65/357
pharmacology, 49/602
phase shift, 49/571
phencyclidine intoxication, 65/54
phenethylamine intoxication, 65/51
phenylketonuria, 42/611, 59/75
phenytoin intoxication, 65/500
pheochromocytoma, 39/498, 501, 42/764, 48/427
phosphorus intoxication, 6/823
physiologic, *see* Physiologic tremor
porphyria, 49/589
posttraumatic, *see* Posttraumatic tremor
postural, *see* Postural tremor
precentral cortex, 49/597
primary writing, *see* Primary writing tremor
procainamide intoxication, 37/452, 65/453
progressive pallidal atrophy (Hunt-Van Bogaert),
   42/247
propafenone intoxication, 65/457
propranolol, 37/446
pyramidotomy, 49/597

pyrethroid intoxication, 64/214
recording method, 6/809-811
reflex sympathetic dystrophy, 61/122, 128
Refsum disease, 21/200
respiration, 63/482
respiratory encephalopathy, 63/414
resting, *see* Resting tremor
Roussy-Lévy syndrome, 21/171, 175, 42/108, 49/588
rubral, *see* Rubral tremor
selective serotonin reuptake inhibitor, 65/316, 320
selective serotonin reuptake inhibitor intoxication, 65/323
sertraline intoxication, 65/320
Shapiro syndrome, 50/167
Shy-Drager syndrome, 1/474
spasmodic torticollis, 42/264
spinocerebellar degeneration, 42/190
spinopontine degeneration, 21/390
St. Louis encephalitis, 56/138
Stargardt disease, 42/420
static, *see* Postural tremor *and* Static tremor
stereotaxic thalamotomy, 49/597
stereotaxic ventral intermedius thalamotomy, 49/603
striatopallidodentate calcification, 6/707, 49/422
Sydenham chorea, 49/363
systemic lupus erythematosus, 55/377
task specific, *see* Task specific tremor
temporal lobe tumor, 17/292
thalamic, 2/481
thalamic dementia, 21/590, 592
thalamic ventralis intermedius nucleus, 49/599
thalamogeniculate artery, 2/481
thalamotomy, 74/473
thalamus, 2/485, 49/597
thallium intoxication, 64/325
thyrotoxic, *see* Thyrotoxic tremor
tocainide, 63/192
tocainide intoxication, 65/455
transient ischemic attack, 53/263
treatment, 49/597
tricyclic antidepressant, 65/316, 318
Troyer syndrome, 42/194
tryptophanuria, 42/633
tsukubaenolide intoxication, 65/558
undetermined significant monoclonal gammopathy, 63/399
uremia, 49/589
uremic encephalopathy, 63/504
valproic acid intoxication, 65/504
ventro-oral posterior nucleus, 6/178
vidarabine intoxication, 65/534

vocal, *see* Vocal tremor
Wackel, *see* Wackeltremor
water intoxication, 64/241
Willvonseder syndrome, 42/194
Wilson disease, 27/384, 42/270, 49/225, 589
X-linked optic atrophy, 42/417
Tremor dysarthria ataxia syndrome
Huntington chorea, 49/293
Tremor generation
Parkinson disease, 49/100
Tremor related parkinsonism
essential, *see* Essential tremor related parkinsonism
Tremorine
antitremor agent, 6/224
Trench foot
see also Immersion foot
frostbite, 8/342
nerve injury, 51/136
Trendelenburg sign
gluteal nerve, 2/43
topographical diagnosis, 2/43
Trepang intoxication
food, 37/66
holothurioidea, 37/66
preparation, 37/66
Treponema pallidum
see also Syphilis
bacterial meningitis, 52/4
history, 33/338
human T-lymphotropic virus type I associated myelopathy, 56/538
Hutchinson teeth, 33/371
neurosyphilis, 33/338, 52/273, 282
tropical spastic paraplegia, 56/538
Treponema pallidum immobilization test
syphilis, 33/343, 52/276
Treponema pallidum syndrome
clinical features, 31/217
diagnosis, 31/218
immune response, 31/218
mental deficiency, 31/212
placental role, 31/217
seizure, 31/212, 218
subclinical, 31/218
treatment, 31/218
Tretinoin, *see* Retinoic acid
Triad syndrome, *see* Prune belly syndrome
Triamcinolone
see also Corticosteroid
steroid myopathy, 41/252
Triangular erythrocyte
thrombotic thrombocytopenic purpura, 9/600, 607

Triatoma
  Chagas disease, 52/345
Triatoma infestans
  Chagas disease, 75/387
Triazinate
  antimetabolite, 39/104
  chemotherapy, 39/104
  neurologic toxicity, 39/104
Triazolam
  antiepileptic agent, 65/497
  restless legs syndrome, 51/548
Tributyl phosphorotrithioite
  organophosphate intoxication, 64/155
Tricarboxylic acid cycle
  brain, 27/14-17
  citrate synthase, 27/14-16
  defect, 66/396
  fumarate hydratase deficiency, 66/396, 420
  GABA, 27/15, 17
  histotoxic hypoxia, 9/621
  isocitrate dehydrogenase, 27/16
  mitochondrial disease, 66/395
  α-oxoglutarate dehydrogenase, 27/16
  oxoglutarate dehydrogenase deficiency, 66/654
  pyruvate carboxylase, 66/418
  pyruvate dehydrogenase, 27/14
  pyruvate metabolism, 66/413
  schematic representation, 66/414
  succinate dehydrogenase, 27/16
Tricarboxylic acid cycle deficiency
  gene, 66/419
  mitochondrial disease, 62/503
Triceps brachii reflex
  diagnostic value, 1/243
Triceps surae reflex, see Achilles reflex
Triceps transfer
  brachial plexus, 51/152
Trichinella
  helminthiasis, 52/563
  inflammatory myopathy, 62/373
  nematode, 52/563
  polymyositis, 62/373
Trichinellosis, see Trichinosis
Trichinosis, 35/267-287
  acute polymyositis, 41/58, 373
  ataxia, 52/566
  benzimidazole, 52/573
  brain granuloma, 52/572
  brain hemorrhage, 52/565, 572
  brain infarction, 52/565, 572
  capillary thrombosis, 52/572
  cardiac involvement, 35/276
  chemosis, 52/567

  clinical features, 35/268-276, 52/565
  control, 35/285-287
  corticosteroid, 52/573
  CSF, 35/277, 52/569
  deafness, 52/566
  delirium, 52/566
  diagnosis, 52/568
  ECG, 52/568
  EEG, 35/277, 52/569
  EMG, 35/277, 52/568
  encephalitis, 52/565, 569
  encephalopathy, 52/511, 569
  eosinophilia, 52/565, 569, 63/371
  eosinophilia myalgia syndrome, 64/261
  epidemiology, 35/213, 267, 52/563
  epilepsy, 52/511, 566
  exophthalmos, 52/567
  gastrointestinal symptom, 52/565
  helminthiasis, 35/213
  Herxheimer reaction, 52/574
  history, 35/267
  immunodiagnosis, 52/570
  immunologic test, 35/281-284
  immunology, 52/572
  intracranial hypertension, 52/566
  laboratory evaluation, 35/276-282
  laboratory test, 35/276-282
  mebendazole, 52/574
  meningitis, 52/566
  microglia nodule, 52/572
  mortality, 52/565
  motoneuron disease, 52/566, 569
  multiple cranial neuropathy, 52/566
  muscle biopsy, 52/568
  myalgia, 40/325, 52/565
  myocarditis, 52/571
  myoglobinuria, 52/566
  myopathy, 52/571
  nerve lesion, 7/89
  neuropathology, 52/570
  neuropathy, 35/275, 52/566, 571
  ocular manifestation, 35/275
  ophthalmoplegia, 52/567
  optic neuritis, 52/567
  parasitology, 35/268, 52/564
  pathology, 35/282
  prevalence, 52/564
  prevention, 35/285-287, 52/574
  prognosis, 35/273-276
  progressive bulbar palsy, 59/225
  radiation, 52/568
  sequela, 52/567
  serodiagnosis, 52/570

skin test, 35/280, 52/570
spastic paraplegia, 59/436
spinal cord lesion, 52/567
subungual splinter hemorrhage, 52/565
superior sagittal sinus, 52/572
thiabendazole, 52/573
treatment, 35/284, 52/573
Trichinella spiralis, 52/563
Trichinous myositis
  inflammatory myopathy, 40/51, 325
Trichlorfon, *see* Metrifonate
Trichloroethanol
  ataxia, 65/329
  metabolite, 65/329
  neurotoxin, 65/329
Trichloroethanol intoxication
  ataxia, 65/329
  disulfiram polyneuropathy, 51/309
  metabolism, 65/329
  nausea, 65/329
Trichloroethylene
  bulbar paralysis, 51/280
  nerve conduction, 7/174
  neuropathy, 7/522
  neurotoxicity, 51/279
  neurotoxin, 5/299, 7/522, 36/458, 51/68, 279,
    64/9, 40, 44
  organic solvent intoxication, 64/40, 44
  toxic neuropathy, 51/279, 64/9
  toxic polyneuropathy, 51/68
  trigeminal neuralgia, 5/299
  trigeminal neuropathy, 64/9
Trichloroethylene intoxication, 36/458-463
  animal study, 36/462
  clinical features, 36/458-461
  epidemiology, 36/457
  history, 36/458
  psychophysiology, 36/461
  symptom, 64/9
Trichloroethylene polyneuropathy
  axonal demyelination, 51/280
  facial nerve, 51/280
  multiple cranial neuropathy, 51/280
  neuron degeneration, 51/280
  trigeminal neuropathy, 51/280
2,4,5-Trichlorophenoxy acetic acid
  polychlorinated dibenzodioxin intoxication,
    64/205
Trichoanesthesia
  terminology, 1/89
Trichodontid intoxication
  vitamin A intoxication, 37/87
Trichoepithelioma
  multiple, *see* Multiple trichoepithelioma
Trichoepithelioma papulosum multiplex, *see*
    Multiple trichoepithelioma
Trichomegaly
  intrauterine growth retardation, 46/94
  mental deficiency, 43/294, 46/94
Trichophyton
  fungal CNS disease, 35/571, 52/496
Trichophytosis
  encephalitis, 35/571, 52/497
  meningitis, 35/571, 52/496
Trichopoliodystrophy
  autonomic nervous system, 29/287, 75/22, 238
  birth incidence, 42/584
  brain atrophy, 14/784, 42/584
  case report, 29/284
  cerebellar granular cell loss, 14/784, 42/584
  ceruloplasmin, 42/584
  clinical features, 29/286-290
  complex IV deficiency, 41/212
  convulsion, 29/286
  copper, 29/279-286, 294-300, 42/584
  corpus callosum agenesis, 50/61
  decerebration, 14/784
  epilepsy, 72/222
  feeding disorder, 42/584
  gastrointestinal symptom, 29/286
  genetics, 29/279
  giant axonal neuropathy, 60/82
  growth and development, 29/287
  growth retardation, 14/784
  hair, 29/288, 42/584
  hereditary congenital cerebellar atrophy, 60/289
  hypocupremia, 46/61
  hypothermia, 42/584
  laboratory finding, 29/290-292
  lactic acidosis, 41/212
  mental deficiency, 46/61
  microcephaly, 30/508
  micrognathia, 29/287
  mitochondrial disease, 62/497
  neuronal heterotopia, 42/584
  optic atrophy, 14/784
  pathology, 29/292-294
  pathophysiology, 29/294-300
  pili torti, 14/784, 29/288, 42/584
  prognosis, 29/290
  Purkinje cell loss, 14/784, 42/584
  secondary neuroaxonal dystrophy, 49/408
  seizure, 29/286, 42/584, 46/61
  treatment, 29/300
  trichorrhexis nodosa, 42/584
  X-linked recessive ataxia, 60/640

Trichorrhexis nodosa
  see also Netherton syndrome
  argininosuccinic aciduria, 21/577
  hair brain syndrome, 43/297
  hyperammonemia, 21/577
  trichopoliodystrophy, 42/584
Trichosporon
  fungal CNS disease, 52/497
Tri-*o*-cresyl phosphate
  amyotrophic lateral sclerosis, 21/33, 22/310
  ataxic neuropathy, 56/526
  characteristics, 64/95
  distal axonopathy, 64/11
  experimental myopathy, 40/154
  methanol, 64/95
  nerve conduction, 7/173
  neuropathy, 7/521, 544
  neurotoxic esterase, 37/475, 51/284
  neurotoxin, 37/471, 475, 40/151, 51/68, 285,
    64/10, 152, 160
  organophosphate, 64/152
  organophosphate induced delayed polyneuropathy,
    64/152, 175
  organophosphate intoxication, 37/475, 554,
    64/152, 155
  toxic neuropathy, 51/285, 64/10, 13
  toxic polyneuropathy, 51/68
  tropical myeloneuropathy, 56/526
  use, 64/95
Tri-*o*-cresyl phosphate intoxication
  action site, 64/10
  amyotrophy, 22/28
  biochemical, 37/474
  clinical features, 37/472
  esterase, 37/475
  experimental, 37/473
  history, 37/471
  mechanism of damage, 37/475
  neurotoxicity, 37/475
  organophosphate induced delayed polyneuropathy,
    64/152
  pathology, 37/473
  pathomechanism, 64/11
  polyneuropathy, 64/160
  spinal cord, 64/160
Tri-*o*-cresyl phosphate paralysis
  tri-*o*-cresyl phosphate polyneuropathy, 51/285
Tri-*o*-cresyl phosphate polyneuropathy
  ataxia, 51/285
  axonal degeneration, 51/285
  exposure source, 51/285
  motor neuropathy, 51/285
  spasticity, 51/285

tri-*o*-cresyl phosphate paralysis, 51/285
Tricyclic antidepressant
  adrenergic receptor affinity, 65/315
  amitriptyline, 65/312
  ataxia, 65/316-318
  autonomic nervous system, 65/321
  behavior disorder, 46/595
  clomipramine, 65/312
  delirium, 65/316
  depression, 46/448
  desipramine, 65/312
  dibenzodiazepine structure, 65/312
  dopamine reuptake, 65/315
  dopaminergic receptor affinity, 65/315
  dosulepin, 65/312
  doxepin, 65/312
  drug induced polyneuropathy, 51/303
  dystonia, 65/316
  EEG, 65/318
  elimination half-life, 65/314
  epilepsy, 65/316-318
  extrapyramidal effect, 65/316
  headache, 65/316
  histaminic receptor affinity, 65/315
  imipramine, 65/312
  intoxication treatment, 65/323
  lofepramine, 65/312
  memory, 65/319
  metabolism, 65/313
  mode of action, 65/312
  muscarinic receptor affinity, 65/315
  myoclonus, 38/582
  neuropathy, 65/316
  neurotoxicity, 51/303
  neurotoxin, 51/303, 65/311
  noradrenalin reuptake, 65/315
  nortriptyline, 65/312
  orthostatic hypotension, 65/316
  pain, 69/47
  pharmacokinetics, 65/313
  plasma level, 65/318
  polyneuropathy, 65/316
  protriptyline, 65/312
  receptor affinity, 65/315
  serotonin reuptake, 65/315
  serotonin reuptake inhibition, 65/313
  sleep, 65/318
  steady state concentration, 65/314
  tremor, 65/316, 318
  trimipramine, 65/312
  visual accommodation, 65/316
  withdrawal symptom, 65/319
Tricyclic antidepressant intoxication, 37/315-319

acute, 37/315-317
allergic reaction, 37/319
ataxia, 65/321
cardiovascular complication, 37/318
clinical picture, 37/315-317
CNS, 37/317
coma, 65/321
delirium, 65/321
epilepsy, 65/321
fatality rate, 65/324
hallucination, 65/321
hypersensitivity, 37/319
mydriasis, 65/321
myoclonus, 65/321
relative toxicity, 65/324
Tridacna maxima intoxication
bivalves, 37/64
ciguatoxin, 37/64
death, 37/64
ingestion, 37/64
Tridione, *see* Trimethadione
Trientine, *see* Triethylenetetramine
Triethylenetetramine
Wilson disease, 49/235
Triethyltin
brain edema, 28/375
CNS, 28/427, 51/274
experimental demyelination, 47/564
myelin vacuolation, 51/274
neurotoxin, 16/196, 36/280, 284, 286, 47/564,
51/68, 274, 64/10
toxic neuropathy, 64/10
toxic polyneuropathy, 51/68
Triethyltin intoxication
action site, 64/10
brain edema, 16/179, 196
clinical features, 36/280-282
CSF, 36/282
cytotoxic brain edema, 11/155
diagnosis, 36/282
EEG, 36/282
experimental, 36/284-288, 40/157
history, 36/279
leukoencephalopathy, 47/564
neuroradiology, 36/282
pathology, 36/283
source, 36/279
spongy glial dystrophy, 21/66
treatment, 36/283
Triethyltin sulfate
experimental myopathy, 41/5
Trifluoperazine
chemical formula, 65/275

neuroleptic agent, 65/275
phenothiazine, 65/275
toxic myopathy, 62/601
Triflupromazine
chemical formula, 65/275
neuroleptic agent, 65/275
phenothiazine, 65/275
Trigeminal antidromic wave
cranial neuralgia, 5/287, 289, 291
Trigeminal artery
carotid cavernous fistula, 24/407
primitive, *see* Primitive trigeminal artery
Trigeminal descending spinal tract lesion
cervicomedullary injury, 24/163-167, 25/147
syringobulbia, 50/444
Trigeminal ganglion
headache, 48/108-110
horseradish peroxidase, 48/108
middle cerebral artery, 48/109
middle ear, 48/333
miosis, 48/333
paratrigeminal syndrome, 48/332
ptosis, 48/333
substance P, 48/110
wheat germ agglutinin, 48/110
Trigeminal ganglionectomy
headache, 48/109
Trigeminal injection
history, 5/306, 307
Trigeminal nerve
acoustic neuroma, 42/756
buccopharyngeal herpes, 7/483
cerebellar tumor, 17/711
cervicomedullary injury, 24/164
cluster tic syndrome, 48/268
cold face test, 74/606, 622
congenital anesthesia, 30/404
congenital anosmia, 42/357
congenital paralysis, 30/403
corneal reflex, 2/56
corneomandibular reflex, 2/56
diabetes mellitus, 27/119
embryology, 30/397
facial diplegia, 30/404
Gradenigo syndrome, 2/91
headache, 48/107, 274
hereditary multiple recurrent mononeuropathy,
51/554
herpes zoster, 34/171, 56/232
herpes zoster ganglionitis, 51/181
injection of branch, 5/388
leprosy, 33/436
leprous neuritis, 51/220

lesion cause, 2/58
Möbius syndrome, 30/404
neuritis, 5/272
ota nevus, 43/37
pathophysiology, 2/54-59
pontine infarction, 12/16
posttraumatic syringomyelia, 26/125
recording, 5/288, 290, 293
root section, 5/392, 396
Sturge-Weber syndrome, 43/47
subarachnoid hemorrhage, 55/18
superior cerebellar artery syndrome, 12/20
tear, 74/427
traumatic carotid cavernous sinus fistula, 57/353
Willis circle, 48/108
Trigeminal nerve agenesis
abducent nerve agenesis, 50/214
alopecia, 50/214
anophthalmia, 50/214
ataxia, 50/214
brain stem hypoplasia, 50/214
cerebellar hypoplasia, 50/214
cheiloschisis, 50/214
congenital facial anesthesia, 50/214
corneal ulceration, 50/214
cranial nerve agenesis, 50/213
facial nerve agenesis, 50/214
focal dermal dysplasia, 50/214
Goldenhar syndrome, 38/411, 50/214
hyperreflexia, 50/214
hypertelorism, 50/214
hypoglossal nerve abnormality, 50/213
imperforate anus, 50/214
internuclear ophthalmoplegia, 50/214
mandibular hypoplasia, 50/214
masseter muscle, 50/214
mental deficiency, 50/214
Möbius syndrome, 50/213
occipital encephalocele, 50/214
palatoschisis, 50/214
pontine lesion, 50/214
preauricular tag, 50/214
vertebral abnormality, 50/214
Trigeminal nerve fiber
spreading depression, 48/113
Trigeminal nerve lesion
corneal reflex, 1/239
reflex sneezing, 2/57
Trigeminal nerve palsy
aqueduct stenosis, 50/312
Trigeminal neuralgia
see also Cranial neuralgia and Facial pain
aconitine, 5/299

alcohol injection, 5/388
amyloidosis, 8/9, 42/353
anesthesia dolorosa, 48/454
aneurysm, 48/450
annual incidence, 42/353
anticonvulsant, 5/300, 45/230
antidromic potential, 5/287, 289, 291
arsenic, 5/299
arteriovenous malformation, 12/248
associated disease, 5/297
atypical, 5/276
autonomic nervous system, 75/298
bilateral, 5/297, 399, 48/453
carbamazepine, 5/383
cardiovascular agent intoxication, 37/429
central trigeminal pathway, 5/275
cerebellopontine angle, 48/450
cerebellopontine angle tumor, 5/273
chronic migrainous neuralgia, 15/330, 48/250, 254
clinical features, 5/296, 310
cluster headache, 48/224
Dandy operation, 5/306, 396
decompression, 48/449
definition, 48/10
diagnosis, 5/3
differential diagnosis, 5/316
digoxin, 63/195
distribution of divisions, 5/297
drug treatment, 5/382
electrode, 48/454
etiology, 5/298, 48/450
familial, 5/311
foramen ovale, 48/453
gangliectomy, 5/301-305
gasserian ganglion, 5/273, 42/352
gelsemium, 5/299
genetics, 5/297
glycerol injection, 48/454
glycoside intoxication, 37/429
Gradenigo syndrome, 2/313, 16/210, 48/302
history, 5/296, 298-300, 305, 48/449
hydrocyanic acid, 5/299
ideal operation, 48/451
incidence, 5/308
infection, 48/450
intermedius neuralgia, 48/487
intramedullary trigeminal tractotomy, 5/397
Kirschner coagulation, 5/399
leprosy, 51/220
leprous neuritis, 51/220
main operation, 5/387
Meckel cave, 48/452
medullary tractotomy, 48/449

metal treatment, 5/299
middle fossa section, 48/452
migrainous, *see* Migrainous trigeminal neuralgia
motor root damage, 48/454
multiple sclerosis, 5/309, 311, 9/168, 42/352, 47/140
neurocysticercosis, 5/273
neurophysiology, 5/281
nociceptor afferent, 45/230
nucleus subcaudalis, 5/287
nystagmus, 5/312
osteitis deformans (Paget), 38/364, 50/399
pain, 45/230
pain distribution, 5/312
pain perception, 45/228
pain in time, 5/313
paroxysmal reflex, 5/272
pathology, 5/273, 298
pathophysiology, 5/272
percutaneous microcompression, 48/454
percutaneous thermocoagulation, 48/454
petrous apex, 5/273
pontine arteriovenous malformation, 2/258
posterior fossa microvascular decompression, 48/455
posterior fossa tumor, 5/312
prevalence, 5/308
prodromata, 5/314
progressive hemifacial atrophy, 14/777, 31/253, 43/408
quality of pain, 5/314
season, 5/310
sex ratio, 5/311, 42/353
side incidence, 5/297
sinus pain, 5/330
Sjöqvist operation, 5/397
Spiller-Frazier operation, 5/397
strychnine, 5/299
subtemporal rhizotomy, 5/303-305
surgery, 5/300, 386
surgical aspect, 48/449-456
surgical indication, 48/455
temporomandibular cephalalgia, 48/415
thalamic syndrome (Dejerine-Roussy), 5/275
thalamus, 5/275
thermocoagulation, 48/453
transoccipital rhizotomy, 5/305
treatment, 5/299, 399
trichloroethylene, 5/299
trigeminal rhizotomy, 48/452
tuberculoma, 18/418
tumor, 48/450
vitamin B1, 5/299

vitamin B12 deficiency, 5/300
Trigeminal neurilemoma
pituitary adenoma, 17/412
Trigeminal neurinoma, 68/480, 537
cranial neuralgia, 5/273
skull base, 68/466
vestibular neurinoma, 68/425
Trigeminal neuropathy
bilateral sensory, *see* Bilateral sensory trigeminal neuropathy
isolated, *see* Sensory trigeminal neuropathy
polyarteritis nodosa, 51/452, 55/356
sensory, *see* Sensory trigeminal neuropathy
Sjögren syndrome, 39/427, 71/74, 85
superior petrosal sinus thrombosis, 54/398
systemic lupus erythematosus, 55/377
trichloroethylene, 64/9
trichloroethylene polyneuropathy, 51/280
Whipple disease, 51/184, 52/138
Trigeminal neurotomy
blood pressure, 48/108
Trigeminal pain
gasserian ganglion, 5/273
headache, 48/281
Trigeminal pathway
cluster headache, 48/268
Trigeminal reflex
migraine, 48/113
sneezing, 2/57
Trigeminal rhizotomy
trigeminal neuralgia, 48/452
Trigeminal root
Biemond syndrome, 21/378
chronic benign trigeminal paresis, 2/59
compression, 5/274
history, 5/300, 305
rhizotomy, 5/303-306
surgery, 5/300, 305
surgical avulsion, 5/301
Trigeminal schwannoma, *see* Trigeminal neurilemoma
Trigeminal sensory deficit
Biemond syndrome, 60/657
neuronal intranuclear hyaline inclusion disease, 60/657
Trigeminal sensory neuropathy
scleroderma, 71/112
Trigeminal spinal nucleus
electrocoagulation, 5/306
Trigeminal tractotomy
intramedullary, *see* Intramedullary trigeminal tractotomy
Trigeminal vein

534

anatomy, 11/54
Trigemino-oculomotor synkinesis
EMG, 42/320
Marcus Gunn phenomenon, 42/320
neurophysiologic test, 42/320
Trigeminopupillary reflex
headache, 75/283
Trigeminovascular system
brain blood vessel, 48/112
cerebrovascular disease, 75/364
headache, 48/107
vascular headache, 48/107-113
Triglidae intoxication
family, 37/78
venomous spine, 37/78
Triglyceride
see also Lipid
antihypertensive agent, 63/83
Bassen-Kornzweig syndrome, 42/512
carnitine deficiency, 43/175
carnitine palmitoyltransferase deficiency, 43/177
cytoplasmic excess, 40/36
plaque, 9/313
serum, 10/266
sudanophilia, 10/28
total, 10/266
Wolman disease, 10/352, 504, 546, 14/779
Triglyceride storage disease
congenital ichthyosis, 62/496
lipid metabolic disorder, 62/496
Trigonocephaly, 43/486-488
see also Craniosynostosis and Metopic suture
arhinencephaly, 43/486
chromosomal aberration, 43/487
craniosynostosis, 42/20, 50/113, 117
holoprosencephaly, 30/451, 43/486, 50/235
mental deficiency, 30/228, 43/486, 50/127
olfactory bulb agenesis, 43/486
9p partial monosomy, 43/512, 50/578
premature closure syndrome, 30/227
prevalence, 43/486
11q partial monosomy, 43/522
13q partial monosomy, 43/526
r(9) syndrome, 43/513
r(13) syndrome, 50/584
terminology, 30/222
Trihexosyl ceramidase deficiency, see Fabry disease
Trihexosyl ceramide galactosidase
Fabry disease, 21/51
Trihexyphenidyl
extrapyramidal disorder, 6/827
motoneuron disease, 59/467
Wilson disease, 49/235

Trihydroxycholestanemia
facial dysmorphia, 60/666
Trilaminar myopathy
classification, 40/288
congenital myopathy, 41/22, 62/332
double ring myopathy, 62/341
facial weakness, 62/334
histopathology, 62/341
ultrastructure, 40/80
Trilateral retinoblastoma
primitive neuroectodermal tumor, 68/224
Trimeperidine
chemistry, 37/366
Trimethadione
antiepileptic agent, 37/200, 65/497
epilepsy, 15/641
side effect, 15/718
Trimethadione intoxication
drug interaction, 37/200
hemeralopia, 37/202
metabolite, 37/200
Trimethadione syndrome
fetal, see Fetal trimethadione syndrome
Trimethoprim
ataxia, 65/472
chemical meningitis, 56/130
headache, 65/472
neurotoxicity, 71/368
neurotoxin, 65/472
paracoccidioidomycosis, 35/537
2,4,5-Trimethoxyamphetamine
amphetamine derivative, 65/50
chemical derivation, 65/50
designer drug, 65/364
hallucinogenic agent, 65/42
phenethylamine derivative, 65/50
3,4,5-Trimethoxyamphetamine
see also Hallucinogenic agent
hallucinogenic agent, 65/42
3,4,5-Trimethoxyamphetamine intoxication
hallucination, 65/50
3,4,5-Trimethoxyphenethylamine, see Mescaline
Trimethylamine
osmoregulation, 63/555
Trimethylaminuria
bleeding tendency, 29/231
platelet, 29/231
splenomegaly, 29/231
Turner syndrome, 29/231
2-Trimethylammoniumethyl
methylphosphonofluoridate iodide intoxication
chemical classification, 37/550
organophosphorus compound, 37/550

2-Trimethylammonium-1-methylethyl
    methylphosphonofluoridate iodide intoxication
  chemical classification, 37/550
  organophosphorus compound, 37/550
3-Trimethylammoniumpropyl
    methylphosphonofluoridate iodide intoxication
  chemical classification, 37/550
  organophosphorus compound, 37/550
Trimethylnitrosourea
  brain tumor carcinogenesis, 17/16
Trimipramine
  tricyclic antidepressant, 65/312
Triorthocresyl phosphate, *see* Tri-*o*-cresyl phosphate
Triparanol
  animal model, 40/548
  cataract, 40/514
  experimental myotonia, 40/155
  iniencephaly, 50/135
  myotonia, 40/514
Tripelennamine
  opiate intoxication, 65/357
Triperinol intoxication
  myotonic syndrome, 62/264
Triphosphoinositide
  biochemistry, 9/10
  chemistry, 10/238
Triphosphoinositide phosphodiesterase
  chemistry, 10/239
Triphosphoinositide phosphomonoesterase
  white matter, 9/16
Triphosphopyridine nucleotide diaphorase
  GM_2 gangliosidosis, 10/291
Triple Y syndrome, *see* Sex chromosomes
Triploidy
  see also Chromosome number abnormality
  birth incidence, 43/564
  cerebellar atrophy, 43/566
  cheilopalatoschisis, 43/564-566
  corpus callosum agenesis, 43/566
  definition, 50/540
  dermatoglyphics, 43/566
  encephalocele, 43/566
  heart disease, 43/566
  holoprosencephaly, 43/566
  macroglossia, 43/564-566
  myelomeningocele, 43/566
  pathogenesis, 50/555
  placental enlargement, 43/564, 566
  polyhydramnios, 43/564, 566
  polyploidy, 50/555
  renal abnormality, 43/566
  spina bifida, 43/564-566
  syndactyly, 43/564-566

Triploidy syndrome
  hydrocephalus, 50/289
Tripneustes gratilla intoxication
  Echinodermata, 37/68
Trismus
  carbon monoxide intoxication, 49/474
  cause, 1/285
  diazoxide intoxication, 37/433
  hyponatremia, 63/545
  infantile Gaucher disease, 42/438, 66/124
  juvenile Gaucher disease, 42/438
  pontine hemorrhage, 2/253
  progressive supranuclear palsy, 22/219
  sea snake intoxication, 37/10, 92
  tetanus, 41/304, 52/230, 65/217
Trisomies, *see* Chromosome
Trisymptom complex, *see* Behçet syndrome
Trisymptomatic syndrome, *see* Behçet syndrome
Tritanomalopia
  primary pigmentary retinal degeneration, 13/190
Tritanomaly, *see* Tritanomalopia
Trivalent arsenic
  neurotoxicity, 51/265
tRNA molecule
  adaptor, 66/76
Trochlear nerve
  clinical examination, 57/136
  diabetes mellitus, 27/117
  doll head eye phenomenon, 2/306
  embryology, 30/396
  function, 2/305
  Garcin syndrome, 2/308
  lesion site, 2/305-310
  neurinoma, 68/536
  subarachnoid hemorrhage, 55/17
  symptom, 2/305-310
Trochlear nerve abnormality
  see also Congenital ptosis, Marcus Gunn
    phenomenon *and* Stilling-Türk-Duane
    syndrome
  congenital ptosis, 30/401
  Marcus Gunn phenomenon, 30/401
  Stilling-Türk-Duane syndrome, 30/403
Trochlear nerve agenesis
  cranial nerve agenesis, 50/212
  nuclear aplasia, 50/212
Trochlear nerve injury
  see also Ocular imbalance
  anatomy, 24/65
  brain injury, 24/65
  fascicular lesion, 24/78
  head injury, 24/65
  history, 24/59

incidence, 24/77, 83
mechanism, 24/65
nuclear lesion, 24/78
orbital injury, 24/86
physiology, 24/65
prognosis, 24/84
sign, 24/65
skull base fracture, 24/65
skull fracture, 24/65
topical diagnosis, 24/80, 83
treatment, 24/66
Trochlear nerve paralysis
allergic granulomatous angiitis, 51/453
clinical picture, 1/610
congenital type, 2/308
nuclear lesion, 2/309
Parinaud syndrome, 2/304
root lesion, 2/308
skew deviation, 2/309
traumatic carotid cavernous sinus fistula, 57/346
Trolard vein
anatomy, 11/47
petrosal sinus, 11/54
radioanatomy, 11/106
superior longitudinal sinus occlusion, 54/397
Trömner reflex
method, 1/244
Tropheryma whippelii
Whipple disease, 70/241
Trophic disorder
autonomic polyneuropathy, 51/481
classification, 1/481
CNS, 1/481-488
hemiplegia, 1/487
muscle absence, 1/481
nail, 1/486
nerve lesion, 7/260
peripheral motoneuron lesion, 1/227
peripheral nervous system lesion, 1/485-487
postnatal, 1/484
tabes dorsalis, 33/365
Takayasu disease, 12/405
Trophic function
motoneuron, 1/59
nerve, 7/89
nerve cell, 1/59
spinal cord, 1/59
Trophic interaction
see also Neuromuscular transmission
chloride conductance, 40/555
experimental model, 40/138-143
Trophic ulcer
facial spasm, 42/315

hereditary amyloid polyneuropathy, 42/518
hereditary motor and sensory neuropathy type I,
42/339
hereditary motor and sensory neuropathy type II,
42/339
Muckle-Wells syndrome, 42/396
neurogenic acro-osteolysis, 42/293-295
sensory radicular neuropathy, 42/351
spastic paraplegia, 42/180
Trophotropic reaction
autonomic nervous system, 74/17
Trophotropic system
emotion, 3/320
Tropical ataxic myelopathy
cause, 65/30
cyanide intoxication, 65/29
cyanogenic glycoside intoxication, 65/29
Tropical ataxic neuropathy
see also Anterolateral myelopathy
ataxia, 7/640, 51/322, 65/29
burning feet syndrome, 51/322
cassava intake, 7/641, 28/26, 30, 64/12
cyanide intoxication, 65/29
cyanogenic glycoside, 64/12
cyanogenic glycoside intoxication, 7/642,
36/522-524, 65/29
deafness, 7/640, 65/29
deficiency neuropathy, 7/640, 51/322
dementia, 65/29
dermatitis, 7/640, 51/322
dysentery, 51/323
enterovirus infection, 51/323
familial spastic paraplegia, 59/306
Jamaican neuropathy, 7/640, 51/323
malnutrition, 7/640, 28/26, 30
multivitamin deficiency, 51/323
myelopathy, 7/641, 51/322
neuritis, 70/411
optic atrophy, 65/29
optic neuropathy, 51/322
paresthesia, 70/411
pellagra, 7/641, 51/323
spinal ataxia, 7/640
tropical spastic paraplegia, 51/324
vitamin B1 deficiency, 28/55
Tropical disease
epilepsy, 72/146
recurrent nonhereditary multiple cranial
neuropathy, 51/570
Tropical myeloneuropathy
acute hemorrhagic conjunctivitis, 56/526
angiostrongyliasis, 56/526
cassava intake, 7/641, 56/526

clioquinol, 56/526
cyanide intoxication, 56/525
filariasis, 56/526
herpes simplex virus type 2, 56/526
hydatidosis, 56/526
Lathyrus sativus, 56/525
neurocysticercosis, 56/526
pellagra, 56/526
schistosomiasis, 56/526
tri-*o*-cresyl phosphate, 56/526
tropical spastic paraplegia, 56/525
vitamin B5, 56/526
vitamin B6, 56/526
vitamin B12 deficiency, 56/526
vitamin Bc, 56/526
vitamin E, 56/526
Tropical myositis, *see* Tropical polymyositis
Tropical neurology
  introduction, 35/1-21
  mycosis, 35/15-21
  parasitic disease, 35/1-15
Tropical polymyositis
  acquired immune deficiency syndrome, 62/373
  Clostridium welchii, 41/51
  nuclear magnetic resonance, 62/373
  Staphylococcus aureus, 62/373
  Streptococcus, 62/373
  Yersinia, 62/373
Tropical spastic paraplegia
  see also Human T-lymphotropic virus type I
    associated myelopathy
  age, 56/534
  amyotrophic lateral sclerosis, 59/451-453
  ataxia, 56/536
  brain stem auditory evoked potential, 56/538
  cassava intake, 65/30
  cause, 56/529
  chronic myelitis, 56/529
  clinical features, 56/535
  CSF, 56/527, 538
  cyanide intoxication, 65/30
  cyanogenic glycoside intoxication, 65/30
  deafness, 56/536
  definition, 56/529
  differential diagnosis, 65/6
  epidemiology, 56/533
  evoked potential, 56/537
  familial aggregation, 56/535
  familial spastic paraplegia, 59/306
  flower lymphocyte, 56/538
  geography, 56/534, 536
  history, 56/535
  human T-lymphotropic virus type I, 56/525

incidence, 56/531
laboratory examination, 56/538
lymphocytic meningitis, 56/538
monoclonal IgG, 56/527
nerve conduction, 56/537
neuroepidemiology, 56/531
neuropathology, 56/531-533, 536
onset, 56/535
pathogenesis, 56/531
plasma thiocyanate, 65/31
prevalence, 56/531
prevention, 56/538
progression, 56/535
race, 56/535
retrobulbar neuropathy, 56/536
sex ratio, 56/531
symptom, 65/31
treatment, 56/538
Treponema pallidum, 56/538
tropical ataxic neuropathy, 51/324
tropical myeloneuropathy, 56/525
viral meningitis, 56/538
Tropical splenomegaly
  visceral, *see* Visceral leishmaniasis
Tropical sprue
  clinical picture, 7/628
  deficiency neuropathy, 7/628, 51/329
  epidemiology, 7/628
  malabsorption syndrome, 7/628, 28/235
  neuropathy, 28/236
  vitamin B12 deficiency, 7/629
  vitamin Bc deficiency, 7/629
Troponin
  calcium release, 40/553
  contractile protein, 40/130
  excitation contraction coupling, 40/130
  mitochondria, 40/553
  sarcoplasmic reticulum, 40/553
Trousseau sign
  hyperventilation syndrome, 38/326, 63/432
  hypoparathyroidism, 27/300, 41/248
  muscle cramp, 40/330
  pseudohypoparathyroidism, 42/621, 63/560
  renal insufficiency, 27/336
  tetany, 41/302, 63/560
Troxidone, *see* Trimethadione
Troyer syndrome
  ataxia, 42/194
  Babinski sign, 42/193
  choreoathetosis, 42/194
  contracture, 42/193
  familial spastic paraplegia, 59/374
  gene frequency, 42/194

heterozygote frequency, 42/194
hyperreflexia, 42/193
muscular atrophy, 42/193
psychomotor retardation, 42/193
spastic paraplegia, 42/193
speech disorder in children, 42/193
spinal muscular atrophy, 59/374
talipes, 42/193
tremor, 42/194
TRPR, see Trypsin resistant protein residue
True porencephaly, 30/681-691
associated malformation, 30/682, 685
clinical features, 30/685-690, 50/359
diagnosis, 30/690
EEG, 30/690
epilepsy, 30/686
familial incidence, 30/685
hydranencephaly, 30/684
mental deficiency, 30/681, 685-687
optic atrophy, 30/685
pathogenesis, 30/683, 50/357
pathology, 30/682, 686-690
Truncal ataxia
archicerebellar syndrome, 2/418
definition, 1/314
kuru, 56/556
Lafora progressive myoclonus epilepsy, 27/173
late infantile metachromatic leukodystrophy,
66/167
multiple sclerosis, 47/56
Truncal nerve
topographical diagnosis, 2/38
Truncus arteriosus communication
holoprosencephaly, 50/237
Trypan blue
anencephaly, 30/184
cephalocele, 50/98
spina bifida, 32/528, 531
teratogenic agent, 32/169
Trypanosoma
African trypanosomiasis, 35/67, 74, 52/339
inflammatory myopathy, 62/373
polymyositis, 62/373
Trypanosoma brucei
see also African trypanosomiasis
CNS invasion, 52/339
geography, 52/339
tsetse fly, 52/339
Trypanosoma cruzi
see also Chagas disease
acquired immune deficiency syndrome, 71/305
biology, 52/345
Chagas disease, 75/388

CNS, 31/221
epilepsy, 72/170
intestinal pseudo-obstruction, 51/494
intrauterine infection, 31/221
Trypanosomiasis
African, see African trypanosomiasis
South American, see Chagas disease
Tryparsamide
arsenic intoxication, 9/648, 64/284
encephalomyelopathy, 9/648
papilledema, 9/648
retrobulbar neuritis, 9/648
Trypsin resistant protein residue
see also Proteolipid
CNS, 9/32
nerve, 9/32
proteolipid, 9/32
Tryptamine
hallucinogenic agent intoxication, 37/329, 331
platelet MAO, 48/94
Tryptamine derivative intoxication
hallucination, 37/329, 331
Tryptoline
designer drug, 65/398
1-methyl-4-phenyl-1,2,3,6-tetrahydropyridine
intoxication, 65/399
neurotoxin, 65/398
pharmacology, 65/399
Tryptophan
see also Eosinophilia myalgia syndrome
acquired hepatocerebral degeneration, 49/216
biochemistry, 7/576
catabolic pathway, 64/264
daily intake, 64/250
DMAB nitrous acid, 9/24
Down syndrome, 31/433-435
eosinophilia, 63/371
eosinophilia myalgia syndrome, 63/375
glycosaminoglycanosis, 10/432
Hartnup disease, 7/577, 21/564, 29/163
insomnia, 64/250
migraine, 48/86, 197
myelin sheath, 9/24
pellagra, 7/574, 28/59, 71, 38/648, 46/336
phenylketonuria, 28/128, 29/31-33
primary pigmentary retinal degeneration, 13/196
staining, 9/24
toxic myopathy, 62/597
Tryptophan containing fragment
myelin basic protein, 47/487
Tryptophan hydroxylase
migraine, 48/86
Tryptophan load test

vitamin B6 deficiency, 70/426
Tryptophan metabolism
  biochemistry, 29/156-158
  Hartnup disease, 29/156
  hydroxykynureninuria, 29/223
  schema, 42/634
  xanthurenic aciduria, 42/636
Tryptophan nicotinic acid metabolism
  epilepsy, 42/716
  vitamin B6, 28/111-114
Tryptophanuria
  ataxia, 42/633
  cutaneous pigmentation, 42/633
  dwarfism, 42/633
  Hartnup disease, 42/146
  mental deficiency, 42/633
  photosensitivity, 42/633
  tremor, 42/633
Tsai area, see Ventral tegmental area
Tsetse fly
  African trypanosomiasis, 35/67-69, 52/339
  Trypanosoma brucei, 52/339
TSH, see Thyrotropic hormone
Tsukubaenolide
  neuropathy, 69/461, 472
Tsukubaenolide intoxication
  akinetic mutism, 65/558
  encephalopathy, 65/558
  epilepsy, 65/558
  headache, 65/558
  insomnia, 65/558
  tremor, 65/558
  vertigo, 65/558
Tuber
  ataxia telangiectasia, 14/74
  cortical, see Cortical tuber
  periventricular, see Periventricular tuber
  tuberous sclerosis, 14/39-43, 354
Tuber cinereum
  anatomy, 2/432
  blood-brain barrier, 53/84, 56/80
  chorea, 6/354
  nucleus, 2/436
Tuberal artery
  syndrome, 2/486
  thalamo, see Thalamotuberal artery
Tubercle
  Benedikt syndrome, 2/298
  olfactory, see Olfactory bulb
Tuberculin
  intrathecal, 9/401
  multiple sclerosis, 9/401, 47/354
Tuberculin test

tuberculous meningitis, 33/218, 52/205
Tuberculoid granuloma
  cheilitis granulomatosa, 8/222
Tuberculoid leprosy
  classification, 51/216
Tuberculoma, 18/413-424
  acquired immune deficiency syndrome, 71/281
  brain angiography, 18/418
  calcification, 18/418
  Cestan-Chenais syndrome, 2/261, 336
  clinical features, 18/417
  CSF fistula, 18/417
  cystic, 18/422
  distribution on site, 18/417
  epidemiology, 16/65
  epilepsy, 15/307
  exophthalmos, 18/418
  geography, 18/416
  hemiballismus, 49/373
  hypophyseal tumor, 18/417
  hypothalamic tumor, 68/71
  infectious striatopallidodentate calcification,
    49/417
  neuropathology, 18/420
  optic chiasm compression, 68/75
  optic pathway tumor, 68/74
  PEG, 18/418
  pituitary tumor, 17/427
  pontine syndrome, 2/259
  prevalence, 18/414
  scalp swelling, 18/417
  sex and age, 18/416
  tegmentum, 17/633
  treatment, 18/423
  trigeminal neuralgia, 18/418
  tuberculous meningitis, 52/198, 208, 55/429
  Wuchereria bancrofti, 18/423
Tuberculoma en plaque
  dura mater, 18/421
Tuberculosis
  acquired immune deficiency syndrome, 71/279,
    281
  acute cerebellar ataxia, 34/625
  adrenal insufficiency, 70/210
  brain, 16/230
  brain abscess, 18/421
  brain vasculitis, 55/429
  carpal tunnel syndrome, 7/295
  cerebrovascular disease, 55/432
  chronic meningitis, 56/645
  congenital brain granuloma, 42/736
  cryptococcosis, 52/433
  CSF, 71/282

deafness, 51/186
diamorphine, 37/392
encephalopathy, 33/207-209, 216
epilepsy, 72/160
ethambutol, 51/294
Guillain-Barré syndrome, 51/186
hemifacial spasm, 51/186
hyperbeta-alaninemia, 29/230
immunopathology, 55/423
intrauterine infection, 31/221
isoniazid neuropathy, 7/516, 533, 582
laboratory, 71/281
Melkersson-Rosenthal syndrome, 8/227, 229, 233
multiple cranial neuropathy, 51/186
nerve lesion, 7/478
nonrecurrent nonhereditary multiple cranial
   neuropathy, 51/571
optic atrophy, 51/186
prognosis, 71/283
progressive multifocal leukoencephalopathy,
   71/400
recurrent laryngeal nerve, 51/186
sarcoidosis, 71/471
spastic paraplegia, 59/435
stroke like syndrome, 55/427
Takayasu disease, 12/414
therapy, 71/283
Tuberculous meningitis, 33/195-234
  see also Mycobacterium tuberculosis
  acquired immune deficiency syndrome, 71/279
  acute inflammatory caseous meningitis, 52/196
  adjuvant treatment, 33/229
  age, 33/210, 52/214
  amino acid, 33/220
  4-aminosalicylic acid, 33/226-228, 52/211-213
  angiography, 33/222-224
  arachnoiditis, 52/216
  arteritis, 55/429
  associated disease, 52/215
  associated pulmonary lesion, 33/217
  basal exudate, 52/197, 207
  basal ganglion infarction, 52/208
  BCG immunization, 52/201
  BCG test, 33/218
  BCG vaccine, 52/215
  blood-brain barrier, 52/205
  blood cerebrospinal bromide partition test, 33/220
  brain angiography, 52/209, 55/424, 429
  brain edema, 52/198
  brain infarction, 52/200, 208, 55/424
  brain ischemia, 52/200
  brain parenchyma, 52/199
  brain scanning, 33/224
  brain vasculitis, 55/424, 429
  bromide partition test, 52/205
  caseation, 52/200
  caseous meningitis, 33/201
  chemotherapy, 71/284
  choroid plexus, 52/198
  choroidal tubercle, 33/216, 52/203
  cisternography, 33/224
  classification, 71/284
  clinical features, 52/201
  coma, 52/203
  computer assisted tomography, 52/207, 215,
    55/429
  cortical encephalitis, 52/209
  corticosteroid, 52/213
  course, 33/212-216, 52/202, 55/431
  cranial nerve palsy, 33/232, 52/216
  CSF, 33/218-221, 52/203, 215, 55/428, 71/282
  CSF block, 52/210
  CSF lymphocyte transformation, 52/207
  cycloserine, 33/225-229, 52/212
  deafness, 52/216
  decerebration, 52/202
  decortication, 52/203
  dementia, 46/387, 52/215
  diagnosis, 33/217-226, 52/203
  diagnostic error, 18/129
  differential diagnosis, 52/205
  disseminated miliary tubercules, 52/196
  dystonia, 6/559
  EEG, 33/225, 52/210, 55/429
  ELISA, 52/206
  encephalopathy, 33/207-209, 216, 52/200
  endocrine sequela, 52/216
  ependymitis, 52/198
  epidemiology, 52/195
  epilepsy, 15/307, 33/213, 232, 52/202, 216,
    55/426, 72/349
  ethambutol, 33/228, 52/211
  ethionamide, 33/228, 52/212
  focal caseous plaque, 52/196
  granuloma, 52/200
  headache, 48/32
  hemagglutination test, 52/205
  hemiballismus, 49/373
  Herxheimer reaction, 52/211
  histopathology, 52/197
  history, 33/195
  hyaluronidase, 52/213
  hydrocephalus, 30/545, 33/206, 52/200, 202, 207,
    214, 55/431
  hyponatremia, 28/499
  immunodiagnosis, 52/205

incidence, 33/196, 210, 52/195
infantile cerebrovascular disease, 54/40
internal carotid artery, 52/198
intracranial calcification, 52/207
intracranial hypertension, 52/202, 215
intrathecal agent, 52/213
intrathecal tuberculin, 33/228
isoniazid, 33/226-228, 52/211
isotope cisternography, 52/209
isotope scan, 33/224
laboratory diagnosis, 55/428
laboratory finding, 52/204
latex agglutination, 52/206
lymphocytic meningitis, 52/205
malnutrition, 33/230
meningeal vein, 52/198
meningitis classification, 52/196
meningitis serosa, 52/203
meningomyelitis, 52/202
mental deficiency, 46/19
microbiology, 33/197, 52/196
middle cerebral artery, 52/198
miliary brain lesion, 52/209
monoclonal antibody, 52/206
mortality, 52/214
moyamoya disease, 53/164
moyamoya pattern, 55/430
multiple cranial neuropathy, 55/426
myelopathy, 33/233, 52/216
myoclonia, 55/426
necrotizing panarteritis, 52/200
neurobrucellosis, 52/594
neurologic symptom, 52/202
neurosurgery, 52/213
nuchal rigidity, 52/201
onset, 33/212, 52/201
ophthalmoplegia, 52/202
opisthotonos, 52/201
optic atrophy, 52/202
optic neuritis, 52/202
pathogenesis, 33/198-200, 52/196
pathology, 33/200-209, 55/424
PEG, 33/221
phenytoin, 52/211
phlebitis, 52/198
plexus choroideus, 52/203
prodromal phase, 33/211
prodromal stage, 52/201
prognosis, 33/230-234, 52/214, 55/431, 71/283
proliferative meningitis, 33/201, 52/196
pulmonary lesion, 33/217, 52/203
pyrazinamide, 33/228, 52/212
racial preponderance, 52/195

radiculomyelitis, 52/202
recurrent nonhereditary multiple cranial
    neuropathy, 51/570
relapse, 33/233
retrobulbar neuritis, 52/212
rifampicin, 33/228, 52/211
sago grain lesion, 52/210
seasonal incidence, 33/197
sequela, 33/231, 52/215
sighing respiration, 1/667
sign, 33/211-216
skull radiation, 33/220, 52/207
soluble antigen fluorescent antibody, 52/206
spinal, 33/234-238
spinal block, 33/233, 55/431
spinal cord tumor, 32/362
streptomycin, 33/226-229, 52/211-213
subdural effusion, 52/216
surgery, 33/229
symptom, 33/211-216, 55/424
symptomatic dystonia, 6/559
thiacetazone, 52/212
thrombosis, 52/199
transverse myelitis, 52/202
treatment, 33/226-230, 52/211, 55/431, 71/283
tuberculin test, 33/218, 52/205
tuberculoma, 52/198, 208, 55/429
vasculitis, 52/199
Tuberculous neuritis
    symptom, 51/186
Tuberculous pachymeningitis
    otitis media, 33/161
Tuberculous spondylitis, 33/238-251
    age, 33/238
    clinical features, 33/242-244
    etiology, 33/240
    history, 33/238
    incidence, 33/238-240
    myelography, 33/236
    paraplegia, 33/241-244
    pathogenesis, 33/240
    pigmented paravenous retinochoroidal atrophy,
        13/268
    prognosis, 33/250
    sex incidence, 33/239
    spinal cord compression, 19/370
    spinal tumor syndrome, 33/243
    spinal X-ray, 33/244-246
    treatment, 33/247-250
Tuberohypophyseal tract
    anatomy, 2/437
Tuberoinfundibular nucleus
    anatomy, 2/436

neurosecretion, 2/437
Tuberoinfundibular system
  arcuate nucleus, 74/312
  stress, 74/312
Tuberothalamic artery
  brain lacunar infarction, 54/236
Tuberous sclerosis, 14/340-381
  absence, 50/372
  amyotrophy, 14/349
  angiomyolipoma, 50/371
  aorta aneurysm, 63/48
  aorta coarctation, 14/381, 50/373
  arthrogryposis multiplex congenita, 14/496
  ash leaf spot, 50/371
  associated disease, 14/496
  associated malformation, 31/6
  astrocytoma, 14/46, 18/300, 42/732, 50/371
  astrocytosis, 14/42
  ataxia telangiectasia, 14/74, 279
  athetosis, 6/457, 14/349
  atypical ocular lesion, 14/624
  autosomal dominant, 50/371
  biochemistry, 14/496
  birth incidence, 43/50
  blood group ABO, 50/371
  bone cyst, 14/51
  brain biopsy, 10/686
  brain cortex heterotopia, 50/248
  brain embolism, 55/163
  brain infarction, 55/163
  brain tumor, 14/45, 380, 18/297, 299-303, 322
  café au lait spot, 14/343
  calcification, 14/45-47
  cardiac lesion, 14/50
  cause, 14/377
  cerebellum, 14/47, 366
  cerebrovascular disease, 53/43, 55/163
  cheilognathopalatoschisis, 14/102
  chordoma, 50/373
  choroid hamartoma, 50/372
  classification, 14/11
  clinical features, 31/6-9, 50/370
  colloid body, 13/75
  colon epithelioma, 14/51
  computer assisted tomography, 50/371
  congenital cardiopathy, 14/50
  congenital malformation, 30/101
  convulsion, 14/108
  corpora amylacea, 14/360
  corpus callosum agenesis, 30/286
  cortical, 14/39
  craniopharyngioma, 18/134
  cutaneous dysplasia, 14/101

  cutaneous hypopigmentation, 50/371
  cutis verticis gyrata, 14/496
  dental pit, 50/374
  dermatology, 14/572-580
  diagnosis, 31/9
  differential diagnosis, 13/251, 14/579
  diffuse glioblastosis, 18/77
  digestive tract, 14/377
  Down syndrome, 14/350
  drusen, 14/623
  dysraphia, 14/398
  EEG, 14/347
  endocrinology, 14/744-746
  epidemiology, 30/159
  epilepsy, 14/108, 346, 31/3, 6-10, 43/49,
    72/129-131, 349
  features, 31/4
  fibroma, 43/49
  fibrous dysplasia, 14/199
  flexor spasm, 14/347
  forme fruste, 14/349-352
  general concept, 14/86
  genetics, 14/493-496, 16/30, 30/159, 31/5, 67/6
  giant cell, 14/42, 357-360
  giant cell astrocytoma, 14/46
  gonadal lesion, 14/377
  hamartoma, 18/300
  heart disease, 14/372
  heart rhabdomyoma, 50/373, 63/94, 103
  hemihypertrophy, 22/549
  hereditary, 14/350-352, 494
  heterophenia, 14/351
  heterotopia, 14/45, 47, 354, 360, 366
  history, 14/572
  hydrocephalus, 31/2, 6
  hypernephroma, 14/686
  hypsarrhythmia, 14/347, 43/49, 50/372
  incidence, 50/371
  infantile spasm, 14/347, 43/49
  interstitial lung disease, 50/371
  intracerebral calcification, 43/50
  intracranial calcification, 14/356, 379
  iris coloboma, 14/496
  iris hypopigmentation, 50/372
  ivory vertebra, 14/375
  kidney tumor, 14/370
  Klippel-Trénaunay syndrome, 14/398
  Koenen fibroma, 31/7
  Koenen tumor, 14/49, 343
  laboratory finding, 14/353
  laminar brain atrophy, 14/47
  late cortical cerebellar atrophy, 21/477
  liver lesion, 14/376

macrencephaly, 14/102, 31/2, 6, 42/41

melanin, 14/102

meningioma, 14/368, 42/744

mental deficiency, 14/348, 31/6, 8, 43/49, 46/43, 50/371

micrencephaly, 14/102, 367

microcephaly, 31/2, 6

microglia, 14/42

microglia cell, 14/42

microgyria, 14/42

moyamoya disease, 42/751

mucous membrane, 14/577

mucoviscidosis, 14/496

multiple trichoepithelioma, 14/110

mutation rate, 14/496, 43/50

myoclonic epilepsy, 14/347, 496

neurinoma, 14/8

neurocutaneous syndrome, 14/101

neurofibromatosis, 14/6, 368

neurofibromatosis type I, 14/133

neurology, 14/349

neuropathology, 14/38-53, 353-369

nosology, 14/573

ocular lesion, 14/343

ophthalmology, 14/619

optic atrophy, 13/81, 14/346

optic nerve tumor, 17/353

pancreas adenoma, 14/51

papilledema, 14/345

parkinsonism, 14/350

partial albinism, 14/496

partial gigantism, 14/102

pathogenesis, 14/377

pathognomonic lesion, 14/573-578

pathologic laughing, 14/347

pathology, 31/5

periungual fibroma, 50/371

periventricular tuber, 50/373

periventricular tumor, 14/356-359

phakoma, 14/343-345, 620

phakomatosis, 14/19

pheochromocytoma, 14/51

piloid gliosis, 14/356

pleiotropism, 14/8, 351

poliosis circumscripta, 14/574

precocious puberty, 14/745

prevalence, 14/351, 493, 43/50

prognosis, 14/381, 580

progressive myoclonus epilepsy, 14/496

pseudohypernephroma, 14/50, 60

psychiatric disorder, 14/760-762

psychosis, 14/349

pulmonary artery stenosis, 14/51

pulmonary cyst, 14/51, 373

pulmonary fibromyoma, 14/373

pulmonary interstitial fibrosis, 50/373

pulmonary lesion, 14/51

pulmonary myomatosis, 14/51

race, 14/353

radiologic diagnosis, 14/379

renal angiolipoma, 50/373

renal cyst, 14/50

renal insufficiency, 50/371

renal lesion, 14/50

reproductive fitness, 43/50

reticular gliosis, 14/355

retinal phakoma, 43/49

retinal tumor, 14/47

rhabdomyoma, 14/51, 371, 687, 50/371

sebaceous adenoma, 14/11, 48, 341, 572, 31/6, 43/9, 49, 50/371

seizure, 50/371-373

shagreen patch, 31/7

shagreen skin, 14/102, 342, 576, 43/49, 50/371

skeletal lesion, 14/374-376

skin lesion, 14/48, 341, 573

spina bifida, 14/368, 398, 31/6

spinal cord lesion, 14/47

spleen lesion, 14/376

spongioblastoma, 14/46, 102, 350, 365, 380, 18/300

Sturge-Weber syndrome, 14/47, 532, 578

subarachnoid hemorrhage, 55/29

subependymal glomerular astrocytoma, 18/14

subependymal nodule, 14/349, 363, 379

subungual fibroma, 50/372

supratentorial brain tumor, 18/322

synopsis, 14/2-4

tache de bougie nodule, 14/44

testicular heterotopia, 14/51

tissue culture, 17/77

treatment, 14/379, 580, 31/9

tuber, 14/39-43, 354

tuber type, 14/39-41

tumor, 18/3

twin study, 14/352

uterine leiomyoma, 14/51

ventricular, 14/43

ventricular tumor, 8/133

visceral lesion, 14/49, 369-374, 682-691

vitiligo, 14/342

Vogt triad, 68/291

Von Hippel-Lindau disease, 14/251

Tuberous sclerosis complex
  bone cyst, 68/292
  confetti skin lesion, 68/292

diagnosis, 68/292
facial angiofibroma, 68/292
forhead plaque, 68/292
giant cell astrocytoma, 68/293
gingival fibroma, 68/292
pathogenesis, 68/293
pulmonary lymphangiomyomatosis, 68/292
rectal polyp, 68/292
renal angiomyolipoma, 68/292
renal cyst, 68/292
retinal astrocytoma, 68/292
retinal hamartoma, 68/292
shagreen patch, 68/292
Tübingen, Germany
neurology, 1/19
D-Tubocurarine
arthrogryposis multiplex congenita, 30/123
chondrodystrophic myotonia, 41/299
malignant hyperthermia, 41/268, 63/439
myasthenia gravis, 41/96, 108, 127
teratogenic agent, 30/123
Tubular aggregate
muscle fiber, 40/33, 112, 226, 229
myotonic dystrophy, 41/154
nerve fiber, 41/154, 161
periodic paralysis, 40/229, 41/154, 162, 62/465
transverse tubule, 40/33, 112, 226, 229
ultrastructure, 40/112
Tubular aggregate myopathy
clinical features, 62/353
congenital myopathy, 62/332
enzyme histochemistry, 62/352
facial weakness, 62/334
Tubular necrosis
acute, see Acute tubular necrosis
spinal cord injury, 25/80, 84
Tubular vision
agnosia, 4/26
Tubulin
axon regeneration, 61/483
Down syndrome, 50/530
spinal cord repair, 61/483
Tubulofilament
dermatomyositis, 62/377
inclusion body myositis, 62/377
polymyositis, 62/377
Tubulomembranous inclusion myopathy
congenital myopathy, 62/332
facial weakness, 62/334
rimmed muscle fiber vacuole variant, 62/352
Tuffstein granule
cerebroside sulfatase, 21/51
metachromatic leukodystrophy, 21/51, 51/368,

372, 60/126
Tuffstone body, see Tuffstein granule
Tuft
hair, see Hair tuft
Tularemia
acquired myoglobinuria, 62/577
Tumeur à myéloplaxe
see also Giant cell tumor
aneurysmal bone cyst, 19/170
Tumeur royale
classification, 8/445, 14/38
neurofibromatosis, 14/30
neurofibromatosis type I, 14/30, 32
type, 8/445
Tumor
acoustic nerve, see Acoustic nerve tumor
ACTH secreting, see ACTH secreting tumor
amyloid, see Amyloid tumor
anterior condylar canal syndrome, 2/99
Antoni type A, see Antoni type A tumor
Antoni type B, see Antoni type B tumor
aorta body, see Aorta body tumor
atypical teratoid/rhabdoid, see Atypical
teratoid/rhabdoid tumor
Bailey, see Oligodendroglioma
barrel shaped vertebra, 20/30
basal ganglion, see Basal ganglion tumor
Benedikt syndrome, 2/298
benign spinal chondrogenic, see Benign spinal
chondrogenic tumor
benign spinal osteogenic, see Benign spinal
osteogenic tumor
bony, see Bony tumor
brachial plexus, 2/143
brain, see Brain tumor
brain edema, 16/176
brain embolism, 16/6
brain stem, see Brain stem tumor
brain stem encephalitis, 18/407
brain vascular, see Brain vascular tumor
brain ventricle, see Lateral ventricle tumor
Brown hyperparathyroid, see Brown
hyperparathyroid tumor
callosal, see Corpus callosum tumor
carcinoid, see Carcinoid tumor
carotid body, see Carotid body tumor
carpal tunnel syndrome, 7/294
cauda equina, see Cauda equina tumor
cavernous sinus, see Cavernous sinus tumor
cerebellar, see Cerebellar tumor
cerebellar astrocytoma, 17/709, 715
cerebellar hemangioblastoma, 17/715
cerebellopontine angle, see Cerebellopontine

angle tumor

cerebral, *see* Brain tumor

choroid plexus, *see* Choroid plexus tumor

ciliary body, *see* Ciliary body tumor

classic primitive neuroectodermal, *see* Classic primitive neuroectodermal tumor

CNS, *see* CNS tumor

congenital brain, *see* Congenital brain tumor

congenital spinal cord, *see* Congenital spinal cord tumor

conjunctival, *see* Conjunctival tumor

corpus callosum, *see* Corpus callosum tumor

corpus callosum ataxia, 1/343, 17/508

cranial vault, *see* Cranial vault tumor

dermoid, *see* Dermoid tumor

desmoid, *see* Desmoid tumor

diencephalic, *see* Diencephalic tumor

dysembryoplastic neuroepithelial, *see* Dysembryoplastic neuroepithelial tumor

ear, *see* Ear tumor

ectopic hormone production, 75/542

endodermal sinus, *see* Endodermal sinus tumor

epidermoid, *see* Epidermoid tumor

epidural, *see* Epidural tumor

epilepsy, 69/9, 72/113

exophthalmos, 68/77

experimental brain, *see* Experimental brain tumor

extramedullary spinal, *see* Extramedullary spinal tumor

facial nerve injury, 24/109

facial paralysis, 8/282

follitropin/luteinizing hormone secreting, *see* Follitropin/luteinizing hormone secreting tumor

frontal lobe, *see* Frontal lobe tumor

ganglion cell, *see* Ganglion cell tumor

genital tract, *see* Genital tract tumor

germ cell, *see* Germ cell tumor

giant, *see* Giant tumor (Elsberg)

giant cell, *see* Giant cell tumor

glomus, *see* Glomus tumor

glomus jugulare, *see* Glomus jugulare tumor

growth hormone secreting, *see* Growth hormone secreting tumor

headache, 48/5

hemiplegia, 1/192

hourglass, *see* Hourglass tumor

hyperhidrosis, 75/530

hypophyseal, *see* Hypophyseal tumor

hypothalamic, *see* Hypothalamic tumor

hypothalamic syndrome, 2/450

hypothalamus, *see* Hypothalamus tumor

infratentorial, *see* Infratentorial tumor

intracerebral hematoma, 11/664

intracranial, *see* Intracranial tumor

intramedullary spinal, *see* Intramedullary spinal tumor

intraspinal, *see* Intraspinal tumor

jaw, 5/227

Koenen, *see* Koenen tumor

lacrimal gland, *see* Lacrimal gland tumor

lateral ventricle, *see* Lateral ventricle tumor

Leydig cell testicular, *see* Leydig cell testicular tumor

Lindau, *see* Hemangioblastoma

lung, *see* Lung tumor

malignant peripheral nerve sheath, *see* Malignant peripheral nerve sheath tumor

medial thalamic syndrome, 2/485

mediastinal, *see* Mediastinal tumor

medulla oblongata, 67/160

meningeal, *see* Meningeal tumor

mesencephalic, *see* Mesencephalic tumor

metastatic, *see* Metastatic tumor

metastatic brain, *see* Brain metastasis

metastatic cardiac, *see* Metastatic cardiac tumor

midbrain, *see* Midbrain tumor

mixed neuronoglial, *see* Mixed neuronoglial tumor

mixed parotid, *see* Mixed parotid tumor

multiple brain, *see* Multiple brain tumor

multiple spinal, *see* Multiple spinal tumor

muscle, *see* Muscle tumor

nasopharyngeal, *see* Nasopharyngeal tumor

nerve, *see* Nerve tumor

nerve sheath, *see* Nerve sheath tumor

nervous system, *see* Nervous system tumor

neural, *see* Neural tumor

neurologic intensive care, 55/204

neuronal, *see* Neuronal tumor

neuronoglial, *see* Neuronoglial tumor

nongerminomatous germ cell, *see* Nongerminomatous germ cell tumor

occipital lobe, *see* Occipital lobe tumor

occipitoparietal, *see* Occipitoparietal tumor

optic atrophy, 13/68

optic chiasm, *see* Optic chiasm tumor

optic nerve, *see* Optic nerve tumor

optic pathway, *see* Optic pathway tumor

orbital, *see* Orbital tumor

Pancoast, *see* Pancoast tumor

pancreas, *see* Pancreas tumor

paraneoplastic syndrome, 69/375

parietal lobe, *see* Parietal lobe tumor

peroneal nerve, 70/46

pineal, *see* Pineal tumor

pituitary, *see* Pituitary tumor

pontine, *see* Pontine tumor

pontine syndrome, 2/259

posterior fossa, *see* Posterior fossa tumor

prefrontal syndrome, 17/239-241

primary cardiac, *see* Primary cardiac tumor

primitive neuroectodermal, *see* Primitive neuroectodermal tumor

radiation induced, *see* Radiation induced tumor

RES, *see* RES tumor

RES brain, *see* RES brain tumor

reticulosarcomatosis, 18/241

retinal, *see* Retinal tumor

retractile, *see* Retractile tumor

retro-orbital, *see* Retro-orbital tumor

rhabdoid, *see* Rhabdoid tumor

scalp, *see* Scalp tumor

Schmincke, *see* Schmincke tumor

sella turcica, 16/690, 23/178

sellar, *see* Sellar tumor

septum pellucidum, *see* Septum pellucidum tumor

sinus pain, 5/330

skull, *see* Skull tumor

skull base, *see* Skull base tumor

sphenoid bone, 13/82

spinal, *see* Spinal tumor

spinal column, *see* Spinal column tumor

spinal cord, *see* Spinal cord tumor

spinal epidural, *see* Spinal epidural tumor

spinal Lindau, *see* Hemangioblastoma

spinal subdural, *see* Spinal subdural tumor

spondylotic radiculopathy, 19/85

sternocleidomastoid, *see* Sternocleidomastoid tumor

subarachnoid hemorrhage, 12/118

subependymal giant cell, *see* Subependymal giant cell tumor

suprasellar pearly, *see* Suprasellar pearly tumor

supratentorial brain, *see* Supratentorial brain tumor

supratentorial midline, *see* Supratentorial midline tumor

sympathetic ganglion, 8/459

temporal bone, *see* Temporal bone tumor

temporal lobe, *see* Temporal lobe tumor

teratoid, *see* Teratoid tumor

thalamic, *see* Thalamic tumor

thalamic dementia, 2/484

third ventricle, *see* Third ventricle tumor

thyrotropic hormone secreting, *see* Thyrotropic hormone secreting tumor

trigeminal neuralgia, 48/450

tuberous sclerosis, 18/3

turban, *see* Turban tumor

vagus nerve, 75/529

vascular, *see* Vascular tumor

vasculitis, 71/158

vasoformative, *see* Vasoformative tumor

vertebral, 69/58

vertebral column, *see* Vertebral column tumor

vertigo, 2/361

Wilms, *see* Wilms tumor

yolk sac, *see* Yolk sac tumor

Tumor cell

anatomic pathway, 71/644

ataxia telangiectasia, 67/31

Caenorhabditis elegans, 67/29

cancer gene, 67/18

CSF, 67/220

CSF cytology, 16/375, 378, 381, 383, 389

cycle, 67/25

cycle checkpoint, 67/26

cycline, 67/25

DNA damage, 67/28

gene mutation, 67/35

growth hormone releasing factor receptor, 67/23

guanine nucleotide binding protein, 67/23

mutator gene, 67/21

neuroblastoma, 67/21, 34

neurofibromatosis type I, 67/31

neurofibromatosis type II, 67/33

oncogene, 67/18

peptide, 67/35

phase, 67/25

primary CNS lymphoma, 68/259

programmed cell death, 67/29

regulatory pathway, 67/17

restriction point, 67/26

signal transduction pathway, 67/22

spindle checkpoint, 67/28

suppressor gene, 67/19

TP53 gene, 67/20

Von Hippel-Lindau disease, 67/34

Wilms tumor, 67/21

Tumor cyst

low grade glioma, 68/50

Tumor growth lesion

lumbosacral plexus, 51/165

Tumor infiltrating lymphocyte

astrocytoma, 67/295

Tumor necrosis factor

critical illness polyneuropathy, 51/584

headache, 65/562

Tumor necrosis factor-α

brain tumor, 67/298

muscle tissue culture, 62/92

Tumor necrosis factor intoxication

exploding headache, 65/562
Tumor neovascularization
  brain tumor, 67/302
Tumor sign
  direct, *see* Direct tumor sign
  indirect, 20/213, 230
  spinal neurinoma, 20/258
Tumor suppressor gene
  brain metastasis, 69/108
Tumoral calcinosis, *see* Lipocalcinogranulomatosis
Tumorigenesis
  signal transduction pathway, 67/22
Tuna intoxication
  vitamin A intoxication, 37/87
Tunbridge-Paley disease
  Alström-Hallgren syndrome, 22/507
  Friedreich ataxia, 22/507
  hereditary hearing loss, 22/507
  hypogonadism, 22/508
  juvenile diabetes mellitus, 22/507
  Laurence-Moon-Biedl syndrome, 22/508
Tunicata intoxication
  chordata, 37/68
Tuning fork test
  acoustic nerve tumor, 17/676
  postural sensibility, 2/683
  Rinne test, 2/74
  Schwabach test, 2/73
Turban tumor
  cranial vault, 17/124
  soft tissue, 17/124
Turcot syndrome, *see* Glioma polyposis syndrome
Turin, Italy
  neurology, 1/9
Türk-Stilling-Duane syndrome, *see*
  Stilling-Türk-Duane syndrome
Turkey
  neurology, 1/11
Turmschädel, *see* Oxycephaly
Turner syndrome
  see also Gonadal dysgenesis, Noonan syndrome
    and Ullrich-Turner syndrome
  amenorrhea, 50/543
  aneuploidy, 50/543, 566
  antimongoloid eyes, 50/543
  aorta aneurysm, 63/50
  aorta coarctation, 50/544
  associated malformation, 31/500
  atlas, 32/82
  Bardet-Biedl syndrome, 13/400
  behavior, 31/500, 50/545
  birth incidence, 43/551
  brachymetacarpalism, 14/466

chromosomal abnormality, 30/250
clinical features, 31/496, 498-502, 50/544
CNS, 31/500
congenital hip dislocation, 14/108
cubitus valgus, 50/543
cytogenetics, 31/497
definition, 31/495, 50/543
dementia, 14/108
dermatoglyphics, 31/499, 43/550
developmental dyslexia, 4/379
diabetes mellitus, 50/544
differential diagnosis, 14/14, 31/503
digital abnormality, 43/550
Duchenne muscular dystrophy, 40/352
EEG, 50/545
endocrine function, 31/501
epicanthus, 43/550, 50/543
epidemiology, 31/497
face, 31/499
familial incidence, 31/497
Gerstmann syndrome, 50/545
gonadal dysgenesis, 14/109, 43/550
Guérin-Stern syndrome, 14/109
Hashimoto thyroiditis, 50/544
heart disease, 43/550
history, 31/496, 50/543
hypergonadotropic hypogonadism, 50/544
hypertelorism, 14/109, 30/248-250, 43/550,
  50/543
hypogonadism, 43/550, 50/544
infertility, 50/543
intelligence, 31/500, 50/544
keloid, 50/543
Klippel-Feil syndrome, 14/109
lymphedema, 43/550, 50/544
mental deficiency, 31/500, 46/74
microgyria, 50/545
mosaicism, 31/502, 50/544
multiple interstitial telangiectasia, 50/544
multiple pigmented nevi, 50/543
nail hypoplasia, 43/550, 50/543
neurocutaneous syndrome, 14/101
neurofibromatosis type I, 14/492
neurosensory hearing loss, 50/545
Nielsen syndrome, 14/111
nondisjunction, 31/497
Noonan-Ehmke syndrome, 31/503, 46/75, 50/546
osteoporosis, 50/544
otitis media, 50/543
palate, 50/543
pathogenesis, 31/497, 50/543
prevention, 50/546
pterygium, 14/109

pterygium colli, 30/249, 43/550, 50/543
ptosis, 50/543, 545
radiologic finding, 31/500
renal abnormality, 43/550
Rossi syndrome, 14/14, 109
sex chromosomes, 43/550-552, 46/74, 50/543
sexual infantilism, 50/544
short metacarpal, 50/543
short metatarsal, 50/544
short stature, 30/249, 31/499, 43/550, 50/544
treatment, 31/502, 50/546
trimethylaminuria, 29/231
twinning, 31/498
Ullmann syndrome, 14/109
Turning behavior
amphetamine, 45/167
apomorphine, 45/167
dopamine, 45/167
GABA, 45/167
neuropharmacology, 45/167
noradrenalin, 45/167
serotonin, 45/167
Turricephaly, *see* Oxycephaly
Turtle intoxication, 37/89-91
Galapagos, 37/89
green, 37/89
hawksbill, 37/89
leathery, 37/89
loggerhead, 37/89
softshell, 37/89
Tussive epilepsy, *see* Cough syncope
Tussive syncope, *see* Cough syncope
Twilight state
absence status, 15/595
consciousness, 3/115
dysphrenia hemicranica, 5/79
epilepsy, 15/595
foamy spheroid body, 59/291
migraine, 48/161
temporal lobe epilepsy, 2/714
Twin study
brain tumor, 16/29
familial gliomatosis-glioblastomatosis, 14/499
intelligence, 3/303
Krabbe-Bartels disease, 14/528
multiple sclerosis, 9/90-92, 47/292-295
neurocutaneous melanosis, 14/525
neurofibromatosis type I, 14/134, 489
sexual deviation, 43/215
spinal epidural cyst, 20/143
telangiectasia, 14/520
tuberous sclerosis, 14/352
Twin zygosity

multiple sclerosis, 47/292-294
Twining line
basilar impression, 32/19
Twins
anencephaly, 30/185, 194, 50/79
axial mesodermal dysplasia spectrum, 50/517
cavum septi pellucidi, 30/326
CNS malformation, 30/151
congenital fiber type disproportion, 62/356
cri du chat syndrome, 31/569
dizygotic, *see* Dizygotic twins
Down syndrome, 30/161, 31/372, 393
dyssynergia cerebellaris myoclonica, 21/509
glioblastoma multiforme, 18/52
handedness, 4/252
holocardiac, *see* Holocardiac twins
infantile spinal muscular atrophy, 22/89
Kii Peninsula amyotrophic lateral sclerosis, 22/371
migraine, 48/19, 23, 28
monozygotic, *see* Monozygotic twins
multiple sclerosis, 9/101
syringomyelia, 50/451
Wohlfart-Kugelberg-Welander disease, 22/69
zygotic, *see* Zygotic twins
Twisted hair, *see* Pili torti
Twisted neurotubule
ultrastructure, 21/57
Two point discrimination
terminology, 1/105
test parameter, 1/88
Tybamate
anxiolytic agent, 37/354
structural formula, 37/354
Tympanic membrane rupture
electrical injury, 61/195
Tympanic neurectomy
motoneuron disease, 59/467
Tympanic plexus
facial nerve, 48/488
Tympanic plexus neuralgia
dysautonomia, 22/254
glossopharyngeal neuralgia, 48/490
Type I fiber, *see* Muscle fiber type I
Type II fiber, *see* Muscle fiber type II
Typhoid fever
acquired myoglobinuria, 62/577
autoscopia, 4/232, 45/386
brain embolism, 55/163
brain infarction, 55/163
cerebrovascular disease, 55/163
Guillain-Barré syndrome, 51/187
mononeuritis, 51/187

β-naphthol, 13/224
nerve lesion, 7/482
neuritis, *see* Typhoid neuritis
Parkinson disease, 6/245
pellagra, 28/93
polyneuritis, 51/187
Typhoid neuritis
  symptom, 51/187
Typhoid vaccination
  neuropathy, 7/482, 8/86
Typhus
  African tick-borne, *see* African tick-borne typhus
  brain infarction, 53/162
  brain ischemia, 53/162
  eastern hemisphere tick-borne, *see* Eastern
    hemisphere tick-borne typhus
  endemic, *see* Endemic typhus
  flea-borne, *see* Flea-borne typhus
  Indian tick-borne, *see* Indian tick-borne typhus
  Kenyan tick-borne, *see* Kenyan tick-borne typhus
  louse-borne, *see* Louse-borne typhus
  mite-borne, *see* Mite-borne typhus
  murine, *see* Murine typhus
  North Asian tick-borne, *see* North Asian
    tick-borne typhus
  Queensland tick-borne, *see* Queensland tick-borne
    typhus
  recrudescent, *see* Recrudescent typhus
  scrub, *see* Scrub typhus
  tick-borne, *see* Tick-borne typhus
Typus degenerativus Amstelodamensis, *see* De
  Lange syndrome
Tyramine
  dopamine β-mono-oxygenase deficiency, 59/163
  headache, 48/97, 445
  migraine, 48/20, 92-94, 97
  orthostatic hypotension, 22/270
  progressive dysautonomia, 59/151
  pure autonomic failure, 22/240
Tyramine sulfate
  migraine, 48/94
Tyrosinase
  copper enzyme, 49/229
  Wilson disease, 6/120
Tyrosinase negative
  oculocutaneous albinism, 43/4
Tyrosinase positive
  oculocutaneous albinism, 43/5
Tyrosine
  albinism, 14/106
  GM2 gangliosidosis, 10/296
  Millon reaction, 9/24
  phenylketonuria, 28/129, 29/31-33, 42/610

primary pigmentary retinal degeneration, 13/196
Tyrosine hydroxylase
  autonomic nervous system, 75/233
  Huntington chorea, 49/75
  inhibition, 75/239
  Parkinson disease, 49/123
  parkinsonism, 42/256
  progressive supranuclear palsy, 46/302, 49/251
  taurine deficiency, 42/256
Tyrosine transaminase
  tyrosinemia, 42/635
Tyrosinemia
  biochemical finding, 29/215
  cystathionine synthetase deficiency, 42/635
  cytosol tyrosine aminotransferase, 42/635
  Fanconi syndrome, 29/172
  hepatorenal tubular dysfunction, 29/214-217
  hereditary, *see* Hereditary tyrosinemia
  4-hydroxyphenylpyruvic acid oxidase, 42/635
  hyperkeratosis, 42/635
  hypermethioninemia, 29/119
  methionine activating enzyme, 42/635
  neonatal transient, 29/38-42, 219-221
  tyrosine transaminase, 42/635
  tyrosinosis, 42/634
  without hepatorenal tubular dysfunction,
    29/216-219
Tyrosinemia type II
  mental deficiency, 42/635
  microcephaly, 42/635
Tyrosinosis
  chronic liver disease, 29/38, 41
  hepatorenal tubular dysfunction, 29/214-217
  oculocutaneous, *see* Oculocutaneous tyrosinosis
  tyrosinemia, 42/634
Tyrosyluria
  biochemistry, 29/215
  clinical features, 29/214
  diagnosis, 29/215
  enzyme defect, 29/215
  genetics, 29/216
  pathology, 29/215
  renal tubular dysfunction, 29/214-217
  treatment, 29/217

Ubidecarenone
  ataxia, 62/504
  creatine kinase, 62/504
  epilepsy, 62/504
  exercise intolerance, 62/504
  lactic acidosis, 62/504
  multiple neuronal system degeneration, 62/504
  muscle biopsy, 62/504, 66/423

muscle weakness, 62/504
myoglobinuria, 62/504
Ubidecarenone deficiency
encephalomyopathy, 66/423
exercise intolerance, 66/423
mental deficiency, 66/423
myoglobinuria, 66/423
seizure, 66/423
Ubiquinol cytochrome c reductase, *see* Complex III
Ubiquinone
serum creatine kinase, 40/343
Ubiquinone 10, *see* Ubidecarenone
Ubiquinone deficiency
myoglobinuria, 62/558
UDP galactose, *see* Uridine diphosphogalactose
UDP-galactose-4-epimerase, *see* Uridine
diphosphate glucose-4-epimerase
UDP glucose, *see* Uridine diphosphoglucose
UDPG-glycogen synthetase, *see* Uridine
diphosphoglucose-glycogen synthetase
Uhrmacher tic
facial nerve, 7/334
UICC classification
astrocytoma, 18/3
brain tumor, 16/9
nomenclature, 16/8
Ulcer
aphthous, *see* Aphthous ulcer
intestine, *see* Intestinal ulcer
nasal, *see* Nasal ulcer
neurogenic, *see* Neurogenic ulcer
neuropathic, *see* Neuropathic ulcer
peptic, *see* Peptic ulcer
perforating foot, *see* Perforating foot ulcer
perforating skin, *see* Perforating skin ulcer
stomach, *see* Stomach ulcer
stress, *see* Stress ulcer
trophic, *see* Trophic ulcer
Ulceration
actiniaria intoxication, 37/53
brain arteriosclerosis, 53/94
corneal, *see* Corneal ulceration
hydrozoa intoxication, 37/38, 40
neurogenic acro-osteolysis, 42/293
sea anemone intoxication, 37/53
Ulcerative colitis
autoimmune disease, 41/102
autonomic nervous system, 75/653
brain infarction, 53/163
brain ischemia, 53/163
chronic, *see* Chronic ulcerative colitis
Hermansky-Pudlak syndrome, 43/6
Ulegyria

see also Brain cortex heterotopia
cerebral hemiatrophy, 50/204, 206
cerebral palsy, 46/14
congenital cerebellar atrophy, 21/460
epilepsy, 72/109, 112
mental deficiency, 46/14
microgyria, 30/479, 50/254
phenylketonuria, 29/33
Ullmann syndrome, 14/446-453
angiomatosis, 14/72, 447
cavernous angioma, 14/450
cerebellar angioma, 14/72
definition, 14/449
dysgenic cyst, 14/448
features, 14/14
hereditary hemorrhagic telangiectasia (Osler),
14/72, 448
history, 14/447
nosology, 14/447, 452
pathology, 14/449
phakomatosis, 31/26
Sturge-Weber syndrome, 14/448
telangiectasia, 14/450
Turner syndrome, 14/109
Von Hippel-Lindau disease, 14/447
Ullrich disease, *see* Congenital atonic sclerotic
muscular dystrophy
Ullrich-Feichtiger syndrome
Gruber syndrome, 14/115, 31/244
Patau syndrome, 14/121
Smith-Lemli-Opitz syndrome, 31/244
Ullrich-Fremerey-Dohna syndrome
ectodermal dysplasia, 14/112
Ullrich syndrome
asymmetric, 14/109
Ullrich-Turner syndrome
see also Turner syndrome
gonadal dysgenesis, 14/109
Ulnar fracture
nerve injury, 70/33
Ulnar nerve
anesthesia, 1/92
brachial plexus, 2/146
claw hand, 2/36, 146
compression, 7/162
compression neuropathy, 51/98
dorsal branch entrapment, 7/306
entrapment elbow, 7/306, 328
entrapment wrist, 7/303, 328
lepromatous sensory polyneuritis, 51/220
lesion site diagram, 2/34
nerve lesion topography, 2/32-38
occupational lesion, 7/327

551

plexus lesion, 2/146
rheumatoid arthritis, 38/483
tardy ulnar paralysis, 7/434, 440
wrist fracture, 70/34
Ulnar nerve grafting
vascularized, *see* Vascularized ulnar nerve grafting
Ulnar nerve lesion
heart transplantation, 63/181
iatrogenic neurological disease, 63/181
Ulnar neuropathy
bacterial endocarditis, 52/300
rheumatoid arthritis, 71/18
uremic polyneuropathy, 63/511
Ulnar paralysis
compression neuropathy, 51/87
Froment sign, 2/37
tardy, *see* Tardy ulnar paralysis
Ultrasonography
Arnold-Chiari malformation type II, 50/408
corpus callosum agenesis, 50/154, 156
Dandy-Walker syndrome, 50/329
parietal cephalocele, 50/108
parietal encephalocele, 50/108
Ultrasound
aspirator, 67/254
brain stem death, 57/464
brain tumor, 67/239
head injury, 57/165
nerve conduction, 7/132
nerve injury, 51/139
radiodiagnosis, 57/165
schizencephaly, 50/201
syringomyelia, 50/450
Ultrasound diagnosis
see also B-mode sonography *and* Doppler sonography
anencephaly, 50/86
brain arteriovenous malformation, 54/35
brain hemorrhage, 54/35
brain infarction, 54/35
intracranial hemorrhage, 54/35
pediatric cerebrovascular disease, 54/35
Ultraviolet light
Cockayne-Neill-Dingwall disease, 13/432
hypersensitivity, 13/432
Umbilical hernia
holoprosencephaly, 50/237
Hurler-Hunter disease, 10/436
Patau syndrome, 31/507, 50/558
Uncal vein
anatomy, 11/48
Uncinate fasciculus
dentatorubropallidoluysian atrophy, 49/442

Uncinate fit
hemianopia, 2/586
medial hippocampal lesion, 2/702
temporal lobe lesion, 2/586, 616
visual hallucination, 2/616
Uncovertebral joint
anatomy, 19/323
vertebral exostosis, 7/447
Uncus
apnea, 63/481
respiration, 63/481
Undescended testis, *see* Cryptorchidism
Undetermined significant monoclonal gammopathy
ataxia, 63/399
axonal neuropathy, 63/402
borderline polyneuropathy, 47/619
chondroitin, 63/402
chronic inflammatory demyelinating polyneuropathy, 63/402
ciclosporin, 63/403
course, 63/399, 402
CSF, 63/399
demyelination, 63/399
distal polyneuropathy, 63/402
EMG, 63/399
features, 63/398
ganglioside, 63/402
glucuronylparagloboside sulfate, 63/401
glycoprotein Po, 63/401
glycosaminoglycan, 63/402
Guillain-Barré syndrome, 63/399, 402
HNK-1/L-2 epitope, 63/400
Ig, 63/403
IgA, 63/402
IgA type, 63/398
IgG, 63/402
IgM, 63/402
IgM binding, 63/400
IgM-kappa, 63/400
IgM type, 63/398
immunohistochemistry, 63/400
immunopathology, 63/400
immunosuppressive therapy, 63/403
incidence, 63/398
lower motoneuron disease, 63/402
lysosome associated membrane protein 1, 63/400
motoneuron disease, 63/402
motor conduction velocity, 63/399
myelin associated glycoprotein, 63/399
myelin associated glycoprotein IgM antibody, 63/401
myelin lamella widening, 63/399
myelin sheath, 63/399

nerve biopsy, 63/399, 402
nerve cell adhesion molecule, 63/400
nerve fiber edema, 63/399
nerve ultrastructure, 63/399
neuropathology, 63/399
paraproteinemia, 63/399
plasma cell dyscrasia, 63/391
plasmapheresis, 63/403
polyneuropathy, 63/398, 402
postural tremor, 63/399
radiculoneuropathy, 63/399
sensorimotor polyneuropathy, 63/399
sensory action potential, 63/399
sex ratio, 63/399
tenascin, 63/400
treatment, 63/403
tremor, 63/399
visual evoked response, 63/399
Waldenström macroglobulinemia, 47/619
Undulant fever, see Brucellosis
Unicentric plaque
senile plaque, 56/555
Uniform muscle fiber type I myopathy
congenital myopathy, 62/332
facial weakness, 62/334
muscle fiber type I predominance, 62/354
symptom, 62/354
Unilateral agraphia
disconnection syndrome, 45/103
Unilateral anesthesia, see Hemianesthesia
Unilateral cervical facet interlocking
see also Cervical vertebral column injury and
Spinal injury
cause, 25/329
cervical spine, 25/136
clinical features, 25/335
experimental, 25/3
pathology, 25/331
radiology, 25/332, 336-338
treatment, 25/338
Unilateral elevator palsy
see also Internuclear ophthalmoplegia and Medial
longitudinal fasciculus
localization, 2/338
pseudoptosis, 2/338
ptosis, 2/338
testing method, 2/338
Unilateral epileptic seizure, 15/235-245
see also Nonconvulsive seizure
atonic, 15/242
classification, 15/107
clonic, 15/235, 238
infantile spasm, 15/241

massive myoclonus, 15/241
physiology, 15/242
tonic, 15/236
tonic clonic, 15/236
Unilateral exophthalmos
frontal lobe tumor, 18/327
Unilateral hemimelic ichthyosiform erythromelia
psoriasis, 43/416
Unilateral inattention, see Visuospatial agnosia
Unilateral paramedian pontine syndrome
combined form, 12/18
pontine hemiplegia, 12/16
symptom, 12/17
Unilateral pontine hemorrhage
Babinski sign, 12/40
conjugate deviation, 12/40
flaccid paralysis, 12/40
fourth ventricle, 12/40
hemiplegia, 12/40
hyperreflexia, 12/40
spastic hemiparesis, 12/40
Unilateral pulsating tinnitus
brain embolism, 53/207
brain infarction, 53/207
cerebrovascular disease, 53/207
intracranial hemorrhage, 53/207
transient ischemic attack, 53/207
Unilateral spatial agnosia, see Visuospatial agnosia
Unilateral spatial neglect, see Visuospatial agnosia
Unilateral tactile anomia
disconnection syndrome, 45/103
Unilateral transverse spinal lesion, see
Brown-Séquard syndrome
Uninucleated epithelioid cell
globoid cell leukodystrophy, 66/10
Unipolar affective disorder
genetics, 43/209, 46/472
Unipolar spongioblastoma, see Spongioblastoma
United Kingdom
head injury epidemiology, 57/5
multiple sclerosis, 47/399
neurology, 1/6
United States
head injury epidemiology, 57/6, 396
neurology, 1/5
Unithiol
arsenic intoxication, 64/289
bismuth intoxication, 64/344
lead intoxication, 64/439
Minamata disease, 64/416
Universal insensitivity
Lissauer tract absence, 1/110
University of Pennsylvania, 1/5

Unmyelinated nerve fiber
  electron microscopy, 7/9, 73
  nerve, 51/3, 7
  neurinoma *vs* neurofibrome, 14/29
  Schwann cell, 7/10, 12, 24, 73
  Schwann cell process, 51/17
  structure, 7/9
Unsaturated fatty acid
  peracetic acid Schiff staining, 9/25
  performic acid Schiff staining, 9/25
  plasma staining, 9/25
  poly, *see* Polyunsaturated fatty acid
Unsaturated sterol
  perchloric acid naphthoquinone method, 9/26
Unsteadiness
  gait, *see* Gait disturbance
Unverricht-Lundborg disease, *see*
    Unverricht-Lundborg progressive myoclonus
    epilepsy
Unverricht-Lundborg progressive myoclonus
    epilepsy
  ataxia, 42/705
  Baltic myoclonus epilepsy, 49/616
  behavior disorder, 42/705
  brain biopsy, 10/685
  dementia, 15/334, 42/705-708, 46/400, 49/617
  dyssynergia cerebellaris myoclonica, 49/616
  genetics, 27/173
  laboratory finding, 73/298
  Lafora body, 27/176-180, 184-187, 42/706
  Lafora progressive myoclonus epilepsy, 73/293
  myoclonic epilepsy, 60/662
  myoclonus, 15/334, 27/172
  neurologic finding, 73/297
  neuropathology, 27/175
  semantics, 49/616
Upbeating nystagmus
  cerebellar vermis atrophy, 60/656
  hereditary cerebellar ataxia, 60/656
Upper airway sleep apnea syndrome, *see* Sleep
    apnea syndrome
Upper eyelid retraction
  Gerstmann-Sträussler-Scheinker disease, 60/656
  spino-olivocerebellar degeneration, 60/655
Upper eyelid spasm
  eye movement disorder, 2/519
Upper lip hypoplasia
  fetal alcohol syndrome, 50/279
Upper motoneuron disease
  agraphia, 45/464
  ataxia, 60/657
  deafness, 60/657
  mental deficiency, 60/657

Upper motoneuron sign
  adult polyglucosan body disease, 62/484
Upper pontine tegmentum syndrome, *see*
    Cestan-Chenais syndrome *and* Foville syndrome
Upstream promotor element
  gene transcription, 66/71
Upward gaze paralysis
  deficiency neuropathy, 51/326
  Von Gräfe sign, 2/338
Uranitis granulomatosis
  Melkersson-Rosenthal syndrome, 8/213
Uranoscopidae intoxication
  family, 37/78
  picture, 37/73
  venomous spine, 37/78
Uranoscopus intoxication
  family, 37/78
  venomous spine, 37/78
Uranoscopus japonicus Houttyn intoxication
  picture, 37/73
Urbach-Wiethe disease
  alopecia, 42/595
  clinical features, 38/5
  congenital indifference to pain, 14/777
  diabetes mellitus, 42/595
  drusen, 13/251
  epilepsy, 14/777
  Grönblad-Strandberg syndrome, 14/114
  intracerebral calcification, 42/595
  intracranial calcification, 14/777, 38/5
  mental deficiency, 14/777, 42/596
  pathology, 38/5
  seizure, 42/595
  short stature, 42/595
  striatopallidodentate calcification, 49/424
Urea
  CSF pressure, 16/361
  hypertonic, *see* Hypertonic urea
  intracranial pressure, 23/212
  spinal cord injury, 26/382, 387, 401
Urea cycle
  disorder, 29/87-106, 42/560
  Krebs-Henseleit, *see* Krebs-Henseleit urea cycle
Urea synthesis
  ammonia, 29/87
Uremia
  see also Renal insufficiency
  amaurosis, 27/334
  anorexia, 63/504
  asterixis, 63/504
  autonomic nervous system, 75/31
  autonomic polyneuropathy, 51/479
  blood-brain barrier, 28/391

brain edema, 9/639
central pontine myelinolysis, 47/583
Cheyne-Stokes respiration, 1/497
chronic axonal neuropathy, 51/531
coma, 63/504
confusion, 63/504
Creutzfeldt-Jakob disease, 56/546
demyelination, 9/639, 47/621
dialysis, 63/504, 523
dialysis dementia, 46/398
dialysis risk, 63/526
diamorphine addiction, 55/517
disorientation, 63/504
fatigue, 63/504
hypertensive encephalopathy, 63/504
lassitude, 63/504
metabolic neuropathy, 8/1, 51/68
metallic taste, 63/504
β2-microglobulin amyloidosis, 63/528
multiple myeloma, 20/11, 63/392
nausea, 63/504
nerve conduction, 7/166
ophthalmoscopy, 11/214
peripheral demyelination, 47/583
restless legs syndrome, 8/316, 51/545
sensorimotor polyneuropathy, 63/510
sleep rhythm, 63/504
spongiform cortical encephalopathy, 9/639
spongiform white matter degeneration, 47/583
subarachnoid hemorrhage, 12/114
symptom, 63/504
tremor, 49/589
vascular occlusion, 9/639
vomiting, 63/504
Uremic encephalopathy, 27/321-344
  apathy, 63/504
  aphasia, 63/505
  apraxia, 63/505
  asterixis, 63/504
  blood-brain barrier, 27/340, 28/391
  brain aluminum, 63/516
  brain atrophy, 63/506
  brain blood flow, 63/506
  brain calcium, 63/515
  brain oxygen consumption, 63/506
  brain sodium, 63/515
  brain stem auditory evoked potential, 63/508
  branched chain aminoacid level, 63/514
  carpopedal spasm, 63/505
  clinical features, 63/504
  cognitive function, 63/509
  coma, 63/504
  computer assisted tomography, 63/506

confusional state, 63/504
convulsion, 27/330-332
course, 63/507
cranial nerve sign, 27/334
creatinine, 63/514
CSF, 63/507
decorticate posture, 63/505
decortication, 63/505
delirium, 46/542, 63/504, 506
dementia, 46/398, 63/504
dialysis, 63/505
dysarthria, 63/504
EEG, 27/332-334, 63/505, 507, 515
EMG, 63/505
epilepsy, 63/504
fatigue, 63/504
GABA, 63/514
glycine, 63/514
guanidinosuccinic acid, 63/514
hallucination, 63/504
hemiparesis, 63/505
hyperparathyroidism, 63/515
hypertensive encephalopathy, 54/211
impaired concentration, 63/504
inositol, 63/514
insomnia, 63/504
MAO, 63/514
mental depression, 63/506
mental status abnormality, 27/322-326
methylguanidine, 63/514
muscle cramp, 27/336, 63/505
myoclonus, 63/504
parathyroid hormone, 63/514-516
pathogenesis, 27/326
peripheral neuritis, 27/338-340
phenytoin, 63/505
psychiatric disorder, 63/506
restless legs, 63/504
stupor, 63/504
survey, 63/503
tetany, 63/504
treatment complication, 27/341-344
tremor, 63/504
visual evoked response, 63/508
Uremic leukoencephalopathy, 47/583
  spongiform leukoencephalopathy, 9/639
Uremic neuropathy
  autonomic involvement, 75/699
  autonomic nervous system, 27/338
  axonal degeneration, 7/166, 8/2, 27/339
  burning feet syndrome, 27/338
  clinical features, 8/2
  creatinine, 7/554, 27/339

555

electron microscopy, 8/2
EMG, 7/166, 27/340
histology, 7/166
motor nerve conduction, 7/166, 8/1, 27/338, 340
motor nerve conduction velocity, 7/166
nerve conduction velocity, 7/554
prognosis, 8/3
renal dialysis, 7/554, 8/1
renal transplantation, 8/1
restless legs syndrome, 27/338
segmental demyelination, 7/166, 8/2, 27/339
sensory nerve conduction, 7/166, 8/1, 27/340
treatment, 8/3
wallerian degeneration, 7/166, 8/2
Uremic polyneuropathy
see also Hemodialysis *and* Renal insufficiency
    polyneuropathy
amyotrophy, 51/357
apathy, 51/356
autonomic neuropathy, 63/511
autonomic polyneuropathy, 63/511
axonal degeneration, 51/359, 63/511
biochemistry, 63/513
brain calcium, 63/515
brain hemorrhage, 63/513
brain sodium, 63/515
branched chain aminoacid level, 63/514
carpal tunnel syndrome, 63/511
clinical features, 51/356
coma, 51/355
cranial neuropathy, 63/512
creatinine, 63/514
creatinine clearance, 51/360
CSF protein, 51/355, 360
deafness, 63/512
dialysis mononeuropathy, 63/511
differential diagnosis, 63/510
EEG, 63/515
EMG, 51/356, 361
etiology, 51/361
GABA, 63/514
glycine, 63/514
guanidine, 63/514
guanidinosuccinic acid, 63/514
Guillain-Barré syndrome, 51/356, 360
hearing loss, 63/512
hemodialysis, 63/511
history, 51/355
hyperparathyroidism, 63/515
incidence, 51/357
inositol, 51/361, 63/514
intermittent hemodialysis, 51/355, 359
MAO, 63/514

metabolic acidosis, 63/512
methylguanidine, 51/361, 63/514
β2-microglobulin, 63/511
miosis, 63/512
muscle cramp, 51/356
nerve conduction velocity, 63/510
neuropathology, 63/511
nystagmus, 63/512
orthostatic hypotension, 63/511
paranodal demyelination, 51/359
parathyroid hormone, 51/362, 63/513-515
paresthesia, 51/356
pathophysiology, 51/360, 63/512
peritoneal dialysis, 51/360
plasma osmolality, 63/512
polyamine, 51/361
prevalence, 63/510
pruritus, 51/356
restless legs syndrome, 51/356
segmental demyelination, 63/511
sensorimotor polyneuropathy, 51/357, 63/510
sex ratio, 63/510
sexual impotence, 63/511
sodium-potassium adenosine triphosphatase,
    51/361
subdural hematoma, 63/513
survey, 63/503
sweating, 51/357
tetraplegia, 51/360
transketolase, 51/361
transplantation, 51/360
treatment, 51/359
ulnar neuropathy, 63/511
Valsalva maneuver, 63/511
vitamin B1, 51/361
vitamin B1 deficiency, 51/361
Ureter
double, *see* Double ureter
Urethan
carbamate intoxication, 64/184, 187
multiple myeloma, 18/256, 20/14, 17
vertebral myeloma, 20/17
Urethra
bulbocavernous muscle, 1/361
innervation, 75/669
nerve, 74/112
Urethra atresia
caudal aplasia, 50/512
renal agenesis syndrome, 50/512
Urethral sphincter resection
urinary tract, 26/411, 425
Uric acid
congenital hyperphosphatasia, 31/259

hereditary myoclonus ataxia syndrome, 21/511
Lesch-Nyhan syndrome, 6/456, 29/266, 46/66,
  60/19
myoclonus, 21/511
Uridine diphosphate glucose-4-epimerase
  alcohol intoxication, 64/110
Uridine diphosphate glucuronosyltransferase
  Crigler-Najjar syndrome, 42/540
Uridine diphosphogalactose
  biochemistry, 9/9
  ceramide, 10/250
Uridine diphosphogalactose N-acetylglucosamine
    pyrophosphorylase
  metachromatic leukodystrophy, 10/48
Uridine diphosphoglucose
  biochemistry, 9/9
Uridine diphosphoglucose-glycogen synthetase
  glycogen storage disease, 40/3
  histochemistry, 40/3
Uridine diphosphoglucose-glycogen
    transglucosylase
  metachromatic leukodystrophy, 10/48
Uridine monophosphate kinase
  Hemophilus influenzae meningitis, 52/62
Urilophidae intoxication
  elasmobranche, 37/70
  wound, 37/71
Urinary arylsulfatase
  metachromatic leukodystrophy, 10/58
Urinary bladder
  autonomic nervous system, 74/14
  Chagas disease, 75/398
  spina bifida, 50/493
Urinary drainage
  external, 26/423
Urinary hesitancy
  dimethylaminopropionitrile polyneuropathy,
    51/282
  Parkinson disease, 59/138
Urinary homoprotocatechuic acid
  chorea-acanthocytosis, 49/330
Urinary incontinence
  anterior cerebral artery syndrome, 53/342
  bismuth intoxication, 64/338
  carbamate intoxication, 64/190
  carbon monoxide intoxication, 64/32
  Charlevoix-Saguenay spastic ataxia, 60/454, 664
  chlormethine intoxication, 64/233
  cingulate gyrus lesion, 2/767
  concentric sclerosis, 47/414
  dentatorubropallidoluysian atrophy, 60/664
  diabetes mellitus, 27/115
  Ferguson-Critchley syndrome, 59/319

frontal lobe syndrome, 55/138
hereditary spastic ataxia, 60/664
hydrocephalic dementia, 46/328
lathyrism, 65/3
lumbar vertebral canal stenosis, 20/706
multiple neuronal system degeneration, 75/162
multiple sclerosis, 9/171, 414, 47/58
neuronal intranuclear hyaline inclusion disease,
  60/664
orthochromatic leukodystrophy, 42/500
orthostatic hypotension, 43/66, 63/143
osteitis deformans (Paget), 38/366, 46/400
paracentral lobule syndrome, 53/345
paraplegia, 61/291
Parkinson disease, 49/94, 59/138
prazosin, 65/448
sacrococcygeal chordoma, 19/289
Shy-Drager syndrome, 1/474
spinal cord injury, 61/291
tabes dorsalis, 33/363, 52/280
Urinary retention
  acrylamide intoxication, 64/68
  ciclosporin intoxication, 63/537
  diabetes mellitus, 27/115
  diabetic autonomic visceral neuropathy, 8/37
  spinal cord injury, 75/576
  spinal shock, 2/188
Urinary sphincter, 1/356-398
  disorder, 1/356
  primary lateral sclerosis, 59/172
  reflex, 1/254
  resection, 26/411, 425
Urinary tract
  see also Bladder
  anal reflex, 26/416
  autonomic dysfunction, 75/670
  autonomic nervous system, 75/571
  bladder neck resection, 26/411
  bladder nerve supply, 26/412, 417, 419
  bulbocavernosus reflex, 26/416
  EMG, 61/298
  follow-up, 26/432
  history, 26/409
  infection, 26/415, 426-432
  lower, see Lower urinary tract
  manual expression, 26/409
  nerve, 74/112
  paraplegia management, 26/208-211
  pH, 26/432
  pharmacologic test, 61/297
  radiodiagnosis, 26/269-274
  radiology, 26/424
  sacral evoked potential, 61/298

spinal cord injury, 26/208, 269-274, 409-436
  upper urinary tract, 26/273
  urethral pressure profilometry, 61/297
  urethral sphincter resection, 26/411, 425
  uroflow measurement, 61/297
Urinary tract defect
  hyperprolinemia type I, 29/133
  iniencephaly, 50/131
  prune belly syndrome, 43/466, 50/515
Urine chemistry
  spinal cord injury, 26/387
Urine copper
  Wilson disease, 27/399, 49/231
Urine disease
  Oasthouse, *see* Oasthouse urine disease
Urochordata intoxication, *see* Tunicata intoxication
Urogenital agenesis
  aplasia, 14/529
  Sturge-Weber syndrome, 14/512
Urogenital tract
  autonomic nervous system, 74/116
Urokinase, *see* Tissue plasminogen activator
Uronic acid
  biochemical features, 10/431
Uroporphyrinogen I synthetase
  acute intermittent porphyria, 42/618
Urticaria
  acute, *see* Acute urticaria
  bee sting intoxication, 37/107
  dominant, *see* Dominant urticaria
  ethchlorvynol intoxication, 37/354
  heat injury, 8/342
  Henoch-Schönlein purpura, 38/71
  ichthyohaemotix fish intoxication, 37/87
  neptunea intoxication, 37/64
  neuropathy, 5/234
  penicillamine intoxication, 27/405, 49/235
  rhizostomeae intoxication, 37/53
  sensorineural deafness, 42/396
Urticaria pigmentosa
  abdominal pain, 14/790
  convulsive seizure, 14/790
  dizziness, 14/790
  headache, 14/790
  histamine, 14/789
  hypotension, 14/790
  nausea, 14/790
  precordial pain, 14/790
  syncope, 14/790
Uruguay
  neurology, 1/12
US Veterans Administration Classification
  epilepsy, 15/13

Usher syndrome, 13/441-448
  ataxia, 42/392
  autosomal recessive, 30/409
  cataract, 60/653
  chorioretinal degeneration, 13/40
  Cockayne-Neill-Dingwall disease, 13/438
  congenital deafness, 50/218
  Corti organ degeneration, 43/392
  deaf mutism, 13/241
  deafness, 13/441, 60/657
  Flynn-Aird syndrome, 22/520
  gene frequency, 42/392
  hereditary deafness, 13/441, 30/409, 50/218
  hereditary hearing loss, 22/499
  mental deficiency, 22/501, 60/664
  optic atrophy, 13/40
  pleiotropism, 13/447
  prevalence, 13/445, 42/392
  psychiatric disorder, 60/665
  Refsum disease, 21/215, 60/235, 66/497
  secondary pigmentary retinal degeneration,
    13/253, 443, 60/654, 722
  Sjögren-Larsson syndrome, 13/479
  Sturge-Weber syndrome, 14/228
  vestibular symptom, 13/443
Ustilago
  fungal CNS disease, 35/572, 52/498
Ustilagomycosis
  meningitis, 35/572, 52/498
Uterine dysfunction
  benign intracranial hypertension, 42/740
Uterine epilepsy
  Prichard classification, 15/4
Uterine leiomyoma
  tuberous sclerosis, 14/51
Uterus bicornis
  holoprosencephaly, 50/237
  Patau syndrome, 31/508, 50/558
Uterus cancer
  opsoclonus, 60/657
Uterus duplex
  Bardet-Biedl syndrome, 13/394
Utriculitis
  positional nystagmus, 2/371
Uvea melanoma
  levodopa, 60/736
  Parkinson disease, 60/736
Uveitis
  see also Vogt-Koyanagi-Harada syndrome
  anterior, *see* Anterior uveitis
  Freund adjuvant, 9/506
  headache, 5/206
  human T-lymphotropic virus type I associated

myelopathy, 59/450
Kawasaki syndrome, 56/638
Köhlmeier-Degos disease, 55/278
multiple sclerosis, 9/115, 176, 47/338
neurosyphilis, 52/274
syphilis, 52/274
Vogt-Koyanagi-Harada syndrome, 56/611
Uveoencephalitic syndrome, *see*
    Vogt-Koyanagi-Harada syndrome
Uveoencephalitis, *see* Vogt-Koyanagi-Harada
    syndrome
Uveomeningitis
    see also Vogt-Koyanagi-Harada syndrome
    chronic meningitis, 56/645
    features, 56/611
Uveomeningoencephalitic syndrome, *see*
    Vogt-Koyanagi-Harada syndrome
Uveo-otocutaneous syndrome, *see*
    Vogt-Koyanagi-Harada syndrome
Uvula
    boundary, 2/399
    sphenopalatine neuralgia, 48/477

VA composite rating
    epilepsy, 72/419
VA Neurotoxicity Scale
    epilepsy, 72/430
VA Systemic Toxicity Scale
    epilepsy, 72/430
Vaccination
    diphtheria, 52/228
    diphtheritic neuropathy, 8/91
    influenza, *see* Influenza vaccination
    meningococcal meningitis, 33/29, 52/21, 31
    multiple sclerosis, 47/339
    pertussis, *see* Pertussis vaccination
    rabies, *see* Rabies vaccination
    swine influenza, *see* Swine influenza vaccination
    TAB, *see* TAB vaccination
    typhoid fever, *see* Typhoid vaccination
    vaccinia, 31/220
Vaccine
    African trypanosomiasis, 52/343
    bacterial meningitis, 52/15
    diphtheria, 52/235
    diphtheria pertussis tetanus, *see* Diphtheria
      pertussis tetanus vaccine
    epilepsy, 72/233
    Hemophilus influenzae meningitis, 52/67
    Margulis, *see* Margulis vaccine
    paratyphoid, *see* Paratyphoid vaccine
    Pasteur rabies, *see* Pasteur rabies vaccine
    pertussis, *see* Pertussis vaccine

poliomyelitis, 34/120-124
poliovirus, *see* Poliovirus vaccine
rabies, *see* Rabies vaccine
rubella, *see* Rubella vaccine
smallpox, *see* Smallpox vaccine
tetanus, 52/235
Vaccinia
    fetal infection, 31/220
    immune mediated encephalomyelitis, 34/412-414
    lymphocytic choriomeningitis, 9/536
    measles, 9/545
    multiple sclerosis, 34/438
    neuralgic amyotrophy, 9/550
    teratogenic agent, 34/378
    vaccination, 31/220
Vaccinia vaccine, *see* Cowpox vaccine
Vaccinia virus, *see* Cowpox virus
Vaccinogenic neuropathy, 8/86-92
    neuralgic amyotrophy, 8/86, 51/172
Vacor
    autonomic nervous system, 75/39
Vacor intoxication
    autonomic nervous system, 75/509
    diabetic polyneuropathy, 51/512
VACTERL syndrome
    see also Caudal aplasia
    cardiac defect, 50/513
    cerebrohepatorenal syndrome, 50/514
    Edwards syndrome, 50/514
    Goldenhar syndrome, 50/514
    Gruber syndrome, 50/514
    hydrocephalus, 50/514
    incidence, 50/514
    limb reduction defect, 50/513
    müllerian aplasia, 50/514
    Patau syndrome, 50/514
    polydactyly, 50/513
Vacuolar myelopathy, *see* Subacute combined spinal
    cord degeneration
Vacuolated lymphocyte
    aspartylglycosaminuria, 42/527
    Batten disease, 10/576-578
    fucosidosis, 42/550
    juvenile neuronal ceroid lipofuscinosis, 42/465
    mannosidosis, 42/597
    neuronal ceroid lipofuscinosis, 60/674
    Niemann-Pick disease, 60/674
    Salla disease, 43/306, 51/382, 59/360
Vacuolation
    lymphocyte, *see* Lymphocyte vacuolation
    myelin, *see* Myelin vacuolation
Vacuole
    autophagic, *see* Autophagic vacuole

muscle fiber, *see* Rimmed muscle fiber vacuole
nerve fiber, 41/148, 153, 160, 178, 187, 191, 197
neuronal mitochondrial, *see* Neuronal
  mitochondrial vacuole
Remak cell, *see* Remak cell vacuole
rimmed muscle fiber, *see* Rimmed muscle fiber
  vacuole
Schwann cell, *see* Schwann cell vacuole
Vacuum headache
  differential diagnosis, 5/218
Vagal body
  nerve tumor, 8/493
Vagal neuralgia, 5/362-367
  dysautonomia, 22/253
  form, 5/363
  glossopharyngeal neuralgia, 22/253
  surgical treatment, 5/400
Vagal paraganglioma
  autonomic nervous system, 75/232
  vagus nerve, 75/232
Vagina
  autonomic nervous system, 2/110
  double, *see* Double vagina
Vagina septa
  Bardet-Biedl syndrome, 13/394
Vaginal atresia
  Edwards syndrome, 50/561
Vagotomy
  apneustic center, 63/479
  diarrhea, 75/653
  malabsorption, 75/653
Vagotonia
  body type, 22/252
Vagovagal reflex
  circuit, 75/625
  digestive neural reflex, 75/624
  dorsal, 75/624
Vagus nerve
  abnormality, 30/410
  acid secretion, 75/620
  anatomy, 2/77-79
  autonomic nervous system, 2/110, 74/62, 619,
    75/428, 529
  blood flow, 75/620
  bradycardia, 75/428
  congenital larynx stridor, 30/410
  embryology, 30/398-400
  epilepsy, 75/355
  exercise, 75/438
  heart rate, 74/619
  neurinoma, 68/540
  orthostatic hypotension, 63/144
  pain, 5/209

petrobasilar suture syndrome, 2/313
respiration, 1/655, 74/619
respiratory center, 63/431
subarachnoid hemorrhage, 55/18
tumor, 75/529
vagal paraganglioma, 75/232
vasodilatation, 75/428
vocal cord paralysis, 30/411
Vagus nerve agenesis
  cardiovascular abnormality, 50/219
  cervical meningocele, 50/219
  cranial nerve agenesis, 50/219
  dysphagia, 50/219
  esophagus abnormality, 50/219
  genetics, 50/219
  Möbius syndrome, 50/219
Vagus nerve dorsal nucleus
  Parkinson disease, 49/110
Vagus nerve injury
  Sicard-Collet syndrome, 24/179
Vagus stimulation
  CSF pressure, 19/113
Vail neuralgia, *see* Sphenopalatine neuralgia
Valine
  GM2 gangliosidosis, 10/296
Valine metabolism
  3-hydroxyisobutyric aciduria, 66/653
Valine transamination
  hypervalinemia, 42/574
Valium, *see* Diazepam
Valproic acid
  anencephaly, 50/80
  antiepileptic agent, 65/497
  carnitine deficiency, 62/494
  cognitive impairment, 72/395
  dyssynergia cerebellaris myoclonica, 60/603
  epilepsy, 73/357
  hemiballismus, 49/376
  metabolic pathway, 65/130
  neurotoxin, 37/200, 202, 49/217, 65/115, 125,
    497, 503
  Reye syndrome, 49/217, 56/159
  toxic encephalopathy, 49/217
Valproic acid induced hepatic necrosis
  acetyl-coenzyme A, 65/132
  acetylsalicylic acid, 65/127
  acyl-coenzyme A dehydrogenase, 65/126
  Alpers disease, 65/128
  anorexia, 65/125
  carnitine, 65/132
  carnitine depletion theory, 65/132
  coenzyme A depletion theory, 65/133
  coma, 65/125

diagnosis, 65/128
epidemiology, 65/127
epilepsy, 65/125
fatty acid hypothesis, 65/131
fatty acid metabolism, 65/132
fatty liver degeneration, 65/127
histopathology, 65/127
Huttenlocher variant, 65/128
hyperammonemia, 65/125
hyperbilirubinemia, 65/126
hyperventilation, 65/125
hypoglycemia, 65/126
hypoglycin, 65/126
jaundice, 65/125
lethargy, 65/125
3-methylenecyclopropylacetic acid, 65/126
mitochondria, 65/119, 127
mortality, 65/129
outcome, 65/129
pathogenesis, 65/131
pathomechanism, 65/129
4-pentenoic acid, 65/126
2-propyl-4-pentenoic acid, 65/126
Reye syndrome, 65/115
screening, 65/128
spontaneous hemorrhage, 65/125
symptom, 65/125
treatment, 65/129
ultrastructure, 65/127
Valproic acid intoxication
    asterixis, 65/504
    ataxia, 65/504
    blood ammonia, 65/504
    camptocormia, 65/505
    carbamoylphosphate synthase deficiency, 65/504
    carnitine deficiency, 65/504
    cognitive effect, 65/504
    dementia, 65/504
    digestive system, 37/203
    drowsiness, 37/202
    drug induced systemic lupus erythematosus,
        65/505
    EEG, 65/504
    extrapyramidal disorder, 37/202
    lipid metabolic disorder, 62/493
    myoclonus, 65/505
    nystagmus, 65/504
    parkinsonism, 65/504
    reticular activating system, 37/203
    stupor, 65/504
    systemic lupus erythematosus, 65/505
    tremor, 65/504
Valsalva maneuver
    autonomic nervous system, 74/615
    blood pressure, 74/615
    blood pressure overshoot, 1/471
    cardiovascular function, 74/597
    CSF pressure, 19/110
    exertional headache, 48/376
    ganglion blockade, 74/169
    heart rate, 74/615, 618
    hereditary neuropathy, 60/261
    method, 63/157
    migraine, 48/419
    orthostatic hypotension, 63/157
    polyneuropathy, 60/261
    pure autonomic failure, 22/233
    reflex vasomotor response, 2/114
    spinal cord injury, 61/445
    subarachnoid hemorrhage, 12/125
    tetraplegia, 61/445
    transverse spinal cord lesion, 61/445
    uremic polyneuropathy, 63/511
Valsalva ratio
    age, 74/598
    diabetic neuropathy, 74/604
    orthostatic hypotension, 63/157
Valsalva response
    amyloid polyneuropathy, 51/416
Valve disease
    heart, see Cardiac valvular disease
Valve mechanism
    hydromyelia, 50/428
Valvular heart disease
    antiphospholipid antibody, 63/330
    autonomic nervous system, 75/447
VAMP, see Synaptobrevin
Vampire bat
    rabies, 34/251, 254, 56/385
    rabies paralysis, 34/254
Vampire hormone
    darkness, 74/499
    light, 74/499
    melatonin, 74/499
    pineal gland, 74/499
Van Bogaert-Bertrand-Canavan disease, see CNS
        spongy degeneration
Van Bogaert-Bertrand classification
    diffuse sclerosis, 10/13
Van Bogaert classification
    diffuse sclerosis, 10/13, 15
Van Bogaert disease, see Neurocutaneous melanosis
Van Bogaert-Scherer-Epstein disease, see
        Cerebrotendinous xanthomatosis
Van Buchem disease, 38/406-410
    see also Hyperphosphatasemia

alkaline phosphatase, 31/258, 38/410, 43/411
autosomal recessive, 31/258
blindness, 43/410
cranial hyperostosis, 38/407, 43/410
diagnosis, 38/410
facial paralysis, 31/258, 38/408, 43/410
genetics, 38/410
headache, 31/258, 38/409, 43/410
hereditary cranial nerve palsy, 60/43
hereditary progressive cochleovestibular atrophy,
    60/774
jaw, 38/407
nonrecurrent nonhereditary multiple cranial
    neuropathy, 51/571
optic atrophy, 13/81, 31/258, 38/408, 43/410
progressive cochleovestibular atrophy, 22/494
sclerosteosis, 31/258
sensorineural deafness, 31/258, 38/408, 43/410
treatment, 38/410
Van den Bosch syndrome
    chorioretinal degeneration, 13/41
    optic atrophy, 13/41
Van der Hoeve-Halbertsma-Waardenburg-Gualdi
    syndrome, see Klein-Waardenburg syndrome
Van der Hoeve-Halbertsma-Waardenburg syndrome,
    see Klein-Waardenburg syndrome
Van der Hoeve-Waardenburg-Klein syndrome, see
    Klein-Waardenburg syndrome
Van Gehuchten, P., 1/10
Van Laere syndrome, see Brown-Vialetto-Van Laere
    syndrome
Van Leeuwenhoek, A., 2/10
Van Meekeren-Danlos-Ehlers syndrome, see
    Ehlers-Danlos syndrome
Van Valkenburg, C.T., 1/10
Vanillin
    relaxation, 74/365
Vanillylmandelic acid, see Vanilmandelic acid
Vanilmandelic acid
    dystonia, 42/214
    excretion, 8/478
    hereditary progressive diurnal fluctuating
        dystonia, 42/214
    hereditary sensory and autonomic neuropathy type
        III, 1/478, 21/109, 60/25, 28
    neuroblastoma, 8/478
    pheochromocytoma, 8/478, 481, 39/499, 42/765
Vapor
    mercury, see Mercury vapor
Vaquez disease
    athetosis, 6/458
Variable degenerates chorea, see Chorée variable
    des dégénérés

Varicella encephalitis, see Varicella zoster virus
    meningoencephalitis
Varicella zoster
    see also Herpes zoster
    acquired immune deficiency syndrome, 71/268
    acute cerebellar ataxia, 34/162-167, 622, 627
    classification, 34/161
    CNS infection, 69/448
    colchicine, 51/180
    epidemiology, 34/170
    history, 34/161
    immune mediated encephalomyelitis, 34/416-418
    morphology, 34/161
    myelitis, 34/167
    neuralgia, 5/323-325
    neurologic complication, 34/161-177
    postherpetic neuralgia, 51/180
    radiculitis, 71/268
    sensory radiculitis, 51/180
    spastic paraplegia, 59/436
Varicella zoster neuritis, see Herpes zoster
    ganglionitis
Varicella zoster virus
    see also Herpes zoster virus
    acquired immune deficiency syndrome, 56/514
    acute postinfectious disseminated
        encephalomyelitis, 56/229
    allergic granulomatous angiitis, 55/436
    brain vasculitis, 56/237
    cerebrovascular disease, 55/436
    congenital, see Chickenpox
    cranial nerve, 56/479
    deafness, 56/107
    DNA virus, 56/229
    facial paralysis, 56/229
    granulomatous arteritis, 56/237
    Guillain-Barré syndrome, 56/236
    hepatic encephalopathy, 56/149
    herpes virus, 56/8
    herpes virus family, 56/229
    herpes zoster, 56/229
    herpes zoster oticus, 56/118
    immunocompromise, 56/479
    immunocompromised host, 56/478
    interferon, 56/231
    neurotropic virus, 56/229
    Oka strain, 56/231
    optic neuritis, 56/229
    polyneuritis, 56/229
    postherpetic neuralgia, 5/323, 56/229
    postinfectious encephalomyelitis, 47/326
    reactivation, 56/231
    Reye syndrome, 27/363, 34/168, 56/149, 161, 229,

238
structure, 56/230
vasculitis, 56/480
vertigo, 56/107
virus dissemination, 56/231
Varicella zoster virus encephalitis
brain hemorrhagic infarction, 54/72
malignant lymphoma, 56/235
mortality, 56/479
Varicella zoster virus infection
animal model, 56/240
brain vasculitis, 55/436
cerebellar ataxia, 55/436
histology, 55/436
immunology, 56/241
insidious stroke, 55/437
latency, 56/241
neurologic symptom, 55/436
neurology, 56/229-243
pathogenesis, 56/239
Reye syndrome, 9/549, 33/293, 34/169, 49/217,
    56/163, 238, 63/438
spread, 56/240
viral meningitis, 55/436
Varicella zoster virus meningitis
CSF, 55/437
treatment, 55/437
Varicella zoster virus meningoencephalitis
acute postinfectious disseminated
    encephalomyelitis, 56/229
clinical features, 56/235
Cowdry A inclusion body, 56/235
CSF, 56/235
demyelination, 56/236
epilepsy, 56/235
headache, 56/235
hemorrhagic encephalitis, 56/235
hypoglycorrhachia, 56/236
neuropathology, 56/235
perivenous encephalomalacia, 56/235
Varicella zoster virus myelitis
ascending myelopathy, 56/479
Brown-Séquard syndrome, 56/236
CSF, 56/236
gangliomyelopathy, 56/236
neurologic complication, 56/229
neuropathology, 56/236
spastic paraplegia, 56/236
Varices
brain arteriovenous malformation, 12/234
Klippel-Trénaunay syndrome, 14/116, 393
Varicocele
spinal cord, 14/431

Varicosis medullae spinalis, 20/448
nomenclature, 20/481
pathologic anatomy, 20/440
spinal arteriovenous malformation, 20/481
Variegate porphyria, see Porphyria variegata
Variola virus
deafness, 56/107
neurotropic virus, 56/5
vertigo, 56/107
Varix
brain, see Brain varix
brain venous system, 12/234
venous malformation, 12/234
Varolius
history, 2/10
Vasa nervorum
adrenergic, 75/703
temporal arteritis, 48/314
Vasa nervorum disease
amyloidosis, 12/658
arteriosclerosis, 12/654
blood vessel, 12/645
diabetes mellitus, 12/655-658
hemorrhage, 12/655
ischemia, 12/651
peripheral nerve, 12/644-661
polyarteritis nodosa, 8/125, 161, 12/659
thromboangiitis obliterans, 12/655
Vascular autoregulation
brain edema, 23/138, 142, 144
brain injury, 23/138, 142, 144, 153
Vascular cell adhesion molecule 1
brain metastasis, 69/111
Vascular dementia, 46/353-365
allergic granulomatous angiitis, 46/359
aorta arch aneurysm, 46/359
arteriosclerosis, 46/357
arteriovenous malformation, 46/360
Binswanger disease, 46/359
borderzone infarction, 46/356
brain aneurysm, 46/360
brain anoxia, 46/357
brain atrophy, 46/364
brain blood flow, 46/362-364
brain hypoxia, 46/357
brain lacunar infarction, 46/355
carotid artery occlusion, 46/356
classification, 46/353
clinicopathologic correlation, 46/364
computer assisted tomography, 46/361
EEG, 46/360
multi-infarct dementia, 46/353-355
polyarteritis nodosa, 46/359

psychological testing, 46/364
regional brain blood flow, 46/362-364
spinal arachnoiditis, 46/360
subarachnoid hemorrhage, 46/360
systemic lupus erythematosus, 46/359
Takayasu disease, 46/357-359, 358
temporal arteritis, 46/357, 359
vasculitis, 46/359
vertebrobasilar insufficiency, 46/357
watershed infarction, 46/356
Vascular disease
  Benedikt syndrome, 2/298
  brain tumor, 69/413
  collagen, *see* Collagen vascular disease
  headache, 48/274
  hyperglycemia, 27/94
  mixed collagen, *see* Mixed connective tissue
    disease
  occlusive, *see* Occlusive vasculopathy
  Sjögren syndrome, 71/79
  spinal cord, *see* Spinal cord vascular disease
Vascular distensibility, 2/118
Vascular ectasia
  arterial hypertension, 63/84
  Ekbom other syndrome, 63/85
Vascular endothelial growth factor, *see*
  Vasculotropin
Vascular epilepsy, 72/209
  brain ischemia, 15/302
Vascular fibrosis
  multiple sclerosis, 47/246
Vascular headache
  cluster variant syndrome, 48/270
  coitus, 48/380
  heart transplantation, 63/179
  muscle contraction headache, 5/20
  psychiatric headache, 48/355
  trigeminovascular system, 48/107-113
Vascular hypertension
  see also Arterial hypertension
  acute paroxysmal, 5/153
  arterial pressure, 5/5
  facial paralysis, 8/283
  headache, 5/138, 149, 153
  migraine, 5/54, 56
  papilledema, 16/282, 291
  pontine hemorrhage, 12/49
  Raynaud phenomenon, 8/345
  stroke management, 12/453
Vascular hypotension
  see also Arterial hypotension
  headache, 5/154
  pure autonomic failure, 22/231-242

Vascular injury, *see* Arterial vascular injury
Vascular malformation
  brain, 18/270
  cryptic hemangioma, 12/104
  epilepsy, 72/115, 359
  headache, 48/287
  hydrocephalus, 30/543, 31/137, 145, 177, 193
  Parkes Weber syndrome, 14/396
  progressive bulbar palsy, 59/375
  spinal angiography, 19/231
  spinal cord, *see* Spinal cord vascular malformation
  spinal epidural angioma, 20/441
  spinal phlebography, 19/239
  spontaneous cerebellar hemorrhage, 12/54
  subarachnoid hemorrhage, 12/104, 110
Vascular mineralization
  calcium, 11/141
  iron, 11/141
  striatopallidodentate calcification, 11/141
Vascular myelopathy
  see also Myelomalacia
  anatomy, 12/478, 492
  blood flow decrease, 12/498
  progressive senile, *see* Progressive senile vascular
    myelopathy
  syndrome, 12/498
  transverse syndrome, 12/498
  vascular obstruction, 12/497
  venous origin, 12/501
  vertebral artery injury, 12/501
Vascular neuropathy
  nerve conduction, 7/165
Vascular permeability factor, *see* Vasculotropin
Vascular pseudopapillitis, *see* Anterior ischemic
  optic neuropathy
Vascular resistance
  acute hypoxia, 75/263, 266
  brain blood flow, 2/117, 11/124, 53/107
Vascular response
  adrenalin, 74/168
  ice cream headache, 5/190
  poliomyelitis, 34/107
Vascular spasm
  anterior cerebral artery syndrome, 53/341
  brain, *see* Brain vasospasm
  brain aneurysm, 12/209
  brain infarction, 53/163
  brain injury, 23/164, 180
  brain ischemia, 53/163
  brain microcirculation, 53/83
  carotid artery occlusion, 26/47
  hypertensive encephalopathy, 11/566, 571
  mechanism, 1/469

subarachnoid hemorrhage, 53/83
transient ischemic attack, 11/522, 53/264
Vascular striatopallidodentate calcification
brain aneurysm, 49/417
brain angioma, 49/417
epidural hematoma, 49/417
intracerebral hematoma, 49/417
subdural hematoma, 49/417
Vascular tumor
angioreticuloma, 20/449
brain tumor, 68/269
CNS tumor, 68/269
spinal cord, see Spinal vascular tumor
spinal cord compression, 19/362
survey, 68/269
Vascular vitreous veil
congenital, see Sex linked idiopathic retinoschisis
Vascularized ulnar nerve grafting
brachial plexus, 51/149
Vasculitic neuropathy
paraneoplastic syndrome, 69/374, 380
scleroderma, 71/112
Vasculitis, 71/7
see also Angiitis
allergic, see Allergic granulomatous angiitis
allergic granulomatous angiitis, 71/8, 155
anoxic ischemic leukoencephalopathy, 47/541
antibody therapy, 71/162
anticoagulation, 71/162
anti-inflammatory agent, 71/161
antiprostaglandin, 71/162
autoantibody, 71/152
autonomic nervous system, 75/11
azathioprine, 71/161
Behçet syndrome, 71/8
brain, see Brain vasculitis
brain lacunar infarction, 54/240
candidiasis, 52/401
chronic axonal neuropathy, 51/531
ciclosporin, 71/161
classification, 71/191
CNS granulomatous, see CNS granulomatous
vasculitis
coagulation, 71/153
corticosteroid, 71/160
cranial, see Cranial vasculitis
cryoglobulinemia, 71/8
cryoglobulinemic polyneuropathy, 51/434
CSF examination, 11/239
cyclophosphamide, 71/160
cytokine, 71/163
dapsone, 71/162
encephalopathy, 45/118

epilepsy, 46/359
etiology, 46/359
giant cell arteritis, 71/8
Goodpasture syndrome, 71/8
granulomatous, see Allergic granulomatous
angiitis
headache, 48/285
Henoch-Schönlein purpura, 71/8
history, 71/149
human T-lymphotropic virus type I associated
myelopathy, 59/450
hypereosinophilic syndrome, 63/373
idiopathic, see Idiopathic vasculitis
Ig, 71/161
IgG2 deficiency, 56/468
immune complex, see Immune complex vasculitis
immunopathogenic mechanism, 71/150
immunotherapy, 71/162
infection, 71/157
inflammation, 71/153
ischemic, see Ischemic vasculitis
Kawasaki syndrome, 56/641, 71/8
leprosy, 51/216
leukapheresis, 71/161
leukocyte, 71/150
leukocytoclastic, see Leukocytoclastic vasculitis
lupus, see Lupus vasculitis
meningococcal meningitis, 33/23, 52/26
methotrexate, 71/161
microscopic polyangiitis, 71/8
myelomalacia, 63/51
necrotizing, see Necrotizing vasculitis
neurobrucellosis, 52/585
neuroradiography, 71/158
neurosyphilis, 52/277
papilledema, 16/287
plasma exchange, 71/161
polymyositis, 41/56, 62/373
retinal, see Retinal vasculitis
rheumatoid, see Rheumatoid vasculitis
secondary, 71/157
spinal artery injury, 26/67
spinal cord injury, 26/67
stroke, 46/359
survey, 71/149, 191
systemic, see Systemic vasculitis
systemic lupus erythematosus, 71/38
Takayasu disease, 71/8
therapy, 71/159
toxin, 71/158
transient ischemic attack, 53/266
tuberculous meningitis, 52/199
tumor, 71/158

varicella zoster virus, 56/480
vascular dementia, 46/359
vertigo, 55/133
Wegener granulomatosis, 8/112, 71/8, 173
Vasculopathy
    Castleman disease, 63/395
    diamorphine intoxication, 65/356
    multiple myeloma, 63/395
    opiate intoxication, 65/356
    osteosclerotic myeloma, 63/395
    POEMS syndrome, 63/395
    radiotherapy, 39/52
    scleroderma, 71/105, 108
    systemic lupus erythematosus, 55/371
Vasculotropin
    brain metastasis, 69/113
Vasoactive intestinal peptide
    Alzheimer disease, 46/268
    basal ganglion, 49/33
Vasoactive intestinal polypeptide
    autonomic nervous system, 75/85
    brain blood flow, 53/63
    function, 74/475
    gastrointestinal disease, 39/451
    headache, 48/109
    Huntington chorea, 49/258
    hypothalamus, 74/482
    migraine, 48/99
    penis erection, 61/314
    sexual behavior, 74/498
    substance P, 48/99
    suprachiasmatic nucleus, 74/475, 477
Vasoactive intestinal polypeptide neuron
    age, 74/495
    biological rhythm, 74/495
    female, 74/495
    sex, 74/495
Vasoconstriction
    electrical injury, 74/321
    ephedrine, 75/723
    ergot intoxication, 65/63
    ergotamine, 48/185
    Godin syndrome, 5/330
    Guttmann sign, 2/190, 202
    headache, 5/190, 48/72
    ice cream headache, 5/190
    local anesthetic agent, 65/427
    mechanism, 1/469
    migraine, 48/72, 126
    muscle, 75/110
    orthostatic hypotension, 75/723
    Raynaud disease, 22/261
    respiration, 1/440

respiration reflex, 1/440
Vasoconstrictor neuron
    classification, 74/24
    muscle, see Muscle vasoconstrictor neuron
    visceral, see Visceral vasoconstrictor neuron
Vasoconstrictor pathway
    autonomic nervous system, 74/23
Vasodepressor syncope, see Vasovagal syndrome
Vasodilatation
    autonomic nervous system, 75/428
    bradykinin, 48/97
    cluster headache, 5/19
    hand, 1/438
    hypoxia, 75/264
    mechanism, 1/469
    papaverine intoxication, 37/456
    procainamide intoxication, 37/452
    Raynaud disease, 8/322, 22/261, 43/70
    skin radiation, 1/438
    substance P, 48/98
    syncope, 75/207
    therapeutic, 11/356, 512
    vagus nerve, 75/428
Vasodilator agent
    amyl nitrite intoxication, 37/453
    arterial hypertension, 54/209
    brain infarction, 11/356
    chronic muscle contraction headache, 48/350
    glyceryl trinitrate, 37/453
    nitrite, 37/453
    papaverine, 37/454-456
    thermoregulation, 75/61
Vasodilator neuron
    autonomic nervous system, 74/32
Vasoformative tumor
    tissue culture, 17/85
Vasogenic brain edema, 47/531-535
    anoxic ischemic leukoencephalopathy, 47/533
    blood-brain barrier, 53/138
    brain infarction, 11/155, 53/138
    brain ischemia, 53/138
    CNS, 28/426
    dexamethasone, 47/535
    experimental, 47/535, 63/213
    malignant hypertension, 11/155
    mechanical injury, 57/261
    metabolic encephalopathy, 57/213
    pathogenesis, 67/75
    toxic encephalopathy, 57/213
Vasomotor activity
    control, 26/259
    newborn, 1/468, 472, 475
Vasomotor control

newborn, 75/106
spinal cord injury, 26/259
spinal shock, 26/259
Vasomotor disorder
ECG, 24/579
lightning injury, 23/695
paralysis, 7/272
reflex, *see* Reflex vasomotor response
traumatic vegetative syndrome, 24/578
Vasopressin
accidental injury, 74/324
age, 74/495
Alzheimer disease, 46/268
annual rhythm, 74/490
biological rhythm, 74/473, 495
brain injury, 23/114, 28/499
brain stem death, 57/484
catamenial migraine, 5/232
circadian rhythm, 74/488
CSF, 74/487
epilepsy, 15/731
Huntington chorea, 49/258
medulla, 74/154
menstrual cycle, 74/491
motion sickness, 74/358
neurogenic arterial hypertension, 63/230
orthostatic hypotension, 63/145
Parkinson disease, 49/120
postlumbar puncture syndrome, 61/156
progressive dysautonomia, 59/151
secretion, *see* Inappropriate antidiuretic hormone
secretion
sexual behavior, 74/498
sodium metabolism, 28/498
subarachnoid hemorrhage, 28/499
suprachiasmatic nucleus, 74/477, 488, 495
toxic myopathy, 62/601
traumatic intracranial hematoma, 57/281
Vasopressin neuron
biological rhythm, 74/471
depression, 74/506
homosexual man, 74/501
Vasospastic disease
traumatic, *see* Vibration neuropathy
Vasovagal syncope
lumbar puncture, 61/165
migraine, 48/85
vertigo, 2/359
Vasovagal syndrome
see also Swallow syncope *and* Syncope
carotid sinus, *see* Carotid sinus syndrome
cervical spondylosis, 7/452
dysautonomia, 22/263

symptom, 2/359
tussive, 1/499
Vasovegetative lability
thalamic syndrome (Dejerine-Roussy), 2/482
VATER association
cardiofacial syndrome, 42/308
Vater-Pacini corpuscle
spinal cord enteric cyst, 32/438
VATER syndrome
see also Caudal aplasia
anal atresia, 50/513
esophagus atresia, 50/513
polydactyly, 50/513
radial aplasia, 50/513
renal malformation, 50/513
tracheoesophageal fistula, 50/513
vertebral malformation, 50/513
Vault tumor
cranial, *see* Cranial vault tumor
VCAM-1, *see* Vascular cell adhesion molecule 1
VDRL test, *see* Venereal Disease Research
Laboratory test
Vecuronium
intensive care, 71/536
multiple organ failure, 71/536
Vegetarian
deficiency neuropathy, 51/322
vitamin B12 deficiency, 51/339
Vegetative crisis
thalamic syndrome (Dejerine-Roussy), 2/482, 486
Vegetative dystonia syndrome
ionizing radiation, 74/322
stress, 74/322
Vegetative nervous function
blood flow, 1/466
intracranial pressure, 23/208
Vegetative nervous system, *see* Autonomic nervous
system
Vegetative state
brain stem death, 57/473
cardiac arrest, 63/218
Glasgow Outcome Scale, 57/132
persistent, *see* Persistent vegetative state
Vegetative syndrome
traumatic, *see* Traumatic vegetative syndrome
VEGF, *see* Vasculotropin
Vein
angular, *see* Angular vein
anterior cardinal, *see* Anterior cardinal vein
anterior vestibular, *see* Anterior vestibular vein
central cerebral, *see* Central cerebral vein
cerebral, *see* Cerebral vein
choroidal, *see* Choroidal vein

deep middle cerebral, *see* Deep middle cerebral vein

descending cerebral, *see* Descending cerebral vein

descending medial occipital, *see* Descending medial occipital vein

diploic, *see* Diploic vein

dorsal callosal, *see* Dorsal callosal vein

emissary, *see* Emissary vein

ethmoidal, *see* Ethmoidal vein

external jugular, *see* External jugular vein

frontal diploic, *see* Frontal diploic vein

frontopolar, *see* Frontopolar vein

Galen, *see* Galen vein

hippocampal, *see* Hippocampal vein

hypoglossal, *see* Hypoglossal vein

inferior cerebral, *see* Inferior cerebral vein

inferior choroidal, *see* Inferior choroidal vein

inferior frontal, *see* Inferior frontal vein

inferior ophthalmic, *see* Inferior ophthalmic vein

inferior posterior cerebellar, *see* Inferior posterior cerebellar vein

inferior ventricular, *see* Inferior ventricular vein

internal auditory, *see* Internal auditory vein

internal cerebral, *see* Internal cerebral vein

internal jugular, *see* Internal jugular vein

interpeduncular, *see* Interpeduncular vein

intracranial infection, 24/218

Labbé, *see* Labbé vein

lateral mesencephalic, *see* Lateral mesencephalic vein

meningeal, *see* Meningeal vein

middle frontal, *see* Middle frontal vein

middle meningeal, *see* Middle meningeal vein

occipital diploic, *see* Occipital diploic vein

ophthalmic, *see* Ophthalmic vein

orbital, *see* Orbital vein

parietal, *see* Parietal vein

parietal emissary, *see* Parietal emissary vein

petrosal, *see* Petrosal vein

pontine, *see* Pontine vein

posterior parietal, *see* Posterior parietal vein

precentral cerebellar, *see* Precentral cerebellar vein

primitive maxillary, *see* Primitive maxillary vein

Rolando, *see* Rolando vein

Rosenthal, *see* Rosenthal vein

septal, *see* Septal vein

spinal cord, *see* Spinal cord vein

superficial middle cerebral, *see* Superficial middle cerebral vein

superior frontal, *see* Superior frontal vein

superior ophthalmic, *see* Superior ophthalmic vein

superior petrosal, *see* Superior petrosal vein

superior posterior cerebellar, *see* Superior posterior cerebellar vein

thalamostriate, *see* Thalamostriate vein

trigeminal, *see* Trigeminal vein

Trolard, *see* Trolard vein

uncal, *see* Uncal vein

ventral diencephalic, *see* Ventral diencephalic vein

ventral myelencephalic, *see* Ventral myelencephalic vein

ventrolateral cerebellopontine, *see* Ventrolateral cerebellopontine vein

vertebral phlebography, 11/108

Vein aneurysm
  Galen, *see* Galen vein aneurysm

Vein calcification
  white matter, *see* White matter vein calcification

Vein of Galen, *see* Galen vein

Velocity
  motor conduction, *see* Motor conduction velocity
  motor nerve conduction, *see* Motor nerve conduction velocity
  nerve conduction, *see* Nerve conduction velocity
  sensory conduction, *see* Sensory conduction velocity

Velopalatine myoclonus
  pontine syndrome, 2/240

Venae capitis lateralis
  anatomy, 11/56

Venae capitis medialis
  anatomy, 11/56

Venereal Disease Research Laboratory test
  syphilis, 33/341, 52/276

Venerupis semidecussata intoxication
  abdominal pain, 37/64
  anorexia, 37/64
  blood clotting, 37/64
  constipation, 37/64
  headache, 37/64
  ingestion, 37/64
  mortality, 37/64
  nausea, 37/64
  petechial hemorrhage, 37/64
  vomiting, 37/64

Venezuelan equine encephalitis
  acute viral encephalitis, 34/73, 56/134, 137
  alphavirus, 56/12
  Culex mosquito, 56/137
  geography, 56/137
  headache, 56/137
  myalgia, 56/137
  teratogenic agent, 34/378

Venezuelan equine virus
  experimental acute viral myositis, 56/195

Venom
  arthropod, *see* Arthropod venom
  arthropod envenomation, 65/198
  bee, *see* Bee venom
  black widow spider venom intoxication, 41/105
  brown spider, *see* Brown spider venom
  cobra, *see* Cobra snake venom
  coral snake intoxication, 37/13
  crotoxin intoxication, 37/17
  cubomedusan intoxication, 37/48-50
  elapid intoxication, 37/12-16
  elapid snake, *see* Elapid snake venom
  envenomation, 37/12-16
  krait intoxication, 37/12, 15
  Lactrodectus, *see* Lactrodectus venom
  mamba intoxication, 37/13, 15
  molluscs intoxication, 37/60-62
  neuropharmacology, 37/13-16
  neurotoxin, 37/13
  octopus intoxication, 37/62
  saxitoxin intoxication, 37/56
  scorpion, *see* Scorpion venom
  sea snake intoxication, 37/56
  snake, *see* Snake venom
  stingray intoxication, 37/72
  stonefish intoxication, 37/69, 76
  tick, *see* Tick venom
  tiger snake, 37/13, 16
  viper, 37/16
  wasp, *see* Wasp venom
  zebrafish intoxication, 37/75
Venom X
  neurotoxin, 64/174
  organophosphate, 64/174
Venomous fish intoxication
  pisces, 37/69
Venomous snake
  Crotalidae, 65/177
  Elapidae, 65/177
  Hydrophiidae, 65/177
  Viperidae, 65/177
Venous angioma
  angiomatous malformation, 14/52
  brain, 18/272
  brain arteriovenous malformation, 53/35
  cerebellar hemorrhage, 12/55
  Cobb syndrome, 14/441
  epilepsy, 72/213
  pial, *see* Leptomeningeal venous angioma
  subarachnoid hemorrhage, 12/104, 55/2
Venous brain infarction, *see* Brain venous infarction
Venous drainage obstruction
  benign intracranial hypertension, 16/152

Venous lacuna
  anatomy, 11/52
Venous malformation
  see also Brain arteriovenous malformation
  calvaria, 12/236
  cardiomegaly, 12/232, 236
  dura mater, 12/236
  falx, 12/236
  Galen vein, 12/236
  scalp, 12/236
  Sturge-Weber syndrome, 12/234
  tentorium, 12/236
  varix, 12/234
  Von Hippel-Lindau disease, 12/236
Venous occlusion
  bacterial meningitis, 52/14
  Behçet syndrome, 56/595
  brain tumor, 69/421
  intracranial, *see* Intracranial venous occlusion
  ophthalmoscopy, 11/213
Venous plexus
  basilar, *see* Basilar venous plexus
  dural, *see* Dural venous plexus
  epidural, *see* Epidural venous plexus
Venous sinus
  anatomy, 11/52
  radioanatomy, 11/111
  topography, 11/111
Venous sinus occlusion
  age, 54/429
  carotid angiography, 54/442
  child, 54/426
  clinical features, 54/429
  cortical infarction, 54/429
  CSF, 54/441
  dehydration, 54/426
  EEG, 54/441
  epilepsy, 54/429
  erythrocyte sedimentation rate, 54/441
  fever, 54/429
  functional anatomy, 54/428
  gestation, 54/427
  headache, 54/429
  hemiplegia, 54/429
  incidence, 54/425
  India, 54/426
  infant, 54/426
  intracranial hypertension, 54/429
  meningeal infection, 54/426
  monoplegia, 54/429
  mortality, 54/427
  nuclear magnetic resonance, 54/445
  oral contraceptive agent, 54/426

pathology, 54/425, 437
pregnancy, 54/425
prognosis, 54/448
puerperal psychosis, 54/429
sinography, 54/442
treatment, 54/445
youth, 54/425
Venous sinus thrombophlebitis, 33/162-182
see also Brain thrombophlebitis
anticoagulant, 33/181
computer assisted tomography, 33/166
Crowe test, 33/165
CSF, 33/165
diagnosis, 33/165-167
incidence, 12/422
Queckenstedt test, 33/165
Tobey-Ayer test, 33/165
treatment, 33/181
Venous sinus thrombosis
see also Cavernous sinus thrombosis, Sagittal
   sinus thrombosis, Superior longitudinal sinus
   thrombosis *and* Transverse sinus thrombosis
adult, 12/434
adult cause, 12/434
adult histology, 12/434
akinetic mutism, 54/432
anemia, 12/431
angiography, 54/405
arteritis, 54/411
asparaginase, 63/345
bacterial meningitis, 55/420
Behçet syndrome, 71/224
benign intracranial hypertension, 54/411
brain collateral blood flow, 54/397
brain edema, 63/417
brain injury, 24/369
brain tumor headache, 69/24
caput medusae, 54/397
cause, 54/407
cerebrovascular disease, 53/31, 55/418
clinical features, 12/436, 54/430
clinical finding, 54/396
coexisting state, 54/409
coma, 54/395
computer assisted tomography, 54/65, 400, 443
cord sign, 54/402
cortical blindness, 54/432
cranial nerve palsy, 54/431
CSF, 54/399
deep leg vein thrombosis, 54/432
dense delta sign, 54/401
diagnosis, 54/399, 440
diffuse encephalopathy, 54/432

diffuse sclerosis, 12/433
digital venous angiography, 54/407
EEG, 54/399
epidemiology, 54/395
epilepsy, 54/395, 72/212, 215
epistaxis, 54/397
gestation, 54/427
Griesinger sign, 54/431
head injury, 24/369
headache, 5/154, 54/395, 429
heparin, 63/306
history, 12/423
hydrocephalus, 12/427
hypercoagulable state, 54/434
hypernatremia, 63/554
incidence, 12/422
infant, 12/429
infection, 12/431
intracranial hypertension, 54/395
intravascular consumption coagulopathy, 63/345
jugular venography, 54/407
Köhlmeier-Degos disease, 55/277
leukemia, 63/345
malignant lymphoma, 55/489
mechanical factor, 54/434
mortality, 54/427
negative delta sign, 54/402
nuclear magnetic resonance, 54/404, 445
otitic hydrocephalus, 54/431
paroxysmal nocturnal hemoglobinuria, 55/467
pathogenesis, 12/424, 54/434
polycythemia, 54/435
pregnancy, 54/425
prognosis, 12/442, 54/414, 448
pseudostroke, 54/432
puerperal psychosis, 54/432
radiology, 12/439
radionuclide scan, 54/400
septic origin, 52/172
sickle cell trait, 55/513
space occupying lesion, 16/235-237
Staphylococcus aureus, 54/431
Sturge-Weber syndrome, 55/446
subdural effusion, 24/333
superior longitudinal sinus, 12/427
symptom, 54/395
time at onset, 54/432
transverse sinus, 12/425
traumatic, *see* Traumatic venous sinus thrombosis
treatment, 12/443, 54/415, 445
type, 12/438
youth, 54/425
Venous thrombosis

brain arteriovenous malformation, 54/364
central retinal, *see* Central retinal venous
    thrombosis
congenital heart disease, 63/3
Guillain-Barré syndrome, 51/257
vitamin B$_{12}$ intoxication, 28/213
Venous tone
    autonomic involvement, 1/468
Ventilation
    see also Respiration
    assisted, *see* Assisted ventilation
    autonomic nervous system, 74/249
    blood gas analysis, 24/614
    brain injury, 24/612
    head injury, 57/229
    increase, 74/578
    increased, 74/587
    peripheral chemoreceptor, 74/587
    sleep, 74/533
Ventilator
    artificial, *see* Artificial ventilator
    brain stem death, 57/472
Ventilatory failure
    Guillain-Barré syndrome, 51/246
Ventral diencephalic vein
    anatomy, 11/60
Ventral globus pallidus
    see also Basal ganglion tier II
    amygdaloid nucleus, 49/26
    component, 49/22
    dorsomedial thalamic nucleus, 49/26
    function, 49/25
    neuroanatomy, 49/20
    neuroaxonal dystrophy, 49/392
    olfactory bulb, 49/20
    thalamic ventral lateral nucleus-anterior ventral
        nucleus complex, 49/26
Ventral horn inflammation
    carcinomatous neuromyopathy, 22/31
Ventral intermedius nucleus
    thalamic, *see* Thalamic ventralis intermedius
        nucleus
Ventral intermedius thalamotomy
    stereotaxic, *see* Stereotaxic ventral intermedius
        thalamotomy
Ventral lateral nucleus
    thalamic, *see* Thalamic ventral lateral nucleus
Ventral lateral thalamotomy
    ballismus, 49/77
Ventral myelencephalic vein
    anatomy, 11/60-62
Ventral neostriatum
    see also Basal ganglion tier III

component, 49/22
function, 49/25
neuroanatomy, 49/20
olfactory bulb, 49/20
Ventral pallidum, *see* Ventral globus pallidus
Ventral pontine syndrome
    classification, 12/22
Ventral spinal nerve root
    efferent, 2/160-165
    segment pointer muscle, 2/160-162
Ventral spinal nerve root avulsion
    see also Spinal nerve root avulsion
    mechanics, 7/407
Ventral spinocerebellar tract
    anterolateral funiculus, 45/232
Ventral striatum, *see* Ventral neostriatum
Ventral tegmental area
    akinesia, 49/67
    dopaminergic system, 49/9
    mesolimbic dopamine system, 49/50
    Parkinson disease, 49/110
Ventral thalamic peduncle
    anatomy, 6/33
Ventralis oralis nucleus
    lateral thalamic nuclear group, 2/472
    thalamic, *see* Thalamic ventral oral posterior
        nucleus
Ventricular cyst
    congenital ichthyosis, 43/406
Ventricular dilatation
    Alzheimer disease, 42/277
    amyotrophic dystonic paraplegia, 42/199
    Berardinelli-Seip syndrome, 42/592
    cerebellar dysarthria, 42/137
    cerebellar vermis agenesis, 42/4
    cerebro-oculofacioskeletal syndrome, 43/341
    Cockayne-Neill-Dingwall disease, 43/350
    Coffin-Lowry syndrome, 43/239
    corpus callosum agenesis, 42/10
    cri du chat syndrome, 50/275
    dermatoleukodystrophy, 42/488
    formiminotransferase deficiency, 42/546
    Fukuyama syndrome, 43/92
    hereditary benign chorea, 42/203
    hereditary essential tremor, 42/269
    Huntington chorea, 42/226
    hyperargininemia, 42/562
    hypertelorism, 43/417
    mental deficiency, 43/277
    myotonic dystrophy, 43/152
    4p partial monosomy, 43/497
    4p partial trisomy, 43/500
    Pena-Shokeir syndrome type I, 43/438

presenile dementia, 42/288
  Sjögren-Larsson syndrome, 43/307
  spina bifida, 32/532
Ventricular diverticulum
  arachnoid cyst, 31/125
  CSF, 31/125
  hydrocephalus, 50/290
Ventricular drainage
  brain, *see* Brain ventricle drainage
  head injury, 57/227, 232
  intracranial pressure, 57/227, 232
Ventricular enlargement
  bacterial meningitis, 52/8
  brain, *see* Brain ventricle enlargement
  subarachnoid hemorrhage, 12/152
Ventricular fibrillation
  mitral valve prolapse, 63/21
  paroxysmal, *see* Paroxysmal ventricular
    fibrillation
Ventricular hemorrhage
  brain, *see* Brain ventricle hemorrhage
  brain arteriovenous malformation, 54/374, 376
  hydrocephalus, 30/579
  subarachnoid hemorrhage, 12/167
Ventricular septal defect, *see* Cardiac ventricular
    septum defect
Ventricular system
  brain, *see* Brain ventricular system
  brain abscess, 33/118
  CNS, 30/32
  happy puppet syndrome, 31/309
Ventricular system atresia
  iniencephaly, 50/131
Ventricular tachycardia
  paroxysmal, *see* Paroxysmal ventricular
    tachycardia
Ventricular vein
  inferior, *see* Inferior ventricular vein
Ventriculoatrial shunt
  aqueduct stenosis, 50/314
  chronic subdural hematoma, 24/318
  hydranencephaly, 50/351
  hydrocephalic dementia, 46/329
  pinealoma, 18/365, 42/767
Ventriculocele
  cerebellar agenesis, 50/177
Ventriculocisternostomy
  aqueduct stenosis, 50/314
  hydrocephalus, 30/568
Ventriculography
  acute subdural hematoma, 24/286
  brain abscess, 33/124
  cerebellar hemorrhage, 12/59

  cerebellar tumor, 17/713
  chronic subdural hematoma, 24/312
  craniopharyngioma, 18/551
  craniovertebral region, 32/66
  frontal lobe tumor, 17/267
  hydrocephalus, 30/550-555
  metrizamide, *see* Metrizamide ventriculography
  midline shift, 57/217
  neurocysticercosis, 35/307
  pineal tumor, 17/660
  pinealoma, 18/365, 42/767
  supratentorial brain abscess, 33/124
  supratentorial brain tumor, 18/336
  third ventricle tumor, 17/465, 467, 469
Ventriculomegaly
  congenital heart disease, 63/11
Ventriculoperitoneal shunt
  aqueduct stenosis, 50/314
  Burkitt lymphoma, 63/352
  cranial subdural empyema, 52/170
Ventriculostium
  pia, 31/126
  subarachnoid space, 31/126
Ventriculostomy
  aqueduct stenosis, 50/314
  terminal, *see* Terminal ventriculostomy
Ventrobasal complex
  pain, 45/236
Ventrolateral cerebellopontine vein
  radioanatomy, 11/113
  topography, 11/113
Ventromedial hypothalamic nucleus
  anatomy, 2/434-436
Verapamil
  calcium antagonist, 65/440
  dystonia, 63/194
  encephalopathy, 63/194
  migraine, 48/200
  oculogyric crisis, 63/194
Veratrum alkaloid
  myasthenia gravis, 41/132
Verbal anosognosia
  body scheme disorder, 4/218, 45/378
  hallucination, 45/378
Verbal aphasia, *see* Motor aphasia
Verbal apraxia
  developmental dyspraxia, 4/459
  speech disorder in children, 4/459
Verbal auditory agnosia
  behavior disorder in children, 43/228, 72/9
  developmental dysphasia, 46/140
  EEG, 43/229
  Landau-Kleffner syndrome, 72/9, 73/282

572

seizure, 43/229
sex ratio, 43/229
speech disorder in children, 4/428, 43/228
stroke, 43/229
temporal lobe lesion, 43/229
Verbal comprehension
    brain hemisphere, 45/103
Verbal hallucination
    see also Auditory hallucination
    polyglot, 45/356
    word deafness, 45/357
Verbal memory
    temporal lobectomy, 45/44
    Wernicke-Korsakoff syndrome, 45/198
Verbal paraphasia
    agraphia, 4/169
    conceptual defect, 4/92
    semantic defect, 4/92
Verbal skill
    physical rehabilitation, 61/476
Verger-Dejerine syndrome
    parietal lobe syndrome, 45/65
Verhaart progressive supranuclear palsy, *see*
        Progressive supranuclear palsy
Verine
    chemistry, 37/365
Vermian aplasia, *see* Cerebellar vermis aplasia
Vermis cerebelli, *see* Cerebellar vermis
Vermouth
    neuropathy, 7/521
Vernet
    signe du rideau, *see* Signe du rideau de Vernet
Vernet-Sargnon syndrome, *see* Sicard-Collet
        syndrome
Vernet syndrome
    see also Jugular foramen syndrome
    nonrecurrent nonhereditary multiple cranial
        neuropathy, 51/571
    skull base tumor, 17/182, 68/469
Verocay nodule
    Antoni type A tumor, 14/22, 27
Verrucose brain dysplasia, *see* Status verrucosus
Verrucose retinal dysplasia
    status verrucosus, 50/203
Verruga peruana, 34/659-672
    choroid plexus, 34/667
    clinical features, 34/661-664
    CSF, 34/665
    diagnosis, 34/672
    encephalitis, 34/664
    ependymitis, 34/665
    etiology, 34/660
    laboratory finding, 34/664

leptomeninges, 34/665
meningoencephalitis, 34/664
pathogenesis, 34/661, 669-672
pathology, 34/665-669
prognosis, 34/672
sandfly, 34/659
spastic paraplegia, 59/436
treatment, 34/672
vascular lesion, 34/665
Versenate, *see* Edetic acid
Vertebra
    articular process, 25/258, 361, 365
    barrel shaped, *see* Barrel shaped vertebra
    blind, *see* Blind vertebra
    coin-shaped, *see* Coin-shaped vertebra
    eggshell, *see* Eggshell vertebra
    hemangiopericytoma, 20/443
    hemi, *see* Hemivertebra
    occipital, *see* Occipital vertebrae
    one eyed, *see* One eyed vertebra
    ossification center, 20/21
Vertebra plana, 19/166
    eosinophilic granuloma, 20/30
Vertebral abnormality
    see also Hemivertebra
    Aarskog syndrome, 43/314
    achondroplasia, 30/100, 43/316
    acrocephalosyndactyly type V, 43/320
    Aicardi syndrome, 42/696
    anencephaly, 43/12
    anterior sacral meningocele, 32/193, 206
    blepharoptosis, 43/488
    brain collateral blood flow, 11/173
    cardiofacial syndrome, 42/308
    caudal aplasia, 42/50
    centrofacial lentiginosis, 43/26
    cerebrocostomandibular syndrome, 43/237
    cleidocranial dysplasia, 43/349
    Coffin-Lowry syndrome, 43/238
    Conradi-Hünermann syndrome, 43/346
    corpus callosum agenesis, 42/7
    diastematomyelia, 42/24
    diastrophic dwarfism, 43/384
    Down syndrome, 43/544
    dwarfism, 43/381, 385
    Faulk-Epstein-Jones syndrome, 30/100
    fetal face syndrome, 43/400
    Forney syndrome, 30/101
    frontometaphyseal dysplasia, 31/257, 43/402
    fucosidosis, 42/550
    glycosaminoglycanosis type III, 42/456
    Goldenhar syndrome, 43/443
    hyperostosis corticalis deformans, 43/409

intraspinal neurenteric cyst, 32/426-429
Klippel-Feil syndrome, 30/100
Kuskokwim arthrogryposis, 43/85
meningocele, 32/193, 206
mental deficiency, 43/279, 288
multiple nevoid basal cell carcinoma syndrome,
    43/31
muscular atrophy, 42/82
myotonic myopathy, 43/165
osteitis deformans (Paget), 20/782, 43/450
4p partial monosomy, 43/497
20p partial trisomy, 43/539
spina bifida, 32/540
spondyloepiphyseal dysplasia, 43/478-482
spondylometaphyseal dysplasia, 43/482
Sprengel deformity, 43/483
Stickler syndrome, 43/484
Taybi-Linder syndrome, 43/385
thanatophoric dwarfism, 43/388
trigeminal nerve agenesis, 50/214
Vertebral agenesis
    r(13) syndrome, 50/585
Vertebral angioma, *see* Spinal angioma
Vertebral arterial malformation
    congenital abnormality, 12/289-335
Vertebral arteritis
    temporal arteritis, 48/322
Vertebral artery
    anatomy, 11/24, 25/233, 239
    anomalous origin, 11/173, 12/335
    anterior spinal branch, 11/25
    birth incidence, 25/168, 26/51
    brain aneurysm, 12/96
    cervical migraine, 5/196
    cervical vertebral column injury, 25/263, 61/33
    cervicomedullary injury, 24/145, 156-159, 168
    compression, 12/7, 19/153
    congenital absence, 32/53
    congenital aneurysm, 31/156
    cranial nerve, 26/55
    craniospinal injury, 25/263
    ectasia, 53/380
    ectodermal dysplasia, 53/380
    embryology, 11/24, 35
    endarterectomy, 12/11
    extracranial segment, 11/88
    fibromuscular dysplasia, 11/373, 53/378, 55/283,
        286
    football injury, 24/157-160
    head injury, 24/145, 156, 57/52
    headache, 5/132
    hypoplastic, 11/173, 176
    insufficiency, 12/592

intracranial branch, 53/379
intracranial segment, 11/90
ischemia, 5/199
lesion, 12/592
occlusive disease, 12/1
pain, 48/274
posterior spinal branch, 11/25
radioanatomy, 11/88-98
reserve loop, 11/90
segment, 53/374
spinal injury, 26/55-62
spondylotic compression, 53/378
syndrome, 7/453
temporal arteritis, 48/323, 53/379
transient ischemic attack, 12/9-11
traumatic insufficiency, 24/157
traumatic thrombosis, *see* Vertebral artery
    thrombosis
variation, 53/375
Vertebral artery aneurysm
    lateral medullary infarction, 12/96
Vertebral artery angiography, 11/65-98, 102-115
    brain aneurysm neurosurgery, 12/207
    brain tumor, 16/632
    cervical vertebral column injury, 25/304
    fibromuscular dysplasia, 11/373
    peduncular hallucination, 45/354
    transient global amnesia, 45/208, 214
Vertebral artery arteriosclerosis
    asymptomatic carotid artery stenosis, 53/376
    ataxia, 53/376
    headache, 53/376
    intracranial, 53/379
    neck motion, 53/377
    plaque site, 53/375
    symptom, 53/376
    treatment, 53/376
    vertigo, 53/376
    visual impairment, 53/376
Vertebral artery injury
    angiography, 26/58
    arteriovenous fistula, 53/377
    autopsy, 26/57
    birth, 26/52-54
    chiropractic manipulation, 53/377
    clinical features, 26/61
    hyperextension, 26/57, 60
    rotation injury, 26/58-60
    spine manipulation, 26/61
    thrombosis, 26/57-62
    vascular myelopathy, 12/501
Vertebral artery occlusion
    bilateral, 53/385

cerebellar infarction, 53/380
craniovertebral region, 32/55
lateral medullary infarction, 53/380
lateral medullary syndrome, 5/132
myelomalacia, 55/100
oral contraceptive agent, 53/379
sickle cell anemia, 55/506
symptom, 53/385
vertebral column injury, 61/527
Vertebral artery stenosis
  angiography, 12/11
  anticoagulant, 12/11
  carotid artery occlusion, 12/14, 23
  Doppler sonography, 54/1
  embolism, 12/11
  management, 12/11
Vertebral artery thrombosis
  see also Cerebrovascular disease
  atlas fracture, 25/263
  injury, 24/159
  temporal arteritis, 48/315
  traumatic, 24/159
Vertebral body defect
  intraspinal neurenteric cyst, 31/88, 32/426-429
  vertebral column, 32/137, 147
Vertebral body fracture
  see also Vertebral fracture
  cervical vertebral column injury, 61/23
Vertebral bony spur
  diastematomyelia, 50/435, 437, 439
  diplomyelia, 50/437, 439
Vertebral canal
  anatomy, 26/5, 34
  cervical, *see* Cervical spinal canal
  deformation during movement, 25/236, 26/11
  diameter, 19/141, 20/638
  Elsberg sign, 20/210
  epidermoid cyst, 31/89
  functional anatomy, 25/235
  interlocking, 25/334
  interpedicular diameter, 19/143-149
  measurement, 19/141
  multiple meningioma, 20/233
  radiographic enlargement, 19/141
  sagittal diameter, 19/145
  sagittal infant, 19/145
  stenosis, *see* Vertebral canal stenosis
  transverse diameter, 19/143-149
  traumatic deformation, 25/27-42
Vertebral canal dilation
  pilonidal sinus, 42/47
Vertebral canal stenosis, 20/617-801
  see also Cauda equina compression *and* Cauda

equina intermittent claudication
  achondroplasia, 19/85, 43/316
  ap diameter, 50/474
  Babin cloth peg sign, 50/466
  Babin index, 50/466
  cauda equina intermittent claudication, 19/149,
    20/617
  cervical region, 20/617
  cervical spondylarthrosis, 42/58
  cervical spondylarthrotic myelopathy, 50/469
  cervical stenosis, 50/468, 470
  combined stenosis, 50/466
  congenital, 50/465
  constitutional, 50/465
  craniovertebral stenosis, 50/468
  definition, 20/720
  dysostotic stenosis, 50/465
  dysplastic stenosis, 50/465
  Elsberg sign, 50/466
  epidemiology, 50/474
  four radiologic signs, 50/472
  genetics, 50/474
  hereditary multiple exostosis, 20/637
  Jones-Thompson index, 50/466
  lumbar, *see* Lumbar vertebral canal stenosis
  lumbar intervertebral disc prolapse, 20/597
  lumbar stenosis, 50/465, 471, 474
  management, 50/475
  osteochondrodystrophy, 20/637
  pathomorphology, 50/470, 472
  pedicular arch hyperplasia, 50/474
  posterior arch hypertrophy, 50/474
  rheumatoid disease, 50/474
  spondyloepiphyseal dysplasia, 43/480
  symptom, 50/468, 471
  syndrome, 20/598
  thoracic region, 20/618
  thoracic stenosis, 50/471
  traumatic, 61/42
Vertebral column, 32/131-157
  anencephaly, 50/83
  aneurysmal bone cyst, 19/300, 20/29, 442
  angioma, *see* Spinal angioma
  angiosarcoma, 20/35
  anterior element, 61/7, 17
  bamboo, *see* Bamboo spine
  benign osteoblastoma, 20/442
  biomechanics, 61/3-19
  bone disease, 19/166
  bone dysplasia, 32/138
  bone grafting, 32/154
  cervical, *see* Cervical spine *and* Cervical vertebral
    column

cervical spondylosis, 26/98
chloroma, 18/252
chordoma, 20/39
classification, 32/133-152
clinical features, 32/139-141
CNS, 32/141
compression fracture, 25/125
compression injury, 25/67
congenital lordosis, 32/148
conventional tomography, 19/139
coordinate system, 61/3
craniocervical region, 32/41
craniovertebral, *see* Craniovertebral region
diagnosis, 32/152
duplication, 32/138
epidemiology, 32/131
Ewing sarcoma, 20/34
experimental injury, *see* Spine experimental injury
generalized bone dysplasia, 32/138
giant cell tumor, 20/443
growth arrest, 32/156
hereditary multiple exostosis, 19/313-319
heterotopic endometrium, 20/443
injuries, *see* Spinal injury
joint disease, 19/166
kinematics, 61/3
Klippel-Feil syndrome, 32/111, 143
kyphosis, 32/147
lumbar, *see* Lumbar vertebral column
meningioma, *see* Spinal meningioma
multiple myeloma, 20/9-18, 39/139
myxosarcoma, 20/33
neural arch defect, 32/137
neurinoma, *see* Spinal neurinoma
neurofibroma, *see* Spinal neurofibroma
neurofibromatosis type I, 32/141-143
osteoid osteoma, 20/25
pain, 32/140
posterior element, 61/7, 17
rachischisis, 32/177
railway, *see* Railway spine
reticulosarcoma, 20/26, 33
reticuloxanthoma, 20/26
retroflexion, *see* Spine retroflexion
sagittal diameter, 61/10
scapula elevation, 32/144
scoliosis, 32/132, 148, 151, 177
segmentation failure, 32/133-137
solitary myeloma, 20/9, 35-39
spinal cord teratoma, 19/363
stab wound, 25/52
stability, *see* Clinical vertebral column stability
thoracic, *see* Thoracic vertebral column

three column concept, 61/95
traction, *see* Spine traction
traumatic, *see* Traumatic spine deformity
treatment, 32/153-157
vertebral body defect, 32/137, 147
Wolter typing, 61/95
Vertebral column injury
angiography, 61/515
burst fracture, 61/519
canal encroachment, 61/91
cause, 61/21
cervical, *see* Cervical vertebral column injury
cervical ligament, 61/517
child, *see* Juvenile vertebral column injury
chronic backache, 61/532
computer assisted tomography, 61/515
data base, 61/21
dura mater rupture, 61/96
facet fracture, 61/523
hangman fracture, 61/519
interlocking facet, 61/524
intervertebral disc prolapse, 61/522
ligament rupture, 61/522
myelography, 61/515
nuclear magnetic resonance, 61/515
odontoid fracture type, 61/518
pathology, 61/21
perched facet, 61/524
Pillar fracture, 61/522
radiology, 61/515
segmental frequency distribution, 61/22
segmental pathology difference, 61/22
tethered cord, 61/529
thoracolumbar, *see* Thoracolumbar vertebral
  column injury
three column concept, 61/91
vertebral artery occlusion, 61/527
vertebral subluxation, 61/522
vertebral teardrop fracture, 61/519
vertebral teardrop fracture type, 61/518
Vertebral column syndrome
hourglass tumor, 20/298
spinal meningioma, 20/207
spinal neurinoma, 20/257
Vertebral column tumor
see also Spinal tumor
angiography, 20/23
benign skeletal tumor, 20/24
bronchopulmonary carcinoma, 20/43
chondroma, 20/25
chondrosarcoma, 20/32
classification, 20/20
clinical features, 20/20-22

death course, 20/236
death rate, 20/237
dysphagia, 20/30
eosinophilic granuloma, 20/30
epidural, *see* Spinal epidural tumor
etiology, 20/20
immediate postoperative period, 20/819
intervertebral disc, 20/21, 33
laboratory, 20/24
mammary carcinoma, 20/43
myelography, 20/23
osteoblastoma, 20/26
osteolytic type, 20/22
osteosarcoma, 20/32
primary tumor, 20/19-40
proliferative type, 20/23
prostatic carcinoma, 20/42
rehabilitation, 20/819-845
renal carcinoma, 20/43
roentgenology, 20/22
secondary tumor, 20/40-54
site and malignancy, 20/21
skin care, 20/820
thyroid carcinoma, 20/43
treatment, 20/50
Vertebral compression
    see also Spinal cord compression
    en galette, 20/22
    myeloma, 20/11
    spinal cord compression, 19/365
    spinal histiocytosis X, 19/365
Vertebral compression fracture, *see* Compression
    fracture
Vertebral dissecting aneurysm
    angiography, 54/278
    bilateral, 54/278
    brain infarction, 53/39
    cerebellar infarction, 54/281
    cerebrovascular disease, 53/39
    cervical pain, 53/205
    chiropractic manipulation, 53/377, 54/276
    fibromuscular dysplasia, 54/277
    head rotation, 54/276
    headache, 54/279
    histology, 54/281
    homocystinuria, 54/282
    intracranial, 53/380
    Marfan syndrome, 54/282
    migraine, 54/277
    neck trauma, 53/377
    polyarteritis nodosa, 54/282
    sex ratio, 54/277
    spontaneous, 53/39

    subarachnoid hemorrhage, 54/279
    syphilitic arteritis, 54/282
    thunderclap headache, 54/277
    trauma, 53/39
    treatment, 54/283
Vertebral enchondroma
    spinal cord compression, 19/367
Vertebral end plate injury
    3 types, 25/508
Vertebral exostosis
    uncovertebral joint, 7/447
Vertebral experimental injury
    acceleration, 25/4
    cervical spine, 25/3
    intervertebral disc, 25/1
    ligament, 25/2
    model, 25/1
    spinal unit, 25/2
Vertebral facet
    cervical, *see* Cervical vertebral facet
Vertebral fracture
    see also Vertebral body fracture *and* Vertebral
        wedge compression fracture
    back pain, 19/86
    burst type, 61/78
    cervical vertebral column injury, 61/28
    compression type, 61/77
    flexion type, 61/77
    unstable, 70/30
    vertebral column concept, 61/90
Vertebral fracture instability
    criteria, 26/287
Vertebral fusion, 14/459
    ankylosing spondylitis, 26/182
    blepharophimosis, 43/488
    bony, *see* Bony fusion
    cervical spondylotic myelopathy, 26/108
    contracture, 43/488
    diastematomyelia, 50/436
    diplomyelia, 50/436
    lactate dehydrogenase, 43/489
    ptosis, 43/488
    spinal deformity, 26/171
    spinal injury, 26/213
    spondylotic radiculopathy, 26/107
Vertebral hemangioma
    pathologic anatomy, 20/438-441
    plain X-ray, 20/453
    pregnancy, 20/441
    spastic paraplegia, 59/438
    spinal cord compression, 19/368
Vertebral hydatid disease
    echinococcosis, 35/194-198

Vertebral hypoplasia
  kyphoscoliosis, 42/61
Vertebral interpedicular widening
  diastematomyelia, 50/439
  diplomyelia, 50/439
Vertebral joint ganglion
  spinal cord compression, 20/128
  spinal epidural cyst, 20/163
Vertebral lesion
  cancer, 69/57
  spinal cord compression, 19/364
Vertebral malformation
  VATER syndrome, 50/513
Vertebral manipulation, *see* Spine manipulation
Vertebral metastasis, 20/415-430
  age, 20/416
  bronchial carcinoma, 20/423
  carcinoma, 20/415
  case series, 20/415
  chemotherapy, 20/421
  clinical features, 20/418
  cordotomy, 20/421
  CSF, 20/419
  decompressive laminectomy, 20/420
  diagnosis, 71/631
  dissemination, 20/426
  dorsal horn pain, 20/427
  EMG, 20/420
  fibroplastic type, 20/188
  hypernephroma, 20/423
  location, 20/416
  mammary carcinoma, 20/424
  melanoma, 20/415-417
  motor symptom, 20/428
  myelography, 20/420
  origin, 20/417, 423
  pain, 20/418, 427, 69/61
  pain pattern, 69/61
  pathogenesis, 71/631
  prognosis, 20/422
  prostatic carcinoma, 20/425
  radicular pain, 20/427
  radiology, 20/420
  radiotherapy, 20/421
  renal carcinoma, 20/423
  sarcoma, 20/415, 425
  sex ratio, 20/415
  spastic paraplegia, 59/437
  sphincter disturbance, 20/418
  spinal cord compression, 19/369
  surgical treatment, 20/420
  thyroid carcinoma, 20/424
  treatment, 20/429, 71/632

Wolfson syndrome, 20/428
Vertebral myeloma
  history, 20/9
  macroglobulinemia, 20/11
  melphalan, 20/14, 17
  postmenopausal osteoporosis, 20/12
  roentgenology, 20/12
  senile osteoporosis, 20/12
  urethan, 20/17
Vertebral osteolysis
  radiography, 19/166
  spinal cord compression, 19/166
Vertebral osteoma
  spinal cord compression, 19/367
Vertebral osteomyelitis
  diamorphine intoxication, 65/355
  opiate intoxication, 65/355
  spinal cord abscess, 52/190
  spinal cord compression, 19/369
  spinal epidural empyema, 52/73, 186
  spinal subdural empyema, 52/189
Vertebral osteosclerosis, 19/170
Vertebral pain
  extramedullary tumor, 19/54
Vertebral pedicle
  congenital absence, 19/163
  metastatic carcinoma, 20/12
  myeloma, 20/12
Vertebral phlebography
  ascending occipital cerebral vein, 11/108
  diagram, 11/109
  spinal neurinoma, 20/277
  vein, 11/108
Vertebral scalloping
  fibrous dysplasia, 14/193
  metastatic carcinoma, 20/13
  multiple myeloma, 20/13
  neurofibromatosis type I, 14/158
  osteitis deformans (Paget), 14/193
  spinal tumor, 19/63
Vertebral subluxation
  glycosaminoglycanosis type IH, 66/285
  ischemic myelopathy, 63/43
  Pillar fracture, 61/522
  vertebral column injury, 61/522
Vertebral teardrop fracture
  cervical segment, 61/29
  cervical vertebral column injury, 25/346-349
  cervicomedullary injury, 24/163
  disc lesion, 61/519
  extension injury, 61/9
  flexion injury, 61/9
  hyperextension type, 61/519

hyperflexion type, 61/519
ligament lesion, 61/519
radiographic criteria, 61/8
type, 61/518
vertebral column injury, 61/519
Vertebral torsion
scoliosis, 50/417
Vertebral wedge compression fracture
see also Vertebral fracture
thoracolumbar vertebra, 61/35
thoracolumbar vertebral column injury, 61/35
Vertebrata intoxication
chordata, 37/68
Vertèbre d'ivoire, see Ivory vertebra
Vertebrobasilar aneurysm
brain aneurysm neurosurgery, 55/69
frequency, 2/257
Vertebrobasilar arterial system
anatomy, 12/1
transient ischemic attack, 12/2
Willis circle, 12/24
Vertebrobasilar artery thrombosis
streptokinase, 54/181
tissue plasminogen activator, 54/181
Vertebrobasilar insufficiency, 12/1-12
see also Vertebrobasilar system
ankylosing spondylitis, 38/513
anticoagulant, 53/385
ataxia, 53/262
Blalock-Taussig anastomosis, 63/5
cerebrovascular disease, 55/132
drop attack, 2/281, 360, 53/262
EEG, 54/158
EEG fast rhythm, 54/158
hemiballismus, 49/373
labyrinthine ischemia, 55/132
periodic ataxia, 21/570
spondylotic compression, 53/378
streptokinase, 53/385
temporal arteritis, 55/344
thalamic hemorrhage, 54/309
thrombosis, 2/245, 251
vascular dementia, 46/357
vertigo, 2/360, 55/132
Vertebrobasilar stenosis
drop attack, 12/23
Vertebrobasilar system
see also Vertebrobasilar insufficiency
anatomy, 11/210
anticoagulant, 53/385
antithrombocytic agent, 53/385
basilar artery supply area, 53/387
embryology, 32/45

lateral long type, 11/28
lateral penetrating artery, 11/28
lateral short type, 11/28
median penetrating branch, 11/28
sickle cell anemia, 55/509
transient ischemic attack, 53/261
Vertebrobasilar system syndrome, 53/371-397
abulia, 53/395
amnesia, 53/392, 395
anterior inferior cerebellar artery syndrome,
53/396
arteriosclerosis, 53/371, 375
ataxia, 53/373, 376, 390, 394, 397
ataxic hemiparesis, 53/394
basilar apex syndrome, 53/392
behavior, 53/395
brain lacunar infarction, 53/394
cerebellar infarction, 53/380
cognition, 53/395
Collier sign, 53/391
computer assisted tomography, 53/392
corectopia iridis, 53/391
diplopia, 53/373
dysarthria, 53/394
facial paresthesia, 53/390, 394
fibromuscular dysplasia, 53/378
headache, 53/373, 376
hemichorea, 53/397
Horner syndrome, 53/397
hyperconvergence, 53/391
hypersomnia, 53/395
internuclear ophthalmoplegia, 53/397
ischemic arm exercise, 53/373
lateral medullary infarction, 53/380
meiosis, 53/390
nuclear magnetic resonance, 53/392
nystagmus, 53/390, 394
ocular bobbing, 53/390
oral contraceptive agent, 53/379
palatal myoclonus, 53/390
peduncular hallucination, 53/392
perioral paresthesia, 53/373
pontine infarction, 53/388
pseudoabducens nerve sign, 53/391
ptosis, 53/390
pure motor hemiplegia, 53/394
pure sensory stroke, 53/395
skew deviation, 53/391, 397
somnolence, 53/391
streptokinase, 53/385
superior cerebellar artery, 53/397
symptom, 53/376
temporal arteritis, 53/379

treatment, 53/376
verbal perseveration, 53/395
vertical gaze palsy, 53/391, 395
vertical nystagmus, 53/397
vertigo, 53/373, 376
visual impairment, 53/373, 376
Vertebrobasilar transient ischemic attack
  brain stem auditory evoked potential, 54/162, 164
  computer assisted tomography, 54/53
  EOG, 54/164
  subclavian steal syndrome, 12/11
Vertebrocerebellar artery (Critchley-Schuster), *see*
  Posterior inferior cerebellar artery
Vertebrogenic syndrome
  lumbar vertebral canal stenosis, 20/681
Vertical axis sparing
  occipital lobe tumor, 17/321
  visual field, 17/321
Vertical gaze palsy
  Bell sign, 2/338
  brain atrophy, 42/188
  brain embolism, 53/212
  brain infarction, 53/212
  brain lacunar infarction, 53/395, 54/245
  cause, 1/602
  cerebellar atrophy, 42/188
  cerebrovascular disease, 53/212, 55/123
  climbing eye movement, 2/338
  Darkschewitsch nucleus, 2/338
  double athetosis, 49/384
  Huntington chorea, 49/280
  intracranial hemorrhage, 53/212
  kernicterus, 49/384
  lesion site, 53/212
  Marinesco-Sjögren syndrome, 60/343, 656
  mesencephalic infarction, 53/391
  1-methyl-4-phenyl-1,2,3,6-tetrahydropyridine
    intoxication, 65/372
  muscular atrophy, 42/188
  olivopontocerebellar atrophy (Dejerine-Thomas),
    60/656
  olivopontocerebellar atrophy (Wadia-Swami),
    60/656
  pallidoluysionigral degeneration, 49/451, 456
  pallidoluysionigrothalamic degeneration, 49/456
  Parinaud syndrome, 2/337
  progressive supranuclear palsy, 22/219, 42/259,
    49/239
  spinocerebellar ataxia, 42/188
  thalamic infarction, 53/395
  transient ischemic attack, 53/212
  vertebrobasilar system syndrome, 53/391, 395
Vertical meridian

cyclopic vision, 3/31
visual field, 3/31
Vertical nystagmus, 42/158-161
  brain embolism, 53/213
  brain infarction, 53/213
  brain stem lesion, 16/322
  cerebrovascular disease, 53/213
  hypomagnesemia, 63/562
  intracranial hemorrhage, 53/213
  late cerebellar atrophy, 42/135
  oscilloscopsia, 42/159
  transient ischemic attack, 53/213
  vertebrobasilar system syndrome, 53/397
  visual disorder, 42/159
Vertical ocular myoclonus
  lesion site, 53/213
Vertical one and half syndrome
  brain embolism, 53/212
  brain infarction, 53/212
  cerebrovascular disease, 53/212
  intracranial hemorrhage, 53/212
  lesion site, 53/212
  transient ischemic attack, 53/212
Vertical supranuclear ophthalmoplegia
  β-*N*-acetylhexosaminidase A deficiency, 60/657
  brain atrophy, 42/605
  cerebellar ataxia, 42/605
  β-galactosidase, 42/606
  hepatosplenomegaly, 42/605
  neurovisceral storage disease, 42/604-606
  Niemann-Pick disease type C, 60/657
  sea blue histiocyte, 42/605
  seizure, 42/605
  sphingomyelinase, 42/605
Vertiginous illusion
  temporal lobe epilepsy, 73/72
Vertiginous neuraxitis, *see* Vestibular neuronitis
Vertigo, 2/358-390
  see also Dizziness
  acetylsalicylic acid intoxication, 65/460
  acoustic nerve tumor, 17/670
  adenovirus, 56/107
  alcohol intoxication, 64/112, 115
  alternobaric, *see* Alternobaric vertigo
  anterior inferior cerebellar artery syndrome,
    55/132
  aprindine intoxication, 65/455
  ataxic hemiparesis, 54/248
  aural origin, 2/362-365
  barbiturate intoxication, 65/332
  Barré syndrome, 2/219
  basilar artery migraine, 48/136
  basilar ischemia, 12/4

580

benign paroxysmal positional nystagmus, 2/370
benign recurrent, 48/157
benzodiazepine intoxication, 65/338, 343
Binswanger disease, 54/222
blood hyperviscosity, 55/484
brain abscess, 52/152
brain aneurysm, 55/134
brain arteriovenous malformation, 55/134
brain embolism, 53/207
brain infarction, 53/207
brain injury, 23/11
brain tumor, 16/129
buspirone intoxication, 65/344
calcium antagonist intoxication, 65/441
carbamazepine intoxication, 65/505
caulerpa intoxication, 37/97
caulerpicin intoxication, 37/97
cause, 53/235, 240
cerebellar hemorrhage, 55/91
cerebellar migraine, 5/78, 48/157
cerebellar tumor, 17/710
cerebellopontine angle syndrome, 2/94
cerebellopontine angle tumor, 2/377
cerebrovascular disease, 53/207
chelonia intoxication, 37/89
chlorophyceae intoxication, 37/97
cholesteatoma, 2/362
cinchonism, 65/452
classic migraine, 48/157
cluster, see Cluster vertigo
cluster headache, 48/270
Cogan syndrome type II, 2/360, 55/133
common migraine, 48/157
congenital nystagmus, 2/361
Coxsackie virus, 56/107
cycloserine, 65/481
cycloserine intoxication, 65/481
cytomegalovirus, 56/107
decaborane intoxication, 64/360
decompression illness, 63/419
diborane intoxication, 64/359
dipeptidyl carboxypeptidase I inhibitor
    intoxication, 65/445
diuretic agent intoxication, 65/440
encainide intoxication, 65/456
encephalomyocarditis virus, 56/107
epidemic, see Vestibular neuronitis
epilepsy, 2/359
Epstein-Barr virus, 56/107
ethylene glycol intoxication, 64/123
felbamate intoxication, 65/512
flecainide intoxication, 65/456
furosemide, 65/440

gabapentin intoxication, 65/512
green seaweed intoxication, 37/97
gyromitrine intoxication, 65/36
head injury, 24/121, 126
hepatitis virus, 56/107
herpes simplex virus, 56/107
herpes zoster oticus, 2/360
heterocyclic antidepressant, 65/320
hyperventilation syndrome, 63/433
hyperviscosity syndrome, 55/484
influenza virus, 56/107
infratentorial lesion, 2/359
intracranial hemorrhage, 53/207
isoniazid intoxication, 65/480
labyrinthine hemorrhage, 55/133
lamotrigine intoxication, 65/512
lateral medullary infarction, 53/381
legionellosis, 52/254
lidocaine intoxication, 65/454
lymphocytic choriomeningitis virus, 56/107
marine toxin intoxication, 65/153
mefloquine intoxication, 65/485
Meniere disease, 2/362-364, 42/402
methyldopa, 65/448
methyl salicylate intoxication, 37/417
metronidazole intoxication, 65/487
migraine, 2/359, 5/47, 48/38, 156, 55/132
multiple sclerosis, 2/359, 9/168, 47/56
mumps virus, 56/107
neurosyphilis, 52/278
nonsteroid anti-inflammatory agent intoxication,
    65/416, 563
organotin intoxication, 64/138
orientation, 3/206
osteitis deformans (Paget), 2/362, 38/364
otosclerosis, 2/365
parainfluenza virus, 56/107
paralytic shellfish poisoning, 65/153
parietal lobe epilepsy, 73/101
pellagra, 28/86
pentaborane intoxication, 64/359
perilabyrinthitis, 2/365
periodic, see Periodic vertigo
periodic vestibulocerebellar ataxia, 42/116
phenytoin intoxication, 65/500
platybasia, 2/362
poliomyelitis, 56/107
polycythemia, 63/250
polycythemia vera, 55/469, 63/250
pontine infarction, 2/241, 12/15, 22
posterior inferior cerebellar artery occlusion,
    55/90
posttraumatic syndrome, 24/686

procainamide intoxication, 37/452
propafenone intoxication, 65/457
pseudo-Meniere syndrome, 7/485
psychogenic cause, 2/359
quinidine intoxication, 37/448
Ramsay Hunt syndrome, 2/360
recurrent, 48/138
reserpine intoxication, 37/438
rubella virus, 56/107
salicylic acid intoxication, 37/417
sodium salicylate intoxication, 37/417
St. Louis encephalitis virus, 56/107
subarachnoid hemorrhage, 55/134
subclavian steal syndrome, 53/373
superior cerebellar artery syndrome, 12/20
supratentorial lesion, 2/359
syringobulbia, 2/361
temporal bone fracture, 24/121, 125
temporal lobe tumor, 17/293
tetrodotoxin intoxication, 65/156
tick-borne encephalitis virus, 56/107
tocainide intoxication, 65/455
transient ischemic attack, 12/4, 53/207, 263
transverse sinus thrombosis, 52/176
traumatic vegetative syndrome, 24/577
trazodone intoxication, 65/320
tsukubaenolide intoxication, 65/558
tumor, 2/361
varicella zoster virus, 56/107
variola virus, 56/107
vasculitis, 55/133
vasovagal syncope, 2/359
vertebral artery arteriosclerosis, 53/376
vertebrobasilar insufficiency, 2/360, 55/132
vertebrobasilar system syndrome, 53/373, 376
vestibular neurinoma, 68/425
vestibular neuronitis, 2/364, 56/116
vigabatrin intoxication, 65/511
viral infection, 56/105
vitamin A intoxication, 37/96, 65/568
Vogt-Koyanagi-Harada syndrome, 2/360
Von Hippel-Lindau disease, 14/115
Waldenström macroglobulinemia, 63/398
western equine encephalitis virus, 56/107
yellow fever virus, 56/107
Very late antigen 4
  brain metastasis, 69/111
Very long chain acyl-coenzyme A dehydrogenase
  inner mitochondrial membrane system, 66/406
  membrane long chain acyl-coenzyme A
    dehydrogenase, 66/409
Very long chain acyl-coenzyme A dehydrogenase
  deficiency

inner mitochondrial membrane system, 66/407
Very long chain acyl-coenzyme A dehydrogenase
  gene
  inner mitochondrial membrane system, 66/407
Very long chain acyl-coenzyme A dehydrogenase
  protein
  inner mitochondrial membrane system, 66/407
Very long chain fatty acid
  adrenoleukodystrophy, 47/597, 51/384
  biochemistry, 66/519
  peroxisomal acetyl-coenzyme A acyltransferase
    deficiency, 66/518
  peroxisomal disease, 60/674
Very long chain fatty acid β-oxidation
  adrenoleukodystrophy, 60/169
Very low density lipoprotein
  Bassen-Kornzweig syndrome, 63/275
  brain arteriosclerosis, 53/97
  clofibrate, 41/272
  hyperlipoproteinemia type III, 66/541
Vesalius, A., 1/3, 2/8
  history, 2/10
Vesical plexus
  anatomy, 1/361
Vesicle associated membrane protein, see
  Synaptobrevin
Vesicourethral neuropathy
  bladder capacity, 61/302
  bladder compliance, 61/302
  bladder contractility, 61/303
  bladder neck resection, 61/304
  bladder reflex activity, 61/301
  bladder sensation, 61/302
  classification, 61/291
  complication, 61/308
  Crede expression, 61/300
  cutaneous urinary diversion, 61/307
  detrusor dyssynergia, 61/296
  detrusor function, 61/296
  detrusor stimulation, 61/305
  external sphincterotomy, 61/304
  functional diagnosis, 61/299
  hyperactive bladder, 61/302
  hyperactive urethra, 61/302
  hypoactive bladder, 61/302
  hypoactive urethra, 61/302
  infection, 61/308
  intramural ureter obstruction, 61/308
  investigation, 61/296
  lesion level, 61/291, 295
  mixed type, 61/295
  pharmacologic test, 61/297
  pharmacotherapy, 61/303

residual urine volume, 61/296
sacral anterior root stimulation, 61/305
sacral nerve root electrostimulation, 61/303
sacral type, 61/295
suprasacral type, 61/295
surgical treatment, 61/303
ultrasound scan, 61/296
urethral outflow, 61/305
urethral pressure profilometry, 61/297
urethral relaxation, 61/305
urethral resistance enhancement, 61/307
urethral resistance restoration, 61/306
urinary calculi, 61/308
urodynamic study, 61/296
uroflow measurement, 61/297
vesicoureteric reflux, 61/308
videocystography, 61/296
Wall stent, 61/305
Vesicular stomatitis virus
experimental acute viral myositis, 56/195
neurotropism, 56/35
Vestibular apparatus
radiation injury, 7/394, 398
Vestibular complex syndrome
terminology, 2/417
Vestibular disorder
Friedreich ataxia, 21/337, 387
hereditary motor and sensory neuropathy type I,
21/284, 308
hereditary motor and sensory neuropathy type II,
21/284, 308
hereditary spastic ataxia, 21/383-389
temporal lobe tumor, 17/289
Vestibular function
acoustic nerve tumor, 17/676
autonomic nervous system, 74/351
caloric test, 16/324
examination sheet, 2/388
Friedreich ataxia, 60/658
gait test, 2/366
late onset ataxia, 60/658
motion sickness, 74/351
Northern Swedish recessive ataxia, 60/658
optokinetic nystagmus, 2/706
progressive spinal muscular atrophy, 22/40
spino-olivocerebellar degeneration, 60/658
stance, 2/366
survey, 74/351
temporal lobe epilepsy, 2/706
test, 2/365-390
tone, 16/302
Wernicke-Korsakoff syndrome, 28/248
Vestibular nerve atrophy

hereditary sensory and autonomic neuropathy type
I, 60/9
Vestibular neuraxitis, see Vestibular neuronitis
Vestibular neurinoma, 68/480
audiometry, 68/426
bilateral, 68/453
brain stem audiometry, 68/427
clinical features, 68/423
cochlear nerve, 68/442
dysequilibrium, 68/446
elderly patient, 68/452
epidemiology, 68/423
facial nerve, 68/441, 443
genetics, 68/435
growth rate, 68/437
headache, 68/425, 446
hearing, 68/424, 445
history, 68/421
imaging, 68/428
laboratory, 68/426
nomenclature, 68/421
pathology, 68/433
radiotherapy, 68/452
screening, 68/432
skull base, 68/466
stapedial reflex testing, 68/427
stereotactic radiosurgery, 68/448
surgery, 68/437, 439, 447, 451
symptom, 67/159
tinnitus, 68/424, 444
trigeminal neurinoma, 68/425
vertigo, 68/425
Vestibular neuronitis
animal experiment, 56/117
clinical features, 56/116
diagnosis, 56/117
features, 2/377, 7/484
Koch postulate, 56/105
management, 56/117
nausea, 56/117
symptom, 2/364
temporal bone pathology, 56/117
vertigo, 2/364, 56/116
viral infection, 56/105
vomiting, 56/117
Vestibular neuropathy, see Vestibular neuronitis
Vestibular nystagmus
clinical features, 1/588
progressive supranuclear palsy, 49/241
rotation, see Rotation vestibular nystagmus
vestibular test, 2/366
Vestibular schwannoma, see Vestibular neurinoma
Vestibular syndrome, 2/358-390

complex, 2/417
optokinetic nystagmus, 2/367
Vestibular system, 16/301-337
  autonomic nervous system, 74/262
  basal ganglion, 16/333
  brain stem tumor, 16/321
  carbamazepine intoxication, 37/203
  cerebellum, 16/302
  cerebral location test, 16/328
  childhood psychosis, 46/491
  coordinate system, 3/28
  diazepam intoxication, 37/203
  exercise, 74/262
  frontal lobe, 16/331
  gait, 16/315
  mechanism, 2/365
  mesuximide intoxication, 37/203
  nerve function, 16/315
  nystagmus, 16/303
  optomotor system, 16/331
  phasic element, 2/365
  phenobarbital intoxication, 37/203
  phenytoin intoxication, 37/203
  primidone intoxication, 37/203
  rotation nystagmus, 16/322
  schizophrenia, 46/491
  seesaw nystagmus, 16/323
  temporal lobe, 16/328
  test, 16/321
  tone, 16/302
  tonic element, 2/365
Vestibulocerebellar syndrome, *see* Archicerebellar
    syndrome
Vestibulocochlear nerve, *see* Statoacoustic nerve
Vestibulocochlear nerve injury
  clinical examination, 57/138
  head injury, 57/138
Vestibulo-ocular disorder
  intracranial hypertension, 16/133
Vestibulo-ocular reflex
  brain stem death, 57/473, 476
  multiple sclerosis, 47/133
Vialetto disease, *see* Familial bulbopontine paralysis
Vialetto-Van Laere syndrome, *see*
    Brown-Vialetto-Van Laere syndrome
Vibration
  brain concussion, 23/95
  ejaculation, 61/322
  H-reflex, 1/64
  head injury, 23/95
  immersion foot, 75/695
  lesion, 7/337
  nerve injury, 51/134

sensation, 1/95
Vibration neuropathy
  acceleration, 51/134
  autonomic dysfunction, 51/134
  chain saw, 51/135
  compression neuropathy, 51/87
  diagnosis, 51/136
  exposure, 51/135
  IgM paraproteinemic polyneuropathy, 51/437
  induced, 75/698
  resonance frequency, 51/134
  smoking, 51/135
  stress, 74/317
  trauma, 74/317
  vibration frequency, 51/134
  widespread tissue lesion, 7/337
Vibration tool, *see* Vibration neuropathy
Vibration white finger, *see* Vibration neuropathy
Vibrator, *see* Electrovibrator
Vicia sativa
  cyanoalanine, 65/11
Vidarabine
  antineoplastic agent, 65/528
  Creutzfeldt-Jakob disease, 46/294
  herpes simplex, 34/154
  herpes simplex virus encephalitis, 34/154, 56/219
  neurotoxin, 65/528, 534
  postherpetic neuralgia, 56/234
  toxic myopathy, 62/609
Vidarabine intoxication
  myoclonus, 65/534
  tremor, 65/534
Vidian artery
  anastomosis, 12/308
Vidian nerve
  anatomy, 5/351
  autonomic nervous system, 75/291
  crocodile tear syndrome, 48/480
  sphenopalatine neuralgia, 48/479
Vidian neuralgia, *see* Sphenopalatine neuralgia
Vienna, Austria
  neurology, 1/10
Vigabatrin
  antiepileptic agent, 65/496
  epilepsy, 73/360
  neurotoxin, 63/496, 65/511
  toxic encephalopathy, 65/511
Vigabatrin intoxication
  akathisia, 65/511
  ataxia, 65/511
  diplopia, 65/511
  headache, 65/511
  hyperkinesia, 65/511

mental depression, 65/511
myelin vacuolation, 65/511
pathologic laughing, 65/511
psychosis, 65/511
vertigo, 65/511
Vigilance
  amphetamine intoxication, 65/257
  attention, 3/142-144, 155, 168, 173
  sleep, 3/86, 90
  terminology, 3/52
  thalamic syndrome (Dejerine-Roussy), 2/482
Villaret syndrome
  clinical picture, 2/100
  Collet-Bonnet syndrome, 2/101
  Horner syndrome, 2/100
  multiple cranial neuropathy, 2/225
  nonrecurrent nonhereditary multiple cranial
    neuropathy, 51/571
  paratrigeminal syndrome, 2/314
  radiography, 2/225
  Sicard-Collet syndrome, 2/100
  skull base, 67/147
  skull base tumor, 17/182
  thalamic syndrome (Dejerine-Roussy), 2/482
Viloxazine
  migraine, 48/196
Vilyuisk virus
  picornavirus, 34/140
Vinblastine
  antineoplastic agent, 65/528
  chemotherapy, 39/110, 67/284
  distal axonopathy, 64/13
  iniencephaly, 50/135
  leukemia, 39/17
  myopathy, 40/555
  neurofibrillary tangle, 46/271
  neurologic toxicity, 39/110
  neuropathy, 69/460
  neurotoxin, 39/17, 94, 107, 110, 64/13, 65/528,
    535
  toxic neuropathy, 64/13
Vinblastine intoxication
  sensorimotor polyneuropathy, 65/535
  symptom, 64/13
Vinca alkaloid
  adverse effect, 69/493
  chemotherapy, 67/284
  human immunodeficiency virus infection, 71/385
  neuropathy, 69/459, 463
  neurotoxicity, 71/368
  side effect, 71/385
Vincent, C., 1/17
Vincristine

antineoplastic agent, 65/528
autonomic nervous system, 75/39, 531
autonomic polyneuropathy, 63/358
brain tumor, 18/519
cerebellar medulloblastoma, 18/190
chemotherapy, 39/106-110, 67/284
chronic axonal neuropathy, 51/531
confusional state, 63/358
distal axonopathy, 64/13
drug induced polyneuropathy, 51/294, 299
epilepsy, 63/359
experimental myopathy, 40/154
hallucination, 63/359
Hodgkin lymphoma, 63/358
inappropriate antidiuretic hormone secretion,
    63/359
intrathecal chemotherapy, 61/151
lethargy, 63/359
leukemia, 39/16, 63/358
lymphocytic leukemia, 63/358
microtubulin binding, 63/358
myeloencephalopathy, 61/151, 63/359
myeloid leukemia, 63/358
myotonia, 40/548
neuroblastoma, 8/472
neurofibrillary tangle, 46/271
neuropathy, 7/545, 69/460, 75/496, 531
neurotoxicity, 39/106-110, 51/294, 63/358
neurotoxin, 7/545, 39/16, 94, 106-110, 119,
    40/151, 339, 51/68, 294, 299, 478, 531, 61/151,
    62/606, 609, 63/152, 358, 64/10, 13, 65/528,
    534, 67/363
non-Hodgkin lymphoma, 63/358
sarcoplasmic reticulum, 40/154
sensorimotor polyneuropathy, 63/358
toxic myopathy, 62/600, 606
toxic neuropathy, 51/299, 64/10, 13
toxic polyneuropathy, 51/68
transverse tubule, 40/154
Vincristine encephalopathy
  Alzheimer neurofibrillary change, 21/57
Vincristine intoxication
  action site, 64/10
  age, 65/535
  autonomic neuropathy, 65/535
  autonomic polyneuropathy, 51/478
  axonal transport, 65/535
  cranial neuropathy, 65/528
  EMG, 65/535
  encephalopathy, 65/528, 534
  myoglobinuria, 40/339
  orthostatic hypotension, 63/152
  polyneuropathy, 65/528, 534

rhabdomyolysis, 40/339
sensorimotor polyneuropathy, 65/535
symptom, 64/13
Vincristine myopathy
differential diagnosis, 40/438
experimental, 40/154
spheromembranous body, 62/609
Vincristine polyneuropathy
abdominal pain, 51/299
autonomic neuropathy, 51/299
axonal degeneration, 51/299
bladder function, 51/299
constipation, 51/299
cranial nerve palsy, 51/299
inappropriate antidiuretic hormone secretion,
51/299
microtubule, 51/299
multiple cranial neuropathy, 51/299
ophthalmoplegia, 51/299
optic atrophy, 51/299
sexual impotence, 51/299
Vindesine
antineoplastic agent, 65/528
chemotherapy, 39/110-114
neurologic toxicity, 39/110-114
neuropathy, 69/460
neurotoxin, 39/94, 107, 110, 65/528, 535
Vindesine intoxication
polyneuropathy, 65/535
sensorimotor polyneuropathy, 65/535
Vinorelbine, *see* Navelbine
Vinyl chloride
angiosarcoma, 68/279
Vinyl ether
anesthetic agent, 37/412
Vinyl ether intoxication
dog, 37/412
hypocapnia, 37/412
inhalation anesthetics, 37/412
seizure, 37/412
Violence
cardiac arrest, 63/217
migraine, 48/161
VIP, *see* Vasoactive intestinal peptide
Viper bite
acanthocytosis, 63/289
Viral antibody
CSF, *see* CSF virus antibody
Epstein-Barr, *see* Epstein-Barr virus antibody
multiple sclerosis, 47/363
Viral antigen
cellular response, 47/356
multiple sclerosis, 47/356

Viral antigenicity
mechanism, 56/57
Viral disease
animal, *see* Animal viral disease
autonomic nervous system, 75/407, 411
border, *see* Border virus disease
Guillain-Barré syndrome, 51/242
myositis, 62/369
slow, *see* Slow virus disease
survey, 75/407
Viral encephalitis
acquired immune deficiency syndrome, 56/490
acute, *see* Acute viral encephalitis
agammaglobulinemia, 56/327
canine distemper virus, 56/587
Coxsackie virus A2, 56/326
Coxsackie virus A4-A9, 56/326
Coxsackie virus B1-B6, 56/326
CSF, 34/66
dementia, 46/388
differential diagnosis, 56/327
ECHO virus 2-4, 56/326
ECHO virus 7, 56/326
ECHO virus 9, 56/326
ECHO virus 11, 56/326
ECHO virus 14, 56/326
ECHO virus 17-19, 56/326
ECHO virus 25, 56/326
enterovirus 71, 56/326
enterovirus 72, 56/326
epilepsy, 15/308
heart transplantation, 63/180
hydranencephaly, 50/337
JHM virus, 56/441
lymphocytic choriomeningitis, 56/360
mental deficiency, 46/19
poliovirus type 1-3, 56/326
porcine hemagglutinating encephalomyelitis virus,
56/441
Viral encephalopathy
myoclonus, 49/618
Viral gene delivery
brain tumor, 67/299
Viral hepatitis
hepatic polyneuropathy, 7/486
Viral infected motoneuron disorder
WFN classification, 59/5
Viral infection, 56/51-61, 65-75, 79-87, 93-102
acoustic neuritis, 56/105
acquired myoglobinuria, 62/572
amyotrophic lateral sclerosis, 59/172-174, 201
amyotrophy, 22/32
animal model, 34/43-54

antibody dependent cytotoxicity, 56/60, 65
antibody production, 56/57
antigen presenting cell, 56/66
antigen specificity, 56/73
aqueduct stenosis, 50/306
Arthus reaction, 56/65
astrocyte, 56/66
autoimmune disease, 56/60
B-lymphocyte, 56/66
blood-brain barrier, 56/79
brain blood flow, 56/79
brain edema, 56/86
brain embolism, 55/163
brain infarction, 55/163
brain metabolism, 56/79
cell surface receptor, 56/34
cellular immunity, 56/65
cerebellar agenesis, 50/178
cerebrovascular disease, 55/163
chorea, 49/554
choreoathetosis, 49/385
cluster headache, 48/220
CNS pathway, 34/18-27
complement, 56/58
complement alternate path, 56/59
complement classic path, 56/59
complement fixation, 56/95
computer assisted tomography, 56/86
congenital deafness, 56/106
CSF electrophoresis, 56/84
CSF virus antibody, 56/100
cytotoxic T-lymphocyte, 56/65, 68
deafness, 56/105
delayed hypersensitivity, 56/65, 70
diagnostic method, 56/93
disseminated encephalomyelitis, 47/474
ELISA, 56/96
endothelial cell, 56/84
enterovirus, 56/307
entry, 34/17
Fab fragment, 56/51
Fc fragment, 56/51
Fc receptor, 56/60
Guillain-Barré syndrome, 47/481
heavy chain, 56/52
hematogenous spread, 56/84
hepatitis B virus, 56/295
HLA system, 56/34, 66
host immune response, 56/51
humoral immune response, 56/51
hydrocephalus, 50/289
hydromyelia, 50/427
hypothalamic syndrome, 2/450

idiopathic postural orthostatic tachycardia
    syndrome, 74/327
Ig, 56/51
IgA antibody, 56/95
IgG antibody, 56/95
IgM antibody, 56/95
immune mediated syndrome, see Immune
    mediated syndrome related to virus infection
immune response, 34/39-44, 55, 56/71
immunocompromised host, 56/467
immunofluorescence test, 56/96
immunologic aspect, 34/39-57
immunopathological process, 56/74
immunoperoxidase labeling, 56/97
interferon, 56/73
intrathecal gammaglobulin synthesis, 56/83
kappa light chain, 56/52
Koch phenomenon, 56/65
Kupffer cell, 56/66
laboratory diagnosis, 56/94
labyrinthitis, 56/105
lambda light chain, 56/52
Langerhans cell, 56/66
latex particle agglutination, 56/100
lymphocyte subpopulation, 56/67
lymphokine, 56/65, 70
major histocompatibility complex, 56/70
measurement, 34/39-44, 55
multiple sclerosis, 34/56
myasthenia gravis, 62/404
neurotropism, 56/34
nonspecific cellular process, 56/73
nuclear magnetic resonance, 56/86
nucleic acid hybridization, 56/99
opsonization, 56/58
organ transplantation, 56/472
paraneoplastic polyneuropathy, 51/470
pathogenesis, 34/15-32, 56/25
peripheral nerve, 56/83
picornavirus, 56/11, 307
poliomyelitis, 42/653, 56/307
polyclonal B-cell activation, 56/61
progressive multifocal leukoencephalopathy,
    34/307, 47/503, 511-513
progressive muscular atrophy, 59/19
progressive rubella panencephalitis, 34/56, 331,
    56/406
progressive spinal muscular atrophy, 22/21, 33
radioimmunoassay, 56/96
Reye syndrome, 9/549, 34/80, 49/217, 56/150,
    65/116, 134
sensorineural deafness, 56/105
serology, 56/95

solid phase IgM assay, 56/96
solid phase immunoassay, 56/96-98
spread, 34/27-29
subacute sclerosing panencephalitis, 34/55,
  47/215
T-lymphocyte, 56/66
T-suppressor lymphocyte, 56/71
tick bite intoxication, 37/111
ultrastructure, 56/95
vertigo, 56/105
vestibular neuronitis, 56/105
viral neutralization, 56/58
viral persistence, 56/61
virus entry, 56/80
virus entry into CNS, 56/83
virus isolation, 56/94
virus specific cytotoxic response, 56/68
X-linked agammaglobulinemia, 56/72
Viral leukoencephalitis
  progressive multifocal leukoencephalopathy,
    9/126
Viral meningitis, 56/125-131
  see also Meningitis serosa
  acquired immune deficiency syndrome, 56/490
  adenovirus, 56/283
  arbovirus, 56/126, 128
  arenavirus, 56/16
  benign recurrent, see Mollaret meningitis
  bunyavirus, 56/16
  clinical features, 56/125
  Coxsackie virus, 56/11, 128, 307
  Coxsackie virus A1-A11, 56/326
  Coxsackie virus A9, 56/129
  Coxsackie virus A14, 56/326
  Coxsackie virus A16-A18, 56/326
  Coxsackie virus A22, 56/326
  Coxsackie virus A24, 56/326
  Coxsackie virus B1-B6, 56/326
  Coxsackie virus B2, 56/129
  Coxsackie virus B3, 56/129
  Coxsackie virus B5, 56/129
  CSF cytology, 56/126
  CSF protein, 56/126
  differential diagnosis, 56/326
  ECHO virus, 56/11, 128, 307
  ECHO virus 1-7, 56/326
  ECHO virus 6, 56/129
  ECHO virus 9, 56/129, 326
  ECHO virus 11-25, 56/326
  ECHO virus 20, 56/129
  ECHO virus 27, 56/326
  ECHO virus 30-33, 56/326
  enterovirus, 56/126, 128, 307

enterovirus 70, 56/307
enterovirus 71, 56/307, 326
epidemiology, 56/127
Epstein-Barr virus, 56/126
flavivirus, 56/12
headache, 56/125
herpes simplex virus, see Herpes simplex virus
  meningitis
herpes simplex virus type 2, 56/130
human T-lymphotropic virus type I associated
  myelopathy, 56/538
Kawasaki syndrome, 56/639
Lassa fever virus, 56/16
Leptospira, 56/126, 128
leptospirosis, 56/130
lymphocytic choriomeningitis, 56/129
lymphocytic choriomeningitis virus, 56/126, 128
Mollaret meningitis, 56/131
mumps meningitis, 56/126, 430
mumps virus, 56/126, 128
Mycoplasma pneumoniae, 56/130
paramyxovirus, 56/14
pathology, 56/127
poliomyelitis, 56/11, 128, 325
poliovirus type 1-3, 56/326
prognosis, 56/131
recurrent, 56/131
season, 56/127
tropical spastic paraplegia, 56/538
varicella zoster virus infection, 55/436
Vogt-Koyanagi-Harada syndrome, 56/611
Wallgren syndrome, 56/359
Viral meningoencephalitis
  experimental, see Experimental viral
    meningoencephalitis
Viral myositis
  see also Bornholm disease
  acute, see Acute viral myositis
  Bornholm disease, 41/54
  Coxsackie virus, 41/20, 53, 73, 377
  dermatomyositis, 62/381
  inclusion body myositis, 62/381
  inflammatory myopathy, 41/53
  myalgia, 41/53
  myoglobinuria, 41/275
  myxovirus, 41/71
  picornavirus, 41/71, 62/381
  polymyositis, 62/381
Viral nuchae myositis, see Bornholm disease
Viral protein HA
  influenza virus, 56/34
  neurotropism, 56/34
  neurovirulence, 56/39

Viral protein HA$_1$
  exopeptidase cleavage, 56/28
Viral protein HA$_2$
  exopeptidase cleavage, 56/28
Viral protein VP1
  neurotropism, 56/36
Viral protein VP3
  neurotropism, 56/36
Viral protein VP4
  acidic ph, 56/28
Viral protein VP5
  trypsin, 56/28
Virchow-Robin perivascular space
  amyloid, 46/257
Viremia
  acquired immune deficiency syndrome, 56/511
  immunocompromised host, 56/468
  neurotropic virus, 56/30
  poliomyelitis, 34/94, 103, 108, 122
  subarachnoid hemorrhage, 55/20
Virgin hand
  congenital tactile agnosia, 4/34
  developmental disorder, 4/467
  developmental dyspraxia, 4/467
  parietal lobe lesion, 4/467
Virilization
  Morgagni-Stewart-Morel syndrome, 43/411
Virology
  animal, 34/1-13
  chemical composition, 34/2-4
  classification, 34/4-9
  cytomegalovirus, 56/263
  definition, 34/1
  DNA virus, 34/7-9
  fundamental aspect, 34/1-13
  lipid, 34/4
  morphology, 34/2
  multiple sclerosis, 34/435-446, 47/319-332, 357
  multiplication, 34/11
  nucleic acid, 34/3
  progressive multifocal leukoencephalopathy,
    71/412
  protein, 34/3
  RNA virus, 34/4-7
Virus
  Akabane, see Akabane virus
  alpha, see Alphavirus
  aphtho, see Aphthovirus
  arena, see Arenavirus
  ataxia telangiectasia, 60/371
  avian herpes, see Avian herpes virus
  BK, see BK virus
  bluetongue, see Bluetongue virus

bovine rinderpest, see Bovine rinderpest virus
bovine viral diarrhea mucosal disease, see Bovine
  viral diarrhea mucosal disease virus
brain tumor, 16/40
brain tumor carcinogenesis, 16/34, 40
brain tumor origin, 16/40
bunya, see Bunyavirus
California encephalitis, see California encephalitis
  virus
canine distemper, see Canine distemper virus
Chikungunya, see Chikungunya virus
chromosomal aberration, 30/103
Colorado tick fever, see Colorado tick fever virus
corona, see Coronavirus
cowpox, see Cowpox virus
Coxsackie, see Coxsackie virus
cytomegalo, see Cytomegalovirus
dengue, see Dengue virus
dermatomyositis, 40/264
distemper, see Distemper virus
DNA, see DNA virus
Duvenhage, see Duvenhage virus
ECHO, see ECHO virus
encephalomyelitis, see Encephalomyelitis virus
encephalomyocarditis, see Encephalomyocarditis
  virus
enteric cytopathic human organ, see ECHO virus
entero, see Enterovirus
epilepsy, 72/152
Epstein-Barr, see Epstein-Barr virus
flavi, see Flavivirus
foot and mouth disease, see Foot and mouth
  disease virus
gammaherpes, see Gammaherpes virus
hepadna, see Hepadnavirus
hepatitis, see Hepatitis virus
hepatitis B, see Hepatitis B virus
hepetitis A, see Hepatitis A virus
herpes, see Herpes virus
herpes simplex, see Herpes simplex virus
herpes zoster, see Herpes zoster virus
hog cholera, see Hog cholera virus
human immunodeficiency, see Human
  immunodeficiency virus
influenza, see Influenza virus
influenza A, see Influenza A virus
influenza A$_2$, see Influenza A$_2$ virus
influenza B, see Influenza B virus
Jamestown Canyon, see Jamestown Canyon virus
JC, see JC virus
JHM, see JHM virus
junin, see Junin virus
La Cross, see La Cross virus

lactate dehydrogenase elevating, *see* Lactate dehydrogenase elevating virus

Lassa fever, *see* Lassa fever virus

lenti, *see* Lentivirus

lymphocytic choriomeningitis, *see* Lymphocytic choriomeningitis virus

Lyssa, *see* Lyssavirus

Machupo, *see* Machupo virus

Marek disease, *see* Marek disease virus

mastadeno, *see* Mastadenovirus

Mayaro, *see* Mayaro virus

measles, *see* Measles virus

mengo, *see* Mengovirus

mokola, *see* Mokola virus

molluscum contagiosum, *see* Molluscum contagiosum virus

Moloney murine leukemia, *see* Moloney murine leukemia virus

monkey pox, *see* Monkey poxvirus

multiple sclerosis, 9/120, 47/215

mumps, *see* Mumps virus

murine leukemia, *see* Murine leukemia virus

murine neurotropic retro, *see* Murine neurotropic retrovirus

murine sarcoma, *see* Murine sarcoma virus

nairo, *see* Nairovirus

neural tumor, 71/418

neurotropic, *see* Neurotropic virus

Newcastle disease, *see* Newcastle disease virus

oncogenic, *see* Oncogenic virus

O'nyong-nyong, *see* O'nyong-nyong virus

orbi, *see* Orbivirus

Orf, *see* Orf virus

orthomyxo, *see* Orthomyxovirus

orthopox, *see* Orthopoxvirus

ortiiomyxo, *see* Ortiiomyxovirus

panleukopenia, *see* Panleukopenia virus

papilloma, *see* Papillomavirus

papova, *see* Papovavirus

parainfluenza, *see* Parainfluenza virus

parainfluenza type 1, *see* Parainfluenza virus type 1

parainfluenza type 3, *see* Parainfluenza virus type 3

paramyxo, *see* Paramyxovirus

parapox, *see* Parapoxvirus

parvo, *see* Parvovirus

Phlebotomus fever, *see* Phlebotomus fever virus

picorna, *see* Picornavirus

polio, *see* Poliovirus

polymyositis, 40/264

polyoma, *see* Polyomavirus

porcine hemagglutinating encephalomyelitis, *see* Porcine hemagglutinating encephalomyelitis virus

Powassan, *see* Powassan virus

pox, *see* Poxvirus

rabies, *see* Rabies virus

reo, *see* Reovirus

replication, 56/3

respiratory syncytial, *see* Respiratory syncytial virus

retro, *see* Retrovirus

Reye syndrome, 70/268

rhino, *see* Rhinovirus

RNA, *see* RNA virus

rocio, *see* Rocio virus

Ross river, *see* Ross river virus

rota, *see* Rotavirus

Rous sarcoma, *see* Rous sarcoma virus

rubella, *see* Rubella virus

rubi, *see* Rubivirus

saimiri-ateles, *see* Saimiri-ateles virus

secondary pigmentary retinal degeneration, 13/216

Semliki forest, *see* Semliki forest virus

simian T-cell tropic retro, *see* Simian T-cell tropic retrovirus

Sindbis, *see* Sindbis virus

slow, *see* Slow virus

snowshoe hare, *see* Snowshoe hare virus

spuma, *see* Spumavirus

St. Louis encephalitis, *see* St. Louis encephalitis virus

subacute myelo-optic neuropathy like, *see* Subacute myelo-optic neuropathy like virus

subacute sclerosing panencephalitis, *see* Subacute sclerosing panencephalitis virus

superior laryngeal neuralgia, 48/498

SV5, *see* Simian virus 5

SV40, *see* Simian virus 40

Tacaribe, *see* Tacaribe virus

Tahyna, *see* Tahyna virus

Tamiami, *see* Tamiami virus

Tana pox, *see* Tana poxvirus

Theiler, *see* Theiler virus

tick-borne encephalitis, *see* Tick-borne encephalitis virus

tick-borne encephalitis flavi, *see* Tick-borne encephalitis flavivirus

Toga, *see* Togavirus

transcription, 56/3

varicella zoster, *see* Varicella zoster virus

variola, *see* Variola virus

Venezuelan equine, *see* Venezuelan equine virus

vesicular stomatitis, *see* Vesicular stomatitis virus

Vilyuisk, *see* Vilyuisk virus
visna, *see* Visna virus
visna like, *see* Visna like virus
visna-maedi, *see* Visna-maedi virus
West Nile, *see* West Nile virus
western equine encephalitis, *see* Western equine
   encephalitis virus
Yaba pox, *see* Yaba poxvirus
yellow fever, *see* Yellow fever virus
Virus encephalomyelitis
   Theiler, *see* Theiler virus encephalomyelitis
Virus induced demyelination
   multiple sclerosis, 47/327
Virus like particle
   cultured muscle, 40/192
   inclusion body myositis, 40/52, 284, 322
   muscle fiber, 40/109
   muscle tissue culture, 40/192
   myelinoclastic diffuse sclerosis, 9/481
   ultrastructure, 40/109
Virus taxonomy
   classification criteria, 56/2
   DNA virus, 56/2
   RNA virus, 56/2
Visceral brain
   emotion, 2/715, 719, 3/320
Visceral cancer
   neurofibromatosis type I, 50/369
Visceral cyst
   Von Hippel-Lindau disease, 50/374
Visceral leishmaniasis
   mosquito, 35/157
   treatment, 35/158
Visceral lesion
   adrenal adenoma, 14/51
   clinical features, 14/669-682
   neurofibromatosis, 14/669-682
   phakomatosis, 14/668-730
   Sturge-Weber syndrome, 14/696-699
   tuberous sclerosis, 14/49, 369-374, 682-691
   Von Hippel-Lindau disease, 14/60, 242, 247,
      691-696
Visceral migraine sine hemicrania
   cardiac symptom, 48/157
Visceral myopathy
   familial type, 51/493
   intestinal pseudo-obstruction, 51/492, 70/324
   ophthalmoplegia, 51/493
   sporadic, *see* Sporadic visceral myopathy
Visceral neuropathy
   see also Autonomic polyneuropathy
   familial, *see* Familial visceral neuropathy
   intestinal pseudo-obstruction, 51/492, 70/324

   sporadic, 70/327
   Wolman disease, 51/379
Visceral pain
   hereditary sensory and autonomic neuropathy type
      II, 60/11
   nociceptive, 69/32
   paraplegia, 26/494
   property, 1/131
Visceral sensation
   property, 1/99
Visceral vasoconstrictor neuron
   autonomic nervous system, 74/26
Viscerocranium
   malformation, 32/180
   rachischisis, 32/180
Viscerocystic retinoangiomatosis, *see* Von
      Hippel-Lindau disease
Viscerogenic spinal reflex
   referred pain, 2/175
Visceromegaly
   Beckwith-Wiedemann syndrome, 46/82
   multiple myeloma polyneuropathy, 51/430
   Waldenström macroglobulinemia, 63/396
Vision
   see also Eye
   autonomic nervous system, 74/399
   blurred, *see* Visual impairment
   carotid cavernous fistula, 24/415, 422
   cinematographic, *see* Cinematographic vision
   color, *see* Color vision
   congenital retinal blindness, 22/530
   Creutzfeldt-Jakob disease, 46/291
   cyclopic, *see* Cyclopic vision
   disorder, *see* Visual disorder
   gaze, 2/666
   handedness, 4/323
   Minamata disease, 36/74
   mosaic, *see* Mosaic vision
   organic mercury intoxication, 36/74
   oxoglutarate dehydrogenase deficiency, 62/503
   perception, 45/521
   phylogeny, 2/542
   reticular formation, 2/533
   retinovestibular hereditary spastic ataxia, 21/384,
      386
   sensory extinction, 3/191
   tubular, *see* Tubular vision
   zoom, *see* Zoom vision
Vision loss
   clinical examination, 57/135
   Harboyan syndrome, 42/368
   infantile optic atrophy, 42/408
   monocular, *see* Monocular vision loss

optic chiasm syndrome, 18/354
Stargardt disease, 13/262
sudden, *see* Acute amaurosis
Vision spéculaire, *see* Autoscopia
Visna
  animal model, 56/453
  animal viral disease, 9/681, 34/294
  antibody, 56/61
  demyelination, 9/681, 56/583
  lentivirus, 56/16
  multicystic necrotizing myelitis, 56/462
  multiple sclerosis, 9/124, 47/430
  neuropathology, 56/459
  retrovirus, 56/16, 453, 459, 529
  retrovirus infection, 56/583
  Scherer acute multiple sclerosis, 9/671
  sheep, 9/681
  slow virus disease, 9/681, 34/294, 56/459, 544
Visna infection
  immunosuppression, 56/462
Visna like virus
  multiple sclerosis, 47/323
Visna-maedi disease
  animal model, 56/583
  lentivirus, 56/16, 460
  retrovirus, 56/16, 529
  slow virus disease, 9/681, 56/583
Visna-maedi virus
  human immunodeficiency virus, 56/584
Visna virus
  demyelination, 56/583
  human immunodeficiency virus, 56/463
  lentivirus family, 56/583
Visual acuity
  agnosia, 45/334-336
  Alström-Hallgren syndrome, 13/461
  Behr disease, 13/89
  cardiovascular agent intoxication, 37/426
  central retinal venous thrombosis, 16/275
  dominant infantile optic atrophy, 13/112
  glycoside intoxication, 37/426, 429
  hexachlorophene intoxication, 37/486
  homocarnosinase deficiency, 42/558
  hyaloidoretinal degeneration (Wagner), 13/274
  oculocutaneous albinism, 43/3
  papillomacular lesion, 13/51
  Snellen chart, 13/9
  Stargardt disease, 13/129, 270
Visual adaptation
  Behr disease, 13/89
  curve, 13/50, 159
  dominant infantile optic atrophy, 13/116
  hereditary adult foveal dystrophy (Behr), 13/89

homonymous hemianopia, 2/584
hyaloidotapetoretinal degeneration
  (Goldmann-Favre), 13/37
schema, 13/15
Stargardt disease, 13/129
Visual agnosia, 4/14-32
  agnostic alexia, 2/656, 4/16, 95
  alexia, 4/116, 122, 45/439
  Alzheimer disease, 46/251
  amnesia, 4/16
  anatomic localization, 4/20-22, 298-301
  attention, 2/651
  attention disorder, 4/18
  Balint syndrome, 2/602, 45/55
  body scheme disorder, 2/668
  cerebral dominance, 4/20, 289-301
  Charcot-Wilbrand syndrome, 2/654
  classification, 4/15-18, 20
  clinical features, 4/16-20
  color, 2/649
  definition, 2/597, 601
  examination, 4/18, 22
  eye movement, 4/22, 30
  Gestaltzerfall, 2/655
  imperception, 2/601, 4/14
  Klüver-Bucy syndrome, 45/260
  left posterior cerebral artery, 53/412
  music type, 2/607
  neuropsychology, 45/13
  object, 4/5, 15, 20, 300, 45/14
  occipital lobe lesion, 2/541, 4/20-22
  occipital lobe syndrome, 45/53-56
  occipital lobe tumor, 17/325, 327, 331, 67/156
  parietal lobe lesion, 2/587, 4/20-22
  pathopsychophysiogenesis, 4/24-29
  posterior cerebral artery syndrome, 53/412
  prosopagnosia, 2/602, 655, 657, 4/17, 45/14
  simultanagnosia, 45/55
  spatial, 3/218, 224, 4/16
  status, 2/658
  temporal lobe lesion, 4/20-22
  test, 4/18
  type, 2/603, 655
  visual localization by representation, 4/18
  visual memory, 45/14
Visual allesthesia
  cerebral dominance, 3/29
  definition, 2/612
  visual perseveration, 45/366
Visual amnesia
  memory disorder, 4/16
Visual association area
  anatomy, 2/643

Klüver-Bucy syndrome, 2/602
Visual association cortex
  inferior, *see* Inferior visual association cortex
Visual attention
  Balint syndrome, 2/667, 45/409
  conjugate eye movement, 3/25
  constructional apraxia, 4/61
  fixation, 2/651
  optomotor ataxia, 2/340
  spatial orientation disorder, 3/214, 226
Visual blurring, *see* Visual impairment
Visual clarity
  abnormal, *see* Abnormal visual clarity
Visual cortex
  alexia, 45/12
  basal dendritic tree, 2/540
  binocular interaction, 2/538
  column, 2/539
  complex cell, 2/538, 644
  component, 2/643
  extent, 2/536
  function, 2/539
  Gennari-Bamberger stripe, 2/538
  hypercomplex cell, 2/644
  injury, *see* Visual cortex injury
  lamination, 2/536
  macular representation, 2/560
  Meynert giant pyramidal cell, 2/537
  Meynert solitary cell, 2/537
  movement perception, 2/540
  ocular inhibition, 2/539
  optic organization, 2/539
  orientational selection, 2/540
  quadrantanopia, 2/558
  receptive field, 2/541
  retinotopic representation, 2/559
  simple cell, 2/538, 644
  synapse, 2/538
  traumatic blindness, 24/28, 48-52
  visual field defect, 2/592
Visual cortex injury
  homonymous hemianopia, 24/47
  optokinetic nystagmus, 24/47
  penetrating, 24/47
  quadrantanopia, 24/47
  sign, 24/47
  vascular supply, 24/31
Visual discrimination
  Klüver-Bucy syndrome, 2/603
  Parkinson dementia, 46/374
Visual disorder
  α-amino-n-butyric acid, 13/339
  apraxia, 2/597

aqueduct stenosis, 50/310
Batten disease, 10/604, 606
central, *see* Central visual disturbance
developmental dyslexia, 4/384
digoxin intoxication, 65/450
Divry-Van Bogaert syndrome, 42/725
eclampsia, 39/546
ethylhydrocupreine, 13/220
head injury, 46/617-620
hypertensive encephalopathy, 11/555
migraine, 5/50, 48/6, 18, 125, 127, 162
neptunea intoxication, 37/64
optic pathway, 68/76
parietal lobe tumor, 67/153
pellagra, 7/571
pheochromocytoma, 39/498, 500
pituitary adenoma, 17/385-387
primary pontine hemorrhage, 12/39
quinidine intoxication, 37/448
scotoma, 48/127
temporal lobe tumor, 17/290
thalassemia major, 42/629
vertical nystagmus, 42/159
Visual disorientation
  agnosia, 45/335
  body scheme disorder, 2/668
  classification, 2/612
  pseudoagnosia, 45/336
  visual sensory deficit, 45/335
Visual dreaming
  Charcot-Wilbrand syndrome, 2/654
  content, 2/654
  element, 17/328
  experience, 3/88
  occipital lobe tumor, 17/328
Visual evoked potential, *see* Visual evoked response
Visual evoked response
  see also Evoked cortical response
  amaurosis fugax, 53/318
  average, 2/544
  Bassen-Kornzweig syndrome, 51/327, 63/275
  blindness, 2/545
  brain death, 24/731
  brain embolism, 53/232
  brain infarction, 53/232
  brain stem infarction, 54/160
  cerebrovascular disease, 53/232
  Charlevoix-Saguenay spastic ataxia, 60/455
  chorea-acanthocytosis, 63/282
  cortical blindness, 45/50
  Down syndrome, 50/529
  EEG, 54/160, 57/197
  familial spastic paraplegia, 59/311

Friedreich ataxia, 60/321
head injury, 57/197
hereditary paroxysmal kinesigenic
   choreoathetosis, 49/354
Huntington chorea, 49/289
hyperactivity, 46/184
infantile optic atrophy, 42/408
intracranial hemorrhage, 53/232
lead polyneuropathy, 64/437
long latency wave, 2/546
migraine, 48/76
multiple sclerosis, 47/133-135, 142
neurosyphilis, 52/278
optic nerve injury, 24/37
optic neuritis, 47/54
pattern, *see* Pattern visual evoked response
pattern shift, *see* Pattern shift visual evoked
   response
technique, 13/22
thalamic infarction, 54/160
transient ischemic attack, 53/232
undetermined significant monoclonal
   gammapathy, 63/399
uremic encephalopathy, 63/508
Visual extinction, *see* Visual inattention
Visual fiber
   decussation, 2/550
   geniculocalcarine tract, 2/557
   optic chiasm, 2/550-554
   optic nerve, 2/548
   optic tract, 2/554-557
   retina, 2/547
Visual field
   abnormal clarity, 17/331
   basilar ischemia, 12/3
   Behr disease, 13/89
   campimetry, 13/13
   centrocecal scotoma, 13/113
   chromophobe adenoma, 17/385-387
   chronic mercury intoxication, 64/370
   craniopharyngioma, 18/511
   dominant infantile optic atrophy, 13/113
   epilepsy, 17/322
   intracranial hypertension, 16/133
   macular sparing, 17/316, 320
   mercury polyneuropathy, 51/272
   Minamata disease, 64/413
   neuromyelitis optica, 9/428
   occipital lobe tumor, 17/314-317, 319, 339-342
   perimetry, 13/12
   primary pigmentary retinal degeneration,
      13/160-165
   sieve scotoma, 13/114

Stargardt disease, 13/129, 270
   torsion, 17/321
   vertical axis sparing, 17/321
   vertical meridian, 3/31
   visual hallucination, 17/322
Visual field constriction
   Minamata disease, 64/415
Visual field defect
   see also Scotoma
   agnosia, 45/334
   alexia, 45/438
   anterior ischemic optic neuropathy, 55/114
   arcuate defect, 2/563
   cause, 13/50
   conductive element lesion, 2/562
   confrontation method, 2/563
   deferoxamine, 63/534
   denial, 2/670, 672
   head injury outcome, 57/374
   homonymous paracentral scotoma, 2/585
   insight, 2/672
   junction scotoma, 2/567
   lateral geniculate nucleus, 2/577
   middle cerebral artery syndrome, 53/363
   occipital lobe, 2/588, 17/314-317, 319
   optic chiasm, 2/570
   optic tract, 2/576
   percipient element lesion, 2/562
   pituitary abscess, 52/152
   posterior cerebral artery syndrome, 11/306, 53/410
   retinal defect, 2/561-567
   subarachnoid hemorrhage, 12/136
   temporal crescent, 2/596
   temporal lobe, 2/587
   topical diagnosis, 2/561
   transient ischemic attack, 12/3
   visual cortex, 2/592
   visual hallucination, 45/356
   visuospatial agnosia, 45/169
Visual fixation
   abnormality, 2/666
   attention, 2/651
   Balint syndrome, 45/406
   ENG, 2/378, 380
   fixation defect, 2/340
   impairment, 1/599
   ocular ataxia, 1/311
   optomotor ataxia, 2/340
   spatial orientation disorder, 3/214, 226
Visual function
   cerebral dominance, 4/277
   hereditary adult foveal dystrophy (Behr), 13/89
   optic atrophy, 13/49

rehabilitation, 55/245
Visual hallucination
  allesthesia, 45/356
  anatomoclinical correlation, 46/567
  auditory hallucination, 2/709
  basilar artery insufficiency, 46/567
  brain angiography, 55/146
  brain embolism, 53/207
  brain infarction, 53/207, 55/145
  brain stem infarction, 55/146
  cause, 2/614-617
  cerebrovascular disease, 53/207, 55/145
  Charles-Bonnet syndrome, 55/146
  ciclosporin, 63/10
  deafferentation state, 46/567
  déjà vu, 45/356
  delirium, 46/526
  N,N-dimethyltryptamine intoxication, 65/45
  dreamy state, 2/616, 45/356
  drug intoxication, 55/146
  epilepsy, 2/646, 17/322, 323, 55/146
  hemianopic field, 2/616
  intracranial hemorrhage, 53/207
  levodopa intoxication, 46/593
  lilliputian hallucination, 2/617
  macropsia, 45/356
  micropsia, 45/356
  migraine, 45/510, 55/146
  morphopsia, 45/356
  occipital lobe infarction, 55/146
  occipital lobe tumor, 17/322, 331
  peduncular, 2/617
  peduncular hallucinosis, 2/617, 46/419
  posterior cerebral artery syndrome, 53/411
  right occipital lesion, 4/5
  schizophrenia, 46/562
  scotoma, 45/356
  temporal lobe, 2/709, 45/356
  temporal lobe tumor, 17/290
  transient ischemic attack, 53/207
  uncinate fit, 2/616
  visual field, 17/322
  visual field defect, 45/356
Visual hemifield
  disconnection syndrome, 45/101
Visual image persistence, see Visual perseveration
Visual impairment
  Argentinian hemorrhagic fever, 56/366
  carbon disulfide intoxication, 64/8
  clioquinol intoxication, 64/8
  cluster headache, 48/222
  conidae intoxication, 37/60
  cyanogenic glycoside intoxication, 64/8
  disopyramide intoxication, 65/453
  disulfiram intoxication, 64/8
  empty sella syndrome, 17/431
  ethambutol intoxication, 64/8
  ethchlorvynol intoxication, 37/354
  ethionamide intoxication, 65/481
  flecainide intoxication, 65/456
  frontometaphyseal dysplasia, 38/410
  Lafora progressive myoclonus epilepsy, 15/385, 27/173
  lidocaine intoxication, 37/451
  methanol intoxication, 36/352, 64/8, 99
  Minamata disease, 36/105
  myotonic hyperkalemic periodic paralysis, 62/462
  neuromyelitis optica, 9/428, 47/397
  organophosphate intoxication, 37/554, 64/167
  orthostatic hypotension, 63/143
  polycythemia, 63/250
  polycythemia vera, 63/250
  pontine infarction, 12/15, 22
  progressive supranuclear palsy, 22/219, 49/240
  propafenone intoxication, 65/457
  quinine intoxication, 13/224, 64/8
  subclavian steal syndrome, 53/373
  thallium intoxication, 64/8
  tocainide intoxication, 65/455
  vertebral artery arteriosclerosis, 53/376
  vertebrobasilar system syndrome, 53/373, 376
  vitamin B$_{12}$ deficiency, 63/254
  Waldenström macroglobulinemia, 63/396
Visual inattention
  Balint syndrome, 45/410
  testing method, 2/584
  unilateral, 2/584
Visual integration
  abnormality, 2/708-711
Visual localization by representation
  visual agnosia, 4/18
Visual memory
  agnosia, 45/344
  occipital lobe tumor, 17/326, 328
  parietal lobe, 45/359
  visual agnosia, 45/14
Visual object agnosia
  definition, 2/606
  description, 2/656
  features, 45/14
  Gerstmann syndrome, 2/656
Visual pathway
  see also Optic tract
  arterial supply, 55/116
  retinotopic organization, 2/545
Visual perception

blanking, 2/545
brain tumor, 67/396
cognitive measurement, 45/521
constructional apraxia, 45/492-494
dissociation, *see* Dissociated visual perception
Gestaltzerfall, 2/662
interaction, 2/545
movement, 2/665
physiology, 2/643
pulvinar, 2/533
Visual perception disorder
depersonalization, 2/709
derealization, 2/709
disconnection syndrome, 2/597, 4/17
Klüver-Bucy syndrome, 2/708, 716
penetrating head injury, 24/491
traumatic psychosyndrome, 24/549
Visual perseveration
Bell sign, 45/365
cardiac catheterization, 63/176
delirium tremens, 45/365
description, 2/617, 664
epilepsy, 45/509
hallucination, 45/364-366
hallucination classification, 45/56
hallucinosis, 46/562
illusory visual speed, 2/617
mescaline, 45/365
migraine, 45/510, 48/162, 166
occipital lobe, 45/364-366
occipital lobe tumor, 17/331
parietal lobe, 45/360
perception, 45/351-369
posterior cerebral artery syndrome, 53/411
temporal lobe epilepsy, 73/71
visual allesthesia, 45/366
Visual processing
rehabilitation, 46/618
Visual recognition
occipital lobe tumor, 17/327
Visual seizure
occipital spike, 73/25
Visual sensory deficit
agnosia, 45/333-336
monochromatism, 45/335
pseudoagnosia, 4/25
shape perception, 45/334
visual disorientation, 45/335
Visual sensory system, 2/506-625
anatomy, 2/506-550
Henle fiber layer, 2/510
Klüver-Bucy syndrome, 2/602
opsoclonus, 2/534

topical diagnosis, 2/561-597
Visual space
absolute localization, 2/685
disorder, 24/491, 549
distortion, 3/28
estimation of distance, 2/686
inversion, 3/29
recognition of length and size, 2/686
relative localization, 2/686
Visual speed
illusory, *see* Illusory visual speed
Visual stimulation
epilepsy, 15/442
eye movement, 1/576
eyelid, 1/614
reflex, 1/614
startle epilepsy, 15/441
Visuoconstructive disturbance
constructional apraxia, 45/495-500
Visuomotor apraxia, *see* Oculomotor apraxia
Visuomotor cortex, *see* Parastriate cortex
Visuosensory band (Elliot Smit)
anatomy, 2/643
Visuospatial agnosia
Anton-Babinski syndrome, 3/29
aphasia, 45/498
attention disorder, 3/194, 4/215
body scheme disorder, 4/215
cancellation test, 45/168
cerebral dominance, 4/298, 45/174
cerebral localization, 3/25
clinical features, 4/16
consciousness, 3/56
constructional apraxia, 4/77, 45/492, 498, 504
epilepsy, 45/171
gaze defect, 45/170
hemianopia, 45/168
hemispace, 45/169
hemispatial attention deficit, 45/170
line bisection task, 45/168
neglect, 45/16
neglect syndrome, 45/167-172
paralexia, 45/168
parietal lobe, 45/172
parietal lobe syndrome, 45/73, 78, 80
pathophysiology, 45/168-172
reaction time, 45/169
representational map hypothesis, 45/169-172
space perception, 45/411
spatial orientation disorder, 3/30, 216, 218, 224,
    45/411
thalamus, 45/170
visual field defect, 45/169

Visuospatial disorder, *see* Spatial orientation
   disorder
Visuospatial inattention
  psychometry, 45/520
Visuospatial neglect, *see* Visuospatial agnosia
Vitamin A, 28/43-46
  acid maltase deficiency, 41/180
  Alopex lagopus intoxication, 37/96
  Arnold-Chiari malformation type II, 50/404
  benign intracranial hypertension, 16/159, 67/111
  Canis familiaris intoxication, 37/96
  cephalocele, 50/98
  erignathus barbatus intoxication, 37/96
  function, 28/44
  halichoerus grypus intoxication, 37/96
  leukodystrophy, 10/58
  metachromatic leukodystrophy, 10/58, 47/593
  phocca groenlandica intoxication, 37/96
  primary pigmentary retinal degeneration, 13/193,
   202
  spina bifida, 32/528
  teratogenic agent, 32/169, 172
  tharlarctos maritimus intoxication, 37/96
Vitamin A deficiency
  Bassen-Kornzweig syndrome, 51/326, 63/277
  benign intracranial hypertension, 28/43
  clinical features, 28/43
  congenital anosmia, 42/357
  deficiency neuropathy, 51/326
  etiology, 28/43
  malabsorption syndrome, 70/233
  pellagra, 28/86
  secondary pigmentary retinal degeneration,
   60/736
Vitamin A intoxication
  abdominal pain, 37/96
  acute type, 65/567
  anencephaly, 30/182, 185, 188, 192, 50/28
  Balaenoptera borealis intoxication, 37/95
  benign intracranial hypertension, 28/45, 37/27, 96,
   65/568
  canis familiaris, 37/96
  carotenoid, 65/567
  cetacea intoxication, 37/94-97
  chronic type, 65/568
  clinical features, 29/44-46, 37/95-97
  craniofacial abnormality, 65/569
  CSF pressure, 37/96
  desquamation, 37/96
  diagnosis, 28/46
  diarrhea, 37/96
  diplopia, 37/96
  drowsiness, 37/96

  exophthalmos, 37/96
  fox intoxication, 37/96
  grouper intoxication, 37/87
  hatahata intoxication, 37/87
  headache, 37/96, 65/568
  hypercalcemia, 65/569
  ichthyohepatoxic fish intoxication, 37/87
  intracranial hypertension, 65/568
  irritability, 37/96
  isotretinoin, 65/569
  Japanese sandfish intoxication, 37/87
  mammalia intoxication, 37/94-97
  nausea, 37/96, 65/568
  papilledema, 28/45, 37/96
  physeter catodon intoxication, 37/95
  polar bear intoxication, 37/96
  polyarthralgia, 37/96
  sea bass intoxication, 37/87
  serranid, 37/87
  serranid intoxication, 37/87
  sign, 37/96
  snapper intoxication, 37/87
  sparid intoxication, 37/87
  symptom, 37/96
  tetracycline, 65/568
  trichodontid, 37/87
  trichodontid intoxication, 37/87
  tuna, 37/87
  tuna intoxication, 37/87
  vertigo, 37/96, 65/568
  vomiting, 37/96, 65/568
  whale intoxication, 37/95
  wolf intoxication, 37/96
Vitamin B complex
  burning feet syndrome, 28/14
Vitamin B1
  see also Beriberi
  alcohol withdrawal delirium, 46/541
  antibiotic agent, 7/632
  antimetabolite, 7/568
  antivitamin, 7/567
  biochemistry, 7/569
  chronic axonal neuropathy, 51/531
  curare like effect, 28/211
  deficiency neuropathy, 51/324
  dialysis, 63/528
  dialysis encephalopathy, 63/528
  fluorouracil intoxication, 65/532
  glycogen storage disease, 7/570
  hereditary periodic ataxia, 21/569
  lethal dose, 28/210
  pyruvate carboxylase deficiency, 29/213
  pyruvate decarboxylase deficiency, 41/428,

62/502
subacute necrotizing encephalomyelopathy, 28/353-356
trigeminal neuralgia, 5/299
uremic polyneuropathy, 51/361
Wernicke encephalopathy, 45/196
Wernicke-Korsakoff syndrome, 45/195, 47/155
Vitamin B$_1$ deficiency, 28/49-56
see also Beriberi
age, 51/331
alcohol, 47/155, 51/331
alcoholic neuropathy, 7/166
autonomic nervous system, 75/491
autonomic neuropathy, 28/5
beriberi, 7/560, 636, 28/2-4, 6-9, 50, 51/331
burning feet syndrome, 28/13
carbohydrate metabolism, 28/3
celiac disease, 51/331
central pontine myelinolysis, 28/311
Changi Camp syndrome, 7/560, 568
Chastek paralysis, 7/561, 568
clinical features, 51/331
cocarboxylase loading test, 51/332
deficiency neuropathy, 7/567, 51/330
delirium, 46/543
dementia, 46/399
EMG, 51/332
features, 51/325
hoarse voice, 51/332
hyperthyroidism, 51/331
instant food, 51/331
late cortical cerebellar atrophy, 21/489
malabsorption syndrome, 70/229
malnutrition, 28/2-9, 15-17
memory disorder, 27/464
mental deficiency, 46/18
neuropathology, 51/332
neuropathy, 7/567, 28/4-9, 49-55
pellagra, 28/85
polyneuropathy, 7/552
pregnancy, 51/331
transketolase, 51/332
tropical ataxic neuropathy, 28/55
uremic polyneuropathy, 51/361
Wernicke encephalopathy, 7/561, 28/15-17
Wernicke-Korsakoff syndrome, 28/243, 267, 45/197
Vitamin B$_1$ intoxication
animal experiment, 28/210
convulsion, 65/572
epilepsy, 65/572
headache, 65/572
paralysis, 65/572

Vitamin B$_1$ response
clinical features, 29/61
maple syrup urine disease, 29/53, 61, 80
treatment, 29/80
Vitamin B$_1$ transferase inhibitor
subacute necrotizing encephalomyelopathy, 42/626
Vitamin B$_2$
cheilosis, 28/200
experimental deficiency, 7/578, 28/205
hypoglycin intoxication, 65/105
Jamaican vomiting sickness, 65/105
lethal dose, 28/210
neuroleptic agent, 28/206
neurologic disorder, 28/205
phenothiazine, 28/206
requirement, 28/200
retardation, 7/578
retrobulbar neuropathy, 28/20
triad, 7/577
Vitamin B$_2$ deficiency
animal experiment, 28/206
ataxic neuropathy, 28/206, 51/332
burning feet syndrome, 28/13, 200, 51/332
cyanide intoxication, 51/332
deficiency neuropathy, 7/577, 51/332
hyperadrenocorticism, 51/332
malnutrition, 51/332
optic neuropathy, 51/332
pellagra, 28/205
phenothiazine, 51/332
retrobulbar neuropathy, 51/332
staple food, 51/332
thallium polyneuropathy, 51/332
Vitamin B$_2$ intoxication
phosphate, 28/214
Vitamin B$_3$
see also Vitamin PP
aura, 5/379
biochemistry, 28/59
brain infarction, 53/421
brain vasospasm, 11/125
clinical features, 51/333
deficiency neuropathy, 7/572, 576, 51/325
derivative, 7/572
lethal dose, 28/210
vitamin B$_6$, 28/111-114
Vitamin B$_3$ deficiency
see also Pellagra
alcoholic dementia, 46/336
burning feet syndrome, 28/13
deficiency neuropathy, 7/576, 51/333
delirium, 46/543

dementia, 51/333
dermatitis, 51/333
diarrhea, 51/333
endemic pellagra, 51/333
features, 51/325
Hartnup disease, 7/577
isoleucine, 51/333
leucine, 51/333
maize, 28/64, 51/333
malnutrition, 28/18
mental deficiency, 46/18
millet jowar, 51/333
neuropathy, 7/552
pellagra, 7/572
polyneuropathy, 51/333
retrobulbar neuropathy, 28/20
Vitamin B3 deficiency encephalopathy
polyneuropathy, 46/337
Wernicke-Korsakoff syndrome, 46/337
Vitamin B3 intoxication
animal experiment, 28/213
Vitamin B4 deficiency
deficiency neuropathy, 7/576
dementia, 28/201
Vitamin B5
burning feet syndrome, 28/207
deficiency neuropathy, 7/578, 51/325
lethal dose, 28/210
Melkersson-Rosenthal syndrome, 8/235
neuropathy, 7/553
nontoxic, 28/214
requirement, 28/200
tropical myeloneuropathy, 56/526
Vitamin B5 deficiency
burning feet syndrome, 28/200, 51/341
features, 51/325
polyneuropathy, 51/341
Vitamin B6, 28/105-132
acute intermittent porphyria, 28/130
aminoaciduria, 28/117-123
axonopathy, 51/264
celiac disease, 28/232
contraceptive, 28/117-120
cystathioninuria, 28/117-120, 29/116-119
daily requirement, 28/107
deficiency neuropathy, 51/325
dependence, 28/110-114
depression, 28/131
Down syndrome, 28/129
drug induced seizure, 28/114-117
epilepsy, 15/63, 316, 28/109-116, 114-117
GABA, 28/113
homocystinuria, 28/120-123, 29/115, 55/332

infantile myoclonic spasm, 28/130
isoniazid, 51/297
isoniazid leukoencephalopathy, 47/565
lethal dose, 28/210
levodopa, 28/124-128
Melkersson-Rosenthal syndrome, 8/235
neuronopathy, 51/264
neuropathy, 7/553, 70/432
oral contraceptive agent, 28/131
Parkinson disease, 28/123-129
parkinsonism, 28/123-129
pellagra, 28/71, 73
penicillamine, 28/116, 51/297
phenylketonuria, 28/128
pregnancy toxemia, 28/131
requirement, 28/107
seizure, 70/431
structure, 28/105
sulfur containing amino acid, 28/117, 122
tropical myeloneuropathy, 56/526
tryptophan nicotinic acid metabolism, 28/111-114
vitamin B3, 28/111-114
vitamin B6 deficiency, 42/716, 51/297, 334
xanthurenic aciduria, 42/636
Vitamin B6 abuse
ataxia, 51/298
vitamin B6 deficiency, 51/335
Vitamin B6 deficiency
acrodynia, 7/581
alcoholic dementia, 46/336
amino acid, 70/423
anemia, 51/297
antagonist, 51/334
antimetabolite, 7/581, 583
ataxia, 51/335
carbon disulfide intoxication, 64/24
carpal tunnel syndrome, 70/432
celiac disease, 28/232
clinical features, 51/334
convulsion, 42/716, 51/297
deficiency neuropathy, 7/580, 51/333
diabetic neuropathy, 70/432
dietary, 70/426
experimental deficiency, 7/586
features, 51/325
hydralazine, 63/87, 70/430
hydralazine neuropathy, 42/467, 51/297, 334
hydralazine polyneuropathy, 63/87
isoniazid, 42/468, 47/565, 51/297, 334
isoniazid neuropathy, 7/516, 533, 582, 42/648
isoniazid polyneuropathy, 51/334
kynureninase, 51/333
Lhermitte sign, 51/335

malabsorption syndrome, 70/229
mental deficiency, 42/716
motor polyneuritis, 7/582
neuropathology, 51/334
neuropathy, 7/580, 70/427
neurotoxicity, 51/294
pathology, 7/583
pellagra, 46/336
penicillamine, 51/297
penicillamine intoxication, 27/405, 49/235
phenelzine, 70/431
psychomotor retardation, 42/716
seizure, 28/109, 42/716, 51/334
serine palmitoyltransferase, 51/297
sphingomyelin synthesis, 51/297
status epilepticus, 42/716
survey, 70/423
tryptophan load test, 70/426
vitamin B6, 42/716, 51/297, 334
vitamin B6 abuse, 51/335
Vitamin B6 deficiency polyneuropathy
antihypertensive agent, 63/87
neuropathy, 51/298
sensory neuropathy, 51/298
Vitamin B6 dependency
convulsion, 42/690, 716
glutamate decarboxylase, 42/716
isoniazid intoxication, 65/480
mental deficiency, 42/716
prenatal onset epilepsy, 42/690
Vitamin B6 dependent xanthurenic aciduria
mental deficiency, 29/224
Vitamin B6 intoxication
animal experiment, 28/212
autonomic polyneuropathy, 51/479
axonal degeneration, 51/298, 65/572
demyelination, 65/572
ganglioradiculopathy, 65/572
gasserian ganglion, 65/572
homocystinuria, 65/572
mechanism, 51/299
muscle weakness, 65/572
nerve biopsy, 65/572
oxalosis, 65/572
pain, 65/572
paresthesia, 65/572
polyneuropathy, 65/480, 572
sensory ataxia, 51/298
sensory neuropathy, 65/572
sensory polyneuropathy, 65/572
Vitamin B6 metabolism
cycloserine, 70/430
cystathioninuria, 28/118

hydrazine, 70/430
penicillamine, 70/430
structure, 28/105-109
survey, 28/105, 70/425
xanthurenic aciduria, 42/635
Vitamin B6 overdose
autonomic nervous system, 75/502
Vitamin B6 status
measurement, 70/425
Vitamin B12
absorption, 28/186, 187
alcoholic amblyopia, 28/174
amblyopia, 28/173-178
child, 28/169
chronic axonal neuropathy, 51/531
congenital retinal blindness, 28/175
CSF, see CSF vitamin B12
cyanide intoxication, 65/32
cyanogenic glycoside intoxication, 36/518, 521,
65/32
deficiency neuropathy, 51/325
dependent biochemical reaction, 28/142
fatty acid, 70/381
features, 51/325
homocystinuria, 29/115, 55/331
intrinsic factor, 28/146
Leber optic atrophy, 28/175
lethal dose, 28/210
malabsorption, 7/596
megaloblastic madness, 38/21
metabolism, 70/369
methionine, 70/378
methionine synthesis, 28/144
methylmalonic acidemia, 42/601, 65/574
multiple sclerosis, 9/322, 368, 400, 47/155
nucleic acid metabolism, 28/144-146
optic atrophy, 28/175
pernicious anemia, 7/587, 592
pernicious psychosis, 7/587, 592
serum, 9/322
side effect, 28/212
structure, 28/142
subacute combined spinal cord degeneration,
7/587
tobacco amblyopia, 28/174
vitamin Bc, 28/144-147
Vitamin B12 binding protein
vitamin B12 deficiency, 51/337
Vitamin B12 deficiency
achlorhydria, 51/339
acquired immune deficiency syndrome, 70/387
apathy, 63/254
ataxia, 63/254, 70/377

autonomic nervous system, 75/491
autonomic neuropathy, 51/340
carcinoma, 38/650
cause, 28/148-150
celiac disease, 28/232
cerebral symptom, 70/383
Crohn disease, 7/620
cycloleucine, 51/339
cystic fibrosis, 51/339
deficiency neuropathy, 7/587, 51/327, 335
delirium, 46/543
dementia, 46/399
differential diagnosis, 7/553
diphyllobothrium latum, 7/631, 51/339
diphyllobothrium latum myelopathy, 7/631
dorsal column, 70/384
EMG, 51/340
endogenous perniciosa, 7/591
fatty acid, 70/381
gastrectomy, 51/339
history, 70/367
homocystinuria, 55/325
ileal receptor, 51/338
immunologic cause, 70/373
intrinsic factor, 51/337, 70/373
leukoencephalopathy, 47/584
malabsorption, 70/373
malabsorption syndrome, 70/233
malignancy, 38/650
malnutrition, 28/22-24
mental change, 51/340
metabolism, 51/337
methylmalonyl CoA carbamyl mutase deficiency,
    29/204
methylmalonyl CoA mutase, 51/338
methyltetrahydrofolate-homocysteine transferase,
    51/338
micturition, 51/341
muscular atrophy, 51/341
myxedema, 70/373
neuroglia relationship, 28/412
neuropathology, 51/341
neuropathy, 7/587-594, 28/179-184, 70/380
nitrous oxide, 70/389
optic neuropathy, 70/382
orthostatic hypotension, 51/341, 63/152
pain, 70/377
pancreatic insufficiency, 51/339
paresthesia, 70/377
pathophysiology, 70/373
pernicious anemia, 38/19, 42/609, 51/338, 340,
    70/371, 373
poliosis, 60/667

polyneuropathy, 7/553, 51/340
retrobulbar neuropathy, 28/24
Schilling test, 7/553
sexual impotence, 51/341
spinal cord, 70/384
spinal lesion, 70/370
subacute combined spinal cord degeneration,
    46/399, 51/339, 63/254, 70/375
subtle vitamin B_{12} deficiency, 70/394
survey, 70/367
symptom, 70/373
transcobalamin, 51/338
treatment, 51/341, 70/395
trigeminal neuralgia, 5/300
tropical myeloneuropathy, 56/526
tropical sprue, 7/629
vegetarian, 51/339
visual impairment, 63/254
vitamin B_{12} binding protein, 51/337
vitamin B_c, 70/380
Vitamin B_{12} deficiency anemia
    see also Megaloblastic anemia
    biochemistry, 63/253
    diagnostic test, 63/253
    gastric resection, 63/254
    hereditary vitamin B_{12} metabolism defect, 63/254
    ileal resection, 63/254
    Ismerlund-Grasbeck disease, 63/254
    nitrous oxide, 63/254
    subacute combined spinal cord degeneration,
        63/254
    tapeworm infection, 63/254
Vitamin B_{12} depletion
    malabsorption syndrome, 70/229
Vitamin B_{12} intoxication
    arterial thrombosis, 28/213
    hypoglycemia, 28/212
    methylmalonic acidemia, 65/574
    venous thrombosis, 28/213
    vitamin B_{12}a, 65/574
Vitamin B_{12} neuropathy
    see also Pernicious anemia
    animal, 7/594
    chemistry, 7/590
    demyelination, 7/593
    subacute combined spinal cord degeneration,
        7/587, 592
Vitamin B_{12}a
    tobacco amblyopia, 28/175
    vitamin B_{12} intoxication, 65/574
Vitamin B_c
    anemia, 28/200
    anticonvulsant, 28/203-205

antiepileptic agent, 65/516
antiepileptic treatment, 15/729
celiac disease, 28/232
chronic axonal neuropathy, 51/531
congenital defect, 28/203
Crohn disease, 7/620
CSF, *see* CSF vitamin B$_C$
deficiency neuropathy, 51/325
dementia, 28/201
homocystinuria, 55/325
lethal dose, 28/210
megaloblastic madness, 38/21
neuropathy, 7/553, 28/200
phenytoin intoxication, 65/503
psychiatric disorder, 28/202
requirement, 28/200
tropical myeloneuropathy, 56/526
vitamin B12, 28/144-147
vitamin B12 deficiency, 70/380
Vitamin B$_C$ deficiency
   alcoholism, 51/336
   aminopterin syndrome, 46/78
   amnesia, 63/255
   anticonvulsant, 28/203-205, 46/24
   antiepileptic agent intoxication, 37/208-210
   celiac disease, 28/232, 51/336
   clinical features, 51/332, 336
   Crohn disease, 51/336
   CSF, 51/336
   deficiency neuropathy, 51/335
   delirium, 46/543
   dementia, 46/399, 63/255
   dihydrofolate, 51/336
   dihydrofolate reductase, 51/336
   features, 51/325
   gastrectomy, 51/336
   homocystinuria, 51/336, 55/325
   insomnia, 63/255
   Kearns-Sayre-Daroff-Shy syndrome, 51/336
   malabsorption syndrome, 70/230
   malnutrition, 28/24
   megaloblastic anemia, 38/23, 51/336, 63/253
   5,10-methylene-*N*-methyltetrahydrofolate
      reductase deficiency, 51/336
   methyltetrahydrofolate, 51/335
   neuropathy, 28/200-202
   phenytoin, 51/336
   phenytoin intoxication, 42/641
   phenytoin polyneuropathy, 51/308
   polyneuropathy, 51/336
   restless legs syndrome, 51/336, 546
   subacute combined spinal cord degeneration,
      63/255

symptom, 28/200
tea and toast diet, 51/336
tropical sprue, 7/629
Vitamin B$_C$ deficiency anemia
   subacute combined spinal cord degeneration,
      63/255
Vitamin B$_C$ intoxication
   anaphylactoid reaction, 28/214
   animal experiment, 28/214
   animal status epilepticus, 65/573
   anxiety, 65/574
   dyssomnia, 28/214
   hypomania, 65/574
   insomnia, 65/574
   mental change, 28/214
Vitamin B$_C$ transport defect
   ataxia, 43/278
   mental deficiency, 43/278
   seizure, 43/278
Vitamin B$_t$
   fatty acid oxidation disorder, 66/399
Vitamin B$_W$
   animal experiment, 28/215
   apocarboxylase, 66/418
   epilepsy, 72/222
   lethal dose, 28/210
   neurotropic virus, 56/26
   pyruvate carboxylase deficiency, 42/586, 66/418
   requirement, 28/200
Vitamin B$_W$ deficiency
   deficiency neuropathy, 7/579
   experimental deficiency, 28/200, 207
   myoclonus, 49/618
Vitamin B$_W$ metabolism
   defect, 66/396
   pyruvate carboxylase deficiency, 66/396
Vitamin C
   aluminum intoxication, 64/275
   animal experiment, 28/215
   cyanide intoxication, 65/32
   cyanogenic glycoside intoxication, 65/32
   cystinosis, 29/179
   dopamine, 74/145
   free radical scavengers, 64/19
   lethal dose, 28/210, 215
   noradrenalin, 74/145
   pharmacology, 74/145
   pressure sore, 61/350
   requirement, 28/200
   restless legs syndrome, 8/318
   side effect, 28/215
Vitamin C deficiency
   animal experiment, 28/200, 209

bleeding tendency, 28/200, 208
Möller-Barlow disease, 28/200, 208
psychiatry, 28/200, 208
scurvy, 28/200, 208, 38/71
symptom, 28/200, 208
Vitamin D
    epilepsy, 15/730
    hyperparathyroidism, 41/247
    osteomalacia, 41/247
    parathyroid disease, 27/284
Vitamin D deficiency
    deficiency neuropathy, 51/326
    limb girdle syndrome, 40/438, 62/188
    postgastrectomy syndrome, 28/233, 51/327
Vitamin D2
    pseudohypoparathyroidism, 42/578
Vitamin D3 intoxication
    1,25-dihydroxyvitamin D3, 65/569
    epidemic, 65/569
    hypercalcemia, 65/569
    mental depression, 65/569
    metabolite, 65/569
    parvalbumin, 65/569
Vitamin deficiency
    autonomic nervous system, 75/492
    metabolic neuropathy, 51/68
Vitamin E
    Bassen-Kornzweig syndrome, 63/277
    chronic axonal neuropathy, 51/531
    lipopigment dystrophy, 21/67
    malabsorption syndrome, 70/230
    motoneuron disease, 22/31
    secondary pigmentary retinal degeneration,
        13/194
    tropical myeloneuropathy, 56/526
Vitamin E deficiency
    autonomic nervous system, 75/492
    Bassen-Kornzweig syndrome, 51/326, 394,
        63/276, 70/443
    blind loop syndrome, 63/277
    cerebellum, 70/448
    CNS, 70/446
    cystic fibrosis, 63/277
    deficiency neuropathy, 51/326
    diagnosis, 70/449
    EMG, 70/445
    ERG, 70/446
    excessive, 70/450
    experimental myopathy, 40/170
    limb girdle syndrome, 62/188
    lipid peroxidation, 40/171
    lipopigment, 21/58
    lipopigment dystrophy, 21/67

malabsorption state, 70/442
malabsorption syndrome, 70/228, 444
memory disorder, 27/464
muscle, 70/448
muscular atrophy, 70/228
myopathology, 70/447
nerve conduction, 7/166, 70/445
neuroaxonal dystrophy, 6/626, 49/394
neuropathology, 63/277
neuropathy, 60/660
obstructive liver disease, 63/278
oxidative phosphorylation, 40/171
peripheral nerve, 70/447
posterior column, 70/448
retina, 70/448
Rottnest quokka, 40/171
secondary neuroaxonal dystrophy, 49/408
somatosensory evoked potential, 70/445
steatorrhea, 60/668
survey, 70/439
symptom, 70/442
Vitamin E intoxication
    anticoagulant, 65/570
    lipid peroxidation, 65/570
    necrotizing myopathy, 65/570
    α-tocotrienol, 65/569
Vitamin E replacement
    method, 70/450
Vitamin G, see Vitamin B2
Vitamin K
    epilepsy, 15/730
Vitamin K deficiency
    Bassen-Kornzweig syndrome, 51/326
    brain matrix hemorrhage, 53/30
    cerebrovascular disease, 55/476
    coagulopathy, 63/309
    mitral valve prolapse, 9/607
Vitamin K1 deficiency
    subarachnoid hemorrhage, 12/126
Vitamin K2
    bilirubin, 6/492
Vitamin K2 deficiency
    subarachnoid hemorrhage, 12/126
Vitamin K3 intoxication
    hemolytic anemia, 65/570
    hyperbilirubinemia, 65/570
    kernicterus, 65/570
Vitamin overdose
    autonomic nervous system, 75/502
Vitamin PP
    see also Vitamin B3
    acquired myoglobinuria, 62/577
    biochemistry, 7/576

Huntington chorea, 49/298
  pellagra, 7/572, 574, 28/59, 71, 46/336
Vitamin PP deficiency
  Hartnup disease, 21/564
  pellagra, 7/576, 28/18, 59, 68, 38/648, 46/336
Vitamin PP intoxication
  acanthosis nigricans, 65/573
  amblyopia, 65/573
  panic attack, 65/573
Vitamins
  ascorbic acid, see Vitamin C
  biotin, see Vitamin B$_W$
  L-carnitine, see Vitamin B$_t$
  cyanocobalamin, see Vitamin B$_{12}$
  farnoquinone, see Vitamin K$_2$
  folate, see Vitamin B$_c$
  folic acid, see Vitamin B$_c$
  nicotinamide, see Vitamin B$_3$
  nicotinic acid, see Vitamin PP
  pantothenic acid, see Vitamin B$_5$
  pyridoxine, see Vitamin B$_6$
  retinol, see Vitamin A
  riboflavin, see Vitamin B$_2$
  tetrahydrofolic acid, see Vitamin B$_c$
  thiamine, see Vitamin B$_1$
  α-tocopherol, see Vitamin E
Vitelliform foveal dystrophy
  adult type, 13/266
  congenital, see Vitelliruptive macula degeneration
  differential diagnosis, 13/133
  EOG, 13/267
  infantile type, 13/266
  juvenile, see Stargardt disease
Vitelliform macula cyst
  congenital, see Vitelliruptive macula degeneration
Vitelliruptive macula degeneration
  dominant inheritance, 13/266
  primary pigmentary epithelium degeneration,
    13/29
Vitiligo
  Eaton-Lambert myasthenic syndrome, 62/421
  Maffucci-Kast syndrome, 43/29
  Maffucci syndrome, 14/119
  Mollaret meningitis, 34/549, 56/631
  secondary pigmentary retinal degeneration,
    13/186
  tuberous sclerosis, 14/342
  Vogt-Koyanagi-Harada syndrome, 56/611, 622
Vitreoretinal degeneration
  autosomal recessive, see Hyaloidotapetoretinal
    degeneration (Goldmann-Favre)
  category, 13/36
  hereditary, see Hereditary vitreoretinal

degeneration
  hereditary retinal detachment, 13/37
  hyaloidoretinal degeneration (Wagner), 13/37
  sex linked idiopathic retinoschisis, 13/37
Vitreoretinal traction disease
  Stargardt disease, 13/134
Vitreotapetoretinal dystrophy
  sex linked juvenile retinoschisis, 13/139
Vitreous hemorrhage disease
  Eales, see Eales vitreous hemorrhage disease
Vitreous opacity
  hereditary amyloid polyneuropathy type 1,
    42/518, 51/424
  hereditary amyloid polyneuropathy type 2,
    42/520, 51/423
  hyaloidoretinal degeneration (Wagner), 13/178
  hyaloidotapetoretinal degeneration
    (Goldmann-Favre), 13/178
VLA-4, see Very late antigen 4
Vocal amusia
  frontal lobe lesion, 45/488
  musical dyslexia, 45/486
Vocal cord
  ataxia, 1/314
Vocal cord paralysis
  accessory nerve, 30/411
  alcoholic polyneuropathy, 51/316
  brachial neuritis, 42/303
  bulbopontine paralysis, 42/96
  congenital, see Vagus nerve agenesis
  Fazio-Londe disease, 42/95
  glossopharyngeal nerve, 30/411
  multiple neuronal system degeneration, 59/138
  progressive bulbar palsy, 59/219
  spinal muscular atrophy, 59/373
  syringobulbia, 50/445
  Tapia syndrome, 2/102
  tetanus, 65/218
  vagus nerve, 30/411
Vocal fold
  autonomic nervous system, 74/388
Vocal tremor
  task specific tremor, 49/592
Vocalization
  Gilles de la Tourette syndrome, 6/788, 49/629
Vocational planning
  future, 41/483
Vogt, C., 1/9, 27
Vogt, H., 10/214-217
Vogt-Hopf syndrome, see Chorea-acanthocytosis
Vogt-Koyanagi disease, see Vogt-Koyanagi-Harada
    syndrome
Vogt-Koyanagi-Harada syndrome, 34/513-540,

56/611-624
see also Uveitis *and* Uveomeningitis
age distribution, 56/612
alopecia, 56/611, 622
ataxia, 56/621
Behçet syndrome, 56/623
chronic meningitis, 56/645
classification, 56/612
clinical course, 56/612
clinical features, 34/6514-528
convalescent type, 56/613
course, 34/528
cranial nerve palsy, 56/621
CSF, 34/527, 56/621
CSF pleocytosis, 56/611
Dalen-Fuchs nodule, 56/613
deafness, 56/613, 622
depigmentation, 56/613
dermatologic features, 34/527
diagnosis, 34/529-532
diagnostic criteria, 56/613
differential diagnosis, 56/623
dysacusis, 56/611
EEG, 34/526, 56/621
epilepsy, 56/620
etiology, 34/534-536, 56/623
granulomatous inflammation, 56/618
histopathology, 56/615
history, 56/611
immunology, 56/623
incidence, 56/612
initial symptom, 56/614
meningitis, 56/611
Mollaret meningitis, 34/549, 56/629
multiple sclerosis, 56/623
neurologic manifestation, 34/518-527
neurologic sequela, 34/529
neurology, 56/619
neuropathology, 34/533
ocular manifestation, 34/515-518
ocular pathology, 34/532
ocular sequela, 34/529
oculodermal depigmentation, 56/612
optic atrophy, 13/83
paraplegia, 56/621
pathology, 34/532-534
poliosis, 56/622
presenting symptom, 34/514
prevalence, 34/514
recovery type, 56/612
recurrent type, 56/612
retinal ablation, 56/611
sarcoidosis, 56/623

sequela, 34/529
spastic paraplegia, 59/436
Sugiura sign, 56/614
sympathetic ophthalmia, 56/612
sympathetic ophthalmitis, 34/537-540
tinnitus, 56/622
transverse myelitis, 56/620
treatment, 34/536, 56/623
ultrastructure, 56/617
uveitis, 56/611
vertigo, 2/360
viral meningitis, 56/611
vitiligo, 56/611, 622
Vogt-Koyanagi syndrome, *see*
    Vogt-Koyanagi-Harada syndrome
Vogt, O., 1/9
Vogt striatal athetotic syndrome
    posthemiplegic athetosis, 49/386
Vogt triad
    neurofibromatosis, 68/291
    tuberous sclerosis, 68/291
Vogt-Waardenburg syndrome
    hypertelorism, 30/246
    skull dysplasia, 30/246
Voice
    autonomic nervous system, 74/388
    cerebral palsy, 46/30
    high pitched, *see* High pitched voice
    hoarse, *see* Hoarse voice
    Huntington chorea, 49/277
    nasal, *see* Nasal voice
Volition
    migraine, 48/162
Voltage clamp
    study, 47/15, 19
Volume pressure response
    brain injury, 23/140
    brain microcirculation, 23/140, 142
    intracranial pressure, 23/217
    intracranial pressure monitoring, 57/219
Volume ratio
    axon-perikaryon, *see* Axon-perikaryon volume
    ratio
Vomeronasal organ
    human, 74/376
    odorant, 74/376
    olfactory system, 74/376
    pheromone, 74/376
Vomiting
    acute mercury intoxication, 64/375
    alcohol intoxication, 64/112
    apistus intoxication, 37/76
    attack, 5/46

autonomic nervous system, 74/74
Balaenoptera borealis intoxication, 37/94
basilar artery migraine, 48/138
brain embolism, 53/206
brain infarction, 53/206
brain injury, 23/12
brain tumor, 16/129, 67/143
buspirone intoxication, 65/344
carbamate intoxication, 64/186
cardiovascular agent intoxication, 37/426
cat scratch disease, 52/127
cephalopoda intoxication, 37/62
cerebellar tumor, 17/710
cerebrovascular disease, 53/206
chelonia intoxication, 37/89
chloral hydrate intoxication, 37/348
cinchonism, 65/452
cluster headache, 48/222
congestin intoxication, 37/37
crab intoxication, 37/58
cubomedusae intoxication, 37/43
digoxin intoxication, 65/450
echinoidea intoxication, 37/67
elfin face syndrome, 31/317
episodic, *see* Episodic vomiting
ethchlorvynol intoxication, 37/354
far eastern tick-borne encephalitis, 56/139
fish roe intoxication, 37/86
fluroxene intoxication, 37/412
fox intoxication, 37/96
fumarate hydratase deficiency, 66/420
glycoside intoxication, 37/426
gram-negative bacillary meningitis, 52/119
gymnapistes intoxication, 37/76
headache, 48/32, 75/286
hereditary periodic ataxia, 21/564
hereditary sensory and autonomic neuropathy type
    III, 1/475, 21/109, 60/26
hexachlorophene intoxication, 37/486, 65/476
hyperpipecolatemia, 29/222
hypervalinemia, 42/574
ichthyohaemotix fish intoxication, 37/87
ichthyohepatoxic fish intoxication, 37/87
intracranial hemorrhage, 53/206
Jamaican vomiting sickness, 37/512, 65/79
Köhlmeier-Degos disease, 39/437
lateral medullary infarction, 53/381
loxosceles intoxication, 37/112
lymphocytic choriomeningitis, 9/197, 56/359
marine toxin intoxication, 37/33, 65/153
meningococcal meningitis, 33/23, 52/24
merostomata intoxication, 37/58
methyl salicylate intoxication, 37/417

migraine, 48/3, 8, 35, 117, 125, 156, 158
motion sickness, 74/354
neptunea intoxication, 37/64
neuropathy, 8/14
nicotine intoxication, 36/427, 65/261
nothesthes intoxication, 37/76
occipital headache, 48/137
octopoda intoxication, 37/62
ostrea gigas intoxication, 37/64
oxoglutarate dehydrogenase deficiency, 62/503
paraldehyde intoxication, 37/349
paralytic shellfish poisoning, 37/33, 65/153
pediatric head injury, 57/329
pellagra, 6/748, 38/648
periodic ataxia, 60/647
phenylketonuria, 29/30
pheochromocytoma, 39/500, 48/427
physiology, 1/505
polar bear intoxication, 37/96
primary pontine hemorrhage, 12/39
procainamide intoxication, 37/452
propionic acidemia, 27/72, 28/520, 29/208
pterois volitans intoxication, 37/75
Reye syndrome, 9/549, 27/363, 29/332, 31/168,
    34/80, 169, 56/150, 238, 63/438, 65/116
salicylic acid intoxication, 37/417
scorpaena intoxication, 37/76
scorpaenodes intoxication, 37/76
scorpaenopsis intoxication, 37/76
sea snake intoxication, 37/92
sebastapistes intoxication, 37/76
sodium salicylate intoxication, 37/417
stingray intoxication, 37/72
subarachnoid hemorrhage, 55/14
subdural hematoma, 57/337
superior cerebellar artery syndrome, 12/20
transient ischemic attack, 53/206
traumatic vegetative syndrome, 24/577
uremia, 63/504
venerupis semidecussata intoxication, 37/64
vestibular neuronitis, 56/117
vitamin A intoxication, 37/96, 65/568
wolf intoxication, 37/96
zinc intoxication, 36/337
zozymus aeneus intoxication, 37/56
Vomiting sickness
    Jamaican, *see* Jamaican vomiting sickness
Vomiting syncope
    dysautonomia, 22/264
Von Baló disease, *see* Concentric sclerosis
Von Baló, J., 10/10
Von Bechterew, V.M., 1/11
Von Bergmann, E., 1/17

Von Economo, C., 1/10
Von Economo disease, *see* Encephalitis lethargica
Von Economo lethargic encephalitis, *see* Encephalitis lethargica
Von Frankl-Hochwart, L., 1/10
Von Gierke disease, *see* Glycogen storage disease type I
Von Gräfe-Liebreich-Hallgren syndrome, *see* Usher syndrome
Von Gräfe-Lindenov syndrome, *see* Lindenov-Hallgren syndrome
Von Gräfe sign
  Dalrymple sign, 2/318
  description, 2/294
  hyperthyroidism, 2/318
  myotonia congenita, 2/324
  pseudo, *see* Pseudo-Von Gräfe sign
  upward gaze paralysis, 2/338
Von Gräfe-Sjögren syndrome
  Flynn-Aird syndrome, 14/113
Von Gräfe syndrome, *see* Progressive external ophthalmoplegia
Von Hippel-Czermak disease, *see* Von Hippel-Lindau disease
Von Hippel-Lindau disease, 14/241-255
  adiadochokinesia, 14/115
  associated lesion, 14/647
  ataxia telangiectasia, 14/252
  autosomal dominant, 50/374
  Babinski sign, 42/156
  brain, 12/236
  brain arteriovenous malformation, 31/199
  cerebellar cyst, 14/53, 242-244, 16/31
  cerebellar hemangioblastoma, 14/13, 242-244, 18/286, 401, 31/16, 50/375, 68/294
  cerebellar hypoplasia, 42/18
  cerebellar tumor, 43/21
  chromosome, 67/57
  clinical features, 14/246, 31/17, 50/374
  CNS tumor, 30/101, 67/54
  consciousness, 14/115
  dementia, 43/21
  diagnosis, 68/294
  dizziness, 14/115
  endocrinology, 14/746
  epidemiology, 30/159, 31/1, 68/274
  epididymal cystadenoma, 50/375
  epididymis, 14/13, 245
  epilepsy, 50/374
  erythrocytosis, 31/18
  forme fruste, 14/245
  gait disturbance, 14/115
  general concept, 14/88
  genetics, 14/242, 502-504, 16/31, 30/159, 31/16
  glaucoma, 14/647, 50/375
  globoid cell leukodystrophy, 14/7
  Gruber syndrome, 14/60, 252, 505
  headache, 14/115
  hemangioblastoma, 14/242, 18/401, 20/449, 68/294
  hemihypertrophy, 22/549
  hereditary, 14/503
  history, 14/2-4, 241
  hypernephroma, 14/242, 245, 249, 695, 16/31
  hyperreflexia, 42/156
  kidney lesion, 14/13, 60, 245
  Klippel-Trénaunay syndrome, 14/398, 532
  Krabbe-Bartels disease, 14/7
  laboratory finding, 14/249
  lipomatosis, 14/532
  mental deficiency, 46/44, 50/374
  molecular biology, 68/275
  mutation rate, 14/503
  neurocutaneous syndrome, 14/101
  neurofibromatosis, 14/6, 251, 532
  neurofibromatosis type I, 14/133, 568
  neuropathology, 14/53-61, 243
  nuchal rigidity, 14/115
  ophthalmology, 14/644-649
  orthostatic hypotension, 14/255
  pancreas lesion, 14/13, 60, 245
  pancreatic cyst, 14/13, 242, 694, 16/31, 50/375
  papilledema, 43/21, 60/655
  pathology, 31/16
  phakomatosis, 14/19
  pheochromocytoma, 14/60, 249, 252, 747, 750, 39/503, 42/765, 43/21, 50/375
  pneumothorax, 14/255
  polycythemia, 14/249, 43/21, 60/674
  polycythemia vera, 14/249, 701
  postlateral myelopathy, 42/156
  Potter syndrome type II, 14/118
  prevalence, 14/502
  pseudoxanthoma cell, 14/244
  psychiatric disorder, 14/766, 60/665
  renal carcinoma, 43/21, 50/374
  renal cyst, 16/31
  retinal angioma, 14/115, 16/31
  retinal hamartoma, 50/374
  retinal hemangioblastoma, 14/13, 245, 248, 645, 43/21
  retinal tumor, 31/16
  roentgenology, 14/250
  secondary pigmentary retinal degeneration, 60/735
  spastic paraplegia, 42/156, 43/21

spastic tetraplegia, 43/21
spinal cord hemangioblastoma, 14/59, 244, 248,
    50/375, 68/498, 507, 575
spinal cord tumor, 16/31, 43/21
Sturge-Weber syndrome, 14/7, 13, 251-253, 532
syringomyelia, 14/13, 31/17, 32/287, 43/21
tonsillar herniation, 14/115
treatment, 14/253
tuberous sclerosis, 14/251
tumor cell, 67/34
Ullmann syndrome, 14/447
venous malformation, 12/236
vertigo, 14/115
visceral cyst, 50/374
visceral lesion, 14/60, 242, 247, 691-696
Zinn circle, 2/517
Von Leyden, E., 1/8
Von Möllendorff Hemikranie
    cluster headache, 48/217
Von Monakow, K., 1/2, 10
Von Recklinghausen disease, *see* Neurofibromatosis
    type I
Von Recklinghausen neurofibromatosis, *see*
    Neurofibromatosis type I
Von Romberg classification
    nerve disease, 7/200
    pathology, 7/200
Von Romberg disease, *see* Hemifacial atrophy *and*
    Progressive hemifacial atrophy
Von Romberg, M.H., 1/4, 8
Von Romberg sign
    astasia abasia, 1/311, 345, 2/420
    ataxia telangiectasia, 60/364
    Biemond syndrome, 42/165
    labyrinthine disease, 1/334
    late cerebellar atrophy, 42/135
    multiple hamartoma syndrome, 42/755
    posterior column ataxia, 42/165
    tabes dorsalis, 52/280
Von Romberg test
    equilibrium, 1/344
    sensory neuropathy, 42/348
Von Stein test
    ataxia, 1/335
Von Strümpell-Lorrain disease, *see* Familial spastic
    paraplegia
Von Weizsäcker, V., 1/2, 8, 28
Von Willebrand disease
    bleeding time, 63/312
    cerebrovascular disease, 53/43, 55/472, 475
    classification, 63/311
    clotting time, 63/312
    coagulopathy, 63/309, 311

head trauma, 63/312
hemostasis, 63/302
treatment, 63/312
type I, 63/311
Von Winiwarter-Buerger disease, *see*
    Thromboangiitis obliterans
VP 16-213, *see* Etoposide
Vrolik disease
    osteogenesis imperfecta, 22/494
Vulpian-Bernhardt disease, *see* Scapulohumeral
    spinal muscular atrophy
Vulpian-Bernhardt scapulohumeral muscular
    atrophy, *see* Scapulohumeral spinal muscular
    atrophy
Vulpian, E.F.A., 1/7
VX, *see* O-Ethyl
    S-[2-(diisopropylamino)ethyl]methylphosphono
    thionate

W index
    dystopia canthorum, 43/51
W syndrome
    antimongoloid eyes, 43/271
    hyperreflexia, 43/271
    mental deficiency, 43/271
    seizure, 43/271
Waardenburg-Shah syndrome
    congenital megacolon, 75/635
Waardenburg syndrome, *see* Klein-Waardenburg
    syndrome
Wackeltremor
    definition, 2/481
Wada test
    aphasia, 4/88
    cerebral dominance, 4/2, 89, 266, 318-320
    epilepsy, 72/399
Wadia sign
    brain arteriovenous malformation, 31/188
Wadia-Swami olivopontocerebellar atrophy, *see*
    Olivopontocerebellar atrophy (Wadia-Swami)
Wadia-Swami syndrome, *see* Olivopontocerebellar
    atrophy (Wadia-Swami)
Wadia test
    cerebrovascular disease, 11/212
Wagner hereditary vitreoretinal degeneration, *see*
    Hereditary vitreoretinal degeneration
Wagner-Meissner corpuscle
    neurofibroma, 8/444
Wagner Von Jauregg, J., 1/10
Wakefulness
    ascending reticular activating system, 3/171
    delirium, 46/529
    encéphalé isolé, 2/280

multiple neuronal system degeneration, 59/159
progressive dysautonomia, 59/159
Waking
  attention, 3/171-181
  conscious experience, 3/171-181
  EEG criteria, 1/490
  epilepsy, 15/457
  mechanism, 3/64
  neurophysiology, 3/171-181
  REM cycle, 1/492
  terminology, 3/52
Waldenström macroglobulinemia
  amyloid deposition, 63/397
  anoxic ischemic leukoencephalopathy, 47/541
  aphasia, 63/396
  ataxia, 63/396, 398
  brain hemorrhage, 55/486, 63/396
  brain infarction, 55/486, 63/396
  chronic axonal neuropathy, 51/531
  chronic inflammatory demyelinating
    polyradiculoneuropathy, 51/530
  classification, 69/290
  clinical features, 63/396
  coma, 63/396, 398
  confusion, 63/396
  confusional state, 63/396
  cryoglobulinemia, 39/181
  CSF, 18/246
  deafness, 55/486, 63/398
  demyelinating neuropathy, 47/616, 619, 63/397
  diplopia, 63/397
  dizziness, 63/398
  EEG, 63/398
  EMG, 63/397
  encephalopathy, 55/486, 63/396
  facial paralysis, 63/396
  fatigue, 63/396
  Guillain-Barré syndrome, 63/397
  headache, 55/486, 63/396, 398
  hearing loss, 63/396
  hemiplegia, 63/396
  hyperviscosity syndrome, 55/485, 63/397
  IgM paraproteinemia, 63/397
  immunocompromised host, 56/470
  lethargy, 63/396
  mental change, 63/396
  monoclonal gammopathy, 63/396
  monoclonal protein, 63/397
  mononeuritis multiplex, 63/397
  motoneuron disease, 63/397
  myelin lamella widening, 63/399
  myelin sheath, 63/399
  myelopathy, 63/396

  neurologic complication, 63/396
  papilledema, 63/398
  paraplegia, 63/396
  paraproteinemia, 69/290, 306
  paraproteinemic polyneuropathy, 51/429
  plasma cell dyscrasia, 63/391
  primary amyloidosis, 51/415
  psychosis, 63/396
  reticuloendothelioma, 18/235
  reticulosarcoma, 18/237
  retinal hemorrhage, 63/396
  retinal vessel tortuosity, 63/397
  segmental demyelination, 51/71
  sensorimotor neuropathy, 63/397
  sensorimotor polyneuropathy, 63/397
  sensory neuropathy, 63/397
  sensory polyneuropathy, 63/397
  sign, 55/485
  spastic paraparesis, 63/396
  spinal cord compression, 63/396
  spinal muscular atrophy, 63/396
  subarachnoid hemorrhage, 12/110, 55/485, 63/396
  survey, 71/436
  symptom, 55/485, 71/436
  tetanus, 33/532
  tetraplegia, 63/396
  treatment, 63/398
  undetermined significant monoclonal
    gammopathy, 47/619
  vertigo, 63/398
  visceromegaly, 63/396
  visual impairment, 63/396
  weakness, 63/396
  weight loss, 63/396
Waldenström macroglobulinemia polyneuropathy
  autoimmune disease, 41/82
  cauda equina syndrome, 51/432
  classification, 18/235
  clinical features, 39/190, 51/432, 63/397
  deafness, 51/432
  demyelinating neuropathy, 51/433
  EMG, 51/432
  encephalopathy, 39/191, 532, 51/432
  history, 39/189
  hyperviscosity syndrome, 39/532
  IgM macroglobulin, 51/432
  multiple cranial neuropathy, 51/432
  myelopathy, 39/192, 532
  nerve conduction velocity, 51/433
  neuropathy, 39/193
  pathogenesis, 39/194-196, 51/433
  pathology, 51/433
  RES tumor, 18/237

reticular system, 18/246
retinopathy, 39/190
sensorimotor neuropathy, 51/432
stroke, 39/191
subarachnoid hemorrhage, 12/110, 39/191, 532,
    51/432
treatment, 39/196, 51/433
ultrastructure, 51/433
Waldeyer neuron theory
  history of neurology, 1/46
Walker, A.E., 1/38
Walker syndrome, *see* HARD syndrome *and*
    HARDE syndrome
Walker-Warburg syndrome, *see* HARD syndrome
    and HARDE syndrome
Walking, *see* Gait
Wallenberg, A., 1/9
Wallenberg artery
  syncopic cervical syndrome, 7/452
Wallenberg syndrome, *see* Lateral medullary
    infarction
Wallerian degeneration
  acetonylacetone intoxication, 64/89
  acetyl coenzyme A synthetase, 7/215
  acid phosphatase, 7/215, 9/36
  adenosine triphosphatase, 9/37
  adenosine triphosphate, 7/215
  aliphatic esterase, 7/214, 9/37
  alkaline phosphatase, 7/215
  aluminum intoxication, 64/276
  axonal lysosome, 9/36
  axoplasm, 9/36
  Bi-Col procedure, 9/35
  Binswanger disease, 9/590
  butyrylcholinesterase, 7/215
  chemistry, 7/212
  choline acetylase, 7/214
  choline kinase, 7/215
  cholinephosphotransferase, 7/215
  cholinesterase, 7/214
  CNS myelin, 9/36
  CNS-nerve myelin difference, 9/36
  complex IV, 9/37
  compression neuropathy, 51/90
  DPN diaphorase, 9/37
  elastase, 51/33
  enzyme, 7/214
  experimental allergic encephalomyelitis, 9/38
  fructose-bisphosphate aldolase, 7/217
  fumarate hydratase, 7/217
  β-galactosidase, 7/215, 9/36
  globule ellipsoid stage, 9/34
  glucose-6-phosphate dehydrogenase, 7/217

β-glucosidase, 9/36
  β-glucuronidase, 7/215, 9/36
  glycerol-3-phosphate dehydrogenase, 7/217
  hereditary sensory and autonomic neuropathy type
    I, 21/75
  history, 7/198, 202
  hypertrophic interstitial neuropathy, 21/162
  idiopathic facial paralysis, 2/68
  incisure frequency, 47/4
  indoxyl esterase, 9/36
  isocitrate dehydrogenase, 7/215
  isoniazid, 47/565
  isoniazid neuropathy, 7/174
  lactate dehydrogenase, 7/215, 9/37
  leucine aminopeptidase, 9/36
  lightning injury, 7/212
  lipid, 7/212
  macrophage, 51/64
  malate dehydrogenase, 9/37
  Marchi staining, 9/35
  multiple sclerosis, 47/222
  myelin constituent, 51/33
  nerve, 7/153
  nerve conduction, 7/153
  nerve myelin, 9/36
  neuropathology, 7/202-219, 51/63
  neutral proteinase, 7/215
  nucleic acid, 7/213
  5'-nucleotidase, 7/215
  organophosphate induced delayed neuropathy,
    64/229
  P_0 protein, 51/32
  peptidase, 7/215
  plasminogen, 51/33
  polyarteritis nodosa, 12/659, 51/453
  proteinase, 9/36
  purine nucleoside phosphorylase, 7/215, 9/36
  rabies, 56/395
  rate, 9/36
  reduced nicotinamide adenine dinucleotide
    phosphate dehydrogenase, 9/37
  satellitosis, 9/37
  second stage, 9/35
  spinal cord injury, 61/480
  stage, 9/34-37
  succinate dehydrogenase, 9/37
  sudanophilic stage, 9/36
  time scale, 51/64
  ultrastructure, 7/208, 9/36, 51/63
  uremic neuropathy, 7/166, 8/2
  Wegener granulomatosis, 51/451
Wallgren syndrome
  lymphocytic choriomeningitis virus, 56/359

viral meningitis, 56/359
Walsh syndrome
  deafness, 13/326
Walshe, F.M.R., 1/2, 23, 26, 31, 35
Walter, W.G., 1/24
Warburg syndrome, *see* HARD syndrome *and*
    HARDE syndrome
Ward-Gorlin-Goltz syndrome, *see* Multiple nevoid
    basal cell carcinoma syndrome
Ward syndrome, *see* Multiple nevoid basal cell
    carcinoma syndrome
Warfarin
  anticoagulant, 63/306
  anticoagulant overdose, 65/460
  Conradi-Hünermann syndrome, 31/275
  Dandy-Walker syndrome, 50/333
  hemostasis, 63/306
  radiation myelopathy, 61/210
  teratogenic agent, 30/123, 31/77
Warfarin syndrome
  fetal, *see* Fetal warfarin syndrome
WARG syndrome
  aniridia, 50/581
  chromosome deletion syndrome, 50/581
  genital abnormality, 50/581
  mental deficiency, 50/581
  11p interstitial deletion, 50/581
  Wilms tumor, 50/581
Warm spring
  upper extremity, 8/405
  wrist extension, 8/405
Warsaw, Poland
  neurology, 1/11
Warts
  cortical, *see* Status verrucosus
Washington Psychosocial Seizure Inventory
  America, 72/412
  Asia, 72/411
  Australia, 72/412
  epilepsy, 72/409
  Europe, 72/411
  Scandinavia, 72/412
  standardization, 72/410
Wasp venom
  neurotoxin, 62/601, 65/194
  polyradiculoneuropathy, 65/198
  toxic myopathy, 62/601, 612
  transient ischemic attack, 65/198
Wasting
  distal muscle, *see* Distal muscular atrophy
  muscle, *see* Muscular atrophy
Water content
  brain, *see* Brain water content

Water intoxication
  argipressin, 64/240
  ataxia, 64/241
  baroreceptor, 64/240
  beer potomania, 64/244
  behavior, 63/547
  brain edema, 63/546, 64/240
  carbamazepine, 64/244
  carbamazepine intoxication, 65/506
  clinical features, 64/239, 241
  coma, 63/547, 64/241
  cyclophosphamide, 64/244
  daily water intake, 64/242
  death, 63/547
  definition, 64/240
  delirium, 64/241
  EEG, 64/241
  epilepsy, 64/241
  headache, 64/241
  healthy adult, 64/243
  healthy infant, 64/243
  hemiparesis, 64/241
  history, 64/239
  hyponatremia, 28/482, 63/545, 64/239
  metabolic encephalopathy, 64/240
  mortality, 63/547
  muscle cramp, 51/356
  myoclonia, 64/241
  nystagmus, 64/241
  osmolality, 63/545
  oxcarbazepine intoxication, 65/507
  oxytocin, 64/244
  physiology, 64/240
  plasma osmolality, 64/240
  polydipsia, 64/242
  postprostatectomy, 64/243
  rigidity, 64/241
  status epilepticus, 63/547
  symptom, 63/545
  synonym, 64/242
  systemic brain infarction, 11/463
  terminology, 64/241
  treatment, 64/244
  tremor, 64/241
  water balance, 64/240
  water type intoxication, 64/244
Water metabolism, *see* Electrolyte balance *and* Salt
    and water balance
Water retention
  headache, 48/436
  hydralazine intoxication, 37/431
Waterhouse-Friderichsen syndrome
  see also Fulminant meningococcemia

meningococcal meningitis, 33/23, 52/27, 30
Watershed infarction
  brain embolism, 53/314
  brain infarction, 53/110, 159, 314
  brain ischemia, 53/159
  carotid system syndrome, 53/314
  cause, 54/56
  computer assisted tomography, 54/55, 64
  facial paralysis, 53/210
  vascular dementia, 46/356
Wave and spike seizure
  multiple, 15/8
Wax like rigidity, see Plastic rigidity
WB 4101, see
    2-{[2-(2,6-Dimethoxyphenoxy)ethyl]aminomet
    hyl}-1,4-benzodioxan
Weakness
  acetonylacetone intoxication, 64/86
  arthrogryposis, 43/78
  beriberi, 28/8
  bulbospinal muscular atrophy, 59/43
  cardiovascular agent intoxication, 37/426
  cataplexy, 42/711
  central pontine myelinolysis, 28/292
  chrysaora intoxication, 37/51
  complex II deficiency, 66/422
  Duchenne muscular dystrophy, 40/358-360
  eye convergence, 24/75
  facial, see Facial weakness
  facial paralysis, 40/304, 422
  glycoside intoxication, 37/426
  hyperaldosteronism, 41/250
  hypermagnesemia, 63/563
  juvenile metachromatic leukodystrophy, 66/168
  Köhlmeier-Degos disease, 39/437
  late infantile metachromatic leukodystrophy,
    29/357
  motor, see Motor weakness
  muscle, see Muscle weakness
  neck muscle, 40/308
  neonatal myasthenia gravis, 41/100
  palatal, see Palatal weakness
  paralytic shellfish poisoning, 37/33
  pellagra, 7/573, 575, 28/90, 46/336
  pheochromocytoma, 39/498, 501, 48/427
  phosphoglucomutase deficiency, 27/222, 41/192
  primary amyloidosis, 51/415
  progressive, see Progressive weakness
  proximal muscle, see Proximal muscle weakness
  puffer fish intoxication, 37/80
  reserpine intoxication, 37/438
  Shy-Drager syndrome, 1/474
  stingray intoxication, 37/72

  tick bite intoxication, 37/111
  tick paralysis, 40/297
  Waldenström macroglobulinemia, 63/396
  zinc intoxication, 36/337
Weaver syndrome
  intrauterine growth acceleration, 46/82
  mental deficiency, 46/82
Webbed neck, see Pterygium colli
Weber disease, see Neurotrophic parietal
    endocarditis
Weber syndrome
  brain lacunar infarction, 54/244
  cause, 2/300
  child, 18/362
  lesion site, 2/296
  localization, 1/187
  pontine infarction, 12/16
  symptom, 2/277
Weber test
  deafness, 2/74
Weber-Von Leyden syndrome, see Weber syndrome
Wechsler Adult Intelligence Scale
  head injury outcome, 57/403
  performance impairment, 57/403
  verbal impairment, 57/403
Wechsler, I.S., 1/2, 14, 30, 35
Wechsler test
  cognition, 45/521
  epilepsy, 15/563
  neuropsychologic defect, 45/521
Weeping spell, see Crying spell
Wegener granulomatosis, 39/343-345
  see also Midline granuloma
  allergic granulomatous angiitis, 51/450
  angiopathic polyneuropathy, 51/446
  antibody, 71/183
  anticytoplasmic antibody, 51/450
  autoimmune disease, 41/87
  brain infarction, 53/162
  brain ischemia, 53/162
  carotid system syndrome, 53/308
  cerebrovascular disease, 53/41, 71/180
  chronic meningitis, 56/644
  classification, 71/4
  clinical features, 39/343, 51/450
  clinical finding, 71/174
  CNS, 39/344
  cranial nerve, 71/177
  CSF, 71/184
  diabetes insipidus, 71/182
  eosinophilia, 63/371
  facial paralysis, 51/451
  generalized form, 51/450

granulomatous CNS vasculitis, 55/396
headache, 48/284
hearing loss, 71/177
histopathology, 71/174
Horner syndrome, 71/179
imaging, 71/185
ischemic optic neuropathy, 51/450
laboratory finding, 71/183
leukoencephalopathy, 9/610-613, 39/344
localized form, 51/450
lymphomatoid granulomatosis, 39/529
mononeuritis multiplex, 39/344, 51/450, 63/423
morbidity, 71/186
multiple cranial neuropathy, 51/450
myelomalacia, 63/55
myelopathy, 71/179
myopathy, 51/451, 71/179
necrotizing arteritis, 71/181
necrotizing vasculitis, 71/178
neurogenic muscular atrophy, 51/451
neurologic symptom, 63/423
neurology, 71/175
neuropathy, 71/176
nonrecurrent nonhereditary multiple cranial
    neuropathy, 51/571
ophthalmoplegia, 71/177
optic nerve, 71/177
optic neuritis, 39/345, 51/451
orbital apex syndrome, 51/450
orbital pseudotumor, 51/450
paratrigeminal syndrome, 48/337
pathergic granulomatosis, 8/116
periarteritis, 8/112
polyarteritis nodosa, 9/610, 55/362
polyneuropathy, 8/128, 63/423
renal lesion, 71/174
seizure, 71/182
sensorimotor polyneuropathy, 51/451
sex ratio, 39/343, 51/450
sinusitis, 8/114, 39/343
survey, 71/173
swayback nose, 51/451
systemic brain infarction, 11/453
temporal arteritis, 55/346, 71/182
treatment, 71/185
vasculitis, 8/112, 71/8, 173
wallerian degeneration, 51/451
Weigert, C., 1/9, 27
Weight
  birth, see Birth weight
  brain, see Brain weight
  brain stem, see Brain stem weight
  cerebellum, see Cerebellum weight

Weight lifting
  headache, 48/124
  rehabilitative sport, 26/527, 532
Weight loss
  acrylamide intoxication, 64/67
  adrenal insufficiency myopathy, 62/536
  allergic granulomatous angiitis, 63/383
  arterial hypertension, 54/206
  atrial myxoma, 63/97
  carbamate intoxication, 64/186
  corticosteroid withdrawal myopathy, 62/536
  dermatomyositis, 62/374
  gastrogenous polyneuropathy, 7/699
  Huntington chorea, 49/261
  idiopathic postural orthostatic tachycardia
      syndrome, 74/327
  inclusion body myositis, 62/374
  obstructive sleep apnea, 63/458
  Parkinson disease, 75/180
  pellagra, 28/71
  pheochromocytoma, 39/498, 42/764
  polymyositis, 38/485, 62/374
  primary amyloidosis, 51/415
  Rett syndrome, 60/637
  Takayasu disease, 55/337
  temporal arteritis, 55/342
  toxic oil syndrome, 63/380
  transverse sinus thrombosis, 54/397
  Waldenström macroglobulinemia, 63/396
Weil disease, see Leptospirosis
Weil-Felix reaction
  Rickettsia, 34/643
Weisenburg, T.H., 1/21
Weiss-Alström syndrome, 13/463-465
  Alström-Hallgren syndrome, 13/463
  chorioretinal degeneration, 13/40
  diabetes mellitus, 13/463
  hypogonadism, 13/463
  introduction, 13/451
  Laurence-Moon-Bardet-Biedl syndrome, 13/463
  muscular hypotonia, 13/463
  obesity, 13/463
  optic atrophy, 13/40
  Prader-Labhart-Willi syndrome, 13/463
Weizsäcker, V. Von, see Von Weizsäcker, V.
Welander distal muscular dystrophy, see Distal
    muscular dystrophy
Welander distal myopathy, see Distal myopathy
Welander muscular dystrophy
  distal muscle weakness, 43/96
Welcker angle
  platybasia, 32/15
  skull radiography, 32/15

Welder ague, *see* Zinc intoxication
Well-being scale
  epilepsy, 72/422
Well leg raising test, *see* Crossed Lasègue test
Werdnig-Hoffmann disease, *see* Infantile spinal
    muscular atrophy
Werlhof essential thrombocytopenic purpura
  leukoencephalopathy, 9/598
Wermer syndrome
  Cushing syndrome, 42/752
  Zollinger-Ellison syndrome, 42/752
Werner syndrome
  alopecia, 43/489
  arteriosclerosis, 43/489
  ataxia telangiectasia, 14/315
  birth incidence, 43/489
  cataract, 43/489
  chorioretinal degeneration, 13/40
  diabetes mellitus, 43/489
  DNA synthesis, 43/480
  dwarfism, 13/436
  ectodermal dysplasia, 14/112
  growth retardation, 43/489
  hereditary connective tissue disease, 39/400-405
  heterozygote frequency, 43/489
  hyperkeratosis, 43/489
  meningioma, 43/489
  muscular atrophy, 43/489
  optic atrophy, 13/40
  osteoporosis, 43/489
  progeria, 39/403
  Refsum disease, 21/218, 60/235, 66/497
  sarcoma, 43/489
Wernicke aphasia, 45/313-317
  acalculia, 4/98, 45/476
  age, 45/304
  aging, 46/211
  agraphia, 45/313, 461
  alexia, 4/130, 45/313, 447
  amusia, 45/487
  anosognosia, 4/98
  apraxia, 45/429
  auditory aphasia, 4/98
  autotopagnosia, 4/98
  conduction aphasia, 4/98, 45/316
  consciousness, 45/115
  cortical deafness, 45/315
  cortical sensory aphasia, 4/98
  depression, 46/428
  diagnosis, 45/115
  finger agnosia, 4/98
  fluent agraphia, 45/460
  global aphasia, 4/98

jargon aphasia, 4/98
  neologistic jargon, 45/314
  paraphasia, 4/98
  parietal lobe lesion, 4/98
  perseveration, 4/98
  phonetic jargon, 45/315
  phonology, 45/308
  pure word deafness, 45/315
  radioisotope diagnosis, 45/298
  reading aloud, 4/98
  reading disorder, 4/98
  right left disorientation, 4/98
  schizophrenia, 45/313
  semantic aphasia, 4/98
  semantic jargon, 45/315
  sensory aphasia, 4/98, 45/313
  somatognosia, 4/98
  syntaxis, 45/307
  temporal lobe, 45/18
  temporal lobe lesion, 4/98
  writing disorder, 4/98
Wernicke disease, *see* Wernicke encephalopathy *and*
    Wernicke-Korsakoff syndrome
Wernicke encephalopathy
  see also Wernicke-Korsakoff syndrome
  alcohol withdrawal delirium, 46/541
  alcoholic polyneuropathy, 51/317
  alcoholism, 70/348, 75/494
  arousal disorder, 45/108
  ataxic tremor, 21/569
  autonomic nervous system, 75/30
  beriberi, 28/4
  central pontine myelinolysis, 28/289, 293, 308,
    47/586
  clinical criteria, 22/180
  confusion, 2/450
  dialysis, 63/523
  dialysis encephalopathy, 63/528
  fornix, 2/772
  hypothalamus, 2/450
  late cerebellar atrophy, 60/584
  leukoencephalopathy, 47/586
  malnutrition, 28/14-19
  memory disorder, 2/450, 3/277
  neuropathy, 28/17
  ophthalmoplegia, 28/17
  orthostatic hypotension, 63/151
  pellagra, 38/648
  protein deficiency, 7/635
  renal insufficiency, 27/324, 327
  thalamus, 2/487
  thalamus degeneration, 21/593
  thermoregulation, 75/60

vitamin B1, 45/196
vitamin B1 deficiency, 7/561, 28/15-17
Wernicke-Korsakoff syndrome, 45/194-196
Wernicke field
  migraine, 5/85
Wernicke, G., 1/5, 8, 20
Wernicke hemianopic reaction, *see* Hemianopic
  pupillary phenomenon
Wernicke-Korsakoff encephalopathy
  malabsorption syndrome, 70/233
Wernicke-Korsakoff syndrome, 28/243-268,
  45/193-201
  see also Memory disorder *and* Wernicke
    encephalopathy
  affect, 45/199
  age at onset, 28/245
  alcoholic dementia, 45/196, 46/335
  alcoholism, 70/348
  amnesia, 28/251-254
  ancillary finding, 28/248
  anterograde amnesia, 45/197
  apathy, 45/194
  arsenic intoxication, 46/392
  ataxia, 28/248, 45/195
  axial amnesia, 3/259
  Binswanger disease, 46/320
  body scheme disorder, 4/217
  brain contusion, 23/427
  brain cortex, 45/196
  brain death, 24/725
  brain infarction, 55/140
  brain injury, 23/11
  carcinoma, 38/648
  cardiac arrest, 63/214, 217
  cause of death, 28/254
  cerebellar degeneration, 28/272
  clinical anatomical correlation, 28/265
  clinical features, 28/245-254
  confabulation, 28/252
  confusional state, 46/341
  consciousness, 3/114, 28/246
  course, 28/249
  CSF, 28/249
  definition, 28/250
  delirium, 46/341
  delusion, 45/194
  dementia, 46/399
  diagnostic criteria, 45/193
  differential diagnosis, 45/199, 53/237
  disorientation, 45/194
  dorsomedial thalamic nucleus, 30/330
  EEG, 28/249
  etiology, 28/267

  euphoria, 45/194
  features, 3/273-283
  fornix, 2/772
  frontal lobe, 45/197
  frontal lobe syndrome, 45/29
  hallucination, 4/331, 45/194
  herpes simplex virus encephalitis, 45/199
  history, 28/243-245
  Huntington chorea, 45/199
  imagery, 45/199
  incidence, 28/245
  intelligence, 45/199
  learning, 27/469
  magnesium deficiency, 45/197
  malignancy, 38/648
  memory, 27/468-470, 46/621
  memory deficit, 45/197-199
  memory disorder, 27/469
  mental disorder, 28/246
  mental function, 28/246
  neurologic abnormality, 28/253
  neuropathologic finding, 28/255-266
  neuropathology, 46/347
  neuropathy, 28/253, 45/195
  neuropsychology, 45/200
  nutritional cerebellar degeneration, 28/272
  nystagmus, 45/194
  occipital lobe syndrome, 45/51
  ocular abnormality, 28/247-249
  pathologic features, 28/254
  pathology, 24/725, 45/196, 46/335
  pellagra, 6/748, 46/336, 399
  perception, 45/199
  phonemic jargon, 45/315
  presenting symptom, 28/245
  primary alcoholic dementia, 46/340-343
  prognosis, 28/252
  recognition, 45/198
  renal insufficiency, 27/324
  retrograde amnesia, 45/197
  sex distribution, 28/245
  sign, 24/725
  subarachnoid hemorrhage, 12/140
  substrate, 3/286
  symptom, 2/450
  thalamus, 45/197
  traumatic amnesia, 24/478
  treatment, 28/268
  verbal memory, 45/198
  vestibular function, 28/248
  vitamin B1, 45/195, 47/155
  vitamin B1 deficiency, 28/243, 267, 45/197
  vitamin B3 deficiency encephalopathy, 46/337

Wernicke encephalopathy, 45/194-196
Wernicke-Mann syndrome
lesion site, 2/301
Wernicke pupil
chiasm, 74/416
Wertheim Salomonson, J.K.A., 1/10
West Nile encephalitis
acute viral encephalitis, 34/76
West Nile virus
experimental acute viral myositis, 56/195
flavivirus, 56/12
West syndrome, *see* Infantile spasm
Western encephalitis
see also Western equine encephalitis
acute viral encephalitis, 56/134, 136
age, 56/136
Culex tarsalis, 56/136
geography, 56/136
season, 56/136
Western equine encephalitis
see also Western encephalitis
acute viral encephalitis, 34/72
Western equine encephalitis virus
deafness, 56/107
vertigo, 56/107
Western Pacific amyotrophic lateral sclerosis, *see*
Guam amyotrophic lateral sclerosis
Westphal
variant, *see* Juvenile Huntington chorea
Westphal paradoxical contraction, *see* Paradoxical
contraction (Westphal)
Westphal phenomenon, *see* Paradoxical contraction
(Westphal)
Westphal-Strümpell pseudosclerosis, *see*
Creutzfeldt-Jakob disease
Wetherbee ail
see also Amyotrophic lateral sclerosis *and*
Progressive muscular atrophy
muscular atrophy, 42/78-81
Wetterwinkel (Steiner)
multiple sclerosis, 9/221
plaque, 9/252
Weyers-Fülling dentofacial syndrome, *see*
Dentofacial syndrome (Weyers-Fülling)
WFN classification
abnormal immunologic motoneuron disease, 59/6
amyotrophic lateral sclerosis, 59/3
ataxia classification proposal 1985, 60/273
chemical motoneuron disorder, 59/3
congenital motoneuron disorder, 59/4
congenital neuropathy, 60/2
developmental motoneuron disorder, 59/4
endocrinopathic motoneuron disease, 59/6

heavy metal motoneuron disorder, 59/3
hereditary motoneuron disease, 59/1-4
hereditary neuropathy, 60/1-3
hyperactive motoneuron disorder, 59/6
motoneuron disease, 59/1-7
spinal muscular atrophy, 59/1-7
toxic motoneuron disorder, 59/5
traumatic motoneuron disorder, 59/5
unknown etiologic motoneuron disease, 59/6
viral infected motoneuron disorder, 59/5
Whale intoxication
ingestion, 37/94
vitamin A intoxication, 37/95
Wheat germ agglutinin
trigeminal ganglion, 48/110
Wheelchair
daily life activity, 41/491
design for scoliosis, 41/477
motoneuron disease, 59/469
neurologic deficit, 8/398
psychological aspect, 41/473
Wheelchair dash
rehabilitative sport, 26/534
Whip ray intoxication
elasmobranche, 37/70
Whiplash injury
brain injury, 23/96, 98
cervical disc injury, 61/31
cervical disc lesion, 61/32
critique, 7/461
head injury, 23/76, 93
headache, 5/200
posttraumatic headache, 24/508
spine experimental injury, 25/4
symptom, 61/33
tissue lesion, 25/68
vertebral artery insufficiency, 12/594
Whipped cream dispenser neuropathy
organic solvent intoxication, 36/374
Whipple disease
apathy, 52/137
ataxia, 51/184
brain cortex, 70/242
chronic meningitis, 56/644
CNS, 70/241
computer assisted tomography, 52/138
co-trimoxazole, 52/140
CSF, 52/138
CSF protein, 51/184
deficiency neuropathy, 51/328
delirium, 52/139
dementia, 46/389, 52/137
demyelination, 52/139

differential diagnosis, 52/137, 140
EMG, 51/185
encephalopathy, 7/627, 46/389
ependymitis, 52/139
epidemiology, 52/136
epilepsy, 52/137
etiology, 52/136
gastrointestinal disease, 39/459-464
headache, 52/138
HLA-B27, 52/136
hydrocephalus, 52/138
immunopathology, 52/136
Kleine-Levin syndrome, 52/138
laboratory, 70/242
macrophage, 70/239
malabsorption syndrome, 28/237-239, 51/328,
    52/137
meningitis, 52/138
multiple cranial neuropathy, 51/184
muscular atrophy, 52/138
myoclonus, 52/137
myopathy, 51/184, 52/138
neuritis, 51/184
neurologic aspect, 39/495-464
neurologic involvement, 52/135
neurologic manifestation, 28/237-239
neurology, 70/244
neuropathology, 52/139
oculomasticatory myoclonus, 52/137
ophthalmoplegia, 52/137
opsoclonus, 52/137
panencephalitis, 52/139
pathogenesis, 52/136
pathology, 70/242
penicillin, 52/140
polyarthralgia, 52/136
polyneuropathy, 51/184, 52/138
posterior column demyelination, 51/184
prevalence, 52/135
relapse rate, 52/140
retrobulbar neuritis, 52/138
sign, 70/245
sleep inversion, 52/137
subependymal granuloma, 70/243
survey, 70/239
symptom, 52/136, 70/241
Tangier disease, 8/21
tetracycline, 52/140
treatment, 52/140, 70/244
trigeminal neuropathy, 51/184, 52/138
Tropheryma whippelii, 70/241
Whispering dysphonia
    hereditary, see Focal dystonia musculorum

deformans
Whistling face syndrome, see Freeman-Sheldon
    syndrome
White finger disease, see Vibration neuropathy
White hair
    ataxia, 42/230
    Klein-Waardenburg syndrome, 14/115, 30/408,
        43/51
    piebald pigmentation, 43/42
    sensorineural deafness, 43/42
White matter
    amino acid composition, 9/12
    amyotrophic lateral sclerosis, 59/182
    aspartic acid, 9/16
    brain, see Brain white matter
    brain contusion, 23/40
    brain edema, 16/169
    brain lacunar infarction, 55/488
    brain water content, 9/16
    cholesterol, 9/5
    citric acid cycle enzyme, 9/14
    composition, 66/39
    2',3'-cyclic nucleotide-3'-phosphohydrolase,
        9/11, 14
    dementia, 67/380
    development, 9/15-19
    enzyme, 9/14-18
    fatty acid, 9/5
    fool plaque, 9/587
    ganglioside, 9/4
    glia, 9/6
    globoid cell leukodystrophy, 10/85
    glycerophosphate dehydrogenase, 9/16
    glycogen, 9/15
    lipid, 9/1, 6, 10
    minor lipid component, 9/5
    multiple brain infarction, 55/488
    myelination, 9/16
    Niemann-Pick disease type A, 10/493
    phosphoglucomutase, 9/15
    protein, 9/1, 11, 13
    proteolipid, 9/12
    Refsum disease, 21/238
    spina bifida, 32/544
    spongy degeneration, see Spongy cerebral
        degeneration
    triphosphoinositide phosphomonoesterase, 9/16
White matter atrophy
    glycosaminoglycanosis type IH, 66/26
    hydrocephalus, 50/290
White matter degeneration
    brain, see Brain white matter degeneration
    globoid cell leukodystrophy, 42/490

White matter demyelination
  Divry-Van Bogaert syndrome, 55/320
  Gerstmann-Sträussler-Scheinker disease, 60/623
  traumatic encephalopathy, 3/69
White matter disease
  CSF marker, 47/115-119
White matter hemorrhage
  brain fat embolism, 55/188
White matter hyperdensity
  glycosaminoglycanosis, 60/175
White matter injury
  brain, *see* Brain white matter injury
  clumsy hand dysarthria syndrome, 53/318
White matter necrosis
  anoxic leukomalacia, 47/526-528
  congenital heart disease, 63/2
White matter petechia
  brain fat embolism, 55/187
White matter vein calcification
  striatopallidodentate calcification, 49/427
White spirit, *see* Stoddard solvent
White syndrome
  periodic vertigo, 21/569
Whole brain radiation therapy
  brain metastasis, 69/142
Whooping cough, *see* Pertussis
Whorls of cytomembrane
  muscle cell, 40/39
Wicke classification
  diffuse sclerosis, 10/14
  leukodystrophy, 10/14
Widely spaced nippels
  Edwards syndrome, 43/535, 50/275
Wieacker-Wolff syndrome
  distal amyotrophy, 60/659
  oculomotor apraxia, 60/656
Wildervanck syndrome
  cranial nerve, 30/100
  ear abnormality, 43/344
  facial asymmetry, 43/343
  hypertelorism, 30/247
  Klippel-Feil syndrome, 43/343
  mental deficiency, 43/343
  prevalence, 43/344
  sensorineural deafness, 43/343
  sex ratio, 43/344
  spasmodic torticollis, 43/343
  Stilling-Türk-Duane syndrome, 43/343, 50/213
Willi-Prader syndrome, *see* Prader-Labhart-Willi
  syndrome
Williams syndrome
  elfin face, 43/259
  hoarse, 43/259, 46/92

hypercalcemia, 43/260, 46/92
hypertelorism, 30/250
mental deficiency, 43/259, 46/92
microcephaly, 30/508
Willis circle
  anastomosis, 11/178
  anterior cerebral artery, 11/176
  occlusion, 42/749-751
  posterior cerebral artery, 11/176
  trigeminal nerve, 48/108
  vertebrobasilar arterial system, 12/24
Willis circle abnormality
  brain collateral blood flow, 11/174
Willis circle occlusion, *see* Moyamoya disease
Willis classification
  epilepsy, 15/87
Willis cord
  anatomy, 11/52
Willis headache
  brain ischemia, 48/281
Willis, T., 2/10, 8/311
Willvonseder syndrome
  ataxia, 42/194
  ceruloplasmin, 42/194
  copper, 22/227
  dementia, 42/194
  speech disorder, 42/194
  speech disorder in children, 42/194
  tremor, 42/194
Wilms tumor
  aniridia, 50/581, 60/655
  ARG syndrome, 50/582
  Beckwith-Wiedemann syndrome, 43/398
  neurofibromatosis type I, 50/370
  11p interstitial deletion, 50/581, 583
  tumor cell, 67/21
  WARG syndrome, 50/581
Wilson disease, 6/267-277, 42/270-272
  see also Chronic hepatocerebral degeneration *and*
    Hepatic encephalopathy
  acquired, *see* Acquired hepatocerebral
    degeneration
  age at onset, 49/223
  Alzheimer type 1 glia, 6/268
  Alzheimer type 2 glia, 6/269, 292
  Alzheimer type I astrocyte, 9/635
  aminoaciduria, 27/387, 42/271, 49/229, 233
  amyotrophic lateral sclerosis, 59/410
  asterixis, 6/274, 282, 284, 823, 49/225
  athetosis, 6/459, 49/225
  autosomal recessive, 49/223
  basal ganglion, 6/62
  basal ganglion degeneration, 42/270

behavior disorder, 42/271
behavior disorder in children, 42/271
biochemistry, 6/116-127
birth incidence, 42/271
blepharospasm, 49/228
bone disease, 49/223
bone lesion, 6/273
bradykinesia, 49/225
case history, 49/224
cataract, 27/386
catatonia, 49/228
central pontine dystrophy, 9/638
central pontine myelinolysis, 28/293, 312
ceruloplasmin, 21/52, 27/395-399, 42/270, 46/58,
    49/231
ceruloplasmin copper, 49/229
chorea, 49/224
classification, 6/281, 27/199
clinical chemistry, 27/399-402
clinical features, 6/271, 27/381-387, 49/223
CNS, 27/393-399, 28/438
cogwheel rigidity, 49/225
computer assisted tomography, 49/233
copper, 27/393-401, 28/438, 42/270, 49/229
copper deposition, 21/52, 27/393-399
copper intoxication, 27/392
Creutzfeldt-Jakob disease, 6/746
cutis laxa, 49/235
cytochrome, 49/229
dementia, 46/130, 398, 400, 59/410
demyelination, 9/634
diagnosis in sibs, 6/125
differential diagnosis, 6/278
dimercaprol, 27/380, 402, 404
double athetosis, 6/155
drooling, 49/225
drug treatment, 6/840
dysarthria, 6/272, 274, 49/225
dysdiadochokinesis, 49/226
dystonia, 6/552, 27/384, 42/270, 49/224, 227
dystonia musculorum deformans, 42/215, 49/524
dystonic posture, 49/225
edetic acid, 27/404
elastosis performans serpiginosa, 49/235
endocrinopathy, 49/223
epilepsy, 15/340, 27/384
etiology, 6/271
Fanconi syndrome, 27/387, 29/172
free copper, 49/230
gait, 49/228
gallstone, 49/224
gene frequency, 42/271
genetics, 27/406-408, 49/235

glial dystrophy, 21/68
glycosuria, 27/387, 49/233
hematologic abnormality, 27/400
hematuria, 49/229
hemiplegic dystonia, 6/651
hemolysis, 49/223
hepatic manifestation, 27/383, 49/224
hepatosplenomegaly, 42/270
heterozygote detection, 42/271
heterozygote frequency, 42/271
historic aspect, 27/379-381
hyperammonemia, 6/292
hypercalciuria, 49/229
hyperhidrosis, 49/227
hyperphosphaturia, 49/229
incidence, 27/379
incontinence, 49/227
inheritance, 27/379
initial chorea, 6/153, 155
inose type, 27/200
ischemic type, 27/200
jaundice, 49/224
joint disease, 49/223
Kayser-Fleischer ring, 6/267, 270, 274, 27/379,
    386, 42/270, 46/58, 49/224, 228
laboratory test, 49/231
late cortical cerebellar atrophy, 21/477
leukoencephalopathy, 9/634
lip retraction, 49/225
liver copper, 27/399-401
liver disease, 27/383, 42/270
liver function, 27/400
lysyl oxidase, 49/229
mental deficiency, 46/18, 58
mental symptom, 6/274
multiple sclerosis, 47/72
muscle cramp, 49/224
mutism, 49/228
myoclonus, 21/514, 49/618, 60/662
nephrolithiasis, 49/229
neurologic manifestation, 27/384, 49/225
neuropathology, 46/18, 347
nomenclature, 6/268
ocular change, 27/386
onset, 6/271
Opalski cell, 6/268, 289
pallidal degeneration, 6/651
pallidoluysionigral degeneration, 49/459
Parkinson disease, 49/96
parkinsonism, 49/225
pathogenesis, 27/387-399, 49/229
pathohistology, 6/268-270
pathology, 27/200-202, 352

penicillamine, 27/380, 402-404, 49/233
peptiduria, 49/229
periodic ataxia, 21/570
physiology, 6/276
plasma copper, 27/399
postural tremor, 49/589
posture sustenance, 49/227
pregnancy, 27/407
presentation, 27/382
prevalence, 42/271, 49/223
previous jaundice, 49/224
prognosis, 6/275
protein metabolism, 27/386
pseudoulegyric type, 27/200
psychiatric disorder, 27/385, 42/271, 49/223, 228
reduced glomerular blood flow, 49/229
renal dysfunction, 49/229
renal function, 27/400
renal osteodystrophy, 49/229
rigidity, 49/69
salivation, 49/227
secondary neuroaxonal dystrophy, 49/408
serum copper, 9/322, 27/399, 49/231
skeletal system, 49/229
space occupying lesion, 16/242
speech disorder, 42/270
speech disorder in children, 42/270
striatal lesion, 2/499, 6/671
sunflower cataract, 27/386, 49/228
superoxide dismutase, 49/229
supranuclear ophthalmoplegia, 60/656
symptomatic dystonia, 6/553, 49/543
tetanoid chorea, 6/553
tetrathiomolybdic acid, 49/235
tonic postural reflex (Foix-Thévenard), 6/141
treatment, 6/124, 275, 27/400-408, 49/233
tremor, 27/384, 42/270, 49/225, 589
triethylenetetramine, 49/235
trihexyphenidyl, 49/235
tyrosinase, 6/120
urine copper, 27/399, 49/231
vasoglial dystrophy, 21/68
zinc sulfate, 49/235
Wilson, K., 1/6
Wilson pseudosclerosis, see Rigid Huntington chorea
Winged scapula
  Emery-Dreifuss muscular dystrophy, 62/146
  facioscapulohumeral muscular dystrophy, 40/417, 62/162
  limb girdle syndrome, 40/436
  myopathy, 40/312, 422
  scapulohumeral muscular dystrophy, 62/170

scapuloperoneal muscular dystrophy, 42/99
  topographical diagnosis, 2/20
Winiwarter-Buerger disease, see Thromboangiitis obliterans
Winkler, C., 1/10
Wintergreen oil intoxication, see Methyl salicylate intoxication
Wiring
  cervical vertebral column injury, 25/297
Wiskott-Aldrich syndrome
  herpes simplex virus, 56/211
  immunocompromised host, 56/468
Witch hazel
  characteristics, 64/95
  methanol, 64/95
  use, 64/95
Withdrawal delirium
  alcohol, see Alcohol withdrawal delirium
  hypnotic agent, see Hypnotic agent withdrawal delirium
Withdrawal reaction
  body scheme disorder, 45/380
Withdrawal syndrome
  alcoholism, 70/347
  barbiturate intoxication, 65/332
  benzodiazepine intoxication, 65/340
  cocaine intoxication, 65/253
  diamorphine, see Diamorphine withdrawal syndrome
  insomnia, 45/138
  methadone, see Methadone withdrawal syndrome
  myoclonus, 6/166
  nicotine intoxication, 65/263
  opiate, see Opiate withdrawal syndrome
  opiate intoxication, 65/354
Wittmaack-Ekbom disease, see Restless legs syndrome
Witzelsucht
  emotional disorder, 3/357
Wohlfart-Kugelberg-Welander disease, 22/67-77, 42/91-93, 59/81-93, 97
  β-N-acetylhexosaminidase A deficiency, 59/100, 372
  β-N-acetylhexosaminidase B deficiency, 59/372
  acid maltase deficiency, 59/372
  age at onset, 22/68, 59/81
  animal model, 59/91
  atypical case, 62/200
  autopsy, 22/73
  autosomal dominant form, 22/136
  autosomal dominant inheritance, 59/84
  autosomal recessive inheritance, 59/82
  Becker muscular dystrophy, 59/372

birth incidence, 42/91
bulbar paralysis, 22/135
bulbospinal muscular atrophy, 59/91
calf hypertrophy, 59/372
cardiac involvement, 59/87, 372
cardiomyopathy, 59/87
chromosome, 22/71
classification, 59/81, 367
clinical features, 22/69, 40/306, 308, 318, 331, 342, 59/85
contracture, 22/70, 42/91, 59/86
course, 22/77, 59/87
cranial nerve, 59/86
cranial nerve deficit, 22/70
cranial nerve involvement, 59/372
creatine kinase, 22/71
definition, 59/81
diagnosis, 22/76
differential diagnosis, 21/29, 22/15, 40/437, 59/91-93, 367-377
distal myopathy, 62/199
dysarthria, 22/70
dysphagia, 22/70
electrophysiologic study, 59/88
EMG, 22/34, 71
extensor plantar response, 22/71, 59/372
external ophthalmoplegia, 22/141
facioscapulohumeral form, 22/136
familial history, 22/69
fasciculation, 22/70, 42/91, 59/85
Fazio-Londe disease, 22/109
fructose-bisphosphate aldolase, 22/71
genetics, 22/69, 75, 41/439, 59/82-85
geography, 22/68
GM2 ganglioside, 59/93
gynecomastia, 59/88
hereditary spinal muscular atrophy, 59/24
Hirayama disease, 59/93
histochemistry, 22/72, 40/44
histologic differentiation, 22/96
history, 22/67
hyperlipoproteinemia, 59/372
hypertriglyceridemia, 59/372
inclusion body myositis, 62/204
infantile spinal muscular atrophy, 22/74, 82, 88, 91, 95
kyphoscoliosis, 42/91
laboratory study, 59/87
limb girdle syndrome, 40/433, 62/188
main features, 21/24, 28
mental deficiency, 59/93
mode of onset, 22/70
molecular biology, 59/85

mortality, 22/74
motor nerve conduction velocity, 59/89
muscle biopsy, 59/89, 62/200
muscle change, 22/72
muscle cramp, 59/86
muscle stretch reflex, 22/71
muscular atrophy, 22/70
myopathic change, 22/73
myopathy, 42/92, 59/88, 90
neurogenic quadriceps amyotrophy, 59/93
neuropathology, 22/72
nonprogressive spinal muscular atrophy, 59/92
nosology, 22/73
ophthalmoplegia, 22/70, 73, 180, 192
pain, 59/86
pathogenesis, 59/90
pathology, 59/90, 62/200
polymyositis, 59/372
presenting symptom, 22/69
prognosis, 22/77, 59/81
progressive bulbar palsy, 22/129
progressive external ophthalmoplegia, 22/210
progressive external ophthalmoplegia classification, 22/179, 62/290
progressive muscular atrophy, 22/13
progressive spinal muscular atrophy, 22/12, 310, 59/92
proximal muscle weakness, 42/91
reflex, 59/86
respiratory function, 59/87
rimmed muscle fiber vacuole, 62/200, 204
scapuloperoneal spinal muscular atrophy, 59/92
sensory disturbance, 59/87
sensory involvement, 59/372
serum creatine kinase, 59/88
serum esterone, 59/88
sex ratio, 21/28, 22/69, 59/87
spinal muscular atrophy, 22/134, 59/92, 371
subclinical type, 22/73
symptom, 62/199
tongue atrophy, 22/70
treatment, 22/77
twins, 22/69
unusual pedigree, 59/84
X-linked recessive inheritance, 59/84
Wolf-Hirschhorn syndrome, 31/550-561
associated malformation, 31/551, 553, 555-559
behavior, 31/559
CNS, 31/551, 553, 557, 43/497
cytogenetics, 31/550, 552, 561
dermatoglyphics, 31/553-555
epidemiology, 30/161, 31/559
epilepsy, 43/497

external appearance, 31/549-553
genetics, 30/508
hypertelorism, 30/249, 43/497
intelligence, 31/559
Marcus Gunn phenomenon, 31/552
mental deficiency, 43/497, 46/68
parental karyotype, 31/559
skeletal abnormality, 31/558, 43/497
Wolf intoxication
arsenic intoxication, 37/96
brain tumor, 37/96
diarrhea, 37/96
diplopia, 37/96
drowsiness, 37/96
headache, 37/96
ingestion, 37/95
irritability, 37/96
papilledema, 37/96
scurvy, 37/96
vitamin A intoxication, 37/96
vomiting, 37/96
Wolff, H.G., 5/1
Wolff-Parkinson-White syndrome
hyperuricemia, 42/571
Leber hereditary optic neuropathy, 66/428
syringobulbia, 32/276
Wolfgram protein
nerve myelin, 51/27
Wolfram syndrome
opticocochleodentate degeneration, 60/758
Wolfslast-Blumel-Subrahmanyam syndrome
hereditary congenital cerebellar atrophy, 60/293
Wolfson syndrome
alkaline phosphatase, 20/428
erythrocyte sedimentation rate, 20/428
spinal nerve root pain, 20/428
vertebral metastasis, 20/428
Wolman disease
see also Adrenal calcification
acanthocyte, 49/332
acid lipase, 21/50, 51/379
adrenal calcification, 10/504, 546, 14/779, 60/165
anemia, 10/546
Auerbach myenteric plexus, 29/373, 51/379
autonomic polyneuropathy, 51/379
bleeding tendency, 10/546
brain stem, 14/779
cerebellum, 14/779
cholesterol, 10/504, 546, 14/779
cholesterol ester, 10/352, 546
cholesterol ester storage disease, 60/166
cholesterol esterase, 10/546
chromosome 10, 60/159, 166

classification, 10/546
clinical diagnosis, 10/546
clinical features, 60/160
definition, 60/147
dementia, 60/165
enzyme deficiency polyneuropathy, 51/369, 379
foam cell, 10/504, 546
hepatosplenomegaly, 10/504, 546
hyperlipidemia, 60/165
jaundice, 10/546
laboratory study, 60/160
lipidosis, 66/16
neuroacanthocytosis, 60/674, 63/271, 290
neurolipidosis, 10/546
neuropathology, 51/379, 60/160
Niemann-Pick disease type A, 10/504, 546
orthochromatic leukodystrophy, 10/114
polyneuropathy, 60/165
steatorrhea, 51/379
submucosal plexus, 51/379
treatment, 60/160
triglyceride, 10/352, 504, 546, 14/779
visceral neuropathy, 51/379
xanthomatosis, 10/504, 546, 27/251
Wood alcohol intoxication, see Methanol
intoxication
Woods-Pendleton syndrome, see Familial
pallidonigral degeneration
Woods-Schaumburg syndrome
see also Nigrospinodentate degeneration
Ferguson-Critchley syndrome, 59/321
muscular atrophy, 22/171
nigrospinodentate degeneration, 21/401, 575
nuclear ophthalmoplegia, 21/401, 575
spinopontine degeneration, 21/401
Word blindness
lesion site, 2/587
pure, see Pure word blindness
Word deafness
congenital, see Verbal auditory agnosia
pure, see Pure word deafness
speech disorder in children, 4/428
verbal hallucination, 45/357
Word production anomia
anomic aphasia, 45/318
Word selection anomia
anomic aphasia, 45/319
Work
intensity, 74/247
oxygen, 74/247
World Federation of Neurology classification, see
WFN classification
World Health Organization classification

ataxia classification proposal 1989, 60/274
brain tumor, 16/24, 68/166
cerebrovascular disease, 11/184
embryonal brain tumor, 67/3
epilepsy, 15/25
Worried look
9p partial trisomy, 43/515
Wound
knife, *see* Knife wound
scalp injury, 24/215
spinal injury, 37/71
stab, *see* Stab wound
stingray intoxication, 37/71
urilophidae intoxication, 37/71
Wound botulism
clinical features, 51/190
diamorphine intoxication, 65/355
drug abuse, 65/209
entry portal, 51/188
incidence, 65/211
needle abscess, 65/211
opiate intoxication, 65/355
WPSI, *see* Washington Psychosocial Seizure
Inventory
Wrinkled abdomen
prune belly syndrome, 50/515
Wrisberg nerve, *see* Intermedius nerve *and* Medial
cutaneous brachial nerve
Wrist drop
lead polyneuropathy, 64/436
Wrist fracture
median nerve, 70/34
radial nerve, 70/34
ulnar nerve, 70/34
Writers cramp
agraphia, 45/467
athetosis, 49/382
botulinum toxin A, 65/213
continuous muscle fiber activity, 1/290
dystonia, 45/467
dystonia musculorum deformans, 6/528, 42/215,
49/521-523
focal athetosis, 49/383
occupational neurosis, 1/290, 7/338-340
stereotaxic surgery, 49/605
Writing
agraphia, 4/141-143, 45/459
alexia, 4/120
anterior alexia, 45/442
ataxic, 1/328
central alexia, 45/440
cerebral dominance, 4/280
developmental dyslexia, 46/108-113

expression, 45/523
frontal lobe tumor, 67/150
hand, *see* Hand writhing
higher cortical function, 3/382
mirror, *see* Mirror writing
nonfluent agraphia, 45/459
normal process, 46/108
Parkinson disease, 49/90
Writing ability
agraphia, 45/458
Writing tremor
primary, *see* Primary writing tremor
Wry neck
caput obstipum, 6/567
cervical vertebral column injury, 25/311
sprain, 25/311
Wyburn-Mason syndrome, *see*
Bonnet-Dechaume-Blanc syndrome

X-linked adrenoleukodystrophy, 66/447-473
Addison disease, 66/457
adrenal gland, 66/460
adrenal insufficiency, 66/457
adrenomyeloneuropathy, 66/453
asymptomatic, 66/458
bone marrow transplantation, 66/470
childhood, 66/448
clinical aspect, 66/447
computer assisted tomography, 60/672
demyelinating lesion, 66/449-451
diagnosis, 66/468
enzyme deficiency, 66/462
genetic mutation, 66/466
heterozygote, 66/457
history, 66/447
localization, 66/463
mRNA expression, 66/463
muscle, 66/460
myelin, 66/461
neonatal adrenoleukodystrophy, 66/512
neuropathology, 66/459
neuropsychological deficit, 66/449
neuroradiology, 66/458
pathogenesis, 66/471
peripheral nerve, 66/460
peroxisomal disorder, 66/510
posterior cerebral form, 66/449
protein, 66/464, 466
pyramidal tract, 66/455
testis, 66/460
treatment, 66/468
X-linked adrenoleukodystrophy complex
clinical classification, 66/510

X-linked agammaglobulinemia
    immunocompromised host, 56/468
    viral infection, 56/72
X-linked ataxia
    areflexia, 60/660
    dementia, 60/665
    distal amyotrophy, 60/659
    kyphoscoliosis, 60/669
    nerve conduction velocity, 60/670
    papilledema, 60/655
    spasticity, 60/661
X-linked cerebellar ataxia
    clinical features, 42/134
    prevalence, 42/135
X-linked copper malabsorption, *see*
    Trichopoliodystrophy
X-linked dystonia musculorum deformans
    deafness, 49/521
    prevalence, 42/218
X-linked hereditary motor and sensory neuropathy
    hereditary motor and sensory neuropathy type I,
      60/202
    hereditary motor and sensory neuropathy type II,
      60/202
    molecular genetics, 60/202
X-linked hereditary sensory and autonomic
    neuropathy
    clinical features, 60/16
    nerve biopsy, 60/17
    neurophysiologic study, 60/17
X-linked humeroperoneal neuromuscular disease,
    see Emery-Dreifuss muscular dystrophy
X-linked hydrocephalus
    prevalence, 50/288
X-linked hypogammaglobulinemia
    myoclonic epilepsy, 60/663
X-linked ichthyosis
    cataract, 60/653
    corneal dystrophy, 60/652
    sex ratio, 13/477
    X-linked recessive ataxia, 60/640
X-linked inheritance
    chorioretinal degeneration, 13/25
    corpus callosum agenesis, 50/151
    neuroretinal degeneration, 13/25
X-linked lissencephaly
    epilepsy, 72/136
X-linked lymphoproliferative syndrome
    Epstein-Barr virus, 56/250
X-linked mental deficiency
    aneuploidy, 50/566
    gene, 72/136
    Renpenning type, 43/302

X-linked muscular dystrophy
    Becker muscular dystrophy, 40/387-389, 62/117
    Duchenne muscular dystrophy, 40/277, 349-391,
      41/408, 415, 62/117
    Emery-Dreifuss muscular dystrophy, 40/389,
      62/117
    myopathy, 40/277
    scapuloperoneal muscular dystrophy, 22/59,
      40/426
    X-linked myotubular myopathy, 62/117
    X-linked neutropenic cardioskeletal myopathy,
      62/117
X-linked myotubular myopathy
    asphyxia, 62/156
    chromosome Xq28, 62/156
    congenital myopathy, 62/117
    fatal course, 62/156
    X-linked muscular dystrophy, 62/117
X-linked neuromuscular dystrophy
    Foster Kennedy syndrome, 62/117
    hereditary motor and sensory neuropathy type I,
      62/117
    hereditary motor and sensory neuropathy type II,
      62/117
X-linked neutropenic cardioskeletal myopathy
    carnitine, 62/156
    chromosome Xq28, 62/156
    congenital myopathy, 62/117
    X-linked muscular dystrophy, 62/117
X-linked optic atrophy
    areflexia, 42/417
    mental deficiency, 42/417
    rare type, 42/416
    tremor, 42/417
X-linked pigmentary abnormality
    hereditary deafness, 50/218
X-linked recessive ataxia, 60/635-641
    adrenoleukodystrophy, 60/638
    history, 60/635-639
    microcephaly, 60/636
    olivopontocerebellar atrophy, 60/638
    papilledema, 60/636
    pyruvate metabolism, 60/640
    Rett syndrome, 60/636
    review, 60/635-639
    sex chromosome fragility, 60/635
    sideroblastic anemia, 60/638
    spastic paraplegia, 60/639
    trichopoliodystrophy, 60/640
    X-linked ichthyosis, 60/640
X-linked recessive disorder
    Fabry disease, 51/376
    phosphoglycerate kinase deficiency, 62/489

X-linked recessive immunodeficiency
  gram-negative bacillary meningitis, 52/117
X-linked recessive inheritance
  bulbospinal muscular atrophy, 59/43
  congenital ataxic diplegia, 42/122
  corpus callosum agenesis, 50/151
  dyskeratosis congenita, 14/784, 43/392
  renal agenesis syndrome, 50/513
  Wohlfart-Kugelberg-Welander disease, 59/84
X-linked recessive spastic ataxic syndrome
  hereditary congenital cerebellar atrophy, 60/293
X-linked recessive transmission (Menkes), *see*
    Trichopoliodystrophy
X-linked scapuloperoneal syndrome, *see*
    Emery-Dreifuss muscular dystrophy
X-linked spastic paraplegia
  Pelizaeus-Merzbacher disease, 66/571
  proteolipid protein gene, 66/571
X-linked spinocerebellar degeneration
  Babinski sign, 42/191
  muscular atrophy, 42/190
  spinocerebellar ataxia, 42/189-192
X-linked transmission
  spina bifida, 50/479
X-radiation, *see* Radiation
X-ray, *see* Radiation
Xanthelasma
  low density lipoprotein, 10/272
Xanthine
  brain infarction, 53/421
  brain vasospasm, 11/125
Xanthine oxidase deficiency
  xanthinuria, 43/192
Xanthinuria
  creatine kinase, 43/193
  hyporeflexia, 43/193
  mental deficiency, 43/193
  myalgia, 43/192
  myopathy, 43/192
  xanthine oxidase deficiency, 43/192
Xanthoastrocytoma
  pleomorphic, *see* Pleomorphic xanthoastrocytoma
Xanthochromia
  brain embolism, 53/224
  brain infarction, 53/224
  cerebrovascular disease, 53/224
  CSF, 12/144, 54/196
  CSF examination, 11/236
  intracranial hemorrhage, 53/224
  Nonne-Froin syndrome, 7/498, 19/125, 20/459
  spinal cord tumor, 19/62
  transient ischemic attack, 53/224
Xanthodes reynaudi intoxication

  eating, 37/56
Xanthogranuloma
  juvenile, *see* Juvenile xanthogranuloma
  Sturge-Weber syndrome, 50/377
Xanthoma
  cerebrotendinous xanthomatosis, 29/374, 66/600
  choroid plexus, *see* Choroid plexus xanthoma
  cutaneotendinous, *see* Cutaneotendinous
    xanthoma
  glycogen storage disease type I, 27/225
  Hand-Schüller-Christian disease, 27/248
  hereditary hyperlipidemia, 11/462
  histiocytosis X, 27/248
  hyperlipoproteinemia type II, 10/272-274
  hyperlipoproteinemia type IV, 10/276
  hyperlipoproteinemia type V, 10/277
  metabolic ataxia, 21/575
  spinal cholesterosis, 27/250
  tendon, *see* Tendon xanthoma
Xanthoma cell
  pseudo, 14/244
Xanthoma tendinosum
  low density lipoprotein, 10/272
Xanthomatosis, 27/241-251
  brain biopsy, 10/686
  cerebrotendinous, *see* Cerebrotendinous
    xanthomatosis
  cholesterolemic, *see* Cholesterolemic
    xanthomatosis
  choroid plexus, 27/244-246
  demyelination, 27/247
  glycogen storage disease type I, 42/586
  Hand-Schüller-Christian disease, 10/343
  histiocytosis X, 10/539, 27/248
  hyperlipoproteinemia type I, 42/443
  hyperlipoproteinemia type II, 10/272, 42/444
  lipoid granulomatosis, 38/96
  meningioma, 27/246
  neurinoma, 27/246
  neuropathy, 8/18
  Niemann-Pick disease, 10/504, 27/242, 251
  osteoporosis, 70/20
  pathogenesis, 27/241-244
  spinal cholesterosis, 14/779
  systemic reticulosis, 27/249
  Wolman disease, 10/504, 546, 27/251
Xanthomatous deposit
  hepatic insufficiency polyneuropathy, 51/363
Xanthomatous process
  pathogenesis, 27/241-244
Xanthomatous transformation
  choroid plexus, 27/244-247
  demyelination, 27/247

histiocytosis X, 27/248
meningioma, 27/246
neurinoma, 27/246
neurodegenerative disease, 27/246
systemic reticulosis, 27/249
Xanthomyxolipoma, *see* Hibernoma
Xanthurenic aciduria
kynureninase, 42/636
mental deficiency, 42/635
pyridoxal-5-phosphate, 42/636
pyridoxine dependency, *see* Vitamin B6 dependent
xanthurenic aciduria
tryptophan metabolism, 42/636
vitamin B6, 42/636
vitamin B6 metabolism, 42/635
Xenon
brain edema, 23/157
brain injury, 23/157
Xenon computer assisted tomography
brain blood flow, 54/149
brain infarction, 54/65
brain stem death, 57/464
Xeroderma pigmentosum, 38/6-8, 60/335-339
age at onset, 60/335
areflexia, 43/13, 60/336, 660
ataxia, 43/13, 51/399, 60/336
axonal degeneration, 51/399
Bambi leg, 60/336
basal cell carcinoma, 43/12
cerebellar atrophy, 43/13
chorea, 14/788
choreoathetosis, 43/13, 49/385, 51/399, 60/336,
663
clinical features, 60/335
CNS, 31/290, 60/337
computer assisted tomography, 60/337
contracture, 60/669
cranial hyperostosis, 60/337
CSF homovanillic acid, 60/337
cutaneous abnormality, 31/290
cutaneous manifestation, 60/335
deafness, 51/399, 60/336, 657
dementia, 60/335
distal amyotrophy, 60/659
DNA excision repair, 60/339
DNA repair mechanism, 43/13, 60/335, 339
dwarfism, 14/13, 51/399
dystonia, 60/664
EEG, 60/337
EMG, 51/399
endocrine disorder, 43/13
fasciculation, 60/659
Friedreich ataxia, 14/13, 315, 21/348

genetics, 14/529, 530, 31/291
gonadal hypoplasia, 51/399
growth retardation, 43/12
heterozygote detection, 43/13
hypogonadism, 60/668
hyporeflexia, 43/13
laboratory study, 60/336
leukemia, 60/674
lymphoma, 60/674
malignant epithelioma, 60/335
mental deficiency, 14/13, 788, 31/290, 43/12,
46/59, 51/399, 60/664
mental deterioration, 51/399
metabolic polyneuropathy, 51/399
Mibelli angiokeratoma syndrome, 14/529
microcephaly, 30/100, 508, 31/290, 43/12, 51/399,
60/335, 666
muscular atrophy, 60/336
nerve conduction, 60/337
nerve conduction velocity, 60/670
nerve fiber loss, 60/337
neurinoma, 43/12
neurocutaneous syndrome, 14/101
neuropathology, 60/337
neuropathy, 60/660
peripheral nervous system, 60/337
phakomatosis, 14/13, 50/377
photophobia, 43/12
photosensitive dermatitis, 51/399
photosensitivity, 14/788, 60/666
polyneuropathy, 60/336
porphyrin metabolism, 14/13
prenatal diagnosis, 43/13
prognosis, 31/291
pupillary abnormality, 60/336
scoliosis, 60/336
seizure, 43/13
sensorimotor polyneuropathy, 51/399
sensorineural deafness, 43/13
Sjögren-Larsson syndrome, 22/478
skeletal abnormality, 60/336
spastic paraplegia, 60/336
spasticity, 43/13, 51/399, 60/336
spinal muscular atrophy, 59/368, 370, 372
squamous carcinoma, 43/12
9 subgroups, 60/335
substantia nigra degeneration, 60/337
Xerodermic idiocy, *see* Xeroderma pigmentosum
Xerodermic idiocy (De Sanctis-Cacchione), *see*
Xeroderma pigmentosum
Xerophthalmia
parasympathetic insufficiency, 75/117
Xerosis

human T-lymphotropic virus type I associated myelopathy, 59/450

Xerostomia
parasympathetic insufficiency, 75/117

Xiphosura intoxication
Carcinoscorpius rotundicauda, 37/58
tachypleus gigas, 37/58
tachypleus tridentatus, 37/58

XO syndrome, *see* Turner syndrome

XX male
birth incidence, 43/556
gynecomastia, 43/555
HY antigen, 43/556
hypogonadism, 43/555
hypospadias, 43/555
Klinefelter syndrome, 31/484
mental deficiency, 43/556
sex chromosomes, 43/555-557, 46/75

XXX syndrome
aneuploidy, 50/553
birth incidence, 43/553
epilepsy, 31/494
history, 50/553
maternal age, 43/553
mental deficiency, 31/493, 43/552, 46/74, 50/553
pathogenesis, 50/554
polyploidy, 50/553
schizophrenia, 43/552
sex chromosomes, 46/74
speech disorder, 43/552
speech disorder in children, 43/552
symptomatology, 50/554

XXXX syndrome
aneuploidy, 50/554
cataract, 43/554
corpus callosum hypoplasia, 50/555
deafness, 50/554
dermatoglyphics, 43/554
digital abnormality, 43/554
dysmorphic features, 50/554
epicanthus, 43/554
hypergonadotropic hypogonadism, 50/555
hypertelorism, 30/250, 43/554
mental deficiency, 31/493, 43/554, 46/74, 50/553, 555
microcephaly, 43/554
nystagmus, 50/554
polyploidy, 50/553
radioulnar dysostosis, 43/554
radioulnar synostosis, 50/554
sex chromosomes, 46/74
short stature, 43/554
strabismus, 43/554

swayback nose, 43/554
symptomatology, 50/554

XXXXX syndrome
aneuploidy, 50/555
camptodactyly, 50/555
cataract, 43/554
dermatoglyphics, 43/554
digital abnormality, 43/554
epicanthus, 43/554
hypertelorism, 30/250, 43/554
mental deficiency, 31/495, 43/554, 46/74, 50/553, 555
microcephaly, 43/554, 50/555
polyploidy, 50/553
radioulnar dysostosis, 43/554
radioulnar synostosis, 50/555
sex chromosomes, 46/74
short stature, 43/554
strabismus, 43/554
swayback nose, 43/554
symptomatology, 50/555

XXXXY syndrome, *see* Klinefelter syndrome *and* Klinefelter variant XXXXY

XXXY syndrome, *see* Klinefelter syndrome *and* Klinefelter variant XXXY

XXY syndrome, *see* Klinefelter syndrome

XXYY syndrome, *see* Klinefelter syndrome

Xylose
progressive dysautonomia, 59/158

β-Xylosidase
fucosidosis, 42/550

XYY syndrome, 31/486-491
above average height, 50/551
behavior, 31/487-489
choreiform dyskinesia, 50/551
clinical features, 31/487-490
CNS, 31/489
convulsion, 31/489
criminality, 43/205, 46/75, 50/551
cytogenetics, 31/490
EEG, 31/489
endocrine function, 31/489
epidemiology, 31/486
epilepsy, 31/489
ethics, 50/552
history, 50/550
hormonal finding, 50/552
hypotonia, 50/551
intention tremor, 50/551
mental deficiency, 31/487
pathogenesis, 31/487, 50/551
prevalence, 50/551
psychological finding, 50/551

radioulnar synostosis, 50/551
skeletal abnormality, 31/490
stature, 31/487
Sturge-Weber syndrome, 14/235
symptomatology, 50/551

Y syndrome
double, *see* XYY syndrome
Yaba poxvirus
neurotropic virus, 56/5
Yale Scale
spinal cord injury, 61/423
spinal cord injury recovery score, 61/423
transverse spinal cord lesion, 61/423
Yawning
anencephaly, 63/493
autonomic nervous system, 74/584
brain infarction, 63/493
epilepsy, 63/493
hypothalamic syndrome, 2/450
locked-in syndrome, 63/493
migraine, 48/59, 124, 156
reflex physiology, 1/499
sleep disorder, 3/101
sneezing, 63/493
sympathetic nerve, 74/652
synkinesis, 63/493
temporal lobe epilepsy, 63/493
Yeast
cancer, 69/438
CNS infection, 69/438
Yellow fever
flavivirus, 56/12
Yellow fever virus
deafness, 56/107
vertigo, 56/107
Yellow jacket intoxication
brain infarction, 37/107
brain stem, 37/107
hemorrhage, 37/107
neurologic complication, 37/107
Yellow orange tonsil
a-alphalipoproteinemia, 51/396
hereditary neuropathy, 60/265
Tangier disease, 10/547, 60/133
Yellow star thistle intoxication
chewing disease, 6/665
pallidonigral encephalomalacia, 6/665, 49/466, 483
pallidonigral necrosis, 49/483
Yersinia
tropical polymyositis, 62/373
Yo antigen

paraneoplastic syndrome, 69/330, 334
Yohimbine
hypertension, 74/168
pharmacology, 74/168
Yolk sac tumor
spinal cord teratoma, 32/361
Yucheng disease
organochlorine insecticide intoxication, 64/207
polychlorinated biphenyl intoxication, 64/207
Yusho disease
organochlorine insecticide intoxication, 64/207
polychlorinated biphenyl intoxication, 64/207

Z band
calcavin, 40/157
central core myopathy, 40/215, 43/81
dimethyl-*p*-phenylenediamine, 40/157
multicore myopathy, 43/121
muscle fiber, 62/7, 26
myopathy, 62/7
nemaline myopathy, 41/10, 43/122
neuromuscular disease, 62/7
plasmocid, 40/157
rod body, 40/240
target fiber, 40/215
tenotomy, 40/144
ultrastructure, 40/63
Z band plaque myopathy
congenital myopathy, 62/332
respiratory dysfunction, 62/344
Z disc streaming
muscle fiber, 40/24, 81, 151, 157, 215
Zackai syndrome
microcephaly, 30/100
Zalcitabine
neuropathy, 71/370
neurotoxicity, 71/368
Zarontin, *see* Ethosuximide
Zebra body
congenital weakness, 40/248
glycosaminoglycanosis, 10/379, 382
glycosaminoglycanosis type II, 10/445
honeycomb myopathy, 62/353
metachromatic leukodystrophy, 51/368, 60/126
muscle fiber, 40/113
normal location, 40/248
peripheral nerve, 22/212
progressive external ophthalmoplegia, 22/194, 212
structural abnormality, 40/247
Zebra body myopathy
congenital myopathy, 41/21, 62/332
Zebrafish intoxication

venom, 37/75
Zellweger like syndrome
    peroxisomal disorder, 66/510
Zellweger syndrome, *see* Cerebrohepatorenal
        syndrome
Zidovudine
    acquired immune deficiency syndrome, 56/492
    human immunodeficiency virus infection, 71/377
    mitochondrial DNA depletion, 62/511
    neuropathy, 71/373
    neurotoxicity, 71/368
    polymyositis, 62/372, 382
    toxic myopathy, 62/597, 600, 602
Zidovudine myopathy
    ragged red fiber, 62/383
Ziehen-Oppenheim syndrome, *see* Dystonia
        musculorum deformans
Zimeldine
    drug induced polyneuropathy, 51/304
    Guillain-Barré syndrome, 51/304
    migraine, 48/196
    neurotoxicity, 51/304
    neurotoxin, 51/304, 65/311
    toxic myopathy, 62/597
Zimeldine polyneuropathy
    influenza attack, 51/304
    myalgia, 51/304
Zimelidine, *see* Zimeldine
Zinc
    CNS, 28/438
    disulfiram polyneuropathy, 51/309
    element data, 64/361
    ethambutol polyneuropathy, 51/295
    metalloenzyme, 64/361
    Pick disease, 46/242-244
    pressure sore, 61/350
    pseudocalcium, 6/715, 718
    striatopallidodentate calcification, 6/704, 49/425
    zinc deficiency, 64/361
Zinc deficiency
    Danbolt disease, 36/336
    zinc, 64/361
Zinc intoxication
    ataxia, 36/338
    chill, 64/362
    coma, 36/337
    diarrhea, 36/337, 64/362
    dyspnea, 36/337
    fatality, 64/362
    fever, 64/362
    headache, 64/362
    hypogeusia, 36/335
    incoordination, 36/338

intention tremor, 36/338
    lethargy, 36/337
    leukocytosis, 64/362
    morbidity, 64/362
    muscle spasm, 36/337
    nausea, 36/337
    nyctalopia, 36/335-337
    prevention, 64/363
    sweating, 64/362
    symptom, 64/361
    treatment, 64/363
    vomiting, 36/337
    weakness, 36/337
Zinc phosphide
    toxic myopathy, 62/601
Zinc pyridinethione, *see* Pyrithione zinc
Zinc sulfate
    CSF, 9/335
    Wilson disease, 49/235
Zinn circle
    Von Hippel-Lindau disease, 2/517
Zinsser-Engman-Cole syndrome, *see* Congenital
        dyskeratosis
Zoantharia-zoanthidae intoxication, *see* Palythoa
        vestitus intoxication
Zollinger-Ellison syndrome
    deficiency neuropathy, 7/608
    Wermer syndrome, 42/752
Zolpidem
    benzodiazepine intoxication, 65/334, 342
Zombi phenomenon
    tetrodotoxin intoxication, 65/155
Zona incerta
    anatomy, 6/20, 38
Zona reticulata substantia nigra
    neuroaxonal dystrophy, 49/392
Zona reticulosa substantia nigra
    ventral pallidostriatum, 49/20
Zonisamide
    epilepsy, 73/361
Zoom vision
    hallucination classification, 45/56
Zoöneurology
    demyelinating disease, 9/666-668
Zopiclone
    benzodiazepine intoxication, 65/343
Zoster encephalitis, *see* Varicella zoster virus
        meningoencephalitis
Zozymus aeneus intoxication
    aphasia, 37/56
    collapse, 37/56
    death, 37/56
    nausea, 37/56

paralysis, 37/56
paresthesia, 37/56
vomiting, 37/56
Zuckerkandl organ
    nerve tumor, 8/496
    pheochromocytoma, 8/473, 42/765
Zülch, K.J., 1/27
Zürich, Switzerland
    neurology, 1/10
Zygodactyly
    Capute-Rimoin-Konigsmark syndrome, 14/110
Zygomatic arch absence

Treacher-Collins syndrome, 43/421
Zygomatic bone fracture
    orbital fracture, 23/375, 24/95
    treatment, 24/97
Zygomycetes
    cerebrovascular disease, 55/432
    meningitis, 55/432
Zygomycosis, *see* Mucormycosis
Zygosity
    twins, *see* Twin zygosity
Zygotic twins
    multiple sclerosis, 9/91